# Dive into Oceanography with Trusted Content and Innovative Media



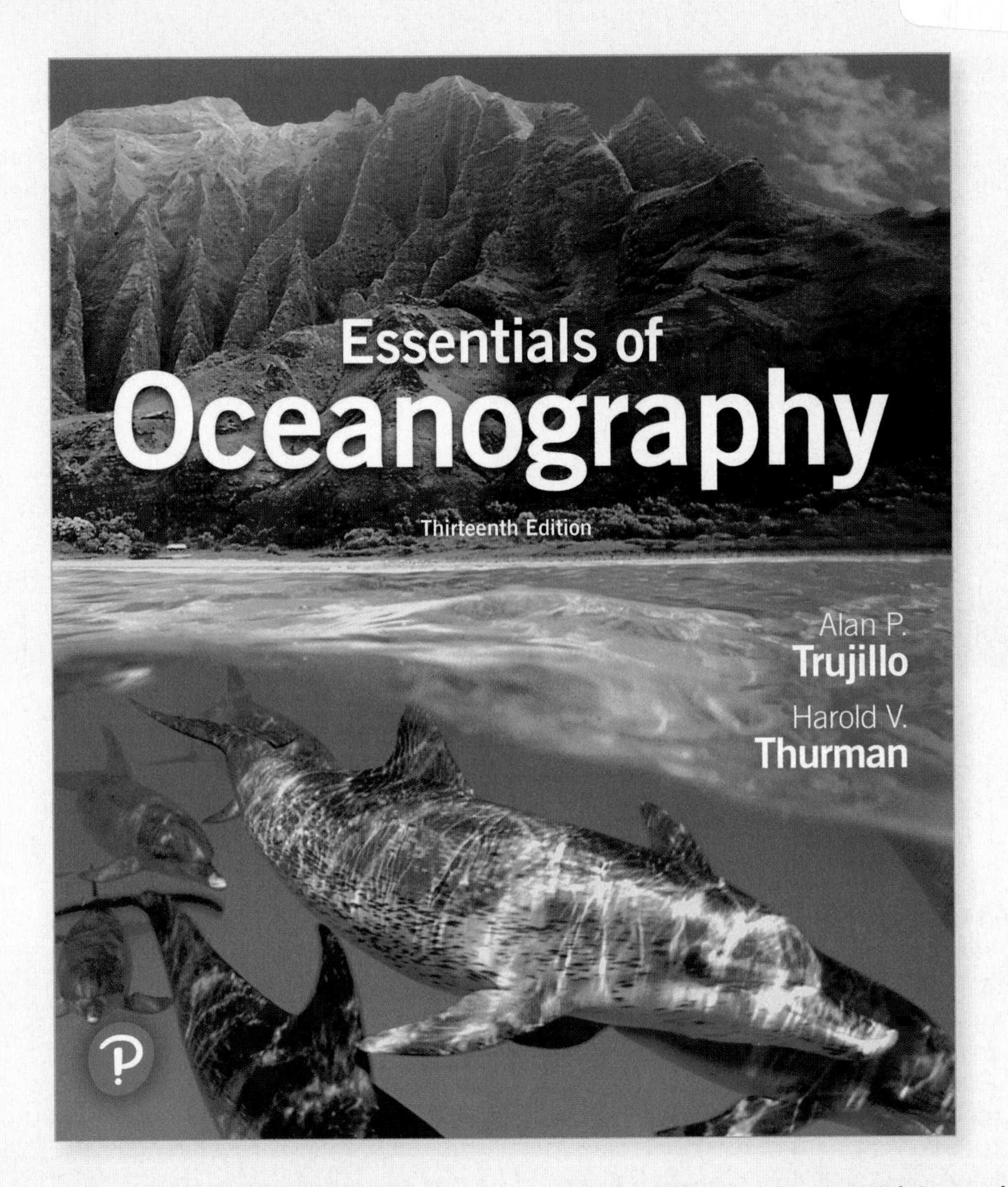


The best-selling brief book in the oceanography market combines dynamic visuals and a student-friendly narrative to bring oceanography to life and inspire students to engage and learn more about the oceans and environments around them. The 13th edition creates an interactive learning experience, providing tightly integrated text and digital offerings that make oceanography approachable and digestible for students. An emphasis on the **process of science** throughout the text provides students with an understanding of how scientists think and work. It also helps students develop the scientific skills of devising experiments and interpreting data.

# Dive into the Process of Science

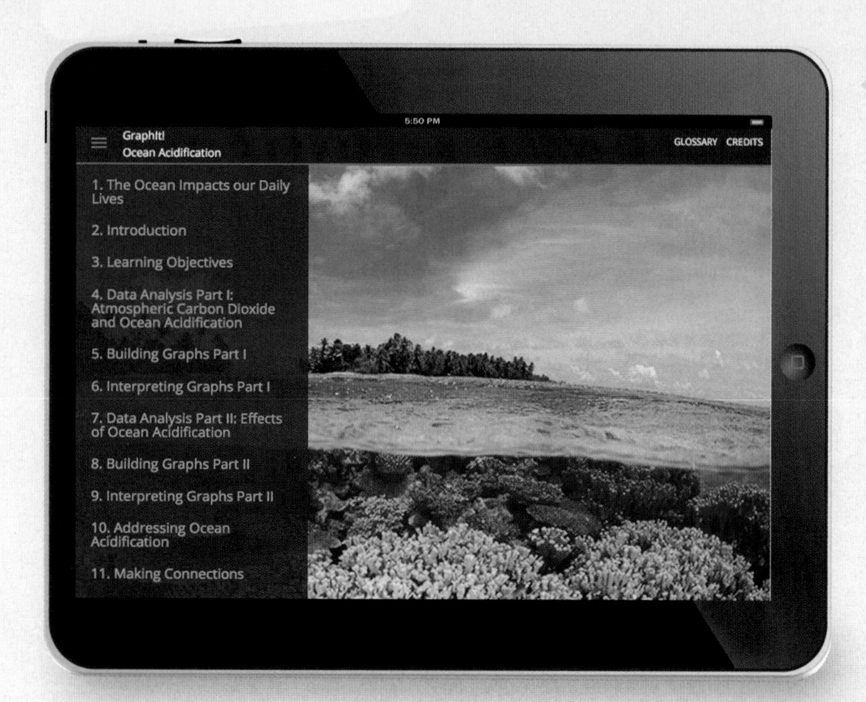


**GraphIt! Activities** are a great way to help your students develop their understanding of graphs and data. These activities use real data to help students build science literacy skills and their understanding of how to analyze and interpret graphs.

**NEW Process of Science** features develop student understanding of how scientists think and work. Each one highlights an area of oceanographic inquiry and explicitly points out the associated background, method, and conclusion of the inquiry.

**NEW Process of Science in Mastering**. Each of the new Process of Science features includes an assignable Mastering coaching activity to enable students to be active participants and develop 21st Century Skills.

## PROCESS OF SCIENCE 4.1: When The Dinosaurs Died: The Cretaceous–Tertiary (K–T) Event

### BACKGROUND

The extinction of the dinosaurs—and about 75% of all plant and animal species on Earth, including many marine species—occurred about 66 million years ago. This extinction marks the boundary between the Cretaceous (K) and Tertiary (T) Periods of geologic time and is known as the **K–T event** or, because of recent changes in the geologic time scale, the *Cretaceous–Paleogene (K–Pg) event*. Did slow climate change lead to the extinction of these organisms, or was it a catastrophic event? Was their demise related to disease, diet, predation, or volcanic activity? Earth scientists have long sought clues to this mystery.

### FORMING A HYPOTHESIS

In 1980, geologist Walter Alvarez, his father, Nobel Physics Laureate Luis Alvarez, and two nuclear chemists, Frank Asaro and Helen Michel, reported that marine deposits collected in northern Italy from the K–T boundary contained an unusual clay layer with high proportions of the metallic element iridium (Ir), an element rare in Earth rocks but much more abundant in meteorites. The high concentrations of iridium suggested minerals in the clay had an extraterrestrial origin. In addition, the clay layer contained shocked quartz grains, indicating an event had occurred with enough force to fracture and partially melt pieces of quartz. Other deposits from the K–T boundary revealed similar features, supporting the hypothesis that Earth experienced an extraterrestrial impact at the same time that the dinosaurs died.

One problem with the impact hypothesis, however, is that dust spewing from volcanic eruptions on Earth could create similar clay deposits enriched in iridium and containing shocked quartz. In fact, at about the same time as the dinosaur extinction, large outpourings of basaltic volcanic rock in India (called the Deccan Traps) and other locations had occurred. Also, if there was a catastrophic meteor impact, where was the crater?

In the early 1990s, the 190-kilometer (120-mile)-wide *Chicxulub* (pronounced "SCHICK-sue-lube") *Crater* off the Yucatán coast in the Gulf of Mexico was identified as a likely candidate because of its structure, age, and size. To create a crater this large, a 10-kilometer (6-mile)-wide object composed of rock and/or ice traveling at speeds up to 72,000 kilometers (45,000 miles) per hour must have slammed into Earth (**Figure 4B**). Such an impact would have created huge waves—estimated to be more than 900 meters (3000 feet) high—that traveled throughout the oceans. In addition, the dust and debris lifted into the atmosphere most likely limited photosynthesis, chilled Earth's surface, and brought about the extinction of the dinosaurs and many other species. Finally, acid rains and global fires may have added to the environmental disaster.

### DEVISING AN EXPERIMENT

Supporting evidence for the meteor impact hypothesis was provided in 1997 by recovering cores of sediment from the sea floor. Previous drilling close to the impact site did not reveal any K–T deposits. Evidently, the impact and resulting huge waves had stripped the ocean floor of its sediment. However, at 1600 kilometers (1000 miles) from the impact site, the telltale sediments from the catastrophe, such as the iridium-rich clay layer, were preserved in sea floor sediments.

### INTERPRETING THE RESULTS

Convincing evidence of the K–T impact from this and other cores collected in 2016 suggests that Earth has experienced many such extraterrestrial impacts over geologic time. Statistics show that an impact the size of the K–T event should occur on Earth about once every 100 million years, severely affecting life on Earth as it did the dinosaurs. This frequency is consistent with the fossil record, which indicates that in the last 500 million years, Earth has experienced five major extinction events.

### THINKING LIKE A SCIENTIST: WHAT'S NEXT?

What kind of evidence would you expect to find in coastal rock sequences that were deposited during the time of the huge waves that were created by the meteor impact?

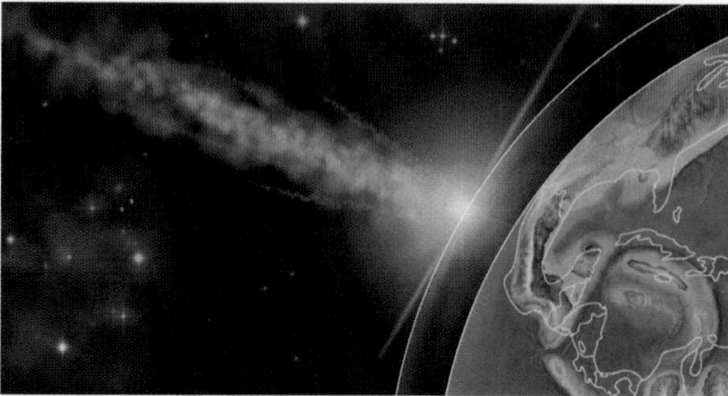

Figure 4B The K–T meteorite impact event.

# with Trusted Content and Dynamic Media

**EXPLORING DATA** ► Using the equation shown in Figure 3.7a, determine the time it takes sound to reach the deepest part of the ocean, the Challenger Deep in the Mariana Trench, which is 11,022 meters (36,161 feet) below sea level. Use the average speed of sound in seawater of 1507 meters (4945 feet) per second.

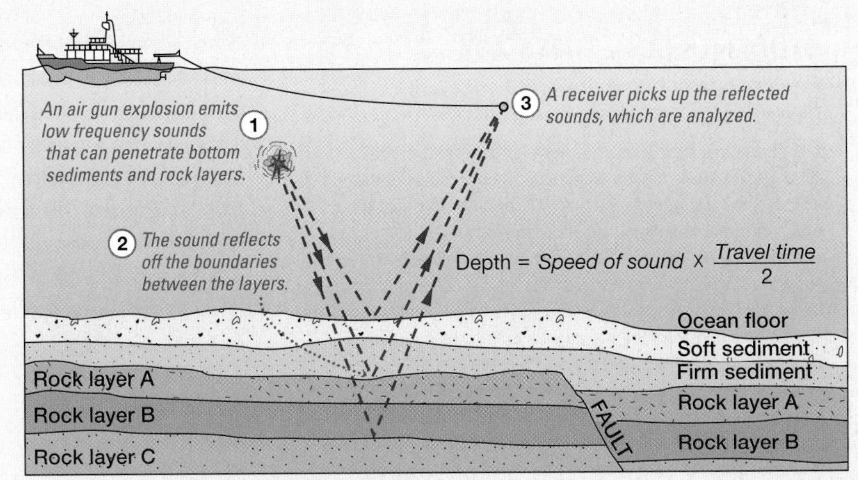


**(a)** A ship conducting seismic profiling. Note that depth can be determined by knowing the speed of sound in seawater and the travel time of the sound.

**NEW Exploring Data Activities** help students engage with graphs and other data-driven features to enable them to practice their data interpretation skills.

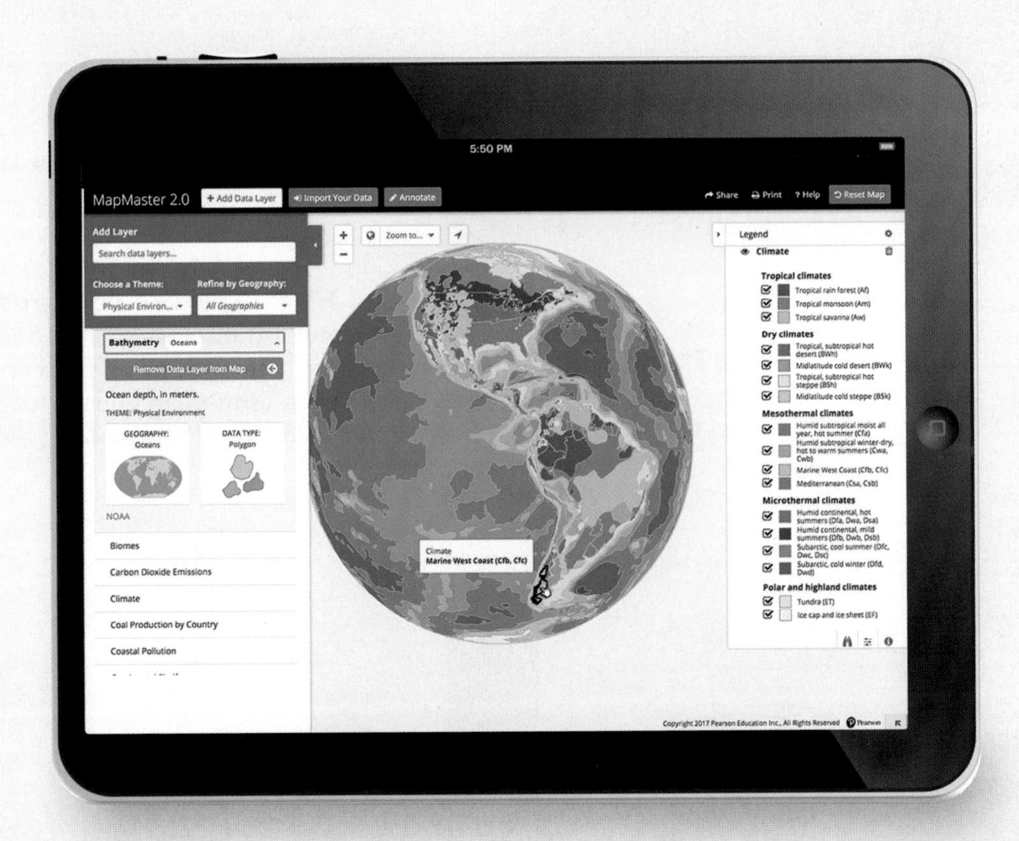


**NEW MapMaster 2.0** is GIS inspired, allowing students to layer various thematic maps to analyze spatial patterns and data at regional and global scales. Now, fully mobile, this tool includes zoom and annotation functionality with hundreds of map layers leveraging recent data from sources such as NOAA, NASA, USGS, United Nations, and CIA. Students can also upload their own data. Students are able to access MapMaster 2.0 in the Study Area on their own and instructors can assign auto-graded activities.

# Dive into Student Engagement

## STUDENTS SOMETIMES ASK . . .

***How can I accept a scientific idea if it's just a theory?***

When most people use the word "theory" in everyday life, it usually means an idea or a guess (such as the all-too-common "conspiracy theory"), but the word has a much different meaning in science. In science, a theory is not a guess or a hunch. It's a well-substantiated, well-supported, well-documented explanation for observations about the natural world. It's a powerful tool that ties together all the facts about something, providing an explanation that fits all the observations and is used to make predictions (for example, what will happen given a certain set of circumstances). In science, a theory is a well-established explanation of how the natural world works. For a scientific theory to exist, scientists have to be *very* sure about it. So, don't discount a scientific idea because it's "just a theory." As famed astrophysicist Neil deGrasse Tyson has stated about the validity of science, *"The good thing about science is that it's true whether or not you believe in it."*

**Students Sometimes Ask features** display common and often entertaining questions posed by real students, like "Why do my fingers get wrinkly when they are in the water for a long time?" and pose scientific explanations.

CREATURE FEATURE 2.1

### Can you help me find my way home?

The **green sea turtle** (*Chelonia mydas*) is not named for its external color, but for the greenish color of its body fat.

Green sea turtles spend many years living at sea in tropical and subtropical oceans but unerringly return to their place of birth to lay eggs on sandy beaches. How do they navigate so precisely? See **Process of Science 2.1**.

**NEW Creature Features** draw student interest by introducing compelling facts about marine organisms in an engaging "Who am I?" format.

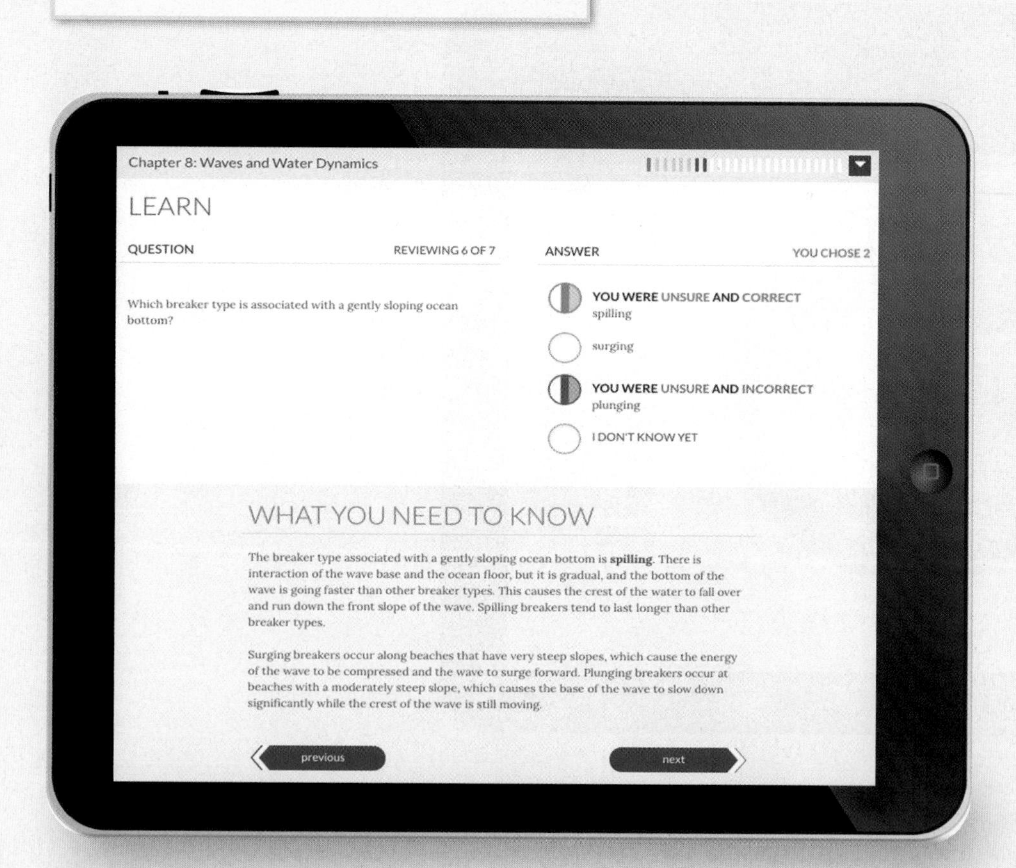


**NEW! Dynamic Study Modules** help students study effectively—and at their own pace—by keeping them motivated and engaged. The assignable modules rely on the latest research in cognitive science, using methods—such as adaptivity, gamification, and intermittent rewards—to stimulate learning and improve retention.

# with Tools that Enhance Learning

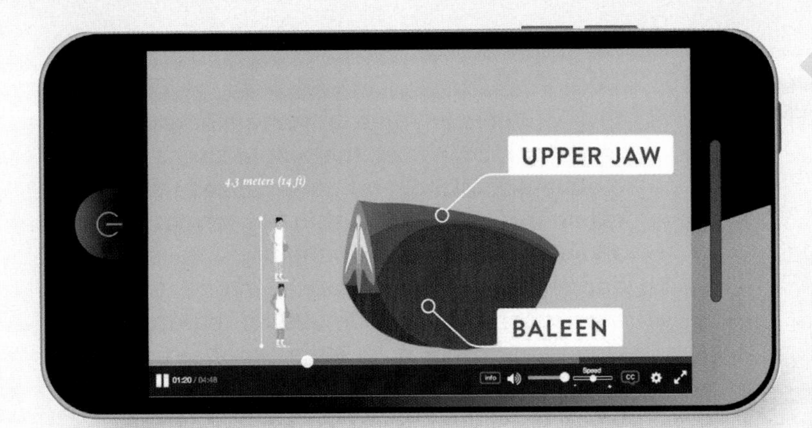


**Oceanography Animations** captivate students with animations that illustrate key concepts in a visually dynamic and engaging way.

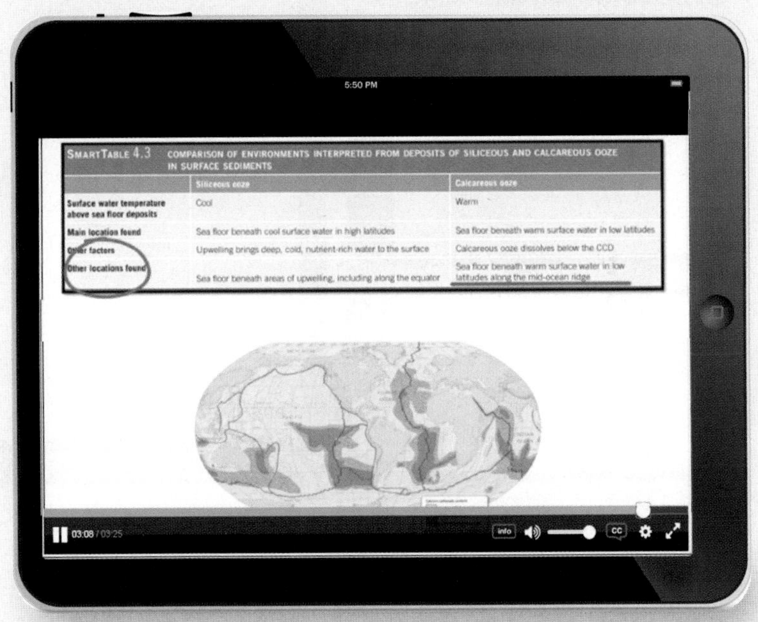

SmartTable 4.3 COMPARISON OF ENVIRONMENTS INTERPRETED FROM DEPOSITS OF SILICEOUS AND CALCAREOUS OOZE IN SURFACE SEDIMENTS

| | Siliceous ooze | Calcareous ooze |
|---|---|---|
| Surface water temperature above sea floor deposits | Cool | Warm |
| Main location found | Sea floor beneath cool surface water in high latitudes | Sea floor beneath warm surface water in low latitudes |
| Other factors | Upwelling brings deep, cold, nutrient-rich water to the surface | Calcareous ooze dissolves below the CCD |
| Other locations found | Sea floor beneath areas of upwelling, including along the equator | Sea floor beneath warm surface water in low latitudes along the mid-ocean ridge |



**SmartFigures** are 3- to 4-minute mini video lessons containing explanations of difficult-to-understand oceanographic concepts and numerical data directed by an oceanography teaching expert and NASA Science Communicator.

The **Student-Centric Approach** enables students to form a path to successful learning. There is a **Recap** feature throughout each chapter, summarizing essential concepts. **Critical Thinking Questions** and **Active Learning Exercises** encourage students to think deeply about and engage with chapter topics.

**ESSENTIAL LEARNING CONCEPTS**

At the end of this chapter, you should be able to:

- ☐ **3.1** Discuss the techniques that are used to determine ocean bathymetry.
- ☐ **3.2** Describe the sea floor features that exist on continental margins.
- ☐ **3.3** Describe the sea floor features that exist in the deep-ocean basins.
- ☐ **3.4** Describe the sea floor features that exist along the mid-ocean ridge.

**RECAP** **Sending pings of sound into the ocean (echo sounding) is a commonly used technique for determining ocean bathymetry. More recently, satellites are being used to map sea floor features.**

CONCEPT CHECK 3.1 ▸ **Discuss the techniques that are used to determine ocean bathymetry.**

1 What is bathymetry? How is it different from topography?

2 Describe how an echo sounder works.

3 Discuss the development of bathymetric techniques, indicating significant advancements in technology.

## ESSENTIAL CONCEPTS REVIEW

### 3.1 ▶ What techniques are used to determine ocean bathymetry?

- *Bathymetry is the measurement of ocean depths and the charting of ocean floor topography.* The varied bathymetry of the ocean floor was first determined using *soundings* to measure water depth. Later, the development of the *echo sounder* gave ocean scientists a more detailed representation of the sea floor.
- Today, much of our knowledge of the ocean floor has been obtained using various *multibeam echo sounders* or *side-scan sonar instruments* (to make detailed bathymetric maps of a small area of the ocean floor), *satellite measurement* of the ocean surface (to produce maps of the world ocean floor), and *seismic reflection profiles* (to examine Earth structure beneath the sea floor).

**Selected Key Terms**

Use the **glossary** at the end of this book to discover the meanings of these Selected Key Terms: **bathymetry, sounding, echo sounder, sonar, seismic reflection profile.**

**Critical Thinking Question**

Describe how satellite measurements of the ocean *surface* allow oceanographers to create a map of the sea *floor*.

**Active Learning Exercise**

Use the Internet to research how a "fish finder" works on modern sport-fishing boats. How do these techniques compare to the sonar techniques described in this textchapter?

# Dive into a Whole New Learning Experience with Pearson eText

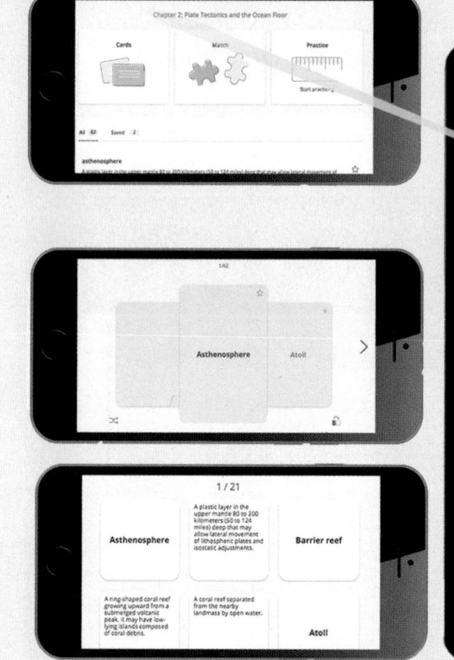


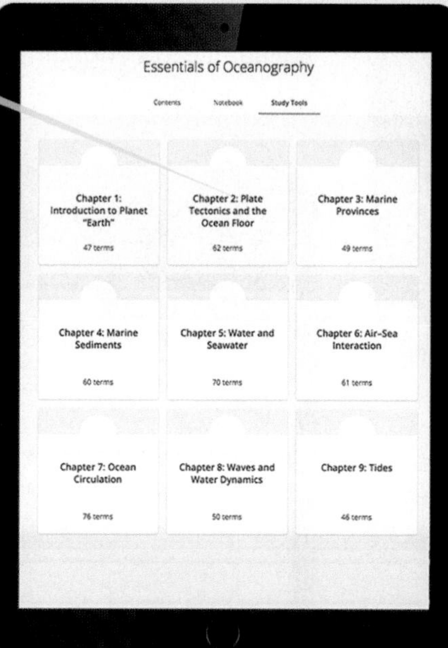


Give students anytime, anywhere access with **Pearson eText**, the simple-to-use, mobile-optimized, and personalized reading experience available within Mastering. It allows students to easily highlight, take notes, and review key vocabulary all in one place—even when offline. Seamlessly integrated videos and other rich media engage students and give them access to the help they need, when they need it.

**"I absolutely love the digital book that came with my main paper book, all the audio and video materials made my course so entertaining, and as a result A+ for the semester!"**

**Pearson eText for Oceanography** has hundreds of videos and animations that bring concepts to life. Instructors are able to highlight key concepts within the eText and share with their students to guide them through the reading and help them grasp key concepts.

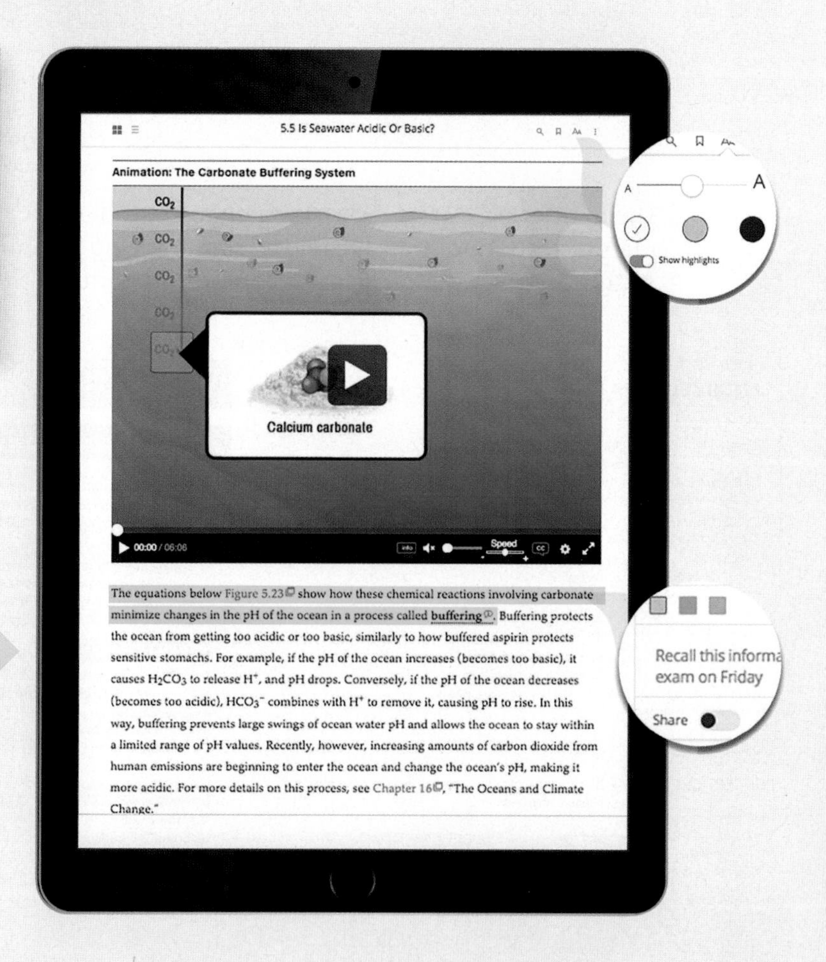

# Essentials of Oceanography

THIRTEENTH EDITION

**Alan P. Trujillo**

DISTINGUISHED TEACHING PROFESSOR
PALOMAR COLLEGE

**Harold V. Thurman**

FORMER PROFESSOR EMERITUS
MT. SAN ANTONIO COLLEGE

Courseware Portfolio Manager: Cady Owens
Director of Portfolio Management: Beth Wilbur
Content Producer: Melanie Field
Managing Producer: Mike Early
Courseware Director, Content Development: Ginnie Simione Jutson
Courseware Sr. Analyst: Barbara Price
Courseware Editorial Assistant: Sherry Wang
Rich Media Content Producer: Christine Hostetler and Mia Sullivan
Full-Service Vendor: Pearson CSC
Copyeditor: Carrie Bell-Hoerth
Compositor: Pearson CSC
Art Coordinator: Kevin Lear, International Mapping
Design Manager: Mark Ong
Interior and Cover Designer: Gary Hespenheide
Rights & Permissions Project Manager: Matt Perry
Rights & Permissions Management: Ben Ferrini
Photo Researcher: Kristin Piljay
Manufacturing Buyer: Stacey Weinberger
Product Marketing Manager: Alysun Burns
Cover Photo Credit: Getty Images/Jim Watt



**Library of Congress Cataloging-in-Publication Data**

Names: Trujillo, Alan P., author. | Thurman, Harold V., author.
Title: Essentials of oceanography / Alan P. Trujillo, Distinquished Teaching Professional, Palomar College, Harold V. Thurman, Former Professor Emeritus, Mt. San Antonio College.
Description: Thirteenth edition. | Hoboken, NJ : Pearson, [2020] | Includes index.
Identifiers: LCCN 2018046627| ISBN 9780134073545 (alk. paper) | ISBN 013489152X (alk. paper) | ISBN 0134251946 | ISBN 0135257581 | ISBN 9780134891521 (student edition) | ISBN 9780134251943 (instructor's review copy) | ISBN 9780135257586 (NASTA)
Subjects: LCSH: Oceanography--Textbooks.
Classification: LCC GC11.2 .T49 2020 | DDC 551.46--dc23 LC record available at https://lccn.loc.gov/2018046627

## Cover Photo Description

**A pod of dolphins swims off the Hawaiian coast.** A pod of pantropical spotted dolphin (*Stenella attenuata*) swims in shallow water off the northern coast of the Hawaiian island of Kauai. This species of dolphin is often found in deeper water throughout the world's temperate and tropical oceans swimming above schools of tuna and as a result were often caught in purse seine nets as bycatch by the tuna fishing industry, which is discussed in Chapter 13, "Biological Productivity and Energy Transfer." In the 1980s, the rise of "dolphin-friendly" tuna capture methods saved millions of these marine mammals in the eastern Pacific Ocean and it is now one of the most abundant dolphin species in the world.

ISBN 10: **0-134-89152-X;** ISBN 13: **978-0-134-89152-1** (Student edition)
ISBN 10: **0-135-20430-5;** ISBN 13: **978-0-135-20430-6** (Loose-Leaf Edition)
ISBN 10: **0-135-25758-1;** ISBN 13: **978-0-135-25758-6** (NASTA)

**www.pearson.com**

# ABOUT THE AUTHORS

**ALAN P. TRUJILLO** Al Trujillo is a Distinguished Teaching Professor in the Earth, Space, and Environmental Sciences Department at Palomar College in San Marcos, California. He received his bachelor's degree in geology from the University of California at Davis and his master's degree in geology from Northern Arizona University, afterward working for several years in industry. Al began teaching at Palomar in 1990. In 1997, he was awarded Palomar's Distinguished Faculty Award for Excellence in Teaching, and in 2005 he received Palomar's Faculty Research Award. He has coauthored *Introductory Oceanography* with Hal Thurman and is a contributing author for the textbooks *Earth* and *Earth Science*. In addition to writing and teaching, Al works as a naturalist and lecturer aboard Holland America Line and natural history expedition vessels for Lindblad Expeditions/National Geographic in Alaska, Iceland, the Sea of Cortez/Baja California, Central/South America, and across the South Pacific Ocean. His research interests include beach processes, sea cliff erosion, and active teaching techniques. He enjoys photography, and he collects sand as a hobby. Al and his wife, Sandy, have two children, Karl and Eva.

**HAROLD V. THURMAN** Hal Thurman's interest in geology led to a bachelor's degree from Oklahoma A&M University, followed by seven years working as a petroleum geologist, mainly in the Gulf of Mexico, where his interest in the oceans developed. He earned a master's degree from California State University at Los Angeles. Hal began teaching at Mt. San Antonio College in Walnut, California, in 1968 as a temporary teacher and taught Physics 1 (a surveying class) and three Physical Geology labs. In 1970, he taught his first class of General Oceanography. It was from this experience that he decided to write a textbook on oceanography and received a contract with Charles E. Merrill Publishing Company in 1973. The first edition of his book *Introductory Oceanography* was released in 1975. Harold authored or coauthored over 20 editions of textbooks that include *Introductory Oceanography*, *Essentials of Oceanography*, *Physical Geology*, *Marine Biology*, and *Oceanography Laboratory Manual*, many of which are still being used today throughout the world. In addition, he contributed to the *World Book Encyclopedia* on the topics of "Arctic Ocean," "Atlantic Ocean," "Indian Ocean," and "Pacific Ocean." Hal Thurman retired in May 1994, after 24 years of teaching, and moved to be closer to family in Oklahoma, then to Florida. Hal passed away at the age of 78 on December 29, 2012. His writing expertise, knowledge about the oceans, and easygoing demeanor are dearly missed.

Dedicated to my wife Sandy, who has taken care of me through thick and thin

—AL TRUJILLO

In memory of Dr. Anthony Trujillo (1926–2017)

# CONTENTS

## BONUS ONLINE CONTENT

# PREFACE

*"The sea, once it casts its spell, holds one in its net of wonder forever."*

—Jacques-Yves Cousteau, oceanographer, underwater videographer, and explorer (circa 1963)

## To the Student

Welcome! You're about to embark on a journey that is far from ordinary. Over the course of this term, you will discover the central role the oceans play in the vast global system, of which you are a part.

This book's content was carefully developed to provide a foundation in science by examining the vast body of oceanic knowledge. This knowledge includes information from a variety of scientific disciplines—geology, chemistry, physics, and biology—as they relate to the oceans. However, no formal background in any of these disciplines is required to successfully master the subject matter contained within this book. Our desire is to have you take away from your oceanography course much more than just a collection of facts. Instead, we want you to develop a fundamental understanding of how the oceans work and why the oceans behave the way that they do.

This book is intended to help you in your quest to know more about the oceans. Taken as a whole, the components of the ocean—its sea floor, chemical constituents, physical components, and life-forms—comprise one of Earth's largest interacting, interrelated, and interdependent systems. Because human activities impact Earth systems, it is important to understand not only how the oceans operate but also how the oceans interact with Earth's other systems (such as its atmosphere, biosphere, and hydrosphere) as part of a larger picture. Thus, this book uses a systems approach to highlight the interdisciplinary relationships among oceanographic phenomena and how those phenomena affect other Earth systems.

### DIVING DEEPER PREFACE.1

### A USER'S GUIDE FOR STUDENTS: HOW TO READ A SCIENCE TEXTBOOK

Have you known someone who could scan a reading assignment or sleep with it under their pillow and somehow absorb all the information? Studies have shown that those people haven't really committed anything to long-term memory. For most of us, it takes a focused, concentrated effort to gain knowledge through reading. Interestingly, if you have the proper motivation and reading techniques, you can develop excellent reading comprehension. What is the best way to read a science textbook such as this one that contains many new and unfamiliar terms?

One common mistake is to approach reading a science textbook as one would read a newspaper, magazine, or novel. Instead, many reading instructors suggest using the SQ4R reading technique, which is based on research about how the brain learns. The SQ4R technique includes these steps:

1. **Survey:** Read the title, introduction, major headings, first sentences, concept statements, review questions, summary, and study aids to become familiar with the content in advance.
2. **Question:** Have questions in mind when you read. If you can't think of any good questions, use the chapter questions as a guide.
3. **Read:** Read flexibly through the chapter, using short time periods to accomplish the task one section at a time (not all in one sitting).
4. **Recite:** Answer the chapter questions. Take notes after each section and review your notes before you move on.
5. **(w)Rite:** Write summaries and/or reflections on what you've read. Write answers to the questions in Step 2.
6. **Review:** Review the text using the strategy in the survey step. Take the time to review your end-of-section notes as well as your summaries.

To help you study most effectively, this textbook includes many study aids that are designed to be used with the SQ4R technique. For example, each chapter includes a list of learning objectives that are tied to the Essential Concepts throughout the chapter; review Concept Check questions embedded at the end of each section; and an Essential Concepts Review that includes a chapter summary, study resources, and critical thinking questions.

Here are some additional reading tips that may seem like common sense and are based on brain-based research, but are often overlooked:

- Don't attempt to do your reading when you are tired, distracted, or agitated.
- Break up your reading into manageable sections. Don't save it all until the last minute.
- Take a short break if your concentration begins to fade. Listen to music, call a friend, have a snack, or drink some water. Then return to your reading.

Remember that every person is different, so experiment with new study techniques to discover what works best for you. In addition, being a successful student is hard work; it is not something one does in his/her spare time. With a little effort in applying the SQ4R reading technique, you will begin to see a difference in what you remember from your reading.

To that end—and to help you make the most of your study time—we focused the presentation in this book by organizing the material around three essential components:

1. **CONCEPTS:** General ideas derived or inferred from specific instances or occurrences (for instance, the concept of density can be used to explain why the oceans are layered)
2. **PROCESSES:** Actions or occurrences that bring about a result (for instance, the process of waves breaking at an angle to the shore results in the movement of sediment along the shoreline)
3. **PRINCIPLES:** Rules or laws concerning the functioning of natural phenomena or mechanical processes (for instance, the principle of sea floor spreading suggests that the geographic positions of the continents have changed through time)

Interwoven within these concepts, processes, and principles are hundreds of photographs, illustrations, real-world examples, and applications that make the material relevant and accessible (and maybe sometimes even entertaining) by bringing science to life.

Ultimately, it is our hope that by understanding how the oceans work, you will develop a new awareness and appreciation of all aspects of the marine environment and its role in Earth systems. To this end, the book has been written for you, a student of the oceans. So enjoy and immerse yourself! You're in for an exciting ride.

—**Al Trujillo**

## To the Instructor

This thirteenth edition of *Essentials of Oceanography* is designed to accompany an introductory college-level course in oceanography taught to students who have no formal background in mathematics or science. As in previous editions, the goal of this edition of the textbook is to clearly present the relationships of scientific principles to ocean phenomena in an engaging and meaningful way. In addition, the content of this book is carefully designed to help students engage with and learn oceanographic material.

This edition has greatly benefited from being thoroughly reviewed by hundreds of students who made numerous suggestions for improvement. Comments by former students about the book include, *"I have really enjoyed the oceanography book we've used this semester. It had just the right mix of graphics, text, and user-friendliness that really held my interest;" "I really liked the videos embedded in the daily chapter quizzes, particularly the SmartFigures done by Laura Faye Tenenbaum. I loved her delivery. Her style helped me understand some complex topics and just made it really digestible. She's so bright and her humor came through just in the right way, kept it lively;"* and *"What I really liked about the book is that it's a welcoming textbook—open and airy. You could almost read it at bedtime like a story because of all the interesting pictures."*

This edition has been reviewed in detail by a host of instructors from leading institutions across the country. Reviewers of the twelfth edition described the text as follows: *"Essentials of Oceanography is a great textbook to introduce oceanography to non-science majors, and it has a lot of great supplemental materials for you and your students;" "Students find it easily understandable. The writing and graphics are excellent; easy to comprehend and remember;" "Your book is truly wonderful. Your writing voice is so excellent, and the fact that you have included so many etymological roots of terms is a real memory aid for students. I'm always stressing to them how, among other things, science is a language, and your book is right in groove with that;"* and *"An excellent introductory oceanography textbook that can be used for courses from two to four credit hours. Easily read, flows well through the chapters and from chapter-to-chapter. Many helpful aids for students as well as ancillaries for instructors. It makes our job easier, and students are happy because they can understand the topics well, leading to higher average grades."*

In 2012, the tenth edition of *Essentials of Oceanography* received a Textbook Excellence Award, called a "Texty," from the Text and Academic Authors Association (TAA). The Texty award recognizes written works for their excellence in the areas of content, presentation, appeal, and teachability. The publisher, Pearson Education, nominated the book for the award, and the textbook was critically reviewed by a panel of expert judges. In 2017, the twelfth edition of *Essentials of Oceanography* received TAA's McGuffey Longevity Award for its long-standing history of publication.

The 16-chapter format of this textbook is designed for easy coverage of the material in a 15- or 16-week semester. For courses taught on a 10-week quarter system, instructors may need to select those chapters that cover the topics and concepts of primary relevance to their course. Chapters are self-contained and can thus be covered in any order. Following the introductory chapter (Chapter 1, which covers the general geography of the oceans; a historical perspective of oceanography; the method behind the process of science; and a discussion of the origin of Earth, the atmosphere, the oceans, and life itself), the four major academic disciplines of oceanography are represented in the following chapters:

- Geological oceanography (Chapters 2–4 and Chapter 10)
- Chemical oceanography (Chapter 5 and Chapter 11)
- Physical oceanography (Chapters 6–9)
- Biological oceanography (Chapters 12–15)
- Interdisciplinary oceanography: Climate change (Chapter 16)

We strongly believe that oceanography is at its best when it links together several scientific disciplines and shows how they are interrelated in the oceans. Therefore, this interdisciplinary approach is a key element of every chapter, particularly Chapter 16, "The Oceans and Climate Change."

## What's New in This Edition?

Changes in this edition are designed to increase the readability, relevance, and appeal of this book. Major changes include the following:

- An emphasis on the process of science, including a new "Process of Science" boxed feature in most chapters that illustrates the scientific method by highlighting an area of oceanographic inquiry and explicitly pointing out how the process of science was used in that particular case; each feature also includes a critical thinking assessment question "Thinking Like a Scientist: What's Next?" so that students gain practice approaching problems scientifically and analytically
- New "Exploring Data" questions added to every chapter; this new feature directs students to engage with data and checks their

## DIVING DEEPER PREFACE.2

## OCEAN LITERACY: WHAT SHOULD PEOPLE KNOW ABOUT THE OCEAN?

The ocean is the defining feature of our planet. Accordingly, there is great interest in developing *ocean literacy*, which means understanding the ocean's influence on humans as well as humans' influence on the ocean. For example, scientists and educators agree that an ocean-literate person:

- Understands the essential principles and fundamental concepts about the functioning of the ocean.
- Can communicate about the ocean in a meaningful way.
- Is able to make informed and responsible decisions regarding the ocean and its resources.

To achieve this goal, ocean educators and experts have developed the **Seven Principles of Ocean Literacy**. The following ideas are what everyone—especially those who successfully pass a college course in oceanography or marine science—should understand about the ocean:

1. Earth has one big ocean with many features.
2. The ocean and life in the ocean shape the features of Earth.
3. The ocean is a major influence on weather and climate.
4. The ocean makes Earth habitable.
5. The ocean supports a great diversity of life and ecosystems.
6. The ocean and humans are inextricably interconnected.
7. The ocean is largely unexplored.

This book is intended to help all people achieve ocean literacy. For more information about the Seven Principles of Ocean Literacy, see **http://oceanliteracy.wp2.coexploration.org/**

understanding by asking data interpretation questions related to data-rich figures, graphs, tables, and maps

- The addition in all chapters of a new "Creature Feature," which uses compelling facts about a marine organism to reinforce the theme of the chapter. Each "Creature Feature's" title is written in an engaging "Who Am I?" format to draw student interest
- Expansion of the discussion of carbon and oxygen in the ocean in Chapter 5, "Water and Seawater," which includes explanation of how the distribution of dissolved gases and pH changes with depth, and their significance
- A thoroughly updated Chapter 16 "The Oceans and Climate Change," introducing a new discussion about the carbon cycle, and describing the most recent findings of the IPCC; the rewrite includes highlights of the 2017 *Climate Science Special Report: Fourth National Climate Assessment*, which was produced at the behest of the U.S. Congress to provide an assessment of the state of science relating to climate change and its physical impacts in the United States; also included in this chapter is a new review of solutions to human-caused greenhouse gas emissions in the atmosphere, and four new or revised "Students Sometimes Ask . . . " questions that address student misconceptions and concerns regarding climate change
- Greater emphasis on the ocean's role in Earth systems
- A stronger learning path that directly links the learning objectives listed at the beginning of each chapter to the end-of-section "Concept Checks," which allow and encourage students to pause and test their knowledge as they proceed through the chapter
- A new active learning pedagogy that divides chapter material into easily digestible chunks, which makes studying easier and assists student learning (cognitive science research shows that the ability to "chunk" information is essential to enhancing learning and memory)
- The inclusion of an array of new SmartFigures and SmartTables, which provide a video explanation of difficult-to-understand oceanographic concepts and numerical data by an oceanography teaching expert
- The addition of one or more "What Did You Learn?" assessment questions to each "Diving Deeper" boxed feature
- Removal of all footnotes; pertinent information from previous footnotes is now contained within the body of the text
- Migration of each chapter's Squidtoons call-out to Mastering Oceanography Study Area as Bonus Web Content
- In all Essential Concept Review (end-of-chapter) materials, the revision of existing "Critical Thinking Questions" and "Active Learning Exercise" questions that can be used for in-class group activities
- The addition of a new "Selected Key Terms" feature in each section's end-of-chapter box that simplifies and replaces the word cloud formerly at the beginning of each chapter and directs students to the glossary at the end of the book to discover the meanings of the most important vocabulary terms that are boldfaced in each section of the text
- Updating of information throughout the text to include technological advances that have resulted in the modernization of oceanographic research and continue to shape the discipline today; for example, space-based oceanographic and atmospheric observations from NASA Earth-observing satellite missions
- Addition of an array of new "Students Sometimes Ask . . . " questions throughout the book
- An enhanced illustration package showcasing new photos, satellite images, and figures to make oceanographic topics more accessible, current, and engaging
- The revision or updating of over half of existing figures and incorporating annotations and labels within key figures that direct student attention and help explain information in storyboard form; this research-proven technique helps students focus on the most relevant information, interpret complex art, and integrate written and visual information

- Standardization of the color scheme and labeling of all figures to make them more appealing and consistent throughout
- Inclusion of more than 70 Web Animations from Pearson's Geoscience Animations Library, which include state-of-the-art computer animations that have been created by Al Trujillo and a panel of geoscience educators
- An enhanced eText, which allows students to review previously learned material with a single click that will place this content side-by-side the page they are currently studying
- Selected Diving Deeper feature boxes have been migrated online to Mastering Oceanography as Bonus Web Content in an effort to reduce the length of the text
- The remaining Diving Deeper features appearing in the book are organized around the following four themes:
  - **HISTORICAL FEATURES**, which focus on historical developments in oceanography that tie into chapter topics
  - **RESEARCH METHODS IN OCEANOGRAPHY**, which highlight how oceanographic knowledge is obtained
  - **OCEANS AND PEOPLE**, which illustrate the interaction of humans and the ocean environment
  - **FOCUS ON THE ENVIRONMENT**, which emphasizes environmental issues that are an increasingly important component of ocean studies
- The former Afterword has been shortened to one page; information about Marine Protected Areas (MPAs) has been moved to Chapter 13 and information about what individuals can do to minimize human impact on the oceans (including former Diving Deeper Aft.1) has been moved to Chapter 16
- All text in the chapters has been thoroughly reviewed and edited by students and oceanography instructors in a continued effort to refine the style and clarity of the writing

Note that a detailed list of specific chapter-by-chapter changes is available at **https://www2.palomar.edu/pages/atrujillo/**

In addition, this edition continues to offer some of the previous edition's most popular features, including the following:

- Scientifically accurate and thorough coverage of oceanography topics
- A series of SmartFigures and SmartTables, which maximize instructional value of the media and help students learn important content
- "Students Sometimes Ask . . . " questions, which present actual student questions along with the authors' answers
- A "Recap" feature that summarizes key points throughout the text, making studying easier
- The continuation of existing "Critical Thinking Questions" and "Active Learning Exercise" questions that can be used for group activities in class in all Essential Concept Review (end-of-chapter) materials
- QR codes embedded in the text that allow students to use their mobile devices to link directly to Mastering Oceanography Animations, SmartFigures and SmartTables, and Web Videos
- QR codes and links to more than 50 hand-picked Web videos that show important oceanographic processes in action
- Use of the international metric system (Système International [SI] units), with comparable English system units in parentheses
- Explanation of word etymons (*etumon* = sense of a word) as new terms are introduced, in an effort to demystify scientific terms by showing what the terms actually mean
- A "Climate Change Connection" icon that alerts students to topics that are related to the overarching theme of global climate change
- Use of **bold print** on key terms, which are defined when they are introduced and are described in the glossary
- A reorganized "Essential Concepts Review" summary at the end of each chapter
- **Mastering Oceanography**, which features chapter-specific Self Study Quizzes, SmartFigures and SmartTables, Oceanography Videos and Animations, Squidtoons, Dynamic Study Modules, and an optional Pearson eText with embedded videos.

## For the Student

- **MASTERING OCEANOGRAPHY** delivers engaging, dynamic learning opportunities—focused on course objectives and responsive to each student's progress—that are proven to help students absorb course material and understand difficult concepts. Mastering Oceanography is a customized learning resource that includes:
  - **Student Study Area**, which is designed to be a one-stop resource for students to acquire study help and serve as a launching pad for further exploration. Content for the site was written by author Al Trujillo and is tied, chapter-by-chapter, to the text. The Student Study Area is organized around a four-step learning pathway:
    1. *Review*, which contains **Essential Concepts** as learning objectives
    2. *Read*, which contains the **eText** and **Bonus Web Content**
    3. *Visualize*, which contains Oceanography Animations, Oceanography Videos, and Smart Figures.
    4. *Test Yourself*, which contains a **Chapter Quiz** that is automatically graded for instant feedback.
  - **Study Tools** such as flashcards and a searchable online glossary to help make the most of students' study time
- **THE PEARSON eTEXT** gives students complete access to a digital version of the text whenever and wherever they have access to the Internet.

## For the Instructor

- **MASTERING OCEANOGRAPHY: CONTINUOUS LEARNING BEFORE, DURING, AND AFTER CLASS** Mastering Oceanography is an online homework, tutorials, and assessments program designed to improve results by helping students quickly master oceanography concepts. Students will benefit from self-paced tutorials that feature immediate wrong-answer feedback and hints that emulate the office-hour experience to help keep them on track. With a wide range of interactive, engaging, and assignable activities, students will be encouraged to actively learn and retain tough course concepts:

- **New Process of Science Coaching Activities** support the text feature that highlights an area of oceanographic inquiry and explicitly point out the associated background, method, and conclusion.
- **New Exploring Data activities** help students actively engage with graphs and other data-driven features and their data interpretation skills.
- **SmartFigures/SmartTables**, which are three- to four-minute mini-lessons that examine and explain the concepts illustrated by a figure or table. Over 90 SmartFigures/SmartTables are assignable in **Mastering**.
- **Oceanography Animations,** which illuminate the most difficult-to-understand topics in oceanography and were created by an expert team of geoscience educators. The animation activities include audio narration, a text transcript, and assignable multiple-choice questions with specific wrong-answer feedback.
- **Video Field Trips** give students fascinating behind-the-scenes experiences at prescribed fire burns, solar energy, and coal-fired power plants, wastewater treatment facilities, landfills, farms, and more. Each Video Field Trip includes assessment questions for easily assignable homework.
- **Visualizing Oceanography Activities** ask students to label art from the text to ensure they are interpreting and understanding figures.
- **Dynamic Study Modules**, which help students study effectively on their own by continuously assessing their activity and performance in real time. Here's how it works: Students complete a set of questions with a unique answer format that also asks them to indicate their confidence level. Questions repeat until the student can answer them all correctly and confidently. Once completed, Dynamic Study Modules explain the concept using materials from the text. These are available as graded assignments prior to class, and accessible on smartphones, tablets, and computers.
- **Learning Catalytics™**, which are an interactive student response tool that uses students' smartphones, tablets, or laptops to engage them in more sophisticated tasks and thinking. Now included with MyLab & Mastering and eText, Learning Catalytics™ enables you to generate classroom discussion, guide your lecture, and promote peer-to-peer learning with real-time analytics.

- **STUDENT PERFORMANCE ANALYTICS** Mastering Oceanography allows an instructor to gain easy access to information about student performance and their ability to meet student learning outcomes. Instructors can quickly add their own learning outcomes, or use publisher-provided ones, to track student performance.
- **INSTRUCTOR MANUAL (DOWNLOAD ONLY)** This resource contains learning objectives, chapter outlines, answers to embedded end-of-section questions, and suggested teaching tips to spice up your lectures.
- **TESTGEN® COMPUTERIZED TEST BANK (DOWNLOAD ONLY)** This resource is a computerized test generator that lets instructors view and edit *Test Bank* questions, transfer questions to tests, and print the test in a variety of customized formats. The *Test Bank* includes over 1200 multiple-choice, matching, and short-answer/essay questions. All questions are tied to the chapter's learning outcomes, include a rating based on Bloom's taxonomy of learning domains (Bloom's 1–6) and contain the section number in which each question's answer can be found.
- **INSTRUCTOR POWERPOINT® PRESENTATIONS (DOWNLOAD ONLY)** Instructor Resource Materials include the following three PowerPoint® files for each chapter so that you can cut down on your preparation time, no matter what your lecture needs:

    1. **EXCLUSIVELY ART:** This file provides all the photos, art, and tables from the text, in order, loaded into PowerPoint® slides.
    2. **LECTURE OUTLINE:** This file averages 50 PowerPoint® slides per chapter and includes customizable lecture outlines with supporting art.
    3. **CLASSROOM RESPONSE SYSTEM (CRS) QUESTIONS:** Authored for use in conjunction with classroom response systems, this PowerPoint® file allows you to electronically poll your class for responses to questions, pop quizzes, attendance, and more.

For more information about these instructor resources, contact your Pearson textbook representative.

## Acknowledgments

Al Trujillo is indebted to many individuals for their helpful comments and suggestions during the revision of this book. I am particularly indebted to Courseware Senior Analyst Dr. Barbara Price of Pearson Education for her encouragement, ideas, and tireless advocacy that she provided to improve the book. It was a pleasure having such a wonderful colleague to work with during the long journey of writing this book.

Many people were instrumental in helping the text evolve from its manuscript stage. My chief liaison at Pearson Education, Courseware Portfolio Manager Cady Owens, suggested many of the new ideas in the book to make it more student-friendly and expertly guided the project. The copy editor at SPi Global did a superb job of editing the manuscript, catching many English and other grammar errors, including obscure errors that had persisted throughout several previous editions. Courseware Director, Content Development Ginnie Jutson kept the book on track by making sure deadlines were met along the way and facilitated the distribution of various versions of the manuscripts. Content Producer for media Christine Hostetler helped create the electronic supplements that accompany this book, including Mastering Oceanography and all of its outstanding features. The art studio International Mapping, and in particular Kevin Lear, did a beautiful job of updating various maps and coming up with creative solutions to improve many of the figures that help tell the story of the content through the art. In addition, Norine Strang coordinated the art house as well as the compositor, who develops the page layout. Art Development Editor Jay McElroy's creative vision helped us develop new illustrations for this edition. The artful design elements of the text, including its color scheme, text wrapping, and end-of-chapter features, was developed by text designer Gary Hespenheide in conjunction with Pearson's Design Manager Mark Ong. New photos were researched and secured by Photo Researcher Kristin Piljay. Last but not least, Content Producer Melanie Field deserves special recognition for her persistence and encouragement during the many long hours of turning the manuscript into the book you see today.

Al Trujillo thanks his students, whose questions provided the material for the "Students Sometimes Ask . . . " sections and whose continued input has proved invaluable for improving the text. Because scientists (and all good teachers) are always experimenting, thanks also for allowing yourselves to be a captive audience with which to conduct my experiments.

Al Trujillo also thanks his patient and understanding family for putting up with his absence during the long hours of preparing "The Book." Finally, appreciation is extended to the chocolate manufacturers Hershey, See's, and Ghirardelli, for providing inspiration. A heartfelt thanks to all of you!

Many other individuals (including dozens of anonymous reviewers) have provided valuable technical reviews for this and previous works. The following reviewers are gratefully acknowledged:

Patty Anderson, *Scripps Institution of Oceanography*; Shirley Baker, *University of Florida*; William Balsam, *University of Texas at Arlington*; Tsing Bardin, *City College of San Francisco*; Tony Barros, *Miami-Dade Community College*; Mark Baskaran, *Wayne State University*; Steven Benham, *Pacific Lutheran University*; Lori Bettison-Varga, *College of Wooster*; Thomas Bianchi, *Tulane University*; David Black, *University of Akron*; Mark Boryta, *Consumnes River College*; Laurie Brown, *University of Massachusetts*; Kathleen Browne, *Rider University*; Jonathan Bryan, *Northwest Florida State College*; Aurora Burd, *Green River Community College*; Nancy Bushell, *Kauai Community College*; Chatham Callan, *Hawaii Pacific University*; Mark Chiappone, *Miami-Dade College–Homestead Campus*; Chris Cirmo, *State University of New York, Cortland*; G. Kent Colbath, *Cerritos Community College*; Thomas Cramer, *Brookdale Community College*; Richard Crooker, *Kutztown University*; Cynthia Cudaback, *North Carolina State University*; Warren Currie, *Ohio University*; Hans Dam, *University of Connecticut*; Dan Deocampo, *California State University, Sacramento*; Jean DeSaix, *University of North Carolina at Chapel Hill*; Richard Dixon, *Texas State University*; Holly Dodson, *Sierra College*; Joachim Dorsch, *St. Louis Community College*; Wallace Drexler, *Shippensburg University*; Walter Dudley, *University of Hawaii*; Iver Duedall, *Florida Institute of Technology*; Debra Duffy, *Tidewater Community College*; Jennifer Duncan, *Palomar College*; Jessica Dutton, *Adelphi University*; Charles Ebert, *State University of New York, Buffalo*; Ted Eckmann, *University of Portland*; Charles Epifanio, *University of Delaware*; Jiasong Fang, *Hawaii Pacific University*; Diego Figueroa, *Florida State University*; Kenneth Finger, *Irvine Valley College*; Catrina Frey, *Broward College*; Jessica Garza, *MiraCosta College*; Sarah Gerken, *University of Alaska, Anchorage*; Benjamin Giese, *Texas A&M University*; Cari Gomes, *MiraCosta College*; Dave Gosse, *University of Virginia*; Carla Grandy, *City College of San Francisco*; John Griffin, *University of Nebraska, Lincoln*; Elizabeth Griffith, *University of Texas at Arlington*; Gary Griggs, *University of California, Santa Cruz*; Ingrid Hendy, *University of Michigan, Ann Arbor*; Amy Hirons, *Nova Southeastern University*; Joseph Holliday, *El Camino Community College*; Mary Anne Holmes, *University of Nebraska, Lincoln*; Timothy Horner, *California State University, Sacramento*; Alan Jacobs, *Youngstown State University*; Rozalind Jester, *Florida SouthWestern State College*; Ron Johnson, *Old Dominion University*; Uwe Richard Kackstaetter, *Metropolitan State University of Denver*; Charlotte Kelchner, *Oakton Community College*; Yong Hoon Kim, *West Chester University of Pennsylvania*; Matthew Kleban, *New York University*; Jessica Kleiss, *Lewis & Clark College*; Eryn Klosko, *State University of New York, Westchester Community College*; Ernest Knowles, *North Carolina State University*; M. John Kocurko, *Midwestern State University*; Lawrence Krissek, *Ohio State University*; Jason Krumholz, *NOAA/University of Rhode Island*; Paul LaRock, *Louisiana State University*; Gary Lash, *State University of New York, Fredonia*; Richard Laws, *University of North Carolina*; Richard Little, *Greenfield Community College*; Stephen Macko, *University of Virginia, Charlottesville*; Chris Marone, *Pennsylvania State University*; Jonathan McKenzie, *Florida SouthWestern State College*; Matthew McMackin, *San Jose State University*; James McWhorter, *Miami-Dade Community College*; Gregory Mead, *University of Florida*; Keith Meldahl, *MiraCosta College*; Nancy Mesner, *Utah State University*; Chris Metzler, *MiraCosta College*; Frank Millero, *University of Miami*; Johnnie Moore, *University of Montana*; P. Graham Mortyn, *California State University, Fresno*; Joy Moses-Hall, *East Carolina University*; Andrew Muller, *Millersville University*; Andrew Muller, *Utah State University*; Daniel Murphy, *Eastfield College*; Jay Muza, *Florida Atlantic University*; Jennifer Nelson, *Indiana University–Purdue University at Indianapolis*; Jim Noyes, *El Camino Community College*; Sarah O'Malley, *Maine Maritime Academy*; B. L. Oostdam, *Millersville University*; William Orr, *University of Oregon*; Joseph Osborn, *Century College*; Angela Osen, *Tarrant County College, Northwest*; Donald Palmer, *Kent State University*; Nancy Penncavage, *Suffolk County Community College*; Curt Peterson, *Portland State University*; Adam Petrusek, *Charles University, Prague, Czech Republic*; Edward Ponto, *Onondaga Community College*; Donald Reed, *San Jose State University*; Randal Reed, *Shasta College*; Robert Regis, *Northern Michigan University*; Erin Rempala, *San Diego City College*; M. Hassan Rezaie Boroon, *California State University, Los Angeles*; Cathryn Rhodes, *University of California, Davis*; James Rine, *University of South Carolina*; Felix Rizk, *Manatee Community College*; Angel Rodriguez, *Broward College*; Sarah Schliemann. *Metropolitan State University of Denver*; Diane Shepherd, *Shepherd Veterinary Clinic, Hawaii*; Beth Simmons, *Metropolitan State College of Denver*; Jill Singer, *State University of New York, Buffalo*; Michael Slattery, *University of Tampa*; Arthur Snoke, *Virginia Polytechnic Institute*; Pamela Stephens, *Midwestern State University*; Dean Stockwell, *University of Alaska, Fairbanks*; Scott Stone, *Fairfax High School, Virginia*; Lenore Tedesco, *Indiana University–Purdue University at Indianapolis*; Shelly Thompson, *West High School*; Craig Tobias, *University of North Carolina, Wilmington*; M. Craig Van Boskirk, *Florida State College at Jacksonville*; Paul Vincent, *Oregon State University*; George Voulgaris, *University of South Carolina*; Bess Ward, *Princeton University*; Jackie Watkins, *Midwestern State University*; Jamieson Webb, *Gulf Coast State College*; Arthur Wegweiser, *Edinboro University of Pennsylvania*; Diana Wenzel, *Seminole State College of Florida*; John White, *Louisiana State University*; Katryn Wiese, *City College of San Francisco*; Raymond Wiggers, *College of Lake County*; John Wormuth, *Texas A&M University*; Memorie Yasuda, *Scripps Institution of Oceanography*

Although this book has benefited from careful review by many individuals, the accuracy of the information rests with the authors. If you find errors or have comments about the text, please contact me.

Al Trujillo
Department of Earth, Space, and Environmental Sciences
Palomar College
1140 W. Mission Rd.
San Marcos, CA 92069
**atrujillo@palomar.edu**
**https://www2.palomar.edu/pages/atrujillo/**

*"If there is magic on this planet, it is contained in water."*
—Loren Eiseley, American educator and natural science writer (1907–1977)

# SmartFigures

# SmartTables

# Introduction To Planet "Earth"

1

## ESSENTIAL LEARNING CONCEPTS

At the end of this chapter, you should be able to:

- ☐ **1.1** Compare the characteristics of Earth's oceans.
- ☐ **1.2** Discuss how early exploration of the oceans was achieved.
- ☐ **1.3** Explain why oceanography is considered an interdisciplinary science.
- ☐ **1.4** Describe the process of science and the nature of scientific inquiry.
- ☐ **1.5** Explain how Earth and the solar system formed.
- ☐ **1.6** Explain how Earth's atmosphere and oceans formed.
- ☐ **1.7** Discuss why life is thought to have originated in the oceans.
- ☐ **1.8** Demonstrate an understanding of how old Earth is.

↑ *Check when completed*

The **oceans**[1] are the largest and most prominent feature on Earth. In fact, they are the single most defining feature of our planet. As viewed from space, our planet is a beautiful blue, white, and brown globe (see this chapter's opening photo). The abundance of liquid water on Earth's surface is a distinguishing characteristic of our home planet, and hidden under the ocean's surface is a submarine landscape that rivals anything on dry land.

So it seems perplexing that our planet is called "Earth" when 70.8% of its surface is covered by oceans. Many early human cultures that lived near the Mediterranean (*medi* = middle, *terra* = land) Sea envisioned the world as being composed of large landmasses surrounded by marginal bodies of water. From their viewpoint, landmasses—not oceans—dominated the surface of Earth. How surprised they must have been when they ventured into the larger oceans of the world. Our planet is misnamed "Earth" because we live on the land portion of the planet. If we were marine animals, our planet would probably be called "Ocean," "Water," "Hydro," "Aqua," or even "Oceanus" to indicate the prominence of Earth's oceans. Let's begin our study of the oceans by examining some of the unique geographic characteristics of our watery world.

## 1.1 ▸ How Are Earth's Oceans Unique?

In all of the planets and moons in our solar system, Earth is the only one that has oceans of liquid water on its surface. No other body in the solar system has a confirmed ocean, but recent satellite missions to other planets have revealed some tantalizing possibilities. For example, the spidery network of dark cracks on Jupiter's moon Europa (**Figure 1.1**) almost certainly betrays the presence of an ocean of liquid water beneath its icy surface. In fact, a recent analysis of the icy blocks that cover Europa's surface indicates that the blocks are actively being reshaped in a process analogous to plate tectonics on Earth. Two other moons of Jupiter, Ganymede and Callisto, may also have liquid oceans of water beneath their cold, icy crust. Yet another possibility for a nearby world with an ocean beneath its icy surface is Saturn's tiny moon Enceladus, which displays geysers of water vapor and ice that have recently been analyzed and, remarkably, contain salt. Recent analysis of the gravity field of Enceladus suggests the presence of a 10-kilometer (6.2-mile) deep saltwater ocean beneath a thick layer of surface ice. Also contained in the geysers' icy spray are tiny mineral grains; in 2015, analysis of these particles by a spacecraft flyby indicated that

***"When you're circling the Earth every 90 minutes, what becomes clearest is that it's mostly water; the continents look like they're floating objects."***

***—Loren Shriver, NASA astronaut (2008)***

[1]Throughout this book, all **bolded** words are key vocabulary terms that are defined in the glossary at the end of this book.

◂ **The blue marble, next generation.** This composite image of satellite data shows Earth's interrelated atmosphere, oceans, and land—including human presence. Its various layers include the land surface, sea ice, ocean, cloud cover, city lights, and the hazy edge of Earth's atmosphere.

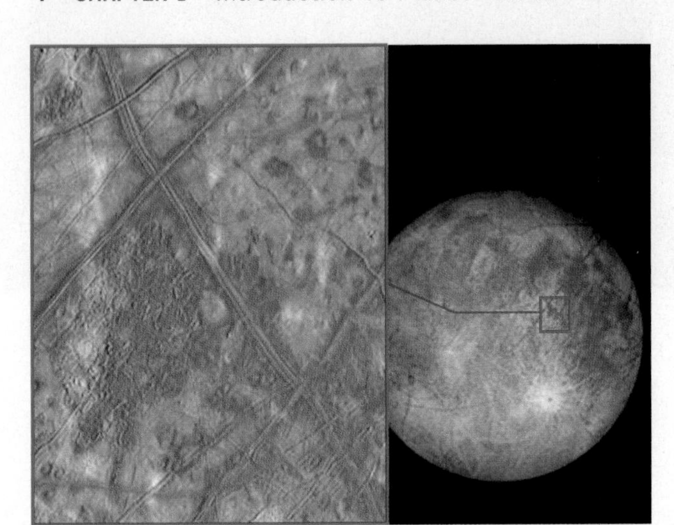

**Figure 1.1 Jupiter's moon Europa.** Scientists speculate that Europa's network of dark cracks is the result of tidal forces on an ocean beneath its icy surface. It's possible that, when Europa's orbit takes it close to Jupiter, the tide of the sea beneath the ice rises higher than normal. If this is so, the constant raising and lowering of the sea may have caused many of the cracks observed on the surface of the moon.

## STUDENTS SOMETIMES ASK . . .

***How can a spacecraft flyby determine if a planetary body has an ocean below its surface?***

Modern spacecraft like NASA's Cassini carry an impressive array of scientific instruments for studying planetary bodies but perhaps the most important of these for detecting subsurface oceans are cameras. Images of geysers of water vapor and ice spurting into space obtained by the Cassini spacecraft give away the presence of liquid subsurface water on Saturn's icy moon Enceladus. Cameras also reveal the presence of fissures and cracks in the icy crust of Jupiter's moon Europa, which may be caused by tidal forces that raise and lower a liquid ocean beneath the surface. The possibility of a liquid underground ocean on Europa was first suggested after a flyby by the Galileo spacecraft detected fluctuations in Europa's magnetic field, indicating the presence of a subsurface conducting fluid—likely a salty ocean.

the dust-sized grains likely form when hot, mineral-laden water from the moon's rocky interior travels upward, coming into contact with cooler water. This evidence of subsurface hydrothermal activity is reminiscent of underwater hot springs in the deep oceans on Earth, a place that may have been key to the development of life on Earth. In 2016, another of Saturn's moons, Dione, was reported to have the telltale sign of a liquid ocean deep beneath its icy surface. And evidence continues to mount that Saturn's giant moon Titan hosts small seas of liquid hydrocarbons, suggesting that Titan may be the only other body in the solar system besides Earth known to have liquid at its surface. Even the dwarf planet Pluto has surface features that imply it too may harbor an ocean beneath its surface. Because these planetary bodies almost certainly have oceans of one sort or another, they are all enticing targets for space missions to search for signs of extraterrestrial life. Still, the fact that our planet has so much water on its surface, *and in the liquid form*, is unique in the solar system.

## Earth's Amazing Oceans

Earth's oceans have had a profound effect on our planet and continue to shape our planet in critical ways. The oceans are essential to all life-forms and are in large part responsible for the development of life on Earth, providing a stable environment in which life could evolve over billions of years. Today, the oceans contain the greatest number of living things on the planet, from microscopic bacteria and algae to the largest life-form alive today (the blue whale). Interestingly, water is the major component of nearly every life-form on Earth, and our own body fluid chemistry is remarkably similar to the chemistry of seawater.

Another unique characteristic of Earth's oceans is that *the volume of the oceans is immense*. The oceans comprise the planet's largest habitat and contain 97.2% of all the water on or near Earth's surface (**Figure 1.2**). The oceans influence climate and weather all over the globe—even in continental areas far from any ocean—through an intricate pattern of currents and heating/cooling mechanisms, some of which scientists are only now beginning to understand. The oceans are also the "lungs" of the planet, taking carbon dioxide gas out of the atmosphere and replacing it with oxygen gas. Scientists have estimated that the oceans supply as much as 70% of the oxygen that humans breathe.

The oceans determine where our continents end and have thus shaped political boundaries and human history. The oceans conceal many features; in fact, the majority of Earth's geographic features are on the ocean floor. Remarkably, there was once more known about the surface of the Moon than about the floor of the oceans! Fortunately, our knowledge of both has increased dramatically over the past several decades.

The oceans also hold many secrets waiting to be discovered, and new scientific discoveries about the oceans are made nearly every day. The oceans are a source of food, minerals, and energy that remains largely untapped. More than half of the world's population lives in coastal areas near the oceans, taking advantage of the mild climate, an inexpensive form of transportation, proximity to food resources, and vast recreational opportunities. Unfortunately, the oceans are also the dumping ground for many of society's wastes. In fact, the oceans are currently showing alarming changes caused by pollution, overfishing, invasive species, and climate change, among other things. All of these and many other topics are contained within this book.

## How Many Oceans Exist on Earth?

The oceans are a common metaphor for vastness. When one examines a world map (**Figure 1.3**), it's easy to appreciate the impressive extent of Earth's oceans. Notice that *the oceans dominate the surface area of the globe*. For those people who have traveled by boat across an ocean (or even flown across one in an airplane), the one thing that immediately strikes them is that the oceans are enormous. Notice, also, that *the oceans are interconnected* and form a single continuous body of seawater, which is why the oceans are commonly referred to as a "world ocean" (singular, not plural). For instance, a vessel at sea can travel from one ocean to another, whereas it is impossible to travel on land from one continent to most others without crossing an ocean.

## The Four Principal Oceans, Plus One

Our world ocean can be divided into four principal oceans plus an additional ocean, based on the shape of the ocean basins and the positions of the continents (Figure 1.3).

**PACIFIC OCEAN** The **Pacific Ocean** is the world's largest ocean, covering more than half of the ocean surface area on Earth (**Figure 1.4b**). The Pacific Ocean is the single largest geographic feature on the planet, spanning more than one-third of Earth's entire surface. The Pacific Ocean is so large that *all* of the continents could fit into the space occupied by it—with room left over! Although the Pacific Ocean is also the deepest ocean in the world (**Figure 1.4c**), it contains many small tropical islands. It was named in 1520 by explorer Ferdinand Magellan's party in honor of the fine weather they encountered while crossing into the Pacific (*paci* = peace) Ocean.

**ATLANTIC OCEAN** The **Atlantic Ocean** is about half the size of the Pacific Ocean and is not quite as deep (Figure 1.4c). It separates the Old World (Europe, Asia, and Africa) from the New World (North and South America). The Atlantic Ocean was named after Atlas, who was one of the Titans in Greek mythology.

**INDIAN OCEAN** The **Indian Ocean** is slightly smaller than the Atlantic Ocean and has about the same average depth (Figure 1.4c). It is mostly in the Southern Hemisphere (south of the equator, or below 0 degrees latitude in Figure 1.3). The Indian Ocean was named for its proximity to the subcontinent of India.

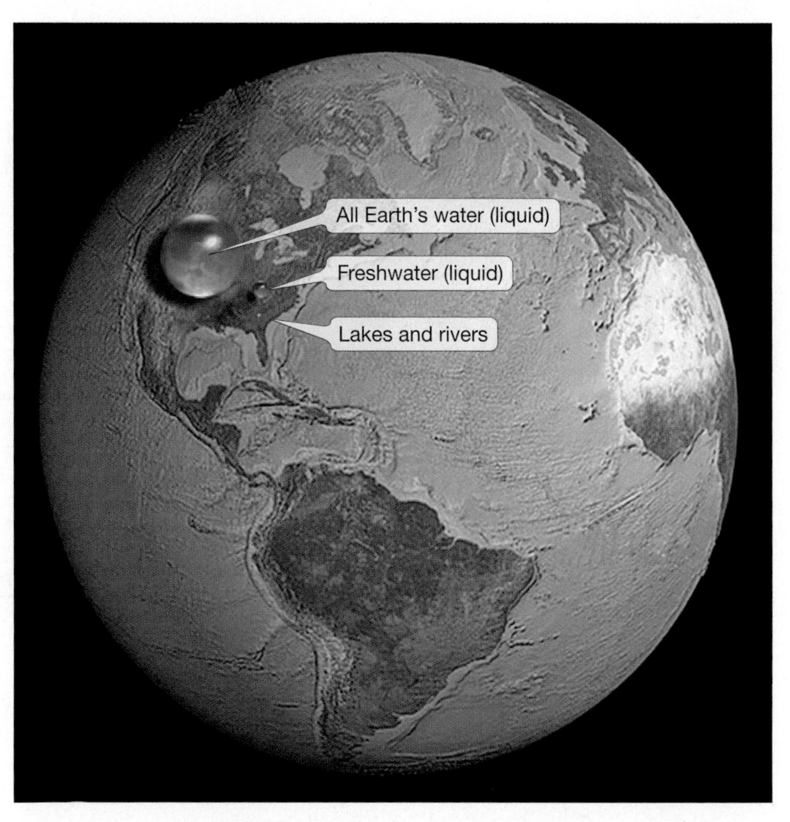


**Figure 1.2 Relative sizes of the spheres of water on Earth.** This image shows all of Earth's liquid water using three blue spheres of proportional sizes. The big sphere is all liquid water in the world, 97% of which is seawater. The next smallest sphere represents a subset of the larger sphere, showing freshwater in the ground, lakes, swamps, and rivers. The tiny speck below it represents an even smaller subset of all the water—just the freshwater in lakes and rivers.

**SmartFigure 1.3 Earth's oceans.** Map showing the four principal oceans plus the Southern Ocean, or Antarctic Ocean. https://goo.gl/BJXqyt

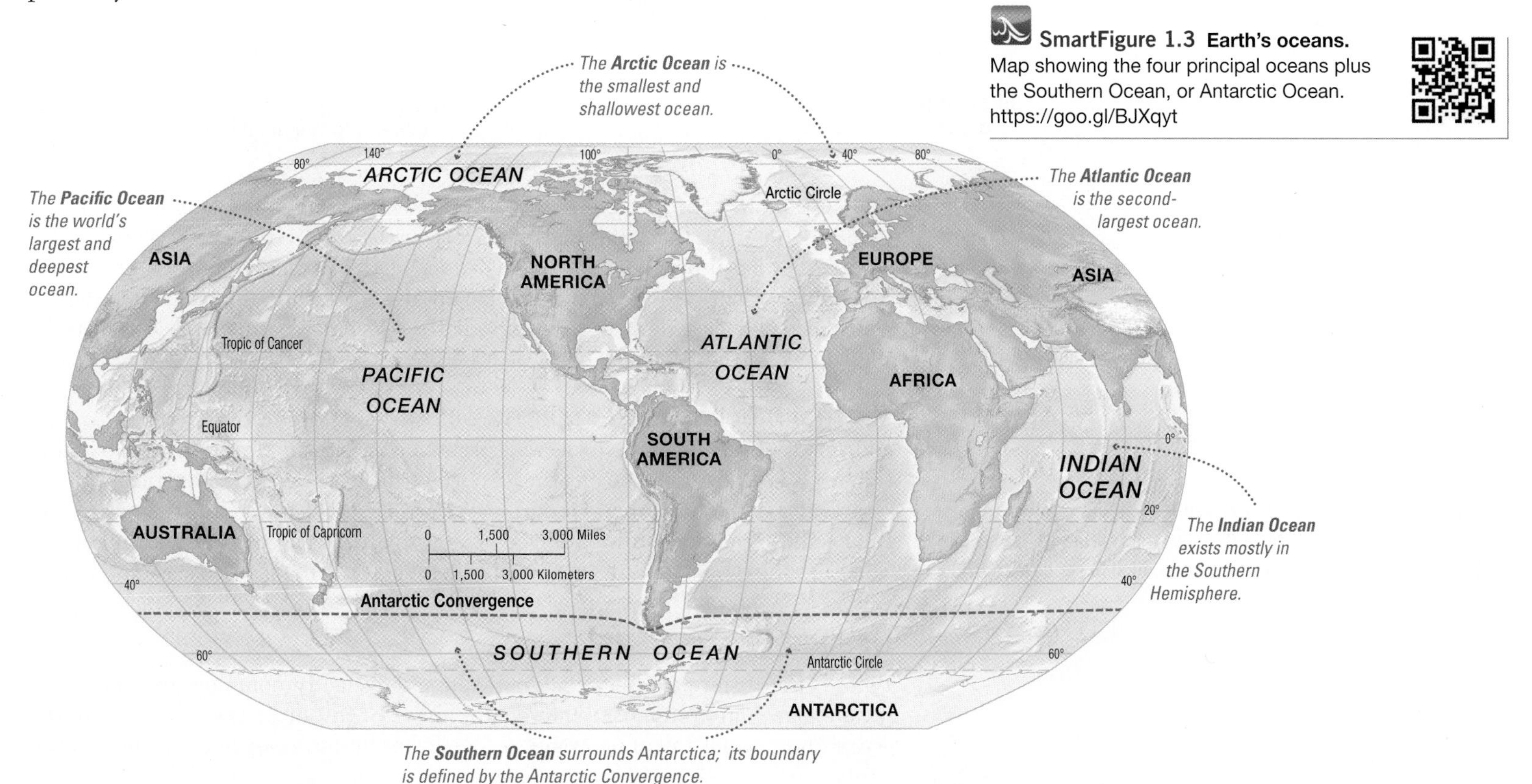

**Figure 1.4 Ocean size and depth.** **(a)** Relative proportions of land and ocean on Earth's surface. **(b)** Relative size of the four principal oceans. **(c)** Average ocean depth. **(d)** Comparing average and maximum depth of the oceans to average and maximum height of land.

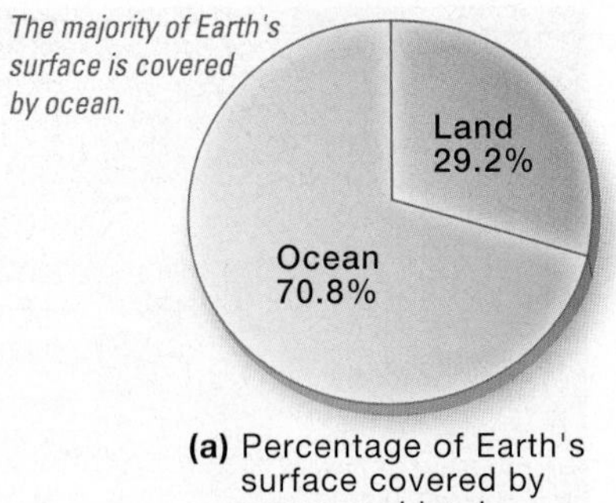


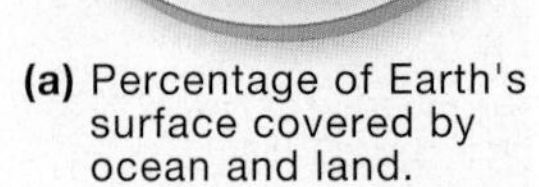
**(a)** Percentage of Earth's surface covered by ocean and land.

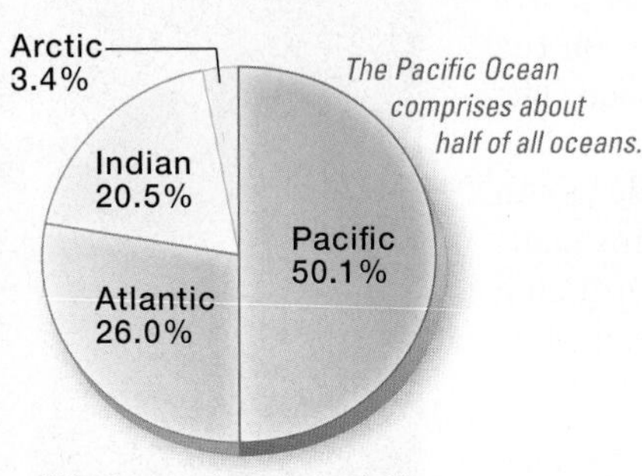


**(b)** Comparing the relative size of each ocean.

1. **Rank the four main world's oceans from largest to smallest, and also from deepest to shallowest.**
2. **Using data, support the argument that the Arctic Ocean technically should be classified as a sea. If you are not sure about the difference between an ocean and a sea, please read on . . . .**

**Animation**
Earth's Water and the Hydrologic Cycle
http://goo.gl/kAo8FC

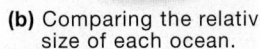

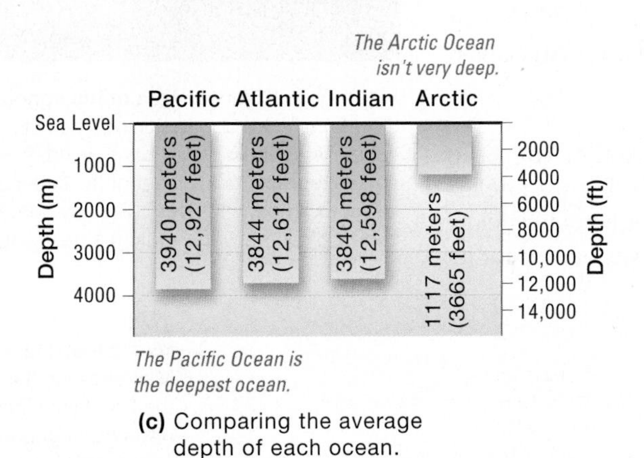


**(c)** Comparing the average depth of each ocean.

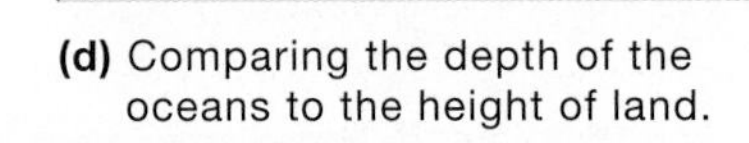


**(d)** Comparing the depth of the oceans to the height of land.

**ARCTIC OCEAN** The **Arctic Ocean** is about 7% the size of the Pacific Ocean and is only a little more than one-quarter as deep as the rest of the oceans (Figure 1.4c). Although it has a permanent layer of sea ice at the surface, the ice is only a few meters thick. The Arctic Ocean was named after its location in the Arctic region, which exists beneath the northern constellation Ursa Major, otherwise known as the Big Dipper, or the Bear (*arktos* = bear).

**SOUTHERN OCEAN, OR ANTARCTIC OCEAN** Oceanographers recognize an additional ocean near the continent of Antarctica in the Southern Hemisphere (Figure 1.3). Defined by the meeting of currents near Antarctica called the Antarctic Convergence, the **Southern Ocean**, or **Antarctic Ocean**, is really the portions of the Pacific, Atlantic, and Indian Oceans south of about 50 degrees south latitude. This ocean was named for its location in the Southern Hemisphere.

**RECAP** **The four principal oceans are the Pacific, Atlantic, Indian, and Arctic Oceans. An additional ocean, the Southern Ocean, or Antarctic Ocean, is also recognized.**

## Oceans versus Seas

What is the difference between an ocean and a sea? In common use, the terms *sea* and *ocean* are often used interchangeably. For instance, a *sea* star lives in the *ocean*, the *ocean* is full of *sea* water, *sea* ice forms in the *ocean*, and one might stroll

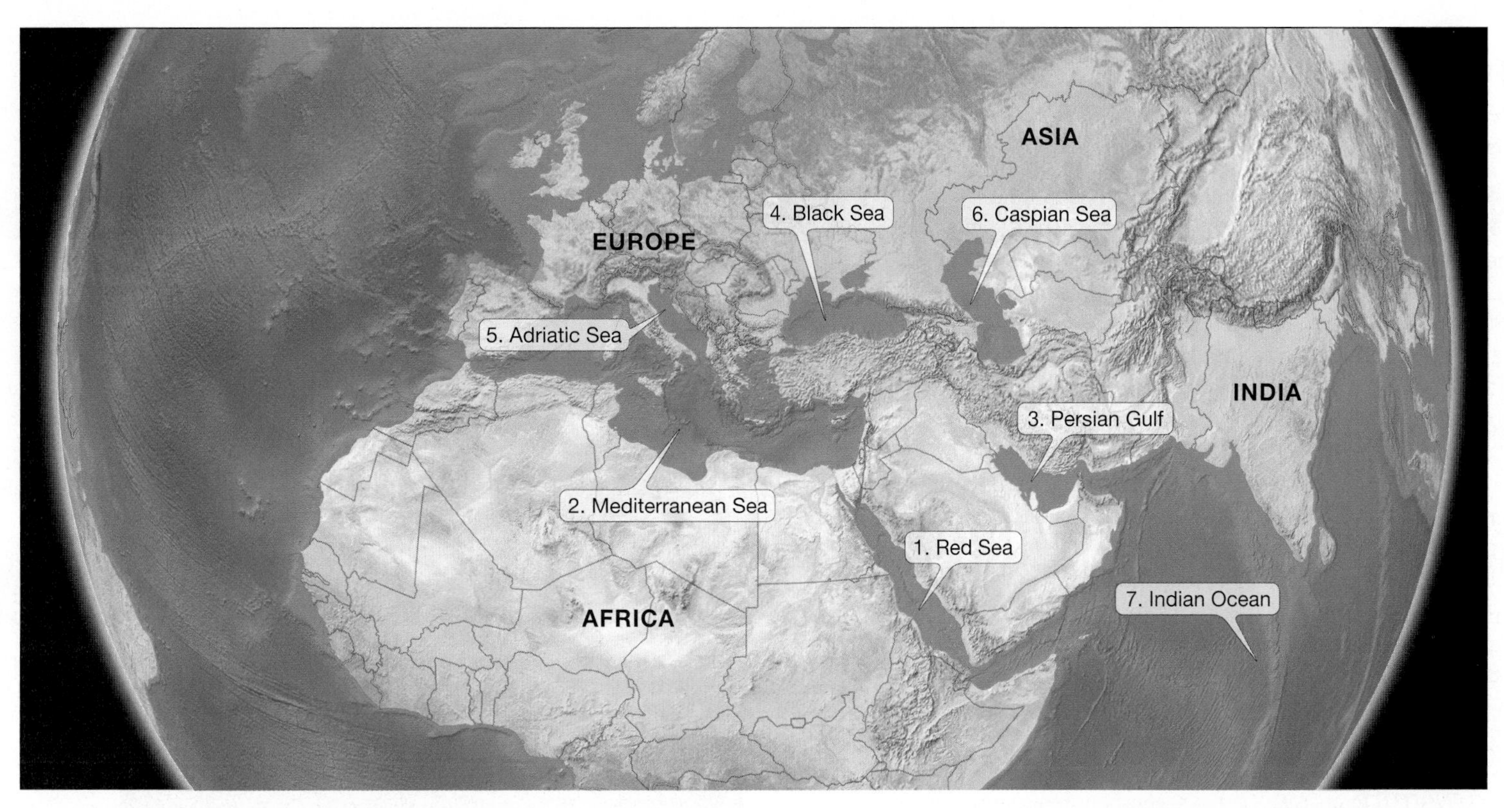


Figure 1.5 **Map of the ancient seven seas.** This map represents the extent of the known world to Europeans before the 15th century.

the *sea* shore while living on *ocean*-front property. Technically, however, a *sea* is defined as follows:

- Smaller and shallower than an ocean (this is why the Arctic Ocean might be more appropriately considered a sea)
- Composed of salt water (although some inland "seas," such as the Caspian Sea in Asia, are actually large lakes with relatively high salinity)
- Somewhat enclosed by land (although some seas, such as the Sargasso Sea in the Atlantic Ocean, are defined by strong ocean currents rather than by land)
- Directly connected to the world ocean

**COMPARING THE OCEANS TO THE CONTINENTS** **Figure 1.4d** shows that the average depth of the world's oceans is 3682 meters[2] (12,080 feet). This means that there must be some extremely deep areas in the ocean to offset the shallow areas close to shore. Figure 1.4d also shows that the deepest depth in the oceans (the Challenger Deep region of the Mariana Trench, which is near Guam) is a staggering 11,022 meters (36,161 feet) below sea level.

How do the continents compare to the oceans? Figure 1.4d shows that the average height of the continents is only 840 meters (2756 feet), illustrating that the average height of the land is not very far above sea level. The highest mountain in the world (the mountain with the greatest height above sea level) is Mount Everest in the Himalaya Mountains of Asia, at 8850 meters (29,035 feet). Even so, Mount Everest is a full 2172 meters (7126 feet) shorter than the Mariana Trench is deep. The mountain with the *greatest total height* from base to top is Mauna Kea on the island of Hawaii in the United States. It measures 4206 meters (13,800 feet) above

## STUDENTS SOMETIMES ASK . . .

***Where are the seven seas?***

"Sailing the seven seas" is a familiar phrase in literature and song, but the origin of the saying is shaded in antiquity. To the ancients, the term "seven" often meant "many," and before the 15th century, Europeans considered these the main seas of the world (**Figure 1.5**):

1. The Red Sea
2. The Mediterranean Sea
3. The Persian Gulf
4. The Black Sea
5. The Adriatic Sea
6. The Caspian Sea
7. The Indian Ocean (notice how "ocean" and "sea" are used interchangeably)

Today, however, more than 100 seas, bays, and gulfs are recognized worldwide, nearly all of them smaller portions of the huge interconnected world ocean.

[2]Throughout this book, metric measurements are used and the corresponding English measurements follow in parentheses. See Mastering Oceanography Appendix I, "Metric and English Units Compared," for conversion factors between the two systems of units.

## STUDENTS SOMETIMES ASK . . .

### *Have humans ever explored the deepest ocean trenches? Could anything live there?*

Humans have indeed visited the deepest part of the oceans—where there is crushing high pressure, complete darkness, and near-freezing water temperatures—and they first did so over half a century ago! In January 1960, U.S. Navy Lt. Don Walsh and explorer Jacques Piccard descended to the bottom of the Challenger Deep region of the Mariana Trench in the *Trieste*, a deep-diving bathyscaphe (*bathos* = depth, *scaphe* = a small ship) (**Figure 1.6**). At 9906 meters (32,500 feet), the men heard a loud cracking sound that shook the cabin. They were unable to see that a 7.6-centimeter (3-inch) Plexiglas viewing port had cracked (miraculously, it held for the rest of the dive). More than five hours after leaving the surface, they reached the bottom, at 10,912 meters (35,800 feet)—a record depth for human descent. They did observe some small organisms that are adapted to life in the deep: a flatfish, a shrimp, and some jellies.

In 2012, film icon James Cameron made a historic solo dive to the Mariana Trench in his submersible *DEEPSEA CHALLENGER* (**Figure 1.7**). On the seven-hour round-trip voyage, Cameron spent about three hours at the deepest spot on the planet to take photographs and collect samples for scientific research. Other notable voyages to the deep ocean in submersibles are discussed in Mastering Oceanography **Web Diving Deeper 1.3**.

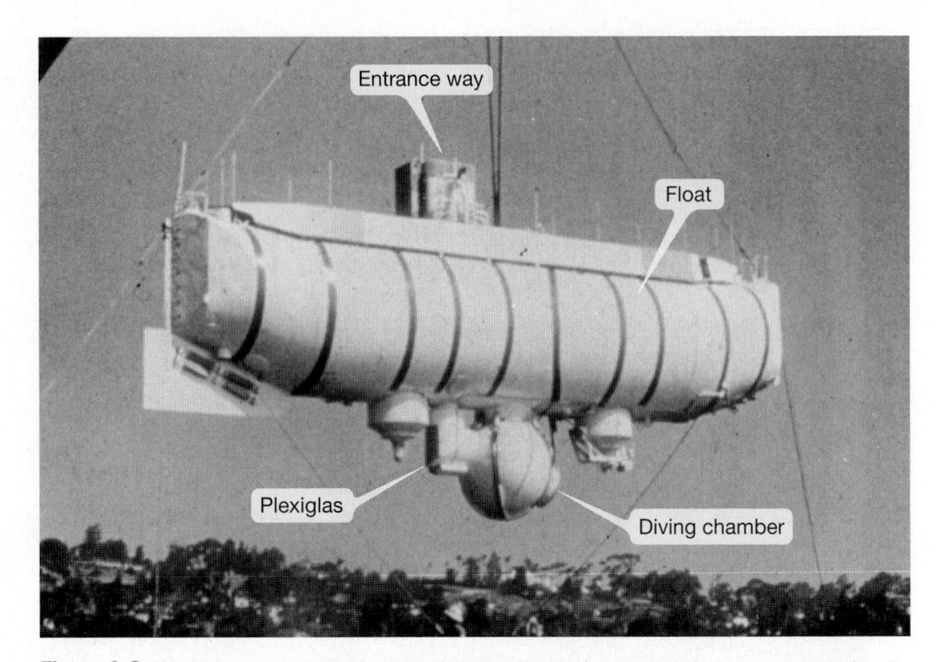


Figure 1.6 **The U.S. Navy's bathyscaphe *Trieste*.** The *Trieste* suspended on a crane before its record-setting deep dive in 1960. The 1.8-meter (6-foot) diameter diving chamber (the round ball below the float) accommodated two people and had steel walls 7.6 centimeters (3 inches) thick.

## CREATURE FEATURE 1.1

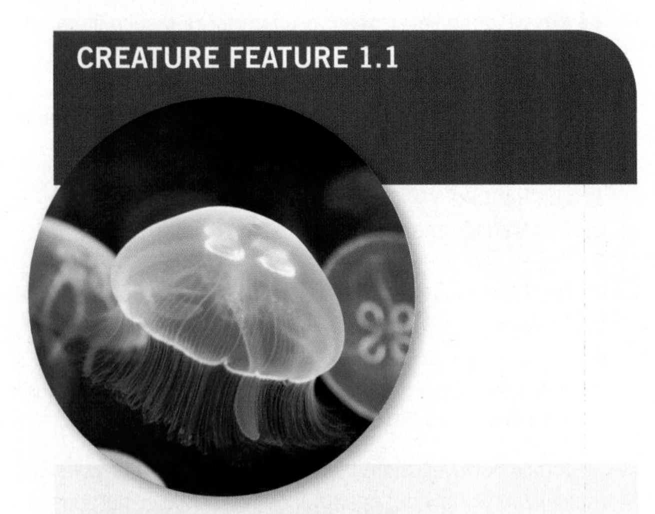

### I light up the deep sea!

**Jellies** (often incorrectly called *jellyfish*) are gelatinous marine organisms that live throughout the oceans.

Jellies are primitive, free-drifting marine organisms that use stinging cells to capture food. About half of all jellies are capable of bioluminescence, which is the ability to produce light; jellies thus are bright spots in the vast darkness of the deep sea.

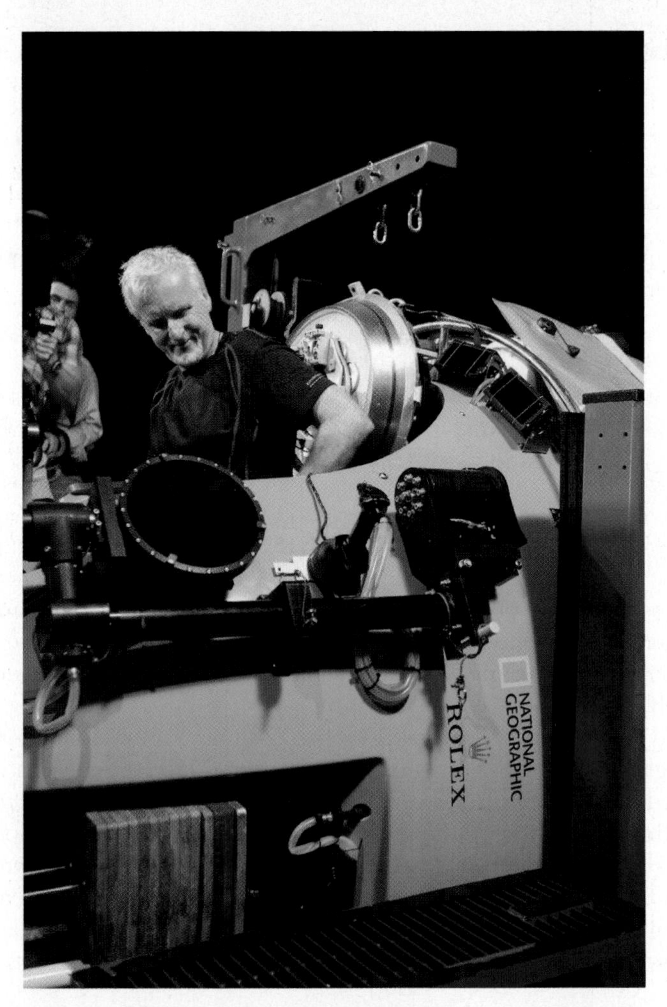


Figure 1.7 **James Cameron emerges from the submersible *DEEPSEA CHALLENGER* after his solo dive to the Mariana Trench.** In 2012, famous moviemaker James Cameron completed a record-breaking solo dive to the bottom of the Mariana Trench, becoming only the third human to visit the deepest spot on Earth.

sea level and 5426 meters (17,800 feet) from sea level down to its base, for a total height of 9632 meters (31,601 feet). The total height of Mauna Kea is 782 meters (2566 feet) higher than Mount Everest, but it is still 1390 meters (4560 feet) shorter than the Mariana Trench is deep. Therefore, no mountain on Earth is taller than the Mariana Trench is deep.

**RECAP** **The deepest part of the ocean is the Mariana Trench in the Pacific Ocean. It is 11,022 meters (36,161 feet) deep and has been visited only twice by humans: once in 1960 and more recently in 2012.**

**CONCEPT CHECK 1.1 ▸ Compare the characteristics of Earth's oceans.**

1 How did the view of the ocean by early Mediterranean cultures influence the naming of planet Earth?

2 Although the terms *ocean* and *sea* are sometimes used interchangeably, what is the technical difference between an ocean and a sea?

3 Where is the deepest part of the ocean? How deep is it, and how does it compare to the height of the tallest mountain on Earth?

# 1.2 ▸ How Was Early Exploration of the Oceans Achieved?

As a testimony to the human spirit, early cultures explored the ocean's furthest reaches in spite of the risks caused by crossing vast expanses of open ocean. Over time, humans developed technology that allowed entire civilizations to safely travel across even the largest oceans. For example, today we can cross even the Pacific Ocean in less than a day by airplane. Even so, much of the deep ocean remains out of reach and woefully unexplored. In fact, the surface of the Moon has been mapped more accurately than most parts of the sea floor. Nonetheless, new technologies employed on land and sea, and Earth-observing satellites orbiting at great distances, are being used to gain knowledge about our watery home at an unprecedented rate.

## Early History

Humankind probably first viewed the oceans as a source of food. Archeological evidence suggests that when boat technology was developed about 40,000 years ago, people probably traveled the oceans. Most likely, their vessels were built to move upon the ocean's surface and transport oceangoing people to new fishing grounds. The oceans also provided an inexpensive and efficient way to move large and heavy objects, facilitating trade and interaction between cultures.

**PACIFIC NAVIGATORS** The peopling of the Pacific Islands (Oceania) is somewhat perplexing because there is no anthropological evidence that people actually evolved on these islands—in other words, they had to come from elsewhere. Their presence required travel over hundreds or even thousands of kilometers of open ocean from the continents (probably in small vessels of that time—double canoes, outrigger canoes, or balsa rafts), as well as remarkable navigation skills (see Mastering Oceanography **Web Diving Deeper 1.4**). The islands in the Pacific Ocean are widely scattered, so it is likely that only a fortunate few of the voyagers made landfall and that many others perished during voyages. **Figure 1.8** shows the three major inhabited island regions in the Pacific Ocean: Micronesia

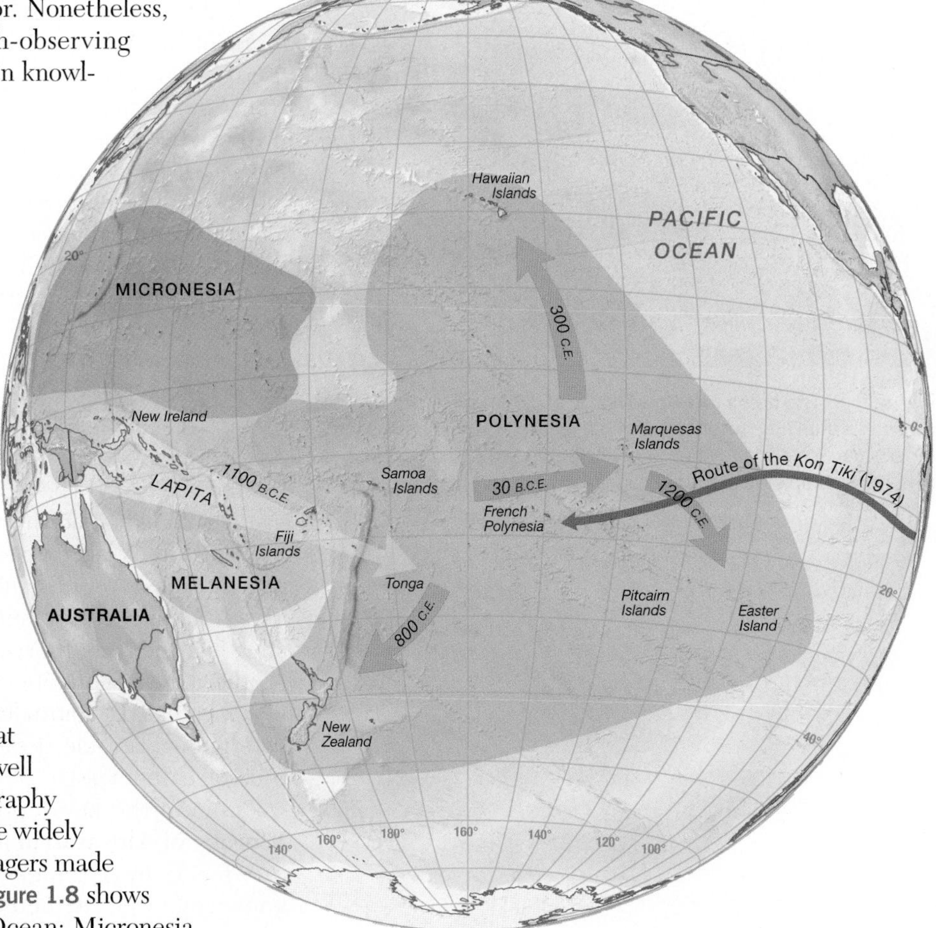


Figure 1.8 **The peopling of the Pacific islands.** The major island groups of the Pacific Ocean are Micronesia (brown shading), Melanesia (*peach shading*), and Polynesia (*green shading*). The "Lapita people" present in New Ireland 5000–4000 B.C.E. can be traced to Fiji, Tonga, and Samoa by 1100 B.C.E. (*yellow arrow*). Green arrows show the peopling of distant islands throughout Polynesia. The 1947 route of Thor Heyerdahl's balsa raft *Kon Tiki* is also shown (*red arrow*).

(*micro* = small, *nesia* = islands); Melanesia (*mela* = black, *nesia* = islands); and Polynesia (*poly* = many, *nesia* = islands), which covers the largest area.

No written records of Pacific human history have been found prior to the arrival of Europeans in the 16th century. Nevertheless, the movement of Asian peoples into Micronesia and Melanesia is easy to imagine, because distances between islands are relatively short. In Polynesia, however, large distances separate island groups, which must have presented great challenges to ocean voyagers. Easter Island, for example, at the southeastern corner of the triangular-shaped Polynesian Islands region, is more than 1600 kilometers (1000 miles) from Pitcairn Island, the next nearest island. Clearly, a voyage to the Hawaiian Islands must have been one of the most difficult because Hawaii is more than 3000 kilometers (2000 miles) from the nearest inhabited islands, the Marquesas Islands (Figure 1.8).

Archeological evidence suggests that humans from New Guinea may have occupied New Ireland as early as 4000 or 5000 B.C.E. However, there is little evidence of human travel farther into the Pacific Ocean before 1100 B.C.E. By then, the *Lapita people,* a group of early settlers who produced a distinctive type of pottery, had traveled on to Fiji, Tonga, and Samoa (Figure 1.8, *yellow arrow*). From there, Polynesians sailed on to the Marquesas (about 30 B.C.E.), which appear to have been the starting point for voyages to other islands in the far reaches of the Pacific (Figure 1.8, *green arrows*), including the Hawaiian Islands (about 300 C.E.) and New Zealand (about 800 C.E.). (Recently, a combination of genetic, linguistic, and archaeological evidence has suggested that the forebears of the Lapita people—and thus Polynesians—originated in Taiwan, just off the coast of China.) Surprisingly, new genetic research suggests that Polynesians populated Easter Island relatively recently, about 1200 C.E.

Figure 1.9 **The balsa raft *Kon Tiki*.** In 1947, Thor Heyerdahl sailed this authentic wooden balsa raft named *Kon Tiki* from South America to Polynesia to show that ancient South American cultures may have completed similar voyages.

Despite the obvious Polynesian backgrounds of the Hawaiians, the Maori of New Zealand, and the Easter Islanders, an adventurous biologist/anthropologist named **Thor Heyerdahl** proposed that voyagers from South America may have reached islands of the South Pacific before the coming of the Polynesians. To prove his point, in 1947 he sailed the ***Kon Tiki***—a balsa raft designed like those that were used by South American navigators at the time of European discovery (**Figure 1.9**)—from South America to the Tuamotu Islands, a journey of more than 11,300 kilometers (7000 miles) (Figure 1.8, *red arrow*). Although the remarkable voyage of the *Kon Tiki* demonstrates that early South Americans could have traveled to Polynesia just as easily as early Asian cultures, anthropologists can find no evidence of such a migration. Further, comparative DNA studies show a strong genetic relationship between the peoples of Easter Island and Polynesia but none between these groups and natives in coastal North or South America.

**EUROPEAN NAVIGATORS** The first Mediterranean people known to have developed the art of navigation were the **Phoenicians**, who lived at the eastern end of the Mediterranean Sea, in the present-day area of Egypt, Syria, Lebanon, and Israel. As early as 2000 B.C.E., they investigated the Mediterranean Sea, the Red Sea, and the Indian Ocean. The first recorded circumnavigation of Africa, in 590 B.C.E., was made by the Phoenicians, who also sailed as far north as the British Isles.

The Greek astronomer-geographer **Pytheas** sailed northward in 325 B.C.E. using a simple yet elegant method for determining latitude (one's position north or south) in the Northern Hemisphere. His method involved measuring the angle between an observer's line of sight to the North Star and line of sight to the northern horizon. (Note that Pytheas's method of determining latitude is featured in Appendix III, "Latitude and Longitude on Earth.") Despite Pytheas's method for determining latitude, it was still impossible to accurately determine longitude (one's position east or west).

One of the key repositories of scientific knowledge at the time was the **Library of Alexandria** in Alexandria, Egypt, which was founded in the 3rd century B.C.E. by *Alexander the Great*. It housed an impressive collection of written knowledge that attracted scientists, poets, philosophers, artists, and writers who studied and researched there. The Library of Alexandria soon became the intellectual capital of the world, featuring history's greatest accumulation of ancient writings.

As long ago as 450 B.C.E., Greek scholars became convinced that Earth was round, using lines of evidence such as the way ships disappeared beyond the horizon and the shadows of Earth that appeared during eclipses of the Moon. This inspired the Greek **Eratosthenes** (pronounced "AIR-uh-TOS-thuh-neez") (276–192 B.C.E.), the second librarian at the Library of Alexandria, to cleverly use the shadow of a stick in a hole in the ground and elementary geometry to determine Earth's circumference. His value of 40,000 kilometers (24,840 miles) compares remarkably well with the true value of 40,032 kilometers (24,875 miles) known today.

An Egyptian-Greek geographer named **Claudius Ptolemy** (c. 85 C.E.–c. 165 C.E.) produced a map of the world in about 150 C.E. that represented the extent of Roman knowledge at that time (**Figure 1.10**). The map not only included the continents of Europe, Asia, and Africa, as did earlier Greek maps, but it also included vertical lines of longitude and horizontal lines of latitude, which had been developed by Alexandrian scholars. Moreover, Ptolemy showed the known seas to be surrounded by land, much of which was as yet unknown and proved to be a great enticement to explorers.

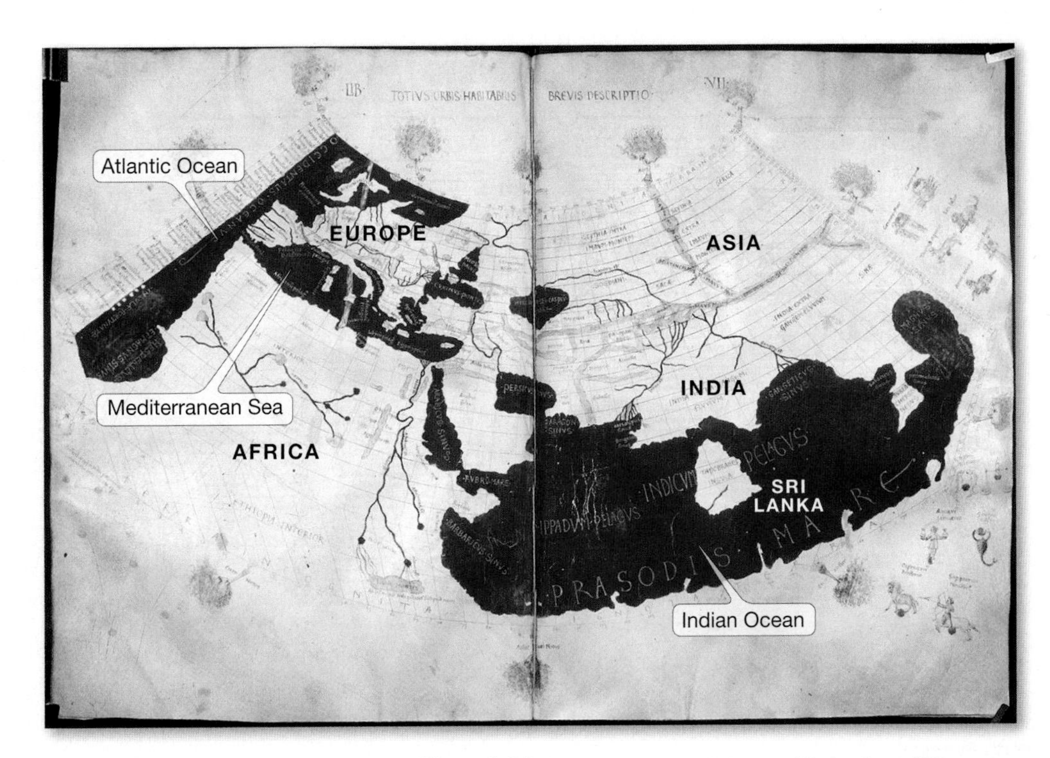


**Figure 1.10 Ptolemy's map of the world.** In about 150 C.E., an Egyptian-Greek geographer named Claudius Ptolemy produced this map of the world that showed the extent of Roman geographic knowledge. Note the use of a coordinate system on land, similar to latitude and longitude used today.

Ptolemy also introduced an (erroneous) update to Eratosthenes's surprisingly accurate estimate of Earth's circumference. Ptolemy wrongly depended on flawed calculations and an overestimation of the size of Asia, and as a result, he determined Earth's circumference to be 29,000 kilometers (18,000 miles), which is about 28% too small. Nearly 1500 years later, Ptolemy's error caused explorer Christopher Columbus to believe he had encountered parts of Asia rather than a new world.

## The Middle Ages

After the destruction of the Library of Alexandria in 415 C.E. (in which all of its contents were burned) and the fall of the Roman Empire in 476 C.E., the achievements of the Phoenicians, Greeks, and Romans were mostly lost. Some of the knowledge, however, was retained by the *Arabs,* who controlled northern Africa and Spain. The Arabs used this knowledge to become the dominant navigators in the Mediterranean Sea area and to trade extensively with East Africa, India, and Southeast Asia. The Arabs were able to trade across the Indian Ocean because they had learned how to take advantage of the seasonal patterns of monsoon winds. During the summer, when monsoon winds blow from the southwest, ships laden with goods would leave the Arabian ports and sail eastward across the Indian Ocean. During the winter, when the trade winds blow from the northeast, ships would return west. (More details about Indian Ocean monsoons can be found in Chapter 7, "Ocean Circulation.")

Meanwhile, in the rest of southern and eastern Europe, Christianity was on the rise. Scientific inquiry counter to religious teachings was actively suppressed, and the knowledge gained by previous civilizations was either lost or ignored. As a result, the Western concept of world geography degenerated considerably during these so-called *Dark Ages*. For example, one notion envisioned the world as a disk with Jerusalem at the center.

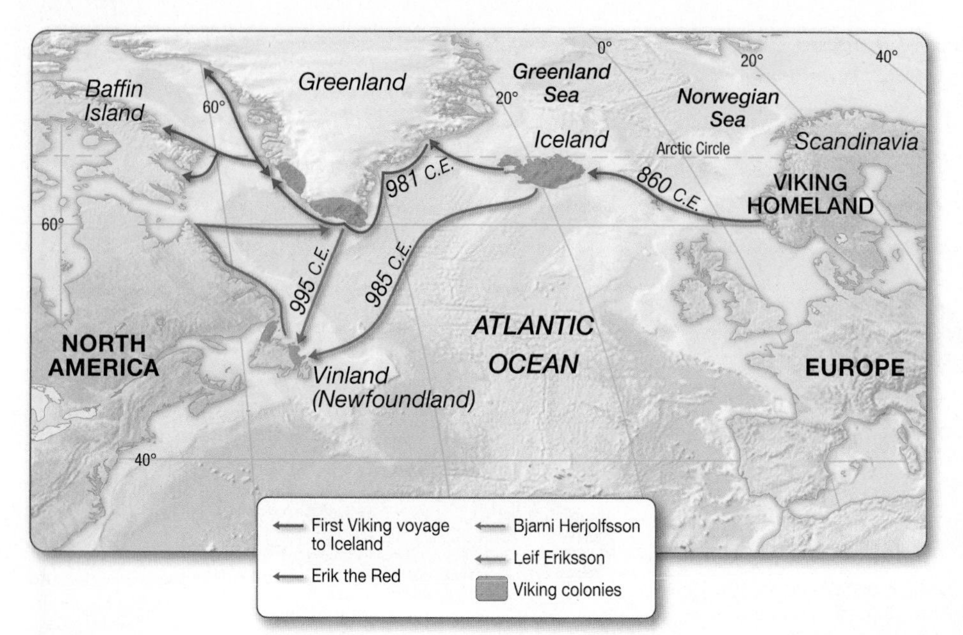


**Figure 1.11 Viking colonies in the North Atlantic.** Map showing the routes and dates of Viking explorations and the locations of the colonies that were established in Iceland, Greenland, and parts of North America.

In northern Europe, the **Vikings** of Scandinavia, who had excellent ships and good navigation skills, actively explored the Atlantic Ocean (Figure 1.11). Late in the 10th century, aided by a period of worldwide climatic warming, the Vikings colonized Iceland. In about 981 C.E., **Erik "the Red" Thorvaldson** sailed westward from Iceland and discovered Greenland. He may also have traveled further westward to Baffin Island. He returned to Iceland and led the first wave of Viking colonists to Greenland in 985 C.E. **Bjarni Herjólfsson** sailed from Iceland to join the colonists, but he sailed too far southwest and is thought to be the first Viking to have seen what is now called Newfoundland. Bjarni did not land but instead returned to the new colony at Greenland. **Leif Eriksson**, son of Erik the Red, became intrigued by Bjarni's stories about the new land Bjarni had seen. In 995 C.E., Leif bought Bjarni's ship and set out from Greenland for the land that Bjarni had seen to the southwest. Leif spent the winter in that portion of North America and named the land *Vinland* (now Newfoundland, Canada) after the grapes that were found there. Climatic cooling and inappropriate farming practices for the region caused these Viking colonies in Greenland and Vinland to struggle and die out by about 1450.

## The Age of Discovery in Europe

The 30-year period from 1492 to 1522 is known as Europe's **Age of Discovery**. During this time, Europeans explored the continents of North and South America, and the globe was circumnavigated for the first time. As a result, Europeans learned the true extent of the world's oceans and that human populations existed elsewhere on newly "discovered" continents and islands with cultures vastly different from those familiar to European voyagers.

Why was there such an increase in ocean exploration during Europe's Age of Discovery? One reason was that Sultan Mohammed II had captured Constantinople (the capital of eastern Christendom) in 1453, a conquest that isolated Mediterranean port cities from the riches of India, Asia, and the East Indies (modern-day Indonesia). As a result, the Western world had to search for new eastern trade routes by sea.

The Portuguese, under the leadership of **Prince Henry the Navigator** (1392–1460), led a renewed effort to explore outside Europe. The prince established a marine institution at Sagres to improve Portuguese sailing skills. The treacherous journey around the tip of Africa was a great obstacle to an alternative trade route. Cape Agulhas (at the southern tip of Africa) was first rounded by **Bartholomeu Diaz** in 1486. He was followed in 1498 by **Vasco da Gama**, who continued around the tip of Africa to India, thus establishing a new eastern trade route to Asia.

Meanwhile, the Italian navigator and explorer **Christopher Columbus** was financed by Spanish monarchs to find a new route to the East Indies across the Atlantic Ocean. During Columbus's first voyage in 1492, he sailed west from Spain and made landfall after a two-month journey (Figure 1.12). Columbus believed that he had arrived in the East Indies somewhere near India, but Earth's circumference had been substantially underestimated, so he was unaware that he had actually arrived in uncharted territory in the Caribbean. Upon his return to Spain and the announcement of his discovery, additional voyages were planned. During the next 10 years, Columbus made three more trips across the Atlantic.

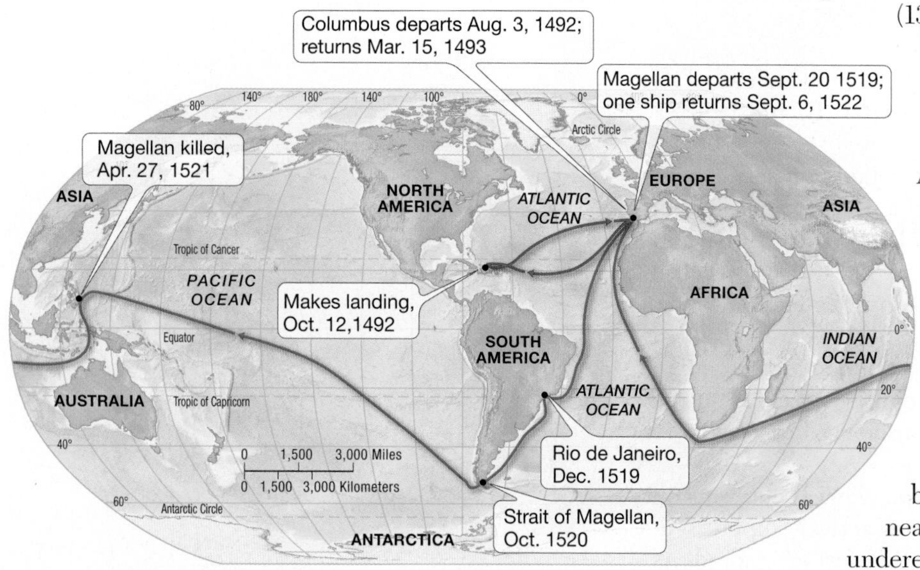


**Figure 1.12 Voyages of Columbus and Magellan.** Map showing the dates and routes of Columbus's first voyage and the first circumnavigation of the globe by Magellan's party.

Even though Christopher Columbus is widely credited with discovering North America, he never actually set foot on the continent. (For more information about the voyages of Columbus, see Diving Deeper 6.1 in Chapter 6, "Air–Sea Interaction.") Still, his journeys inspired other navigators to explore the "New World." For example, in 1497, only five years after Columbus's first voyage, the Italian navigator and explorer **Giovanni Caboto**, who was also known as **John Cabot**, landed somewhere on the northeastern coast of North America. Later, Europeans first saw the Pacific Ocean in 1513, when **Vasco Núñez de Balboa** attempted a land crossing of the Isthmus of Panama and sighted a large ocean to the west from atop a mountain.

The culmination of the Age of Discovery was a remarkable circumnavigation of the globe initiated by **Ferdinand Magellan** (Figure 1.12). Magellan left Spain in September 1519, with five ships and 280 sailors. He crossed the Atlantic Ocean, sailed down the eastern coast of South America, and traveled through a passage to the Pacific Ocean at 52 degrees south latitude, now named the Strait of Magellan in his honor. About a month after landing in the Philippines in March 1521, Magellan was killed in a fight with the inhabitants of these islands. **Juan Sebastian del Caño** completed the circumnavigation by taking the last of the ships, the *Victoria*, across the Indian Ocean, around Africa, and back to Spain in 1522. After three years, just one ship and 18 men completed the voyage.

Following these expeditions, the Spanish initiated many other voyages to take gold from the Aztec and Inca cultures in Mexico and South America. The English and Dutch, meanwhile, used smaller, more maneuverable ships to rob the gold from bulky Spanish galleons, which resulted in many confrontations at sea. The maritime dominance of Spain ended when the English defeated the Spanish Armada in 1588. With control of the seas, the English thus became the dominant world power—a status they retained until early in the 20th century.

## The Beginning of Voyaging for Science

The English realized that increasing their scientific knowledge of the oceans would help maintain their maritime superiority. For this reason, Captain **James Cook** (1728–1779), an English navigator and prolific explorer (**Figure 1.13**), undertook three voyages of scientific discovery with the ships *Endeavour*, *Resolution*, and *Adventure* between 1768 and 1779. He searched for the continent Terra Australis ("Southern Land," or Antarctica) and concluded that it lay beneath or beyond the extensive ice fields of the southern oceans if it existed at all. Cook also mapped many islands previously unknown to Europeans, including the South Georgia, South Sandwich, and Hawaiian Islands. During his last voyage, Cook searched for the fabled "Northwest Passage" from the Pacific Ocean to the Atlantic Ocean and stopped in Hawaii, where he was killed in a skirmish with native Hawaiians.

Cook's expeditions added greatly to the scientific knowledge of the oceans. He determined the outline of the Pacific Ocean and was the first person known to cross the Antarctic Circle in his search for Antarctica. Cook initiated systematic sampling of subsurface water temperatures, measuring winds and currents, taking *soundings* (which are depth measurements that, at the time, were taken by lowering a long rope with a weight on the end to the sea floor), and collecting data on coral reefs. Cook also discovered that a shipboard diet containing the German staple sauerkraut prevented his crew from contracting scurvy, a disease that incapacitated sailors. Scurvy is caused by a vitamin C deficiency, and the cabbage used to make sauerkraut contains large quantities of vitamin C. Prior to Cook's discovery about preventing scurvy, the malady claimed more lives than all other types of deaths at sea, including contagious disease,

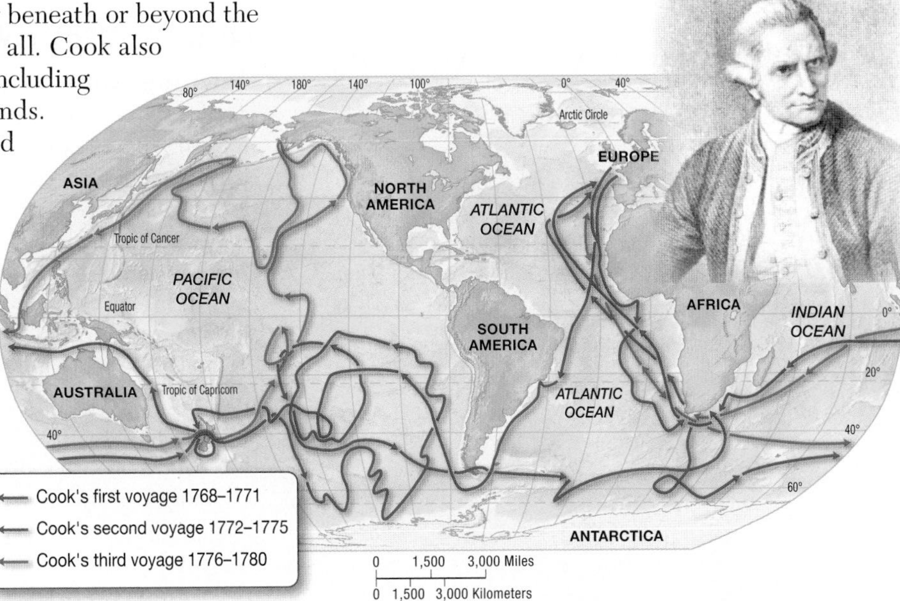


**Figure 1.13 Captain James Cook (1728–1779) and his voyages of exploration.** Routes taken by Captain James Cook (*inset*) on his three scientific voyages, which initiated scientific exploration of the oceans. Cook was killed in 1779 in Hawaii during his third voyage.

gunfire, and shipwreck. In addition, Cook made possible the first accurate maps of Earth's surface by proving the value of a new—and controversial— scientific device for determining longitude, which was a precision timepiece invented by English cabinetmaker John Harrison (see Mastering Oceanography **Web Diving Deeper 1.4**).

## History of Oceanography . . . To Be Continued

Much has changed since the early days of studying the oceans, when scientists used buckets, nets, and lines deployed from ships. And yet, some things remain the same. For example, going to sea aboard ships continues to be a mainstay of ocean science. Also, even though efforts to monitor the ocean are getting bigger and more sophisticated, vast swaths of the marine world remain unknown.

Today, oceanographers employ many high-technology tools, such as state-of-the-art research vessels that routinely use sonar to map the sea floor, remotely operated data collection devices, drifting buoys, robotics, sea floor observation networks, sophisticated computer models, and Earth-orbiting satellites. Many of these tools are featured throughout this book. Further, additional events in the history of oceanography can be found as Diving Deeper features in subsequent chapters. These boxed features are identified by the "Historical Feature" theme, and each introduces an important historical event that is related to the subject of that particular chapter.

**RECAP** **The ocean's large size did not prohibit early explorers from venturing into all parts of the ocean for discovery, trade, or conquest. Voyaging for science began relatively recently, and many parts of the ocean remain unknown.**

**CONCEPT CHECK 1.2 ▶ Discuss how early exploration of the oceans was achieved.**

1 While the Arabs dominated the Mediterranean region during the Middle Ages, what were the most significant ocean-related events taking place in northern Europe?

2 Describe the important events in oceanography that occurred during the Age of Discovery in Europe.

3 List some of the major achievements of Captain James Cook.

### STUDENTS SOMETIMES ASK . . .

***What is NOAA? What is its role in oceanographic research?***

NOAA (pronounced "NO-ah") stands for National Oceanic and Atmospheric Administration and is the branch of the U.S. Department of Commerce that oversees oceanographic research. Scientists at NOAA work to ensure wise use of ocean resources through the National Ocean Service, the National Oceanographic Data Center, the National Marine Fisheries Service, and the National Sea Grant Office. Other U.S. government agencies that work with oceanographic data include the U.S. Naval Oceanographic Office, the Office of Naval Research, the U.S. Coast Guard, and the U.S. Geological Survey (coastal processes and marine geology). The NOAA Website is at **www.noaa.gov**. In 2013, federal officials developed the National Ocean Policy Implementation Plan, which proposes moving NOAA to the Department of the Interior so that agencies dealing with natural resources would all be grouped within the same department.

# 1.3 ▶ What Fields of Science Does Oceanography Include?

**Oceanography** (*ocean* = the marine environment, *graphy* = description of) is literally the description of the marine environment. Although the term was first coined in the 1870s, at the beginning of scientific exploration of the oceans, this definition does not fully portray the extent of what oceanography encompasses: Oceanography does much more than just *describe* marine phenomena. Oceanography could be more accurately called the scientific study of all aspects of the marine environment. Hence, the field of study called oceanography could (and maybe *should*) be called oceanology (*ocean* = the marine environment, *ology* = the study of). However, the science of studying the oceans has traditionally been called *oceanography*. It is also called *marine science* and includes the study of the water of the ocean, the life within it, and the (not so) solid Earth beneath it.

Since prehistoric time, people have used the oceans as a means of transportation and as a source of food. Ocean processes, on the other hand, have been studied using technology only since the 1930s, beginning with the search for offshore petroleum and then expanding greatly during World War II with seafaring nations' interest in ocean warfare. The recognition of the importance of marine problems by governments, their readiness to make money available for research, the growth in the number of ocean scientists at work, and the increasing sophistication of scientific equipment have all made it feasible to study the ocean on a scale and to a degree of complexity never before attempted nor even possible.

Consider, for example, the logical assumption that those who make their living fishing in the ocean will go where the physical processes of the oceans offer good fishing (for more details about this topic, see Chapter 13, "Biological Productivity and Energy Transfer"). How ocean geology, chemistry, and physics work together with biology to create good fishing grounds has been more or less a mystery until only recently, when scientists from those disciplines began to investigate the oceans with new technology. One insight from these studies was the realization of how much of an impact humans are beginning to have on the ocean. As a result, much recent research has been concerned with documenting human impacts on the ocean.

Oceanography is traditionally divided into different academic disciplines (or subfields) of study. The four main disciplines of oceanography that are covered in this book are as follows:

- *Geological oceanography*, which is the study of the structure of the sea floor and how the sea floor has changed through time; the creation of sea floor features; and the history of sediments deposited on it
- *Chemical oceanography*, which is the study of the chemical composition and properties of seawater, how to extract certain chemicals from seawater, and the effects of pollutants
- *Physical oceanography*, which is the study of waves, tides, and currents; the ocean–atmosphere relationship that influences weather and climate; and the transmission of light and sound in the oceans
- *Biological oceanography*, which is the study of the various oceanic life-forms and their relationships to one another, their adaptations to the marine environment, and developing sustainable methods of harvesting seafood

Other disciplines include ocean engineering, marine archaeology, and marine policy. Because the study of oceanography often examines in detail all the different disciplines of oceanography, it is frequently described as being an **interdisciplinary science**, or one covering all the disciplines of science as they apply to the oceans (**Figure 1.14**). In essence, this is a book about *all* aspects of the oceans.

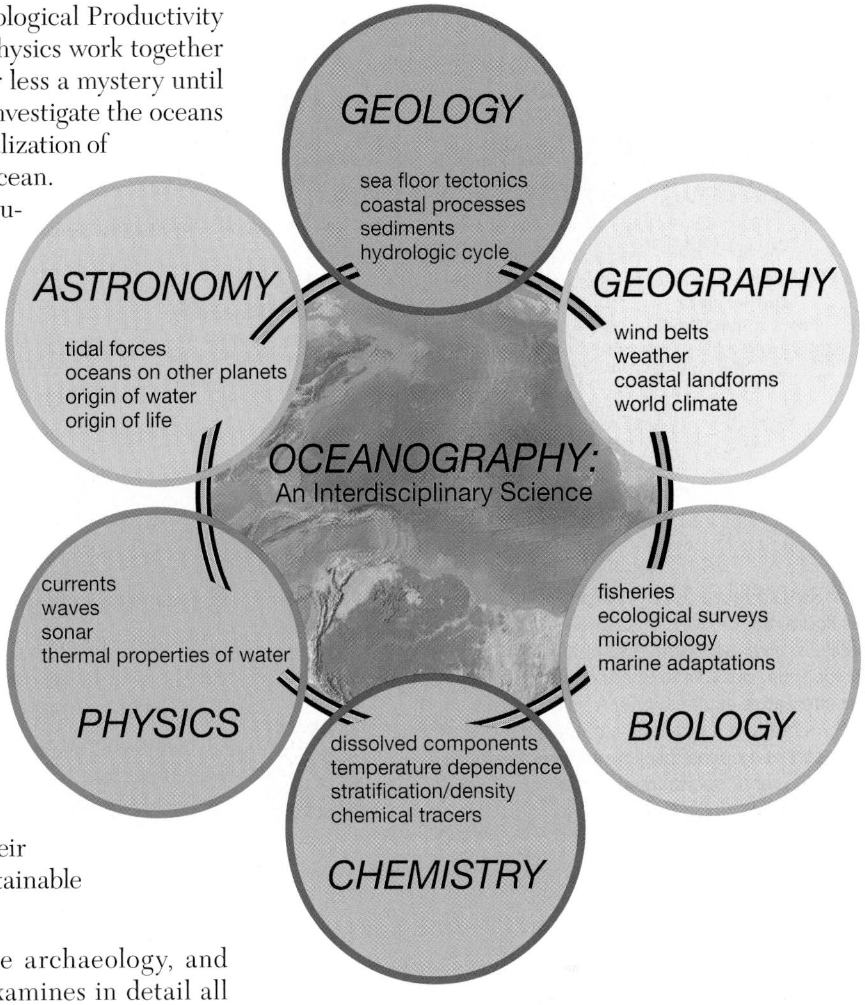

Figure 1.14 **A Venn diagram showing the interdisciplinary nature of oceanography.** Oceanography is an interdisciplinary science that overlaps into many scientific disciplines.

**CONCEPT CHECK 1.3 ▸ Explain why oceanography is considered an interdisciplinary science.**

1 What was the impetus for studying ocean processes that led to the great expansion of the science of oceanography?

2 What are the four main disciplines or subfields of study in oceanography? What other marine-related disciplines exist?

3 What does it mean when oceanography is called an interdisciplinary science?

**RECAP A broad range of interdisciplinary science topics from the diverse fields of geology, chemistry, physics, and biology are included in the study of oceanography.**

## 1.4 ▸ What Is the Process of Science and the Nature of Scientific Inquiry?

In modern society, scientific studies are increasingly used to substantiate the need for political, community, or personal action. However, there is often little understanding of how science operates. For instance, how certain are we about a particular scientific theory? How are facts different from theories?

The overall goal of science is to discover underlying patterns in the natural world and then to use this knowledge to make predictions about what should or should not be expected to happen given a certain set of circumstances. Scientists develop explanations about the causes and effects of various natural

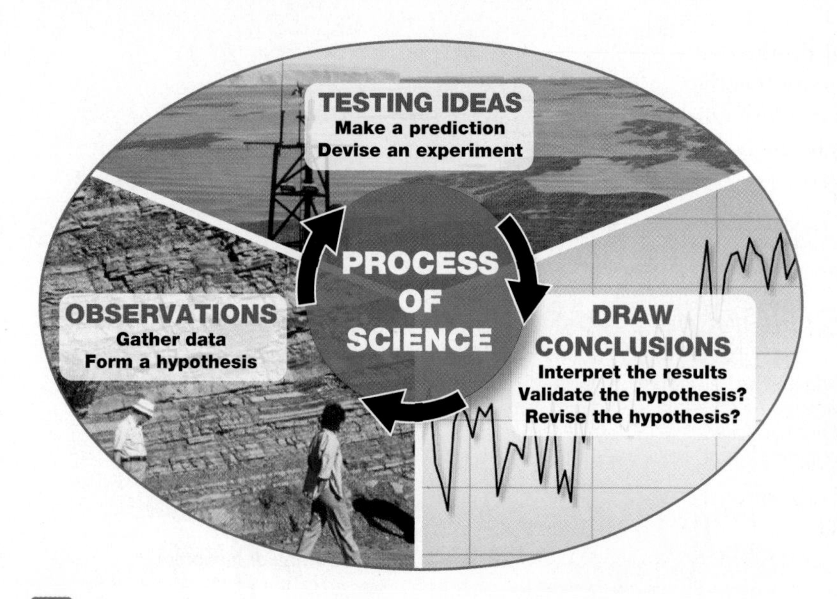


SmartFigure 1.15 **The process of science.** As shown in this figure, the process of science is iterative. After analyzing the results of an experiment, a scientist will often revise a hypothesis, decide more observations are necessary, talk to colleagues, and test alternative explanations. A hallmark of a good study is not that it provides a pat answer to a problem but that it opens the door to new questions and a deeper understanding of the natural world. https://goo.gl/cNH567

phenomena (such as why Earth has seasons or what the structure of matter is). This work is based on an assumption that all natural phenomena are controlled by understandable physical processes, and that the same physical processes operating today have been operating throughout time. Consequently, science has demonstrated remarkable power in allowing scientists to describe the natural world accurately, to identify the underlying causes of natural phenomena, and to better predict future events that rely on natural processes.

Science supports the explanation of the natural world that best explains all available observations. The investigation of natural phenomena using scientific principles is formalized into what is called the **process of science**, which is the way scientists ask questions, collect and evaluate evidence, and draw conclusions. As Figure 1.15 shows, *testing ideas* is central to the process of science: only with evidence that meets the high standards of scientific scrutiny can you separate science from pseudoscience, fact from fiction.

## Observations

Although the process of science is actually cyclic or *iterative* (meaning that, based on the results of one test you go back to the start and run another test) you could say that the process of science begins with **observations**. Observations are events or phenomena we can detect with our senses. They are things we can manipulate, measure, see, touch, hear, taste, or smell, often by experimenting with them directly or by using sophisticated tools (such as a microscope or telescope) to sense them. Observations can occur in the controlled environment of a scientific experiment, or they can happen by chance, such as when a diver on vacation notices an unusual die-off of coral on a reef. When observations are repeated and documented, they become *data*—information about our natural world that then leads scientists to form hypotheses.

## Hypotheses

As observations are being made, the human mind attempts to sort out the observations in a way that reveals some underlying order or pattern in the observations or phenomena. This sorting process—which involves a lot of trial and error—seems to be driven by a fundamental human urge to make sense of our world. This is how **hypotheses** (*hypo* = under, *thesis* = an arranging) are made. A hypothesis is more than an informed or educated guess. A hypothesis is a tentative, testable statement about the general nature of the phenomena being observed. In other words, a hypothesis is an initial idea of how or why things happen in nature. For example, there were several hypotheses about why whales *breach* (that is, why whales sometimes leap entirely out of water) until a group of scientists made headway by systematically studying it (see Process of Science 1.1).

Scientists often use the *multiple working hypotheses* to guide them as they collect data. Only through testing of hypotheses and collection of data does a single hypothesis begin to make sense. That's why testing is an iterative process. Further, a defining property of a hypothesis is its ability to make *predictions*. For example, when testing a hypothesis, a scientist might note that if a certain hypothesis is true, then another thing should be happening. Note that if a hypothesis cannot be tested, it is not scientifically useful, no matter how interesting it might seem.

## Testing

Hypotheses are used to understand certain occurrences that lead to further research and the refinement of those hypotheses. For instance, if we are trying to

## PROCESS OF SCIENCE 1.1: Why Do Whales Leap?

### BACKGROUND

Most species of whales have been observed at times to leap out of the water, a spectacular display called "breaching" (**Figure 1D**). But why do whales breach? It has long been known that whales communicate with each other over long distances using underwater vocalizations, such as clicks, whistles, and even elaborate songs. Whales have also been observed to use pectoral (side) fin and tail (fluke) slaps on the water surface, presumably to communicate with other whales nearby.

### FORMING A HYPOTHESIS

Scientists suspected that breaching was an additional form of surface communication, but how breaching fit in with whales' other ways to communicate has been a mystery until recently. Indeed, many have wondered why such large, bulky creatures would expend so much energy to leap out of the water at all.

### DEVISING AN EXPERIMENT

If breaching is a form of whale communication, then whales should breach in predictable situations and with a defined purpose. In 2010 and 2011, a scientific team examined this idea by observing 94 different groups of humpback whales (*Megaptera novaeangliae*) migrating along the coast of Australia. The migratory route is close to land, so scientists were able to view the whales from shore to ensure that the team did not disturb the normal behavior of the whales. The scientists meticulously observed and recorded the whales' behavior and listened to their sounds using underwater microphones. After several weeks of monitoring the social and environmental contexts of how whales interacted within their group—and with other groups in the vicinity—the scientists began to notice some patterns.

**Figure 1D** **A breaching humpback whale (*Megaptera novaeangliae*).**

### INTERPRETING THE RESULTS

The study showed that distance between groups was the main variable in predicting breaching behavior. The researchers found that groups of whales traveling close to each other engaged in fin- and fluke-slaps more often than breaching. As the groups moved farther away from each other or split, or if ocean conditions were louder, they tended to breach more often. This makes sense if breaching is a form of communication—the noise from a whale's body striking the water during a breach creates a percussion sound that, like a loud hammer, can be heard over longer distances or louder conditions than the quieter fin-and fluke-slaps. The scientists in this study concluded that breaching is indeed related to whale communication, with whale groups using breaching to communicate when they are farther apart.

### THINKING LIKE A SCIENTIST: WHAT'S NEXT?

Other than communication, several alternative explanations for whale breaching have included dislodging of external parasites, mating rituals, territorial displays, or even just for play. Can you think of others? Choose one of these possible explanations and design an experiment to test your hypothesis for whale breaching.

understand why some large sharks attack people, one hypothesis could be that people floating at the surface are mistaken for their food source—which is normally seals and sea lions. Another hypothesis is that sharks are territorial and so they are defending their space. A careful study would have to examine the types and occurrences of shark attacks on people to compile data, which would either support one or the other of the hypotheses, or cause them to be reconsidered and modified. If observations clearly suggest that a hypothesis is incorrect (that is, the hypothesis is *falsified*), then it must be dropped, and other alternative explanations of the facts must be considered. Only after much testing and experimentation—usually done by many experimenters using a wide variety of repeatable tests—does a hypothesis gain validity.

## STUDENTS SOMETIMES ASK . . .

***How can I accept a scientific idea if it's just a theory?***

When most people use the word "theory" in everyday life, it usually means an idea or a guess (such as the all-too-common "conspiracy theory"), but the word has a much different meaning in science. In science, a theory is not a guess or a hunch. It's a well-substantiated, well-supported, well-documented explanation for observations about the natural world. It's a powerful tool that ties together all the facts about something, providing an explanation that fits all the observations and is used to make predictions (for example, what will happen given a certain set of circumstances). In science, a theory is a well-established explanation of how the natural world works. For a scientific theory to exist, scientists have to be *very* sure about it. So, don't discount a scientific idea because it's "just a theory." As famed astrophysicist Neil deGrasse Tyson has stated about the validity of science, *"The good thing about science is that it's true whether or not you believe in it."*

## STUDENTS SOMETIMES ASK . . .

***If a theory is proven again and again, does it become a law?***

No, but that's a common misconception. In science, we collect facts, or observations, we use natural laws to describe them (often using mathematics), and we use a theory to explain them. Natural laws are typically conclusions based on repeated scientific experiments and observations over many years and which have become accepted universally within the scientific community. For example, the *law of gravity* is a description of the force; then there is the *theory of gravitational attraction,* which explains why the force occurs. Theories don't get "promoted" to a law by an abundance of proof, and so a theory never becomes a law. They're really two separate things.

**RECAP** **Science supports the explanation of the natural world that best explains all available observations. Because new observations can modify existing theories, scientific ideas are constantly being refined and updated.**

## Theory

A **theory** (*theoria* = a looking at) is different from a hypothesis in that it provides a much broader explanation of some aspect of the natural world that incorporates facts, scientific laws (which are quantifiable generalizations about the physical universe, such as Newton's law of gravity), and logical inferences, as well as tested hypotheses. A theory is not a guess or a hunch. Rather, it is a well-substantiated understanding of the natural world that develops from extensive observation, experimentation, and creative reflection. Successful theories do not include numerous special cases or exceptions. Examples of prominent, well-accepted theories that are held with a very high degree of confidence include biology's theory of evolution (which is discussed later in this chapter) and geology's theory of plate tectonics (which is covered in the next chapter).

In science, theories are formalized only after many years of testing and verification. Thus, scientific theories have been rigorously scrutinized to the point where most scientists agree that they are the best explanation of certain observable facts. Nonetheless, scientists recognize that a theory is never really "proven." The search for new observations and new questions—as well as the development of new technology ensures the continuation of the cycle that is the process of science.

## Theories and the Truth

We've seen how the process of science is used to develop theories, but does science ever arrive at the undisputed "truth"? Science never reaches an absolute truth because we can never be certain that we have all the observations, especially considering that new technology will be available in the future to examine phenomena in different ways. Notice that there is no end point to the process depicted here. New observations are always possible, so the nature of scientific truth is subject to change. Therefore, it is more accurate to say that science arrives at a conclusion that is *most likely* true, based on the available observations.

It is not a downfall or weakness of science that scientific ideas are modified as more observations are collected. In fact, the opposite is true. Science is a process that depends on reexamining ideas as new observations are made. Thus, science progresses when new observations yield new hypotheses and modification of theories. As a result, science is littered with hypotheses that have been abandoned in favor of later explanations that fit new observations. One of the best known is the idea that Earth was at the center of the universe, a proposal that was supported by the apparent daily motion of the Sun, Moon, and stars around Earth. At the time, this was a very reasonable idea—an obvious "truth"— to anyone who looked up at the sky.

The statements of science should never be accepted as the "final truth." Over time, however, they generally form a sequence of increasingly more accurate statements. Theories are the endpoints in science and do not turn into facts through accumulation of evidence. Nevertheless, the data can become so convincing that the accuracy of a theory is no longer questioned. For instance, the *heliocentric* (*helios* = sun, *centric* = center) *theory* of our solar system states that Earth revolves around the Sun rather than vice versa. Such concepts are supported by such abundant observational and experimental evidence that they are no longer questioned in science and are considered to be scientific facts.

Is the process of science as formal as Figure 1.15 suggests? Actually, the work of scientists is much less formal and is not always done in a clearly logical and systematic manner. In reality, the process of science is a rich and complex process that is not always so methodical. (For a more detailed look at how the process of science works, see https://undsci.berkeley.edu/article/howscienceworks_01.) Like detectives analyzing a crime scene, scientists use ingenuity and serendipity, visualization of models, synthesizing ideas, and sometimes even follow hunches in order to unravel the mysteries of nature.

A final word about theories and scientific truth must take into account the essential role of peer review in verifying scientific ideas. Once scientists make a discovery, their goal is to get word of their results out to the scientific community. This is typically done via a published paper, but a draft of the manuscript is first checked by other experts to see if the work has been conducted according to proper scientific protocols and if the conclusions are valid. Normally, changes or corrections are suggested, and the paper is revised before it is published. This process helps weed out inaccurate or poorly formed ideas. Peer review is the final test of a scientific idea: if the evidence and conclusions of a study meet the strict standards of the scientific community, then it is ready to be shared.

**CONCEPT CHECK 1.4 ▸ Describe the process of science and the nature of scientific inquiry.**

1 Describe the steps involved in the process of science, starting with an observation about the natural world.

2 What is the difference between a hypothesis and a theory?

3 Briefly comment on the phrase "scientific certainty." Is it an oxymoron (a combination of contradictory words), or are scientific theories considered to be the absolute truth?

4 Can a theory ever be so well established that it becomes a fact? Explain.

## 1.5 ▸ How Were Earth and the Solar System Formed?

Earth is the third of eight major planets in our **solar system** that revolve around the Sun (**Figure 1.16**). Note that Pluto, which used to be considered the ninth planet in our solar system, was reclassified by the International Astronomical Union as a "dwarf planet" in 2006, along with other similar bodies. Evidence suggests that the Sun and the rest of the solar system formed about 5 billion years ago from a huge cloud of gas and space dust called a **nebula** (*nebula* = a cloud). Astronomers base this hypothesis on the orderly nature of our solar system and the consistent age of meteorites (pieces of the early solar system). Using sophisticated telescopes, astronomers have also been able to observe distant nebula and planetary systems in various stages of formation elsewhere in our galaxy (**Figure 1.17**). In addition, nearly 4000 planets have been discovered outside our solar system—including several that are about the size of Earth—by detecting the telltale wobble of distant stars or slight changes in the emitted light of remote stars, such as the decrease in brightness as planets pass in front of them.

**(a)** Features and relative sizes of the Sun and the eight major planets of the solar system.

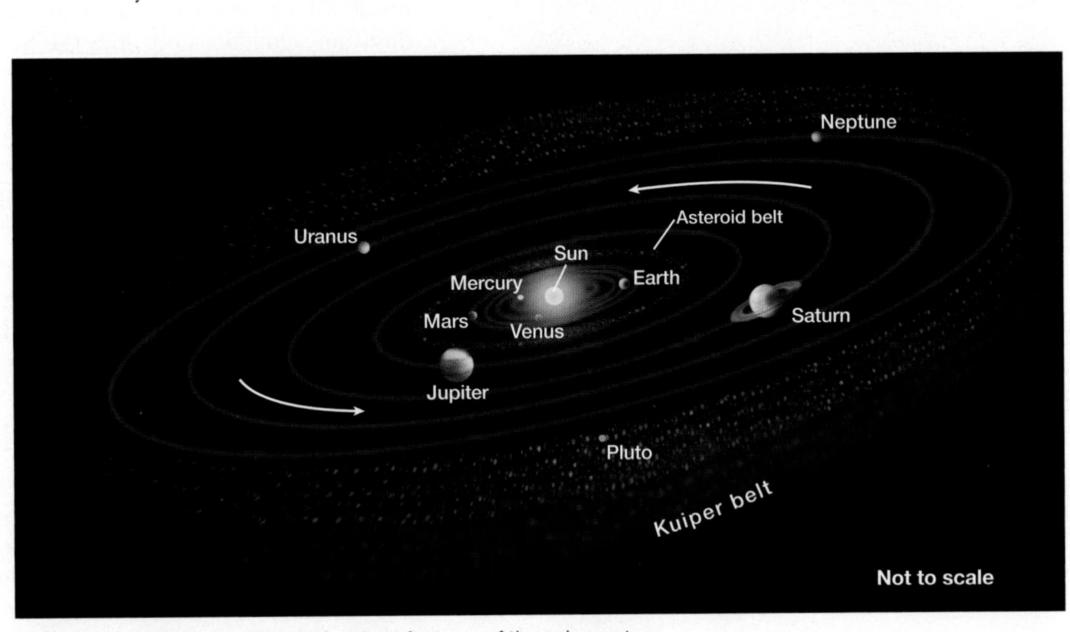


**(b)** Orbits and relative positions of various features of the solar system.

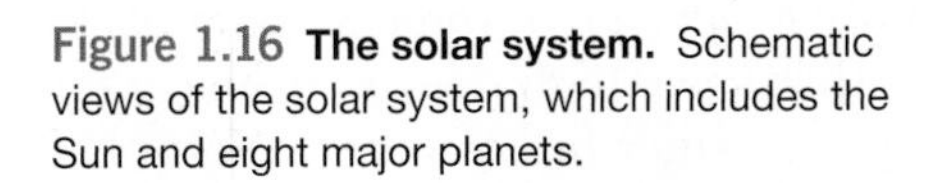

**Figure 1.16 The solar system.** Schematic views of the solar system, which includes the Sun and eight major planets.

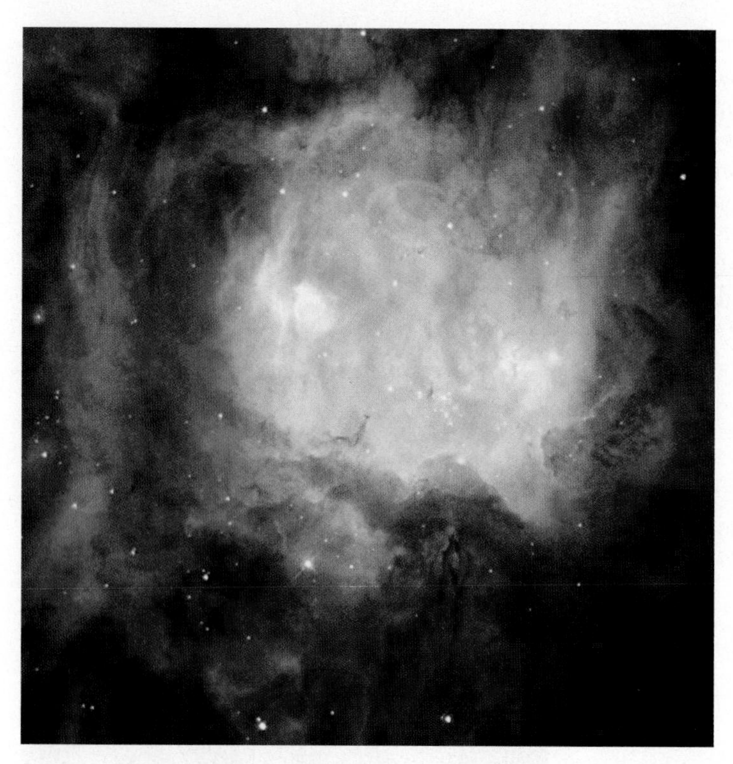

Figure 1.17 **The Ghost Head Nebula.** NASA's Hubble Space Telescope image of the Ghost Head Nebula (NGC 2080), which is a site of active star formation.

## The Nebular Hypothesis

According to the **nebular hypothesis** (Figure 1.18), all bodies in the solar system formed from an enormous cloud composed mostly of hydrogen and helium, with only a small percentage of heavy elements. As this huge accumulation of gas and dust revolved around its center, it began to contract under its own gravity, becoming hotter and denser, eventually forming the Sun.

As the nebular matter that formed the Sun contracted, small amounts of it were left behind in swirling eddies, which are similar to small whirlpools in a stream. The material in these eddies was the beginning of the **protoplanets** (*proto* = original, *planetes* = wanderers) and their orbiting satellites, which later consolidated into the present planets and their moons.

## Proto-Earth

**Proto-Earth** looked very different from Earth today. Its size was larger than today's Earth, and there were neither oceans nor any life on the planet. In addition, the structure of the deep proto-Earth is thought to have been *homogenous* (*homo* = alike, *genous* = producing), which means that it had a uniform composition throughout. The structure of proto-Earth changed, however, as its heavier constituents sank toward the center to form a heavy core.

During this early stage of formation, many meteorites and comets from space bombarded proto-Earth (Figure 1.19). In fact, a leading theory states that the Moon was born in the aftermath of a titanic collision between a Mars-sized planet named *Theia* and proto-Earth. While most of Theia was swallowed up and incorporated into the magma ocean it created on impact, the collision also flung a small world's worth of vaporized and molten rock into orbit. Over time, this debris coalesced into a sphere and created Earth's orbiting companion, the Moon.

During this early formation of the protoplanets and their satellites, the Sun condensed into a body so massive and hot that pressure within its core initiated the process of **thermonuclear fusion** (*thermo* = hot, *nucleos* = a little nut; *fusus* = melted). Thermonuclear fusion occurs when temperatures reach tens of millions of degrees and hydrogen **atoms** (*a* = not, *tomos* = cut) combine to form helium atoms, releasing enormous amounts of energy. (Thermonuclear fusion in stars also creates larger and more complex elements, such as carbon. It is interesting to note that as a result, all matter—even the matter that comprises our bodies—originated as stardust long ago.) Not only does the Sun emit light, it also emits *ionized* (electrically charged) particles that make up the *solar wind*. During the early stages of formation of the solar system, this solar wind blew away the nebular gas that remained from the formation of the planets and their satellites.

The protoplanets closest to the Sun (including Earth) also lost their initial atmospheres (mostly hydrogen and helium), blown away by the bombardment by ionized solar radiation. At the same time, these rocky protoplanets were gradually cooling, causing them to contract and drastically shrink in size. As the protoplanets continued to contract, another source of heat was produced deep within their cores from the spontaneous disintegration of atoms, called *radioactivity* (*radio* = ray, *acti* = to cause).

## Density and Density Stratification

**Density**, which is an extremely important physical property of matter, is defined as mass per unit volume. In common terms, an easy way to think about density is that it is a measure of *how heavy something is for its size*. For instance, an object that has a low density is light for its size (like a dry sponge, foam packing, or a surfboard). Conversely, an object that has a high density is heavy for its size (like cement, most metals, or a large container full of water). Note that density has nothing to do with the *thickness* of an object; some objects (like a

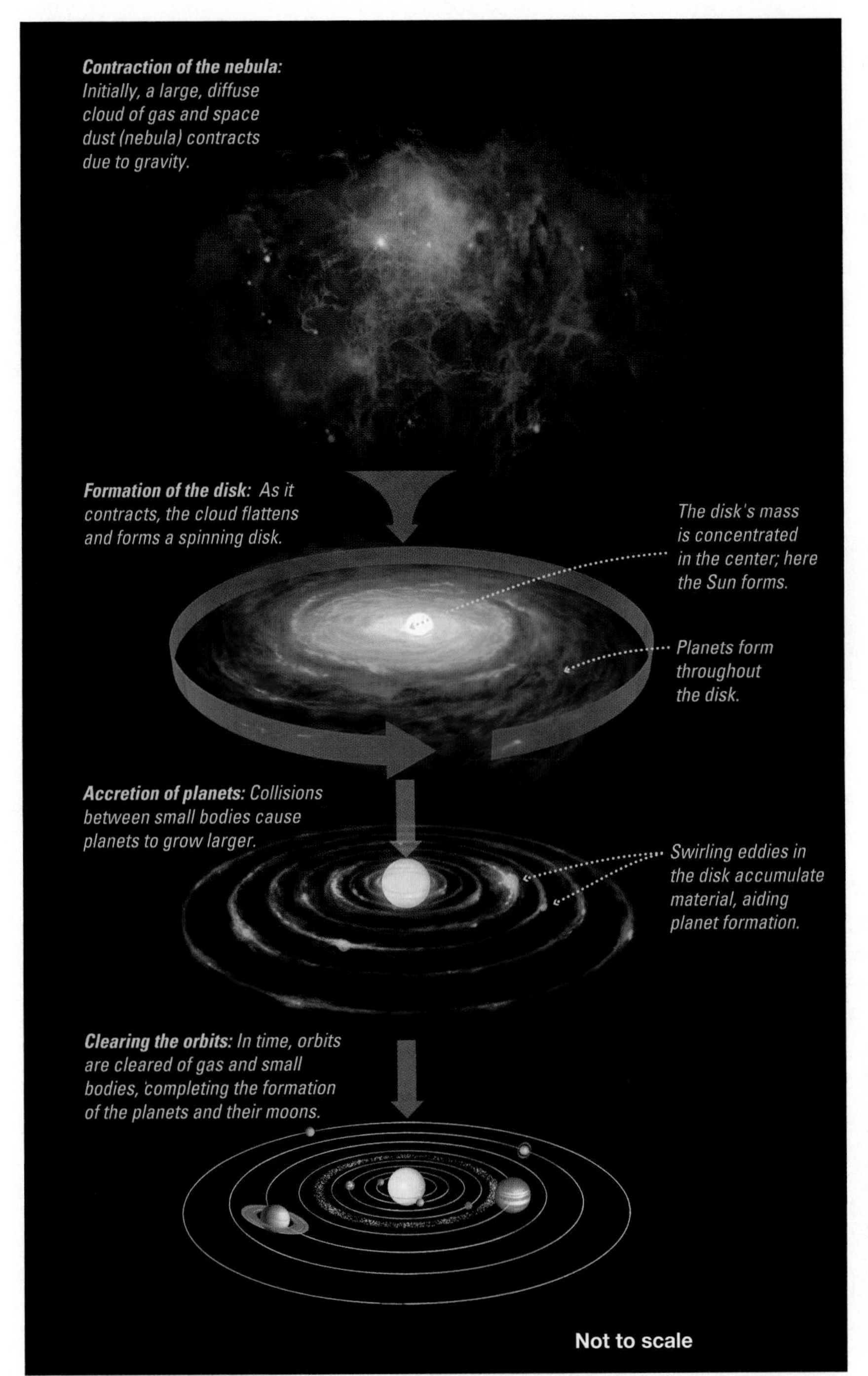

Contraction of the nebula: Initially, a large, diffuse cloud of gas and space dust (nebula) contracts due to gravity.
Formation of the disk: As it contracts, the cloud flattens and forms a spinning disk.
The disk's mass is concentrated in the center; here the Sun forms.
Planets form throughout the disk.
Accretion of planets: Collisions between small bodies cause planets to grow larger.
Swirling eddies in the disk accumulate material, aiding planet formation.
Clearing the orbits: In time, orbits are cleared of gas and small bodies, completing the formation of the planets and their moons.
Not to scale

SmartFigure 1.18 **The nebular hypothesis of solar system formation.** According to the nebular hypothesis, our solar system formed from the gravitational contraction of an interstellar cloud of gas and space dust called a *nebula*. https://goo.gl/FoY7Yt

**Animation**
The Nebular Hypothesis of Solar System Formation
http://goo.gl/KObsRK

Figure 1.19 **Proto-Earth.** An artist's conception of what Earth may have looked like early in its development.

stack of foam packing) can be thick but have low density. In reality, density is related to molecular packing, with higher packing of molecules into a certain space resulting in higher density. Density is an extremely important concept that will be discussed in many other chapters in this book. For example, the density of Earth's layers dramatically affects their locations within Earth (Chapter 2), the density of air masses affects their positions in the atmosphere and other properties (Chapter 6), and the density of water masses determines how deep in the ocean they are found and how they move (Chapter 7).

On the early Earth, heat generated at the surface by the bombardment of space debris and heat released internally by the decay of radioactive elements was so intense that Earth's surface became molten. Once Earth became a ball of hot liquid rock, the elements were able to segregate according to their densities in a process called **density stratification** (*strati* = a layer, *fication* = making), which occurs because of *gravitational separation*. The highest-density materials (primarily iron and nickel) concentrated in the core, whereas progressively lower-density components (primarily rocky material) formed concentric spheres around the core. If you've ever noticed how oil-and-vinegar salad dressing settles out into a lower-density top layer (the oil) and a higher-density bottom layer (the vinegar), then you've seen how density stratification causes separate layers to form.

**EXPLORING DATA ▼**

**1. Comparing the two halves of this figure, explain why the boundaries on the two sides don't always align.**

**2. Other than the center of Earth and Earth's surface, what is the one boundary that actually does match up?**

Composition: Light silicate materials
Ocean
Crust
Property: Rigid
100 km (60 mi)
Lithosphere
Property: Plastic
700 km (430 mi)
Asthenosphere
Composition: Silicate materials rich in iron and magnesium
Mantle
Property: Rigid
Mesosphere
2885 km (1800 mi)
Earth's layers classified by chemical composition
Earth's layers classified by physical properties
Composition: Iron with nickel and sulfur
Property: Liquid
Outer core
5155 km (3200 mi)
Core
Property: Rigid
Inner core
6371 km (3960 mi)

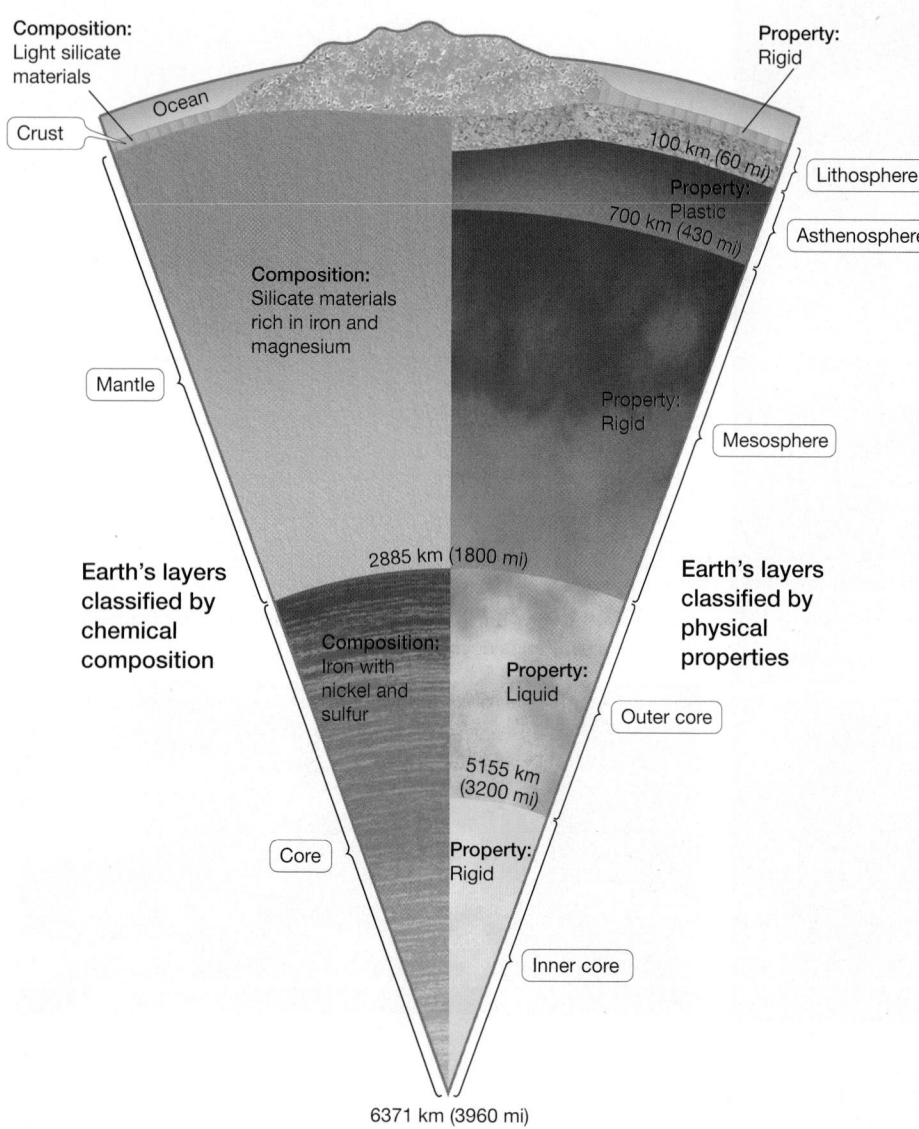

**SmartFigure 1.20 Comparison of Earth's chemical composition and physical properties.** A cross-sectional view of Earth, showing Earth's layers classified by chemical composition along the left side of the diagram. For comparison, Earth's layers classified by physical properties are shown along the right side of the diagram. Layers near the surface are enlarged for clarity.
https://goo.gl/JjglcZ

## Earth's Internal Structure

As a result of density stratification, Earth became a layered sphere based on density, with the highest-density material found near the center of Earth and the lowest-density material located near the surface. Let's examine Earth's internal structure and the characteristics of its layers.

**CHEMICAL COMPOSITION VERSUS PHYSICAL PROPERTIES** The cross-sectional view of Earth in **Figure 1.20** shows that Earth's inner structure can be subdivided according to its chemical composition (the chemical makeup of Earth materials) or its physical properties (how the rocks respond to increased temperature and pressure at depth).

**CHEMICAL COMPOSITION** Based on chemical composition, Earth consists of three layers: the **crust**, the **mantle**, and the **core** (Figure 1.20). If Earth were reduced to the size of an apple, then the crust would be its thin skin. It extends from the surface to an average depth of about 30 kilometers (20 miles). The crust is composed of relatively low-density rock, consisting mostly of various *silicate minerals* (common rock-forming minerals with silicon and oxygen). There are two types of crust—oceanic and continental—that will be discussed in the next section.

Immediately below the crust is the mantle. It occupies the largest volume of the three layers and extends to a depth of about 2885 kilometers (1800 miles). The mantle is composed of relatively high-density iron and magnesium silicate rock.

Beneath the mantle is the core. It forms a large mass from 2885 kilometers (1800 miles) to the center of Earth at 6371 kilometers (3960 miles). The core is composed of even higher-density metal (mostly iron and nickel).

**PHYSICAL PROPERTIES** Based on physical properties, Earth is composed of five layers (Figure 1.20): the **inner core**, the **outer core**, the **mesosphere** (*mesos* = middle, *sphere* = ball), the **asthenosphere** (*asthenos* = weak, *sphere* = ball), and the **lithosphere** (*lithos* = rock, *sphere* = ball).

The lithosphere is Earth's cool, rigid, outermost layer. It extends from the surface to an average depth of about 100 kilometers (62 miles) and includes the crust plus the topmost portion of the mantle. The lithosphere is *brittle* (*brytten* = to shatter), meaning that it will fracture when force is applied to it. As will be discussed in Chapter 2, "Plate Tectonics and the Ocean Floor," the plates involved in plate tectonic motion are the plates of the lithosphere.

Beneath the lithosphere is the asthenosphere. The asthenosphere is *plastic* (*plasticus* = molded), meaning that it will flow when a gradual force is applied to it. It extends from about 100 kilometers (62 miles) to 700 kilometers (430 miles) below the surface, which is the base of the upper mantle. At these depths, it is hot enough to partially melt portions of most rocks.

Beneath the asthenosphere is the mesosphere. The mesosphere extends to a depth of about 2885 kilometers (1800 miles), which corresponds to the middle and lower mantle. Although the asthenosphere deforms plastically, the mesosphere is rigid because of the increased pressure at these depths.

Beneath the mesosphere is the core. The core consists of the outer core, which is liquid and capable of flowing, and the inner core, which is rigid and does not flow. Again, the increased pressure at the center of Earth keeps the inner core from flowing.

## STUDENTS SOMETIMES ASK . . .

***How do we know about the internal structure of Earth?***

You might suspect that the internal structure of Earth has been sampled directly. Although attempts are currently underway, the truth is humans have never penetrated beneath Earth's crust! Instead, the internal structure of Earth is determined by analyzing earthquakes that send vibrations through the deep interior of our planet. These vibrations are called *seismic waves*, which change their speed and are bent and reflected as they move through zones having different properties. For example, seismic waves travel more slowly through areas of hotter rock and speed up thorough colder rock. An extensive network of monitoring stations around the world detects and records these vibrations. The data are analyzed and used to determine the structure and properties of the deep Earth and how they change over time. In fact, repeated analysis of seismic waves that pass through Earth has allowed researchers to construct a detailed three-dimensional model of Earth's interior—similar to an MRI in medical technology—which reveals the inner workings of our planet (**Figure 1.21**).

**Animation**
How Seismic Waves Reveal Earth's Internal Layers
http://goo.gl/76mJf8

Areas of hotter rock (red shading) cause seismic waves to slow down.
Areas of cooler rock (green shading) cause seismic waves to speed up.

**Figure 1.21 Determining the internal structure of Earth.** By analyzing how various seismic waves travel through Earth, scientists are able to map Earth's complex inner structure.

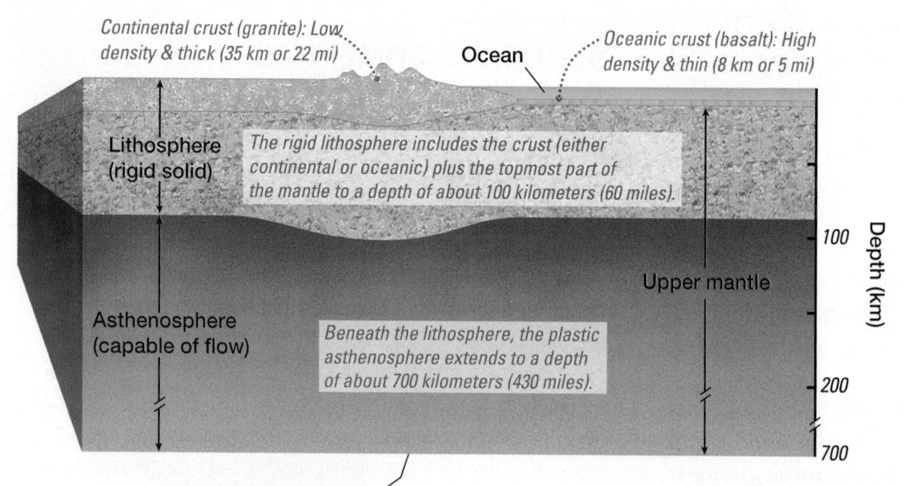


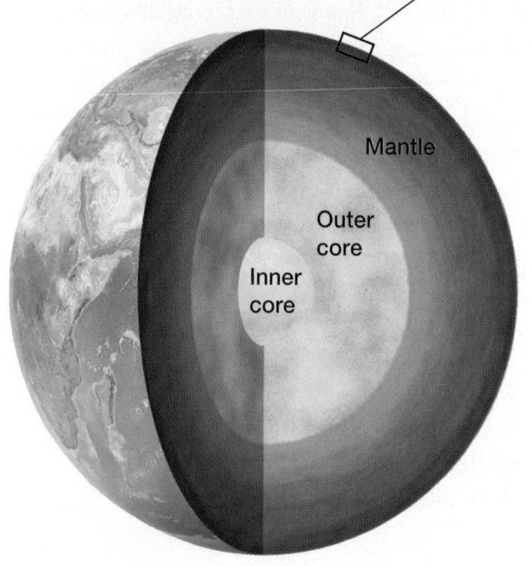


Figure 1.22 **Internal structure of Earth showing an enlargement of layers close to the surface.**

SmartTable 1.1 **Comparing oceanic and continental crust.**
https://goo.gl/EneJOl

**NEAR THE SURFACE** The top portion of **Figure 1.22** shows an enlargement of Earth's layers closest to the surface.

***Lithosphere*** The lithosphere is a relatively cool, rigid shell that includes all the crust and the topmost part of the mantle. In essence, the topmost part of the mantle is attached to the crust, and the two act as a single unit, approximately 100 kilometers (62 miles) thick. The expanded view in Figure 1.22 shows that the crust portion of the lithosphere is further subdivided into oceanic crust and continental crust, which are compared in **Table 1.1**.

***Oceanic versus Continental Crust*** **Oceanic crust** underlies the ocean basins and is composed of the igneous rock **basalt**, which is dark colored and has a relatively high density of about 3.0 grams per cubic centimeter. (Water has a density of 1.0 grams per cubic centimeter. Thus, basalt with a density of 3.0 grams per cubic centimeter is three times denser than water.) The average thickness of the oceanic crust is only about 8 kilometers (5 miles). Basalt originates as molten magma beneath Earth's crust (typically from the mantle), some of which comes to the surface during underwater sea floor eruptions.

**Continental crust** is composed primarily of the lower-density and lighter-colored igneous rock **granite**. (At the surface, continental crust is often covered by a relatively thin layer of surface sediments. Below these, granite can be found.) It has a density of about 2.7 grams per cubic centimeter. The average thickness of the continental crust is about 35 kilometers (22 miles) but may reach a maximum of 60 kilometers (37 miles) beneath the highest mountain ranges. Most granite originates beneath the surface as molten magma that cools and hardens within Earth's crust. No matter which type of crust is at the surface, it is all part of the lithosphere.

***Asthenosphere*** The asthenosphere is a relatively hot, plastic region beneath the lithosphere. It extends from the base of the lithosphere to a depth of about 700 kilometers (430 miles) and is entirely contained within the upper mantle. The asthenosphere can deform without fracturing if a force is applied slowly. This means that it has the ability to flow but has high **viscosity** (*viscosus* = sticky). Viscosity is a measure of a substance's resistance to flow. (Substances that have high viscosity—a high resistance to flow—include toothpaste, honey, tar, and Silly Putty; a common substance that has low viscosity is water. Note that a substance's viscosity often changes with temperature. For instance, as honey is heated, it flows more easily.) Studies indicate that the high-viscosity asthenosphere is flowing slowly through time; this has important implications for the movement of lithospheric plates.

**ISOSTATIC ADJUSTMENT** **Isostatic adjustment** (*iso* = equal, *stasis* = standing)—the vertical movement of crust—is the result of the buoyancy of Earth's lithosphere as it floats on the denser, plastic-like asthenosphere below. **Figure 1.23**, which shows a container ship floating in water, provides an example of isostatic adjustment. It shows that an empty ship floats high in the water. Once the ship is loaded with cargo, though, the ship undergoes isostatic adjustment and floats lower in the water (but hopefully won't sink!). When the cargo is unloaded, the ship isostatically adjusts itself and floats higher again.

**SmartTable 1.1 Comparing oceanic and continental crust**

| | Oceanic crust | Continental crust |
|---|---|---|
| Main rock type | Basalt (dark-colored igneous rock) | Granite (light-colored igneous rock) |
| Density (grams per cubic centimeter) | 3.0 | 2.7 |
| Average thickness | 8 kilometers (5 miles) | 35 kilometers (22 miles) |

**EXPLORING DATA ▲**

**Although the density difference between granite and basalt seems small, the units reported in Table 1.1 are for a very small volume, so let's examine a larger volume. Imagine a typical swimming pool, which has a volume of about 400 cubic meters. If that swimming pool was filled with liquid granite, how much would it weigh? Imagine a second identical pool that was filled with liquid basalt. How much would that pool weigh? And, what is the difference in weight (metric tons) between the two pools?**

Similarly, both continental and oceanic crust float on the denser mantle beneath. Oceanic crust is denser than continental crust, however, so oceanic crust floats lower in the mantle because of isostatic adjustment. Oceanic crust is also thin, which creates low areas for the oceans to occupy. Areas where the continental crust is thickest (such as large mountain ranges on the continents) float higher than continental crust of normal thickness, also because of isostatic adjustment. These mountains are similar to the top of a floating iceberg—they float high because there is a very thick mass of crustal material beneath them, plunged deeper into the asthenosphere. Thus, tall mountain ranges on Earth are composed of a great thickness of crustal material sometimes referred to as a root, which in essence keeps them buoyed up.

Areas that are exposed to an increased or decreased load experience isostatic adjustment. For instance, during the most recent ice age (which occurred during the Pleistocene Epoch between about 1.8 million and 10,000 years ago), massive ice sheets alternately covered and exposed northern regions such as Scandinavia and northern Canada. The additional weight of ice several kilometers thick caused these areas to isostatically adjust themselves lower in the mantle. Since the end of the most recent ice age, the reduced load on these areas caused by the melting of ice caused these areas to rise and experience **isostatic rebound**, which continues today. The rate at which isostatic rebound occurs gives scientists important information about the properties of the upper mantle.

Further, isostatic adjustment provides additional evidence for the movement of Earth's tectonic plates. Because continents isostatically adjust themselves by moving *vertically* they must not be firmly fixed in one position on Earth. As a result, the plates that contain these continents should certainly be able to move *horizontally* across Earth's surface. This remarkable idea will be explored in more detail in the next chapter.

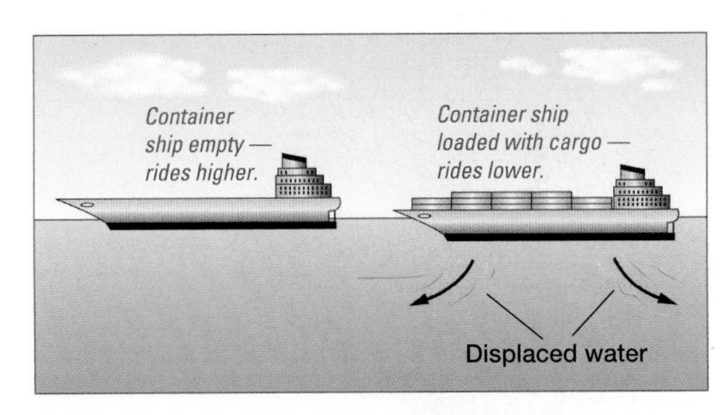


**Figure 1.23 A container ship experiences isostatic adjustment.** A ship will ride higher in water when it is empty and will ride lower in water when it is loaded with cargo, illustrating the principle of isostatic adjustment.

**Animation**
Isostatic Adjustment
https://goo.gl/esrK8U

**RECAP Earth has differences in composition and physical properties that create layers such as the brittle lithosphere and the plastic asthenosphere, which is capable of flowing slowly over time.**

## CONCEPT CHECK 1.5 ▸ Explain how Earth and the solar system formed.

1 Discuss the origin of the solar system using the nebular hypothesis.

2 How was proto-Earth different from Earth today?

3 What is density stratification, and how did it change proto-Earth?

4 What are some differences between the lithosphere and the asthenosphere?

Water vapor and other gases
Early in Earth's history, volcanic activity released large amounts of water vapor into the atmosphere.
Water vapor and other gases
Water vapor condensed into clouds.
Liquid water fell to Earth's surface, where it accumulated in low areas and over time formed the oceans.

Figure 1.24 **Formation of Earth's oceans.**

**Animation**
Formation of Earth's Oceans
https://goo.gl/gCXrDg

# 1.6 ▶ How Were Earth's Atmosphere and Oceans Formed?

The formation of Earth's atmosphere is related to the formation of the oceans; both are a direct result of density stratification.

## Origin of Earth's Atmosphere

Where did the atmosphere come from? As previously mentioned, Earth's initial atmosphere consisted of leftover gases from the nebula, but those particles were blown out to space by the Sun's solar wind. After that, a second atmosphere was most likely expelled from inside Earth by a process called **outgassing**. During the period of density stratification, the lowest-density material contained within Earth was composed of various gases. These gases rose to the surface and were expelled to form Earth's early atmosphere.

What was the composition of these atmospheric gases? They are believed to have been similar to the gases emitted from volcanoes, geysers, and hot springs today: mostly water vapor (steam), with small amounts of carbon dioxide, hydrogen, and other gases. The composition of this early atmosphere was not, however, the same composition as today's atmosphere. The composition of the atmosphere changed over time because of the influence of life (as will be discussed shortly) and possibly because of changes in the mixing of material in the mantle.

## Origin of Earth's Oceans

Where did the oceans come from? Similarly, their origin is linked directly to the origin of the atmosphere. Because outgassing releases mostly water vapor, this was the primary source of water on Earth, including supplying the oceans with water. **Figure 1.24** shows that as Earth cooled, the water vapor released to the atmosphere during outgassing condensed, fell to Earth, and accumulated in low areas. Evidence suggests that by at least 4 billion years ago, most of the water vapor from outgassing had accumulated to form the first permanent oceans on Earth.

Recent research, however, suggests that not all water came from inside Earth. Comets, which are composed of about half water, were once widely held to be the source of Earth's oceans. During Earth's early development, space debris left over from the origin of the solar system bombarded the young planet, and there could have been plenty of water supplied to Earth in this way. However, spectral analyses of the chemical composition of three comets—Halley, Hyakutake, and Hale-Bopp—during near-Earth passes they made in 1986, 1996, and 1997, respectively, revealed a crucial chemical difference between the hydrogen in comet ice and that in Earth's water. In 2014, the European Space Agency's Rosetta spacecraft reached the orbit of a comet to gather data on its ice. Although the lander sent to the comet's surface failed to send back data, the orbiter was able to analyze the comet's ice and determined that it, too, did not chemically match the water in Earth's oceans. If similar comets supplied large quantities of water to Earth, much of Earth's water would still exhibit the telltale type of hydrogen identified in these comets.

Even though comet ice doesn't match the chemical signature of Earth's water, there are a variety of small bodies in the solar system that could have supplied water to Earth. For example, recent analysis of a comet from the Kuiper Belt (an icy debris disk in the outer solar system that includes Pluto) indicates it *does* contain water with nearly the correct type of hydrogen that is found in Earth's water. In addition to Kuiper Belt objects, asteroids—rocky bodies that contain ice and orbit the Sun between Mars and Jupiter—also have a similar type of hydrogen and thus could have contributed water to an early Earth. These findings point to an emerging picture of a complex and dynamic evolution of the early solar system. Although it seems likely that most of Earth's water was derived from outgassing, other sources of water may have contributed to Earth's oceans as well.

**THE DEVELOPMENT OF OCEAN SALINITY** The relentless rainfall that landed on Earth's rocky surface dissolved many elements and compounds and carried them into the newly forming oceans. Even though Earth's oceans have existed since early in the formation of the planet, their chemical composition must have changed. This is because the high carbon dioxide and sulfur dioxide content in the early atmosphere would have created a very acidic rain, capable of dissolving greater amounts of minerals in the crust than occurs today. In addition, volcanic gases such as chlorine became dissolved in the atmosphere. As rain fell and washed to the ocean, it carried some of these dissolved compounds, which accumulated in the newly forming oceans. (Note that some of these dissolved components were removed or modified by chemical reactions between ocean water and rocks on the sea floor.) Eventually, a balance between inputs and outputs was reached, producing an ocean with a chemical composition similar to today's oceans. Further aspects of the oceans' salinity are explored in Chapter 5, "Water and Seawater."

**RECAP** **Originally, Earth had no oceans. The oceans (and atmosphere) came from inside Earth as a result of outgassing and were present by at least 4 billion years ago.**

**CONCEPT CHECK 1.6 ▶ Explain how Earth's atmosphere and oceans formed.**

1 Describe the origin of Earth's oceans.

2 Describe the origin of Earth's atmosphere. How is its origin related to the origin of Earth's oceans?

3 Have the oceans always been salty? Why or why not?

## 1.7 ▶ Did Life Begin in the Oceans?

The fundamental question of how life began on Earth has puzzled humankind since ancient times, and has recently received a great amount of scientific study. The evidence required to understand our planet's prebiotic environment and the events that led to first living systems is scant and difficult to decipher. Still, the inventory of current views on life's origin reveals a broad assortment of opposing positions. One recent hypothesis is that the organic building blocks of life may have arrived embedded in meteors, comets, or cosmic dust. Alternatively, life may have originated around hydrothermal vents—hot springs—on the deep-ocean floor. Yet another idea is that life originated in certain minerals that acted as chemical catalysts within rocks deep below Earth's surface.

According to the fossil record on Earth, the earliest-known life-forms were primitive bacteria that lived in sea floor rocks about 3.5 billion years ago. Unfortunately, Earth's geologic record for these early times is so sparse and the rocks are so deformed by Earth processes that the rocks no longer reveal life's precursor molecules. In addition, there is no direct evidence of Earth's environmental conditions (such as its temperature, ocean acidity, or the exact composition of the atmosphere) at the time of life's origin. Still, it is clear that the basic building blocks for the development of life were available from materials already present on the early Earth. And the presence of oceans on Earth was critical because this is the most likely place for these basic materials to interact and produce life.

### The Importance of Oxygen to Life

Oxygen, which comprises almost 21% of Earth's present atmosphere, is essential to human life for two reasons. First, our bodies need oxygen to "burn" (*oxidize*) food, releasing energy to our cells. Second, oxygen in the upper atmosphere in the form of *ozone* (*ozone* = to smell; ozone gets its name because of its pungent, irritating odor) protects the surface of Earth from most of the Sun's harmful ultraviolet radiation (which is why the atmospheric ozone hole over Antarctica has generated such concern).

### STUDENTS SOMETIMES ASK . . .

***Have the oceans always been salty? Are the oceans growing more or less salty through time?***

It is likely that the oceans have always been salty because wherever water comes in contact with the rocks of Earth's crust, some of the minerals dissolve. This is the source of salts in the oceans, whether from stream runoff or dissolving directly from the sea floor. Today, new minerals are forming on the sea floor at the same rate as dissolved materials are added. Thus, the salt content of the ocean is in a "steady state," meaning that it is not increasing or decreasing.

Interestingly, these questions can also be answered by studying the proportion of water vapor to chloride ion, $Cl^-$, in ancient marine rocks. Chloride ion is important because it forms part of the most common salts in the ocean (for example, sodium chloride, potassium chloride, and magnesium chloride). Also, chloride ion is produced by outgassing, like the water vapor that formed the oceans. Currently, there is no indication that the ratio of water vapor to chloride ion has fluctuated throughout geologic time, so it can be reasonably concluded that the oceans' salinity has been relatively constant through time.

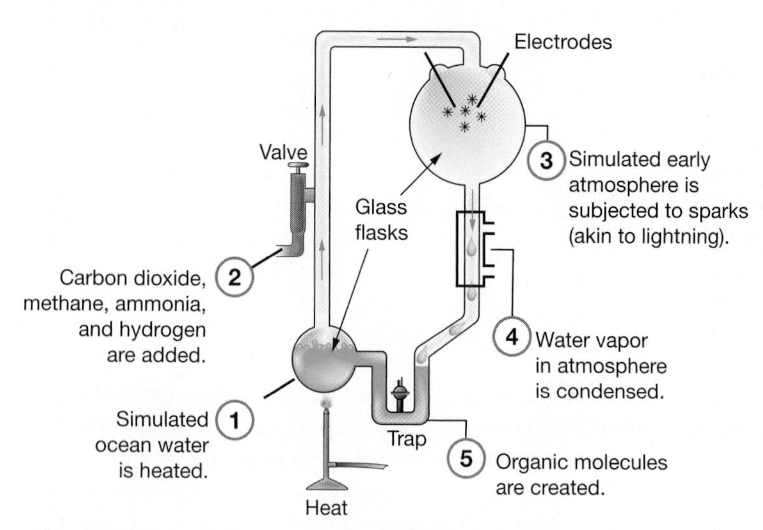


**(a)** Laboratory apparatus used by Stanley Miller to simulate the conditions of the early atmosphere and the oceans. The experiment produced various organic molecules and suggests that the basic components of life were created in a "prebiotic soup" in the oceans.

**(b)** Stanley Miller in 1999, with his famous apparatus in the foreground.

SmartFigure 1.25 **Creation of organic molecules.**
https://goo.gl/qLpjYn

Evidence suggests that Earth's early atmosphere (the product of outgassing) was different from Earth's initial hydrogen–helium atmosphere and different from the mostly nitrogen–oxygen atmosphere of today. The early atmosphere probably contained large percentages of water vapor and carbon dioxide and smaller percentages of hydrogen, methane, and ammonia, but very little free oxygen (oxygen that is not chemically bound to other atoms). Why was there so little free oxygen in the early atmosphere? Oxygen may well have been outgassed, but oxygen and iron have a strong affinity for each other. (As an example of the strong affinity of iron and oxygen, consider how common rust—a compound of iron and oxygen—is on Earth's surface.) As a result, iron in Earth's early crust would have reacted with the outgassed oxygen immediately, removing it from the atmosphere.

Without oxygen in Earth's early atmosphere, moreover, there would have been no ozone layer to block most of the Sun's ultraviolet radiation. The lack of a protective ozone layer may, in fact, have played a crucial role in several of life's most important developmental milestones.

## Stanley Miller's Experiment

In 1952, **Stanley Miller** (**Figure 1.25b**)—then a 22-year-old graduate student of chemist Harold Urey at the University of Chicago—conducted a laboratory experiment that had profound implications about the development of life on Earth. In Miller's experiment, he exposed a mixture of carbon dioxide, methane, ammonia, hydrogen, and water (the components of the early atmosphere and ocean) to ultraviolet light (from the Sun) and an electrical spark (to imitate lightning) (**Figure 1.25a**). By the end of the first day, the mixture turned pink, and after a week it was a deep, muddy brown, indicating the formation of a large assortment of organic molecules, including amino acids—which are the basic components of life—and other biologically significant compounds.

Miller's now-famous laboratory experiment of a simulated primitive Earth in a bottle—which has been duplicated and confirmed numerous times since—demonstrated that vast amounts of organic molecules could have been produced in Earth's early oceans, often called a "prebiotic soup." This prebiotic soup, perhaps spiced by extraterrestrial molecules aboard comets, meteorites, or interplanetary dust, was fueled by raw materials from volcanoes, certain minerals in sea floor rocks, and undersea hydrothermal vents. On early Earth, the mixture was energized by lightning, cosmic rays, and the planet's own internal heat, and it is thought to have created life's precursor molecules about 4 billion years ago.

Exactly how these simple organic compounds in the prebiotic soup assembled themselves into more complex molecules—such as proteins and DNA—and then into the first living entities remains one of the most tantalizing questions in science. Research suggests that with the vast array of organic compounds available in the prebiotic soup, several kinds of chemical reactions led to increasingly elaborate molecular structures. In fact, research suggests that small, simple molecules could have acted as templates, or "molecular midwives," in helping the building blocks of life's genetic material form long chains and thus may have assisted in the formation of longer, more elaborate molecular complexes. Among these complexes, some began to carry out functions associated with the basic molecules of life. As the products of one generation became the building blocks for another, even more complex molecules, or polymers, emerged over many generations that could store and transfer information. Such genetic polymers ultimately became encapsulated within cell-like membranes that were also present in Earth's primitive broth. The resulting cell-like complexes thereby housed self-replicating molecules capable of multiplying—and hence evolving—genetic information. Many specialists consider this emergence of genetic replication to be the true origin of life. Moreover, Miller's experiment demonstrates that simple chemicals under the right conditions could give rise to more complex chemical compounds that may lead to life-like behavior on other worlds.

**RECAP Organic molecules were produced in a simulation of Earth's early atmosphere and ocean, suggesting that life most likely originated in the oceans.**

# THE VOYAGE OF HMS *BEAGLE*: HOW IT SHAPED CHARLES DARWIN'S THINKING ABOUT THE THEORY OF EVOLUTION

*"Nothing in biology makes sense except in the light of evolution."*
*—Geneticist Theodosius Dobzhansky (1973)*

To help explain how biologic processes operating in nature were responsible for producing the many diverse and remarkable species on Earth, the English naturalist **Charles Darwin** (1809–1882) proposed the *theory of evolution* by natural selection, which he referred to as "common descent with modification." Many of the observations upon which he based the theory were made aboard the vessel HMS *Beagle* during its famous expedition from 1831 to 1836 that circumnavigated the globe (**Figure 1E**).

Darwin became interested in natural history during his student days at Cambridge University, where he was studying to become a minister. Because of the influence of John Henslow, a professor of botany, he was selected to serve as an unpaid naturalist on HMS *Beagle*. The *Beagle* sailed from Devonport, England, on December 27, 1831, under the command of Captain Robert Fitzroy. The major objective of the voyage was to complete a survey of the coast of Patagonia (Argentina) and Tierra del Fuego and to make chronometric measurements. The voyage allowed the 22-year-old Darwin—who was often seasick—to disembark at various locations and study local plants and animals. What particularly influenced his thinking about evolution were the discovery of fossils in South America, the different tortoises throughout the Galápagos Islands, and the identification of 15 closely related species of Galápagos finches. These finches differ greatly in the configuration of their beaks (Figure 1E, *left inset*), which are suited to their diverse feeding habitats. After his return to England, Darwin noted the adaptations of finches and other organisms living in different habitats and concluded that all organisms change slowly over time as products of their environment.

Darwin recognized the similarities between birds and mammals and reasoned that they must have evolved from reptiles. Patiently making observations over many years, he also noted the similar skeletal framework of species such as bats, horses, giraffes, elephants, porpoises, and humans, which led him to establish relationships between various groups. Darwin suggested that the differences between species were the result of adaptation over time to different environments and modes of existence.

In 1858, Darwin hastily published a summary of his ideas about natural selection because fellow naturalist *Alfred Russel Wallace*, working half a world away cataloguing species in what is now Indonesia, had independently discovered the same idea. A year later, Darwin published his remarkable masterwork *On the Origin of Species by Means of Natural Selection* (Figure 1E, *right inset*), in which he provided extensive and compelling evidence that all living beings—including humans—have evolved from a common ancestor. At the time, Darwin's ideas were highly controversial because they stood in stark conflict with what most people believed about the origin of humans. Darwin also produced important publications on subjects as diverse as barnacle biology, carnivorous plants, and the formation of coral reefs.

Over 150 years later, Darwin's theory of evolution is so well established by evidence and reproducible experiment that it is considered a landmark influence in the scientific understanding of the underlying biologic processes operating in nature. Discoveries made since Darwin's time—including genetics and the structure of DNA—confirm how the process of evolution works. For example, the sequencing of the genomes of all 15 species of Darwin's finches was published in 2015, confirming Darwin's ideas about their evolutionary history.

It is interesting to note that most of Darwin's ideas have been so thoroughly accepted by scientists that they are now the underpinnings of the modern study of biology. That's why the name *Darwin* is synonymous with evolution. In 2009, to commemorate Darwin's birth and his accomplishments, the Church of England even issued this formal apology to Darwin: *"The Church of England owes you an apology for misunderstanding you and, by getting our first reaction wrong, encouraging others to misunderstand you still."*

## What Did You Learn?

Describe the three different types of organisms that Charles Darwin observed during his voyage on the *Beagle* that influenced his thinking about the theory of evolution.

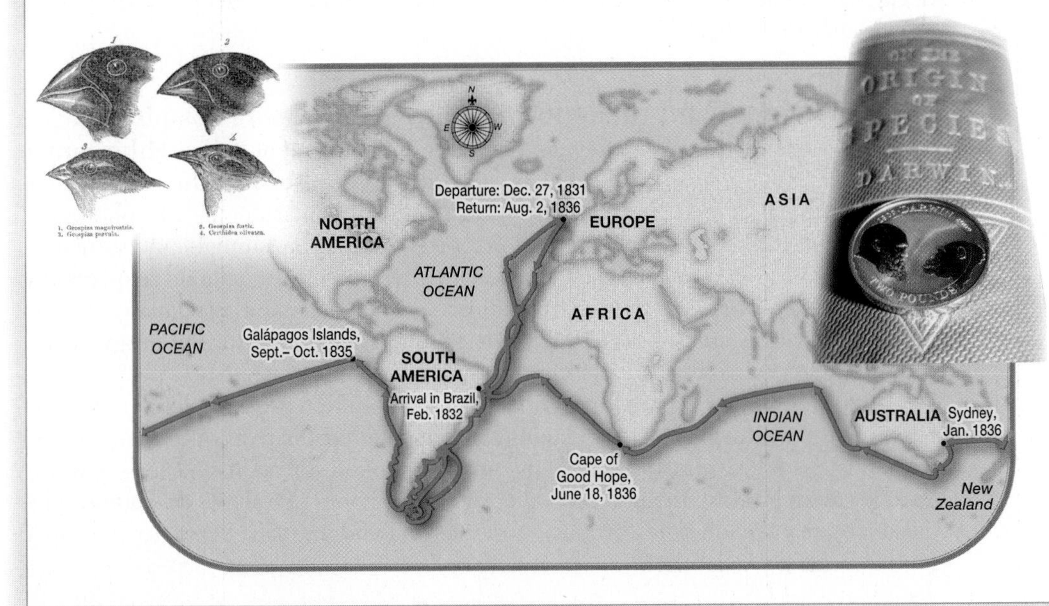


**Figure 1E** **Charles Darwin's legacy: Galápagos finches, route of the HMS *Beagle*, and *On the Origin of Species*.** Map showing the route of the HMS *Beagle*, beak differences in Galápagos finches (*left inset*) that greatly influenced Charles Darwin, and the British two-pound coin commemorating Darwin and his masterwork *On the Origin of Species* (*right inset*).

## STUDENTS SOMETIMES ASK . . .

***I've heard of the discovery of other planets outside of our solar system. Could any of them contain life?***

Outside our solar system, about 4000 exoplanets have been discovered orbiting other star systems, including a few rocky exoplanets that are Earth-sized and may be orbiting their Sun-like stars at just the right distance for water to remain liquid, potentially sustaining life. Astronomers are able to detect if these exoplanets have water or not by analyzing specific frequencies of light. New discoveries of exoplanets are a frequent occurrence, suggesting that there could be hundreds to billions of Earth-like worlds in the vastness of the galaxy. And if our own solar system is any example, salty oceans with the ingredients for life—whether liquid on the surface or deep beneath a world's icy exterior—may be common. However, most of these exoplanets are many light-years away, so we may never know if any of them contain life.

## Evolution and Natural Selection

Every living organism that inhabits Earth today is the result of **evolution** by the process of **natural selection**, which has been occurring since life first existed on Earth. Evolution in organisms is observed when the appearance and characteristics of individuals within populations change, or *evolve*, over time. This happens because naturally-occurring *mutations* in an organism's DNA can sometimes give an individual a survival advantage, and if the helpful mutation is passed down to the next generation, eventually the mutation may become common. This change in the genetic makeup of a population over time is driven by natural selection—the idea that organisms with traits best-suited to a certain environment survive and reproduce at higher rates than individuals lacking those traits. New traits in organisms that arise in response to environmental changes are called *adaptations*. New **species** (*species* = a kind) emerge when the accumulated genetic changes in a population reach a certain threshold (see **Diving Deeper 1.1** on page 29). Evolution by the process of natural selection has been the driving force that allows organisms to inhabit increasingly numerous environments on Earth.

For most organisms, evolution occurs very slowly over long periods of time. As we shall see, when species adapt to Earth's various environments, they can also modify the environments in which they live. This modification can be localized or nearly global in scale. For example, when plants emerged from the oceans and inhabited the land, they changed Earth from a harsh and bleak landscape as barren as that of the Moon to one that is green and lush.

## Plants and Animals Evolve

The very earliest forms of life were probably **heterotrophs** (*hetero* = different, *tropho* = nourishment). Heterotrophs require an external food supply, which was abundantly available in the form of nonliving organic matter in the ocean around them. **Autotrophs** (*auto* = self, *tropho* = nourishment), which can manufacture their own food supply, evolved later. The first autotrophs were probably similar to present-day **anaerobic** (*an* = without, *aero* = air) bacteria, which live without atmospheric oxygen. They may have been able to derive energy from inorganic compounds at deep-water hydrothermal vents using a process called **chemosynthesis** (*chemo* = chemistry, *syn* = with, *thesis* = an arranging). (More details about chemosynthesis are discussed in Chapter 15, "Animals of the Benthic Environment.") In fact, the detection of microbes deep within the ocean crust as well as the discovery of 3.2-billion-year-old microfossils of bacteria from deep-water marine rocks support the idea of life's origin on the deep-ocean floor in the absence of sunlight.

**PHOTOSYNTHESIS AND RESPIRATION** Eventually, more complex single-celled autotrophs evolved. They developed a green pigment called **chlorophyll** (*chloro* = green, *phyll* = leaf), which captures the Sun's energy through cellular **photosynthesis** (*photo* = light, *syn* = with, *thesis* = an arranging). In photosynthesis (**Figure 1.26**), plant and algae cells capture energy from sunlight and store it as sugars, releasing oxygen gas as a by-product. Alternatively, in cellular **respiration** (*respirare* = to breathe) (Figure 1.26), animals who consume the sugars produced by photosynthesis combine them with oxygen, releasing the stored energy of the sugars to carry on cellular tasks important for various life processes.

Figure 1.26 shows that photosynthesis and respiration are complementary processes, with photosynthesis producing what is needed for respiration (sugar and oxygen gas), and respiration producing what is needed for photosynthesis (carbon dioxide gas and water). In fact, the cyclic nature of Figure 1.26 shows that autotrophs (algae and plants) and heterotrophs (most bacteria and animals) began to develop a mutual need for each other.

The oldest fossilized remains of organisms are primitive photosynthetic bacteria recovered from rocks formed on the sea floor about 3.5 billion years ago. However, the oldest rocks containing iron oxide (rust)—an indicator of an oxygen-rich atmosphere—did not appear until about 2.45 billion years ago. This indicates that photosynthetic organisms needed about a billion years to develop and begin producing abundant free oxygen in the atmosphere. Another possible scenario is that a large amount of oxygen-rich (ferric) iron sank to the base of the mantle, where it was heated by the core and subsequently rose as a plume to the ocean floor, releasing large amounts of oxygen through outgassing about 2.5 billion years ago.

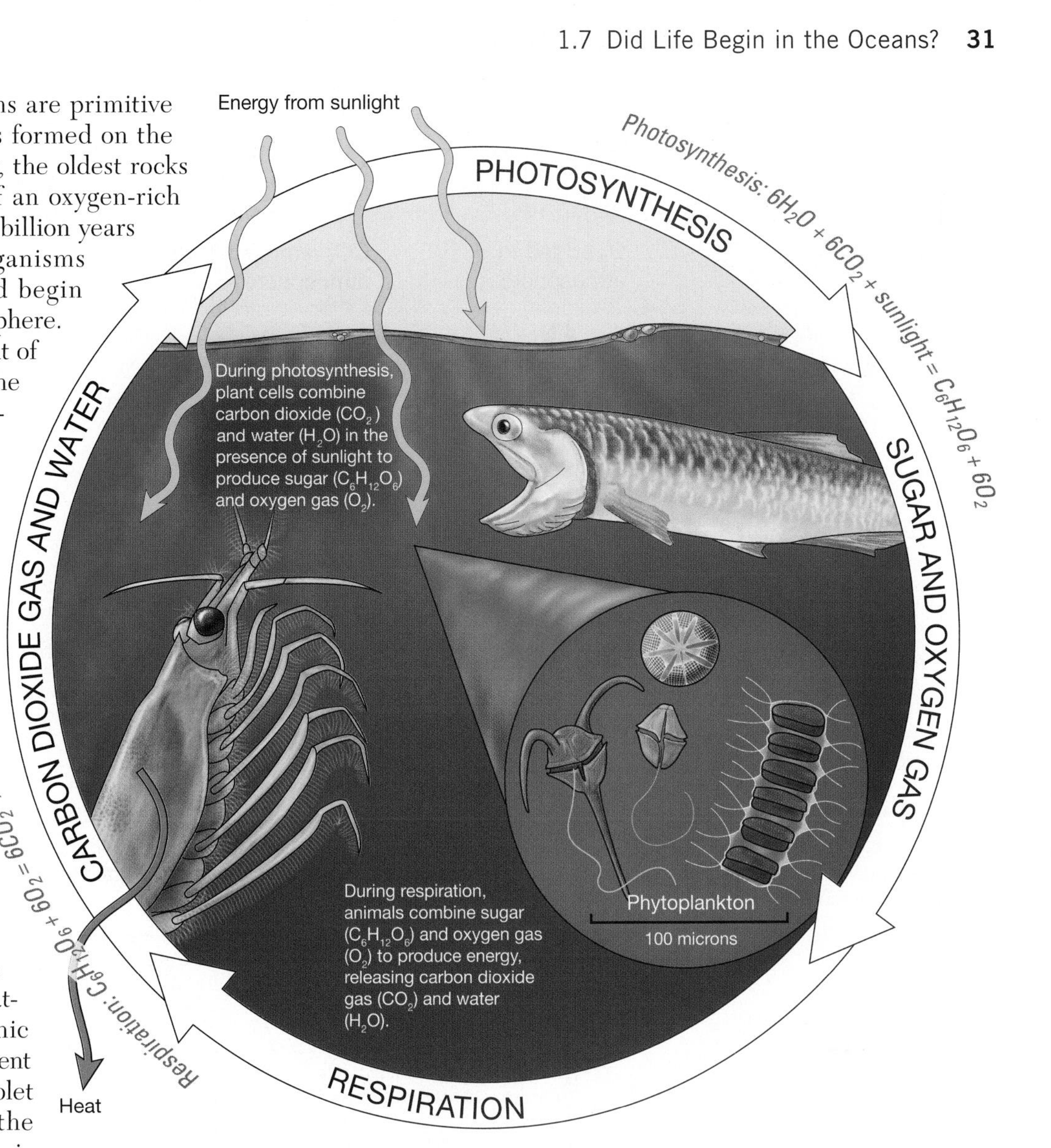
Energy from sunlight
PHOTOSYNTHESIS
Photosynthesis: $6H_2O + 6CO_2$ + sunlight = $C_6H_{12}O_6 + 6O_2$
SUGAR AND OXYGEN GAS
During photosynthesis, plant cells combine carbon dioxide ($CO_2$) and water ($H_2O$) in the presence of sunlight to produce sugar ($C_6H_{12}O_6$) and oxygen gas ($O_2$).
CARBON DIOXIDE GAS AND WATER
Respiration: $C_6H_{12}O_6 + 6O_2 = 6CO_2 + 6H_2O$ + energy
During respiration, animals combine sugar ($C_6H_{12}O_6$) and oxygen gas ($O_2$) to produce energy, releasing carbon dioxide gas ($CO_2$) and water ($H_2O$).
Phytoplankton
100 microns
RESPIRATION
Heat

**SmartFigure 1.26 Photosynthesis and respiration are cyclic and complimentary processes that are fundamental to life on Earth.** https://goo.gl/SsyVda

**THE GREAT OXIDATION EVENT/OXYGEN CRISIS** Based on the chemical makeup of certain rocks, Earth's atmosphere became oxygen-rich about 2.45 billion years ago—called the *great oxidation event*—and fundamentally changed Earth's ability to support life. Particularly for anaerobic bacteria, which had grown successfully in an oxygen-free world, all this oxygen was nothing short of a catastrophe! This is because the increased atmospheric oxygen caused the ozone concentration in the upper atmosphere to build up, thereby shielding Earth's surface from ultraviolet radiation—and effectively eliminating anaerobic bacteria's food supply of organic molecules. (Recall that Stanley Miller's experiment created organic molecules but needed ultraviolet light.) In addition, oxygen (particularly in the presence of light) is highly reactive with organic matter. When anaerobic bacteria are exposed to oxygen and light, they are killed instantaneously. By 1.8 billion years ago, the atmosphere's oxygen content had increased to such a high level that it began causing the extinction of many anaerobic organisms. Nonetheless, descendants of such bacteria survive on Earth today in isolated microenvironments that are dark and free of oxygen, such as deep in soil or rocks, in landfills, and inside other organisms.

Although oxygen is very reactive with organic matter and can even be toxic, it also yields nearly 20 times more energy than anaerobic respiration—a fact that some organisms exploited. For example, blue-green algae, which are also known as *cyanobacteria* (*kuanos* = dark blue), adapted to and thrived in this new oxygen-rich environment. In doing so, they altered the composition of the atmosphere.

**CHANGES TO EARTH'S ATMOSPHERE** The development and successful evolution of photosynthetic organisms are greatly responsible for the world as we know it today (**Figure 1.27**). By the trillions upon trillions, these microscopic organisms transformed the planet by capturing the energy of the Sun to make food and releasing oxygen as a waste product. By this process, these organisms reduced the high amount of carbon dioxide in the early atmosphere and gradually replaced it with free oxygen. This created a third and final atmosphere on Earth: one that is oxygen rich (about 21% today). Little by little, these tiny organisms turned the atmosphere into

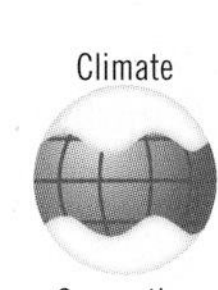
Climate
Connection

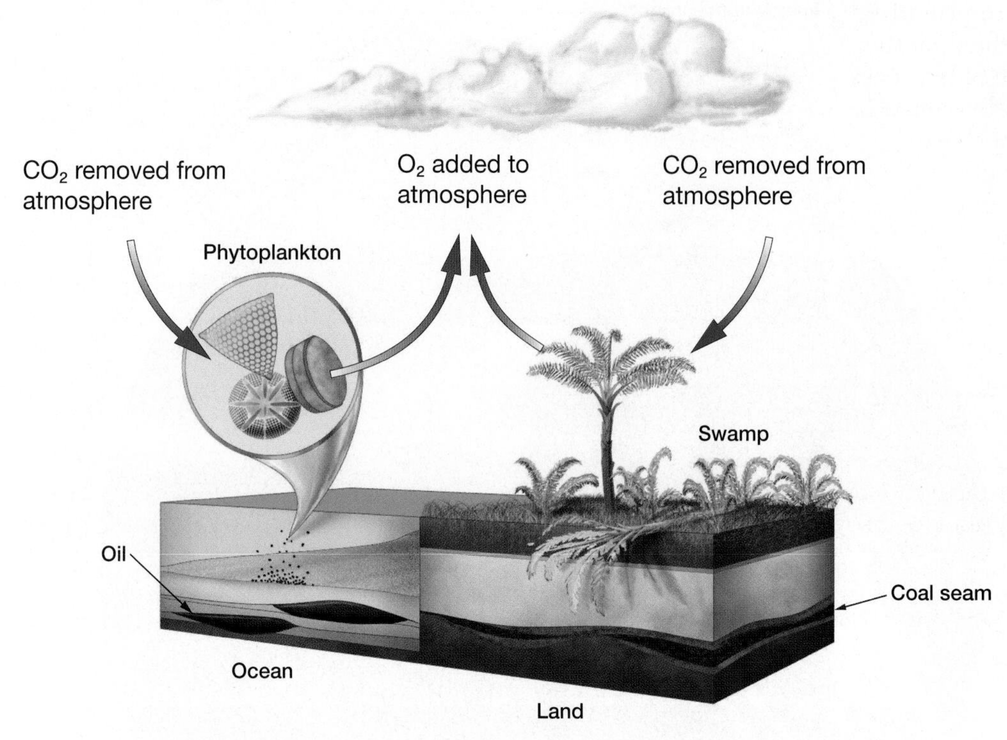


Figure 1.27 **The effect of plants on Earth's environment.** As microscopic photosynthetic cells (*inset*) became established in the ocean, Earth's atmosphere was enriched in oxygen and depleted of carbon dioxide. As organisms died and accumulated on the ocean floor, some of their remains were converted to oil and gas. The same process occurred on land, sometimes producing coal.

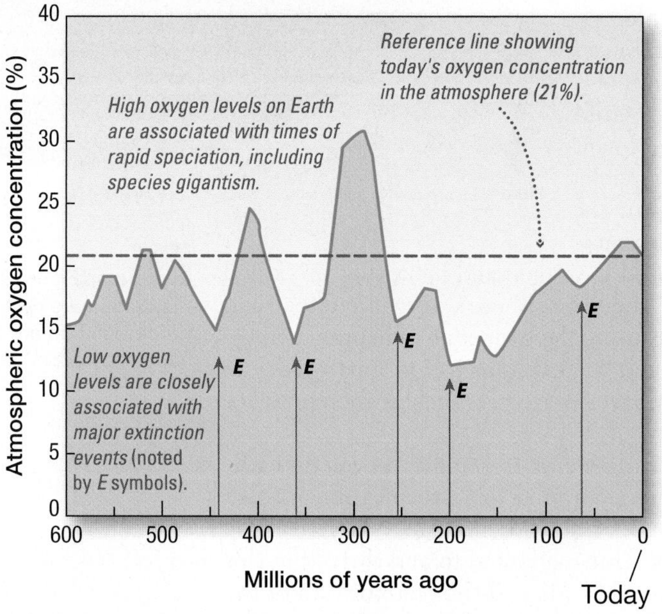

breathable air, opening the way to the diversity of life that followed.

The graph in **Figure 1.28** shows how the concentration of atmospheric oxygen has varied during the past 600 million years. When atmospheric oxygen concentrations are high, organisms thrive, and rapid speciation occurs. At such times in the past, insects grew to gargantuan proportions, reptiles took to the air, and the forerunners of mammals developed a warm-blooded metabolism. More oxygen was dissolved in the oceans, too, and so marine biodiversity increased. At other times when atmospheric oxygen concentrations fell precipitously, biodiversity was smothered. In fact, some of the planet's worst mass extinctions are associated with sudden drops in atmospheric oxygen.

The remains of ancient plants and animals buried in oxygen-free environments have become the oil, natural gas, and coal deposits of today. These deposits, which are called *fossil fuels*, provide more than 90% of the energy humans consume to power modern society. In essence, humans depend not only on the food energy stored in today's plants but also on the energy stored in plants during the geologic past—in the form of fossil fuels.

Because of increased burning of fossil fuels for home heating, industry, power generation, and transportation during the industrial age, the atmospheric concentration of carbon dioxide and other gases that help warm the atmosphere has increased, too. Scientists understand now that these human-generated emissions are increasing global warming and are causing serious environmental problems that will likely get worse in the future. This phenomenon is referred to as the atmosphere's *enhanced greenhouse effect* and is discussed in Chapter 16, "The Oceans and Climate Change."

**EXPLORING DATA ▶**

**1. Based on the graph, what is the average interval in millions of years between Earth's major extinction events (labeled *E*)? Based on the last extinction event, is Earth overdue for one? Explain.**

**2. What is the relationship between atmospheric oxygen concentration and extinction events? Speculate about why this pattern occurs.**

Atmospheric oxygen concentration (%)
40
35
30
25
20
15
10
5
0
Reference line showing today's oxygen concentration in the atmosphere (21%).
High oxygen levels on Earth are associated with times of rapid speciation, including species gigantism.
Low oxygen levels are closely associated with major extinction events (noted by *E* symbols).
E
E
E
E
E
600 500 400 300 200 100 0
Millions of years ago
Today

Figure 1.28 **Atmospheric oxygen concentration.** Graph showing how the concentration of oxygen in Earth's atmosphere has varied during the past 600 million years, including major extinction events (labelled *E*).

**RECAP Life on Earth has evolved over time and changed Earth's environment. For example, abundant photosynthetic organisms created today's oxygen-rich atmosphere.**

**CONCEPT CHECK 1.7 ▶ Discuss why life is thought to have originated in the oceans.**

1 How does the presence of oxygen in our atmosphere help reduce the amount of ultraviolet radiation that reaches Earth's surface?

2 What was Stanley Miller's experiment, and what did it help demonstrate?

3 Earth has had three atmospheres (initial, early, and present). Describe the composition and origin of each one.

# 1.8 ▶ How Old Is Earth?

Science's ideas about planetary and biological evolution are dependent on the knowledge that Earth is very old, because the processes at work are very slow. Thus, having an accurate estimate of Earth's age is fundamental to the acceptance of these ideas by the science community. But how can Earth scientists tell how old a rock is? It can be a difficult task to tell if a rock is thousands, millions, or even billions of years old—unless the rock contains telltale fossils. Fortunately, Earth scientists can determine how old most rocks are by using the radioactive materials contained within rocks. In essence, this technique involves reading a rock's internal "rock clock."

## Radiometric Age Dating

Most rocks on Earth (as well as those from outer space) contain small amounts of radioactive materials such as uranium, thorium, and potassium. These radioactive materials spontaneously break apart or decay into atoms of other elements. Radioactive materials have a characteristic **half-life**, which is the time required for one-half of the atoms in a sample to decay to other atoms. The older the rock is, the more radioactive material will have been converted to decay product. Analytical instruments can accurately measure the amount of radioactive material and the amount of resulting decay product in rocks. By comparing these two quantities, the age of the rock can thus be determined. Such dating is referred to as **radiometric age dating** (*radio* = radioactivity, *metri* = measure) and is an extremely powerful tool for determining the age of rocks.

Figure 1.29 shows an example of how radiometric age dating works. It shows how uranium 235 decays into lead 207 at a rate where one-half of the atoms turn into lead every 704 million years. By counting the number of each type of atom in a rock sample, one can tell how long it has been decaying (as long as the sample does not gain or lose atoms). Using uranium and other radioactive elements and applying this same technique, hundreds of thousands of rock samples have been age-dated from around the world.

## The Geologic Time Scale

The ages of rocks on Earth are shown in the **geologic time scale** (Figure 1.30; see also Mastering Oceanography Web Diving Deeper 1.2), which lists the names of the geologic time periods as well as important advances in the development of life-forms on Earth. Initially, the divisions between geologic periods were based on major extinction episodes as recorded in the fossil record. As radiometric age dates became available, they were also included on the geologic time scale. The oldest known rocks on Earth, for example, are about 4.3 billion years old, and the oldest known crystals within terrestrial rocks have been dated at up to 4.4 billion years old. (The discovery of crystals this old implies that significant continental crust must have formed on Earth early on, perhaps by nearly 4.5 billion years ago.) No rocks

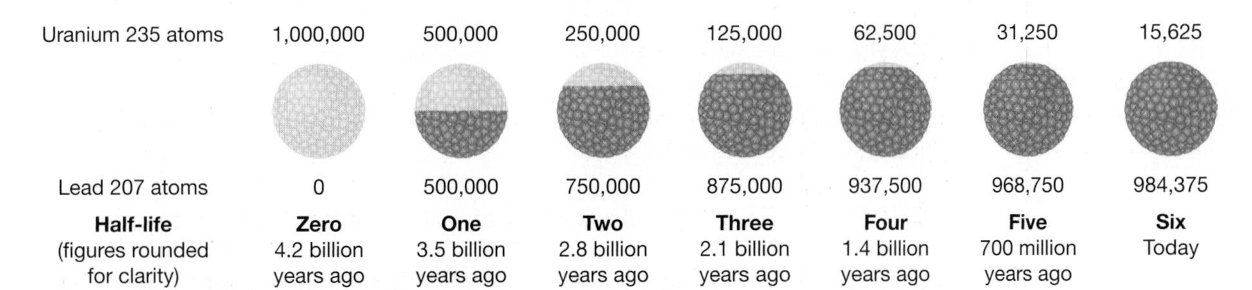


Figure 1.29 **Radiometric age dating.** During one half-life, half of all radioactive uranium 235 atoms decay into lead 207. With each successive half-life, half of the remaining radioactive uranium atoms convert to lead. By counting the number of each type of atom in a rock sample, the rock's age can be determined.

**Animation**
Radioactive Decay
http://goo.gl/iMQIID

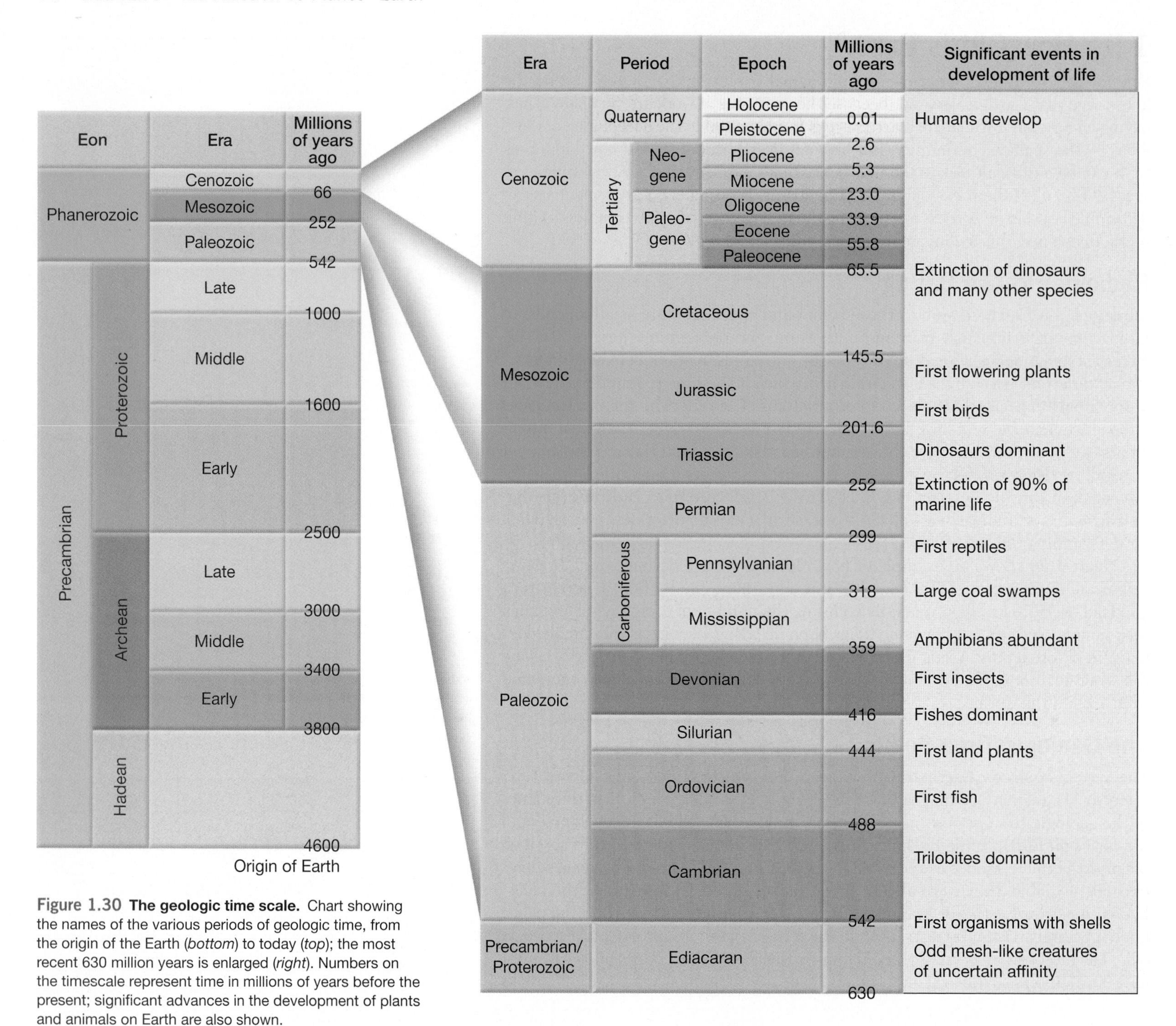


**Figure 1.30 The geologic time scale.** Chart showing the names of the various periods of geologic time, from the origin of the Earth (*bottom*) to today (*top*); the most recent 630 million years is enlarged (*right*). Numbers on the timescale represent time in millions of years before the present; significant advances in the development of plants and animals on Earth are also shown.

**RECAP Earth scientists can accurately determine the age of most rocks by analyzing their radioactive components, some of which indicate that Earth is 4.6 billion years old.**

older than this have been found because few likely survived Earth's molten youth, a time when Earth was being bombarded by meteorites. However, radiometric dating of space rocks left over from the formation of the solar system indicates Earth is about 4.6 billion years old.

## CONCEPT CHECK 1.8 ▸ Demonstrate an understanding of how old Earth is.

1 Describe how the half-life of radioactive materials can be used to determine the age of a rock through radiometric age dating.

2 What is the age of Earth? Describe the major events that demark the boundaries between these time periods: (a) Precambrian/Proterozoic, (b) Paleozoic/Mesozoic, (c) Mesozoic/Cenozoic.

# ESSENTIAL CONCEPTS REVIEW

## 1.1 ▶ How are Earth's oceans unique?

- *Water covers 70.8% of Earth's surface.* The world ocean is a *single interconnected body of water*, which is large in size and volume. It can be divided into *four principal oceans* (the Pacific, Atlantic, Indian, and Arctic Oceans), plus an additional ocean (the Southern Ocean, or Antarctic Ocean). Even though there is a technical distinction between a *sea* and an *ocean*, the two terms are used interchangeably. In comparing the oceans to the continents, it is apparent that *the average land surface does not rise very far above sea level* and that *there is not a mountain on Earth that is as tall as the ocean is deep.*

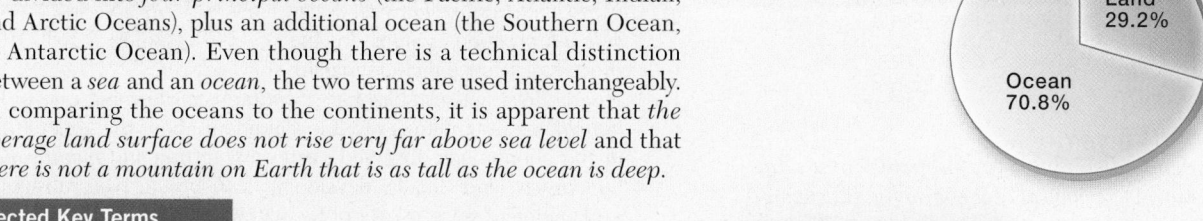

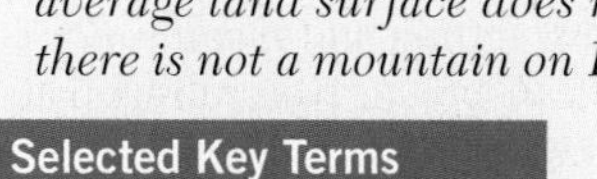

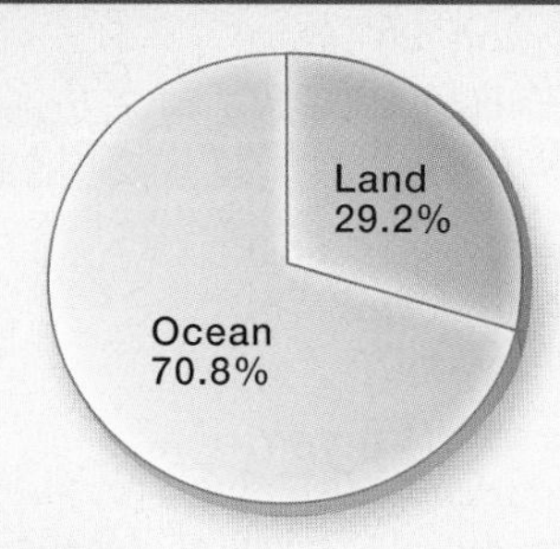


### Selected Key Terms

Use the **glossary** at the end of this book to discover the meanings of these Selected Key Terms: **ocean, Pacific Ocean, Atlantic Ocean, Indian Ocean, Arctic Ocean, Southern (Antarctic) Ocean.**

### Critical Thinking Question

NASA has discovered a new planet that has an ocean. Using today's technology, how would you propose studying that ocean, all that's in it, and the sea floor beneath it? Assume an unlimited budget.

### Active Learning Exercise

If all Earth's glaciers melted, sea level would rise by about 70 meters (230 feet). Since the average height of the continents is only 840 meters (2756 feet), a rise in sea level of this magnitude would seriously impact human activities, especially in low-lying areas. Based on your knowledge of worldwide geography, which areas of the globe would most likely be affected? Be sure to include major population centers that would be under water. Assess these impacts, and discuss as a group.

## 1.2 ▶ How was early exploration of the oceans achieved?

- In the Pacific, *people who populated the Pacific Islands may have been the first great navigators*. In the Western world, the *Phoenicians* were making remarkable voyages as well. Later the *Greeks*, *Romans*, and *Arabs* made significant contributions and advanced oceanographic knowledge. During the Middle Ages, the *Vikings* colonized Iceland and Greenland and made voyages to North America.
- The *Age of Discovery* in Europe renewed the Western world's interest in exploring the unknown. It began with the voyage of *Christopher Columbus* in 1492 and ended in 1522 with the first circumnavigation of Earth by a voyage initiated by *Ferdinand Magellan*. *Captain James Cook* was one of the first to explore the ocean for scientific purposes.

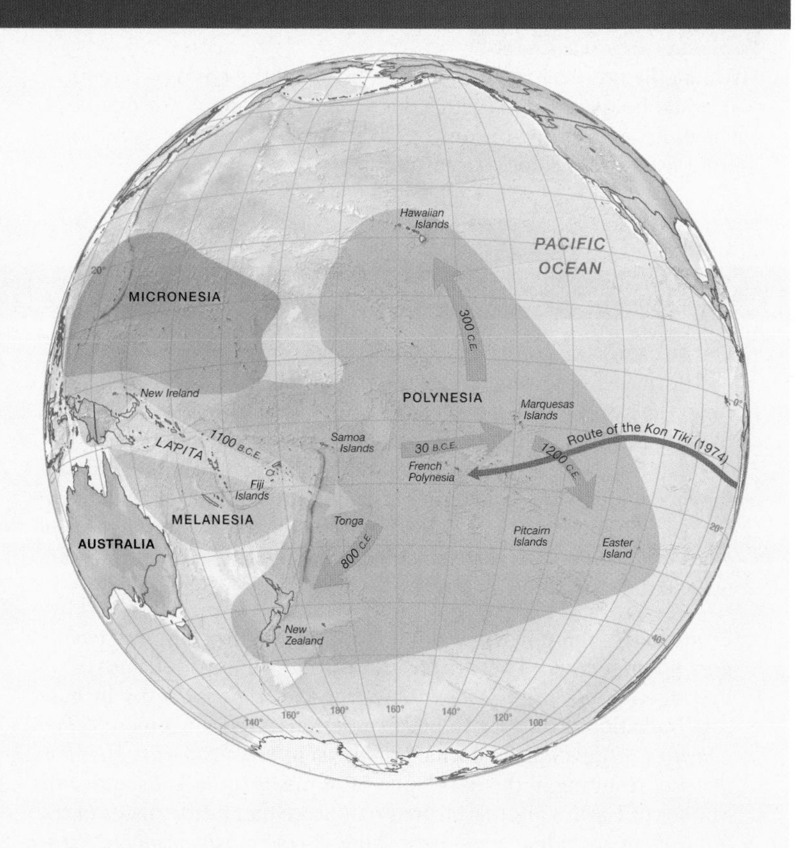


### Selected Key Terms

Use the **glossary** at the end of this book to discover the meanings of these Selected Key Terms: **latitude, longitude, Age of Discovery.**

### Critical Thinking Question

Discuss the technological advantages that allowed seafaring Arabs during the Middle Ages to dominate the Mediterranean Sea and trade with East Africa, India, and southeast Asia.

### Active Learning Exercise

Make a list of the 10 essential items you'd need to take with you on a month-long boat expedition to study the ocean (exclude clothes, personal items, and food). Compare and discuss your list with another student in class. How would your list of 10 essential items be different if you created it during the beginning of voyaging for science in the 1700s?

## 1.3 ▶ What fields of science does oceanography include?

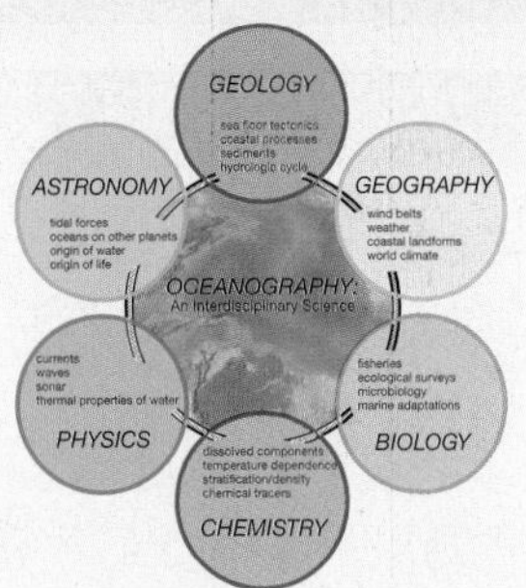


- *Oceanography, or marine science, is the scientific study of all aspects of the marine environment.* During World War II, *a tactical advantage was gained by studying ocean processes*, leading to great advances in technology and the ability to observe and study the oceans in more detail. *Today, much study is focused on human impacts on the ocean.*
- *Oceanography is traditionally divided into four academic disciplines (or subfields) of study.* These four disciplines are: (1) *geological oceanography*, (2) *chemical oceanography*, (3) *physical oceanography*, and (4) *biological oceanography*. *Oceanography is frequently described as being an interdisciplinary science* because it encompasses all the different disciplines of science as they apply to the oceans.

### Selected Key Terms

Use the **glossary** at the end of this book to discover the meanings of these Selected Key Terms: **oceanography, interdisciplinary science.**

### Critical Thinking Question

One of today's most pressing problems is trash in the oceans. Describe how information from at least two different disciplines in oceanography contribute to understanding or potentially solving this problem.

### Active Learning Exercise

With another student in class, make a list of all the types of careers you would be qualified for with a degree in oceanography or marine science. As an example of someone who works in oceanography or marine science, consider your instructor.

## 1.4 ▶ What is the process of science and the nature of scientific inquiry?

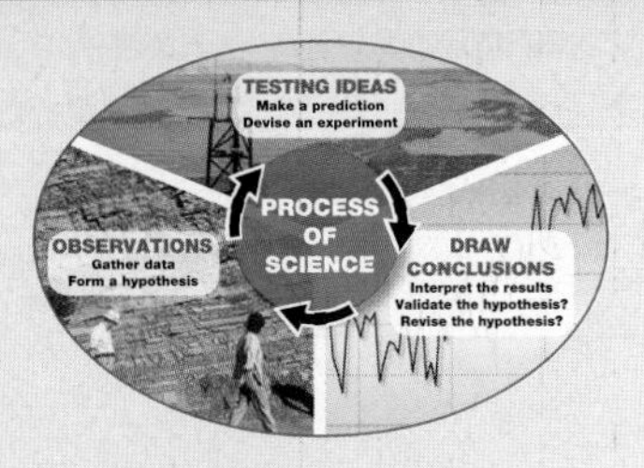


- The *process of science* is used to understand the occurrence of physical events or phenomena and can be stated as *science supports the explanation of the natural world that best explains all available observations*. Steps in the process of science include making *observations* and establishing *scientific facts*; forming one or more *hypotheses* (a tentative, testable statement about the general nature of the phenomena observed); extensive *testing* and *modification of hypotheses*; and, finally, developing a *theory* (a well-substantiated explanation of some aspect of the natural world that can incorporate facts, laws, logical inferences, and tested hypotheses). Science never arrives at the absolute "truth"; rather, *science arrives at what is probably true* based on the available observations and can *continually change because of new observations*.

### Selected Key Terms

Use the **glossary** at the end of this book to discover the meanings of these Selected Key Terms: **process of science, observation, hypothesis, theory.**

### Critical Thinking Question

What is the difference between a fact and a theory? Can either (or both) be revised?

### Active Learning Exercise

With another student in class, discuss if you believe nature is simple enough for humans to truly understand. Give reasons why or why not. If not, do you think it is still reasonable for scientists to make this assumption in applying the process of science in their work?

## 1.5 ▶ How were Earth and the solar system formed?

- Our *solar system*, consisting of *the Sun and eight major planets*, probably formed from a *huge cloud of gas and space dust* called a *nebula*. According to the *nebular hypothesis*, the nebular matter contracted to form the Sun, and the planets were formed from eddies of material that remained. The Sun, composed of hydrogen and helium, was massive enough and concentrated enough to emit *large amounts of energy* from fusion. The Sun also emitted *ionized particles that swept away any nebular gas* that remained from the formation of the planets and their satellites.
- *Proto-Earth*, more massive and larger than Earth today, *was molten and homogenous*. The *initial atmosphere*, composed mostly of hydrogen and helium, *was later driven off into space* by intense solar radiation. Proto-Earth began a period of rearrangement called *density stratification* and formed a *layered internal structure based on density*, resulting in the development of the *crust*, *mantle*, and *core*. Studies of Earth's internal structure indicate that brittle plates of the *lithosphere* are riding on a plastic, high-viscosity *asthenosphere*. Near the surface, the lithosphere is composed of *continental* and *oceanic crust*. Continental crust consists mostly of granite and oceanic crust consists mostly of basalt. *Continental crust is lower in density, lighter in color, and thicker than oceanic crust.* Both types of crust float isostatically on the denser mantle below.

### Selected Key Terms

Use the **glossary** at the end of this book to discover the meanings of these Selected Key Terms: **nebular hypothesis, density stratification, lithosphere, asthenosphere, oceanic crust – basalt, continental crust – granite.**

### Critical Thinking Question

Describe how the chemical composition of Earth's interior differs from its physical properties. Include specific examples.

### Active Learning Exercise

The nebular hypothesis of solar system formation is a scientific hypothesis. Based on your understanding of the process of science, describe to another student in class how sure of this hypothesis you think scientists really are. Why would scientists have this level of certainty?

## 1.6 ▶ How were Earth's atmosphere and oceans formed?

- *Outgassing produced an early atmosphere* rich in water vapor and carbon dioxide. Once Earth's surface cooled sufficiently, the *water vapor condensed and accumulated to give Earth its oceans. Rainfall on the surface dissolved compounds that,* when carried to the ocean, *made it salty*.

### Selected Key Terms

Use the **glossary** at the end of this book to discover the meaning of this Selected Key Term: **outgassing.**

### Critical Thinking Question

Compare the two ways in which Earth was supplied with enough water to have an ocean. Which is likely to have contributed most of the water on Earth?

### Active Learning Exercise

With another student in class, describe in your own words how Earth's oceans became salty.

## 1.7 ▶ Did life begin in the oceans?

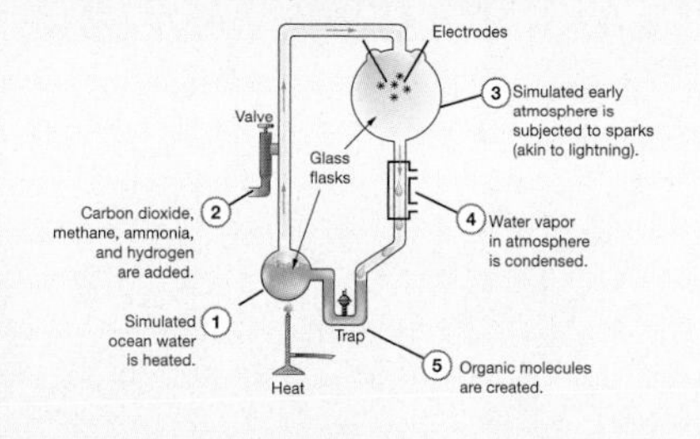


- *Life is thought to have begun in the oceans. Stanley Miller's experiment* showed that ultraviolet radiation from the Sun and hydrogen, carbon dioxide, methane, ammonia, and inorganic molecules from the oceans may have combined to produce *organic molecules such as amino acids*. Certain combinations of these molecules eventually produced *heterotrophic organisms* (which cannot make their own food) that were probably similar to present-day anaerobic bacteria. Eventually, *autotrophs evolved* that had the ability to make their own food through *chemosynthesis*. Later, some cells developed *chlorophyll*, which made *photosynthesis* possible and led to the *development of plants*.
- *Photosynthetic organisms altered the environment* by extracting carbon dioxide from the atmosphere and also by releasing free oxygen, thereby creating today's *oxygen-rich atmosphere*. Eventually, both *plants and animals evolved* into forms that could survive on land.

### Selected Key Terms

Use the **glossary** at the end of this book to discover the meanings of these Selected Key Terms: **evolution, natural selection, species, photosynthesis, respiration.**

### Critical Thinking Question

How would you answer the accusation, made by some religious groups, that scientific theories such as Stanley Miller's theory on the origin of life on Earth are inherently weak because it is a historic event that no one actually observed? Please explain your answer in detail.

### Active Learning Exercise

With another student in class, discuss which of these two statements has more validity: (1) the greatest environmental crisis of all time was the build-up of toxic oxygen in Earth's atmosphere 2 billion years ago or (2) humans are causing the greatest environmental crisis of all time.

## 1.8 ▶ How old is Earth?

| Era | Period | | Epoch | Millions of years ago |
|---|---|---|---|---|
| Cenozoic | Quaternary | | Holocene | 0.01 |
| | | | Pleistocene | 2.6 |
| | Tertiary | Neo-gene | Pliocene | 5.3 |
| | | | Miocene | 23.0 |
| | | Paleo-gene | Oligocene | 33.9 |
| | | | Eocene | 55.8 |
| | | | Paleocene | 65.5 |

- *Radiometric age dating* is used to determine the age of most rocks. Information from extinctions of organisms and from age dating rocks comprises the *geologic time scale*, which indicates that Earth has experienced a long history of changes since *its origin 4.6 billion years ago.*

### Selected Key Terms

Use the **glossary** at the end of this book to discover the meanings of these Selected Key Terms: **half-life, radiometric age dating, geologic time scale.**

### Critical Thinking Question

Explain how radiometric age dating works. Why does the parent material never totally disappear completely, even after many half-lives?

### Active Learning Exercise

Working as a team, construct a representation of the geologic time scale, using an appropriate quantity of any substance (other than dollar bills or toilet paper, which are used as examples in Mastering Oceanography Web Diving Deeper 1.2). Be sure to indicate some of the major changes that have occurred on Earth since its origin, such as "Origin of Earth," "Origin of oceans," "Earliest known life-forms," "Oxygen-rich atmosphere first occurs," "First organisms with shells," "Dinosaurs die out," and "Age of humans."

## Mastering Oceanography

Looking for additional review and test prep materials? With individualized coaching on the toughest topics of the course, Mastering Oceanography offers a wide variety of ways for you to move beyond memorization and deeply grasp the underlying processes of how the oceans work. Visit the Study Area in **www.masteringoceanography.com** to find practice quizzes, study tools, and multimedia that will improve your understanding of this chapter's content. Sign in today to access the following features: Self Study Quizzes, SmartFigures, Oceanography Videos and Animations, Squidtoons, Dynamic Study Modules, and an optional Pearson eText with embedded videos.

# Plate Tectonics and the Ocean Floor

# 2

## ESSENTIAL LEARNING CONCEPTS

At the end of this chapter, you should be able to:

- ☐ **2.1** Evaluate the evidence that supports continental drift.
- ☐ **2.2** Summarize the evidence that supports plate tectonics.
- ☐ **2.3** Discuss the origin and characteristics of features that occur at plate boundaries.
- ☐ **2.4** Show how plate tectonics explains the origin of features not easily explained by other processes.
- ☐ **2.5** Describe how Earth has changed in the past and predict how it will look in the future.

↑ *Check when completed*

Each year at various locations around the globe, several thousand earthquakes and dozens of volcanic eruptions occur, both of which indicate how remarkably dynamic our planet is. These events have occurred throughout history, constantly changing the surface of our planet, yet only a little over 50 years ago, most scientists believed the continents were stationary over geologic time. Since that time, a bold new theory has been advanced that helps explain surface features and phenomena on Earth, including:

- The worldwide locations of volcanoes, faults, earthquakes, and mountain building
- Why mountains on Earth haven't all eroded away during the billions of years Earth has existed
- The origin of most landforms and ocean floor features
- How the continents and ocean floor formed and why they are different
- The continuing development of Earth's surface
- The distribution of past and present life on Earth

This revolutionary new theory is called **plate tectonics** (*plate* = plates of the lithosphere; *tekton* = to build), or "the new global geology." According to the theory of plate tectonics, the outermost portion of Earth is composed of a patchwork of thin, rigid plates that move horizontally with respect to one another, like icebergs floating on water. (Note that these thin, rigid plates are pieces of the *lithosphere*, which comprises Earth's outermost layer and contain oceanic and/or continental crust, as described in Chapter 1.) As a result, the continents are mobile and move about on Earth's surface, controlled by forces deep within Earth.

The interaction of these plates as they move builds features of Earth's crust (such as mountain belts, volcanoes, and ocean basins). For example, the tallest mountain range on Earth is the Himalaya Mountains that extend through India, Nepal, and Bhutan. This mountain range contains rocks that were deposited millions of years ago in a shallow, low-lying sea, providing testament to the power and persistence of plate tectonic activity.

Plate tectonics is extensively supported by data from a variety of sciences, including geological, chemical, physical, and biological sources. Yet the original idea that led to this theory—continental drift—wasn't accepted by many scientists when it was first introduced. In fact, the development of the theory of plate tectonics is a classic example of the process of science: how a seemingly implausible idea, when faced with a preponderance of evidence to support it, became one of the central tenets of a revolutionary theory that now forms the basis of our understanding of fundamental Earth processes.

***"It is just as if we were to refit the torn pieces of a newspaper by matching their edges and then check whether the lines of print run smoothly across. If they do, there is nothing left but to conclude that the pieces were in fact joined in this way."***

—***Alfred Wegener,*** **The Origins of Continents and Oceans** ***(1915)***

◂ **Tall mountains created by tectonic uplift.** Tall coastal mountains such as these in Glacier Bay National Park in southeastern Alaska have been uplifted by plate tectonic processes. The rocks that compose these mountains have come from distant areas of the world, include parts of the sea floor such as coral reefs, and have been elevated thousands of meters above sea level.

Figure 2.1 **Alfred Wegener, circa 1912–1913.** Alfred Wegener (1880–1930), shown here in his research station in Greenland, developed the idea of continental drift. He was one of the first scientists to use multiple lines of evidence to suggest that continents are mobile. Wegener perished in 1930 while trying to establish a year-round meteorological station atop the Greenland ice sheet.

## 2.1 ▶ What Evidence Supports Continental Drift?

**Alfred Wegener** (Figure 2.1), a German meteorologist and geophysicist, was the first to advance the idea of mobile continents in 1912. He envisioned that the continents were slowly drifting across the globe and called his idea **continental drift**. Wegener first published his ideas in *The Origins of Continents and Oceans* in 1915, but the book did not attract much attention until it was translated into English, French, Spanish, and Russian in 1924. From that point until his death in 1930, Wegener's drift hypothesis received much hostile criticism—and sometimes open ridicule—from the scientific community because of flaws in the mechanism he proposed for the movement of the continents. Lacking knowledge of Earth's internal structure, Wegener suggested that the continents plowed through the rock of the ocean basins to reach their present-day positions, propelled by a combination of the gravitational attraction of Earth's equatorial bulge and tidal forces from the Sun and Moon. Further, he proposed that the leading edges of the continents deformed into mountain ridges because of the drag imposed by the sea floor. Scientists rejected the idea as too fantastic and contrary to the laws of physics. Although Wegner was wrong about *how* the continents moved, he was correct that, over time, the continents *did* move, and he was one of the first scientists to use multiple lines of evidence to show this. Let's examine the evidence that led Wegener to formulate the idea of drifting continents.

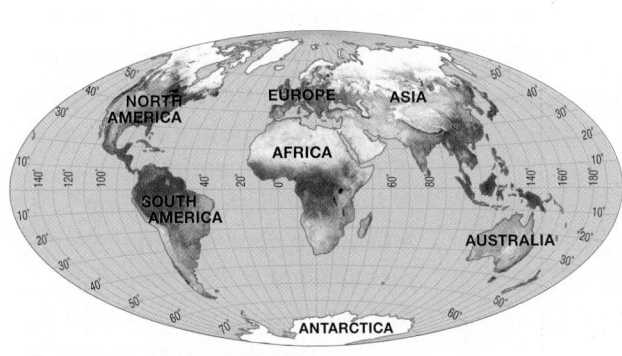

(a) The positions of the continents today.

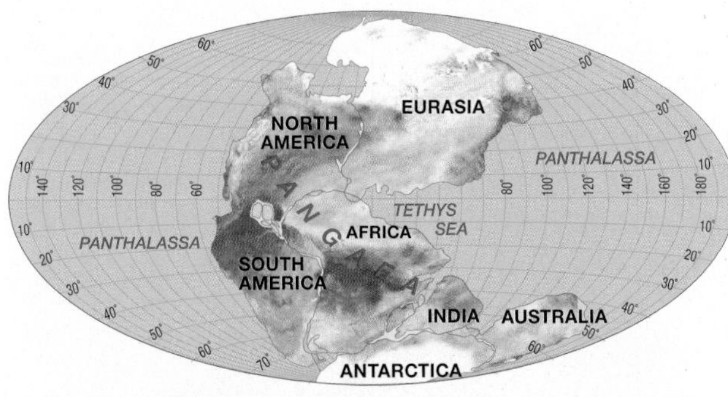


(b) The positions of the continents about 200 million years ago, showing the supercontinent of Pangaea and the single large ocean, Panthalassa.

Figure 2.2 **Reconstruction of Pangaea.**

## PROCESS OF SCIENCE: Why Do the Edges of the Continents Appear to Fit Together?

### FORMING A HYPOTHESIS

The idea that continents—particularly South America and Africa—fit together like pieces of a jigsaw puzzle originated with the development of reasonably accurate world maps. As far back as 1620, Sir Francis Bacon wrote about how the continents appeared to fit together. However, little significance was given to this idea until 1912, when Wegener used the shapes of matching shorelines on different continents as a supporting piece of evidence for continental drift. Wegener suggested that during the geologic past, the continents collided to form a large landmass, which he named **Pangaea** (*pan* = all, *gaea* = Earth) (**Figure 2.2**). Further, a huge ocean, called **Panthalassa** (*pan* = all, *thalassa* = sea), surrounded Pangaea. Panthalassa included several smaller seas, including the **Tethys Sea** (*Tethys* = a Greek sea goddess). Wegener's evidence indicated that as Pangaea began to split apart, the various continental masses started to drift toward their present geographic positions.

### DEVISING AN EXPERIMENT

Wegener's attempt at matching shorelines revealed considerable areas of crustal overlap and large gaps. Some, but not all, of the differences could be explained by material deposited by rivers or eroded from coastlines. What Wegener didn't know at the time was that the shallow parts of the ocean floor close to shore are underlain by materials similar to those beneath continents, and those submerged regions represent the true edge of the continents. In the early 1960s, Sir Edward Bullard and two associates tested the fit of the continents using a computer program. (**Figure 2.3**). Instead of using the shorelines of the continents as Wegener had done, Bullard achieved the best fit (for example, with minimal overlaps or gaps) by mapping the ocean floor contours at a depth of 2000 meters (6560 feet) below sea level. This depth corresponds to halfway between the shoreline and the deep-ocean basins; as such, it represents the true continental margin. By using this depth, the continents fit together remarkably well.

### INTERPRETING THE RESULTS

As we shall see in the following sections, Wegener looked for other evidence that the continents had once been joined by looking at the data available at the time. For example, continental drift predicted matching fossil and rock sequences on opposite sides of the Atlantic Ocean. When these were found, in Wegener's words, "It is just as if we were to refit the torn pieces of a newspaper by matching their edges and then check whether the lines of print run smoothly across. If they do, there is nothing left but to conclude that the pieces were in fact joined in this way." However, it wouldn't be until decades after Wegener's death that advancing technology and new observations from an unexpected source—the sea floor— vindicated Wegener's hypothesis that the continents were mobile. Until that time, the observations supporting continental drift remained a perplexing oddity.

### THINKING LIKE A SCIENTIST: WHAT'S NEXT?

Suppose you are a scientist in the 1920s and you are a passionate advocate of Wegener's hypothesis of continental drift. What kinds of data would you look for that would falsify the idea that the continents had once been joined? Remember, if you accept Wegener's idea of continental drift, think of a very definitive, observable, logical test, because if you *cannot falsify* a hypothesis with the available data, the hypothesis must stand. As a hint, reading through the next headings in this section will likely give you some ideas.

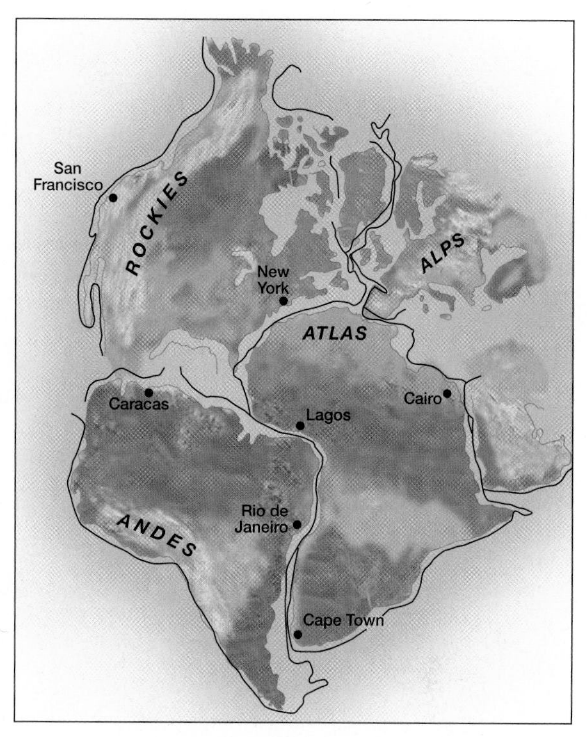


Figure 2.3 **An early computer fit of the continents.** Map showing the 1960s fit of the continents using a depth of 2000 meters (6560 feet) (*black lines*), which is the true edge of the ocean basin. The results indicate a remarkable match, with few overlaps and minimal gaps. Note that the present-day shorelines of the continents are shown with blue lines.

## Testing the Idea: Matching Sequences of Rocks and Mountain Chains

If the continents were once joined together, as Wegener had hypothesized, then evidence should appear in rock sequences that were originally continuous but are now separated by large distances. To test the idea of drifting continents, geologists began comparing the rocks along the edges of continents with rocks found in adjacent positions on matching continents. They wanted to see if the rocks had similar types, ages, and structural styles (the type and degree of deformation). In some areas, younger rocks had been deposited during the millions of years since the continents separated, covering the rocks that held the key to the history of the continents. In other areas, the rocks had been eroded away. Nevertheless, in many other areas, the key rocks were present.

Moreover, these studies showed that many rock sequences from one continent were identical to rock sequences on an adjacent continent—although the two were separated by an ocean. In addition, mountain ranges that terminated abruptly at the edge of a continent continued on another continent across an ocean basin, with identical rock sequences, ages, and structural styles. **Figure 2.4** shows, for example, how similar rocks from the Appalachian Mountains in North America match up with identical rocks from the British Isles and the Caledonian Mountains in Europe.

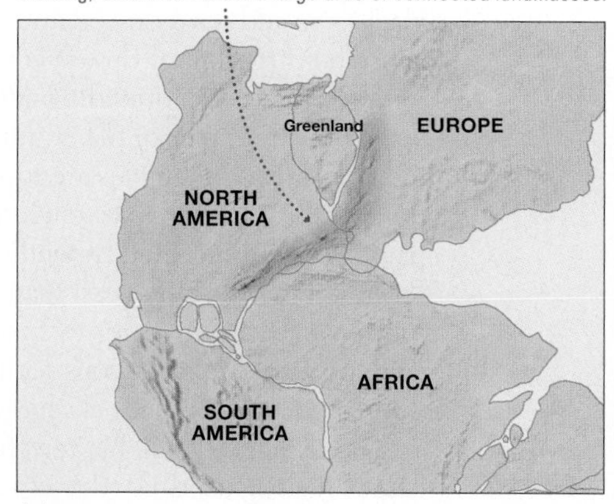


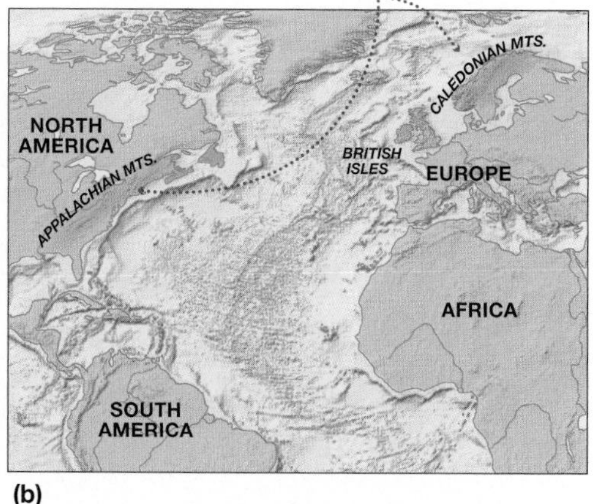


Figure 2.4 **Matching mountain ranges across the North Atlantic Ocean.**

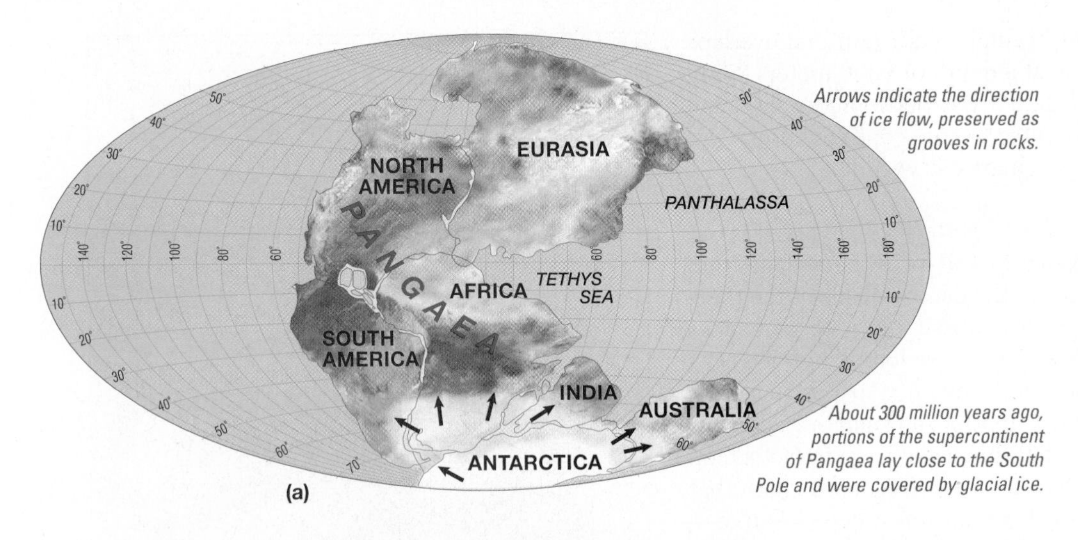
Arrows indicate the direction of ice flow, preserved as grooves in rocks.
NORTH AMERICA
EURASIA
PANTHALASSA
PANGAEA
TETHYS SEA
AFRICA
SOUTH AMERICA
INDIA
AUSTRALIA
ANTARCTICA
About 300 million years ago, portions of the supercontinent of Pangaea lay close to the South Pole and were covered by glacial ice.
(a)

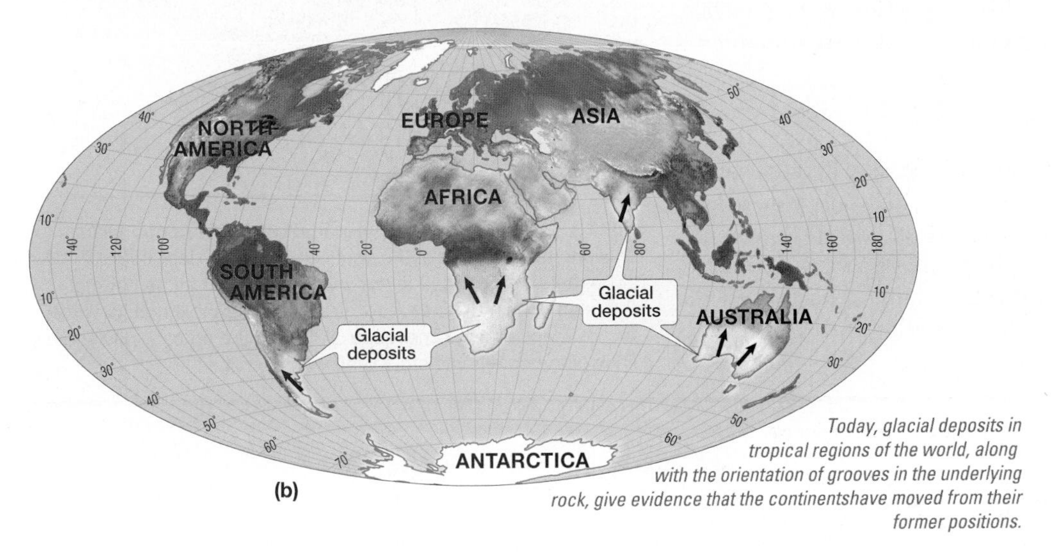
NORTH AMERICA
EUROPE
ASIA
AFRICA
SOUTH AMERICA
Glacial deposits
Glacial deposits
AUSTRALIA
ANTARCTICA
Today, glacial deposits in tropical regions of the world, along with the orientation of grooves in the underlying rock, give evidence that the continentshave moved from their former positions.
(b)

Figure 2.5 **Ice age on Pangaea.**

Wegener noted the similarities in rock sequences on both sides of the Atlantic and used the information as a supporting piece of evidence for continental drift. He suggested that mountains such as those seen on opposite sides of the Atlantic formed during the collision when Pangaea was formed. Later, when the continents split apart, once-continuous mountain ranges were separated. Confirmation of this idea exists in a similar match with mountains extending from South America through Antarctica and across Australia.

## More Observations: Glacial Ages and Other Climate Evidence

Wegener also noticed the occurrence of past glacial activity in areas that are now tropical and suggested that it, too, provided supporting evidence for drifting continents. Currently, the only places in the world where thick continental *ice sheets* occur are in the polar regions of Greenland and Antarctica. However, evidence of ancient glaciation is found in the lower-latitude regions of South America, Africa, India, and Australia.

These deposits, which have been dated at 300 million years old, indicate one of two possibilities: (1) There was a global **ice age** at that time, and even tropical areas were covered by thick ice, or (2) some continents that are now in tropical areas were once located much closer to one of the poles. It is unlikely that the entire world was covered by ice 300 million years ago because coal found in North America and Europe today are from the same geologic age as the glacial deposits, and we know that coal formed in vast semitropical swamps. Thus, a reasonable conclusion is that some of the continents must have been closer to the poles than they are today.

**Animation**
Breakup of Pangaea
http://goo.gl/egACqz

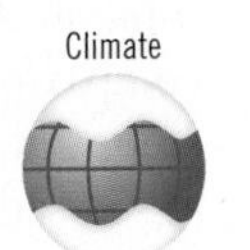
Climate
Connection

Another type of glacial evidence indicates that certain continents have moved from more polar regions during the past 300 million years. When glaciers flow, they move and abrade the underlying rocks, leaving grooves that indicate the direction of flow. The arrows in **Figure 2.5a** show how the glaciers would have flowed away from the South Pole on Pangaea 300 million years ago. The direction of flow is consistent with the grooves found on many continents today (**Figure 2.5b**), providing additional evidence for drifting continents.

Many examples of plant and animal fossils indicate very different climates than today. Two such examples are fossil palm trees in Arctic Spitsbergen and coal deposits in Antarctica. Earth's past environments can be interpreted from these rocks because plants and animals need specific environmental conditions in which to live. Corals, for example, generally need seawater above 18 degrees centigrade (°C), or 64 degrees Fahrenheit (°F), in order to survive. When fossil corals are found in areas that are cold today, two explanations seem most plausible: (1) Worldwide climate has changed dramatically or (2) the rocks have moved from their original location.

As explained in Chapter 16, "The Oceans and Climate Change,"natural processes have caused Earth's climate to change in the geologic past. Although dramatic shifts in Earth's climate might help explain the occurrence of fossils that seem out of place today (such as tropical fossils in polar locations), the distribution of these fossils could also be explained by drifting continents. Unaware of the changes in Earth's climate that are known by Earth scientists today, Wegener suggested that the out-of-place fossils as well as other climate evidence provided support for the slow movement of the continents and added another item to a growing list of evidence.

## Evidence from Fossils: Distribution of Organisms

To add credibility to his argument for the existence of the supercontinent of Pangaea, Wegener cited documented cases of several fossil organisms found on different landmasses that could not have crossed the vast oceans presently separating the continents. For example, the fossil remains of ***Mesosaurus*** (*meso* = middle, *saurus* = lizard), an extinct, presumably aquatic reptile that lived about 250 million years ago, are located only in eastern South America and western Africa (**Figure 2.6**). If *Mesosaurus* had been strong enough to swim across an ocean, why aren't its remains more widely distributed?

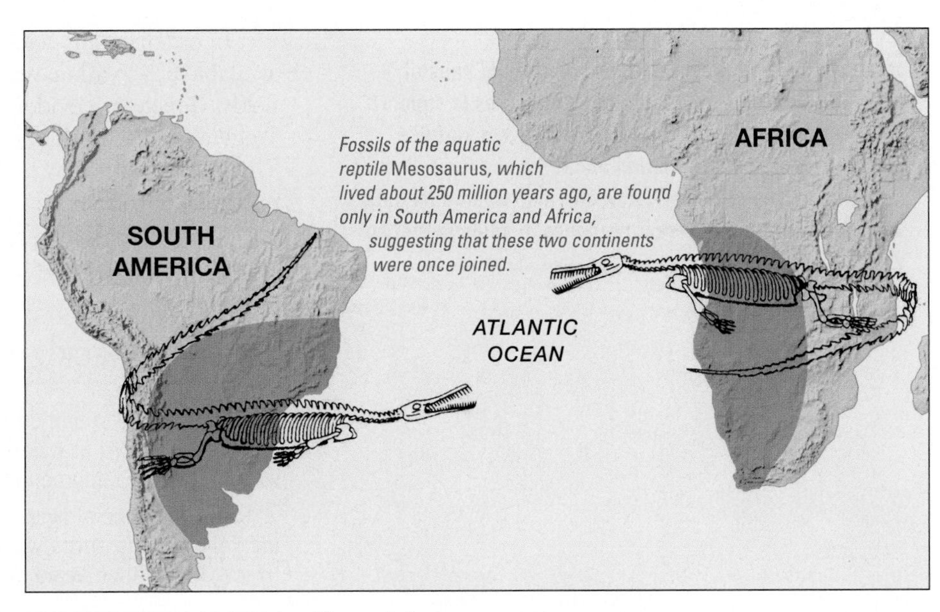


**Figure 2.6** **Fossils of *Mesosaurus*.**

Wegener's idea of continental drift provided an elegant solution to this problem. He suggested that the continents were closer together in the geologic past, so *Mesosaurus* didn't have to be a good swimmer to leave remains on two different continents. Later, after *Mesosaurus* became extinct, the continents moved to their present-day positions, and a large ocean now separates the once-connected landmasses. Other examples of similar fossils on different continents include those of plants, which would have had a difficult time traversing a large ocean.

Before continental drift, several ideas were proposed to help explain the curious pattern of these fossils, such as the existence of island stepping stones or a land bridge. It was even suggested that at least one pair of land-dwelling *Mesosaurus* survived the arduous journey across several thousand kilometers of open ocean by rafting on floating logs. However, there is no evidence to support the idea of island stepping stones or a land bridge, and the idea of *Mesosaurus* rafting across an ocean seems implausible.

Wegener also cited the distribution of present-day organisms as evidence to support the concept of drifting continents. For example, modern organisms with similar ancestries clearly had to evolve in isolation during the past few million years. Most obvious of these are the Australian marsupials (such as kangaroos, koalas, and wombats), which have a distinct similarity to the marsupial opossums found in the Americas.

## Objections to the Continental Drift Model

As compelling as his evidence may seem today, Wegener was unable to convince many of his fellow Earth scientists of the validity of his ideas. Although his hypothesis was correct in principle, it contained several incorrect details, such as the driving mechanism for continental motion and how continents move across ocean basins. For example, material strength calculations showed that ocean rock was too strong for continental rock to plow through it. Further, analysis of gravitational and tidal forces indicated that they were too small to move the great continental landmasses. Even without an acceptable mechanism, many geologists who studied rocks in South America and Africa accepted continental drift because it was consistent with the observations of rocks they were familiar with. However, North American geologists—most of whom were unfamiliar with these Southern Hemisphere rock sequences—remained highly skeptical.

**RECAP Alfred Wegener used a variety of lines of evidence from different scientific disciplines to support continental drift. However, he did not have a suitable mechanism or any information about the sea floor, and his idea was widely criticized. It wasn't until additional data was available from different branches of science that an explanation emerged about a possible mechanism.**

In order for any scientific viewpoint to gain wide acceptance, it must explain all available observations and have supporting evidence from a wide variety of scientific fields. Conclusive evidence for continental drift would not come until decades after Wegener's death, when more details of the nature of the ocean floor were revealed. That, along with new technology that enabled scientists to determine the original positions of rocks on Earth, allowed scientists to reexamine Wegener's hypothesis and address its flaws. This process of testing, collecting more data, and refining hypotheses, is the quintessential example of the process of science.

**CONCEPT CHECK 2.1 ▸ Evaluate the evidence that supports continental drift.**

1 When did the supercontinent of Pangaea exist? What was the ocean that surrounded the supercontinent called?

2 Regarding glacial ages, why is it unlikely that the entire world was covered by ice 300 million years ago?

3 Cite the lines of evidence Alfred Wegener used to support his idea of continental drift. Why did scientists of the time doubt that continents had drifted?

## 2.2 ▸ What Additional Observations Led to the Theory of Plate Tectonics?

Very little new information about Wegener's continental drift hypothesis was introduced between the time of Wegener's death in 1930 and the early 1950s. This changed when new technology allowed scientists to analyze the way rocks retained the signature of Earth's **magnetic field**, which in turn enabled them to determine the location where rocks first formed at Earth's surface. At the same time, studies of the sea floor using sonar, which was initially used for military purposes during World War II and is still used for ocean exploration today, revealed that the ocean basins were bisected by enormous mountain chains and rimmed by deep trenches. Unexpectedly, the rocks collected from these submerged mountain ridges were relatively young and associated with recent volcanic activity. These surprising observations suggested that the sea floor was geologically active, and scientists hypothesized that mid-ocean ridges were the source of new sea floor created in a process they called sea floor spreading. Ultimately, these two independent observations—continental drift and sea floor spreading—were explained by the same process, which today is encompassed in the theory of plate tectonics.

### Earth's Magnetic Field and Paleomagnetism

Earth's magnetic field, which is shown in **Figure 2.7**, plays a crucial role in guiding navigators, and protects Earth's life-forms from solar storms. The invisible lines of magnetic force that originate within Earth and travel out into space resemble the magnetic field produced by a large bar magnet. (Note that the properties of a magnetic field can be explored easily enough with a bar magnet and some iron particles. Place the iron particles on a table and place a bar magnet nearby. Depending on the strength of the magnet, you should get a pattern resembling that in Figure 2.7a.) Like a bar magnet, Earth's magnetic field has opposite poles (labeled N for North and S for South). As Figure 2.7b shows, because Earth's magnetic lines of force wrap around the planet, a magnetic object aligned parallel to the field will also point into Earth to varying degrees, depending on latitude. The degree to which a magnetic particle points into Earth is called its **magnetic dip**, or *magnetic inclination*. Notice also in Figures 2.7a and 2.7b that Earth's geographic North Pole (the rotational axis) and Earth's magnetic north pole (magnetic north) do not coincide.

### STUDENTS SOMETIMES ASK . . .

***What causes Earth's magnetic field?***

Studies of Earth's magnetic field and research in the field of *magnetodynamics* suggest that convective movement of fluids in Earth's liquid iron–nickel outer core is the cause of Earth's magnetic field. The most widely accepted view is that Earth's magnetic field is created by strong electrical currents generated by a dynamo process resulting from the convective flow of molten iron in Earth's outer core. Earth's magnetic field is so complex that it has only recently been successfully modeled using some of the world's most powerful computers. In our solar system, the Sun and most other planets (and even some planets' moons) also exhibit magnetic fields. Interestingly, recent research based on ancient rocks in South Africa reveal that Earth's magnetic field must have been present by 3.45 billion years ago.

**ROCKS AFFECTED BY EARTH'S MAGNETIC FIELD** **Igneous rocks** (*igne* = fire, *ous* = full of) solidify from molten **magma** (*magma* = a mass) either underground or after volcanic eruptions at the surface that produce **lava** (*lavare* = to wash). Nearly all igneous rocks contain **magnetite**, a naturally magnetic iron mineral. Particles of magnetite in magma align themselves with Earth's magnetic field because magma and lava are fluid. Once molten material is cooled to a certain temperature, however, internal magnetite particles are frozen into position, thereby recording the angle of Earth's magnetic field at that place and time. In essence, grains of magnetite serve as tiny compass needles that record the strength and orientation of Earth's magnetic field. Unless the rock is heated to the temperature where magnetite grains are again mobile, these magnetite grains contain information about the magnetic field where the rock originated, regardless of where the rock subsequently moves.

Magnetite is also deposited in sediments. As long as the sediment is surrounded by water, the magnetite particles can align themselves with Earth's magnetic field. After sediment is buried and solidifies into **sedimentary rock** (*sedimentum* = settling), the particles are no longer able to realign themselves if they are subsequently moved. Thus, magnetite grains in sedimentary rocks also contain information about the magnetic field where the rock originated. Although other rock types have been used successfully to reveal information about Earth's ancient magnetic field, the most reliable ones are igneous rocks that have high concentrations of magnetite such as **basalt**, which is the rock type that comprises oceanic crust.

**PALEOMAGNETISM** The study of Earth's ancient magnetic field is called **paleomagnetism** (*paleo* = ancient). Scientists who study paleomagnetism analyze magnetite particles in rocks to determine not only their north–south direction but also their magnetic dip relative to Earth's surface.

**EXPLORING DATA ▼ Using Figure 2.7b, describe the changes to a handheld magnetic dip needle if you carried one to the following locations: (1) the magnetic equator, (2) the magnetic North Pole, (3) halfway between the magnetic equator and the magnetic North Pole, and (4) the magnetic South Pole. Explain why the instrument would experience those changes.**

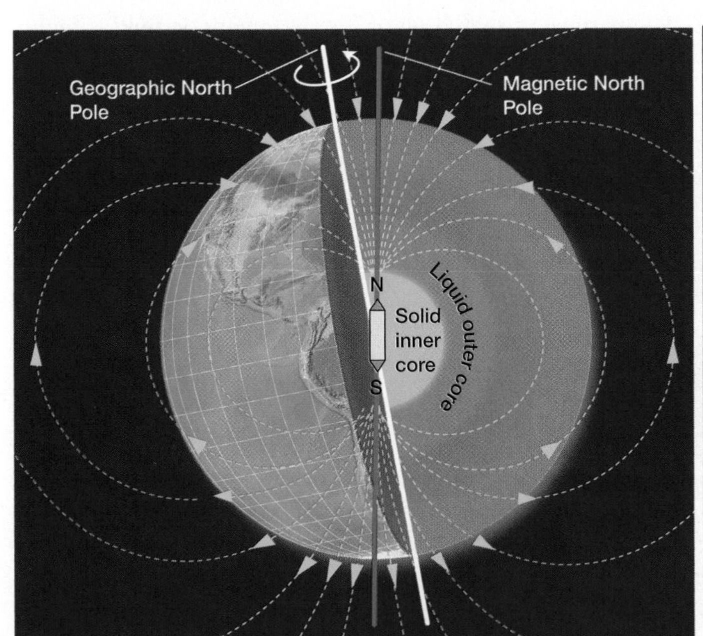


(a) Earth's magnetic field generates invisible lines of magnetic force similar to a large bar magnet. Note that the Geographic North Pole and the Magnetic North Pole are not in exactly the same location.

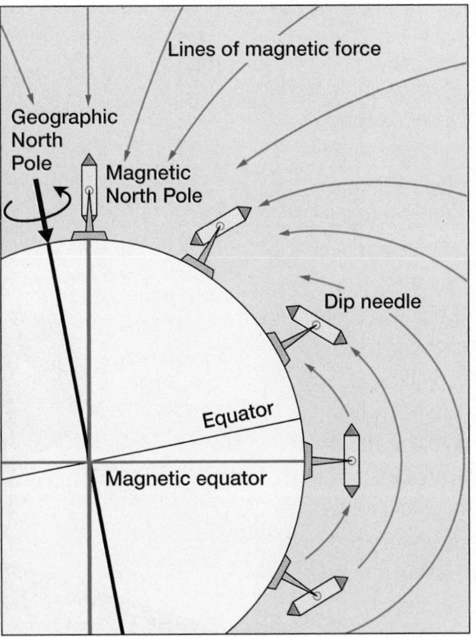


(b) Earth's magnetic field causes a dip needle to align parallel to the lines of magnetic force and change orientation with increasing latitude. Consequently, an approximation of latitude can be determined based on the dip angle.

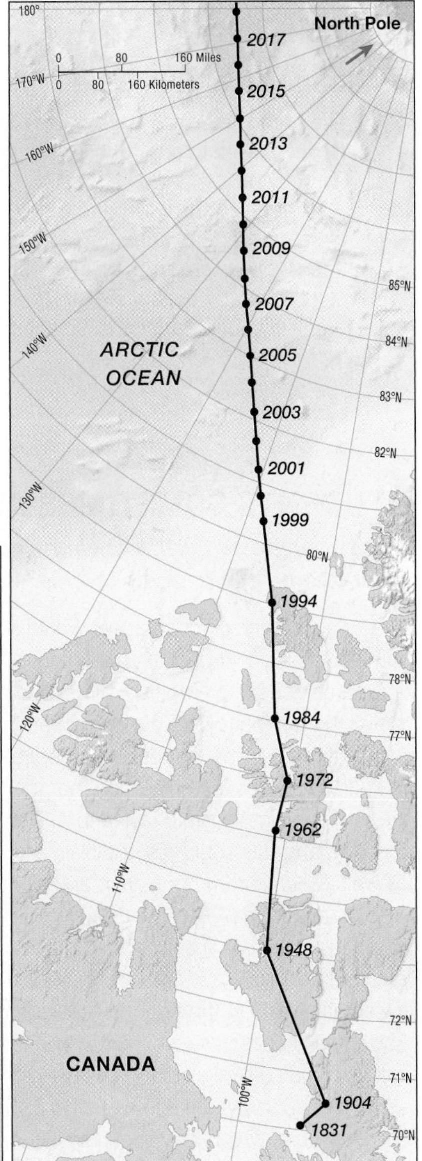


(c) Map showing the location of Earth's north magnetic pole since 1831 (black) and its projected location in the future (green).

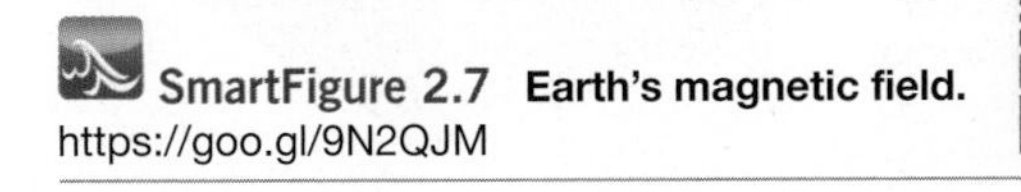

SmartFigure 2.7 **Earth's magnetic field.**
https://goo.gl/9N2QJM

One of the most interesting things about magnetic dip is that it is directly related to latitude on Earth. Figure 2.7b shows that a dip needle does not dip at all at Earth's magnetic equator. Instead, the needle lies horizontal to Earth's surface. At Earth's magnetic north pole, however, a dip needle points straight into the ground. A dip needle at Earth's magnetic south pole is also vertical to the surface, but it points *out* instead of *in*. Thus, magnetic dip increases with increasing latitude, from 0 degrees at the magnetic equator to 90 degrees at the magnetic poles. Because magnetic dip is retained in magnetically oriented rocks, measuring the dip angle reveals the latitude at which the rock initially formed. Done with care, paleomagnetism is an extremely powerful tool for interpreting where rocks first formed. Based on paleomagnetic studies, convincing arguments could finally be made that the continents had drifted relative to one another (**Diving Deeper 2.1**).

**MAGNETIC POLARITY REVERSALS** Magnetic compasses on Earth today follow lines of magnetic force and point toward magnetic north. It turns out, however, that the **polarity** (the north-south orientation of the magnetic field) has reversed itself periodically throughout geologic time. In essence, the north and south magnetic poles *reverse* or *switch* so that magnetic north becomes magnetic south and vice versa. **Figure 2.8** shows how ancient rocks have recorded the switching of Earth's magnetic polarity through time.

Why does Earth's magnetic field switch polarity? Geophysicists who study Earth's magnetic field do not yet fully understand the process of magnetic polarity reversals, but they are in agreement that Earth's rotation causes the electrically conducting liquid iron outer core to generate a self-sustaining magnetic field. Every so often, the flow of liquid iron is disturbed locally and twists part of the magnetic field in the opposite direction, weakening it. What triggers these disturbances is unknown; it may be because of turbulent flow conditions, or it may be just an inevitable consequence of a naturally chaotic system. Interestingly, computer simulations of Earth's core reveal frequent flipping of Earth's magnetic field.

Paleomagnetic studies reveal that 184 major reversals have occurred in the past 83 million years. The pattern of switching of Earth's magnetic field is highly irregular and ranges from 25,000 years to more than 30 million years. Even though the pattern has been described as random, on average a reversal occurs about every

## CREATURE FEATURE 2.1

### Can you help me find my way home?

The **green sea turtle** (*Chelonia mydas*) is not named for its external color, but for the greenish color of its body fat.

Green sea turtles spend many years living at sea in tropical and subtropical oceans but unerringly return to their place of birth to lay eggs on sandy beaches. How do they navigate so precisely? See **Process of Science 2.1**.

## STUDENTS SOMETIMES ASK . . .

***What changes to Earth's environment would occur when the magnetic poles reverse?***

During a reversal, compasses would likely show incorrect directions, and people could have difficulty navigating. The same goes for some fish, birds, and mammals that sense the magnetic field during migrations (see **Process of Science 2.1**). The decrease in strength of the magnetic field also reduces the protection that the field provides for life-forms against cosmic rays and particles coming from the Sun, and this could disrupt low-Earth-orbiting satellites as well as some communication and power grid systems. Also, Earth's *aurora borealis* (the Northern Lights) and its counterpart *aurora australis* (the Southern Lights), which are natural light displays in the sky, might be visible at much lower latitudes. On the bright side, we know that life on Earth has successfully survived previous magnetic reversals, so reversals might not be as dangerous as they are sometimes portrayed (such as in the 2003 science fiction film *The Core*, which is full of scientific inaccuracies).

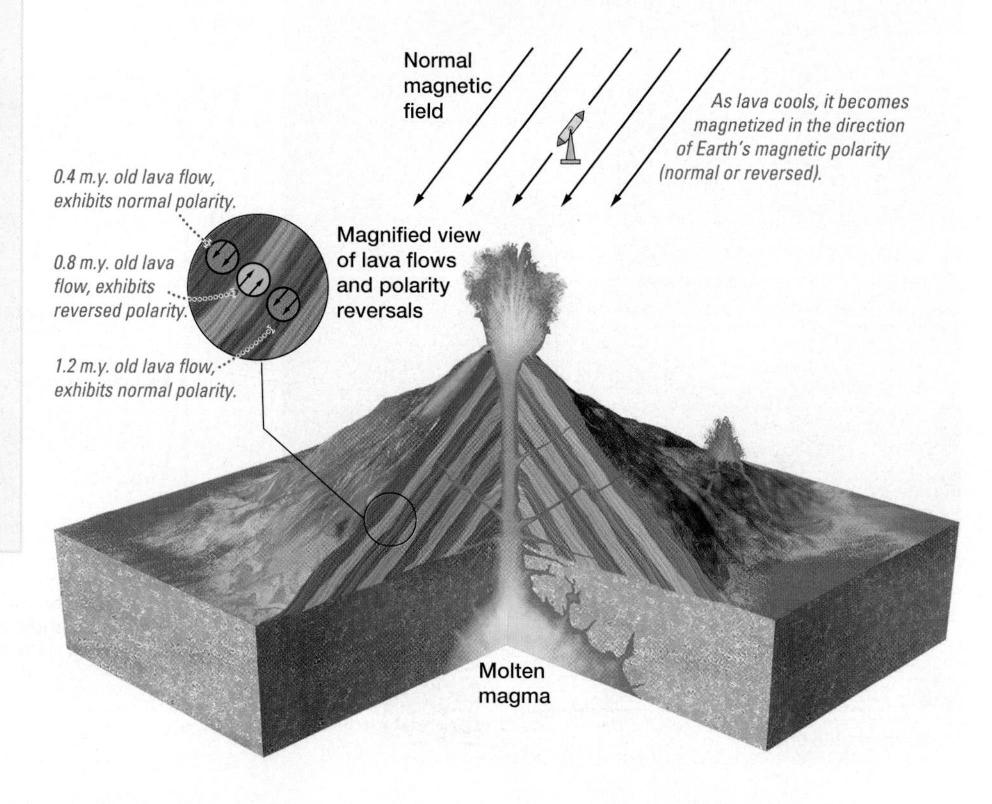


**Figure 2.8 Paleomagnetism preserved in rocks.** The switching of Earth's magnetic polarity through time is preserved in a sequence of rocks such as these lava flows, which are produced successively from the volcano. Note that m.y. = million years.

## PROCESS OF SCIENCE 2.1: Do Sea Turtles Use Earth's Magnetic Field for Navigation?

### BACKGROUND

Biologists have long been intrigued by sea turtles' ability to return to the exact same beach where they hatched when it is time to lay their eggs (**Figure 2A**), even if they have traveled thousands of kilometers and spent decades at sea before mating and returning home. Often these original nesting sites are small, isolated islands. So how do sea turtles navigate through the open ocean, which rarely contains any significant landmarks? Radio tagging of green sea turtles (*Chelonia mydas*) indicates that during their migration, they often travel in an essentially straight-line path to reach their destination. One hypothesis suggests that sea turtles use wave direction to help them steer. However, studies of their migration route reveal that green sea turtles continue along a straight-line path that is independent of wave direction.

### FORMING A HYPOTHESIS

Scientists wondered if the answer might lie in sea turtles' use of *magnetoreception*, which is the ability to sense the magnetic field of Earth. Studies have shown that newly hatched sea turtles can distinguish between different magnetic inclination angles, which in effect allows them to sense latitude. Young sea turtles can also distinguish magnetic field intensity, which gives a rough indication of longitude. By sensing these two magnetic field properties, a sea turtle could potentially determine its position at sea (similar to how global positioning system [GPS] works) and navigate to a tiny island thousands of kilometers away. Such an ability in adult sea turtles would allow them to navigate the ocean with something akin to that of a magnetic compass, in essence pointing the way home. Is that how they do it?

**Figure 2A** **Green sea turtle (*Chelonia mydas*).**

### DEVISING AN EXPERIMENT

Earth's magnetic field is like a large magnet, so if turtles are using that magnet to navigate, could a stronger magnet impair their ability to sense direction? A 2007 study examined green sea turtles and their ability to navigate in the presence of an artificial magnetic field. Nesting female sea turtles were collected from their home beach and moved to a release site 100-120 kilometers (62-75 miles) away. They were divided into three groups: The first group had a strong magnet glued to their heads while they were transported to the release site, which was then removed. This was done in an attempt to "reset" the turtles' magnetic homing ability. The second group had a magnet glued to their heads for the duration of the experiment. The third group was not exposed to magnets and was used as a control group. These turtles had a non-magnetic brass disk attached to their heads. All the turtles were then tagged with a GPS device, released, and then tracked to see if they could navigate their way back to their nesting sites.

### INTERPRETING THE RESULTS

The results show that the magnets did not prevent the turtles from eventually reaching their goals. Significantly, however, the turtles in both groups that had been exposed to the magnets took a more convoluted path back toward the nesting site than did the control turtles. This suggests that magnets adversely affected navigation, and occurred whether the turtles were exposed to the magnets just prior to release or during the entire experiment. The results imply that sea turtles do detect and actively use Earth's magnetic field to navigate while traveling. At the same time, it doesn't rule out the turtles' ability to navigate using non-magnetic processes. Like any good navigator, sea turtles may use other tools, such as olfactory (scent) clues, Sun angles, local landmarks, and oceanographic phenomena to find their way home.

### THINKING LIKE A SCIENTIST: WHAT'S NEXT?

A diverse assortment of other animals may also sense Earth's magnetic field and use it to navigate, including fish, marine mammals, birds, cows, deer, and even humans. Devise an experiment using one of these organisms to determine if it has the ability to orient itself in relation to Earth's magnetic field.

450,000 years or so. The flipping of Earth's magnetic field takes an average of about 5000 years; it can happen in as quickly as 1000 years or as slowly as 20,000 years. Changes in Earth's magnetic polarity are identified in rock sequences by a gradual decrease in the intensity of the magnetic field of one polarity, followed by a gradual increase in the intensity of the magnetic field of opposite polarity. Interestingly, there have been several documented instances of false starts where the weakening of the magnetic field does not lead to a full flip.

**Animation**
Flipping of Earth's Magnetic Field
http://goo.gl/84Hbpo

## USING MOVING CONTINENTS TO RESOLVE AN APPARENT DILEMMA: DID EARTH EVER HAVE TWO WANDERING NORTH MAGNETIC POLES?

The theory of plate tectonics has been very useful in resolving some apparent dilemmas about Earth's history. A classic example of this occurred when magnetic dip data for rocks on various continents were used to determine the ancient position of the magnetic north pole on Earth. Scientists who analyzed the data concluded that Earth's north magnetic pole must be wandering, or moving, through time. Further, the data suggested that rocks on different continents pointed to two different locations for Earth's north magnetic pole.

Figure 2C (*part a, left*) shows the magnetic **polar wandering paths**—sometimes called *polar wandering curves*—for North America and Eurasia. Notice how both paths have a similar shape but, for all rocks older than about 70 million years, the pole determined from North American rocks lies to the west of that determined from Eurasian rocks. From this data, it appeared that Earth had two separate magnetic poles in the geologic past, which would be remarkably different than today, when Earth has a single north magnetic pole. In fact, geophysical data indicates that only one north magnetic pole can exist at any given time and that it is unlikely that its position has changed very much through time because it must remain very closely aligned with Earth's rotational axis. Earth scientists were initially puzzled by these findings until they realized that the discrepancy could be resolved by having a single magnetic pole that remains relatively stationary, while North America and Eurasia moved relative to the pole and relative to each other. It wasn't Earth's magnetic field that was moving; instead, it was the continents themselves that were moving. That's why the magnetic polarity paths are called *apparent* polar wandering paths.

Figure 2C (*part b, right*) shows that when the continents are moved into the positions they occupied when they were part of Pangaea, the two wandering paths match up, providing strong evidence that there were never two magnetic north poles on Earth. A more reasonable conclusion in light of plate tectonics is that the *continents* had moved both toward the north as well as relative to each other throughout geologic time.

### What Did You Learn?

What puzzled sctientists about Earth's ancient magnetic field? How was the apparent dilemma resolved?

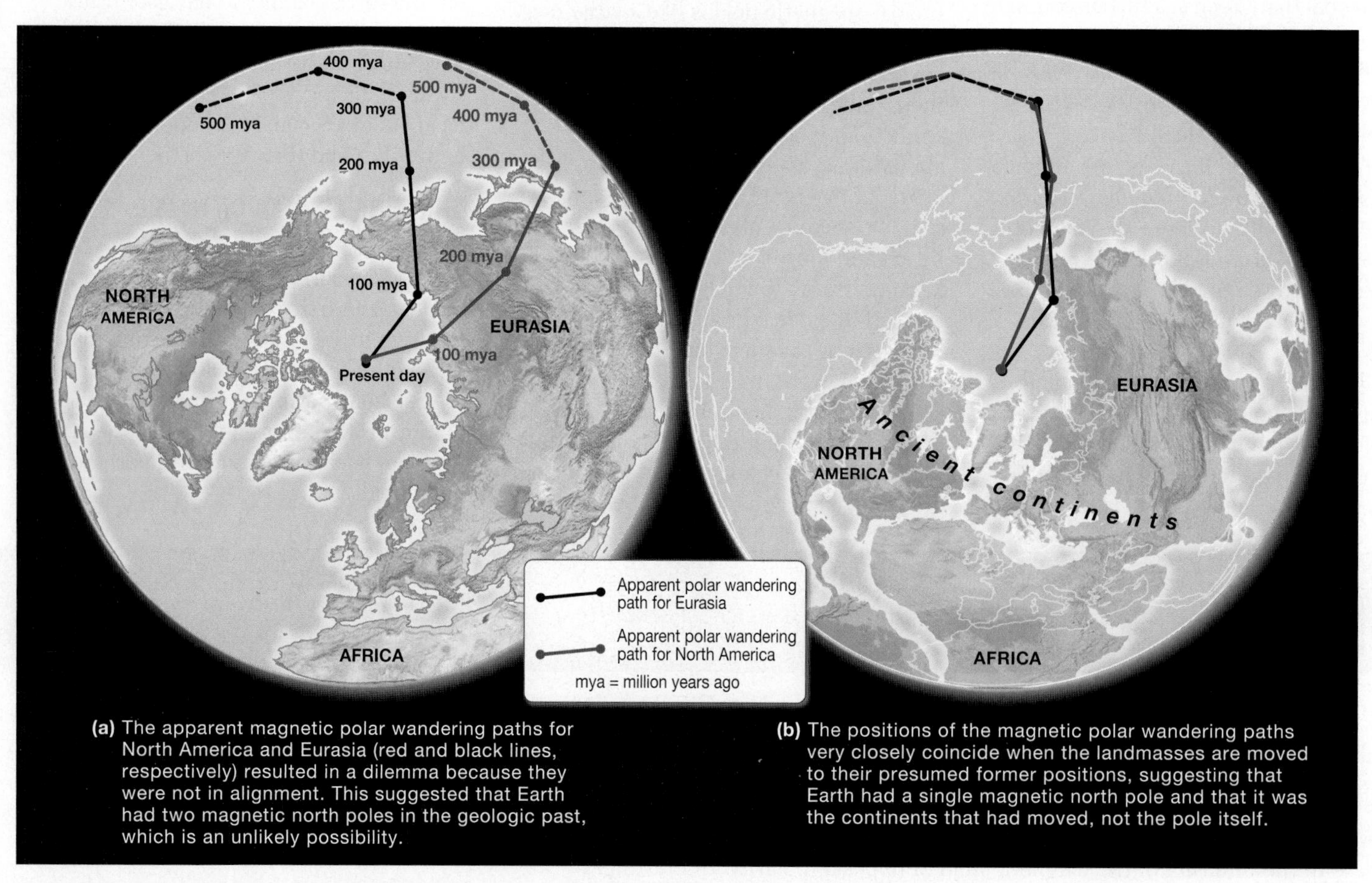


**(a)** The apparent magnetic polar wandering paths for North America and Eurasia (red and black lines, respectively) resulted in a dilemma because they were not in alignment. This suggested that Earth had two magnetic north poles in the geologic past, which is an unlikely possibility.

**(b)** The positions of the magnetic polar wandering paths very closely coincide when the landmasses are moved to their presumed former positions, suggesting that Earth had a single magnetic north pole and that it was the continents that had moved, not the pole itself.

Figure 2B **Apparent polar wandering paths.**

Geologic evidence indicates that Earth's magnetic field has been weakening during the past 2000 years. New satellite analysis reveals that Earth's magnetic field is losing strength at a rate of about 5% per decade, which is more rapidly than previously thought. Geophysicists think that the diminishing strength of Earth's magnetic field may be an indication that Earth's current "normal" polarity may reverse itself. In fact, the last major reversal of Earth's magnetic poles occurred 780,000 years ago, which suggests that the next one is overdue.

**TRUE POLAR WANDERING** Earth's magnetic north pole—which does not coincide with the geographic North Pole—was first located near Boothia Peninsula in the Canadian Arctic in 1831; since that time, it's been migrating northwest by about 50 kilometers (30 miles) each year (Figure 2.7c). If this rate continues, Earth's magnetic pole will be in Siberia by 2050.

**PALEOMAGNETISM AND THE OCEAN FLOOR** Paleomagnetism had certainly proved its usefulness on land, but, up until the mid-1950s, paleomagnetic studies had only been conducted on continental rocks. Would the ocean floor also show variations in magnetic polarity? To test this idea, the U.S. Coast and Geodetic Survey, in conjunction with scientists from Scripps Institution of Oceanography, undertook an extensive deep-water mapping program off Oregon and Washington in 1955. Using a sensitive instrument called a **magnetometer** (*magneto* = magnetism, *meter* = measure), which is towed behind a research vessel, the scientists spent several weeks at sea, moving back and forth in a regularly spaced pattern, measuring Earth's magnetic field and how it was affected by the magnetic properties of rocks on the ocean floor.

When the scientists analyzed their data, they found that the entire surveyed area had a pattern of north–south stripes in a surprisingly regular and alternating pattern of above-average and below-average magnetism. What was even more surprising was that the pattern appeared to be symmetrical with respect to a long mountain range that was fortuitously in the middle of their survey area.

Detailed paleomagnetic studies of this and other areas of the sea floor confirmed a similar pattern of alternating stripes of above-average and below-average magnetism. These stripes are called **magnetic anomalies** (*a* = without, *nomo* = law; an anomaly is a departure from normal conditions). The ocean floor had embedded in it a regular pattern of alternating magnetic stripes, unlike anywhere on land.

Researchers had a difficult time explaining why the ocean floor had such a regular pattern of magnetic anomalies. Nor could they explain how the sequence on one side of the underwater mountain range matched the sequence on the opposite side—in essence, they were a mirror image of each other. To understand how this pattern could have formed, more information was needed about ocean floor features and their origin.

## Sea Floor Spreading and Features of the Ocean Basins

When he was a U.S. Navy captain in World War II, geologist **Harry Hess** (1906–1969), developed the habit of leaving his depth recorder on at all times while his ship was traveling at sea. After the war, compilation of these and many other depth records showed extensive mountain ridges near the centers of ocean basins and extremely deep, narrow trenches at the edges of ocean basins. In 1962, Hess published *History of Ocean Basins*, which contained the idea of **sea floor spreading** and the associated circular movement of rock material in the mantle—**convection cells** (*con* = with, *vect* = carried)—as the driving mechanism (**Figure 2.9**). He suggested that new ocean crust was created at the ridges, split apart, moved away from the ridges, and later disappeared back into the deep Earth at trenches. Mindful of the resistance of North American scientists to the idea of continental drift, Hess referred to his own work as "geopoetry."

As it turns out, Hess's initial ideas about sea floor spreading have been confirmed. The **mid-ocean ridge** (Figure 2.9) is a continuous underwater mountain range that

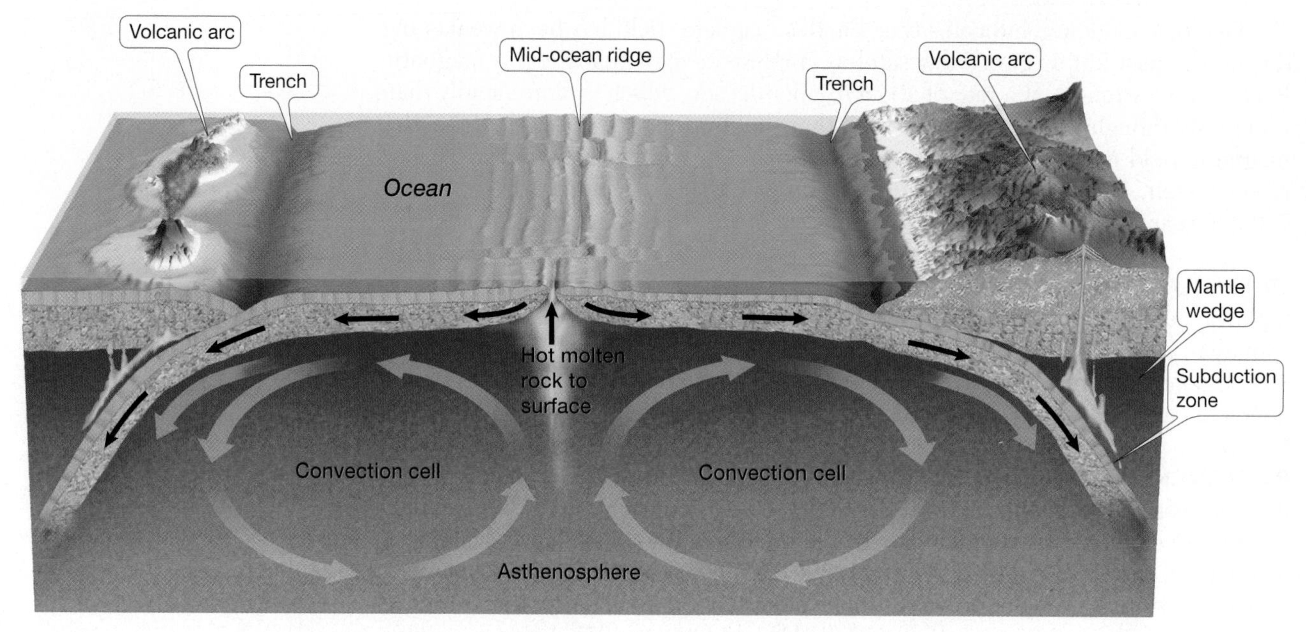


**Figure 2.9 Processes and resulting features of plate tectonics.**

**Animation**
Sea Floor Spreading and Plate Boundaries
http://goo.gl/9iEcQD

winds through every ocean basin in the world and resembles the seam on a baseball. It is entirely volcanic in origin, wraps one-and-a-half times around the globe, and rises more than 2.5 kilometers (1.5 miles) above the surrounding deep-ocean floor. It even rises above sea level in places such as Iceland. New ocean floor forms at the crest, or axis, of the mid-ocean ridge. By the process of sea floor spreading, new ocean floor is split in two and carried away from the axis, replaced by the upwelling of volcanic material that fills the void with new strips of sea floor. Sea floor spreading occurs along the axis of the mid-ocean ridge, which is referred to as a **spreading center**. One way to think of the mid-ocean ridge is as a zipper that is being pulled apart. Thus, Earth's zipper (the mid-ocean ridge) is becoming unzipped!

At the same time, ocean floor is being destroyed at **deep-ocean trenches**. Trenches are the deepest parts of the ocean floor and, on a map of the sea floor, resemble a narrow crease or trough (Figure 2.9). The largest earthquakes in the world occur near these trenches; they are caused by a plate bending downward and slowly plunging back into Earth's interior. This process is called **subduction** (*sub* = under, *duct* = lead), and the sloping area from the trench along the downward-moving plate is called a **subduction zone**.

In 1963, geologists **Frederick Vine** and **Drummond Matthews** of Cambridge University combined the seemingly unrelated pattern of magnetic sea floor stripes with the process of sea floor spreading to explain the perplexing pattern of alternating and symmetric magnetic stripes on the sea floor (**Figure 2.10**). Vine and Matthews interpreted the pattern of above-average and below-average magnetic polarity episodes embedded in sea floor rocks to be caused by Earth's magnetic field alternating between "normal" polarity (like today's magnetic pole position in the north) and "reversed" polarity (with the magnetic pole to the south). They proposed that the pattern could be created when newly formed rocks at the mid-ocean ridge are magnetized with whichever polarity exists on Earth during their formation. As those rocks are slowly moved away from the crest of the mid-ocean ridge, they maintain their original polarity, and subsequent rocks record the periodic switches of Earth's magnetic polarity. The result is an alternating pattern of magnetic polarity stripes that are symmetric with respect to the mid-ocean ridge.

The pattern of alternating reversals of Earth's magnetic field as recorded in the sea floor was the most convincing piece of evidence set forth to support the concept of sea floor spreading—and, as a result, continental drift. However, the continents weren't plowing through the ocean basins as Wegener had envisioned. Instead, the ocean floor was a conveyer belt that was being continuously formed at the mid-ocean

## STUDENTS SOMETIMES ASK . . .

***Figure 2.9 shows that the mantle is moving in large circles. Is the mantle molten?***

No. Because the mantle is often depicted as flowing in convective motion, a common misconception is that the mantle is molten. Seismic studies reveal that the mantle is unambiguously greater than 99% solid, although it does have the ability to flow S-L-O-W-L-Y over time (hence the arrows in the figure). The only places where the mantle is partially molten are (1) underneath the mid-ocean ridge, where release of pressure causes molten material to form; (2) in the mantle wedge above a subducting plate, where water released from the downgoing oceanic plate causes melting; and (3) in isolated mantle plumes, which are discussed later in this chapter. Make no mistake about it: The vast majority of the mantle is composed of hot, solid rock. But even that rock can flow if enough pressure is applied to it. Think of how a blacksmith can deform and shape a red-hot piece of solid iron by using the pressure of repeated hammerings. Imagine how much greater the pressure is inside Earth to cause hot, solid rock to deform and flow!

**Animation**
Convection in a Lava Lamp
http://goo.gl/4SFFCG

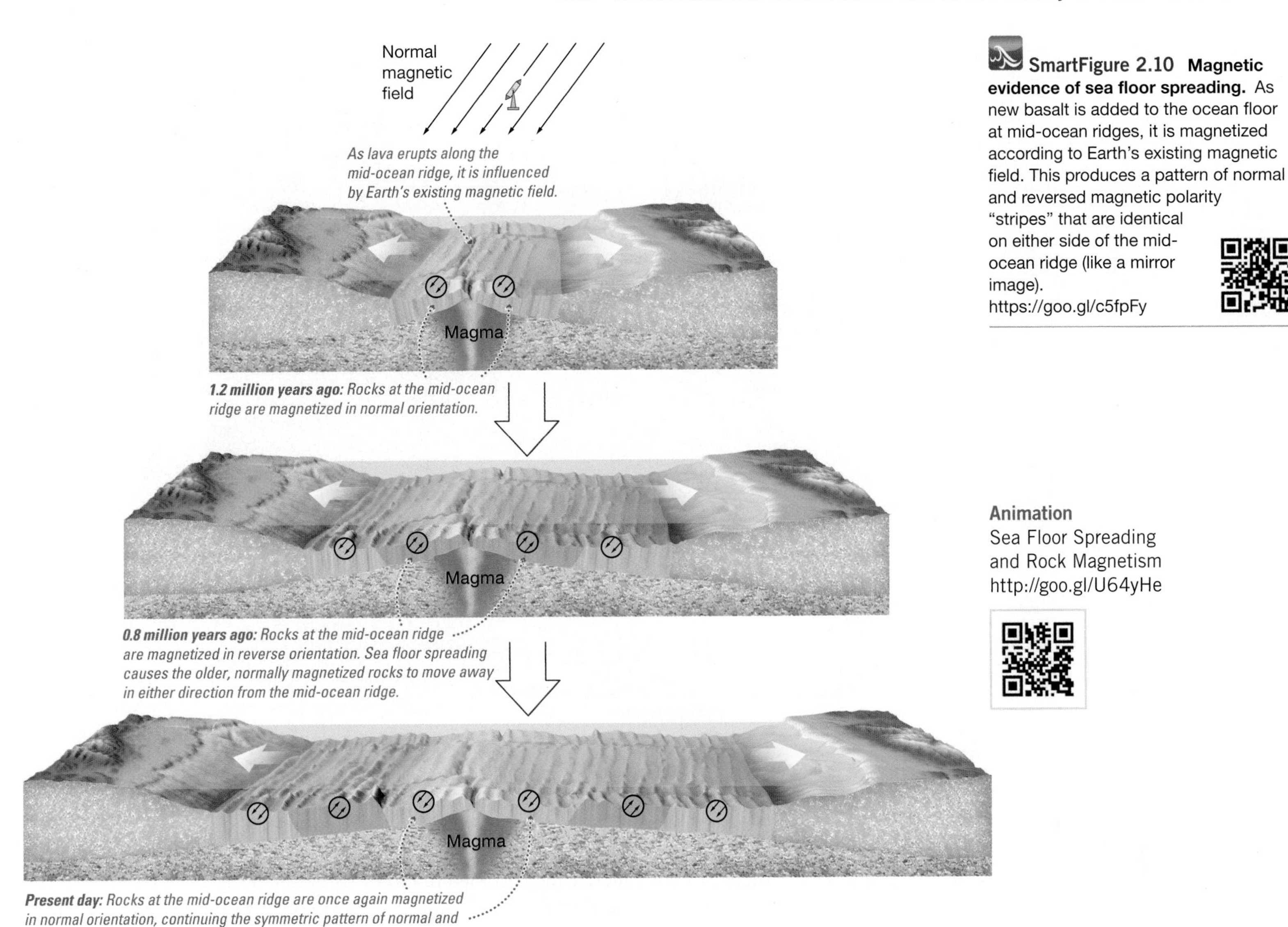
Normal magnetic field
As lava erupts along the mid-ocean ridge, it is influenced by Earth's existing magnetic field.
Magma
1.2 million years ago: Rocks at the mid-ocean ridge are magnetized in normal orientation.
Magma
0.8 million years ago: Rocks at the mid-ocean ridge are magnetized in reverse orientation. Sea floor spreading causes the older, normally magnetized rocks to move away in either direction from the mid-ocean ridge.
Magma
Present day: Rocks at the mid-ocean ridge are once again magnetized in normal orientation, continuing the symmetric pattern of normal and reversed magnetic polarity "stripes" on either side of the mid-ocean ridge.

**SmartFigure 2.10 Magnetic evidence of sea floor spreading.** As new basalt is added to the ocean floor at mid-ocean ridges, it is magnetized according to Earth's existing magnetic field. This produces a pattern of normal and reversed magnetic polarity "stripes" that are identical on either side of the mid-ocean ridge (like a mirror image). https://goo.gl/c5fpFy

**Animation**
Sea Floor Spreading and Rock Magnetism
http://goo.gl/U64yHe

ridge and destroyed at the trenches, with the continents just passively riding along on the conveyer. By the late 1960s, most geologists had changed their stand on continental drift in light of this new evidence, which is a prime example of the process of science.

**RECAP** **The plate tectonic model states that new sea floor is created at the mid-ocean ridge, where it moves outward by the process of sea floor spreading and is destroyed by subduction into ocean trenches.**

## Other Evidence from the Ocean Basins

Even though the tide of scientific opinion had indeed switched to favor a mobile Earth, additional evidence from the ocean floor would further support the ideas of continental drift and sea floor spreading.

**AGE OF THE OCEAN FLOOR** In the late 1960s, an ambitious deep-sea drilling program was initiated to test the existence of sea floor spreading. One of the program's primary missions was to drill into and collect ocean floor rocks for radiometric age dating. If sea floor spreading does indeed occur, then the youngest sea floor rocks would be atop the mid-ocean ridge, and the ages of rocks would increase on either side of the ridge in a symmetric pattern.

The map in **Figure 2.11**, showing the age of the ocean floor beneath deep-sea deposits, is based on the pattern of magnetic stripes verified with thousands of radiometrically age-dated samples. It shows that the ocean floor is youngest along the mid-ocean ridge, where new ocean floor is created, and the age of rocks

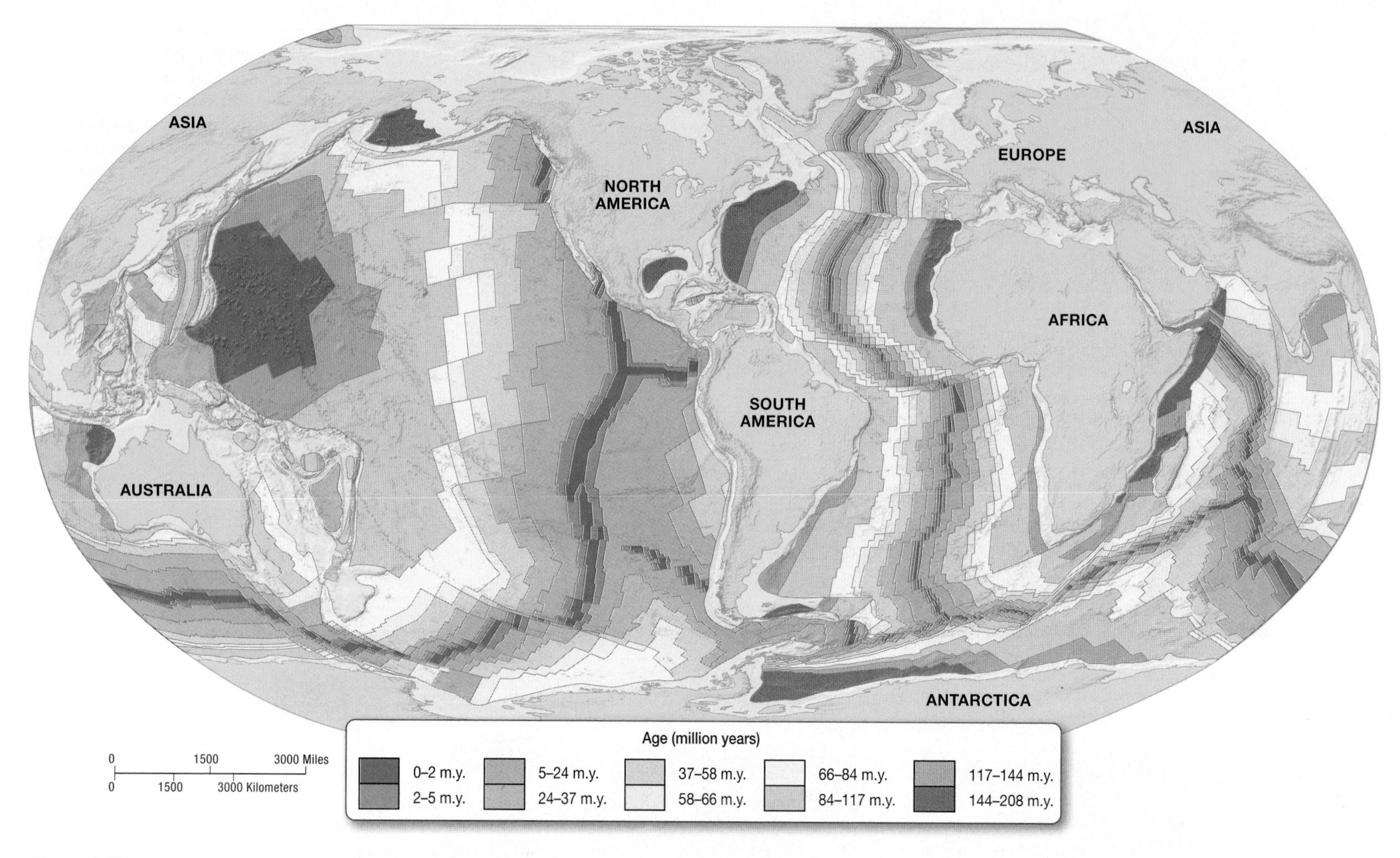


Figure 2.11 **Age of the ocean crust beneath deep-sea deposits.** The youngest rocks (*bright red areas*) are found along the mid-ocean ridge. Farther away from the mid-ocean ridge, the rocks increase linearly in age in either direction. Ages shown are in millions of years before present.

**EXPLORING DATA** ▲

**1. Where, in general, is the world's youngest ocean crust? What sea floor feature is it associated with?**

**2. Identify three locations where the world's oldest ocean crust occurs. Explain why such old ocean crust exists here.**

increases with increasing distance in either direction away from the axis of the ridge. The symmetric pattern of ocean floor ages confirms that the process of sea floor spreading must indeed be occurring.

The Atlantic Ocean has the simplest and most symmetric pattern of age distribution in Figure 2.11. The pattern results from the newly formed Mid-Atlantic Ridge that rifted Pangaea apart. The Pacific Ocean has the least symmetric pattern because many subduction zones surround it. For example, ocean floor east of the East Pacific Rise that is older than 40 million years has already been subducted. The ocean floor in the northwestern Pacific, about 180 million years old, has not yet been subducted. A portion of the East Pacific Rise has even disappeared under North America. The age bands in the Pacific Ocean are wider than those in the Atlantic and Indian Oceans, which suggests that the rate of sea floor spreading is greatest in the Pacific Ocean.

Recall from Chapter 1 that the ocean is at least 4 billion years old. However, the oldest ocean floor is only 180 million years old (or 0.18 billion years old), and the majority of the ocean floor is not even half that old (see Figure 2.11). How could the ocean floor be so incredibly young, while the oceans themselves are so phenomenally old? According to plate tectonic theory, new ocean floor is created at the mid-ocean ridge by sea floor spreading and moves off the ridge to eventually be subducted and remelted in the mantle. In this way, the ocean floor keeps regenerating itself. The floor beneath the oceans today is not the same one that existed beneath the oceans 4 billion years ago.

If the rocks that comprise the ocean floor are so young, why are continental rocks so old? Using radiometric age dating, scientists have determined that the oldest rocks on land are about 4 billion years old. Many other continental rocks approach this age, implying that the same processes that constantly renew the sea floor do not operate on land. Rather, evidence suggests that continental rocks, because of their low density, do not get recycled by the process of sea floor spreading, and thus they remain at Earth's surface for long periods of time.

**HEAT FLOW** The heat from Earth's interior is released to the surface as **heat flow**. Current models indicate that this heat moves to the surface with magma in convective motion. Most of the heat is carried to regions of the mid-ocean ridge spreading centers (see Figure 2.9). Cooler portions of the mantle descend along subduction zones to complete each circular-moving convection cell.

Heat flow measurements show that the amount of heat flowing to the surface along the mid-ocean ridge can be up to eight times greater than the average amount flowing to other parts of Earth's crust. Additionally, heat flow at deep-ocean trenches, where ocean floor is subducted, can be as little as one-tenth the average. Increased heat flow at the mid-ocean ridge and decreased heat flow at subduction zones is what would be expected based on thin crust at the mid-ocean ridge and a double thickness of crust at the trenches (see Figure 2.9).

**WORLDWIDE EARTHQUAKES** *Earthquakes* are sudden releases of energy caused by fault movement or volcanic eruptions. The map in **Figure 2.12a** shows that most large earthquakes occur along ocean trenches, reflecting the energy released during subduction. Other earthquakes occur along the mid-ocean ridge, reflecting the energy released during sea floor spreading. Still others occur along major faults in the sea floor and on land, reflecting the energy released when moving plates contact other plates along their edges. When you examine the two maps in **Figure 2.12**, notice how closely the pattern of major earthquakes matches the locations of plate boundaries. This is because most earthquakes worldwide are created by plates interacting with each other at their margins. In fact, it was the location of earthquakes that first enabled scientists to identify plate boundaries.

## Detecting Plate Motion with Satellites

Since the late 1970s, orbiting satellites have allowed the accurate positioning of locations on Earth. (This technique is also used for navigation by ships at sea; see Diving Deeper 1.2.) One of the predictions of plate tectonic theory is that if the plates are moving, satellite positioning should show this movement over time. The map in **Figure 2.13** shows numerous locations that have been measured in this manner and confirms that regions on Earth are indeed moving in good agreement with the direction and rate of motion predicted by plate tectonics. The successful prediction of how locations on Earth are moving with respect to one another very strongly supports plate tectonic theory.

## The Acceptance of a Theory

The accumulation of lines of evidence such as those mentioned in this section, along with many other lines of evidence in support of moving continents, has convinced scientists of the basic validity of Wegner's concept of continental drift. Since the late 1960s, continental drift and sea floor spreading have been united into a much more encompassing theory known as plate tectonics, which describes the movement of the outermost portion of Earth and the resulting creation of continental and sea floor features. These tectonic plates are pieces of the **lithosphere** (*lithos* = rock, *sphere* = ball) that float on the more fluid **asthenosphere** (*asthenos* = weak, *sphere* = ball) below. (For a discussion of the properties of the lithosphere and asthenosphere, see Chapter 1.)

What forces drive plate motion? Although several mechanisms have been proposed for the force (or forces) responsible for driving this motion, none of them are able to explain all aspects of plate movement. However, scientific studies based on a simple model of lithosphere and mantle interactions suggest that two major

### STUDENTS SOMETIMES ASK . . .

***How fast do plates move, and have they always moved at the same rate?***

Currently, plates move an average of 2 to 12 centimeters (1 to 5 inches) per year, which is about as fast as a person's fingernails grow. A person's fingernail growth is dependent on many factors, including heredity, gender, diet, and amount of exercise, but averages about 8 centimeters (3 inches) per year. This may not sound very fast, but the plates have been moving for millions of years. Over a very long time, even an object moving slowly will eventually travel a great distance. For instance, fingernails growing at a rate of 8 centimeters (3 inches) per year for 1 million years would be 80 kilometers (50 miles) long!

Evidence shows that the plates were moving faster millions of years ago than they are moving today. Geologists can determine the rate of plate motion in the past by analyzing the width of new oceanic crust produced by sea floor spreading, since fast spreading produces more sea floor rock. (By using this relationship and examining Figure 2.11, you should be able to determine whether the Pacific Ocean or the Atlantic Ocean had a faster spreading rate.) Recent studies using this same technique indicate that about 50 million years ago, India attained a speed of 19 centimeters (7.5 inches) per year. Other research indicates that about 530 million years ago, plate motions may have been as high as 30 centimeters (1 foot) per year! What caused these rapid bursts of plate motion? Geologists are not sure why plates moved more rapidly in the past, but greater heat release from Earth's interior is a likely mechanism.

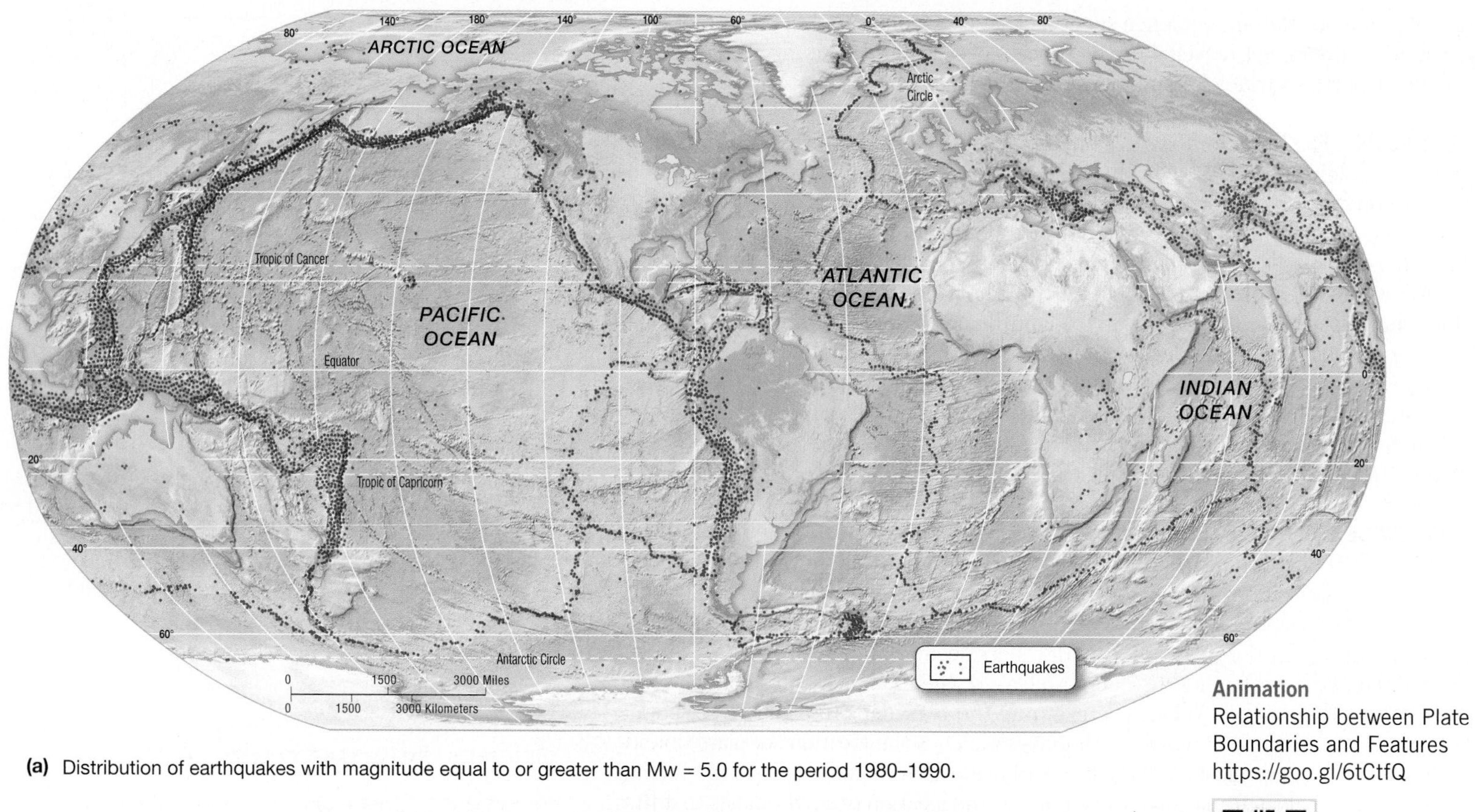


(a) Distribution of earthquakes with magnitude equal to or greater than Mw = 5.0 for the period 1980–1990.

**Animation**
Relationship between Plate Boundaries and Features
https://goo.gl/6tCtfQ

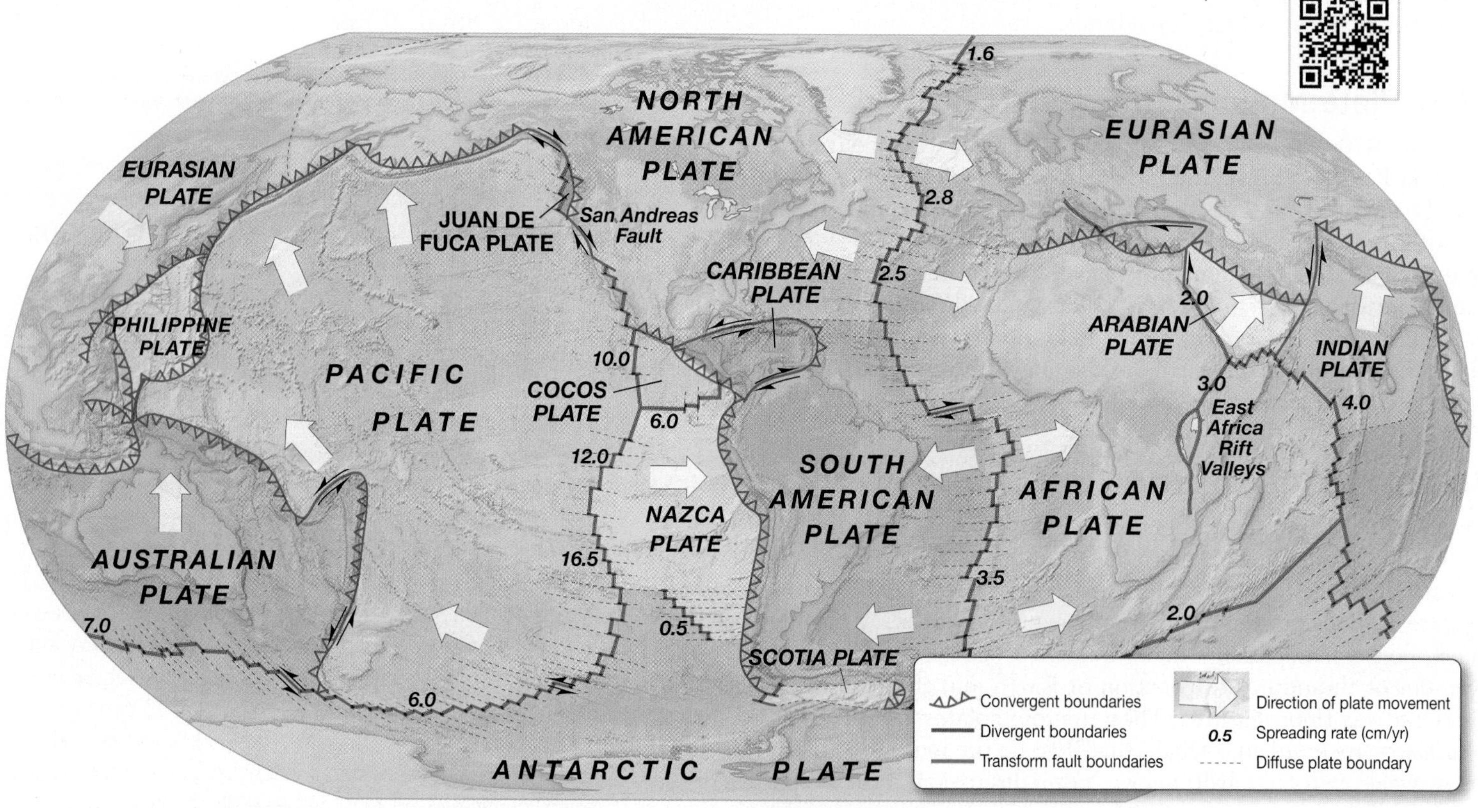


(b) Plate boundaries define the major tectonic plates (shaded), with arrows indicating the direction of motion and numbers representing the rate of motion in centimeters per year.

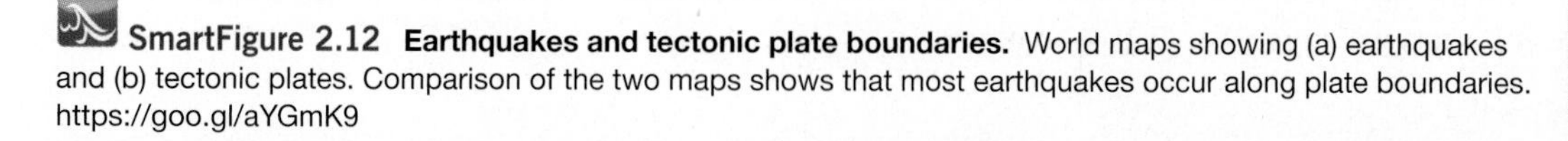

SmartFigure 2.12 **Earthquakes and tectonic plate boundaries.** World maps showing (a) earthquakes and (b) tectonic plates. Comparison of the two maps shows that most earthquakes occur along plate boundaries. https://goo.gl/aYGmK9

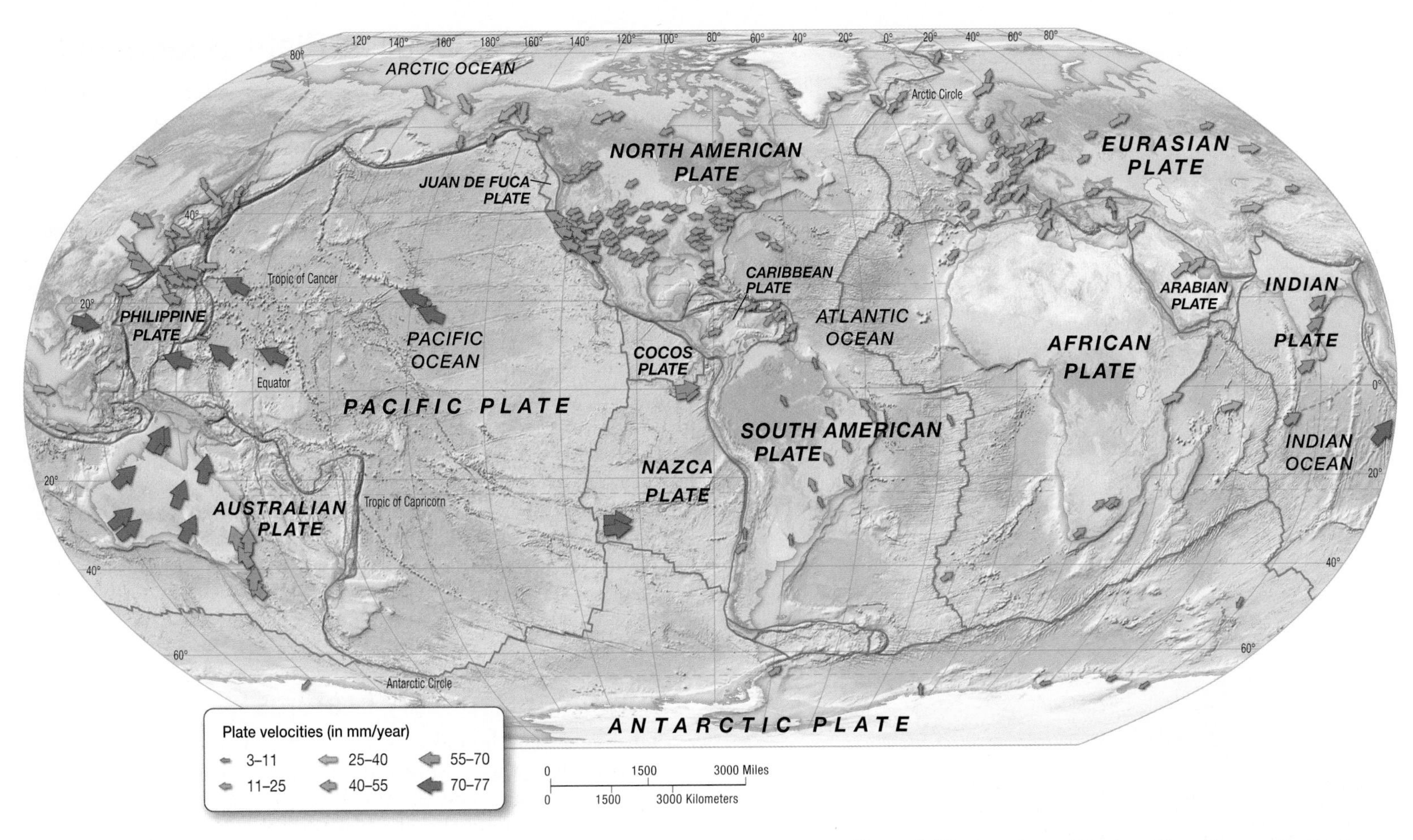


Figure 2.13 **Satellite positioning of locations on Earth.** Arrows show the direction of motion based on repeated satellite measurement of positions on Earth. The rate of plate motion in millimeters per year is indicated with different-colored arrows (*see legend*). Plate boundaries are shown with blue lines and are dashed where uncertain or diffuse.

tectonic forces may act in unison on subducting plates (slabs): (1) *slab pull,* which is generated by the pull of the weight of a plate as it sinks underneath an overlying plate, pulling the rest of the plate behind it, in a similar fashion to how a heavy comforter often slides off a bed onto the floor, and (2) *slab suction,* which is created as a subducting plate drags against the viscous mantle and causes the mantle to flow in toward the subduction zone, thereby sucking in nearby plates, much in the same way pulling a plug from a full bathtub draws floating objects toward its drain. In addition, high-resolution seismic studies have located a weak, partially molten layer at the base of the lithosphere that aids sliding and may reduce the force required for plate subduction. Other modeling studies that include upper mantle viscosity variations suggest that mantle flow differences contribute to either reinforcing or counteracting plate motions. Although researchers continue to model the forces that drive plate motions, these studies are hampered by the inaccessibility and complexity of Earth's mantle.

Since the acceptance of the theory of plate tectonics, much research has focused on understanding various features associated with plate boundaries, both on the sea floor and on land.

**RECAP Many independent lines of evidence, such as the detection of plate motion by satellites, provide strong support for the theory of plate tectonics.**

## CONCEPT CHECK 2.2 ▸ Summarize the evidence that supports plate tectonics.

1 Describe what Earth's magnetic field looks like and how it has changed through time.

2 Describe sea floor spreading and why it was an important piece of evidence in support of plate tectonics.

3 Why does a map of worldwide earthquakes closely match the locations of worldwide plate boundaries?

## 2.3 ▶ What Features Occur at Plate Boundaries?

Plate boundaries—where plates interact with each other—are associated with a great deal of tectonic activity, such as mountain building, volcanic activity, and earthquakes. In fact, the first clues to the locations of plate boundaries were the dramatic tectonic events that occur there, and there is a close correspondence between worldwide earthquakes and plate boundaries. As **Figure 2.12b** shows, Earth's surface is composed of seven major plates, along with many smaller ones. Close examination of Figure 2.12b shows that the boundaries of plates do not always follow coastlines and, as a consequence, nearly all plates contain both oceanic and continental crust. (For a review of the differences between (basaltic) oceanic and (granitic) continental crust, see Chapter 1.) Notice also that about 90% of plate boundaries occur on the sea floor.

**Animation**
Motion at Plate Boundaries
http://goo.gl/LNnG80

There are three types of plate boundaries, as shown in **Figure 2.14**. **Divergent boundaries** (*di* = apart, *vergere* = to incline) are found along oceanic ridges where new lithosphere is being added. **Convergent boundaries** (*con* = together, *vergere* = to incline) are found where plates are moving together and one plate subducts beneath the other. **Transform boundaries** (*trans* = across, *form* = shape) are found where lithospheric plates slowly grind past one another. **Table 2.1** summarizes characteristics, tectonic processes, features, and geographic examples of these plate boundaries.

*The three main types of plate boundaries are...*

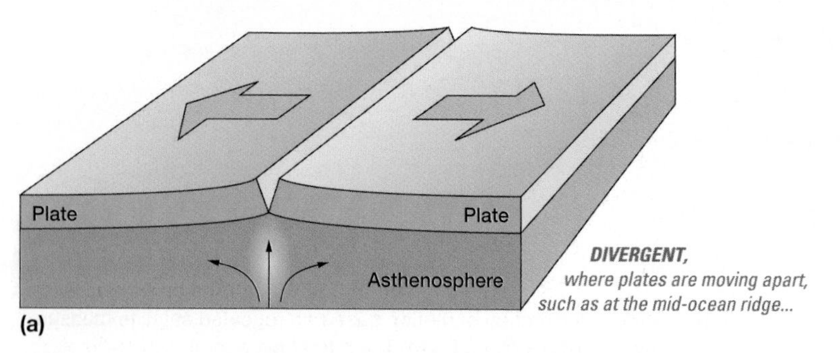
Plate
Plate
Asthenosphere
DIVERGENT,
where plates are moving apart,
such as at the mid-ocean ridge...
(a)

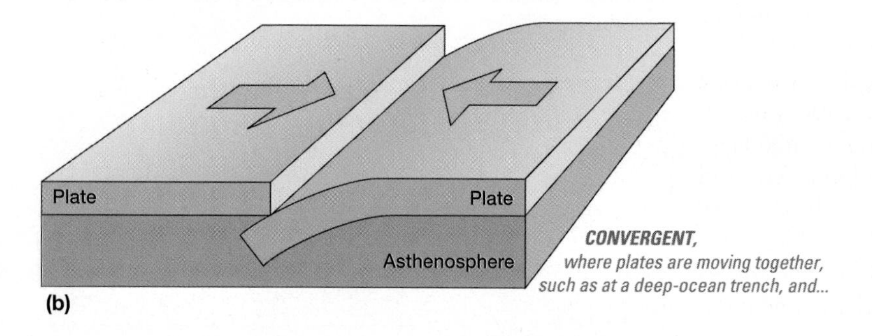
Plate
Plate
Asthenosphere
CONVERGENT,
where plates are moving together,
such as at a deep-ocean trench, and...
(b)

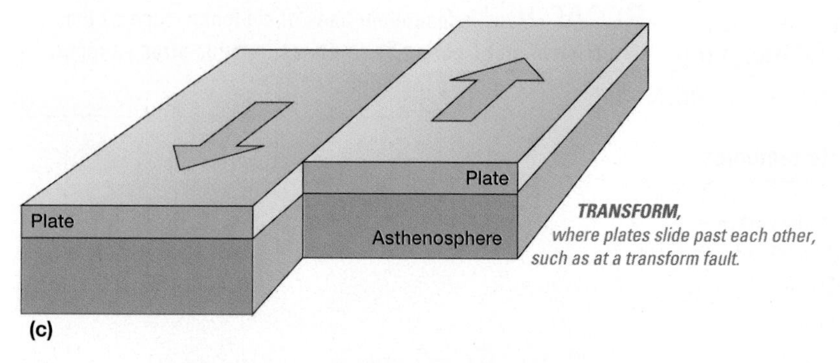
Plate
Plate
Asthenosphere
TRANSFORM,
where plates slide past each other,
such as at a transform fault.
(c)

Figure 2.14 **The three types of lithospheric plate boundaries.**

### Divergent Boundary Features

Divergent plate boundaries occur where two plates move apart, such as along the crest of the mid-ocean ridge, where sea floor spreading creates new oceanic lithosphere (**Figure 2.15**). A common feature along the crest of the mid-ocean ridge is a **rift valley**, which is a central downdropped linear depression (**Figure 2.16**). Pull-apart faults located along the central rift valley show that the plates are *continuously being pulled apart* rather than being pushed apart by the upwelling of material beneath the mid-ocean ridge. Upwelling of magma beneath the mid-ocean ridge is simply filling in the void left by the separating plates of lithosphere. In the process, sea floor spreading produces about 20 cubic kilometers (4.8 cubic miles) of new ocean crust worldwide each year.

**Figure 2.17** shows how the development of a mid-ocean ridge creates an ocean basin. Initially, molten material rises to the surface, causing upwarping and thinning of the crust. Volcanic activity produces vast quantities of high-density basaltic rock. As the plates begin to move apart, a linear rift valley is formed, and volcanism continues. Further splitting apart of the land—a process called **rifting**—and more spreading cause the area to drop below sea level. When this occurs, the rift valley eventually floods with seawater, and a young linear sea is formed. After millions of years of sea floor spreading, a full-fledged ocean basin is created, with a mid-ocean ridge in the middle of the two landmasses.

Two different stages of ocean basin development are shown in the map of East Africa in **Figure 2.18**. First, the rift valleys are actively pulled apart and are at the rift valley stage of formation. Second, the Red Sea is at the linear sea stage. It has rifted apart so far that the land has dropped below sea level. The Gulf of California in Mexico is another linear sea. The Gulf of California and the

**SmartTable 2.1 Characteristics, tectonic processes, features, and examples of plate boundaries**

| Plate boundary | Plate movement | Crust types | Sea floor created or destroyed? | Tectonic process | Sea floor feature(s) | Geographic examples |
|---|---|---|---|---|---|---|
| **Divergent plate boundaries** | Apart | Oceanic–oceanic | New sea floor is created | Sea floor spreading | Mid-ocean ridge; volcanoes; young lava flows | Mid-Atlantic Ridge, East Pacific Rise |
| | | Continental–continental | As a continent splits apart, new sea floor is created | Continental rifting | Rift valley; volcanoes; young lava flows | East African rift valleys, Red Sea, Gulf of California |
| **Convergent plate boundaries** | Together | Oceanic–continental | Old sea floor is destroyed | Subduction | Trench; volcanic arc on land | Peru–Chile Trench, Andes Mountains |
| | | Oceanic–oceanic | Old sea floor is destroyed | Subduction | Trench; volcanic arc as islands | Mariana Trench, Aleutian Islands |
| | | Continental–continental | N/A | Collision | Tall mountains | Himalaya Mountains, Alps |
| **Transform plate boundaries** | Past each other | Oceanic | N/A | Transform faulting | Fault | Mendocino Fault, Eltanin Fault (between mid-ocean ridges) |
| | | Continental | N/A | Transform faulting | Fault | San Andreas Fault, Alpine Fault (New Zealand) |

Red Sea are two of the youngest seas in the world, having been created only a few million years ago. If plate motions continue rifting the plates apart in these areas, they will eventually become large oceans.

SmartTable 2.1 **Characteristics, tectonic process, features, and geographic examples of plate boundaries**
https://goo.gl/Lj7TyM

**OCEANIC RISES VERSUS OCEANIC RIDGES** The rate at which the sea floor spreads apart varies along the mid-ocean ridge and dramatically affects its appearance. Faster spreading, for instance, produces broader and less rugged segments of the global mid-ocean ridge system. This is because fast-spreading segments of the mid-ocean ridge produce vast amounts of rock, which move away from the spreading center at a rapid rate. When compared to rock from a slow-spreading segment of the mid-ocean ridge, rock from a fast-spreading segment has less time to cool, contract, and sink in a process called **subsidence**. As a result, the slope of fast-spreading segments is less steep than the slope of slow-spreading segments. Another distinction is that central rift valleys on slow-spreading segments tend to be larger and better developed (**Figure 2.19**).

The gently sloping and fast-spreading parts of the mid-ocean ridge are called **oceanic rises**. For example, the **East Pacific Rise** (Figure 2.19b) between the Pacific and Nazca Plates is a broad, low, gentle swelling of the sea floor with a small, indistinct central rift valley and has a spreading rate as high as 16.5 centimeters (6.5 inches) per year. (Note that the spreading rate is the total widening rate of an ocean basin resulting from motion of *both* plates away from a spreading center.) Conversely, steeper-sloping and slower-spreading areas of the mid-ocean ridge are called **oceanic ridges**. For instance, the **Mid-Atlantic Ridge** (Figure 2.19a) between the South American and African Plates is a tall, steep, rugged oceanic ridge that has an average spreading rate of 2.5 centimeters (1 inch) per year and stands as much as 3000 meters (10,000 feet) above the surrounding sea floor. Its prominent central rift valley is as much as 32 kilometers (20 miles) wide and averages

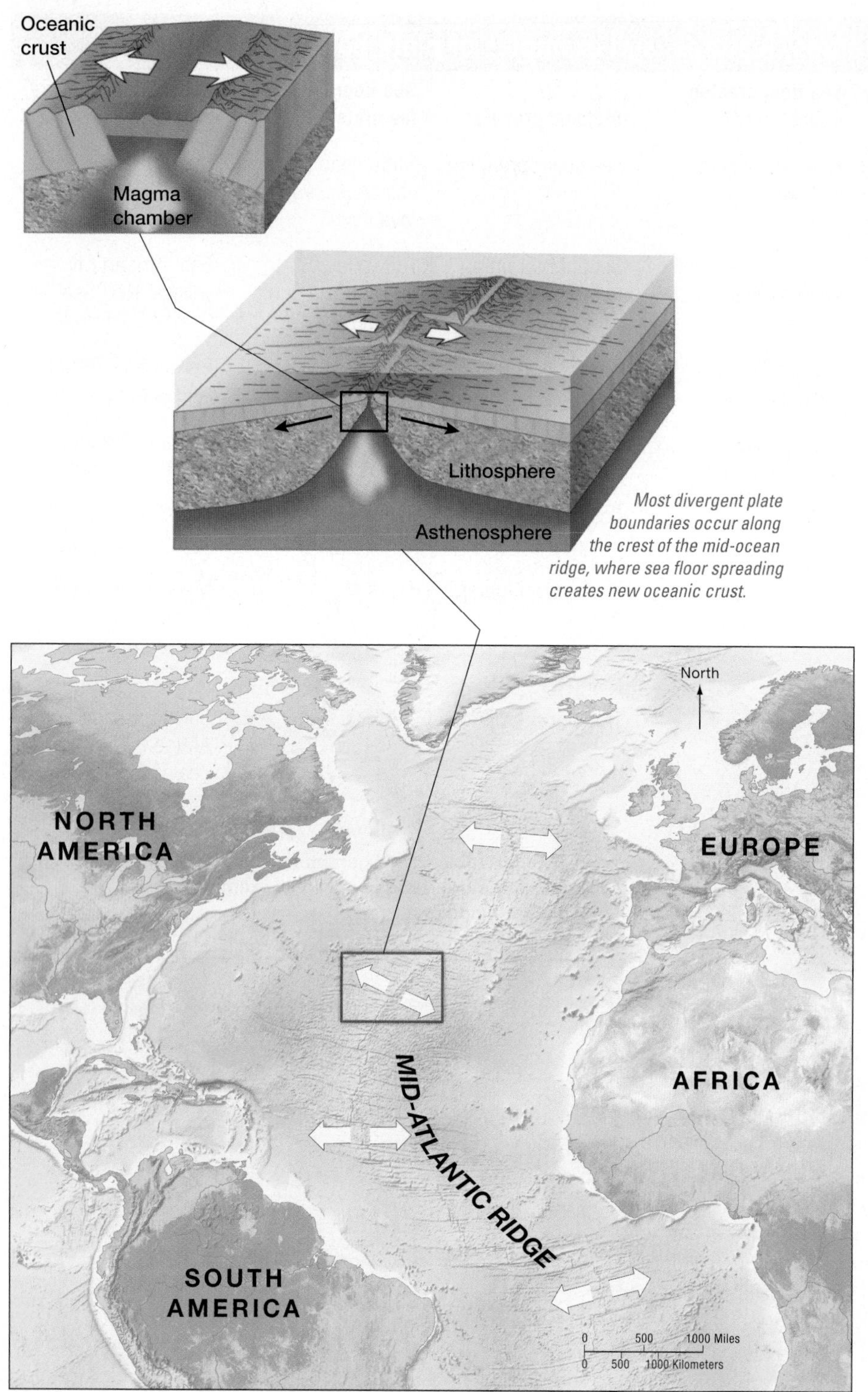


**Figure 2.15** **Divergent boundary at the Mid-Atlantic Ridge.**

**Figure 2.16** **Rift valley in Iceland.** View along a rift valley looking southwest from Laki volcano in Iceland, which sits atop the Mid-Atlantic Ridge (*red dot on inset globe*). The rift valley is marked by the linear row of volcanoes extending from the bottom of the photo to the horizon that are split in half. Note the bus (*red circle*) for scale.

2 kilometers (1.2 miles) deep. Note that the profile views for both oceanic rises and oceanic ridges shown in Figure 2.19 have the exact same scale. Also notice how much more sea floor is produced in 50 million years by the faster rate of spreading along the East Pacific Rise (Figure 2.19b) as compared to the slower rate of spreading along the Mid-Atlantic Ridge (Figure 2.19a).

Recently, a new class of spreading centers called *ultra-slow spreading centers* has also been recognized. These spreading centers, which were discovered along

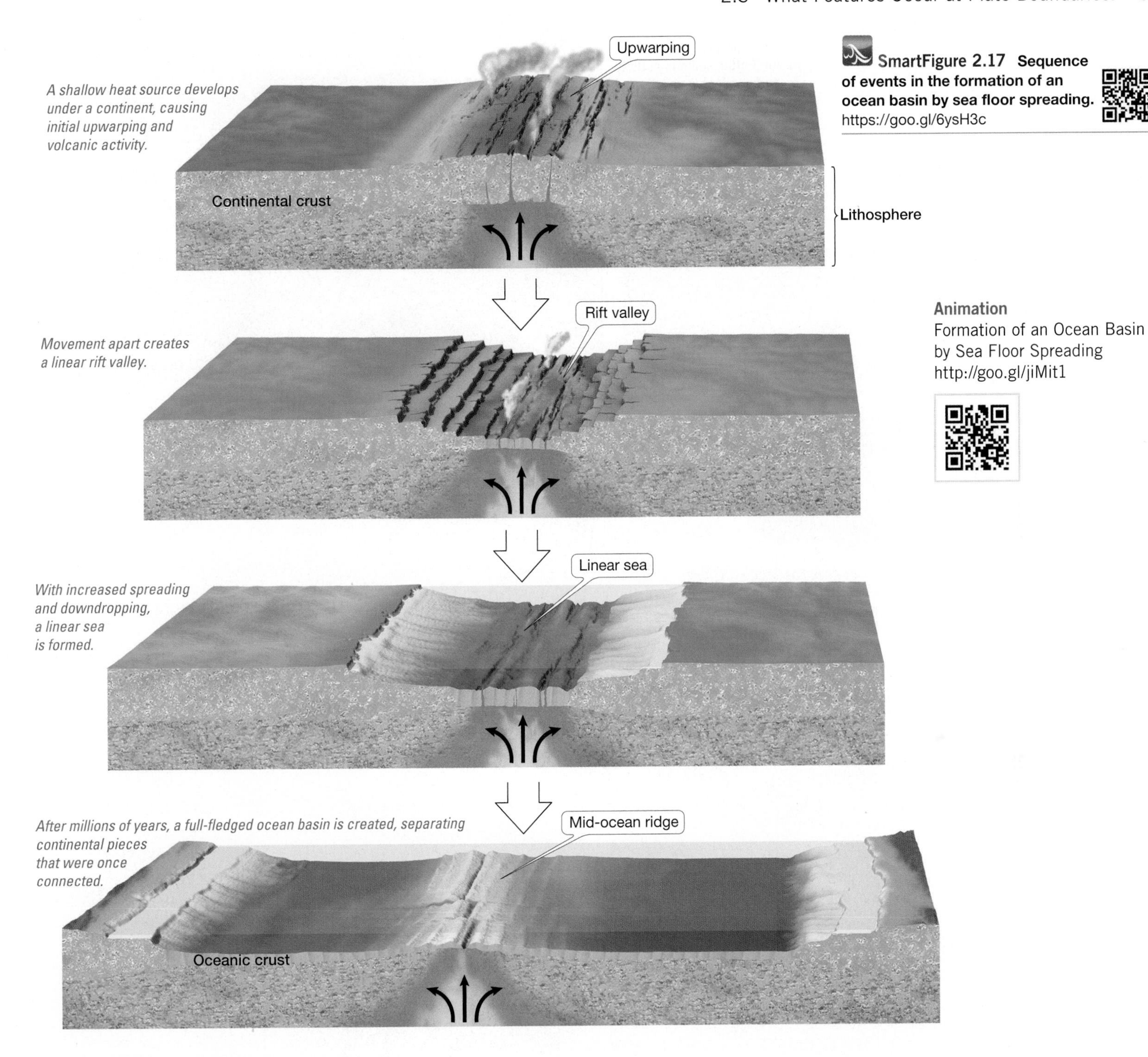

A shallow heat source develops under a continent, causing initial upwarping and volcanic activity.
Upwarping
Continental crust
Lithosphere
Movement apart creates a linear rift valley.
Rift valley
With increased spreading and downdropping, a linear sea is formed.
Linear sea
After millions of years, a full-fledged ocean basin is created, separating continental pieces that were once connected.
Mid-ocean ridge
Oceanic crust

**SmartFigure 2.17 Sequence of events in the formation of an ocean basin by sea floor spreading.**
https://goo.gl/6ysH3c

**Animation**
Formation of an Ocean Basin by Sea Floor Spreading
http://goo.gl/jiMit1

the Southwest Indian and Arctic segments of the mid-ocean ridge, are characterized by spreading rates less than 2 centimeters (0.8 inch) per year, a deep rift valley, and volcanoes that occur only at widely spaced intervals. The ultra-slow ridges are spreading so slowly, in fact, that Earth's mantle itself is exposed on the ocean floor in great slabs of rock between these volcanoes, offering scientists a rare opportunity for study.

**EARTHQUAKES ASSOCIATED WITH DIVERGENT BOUNDARIES** The amount of energy released by earthquakes along divergent plate boundaries is closely related to the spreading rate. The faster the sea floor spreads, the less energy is released

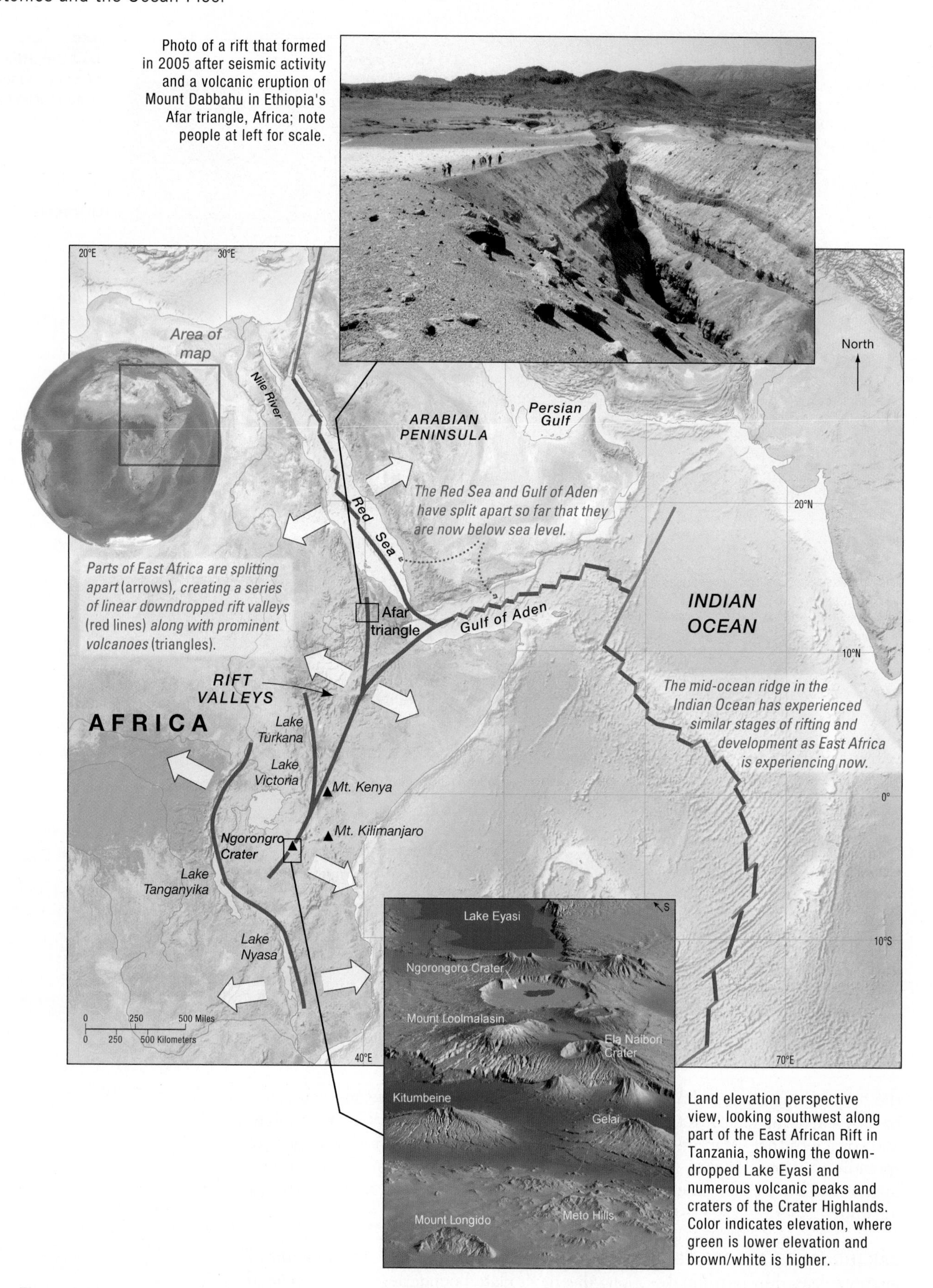

Photo of a rift that formed in 2005 after seismic activity and a volcanic eruption of Mount Dabbahu in Ethiopia's Afar triangle, Africa; note people at left for scale.
20°E
30°E
Area of map
North
Nile River
ARABIAN PENINSULA
Persian Gulf
Red Sea
The Red Sea and Gulf of Aden have split apart so far that they are now below sea level.
20°N
Parts of East Africa are splitting apart (arrows), creating a series of linear downdropped rift valleys (red lines) along with prominent volcanoes (triangles).
Afar triangle
Gulf of Aden
INDIAN OCEAN
10°N
RIFT VALLEYS
The mid-ocean ridge in the Indian Ocean has experienced similar stages of rifting and development as East Africa is experiencing now.
AFRICA
Lake Turkana
Lake Victoria
Mt. Kenya
0°
Mt. Kilimanjaro
Ngorongro Crater
Lake Tanganyika
Lake Nyasa
10°S
0 250 500 Miles
0 250 500 Kilometers
40°E
70°E
S
Lake Eyasi
Ngorongoro Crater
Mount Loolmalasin
Ela Naibori Crater
Kitumbeine
Gelai
Mount Longido
Meto Hills
Land elevation perspective view, looking southwest along part of the East African Rift in Tanzania, showing the down-dropped Lake Eyasi and numerous volcanic peaks and craters of the Crater Highlands. Color indicates elevation, where green is lower elevation and brown/white is higher.

Figure 2.18 **East African rift valleys and associated features.**

*The slow-spreading Mid-Atlantic Ridge is a tall, steep, rugged portion of the mid-ocean ridge with a prominent central rift valley.*

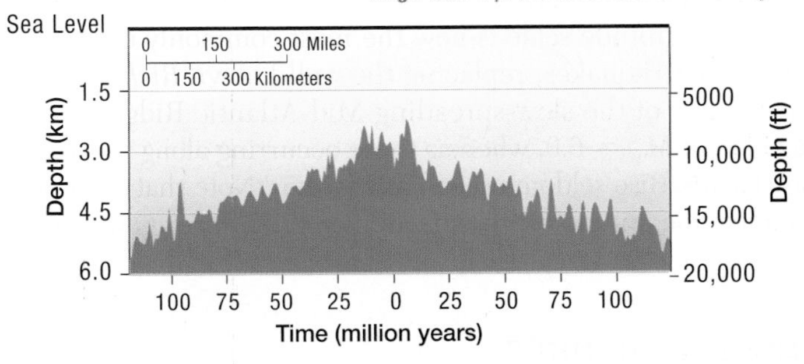


**(a)** Profile view of an oceanic ridge.

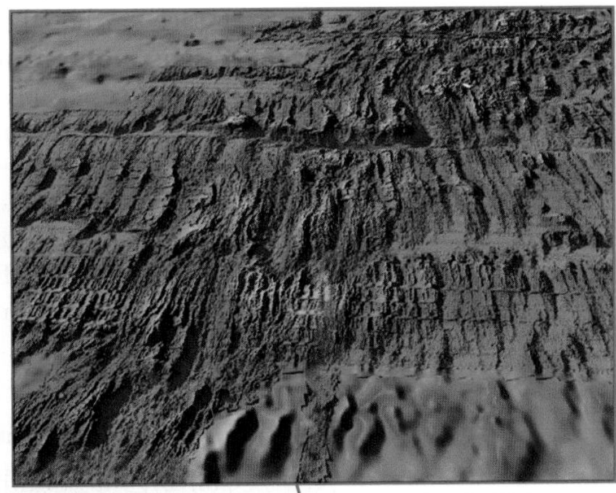

Mid–Atlantic Ridge

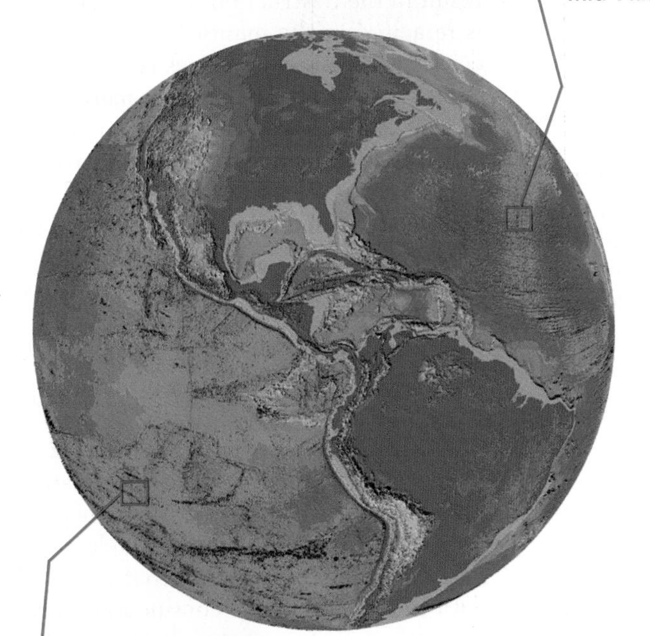

**EXPLORING DATA**

1. Draw a profile view of an oceanic spreading center that has a spreading rate halfway between that of an oceanic ridge and an oceanic rise. Use the same scale as the profiles in Figure 2.19 and label the axes.

2. Next, draw a profile view of an ultra-slow oceanic spreading center. Use the same scale as the profiles in Figure 2.19 and label the axes.

East Pacific Rise

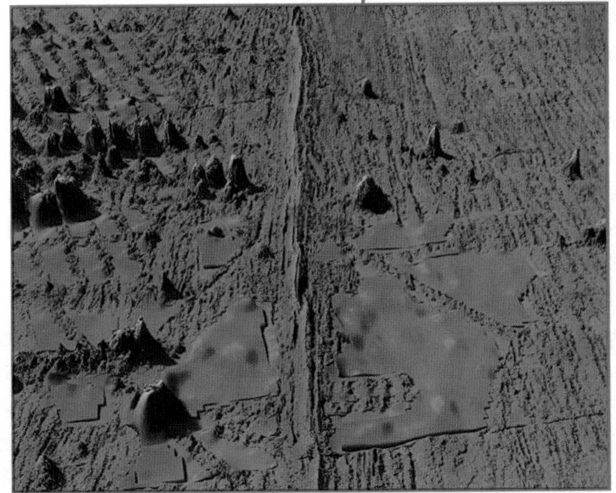

*The fast-spreading East Pacific Rise is a broad, low, gentle swelling of the mid-ocean ridge that lacks a prominent rift valley.*

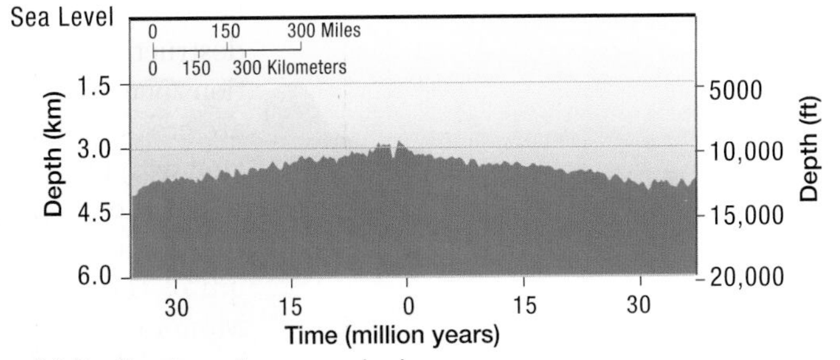


**(b)** Profile view of an oceanic rise.

**SmartFigure 2.19 Comparing oceanic rises and ridges.** Perspective and profile views of the ocean floor based on satellite bathymetry showing differences between oceanic ridges (*part a above*) and oceanic rises (*part b below*). Note that both profile views have the same scale.
https://goo.gl/Zh9QSS

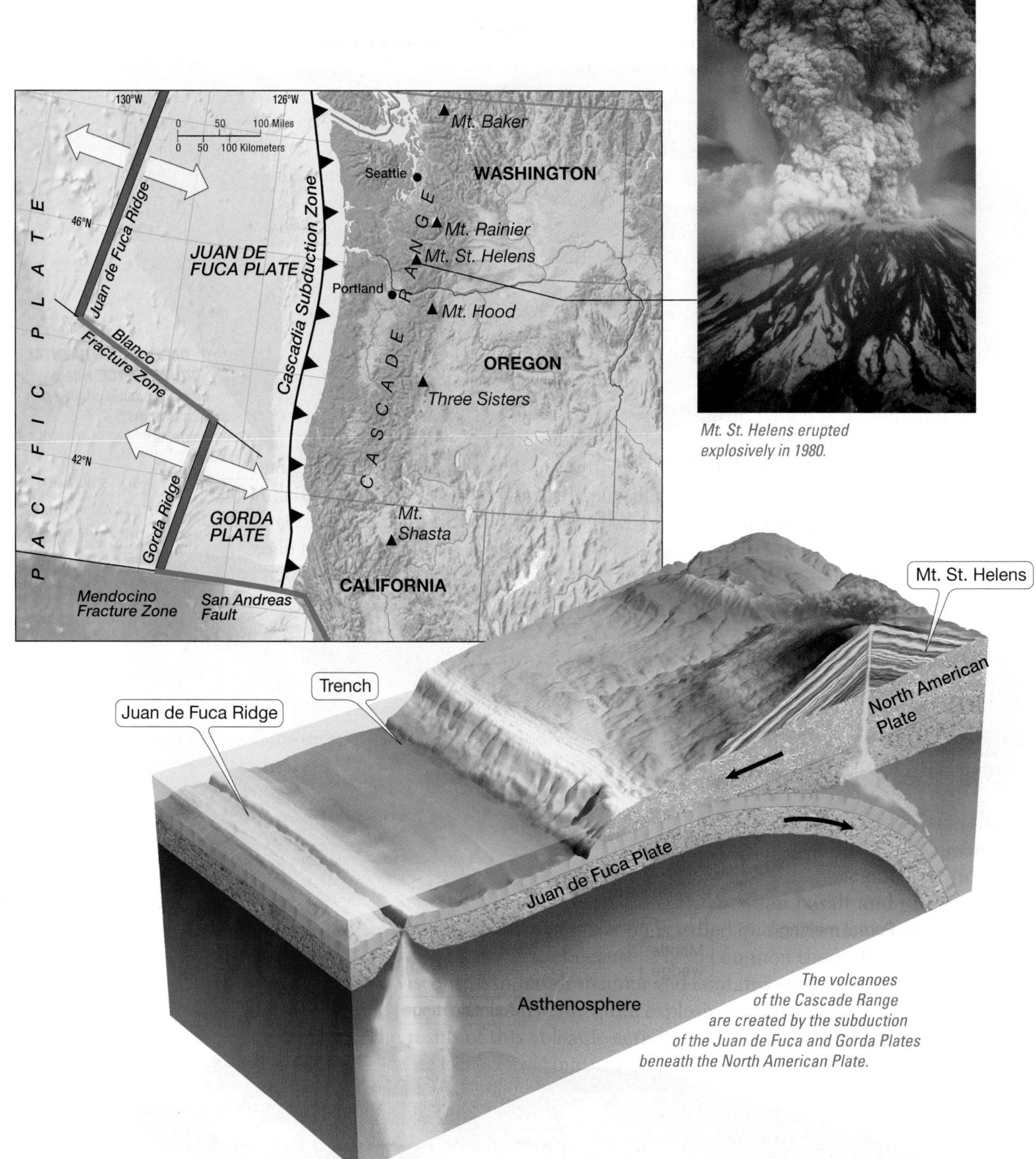


*Mt. St. Helens erupted explosively in 1980.*

*The volcanoes of the Cascade Range are created by the subduction of the Juan de Fuca and Gorda Plates beneath the North American Plate.*

**Animation**
Collapse of Mount St. Helens
goo.gl/IKVQpP

**Figure 2.21 Convergent tectonic activity produces the Cascade Mountains.**

Cascade Range volcanoes of this continental arc have been active within the past 100 years. Most notably, Mount St. Helens erupted in May 1980, killing 62 people.

**OCEANIC–OCEANIC CONVERGENCE** When two oceanic plates converge, the denser oceanic plate is subducted (Figure 2.20b). Typically, the older oceanic plate is denser because it has had more time to cool and contract. This type of convergence produces the deepest trenches in the world, such as the Mariana Trench in the western Pacific Ocean. Similar to oceanic–continental convergence, the subducting oceanic plate becomes heated, releases superheated gases, and partially melts the overlying mantle. This buoyant molten material rises to the surface and fuels the active volcanoes, which occur as an arc-shaped row of volcanic islands that is a type of volcanic arc called an **island arc**. The molten material is mostly basaltic because there is no mixing with granitic rocks from the continents, and the eruptions are not nearly as destructive. Examples of island arc/trench systems are the West Indies' Leeward and Windward Islands/Puerto Rico Trench in the Caribbean Sea and the Aleutian Islands/Aleutian Trench in the North Pacific Ocean.

**CONTINENTAL–CONTINENTAL CONVERGENCE** When two continental plates converge, which one is subducted? You might expect that the older of the two (which is most likely the denser one) will be subducted. Continental lithosphere forms differently than oceanic lithosphere, however, and old continental lithosphere is no denser than young continental lithosphere. It turns out that *neither* subducts because they are both too low in density to be pulled very far down into the mantle. Instead, a tall uplifted mountain range is created by the collision of the two plates (Figure 2.20c). These mountains are composed of folded and deformed sedimentary rocks originally deposited on the sea floor that previously separated the two continental plates. The intervening oceanic crust between the two plates is subducted beneath such mountains as the plates collide. A prime example of continental-continental convergence is the collision of India with Asia (**Figure 2.22**). It began 45 million years ago and has created the Himalaya Mountains, presently the tallest mountains on Earth.

**EARTHQUAKES ASSOCIATED WITH CONVERGENT BOUNDARIES** Both spreading centers and trench systems are characterized by earthquakes, but in different ways. Spreading centers have shallow earthquakes, usually less than 10 kilometers (6 miles) deep. Earthquakes in the trenches, on the other hand, vary from near the surface down to 670 kilometers (415 miles) deep, which are the deepest earthquakes in the world. These earthquakes are clustered in a band about 20 kilometers (12.5 miles) thick that closely corresponds to the location of the subduction zone. In fact, the subducting plate in a convergent plate boundary can be traced below the surface by examining the pattern of successively deeper earthquakes extending from the trench.

Many factors combine to produce large earthquakes at convergent boundaries. The forces involved in convergent-plate-boundary collisions are enormous. Huge lithospheric slabs of rock are relentlessly pushing against each other, and the subducting plate must bend as it dives below the surface. In addition, thick crust associated with convergent boundaries tends to store more energy than the thinner crust at divergent boundaries. Also, mineral structure changes occur at the higher pressures encountered deep below the surface, which are thought to produce changes in volume that lead to some of the most powerful earthquakes in the world. In fact, the largest earthquake ever recorded was the 1960 Chilean earthquake near the Peru–Chile Trench, which had a magnitude of $M_w = 9.5$!

## Transform Boundary Features

A global sea floor map (such as the one inside the front cover of this book) shows that the mid-ocean ridge is offset by many large, elongated features oriented perpendicular (at right angles) to the axis (crest) of the ridge. What causes

**Animation**
Convergent Margins: India–Asia Collision
http://goo.gl/UJhh6H

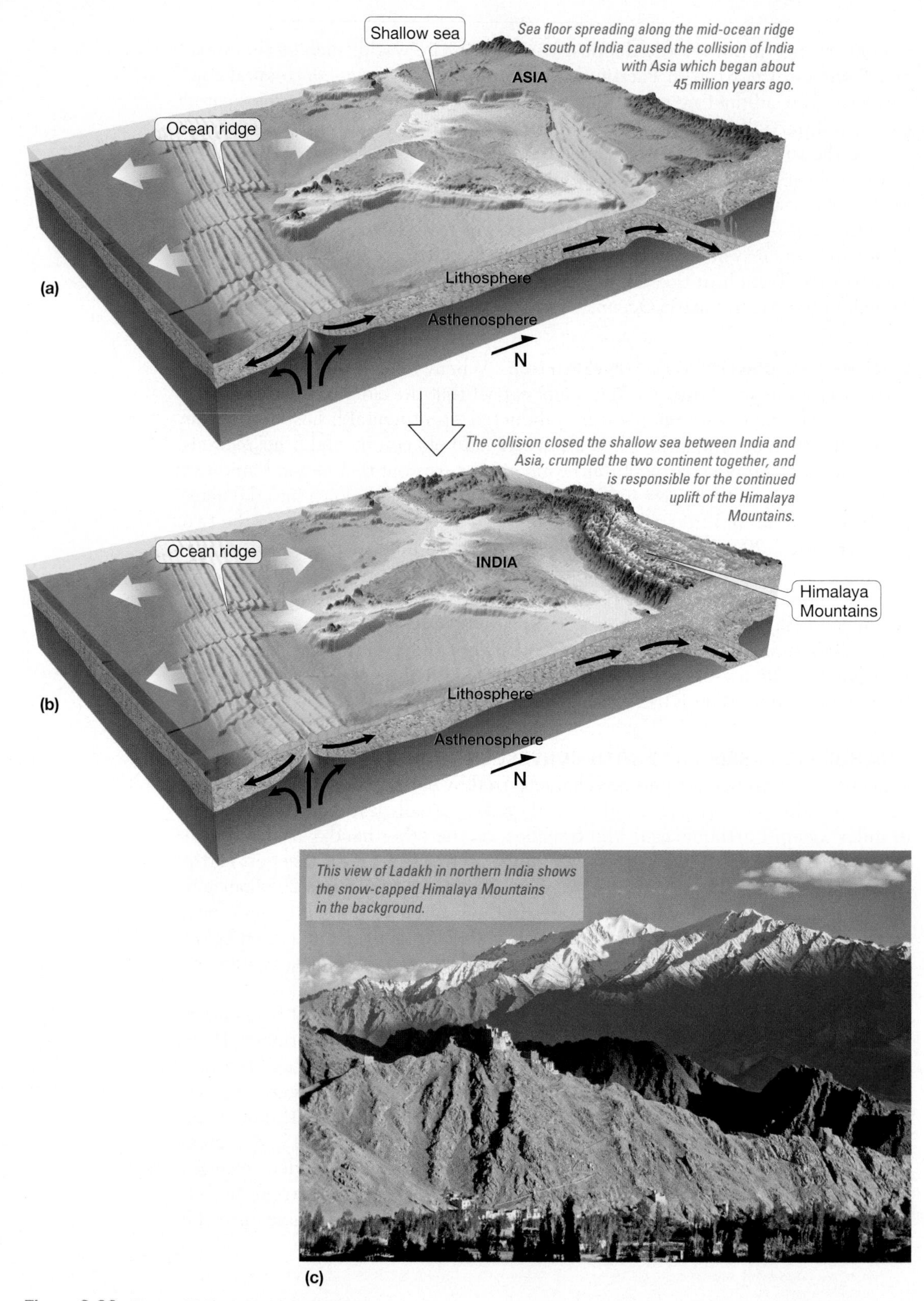

Shallow sea
Sea floor spreading along the mid-ocean ridge south of India caused the collision of India with Asia which began about 45 million years ago.
ASIA
Ocean ridge
(a)
Lithosphere
Asthenosphere
N
The collision closed the shallow sea between India and Asia, crumpled the two continent together, and is responsible for the continued uplift of the Himalaya Mountains.
Ocean ridge
INDIA
Himalaya Mountains
(b)
Lithosphere
Asthenosphere
N
This view of Ladakh in northern India shows the snow-capped Himalaya Mountains in the background.
(c)

Figure 2.22 **The collision of India with Asia.**

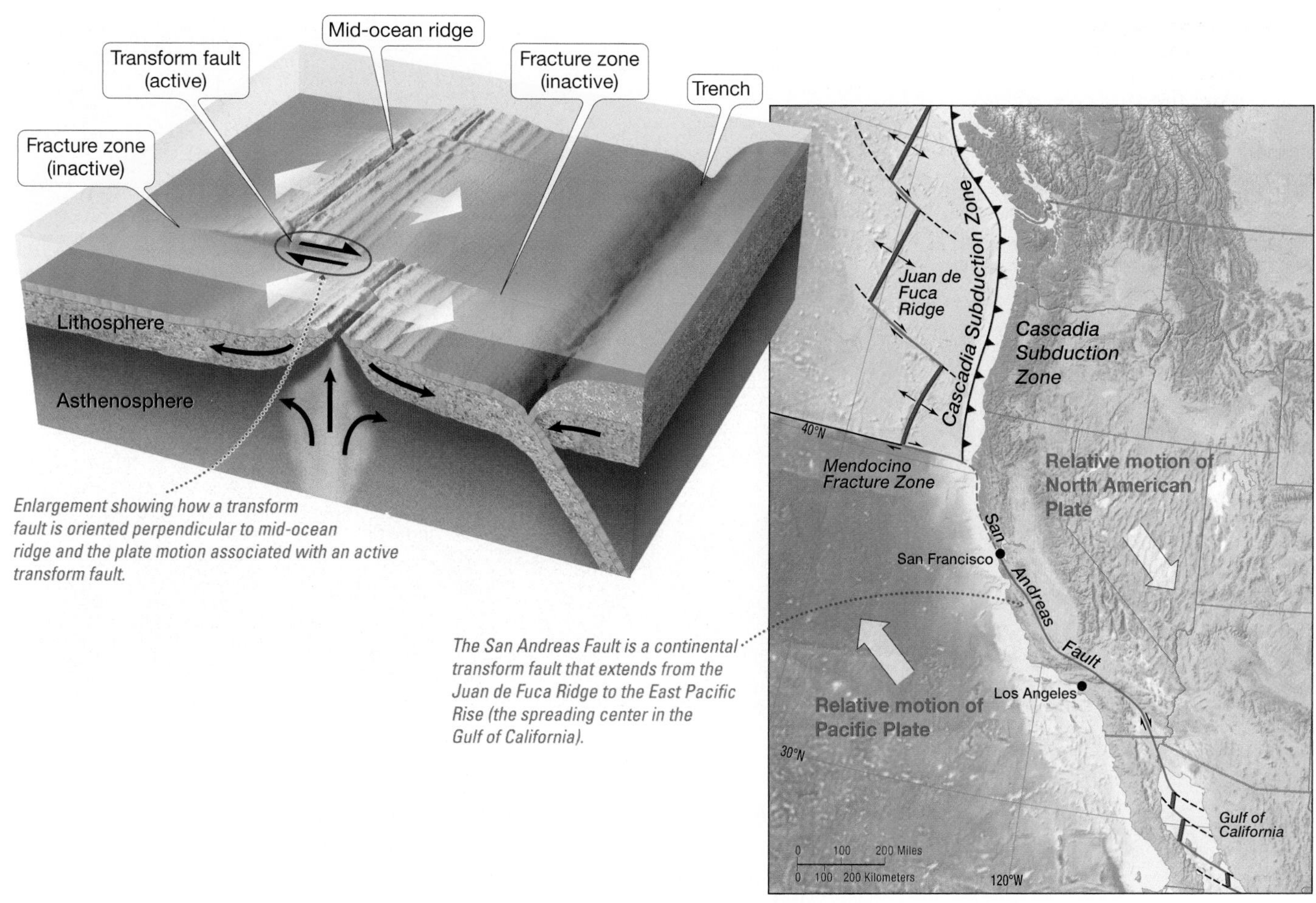


Figure 2.23 **Transform faults.**

**Animation**
Transform Faults
http://goo.gl/B6rQRH

these offsets? They are formed because the movement of lithospheric plates away from a spreading center is always perpendicular to the axis of a mid-ocean ridge, and all parts of a plate must move together. As a result, offsets are oriented perpendicular to the ridge and parallel to each other to accommodate spreading of a linear ridge system on a spherical Earth. In addition, the offsets allow different segments of the mid-ocean ridge to spread apart at different rates. These offsets—called **transform faults**—give the mid-ocean ridge a zigzag appearance. Thousands of these transform faults, some large and some small, dissect the global mid-ocean ridge. In a few instances, transform faults also occur on land.

**OCEANIC VERSUS CONTINENTAL TRANSFORM FAULTS** There are two types of transform faults. The first and most common type occurs wholly on the ocean floor and is called an **oceanic transform fault**. The second type cuts across a continent and is called a **continental transform fault**. Regardless of type, though, transform faults always occur between two segments of a mid-ocean ridge, as shown in Figure 2.23.

## STUDENTS SOMETIMES ASK . . .

***When will California fall off into the ocean?***

Due in part to popular media (such as the 2015 science fiction disaster film *San Andreas*) and because of the fact that California periodically experiences large earthquakes, many people are mistakenly concerned that it will "fall off into the ocean" during a large earthquake along the San Andreas Fault. These earthquakes occur as the Pacific Plate continues to move to the northwest past the North American Plate, at a rate of about 5 centimeters (2 inches) a year. At this rate, Los Angeles (on the Pacific Plate) will be adjacent to San Francisco (on the North American Plate) in just over 12 million years—a time so great that half a million generations of people could live their lives. Although California will never fall into the ocean, people living near this fault should be very aware they are likely to experience a large earthquake within their lifetime.

Figure 2.24 **Aerial view of the San Andreas Fault in California.** The San Andreas Fault cuts through coastal southern and central California and produces many earthquakes. This aerial view of the Carrizo Plain in central California shows the San Andreas Fault as a long linear scar; arrows show relative fault motion.

**EARTHQUAKES ASSOCIATED WITH TRANSFORM BOUNDARIES** The movement of one plate past another—a process called **transform faulting**—produces shallow but often strong earthquakes in the lithosphere. Magnitudes of $M_w = 7.0$ have been recorded along some oceanic transform faults. One of the best-studied faults in the world is California's **San Andreas Fault** (Figure 2.24), a continental transform fault that runs from the Gulf of California through coastal southern and central California past San Francisco and continues offshore parallel to the coast in northern California. Because the San Andreas Fault cuts through continental crust, which is much thicker than oceanic crust, earthquakes are considerably larger than those produced by oceanic transform faults, sometimes up to $M_w = 8.5$.

**RECAP** **The three main types of plate boundaries are divergent (plates moving apart, such as at the mid-ocean ridge), convergent (plates moving together, such as at an ocean trench), and transform (plates sliding past each other, such as at a transform fault).**

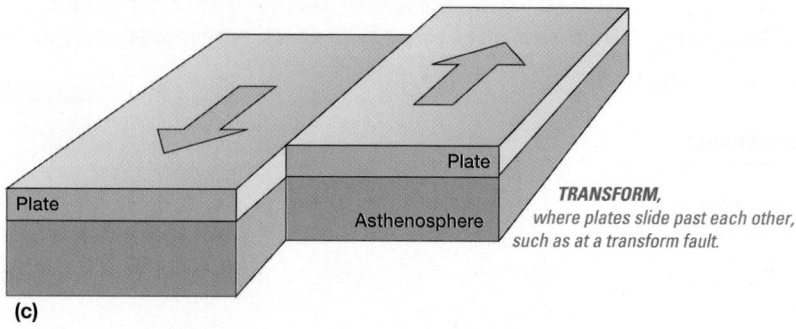

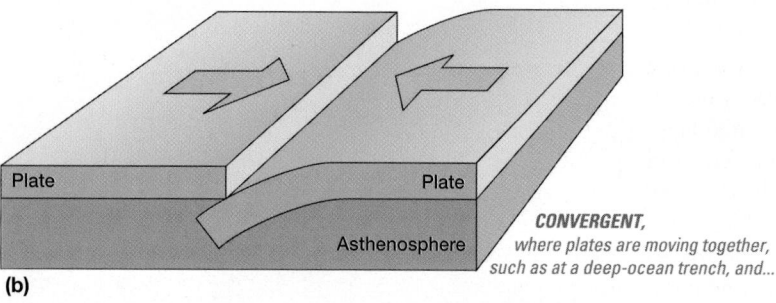

## CONCEPT CHECK 2.3 ▸ Discuss the origin and characteristics of features that occur at plate boundaries.

1 Most lithospheric plates contain both oceanic- and continental-type crust. Use plate boundaries to explain why this is true.

2 Describe the differences between oceanic ridges and oceanic rises. Include in your answer why these differences exist.

3 Using the profile view of the Mid-Atlantic Ridge in Figure 2.19a, calculate its total spreading rate over the past 50 million years (divide total distance by time). Then do a similar calculation for the East Pacific Rise (Figure 2.19b) and compare the two.

4 Convergent boundaries can be divided into three types, based on the type of crust contained on the two colliding plates. Compare and contrast the different types of convergent boundaries that result from these collisions.

5 Describe the differences in earthquake magnitudes that occur between the three types of plate boundaries and explain why these differences occur.

# 2.4 ▶ Testing the Model: Can Plate Tectonics Explain Other Features in the Ocean and on Land?

One of the strengths of plate tectonic theory is how it unifies so many seemingly separate processes and features into a single consistent model. Let's look at a few examples that illustrate how plate tectonic processes can be used to explain the origin of features that, up until the acceptance of plate tectonics, were difficult to explain.

## Hotspots and Mantle Plumes

Although the theory of plate tectonics helped explain the origin of many features near plate boundaries, it did not seem to explain the origin of **intraplate features** (*intra* = within, *plate* = plate of the lithosphere) that are far from any plate boundary. For instance, how can plate tectonics explain volcanic islands near the middle of a plate? Areas of intense volcanic activity that remain in more or less the same location over long periods of geologic time and are unrelated to plate boundaries are called **hotspots**. For example, the continuing volcanism in Yellowstone National Park, located within the North American Plate, and Hawaii, located in the center of the Pacific Plate, is caused by hotspots. Note that a hotspot is different from either a volcanic arc or a mid-ocean ridge (both of which are related to plate boundaries), even though all are marked by a high degree of volcanic activity.

Why is there so much volcanic activity at hotspots? The plate tectonic model infers that hotspot volcanism is caused by the presence of **mantle plumes** (*pluma* = a soft feather), which are vertical tube-shaped areas of hot molten rock that arise from deep within the mantle (**Figure 2.25**). Mantle plumes can be identified by researchers who measure how fast seismic waves from earthquakes travel below ground; the underlying principle is that seismic waves move more slowly through hot rock than cold. Seismic studies suggest that several types of mantle plumes exist: Some come from the core–mantle boundary, while others have a shallower source. Geophysical research reveals that the core–mantle boundary is not a simple, smooth dividing zone but has many regional variations, which has implications for the development of mantle plumes. In addition, new research suggests that

**Animation**
Tectonic Settings of Volcanic Activity
http://goo.gl/RUJyDG

**SmartFigure 2.25 Origin and development of mantle plumes and hotspots.** Schematic cross-sectional views of Earth showing the development of a mantle plume and hotspot according to the plume hypothesis.
https://goo.gl/SWPXyr

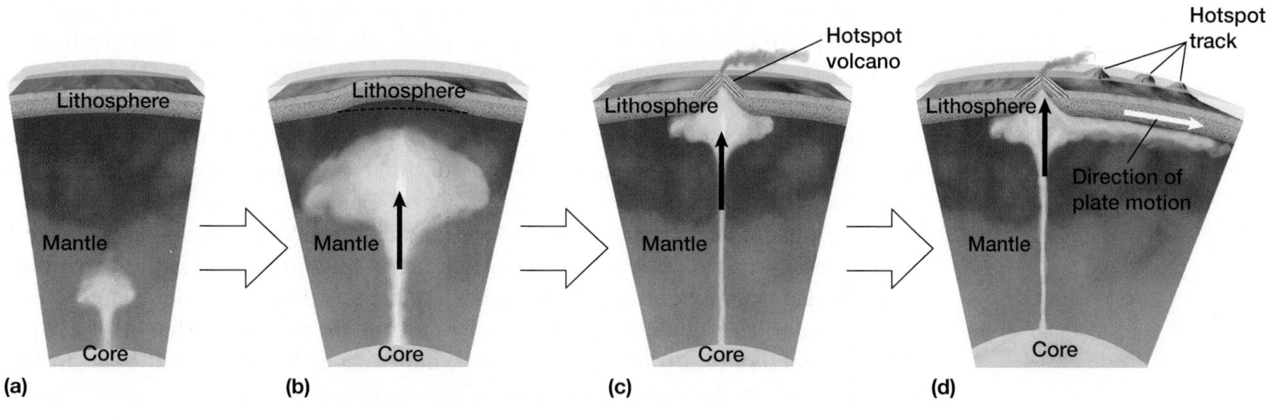
A plume of hot buoyant material detaches from the deep mantle or the core-mantle boundary.
The plume rises more rapidly in its conduit than the plume head can push through the viscous mantle, which inflates the head and elevates Earth's surface.
Decompression near the surface partially melts the plume head, which comes to the surface and creates a hotspot volcano.
The volcano is carried away by plate motion as the plume continues to feed subsequent volcanoes, creating a hotspot track (nematath).
Hotspot volcano
Hotspot track
Lithosphere
Mantle
Core
Direction of plate motion
(a)
(b)
(c)
(d)

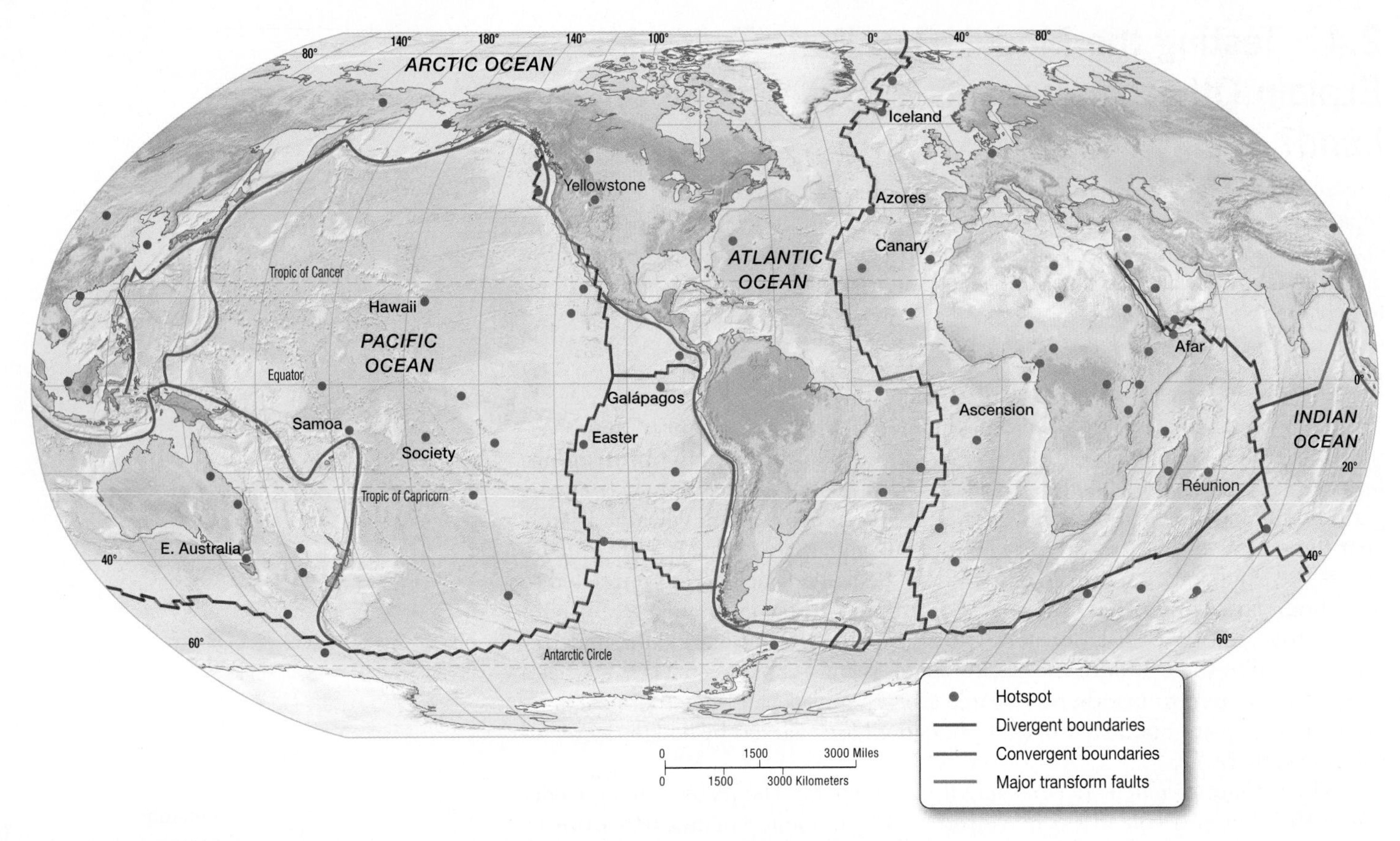


Figure 2.26 **Global distribution of prominent hotspots.** Map showing prominent hotspots, which are shown by red dots; the locations of plate boundaries are also shown. The majority of the world's hotspots are *not* associated with plate boundaries; those that are tend to occur along divergent plate boundaries, where the lithosphere is thin.

the asthenosphere may actually play a more significant role than the core–mantle boundary in the development of hotspots. Because mantle plumes themselves cannot be directly sampled and the thin plume conduits are difficult to resolve using seismic data, their existence has been difficult to confirm. As a result, there is currently vigorous scientific debate regarding mantle plumes and volcanism at hotspots (see, for example, www.mantleplumes.org). In fact, new studies suggest that some mantle plumes are neither deep phenomena nor fixed in position over geologic time, as assumed in the standard plume model.

Worldwide, more than 100 hotspots have been active within the past 10 million years. **Figure 2.26** shows the global distribution of prominent hotspots today. In general, hotspots do not coincide with plate boundaries. Notable exceptions are those that are near divergent boundaries where the lithosphere is thin, such as at the Galápagos Islands and Iceland. In fact, Iceland straddles the Mid-Atlantic Ridge (a divergent plate boundary). It is also directly over a 150-kilometer (93-mile) wide mantle plume, which accounts for its remarkable amount of volcanic activity—so much that it has caused Iceland to be one of the few areas of the global mid-ocean ridge that rise high above sea level.

Throughout the Pacific Plate, many island chains are oriented in a northwestward–southeastward direction. The most intensely studied of these is the **Hawaiian Islands–Emperor Seamount chain** in the northern Pacific Ocean (**Figure 2.27**). What created this chain of more than 100 intraplate volcanoes that stretch over 5800 kilometers (3000 miles)? Further, what caused the prominent bend in the overall direction that occurs in the middle of the chain?

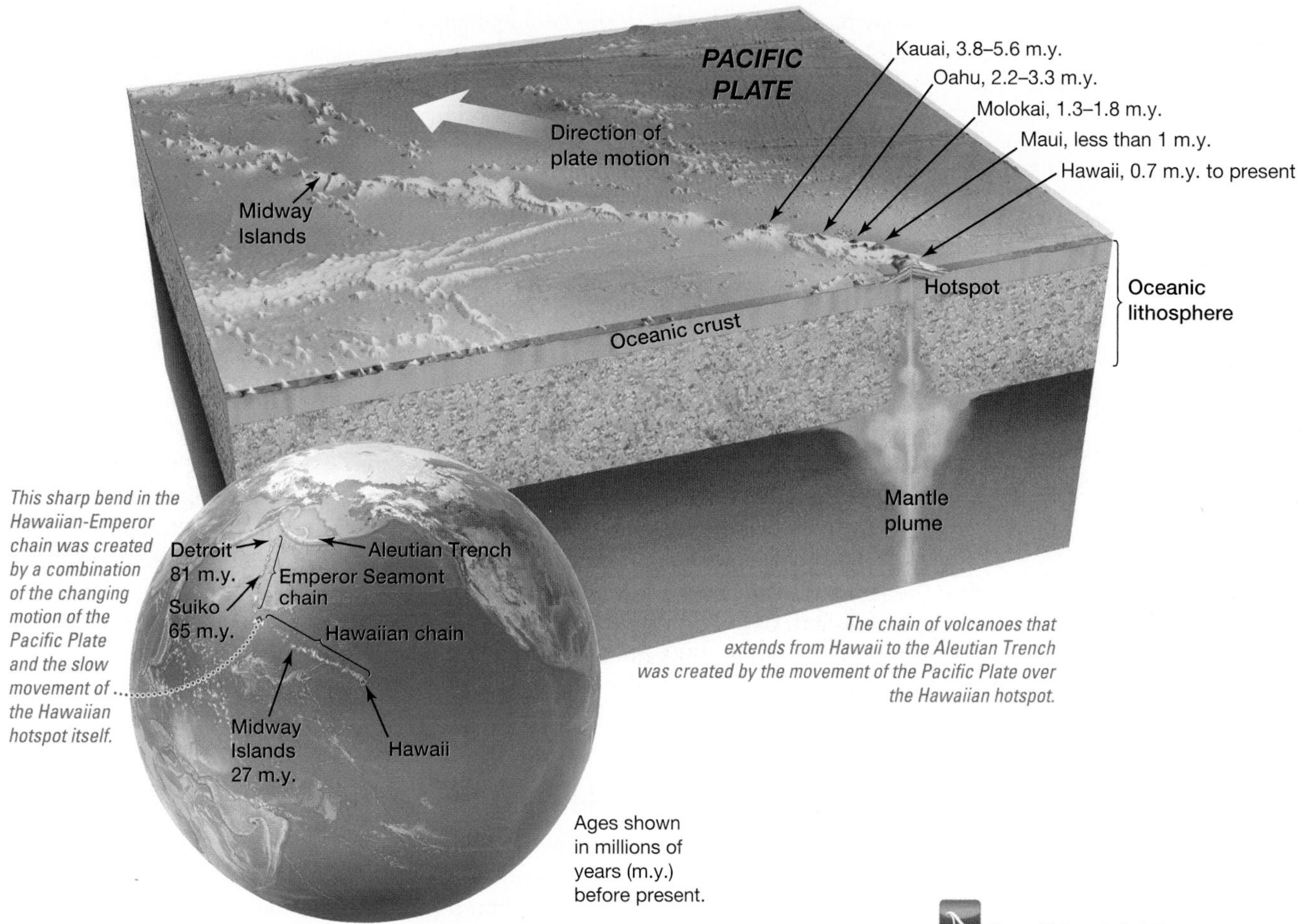


**SmartFigure 2.27 Hawaiian Islands–Emperor Seamount chain.** Schematic diagram showing how movement of the Pacific Plate over the Hawaiian hotspot created the Hawaiian Islands–Emperor Seamount chain that extends from Hawaii to the Aleutian Trench. Numbers represent radiometric age dates in millions of years (m.y.) before present. https://goo.gl/SDwwHP

**Animation**
Hot Spot Volcano Tracks
http://goo.gl/3tpkab

To help answer these questions, let's examine the ages of the volcanoes in the chain. Every volcano in the chain has long since become extinct, except the volcano Kilauea on the island of Hawaii, which is the southeasternmost island of the chain. The age of volcanoes progressively increases northwestward from Hawaii (Figure 2.27). To the northwest, the volcanoes increase in age past Suiko Seamount (65 million years old) to Detroit Seamount (81 million years old), near the Aleutian Trench.

These age relationships suggest that the Pacific Plate has steadily moved northwestward, while the underlying mantle plume has remained relatively stationary. The resulting Hawaiian hotspot created each of the volcanoes in the chain. As the plate moved, it carried the active volcano off the hotspot, and a new volcano began forming, younger in age than the previous one. A chain of extinct volcanoes that is progressively older as one travels away from a hotspot is called a **nematath** (*nema* = thread, *tath* = dung or manure), or a *hotspot track* (see Figure 2.25). Evidence suggests that about 47 million years ago, the Pacific Plate shifted from a northerly to a northwesterly direction. This change in plate motion can account for the bend (large elbow) about halfway through the chain, separating the Hawaiian Islands from the Emperor Seamounts (see Figure 2.27). If this is true, then other hotspot tracks throughout the Pacific Plate should show a similar bend at roughly the same time, but most do not.

Recent research that may help resolve this disparity indicates that hotspots do not remain completely stationary. In fact, several studies have shown that most hotspots move at less than 1 centimeter (0.4 inch) per year, but some, like Hawaii, may have moved faster in the geologic past. Even if Hawaii's hotspot had moved faster in the past, it did not do so in a way that would have created the sharp bend in the Hawaiian–Emperor track seen in Figure 2.27. Moreover, recent plate reconstructions suggest that the observed bend in the Hawaiian–Emperor chain was created by a combination of the changing motion of the Pacific Plate (mainly as a result of changes in plate motions near Australia and Antarctica), the subduction of a plate in the northwest Pacific underneath Asia millions of years ago that altered the direction of mantle flow, and the slow movement of Hawaii's mantle plume itself. In fact, many other hotspot tracks appear to have been at least partially created by motion of their mantle plumes as well. Remarkably, hotspots seem to move in exactly the opposite direction of their overlying plates, so hotspots may still be useful for tracking plate motions.

In the future, what will become of Hawaii—the island that currently resides on the hotspot? Based on the hotspot model, the island will be carried to the northwest, off the hotspot, become inactive, and eventually be subducted into the Aleutian Trench, like all the rest of the volcanoes in the chain to the north of it. In turn, other volcanoes will build up over the hotspot. In fact, a 3500-meter (11,500-foot) volcano named **Loihi** already exists 32 kilometers (20 miles) southeast of Hawaii. Still 1 kilometer (0.6 mile) below sea level, Loihi is volcanically active and, based on its current rate of activity, it should reach the surface sometime between 30,000 and 100,000 years from now. As it builds above sea level, it will become the newest island in the long chain of volcanoes created by the Hawaiian hotspot.

## Seamounts and Tablemounts

Many areas of the ocean floor (most notably on the Pacific Plate) contain tall volcanic peaks that resemble some volcanoes on land. These large volcanoes are called **seamounts** if they are cone-shaped on top, like an upside-down ice cream cone. Some volcanoes are flat on top—unlike anything on land—and these are called **tablemounts**, or **guyots**, after Princeton University's first geology professor, Arnold

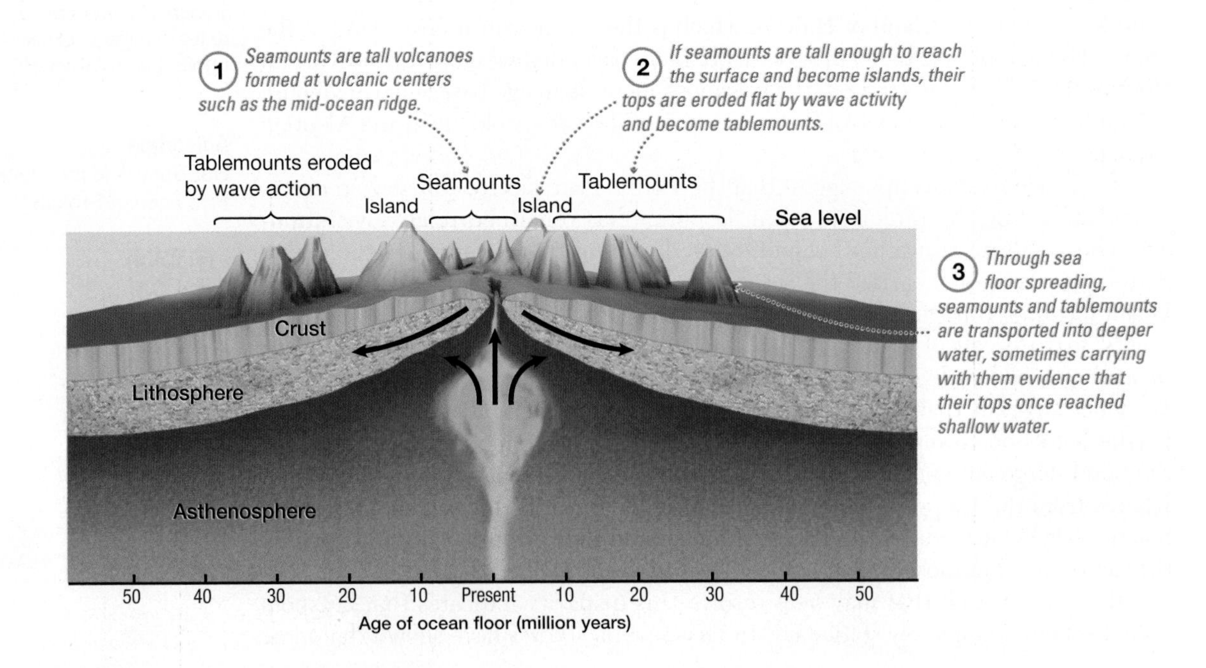
1 Seamounts are tall volcanoes formed at volcanic centers such as the mid-ocean ridge.
2 If seamounts are tall enough to reach the surface and become islands, their tops are eroded flat by wave activity and become tablemounts.
3 Through sea floor spreading, seamounts and tablemounts are transported into deeper water, sometimes carrying with them evidence that their tops once reached shallow water.
Tablemounts eroded by wave action
Island
Seamounts
Island
Tablemounts
Sea level
Crust
Lithosphere
Asthenosphere
50 40 30 20 10 Present 10 20 30 40 50
Age of ocean floor (million years)

**Figure 2.28** **Sequence of events in the formation of seamounts and tablemounts at a mid-ocean ridge.**

**Animation**
Seamounts/Tablemounts and Coral Reef Stages
http://goo.gl/YltBIQ

Guyot. (Note that *guyot* is pronounced "GEE-oh," with a hard g, as in "give.") Until the theory of plate tectonics, it was unclear how the differences between seamounts and tablemounts could have been produced. The theory explains why tablemounts are flat on top and also why the tops of some tablemounts have shallow-water deposits, despite being located in very deep water.

The origin of many seamounts and tablemounts is related to the volcanic activity occurring at hotspots; the origin of others is related to processes occurring at the mid-ocean ridge (**Figure 2.28**). Because of sea floor spreading, active volcanoes (seamounts) occur along the crest of the mid-ocean ridge. Some may be built up so high that they rise above sea level and become islands, at which point wave erosion becomes important. When sea floor spreading has moved the seamount off its source of magma (whether it is a mid-ocean ridge or a hotspot), the top of the seamount can be flattened by waves in just a few million years. This flattened seamount-now a tablemount-continues to be carried away from its source and, after millions of years, is submerged deeper into the ocean. Frequently, tops of tablemounts contain evidence of shallow-water conditions (such as ancient coral reef deposits) that were carried with them into deeper water.

## Coral Reef Distribution

The largest reef system in the world is Australia's **Great Barrier Reef**, home to hundreds of coral species and thousands of other reef-dwelling organisms. This enormous structure lies 40 kilometers (25 miles) or more offshore and averages 150 kilometers (90 miles) in width, and extends for more than 2000 kilometers (1200 miles) along Australia's shallow northeastern coast (**Figure 2.29**). The corals that built the Great Barrier Reef are colonial animals that live in shallow, warm, tropical seawater and produce a hard skeleton of limestone. Once corals are established, they continue to grow upward layer by layer with each new generation attached to the skeletons of its predecessors. Over millions of years, a thick sequence of coral reef deposits may develop, if conditions remain favorable.

As Figure 2.29 shows, the effects of the Indian–Australian Plate moving north toward the equator from colder Antarctic waters are clearly visible in the age, thickness, and structure of the Great Barrier Reef. It is oldest (around 25 million years old) and thickest at its northern end because the northern part of Australia reached water warm enough to grow coral before the southern parts did. More information on the types of coral reefs and their stages of development is in Chapter 15, "Animals of the Benthic Environment."

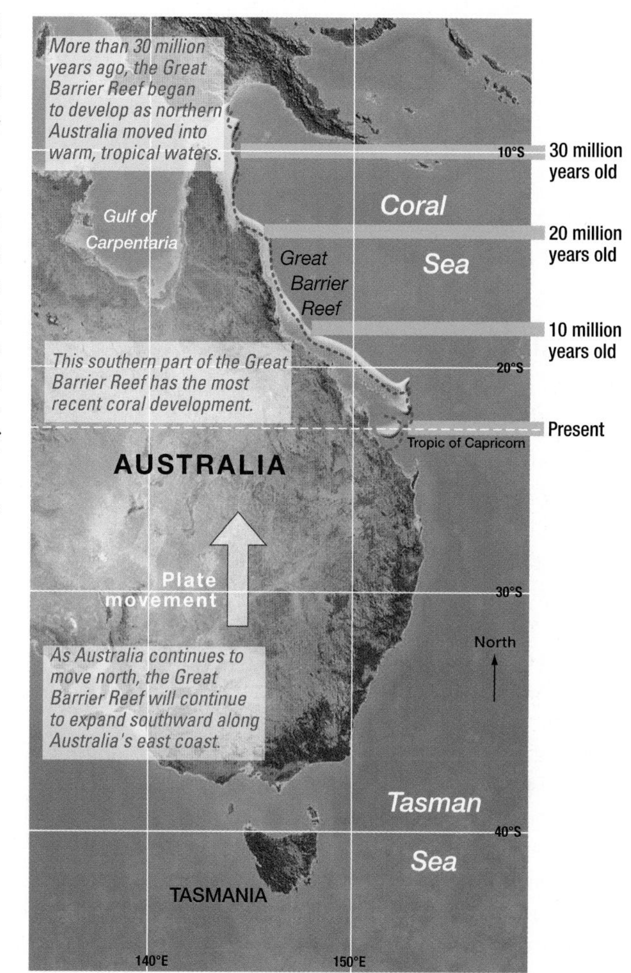


Figure 2.29 **Australia's Great Barrier Reef records plate movement.**

**RECAP Mantle plumes create hotspots at Earth's surface, which produce volcanic chains called nemataths that record the motions of plates.**

**CONCEPT CHECK 2.4 ▸ Show how plate tectonics can be used to explain the origin of features not easily explained by other processes.**

1 How is the age distribution pattern of the Hawaiian Islands–Emperor Seamount chain explained by the position of the Hawaiian hotspot? What could have caused the curious bend in the chain?

2 What are the differences between a mid-ocean ridge and a hotspot?

3 How can plate tectonics be used to help explain the difference between a seamount and a tablemount?

4 Explain how plate movement has affected the age of Australia's Great Barrier Reef.

## 2.5 ▶ How Has Earth Changed in the Past, and How Will It Look in the Future?

One of the most powerful features of any scientific theory is its ability to predict occurrences. Let's examine how plate tectonics can be used to determine the locations of the continents and oceans in the past, as well as what the continents and oceans will look like in the future.

### STUDENTS SOMETIMES ASK . . .

***How long has plate tectonics been operating on Earth? Will it ever stop?***

It's difficult to say with certainty how long plate tectonics has been operating because our planet has been so dynamic since its early history, regularly recycling most of Earth's crust. However, recently discovered ancient volcanic rock sequences uplifted onto Greenland show telltale characteristics of tectonic activity and suggest that plate tectonics has been operating for at least the past 3.8 billion years of Earth's history.

Plate motion has typically been assumed to be an active and continuous process, with new sea floor constantly being formed while old sea floor is being destroyed. Recent research, however, suggests that plates may move more actively at times, then slow down or even stop, and then start up again. The reasons for this intermittent plate motion appear to be related to plate distribution and changes in the amount of heat released from Earth.

Looking into the future, the forces that drive plates will likely decrease until plates no longer move. This is because plate tectonic processes are powered by heat released from within Earth (which is of a finite amount). The erosional work of water, however, will continue to erode Earth's features. What a different world it will be then—an Earth with no earthquakes, no volcanoes, and no mountains. Flatness will prevail!

### The Past: Paleogeography

The study of historical changes of continental shapes and positions is called **paleogeography** (*paleo* = ancient, *geo* = earth, *graphy* = description of). As a result of paleogeographic changes, the size and shape of ocean basins have changed as well.

Figure 2.30 is a series of world maps showing the paleogeographic reconstructions of Earth at 60-million-year intervals. At 540 million years ago, many of the present-day continents are barely recognizable. North America was on the equator and rotated 90 degrees clockwise. Antarctica was on the equator and connected to many other continents.

Between 540 and 300 million years ago, the continents began to come together to form Pangaea. Notice that Alaska had not yet formed. Continents are thought to add material through the process of **continental accretion** (*ad* = toward, *crescere* = to grow). Like adding layers onto a snowball, bits and pieces of continents, islands, and volcanoes are added to the edges of continents and create larger landmasses.

From 180 million years ago to the present, Pangaea separated, and the continents moved toward their present-day positions. North America and South America rifted away from Europe and Africa to produce the Atlantic Ocean. In the Southern Hemisphere, South America and a continent composed of India, Australia, and Antarctica began to separate from Africa.

By 120 million years ago, there was a clear separation between South America and Africa, and India had moved northward, away from the Australia–Antarctica mass, which began moving toward the South Pole. As the Atlantic Ocean continued to open, India moved rapidly northward and collided with Asia about 45 million years ago. Australia had also begun a rapid journey to the north since separating from Antarctica.

One major outcome of global plate tectonic events over the past 180 million years has been the creation of the Atlantic Ocean, which continues to grow as the sea floor spreads along the Mid-Atlantic Ridge. At the same time, the Pacific Ocean continues to shrink due to subduction along the many trenches that surround it and continental plates that bear in from both the east and west.

### The Future: Some Bold Predictions

Using plate tectonics, a prediction of the future positions of features on Earth can be made based on the assumption that the rate and direction of plate motion will remain the same. Although these assumptions may not be entirely valid, they do provide a framework for the prediction of the positions of continents and other Earth features in the future.

Figure 2.31 is a map of what the world may look like 50 million years from now, showing many notable differences from today. For instance, the East African rift valleys may enlarge to form a new linear sea, and the Red Sea may be greatly enlarged if rifting continues to occur there. India may continue to plow into Asia, further uplifting the Himalaya Mountains as India slides to the east. As Australia moves north toward Asia, it may use New Guinea like a snowplow to accrete various islands. North America and South America may continue to move west, enlarging the Atlantic Ocean and decreasing the size of the Pacific Ocean. In addition, several new inland

Climate

Connection

**Animation**
Terrane Formation
http://goo.gl/GQKTXO

**RECAP The geographic positions of the continents and ocean basins are not fixed in time or place. Rather, they have changed in the past and will continue to change in the future.**

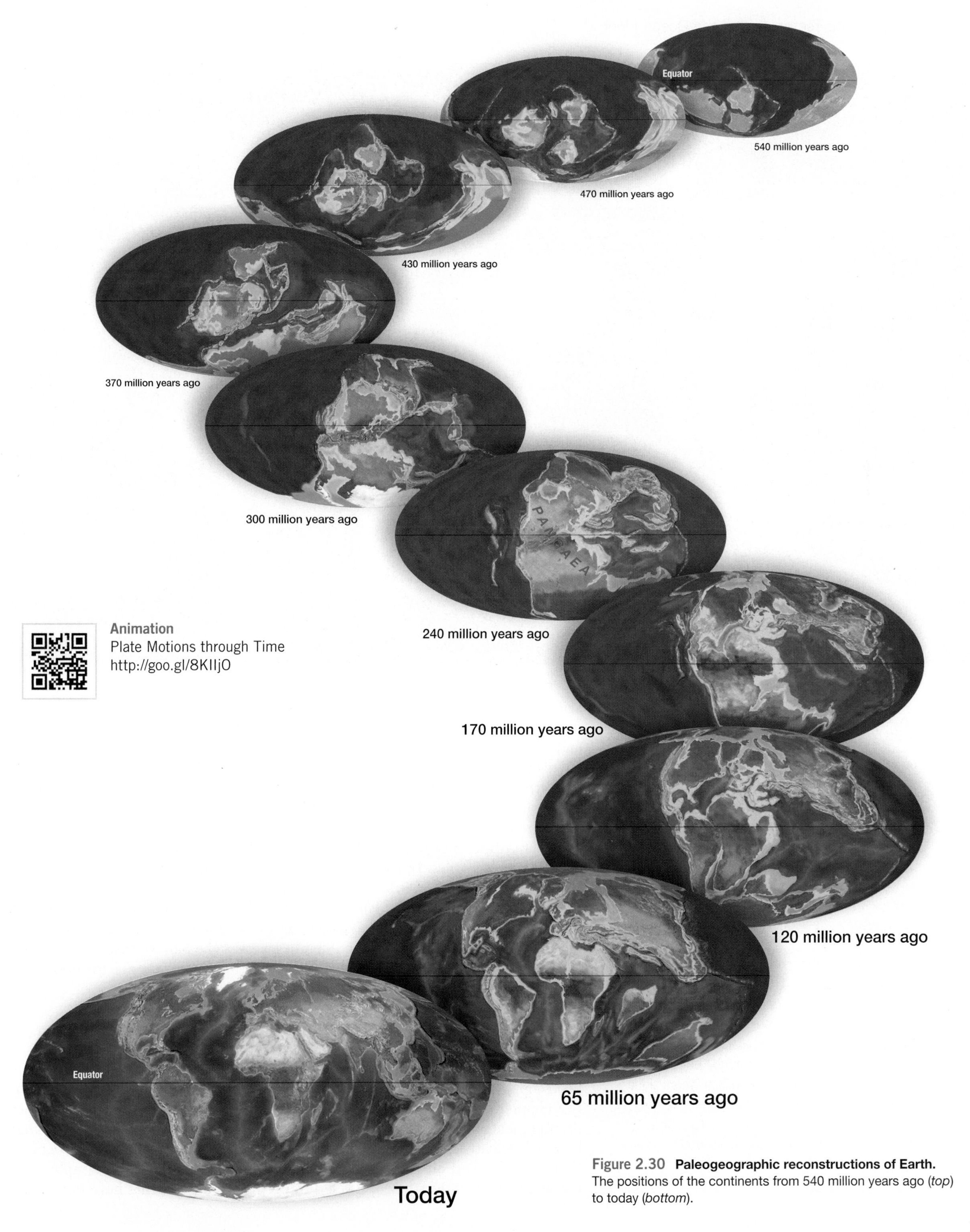


**Animation**
Plate Motions through Time
http://goo.gl/8KIljO

Figure 2.30 **Paleogeographic reconstructions of Earth.** The positions of the continents from 540 million years ago (*top*) to today (*bottom*).

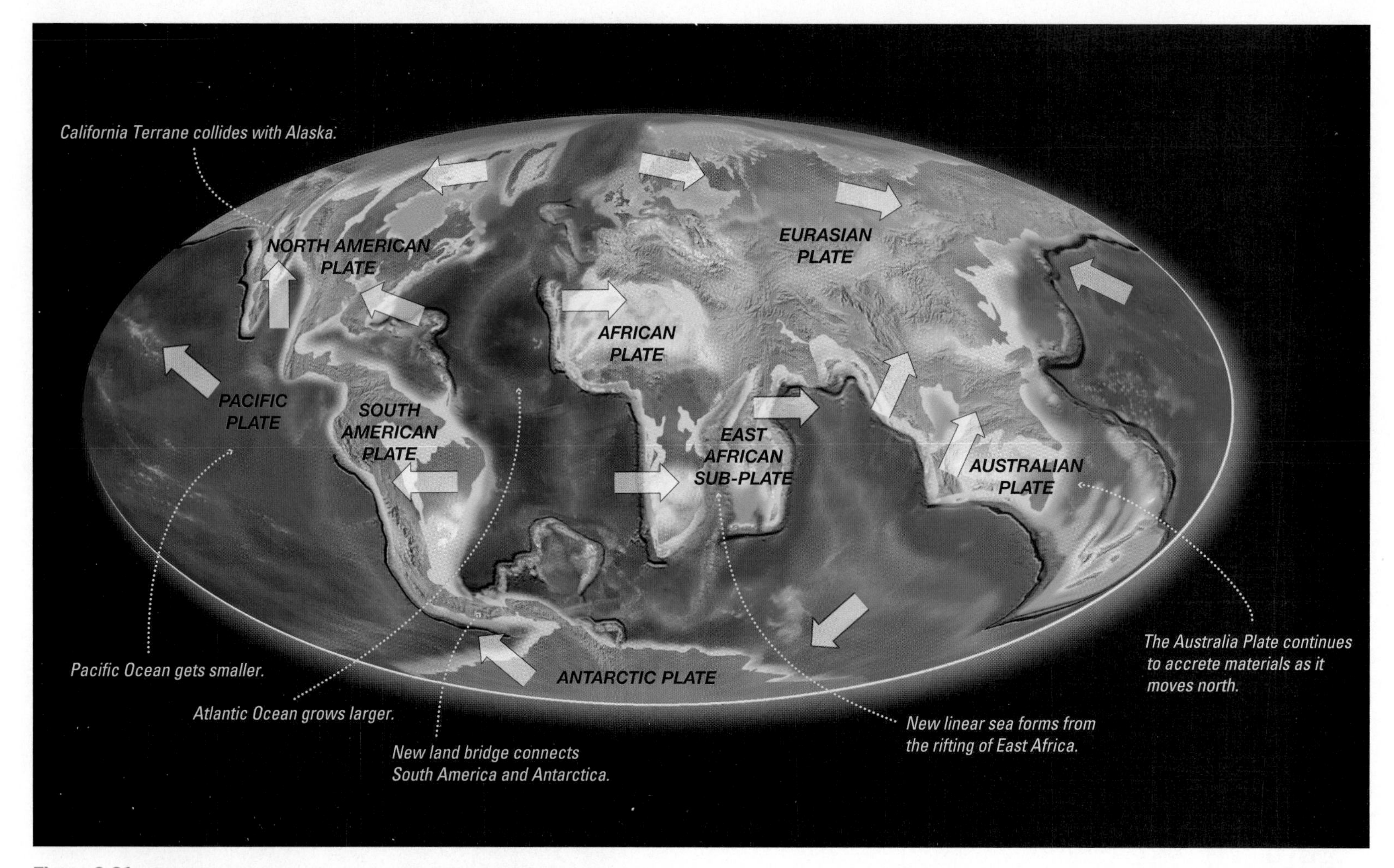

California Terrane collides with Alaska.
NORTH AMERICAN PLATE
EURASIAN PLATE
AFRICAN PLATE
PACIFIC PLATE
SOUTH AMERICAN PLATE
EAST AFRICAN SUB-PLATE
AUSTRALIAN PLATE
ANTARCTIC PLATE
Pacific Ocean gets smaller.
Atlantic Ocean grows larger.
New land bridge connects South America and Antarctica.
New linear sea forms from the rifting of East Africa.
The Australia Plate continues to accrete materials as it moves north.

Figure 2.31 **The world as it may look 50 million years from now.** Based on current plate motions, this map shows the positions of features on Earth 50 million years in the future. Arrows indicate the direction of plate motion.

## STUDENTS SOMETIMES ASK . . .

***Will the continents come back together and form a single landmass anytime soon?***

Yes, it is very likely that the continents will come back together, but not anytime soon. Remember that continents are too low in density to get recycled back into Earth via subduction, so they tend to remain at Earth's surface. When a supercontinent breaks up, the continental pieces can travel only so far across the globe before they collide again on the other side (move your hands around in a spherical shape to see how this works). Research suggests that the continents may form a supercontinent once every 500 million years or so. It has been 200 million years since Pangaea split up, so we only have about 300 million years to establish world peace! Even though that's a long time from now, researchers have already dubbed the new supercontinent "Amasia."

arms of the sea may exist, dramatically affecting world ocean circulation patterns. A new land bridge may exist all the way from North America through Central America and South America to Antarctica; this would dramatically alter present-day ocean circulation, interfere with ocean mixing, and undoubtedly result in climate change.

Other changes are caused by the movement of **terranes** (*terranus* = land), which are fragments of crustal material broken off from one plate and accreted or sutured onto another. Each terrane preserves its own distinctive geologic history that is different from that of the surrounding areas, which is why they are also called *exotic terranes*. In fact, Alaska is built by an accumulation of terranes that moved to present-day Alaska over the past 300 million years from as far away as the equator, bringing with them evidence of their tropical origin. Australia, too, is growing larger as it accumulates terranes as it travels north. Figure 2.31 shows that in the future, the thin sliver of land that lies west of the San Andreas Fault named the *California Terrane* may continue to travel northward and become the next piece that is accreted onto southern Alaska.

## A Predictive Model: The Wilson Cycle

Since its inception by Alfred Wegener nearly 100 years ago, plate tectonics has been supported by a wealth of scientific evidence—some of which is presented in this chapter. Although there are still details to be worked out (such as the exact driving mechanism), the theory of plate tectonics has been universally accepted by Earth scientists today because it helps explain so many features and processes that are observed on Earth (see, for example, Mastering Oceanography Web Diving Deeper 2.2). Further, it has led to predictive models that have been used to successfully understand Earth behavior. One such example is the **Wilson cycle** (**Figure 2.32**), named in honor of geophysicist John Tuzo Wilson for his contribution to the early ideas of plate tectonics. The Wilson cycle uses plate tectonic processes to show the distinctive life cycle of ocean basins during their formation, growth, and destruction over many millions of years.

In the embryonic stage of the Wilson cycle, a heat source beneath the lithosphere creates uplift and begins to split a continent apart. The juvenile stage is characterized by further spreading, downdropping, and the formation of a narrow, linear sea. In the mature stage, an ocean basin is fully developed, and a mid-ocean ridge runs down the middle of it. Eventually, a subduction zone occurs along the continental margin, and the plates come back together, producing the declining stage where the ocean basin shrinks. The terminal stage is marked by the plates coming back together, creating a progressively narrower ocean. Finally, in the suturing stage, the ocean disappears, the continents collide, and tall uplifted mountains are created. Over time, the uplifted mountains erode, and the stage is set for the cycle to repeat.

Not only is plate tectonic activity primarily responsible for the creation of landforms, it also plays a prominent role in the development of ocean floor features—which is the topic of the next chapter. Armed with the knowledge of plate tectonic processes you've gained from this chapter, understanding the history and development of ocean floor features in various marine provinces will be a much simpler task.

| Stage, showing cross-sectional view | Motion | Physiography | Example |
|---|---|---|---|
| EMBRYONIC | Uplift | Complex system of linear rift valleys on continent | East African rift valleys |
| JUVENILE | Divergence (spreading) | Narrow seas with matching coasts | Red Sea |
| MATURE | Divergence (spreading) | Ocean basin with continental margins | Atlantic and Arctic Oceans |
| DECLINING | Convergence (subduction) | Island arcs and trenches around basin edge | Pacific Ocean |
| TERMINAL | Convergence (collision) and uplift | Narrow, irregular seas with young mountains | Mediterranean Sea |
| SUTURING | Convergence and uplift | Young to mature mountain belts | Himalaya Mountains |

SmartFigure 2.32 **The Wilson cycle of ocean basin evolution.** The Wilson cycle depicts the stages of ocean basin development, from the initial embryonic stage of formation to the destruction of the basin as continental masses collide and undergo suturing. https://goo.gl/Z4ugKg

## CONCEPT CHECK 2.5 ▸ Describe how Earth has changed in the past and predict how it will look in the future.

1 Using the paleogeographic reconstructions shown in Figure 2.30, determine when the following events first appear in the geologic record:

a. North America lies on the equator
b. The continents come together as Pangaea
c. The North Atlantic Ocean opens
d. India separates from Antarctica

2 Determine the Wilson cycle stage for each of the following present-day locations, noting the features and processes that support your answers:

a. The Atlantic Ocean
b. The Pacific Ocean
c. The Red Sea
d. The Alps
e. The East African rift valleys
f. Baja California

Then, using the Wilson cycle as a predictive model, describe the sequence of events that will happen in the future to the above locations. Be as detailed as you can.

3 Examine Figures 2.30 and 2.32. In which ocean basin would you expect to find the oldest sea floor? Explain your reasoning.

**RECAP** **The Wilson cycle describes the continuing evolution of ocean basins during their formation, growth, and destruction over millions of years.**

# ESSENTIAL CONCEPTS REVIEW

## 2.1 ▶ What evidence supports continental drift?

- According to the theory of plate tectonics, *the outermost portion of Earth is composed of a patchwork of thin, rigid lithospheric plates that move horizontally* with respect to one another. The idea began as a hypothesis called continental drift proposed by *Alfred Wegener* at the start of the 20th century. He suggested that *about 200 million years ago, all the continents were combined* into one large continent (*Pangaea*) surrounded by a single large ocean (*Panthalassa*).
- *Many lines of evidence were used to support the idea of continental drift,* including the similar shape of nearby continents, matching sequences of rocks and mountain chains, glacial ages and other climate evidence, and the distribution of fossil and present-day organisms. Although this evidence suggested that continents have drifted, other incorrect assumptions about the mechanism involved caused *many geologists and geophysicists to discount this hypothesis* throughout the first half of the 20th century.

### Selected Key Terms

Use the **glossary** at the end of this book to discover the meanings of these Selected Key Terms: **continental drift, Pangaea, Panthalassa.**

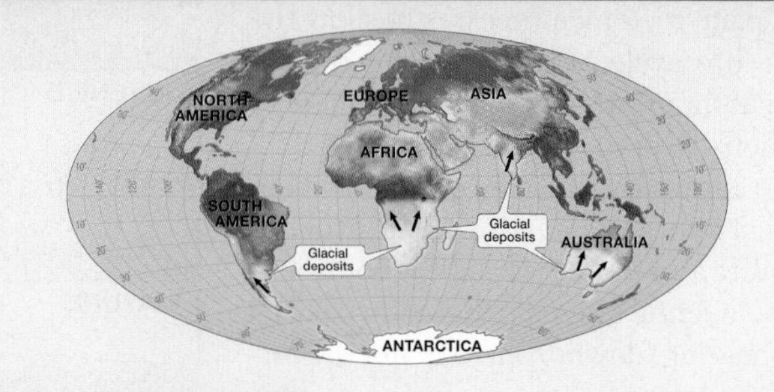


### Critical Thinking Question

If you could travel back in time with three illustrations from this chapter to help Alfred Wegener convince the scientists of his day that continental drift does indeed exist, what would they be, and why?

### Active Learning Exercise

Create two teams to debate the evidence for and against continental drift. Use only knowledge of Earth processes that was available prior to the 1930s.

## 2.2 ▶ What observations led to the theory of plate tectonics?

- *More convincing evidence for drifting continents was introduced in the 1960s, when paleomagnetism*—the study of Earth's ancient magnetic field—*was developed*. Paleomagnetic studies of continental rocks displayed an anomaly scientists called *polar wandering* that could only be explained if the continents had moved relative to each other over geologic time.
- *Harry Hess advanced the idea of sea floor spreading. New sea floor is created at the crest of the mid-ocean ridge* and moves apart in opposite directions and is *eventually destroyed by subduction into an ocean trench*. This helps explain the pattern of magnetic stripes on the sea floor and why sea floor rocks increase linearly in age in either direction from the axis of the mid-ocean ridge.
- Other supporting evidence for plate tectonics includes *oceanic heat flow measurements*, the *pattern of worldwide earthquakes,* and, more recently, *the detection of plate motion by accurate positioning of locations on Earth using satellites*. The combination of evidence has *convinced geologists of Earth's dynamic nature* and helped advance the idea of continental drift into the more encompassing plate tectonic theory.

### Selected Key Terms

Use the **glossary** at the end of this book to discover the meanings of these Selected Key Terms: **paleomagnetism, polarity, magnetic anomalies, sea floor spreading, convection cell, mid-ocean ridge, trench, subduction.**

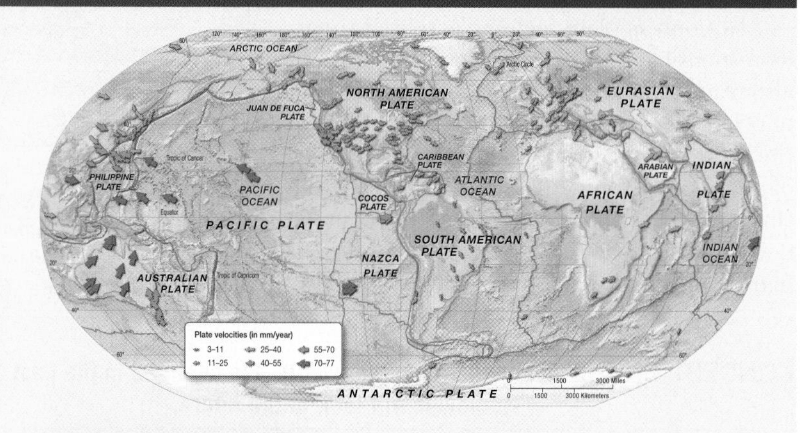


### Critical Thinking Question

If the sea floor didn't exhibit any magnetic polarity reversals, what would that indicate about the age of Earth's ocean basins?

### Active Learning Exercise

A recent discovery suggests that Jupiter's moon Europa is composed of thin, brittle slabs of water ice that undergo plate tectonics, much like Earth's rocky lithospheric plates. Research this discovery on the Internet and describe the evidence for the existence of plate tectonic processes on Europa.

## 2.3 ▶ What features occur at plate boundaries?

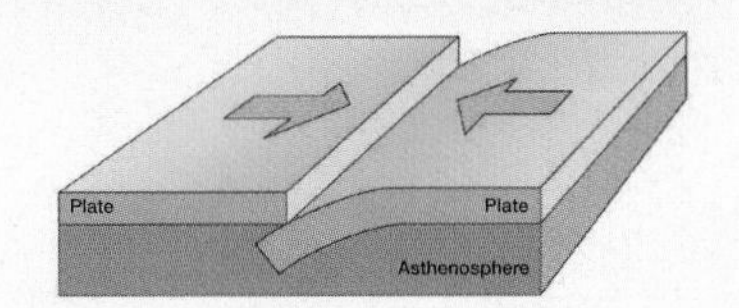


- *As new crust is added to the lithosphere at the mid-ocean ridge (divergent boundaries* where plates move apart), the opposite ends of the *plates are subducted into the mantle at ocean trenches* or beneath continental mountain ranges such as the Himalaya Mountains (*convergent boundaries* where plates come together). In addition, oceanic ridges and rises are offset, and *plates slide past one another along transform faults* (*transform boundaries* where plates slowly grind past one another).

### Selected Key Terms

Use the **glossary** at the end of this book to discover the meanings of these Selected Key Terms: **divergent boundary, convergent boundary, transform boundary, volcanic arc, transform fault.**

### Critical Thinking Question

Using Figure 2.12, analyze and describe the tectonic setting that contributed to these natural disasters: (1) the 1883 eruption of Krakatoa, Indonesia; (2) the 2010 Haitian earthquake; and (3) the 2011 earthquake and tsunami in northeastern Japan.

### Active Learning Exercise

With another student in class, list and describe the three types of plate boundaries. Include in your discussion any sea floor features that are related to these plate boundaries and include a real-world example of each. Construct a map view and cross section showing each of the three types of plate boundaries, including the direction of plate movement and associated features.

## 2.5 ▶ How has Earth changed in the past, and how will it look in the future?

- *The positions of various sea floor and continental features have changed in the past, continue to change today, and will look very different in the future.*
- A predictive working model of plate tectonics is the *Wilson cycle*, which describes the *evolution of ocean basins during their formation, growth, and destruction* over millions of years.

### Selected Key Terms

Use the **glossary** at the end of this book to discover the meanings of these Selected Key Terms: **paleogeography, continental accretion, terrane, Wilson cycle.**

### Critical Thinking Question

Assume that you travel at the same rate as a fast-moving continent—at a rate of 10 centimeters (2.5 inches) per year. Calculate how long it would take you to travel from your present location to a nearby large city. Also, calculate how long it would take you to travel across the United States from the East Coast to the West Coast.

### Active Learning Exercise

You and two of your fellow classmates are colonists on an Earth-sized planet orbiting within the habitable zone of a distant star. As a group, choose one of the following scenarios for your planet: (1) it has extremely active tectonics, (2) it exhibits Earth-like tectonic activity, or (3) it is tectonically dead. Then, based on your planet's chosen level of tectonic activity, describe what your planet looks like, including details about various landforms that would be visible.

## 2.4 ▶ Testing the model: Can plate tectonics explain other features in the ocean and on land?

- *Tests of the plate tectonic model indicate that many features and phenomena provide support for shifting plates.* These include mantle plumes and their associated hotspots that record the motion of plates past them, the origin of flat-topped tablemounts, and the distribution of coral reefs.

### Selected Key Terms

Use the **glossary** at the end of this book to discover the meanings of these Selected Key Terms: **hotspot, mantle plume, nematath, seamount, tablemount.**

### Critical Thinking Question

Describe the differences in origin between the Aleutian Islands (Alaska) and the Hawaiian Islands. Provide evidence to support your explanation.

### Active Learning Exercise

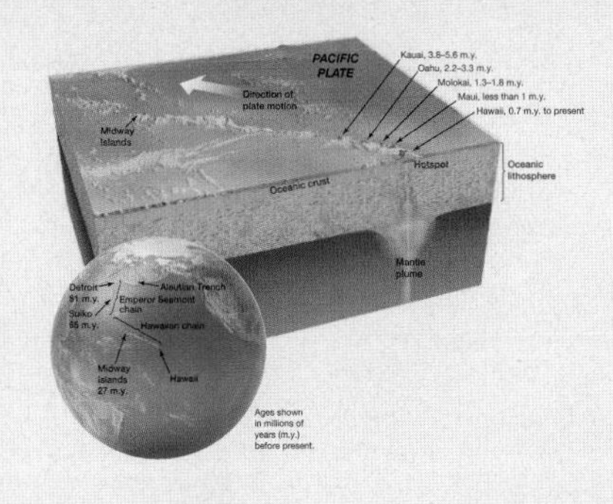


In pairs, investigate the idea that a mantle plume underlies Yellowstone National Park. Report to the class what evidence you have discovered. Using your understanding about plate tectonics, assess the implications for the future of this region.

## Mastering Oceanography

Looking for additional review and test prep materials? With individualized coaching on the toughest topics of the course, Mastering Oceanography offers a wide variety of ways for you to move beyond memorization and deeply grasp the underlying processes of how the oceans work. Visit the Study Area in **www.masteringoceanography.com** to find practice quizzes, study tools, and multimedia that will improve your understanding of this chapter's content. Sign in today to access the following features: Self Study Quizzes, SmartFigures, Oceanography Videos and Animations, Squidtoons, Dynamic Study Modules, and an optional Pearson eText with embedded videos.

# Marine Provinces

# 3

## ESSENTIAL LEARNING CONCEPTS

At the end of this chapter, you should be able to:

- ☐ **3.1** Discuss the techniques that are used to determine ocean bathymetry.
- ☐ **3.2** Describe the sea floor features that exist on continental margins.
- ☐ **3.3** Describe the sea floor features that exist in the deep-ocean basins.
- ☐ **3.4** Describe the sea floor features that exist along the mid-ocean ridge.

↑ *Check when completed*

What does the shape of the ocean floor look like? For the majority of the time that the oceans have been scientifically studied, the deep ocean floor has been largely unknown. During the early age of discovery of the oceans, for instance, most scientists believed that the ocean floor was completely flat and carpeted with a thick layer of muddy sediment, containing little of scientific interest. Further, it was believed that the deepest parts were somewhere in the middle of the ocean basins. However, as more and more vessels crisscrossed the seas to map the ocean floor, scientists found that the terrain of the sea floor is highly varied and includes deep troughs, ancient volcanoes, submarine canyons, and great mountain chains. For example, the ocean floor contains the largest mountain range on Earth, canyons far grander than the Grand Canyon, towering vertical cliffs three times higher than Yosemite's famous walls, and one of the largest volcanoes in the solar system. It is unlike anything on land and, as it turns out, some of the deepest parts of the oceans are actually very close to land! Even today with all of our technological advances, it seems surprising that roughly 80% of the sea floor lacks detailed topographic measurements. There is still so much to discover about the ocean floor.

As marine geologists and oceanographers began to analyze the features of the ocean floor, they realized that certain features had profound implications not only for the history of the ocean floor but also for the history of Earth. How could all these remarkable features have formed, and how can their origin be explained? Scientists now know that over millions of years, the shape of the ocean basins has changed as continents have moved across Earth's surface in response to forces within Earth's interior. The ocean basins as they presently exist reflect the processes of plate tectonics (the topic of Chapter 2), which helps explain the origin of sea floor features.

At first glance, the ocean floor can be divided into three main provinces or regions, which are characterized by water depth and location. Looking at the chapter-opening image, you can see (1) continental margins that are shallow and close to land (*light purple color on the image*), (2) deep-ocean basins, which are deeper areas further from land (*deeper blue color, mostly flat areas on the image*), and (3) the mid-ocean ridge, which, as we learned in Chapter 2, is a tall volcanic mountain range (*light-blue, shallow feature that zigzags through the center of the ocean basin*). In this textchapter, we'll discuss techniques that are used to determine ocean bathymetry, followed by an examination of the features in each of the three main provinces mentioned above.

***"Could the waters of the Atlantic be drawn off so as to expose to view this great sea-gash which separates the continents, and extends from the Arctic to the Antarctic, it would present a scene most rugged, grand, and imposing."***

***—Matthew Fontaine Maury (1854), the "father of oceanography," commenting about the Mid-Atlantic Ridge***

◂ **Sea floor of the North and South Atlantic Oceans.** The sea floor has many interesting features, some of which are completely different from those on land. Recent improvements in technology have aided exploration of the sea floor and given scientists the ability to create stunning high-resolution images like this one that shows the Mid-Atlantic Ridge.

## 3.1 ▶ What Techniques Are Used to Determine Ocean Bathymetry?

**Bathymetry** (*bathos* = depth, *metry* = measurement) is the measurement of ocean depths and the charting of the shape, or *topography* (*topos* = place, *graphy* = description of) of the ocean floor. Determining bathymetry involves measuring the vertical distance from the ocean surface down to the mountains, valleys, and plains of the sea floor.

### Soundings

The first recorded attempt to measure the ocean's depth was conducted in the Mediterranean Sea in about 85 B.C.E. by a Greek explorer named Posidonius. His mission was to answer an age-old question: How deep is the ocean? Posidonius's crew made a **sounding** by letting out nearly 2 kilometers (1.2 miles) of line before the heavy weight on the end of the line touched bottom. (A *sounding* refers to a probe of the environment for scientific observation and was borrowed from atmospheric scientists, who released probes called soundings into the atmosphere. Ironically, the term does not actually refer to sound; the use of sound to measure ocean depths came later.) For the next 2000 years, voyagers used sounding lines to probe the ocean's depths. The standard unit of ocean depth is the **fathom** (*fathme* = outstretched arms). The term fathom is derived from the method used to bring depth sounding lines back on board a vessel by hand. While hauling in the line, workers counted the number of arm-lengths collected. By measuring the length of the person's outstretched arms, the amount of line taken in could be calculated. Much later, the distance of 1 fathom was standardized to equal exactly 6 feet, which is equal to 1.8 meters.

**EXPLORING DATA** ▼

1. **Notice the strong sound reflection of the deep scattering layer (DSL) at a depth between 350 and 400 meters on the left side of the figure. Explain why a dense concentration of marine organisms would create such a feature on an echo sounding record.**
2. **Using the equation shown in Figure 3.7a, determine the time it takes sound to reach 350 meters depth and return to the surface. Use the average speed of sound in seawater of 1507 meters (4945 feet) per second.**

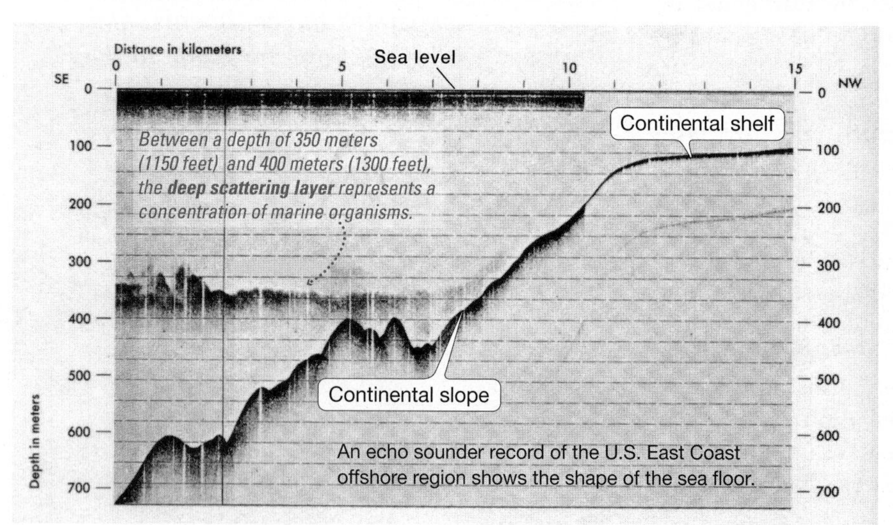


**Figure 3.1 An echo sounder record.** Vertical exaggeration (the amount of expansion of the vertical scale) is 12 times.

The first systematic bathymetric measurements of the oceans were made in 1872 aboard the HMS *Challenger* during its historic three-and-a-half-year voyage. (For more information about the accomplishments of the *Challenger* expedition, see Mastering Oceanography Web Diving Deeper 5.2.) Every so often, *Challenger*'s crew stopped and measured the depth, along with many other ocean properties. These measurements indicated that the deep-ocean floor was not flat but had significant *relief* (variations in elevation), just as dry land does. However, determining bathymetry by making occasional soundings rarely gives a complete picture of the ocean floor. For instance, imagine trying to determine what the surface features on land look like while flying in a blimp at an altitude of several kilometers on a foggy night, using only a long, weighted rope to determine your height above the surface. This is similar to how bathymetric measurements were collected from ships using sounding lines.

### Echo Soundings

The presence of mid-ocean undersea mountains had long been known, but recognition of their full extent into a connected worldwide system had to await the invention and use of the **echo sounder**, or *fathometer*, in the early 1900s. An echo sounder sends a sound signal (called a **ping**) from the ship downward into the ocean, where it produces echoes when it bounces off any density difference, such as marine organisms or the ocean floor (**Figure 3.1**). Water is a good transmitter of sound, and although the speed of sound in seawater varies with salinity, pressure, and temperature, it averages about 1507 meters (4945 feet) per second. Knowing this variable, depth can be determined by calculating the time it takes for the echoes to return. For a continuous echo sounding, it also gives an idea of the corresponding shape of the ocean floor. In 1925,

**Animation**
Sonar and Echolocation
http:/goo.gl/53cu5z

for example, the German vessel *Meteor* used echo sounding to identify the underwater mountain range running through the center of the South Atlantic Ocean.

Echo sounding, however, lacks detail and often gives an inaccurate view of the relief of the sea floor. For instance, the sound beam emitted from a ship 4000 meters (13,100 feet) above the ocean floor widens to a diameter of about 4600 meters (15,000 feet) at the bottom. Consequently, the first echoes to return from the bottom are usually from the closest (highest) peak within this broad area. Nonetheless, most of our knowledge of ocean bathymetry has been provided by the echo sounder.

Because sounds produced by echo sounders bounce off any density difference in the ocean, it was soon discovered that echo sounders could detect and track submarines. During World War II, antisubmarine warfare inspired many improvements in the technology of "seeing" into the ocean using sound.

During and after World War II, there was great improvement in sonar technology. For example, the **precision depth recorder (PDR)**, which was developed in the 1950s, uses a focused high-frequency sound beam to measure depths to a resolution of about 1 meter (3.3 feet). Throughout the 1960s, PDRs were used extensively and provided a reasonably good representation of the ocean floor. From thousands of research vessel tracks, the first reliable global maps of sea floor bathymetry were produced. These maps helped confirm the ideas of sea floor spreading and plate tectonics.

Modern *acoustic* (*akouein* = to hear) instruments that use sound to map the sea floor include *multibeam echo sounders* (which use multiple frequencies of sound simultaneously) and side-scan **sonar** (an acronym for *so*und *n*avigation *a*nd *r*anging). **Seabeam**—the first multibeam echo sounder—made it possible for a survey ship to map the features of the ocean floor along a strip up to 60 kilometers (37 miles) wide. Multibeam systems use sound emitters directed away from both sides of a survey ship, with receivers permanently mounted on the ship's hull. Multibeam instruments emit multiple beams of sound waves, which are reflected off the ocean floor. As the sound waves bounce back with different strengths and timing, computers analyze these differences to determine the depth and shape of the sea floor and whether the bottom is rock, sand, or mud (**Figure 3.2**). In this way, multibeam surveying provides incredibly detailed imagery of the seabed. Because its beams of sound spread out with depth, multibeam systems have resolution limitations in deep water.

In deep water or where a detailed survey is required, **side-scan sonar** can provide enhanced views of the sea floor. A side-scan sonar instrument is towed behind a survey ship and can be lowered to just above the ocean floor to produce a detailed strip map of ocean floor bathymetry (**Figure 3.3**). To maximize its resolution, the

Hull-mounted multibeam instruments emit multiple beams of sound waves, which are reflected off the ocean floor. Receivers collect data that allow oceanographers to determine the depth, shape, and even composition of the sea floor.
As a ship travels back and forth throughout an area, it can produce a detailed image of sea floor bathymetry.



**SmartFigure 3.2 Multibeam sonar.** An artist's depiction of how a survey vessel uses multibeam sonar to map the ocean floor. Colors on the sea floor represent different elevations. https://goo.gl/2oMV73

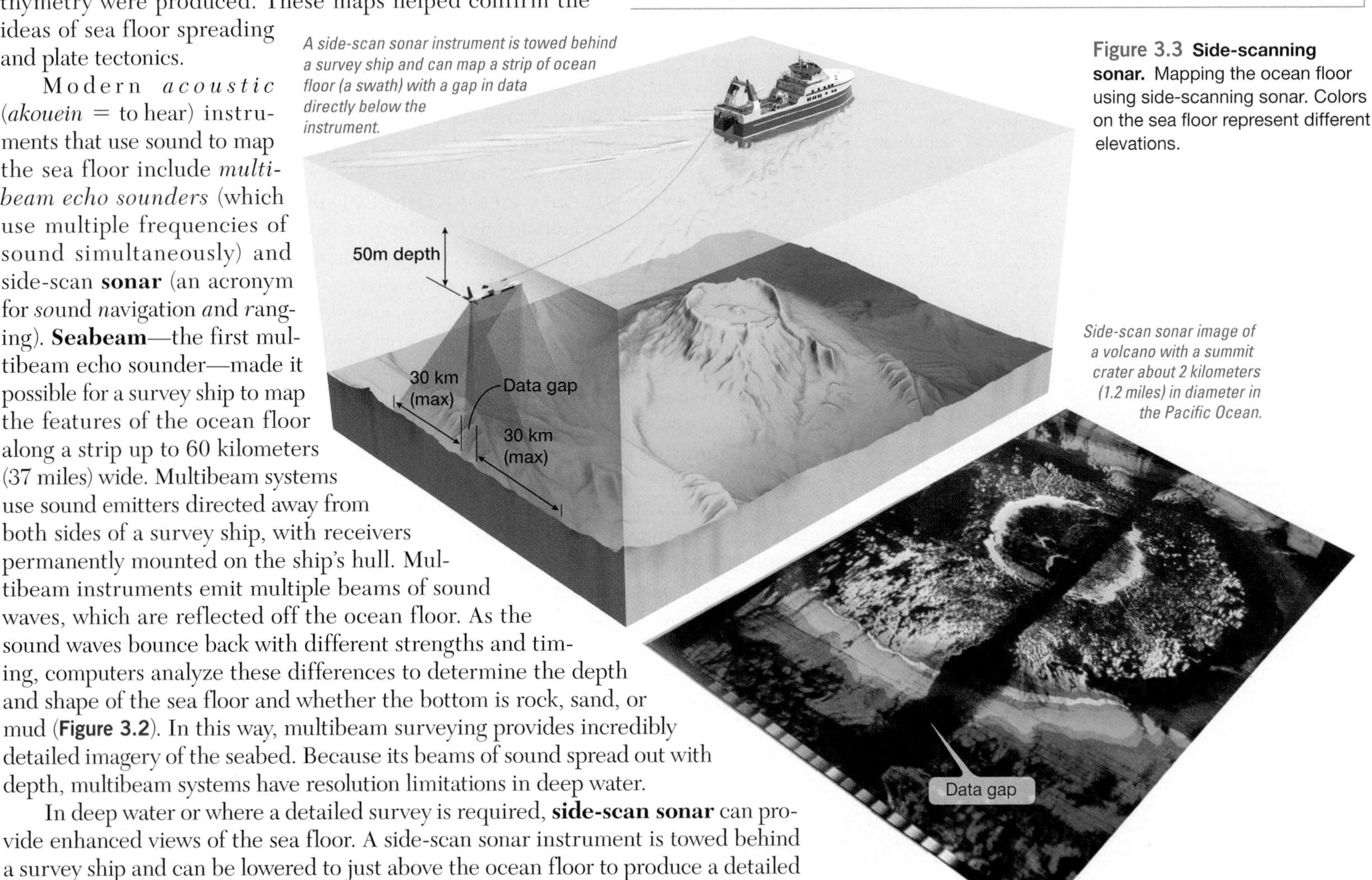


**Figure 3.3 Side-scanning sonar.** Mapping the ocean floor using side-scanning sonar. Colors on the sea floor represent different elevations.

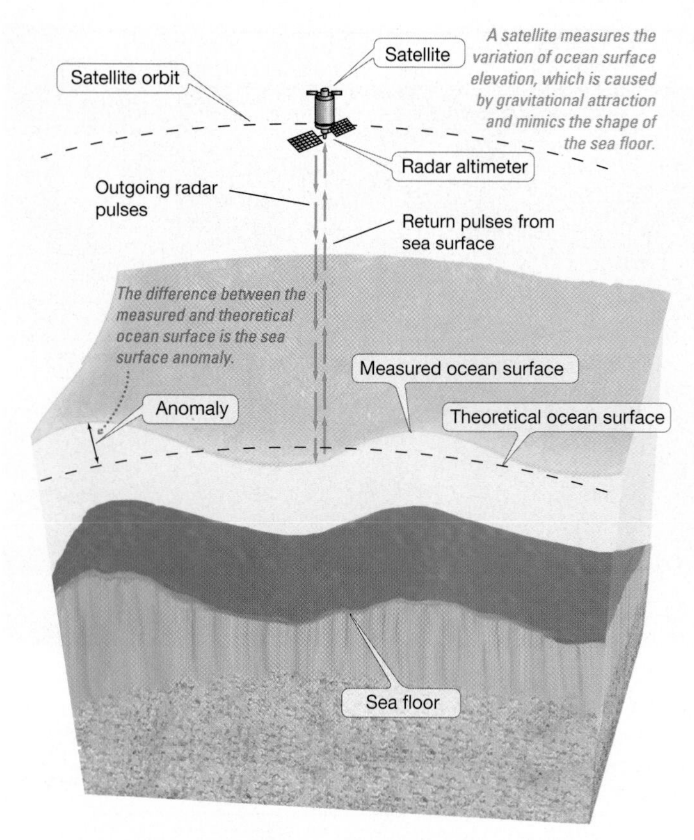


**Figure 3.4 How satellite measurements of the ocean surface are used to map sea floor features.**

side-scan instrument can be lowered on its cable so that it "flies" just above the ocean floor. Underwater robotic vehicles with side-scan sonar that are programmable and navigate independently from a ship can also be used to map the sea floor.

## Using Satellites to Map Ocean Properties from Space

Although multibeam and side-scan sonar produce very detailed bathymetric maps, mapping the sea floor by ship is an expensive and time-consuming process. A research vessel must tediously travel back and forth throughout an area (a process called "mowing the lawn") to produce an accurate map of bathymetric features (see Figure 3.2). Unfortunately, only a small percentage of the ocean floor has been mapped in this way.

An Earth-orbiting satellite, on the other hand, can observe large areas of the ocean at one time. Consequently, satellites are increasingly used to determine ocean properties. A list of recent U.S. oceanographic satellite missions and their objectives are shown in Mastering Oceanography Web Table 3.1. Remarkably, satellite measurements of the ocean surface have been used to make maps of the sea floor. How does a satellite—which orbits at a great distance above the planet and can view only the ocean's *surface*—obtain a picture of the sea *floor*?

The answer lies in the fact that sea floor features directly influence Earth's gravitational field. Deep areas such as trenches correspond to a lower gravitational attraction, and large undersea objects—such as tall volcanoes on the sea floor called **seamounts** (described in Chapter 2)—exert an extra gravitational pull. These differences affect the height of the sea surface directly above these sea floor features, causing the ocean surface to bulge upward and sink downward, mimicking the relief of the ocean floor. A 2000-meter (6500-foot)-high seamount, for example, exerts a small but measurable gravitational pull on the water around it, creating a bulge 2 meters (7 feet) high on the ocean surface. These irregularities are easily detectable by satellites, which use microwave beams to measure sea level to an accuracy of 4 centimeters (1.5 inches). After corrections are made for waves, tides, currents, and atmospheric effects, the resulting pattern of dips and bulges at the ocean surface can be used to indirectly reveal ocean floor bathymetry (**Figure 3.4**). For example, **Figure 3.5** compares two different maps of the same area: one based on bathymetric data from ships (*top*) and the other based on satellite measurements (*bottom*), which shows much higher resolution of sea floor features.

Data from Earth-orbiting satellites such as Geosat, a U.S. Navy satellite, were collected during the 1980s. When this information was declassified, Walter Smith of the National Oceanic and Atmospheric Administration and David Sandwell of Scripps Institution of Oceanography began producing sea floor maps based on the shape of the sea surface. Although the shape of the sea surface is not exactly equivalent to the bathymetry of the sea floor, sea level does mimic the overall shape of the sea floor. The researchers also use depth soundings to calibrate the sea surface height measurements. What is unique about these researchers' maps is that they provide a view of Earth similar to what we could see if we were able to drain the oceans and view the ocean floor directly. For example, their newest high-resolution map of ocean surface gravity (**Figure 3.6**), which was published in 2014,

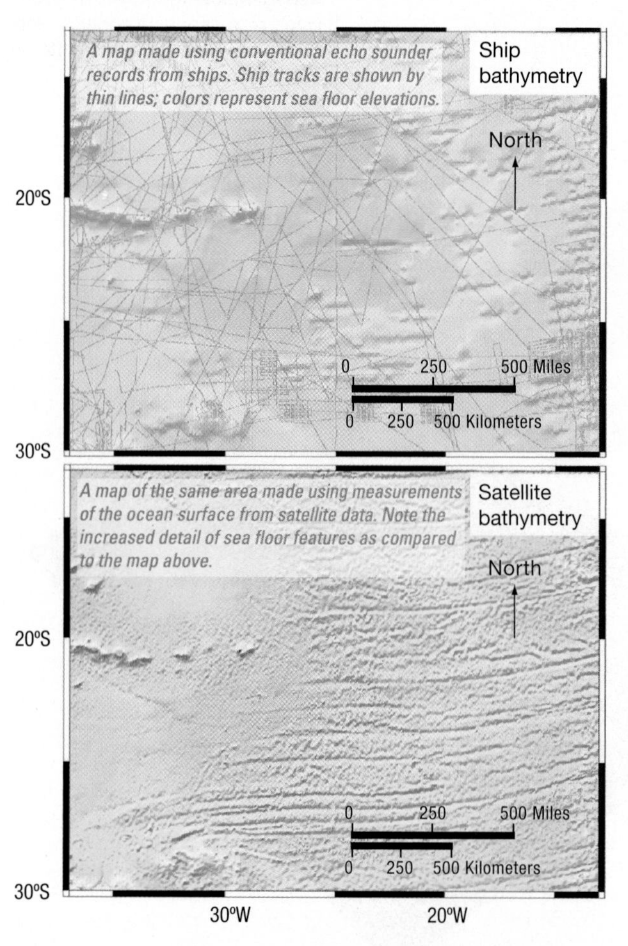


**Figure 3.5 A comparison of echo sounder (ship) and satellite bathymetric maps of the sea floor.** Both bathymetric maps show the same portion of the Brazil Basin in the South Atlantic Ocean. Colors on the sea floor represent different elevations.

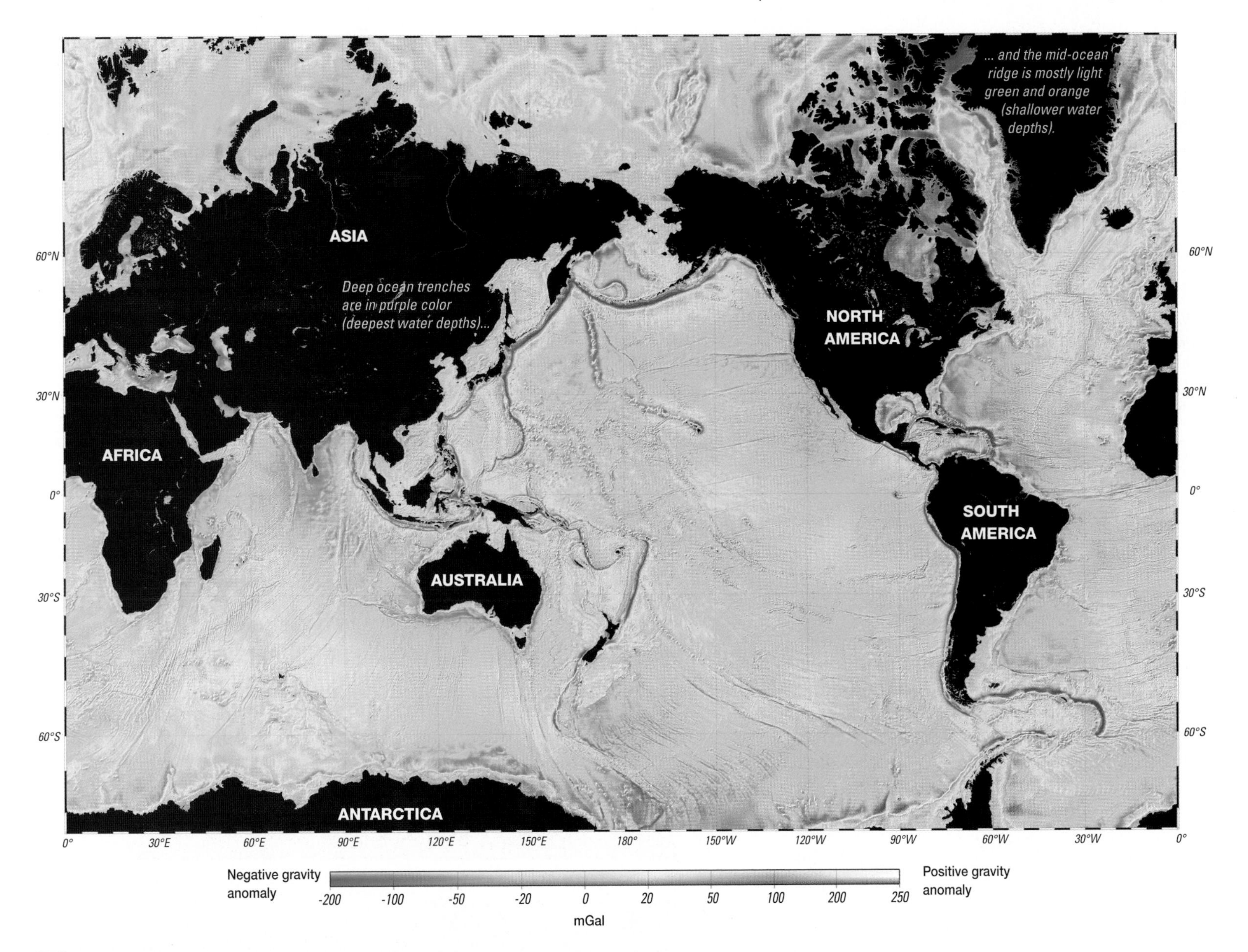


SmartFigure 3.6 **Global sea surface elevation map from satellite data.** This new high-resolution map of the sea floor was produced in 2014 using satellite data of Earth's gravity field, which, when adjusted using measured depths, closely corresponds to ocean depth. Gravity anomalies shown on map are in mGal.
https://goo.gl/bYopVq

uses data mostly from two satellites: CryoSat-2, from the European Space Agency, and Jason-1, from NASA and the French space agency CNES. (In 2002, Jason-1 replaced the TOPEX/Poseidon satellite, which measured ocean surface topography from 1992 through 2005. Jason-1 was decommissioned in 2013 and replaced by updated instruments aboard Jason-2.) This new map of the ocean floor clearly shows the large-scale details of many ocean floor features, such as the mid-ocean ridge, trenches, seamounts, and nemataths (island chains). In fact, the new map delineates ocean bathymetry in areas where research vessels have never conducted sonar surveys and includes many new sea floor features such as thousands of underwater mountains.

## Seismic Reflection Profiles

Oceanographers who want to know about ocean structure beneath the sea floor use strong low-frequency sounds produced by explosions or air guns, as shown in Figure 3.7. These sounds penetrate beneath the sea floor and reflect off the boundaries between different rock or sediment layers, producing **seismic reflection profiles**, which have applications in mineral and petroleum exploration.

**RECAP** **Sending pings of sound into the ocean (echo sounding) is a commonly used technique for determining ocean bathymetry. More recently, satellites are being used to map sea floor features.**

**EXPLORING DATA ▶ Using the equation shown in Figure 3.7a, determine the time it takes sound to reach the deepest part of the ocean, the Challenger Deep in the Mariana Trench, which is 11,022 meters (36,161 feet) below sea level. Use the average speed of sound in seawater of 1507 meters (4945 feet) per second.**

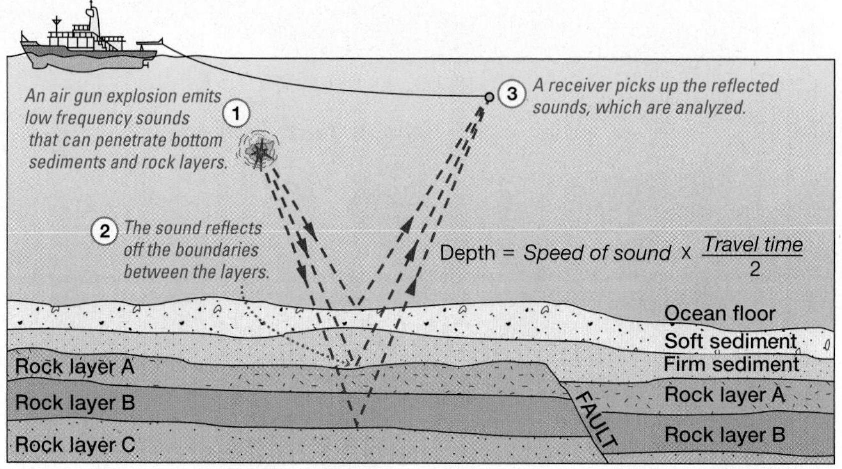

An air gun explosion emits low frequency sounds that can penetrate bottom sediments and rock layers.
1
2
The sound reflects off the boundaries between the layers.
3
A receiver picks up the reflected sounds, which are analyzed.
Depth = Speed of sound x Travel time / 2
Ocean floor
Soft sediment
Firm sediment
Rock layer A
Rock layer B
Rock layer C
FAULT

**(a)** A ship conducting seismic profiling. Note that depth can be determined by knowing the speed of sound in seawater and the travel time of the sound.

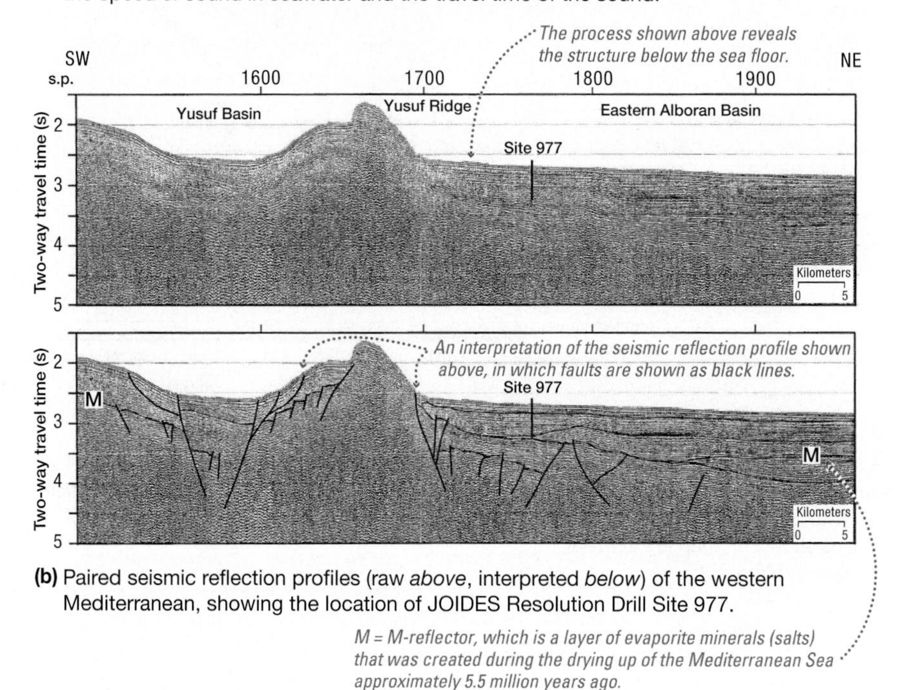

The process shown above reveals the structure below the sea floor.
SW
NE
s.p.
1600
1700
1800
1900
Yusuf Basin
Yusuf Ridge
Eastern Alboran Basin
Site 977
Two-way travel time (s)
2
3
4
5
Kilometers
0
5
An interpretation of the seismic reflection profile shown above, in which faults are shown as black lines.
M

**(b)** Paired seismic reflection profiles (raw *above*, interpreted *below*) of the western Mediterranean, showing the location of JOIDES Resolution Drill Site 977.

*M = M-reflector, which is a layer of evaporite minerals (salts) that was created during the drying up of the Mediterranean Sea approximately 5.5 million years ago.*

Figure 3.7 **Seismic profiling.**

## STUDENTS SOMETIMES ASK . . .

### *What happened to the 2014 Malaysian Airlines flight that vanished after takeoff?*

It's still a mystery. Malaysian Airlines flight MH370 went missing on March 8, 2014, while en route from Kuala Lumpur, Malaysia to Beijing, China. Satellite communications suggest that the flight veered south and ended up running out of fuel and crash-landing in the Indian Ocean west of Australia. Unfortunately, the suspected area of the crash is large, remote, and deep, and the region's rugged sea floor is very poorly explored, all of which has hampered recovery efforts. In the days following the plane's disappearance, large pieces of floating trash were misidentified as pieces of the airplane. To this day, it's unclear exactly what happened to the airplane and all 239 people on board. An extensive sea floor search effort by a multinational team from 2014–2016 found nothing and was called off in 2017. It was the world's largest and most costly (U.S. $56 million) marine search effort to date. In July 2015, a floating piece of a wing washed up on Reunion Island in the Indian Ocean that was later positively identified as belonging to the missing plane. In 2018, a private U.S.-based team began searching the sea floor using ship-based sonar, a small fleet of underwater drones, and data from the oceanographic community, but in spite of valiant efforts to locate it, the remaining wreckage of the plane may never be found.

## CONCEPT CHECK 3.1 ▶ Discuss the techniques that are used to determine ocean bathymetry.

1 What is bathymetry? How is it different from topography?

2 Describe how an echo sounder works.

3 Discuss the development of bathymetric techniques, indicating significant advancements in technology.

# 3.2 ▶ What Features Exist on Continental Margins?

The ocean floor can be divided into three major provinces (**Figure 3.8**): (1) **continental margins**, which are shallow-water areas close to continents, (2) **deep-ocean basins**, which are deep-water areas farther from land, and (3) the **mid-ocean ridge**, which is composed of shallower areas near the middle of an ocean. Plate tectonic processes (discussed in the previous chapters) are integral to the formation of these provinces. Through the process of sea floor spreading, the mid-ocean ridge and deep-ocean basins are created. Elsewhere, as a continent is split apart, new continental margins are formed.

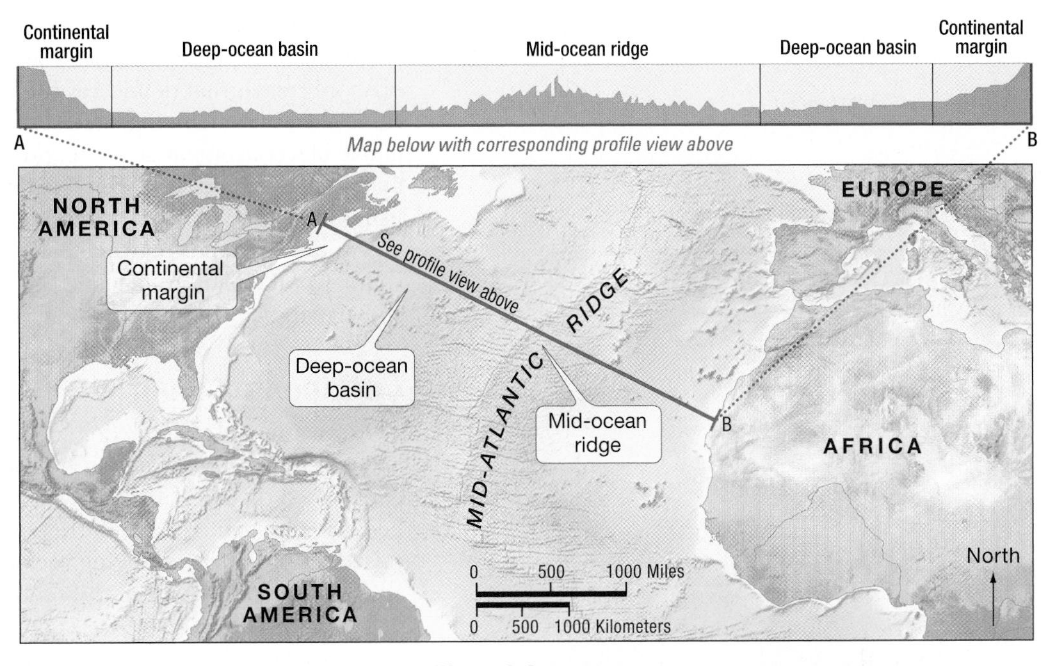


**Figure 3.8 Major regions of the North Atlantic Ocean floor.** Map view (*bottom*) and profile view (*top*), showing that the ocean floor can be divided into three major provinces: continental margins, deep-ocean basins, and the mid-ocean ridge.

## Passive versus Active Continental Margins

Continental margins can be classified as either passive or active, depending on their proximity to plate boundaries. **Passive margins** (**Figure 3.9**, *left*) are embedded within the interior of lithospheric plates and are therefore not in close proximity to any plate boundary. Thus, passive margins usually lack major tectonic activity (such as large earthquakes, eruptive volcanoes, and mountain building).

The East Coast of the United States, where there is no plate boundary, is an example of a passive continental margin. Passive margins are usually produced by rifting of continental landmasses and continued sea floor spreading over geologic time. Features of passive continental margins include the continental shelf, the continental slope, and the continental rise that extends toward the deep-ocean basins (Figures 3.9 and 3.10).

**Active margins** (Figure 3.9, *right*) are associated with lithospheric plate boundaries and are marked by a high degree of tectonic activity. Two types of active margins exist. **Convergent active margins** are associated with oceanic–continental

**SmartFigure 3.9 Diagrammatic view of passive and active continental margins.** Diagrammatic perspective view of typical features across an ocean basin, including a passive continental margin (*left*) and a convergent active continental margin (*right*). Vertical exaggeration is about 10 times.
https://goo.gl/0fnsb6

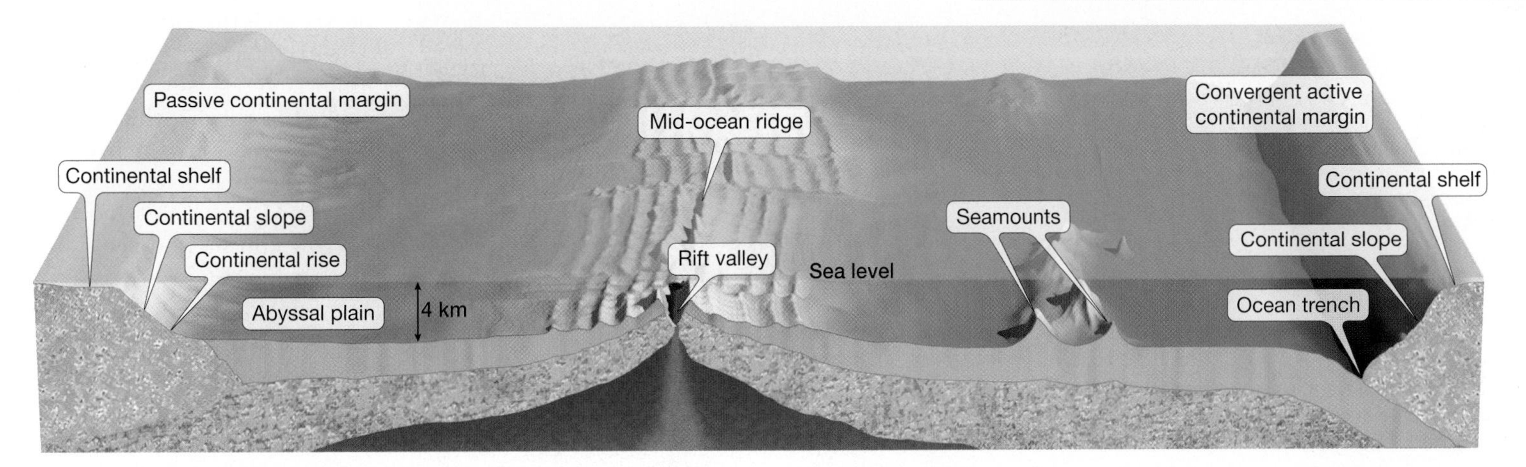

convergent plate boundaries. From the land to the ocean, features include an onshore arc-shaped row of active volcanoes, then a narrow shelf, a steep slope, and an offshore trench that delineates the plate boundary. Western South America, where the Nazca Plate is being subducted beneath the South American Plate, is an example of a convergent active margin. **Transform active margins** are less common and are associated with transform plate boundaries. At these locations, there are usually offshore faults that parallel the main transform plate boundary fault and create linear islands, banks (shallowly submerged areas), and deep basins close to shore. Coastal California along the San Andreas Fault is an example of a transform active margin.

**RECAP** **Passive continental margins lack a plate boundary and have different features than active continental margins, which include a plate boundary (either convergent or transform).**

## Continental Shelf

The **continental shelf** is defined as a generally flat zone extending from the shore beneath the ocean surface to a point at which a marked increase in slope angle occurs, called the **shelf break** (Figure 3.10). It is usually flat and relatively featureless because of marine sediment deposits, but can contain coastal islands, reefs, and raised banks. The underlying rock is granitic continental crust, so the continental shelf is geologically part of the continent. Accurate sea floor mapping is essential for determining the extent of the continental shelf.

Continental shelves vary depending on the local geology and topography. For example, the average width of the continental shelf is about 70 kilometers (43 miles), but it varies from a few tens of meters to 1500 kilometers (930 miles). The broadest shelves occur off the northern coasts of Siberia and North America in the Arctic Ocean. Worldwide, the average depth at which the shelf break occurs is about 135 meters (443 feet). Around Antarctica, however, the shelf break occurs at 350 meters (2200 feet). The average slope of the continental shelf is only about a tenth of a degree, which is similar to the slope given to a large parking lot for drainage purposes.

Climate Connection

Sea level has fluctuated over the history of Earth, causing the shoreline to migrate back and forth across the continental shelf. When colder climates prevailed during the most recent ice age, for example, more of Earth's water was frozen as glaciers on land, so sea level was lower than it is today. During that time, more of the continental shelf was exposed.

The type of continental margin determines the shape and features associated with the continental shelf. For example, the east coast of South America has a broader continental shelf than its west coast. The east coast is a passive margin, which typically has a wider shelf. In contrast, the convergent active margin present along the west coast of South America is characterized by a narrow continental shelf and a shelf break close to shore. For transform active margins such as along California, the presence of offshore faults produces a continental shelf that is not flat. Rather, it is marked

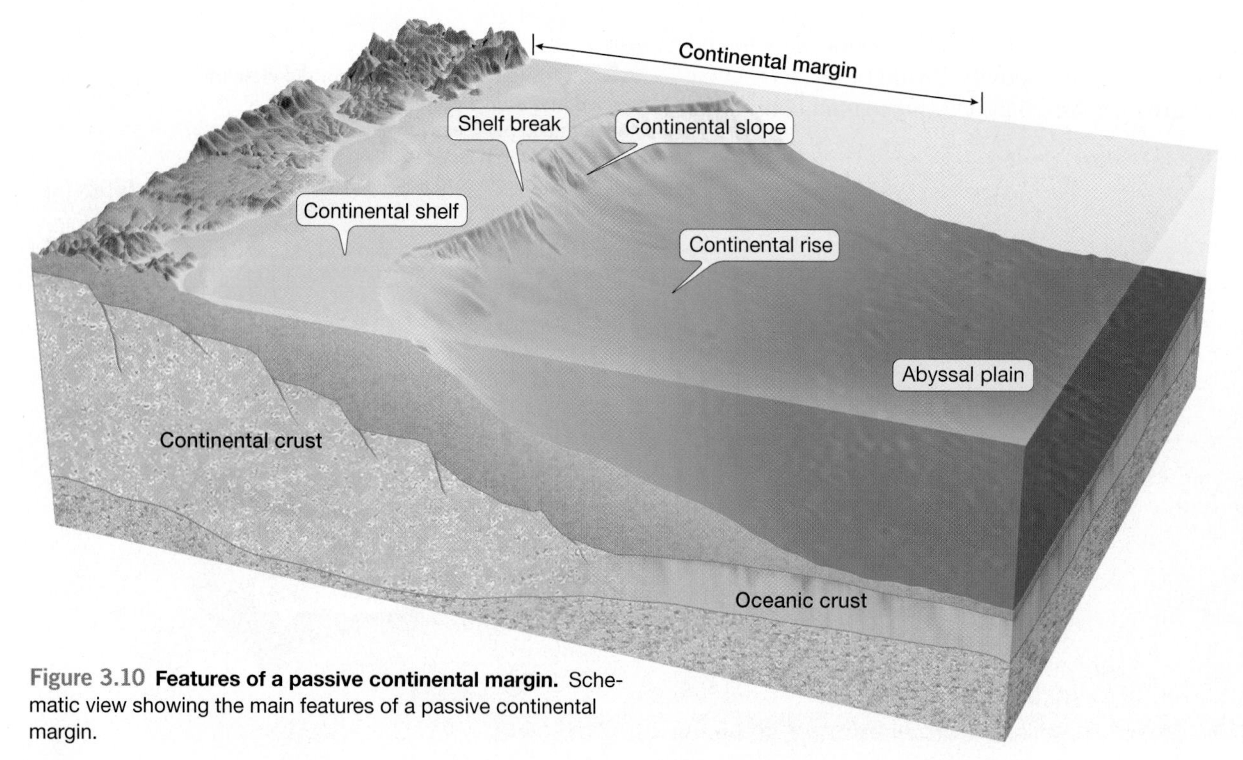


Figure 3.10 **Features of a passive continental margin.** Schematic view showing the main features of a passive continental margin.

by a high degree of relief (islands, shallow banks, and deep basins) called a **continental borderland** (Figure 3.11).

## Continental Slope

The base of the **continental slope**, which lies beyond the shelf break, is where the deep-ocean basins begin. Total relief in this region is similar to that found in mountain ranges on the continents. The break at the top of the slope may be from 1 to 5 kilometers (0.6 to 3 miles) above the deep-ocean basin at its base. Along convergent active margins where the slope descends into submarine trenches, even greater vertical relief is measured. Off the west coast of South America, for instance, the total relief from the top of the Andes Mountains to the bottom of the Peru–Chile Trench is about 15 kilometers (9.3 miles).

Worldwide, the slope of the continental slopes averages about 4 degrees but varies from 1 to 25 degrees (for comparison, a very steep road grade is 8%, or about 5 degrees). A study that compared different continental slopes in the United States revealed that the average slope is just over 2 degrees. Around the margin of the Pacific Ocean, the continental slopes average more than 5 degrees because of the presence of convergent active margins that drop directly into deep offshore trenches. The Atlantic and Indian Oceans, on the other hand, contain many passive margins, which lack plate boundaries. Thus, the amount of relief is lower and slopes in these oceans average about 3 degrees.

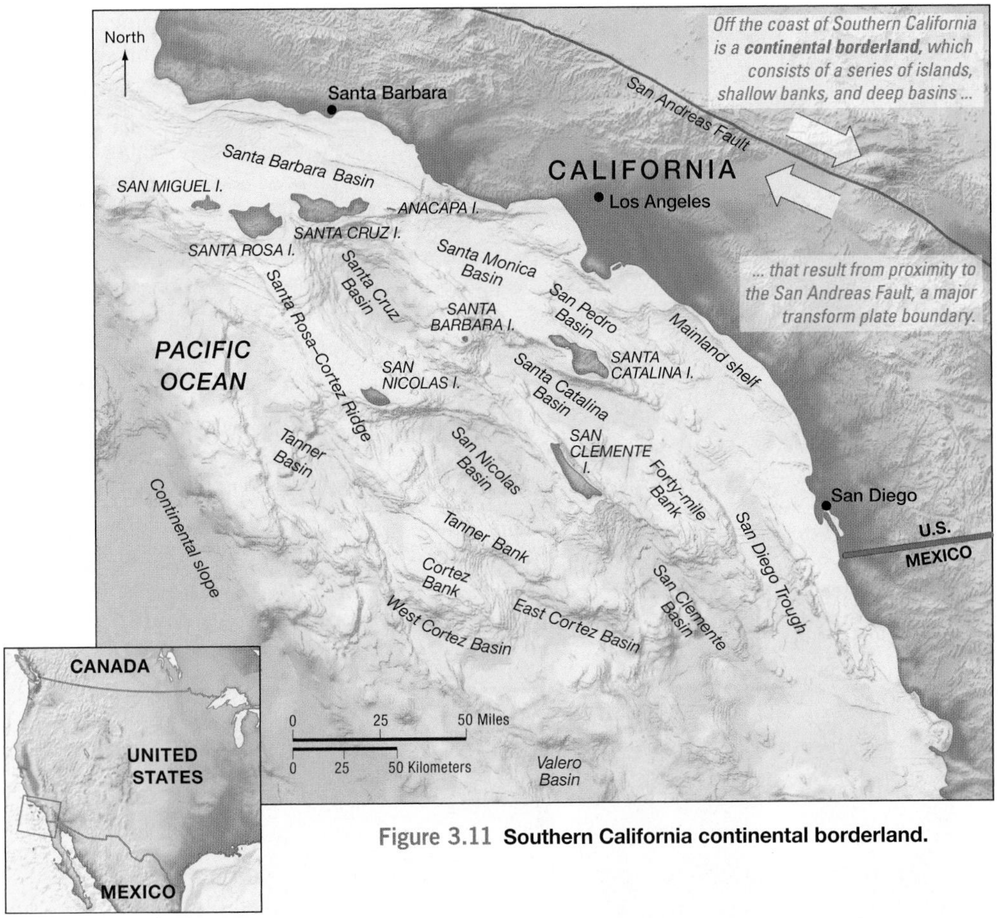

**Figure 3.11 Southern California continental borderland.**

## Submarine Canyons and Turbidity Currents

The continental slope and, to a lesser extent, the continental shelf exhibit **submarine canyons**, which are narrow but deep submarine valleys that are V-shaped in profile view and have branches or tributaries with steep to overhanging walls (Figure 3.12). They resemble canyons formed on land that are carved by rivers and can be quite large. In fact, the Monterey Canyon off California is comparable in size to Arizona's Grand Canyon (Figure 3.13).

How are submarine canyons formed? Initially it was thought that submarine canyons were ancient river valleys created by the erosive power of rivers when sea level was lower and the continental shelf was exposed. Although some canyons are directly offshore from where rivers enter the sea, the majority of them are not. Many, in fact, are confined exclusively to the continental slope. In addition, submarine canyons continue to the base of the continental slope, which averages some 3500 meters (11,500 feet) below sea level. If ancient rivers cut these canyons, then sea level would have had to be some 3500 meters (11,500 feet) lower than it is today, and there is no evidence that sea level has ever been lowered by that much.

Side-scan sonar surveys along the Atlantic coast indicate that the continental slope is dominated by submarine canyons from Hudson Canyon near New York City to Baltimore Canyon in Maryland. Canyons confined to the continental slope are straighter and have steeper canyon floor gradients than those that are cut into the continental shelf. These characteristics suggest the canyons are created on the

**(a)** Turbidity currents move downslope, eroding the continental margin to enlarge submarine canyons. Deep-sea fans are composed of turbidite deposits, which consist of sequences of graded bedding (inset).

**(b)** A diver descends into a submarine canyon in the Red Sea near Dahab, Egypt.

**(c)** Outcrop of layered turbidite deposits that have been tilted and uplifted onto land in California. Each light-colored layer is sandstone that marks the coarser bottom of a graded bedding sequence.

SmartFigure 3.12 **Submarine canyons and turbidity currents.**
https://goo.gl/Bl6Vy3

**Animation**
Turbidity Currents and the Formation of Graded Bedding
http:/goo.gl/CUwQCF

continental slope by some marine process and enlarge into the continental shelf through time.

Both indirect and direct observations of the erosive power of **turbidity currents** (*turbidus* = disordered) (see Mastering Oceanography **Web Diving Deeper 3.3**) have suggested that they are responsible for carving submarine canyons. Turbidity currents are underwater avalanches of muddy water mixed with rocks and other debris. The sediment portion of turbidity currents comes from sea floor materials that move across the continental shelf into the head of a submarine canyon and accumulate there, setting the stage for initiation of a turbidity current. Trigger mechanisms for turbidity currents include shaking by an earthquake, the oversteepening of sediment that accumulates on the shelf, hurricanes passing over the area, and the rapid input of sediment from flood waters. Once a turbidity current is set in motion, the dense mixture of

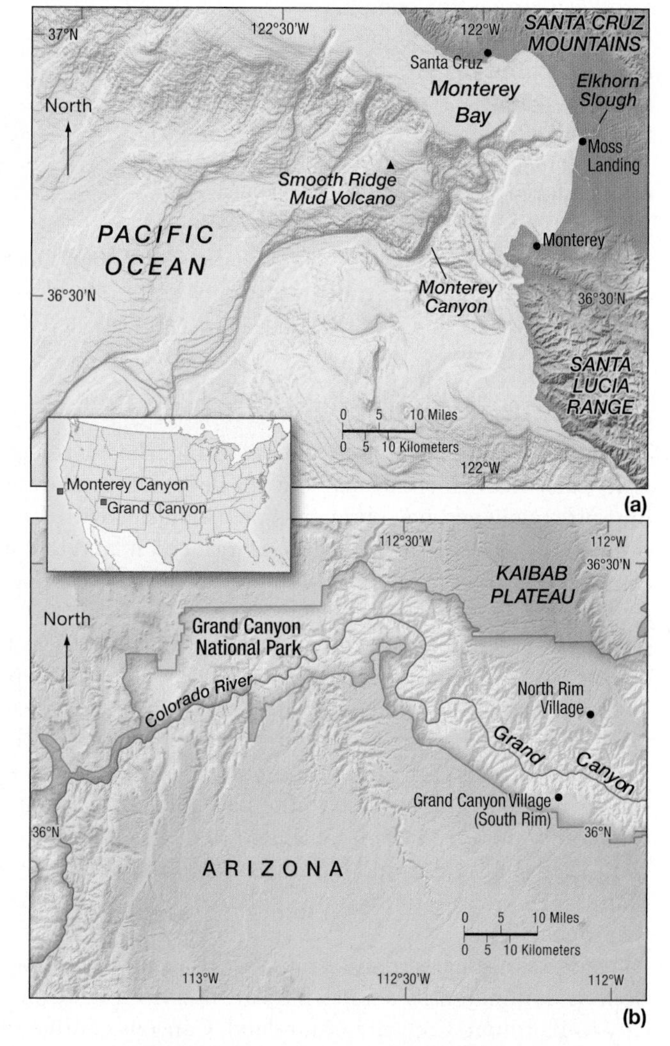


Figure 3.13 **Comparison of the Monterey Submarine Canyon and Arizona's Grand Canyon.** These same-scale maps show that the Monterey Submarine Canyon **(a)** is comparable to Arizona's Grand Canyon **(b)** in terms of length, depth, width, and steepness.

water and debris moves rapidly downslope under the force of gravity and carves the canyon as it goes, resembling a flash flood on land. Turbidity currents are strong enough to transport huge rocks down submarine canyons and cause a considerable amount of erosion over time.

## Continental Rise

The **continental rise** is a transition zone between the continental slope and the deep-ocean floor comprised of a huge submerged pile of debris composed of gravel, sand, mud, and smaller particles. Where did all this debris come from, and how did it get there?

Oceanographers now recognize that the material transported by turbidity currents off the continental shelf is responsible for the creation of continental rises. When a turbidity current moves through and erodes a submarine canyon, it exits through the mouth of the canyon. The slope angle decreases and the turbidity current slows, causing suspended material to settle out in a distinctive type of layering called **graded bedding** that *grades in size upward* (Figure 3.12a, *inset*). As the energy of the turbidity current dissipates, larger pieces settle first, then progressively smaller pieces settle, and eventually even very fine pieces settle out, which may occur weeks or months later.

An individual turbidity current deposits one graded bedding sequence. The next turbidity current may partially erode the previous deposit and then deposit another graded bedding sequence on top of the previous one. After some time, a thick sequence of graded bedding deposits can develop one on top of another. These stacks of graded bedding, which make up the continental rise, are called **turbidite deposits** (Figure 3.12c).

As viewed from above, the deposits at the mouths of submarine canyons are fan-shaped and resemble a fanned-out apron (Figures 3.12a and 3.14). Consequently, these deposits are called **deep-sea fans**, or **submarine fans**. Deep-sea fans create the continental rise when they merge together along the base of the continental slope. Along convergent active margins, however, the steep continental slope leads directly into a deep-ocean trench. Sediment from turbidity currents accumulates in the trench and there is no continental rise.

One of the largest deep-sea fans in the world is the Indus Fan, a passive margin fan that extends 1800 kilometers (1100 miles) south of Pakistan (**Figure 3.14a**). The Indus River carries large amounts of sediment from the Himalaya Mountains to the coast. This sediment eventually makes its way down the submarine canyon and builds the fan, which, in some areas, has sediment that is more than 10 kilometers (6.2 miles) thick! The Indus Fan has a main submarine canyon channel that extends seaward onto the fan but soon divides into several branching distributary channels. These distributary channels are similar to those found on deltas, which form at the mouths of streams. On the lower fan, the surface has a very low slope, and the flow is no longer confined to channels, so it spreads out and forms layers of fine sediment across the fan surface. The Indus Fan has so much sediment, in fact, that it partially buries an active mid-ocean ridge, the Carlsberg Ridge!

### STUDENTS SOMETIMES ASK . . .

***Has anyone ever been caught in and killed by a turbidity current?***

Actually, no. This is mostly because turbidity currents occur in submarine canyons on the continental slope, which are normally so deep that even deep divers don't venture there. However, oceanographic equipment left on the floor of a submarine canyon has frequently been mangled or destroyed by the highly erosive power of turbidity currents. Other times, the equipment is simply swept away by a turbidity current and is never seen again.

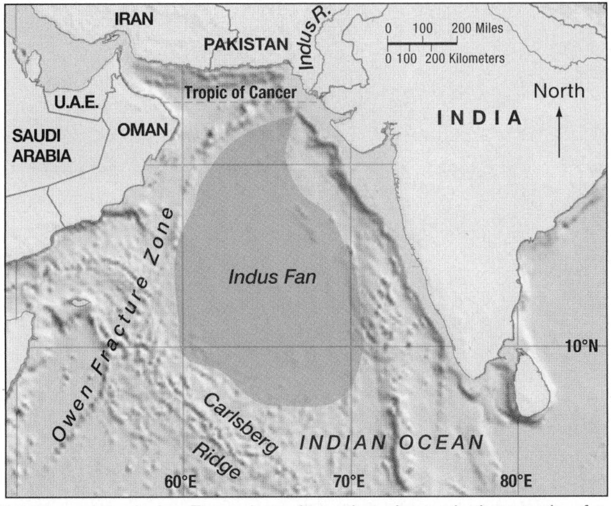


**(a)** Map of the Indus Fan, a large but otherwise typical example of a passive margin fan.

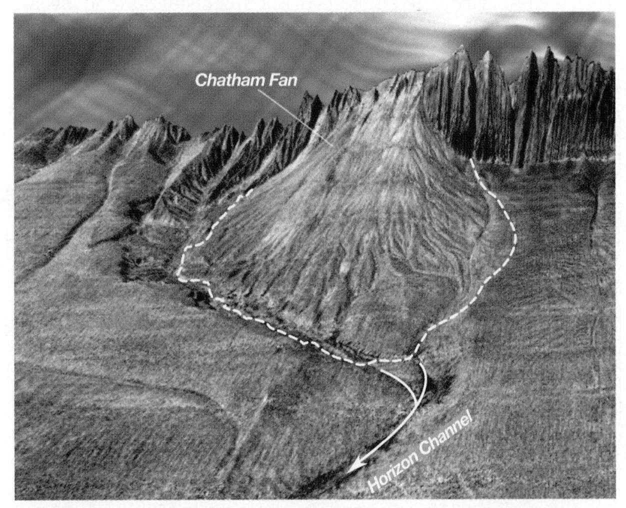


**(b)** Sonar perspective view of southeast Alaska's Chatham Fan, which rises 450 meters (1500 feet) above the surrounding sea floor. Vertical exaggeration is 20 times; view looking northeast.

Figure 3.14 **Examples of deep-sea (submarine) fans.**

**RECAP Turbidity currents are underwater avalanches of muddy water mixed with sediment that move down the continental slope and are responsible for carving submarine canyons.**

### CONCEPT CHECK 3.2 ▸ Describe the sea floor features that exist on continental margins.

1 Describe the major features of a passive continental margin: continental shelf, continental slope, continental rise, submarine canyon, and deep-sea fans.

2 Explain how submarine canyons are created.

3 Explain what graded bedding is and how it forms.

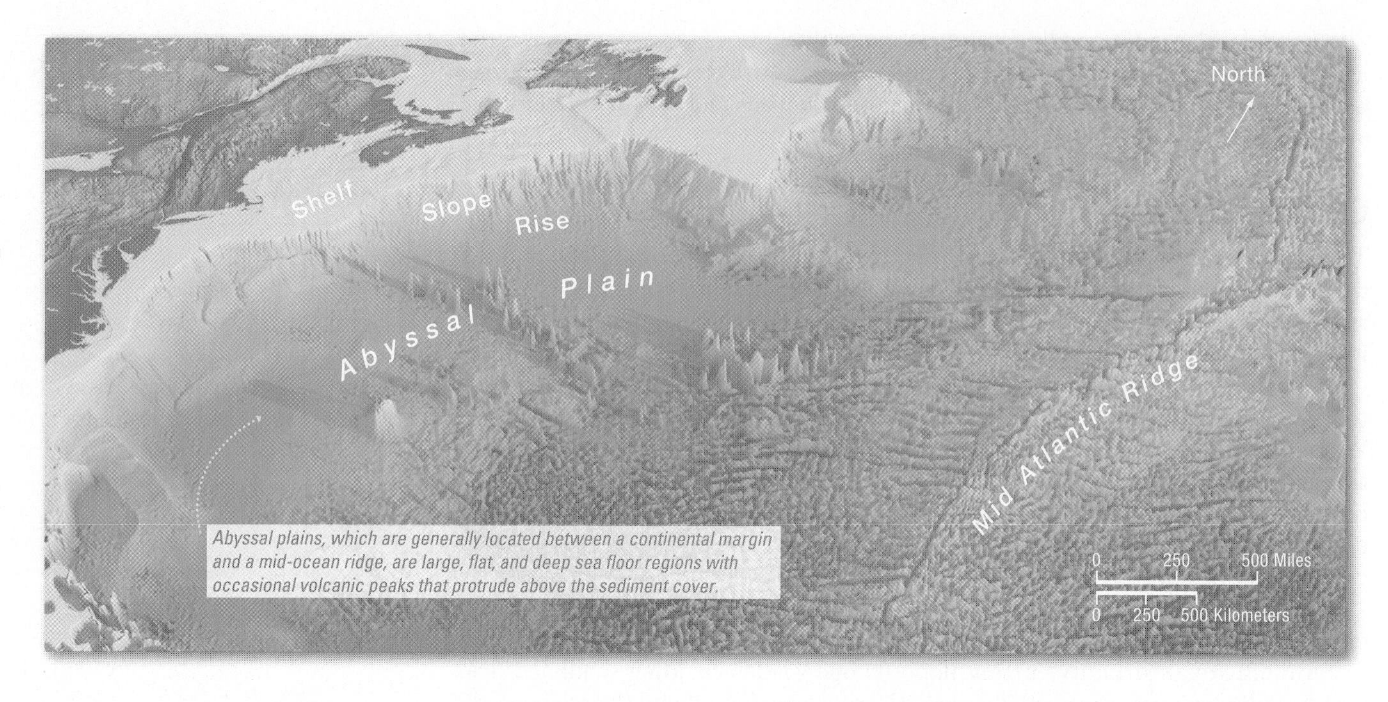


Figure 3.15 **Abyssal plain in the Atlantic Ocean.** False shadow perspective view of the features of the floor of the Atlantic Ocean. Vertical exaggeration is 10 times.

## 3.3 ▸ What Features Exist in the Deep-Ocean Basins?

The deep-ocean floor lies beyond the continental margin province (the shelf, slope, and the rise) and contains a variety of features.

### Abyssal Plains

Extending from the base of the continental rise into the deep-ocean basins are flat depositional surfaces with slopes of less than a fraction of a degree that cover extensive portions of the deep-ocean basins. These **abyssal** (*a* = without, *byssus* = bottom) **plains** average between 4500 meters (15,000 feet) and 6000 meters (20,000 feet) deep. Abyssal plains are large, flat features that cover about one-third of the planet as much as the continents combined. They are not literally bottomless, but they are some of the deepest (and flattest) regions on Earth (Figure 3.15).

Abyssal plains are formed by fine particles of sediment slowly drifting onto the deep-ocean floor. Over millions of years, a thick blanket of sediment is produced by **suspension settling** as fine particles (analogous to "marine dust") accumulate on the ocean floor. With enough time, these deposits cover most irregularities of the deep ocean, as shown in Figure 3.16. In addition, sediment traveling in turbidity currents from land adds to the sediment load.

The type of continental margin determines the distribution of abyssal plains. For instance, few abyssal plains are located in the Pacific Ocean; instead, most occur in the Atlantic and Indian Oceans. The deep-ocean trenches found on the convergent active margins of the Pacific Ocean prevent sediment from moving past the continental slope. In essence, the trenches act like a gutter that traps sediment transported off the land by turbidity currents. On the passive margins of the Atlantic and Indian Oceans, however, turbidity currents travel directly down the continental margin and deposit sediment

Figure 3.16 **Abyssal plain formed by suspension settling.** Paired seismic profile (*above*) and matching drawing (*below*) for a portion of the deep Madeira Abyssal Plain in the eastern Atlantic Ocean, showing how irregular volcanic terrain is buried by sediment.

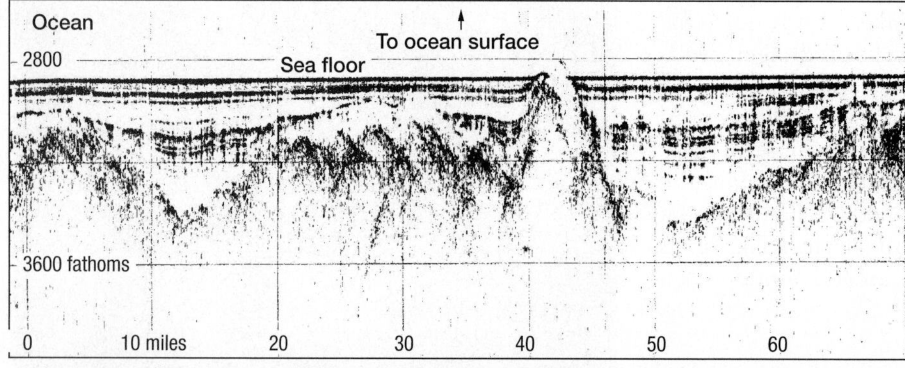


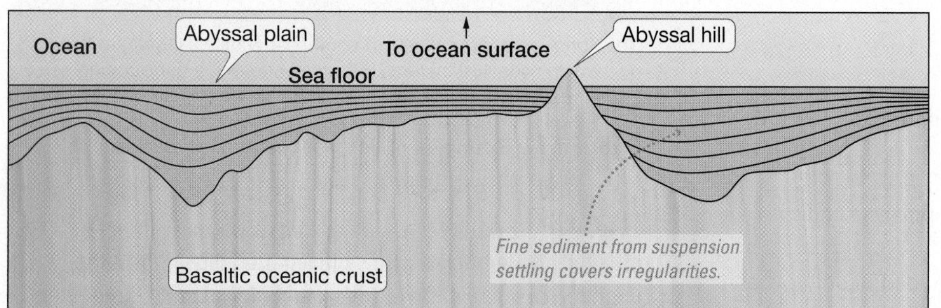

on the abyssal plains. In addition, the distance from the continental margin to the floor of the deep-ocean basins in the Pacific Ocean is so great that most of the suspended sediment settles out before it reaches these distant regions. Conversely, the smaller size of the Atlantic and Indian Oceans does not prevent suspended sediment from reaching their deep-ocean basins.

## Volcanic Peaks of the Abyssal Plains

Poking through the sediment cover of the abyssal plains are a variety of volcanic peaks, which extend to various elevations above the ocean floor. Some extend above sea level to form islands, while others are just below the surface (see Mastering Oceanography Web Diving Deeper 3.2). Those that are below sea level but rise more than 1 kilometer (0.6 mile) above the deep-ocean floor and have a pointy top like an upside-down ice cream cone are called seamounts. Worldwide, scientists estimate that there are at least 125,000 known seamounts, many of which originated at volcanic centers such as hotspots or the mid-ocean ridge. On the other hand, if a volcano has a flattened top, it is called a **tablemount**, or *guyot*. The origin of seamounts and tablemounts is discussed in Chapter 2 (refer to Figure 2.27).

Volcanic features on the ocean floor that are less than 1000 meters (0.6 mile) tall—the minimum height of a seamount—are called **abyssal hills**, or **seaknolls**. Abyssal hills are one of the most abundant features on the planet (several hundred thousand have been identified) and cover a large percentage of the entire ocean basin floor. Many are gently rounded in shape (Figure 3.17), and they have an average height of about 200 meters (650 feet). Most abyssal hills are created by stretching of crust during the creation of new sea floor at the mid-ocean ridge. Interestingly, new research suggests that there is a strong correlation between ice ages and the production of abyssal hills. During ice ages when sea level is lowered, there is less water and therefore less weight overlying the mid-ocean ridge. This reduction in pressure reduces the melting temperature of underground magma, thus makes it easier for rock in the mantle to melt and rise to the surface. The result is an increase in the number of abyssal hills created during ice ages.

In the Atlantic and Indian Oceans, many abyssal hills are found buried beneath abyssal plain sediment. In the Pacific Ocean, the abundance of deep trenches along the margins of the Pacific helps trap land-derived sediment, and so the rate of sediment deposition is lower. Consequently, extensive regions dominated by abyssal hills occur on the Pacific sea floor; these are called **abyssal hill provinces**. Evidence of volcanic activity on the bottom of the Pacific Ocean is particularly widespread. In fact, more than 20,000 volcanic peaks are known to exist on the Pacific sea floor, including the recently-discovered largest single volcano on Earth, Tamu Massif, which is in the northwestern Pacific Ocean and is comparable in size to the largest known volcano in our solar system, Olympus Mons on Mars.

Figure 3.17 **Comparison of an abyssal hill, seamount, and tablemount.** Schematic drawing of the relative size and shape of an abyssal hill (seaknoll), seamount, and tablemount (guyot).

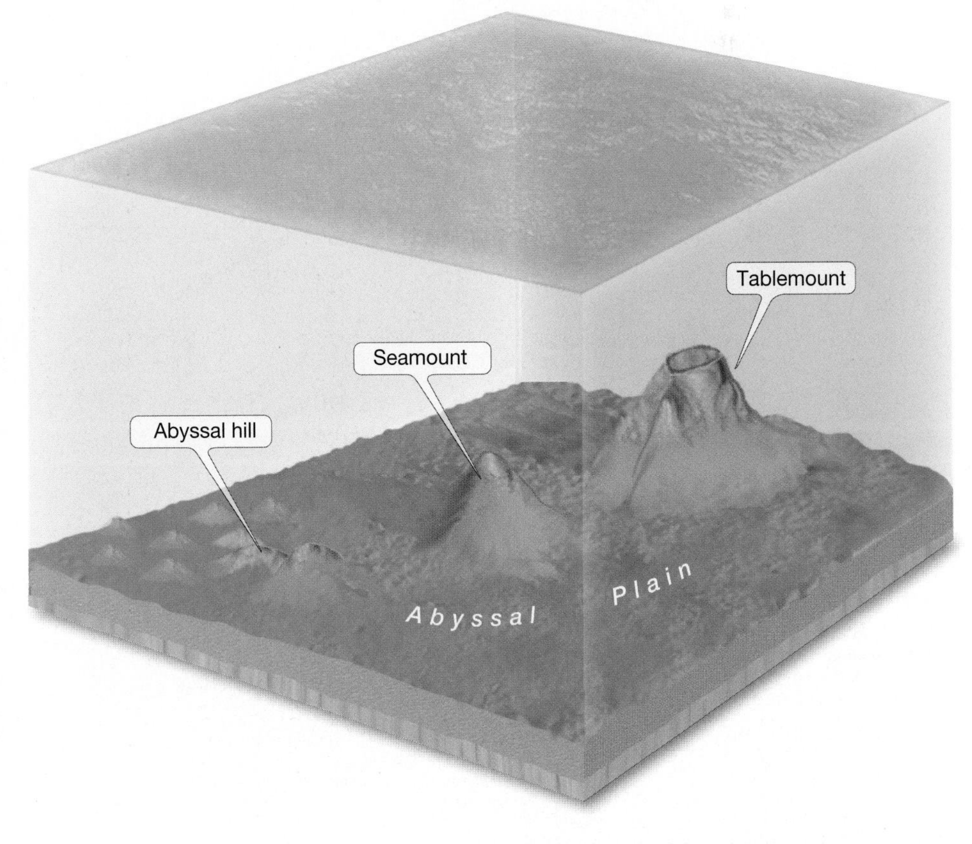


## Ocean Trenches and Volcanic Arcs

Along passive margins, the continental rise commonly occurs at the base of the continental slope and merges smoothly into the abyssal plain. In convergent active margins, however, the slope descends into a long, narrow, steep-sided **ocean trench**. Ocean trenches are deep linear scars in the ocean floor, caused by the collision of two plates along convergent plate margins (as discussed in Chapter 2). The landward side of the trench rises as a **volcanic arc** that may produce islands (such as the islands of Japan, an **island arc**) or a volcanic mountain range along the margin of a continent (such as the Andes Mountains, a **continental arc**).

The deepest portions of the world's oceans are found in these trenches. In fact, the deepest point on Earth's surface—11,022 meters (36,161 feet)—is found in the Challenger Deep area of the Mariana Trench. The majority of ocean trenches are found along the margins of the Pacific Ocean (**Figure 3.18**), and only a few exist in the Atlantic and Indian Oceans.

**THE PACIFIC RING OF FIRE** The **Pacific Ring of Fire** occurs along the margins of the Pacific Ocean. It is home to the majority of Earth's active volcanoes and large earthquakes because of the prevalence of convergent plate boundaries along the Pacific Rim. A part of the Pacific Ring of Fire is South America's western coast, including the Andes Mountains and the associated Peru–Chile Trench. **Figure 3.19** shows a cross-sectional view of this area and illustrates the tremendous amount of relief at convergent plate boundaries where deep-ocean trenches are associated with tall volcanic arcs.

**RECAP** **Deep-ocean trenches and volcanic arcs result from the collision of two plates at convergent plate boundaries; these features occur mostly along the margins of the Pacific Ocean (the Pacific Ring of Fire).**

**EXPLORING DATA** ▼ **Calculate the average depth, width, and length of Pacific Ocean trenches. How do those values compare to the average dimensions of trenches in the Atlantic and Indian Oceans?**

**Selected Pacific Ocean Trenches**

| Name | Depth (km) | Width (km) | Length (km) |
|---|---|---|---|
| Middle America | 6.7 | 40 | 2800 |
| Aleutian | 7.7 | 50 | 3700 |
| Peru–Chile | 8.0 | 100 | 5900 |
| Kermadec–Tonga | 10.0 | 50 | 2900 |
| Kuril | 10.5 | 120 | 2200 |
| Mariana | 11.0 | 70 | 2550 |

**Atlantic Ocean Trenches**

| Name | Depth (km) | Width (km) | Length (km) |
|---|---|---|---|
| South Sandwich | 8.4 | 90 | 1450 |
| Puerto Rico | 8.4 | 120 | 1550 |

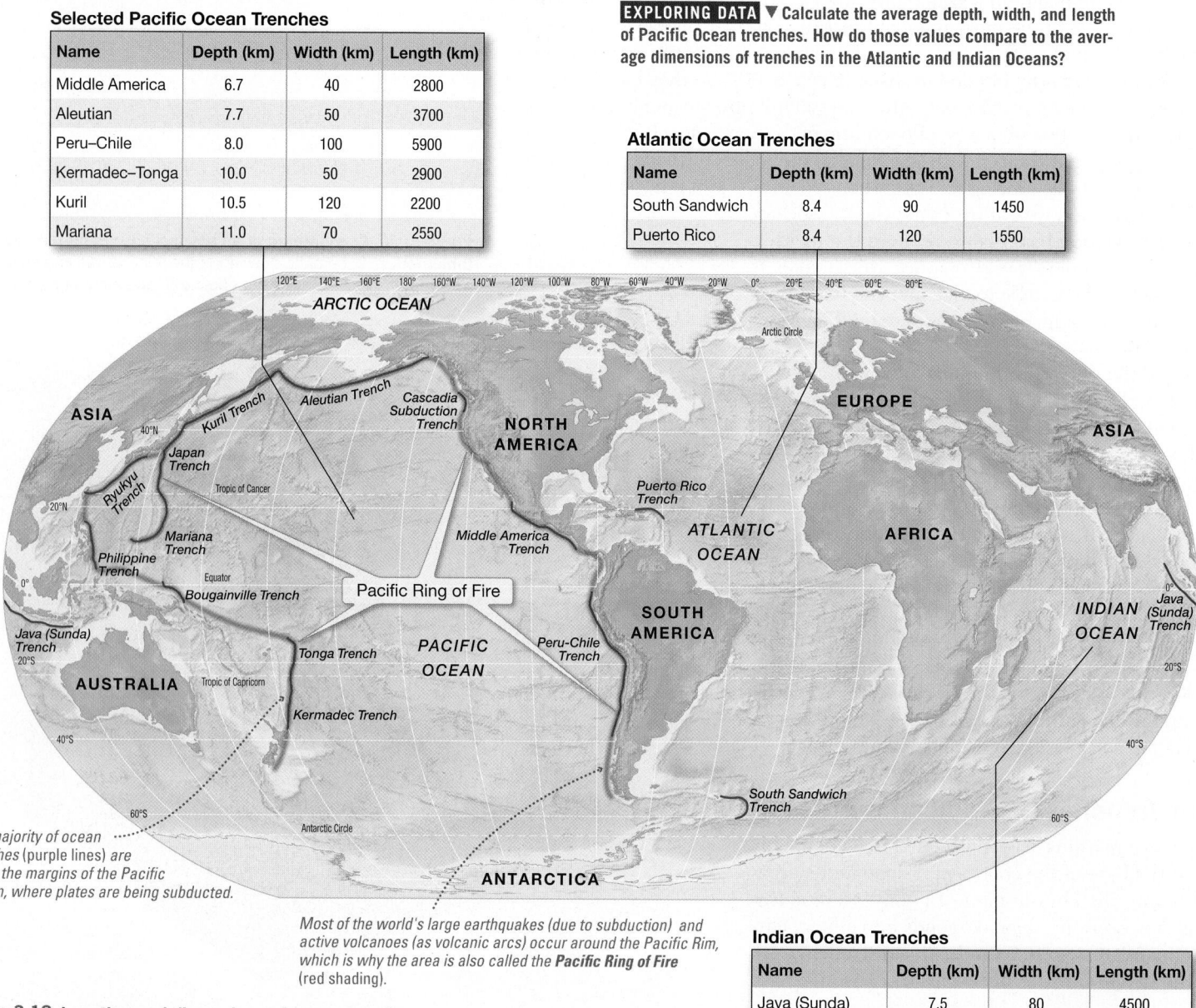

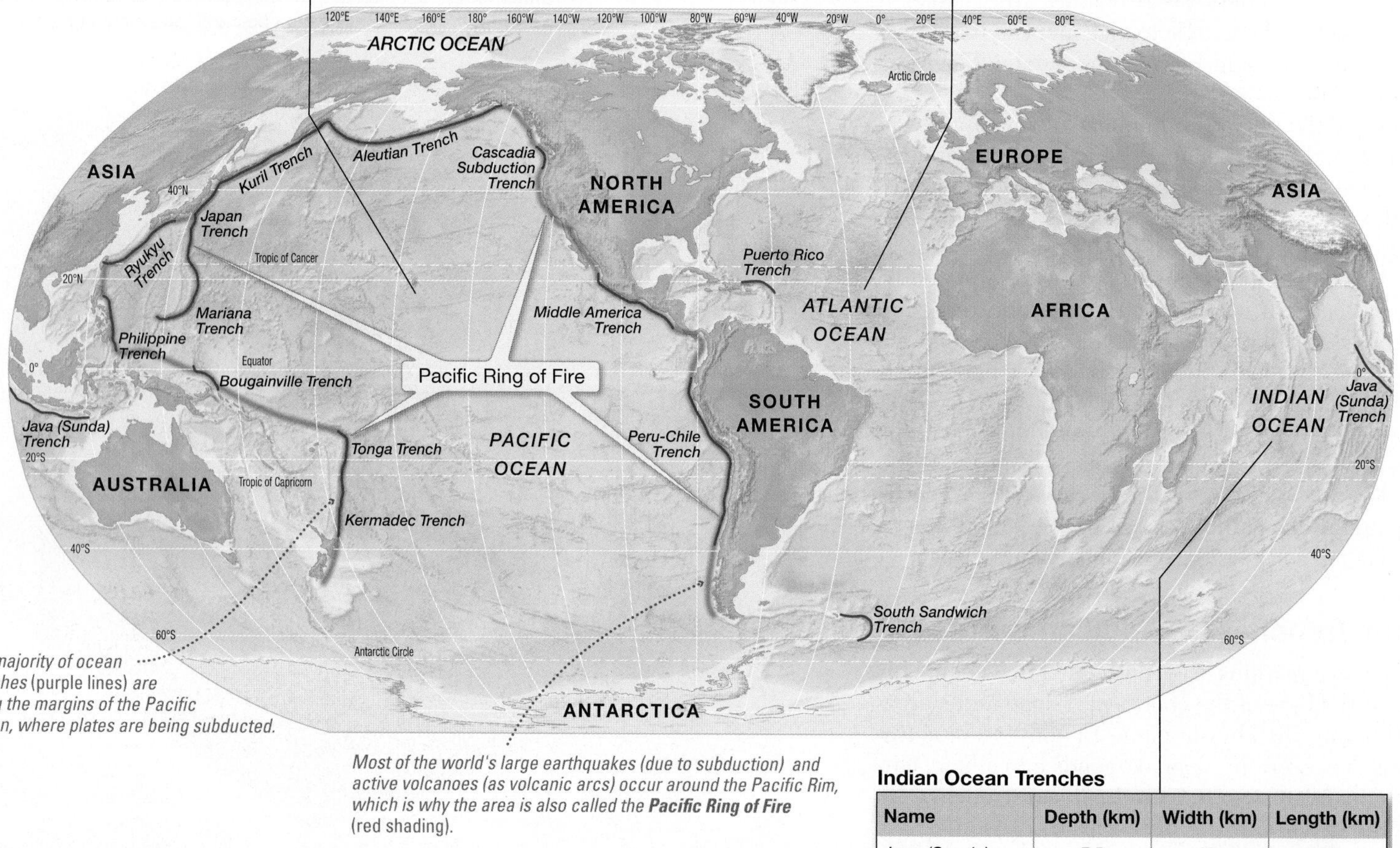

**Indian Ocean Trenches**

| Name | Depth (km) | Width (km) | Length (km) |
|---|---|---|---|
| Java (Sunda) | 7.5 | 80 | 4500 |

*The majority of ocean trenches (purple lines) are along the margins of the Pacific Ocean, where plates are being subducted.*

*Most of the world's large earthquakes (due to subduction) and active volcanoes (as volcanic arcs) occur around the Pacific Rim, which is why the area is also called the **Pacific Ring of Fire** (red shading).*

Figure 3.18 **Location and dimensions of ocean trenches.**

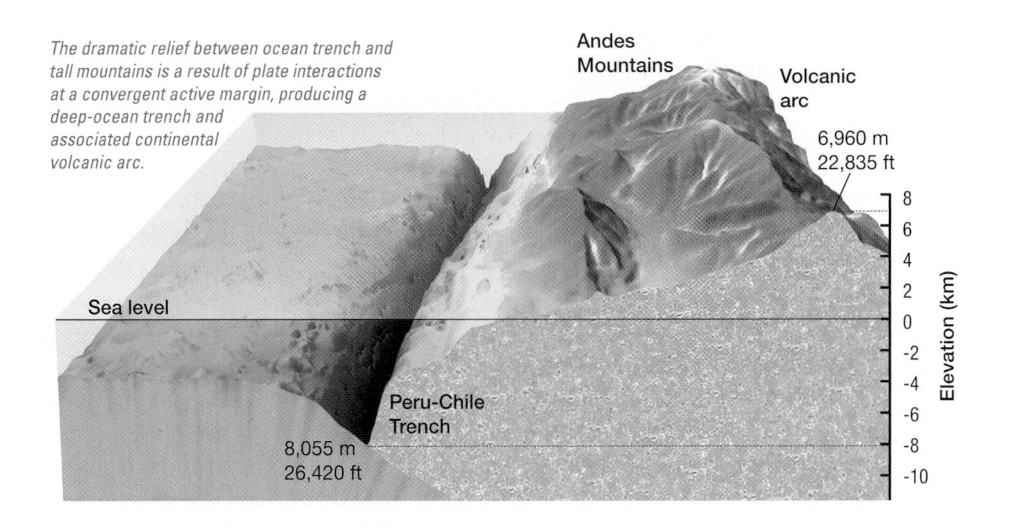


Figure 3.19 **Perspective view of the Peru–Chile Trench and the Andes Mountains.** Over a distance of 200 kilometers (125 miles), there is a dramatic change in elevation of more than 14,900 meters (49,000 feet) from the deep Peru–Chile Trench to the high Andes Mountains. Vertical scale is exaggerated about 15 times.

**CONCEPT CHECK 3.3 ▸ Describe the sea floor features that exist in the deep-ocean basins.**

1 Describe the process by which abyssal plains are created.

2 Discuss the origin of the various volcanic peaks of the abyssal plains: seamounts, tablemounts, and abyssal hills.

3 What are some differences between a submarine canyon and an ocean trench?

CREATURE FEATURE 3.1

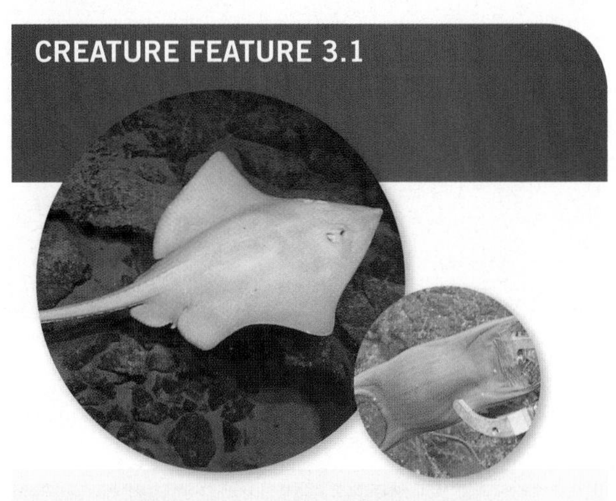

## My babies need the heat!

The **Pacific white skate** (*Bathyraja spinosissima*) is a relative of sharks that inhabits deep water. When mature, a skate deposits one egg (*inset*) in a leathery pouch and leaves it to hatch up to four years later.

Because it takes longer for eggs to hatch in cold water, scientists speculate that Pacific white skates lay their eggs near hydrothermal vents in order to use the natural heat to speed up egg incubation time (see Mastering Oceanography Web Diving Deeper 3.4).

# 3.4 ▸ What Features Exist along the Mid-Ocean Ridge?

The global mid-ocean ridge is a continuous, fractured-looking mountain range that extends through all the world's ocean basins. The portion of the mid-ocean ridge found in the North Atlantic Ocean is called the Mid-Atlantic Ridge (**Figure 3.20**), and it dwarfs all mountain ranges on land. As discussed in Chapter 2, the mid-ocean ridge results from sea floor spreading along divergent plate boundaries (see Figure 2.15 in Chapter 2). The enormous mid-ocean ridge forms Earth's longest mountain chain, extending across some 75,000 kilometers (46,600 miles) of the deep-ocean basin. The width of the mid-ocean ridge averages about 1000 kilometers (620 miles). The mid-ocean ridge is a topographically high feature, extending an average of 2.5 kilometers (1.5 miles) above the surrounding sea floor. The mid-ocean ridge contains only a few scattered islands, such as Iceland and the Azores, where it is exposed above sea level. Remarkably, the mid-ocean ridge covers 23% of Earth's surface.

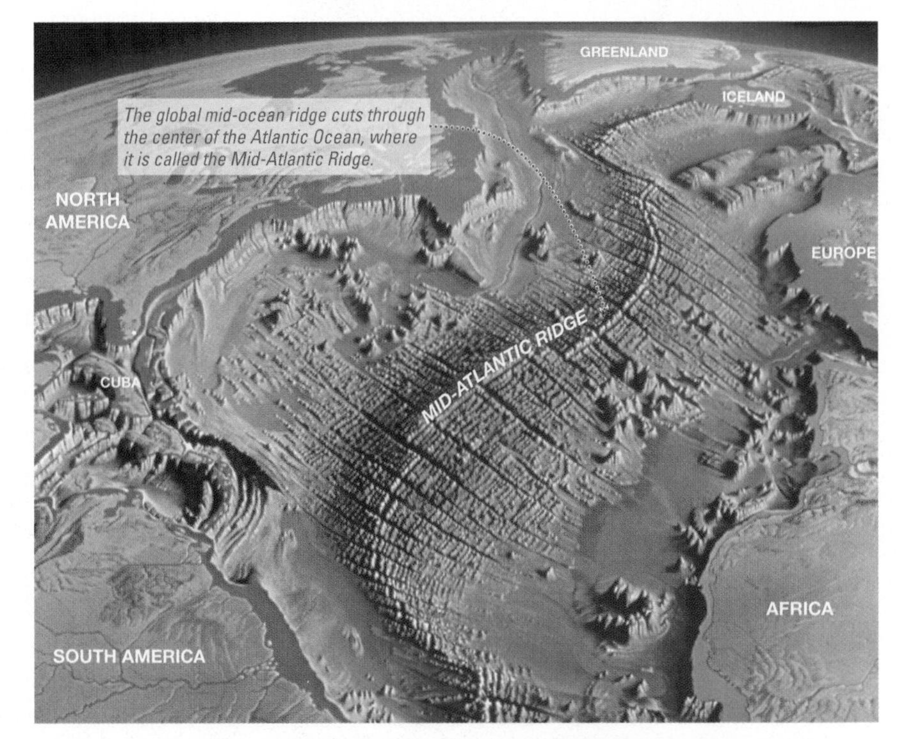


Figure 3.20 **The rugged floor of the North Atlantic Ocean.** This perspective view of the North Atlantic Ocean reveals what the sea floor would look like if the oceans were drained away. Vertical exaggeration is about 20 times.

The mid-ocean ridge is entirely volcanic and is composed of basaltic lavas characteristic of the oceanic crust. Along most of its crest is a central downdropped **rift valley** created by sea floor spreading (rifting) where two plates diverge (see, for example, Figures 2.14 and 2.15). Along the Mid-Atlantic Ridge, for example, is a central rift valley that is as much as 30 kilometers (20 miles) wide and 3 kilometers (2 miles) deep. Here, molten rock presses upward toward the sea floor, setting off earthquakes, creating jets of superheated seawater, and eventually solidifying to form new oceanic crust. Cracks called *fissures* (*fissus* = split) and faults are commonly observed in

**(a)** Perspective view based on sonar mapping of a portion of the East Pacific Rise (*center*) showing volcanic seamount (*left*). Colors represent sea floor elevation; the depth, in meters, is indicated by the color scale along the left margin. Vertical exaggeration is six times.

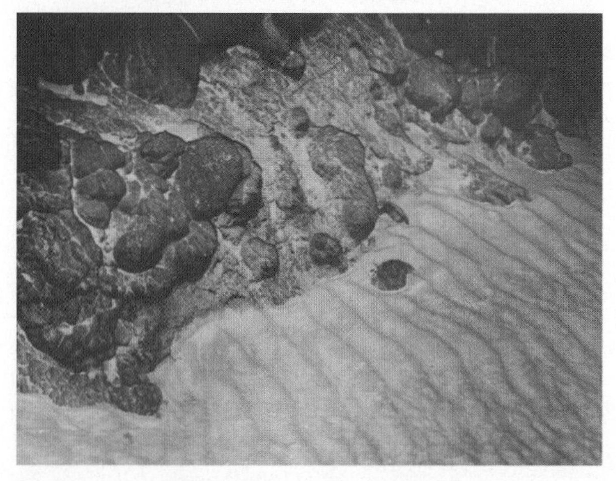

**(b)** Recently formed pillow lava along the East Pacific Rise. Photo shows an area of the sea floor about 3 meters (10 feet) across that also displays ripple marks from deep-ocean currents.

**(c)** Pillow lava that was once on the sea floor but has since been uplifted onto land at Port San Luis, California. Maximum width of a pillow in this photo is 1 meter (3.3 feet).

Figure 3.21 **Mid-ocean ridge volcanoes and pillow lava.**

the central rift valley. Swarms of small earthquakes occur along the central rift valley, caused by the injection of magma into the sea floor or rifting along faults.

Segments of the mid-ocean ridge called **oceanic ridges** have a prominent rift valley and steep, rugged slopes, and **oceanic rises** have slopes that are gentler and less rugged. As explained in Chapter 2, the differences in overall shape are caused by the fact that oceanic ridges (such as the Mid-Atlantic Ridge) spread more slowly than oceanic rises (such as the East Pacific Rise) (see SmartFigure 2.19 in Chapter 2).

## Volcanic Features

Volcanic features associated with the mid-ocean ridge include recent underwater lava flows and tall volcanoes called *seamounts* (Figure 3.21a). In a number of cases, researchers have discovered seamounts that initially formed along the crest of the mid-ocean ridge and have split in two as the plates have spread apart. The most typical underwater volcanic feature associated mid-ocean ridges, however, are **pillow lavas** or **pillow basalts**, which are smooth, rounded lobes of rock that resemble a stack of bed pillows (Figures 3.21b and 3.21c). Pillow basalts form when hot basaltic lava spills onto the sea floor and is exposed to cold seawater. The cold water quickly chills and solidifies the margins of the lava, permitting the viscous flow to assume a rounded shape.

Although most people are not aware of it, frequent volcanic activity is common along the mid-ocean ridge. In fact, 85% of Earth's volcanic activity takes place on the sea floor, and every year about 12 cubic kilometers (3 cubic miles) of molten rock erupts underwater. The amount of erupted lava along the mid-ocean ridge is large enough to fill an Olympic-sized swimming pool every three seconds! Bathymetric studies along the Juan de Fuca Ridge off Washington and Oregon, for example, revealed that 50 million cubic meters (1800 million cubic feet) of new lava had erupted between 1981 and 1987. Subsequent surveys of the area indicated many changes along the mid-ocean ridge, including new volcanic features, recent lava flows, and depth changes of up to 37 meters (121 feet). Interest in the continuing volcanic activity along the Juan de Fuca Ridge has led to the development of a permanent sea floor observation system there (see Mastering Oceanography Web Diving Deeper 2.1). Other parts of the mid-ocean ridge, such as East Pacific Rise, also experience frequent volcanic activity (**Process of Science 3.1**).

## Hydrothermal Vents

Other features in the central rift valley include **hydrothermal vents** (*hydro* = water, *thermo* = heat). Hydrothermal vents are sea floor hot springs created when cold seawater seeps down along cracks and fractures in the ocean crust and approaches an underground magma chamber (Figure 3.22). The water picks up heat and dissolved substances and then works its way back toward the surface through a complex plumbing system, exiting through the sea floor. The temperature of the water that rushes out of a particular hydrothermal vent determines its appearance:

- **Warm-water vents** have water temperatures below 30°C (86°F) and generally emit water that is clear in color.
- **White smokers** have water temperatures from 30° to 350°C (86° to 662°F) and emit water that is white because of the presence of various light-colored compounds, including barium sulfide.
- **Black smokers** have water temperatures above 350°C (662°F) and emit water that is black because of the presence of dark-colored **metal sulfides**, including iron, nickel, copper, and zinc.

Many black smokers spew out of chimney-like structures (Figure 3.22b) that can be up to 60 meters (200 feet) high and were named for their resemblance to factory smokestacks belching clouds of smoke. The dissolved metal particles often come out of solution, or **precipitate** (become a solid), when the hot water mixes with cold seawater. This creates coatings of solid mineral deposits on nearby rocks. Chemical analyses of these deposits reveal that they are composed of various metal sulfides and sometimes even important economic deposits such as silver and gold.

## EARTH'S HYPSOGRAPHIC CURVE: NEARLY EVERYTHING YOU NEED TO KNOW ABOUT EARTH'S OCEANS AND LANDMASSES IN ONE GRAPH

Earth's **hypsographic curve** (*hypos* = height, *graphic* = drawn) (**Figure 3A**), which shows the relationship between the height of the land and the depth of the oceans, is useful for illustrating many things. For example, the bar graph (Figure 3A, *left*) gives the percentage of Earth's surface area at various ranges of elevation and depth. The cumulative curve (Figure 3A, *right*) gives the percentage of surface area from the highest peaks to the deepest depths of the oceans. Together, they show that 70.8% of Earth's surface is covered by oceans, and that the average depth of the ocean is 3729 meters (12,234 feet), while the average height of the land is only 840 meters (2756 feet). This difference can be explained by recalling our discussion of isostasy from Chapter 1, and remembering that the less dense continents float higher in the mantle than the denser oceanic crust, over which the oceans lie (see Figure 1.22 in Chapter 1).

Earth's cumulative hypsographic curve (Figure 3A, *right*) shows five differently sloped segments. On land, the first steep segment of the curve represents tall mountains, while the gentle slope represents low coastal plains (and continues just offshore, representing the shallow parts of the continental margin). The first slope below sea level represents steep areas of the continental margins, and also includes the mountainous mid-ocean ridge. Further offshore, the longest, flattest part of the whole curve represents the deep-ocean basins, followed by the last steep part, which represents ocean trenches.

Interestingly, the shape of Earth's hypsographic curve is evidence of the existence of plate tectonics on Earth. Specifically, the two flat areas and three sloped areas of the curve show that there is a very uneven distribution of area at different depths and elevations. If there were no active mechanism involved in creating such features on Earth, the bar graph portions would all be about the same length, and the cumulative curve would be a straight line. Instead, the variations in the curve suggest that plate tectonics is responsible for producing very dramatic differences in elevation across Earth's surface. For example, the slopes of the curve represent mountains, continental slopes, the mid-ocean ridge, and deep-ocean trenches, all of which are created by plate tectonic processes. Interestingly, analysis of hypsographic curves constructed for other planets and moons using satellite data have been used to determine if plate tectonic processes are actively modifying the surfaces of those worlds or have modified them in the geologic past.

### WHAT DID YOU LEARN?

1. Explain how the shape of Earth's hypsographic curve helps support the existence of plate tectonics on Earth.
2. For a planet that had no plate tectonics, what would its hypsographic curve look like?

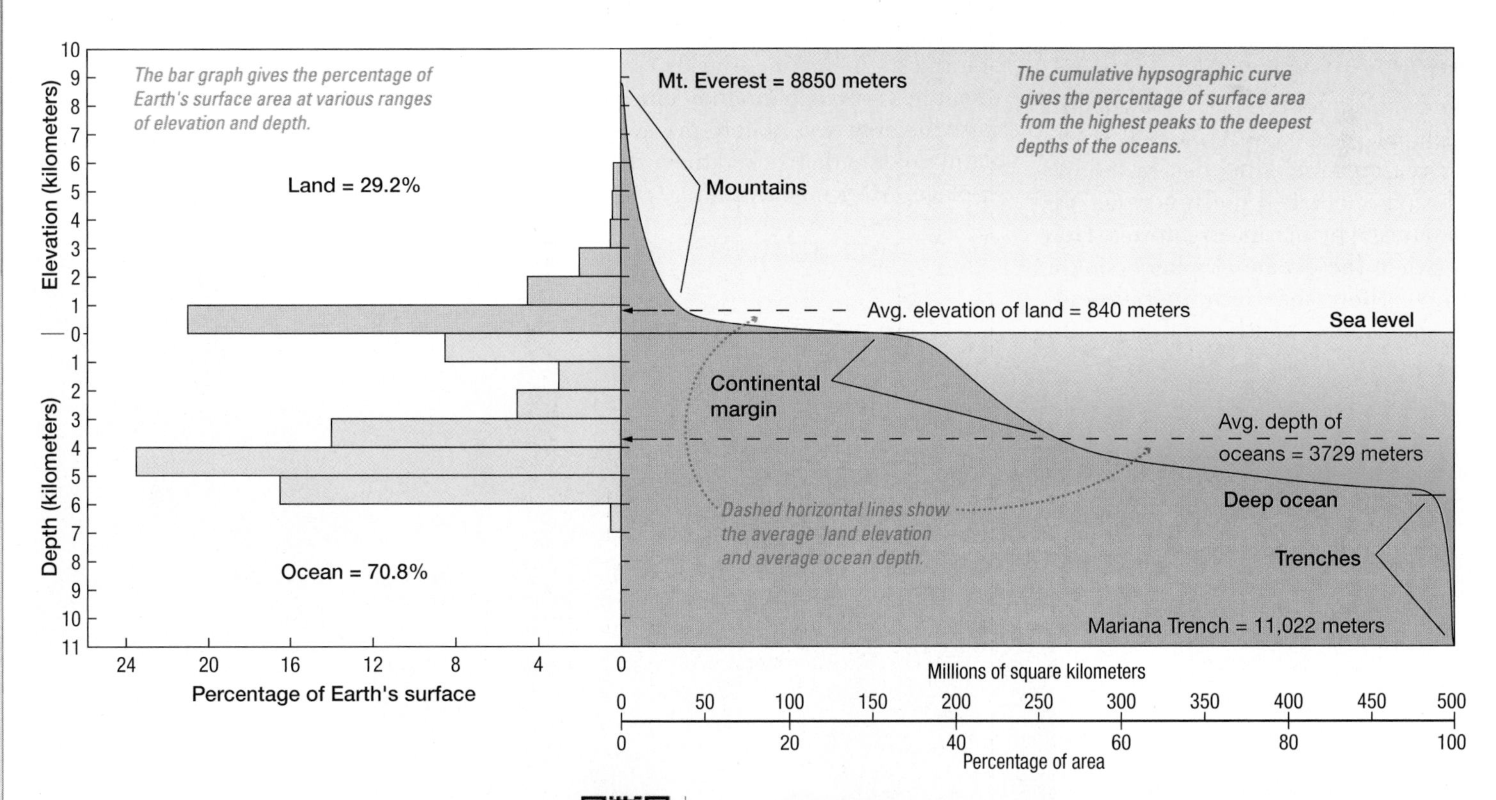

SmartFigure 3A **Earth's hypsographic curve.**
https://goo.gl/NBttvg

**EXPLORING DATA ▲ Compare the average land elevation (height above sea level) versus the average ocean depth (below sea level). Why are the values so dissimilar?**

## PROCESS OF SCIENCE 3.1 Now You See It, Now You Don't: The Unusual Case of the Missing Sea Floor Seismic Equipment

### BACKGROUND

The story begins in 2005, when scientists deployed 12 ocean-bottom seismometers (OBSs) over a few square kilometers of sea floor along an unusually active portion of the East Pacific Rise north of the Galápagos Islands. The OBSs—each about the size and weight of a small refrigerator—were designed to stay on the sea floor for up to a year and record earthquake data that accompanies submarine volcanic events. Researchers returned a year later, thinking that they would simply recover the instruments and send down others. When the research vessel sent a sonar signal to the OBSs to release their weights and use their floats to return to the surface, only four came bobbing up. What happened to the other eight?

### FORMING A HYPOTHESIS

The ocean is a hostile place for scientific equipment, which can be lost at sea for many reasons, including design failures, implosion, electrical malfunctions, and even damage by marine organisms. However, when the oceanographers couldn't retrieve their sea floor instruments, they entertained another, more exciting hypothesis: that the lost instruments had been at least partially buried by a recent underwater lava flow. If this were the case and if they could recover the stuck equipment in water 2.5 kilometers (1.6 miles) deep then they would be the first to have recorded a sea floor volcanic eruption *as it happened*.

### DEVISING AN EXPERIMENT

It took two months, but the scientists returned with a camera-equipped sled that they towed behind their ship to survey the sea floor. They located three OBSs and confirmed that they were indeed embedded in recent lava; the other five OBSs seemed to have vanished and scientists suspected that they were completely buried by lava. Although they tried to nudge and pry the three OBSs loose with the sled, they were thoroughly stuck. Hoping that the stuck OBSs had recorded data while riding out an active sea floor lava flow, the scientists had to wait until a year later, when the tethered robotic vehicle *Jason* was sent down to try to free the instruments. Using *Jason*'s video camera and its mechanical arms controlled remotely from a command center on the ship, the crew was able to pry away large chunks of lava that locked the instruments in place. After much yanking, two of the OBSs finally broke free and rose to the surface, with help from attached floats. Although the researchers attempted to free the third OBS, it was never recovered because it was stuck too tightly in new lava.

### INTERPRETING THE RESULTS

The recovered OBS instruments—both badly scorched from the hot lava (**Figure 3B**)—provided usable data that have given researchers new information about the volcanic processes that occur at the mid-ocean ridge. This and other evidence suggests that the fresh lava had erupted for six hours straight, heating and darkening the water above it and spreading along the ridge for more than 16 kilometers (10 miles). The researchers consider themselves lucky to have fortuitously caught Earth's crust in the very act of ripping itself apart, documenting swarms of undersea earthquakes and culminating in a volcanic eruption that buried their instruments in lava.

### THINKING LIKE A SCIENTIST: WHAT'S NEXT?

Other than instruments to gather earthquake data, what other kinds of instruments would you deploy on a sea floor instrument package to help collect data about an underwater volcanic eruption?

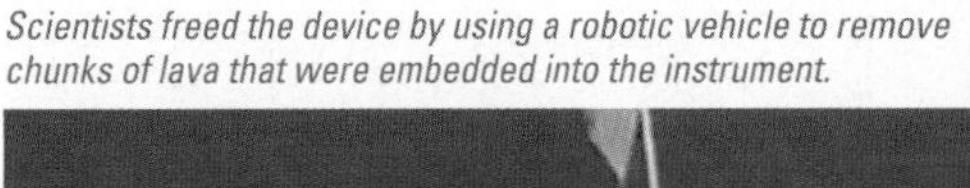

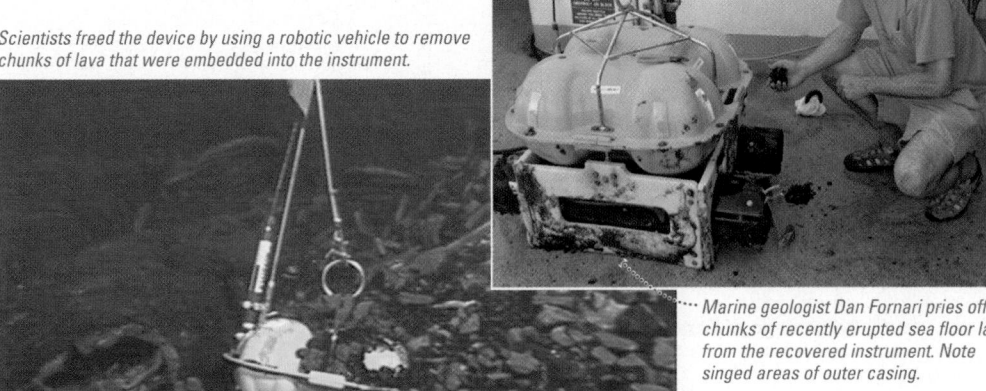

*Scientists freed the device by using a robotic vehicle to remove chunks of lava that were embedded into the instrument.*

*A 2006 sea floor eruption along the East Pacific Rise trapped this and several other OBS instruments in lava. The yellow plastic covering protects glass ball floats that are normally used to raise the instrument to the surface; cable is attached to additional floats above.*

*Marine geologist Dan Fornari pries off chunks of recently erupted sea floor lava from the recovered instrument. Note singed areas of outer casing.*

**Figure 3B An ocean-bottom seismometer (OBS) stuck in lava.**

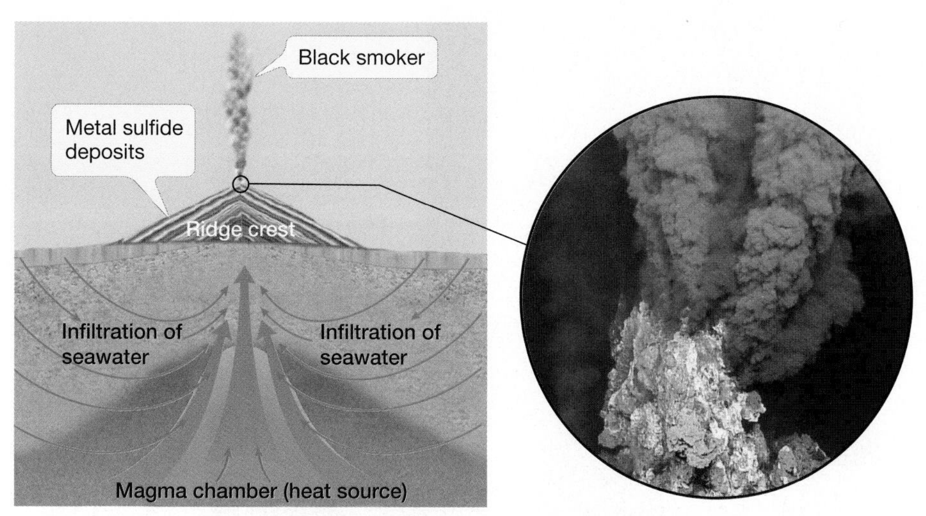


**(a)** Diagram showing hydrothermal circulation along the mid-ocean ridge and the creation of black smokers; photo (inset) showing a close-up view of a black smoker along the East Pacific Rise.

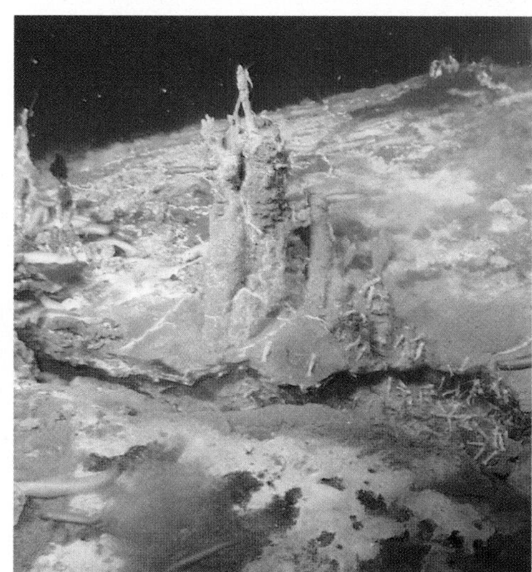

**(b)** Black smoker chimney and fissure at Susu north active site, Manus Basin, western Pacific Ocean. Chimney is about 3 meters (10 feet) tall.

Figure 3.22 **Hydrothermal vents.**

In addition, most hydrothermal vents foster unusual deep-ocean ecosystems that include organisms such as giant tubeworms, large clams, beds of mussels, and many other creatures—most of which were new to science when they were first encountered. These organisms are able to survive in the absence of sunlight because the vents discharge hydrogen sulfide gas, which is metabolized by bacteria and archaeons, microscopic bacteria-like organisms that are a newly discovered domain of life. These microorganisms in turn provide a food source for other organisms in the community. Recent studies of active hydrothermal vent fields indicate that vents have short life spans of only a few years to several decades, which has important implications for the organisms that depend on hydrothermal vents. These interesting biocommunities are discussed in Chapter 15, "Animals of the Benthic Environment."

## Fracture Zones and Transform Faults

The mid-ocean ridge is cut by a number of **transform faults**, which offset spreading zones. Oriented perpendicular to the spreading zones, transform faults give the mid-ocean ridge the zigzag appearance shown in Figure 3.20. As described in Chapter 2,transform faults occur for two reasons: first, to accommodate spreading of a linear ridge system on a spherical Earth and, second, because different segments of the mid-ocean ridge spread apart at different rates.

On the Pacific Ocean sea floor, where scars are less rapidly covered by sediment than in other ocean basins, transform faults are prominently displayed (**Figure 3.23**). Here, they extend for thousands of kilometers away from the mid-ocean ridge and have widths of up to 200 kilometers (120 miles). These extensions, however, are not transform faults. Instead, they are **fracture zones**.

What is the difference between a transform fault and a fracture zone? **Figure 3.24** shows that both run along the same long linear zone of weakness in Earth's crust. In fact, by following the same zone of weakness from one end to the other, it changes from a fracture zone to a transform fault and back again to a fracture zone. A transform fault is a seismically active area that offsets the axis of a mid-ocean ridge. A fracture zone, on the other hand, is a seismically *inactive* area that shows evidence of past transform fault activity. A helpful way to visualize the difference is that transform faults occur *between* offset segments of the mid-ocean ridge, while fracture zones occur *beyond* the offset segments of the mid-ocean ridge.

### STUDENTS SOMETIMES ASK . . .

***What effect does all this volcanic activity along the mid-ocean ridge have at the ocean's surface?***

Sometimes an underwater volcanic eruption is large enough to create what is called a *megaplume* of warm, mineral-rich water that is lower in density than the surrounding seawater and thus rises to the surface. Remarkably, a few research vessels have reported experiencing the effects of a megaplume at the surface while directly above an erupting sea floor volcano! Researchers on board describe bubbles of gas and steam at the surface, a marked increase in water temperature, and the presence of enough volcanic material to turn the water cloudy. In other cases, fishers have reported seeing bobbing, steaming hot "balloons" of lava floating at the water's surface, some the size of refrigerators, which then cool and sink. In terms of warming the ocean, the heat released into the ocean at mid-ocean ridges is probably not very significant, mostly because the ocean is so good at absorbing and redistributing heat.

**Video**
Black Smoker Venting Fluid
https://goo.gl/FK88di

**RECAP The mid-ocean ridge is created by plate divergence and typically includes a central rift valley, faults and fissures, seamounts, pillow basalts, hydrothermal vents, and metal sulfide deposits.**

## STUDENTS SOMETIMES ASK . . .

### *Has anyone seen pillow lava forming?*

Amazingly, yes! In the 1960s, an underwater film crew ventured to Hawaii during an eruption of the volcano Kilauea, where lava spilled into the sea. They braved high water temperatures and risked being burned on the red-hot lava but filmed some incredible footage. Underwater, the formation of pillow lava occurs where hollow channels in the rock (known as *lava tubes*) deliver molten lava directly into the ocean. When hot lava comes into contact with cold seawater, it forms the characteristic smooth and rounded margins of pillow basalt. The divers also cracked open the thin crust on newly formed pillows with a hammer and were able to initiate new lava outpourings. And today, several commercial dive outfitters in Hawaii offer underwater trips to see lava flowing into the sea.

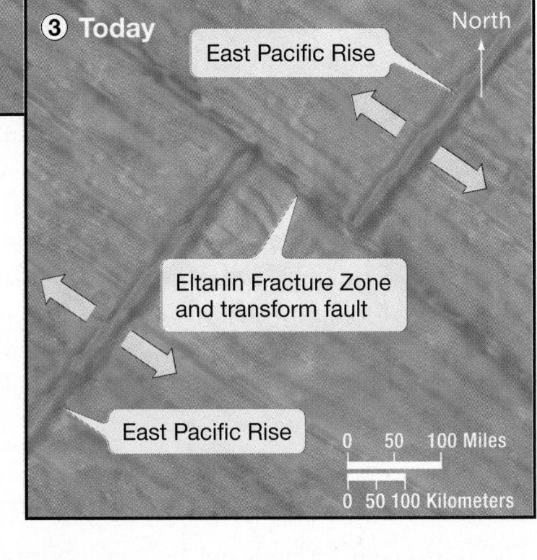


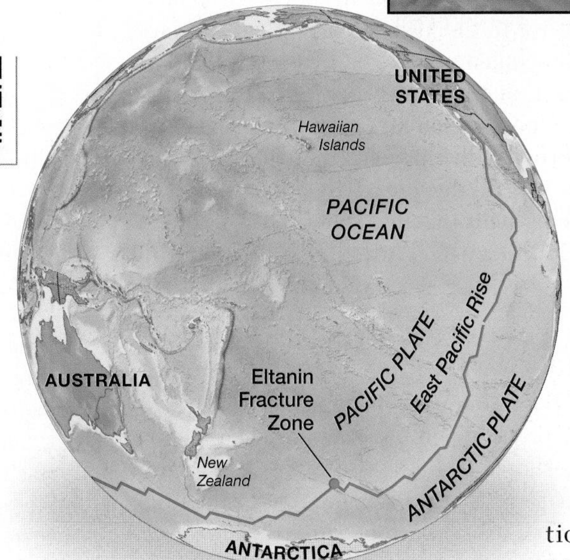


**SmartFigure 3.23 The Eltanin Fracture Zone.** Enlargement of the Eltanin Fracture Zone in the South Pacific Ocean, showing its relationship to the East Pacific Rise and how it has developed through time. Note that the Eltanin Fracture Zone is actually both a fracture zone and a transform fault; the name was given to it before the modern-day understanding of plate tectonic processes.
https://goo.gl/NhqwUN

The relative direction of plate motion across transform faults and fracture zones further differentiates these two features. Across a transform fault, two lithospheric plates are moving in opposite directions. Across a fracture zone (which occurs entirely within a plate), there is no relative motion because the parts of the lithospheric plate cut by a fracture zone are moving in the same direction (Figure 3.24). Transform faults are actual plate boundaries, whereas fracture zones are not. Rather, fracture zones are ancient, inactive fault scars embedded within a plate.

Earthquake activity is also quite different along transform faults, as compared to fracture zones. Along transform faults, two plates are moving in opposite directions, which results in an abundance of earthquakes that are quite shallow (less than 10 kilometers [6 miles] deep). On the other hand, fracture zones separate a *single* plate moving in the same direction; as a result, fracture zones have relatively little seismic activity. **Table 3.1** summarizes the differences between transform faults and fracture zones.

## STUDENTS SOMETIMES ASK . . .

### *If black smokers are so hot, why isn't there steam coming out of them instead of hot water?*

Indeed, black smokers emit water that can be up to four times the boiling point of water at the ocean's surface and hot enough to melt lead. However, at the depth where black smokers are found, the pressure is much higher than at the surface. At these higher pressures, water has a much higher boiling point. Thus, water from hydrothermal vents remains in a liquid state instead of turning into water vapor (steam).

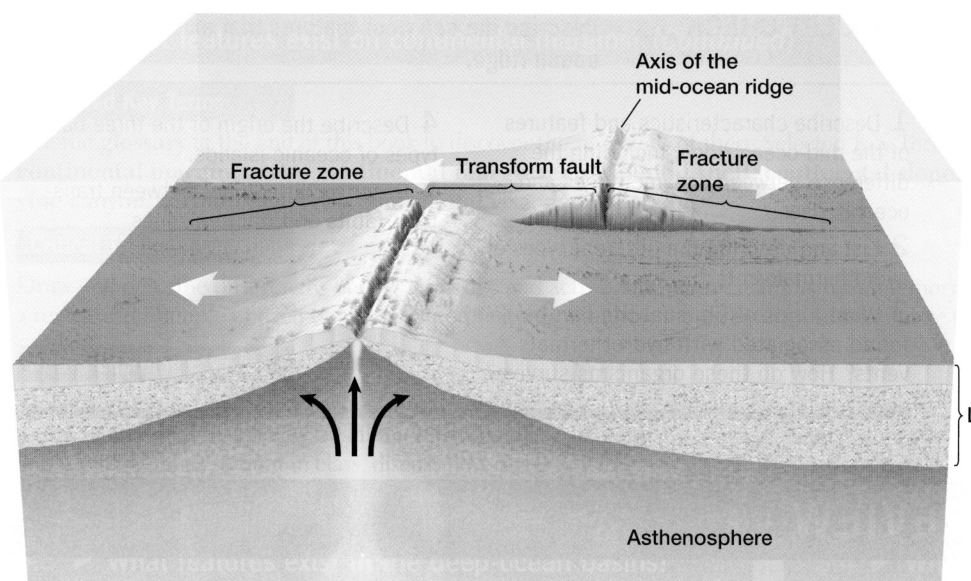


SmartFigure 3.24 **Transform faults and fracture zones.** Transform faults are active transform plate boundaries that occur *between* the segments of the mid-ocean ridge. Fracture zones are inactive intraplate features that occur *beyond* the segments of the mid-ocean ridge. https://goo.gl/1EJnEg

Animation
Transform Faults
http://goo.gl/OOR2Xw

## Oceanic Islands

Some of the most interesting features of ocean basins are islands, which are unusually tall features that reach from the sea floor all the way above sea level. There are three basic types of oceanic islands: (1) islands associated with volcanic activity along the mid-ocean ridge (such as Ascension Island along the Mid-Atlantic Ridge); (2) islands associated with hotspots (such as the Hawaiian Islands in the Pacific Ocean); and (3) islands that are island arcs and associated with convergent plate boundaries (such as the Japanese archipelago in the Pacific Ocean; Figure 3.25). Note that all three types are volcanic in origin. There is an additional type of island that exists: islands that are parts of continents (such as the British Isles off Europe), but these occur close to shore and thus are not considered a true type of oceanic island.

**RECAP** **Transform faults are plate boundaries that occur *between* offset segments of the mid-ocean ridge, while fracture zones are intraplate features that occur *beyond* the offset segments of the mid-ocean ridge.**

Figure 3.25 **A new volcanic island emerges.** The tiny volcanic island of Nishino-shima, south of Japan, was created by successive explosive eruptions and other volcanic activity, most recently in 2017. The island, which is now 2.7 square kilometers (1.04 square miles), is associated with an active convergent plate boundary and continues to grow.

SmartTable 3.1 Comparison between transform faults and fracture zones. https://goo.gl/iMXqfl

**SmartTable 3.1 Comparison between transform faults and fracture zones**

| | **Transform faults** | **Fracture zones** |
|---|---|---|
| **Plate boundary?** | Yes—a transform plate boundary | No—an intraplate feature |
| **Relative movement across feature** | Movement in opposite directions ←/→ | Movement in the same direction ←/← |
| **Earthquakes?** | Many | Few |
| **Relationship to mid-ocean ridge** | Occur *between* offset mid-ocean ridge segments | Occur *beyond* offset mid-ocean ridge segments |
| **Geographic examples** | San Andreas Fault, Alpine Fault, Dead Sea Fault | Mendocino Fracture Zone, Molokai Fracture Zone |

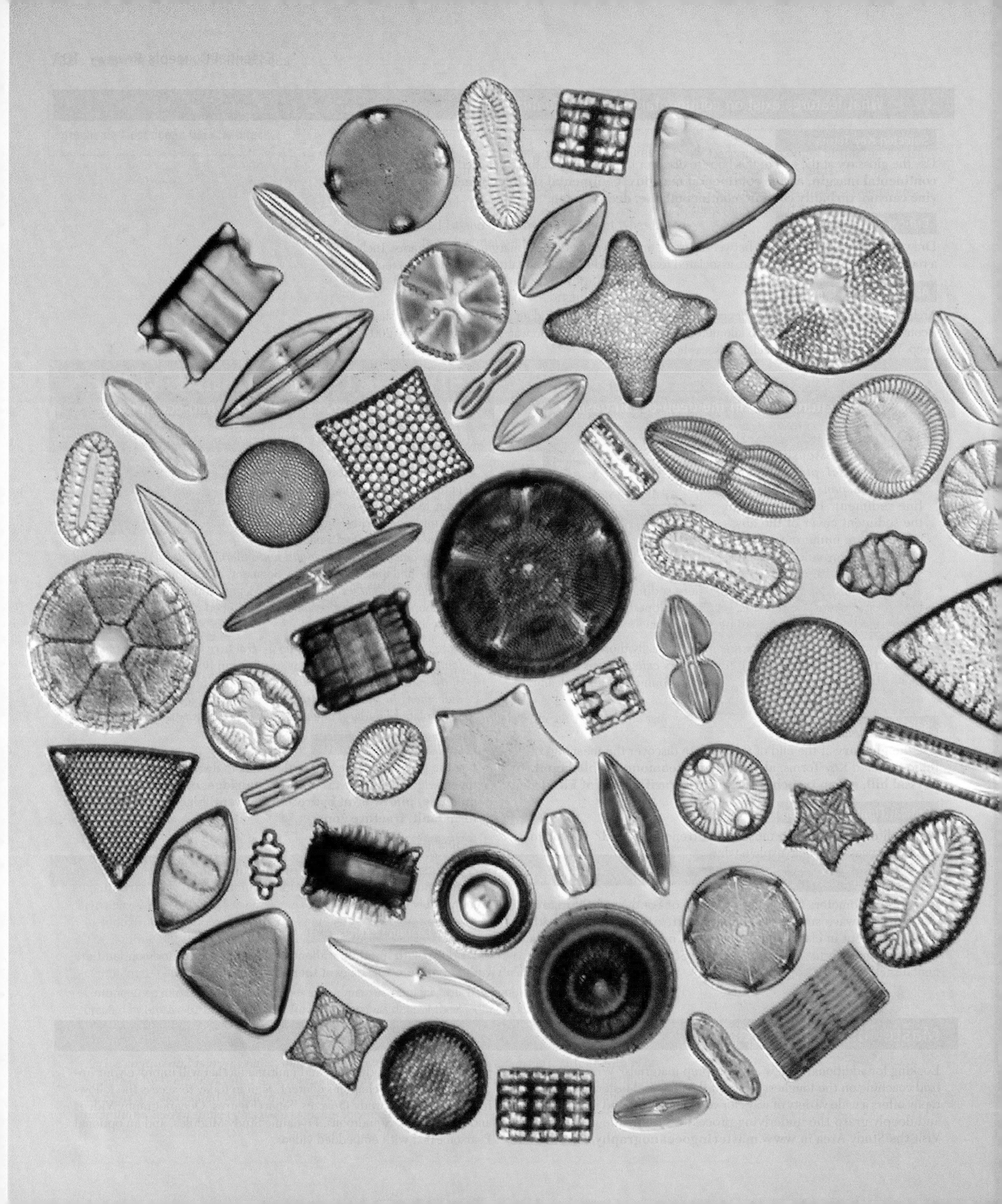

# Marine Sediments

# 4

## ESSENTIAL LEARNING CONCEPTS

At the end of this chapter, you should be able to:

- ☐ **4.1** Demonstrate an understanding of how marine sediments are collected and what historical events they reveal.
- ☐ **4.2** Describe the characteristics of lithogenous sediment.
- ☐ **4.3** Describe the characteristics of biogenous sediment.
- ☐ **4.4** Describe the characteristics of hydrogenous sediment.
- ☐ **4.5** Describe the characteristics of cosmogenous sediment.
- ☐ **4.6** Specify how the distribution of pelagic and neritic deposits is determined by proximity to sediment sources and agents of transport.
- ☐ **4.7** Identify the various resources that marine sediments provide.

↑ *Check when completed*

***"From the sediments the history of the ocean emerged with all its wonders . . . "***

**—*Wolf H. Berger,* Oceans: Reflections on a Century of Exploration *(2009)***

Why are **sediments** (*sedimentum* = settling) interesting to oceanographers? Ocean sediments appear to be little more than fragments of dirt, dust, and other debris that have slowly settled out of the water by the process of **suspension settling** and accumulated on the ocean floor (Figure 4.1). Yet marine sediments are troves of information, revealing much about Earth's history. For example, over millions of years, the thick layers of sediment that accumulate on the ocean floor often contain microscopic fossils that provide clues to the past geographic distributions of marine organisms. Marine sediments are also useful for determining the pattern of ancient ocean circulation, the movement of the sea floor, and even the timing and severity of global extinction events. Further, marine sediments reveal a detailed history of Earth's past climate, thus providing insight into today's climate changes. Remarkably, sediments that accumulate over time on the sea floor comprise a nearly continuous, undisturbed record of Earth's history unlike anything on land. In essence, marine sediments represent Earth's largest museum, with displays of Earth's history dating back millions of years.

Over time, sediments can become *lithified* (*lithos* = stone, *fic* = making)—turned to rock—and form *sedimentary rock*. More than half of the rocks exposed on the continents are sedimentary rocks deposited in ancient ocean environments and uplifted onto land by plate tectonic processes. Perhaps surprisingly, even the tallest mountains on the continents—far from any ocean—contain telltale marine fossils, which indicate that these rocks originated on the ocean floor in the geologic past. For example, the summit of the world's tallest mountain (Mount Everest in the Himalaya Mountains) consists of limestone, which is a type of rock that originated as sea floor deposits.

Climate Connection

Particles of marine sediment come from worn pieces of rocks, the remains of once-living organisms, minerals dissolved in seawater, and even from outer space. Clues to sediment origin are found in its mineral composition and its **texture** (the size and shape of its particles).

This chapter begins with a brief discussion about how marine sediments are collected and the important information they reveal about Earth's history. Then the characteristics, origin, and distribution of the four main types of sediment are examined (Table 4.1). Note that Table 4.1 summarizes much of the content within this chapter and it can be used as a road map of topics to help you organize information as you learn about marine sediments. Mixtures of marine sediment and sediment distribution in the ocean are also considered. Finally, the chapter concludes with a discussion of the resources that marine sediments provide.

For each of the four main types of sediment (*first column*), the table shows important aspects of its composition (*second column*), sources/origin (*third column*), and distribution/main locations found (*fourth column*).

◂ **Microscopic view of arranged diatoms.** The objects in this photomicrograph are diatoms, a type of microscopic marine algae that exists in incredible abundance in the ocean. This image shows various species of diatoms magnified several hundred times and was made by carefully arranging them under a microscope.

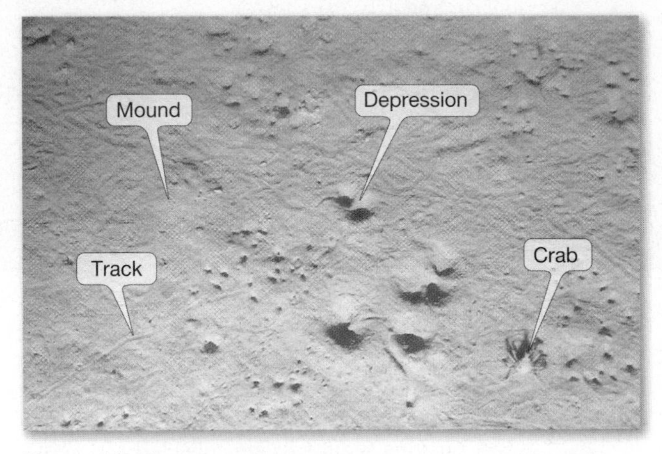


Figure 4.1 **Oceanic sediment.** View of typical deep-ocean floor, which is covered with a thick blanket of fine particles that have slowly settled onto the sea floor by the process of suspension settling. The depressions, mounds, and tracks are made by bottom-dwelling organisms. The crab in the lower right is about 4 inches (10 cm) across.

# 4.1 ▶ How Are Marine Sediments Collected, and What Historical Events Do They Reveal?

One of the difficulties of studying marine sediments is collecting adequate samples from the deep-ocean floor. Until relatively recently, the inaccessibility of the deep ocean has hindered the collection of marine sediments, especially those buried beneath the surface of the sea floor.

## Collecting Marine Sediments

Collecting sediments suitable for analysis from the deep ocean is an arduous process. During early exploration of the oceans, a bucket-like device called a *dredge* was used to scoop up sediment from the deep-ocean floor for analysis. This technique, however, was tedious and had many limitations. For example, it often didn't work correctly, and the dredge came up empty. It also disturbed the sediment and could only gather samples from the surface of the ocean floor. Later, the *gravity corer*—a hollow steel tube with a heavy weight on top—was thrust into the sea floor to collect the first **cores** (cylinders of sediment and rock). Although

**Table 4.1 Classification of marine sediments**

| Type | | Composition | Sources/Origin | | Distribution/Main locations where sediment accumulates |
|---|---|---|---|---|---|
| Lithogenous | Continental margin | Rock fragments<br>Quartz sand<br>Quartz silt<br>Clay | | Rivers; coastal erosion; landslides | Continental shelf |
| | | | | Glaciers | Continental shelf in high latitudes |
| | | | | Turbidity currents | Continental slope and rise; ocean basin margins |
| | Oceanic | Quartz silt<br>Clay | | Wind-blown dust; rivers | Abyssal plains and other regions of the deep-ocean basins |
| | | Volcanic ash | | Volcanic eruptions | |
| Biogenous | Calcium carbonate/ calcite ($CaCO_3$) | Calcareous ooze (microscopic) | Warm surface waters | Coccolithophores (algae)<br>Foraminifers (protozoans) | Low-latitude regions; sea floor above CCD; along mid-ocean ridges and the tops of submarine volcanic peaks |
| | | Shells and coral fragments (macroscopic) | | Macroscopic shell-producing organisms | Continental shelf; beaches |
| | | | | Coral reefs | Shallow low-latitude regions |
| | Silica ($SiO_2 \cdot nH_2O$) | Siliceous ooze | Cold surface waters | Diatoms (algae)<br>Radiolarians (protozoans) | High-latitude regions; sea floor below CCD; upwelling areas where cold, deep water rises to the surface, especially that caused by surface current divergence near the equator |
| Hydrogenous | | Manganese nodules (manganese, iron, copper, nickel, cobalt) | Precipitation of dissolved materials directly from seawater due to chemical reactions | | Abyssal plain |
| | | Phosphorite (phosphorous) | | | Continental shelf |
| | | Oolites ($CaCO_3$) | | | Shallow shelf in low-latitude regions |
| | | Metal sulfides (iron, nickel, copper, zinc, silver) | | | Hydrothermal vents at mid-ocean ridges |
| | | Evaporites (gypsum, halite, other salts) | | | Shallow restricted basins where evaporation is high in low-latitude regions |
| Cosmogenous | | Iron–nickel spherules, tektites (silica glass) | Space dust | | In very small proportions mixed with all types of sediment and in all marine environments |
| | | Iron–nickel meteorites | Meteors | | Localized near meteor impact structures |

the gravity corer could sample below the surface, its depth of penetration was limited. Today, specially designed ships perform **rotary drilling** to collect cores from the deep ocean.

In 1963, the U.S. National Science Foundation began funding a program that borrowed drilling technology from the offshore oil industry to obtain long sections of core from deep below the surface of the ocean floor. The program united four leading oceanographic institutions (Scripps Institution of Oceanography in California; Rosenstiel School of Atmospheric and Oceanic Studies at the University of Miami, Florida; Lamont-Doherty Earth Observatory of Columbia University in New York; and the Woods Hole Oceanographic Institution in Massachusetts) to form the *Joint Oceanographic Institutions for Deep Earth Sampling (JOIDES)*. The oceanography departments of several other leading universities later joined JOIDES.

The first phase of the **Deep Sea Drilling Project (DSDP)** was initiated in 1966, when the specially designed drill ship *Glomar Challenger* was launched. It had a tall drilling rig resembling a steel tower. Cores could be collected by drilling into the ocean floor in water up to 6000 meters (3.7 miles) deep. From the initial cores collected, scientists confirmed the existence of sea floor spreading by documenting that (1) the age of the ocean floor increased progressively with distance from the mid-ocean ridge (see Figure 2.11), (2) sediment thickness increased progressively with distance from the mid-ocean ridge (see Figure 4.24), and (3) Earth's magnetic field polarity reversal patterns recorded in ocean floor rocks were symmetrical on either side of the mid-ocean ridge (see Figure 2.10).

Although the oceanographic research program was initially financed by the U.S. government, it became international in 1975, when West Germany, France, Japan, the United Kingdom, and the Soviet Union also provided financial and scientific support. In 1983, the Deep Sea Drilling Project became the **Ocean Drilling Program (ODP)**, with 20 participating countries under the supervision of Texas A&M University, and a broader objective of drilling the thick sediment layers near the continental margins.

In 1985, the *Glomar Challenger* was decommissioned and replaced by the drill ship *JOIDES Resolution* (**Figure 4.2**). The new ship also has a tall metal drilling rig to conduct *rotary drilling*. The drill pipe is in individual sections of 9.5 meters (31 feet), and sections can be screwed together to make a single string of pipe up to 8200 meters (27,000 feet) long (Figure 4.2). The drill bit, located at the end of the pipe string, rotates as it is pressed against the ocean bottom and can drill up to 2100 meters (6900 feet) into the sea floor. Like twirling a soda straw into a layer cake, the drilling operation crushes the rock around the outside and retains a cylinder of rock (a *core sample*) on the inside of the hollow pipe. A core can then be raised to the surface from inside the pipe, cut in half, and analyzed using state-of-the-art laboratory facilities on board the *Resolution*. Worldwide, more than 2000 holes have been drilled into the sea floor using this method, allowing the collection of cores (**Figure 4.3**) that provide scientists with valuable information about Earth's history, as recorded in sea floor sediments.

In 2003, the ODP was replaced by the **Integrated Ocean Drilling Program (IODP)**, and, in 2013, its name was updated to *International Ocean Discovery Program (IODP): Exploring the Earth Under the Sea*. This new international effort continues over five decades of scientific collaboration that seeks to recover geological data and samples from beneath the ocean floor to study the history and dynamics of Planet Earth. In addition, the program does not

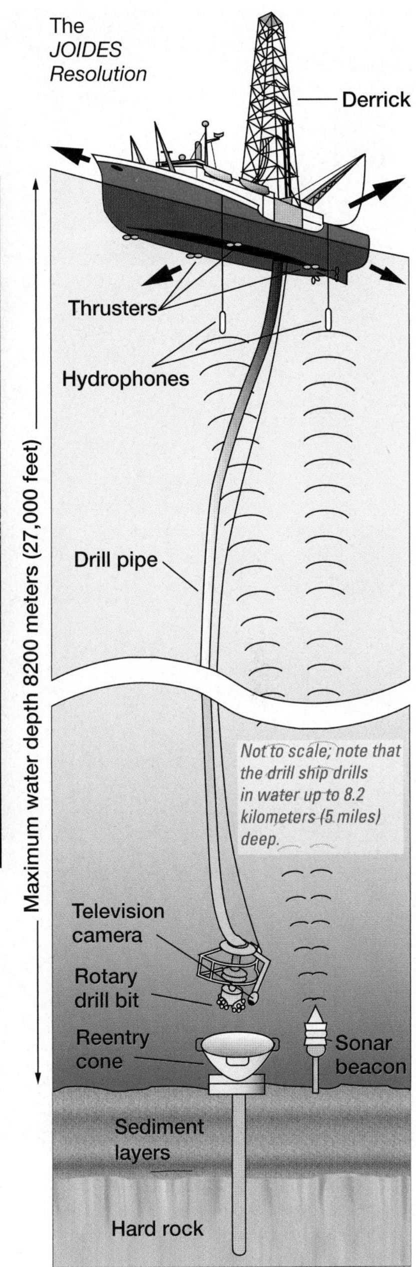


**Figure 4.2 Rotary drilling from the *JOIDES Resolution*.** Using its array of thrusters, the *JOIDES Resolution* (*photo*) can remain in one place at the surface while performing rotary drilling, which is shown diagrammatically (*right*).

Each sediment layer represents a unique event in Earth's history.

Figure 4.3 **An ocean sediment core.** Cylinders of sediment and rock called *cores* are retrieved from the ocean floor and then cut in half for examination. Oldest layers are at the bottom of the core and youngest are at the top.

**RECAP Marine sediments accumulate on the ocean floor and contain a record of Earth's history, including past environmental conditions.**

rely on just one drill ship but uses multiple vessels for exploration. For example, one of the new vessels that began operations in 2007 is a state-of-the-art drill ship named *Chikyu* (which means "Planet Earth" in Japanese) that is capable of drilling deeper at sea than any other science drilling vessel—up to 7000 meters (23,000 feet) into the sea floor. Plans to upgrade the vessel with new drilling technology will allow it to drill even deeper, perhaps all the way through Earth's crust into the mantle. The program's primary mission is to collect cores that will allow scientists to better understand the properties of the deep crust, the microbiology of the deep-ocean floor, Earth's climate change patterns, and earthquake mechanisms.

## Environmental Conditions Revealed by Marine Sediments

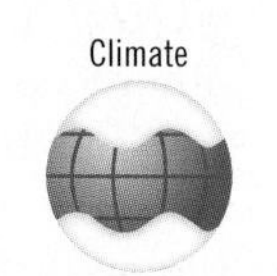


Marine sediments provide a wealth of information about past conditions on Earth. As sediment accumulates on the ocean floor, it preserves the materials—and the conditions of the environment—that existed in the overlying water column. By carefully analyzing cylindrical cores of sediment collected from the sea floor and interpreting them (**Figure 4.4**), Earth scientists can infer past environmental conditions such as sea surface temperature, nutrient supply, abundance of marine life, atmospheric winds, ocean current patterns, volcanic eruptions, major extinction events, changes in Earth's climate, and the movement of tectonic plates. In fact, most of what is known of Earth's past geology, climate, and biology has been learned through studying ancient marine sediments.

## Paleoceanography

The study of how the ocean, atmosphere, and land have interacted in the past to produce changes in ocean chemistry, circulation, biology, and climate is called **paleoceanography** (*paleo* = ancient, *ocean* = the marine environment, *graphy* = description of), a branch of oceanography that relies on sea floor sediments to gain insight into these past changes. Recent paleoceanographic studies, for example, have linked changes in deep-ocean circulation with rapid climate change. In the North Atlantic Ocean, cold, relatively salty water sinks and forms a body of water called *North Atlantic Deep Water.* Water in this deep current circulates through the global ocean, driving deep-ocean circulation and global heat transport, which, in turn, impacts global climate. This is widely viewed as one of the most climatically sensitive regions on Earth, and North Atlantic sea floor sediments from the past several million years have revealed that the region has experienced abrupt changes to its ocean–atmosphere system, triggered by fluctuations of freshwater from melting glaciers. Understanding the timing, mechanisms, and causes of this abrupt climate change is one of the major challenges facing paleoceanography today.

**CONCEPT CHECK 4.1 ▸ Demonstrate an understanding of how marine sediments are collected and what historical events they reveal.**

1 Using Table 4.1, list and describe the characteristics of the four main types of marine sediment.

2 Describe the process of how a drill ship like the *JOIDES Resolution* obtains core samples from the deep-ocean floor.

3 What types of past environmental conditions can be inferred by studying cores of sediment?

# 4.2 ▸ What Are the Characteristics of Lithogenous Sediment?

**Lithogenous sediment** (*lithos* = stone, *generare* = to produce) is derived from preexisting rock material that originates on the continents or islands from erosion, volcanic eruptions, or blown dust. Note that lithogenous sediment is sometimes referred to as **terrigenous sediment** (*terra* = land, *generare* = to produce).

## Origin of Lithogenous Sediment

Lithogenous sediment begins as rocks on continents or islands. Over time, **weathering** agents such as water, temperature extremes, and chemical effects break rocks into smaller pieces, as shown in **Figure 4.5**. When rocks are in smaller pieces, they can be more easily picked up and transported through the process of **erosion**. This eroded material is the basic component of which all lithogenous sediment is composed.

Eroded material from the continents is carried to the oceans by streams, wind, glaciers, and gravity (**Figure 4.6**). Each year, stream flow alone carries about 20 billion metric tons (44 trillion pounds) of sediment to Earth's continental margins; almost 40% is provided by runoff from Asia.

Transported sediment can be deposited in many environments, including bays or lagoons near the ocean, in deltas at the mouths of rivers, along beaches at the shoreline, or further offshore across the continental margin. It can also be carried beyond the continental margin to the deep-ocean basin by turbidity currents, as discussed in Chapter 3.

The greatest quantity of lithogenous material is found around the margins of the continents, where it is constantly moved by high-energy currents along the shoreline and by turbidity currents in deeper water. Lower-energy currents distribute finer components, which settle out onto the deep-ocean basins. Microscopic particles from wind-blown dust or volcanic eruptions can even be carried far out over the open ocean by prevailing winds. These particles are deposited into the ocean either as the wind speed decreases or when they serve as nuclei around which raindrops and snowflakes form, and ultimately settle onto the sea floor as fine layers of sediment.

Figure 4.4 **Examining deep-ocean sediment cores.** Sediment cores reveal interesting aspects of Earth's history such as the past geographic distributions of marine organisms, ocean circulation changes, major extinctions, and Earth's past climate.

## Composition of Lithogenous Sediment

The composition of lithogenous sediment reflects the material from which it was derived. All rocks are composed of discrete crystals of naturally occurring compounds called *minerals*. One of the most abundant, chemically stable, and durable minerals in Earth's crust is **quartz**, composed of silicon and oxygen in the form of $SiO_2$—the same composition as ordinary glass. Quartz is a major component of most rocks. Because quartz is resistant to abrasion, it can be transported long distances and deposited far from its source area. The majority of lithogenous deposits—such as beach sands—are composed primarily of quartz (**Figure 4.7**).

A large percentage of lithogenous particles that find their way into deep-ocean sediments far from continents are transported by prevailing winds, which remove small particles from the continents' subtropical desert regions. The map in **Figure 4.8** shows a close relationship between the location of microscopic fragments

Figure 4.5 **Weathering of a rock outcrop.**

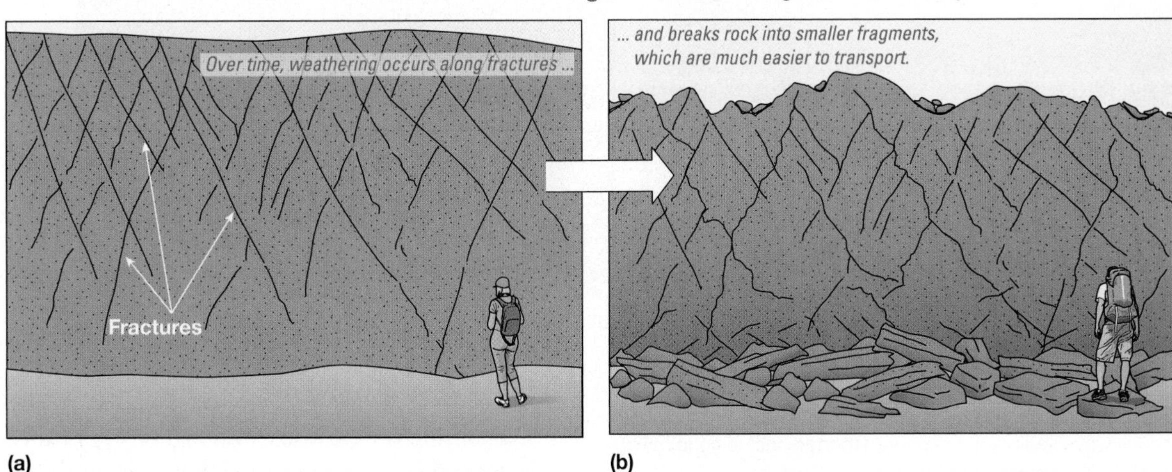

Over time, weathering occurs along fractures ...
Fractures
... and breaks rock into smaller fragments, which are much easier to transport.
(a)
(b)

Light colors are suspended
sediment in the water.

**(a)** Stream: Po River, Italy, which displays a prominent delta and a visible sediment plume in the water.

**(b)** Wind: Dust storm approaching a military base, Australia.

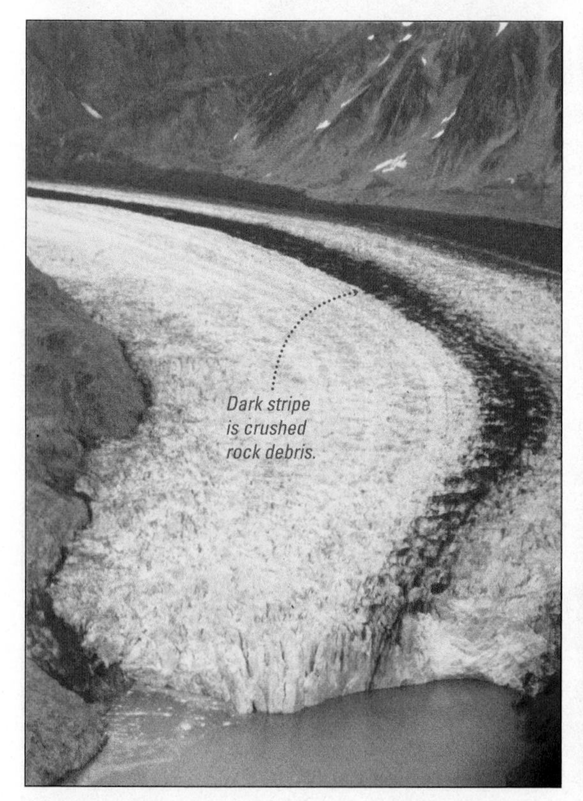
Dark stripe
is crushed
rock debris.

**(c)** Glacier: Riggs Glacier, Glacier Bay National Park, Alaska, which displays a dark stripe of sediment along its length called a *medial moraine*.

**(d)** Gravity, which creates landslides: Del Mar, California.

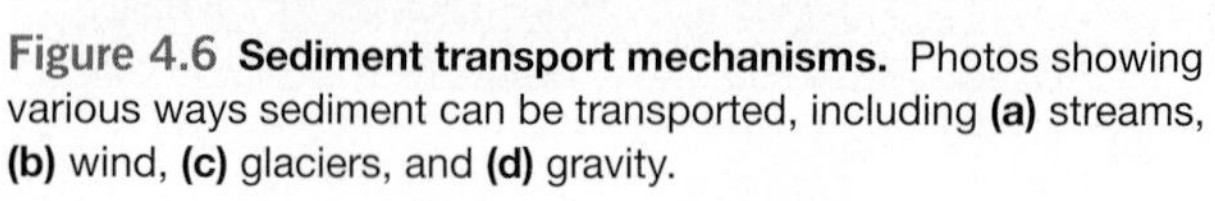

Figure 4.6 **Sediment transport mechanisms.** Photos showing various ways sediment can be transported, including **(a)** streams, **(b)** wind, **(c)** glaciers, and **(d)** gravity.

of lithogenous quartz in the surface sediments of the ocean floor and the strong prevailing winds in the desert regions of Africa, Asia, and Australia. Satellite observations of dust storms (Figure 4.8, *inset*) confirm this relationship. Sediment is not the only item transported by wind. In fact, scientists have documented the transportation of a variety of airborne substances—including viruses, pollutants, and even living insects—from Africa all the way across the Atlantic Ocean to North America.

## Texture of Lithogenous Sediment

Sediment texture describes the physical characteristics of a deposit. (Note that texture is an important concept only for describing lithogenous sediment, not any of the other three types of sediment.) One of the most important components of lithogenous sediment texture is a deposit's **grain size**. (Sediment grains are also known as particles, fragments, or clasts.) The **Wentworth scale of grain size** (Table 4.2) indicates that particles can be classified as boulders (largest), cobbles, pebbles, granules, sand, silt, or clay (smallest). Studies of sediment transport indicate that grain size is proportional to the energy needed to lay down a deposit. For example, deposits laid down where wave action is strong (areas of high energy) are usually composed primarily of larger particles—cobbles and boulders. Fine-grained particles, on the other hand, are deposited where the energy level is low and the current speed is minimal. When clay-sized particles—many of which are flat—are deposited, they tend to stick together by cohesive forces. Consequently, higher-energy conditions than what would be expected based on grain size alone are required to erode and transport clays. In general, however, lithogenous sediment tends to become finer with increasing distance from shore. This relationship is mostly because high-energy transporting agents predominate close to shore and lower-energy conditions exist in the deep-ocean basins.

Figure 4.6 shows some examples of different transporting agents (or processes) that result in deposits with a variety of grain sizes. For example, Figure 4.6a shows a river, which drops coarser sediment close to shore while the finer, suspended materials are deposited further out to sea. Figure 4.6b shows wind deposits, which are generally composed of fine-grained sand that is very uniform (the wind cannot lift any large particles, and so it is said to be a very selective transporting agent). Figures 4.6c and 4.6d shows transporting agents that are not very selective at all (they transport all sizes); as a result, deposits from glaciers and landslides tend to have a wide variety of grain sizes, from boulder- to clay-sized particles.

The texture of lithogenous sediment also depends on its **sorting**. Sorting is a measure of the uniformity of grain sizes and indicates the selectivity of the transportation process. For example, sediments composed of particles that are primarily the same size are well-sorted—such as in coastal sand dunes, where winds (Figure 4.6b) can only pick up a certain size particle. Poorly-sorted deposits, on the other hand, contain a variety of differently sized particles and indicate a transportation process capable of picking up clay- to boulder-sized particles. An example of poorly-sorted sediment is that which is carried by a glacier (Figure 4.6c) and left behind when the glacier melts.

Figure 4.7 **Lithogenous beach sand.** Photomicrograph of well-sorted lithogenous beach sand, which is composed mostly of particles of white quartz plus small amounts of other minerals. This sand, from North Beach, Hampton, New Hampshire, is magnified approximately 23 times.

## Distribution of Lithogenous Sediment

Marine sedimentary deposits can be categorized as either neritic or pelagic. **Neritic deposits** (*neritos* = of the coast) are found on continental shelves and in shallow water near islands; these deposits are generally coarse grained. Alternatively, **pelagic deposits** (*pelagios* = of the sea) are found in the deep-ocean basins and are typically fine-grained. Moreover, lithogenous sediment in the ocean is ubiquitous; at least a small percentage of lithogenous sediment is found nearly everywhere on the ocean floor.

**NERITIC DEPOSITS** Lithogenous sediment dominates most neritic deposits. Lithogenous sediment is derived from rocks on nearby landmasses, consists of coarse-grained deposits, and accumulates rapidly on the continental shelf, slope, and rise. Examples of lithogenous neritic deposits include beach deposits, continental shelf deposits, turbidite deposits, and glacial deposits.

***Beach Deposits*** Beaches are made of whatever materials are locally available. Beach materials are composed mostly of quartz-rich sand that is washed down to the coast by rivers but can also be built of materials having a wide variety of sizes and compositions. This material is transported by waves that crash against the shoreline, especially during storms.

***Continental Shelf Deposits*** At the end of the last ice age (about 10,000 years ago), glaciers melted and sea level rose. As a result, many rivers of the world today drop their sediment in drowned river mouths rather than carry it onto the continental shelf as they did during the geologic past. This explains why, in many areas, the sediments that cover the continental shelf—called *relict* (*relict* = left behind) *sediments*—were deposited from 3000 to 7000 years ago and are not covered by sediments discharged by rivers today. These relict sediments presently cover about 70% of the world's continental shelves. In other areas, deposits of sand ridges on the continental shelves appear to have been formed more recently than the most recent ice age and at present water depths.

Climate Connection

***Turbidite Deposits*** As discussed in Chapter 3, turbidity currents are underwater avalanches that periodically move down the continental slopes and carve submarine canyons. Turbidity currents also carry vast amounts of neritic material. This material spreads out as deep-sea fans, comprises the continental rise, and gradually thins toward the abyssal plains. These deposits are called **turbidite deposits** and are composed of characteristic layering called *graded bedding* (see Figure 3.12).

### STUDENTS SOMETIMES ASK . . .

***How effective is wind as a transporting agent?***

Any material that gets into the atmosphere—including dust from dust storms, soot from forest fires, specks of pollution, and ash from volcanic eruptions—is transported by wind and can be found as deposits on the ocean floor. Every year, wind storms lift an estimated 3 billion metric tons (6.6 trillion pounds) of this material into the atmosphere, where it gets transported around the globe. As much as three-quarters of these particles—mostly dust—come from Africa's Sahara Desert; once airborne, they are carried out across the Atlantic Ocean (see Figure 4.8). Much of this dust falls in the Atlantic, and that's why ships traveling downwind from the Sahara Desert often arrive at their destinations quite dusty. Some of it falls in the Caribbean (where the pathogens it contains have been linked to stress and disease among coral reefs), in Bermuda (where past accumulations have produced the island's red soils), the Amazon (where its iron and phosphorus fertilize nutrient-poor soil), and across the southern United States as far west as New Mexico. The dust also contains bacteria and pesticides—even African desert locusts have been transported alive across the Atlantic Ocean during strong wind storms!

organisms that produce hard parts die, their remains settle onto the ocean floor and can accumulate as biogenous sediment.

Biogenous sediment can be classified as either macroscopic or microscopic. **Macroscopic biogenous sediment** is large enough to be seen without the aid of a microscope and includes shells, bones, and teeth of large organisms. Except in certain tropical beach localities where shells and coral fragments are numerous, this type of sediment is relatively rare in the marine environment, especially in deep water where fewer organisms live. Much more abundant is **microscopic biogenous sediment**, which contains particles so small they can be seen well only through a microscope. Microscopic organisms produce tiny shells called **tests** (*testa* = shell) that begin to sink after the organisms die and continually rain down in great numbers onto the ocean floor. These microscopic tests can accumulate on the deep-ocean floor and form deposits called **ooze** (*wose* = juice). Ooze has the consistency of toothpaste mixed about half and half with water. (As a way to remember this term, imagine walking barefoot across the deep-ocean floor and having the fine sediment *ooze* between your toes). As its name implies, ooze resembles very fine-grained, mushy material. Technically, biogenous ooze must contain at least 30% biogenous test material by weight. What comprises the other part—up to 70%—of an ooze? Commonly, it is fine-grained lithogenous clay that is deposited along with biogenous tests in the deep ocean. By volume, much more microscopic ooze than macroscopic biogenous sediment exists on the ocean floor.

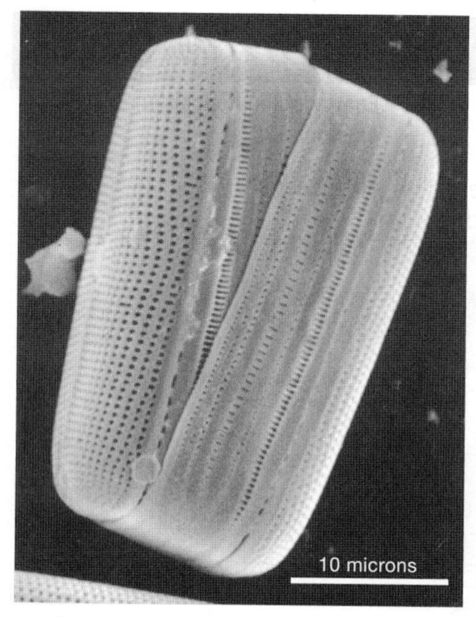


**(a)** Diatom, showing how the two parts of the diatom's test fit together.

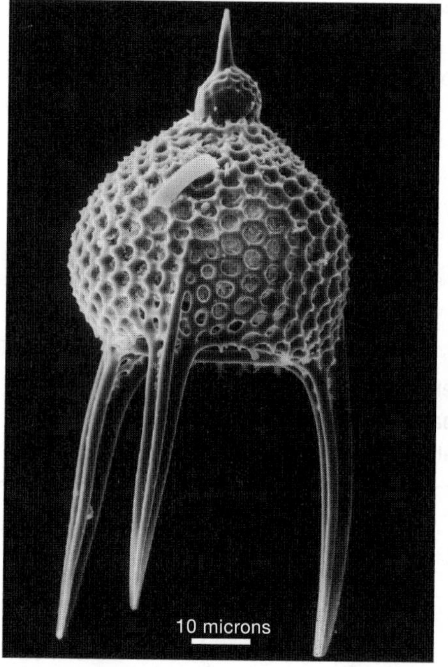


**(b)** Radiolarian.

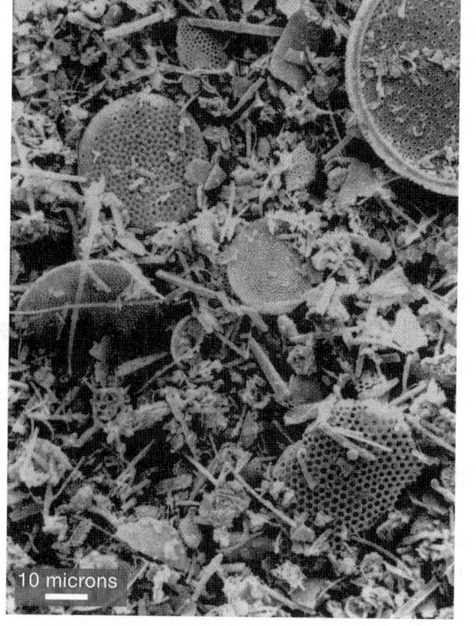


**(c)** Siliceous ooze, showing mostly fragments of diatom tests.

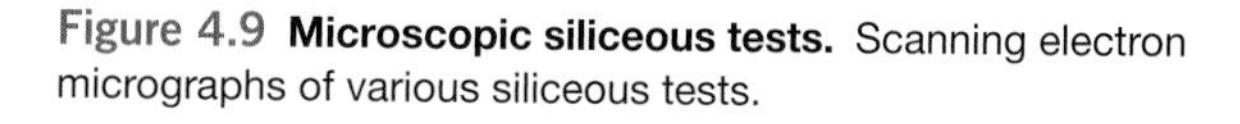

Figure 4.9 **Microscopic siliceous tests.** Scanning electron micrographs of various siliceous tests.

The organisms that contribute to biogenous sediment are chiefly **algae** (*alga* = seaweed) and **protozoans** (*proto* = first, *zoa* = animal). Algae are primarily aquatic, eukaryotic, photosynthetic organisms, ranging in size from microscopic single cells to large organisms like giant kelp. (Note that eukaryotic [*eu* = good, *karyo* = the nucleus] cells contain a distinct membrane-bound nucleus.) Protozoans are any of a large group of single-celled, eukaryotic, usually microscopic organisms that are generally not photosynthetic.

## Composition of Biogenous Sediment

The two most common chemical compounds in biogenous sediment are **calcium carbonate** ($CaCO_3$, which forms the mineral **calcite**) and **silica** ($SiO_2$). Often, the silica is chemically combined with water to produce $SiO_2 \cdot nH_2O$, the hydrated form of silica, which is called *opal*.

**SILICA** Most of the silica in biogenous ooze comes from microscopic algae called **diatoms** (*diatoma* = cut in half) and protozoans called **radiolarians** (*radio* = a spoke or ray).

Because diatoms photosynthesize, they need strong sunlight and are found only within the upper, sunlit surface waters of the ocean. Most diatoms are free-floating, or **planktonic** (*planktos* = wandering). The living organism builds a glass greenhouse out of silica as a protective covering and lives inside. Most species have two parts to their test that fit together like a petri dish or pillbox (**Figure 4.9a**). The tiny tests are perforated with small holes in intricate patterns to allow nutrients to pass in and waste products to pass out. Where diatoms are abundant at the ocean surface, thick deposits of diatom-rich ooze can accumulate below on the ocean floor. When this ooze lithifies, it becomes **diatomaceous earth** (also called diatomite, tripolite, or kieselguhr), which is a lightweight white rock composed of diatom tests and clay (**Diving Deeper 4.1**).

Radiolarians are microscopic single-celled protozoans, most of which are also planktonic. As their name implies, they often have long spikes or rays of silica protruding from their siliceous shell (**Figure 4.9b**). They do not photosynthesize but rely on external food sources such as bacteria and other plankton. Radiolarians typically

## DIATOMS: THE MOST IMPORTANT THINGS YOU HAVE (PROBABLY) NEVER HEARD OF

*"Few objects are more beautiful than the minute siliceous cases of the diatomaceae: were these created that they might be examined and admired under the higher powers of the microscope?"*

*—Charles Darwin (1872)*

Diatoms are microscopic single-celled photosynthetic organisms. Each one lives inside a protective silica test, most of which contain two halves that fit together like a shoebox and its lid. First described with the aid of a microscope in 1702, their tests are exquisitely ornamented with holes, ribs, and radiating spines unique to individual species. The fossil record indicates that diatoms have been on Earth since the Jurassic Period (180 million years ago), and more than 70,000 species of diatoms have been identified.

Diatoms live for a few days to as much as a week, can reproduce sexually or asexually, and occur individually or linked together into long communities. They are found in great abundance floating in the ocean and in certain freshwater lakes but can also be found in many diverse environments, such as on the undersides of polar ice, on the skins of whales, in soil, in thermal springs, and even on brick walls.

When marine diatoms die, their tests rain down and accumulate on the sea floor as siliceous ooze. Hardened deposits of siliceous ooze, called *diatomaceous earth*, can be as much as 900 meters (3000 feet) thick. Diatomaceous earth consists of billions of minute silica tests and has many unusual properties: It is lightweight, has an inert chemical composition, is resistant to high temperatures, and has excellent filtering properties. Diatomaceous earth is used to produce a variety of common products (**Figure 4A**). The main uses of diatomaceous earth include:

- Filters (for refining sugar, separating impurities from wine, straining yeast from beer, and filtering swimming pool water)
- Mild abrasives (in toothpaste, facial scrubs, matches, and household cleaning and polishing compounds)
- Absorbents (for chemical spills, in cat litter, and as a soil conditioner)
- Chemical carriers (in pharmaceuticals, paint, and even dynamite)

Other products from diatomaceous earth include optical-quality glass (because of the pure silica content of diatoms) and space shuttle tiles (because they are lightweight and provide good insulation). Diatomaceous earth is also used as an additive in concrete, a filler in tires, an anticaking agent, a natural pesticide, and even a building stone in the construction of houses.

Further, some estimates suggest that the vast majority of oxygen that all animals breathe is a by-product of photosynthesis by diatoms. In addition, each living diatom contains a tiny droplet of oil. When diatoms die, their tests containing droplets of oil accumulate on the sea floor and are the beginnings of petroleum deposits, such as those found offshore of California.

Given their many practical applications, it is difficult to imagine how different our lives would be without diatoms!

### WHAT DID YOU LEARN?

What are several reasons diatoms are so remarkable? List products that contain or are produced using diatomaceous earth.

**Figure 4A Products containing or produced using diatomaceous earth (diatom *Thalassiosira eccentrica*, inset).**

display well-developed symmetry, which is why they have been described as the "living snowflakes of the sea."

The accumulation of siliceous tests of diatoms, radiolarians, and other silica-secreting organisms produces **siliceous ooze** (Figure 4.9c).

**CALCIUM CARBONATE** Two significant sources of calcium carbonate biogenous ooze are the **foraminifers** (*foramen* = an opening)—close relatives of radiolarians—and microscopic algae called **coccolithophores** (*coccus* = berry, *lithos* = stone, *phorid* = carrying).

Coccolithophores are single-celled algae, most of which are planktonic. Coccolithophores produce thin plates or shields made of calcium carbonate, 20 or 30 of which overlap to produce a spherical test (Figure 4.10a). Like diatoms, coccolithophores photosynthesize, so they need sunlight to live. Coccolithophores are really, *really* small. In fact, coccolithophores are about 10 to 100 times smaller than most diatoms (Figure 4.10b), which is why coccolithophores are often called **nannoplankton** (*nanno* = dwarf, *planktos* = wandering).

When the organism dies, the individual plates (called **coccoliths**) disaggregate and can accumulate on the ocean floor as coccolith-rich ooze. When this ooze lithifies over time, it forms a white deposit called **chalk**, which is used for a variety of purposes (including writing on chalkboards). The White Cliffs of southern England

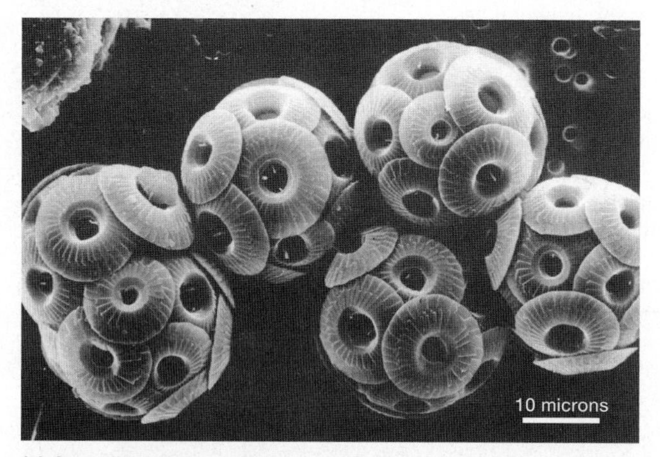


**(a)** Coccolithophores, which resemble tiny spheres.

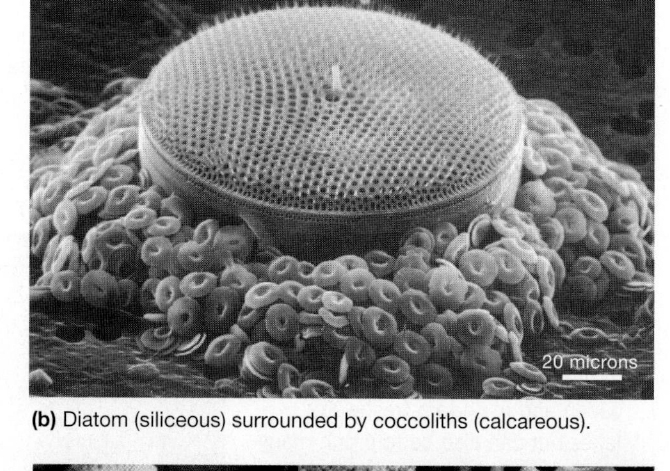


**(b)** Diatom (siliceous) surrounded by coccoliths (calcareous).

**(c)** Foraminifers, which resemble tiny shells found at a beach.

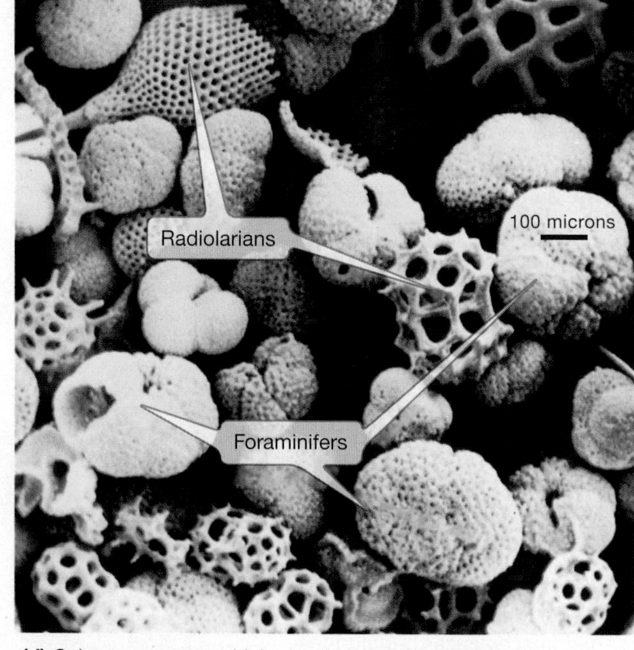


**(d)** Calcareous ooze, which also includes some siliceous radiolarian tests.

Figure 4.10 **Microscopic calcareous tests.** Scanning electron micrographs (*above*) and photomicrographs (*below*) of various calcareous tests.

are composed of hardened, coccolith-rich calcium carbonate ooze, which was deposited on the ocean floor and has been uplifted onto land (Figure 4.11). Deposits of chalk the same age as the White Cliffs are so common throughout Europe, North America, Australia, and the Middle East that the geologic period in which these deposits formed is named the Cretaceous (*creta* = chalk) Period.

Foraminifers are single-celled protozoans, many of which are planktonic, ranging in size from microscopic to macroscopic. They do not photosynthesize, so they must ingest other organisms for food. Foraminifers produce a hard calcium carbonate test in which the organism lives (Figure 4.10c). Most foraminifers produce a segmented or chambered test, and all tests have a prominent opening in one end. Although very small in size, the tests of foraminifers resemble the large shells that one might find at a beach.

Deposits comprised primarily of tests of foraminifers, coccoliths, and other calcareous-secreting organisms are called **calcareous ooze** (Figure 4.10d).

## Distribution of Biogenous Sediment

Biogenous sediment is one of the most common types of pelagic deposits. The distribution of biogenous sediment on the ocean floor depends on three fundamental processes: (1) productivity, (2) destruction, and (3) dilution.

*Productivity* is the number of organisms present in the surface water above the ocean floor. Surface waters with high biologic productivity contain many living and reproducing organisms—conditions that are likely to produce biogenous sediments. Conversely, surface waters with low biologic productivity contain too few organisms to produce biogenous oozes on the ocean floor.

*Destruction* occurs when skeletal remains (tests) dissolve in seawater at depth. In some cases, biogenous sediment dissolves before ever reaching the sea floor; in other cases, it is dissolved before it has a chance to accumulate into deposits on the sea floor.

*Dilution* occurs when the deposition of other sediments decreases the percentage of the biogenous sediment found in marine deposits. For example, other types of sediments can dilute biogenous test material below the 30% necessary to classify it as ooze. Dilution occurs most often because of the abundance of coarse-grained lithogenous material in neritic environments, so biogenous oozes are uncommon along continental margins.

**NERITIC DEPOSITS** Although neritic deposits are dominated by lithogenous sediment, both microscopic and macroscopic biogenous material may be incorporated into lithogenous sediment in neritic deposits. In addition, biogenous carbonate deposits are common in some areas.

***Carbonate Deposits*** **Carbonate** minerals are those that contain $CO_3$ in their chemical formula—such as calcium carbonate, $CaCO_3$. Rocks from the marine environment composed primarily of calcium carbonate are called **limestones**. Most limestones contain fossil marine shells, suggesting a biogenous origin, while other carbonate-containing rocks appear to have formed directly from seawater without the help of any marine organism. Modern environments where calcium carbonate is currently being deposited (such as in the Bahama Banks, Australia's Great Barrier Reef, and the Persian Gulf) suggest that carbonate deposits are formed in shallow, warm-water shelves and around tropical islands as coral reefs and beaches.

Ancient marine carbonate deposits constitute 2% of Earth's crust and 25% of all sedimentary rocks on Earth. In fact, marine limestones form the underlying bedrock of Florida and many Midwestern states, from Kentucky

CREATURE FEATURE 4.1

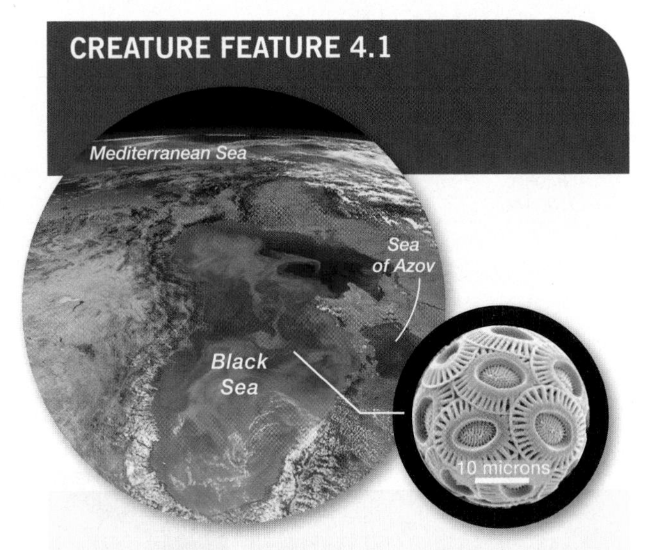


### My buddies and I can be seen from space!

The **coccolithophore** *Emiliania huxleyi* is a microscopic photosynthetic alga that builds its hard shell (test) out of calcite; it is the world's most famous coccolithophore.

Coccolithophores can exist in such great abundance during blooms that they impart a bright blue color to surface waters that can be seen from space; under the right conditions, the tests of coccolithophores form deposits on the sea floor.

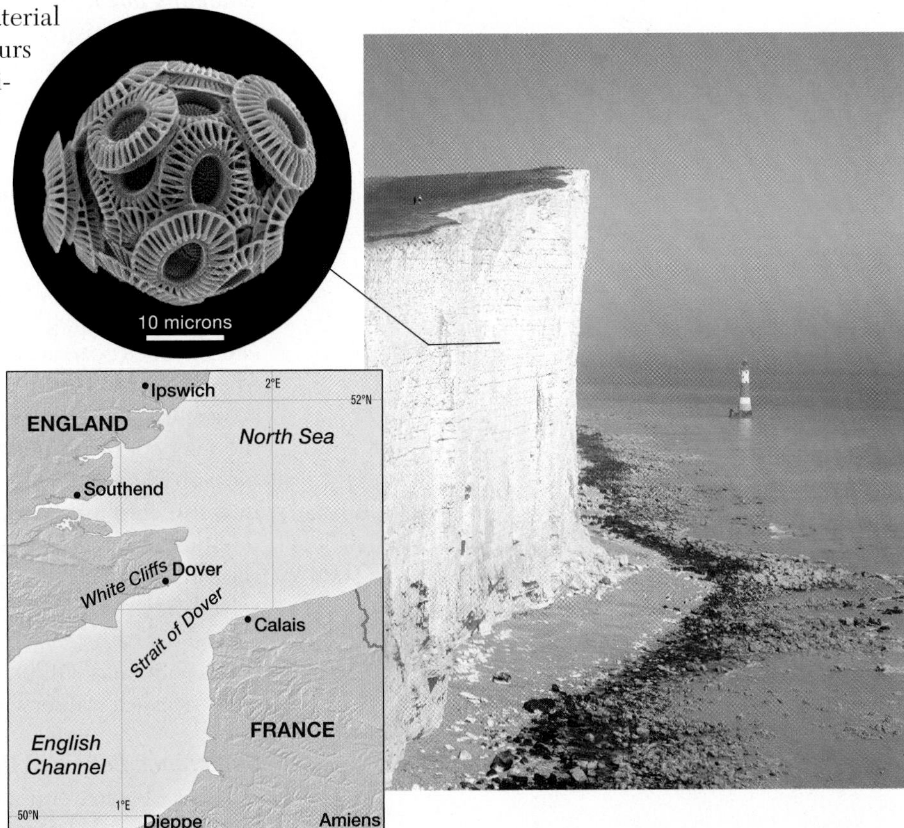


Figure 4.11 **The White Cliffs of southern England.** The White Cliffs near Dover in southern England are composed of chalk, which is hardened coccolith-rich calcareous ooze. Inset shows a colored image of the coccolithophore *Emiliana huxleyi*.

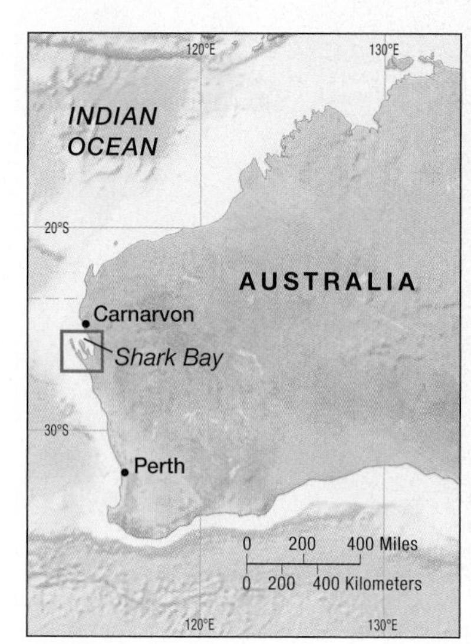


**(a)** Location map of Shark Bay, Australia.

**(b)** Shark Bay stromatolites, which form in high-salinity tidal pools and reach a maximum height of about 1 meter (3.3 feet).

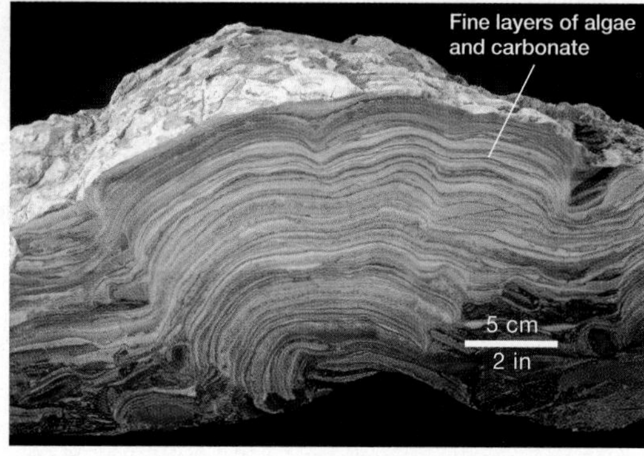


**(c)** Profile view through a stromatolite, showing its internal fine layering.

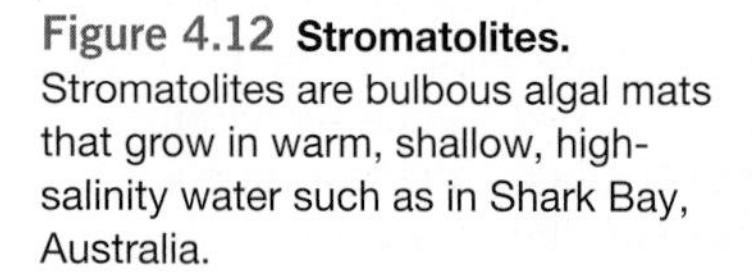

Figure 4.12 **Stromatolites.** Stromatolites are bulbous algal mats that grow in warm, shallow, high-salinity water such as in Shark Bay, Australia.

to Michigan and from Pennsylvania to Colorado. Percolation of groundwater through these deposits has dissolved the limestone to produce sinkholes and, in some cases, spectacular caverns.

***Stromatolites*** **Stromatolites** are lobate structures consisting of fine layers of carbonate that form in specific warm, shallow-water environments such as the high salinity tidal pools in Shark Bay, Western Australia (**Figure 4.12**). Cyanobacteria, which are simple, ancient creatures whose ancestry can be traced back to some of the first photosynthetic organisms on Earth, produce these deposits by trapping fine sediment in mucous mats. Other types of algae produce long filaments that bind carbonate particles together. Like tree rings being added as a tree grows, layer upon layer of these algae colonize the surface, forming a bulbous structure. In the geologic past—particularly from about 1 to 3 billion years ago—conditions were ideal for the development of stromatolites, so stromatolite structures hundreds of meters high can be found in rocks from these ages.

**PELAGIC DEPOSITS** Microscopic biogenous sediment (ooze) is common on the deep-ocean floor because there is so little lithogenous sediment deposited at great distances from the continents that could dilute the biogenous material.

***Siliceous Ooze*** Siliceous ooze contains at least 30% of the hard remains of silica-secreting organisms. When the siliceous ooze consists mostly of diatoms, it is called *diatomaceous ooze*. When it consists mostly of radiolarians, it is called *radiolarian ooze*. When it consists mostly of single-celled silicoflagellates—another type of protozoan—it is called *silicoflagellate ooze*.

The ocean is undersaturated with silica at all depths, which means that any solid particle made up of silica will tend to dissolve in seawater. In fact, if living diatoms, radiolarians, and silicoflagellates were not hard at work creating their silica-containing tests, they would dissolve, too! As a consequence, the destruction of siliceous biogenous particles (that is, the tests of dead organisms drifting to the sea floor), by dissolving in seawater, occurs continuously and slowly at all depths. How can siliceous ooze accumulate on the ocean floor if it is being dissolved? One way is to accumulate the siliceous tests faster than seawater can dissolve them. For instance, many tests sinking at the same time will create a deposit of siliceous ooze on the sea floor below (**Figure 4.13**). An analogy to this is trying to get a layer of sugar to form on the bottom of a cup of hot coffee. If a few grains of sugar are slowly dropped into the cup, a layer of sugar won't accumulate. If a whole bowl full of sugar is dumped into the coffee, however, a thick layer of sugar will form on the bottom of the cup. Similarly, by having more siliceous tests accumulate than are being dissolved, a deposit of siliceous ooze will form on the sea floor. Then, as the deposit is buried beneath other siliceous tests, it is no longer exposed to the dissolving effects of seawater and will be preserved. Thus, siliceous ooze is commonly found in areas below surface waters with high biologic productivity of silica-secreting organisms.

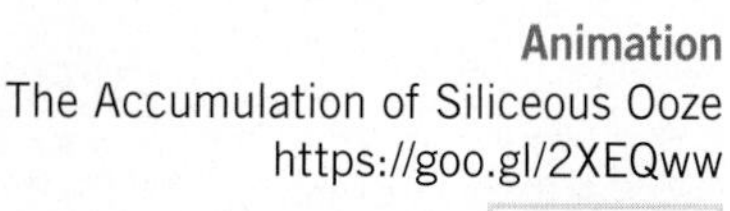

**Animation**
The Accumulation of Siliceous Ooze
https://goo.gl/2XEQww

***Calcareous Ooze and the CCD*** Calcareous ooze contains at least 30% of the hard remains of calcareous-secreting organisms. When it consists mostly of coccolithophores, it is called *coccolith ooze*. When it consists mostly of foraminifers, it is called *foraminifer ooze*. One of the most common types of foraminifer ooze is *Globigerina ooze*, named for a foraminifer that is especially widespread in the Atlantic and South Pacific oceans. Other calcareous oozes include *pteropod oozes* and *ostracod oozes*, named for the marine organisms whose shells are the primary components of the sediment.

The destruction of calcium carbonate (calcite) varies with depth. Calcium carbonate is an unusual solid in that it dissolves more readily in cold water, as opposed to warm water. Also, it dissolves at higher pressure. Accordingly, in the warmer surface and in the shallow parts of the ocean, seawater is generally saturated with calcium carbonate, so calcite does not dissolve. In the deep ocean, however, pressure is greater, the water is colder, and it contains greater amounts of carbon dioxide, which forms carbonic acid—three factors that help speed the dissolution of calcium carbonate.

The depth in the ocean at which the pressure is high enough, the temperature low enough, and the amount of carbon dioxide in seawater great enough to begin dissolving calcium carbonate is called the **lysocline** (*lusis* = a loosening, *cline* = slope). Below the lysocline, calcium carbonate dissolves at an increasing rate with increasing depth until the **calcite compensation depth (CCD)** is reached (**Figure 4.14**). (Because the mineral calcite is composed of calcium carbonate, the *calcite compensation depth* is also known as the *calcium carbonate compensation depth* or the *carbonate compensation depth*. All go by the handy abbreviation CCD.) At the CCD and greater depths, sediment does not usually contain much calcite because it readily dissolves; even the thick tests of foraminifers dissolve within a day or two. In essence, calcite accumulates only near the tops of the tall peaks that rise off the sea floor and extend above the CCD, but dissolves at deeper depths associated with the base of the peaks. This situation creates the marine equivalent of a mountain's "snow line," but with deposits of light-colored calcite on the mountaintop instead of frozen water.

The CCD, on average, is 4500 meters (15,000 feet) below sea level, but depending on the chemistry of the deep ocean, it may be as deep as 6000 meters (20,000 feet) in portions of the Atlantic Ocean or as shallow as 3500 meters (11,500 feet) in the Pacific Ocean. The depth of the lysocline also varies from ocean to ocean but averages about 4000 meters (13,100 feet).

In the geologic past, higher concentrations of carbon dioxide in the atmosphere have led to increased amounts of dissolved carbon dioxide in the ocean, thereby making the ocean more acidic and causing the CCD to rise (see Section 5.8 "What Factors Control the Distribution of Carbon and Oxygen in the Ocean?" in Chapter 5, "Water and Seawater."). Currently, scientists have documented an increase in

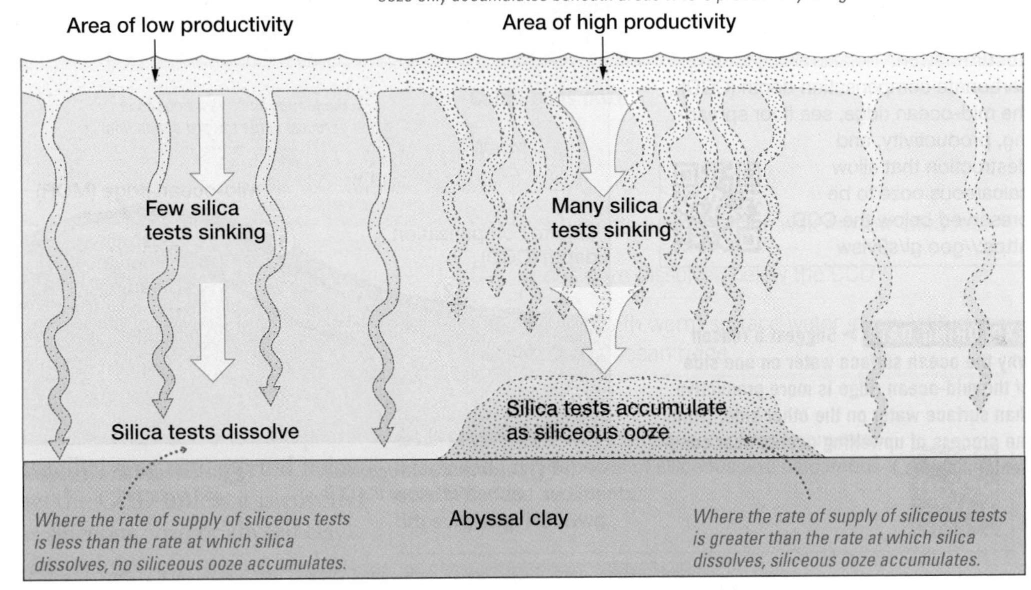
Silica-secreting organisms live in sunlit surface waters; siliceous ooze only accumulates beneath areas where productivity is high.
Area of low productivity
Area of high productivity
Few silica tests sinking
Many silica tests sinking
Silica tests dissolve
Silica tests accumulate as siliceous ooze
Where the rate of supply of siliceous tests is less than the rate at which silica dissolves, no siliceous ooze accumulates.
Abyssal clay
Where the rate of supply of siliceous tests is greater than the rate at which silica dissolves, siliceous ooze accumulates.

SmartFigure 4.13 **Accumulation of siliceous ooze.**
https://goo.gl/4iUEf1

**EXPLORING DATA ▲ What would cause siliceous ooze to accumulate in one area of the sea floor but not in another? Assume the water depth is the same in both places.**

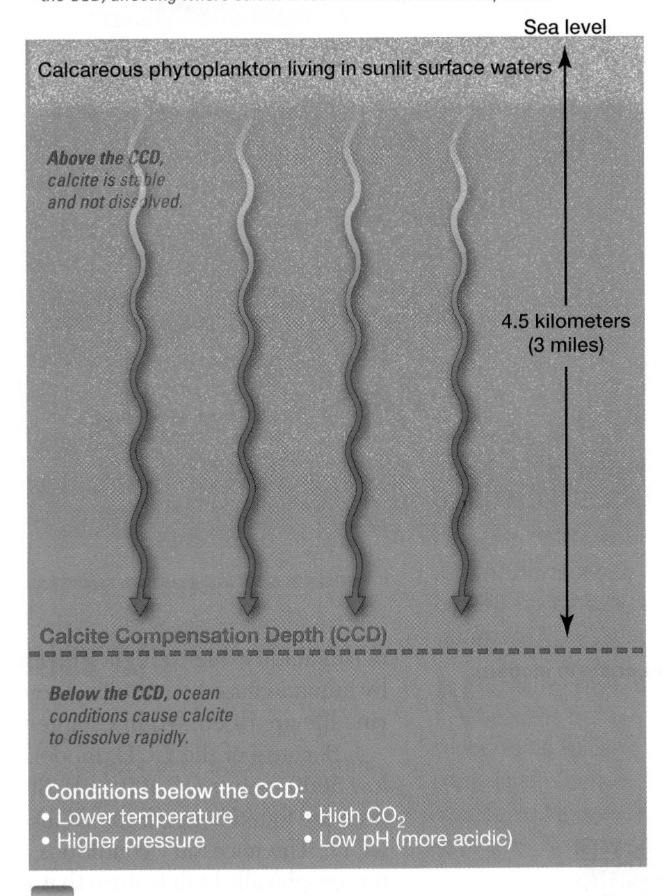
Ocean pressure increases and the properties of seawater change below the CCD, affecting where calcite dissolves and where it is deposited.
Sea level
Calcareous phytoplankton living in sunlit surface waters
Above the CCD, calcite is stable and not dissolved.
4.5 kilometers (3 miles)
Calcite Compensation Depth (CCD)
Below the CCD, ocean conditions cause calcite to dissolve rapidly.
Conditions below the CCD:
• Lower temperature
• Higher pressure
• High $CO_2$
• Low pH (more acidic)

**EXPLORING DATA ◀ Assume a seamount in water 8 kilometers (5 miles) deep rises to within 3.5 kilometers (2.2 miles) of the ocean's surface. To what water depth will calcite accumulate on the seamount's flanks?**

SmartFigure 4.14 **Characteristics of water above and below the calcite compensation depth (CCD).**
https://goo.gl/qbRcsr

dissolved to the solid state). Precipitation usually occurs when there is a *change in conditions*, such as a change in temperature or pressure, or the injection of chemically reactive fluids into seawater, such as happens at hydrothermal vents. In this case, metals such as silver and gold precipitate out of solution to form important mineral deposits. To take a more familiar example, the process is similar to how you would make rock candy. Here, a pan of water is heated and sugar is added. When the water is hot and the sugar dissolved, the pan is removed from the heat, and the sugar water is allowed to cool. The *change in temperature* causes the sugar to become oversaturated, which causes it to precipitate. As the water cools, the sugar precipitates on anything that is put in the pan, such as pieces of string or kitchen utensils.

## Composition and Distribution of Hydrogenous Sediment

Although hydrogenous sediments represent a relatively small portion of the overall sediment in the ocean, they have many different compositions and are distributed in diverse environments of deposition.

**METAL SULFIDES** Deposits of **metal sulfides** are associated with hydrothermal vents and black smokers along the mid-ocean ridge. These deposits contain iron, nickel, copper, zinc, silver, and other metals in varying proportions. Transported away from the mid-ocean ridge by sea floor spreading, these deposits can be found throughout the ocean floor and can even be uplifted onto continents (see Mastering Oceanography **Web Diving Deeper 2.2**). Many important metal ore deposits now found on land originated at deep sea hydrothermal vents.

**(a)** Manganese nodules, including some that are cut in half.

**(b)** Close-up of a baseball-sized manganese nodule cut in half, revealing its central nucleation object and layered internal structure.

**(c)** An abundance of manganese nodules on a portion of the deep South Pacific Ocean floor about 4 meters (13 feet) across.

Figure 4.17 **Manganese nodules.**

**MANGANESE NODULES** **Manganese nodules** are rounded, hard lumps of manganese, iron, and other metals, they typically range from 5 centimeters (2 inches) in diameter to a maximum of about 20 centimeters (8 inches). When cut in half, they often reveal a layered structure formed by precipitation around a central nucleation object (**Figures 4.17a** and **4.17b**). The nucleation object may be a piece of lithogenous sediment, coral, volcanic rock, a fish bone, or a shark's tooth. Manganese nodules lie on sediment in vast expanses of abyssal plains, potentially covering some 60% of the ocean basin at a typical water depth of about 5 kilometers (3.1 miles). Manganese nodules can sometimes occur in concentrations of about 100 nodules per square meter (square yard); in rare cases, they exist in even greater abundance (**Figure 4.17c**), resembling a scattered field of golf ball- to baseball-sized nodules. The formation of manganese nodules requires extremely low rates of lithogenous or biogenous input so that the nodules are not buried.

The major components of these nodules are manganese hydroxide (around 30% by weight) and iron hydroxide (around 20%). The element manganese is important for making high-strength steel alloys. Other accessory metals present in manganese nodules include copper (used in electrical wiring, in pipes, and to make brass and bronze), nickel (used to make stainless steel), and cobalt (used as an alloy with iron to make strong magnets and steel tools). Although the concentration of these accessory metals is usually less than 1%, they can exceed 2% by weight, which may make them attractive sources for these metals in the future.

The origin of manganese nodules has puzzled oceanographers since they were first discovered in 1872 during the voyage of HMS *Challenger*. (For more information about the accomplishments of the *Challenger* expedition, see Mastering Oceanography Web Diving Deeper 5.2.) If manganese nodules are truly hydrogenous and precipitate from seawater, then how can they have such high concentrations of manganese (which occurs in seawater at concentrations often too small to measure accurately)? Furthermore, why are the nodules on *top* of ocean floor sediment and not buried by the constant rain of sedimentary particles?

Unfortunately, nobody has definitive answers to these questions. Perhaps manganese nodules are created by one of the slowest chemical reactions known—on average, they grow at a rate of about 5 millimeters (0.2 inch) per *million years*. Scientific studies suggest that the formation of manganese nodules may be aided by bacteria and an as-yet-unidentified marine organism that intermittently lifts and rotates them. Other studies reveal that the nodules don't form continuously over time but in spurts that are related to specific conditions such as a low sedimentation rate of lithogenous clay and strong deep-water currents. Remarkably, the larger the nodules are, the faster they grow. The origin of manganese nodules is widely considered the most interesting unresolved problem in marine chemistry.

Figure 4.18 **Evaporative salts cover the floor of a seasonally flooded basin.** After seasonal rains at Death Valley, California, the high evaporation rate causes salts (white material) to precipitate out, resulting in this extensive salt flat.

**PHOSPHATES** Phosphorus-bearing compounds (**phosphates**) occur abundantly as coatings on rocks and as nodules on the continental shelf and on banks at depths shallower than 1000 meters (3300 feet). Concentrations of phosphates in such deposits commonly reach 30% by weight and indicate abundant biological activity in surface water above where they accumulate. Because phosphates are valuable as fertilizers, ancient marine phosphate deposits that have been uplifted onto land are extensively mined to supply agricultural needs.

**CARBONATES** The two most important carbonate minerals in marine sediment are calcite and **aragonite**. Both are composed of calcium carbonate ($CaCO_3$), but aragonite has a different crystalline structure that is less stable and transforms into calcite over time. Carbonates are widely used in the construction industry, in the production of cement, and they are commonly used medicinally as calcium supplements or antacids.

As previously discussed, most carbonate deposits are biogenous in origin. However, hydrogenous carbonate deposits can precipitate directly from seawater in tropical climates to form aragonite crystals less than 2 millimeters (0.08 inch) long. In addition, **oolites** (*oo* = egg, *lithos* = rock) are small calcite spheres 2 millimeters (0.08 inch) or less in diameter that have layers like an onion and form in some shallow tropical waters where concentrations of $CaCO_3$ are high. Oolites are thought to precipitate around a nucleus and grow larger as they roll back and forth on beaches by wave action, but some evidence suggests that a type of algae may aid their formation.

**EVAPORITES** Generally, **evaporite minerals** form wherever there are high evaporation rates (dry climates) accompanied by restricted open ocean circulation. One such example is the Mediterranean Sea, which contains thick deposits of evaporites on its floor that suggest that sometime in the geologic past, the sea completely dried up (see Mastering Oceanography Web Diving Deeper 4.1). As water evaporates in these dry areas, the remaining seawater becomes saturated with dissolved minerals, which then begin to precipitate (form a solid). Because they are heavier than seawater, the minerals sink to the bottom or form a white crust of evaporite minerals around the edges of these areas (Figure 4.18). Collectively termed "salts," some evaporite minerals, such as *halite* (common table salt, NaCl),

**RECAP** **Hydrogenous sediment is produced when dissolved materials precipitate out of solution, producing a variety of materials, and are found in localized concentrations on the ocean floor.**

## STUDENTS SOMETIMES ASK . . .

***How are scientists able to identify cosmogenous sediment? I mean, how can they tell that it's extraterrestrial?***

Cosmogenous sediment can be differentiated from other sediment types primarily by its structure but also by its composition. Cosmogenous sediment can be either silicate rock or rich in iron—both of which are common compositions of lithogenous sediment. However, glassy fragments indicative of melting (called tektites) are uniquely cosmogenous, as are iron-rich spherules (see Figure 4.19). Compositionally, cosmogenous particles from outer space typically contain more nickel than those that originate in other ways; most of the nickel in Earth's crust sank below the surface during density stratification early in Earth's history.

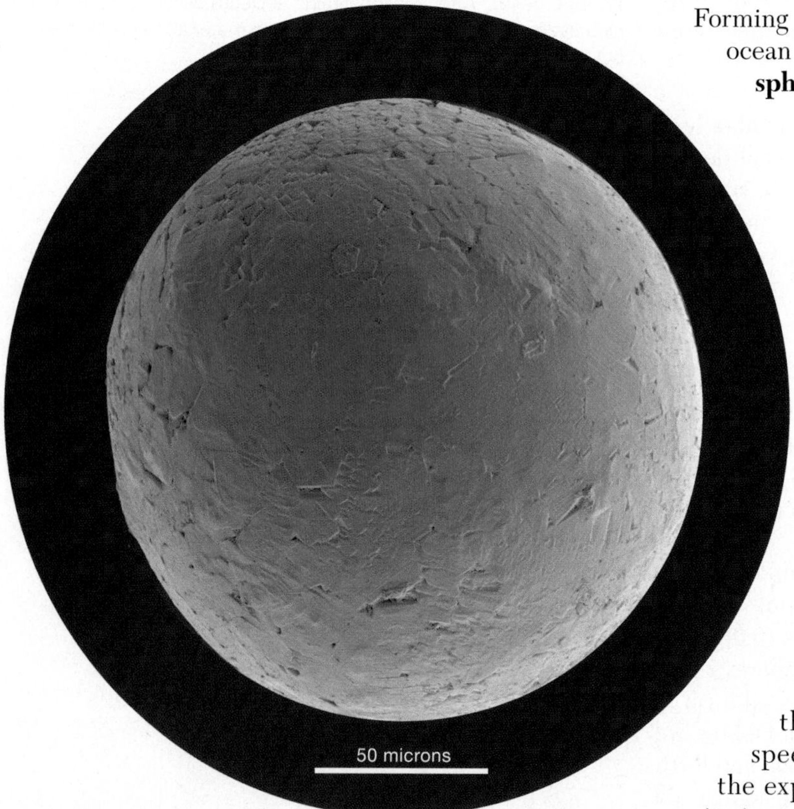

50 microns

**Figure 4.19 Microscopic cosmogenous spherule.** Scanning electron micrograph of an iron-rich spherule of cosmic dust. Note that 50 microns is about half the width of a human hair.

**RECAP Cosmogenous sediment is produced from materials originating in outer space and includes microscopic space dust and macroscopic meteor debris.**

taste salty, and some, such as the calcium sulfate minerals *anhydrite* ($CaSO_4$) and *gypsum* ($CaSO_4 \cdot H_2O$), do not.

**CONCEPT CHECK 4.4 ▸ Describe the characteristics of hydrogenous sediment.**

1 Describe the origin, composition, and distribution of hydrogenous sediment.

2 Describe manganese nodules, including what is currently known about how they form.

## 4.5 ▸ What Are the Characteristics of Cosmogenous Sediment?

**Cosmogenous sediment** (*cosmos* = universe, *generare* = to produce) is derived from extraterrestrial sources.

### Origin, Composition, and Distribution of Cosmogenous Sediment

Forming a very small but important component of the overall sediment on the ocean floor, cosmogenous sediment consists of two main types: microscopic **spherules** and macroscopic **meteor** debris.

Microscopic spherules are small globular masses. Some spherules are composed of silicate rock material and show evidence of being formed by extraterrestrial impact events on Earth or other planets that eject small molten pieces of crust into space. These **tektites** (*tektos* = molten) then rain down on Earth and can form *tektite fields*. Other spherules are composed mostly of iron and nickel (**Figure 4.19**); these form in the asteroid belt between the orbits of Mars and Jupiter when asteroids collide. This material constantly rains down on Earth as a general component of *space dust* or *micrometeorites* captured by Earth's gravity. Although about 90% of micrometeorites are destroyed by frictional heating as they enter the atmosphere, NASA has estimated that as much as 100 metric tons (220,000 pounds) of space dust reach Earth's surface each and every day! The iron-rich space dust that lands in the oceans often dissolves in seawater. Glassy tektites, however, do not dissolve as easily and sometimes comprise minute proportions of various marine sediments.

Macroscopic meteor debris is rare on Earth but can be found associated with meteor impact sites. Evidence suggests that throughout time meteors have collided with Earth at great speeds, and that some larger ones have released energy equivalent to the explosion of multiple large nuclear bombs. To date, over 200 meteorite impact structures have been identified on Earth; most of them are on land, but new ones are being discovered on the ocean floor as well (see **Process of Science 4.1**). The debris from meteors—called **meteorite** material—settles out around the impact site and is either composed of silicate rock material (called *chondrites*) or iron and nickel (called *irons*).

**CONCEPT CHECK 4.5 ▸ Describe the characteristics of cosmogenous sediment.**

1 Describe the origin, composition, and distribution of cosmogenous sediment.

2 Describe the most common types of cosmogenous sediment and give the probable source of these particles.

## PROCESS OF SCIENCE 4.1: When The Dinosaurs Died: The Cretaceous–Tertiary (K–T) Event

### BACKGROUND

The extinction of the dinosaurs—and about 75% of all plant and animal species on Earth, including many marine species—occurred about 66 million years ago. This extinction marks the boundary between the Cretaceous (K) and Tertiary (T) Periods of geologic time and is known as the **K–T event** or, because of recent changes in the geologic time scale, the *Cretaceous–Paleogene (K–Pg) event*. Did slow climate change lead to the extinction of these organisms, or was it a catastrophic event? Was their demise related to disease, diet, predation, or volcanic activity? Earth scientists have long sought clues to this mystery.

### FORMING A HYPOTHESIS

In 1980, geologist Walter Alvarez, his father, Nobel Physics Laureate Luis Alvarez, and two nuclear chemists, Frank Asaro and Helen Michel, reported that marine deposits collected in northern Italy from the K–T boundary contained an unusual clay layer with high proportions of the metallic element iridium (Ir), an element rare in Earth rocks but much more abundant in meteorites. The high concentrations of iridium suggested minerals in the clay had an extraterrestrial origin. In addition, the clay layer contained shocked quartz grains, indicating an event had occurred with enough force to fracture and partially melt pieces of quartz. Other deposits from the K–T boundary revealed similar features, supporting the hypothesis that Earth experienced an extraterrestrial impact at the same time that the dinosaurs died.

One problem with the impact hypothesis, however, is that dust spewing from volcanic eruptions on Earth could create similar clay deposits enriched in iridium and containing shocked quartz. In fact, at about the same time as the dinosaur extinction, large outpourings of basaltic volcanic rock in India (called the Deccan Traps) and other locations had occurred. Also, if there was a catastrophic meteor impact, where was the crater?

In the early 1990s, the 190-kilometer (120-mile)-wide *Chicxulub* (pronounced "SCHICK-sue-lube") *Crater* off the Yucatán coast in the Gulf of Mexico was identified as a likely candidate because of its structure, age, and size. To create a crater this large, a 10-kilometer (6-mile)-wide object composed of rock and/or ice traveling at speeds up to 72,000 kilometers (45,000 miles) per hour must have slammed into Earth (Figure 4B). Such an impact would have created huge waves—estimated to be more than 900 meters (3000 feet) high—that traveled throughout the oceans. In addition, the dust and debris lifted into the atmosphere most likely limited photosynthesis, chilled Earth's surface, and brought about the extinction of the dinosaurs and many other species. Finally, acid rains and global fires may have added to the environmental disaster.

### DEVISING AN EXPERIMENT

Supporting evidence for the meteor impact hypothesis was provided in 1997 by recovering cores of sediment from the sea floor. Previous drilling close to the impact site did not reveal any K–T deposits. Evidently, the impact and resulting huge waves had stripped the ocean floor of its sediment. However, at 1600 kilometers (1000 miles) from the impact site, the telltale sediments from the catastrophe, such as the iridium-rich clay layer, were preserved in sea floor sediments.

### INTERPRETING THE RESULTS

Convincing evidence of the K–T impact from this and other cores collected in 2016 suggests that Earth has experienced many such extraterrestrial impacts over geologic time. Statistics show that an impact the size of the K–T event should occur on Earth about once every 100 million years, severely affecting life on Earth as it did the dinosaurs. This frequency is consistent with the fossil record, which indicates that in the last 500 million years, Earth has experienced five major extinction events.

### THINKING LIKE A SCIENTIST: WHAT'S NEXT?

What kind of evidence would you expect to find in coastal rock sequences that were deposited during the time of the huge waves that were created by the meteor impact?

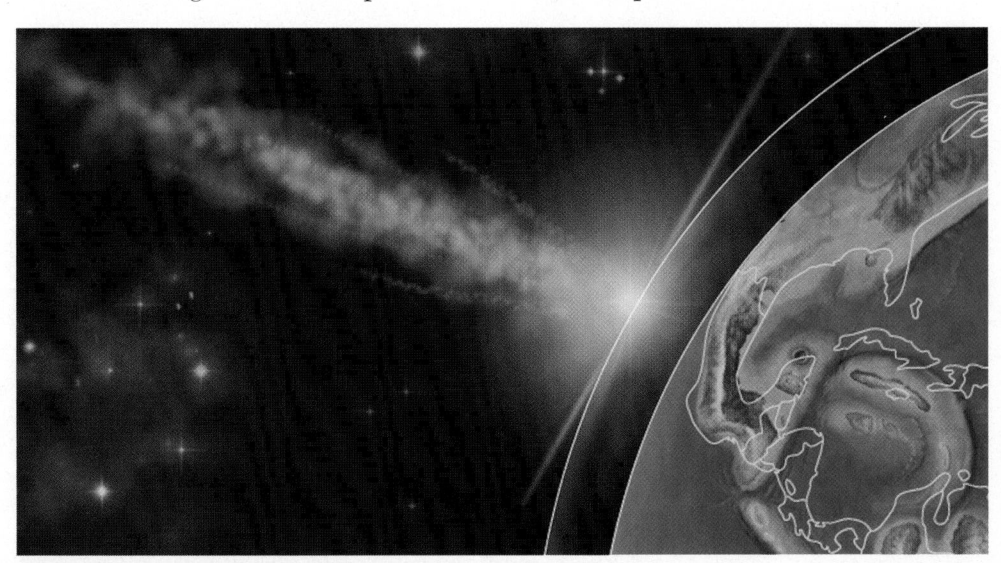

Figure 4B **The K–T meteorite impact event.**

## 4.6 ▶ How Are Pelagic and Neritic Deposits Distributed?

The ocean is a messy place. Lithogenous and biogenous sediment rarely occur as absolutely pure deposits that do not contain other types of sediment. As a result, most marine sediments occur as mixtures.

### Mixtures of Marine Sediment

Mixtures of marine sediments are incredibly common on Earth and are comprised of varying proportions of sediment types. However, in order to classify marine sediments according to a single type, oceanographers note the following characteristics about sediments:

- The abundance of clay-sized lithogenous particles throughout the world and the ease with which they are transported by winds and currents means that these particles are incorporated into every sediment type.
- Most lithogenous sediment contains small percentages of biogenous particles.
- The composition of biogenous ooze includes up to 70% fine-grained lithogenous clays.
- Most calcareous oozes contain some siliceous material and vice versa (see, for example, Figure 4.10d).
- There are many types of hydrogenous sediment.
- Tiny amounts of cosmogenous sediment are mixed in with all other sediment types.

**RECAP** **Although most ocean sediment is a mixture of various sediment types, it is usually dominated by lithogenous, biogenous, or hydrogenous material.**

Although sediment deposits on the sea floor are usually a mixture of different sediment types, typically one type of sediment dominates, which allows the deposit to be classified as primarily lithogenous, biogenous, hydrogenous, or cosmogenous. **Figure 4.20** shows the distribution of sediment across a passive continental margin

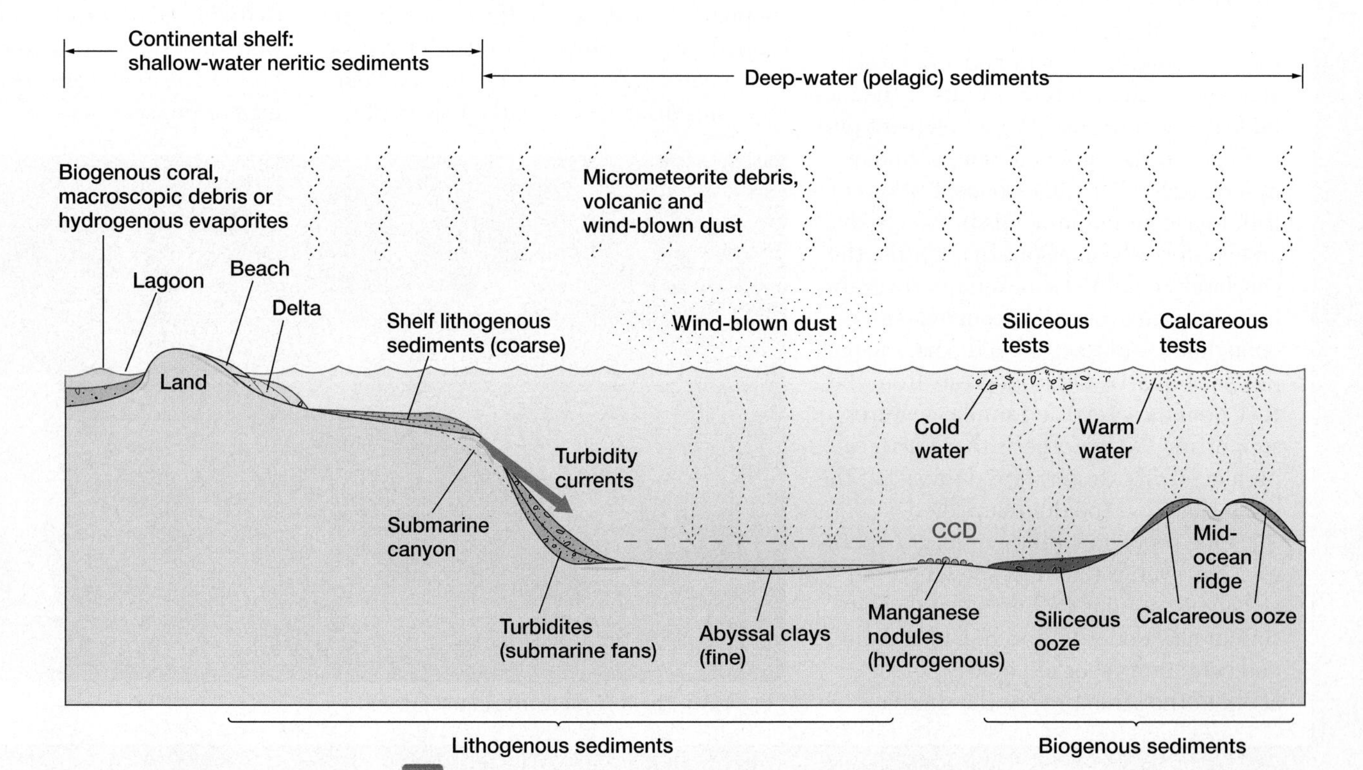


**SmartFigure 4.20 Distribution of sediment across a passive continental margin.** Schematic profile view of various sediment types and their distribution across an idealized passive continental margin and extending out to a mid-ocean ridge. https://goo.gl/07cWpo

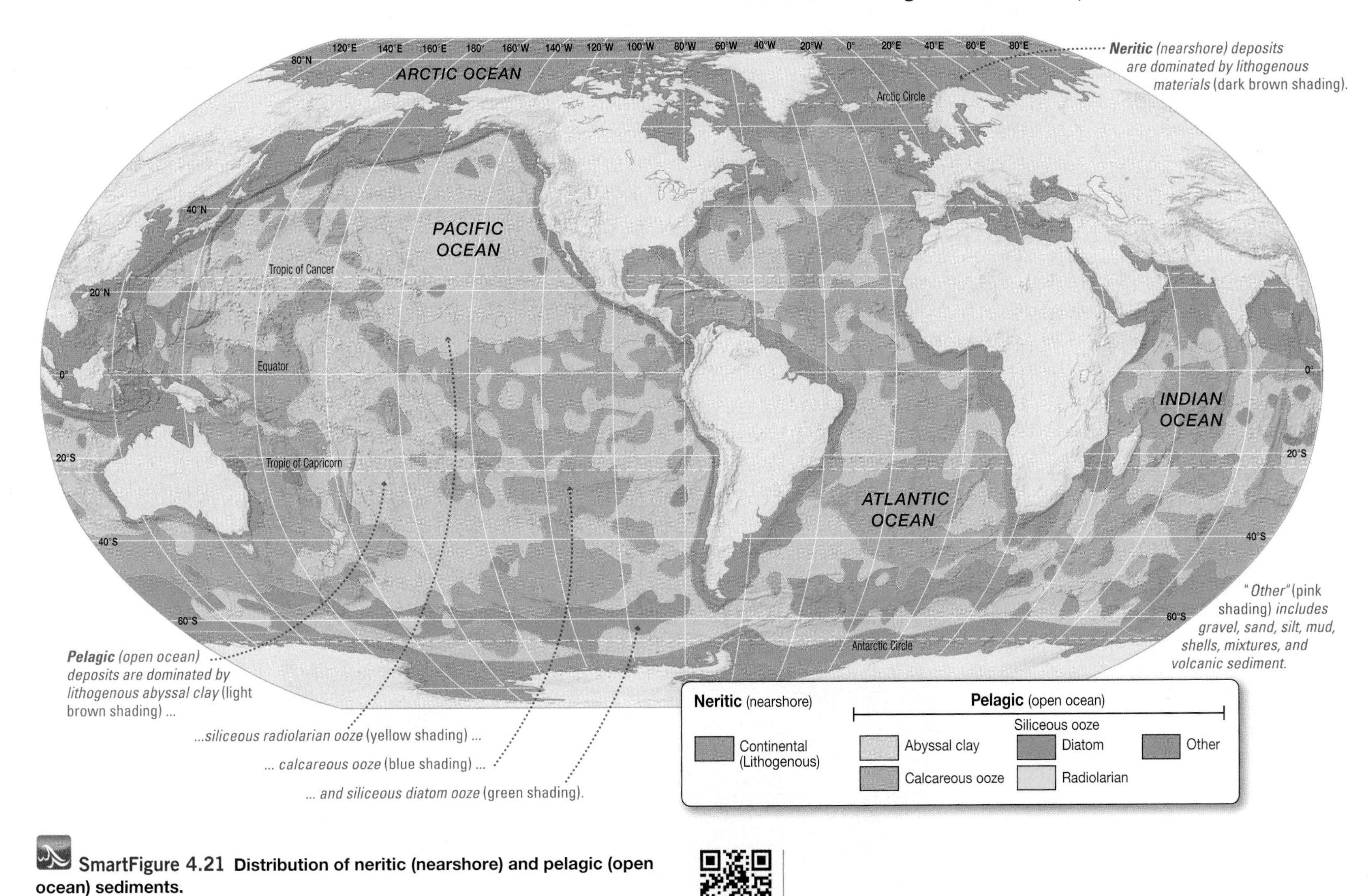
ARCTIC OCEAN
PACIFIC OCEAN
ATLANTIC OCEAN
INDIAN OCEAN
Arctic Circle
Tropic of Cancer
Equator
Tropic of Capricorn
Antarctic Circle
Neritic (nearshore) deposits are dominated by lithogenous materials (dark brown shading).
Pelagic (open ocean) deposits are dominated by lithogenous abyssal clay (light brown shading) ...
...siliceous radiolarian ooze (yellow shading) ...
... calcareous ooze (blue shading) ...
... and siliceous diatom ooze (green shading).
"Other" (pink shading) includes gravel, sand, silt, mud, shells, mixtures, and volcanic sediment.
Neritic (nearshore)
Continental (Lithogenous)
Pelagic (open ocean)
Siliceous ooze
Abyssal clay
Diatom
Other
Calcareous ooze
Radiolarian

**SmartFigure 4.21 Distribution of neritic (nearshore) and pelagic (open ocean) sediments.**
https://goo.gl/0sxYZs

and illustrates how sediment type changes, going from shallow water to deep water, and how mixtures occur.

## Neritic Deposits

Neritic (nearshore) deposits cover about one-quarter of the ocean floor, and pelagic (deep-ocean basin) deposits cover the other three-quarters. The map in **Figure 4.21** shows the distribution of neritic and pelagic deposits in the world's oceans. Coarse-grained lithogenous neritic deposits dominate continental margin areas (*dark brown shading*), which is not surprising because lithogenous sediment is derived from nearby continents. Although neritic deposits usually contain biogenous, hydrogenous, and cosmogenous particles, these constitute only a minor percentage of the total sediment mass.

## Pelagic Deposits

Figure 4.21 shows that pelagic deposits are dominated by biogenous calcareous oozes (*blue shading*), which are found on the relatively shallow deep-ocean areas along the mid-ocean ridge. Biogenous siliceous oozes are found beneath areas of unusually high biological productivity such as the northernmost North Pacific Ocean, surrounding Antarctica (*green shading*, where diatomaceous ooze occurs), and the equatorial Pacific (*yellow shading*, where radiolarian ooze occurs). Fine lithogenous pelagic deposits of abyssal clays (*light brown shading*) are common in deeper areas

## STUDENTS SOMETIMES ASK . . .

***Are there any areas of the ocean floor where no sediment is being deposited?***

Various types of sediment accumulate on nearly all areas of the ocean floor in the same way dust accumulates in all parts of your home (which is why marine sediment is often referred to as "marine dust"). Even the deep-ocean floor far from land receives small amounts of windblown material, microscopic biogenous particles, and space dust. There are some places in the ocean, however, where very little sediment accumulates. A few such places include: (1) the South Pacific Bare Zone east of New Zealand, where a combination of factors limit sediment accumulation, (2) along the continental slope, where there is active erosion by turbidity and other deep-ocean currents, and (3) along the mid-ocean ridge, where the age of the sea floor is so young (because of sea floor spreading) and the rates of sediment accumulation far from land are so slow that there hasn't been enough time for sediments to accumulate.

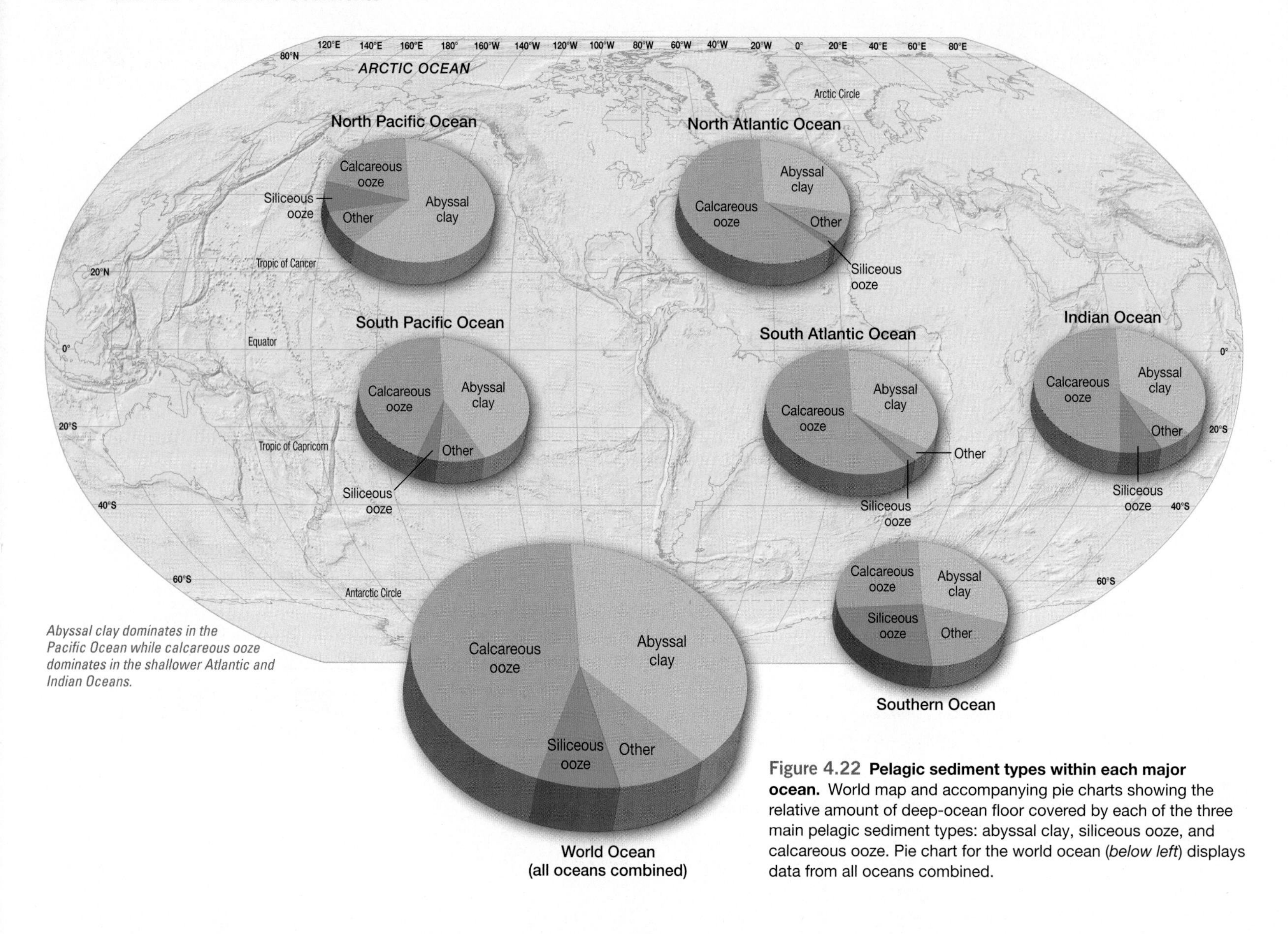


**Figure 4.22 Pelagic sediment types within each major ocean.** World map and accompanying pie charts showing the relative amount of deep-ocean floor covered by each of the three main pelagic sediment types: abyssal clay, siliceous ooze, and calcareous ooze. Pie chart for the world ocean (*below left*) displays data from all oceans combined.

of the ocean basins, such as in the North Pacific. Hydrogenous and cosmogenous sediment comprise only a small proportion of pelagic deposits in the ocean.

Figure 4.22 shows the proportion of each ocean floor that is covered by the pelagic deposits abyssal clay, calcareous ooze, and siliceous ooze. The world ocean (combined) pie chart shows that calcareous ooze is the most dominant sediment worldwide, covering about 45% of the deep-ocean floor. The world ocean pie chart also shows that abyssal clay covers about 38% and siliceous ooze about 8% of the world ocean floor area. If you examine the individual ocean pie charts, they show that the amount of ocean basin floor covered by calcareous ooze decreases in deeper ocean basins because they generally lie beneath the CCD. The dominant oceanic sediment in the deepest basin—the North Pacific—is abyssal clay (see also Figure 4.21). Conversely, calcareous ooze is the most widely deposited sediment in the shallower Atlantic and Indian Oceans. Note that siliceous oozes cover a smaller percentage of the ocean floor because regions of high productivity of organisms that produce silica tests are generally restricted to the equatorial region (for radiolarians) and the high latitudes such as near Antarctica and the far northern Pacific (for diatoms). Table 4.4 shows the average rates of deposition (also called *sedimentation rate*) of selected marine sediments in neritic and pelagic deposits.

**RECAP** **Neritic deposits occur close to shore and are dominated by coarse lithogenous material. Pelagic deposits occur in the deep ocean and are dominated by biogenous oozes and fine lithogenous clay.**

**Table 4.4 Average rates of deposition (sedimentation rate) of selected marine sediments**

| Type of sediment deposit | Average rate of deposition (per 1000 years) | Thickness of deposit after 1000 years equivalent to . . . |
|---|---|---|
| Coarse lithogenous sediment, neritic deposit | 1 meter (3.3 feet) | A meter stick |
| Biogenous ooze, pelagic deposit | 1 centimeter (0.4 inch) | The diameter of a dime |
| Abyssal clay, pelagic deposit | 1 millimeter (0.04 inch) | The thickness of a dime |
| Manganese nodule, pelagic deposit | 0.001 millimeter (0.00004 inch) | A microscopic dust particle |

**EXPLORING DATA ▲ Using the depositional rates shown in Table 4.4, how long would it take to make a deposit 1 meter (3.3 feet) thick of biogenous ooze? A deposit 1 meter (3.3 feet) thick of abyssal clay?**

## Do Sea Floor Sediments Represent Surface Conditions?

Because of their tiny size and the enormous distance to the sea floor, microscopic biogenous tests should take from 10 to 50 years to sink from the ocean surface where the organisms lived to the abyssal depths where biogenous ooze accumulates. During this time, even a sluggish horizontal ocean current of only 0.05 kilometer (0.03 mile) per hour could carry tests as much as 22,000 kilometers (13,700 miles) before they settled onto the deep-ocean floor. Why, then, do biogenous tests on the deep-ocean floor closely reflect the population of organisms living in the surface water directly above? Remarkably, about 99% of the particles that fall to the ocean floor do so as part of *fecal pellets*, which are produced by tiny animals that eat algae and protozoans living in the water column, digest their tissues, and excrete their hard parts. These pellets are full of the remains of algae and protozoans from the surface waters (**Figure 4.23**) and, though still small, are large enough to sink to the deep-ocean floor in only 10 to 15 days. Once fecal pellets settle onto the ocean floor, the organic material in them is quickly consumed by bacteria and other microbes, releasing the indigestible, inorganic hard parts to the sediment.

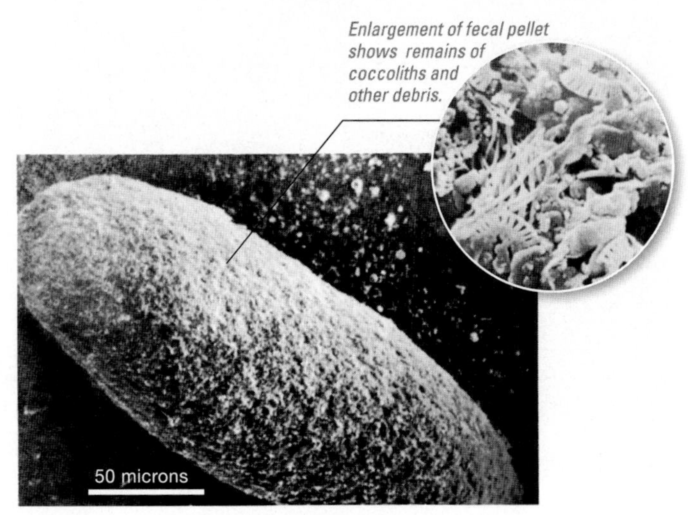

Enlargement of fecal pellet shows remains of coccoliths and other debris.
50 microns

**Figure 4.23 Fecal pellet.** Photomicrograph of a fecal pellet, which is large enough to sink rapidly from the surface to the ocean floor.

## Worldwide Thickness of Marine Sediments

**Figure 4.24** is a map of marine sediment thickness. The map shows that areas of thick sediment accumulation occur on the continental shelves and rises, especially near the mouths of major rivers. The reason sediments in these locations are so thick is because they are close to major sources of lithogenous sediment. Conversely, areas where marine sediments are thinnest are where the ocean floor is young, such as along the crest of the mid-ocean ridge. Since sediments accumulate slowly in the deep ocean and the sea floor is continually being created here, there hasn't been enough time for much sediment to accumulate. However, as the sea floor moves away from the mid-ocean ridge, it gets progressively older and carries a thicker pile of sediments.

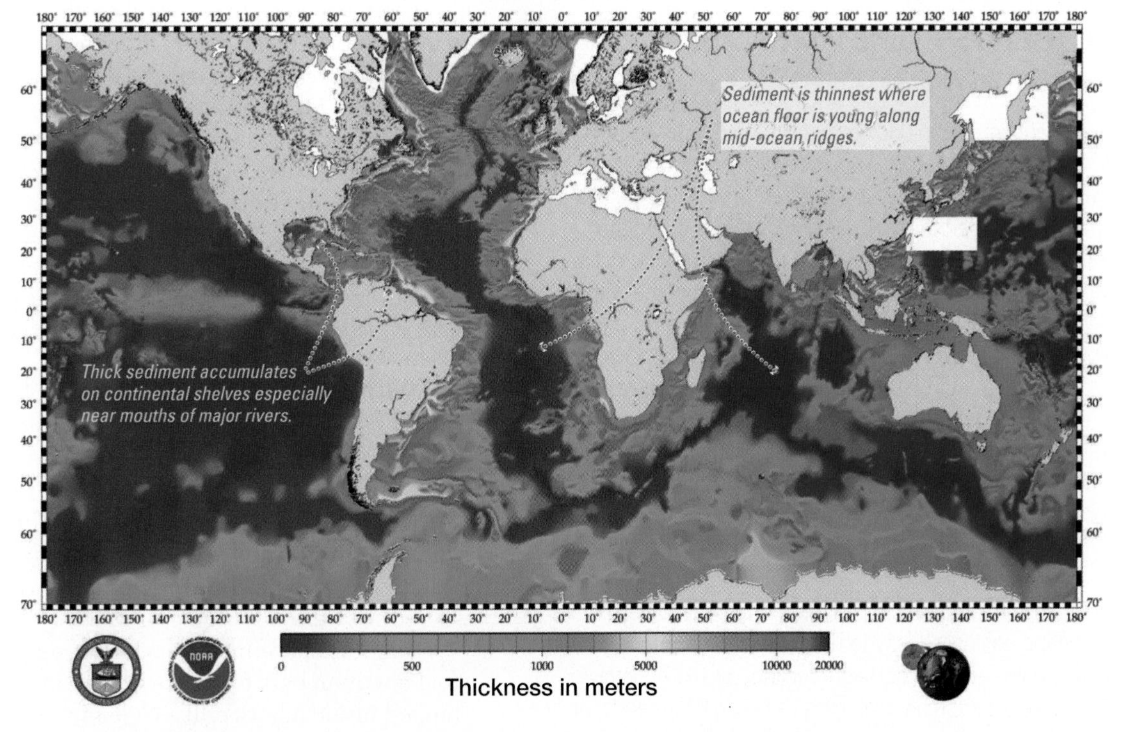


**Figure 4.24 Marine sediment thickness.** Map showing the thickness of sediments in the oceans and marginal seas. Thickness shown in meters; dark blue color represents thinnest sediments, and red represents thickest sediment accumulations. White color indicates no available data.

**CONCEPT CHECK 4.6 ▸ Specify how the distribution of pelagic and neritic deposits is determined by proximity to sediment sources and agents of transport.**

1 Why is it so rare to find a pure marine sediment type? Give some examples of mixtures of sediment.

2 Why is lithogenous sediment the most common neritic deposit? Why are biogenous oozes the most common pelagic deposits?

3 How do fecal pellets help explain why the particles found in the ocean surface waters are closely reflected in the particle composition of the sediment directly beneath? Why is this unexpected?

Figure 4.25 **Offshore oil-drilling platform.** Constructed on tall stilts, drilling platforms are important for extracting petroleum reserves from beneath the continental shelves.

## STUDENTS SOMETIMES ASK . . .

***When will we run out of oil?***

Not anytime soon. However, from an economic perspective, when the world runs completely out of oil—a finite resource—is not as relevant as when production begins to taper off. When this happens, we will run out of the *abundant* and *cheap* oil on which all industrialized nations depend. Current estimates indicate that sometime within the next few decades, more than half of all known and likely-to-be-discovered oil will be gone. Other experts have suggested that petroleum production has already reached its plateau. In fact, several oil-producing countries were once considered to be past the peak of their production, including the United States and Canada, both of which peaked in 1972. However, recent advances in the controversial method of hydraulic fracturing (called *fracking*) have reversed the decline in U.S. oil production, which is now once again near its peak. Still, once the decline begins, it will be increasingly costly to produce oil, and prices will rise dramatically—unless demand declines proportionately or other sources such as coal, extra-heavy oil, tar sands, or gas hydrates become readily available.

# 4.7 ▸ What Resources Do Marine Sediments Provide?

The sea floor is rich in potential mineral and organic resources. Much of these resources, however, are not easily accessible, so their recovery involves technological challenges and high cost. In addition, there are legal concerns about who owns the rights to extract sea floor resources (see Mastering Oceanography **Web Diving Deeper 4.2**). Nevertheless, let's examine some of the most appealing exploration targets.

## Energy Resources

The main energy resources associated with marine sediments are *petroleum* and *gas hydrates*.

**PETROLEUM** The ancient remains of microscopic organisms, buried within marine sediments before they could decompose, are the source of today's **petroleum** (oil and natural gas) deposits. Of the nonliving resources extracted from the oceans, more than 95% of the economic value is in petroleum products.

The percentage of world oil produced from offshore regions has increased from small amounts in the 1930s to more than 30% today. Most of this increase results from continuing technological advancements employed by offshore drilling platforms (**Figure 4.25**). Major offshore reserves exist in the Persian Gulf, in the Gulf of Mexico, off Southern California, in the North Sea, and in the East Indies. Additional reserves are probably located off the north coast of Alaska and in the Canadian Arctic, below Asian seas, and off the coasts of Africa and Brazil. With almost no likelihood of finding major new reserves on land, future offshore petroleum exploration will continue to be intense, especially in deeper waters of the continental margins. However, a major drawback to offshore petroleum exploration is the inevitable oil spills caused by inadvertent leaks or blowouts during the drilling process.

**GAS HYDRATES** **Gas hydrates**, which are also known as *clathrates* (*clathri* = a lattice), are unusually compact chemical structures made of water and natural gas. They form only when high pressures squeeze chilled water and gas molecules into an ice-like solid. Although hydrates can contain a variety of gases—including carbon dioxide, hydrogen sulfide, and larger hydrocarbons such as ethane and propane—**methane hydrates** are by far the most common hydrates in nature. Gas hydrates occur beneath Arctic permafrost areas on land and under the ocean floor, where they were discovered in 1976.

Deep-ocean sediments, where pressures are high and temperatures are low, are ideal environments for water and natural gas combine in such a way that the gas is trapped inside a lattice-like cage of water molecules. Vessels that have drilled into gas hydrates have retrieved cores of mud mixed with chunks or layers of gas hydrate "ice" that fizzle and decompose quickly when exposed to the relatively warm, low-pressure conditions at the ocean surface. Gas hydrates may resemble chunks of ice but ignite when lit by a flame because methane and other flammable gases are released as gas hydrates vaporize (**Figure 4.26**).

Most oceanic gas hydrates are created when bacteria break down organic matter trapped in sea floor sediments, producing methane gas with minor amounts of ethane and propane. These gases can be incorporated into gas hydrates under high-pressure and low-temperature conditions. Most ocean floor areas below 525 meters (1720 feet) provide these conditions, but gas hydrates seem to be confined to continental margin areas, where high productivity surface waters enrich ocean floor sediments below with organic matter.

Studies of the deep-ocean floor reveal that at least 50 sites worldwide may contain extensive gas hydrate deposits. Interestingly, sea floor methane seeps support a rich community of organisms, many of which are species new to science.

The release of methane from the sea floor to the atmosphere can have dramatic effects on global climate. Research suggests that at various times in the geologic past, changes in sea level or sea floor instability have released large quantities of methane, which is the third-most-important greenhouse gas after water vapor and carbon dioxide. In fact, analysis of sea floor sediments off Norway suggests that an abrupt increase in global temperature about 55 million years ago was driven by an explosive release of gas hydrates from the sea floor. Today, a major concern is that recent climate changes could warm ocean waters enough to release additional methane that is trapped beneath the seabed, causing even more warming. Sudden releases of methane hydrates have also been linked to underwater slope failure, which can cause *seismic sea waves*, or *tsunami* (see Chapter 8, "Waves and Water Dynamics").

Climate

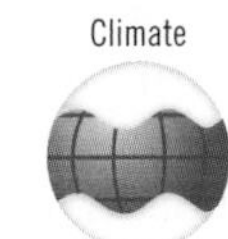

Connection

Some estimates indicate that as much as 20 quadrillion cubic meters (700 quadrillion cubic feet) of methane are locked up in marine sediments containing gas hydrates. This is equivalent to about *twice* as much carbon as Earth's coal, oil, and conventional gas reserves combined (**Figure 4.27**), so gas hydrates may potentially be the world's largest source of usable energy.

In spite of the energy potential that gas hydrates possess, several drawbacks exist. One major drawback in exploiting reserves of gas hydrate is that they rapidly decompose at surface temperatures and pressures. Another problem is that they are typically too dispersed within the sea floor to make collecting them economically feasible. An additional concern is that during commercial extraction of methane hydrates, methane could be accidentally released into the atmosphere, exacerbating fossil fuel-driven climate changes. Although technological advancements may be able to solve many of the specific challenges of safely extracting methane from deposits of gas hydrates, there are additional scientific, engineering, and environmental questions that need to be addressed before commercial operations can produce fuel from hydrates. Nonetheless, a multinational research team has evaluated the economic potential of collecting methane hydrates in the Nankai Trough off Japan and in 2017, China successfully extracted gas hydrates from sedimentary deposits along the continental margin in the South China Sea.

Climate

Connection

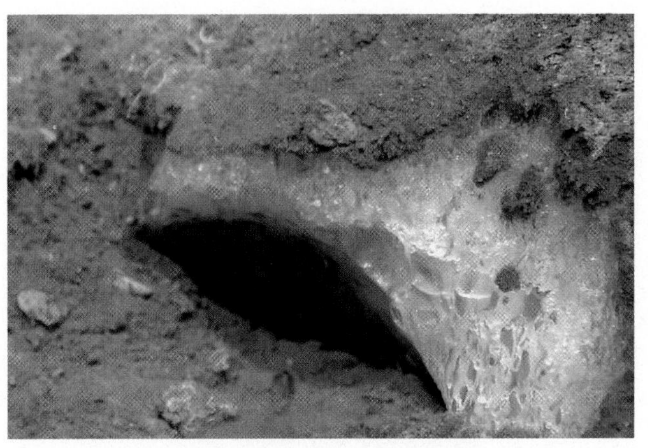

**(a)** Photo of ice-like gas hydrate created from a methane seep on the continental margin, depth = 1055 meters (3500 feet); width of photo is about 1 meter (3.3 feet).

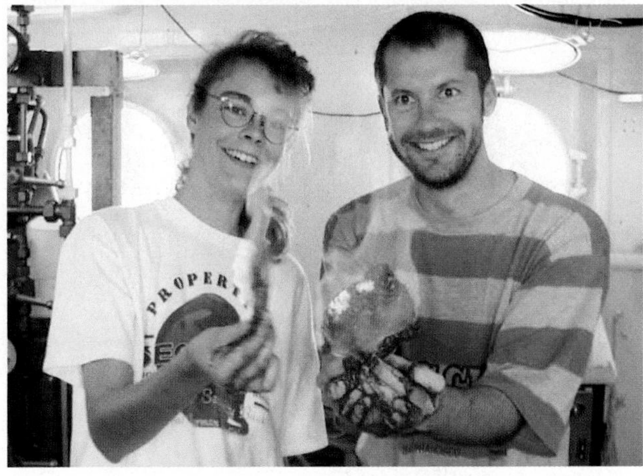

**(b)** Gas hydrates decompose when exposed to surface conditions and release natural gas, which can be ignited.

Figure 4.26 **Gas hydrates.** Gas hydrates are ice-like substances that form in deep-ocean sediments and are composed of natural gas combined with frozen water.

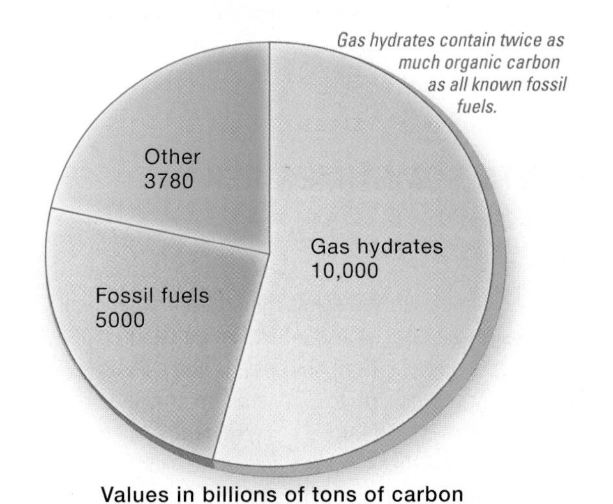


Figure 4.27 **Organic carbon in Earth reservoirs.** Pie chart showing the distribution of various types of organic carbon; "Other" includes sources such as soil, peat, and living organisms.

## Other Resources

Other resources associated with marine sediments include *sand and gravel*, *evaporative salts*, *phosphorite*, *manganese nodules and crusts*, and *rare-earth elements*.

**SAND AND GRAVEL** Sand and gravel, which includes both rock fragments that are washed out to sea and shells of marine organisms, is mined by offshore barges using suction dredges. This material is primarily used as aggregate in concrete, as fill material in grading projects, and on recreational beaches. In terms of economic value, offshore sand and gravel is the second largest sea floor resource behind petroleum.

Offshore deposits are a major source of sand and gravel in New England, New York, and throughout the Gulf Coast. Many European countries, Iceland, Israel, and Lebanon also mine similar offshore deposits.

Some offshore sand and gravel deposits are rich in valuable minerals. Gem-quality diamonds, for example, are recovered from gravel deposits on the continental shelf offshore of South Africa and Australia, where waves rework them during times of lower sea level. Sediments rich in tin have been mined offshore of southeast Asia from Thailand to Indonesia. Platinum and gold have been found in deposits offshore of gold mining areas throughout the world, and some Florida beach sands are rich in titanium. The largest unexplored potential for metallic minerals in offshore sand deposits may exist along the west coast of South America, where rivers have transported Andean metallic minerals.

Figure 4.28 **Mining sea salt.** A salt mining operation at Scammon's Lagoon, Baja California, Mexico. Low-lying areas near the lagoon are allowed to flood with seawater, which evaporates in the arid climate and leaves deposits of salt that are then collected.

**EVAPORATIVE SALTS** When seawater evaporates, the salts increase in concentration until they can no longer remain dissolved, so they precipitate out of solution and form **salt deposits** (Figure 4.28). Extensive sea floor salt deposits indicate that entire seas such as the Mediterranean Sea completely dried up in the geologic past (see Mastering Oceanography Web Diving Deeper 4.1).

The most economically useful salts are *gypsum* and *halite*. Gypsum is used in plaster of Paris to make casts and molds and is the main component in gypsum board (wallboard or sheet rock). Halite—common table salt—is widely used for seasoning, curing, and preserving foods. It is also used to de-ice roads, in water conditioners, in agriculture, and in the clothing industry for dying fabric.

In addition, halite is used in the production of chemicals such as sodium hydroxide (to make soap products), sodium hypochlorite (for disinfectants, bleaching agents, and PVC piping), sodium chlorate (for herbicides, matches, and fireworks), and hydrochloric acid (for use in chemical applications and for cleaning scaled pipes). The manufacture and use of salt is one of the oldest chemical industries. (In fact, an interesting historical note about salt is that part of a Roman soldier's pay was in salt. That portion was called the *salarium*, from which the word *salary* is derived. If a soldier did not earn it, he was said to be not worth his salt.)

**PHOSPHORITE (PHOSPHATE MINERALS)** **Phosphorite** is a sedimentary rock consisting of various phosphate minerals containing the element phosphorus, an important plant nutrient. Consequently, phosphate deposits can be used to produce phosphate fertilizer. Although there is currently no commercial phosphorite mining occurring in the oceans, the marine reserve is estimated to exceed 45 billion metric tons (99 trillion pounds). Phosphorite occurs in the ocean at depths of less than 300 meters (1000 feet) on the continental shelf and slope in regions of upwelling and high productivity.

Some shallow sand and mud deposits contain up to 18% phosphate. Many phosphorite deposits occur as nodules, with a hard crust formed around a nucleus. The nodules may be as small as a sand grain or as large as 1 meter (3.3 feet) in diameter and may contain more than 25% phosphate. For comparison, most land sources of phosphate have been enriched to more than 31% by groundwater leaching. Florida, for example, has large phosphorite deposits and supplies about one-quarter of the world's phosphates.

## STUDENTS SOMETIMES ASK . . .

***Where does Himalayan sea salt come from, and why is it pink?***

Surprisingly, Himalayan sea salt is mined from an ancient evaporite sea salt deposit in the Salt Range region of Pakistan, which is quite far from any ocean. The salt is very pure and typically contains up to 98% sodium chloride. The remainder of the salt is made up of trace minerals, such as iron, potassium, magnesium, and calcium, which give the salt its light pink tint. Because it contains slightly lower amounts of sodium than regular salt, it is touted to be a healthier version of salt.

**MANGANESE NODULES AND CRUSTS** *Manganese nodules* are rounded, hard, golf- to tennis-ball-sized lumps of metals that contain significant concentrations of manganese, iron, and smaller concentrations of copper, nickel, and cobalt, all of which have a variety of economic uses. In the 1960s, mining companies began to assess the feasibility of mining manganese nodules from the deep-ocean floor (Figure 4.29). The map in Figure 4.30 shows that vast areas of the sea floor contain manganese nodules, particularly in the Pacific Ocean.

Technologically, mining the deep-ocean floor for manganese nodules is possible. However, the political issue of determining international mining rights at great distances from land has hindered exploitation of this resource. In addition, environmental concerns about mining the deep-ocean floor have not been fully addressed. Evidence suggests that it takes at least several million years for manganese nodules to form and that their formation depends on a particular set of physical and chemical conditions that probably do not last long at any location. In essence, they are a nonrenewable resource that will not be replaced for a very long time once they are mined.

Of the five metals commonly found in manganese nodules, cobalt is the only metal deemed "strategic" (essential to national security) for the United States. It is required to produce dense, strong alloys with other metals for use in high-speed cutting tools, powerful permanent magnets, and jet engine parts. Currently, the United States must import all of its cobalt from large deposits in southern Africa. However, the United States has considered deep-ocean nodules and **crusts** (hard coatings on other rocks) as a more reliable source of cobalt.

In the 1980s, cobalt-rich manganese crusts were discovered on the upper slopes of islands and seamounts that lie relatively close to shore and within the jurisdiction of the United States and its territories. The cobalt concentrations in these crusts are about one-and-a-half times as rich as the best African ores and at least twice as rich as deep-sea manganese nodules. However, interest in mining these deposits has faded because of lower metal prices from land-based sources.

Figure 4.29 **Mining manganese nodules.** Manganese nodules can be collected by dredging the ocean floor. This metal dredge is shown unloading nodules onto the deck of a ship.

**RARE-EARTH ELEMENTS** *Rare-earth elements*—an assortment of 17 chemically similar metallic elements such as lanthanum and neodymium—are used in a variety of electronic, optical, magnetic, and catalytic applications. For example, rare-earth elements are used in a host of technological gadgets from cell phones and television screens to fluorescent light bulbs and batteries in electric cars. Demand for rare-earth elements has skyrocketed in recent years, with China supplying about 90% of the current world demand.

Over millions of years, deep-sea hot springs associated with the mid-ocean ridge pulled rare-earth elements out of seawater and enriched them in sea floor muds. A recent study of rare-earth elements on the floor of the Pacific Ocean

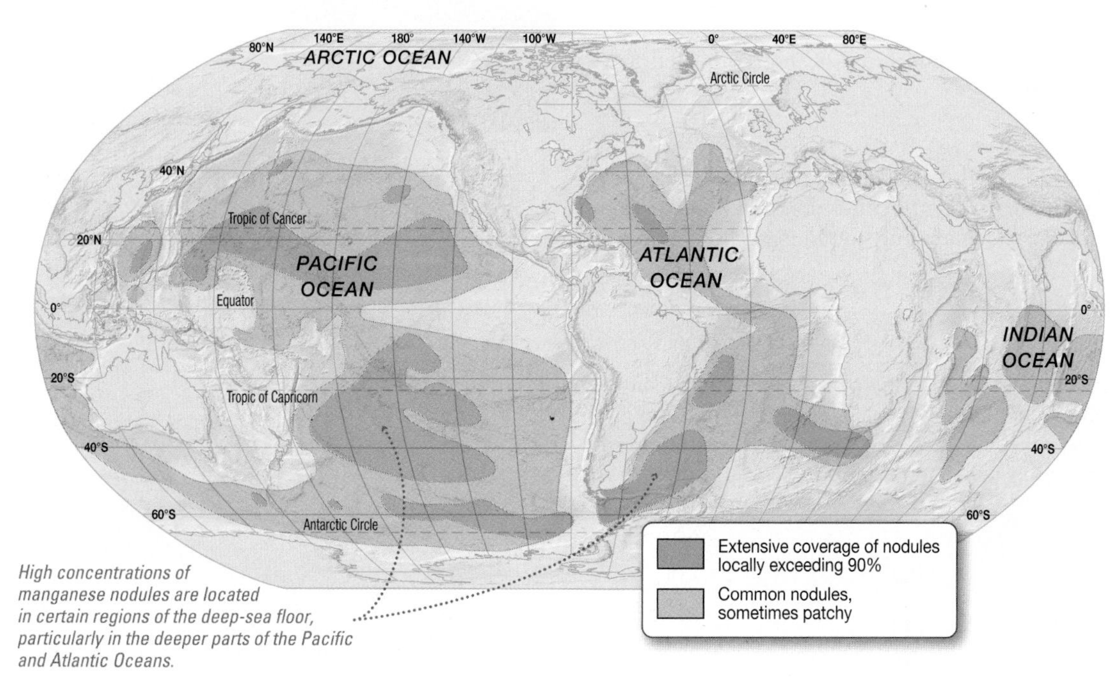


Figure 4.30 **Distribution of manganese nodules on the sea floor.**

**RECAP** **Ocean sediments contain many important resources, including petroleum, gas hydrates, sand and gravel, evaporative salts, phosphorite, manganese nodules and crusts, and rare-earth elements.**

indicated that some locations are particularly enriched. For example, an area of the sea floor near Hawaii measuring 1 square kilometer (0.4 square mile) holds as much as 25,000 metric tons (55 million pounds) of rare-earth elements. Overall, estimates suggest that the ocean floor might hold more rare-earth elements than all the known deposits on land.

**CONCEPT CHECK 4.7 ▸ Identify the various resources that marine sediments provide.**

1 Discuss the present importance and the future prospects for the production of petroleum, sand and gravel, phosphorite, manganese nodules and crusts, and rare-earth elements.

2 What are gas hydrates, where are they found, and why are they important?

# ESSENTIAL CONCEPTS REVIEW

## 4.1 ▶ How are marine sediments collected, and what historical events do they reveal?

- Sediments that accumulate on the ocean floor are *classified by origin* as *lithogenous* (derived from rock), *biogenous* (derived from organisms), *hydrogenous* (derived from water), or *cosmogenous* (derived from outer space).
- The existence of *sea floor spreading was confirmed when the* Glomar Challenger *began the Deep Sea Drilling Project* to sample ocean sediments and the underlying crust, which was continued by the *Ocean Drilling Program's* JOIDES Resolution. Today, the *Integrated Ocean Drilling Program* continues the important work of retrieving sediments from the deep-ocean floor.
- Analysis and interpretation of marine sediments reveal that *Earth has had an interesting and complex history* including *mass extinctions*, the *drying of entire seas*, *global climate change*, and the *movement of tectonic plates*.

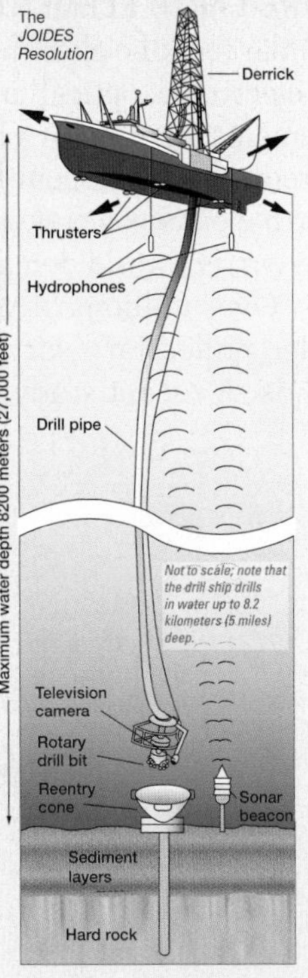


### Selected Key Terms

Use the **glossary** at the end of this book to discover the meanings of these Selected Key Terms: **core, rotary drilling, paleoceanography.**

### Critical Thinking Question

A sediment core is retrieved from the middle of the North Pacific Ocean, about 1000 kilometers (620 miles) south of the Aleutian Islands (Alaska), and at a water depth of 5000 meters (16,400 feet). The core contains coral reef fossils, which are only found in shallow, tropical waters. Develop a hypothesis that could explain this occurrence. Also, develop a test for your hypothesis.

### Active Learning Exercise

It has been said that in the early days of oceanography, collecting marine sediments using a dredge was akin to collecting land samples from a hot air balloon using a bucket—at a height above the ground of several kilometers (a few miles) and at night. With another student in class, evaluate the effectiveness of this type of sample collection (for example, is it representative of the environment being sampled?).

## 4.2 ▶ What are the characteristics of lithogenous sediment?

- *Lithogenous sediments reflect the composition of the rock from which they were derived.* Sediment *texture*—determined in part by the size, sorting, and rounding of particles—is affected greatly by how the particles were transported (by water, wind, ice, or gravity) and the energy conditions under which they were deposited. *Coarse lithogenous material dominates neritic deposits* that accumulate rapidly along the margins of continents, while *fine abyssal clays are found in pelagic deposits.*

### Selected Key Terms

Use the **glossary** at the end of this book to discover the meanings of these Selected Key Terms: **lithogenous sediment, quartz, grain size, sorting, neritic deposit, pelagic deposit, turbidite deposit, abyssal clay.**

### Critical Thinking Question

As part of a research operation you are reviewing a video of the sea floor near the coast of Greenland and see a large boulder lying on top of the sediment. Suggest a way that the boulder could have arrived there.

### Active Learning Exercise

With another student in class, discuss how a deposit with a coarse grain size indicates whether it was deposited by a high- or low-energy transporting agent. Give several examples of various transporting agents that would produce such a deposit.

## 4.3 ▶ What are the characteristics of biogenous sediment?

- *Biogenous sediment consists of the hard remains* (shells, bones, and teeth) *of organisms*. These are composed of either *silica* ($SiO_2$) from diatoms and radiolarians or *calcium carbonate* ($CaCO_3$) from foraminifers and coccolithophores. *Accumulations of microscopic shells* (tests) of organisms must comprise at least 30% of the deposit for it to be classified as *biogenic ooze*.
- *Biogenous oozes are the most common type of pelagic deposits.* The rate of biological productivity, relative to the rates of destruction and dilution of biogenous sediment, determines whether abyssal clay or oozes will form on the ocean floor. *Siliceous ooze* will form only below areas of high biologic productivity of silica-secreting organisms at the surface. *Calcareous ooze* will form only above the *calcite compensation depth (CCD)*—the depth where seawater dissolves calcium carbonate—although it can be covered and transported into deeper water through sea floor spreading.

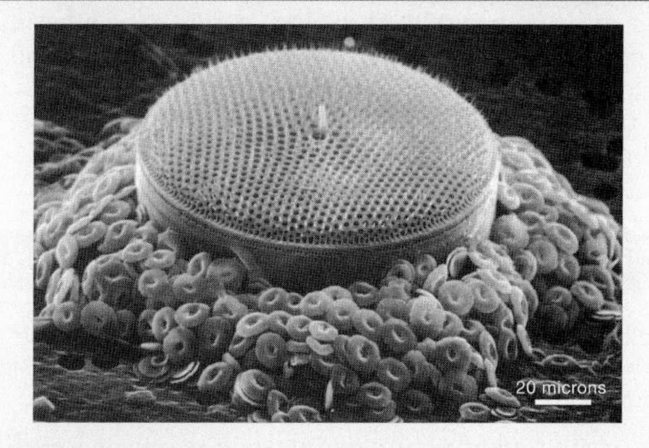


### Selected Key Terms

Use the **glossary** at the end of this book to discover the meanings of these Selected Key Terms: **biogenous sediment, test, ooze, diatom, radiolarian, foraminifer, coccolithophore, calcite compensation depth (CCD).**

### Critical Thinking Question

How do oozes differ from abyssal clay? Discuss how productivity, destruction, and dilution combine to determine whether an ooze or abyssal clay will form on the deep-ocean floor.

### Active Learning Exercise

Working with another student in class, sketch and label two examples of silica-secreting organisms that produce biogenous ooze and two examples of calcareous-secreting organisms that produce biogenous ooze.

## 4.4 ▶ What are the characteristics of hydrogenous sediment?

- *Hydrogenous sediment* includes manganese nodules, phosphates, carbonates, metal sulfides, and evaporites that *precipitate directly from water* or are formed by the interaction of substances dissolved in water with materials on the ocean floor. Hydrogenous sediments represent a relatively small proportion of marine sediment and are distributed in many diverse environments.

### Selected Key Terms

Use the **glossary** at the end of this book to discover the meanings of these Selected Key Terms: **hydrogenous sediment, precipitate, manganese nodule, oolite, metal sulfide, evaporate mineral.**

### Critical Thinking Question

Construct a table that shows the various types of hydrogenous sediment and list both their origin and how they are used by humans.

### Active Learning Exercise

Working with another student in class, design a hypothesis and an associated test to determine if manganese nodules form steadily over time or if they form episodically (in spurts).

## 4.5 ▶ What are the characteristics of cosmogenous sediment?

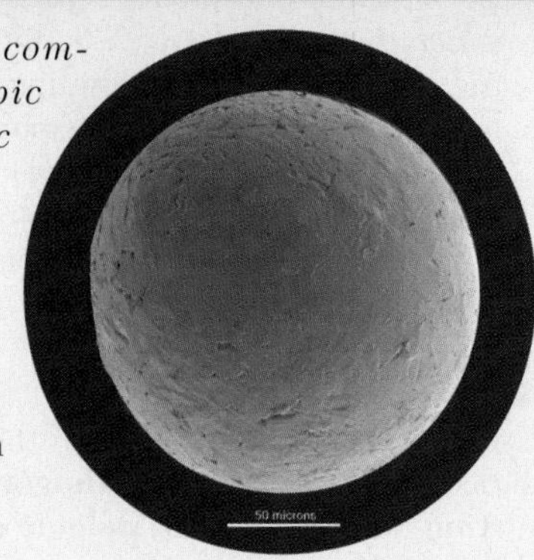


- *Cosmogenous sediment is composed of either macroscopic meteor debris* or *microscopic iron–nickel and silicate spherules* that result from asteroid collisions or extraterrestrial impacts. Minute amounts of cosmogenous sediment are mixed into most other types of ocean sediment.

### Selected Key Terms

Use the **glossary** at the end of this book to discover the meanings of these Selected Key Terms: **cosmogenous sediment, spherule, tektite, meteorite.**

### Critical Thinking Question

Using Mastering Oceanography Web Diving Deeper 4.2, describe what happened on Earth at the Cretaceous–Tertiary (K–T) boundary. What evidence was used to confirm the event, and what environmental effects did it cause?

### Active Learning Exercise

Working with another student in class, discuss why micrometeorites don't form extensive sea floor deposits, even though they are constantly raining down though Earth's atmosphere.

## 4.6 ▶ How are pelagic and neritic deposits distributed?

- Although *most ocean sediment is a mixture of various sediment types*, it is usually dominated by lithogenous, biogenous, hydrogenous, or cosmogenous material.
- *The distribution of neritic and pelagic sediment is influenced by many factors*, including proximity to sources of lithogenous sediment, productivity of microscopic marine organisms, the depth of the ocean floor, and the distribution of various sea floor features. *Fecal pellets* rapidly transport biogenous particles to the deep-ocean floor and cause the composition of sea floor deposits to match the organisms living in surface waters immediately above them.

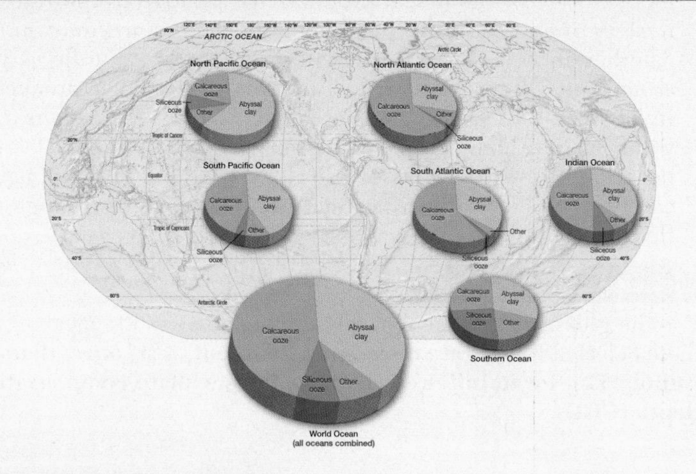


### Selected Key Terms

Use the **glossary** at the end of this book to discover the meanings of these Selected Key Terms: **neritic deposit, pelagic deposit.**

### Critical Thinking Question

Using Figure 4.22, what is the dominant sediment type in the Pacific, Atlantic, and Indian Oceans? Explain why the dominant sediment type differs from ocean to ocean.

### Active Learning Exercise

Working with another student in class, compare neritic and pelagic deposits (for example, describe their location, composition, thickness, and distribution on the sea floor).

## 4.7 ▶ What resources do marine sediments provide?

- *The most valuable nonliving resource from the ocean today is petroleum*, which is recovered from below the continental shelves and used as a source of energy. *Gas hydrates* include vast deposits of ice-like material that may someday be used as a source of energy. Other important resources include *sand and gravel* (including deposits of valuable minerals), *evaporative salts, phosphorite, manganese nodules and crusts*, and *rare-earth elements*.

**Selected Key Terms**

Use the **glossary** at the end of this book to discover the meanings of these Selected Key Terms: **petroleum, gas hydrate, salt deposit, phosphorite, manganese nodule/crust.**

**Critical Thinking Question**

A company wants to mine sea floor minerals. What technological issues would there be for developing a mining operation on the sea floor? Also, evaluate the environmental factors that should be considered before mining materials on the sea floor.

**Active Learning Exercise**

Working with another student in class, research the Internet to make a list of the everyday products that are made from these sea floor deposits: (1) the components found in manganese nodules and crusts and (2) the elements that comprise the rare-earth elements.

## Mastering Oceanography

Looking for additional review and test prep materials? With individualized coaching on the toughest topics of the course, Mastering Oceanography offers a wide variety of ways for you to move beyond memorization and deeply grasp the underlying processes of how the oceans work. Visit the Study Area in **www.masteringoceanography.com** to find practice quizzes, study tools, and multimedia that will improve your understanding of this chapter's content. Sign in today to access the following features: Self Study Quizzes, SmartFigures, Oceanography Videos and Animations, Squidtoons, Dynamic Study Modules, and an optional Pearson eText with embedded videos.

# Water and Seawater

# 5

## ESSENTIAL LEARNING CONCEPTS

At the end of this chapter, you should be able to:

- ☐ **5.1** Specify water's unique chemical properties.
- ☐ **5.2** Discuss water's important physical properties.
- ☐ **5.3** Demonstrate an understanding of what salinity is and how salinity is measured.
- ☐ **5.4** Explain why seawater salinity varies.
- ☐ **5.5** Specify how seawater salinity varies at the surface and with depth.
- ☐ **5.6** Specify how seawater density varies with depth.
- ☐ **5.7** Discuss the acid/base properties of seawater.
- ☐ **5.8** Describe factors that control the distribution of carbon and oxygen in the ocean.

↑ *Check when completed*

***"Chemistry . . . is one of the broadest branches of science, if for no other reason that, when we think about it, everything is chemistry."***

—***Luciano Caglioti,***
**The Two Faces of Chemistry** ***(1985)***

Why are temperature extremes found at places far from the ocean, while areas close to the ocean rarely experience severe temperature variations? The mild climates found in coastal regions are made possible by the unique thermal properties of water. These and other properties of water, which stem from the arrangement of its atoms and how its molecules stick together, give water the ability to store a vast amount of heat and to dissolve almost everything.

Water is so common that we often take its properties for granted, yet it is one of the most peculiar substances on Earth. For example, almost every other liquid contracts as it approaches its freezing point, but water actually *expands* as it freezes. Thus, water stays at the surface as it starts to freeze, and ice floats—a rare property shared by very few other substances. If water's properties followed the pattern of similar chemical compounds, ice would sink and cause all temperate-zone lakes, ponds, rivers, and even oceans to eventually freeze solid from the bottom up; in this scenario, life on Earth as we know it would not exist. Instead, a floating skin of ice forms at water's surface and acts as an insulating cover to protect the organisms that live in the liquid water below.

The chemical properties of water are also essential for sustaining all forms of life. In fact, the primary component of all living organisms is water. The water content of organisms, for instance, ranges from about 65% (humans) to about 90% (most plants) to as much as 95% in jellies. Water is the ideal medium to have within our bodies because it facilitates chemical reactions. Our blood, which serves to transport nutrients and remove wastes within our bodies, is 83% water. The very presence of water on our planet makes life possible, and its remarkable properties make our planet livable.

## 5.1 ▶ Why Does Water Have Such Unusual Chemical Properties?

To understand why water has such unusual properties, let's examine its chemical structure.

### Atomic Structure

**Atoms** (*a* = not, *tomos* = cut) are the basic building blocks of all matter. Every physical substance that we normally come into contact with in our world—chairs, tables, books, people, the air we breathe—is composed of atoms. An atom resembles a microscopic sphere (**Figure 5.1**), and was originally thought to be the smallest form of matter. Additional study has revealed that atoms are composed of even smaller

◂ **Water molecules and the ocean.** The objects shown in this image are water molecules, magnified by many orders of magnitude. Most surface water on Earth is in the ocean; a single droplet of water contains more water molecules than there are grains of sand on a large beach.

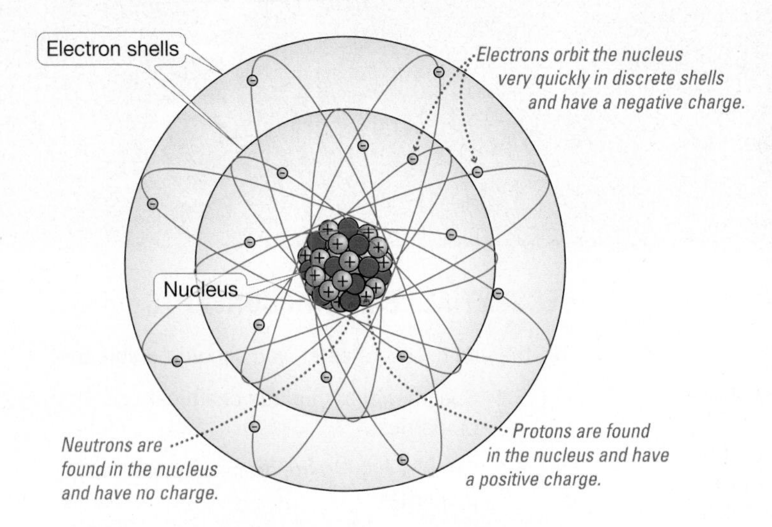


**Figure 5.1 Simplified model of an atom.** An atom consists of a central nucleus composed of protons and neutrons that is encircled by high-speed electrons.

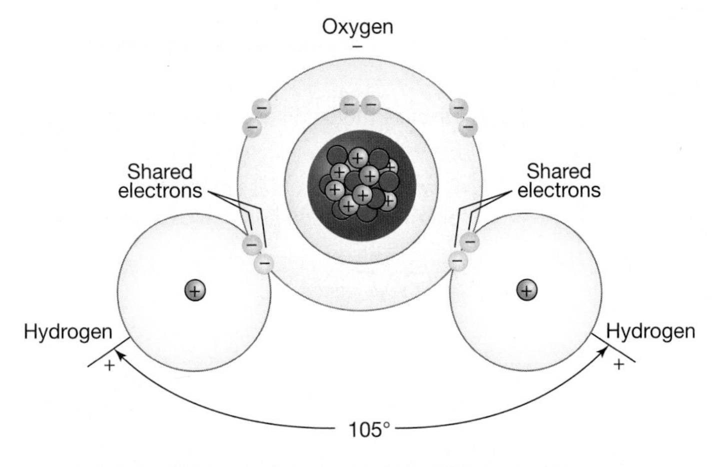


**(a)** Geometry of a water molecule. The oxygen end of the molecule is negatively charged, and the hydrogen regions exhibit a positive charge. Covalent bonds occur between the oxygen and the two hydrogen atoms as electrons are shared.

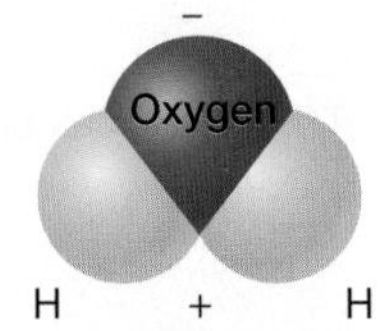


**(b)** A three-dimensional representation of the water molecule.

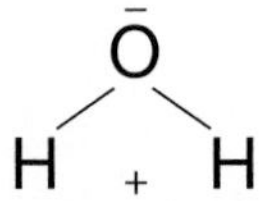


**(c)** The water molecule represented by letters (*H* = hydrogen, *O* = oxygen).

**Figure 5.2 Representations of the water molecule.**

particles, called subatomic particles. (Note that it has been discovered that subatomic particles themselves are composed of a variety of even smaller particles, such as *quarks*, *leptons*, and *bosons*.) As shown in Figure 5.1, the **nucleus** (*nucleos* = a little nut) of an atom is composed of **protons** (*protos* = first) and **neutrons** (*neutr* = neutral) that are bound together by strong forces. Protons have a positive electrical charge, whereas neutrons have no electrical charge. Both protons and neutrons have about the same mass, which is extremely small. Surrounding the nucleus are particles called **electrons** (*electro* = electricity), which have about $^1/_{2000}$ the mass of either protons or neutrons. Electrical attraction between positively charged protons and negatively charged electrons holds electrons in layers, or shells, around the nucleus.

The overall electrical charge of individual atoms is balanced because each atom contains an equal number of protons and electrons. An oxygen atom, for example, has eight protons and eight electrons. Most oxygen atoms also have eight neutrons, which do not affect the overall electrical charge because neutrons are electrically neutral. The number of protons is what distinguishes atoms of the 118 known chemical elements from one another. For example, an oxygen atom (and only an oxygen atom) has eight protons. Similarly, a hydrogen atom (and only a hydrogen atom) has one proton, a helium atom has two protons, and so on (for more details, see Mastering Oceanography Appendix IV, "A Chemical Background: Why Water Has 2 Hs and 1 O"). In some cases, an atom will lose or gain one or more electrons and thus have an overall electrical charge, in which case it is called an **ion** (*ienai* = to go).

## The Water Molecule

A **molecule** (*molecula* = a mass) is a group of two or more atoms held together by mutually shared electrons. It is the smallest piece of a substance that can exist yet still retain the original properties of that substance. When atoms combine with other atoms to form molecules, they share or trade electrons and establish chemical bonds. For instance, the chemical formula for water—$H_2O$—indicates that a water molecule is composed of two hydrogen atoms chemically bonded to one oxygen atom.

**GEOMETRY** Atoms can be represented as spheres of various sizes, and a general rule of thumb is that the more electrons an atom contains, the larger its sphere. It turns out that an oxygen atom (with eight electrons) is about twice the size of a hydrogen atom (with one electron). A water molecule consists of a central oxygen atom covalently bonded to the two hydrogen atoms, which are separated by an angle of about 105 degrees (Figure 5.2a). The **covalent bonds** (*co* = with, *valere* = to be strong) in a water molecule are due to the sharing of electrons between oxygen and each hydrogen atom. They are relatively strong chemical bonds, so a lot of energy is needed to break them.

Figure 5.2b shows a water molecule in a more compact representation, and in Figure 5.2c letter symbols are used to represent the atoms in water (*O* for oxygen, *H* for hydrogen). Instead of water's atoms being in a straight line like most other molecules, *both hydrogen atoms are on the same side of the oxygen atom*. This unusual bend in the geometry of the water molecule is the underlying cause of most of the unique properties of water.

**POLARITY** The bent geometry of the water molecule gives a slight overall negative charge to the end that contains the oxygen atom and a slight overall positive charge to the other side that contains the hydrogen atoms

(Figure 5.2a). This slight separation of charges gives the entire molecule an electrical **polarity** (*polus* = pole, *ity* = having the quality of), so water molecules are **dipolar** (*di* = two, *polus* = pole). Other common dipolar objects are flashlight batteries, car batteries, and bar magnets. In fact, a good way to visualize the polarity of water molecules is to view them as if they contain a tiny, weak bar magnet.

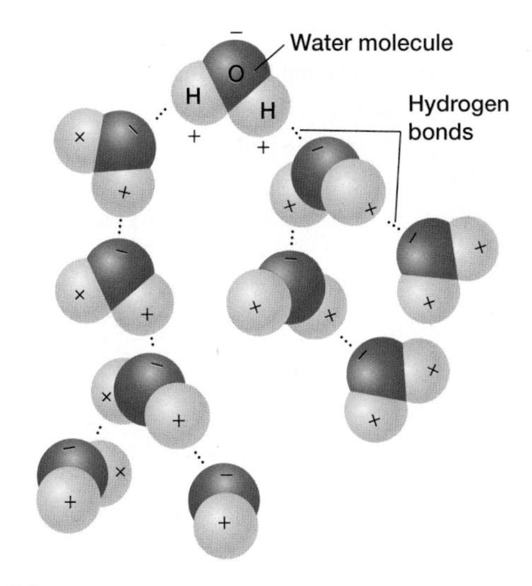


Figure 5.3 **Hydrogen bonding in water.** Dashed lines indicate locations of hydrogen bonds, which occur between water molecules.

**INTERCONNECTIONS OF MOLECULES** If you've ever experimented with bar magnets, you know they have polarity and orient themselves relative to one another such that the positive end of one bar magnet is attracted to the negative end of another. Water molecules have polarity, too, and as a result they orient themselves relative to one another. In water, the positively charged hydrogen area of one water molecule interacts with the negatively charged oxygen end of an adjacent water molecule, forming a **hydrogen bond** (Figure 5.3). The hydrogen bonds between water molecules are much weaker than the covalent bonds that hold the hydrogen and oxygen atoms of water molecules together. In essence, weaker hydrogen bonds form between adjacent water molecules, and stronger covalent bonds occur within water molecules.

Even though hydrogen bonds are weaker than covalent bonds, they are strong enough to cause water molecules to cluster together and exhibit **cohesion** (*cohaesus* = a clinging together). The cohesive properties of water cause it to "bead up" on a waxed surface, such as a freshly waxed car. They also give water its **surface tension**. Water's surface has a thin "skin" that allows a glass to be filled just above the brim without spilling any of the water. Surface tension results from the formation of hydrogen bonds between the outermost layer of water molecules and the underlying molecules. Water's ability to form hydrogen bonds causes it to have the highest surface tension of any liquid except the element mercury. (Mercury is the only metal that is a liquid at normal surface temperatures, which is why it was commonly used in older thermometers; most thermometers in use today have been replaced by digital thermometers, which do not contain toxic mercury.)

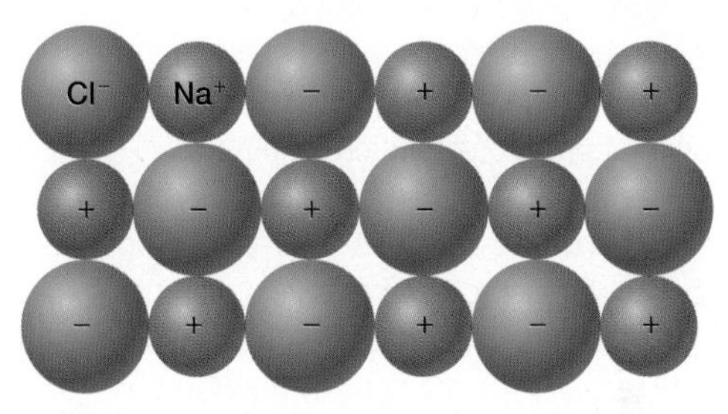


**(a)** Molecular structure of table salt, which is composed of sodium chloride ($Na^+$ = sodium ion, $Cl^-$ = chlorine ion).

**WATER: THE UNIVERSAL SOLVENT** Water molecules are not only attracted to other water molecules but also to other polar chemical compounds. Because of this, water molecules can reduce the attraction between ions of opposite charges in other substances. For instance, ordinary table salt—sodium chloride, NaCl, where sodium is represented by the letters *Na* because the Latin term for sodium is *natrium*—consists of an alternating array of positively charged sodium ions and negatively charged chloride ions (Figure 5.4a). The **electrostatic force of attraction** (*electro* = electricity, *stasis* = standing) between oppositely charged ions produces an **ionic bond** (*ienai* = *to go*). When solid NaCl is placed in water, the electrostatic force of attraction (ionic bonding) between the sodium and chloride ions is reduced by 80 times. This, in turn, makes it much easier for the sodium ions and chloride ions to separate. When the ions separate, the positively charged sodium ions become attracted to the negative ends of the water molecules, the negatively charged chloride ions become attracted to the positive ends of the water molecules (Figure 5.4b), and the salt is dissolved in water. The process by which water molecules completely surround ions is called *hydration* (*hydra* = water, *ation* = action or process).

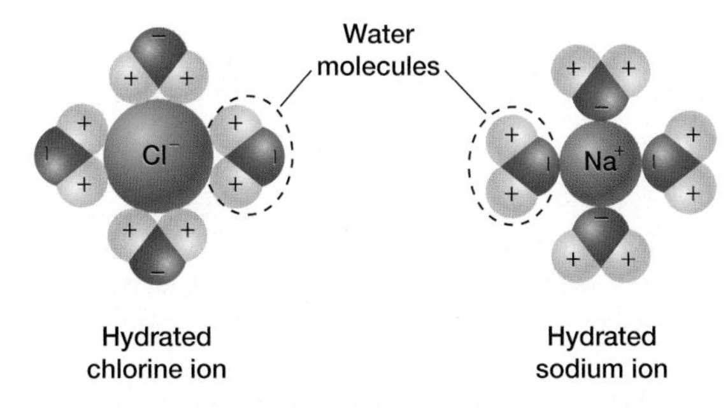


**(b)** As sodium chloride is dissolved, the positively charged ends of water molecules are attracted to the negatively charged $Cl^-$ ion, while the negatively charged ends are attracted to the positively charged $Na^+$ ion.

Figure 5.4 **Water as a solvent.**

Because water molecules interact with other water molecules and other polar molecules, water is able to dissolve nearly everything. If water is such a good solvent, why doesn't oil dissolve in water? As you might have guessed, the chemical structure of oil is remarkably nonpolar. With no positive or negative ends to attract the polar water molecule, oil will not dissolve in water. But given enough time, water can dissolve more substances and in greater quantity than any other known substance. This is why water is called the universal solvent.

It is also why the ocean contains so much dissolved material—an estimated 50 quadrillion metric tons (110 quintillion pounds) of salt—which makes seawater taste salty.

**Animation**
How Salt Dissolves in Water
https://goo.gl/IT8Sd3

**RECAP** **A water molecule has a bend in its geometry, with the two hydrogen atoms on the same side of the oxygen atom. This property gives water its polarity and ability to form hydrogen bonds.**

**CONCEPT CHECK 5.1 ▸ Specify water's unique chemical properties.**

1 Sketch a model of an atom, showing the positions of the subatomic particles: protons, neutrons, and electrons.

2 Describe what condition exists in water molecules to make them dipolar.

3 Sketch several water molecules, showing all covalent and hydrogen bonds. Be sure to indicate the polarity of each water molecule.

4 How does hydrogen bonding produce the surface tension phenomenon of water?

5 Discuss how the dipolar nature of a water molecule makes it such an effective solvent of ionic compounds.

## STUDENTS SOMETIMES ASK . . .

***Why does a water molecule have the unusual shape that it does?***

Based on simple symmetry considerations and charge separations, one might think that a water molecule would have its two hydrogen atoms on opposite sides of the oxygen atom, thus producing a linear 180° shape, like the shape of many other molecules. But water's odd shape, where the hydrogen atoms are separated by only 105°, stems from the fact that the two electrons from the oxygen atom that enter into covalent bonds with hydrogen are positioned at a fixed angle to each other, consistent with the oxygen molecule's pyramid-shaped (tetrahedral) bonding geometry. Ultimately, this is the reason why a water molecule has its unusual bend.

## STUDENTS SOMETIMES ASK . . .

***How many atoms are there in a single drop of water?***

A LOT! This is actually not too difficult to calculate by making a few assumptions and knowing a little bit about chemistry. First, a drop of water is about 50 microliters if dispensed from a dropper. Fifty microliters of water has a mass of 50 milligrams at standard temperature and pressure. Since water is $H_2O$ and thus has a molecular mass of 18.015 grams/mole, 50 milligrams of water contains 2.775 millimoles. According to Avogadro's number (a standard constant in chemistry), one mole of any compound contains $6.022 \times 10^{23}$ molecules, so 2.775 millimoles of water has $1.67 \times 10^{21}$ $H_2O$ molecules. So that means there are 1.67 sextillion molecules of $H_2O$ in a drop of water (that's 1,670,000,000,000,000,000,000). Or, since you asked for atoms, and each $H_2O$ molecule is composed of three atoms, there are about 5 sextillion atoms in a single drop of water!

# 5.2 ▸ What Important Physical Properties Does Water Possess?

Water's important physical properties include its thermal properties (such as water's freezing and boiling points, heat capacity, and latent heats) and how water's thermal expansion and contraction affects its density.

## Water's Thermal Properties

Water exists on Earth as a solid, a liquid, and a gas and has the ability to store and release great amounts of heat. Water's thermal properties influence the world's heat budget and are in part responsible for the development of tropical cyclones, worldwide wind belts, and ocean surface currents.

**HEAT, TEMPERATURE, AND CHANGES OF STATE** Matter around us is usually in one of the three common states: solid, liquid, or gas. (*Plasma* is widely recognized as a fourth state of matter distinct from solids, liquids, and normal gases. Plasma is a gaseous substance in which atoms have been ionized—that is to say, stripped of electrons. Plasma television screens take advantage of the fact that plasmas are strongly influenced by electric currents.) What must happen to change the state of a compound? The attractive forces between molecules in a substance must be overcome if the state of the substance is to be changed from solid to liquid or from liquid to gas. These attractive forces include hydrogen bonds and van der Waals forces. The **van der Waals forces**—named for Dutch physicist Johannes Diderik van der Waals (1837–1923)—are relatively weak interactions that exist between electrically neutral molecules because of the molecules' uneven distribution of charge. Energy must be added to the molecules so they can move fast enough to overcome these attractions.

What form of energy changes the state of matter? Very simply, adding or removing heat causes a substance to change its state of matter. For instance, adding heat to ice cubes causes them to melt, and removing heat from water causes ice to form. Before proceeding, let's clarify the difference between heat and temperature:

- **Heat** is defined as the *amount of energy transferred from one body to another due to a difference in temperature.* Heat is proportional to the average **kinetic energy** (*kinetos* = moving) of the molecules in a body. For example, water can exist as a solid, liquid, or gas, depending on the amount of heat added. Heat may be generated by combustion (a chemical reaction commonly called burning), through other chemical reactions, by friction, or from radioactivity; it can be transferred by conduction, by convection, or by radiation. A **calorie** (*calor* = heat) is the amount of heat required to raise the temperature of 1 gram of water by 1 degree centigrade. (Note that 1 gram [0.035 ounce] of water is equal to about 10 drops.) The familiar "calorie" used to measure the energy content of foods is actually a *kilocalorie*, or 1000 calories. Although the metric unit for thermal energy is the *joule*, calories are directly tied to some of water's thermal properties, as will be discussed in the next section.

- **Temperature** is the *direct measure of the average kinetic energy of the molecules that make up a substance*. The greater the temperature, the greater the kinetic energy of the substance. Temperature changes when heat energy is added to or removed from a substance. Temperature is usually measured in degrees centigrade (°C) or degrees Fahrenheit (°F).

Figure 5.5 shows water molecules in the solid, liquid, and gaseous states. In the *solid state* (ice), water has a rigid structure and does not normally flow over short time scales. Intermolecular bonds are constantly being broken and reformed, but the molecules remain firmly attached. That is, the molecules vibrate with energy but remain in relatively fixed positions. As a result, solids do not conform to the shape of their container.

In the *liquid state* (water), water molecules still interact with each other, but they have enough kinetic energy to flow past each other and take the shape of their container. Intermolecular bonds are being formed and broken at a much greater rate than in the solid state.

In the *gaseous state* (water **vapor**), water molecules no longer interact with one another except during random collisions. Water vapor molecules flow very freely, filling the volume of whatever container they are placed in.

**WATER'S FREEZING AND BOILING POINTS** If enough heat energy is added to a solid, it melts to a liquid. The temperature at which melting occurs is the substance's **melting point**. If enough heat energy is removed from a liquid, it freezes to a solid. The temperature at which freezing occurs is the substance's **freezing point**, which is the same temperature as the melting point (Figure 5.5). For pure water, melting and freezing occur at 0°C (32°F). Note that all melting/freezing/boiling points discussed in this chapter assume a standard sea level pressure of 1 atmosphere (14.7 pounds per square inch).

If enough heat energy is added to a liquid, it converts to a gas. The temperature at which boiling occurs is the substance's **boiling point**. If enough heat energy is removed from a gas, it changes through the process of **condensation** into a liquid. The highest temperature at which condensation occurs is the substance's **condensation point**, which is the same temperature as the boiling point (Figure 5.5). For pure water, boiling and condensation occur at 100°C (212°F).

Both the freezing and boiling points of water are unusually high compared to those of similar chemical substances. As shown in Figure 5.6, if water followed the pattern of other chemical compounds with molecules of similar mass, it should melt at −90°C (−130°F) and boil at −68°C (−90°F). If that were the case, all water on Earth would be in the gaseous state. Instead, water melts and boils at the relatively

## STUDENTS SOMETIMES ASK . . .

***How can it be that water—a liquid at room temperature—can be created by combining hydrogen and oxygen—two gases at room temperature?***

It is true that combining two parts hydrogen gas with one part oxygen gas produces liquid water. This can be accomplished as a chemistry experiment, although care should be taken because much energy is released during the reaction (don't try this at home!). Oftentimes, when combining two elements, the product has very different properties than the pure substances. For instance, combining elemental sodium (Na), a highly reactive metal, with pure chlorine ($Cl_2$), a toxic nerve gas, produces cubes of harmless table salt (NaCl). This is what most people find amazing about chemistry.

**Animation**
Phase Changes of Water
http://goo.gl/gRT6NV

Figure 5.5 **Water in the three states of matter: solid, liquid, and gas.** Diagram showing the three states of matter in which water is found on Earth and the processes associated with changes from one state to another.

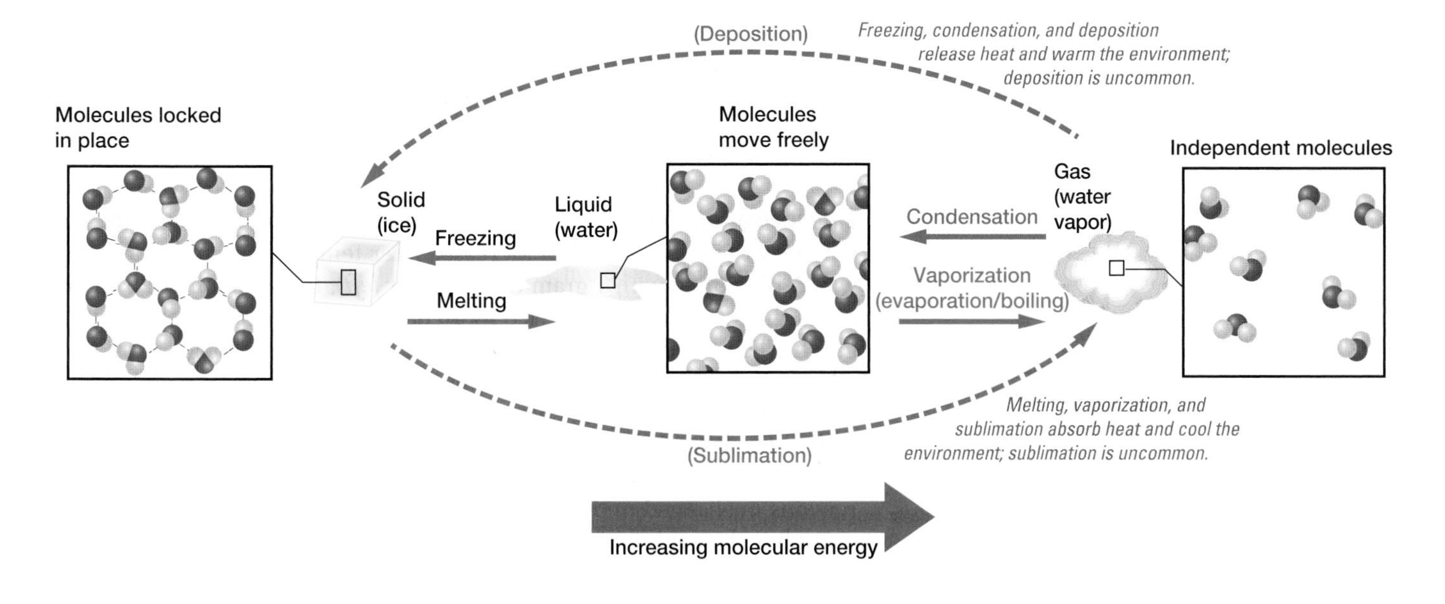

## FOREVER THIRSTY ON A WATER WORLD

The world is on the verge of a water crisis. In fact, more than a third of the world's population already suffers from shortages of drinkable freshwater—a percentage expected to increase to 50% by 2025. Because the human consumption of drinking water is growing even as its supply is dwindling, several countries have begun to use the ocean as a source of freshwater. **Desalination,** which is the salt removal from seawater, can provide freshwater for business, home, and agricultural use.

Currently, there are more than 13,000 desalination plants worldwide, the majority of them very small and located in arid regions of the Middle East, Caribbean, and Mediterranean. These plants produce more than 45 billion liters (12 billion gallons) of freshwater daily. The United States produces only about 10% of the world's desalted water, primarily in Florida.

More than half of the world's desalination plants use *distillation* to purify water, while most of the remaining plants use *membrane processes*.

### DISTILLATION

The process of **distillation** (*distillare* = to trickle) is shown schematically in **Figure 5A**. In distillation, saltwater is boiled, and the resulting water vapor is passed through a cooling condenser, where it condenses and is collected as freshwater. This simple procedure is very efficient at purifying seawater. For instance, distillation of 35‰ seawater produces freshwater with a salinity of only 0.03‰, which is about 10 times fresher than bottled water, so it needs to be mixed with less pure water to make it taste better. Distillation is expensive, however, because it requires large amounts of heat energy to boil the saltwater. Increased efficiency, such as using the waste heat from a power plant, is required to make distillation practical on a large scale. **Solar distillation**, which is also known as **solar humidification**, does not require supplemental heating and has been used successfully in small-scale agricultural experiments in arid regions such as Israel, West Africa, and Peru.

### MEMBRANE PROCESSES

**Reverse osmosis** (*osmos* = to push) may have potential for large-scale desalination. In normal osmosis, water molecules naturally pass through a thin, semipermeable membrane from a freshwater solution to a saltwater solution. In reverse osmosis, water on the salty side is highly pressurized to drive water molecules—but not salt and other impurities—through the membrane to the freshwater side (**Figure 5B**). A significant problem with reverse osmosis is that the membranes are flimsy, become clogged, and must be replaced frequently. Advanced composite materials may help eliminate these problems because they are sturdier, provide better filtration, and last up to 10 years.

### OTHER METHODS OF DESALINATION

Seawater selectively excludes dissolved substances as it freezes—a process called **freeze separation**. As a result, the salinity

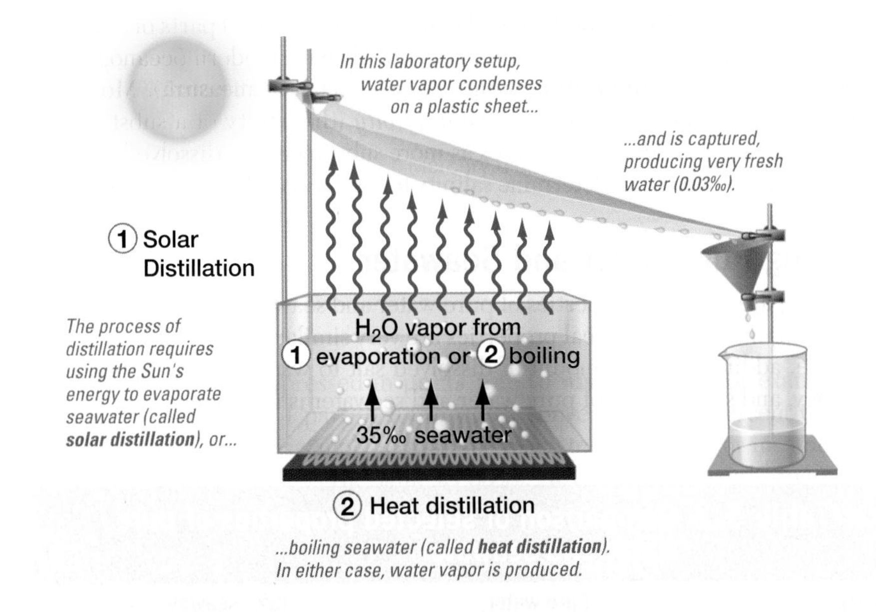
In this laboratory setup, water vapor condenses on a plastic sheet...
...and is captured, producing very fresh water (0.03‰).
① Solar Distillation
The process of distillation requires using the Sun's energy to evaporate seawater (called solar distillation), or...
$H_2O$ vapor from ① evaporation or ② boiling
35‰ seawater
② Heat distillation
...boiling seawater (called heat distillation). In either case, water vapor is produced.

**Figure 5A.** How the process of distillation works to purify seawater.

The dissolved substances in seawater, however, give it slightly different yet important physical properties as compared to pure water. For example, recall that dissolved substances interfere with pure water changing state. The freezing points and boiling points in Table 5.2 show that dissolved substances decrease the freezing point and increase the boiling point of water. Thus, seawater freezes at a temperature of −1.9°C (28.6°F), which is lower than the freezing point of pure water (0°C [32°F]). Similarly, seawater boils at a temperature of 100.6°C (213.1°F), which is higher than the boiling point of pure water (100°C [212°F]). In effect, the salts in seawater extend the range of temperatures in which water is a liquid. This same principle applies to of sea ice (once it is melted) is typically 70% lower than seawater. To make this an effective desalination technique, though, the water must be frozen and thawed multiple times, with the salts washed from the ice between each thawing. Like distillation, freeze separation requires large amounts of energy, so it may be impractical except on a small scale.

Yet another way to obtain freshwater is to melt naturally formed ice. Imaginative thinkers have proposed towing large icebergs to warm coastal waters off countries that need freshwater. Once there, the freshwater produced as the icebergs melt could be collected and pumped ashore. Studies have shown that towing large Antarctic icebergs to arid regions would be technologically feasible and, for certain Southern Hemisphere locations, economically feasible, too.

Today, desalination of seawater provides less than 0.5% of human water needs. Yet as the need for fresh water grows, this percentage will likely grow as well. The technological challenge to produce large scale, low cost freshwater from seawater is considerable, but with oceans of water to tap for a thirsty planet, it is a challenge we may not be able to pass up.

### WHAT DID YOU LEARN?

1. Why is the underlying reason that seawater desalination is so expensive?
2. Describe the two main methods of seawater desalination.

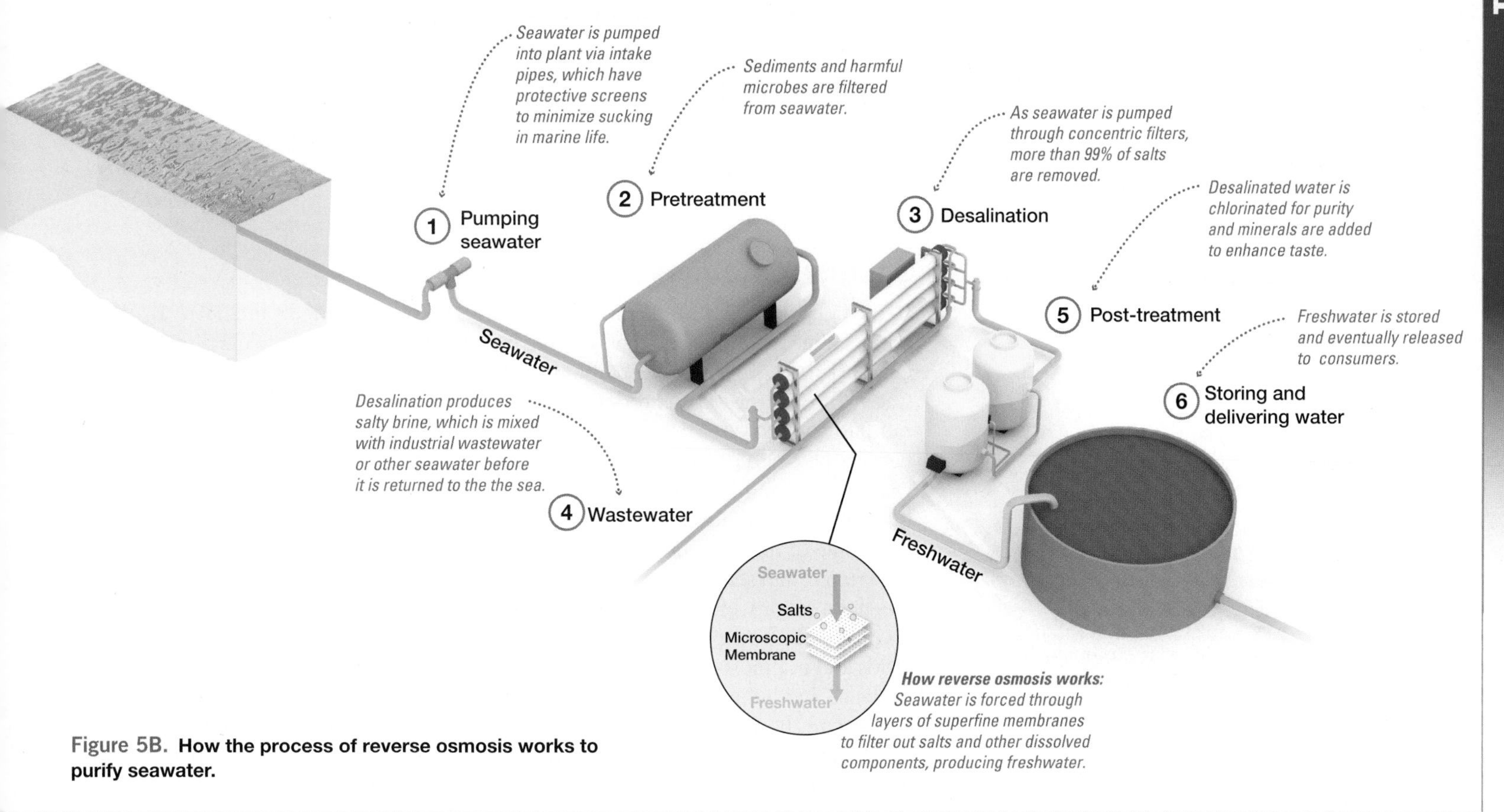


Figure 5B. **How the process of reverse osmosis works to purify seawater.**

antifreeze used in automobile radiators. Antifreeze lowers the freezing point of the water in a radiator and increases the boiling point, thus extending the range over which the water remains in the liquid state. Antifreeze, therefore, protects your radiator from freezing in the winter and from boiling over in the summer.

Density is another property that exhibits small but remarkable differences between pure water and seawater. Recall that density is defined as mass per unit volume. When substances are added to water and dissolved, the volume doesn't change but the water's density increases because more mass has been added per unit volume. Although the difference in density between pure water and seawater seems negligible

**RECAP** **Seawater salinity can be measured using a salinometer and averages 35‰. The dissolved components in seawater give it different yet important physical properties as compared to pure water.**

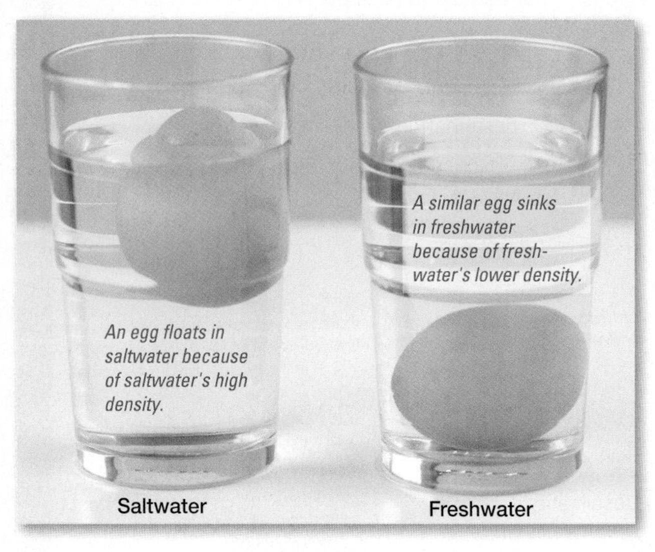

A similar egg sinks in freshwater because of freshwater's lower density.
An egg floats in saltwater because of saltwater's high density.
Saltwater
Freshwater

Figure 5.17 **An egg floats in saltwater but sinks in freshwater.**

## STUDENTS SOMETIMES ASK . . .

***I've seen the labels on electric cords warning against using electrical appliances close to water. Are these warnings because water's polarity allows electricity to be transmitted through it?***

Yes and no. Water molecules are polar, so you might assume that water is a good conductor of electricity. Pure water is a very poor conductor, however, because water molecules are neutral overall and will not move toward the negatively or positively charged pole in an electrical system. If an electrical appliance is dropped into a tub of absolutely pure water, the water molecules will transmit no electricity. Instead, the water molecules will simply orient their positively charged hydrogen ends toward the negative pole of the appliance and their negatively charged oxygen ends toward its positive pole, which tends to neutralize the electric field. Interestingly, it is the dissolved substances that transmit electrical current through water (see Figure 5.16). Even slight amounts, such as those in tap water, allow electricity to be transmitted. That's why there are warning labels on the electric cords of household appliances that are commonly used in the bathroom, such as blow dryers, electric razors, and heaters. That's also why it is recommended to stay out of any water—including a bathtub or shower—during a lightning storm!

(Table 5.2 shows an increase of only 0.028 g/cm$^3$), a simple experiment with an egg in two different glasses of water shows how dramatically small differences in density can affect floating objects (**Figure 5.17**).

Other important properties of seawater (such as its pH and how seawater density varies with depth) are discussed later in this chapter.

**CONCEPT CHECK 5.3 ▸ Demonstrate an understanding of what salinity is and how salinity is measured.**

1 What is the average salinity of seawater? What units are normally used, and why are those units useful?

2 What condition of salinity makes it possible to determine the total salinity of ocean water by measuring the concentration of only one constituent, the chloride ion?

3 In what ways are seawater and pure water similar? How are the two different?

## 5.4 ▸ Why Does Seawater Salinity Vary?

Using salinometers and other techniques, oceanographers have determined that salinity varies from place to place in the oceans. What are the patterns of seawater salinity, and what causes them?

### Salinity Variations

In the open ocean far from land, salinity varies between about 33 and 38‰. In coastal areas, salinity variations can be extreme. In the Baltic Sea, for example, salinity averages only 10‰ because physical conditions create **brackish** (*brak* = salt, *ish* = somewhat) water. Brackish water is produced in areas where freshwater (from rivers and high rainfall) and seawater mix. In the Red Sea, on the other hand, salinity averages 42‰ because physical conditions produce **hypersaline** (*hyper* = excessive, *salinus* = salt) water. Hypersaline water is typical of seas and inland bodies of water that experience high evaporation rates and limited open-ocean circulation.

Some of the most hypersaline water in the world is found in inland lakes, which are often called seas because they are so salty. The Great Salt Lake in Utah, for example, has a salinity of 280‰, and the Dead Sea on the border of Israel and Jordan has a salinity of 330‰. The water in the Dead Sea, therefore, contains 33% dissolved solids and is almost *10 times saltier than seawater*. As a result, hypersaline waters are so dense and buoyant that one can easily float (**Figure 5.18**), even with arms and legs sticking up above water level! Hypersaline waters also taste much saltier than seawater.

Salinity of seawater in coastal areas also varies seasonally. For example, the salinity of seawater off Miami Beach, Florida, varies from about 34.8‰ in October to 36.4‰ in May and June, when evaporation is high. Offshore of Astoria, Oregon, seawater salinity is always extremely low because of the vast freshwater input from the Columbia River. Here, surface water salinity can be as low as 0.3‰ in April and May (when the Columbia River is at its maximum flow rate) and 2.6‰ in October (the dry season, when freshwater input is reduced).

Other types of water have much lower salinity. Tap water, for instance, has salinity somewhere below 0.8‰, and good-tasting tap water is usually below 0.6‰. Salinity of premium bottled water is on the order of 0.3‰, with the salinity often displayed prominently on its label, usually as total dissolved solids (TDS) in units of parts per million (ppm), where 1000 ppm equals 1‰.

### Processes Affecting Seawater Salinity

Processes affecting seawater salinity change either the amount of water ($H_2O$ molecules) or the amount of dissolved substances in the water. Adding more water, for instance, dilutes the dissolved component and lowers the salinity of the sample.

Conversely, removing water increases salinity. Changing the salinity in these ways does not affect the *amount* or the *composition* of the dissolved components, which remain in constant proportions. Let's first examine processes that affect the amount of water in seawater before turning our attention to processes that influence dissolved components.

**PROCESSES THAT DECREASE SEAWATER SALINITY** Table 5.3 summarizes the processes that affect seawater salinity. Precipitation, **runoff** (stream discharge), melting icebergs, and melting sea ice *decrease* seawater salinity by adding more freshwater to the ocean. Precipitation is the way atmospheric water returns to Earth as rain, snow, sleet, and hail. Worldwide, about three-quarters of all precipitation falls directly back into the ocean and one-quarter falls onto land. Precipitation falling directly into the oceans adds freshwater, reducing seawater salinity.

Most of the precipitation that falls on land returns to the oceans indirectly as stream runoff. Even though this water dissolves minerals on land, the runoff is relatively pure water, as shown in Table 5.4. Runoff, therefore, adds mostly water to the ocean, causing seawater salinity to decrease.

*Icebergs* are chunks of ice that have broken free (*calved*) from a glacier when it flows into an ocean or marginal sea and begins to melt. Glacial ice originates as snowfall in high mountain areas, so icebergs are composed of freshwater. When icebergs melt in the ocean, they add freshwater, which is another way in which seawater salinity is reduced.

*Sea ice* forms when ocean water freezes in high-latitude regions and is composed primarily of freshwater. When warmer temperatures return to high-latitude regions

**Figure 5.18 High-salinity water of the Dead Sea allows swimmers to easily float.** The Dead Sea, which has 330‰ salinity (almost 10 times the salinity of seawater), has high density. As a result, it also has high buoyancy that allows swimmers to float easily.

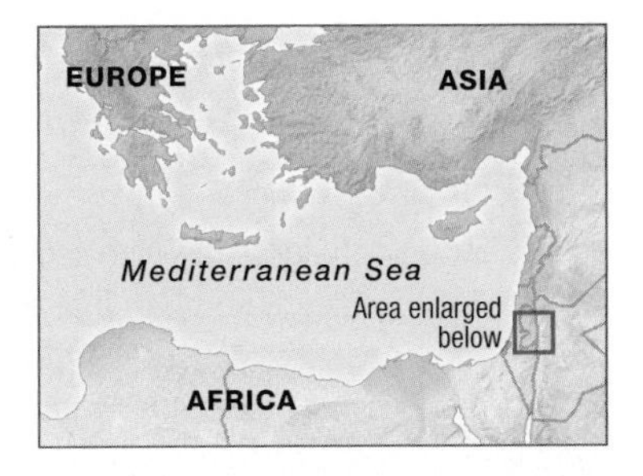


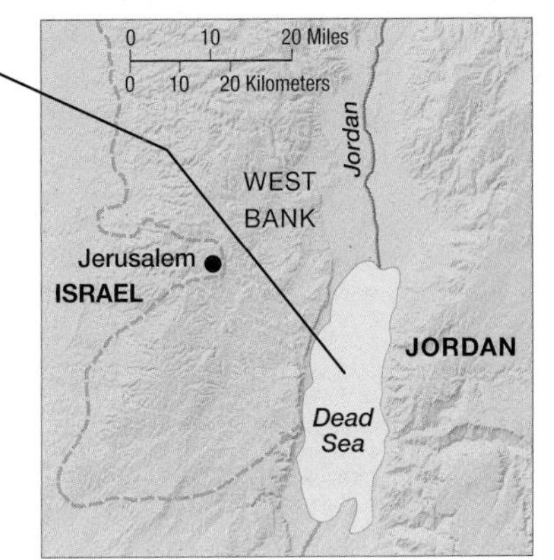


**SmartTable 5.3 Processes that affect seawater salinity.**
https://goo.gl/fZpUXm

## SmartTable 5.3 Processes that affect seawater salinity

| Process | How accomplished | Adds or removes | Effect on salt in seawater | Effect on $H_2O$ in seawater | Salinity increase or decrease? | Source of freshwater from the sea? |
|---|---|---|---|---|---|---|
| Precipitation | Rain, sleet, hail, or snow falls directly on the ocean | Adds very fresh water | None | More $H_2O$ | Decrease | N/A |
| Runoff | Streams carry water to the ocean | Adds mostly fresh water | Negligible addition of salt | More $H_2O$ | Decrease | N/A |
| Icebergs melting | Glacial ice calves into the ocean and melts | Adds very fresh water | None | More $H_2O$ | Decrease | Yes, icebergs from the Antarctic have been towed to South America |
| Sea ice melting | Sea ice melts in the ocean | Adds mostly fresh water and some salt | Adds a small amount of salt | More $H_2O$ | Decrease | Yes, sea ice can be melted and is better than drinking seawater |
| Sea ice forming | Seawater freezes in cold ocean areas | Removes mostly freshwater | 30% of salts in seawater are retained in ice | Less $H_2O$ | Increase | Yes, through multiple freezings, called *freeze separation* |
| Evaporation | Seawater evaporates in hot climates | Removes very pure water | None (essentially all salts are left behind) | Less $H_2O$ | Increase | Yes, through evaporation of seawater and condensation of water vapor, called *distillation* |

## STUDENTS SOMETIMES ASK . . .

***What would happen to a person if he or she drank seawater?***

It depends on the quantity. The salinity of seawater is about four times greater than that of your body fluids. In your body, seawater causes your internal tissues to lose water through *osmosis* (*osmos* = to push), which transports water molecules from higher concentrations (the normal body chemistry of your internal fluids) to areas of lower concentrations (your digestive tract containing seawater). Thus, your natural body fluids would move into your digestive tract and eventually be expelled, causing dehydration.

Don't worry too much if you've inadvertently swallowed some seawater. As a nutritional drink, seawater provides seven important nutrients and contains no fat, cholesterol, or calories. Some people even claim that drinking a small amount of seawater daily gives them good health! However, beware of microbial contaminants in seawater, such as viruses and bacteria that can sometimes be present.

**Table 5.4 Comparison of major dissolved components in streams with those in seawater**

| Constituent | Concentration in streams (parts per million by weight) | Concentration in seawater (parts per million by weight) |
|---|---|---|
| Bicarbonate ion ($HCO_3^-$) | 58.4 | 90 |
| Calcium ion ($Ca^{2+}$) | 15.0 | 400 |
| Silicate ($SiO_2$) | 13.1 | 3 |
| Sulfate ion ($SO_4^{2-}$) | 11.2 | 2700 |
| Chloride ion ($Cl^-$) | 7.8 | 19,200 |
| Sodium ion ($Na^+$) | 6.3 | 10,600 |
| Magnesium ion ($Mg^{2+}$) | 4.1 | 1300 |
| Potassium ion ($K^+$) | 2.3 | 380 |
| **Total (parts per million)** | 119.2 ppm | 34,793 ppm |
| **Total (‰)** | 0.1192‰ | 34.8‰ |

in the summer, sea ice melts in the ocean, adding mostly freshwater with a small amount of salt to the ocean. Seawater salinity, therefore, is decreased.

**PROCESSES THAT INCREASE SEAWATER SALINITY** The formation of sea ice and evaporation *increase* seawater salinity by removing water from the ocean (Table 5.3). Sea ice forms when seawater freezes. Depending on the salinity of seawater and the rate of ice formation, about 30% of the dissolved components in seawater are retained in sea ice. This means that 35‰ seawater creates sea ice with about 10‰ salinity (30% of 35‰ is 10‰). Consequently, the formation of sea ice removes mostly freshwater from seawater, increasing the salinity of the remaining unfrozen water. High-salinity water also has a high density, so it sinks below the surface.

Recall that evaporation is the conversion of water molecules from the liquid state to the vapor state at temperatures below the boiling point. Evaporation removes water from the ocean, leaving its dissolved substances behind. Evaporation, therefore, increases seawater salinity. Worldwide, about 86% of all evaporation occurs in the oceans.

## STUDENTS SOMETIMES ASK . . .

***You mentioned that when seawater freezes, it produces ice with about 10‰ salinity. Once that ice melts, can a person drink it with no ill effects?***

Early Arctic explorers found out the answer to your question by necessity. Some of these explorers who traveled by ship in high-latitude regions became inadvertently or purposely entrapped by sea ice (see, for example, Mastering Oceanography **Web Diving Deeper 7.1**, which describes the remarkable voyage of the *Fram*). Lacking other water sources, they used melted sea ice. Although newly formed sea ice contains little salt, it does trap a significant amount of brine (drops of salty water). Depending on the rate of freezing, newly formed ice may have a total salinity from 4 to 15‰. The more rapidly it forms, the more brine it captures and the higher the salinity. Melted sea ice with salinity this high doesn't taste very good, and it still causes dehydration, but not as quickly as drinking 35‰ seawater does. Over time, however, the brine will trickle down through the coarse structure of the sea ice, so its salinity decreases. By the time it is a year old, sea ice normally becomes relatively pure. Drinking melted sea ice enabled these early explorers to survive.

**THE HYDROLOGIC CYCLE** **Figure 5.19** shows the **hydrologic cycle**, (*hydro* = water, *logos* = study of) which describes the continual movement of water on, above, and below the surface of Earth. The movement of water through various components of the hydrologic cycle involves processes that recycle water among the ocean, the atmosphere, and the continents, illustrating that water is in constant motion between the different components (or *reservoirs*) of the hydrologic cycle. Note that many of the processes of the hydrologic cycle affect seawater salinity. For example, river runoff into the ocean changes seawater salinity in that region. The figure also shows that of Earth's reservoirs, the vast majority of water at or near Earth's surface is contained in the ocean.

In addition, Figure 5.19 shows the average yearly amounts of transfer, or *flux*, of water between various reservoirs.

## Dissolved Components Added to and Removed from Seawater

Seawater salinity is a function of the amount of dissolved components in seawater. Interestingly, dissolved substances do not remain in the ocean forever. Instead, they are cycled into and out of seawater by the processes shown in **Figure 5.20**. These

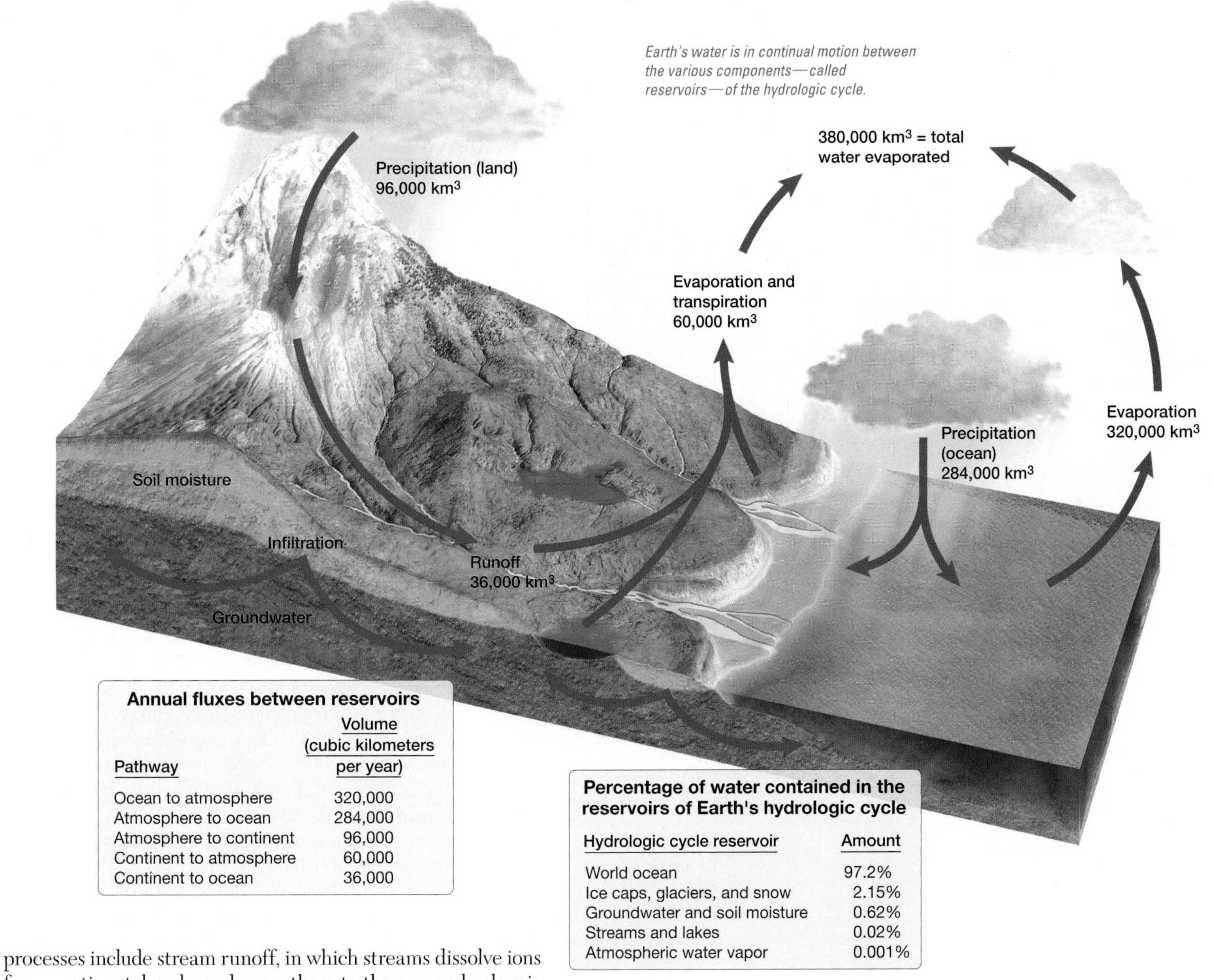


**Annual fluxes between reservoirs**

| Pathway | Volume (cubic kilometers per year) |
|---|---|
| Ocean to atmosphere | 320,000 |
| Atmosphere to ocean | 284,000 |
| Atmosphere to continent | 96,000 |
| Continent to atmosphere | 60,000 |
| Continent to ocean | 36,000 |

**Percentage of water contained in the reservoirs of Earth's hydrologic cycle**

| Hydrologic cycle reservoir | Amount |
|---|---|
| World ocean | 97.2% |
| Ice caps, glaciers, and snow | 2.15% |
| Groundwater and soil moisture | 0.62% |
| Streams and lakes | 0.02% |
| Atmospheric water vapor | 0.001% |

**Figure 5.19 The hydrologic cycle.** Diagrammatic view of Earth's hydrologic cycle. Numbers represent Earth's average yearly flux (volume of water moved between reservoirs) in cubic kilometers. Left table shows average yearly flux between reservoirs; right table shows the percentage of Earth's water in each reservoir.

processes include stream runoff, in which streams dissolve ions from continental rocks and carry them to the sea, and volcanic eruptions, both on the land and on the sea floor. Other sources include the atmosphere (which contributes gases) and biological interactions.

Stream runoff is the primary method by which dissolved substances are added to the oceans. Table 5.4 compares the major components dissolved in stream water with those in seawater. It shows that streams have far lower salinity and a vastly different composition of dissolved substances than seawater. For example, bicarbonate ion ($HCO_3^-$) is the most abundant dissolved constituent in stream water yet is found in only trace amounts in seawater. Conversely, the most abundant dissolved component in seawater is the chloride ion ($Cl^-$), which exists in very small concentrations in streams.

If stream water is the main source of dissolved substances in seawater, why do the components of the two not match each other more closely? One of the reasons is that some dissolved substances stay in the ocean and accumulate over time. **Residence time** is the average length of time that a substance resides in the ocean. Long residence times lead to higher concentrations of the dissolved substance. The sodium ion ($Na^+$), for instance, has a residence time of 260 million years and, as a result, has a high concentration in the ocean. Other elements such as aluminum have a residence time of only 100 years and occur in seawater in much lower concentrations.

**Animation**
Earth's Water and the Hydrologic Cycle
http://goo.gl/3Ra4I2

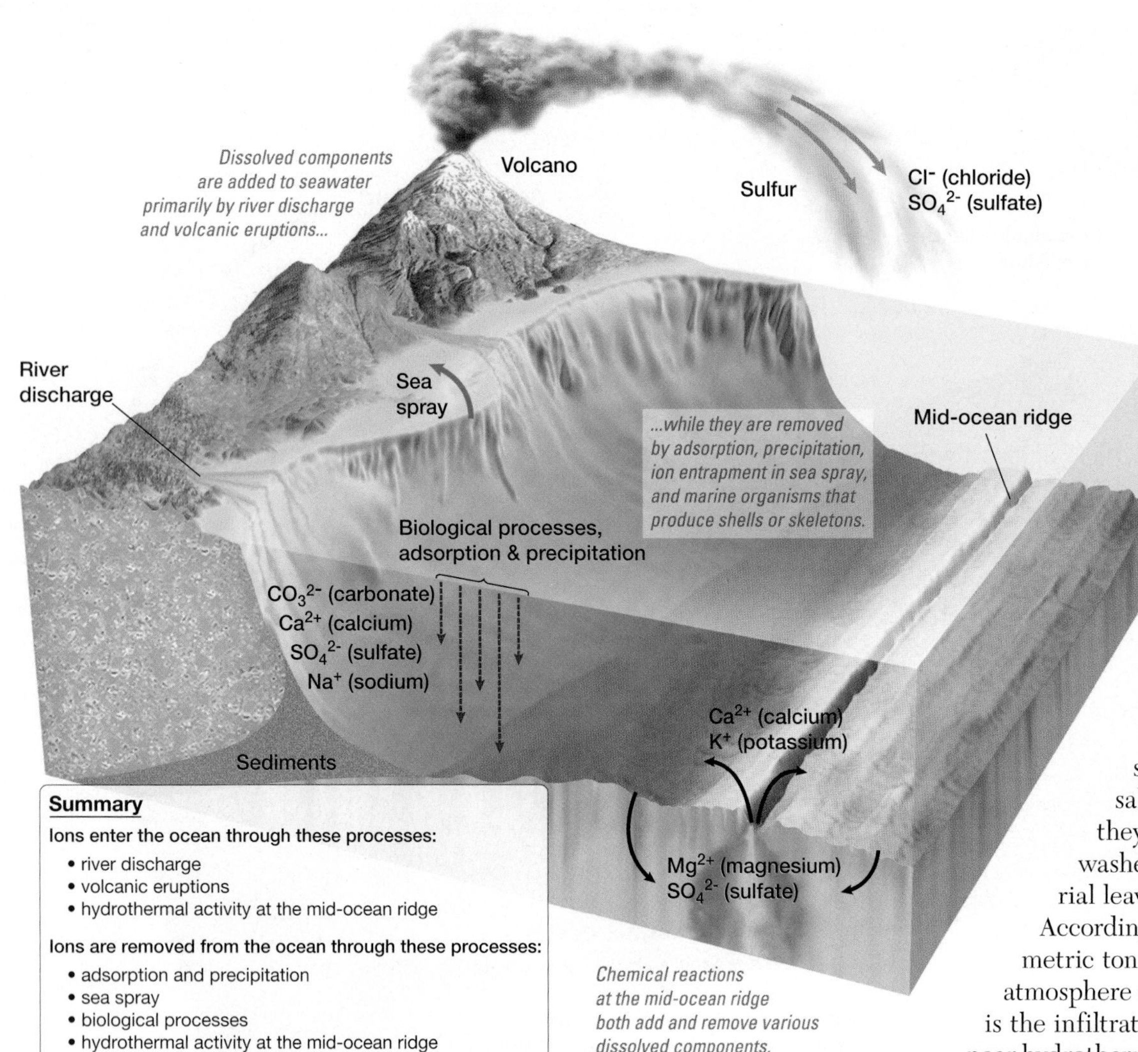

Dissolved components are added to seawater primarily by river discharge and volcanic eruptions...
Volcano
Sulfur
$Cl^-$ (chloride)
$SO_4^{2-}$ (sulfate)
River discharge
Sea spray
...while they are removed by adsorption, precipitation, ion entrapment in sea spray, and marine organisms that produce shells or skeletons.
Mid-ocean ridge
Biological processes, adsorption & precipitation
$CO_3^{2-}$ (carbonate)
$Ca^{2+}$ (calcium)
$SO_4^{2-}$ (sulfate)
$Na^+$ (sodium)
Sediments
$Ca^{2+}$ (calcium)
$K^+$ (potassium)
$Mg^{2+}$ (magnesium)
$SO_4^{2-}$ (sulfate)
Summary
Ions enter the ocean through these processes:
• river discharge
• volcanic eruptions
• hydrothermal activity at the mid-ocean ridge
Ions are removed from the ocean through these processes:
• adsorption and precipitation
• sea spray
• biological processes
• hydrothermal activity at the mid-ocean ridge
Chemical reactions at the mid-ocean ridge both add and remove various dissolved components.

**Figure 5.20 The cycling of dissolved components in seawater.**

Are the oceans becoming saltier through time? This might seem logical since new dissolved components are constantly being added to the oceans and because most salts have long residence times. However, analysis of ancient marine organisms and sea floor sediments suggests that the oceans have not increased in salinity over time, nor have the proportions of dissolved components changed. This must be because the rate at which an element is added to the ocean equals the rate at which it is removed, so the average *amounts* of various elements remain constant (this is called a *steady-state* condition).

Materials added to the oceans are affected by several processes that cycle dissolved substances out of seawater. When waves break at sea, for example, sea spray releases tiny salt particles into the atmosphere, where they may be blown over land before being washed back to Earth. The amount of material leaving the ocean in this way is enormous: According to a recent study, as much as 3.3 billion metric tons (7.3 trillion pounds) of salt enters the atmosphere each year as sea spray. Another example is the infiltration of seawater along mid-ocean ridges near hydrothermal vents (see Figure 5.20), which incorporates magnesium and sulfate ions into sea floor mineral deposits. In fact, chemical studies of seawater indicate that the *entire volume of ocean water* is recycled through this hydrothermal circulation system at the mid-ocean ridge approximately every 3 million years. As a result, the chemical exchange between ocean water and the basaltic crust has a major influence on the composition of ocean water.

Dissolved substances are also removed from seawater in other ways. Calcium, carbonate, sulfate, sodium, and silicon are deposited in ocean sediments within the shells of dead microscopic organisms and animal feces. Vast amounts of dissolved substances can be removed when inland arms of seas dry up, leaving salt deposits called *evaporites* (such as those beneath the Mediterranean Sea; see Mastering Oceanography **Web Diving Deeper 4.1**). In addition, ions dissolved in ocean water are removed by adsorption (physical attachment) to the surfaces of sinking clay and biological particles.

**RECAP Various surface processes either decrease seawater salinity (precipitation, runoff, icebergs melting, or sea ice melting) or increase seawater salinity (sea ice forming and evaporation).**

## CONCEPT CHECK 5.4 ▸ Explain why seawater salinity varies.

1 What physical conditions create brackish water in the Baltic Sea and hypersaline water in the Red Sea?

2 Describe the ways in which dissolved components are added and removed from seawater.

3 List the components (reservoirs) of the hydrologic cycle that hold water on Earth and the percentage of Earth's water in each one. Describe the processes by which water moves among these reservoirs.

# 5.5 ▶ How Does Seawater Salinity Vary at the Surface and with Depth?

From the surface to the ocean depths, the ocean undergoes variations in salinity, temperature, and density that create a layered ocean. This layering affects the mixing of ocean water, the movement of currents, and the distribution of marine life. In this section and the next, we'll explore the variations of properties both at the surface and with depth that cause the ocean to be layered.

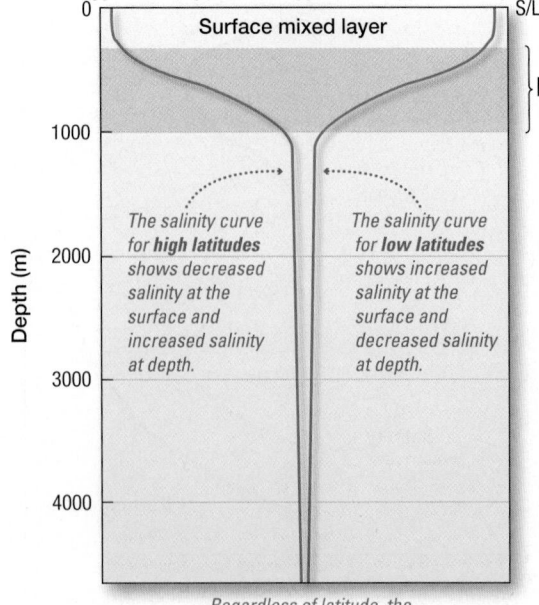

## Surface Salinity Variation

Although seawater salinity at the ocean surface averages 35‰, surface salinity varies depending on the latitude (**Figure 5.21**). The red curve in Figure 5.21 shows temperature, which is low in the high latitudes but steadily increases with latitude all the way to the equator. The green curve in the figure shows salinity, which is lowest at high latitudes, peaks in the lower latitudes near the Tropics of Cancer and Capricorn, and dips near the equator.

Why does surface salinity vary in the pattern shown in Figure 5.21? At high latitudes, abundant precipitation and runoff and the melting of freshwater icebergs all decrease salinity. In addition, cool temperatures limit the amount of evaporation that takes place (which would increase salinity). The formation and melting of sea ice balance each other out in the course of a year and are not a factor in changes in salinity.

In low latitudes, the pattern of Earth's atmospheric circulation (see Chapter 6, "Air–Sea Interaction") causes warm, dry air to descend, so near the Tropics of Cancer and Capricorn evaporation rates are high and salinity increases. In addition, little precipitation and runoff occur to decrease salinity. As a result, the regions near the Tropics of Cancer and Capricorn are the continental and maritime deserts of the world.

Temperatures are warm near the equator, so evaporation rates are high enough to increase salinity. Increased precipitation and runoff partially offsets the high salinity, though. For example, daily rain showers are common along the equator, adding water to the ocean and lowering its salinity.

**EXPLORING DATA** ▼ **Is there relatively high or low seawater salinity in the high latitudes? Is there relatively high or low seawater salinity in the low latitudes? Explain.**

**Figure 5.21 Sea surface salinity and temperature variation with latitude.** Graph showing variation with latitude in sea surface salinity (*green curve*) along with sea surface temperature (*red curve*).

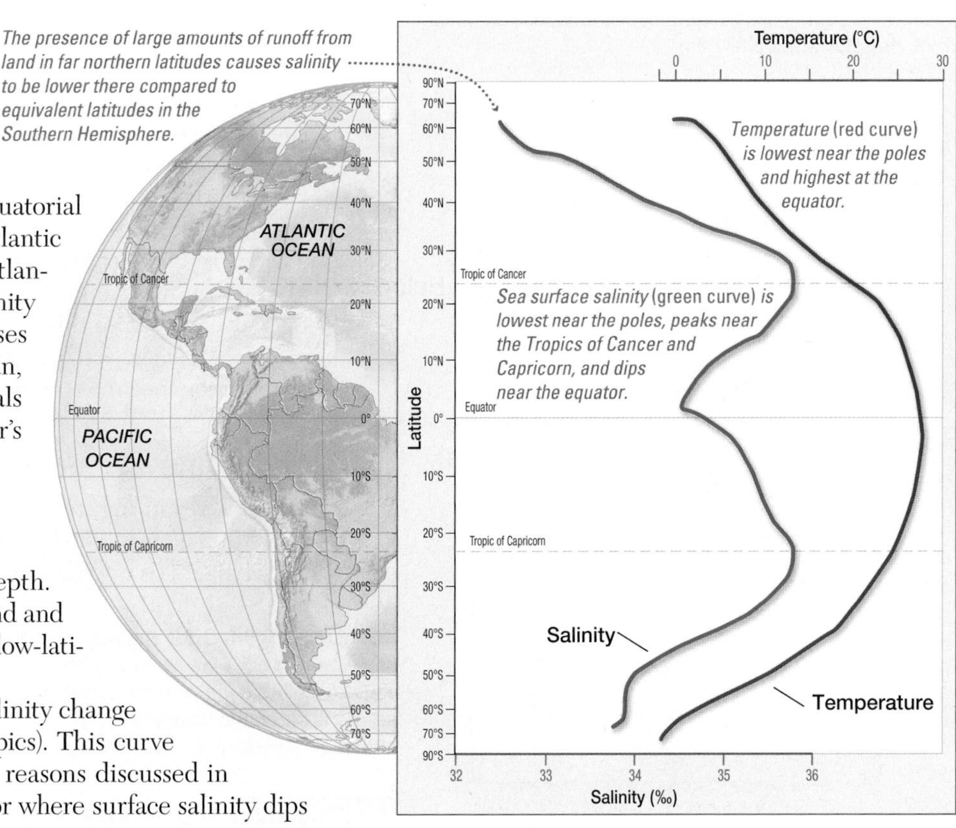


**Figure 5.22** is a map of satellite-collected data that shows how ocean surface salinity varies worldwide. Notice how the overall pattern of the satellite image matches the graph in Figure 5.21. For example, both the graph and the satellite image show high salinity in the subtropics (Figure 5.22, *orange*) and lower salinity in rainy polar regions and equatorial belts (Figure 5.22, *blue*). In addition, notice that the Atlantic Ocean has higher salinity values than the Pacific. The Atlantic Ocean's higher overall salinity is caused by its proximity to land and the associated continental effect. This causes high rates of evaporation in the narrower Atlantic Ocean, particularly in the tropics. The satellite image also reveals an expanse of low-salinity water from the Amazon River's outflow (Figure 5.22, *purple*).

## Salinity Variation with Depth

**Figure 5.23** shows how seawater salinity varies with depth. The graph displays data for the open ocean far from land and shows one curve for high-latitude regions and one for low-latitude regions.

The curve to the right in Figure 5.23 shows the salinity change with depth for low-latitude regions (such as in the tropics). This curve shows *increased* salinity at the surface because of the reasons discussed in the preceding section. Note that even along the equator where surface salinity dips

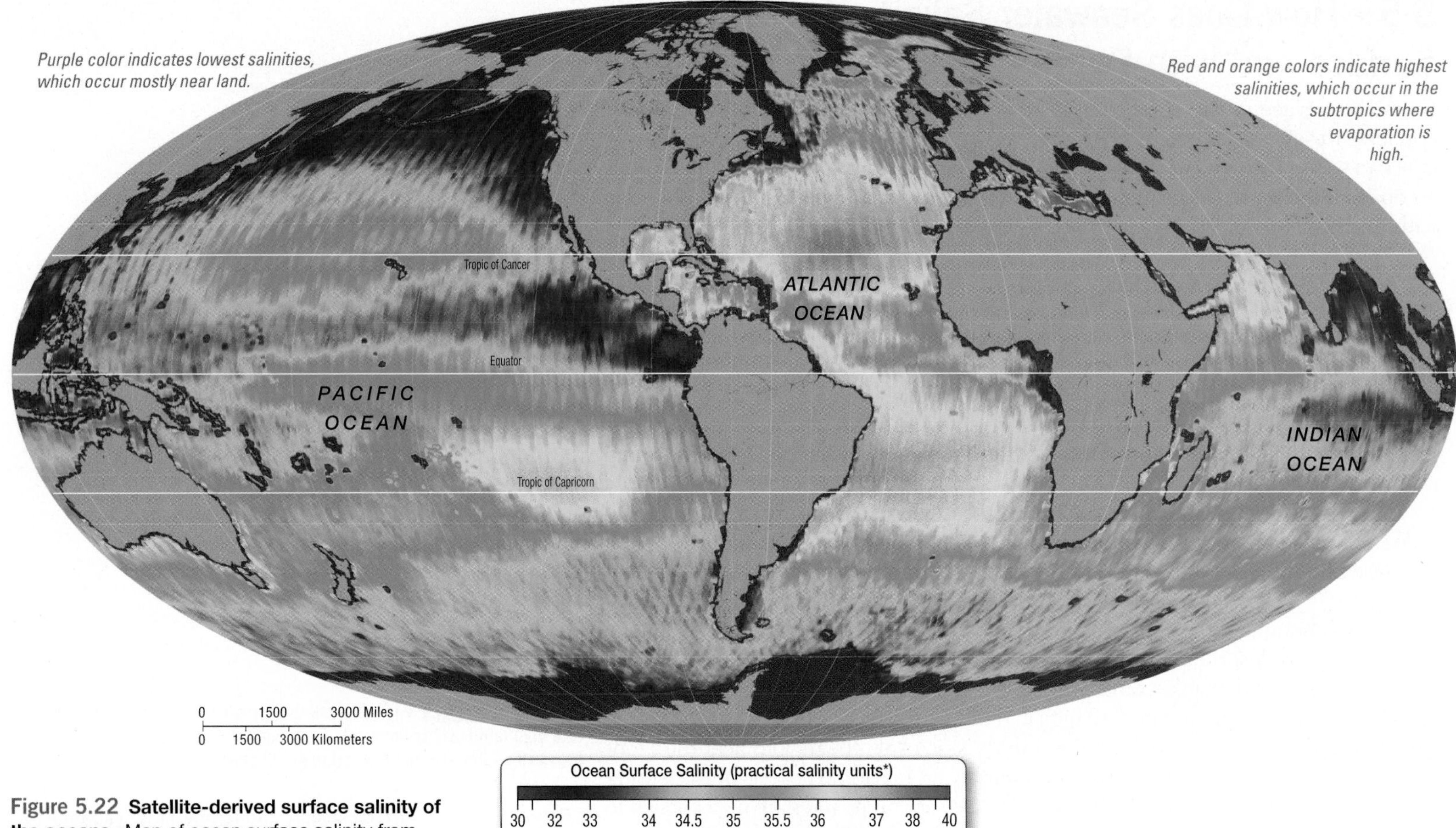


**Figure 5.22 Satellite-derived surface salinity of the oceans.** Map of ocean surface salinity from data collected by the Aquarius satellite during January 2015. Values are in practical salinity units, which are roughly equivalent to parts per thousand (‰); black regions indicate no data; the north-south color striations are an artifact from the satellite's orbital path.

Increasing salinity (‰) ⟶

33 34 35 36 37

S/L

Surface mixed layer

Halocline

Depth (m): 0, 1000, 2000, 3000, 4000

*The salinity curve for **high latitudes** shows decreased salinity at the surface and increased salinity at depth.*

*The salinity curve for **low latitudes** shows increased salinity at the surface and decreased salinity at depth.*

*Regardless of latitude, the salinity at depth is similar.*

(Figure 5.21), the salinity value is still relatively high. Then with increasing depth, the curve changes to a *lower* salinity value for the remainder of the water column.

The curve to the left in Figure 5.23 shows the salinity change with depth for high-latitude regions (such as near Antarctica or in the Gulf of Alaska). This curve shows *decreased* salinity at the surface also because of the reasons discussed in the preceding section. Then with increasing depth, the curve changes to a *higher* salinity value for the remainder of the water column. Note that both curves on the graph show that regardless of latitude, there is a similar, intermediate value of salinity at depth.

These two curves, which together resemble the outline of a wide Champagne glass, show that salinity varies widely at the surface but hardly varies at all in the deep ocean. Why is this so? It occurs because all the processes that affect seawater salinity (precipitation, runoff, melting icebergs, melting sea ice, sea ice forming, and evaporation) occur at the *surface* and thus have no effect on deep water below.

## Halocline

Both curves in Figure 5.23 show a rapid change in salinity between the depths of about 300 meters (980 feet) and 1000 meters (3300 feet). For the low-latitude curve, the change is a *decrease* in salinity. For the high-latitude curve, the change is an

**SmartFigure 5.23 Salinity variation with depth.** Graph showing a vertical profile of high- and low-latitude salinity variation with depth. Horizontal scale is in ‰; vertical scale is depth in meters, with sea level at the top. The layer of rapidly changing salinity is called the halocline.
https://goo.gl/9b1dP7

**EXPLORING DATA ◀ According to SmartFigure 5.23, there is a large amount of salinity variation at the ocean surface, yet only a small amount of salinity variation below 1000 meters (3300 feet). Explain why this is so.**

*increase* in salinity. In both cases, this layer of rapidly changing salinity with depth is called a **halocline** (*halo* = salt, *cline* = slope). Haloclines separate layers of different salinity in the ocean.

**RECAP** **A halocline is a layer of rapidly changing salinity that occurs in both high- and low-latitude regions.**

**CONCEPT CHECK 5.5 ▸ Specify how seawater salinity varies at the surface and with depth.**

1 Why is there low surface salinity in the high latitudes, and why is there higher surface salinity in the low latitudes?

2 Explain the dip in surface salinity values near the equator that is shown in Figure 5.21.

3 Describe the halocline, including where it occurs in the ocean.

## 5.6 ▸ How Does Seawater Density Vary with Depth?

The density of pure water is 1.000 gram per cubic centimeter (g/cm$^3$) at 4°C (39°F). This value serves as a standard against which the density of all other substances can be measured. Seawater contains various dissolved substances that increase its density. In the open ocean, seawater density averages between 1.022 and 1.030 g/cm$^3$ (depending on its salinity). Thus, the density of seawater is 2 to 3% greater than that of pure water. Unlike freshwater, seawater continues to increase in density until it freezes at a temperature of −1.9°C (28.6°F). (Recall that below 4°C [39°F], the density of freshwater actually *decreases*; see Figure 5.12.) At its freezing point, however, seawater behaves in a similar fashion to freshwater: Its density decreases dramatically, which is why sea ice floats, too.

Density is an important property of ocean water because density differences determine the vertical position of ocean water and cause water masses to float or sink, thereby creating deep-ocean currents. For example, if seawater with a density of 1.030 g/cm$^3$ were added to freshwater with a density of 1.000 g/cm$^3$, the denser seawater would sink below the freshwater, initiating a deep current.

### Factors Affecting Seawater Density

The ocean, like Earth's interior, is layered according to density. Low-density water exists near the surface, and higher-density water occurs below. Except for some shallow inland seas with high rates of evaporation that create high-salinity water, the highest-density water is found at the deepest ocean depths. Let's examine how temperature, salinity, and pressure influence seawater density by expressing the relationships using arrows (up arrow = increase; down arrow = decrease):

- As temperature increases (↑), seawater density decreases (↓) (due to thermal expansion; note that a relationship where one variable *decreases* as a result of another variable's *increase* is known as an inverse relationship, in which the two variables are *inversely proportional*).
- As salinity increases (↑), seawater density increases (↑) (due to the addition of more dissolved material).
- As pressure increases (↑), seawater density increases (↑) (due to the compressive effects of pressure).

Of these three factors, only temperature and salinity influence the density of surface water. Pressure influences seawater density only when very high pressures are encountered, such as in deep-ocean trenches. Still, the density of seawater in the deep ocean is only about 5% greater than at the ocean surface, showing that despite tons of pressure per square centimeter, water is nearly incompressible. Unlike air, which can be compressed and put in a tank for use in scuba diving, the molecules in liquid water are already close together and cannot be compressed much more. Therefore, pressure has the least effect on influencing the density of surface water and can largely be ignored.

Temperature, on the other hand, has the greatest influence on surface seawater density because the range of surface seawater temperature is greater than that of salinity. In fact, only in the extreme polar areas of the ocean, where temperatures are low and remain relatively constant, does salinity significantly affect density. Cold water that also has high salinity is some of the highest-density water in the world. The density of seawater—the result of its salinity and temperature—influences currents in the deep ocean because high-density water sinks below less-dense water.

> **RECAP** **Differences in ocean density cause the ocean to be layered. Seawater density increases with decreased temperature, increased salinity, and increased pressure. Temperature has the greatest influence on seawater density.**

**Low latitudes: Comparing temperature and density curves**

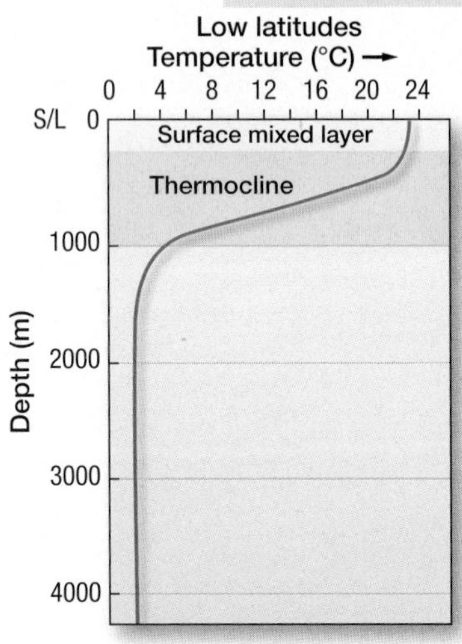


(a) Temperature variation with depth in low-latitude regions. The layer of rapidly changing temperature with depth is the thermocline.

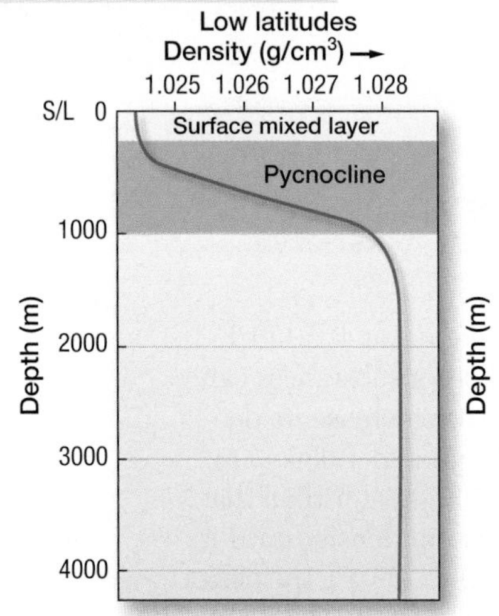


(b) Density variation with depth in low-latitude regions. The layer of rapidly changing density with depth is the pycnocline.

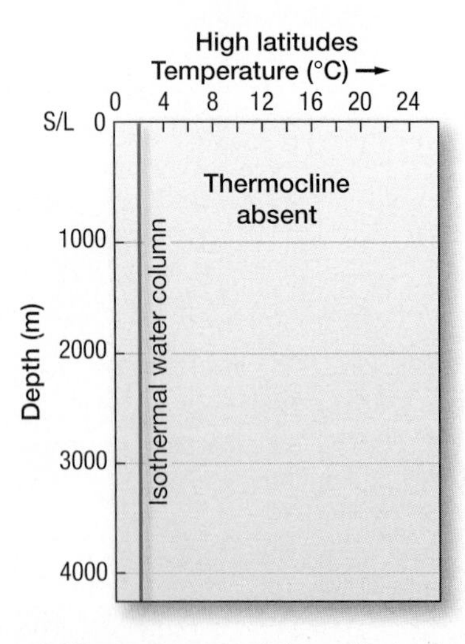


(c) Temperature variation with depth in high-latitude regions. Because the water column is isothermal, there is no thermocline.

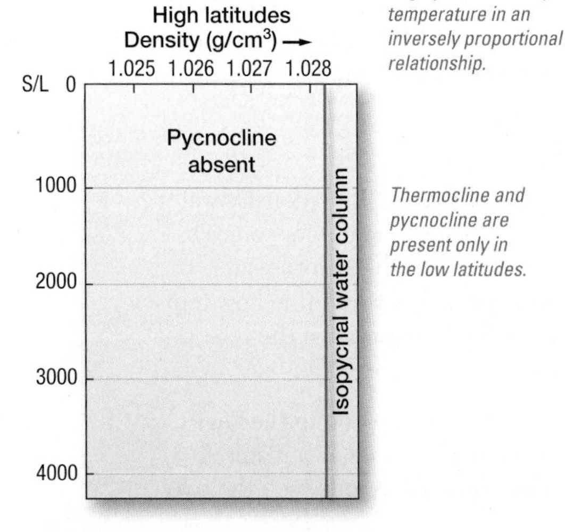


(d) Density variation with depth in high-latitude regions. Because the water column is isopycnal, there is no pycnocline.

**High latitudes: Comparing temperature and density curves**

**SmartFigure 5.24 Comparing typical vertical profile curves for temperature and density in the low and high latitudes.** Paired graphs comparing temperature and density curves in the low latitudes (*a* and *b*, *above*) and temperature and density curves in the high latitudes (*c* and *d*, *below*). Note that the curves are averaged over a year's time. https://goo.gl/1zC18l

## Temperature and Density Variation with Depth

The four graphs in **Figure 5.24** compare the idealized vertical profile curves for temperature and density in both low- and high-latitude regions. Let's examine each graph individually.

Figure 5.24a shows how temperature varies with depth in low-latitude regions, where surface waters are warmed by high Sun angles and constant length of days. However, the Sun's energy does not penetrate very far into the ocean. Surface water temperatures remain relatively constant until a depth of about 300 meters (980 feet) because of good surface mixing mechanisms such as surface currents, waves, and tides. Below about 300 meters (980 feet), the temperature decreases rapidly until a depth of about 1000 meters (3300 feet). Below 1000 meters, the water's low temperature again remains constant down to the ocean floor.

The density curve for low-latitude regions in Figure 5.24b shows that density is relatively low at the surface. Density is low because surface water temperatures are high. (Remember that temperature has the greatest influence on density and temperature is inversely proportional to density.) Below the surface, density also remains constant until a depth of about 300 meters (980 feet) because of good surface mixing. Below about 300 meters (980 feet), the density increases rapidly until a depth of about 1000 meters (3300 feet). Below 1000 meters, the water's low density again remains constant down to the ocean floor.

Figure 5.24c shows how temperature varies with depth in high-latitude regions, where surface waters remain cool year-round and deep-water temperatures are about the same as the surface. The temperature curve for high-latitude regions, therefore, is a straight vertical line, which indicates uniform conditions at the surface and at depth.

The density curve for high-latitude regions (Figure 5.24d) also shows hardly any variation with depth. Density is relatively high at the surface because surface water temperatures are low. Density is high below the surface, too, because water temperature is also low. The density curve for high-latitude regions, therefore, is also represented by a straight vertical line, which indicates uniform conditions at the surface and at depth. These conditions allow cold high-density water to form at the surface, sink, and initiate deep-ocean currents.

One of the most important things to notice about Figure 5.24 is that the two graphs shown in the top portion of Figure 5.24 are related to each other, as are the two graphs shown in the bottom portion of Figure 5.24. If you can imagine folding Figure 5.24 along its vertical dashed line and overlaying the two sets of graphs, you would notice they are identical mirror images of each other. For example, the low latitude temperature graph (Figure 5.24a) is a mirror image of its corresponding density graph (Figure 5.24b). Similarly, the high latitude temperature graph (Figure 5.24c) is a mirror image of its corresponding density graph (Figure 5.24d). Why are the curves mirror images of each other? As discussed previously, temperature is the most important factor that influences seawater density and it operates as an inversely proportional relationship. This is exactly the relationship that is illustrated by the mirror images of these two sets of graphs.

## Thermocline and Pycnocline

Analogous to the halocline (the layer of rapidly changing salinity shown in Figure 5.23), the low-latitude temperature graph in Figure 5.24a displays a curving line that indicates a layer of rapidly changing temperature called a **thermocline** (*thermo* = heat, *cline* = slope). Similarly, the low-latitude density graph in Figure 5.24b displays a curving line that indicates a **pycnocline** (*pycno* = density, *cline* = slope), which is a layer of rapidly changing density. Note that the corresponding high-latitude graphs of temperature (Figure 5.24c) and density (Figure 5.24d) both lack a thermocline and a pycnocline, respectively, because these lines show a constant value with depth (they are straight vertical lines that don't curve). Like a halocline, a thermocline and a pycnocline typically occur between about 300 meters (980 feet) and 1000 meters (3300 feet) below the surface. The temperature difference between water above and below the thermocline can be used to generate electricity (see Mastering Oceanography **Web Diving Deeper 5.1**).

When a pycnocline is established in an area, it presents an incredible barrier to mixing between low-density water above and high-density water below. A pycnocline has a high gravitational stability and thus physically isolates adjacent layers of water. (Note that this is similar to a temperature inversion in the atmosphere, which traps cold, high-density air underneath warm, low-density air.) The pycnocline results from the combined effect of the thermocline and the halocline because temperature and salinity influence density. The interrelationship of these three layers determines the degree of separation between the upper-water and deep-water masses.

The ocean is layered into three distinct water masses based on density. The **mixed surface layer** occurs above a strong permanent thermocline (and corresponding pycnocline; see Figure 5.24). The water is uniform because it is well-mixed by surface currents, waves, and tides. The thermocline and pycnocline occur in a relatively low-density layer called the **upper water**, which is well developed throughout the low and middle latitudes. Denser and colder **deep water** extends from below the thermocline/pycnocline to the deep-ocean floor.

Thermoclines (and corresponding pycnoclines) can occur in other locations, too. Scuba divers, for example, often experience minor thermoclines as they descend into the ocean. Thermoclines can also develop in swimming pools, ponds, and lakes. During the spring and fall, when nights are cool but days can be quite warm, the Sun heats the surface water of the pool, yet the water below the surface can be quite cold. If the pool has not been mixed, a thermocline isolates the warm surface layer from the deeper cold water. The cold water below the thermocline can be quite a surprise for anyone who dives into the pool!

Animation
The Carbonate Buffering System
https://goo.gl/ZsCngq

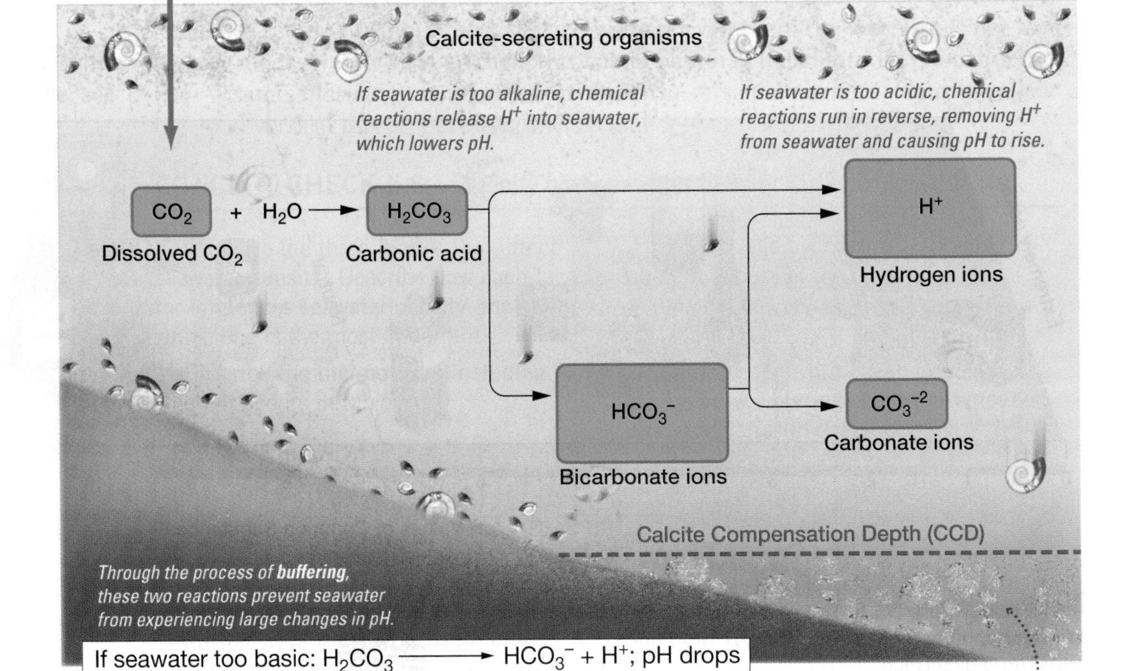


**SmartFigure 5.27 The carbonate buffering system.**
https://goo.gl/d5Gkwp

**RECAP Reactions involving carbonate chemicals serve to buffer the ocean and help maintain its average pH at 8.1 (slightly alkaline, or basic).**

## STUDENTS SOMETIMES ASK . . .

***Why do carbonated beverages burn my throat when I drink them?***

When carbon dioxide gas ($CO_2$) dissolves in water ($H_2O$), its molecules react with water molecules and form carbonic acid ($H_2CO_3$). Carbonic acid is a weak acid, which means most of its molecules are neutral (nonionized) at any given moment. However, a small percentage of those molecules naturally break apart (dissociate) and exist as two fragments: a negatively charged $HCO_3^-$ ion and a positively charged $H^+$ ion. The $H^+$ ions are responsible for acidity—the higher their concentration in a solution, the more acidic that solution. The presence of carbonic acid in carbonated water makes that water acidic—the more carbonated, the more acidic. What you're feeling when you drink a carbonated beverage is the moderate acidity of that beverage irritating your throat.

The equations below Figure 5.27 show how these chemical reactions involving carbonate minimize changes in the pH of the ocean in a process called **buffering**. Buffering protects the ocean from getting too acidic or too basic. In effect, the shells of organisms act as an "antacid" for the deep ocean, analogous to the way commercial antacids use calcium carbonate to neutralize excess stomach acid. For example, if the pH of the ocean increases (becomes too basic), it causes $H_2CO_3$ to release $H^+$, and pH drops. Conversely, if the pH of the ocean decreases (becomes too acidic), $HCO_3^-$ combines with $H^+$ to remove it, causing pH to rise. Since more carbonate shells dissolve as conditions become more acidic, there is always a ready supply of $HCO_3$. In this way, buffering prevents large swings of ocean water pH and allows the ocean to stay within a limited range of pH values. Recently, however, increasing amounts of carbon dioxide from human emissions are beginning to enter the ocean and change the ocean's pH, making it more acidic. For more details on this process, see Chapter 16, "The Oceans and Climate Change."

**CONCEPT CHECK 5.7 ▸ Discuss the acid/base properties of seawater.**

1 Explain the difference between an acidic substance and an alkaline (basic) substance.

2 How does the ocean's buffering system work?

## 5.8 ▸ What Factors Control the Distribution of Carbon and Oxygen in the Ocean?

Of all the substances found in seawater, carbon and oxygen are two of the most interesting—oxygen, because we need it to breathe, and carbon, because of its capacity to form compounds that maintain the ocean's pH, regulate Earth's temperature, and are essential for sustaining all life on the planet. Perhaps surprisingly, the biggest reservoir of carbon on the planet is not found in the biosphere but in the lithosphere, specifically in carbonate-based sedimentary rocks and marine sediments. In fact, estimates put the amount of carbon in rock at more than 100,000,000 times that found in all terrestrial plants. The second largest reservoir of carbon is the oceans, with more than 60 times the amount of carbon found in land plants. It is no wonder that scientists look to the oceans to study where carbon goes and what happens to it. (For a detailed discussion of the carbon cycle, see Chapter 16, "The Oceans and Climate Change.")

## The Solubility and Distribution of Carbon Dioxide in the Ocean

To understand what happens to carbon in the ocean, it is important to know that carbon is comprised of two types: (1) *inorganic carbon*, such as that found in $CO_2$ and the $CaCO_3$-containing shells of marine organisms, and (2) *organic carbon*, such as that found in the soft tissues of living creatures, often represented by the formula $C_6H_{12}O_6$.

Carbon most directly enters the ocean at the sea surface when atmospheric $CO_2$ dissolves in seawater. Here it reacts with water to form carbonic acid, entering the carbonate buffering system as described in Section 5.7 and contributing to the formation of calcium carbonate shells of marine organisms. In the upper sunlit surface waters of the ocean, carbon dioxide is also taken up by marine plants and algae in the production of organic compounds during photosynthesis. The net effect of both processes is to lower dissolved $CO_2$ concentrations in surface waters. As **Figure 5.28** shows, $CO_2$ concentration is low at the surface but increases with increasing depth.

The increase in the concentration of $CO_2$ with depth is due in part to the input of $CO_2$ from the respiration of marine organisms and the decay of dead organic matter, but also because the **solubility** of $CO_2$ increases as seawater gets colder and pressure increases—conditions that occur the deeper you go in the ocean.

A handy way to think of solubility is the tendency of a substance to dissolve in a liquid. Unlike most substances, which typically become *more soluble* as a liquid heats up, the solubility of gases like $CO_2$ *decreases* as the water temperature rises. That is why a cold soft drink that is very fizzy when first opened loses its fizz as it warms up. The carbonation—that is, the carbon dioxide gas dissolved in that beverage—escapes into the air as $CO_2$, and becomes less soluble in the warming liquid. The same analogy applies to the solubility of gases as a function of pressure. When you first open a soft drink and release the pressure inside, there is an initial rush of bubbles. This happens because the $CO_2$ gas dissolved in the liquid is less soluble at atmospheric pressure and so immediately bubbles out. As this analogy suggests, the highest concentration of dissolved $CO_2$ is in seawater that is both cold and under high pressure.

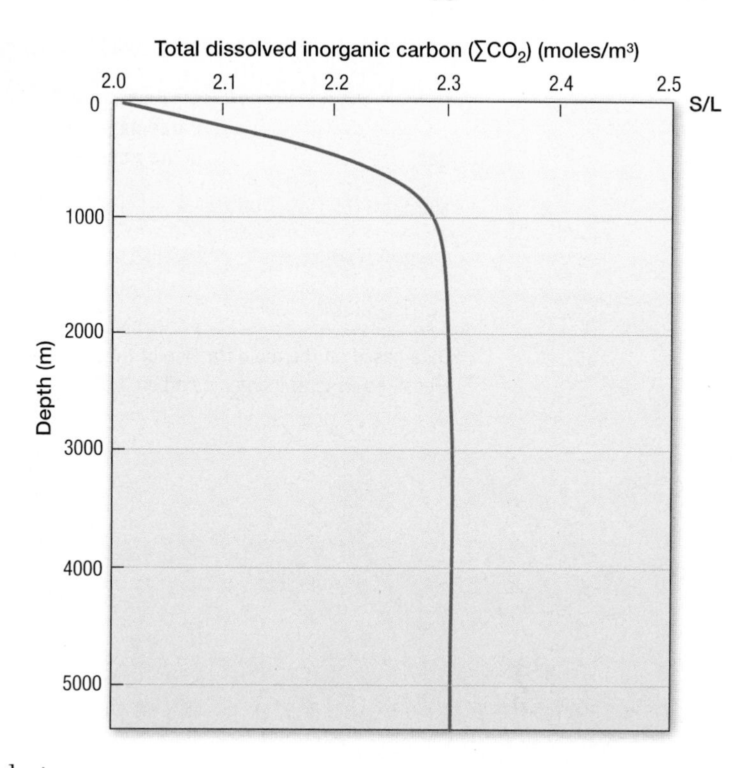


**Figure 5.28 Carbon dioxide concentration in seawater varies with ocean depth.** Carbon dioxide reacts with water and exists in several different forms, known collectively as $\Sigma CO_2$ or *total dissolved inorganic carbon*. $\Sigma CO_2$ consists of free dissolved $CO_2$ gas, carbonic acid ($H_2CO_3$), bicarbonate ($HCO_3^-$), and carbonate ($CO_3^{2-}$). Bicarbonate is by far by most abundant form of dissolved inorganic carbon in the ocean.

**EXPLORING DATA ▲ By examining the change in $\Sigma CO_2$ with depth, identify the depth that corresponds to the bottom of the pycnocline. Hint: recall that the pycnocline is a strong barrier to vertical water mixing in the ocean.**

## The Solubility and Distribution of Dissolved Oxygen in the Ocean

The temperature-pressure-solubility relationship described for $CO_2$ has implications for the distribution of another important gas in the ocean: oxygen. In high latitudes, the concentration of dissolved oxygen in surface waters is greater than that found in low latitudes, due to the increased solubility of $O_2$ gas in the colder water. For example, **Figure 5.29** shows a comparison of the dissolved $O_2$ profiles at two different locations in the Pacific Ocean.

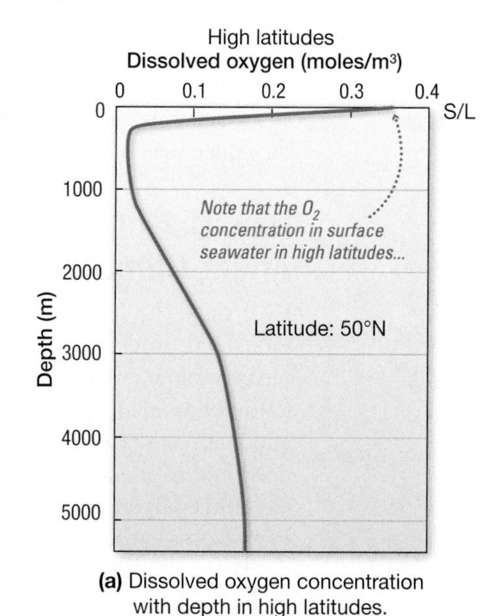


(a) Dissolved oxygen concentration with depth in high latitudes.

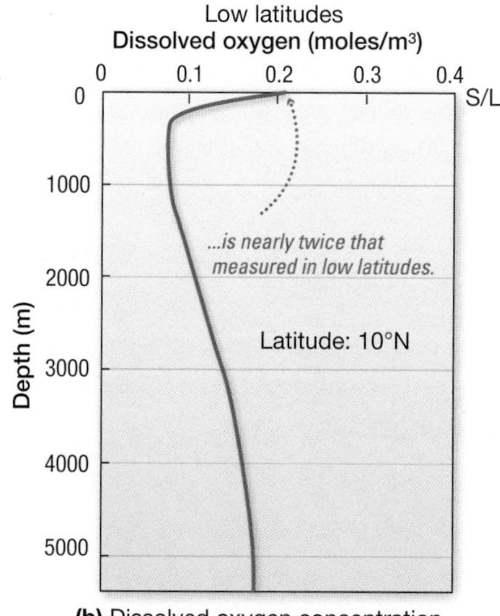


(b) Dissolved oxygen concentration with depth in low latitudes.

**SmartFigure 5.29 Graphs of dissolved oxygen concentration in seawater with depth in the North Pacific, at different latitudes.**
https://goo.gl/s6RL8D

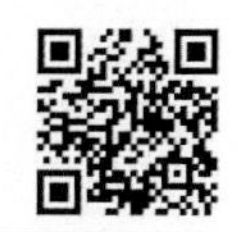

**Table 5.5 Abundances of major gases in the atmosphere and in seawater**

| Gas | Percentage by volume in the atmosphere | Percentage of total gas dissolved in surface seawater[a] | Average concentration in parts per million by mass in the ocean as a whole |
|---|---|---|---|
| Nitrogen ($N_2$) | 78.1 | 63 | 33 |
| Oxygen ($O_2$) | 20.9 | 34 | 6 |
| Carbon dioxide ($CO_2$) | 0.04 | 1.4 | 99[b] |

[a] Data based on the mole fraction of these gases in seawater at equilibrium, after Millero, *Chemical Oceanography*, 4th edition, 2013.
[b] Measured as total inorganic carbon $\Sigma CO_2$, after Broecker, *Chemical Oceanography*, 1974.

Unlike carbon dioxide, however, the major source of oxygen at the sea surface is not exchange with the atmosphere, although excess oxygen can be mixed into the surface ocean by large waves when seas are turbulent, such as during storms. Instead, a significant fraction of oxygen measured in surface seawater is oxygen released by marine photosynthetic organisms, and this causes the oxygen concentration in surface waters to be greater than its concentration in the atmosphere. This means that the oceans are a *net source* of oxygen to the atmosphere; in other words, the oceans supply oxygen to the atmosphere. In fact, estimates suggest that more than half of the world's atmospheric oxygen is produced by marine phytoplankton.

Table 5.5 compares the abundances of major gases in the atmosphere, in surface seawater, and in the oceans as a whole. Note that compared to its abundance in the atmosphere, there is a greater percentage of oxygen in surface seawater because of the vast amount of oxygen produced by phytoplankton during photosynthesis. There is also more carbon dioxide in surface seawater compared to its abundance in the atmosphere. This is because $CO_2$ is highly soluble, readily reacting with water to form other inorganic compounds.

Another way that the concentrations of dissolved $O_2$ and $CO_2$ in seawater differ can be seen by comparing Figures 5.28 and 5.29. At water depths below the sunlit surface, oxygen is consumed and carbon dioxide is produced through the process of respiration, and as a consequence the shape of the profiles of dissolved $CO_2$ and $O_2$ in the top 1000 meters (3300 feet) are the opposite of each other. However, below 1000 meters (3300 feet), both carbon dioxide and oxygen concentrations are fairly constant. This reflects the lower rates of respiration in the deep ocean, a consequence of the fact that very few marine organisms live there. The small rebound in oxygen concentration at depths below the oxygen minimum is because these deep, dense, cold-water masses originated as surface water in high latitudes, where the solubility of oxygen is greater, and therefore the concentration of dissolved oxygen is greater (see Chapter 7, "Ocean Circulation").

## STUDENTS SOMETIMES ASK . . .

***Won't more $CO_2$ in the atmosphere and ocean cause marine plants and algae to grow faster, which will then generate more oxygen for humans to breathe?***

Increases in $CO_2$ in the atmosphere will definitely cause an increase in $CO_2$ in the ocean, but this won't necessarily translate into a bloom of oxygen-producing marine organisms. The availability of other components of seawater, such as nutrients like iron, does more to limit the growth of phytoplankton than the availability of $CO_2$ does. In fact, the increased acidity of the ocean caused by more atmospheric $CO_2$ dissolving in the ocean has been shown to be harmful to most microscopic photosynthetic organisms that build their tiny shells out of $CaCO_3$.

## How Does Dissolved Carbon and Oxygen in the Ocean Affect Climate?

Climate

Connection

Ever since the 1930s when scientists first pointed out that human-caused industrial processes pumping $CO_2$ into the atmosphere had resulted in measurable increases in global temperature, researchers have been trying to understand and predict how carbon dioxide moves between the air, sea, and land. Scientists now know that more than a quarter of the excess $CO_2$ delivered to the atmosphere from the burning of **fossil fuels** (oil, natural gas, and coal) ends up in the ocean, which is called a *carbon sink* because it accumulates excess carbon. However, as global temperatures rise and the oceans warm, this fraction will likely decrease, as the solubility of carbon dioxide in seawater decreases and seawater becomes saturated with $CO_2$. When this happens, more carbon dioxide will stay in the atmosphere, which will only increase the amount of atmospheric global warming (see Chapter 16, "The Oceans and Climate Change").

In addition, warming oceans will also affect the solubility of oxygen in seawater. Indeed, several detailed studies have demonstrated that since 1960, ocean oxygen concentrations have already declined in today's oceans, with more expected in the future. If this trend continues, scientists are concerned that the oceans may suffer from a low oxygen condition called *hypoxia* (*hypo* = under, *oxia* = oxygen), which can broadly impact many types of marine life (for more details about this topic, see Chapter 13, "Biological Productivity and Energy Transfer"). Such changes could have severe detrimental consequences for both fisheries and coastal economies.

**RECAP** **Carbon and oxygen distributions in the ocean are controlled by their concentrations in the atmosphere, solubility in seawater, chemical reactions, biological activities, and the pressure and temperature of seawater.**

### CONCEPT CHECK 5.8 ▸ Describe factors that control the distribution of carbon and oxygen in the ocean.

1 Describe the relationship between solubility, temperature, pressure, and the concentration of gases in seawater.

2 Explain the difference between organic and inorganic carbon and describe a reaction associated with each.

3 Why would global warming cause the ocean to absorb less $CO_2$? What about $O_2$?

# ESSENTIAL CONCEPTS REVIEW

## 5.1 ▸ Why does water have such unusual chemical properties?

- *Water's remarkable properties help make life as we know it possible on Earth.* These properties include the arrangement of its atoms, how its molecules stick together, its ability to dissolve almost everything, and its heat storage capacity.

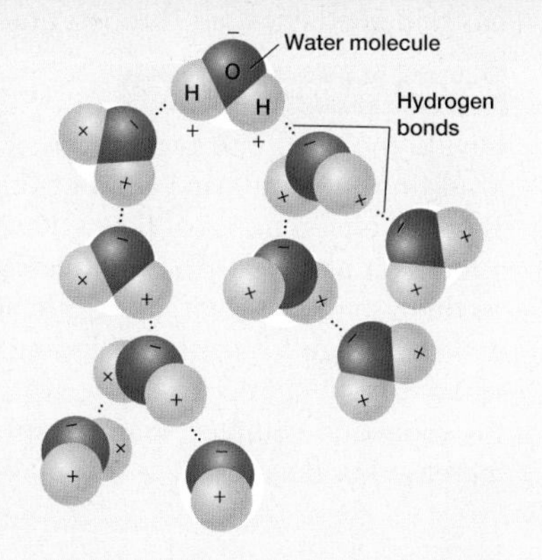


- *The water molecule is composed of one atom of oxygen and two atoms of hydrogen* ($H_2O$). The two hydrogen atoms, which are covalently bonded to the oxygen atom, are *attached to the same side of the oxygen atom* and produce a bend in the geometry of a water molecule. This geometry makes water molecules *polar*, which allows them to form hydrogen bonds with other water molecules or other substances and gives water its remarkable properties. Water, for example, is *the universal solvent* because it can hydrate charged particles (ions), thereby dissolving them.

### Selected Key Terms

Use the **glossary** at the end of this book to discover the meanings of these Selected Key Terms: **atom, nucleus, proton, neutron, electron, molecule, polarity, hydrogen bond, ionic bond.**

### Critical Thinking Question

Using chemical principles, explain why water is considered the universal solvent.

### Active Learning Exercise

Water molecules exhibit polarity. Working with another student in class, discuss what polarity means and come up with a list of common household items that also exhibit polarity.

## 5.2 ▸ What important physical properties does water possess?

- *Water is one of the few substances that exists naturally on Earth in all three states of matter (solid, liquid, and gas).* Hydrogen bonding gives water *unusual thermal properties*, such as a high freezing point (0°C [32°F]) and boiling point (100°C [212°F]), a high heat capacity and high specific heat (1 calorie per gram), a high latent heat of melting (80 calories per gram), and a high latent heat of vaporization (540 calories per gram). Water's high heat capacity and latent heats have important implications in regulating global thermostatic effects.

Crystalline structure is three-dimensional

SOLID

- Like most other chemical substances, *the density of water increases as temperature decreases* and reaches a maximum density at 4°C (39°F). *Below 4°C, however, water density decreases with temperature*, due to the formation of bulky ice crystals. *As water freezes, it expands by about 9%* in volume, so *ice floats* on water.

### Selected Key Terms

Use the **glossary** at the end of this book to discover the meanings of these Selected Key Terms: **heat, temperature, heat capacity, latent heat, marine effect, continental effect.**

### Critical Thinking Question

Explain the differences between the three states of matter, using the arrangement of water molecules and hydrogen bonds in your explanation.

### Active Learning Exercise

Working with another student in class, explain the unusual fact that ice is less dense than liquid water. Be sure to use the terms *thermal contraction*, *water molecules*, and *hydrogen bonds*.

## 5.3 ▶ Why is seawater salty?

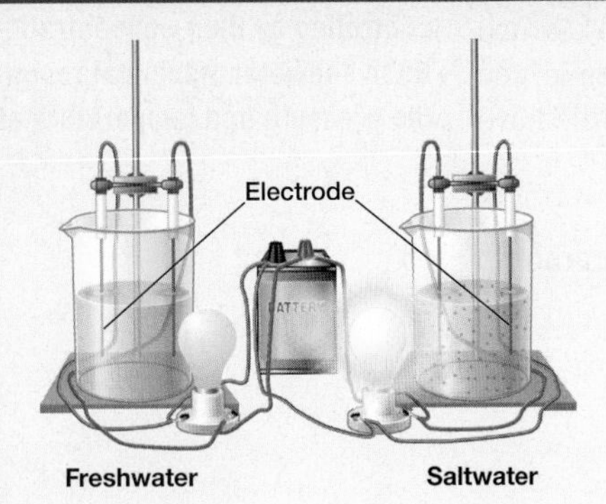


- Dissolved solids make seawater salty. *Salinity is the amount of dissolved solids in ocean water.* It averages about 35 grams of dissolved solids per kilogram of ocean water (*35 parts per thousand* [‰]) but ranges from brackish to hypersaline. Six ions—chloride, sodium, sulfate, magnesium, calcium, and potassium—account for over 99% of the dissolved solids in ocean water. These ions always occur in a *constant proportion* in any seawater sample, so salinity can be determined by measuring the concentration of only one—typically, the chloride ion.
- *The physical properties of pure water and seawater are remarkably similar*, with a few notable exceptions. Compared to pure water, seawater has a *higher pH*, *density*, and *boiling point* (but a *lower freezing point*).
- Although *desalination of seawater* is costly, it *provides freshwater* for business, home, and agricultural use.

### Selected Key Terms

Use the **glossary** at the end of this book to discover the meanings of these Selected Key Terms: **salinity, parts per thousand, principle of constant proportions, salinometer, desalination.**

### Critical Thinking Question

Specify how the principle of constant proportions can be used to determine seawater salinity by measuring only one dissolved component.

### Active Learning Exercise

Working with another student in class, determine the amount of your state sales tax, in parts per thousand.

## 5.4 ▶ Why does seawater salinity vary?

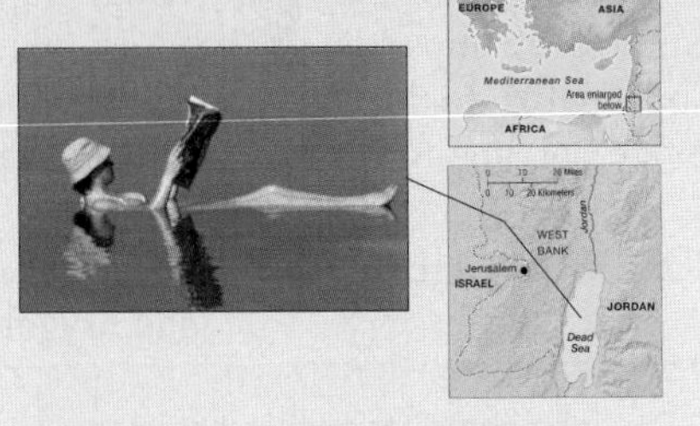


- *Dissolved components in seawater are added and removed by a variety of processes.* Precipitation, runoff, and the melting of icebergs and sea ice add freshwater to seawater and decrease its salinity. The formation of sea ice and evaporation remove freshwater from seawater and increase its salinity.
- The *hydrologic cycle* includes all the reservoirs of water on Earth, including the *oceans*, which contain *97% of Earth's water*. The residence time of various elements indicates how long they stay in the ocean and implies that *ocean salinity has remained constant through time*.

### Selected Key Terms

Use the **glossary** at the end of this book to discover the meanings of these Selected Key Terms: **brackish, hypersaline, hydrologic cycle.**

### Critical Thinking Question

What evidence is used to support the hypothesis that ocean salinity has remained constant through time?

### Active Learning Exercise

Divide the class into two groups. For each group, note that salinity levels in the ocean can vary. For the first group, work with a partner and give one example of how salinity levels might *increase*, including a location along North American where this might occur. For the second group, work with a partner and give one example of how salinity levels might *decrease*, including a location along North American where this might occur. Then, switch partners with the other side of the room and compare examples of how salinity in the ocean may be increased or decreased.

## 5.5 ▶ How does seawater salinity vary at the surface and with depth?

- *The salinity of surface water varies considerably due to surface processes*, with the maximum salinity found near the Tropics of Cancer and Capricorn, and the minimum salinity found in high-latitude regions. Salinity also varies with depth down to about 1000 meters (3300 feet), but below that the salinity of deep water is very consistent. *A halocline is a layer of rapidly changing salinity.*

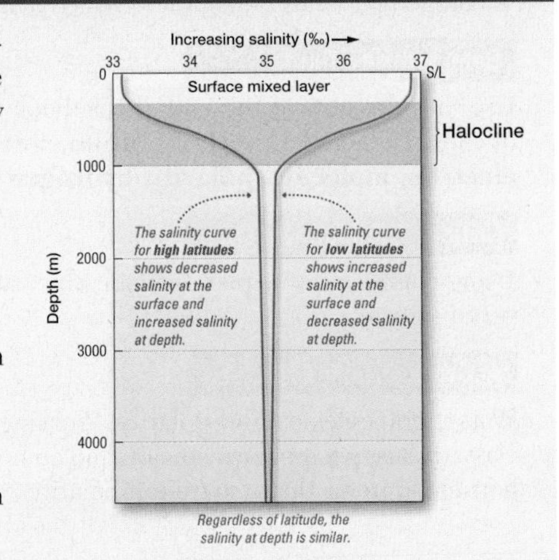

Increasing salinity (‰)
33
34
35
36
37
0
S/L
Surface mixed layer
Halocline
1000
2000
3000
4000
Depth (m)
The salinity curve for high latitudes shows decreased salinity at the surface and increased salinity at depth.
The salinity curve for low latitudes shows increased salinity at the surface and decreased salinity at depth.
Regardless of latitude, the salinity at depth is similar.

### Selected Key Term

Use the glossary at the end of this book to discover the meaning of this Selected Key Term: **halocline.**

### Critical Thinking Question

Using the processes that affect seawater salinity, explain why there is such a large range of salinity variation at the surface but such a narrow range of salinity at depth.

### Active Learning Exercise

Working with another student in class, determine which processes *increase* seawater salinity and which processes *decrease* seawater salinity. For each process, describe how it works on a molecular scale.

## 5.6 ▶ How does seawater density vary with depth?

- *Seawater density increases as temperature decreases and salinity increases*, though temperature influences surface seawater density more strongly than salinity (the influence of pressure is negligible). Temperature and density vary considerably with depth in low-latitude regions, creating a *thermocline (layer of rapidly changing temperature)* and corresponding *pycnocline (layer of rapidly changing density)*, both of which are generally absent in high latitudes.

### Selected Key Terms

Use the **glossary** at the end of this book to discover the meanings of these Selected Key Terms: **thermocline, pycnocline, mixed surface layer, upper water, deep water, isothermal, isopycnal.**

### Critical Thinking Question

Seawater density is inversely proportional to and largely controlled by its temperature. In simple language, explain what this statement means.

### Active Learning Exercise

With another student in class, discuss the following questions for both high and low latitude regions:

1. Is there a pycnocline in this region? Why or why not?
2. Is there a thermocline in this region? Why or why not?
3. Is there a halocline in this region? Why or why not?

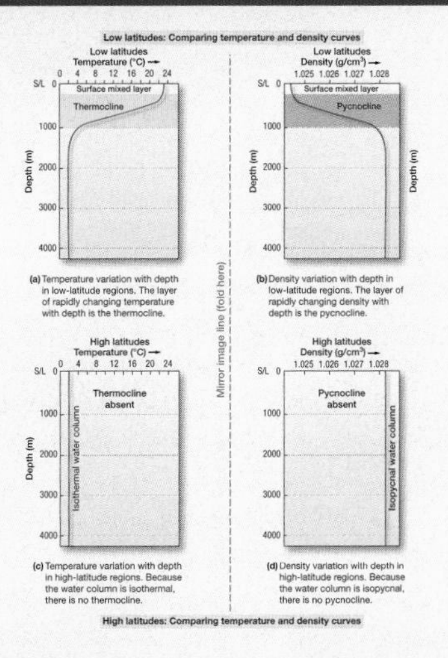


## 5.7 ▶ Is the ocean acidic or basic?

- Ocean surface water has an *average pH of 8.1, making it slightly alkaline*, but the pH of seawater varies both at the surface and at depth. *A natural buffering system* based on the chemical reaction of carbon dioxide in water *exists in the ocean*. This buffering system *regulates any changes in pH*, creating a stable ocean environment.

### Selected Key Terms

Use the **glossary** at the end of this book to discover the meanings of these Selected Key Terms: **acid, base, pH scale, neutral, buffering.**

### Critical Thinking Question

What is the pH of surface waters? Is this value very acidic, slightly acidic, neutral, slightly alkaline, or very alkaline? Also, specify how pH changes with depth, and why those changes occur.

### Active Learning Exercise

Using examples of common household items, work with another student in class to describe the pH scale.

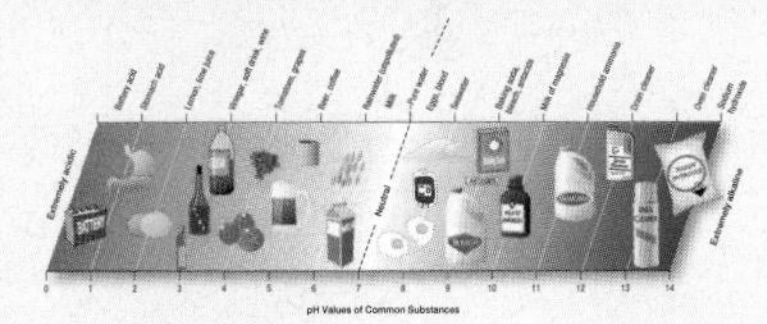

## 5.8 ▶ What factors control the distribution of carbon and oxygen in the ocean?

- *Carbon in the ocean* is regulated by the *concentration of $CO_2$ gas in the atmosphere*, the *solubility of $CO_2$ gas* in seawater, *organic and inorganic reactions*, and the *temperature and pressure* characteristics of surface and deep water. *Oxygen in the ocean* is regulated by similar *physical processes* and also by *biological activities* that both produce and consume oxygen.

### Selected Key Terms

Use the **glossary** at the end of this book to discover the meanings of these Selected Key Terms: **solubility, fossil fuel.**

### Critical Thinking Question

Consider the two dissolved oxygen profiles in Figure 5.29. Think of *two* reasons that could explain why the concentration of oxygen in the surface water in part *b* of the figure is higher than it is in part *a*.

### Active Learning Exercise

Pair up with a classmate and take the position that oxygen is more important in the ocean than carbon, and have your classmate take the opposite position. Debate the issue by describing what happens to oxygen and carbon in seawater, and compare their roles in maintaining conditions necessary for life.

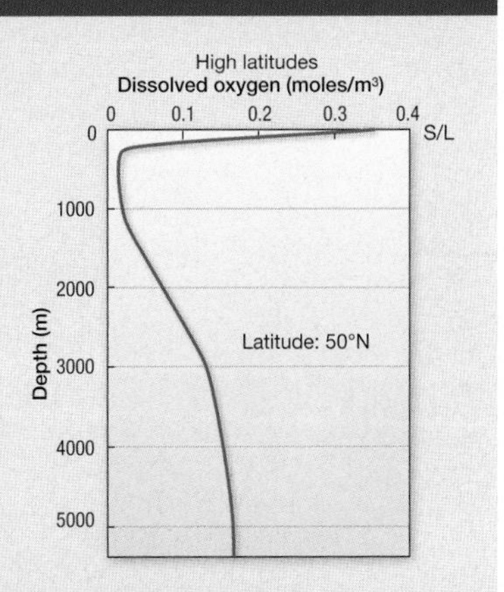


## Mastering Oceanography

www.masteringoceanography.com

Looking for additional review and test prep materials? With individualized coaching on the toughest topics of the course, Mastering Oceanography offers a wide variety of ways for you to move beyond memorization and deeply grasp the underlying processes of how the oceans work. Visit the Study Area in **www.masteringoceanography.com** to find practice quizzes, study tools, and multimedia that will improve your understanding of this chapter's content. Sign in today to access the following features: Self Study Quizzes, SmartFigures, Oceanography Videos and Animations, Squidtoons, Dynamic Study Modules, and an optional Pearson eText with embedded videos.

# Air–Sea Interaction

# 6

## ESSENTIAL LEARNING CONCEPTS

At the end of this chapter, you should be able to:

- ☐ **6.1** Explain variations in solar radiation on Earth, including the cause of Earth's seasons.
- ☐ **6.2** Describe the physical properties of the atmosphere.
- ☐ **6.3** Demonstrate an understanding of the Coriolis effect.
- ☐ **6.4** Explain global atmospheric circulation patterns.
- ☐ **6.5** Describe how the ocean influences global weather phenomena and climate patterns.
- ☐ **6.6** Evaluate the advantages and disadvantages of harnessing winds as a source of energy.

↑ *Check when completed*

One of the most remarkable things about our planet is that the atmosphere and the ocean act as one interdependent system. Observations of the atmosphere–ocean system show that what happens in one causes changes in the other. Further, the two parts of this system are linked by complex feedback loops, some of which reinforce a change and others that counteract the initial change. Surface currents in the oceans, for instance, are a direct result of Earth's atmospheric wind belts. Conversely, certain atmospheric weather phenomena are manifested in the oceans. In order to understand the behavior of the atmosphere and the oceans, their mutual interactions and relationships must be examined.

Solar energy heats the surface of Earth and creates atmospheric winds, which, in turn, drive most of the surface currents and waves in the ocean. Radiant energy from the Sun, therefore, is responsible for motion in the atmosphere and the ocean. In fact, variations in solar radiation drive the global ocean–atmosphere engine, creating pressure and density differences that stir currents and waves in both the atmosphere and the ocean. Recall from Chapter 5 that the atmosphere and ocean use the high heat capacity of water to constantly exchange this energy, shaping Earth's global weather patterns in the process.

Periodic extremes of atmospheric weather, such as droughts and profuse precipitation, are related to periodic changes in oceanic conditions. For instance, it was recognized as far back as the 1920s that El Niño—an ocean event—is tied to catastrophic weather events worldwide. What is as yet unclear, however, is if changes in the ocean produce changes in the atmosphere that lead to El Niño conditions—or vice versa. El Niño–Southern Oscillation events are discussed in Chapter 7, "Ocean Circulation."

Air–sea interactions have important implications in global warming, too. A multitude of recent studies have confirmed that the atmosphere is experiencing unprecedented warming as a result of human-caused emissions of carbon dioxide and other gases that absorb and trap heat in the atmosphere. This atmospheric heat is being transferred to the oceans and has the potential to cause widespread marine ecosystem changes. This issue is discussed in Chapter 16, "The Oceans and Climate Change."

Climate Connection

In this chapter, we'll examine the redistribution of solar heat by the atmosphere and its influence on ocean conditions. First, large-scale phenomena that influence air–sea interactions are studied, and then smaller-scale phenomena are examined.

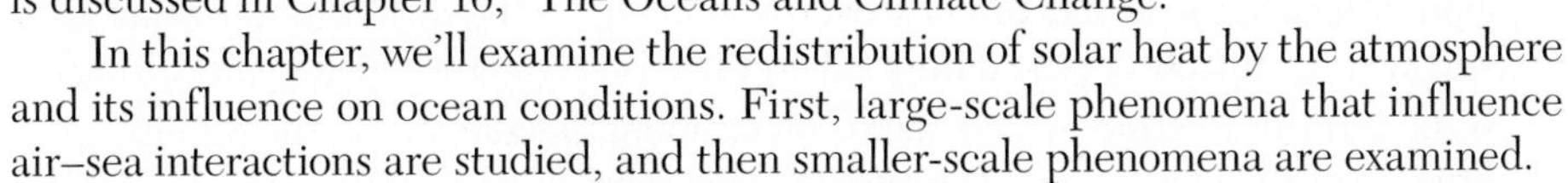

***Down dropt the breeze,***
***the sails dropt down,***
***'Twas sad as sad could be;***
***And we did speak only to break***
***The silence of the sea!***

***Day after day, day after day,***
***We stuck, nor breath nor motion;***
***As idle as a painted ship***
***Upon a painted ocean.***

***—Samuel Taylor Coleridge, about ships getting stuck in the horse latitudes,*** **Rime of the Ancient Mariner** ***(1798)***

◀ **The atmosphere-ocean system.** Although consisting of quite different densities, the atmosphere and the ocean are a linked system: what happens in one is often mirrored in the other, and vice versa. The atmosphere-ocean system also regulates Earth's climate and is impacted by human activities.

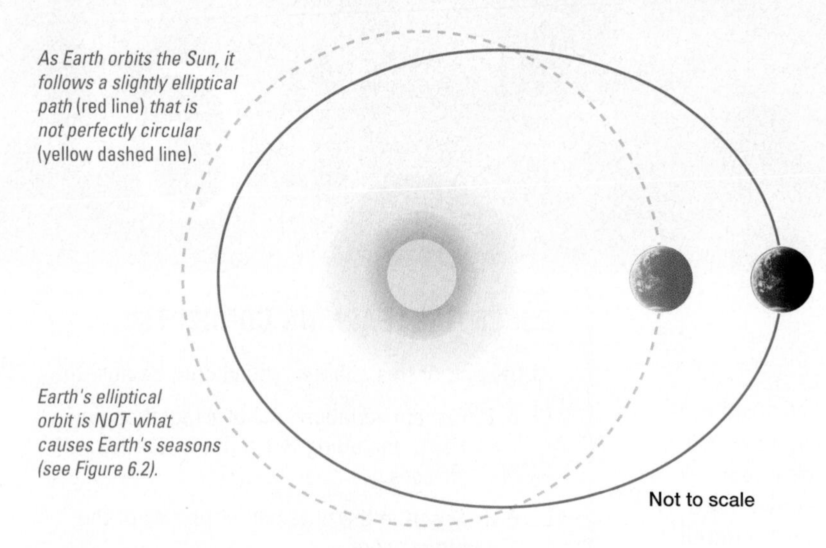
As Earth orbits the Sun, it follows a slightly elliptical path (red line) that is not perfectly circular (yellow dashed line).
Earth's elliptical orbit is NOT what causes Earth's seasons (see Figure 6.2).
Not to scale

**Figure 6.1 Earth has an elliptical orbit, but that's not what causes Earth's seasons.** Comparison between a slightly elliptical orbit (*red line*) and a perfectly circular orbit (*yellow dashed line*). View is directly above Earth's plane of the ecliptic; Earth's elliptical orbit is exaggerated for clarity (not to scale). Note that Earth's elliptical orbit is not the cause of Earth's seasons.

# 6.1 ▸ What Causes Variations in Solar Radiation on Earth?

A variety of factors cause changes in the amount of solar radiation (solar energy) that Earth receives. One of the most striking examples is the daytime–nighttime cycle: The side of Earth facing the Sun (the daytime side) receives a tremendous dose of intense solar radiation, while the nighttime side receives none. An example of a longer-term cycle is the change in seasons.

## What Causes Earth's Seasons?

This seemingly simple question is the source of a common misconception: Even though Earth does revolve around the Sun in an elliptical orbit that varies only slightly from a perfect circle (**Figure 6.1**), Earth's seasons are not caused by Earth's changing distance from the Sun. As will be explained below, Earth's seasons are actually caused by the tilt of Earth's axis.

The surface connecting all points in Earth's orbit is called the **plane of the ecliptic** (**Figure 6.2**). More importantly, Figure 6.2 shows that Earth's axis of rotation is not perpendicular ("upright") relative to the plane of the ecliptic; rather, it tilts at an angle of 23.5 degrees. As a result, *different hemispheres on Earth are tilted more directly toward or away from the Sun* during Earth's yearly orbit (Figure 6.2, *insets*), which is the cause of Earth's seasons (not Earth's elliptical orbit). An interesting consequence of Earth's tilt is that throughout its yearly cycle, Earth's axis *always points in the same direction*, which is toward Polaris, the North Star.

So, the tilt of Earth's rotational axis—not its elliptical orbit—is what causes Earth to have seasons. Let's examine a yearly progression of the seasons from spring, summer, fall, to winter:

- At the **vernal equinox** (*vernus* = spring; *equi* = equal, *noct* = night), which occurs on or about March 21, the Sun is directly overhead along the equator. During this time, all places in the world experience equal lengths of night and day (hence the name *equinox*). In the Northern Hemisphere, the vernal equinox is also known as the *spring equinox*.
- At the **summer solstice** (*sol* = the Sun, *stitium* = a stoppage), which occurs on or about June 21, the Sun reaches its most northerly point in the sky, directly overhead along the **Tropic of Cancer**, at 23.5 degrees north latitude (Figure 6.2, *left inset*). To an observer on Earth, the noonday Sun reaches its northernmost or southernmost position in the sky at this time and appears to pause—hence the term *solstice*—before beginning its next six-month cycle.
- At the **autumnal equinox** (*autumnus* = fall), which occurs on or about September 23, the Sun is directly overhead along the equator again. In the Northern Hemisphere, the autumnal equinox is also known as the *fall equinox*.
- At the **winter solstice**, which occurs on or about December 22, the Sun is directly overhead along the **Tropic of Capricorn**, at 23.5 degrees south latitude (Figure 6.2, *right inset*). In the Southern Hemisphere, the seasons are reversed. Thus, the winter solstice is the time when the Southern Hemisphere is most directly facing the Sun, which is the beginning of the Southern Hemisphere summer.

Because Earth's rotational axis is tilted 23.5 degrees, the Sun's **declination** (angular distance from the equatorial plane) varies between 23.5 degrees north and 23.5 degrees south of the equator on a yearly cycle. As a result, the region between these two latitudes (called the **tropics**) receives much greater annual radiation than polar areas.

Seasonal changes in the angle of the Sun and the length of day profoundly influence Earth's climate. In the Northern Hemisphere, for example, the longest day occurs on the summer solstice and the shortest day on the winter solstice.

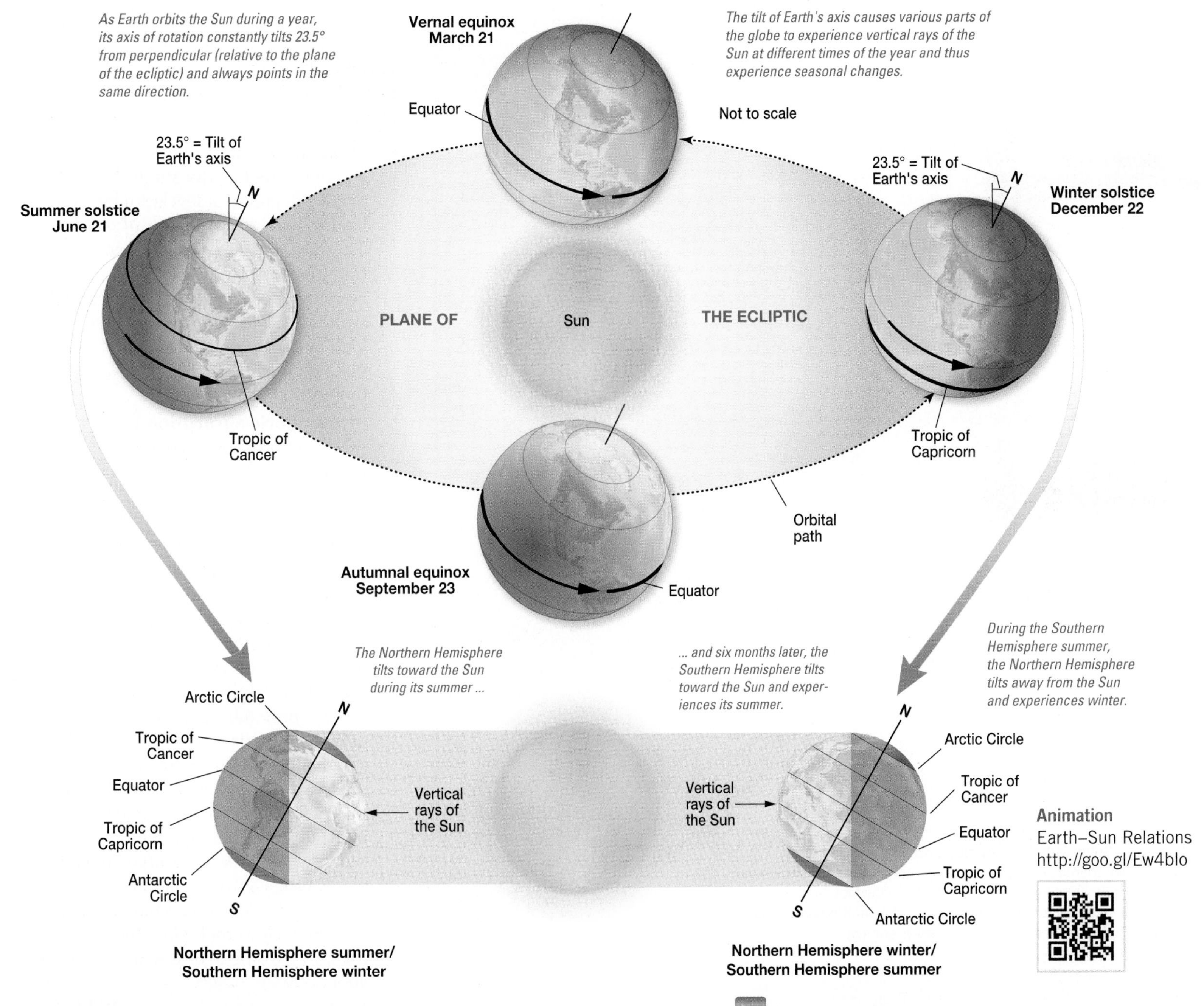

As Earth orbits the Sun during a year, its axis of rotation constantly tilts 23.5° from perpendicular (relative to the plane of the ecliptic) and always points in the same direction.
Vernal equinox
March 21
The tilt of Earth's axis causes various parts of the globe to experience vertical rays of the Sun at different times of the year and thus experience seasonal changes.
Equator
Not to scale
23.5° = Tilt of Earth's axis
N
Summer solstice
June 21
23.5° = Tilt of Earth's axis
N
Winter solstice
December 22
PLANE OF
Sun
THE ECLIPTIC
Tropic of Cancer
Tropic of Capricorn
Orbital path
Autumnal equinox
September 23
Equator
The Northern Hemisphere tilts toward the Sun during its summer ...
... and six months later, the Southern Hemisphere tilts toward the Sun and experiences its summer.
During the Southern Hemisphere summer, the Northern Hemisphere tilts away from the Sun and experiences winter.
Arctic Circle
N
Tropic of Cancer
Equator
Tropic of Capricorn
Antarctic Circle
S
Vertical rays of the Sun
Vertical rays of the Sun
N
Arctic Circle
Tropic of Cancer
Equator
Tropic of Capricorn
Antarctic Circle
S
Northern Hemisphere summer/
Southern Hemisphere winter
Northern Hemisphere winter/
Southern Hemisphere summer
Animation
Earth–Sun Relations
http://goo.gl/Ew4blo

**SmartFigure 6.2 Perspective view of Earth's orbit: Why Earth has seasons.** Earth has seasons not because of its elliptical orbit and varying distance from the Sun; it experiences seasons because of Earth's tilt.
https://goo.gl/x9o96g

Daily heating of Earth also influences climate in most locations. Exceptions to this pattern occur north of the **Arctic Circle** (66.5 degrees north latitude) and south of the **Antarctic Circle** (66.5 degrees south latitude), which at certain times of the year do not experience daily cycles of daylight and darkness. For instance, during the Northern Hemisphere winter, the area north of the Arctic Circle receives no direct solar radiation at all and experiences up to six months of darkness. At the same time, the area south of the Antarctic Circle receives continuous radiation ("midnight Sun"), so it experiences up to six months of light. Half a year later, during the Northern Hemisphere summer (the Southern Hemisphere winter), the situation is reversed.

**RECAP** **Earth's axis is tilted at an angle of 23.5 degrees, which causes the Northern and Southern Hemispheres to take turns "leaning toward" the Sun every six months, and results in the change of seasons.**

## How Latitude Affects the Distribution of Solar Radiation

If Earth were a flat plate in space, with its flat side directly facing the Sun, sunlight would fall equally on all parts of Earth. Earth is spherical, however, so the amount

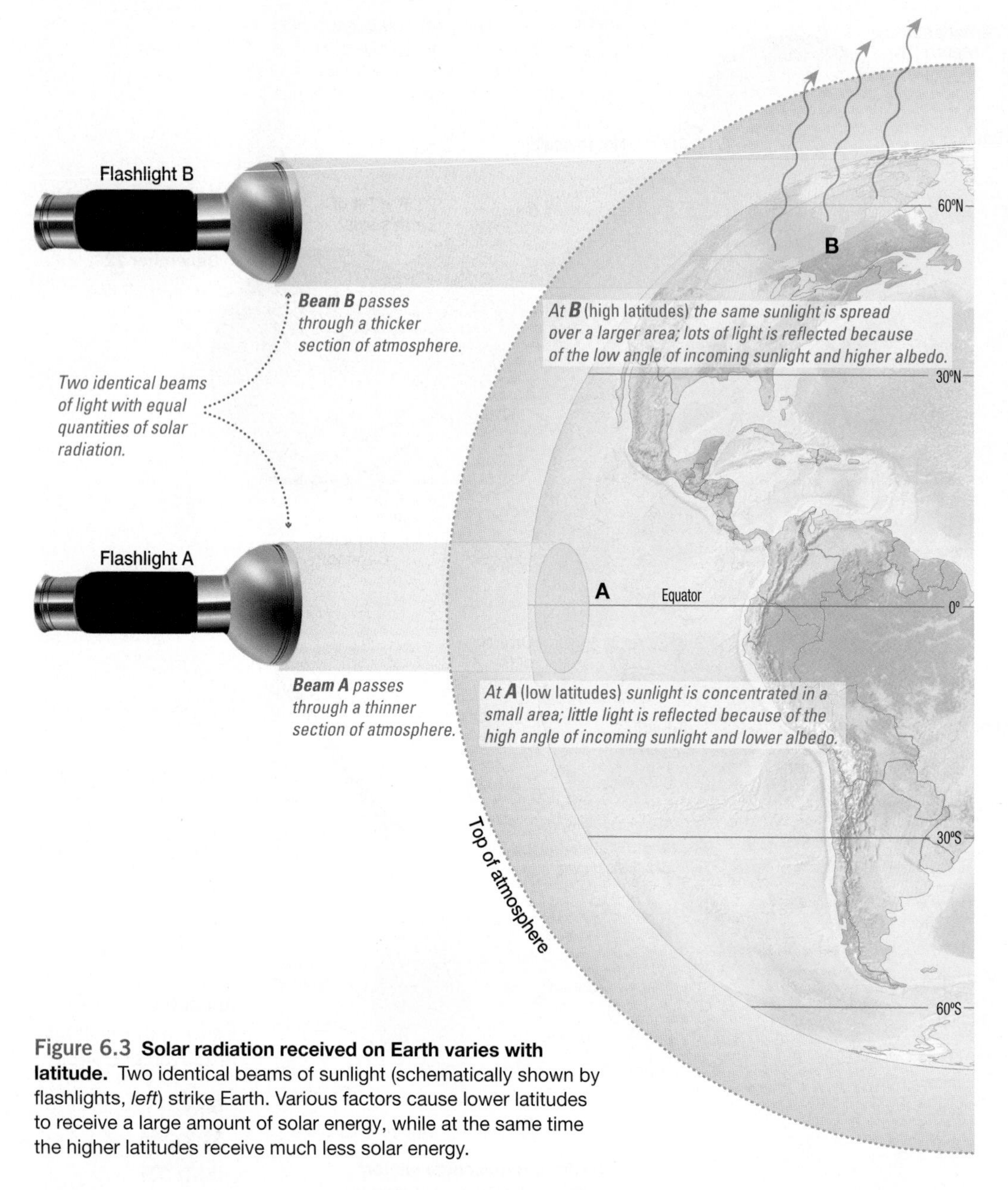

Flashlight B
Beam B passes through a thicker section of atmosphere.
At B (high latitudes) the same sunlight is spread over a larger area; lots of light is reflected because of the low angle of incoming sunlight and higher albedo.
Two identical beams of light with equal quantities of solar radiation.
60°N
B
30°N
Flashlight A
A
Equator
0°
Beam A passes through a thinner section of atmosphere.
At A (low latitudes) sunlight is concentrated in a small area; little light is reflected because of the high angle of incoming sunlight and lower albedo.
30°S
Top of atmosphere
60°S

Figure 6.3 **Solar radiation received on Earth varies with latitude.** Two identical beams of sunlight (schematically shown by flashlights, *left*) strike Earth. Various factors cause lower latitudes to receive a large amount of solar energy, while at the same time the higher latitudes receive much less solar energy.

**Table 6.1 Reflection and absorption of solar energy relative to the angle of incidence on a flat sea**

| Elevation of the Sun above the horizon | 90° | 60° | 30° | 15° | 5° |
|---|---|---|---|---|---|
| Reflected radiation (%) | 2 | 3 | 6 | 20 | 40 |
| Absorbed radiation (%) | 98 | 97 | 94 | 80 | 60 |

and intensity of solar radiation received at higher latitudes are much less than at lower latitudes. The following factors influence the amount of radiation received at low and high latitudes:

- **Solar footprint** Most of the time in the equatorial region, the Sun is directly overhead, and so at low latitudes, sunlight strikes at a *high angle of incidence.* This means solar radiation is concentrated in a relatively small area (area *A* in Figure 6.3). Closer to the poles, sunlight strikes at a *low angle of incidence*, so in high latitudes, the same amount of radiation is spread over a larger area (area *B* in Figure 6.3).
- **Atmospheric absorption** Earth's atmosphere absorbs some radiation, so less radiation reaches Earth's surface at high latitudes, compared to low latitudes, because sunlight must pass through more atmosphere at high latitudes.
- **Albedo** The **albedo** (*albus* = white) of various Earth materials, defined as the percentage of incident radiation that is reflected back to space, varies depending on the material considered. For example, thick sea ice covered by snow reflects back into space as much as 90% of incoming solar radiation, and so has a high albedo. This is one of the reasons a larger proportion of radiation is reflected back into space in ice-covered high latitudes as compared to low latitudes, which lack substantial amounts of ice. Other Earth materials such as ocean, soil, vegetation, sand, and rock have much lower albedo values than ice; the average albedo of Earth's surface is about 30%.
- **Reflection of incoming sunlight** The angle at which sunlight strikes the ocean surface determines how much is absorbed and how much is reflected. If the Sun shines down on a smooth sea from directly overhead, only 2% of the radiation is reflected, but if the Sun is only 5 degrees above the horizon, 40% is reflected back into the atmosphere (Table 6.1). Thus, the ocean reflects more radiation at high latitudes than at low latitudes.

Because of all these reasons, the intensity of radiation at high latitudes is greatly decreased compared with the intensity of radiation received in equatorial regions.

Other factors influence the amount of solar energy that reaches Earth. For example, the amount of radiation received at a particular location on Earth's surface varies *daily* because Earth rotates on its axis, so the surface experiences daylight and darkness each day. In addition, the amount of radiation varies *annually* due to Earth's seasons, as discussed in the previous section.

## Oceanic Heat Flow

Close to the poles, most incoming solar radiation strikes Earth's surface at low angles. In addition, ice has a high albedo, so more energy is reflected back into space than is absorbed. In contrast, between about 35 degrees north latitude and 40 degrees south latitude, sunlight strikes Earth at much higher angles, and more energy is absorbed than is reflected back into space. (Note that this latitudinal range extends farther in the Southern Hemisphere because the Southern Hemisphere has more ocean surface area in the middle latitudes than the Northern Hemisphere does.) The graph in **Figure 6.4** shows how incoming sunlight and outgoing heat combine on a daily basis for a net heat gain in low-latitude oceans and a net heat loss in high-latitude oceans.

Based on Figure 6.4, you might expect that over time the equatorial zone grows progressively warmer and the polar regions grow progressively cooler. The polar regions are always considerably colder than the equatorial zone, but the temperature *difference* remains the same because excess heat is transferred from the equatorial zone to the poles. How is this accomplished? Circulation in both the oceans and the atmosphere transfers the heat.

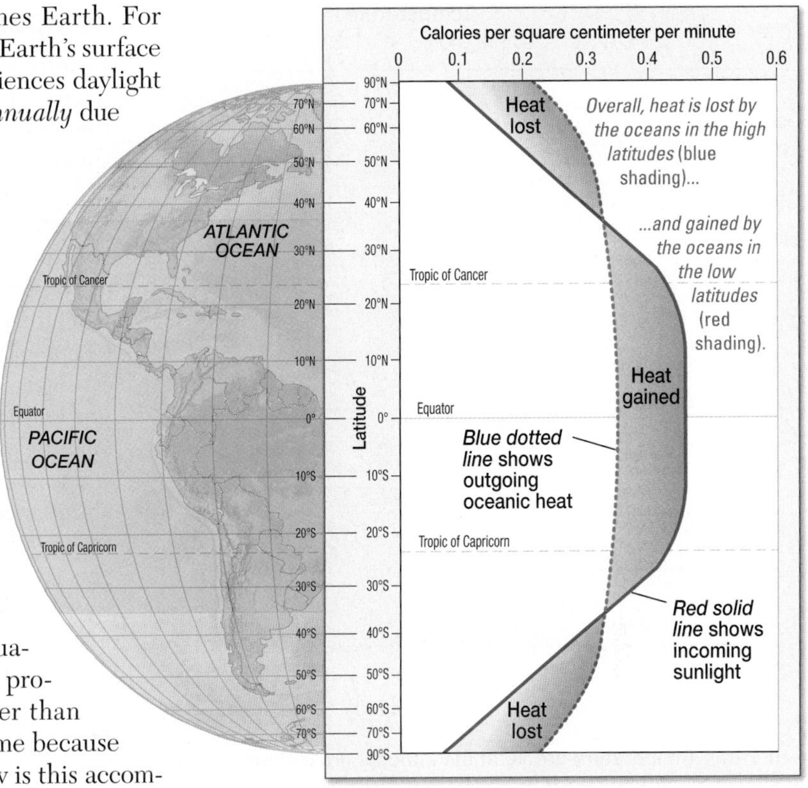

**Figure 6.4 Graph showing the balance between heat gained and heat lost by the oceans.** On average, heat gained and heat lost by the oceans balance each other on a global scale, since the excess heat from low latitudes is transferred to heat-deficient high latitudes by both oceanic and atmospheric circulation.

### CONCEPT CHECK 6.1 ▸ Explain variations in solar radiation on Earth, including the cause of Earth's seasons.

1 Sketch a labeled diagram to explain the cause of Earth's seasons.

2 Along the Arctic Circle, how would the Sun appear during the summer solstice? During the winter solstice?

3 If there is a net annual heat loss at high latitudes and a net annual heat gain at low latitudes, why does the temperature difference between these regions not increase over time?

**RECAP** **Low-latitude regions receive more solar radiation than high-latitude regions, but oceanic and atmospheric circulation transfer heat around the globe.**

# 6.2 ▸ What Physical Properties Does the Atmosphere Possess?

The atmosphere transfers heat and water vapor from place to place on Earth. Within the atmosphere, complex relationships exist among air composition, temperature, density, water vapor content, and pressure. Before we apply these relationships, let's examine the atmosphere's composition and some of its physical properties.

## Composition of the Atmosphere

**Figure 6.5** lists the composition of dry air and shows that the atmosphere consists almost entirely of nitrogen and oxygen. Other gases include argon (an inert gas), carbon dioxide, and others in trace amounts. Although these gases are present in very small amounts, they can trap significant amounts of heat within the atmosphere. For more about how these gases trap heat in the atmosphere, see Chapter 16, "The Oceans and Climate Change."

Climate Connection

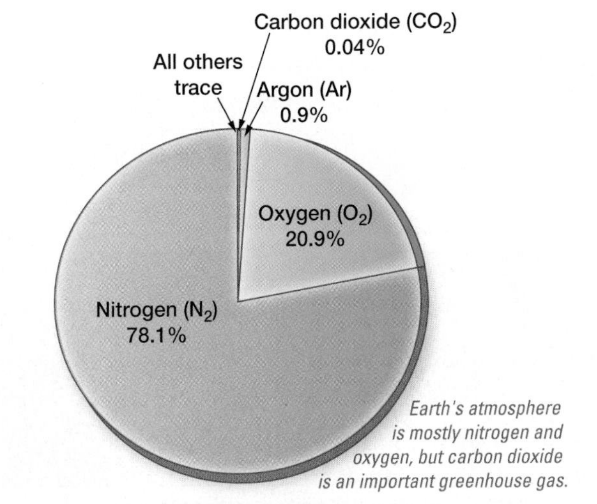


**Figure 6.5 Composition of dry air.** Pie chart showing the composition of dry air (without any water vapor) by volume. Nitrogen and oxygen gas comprise 99% of the total composition of Earth's atmosphere.

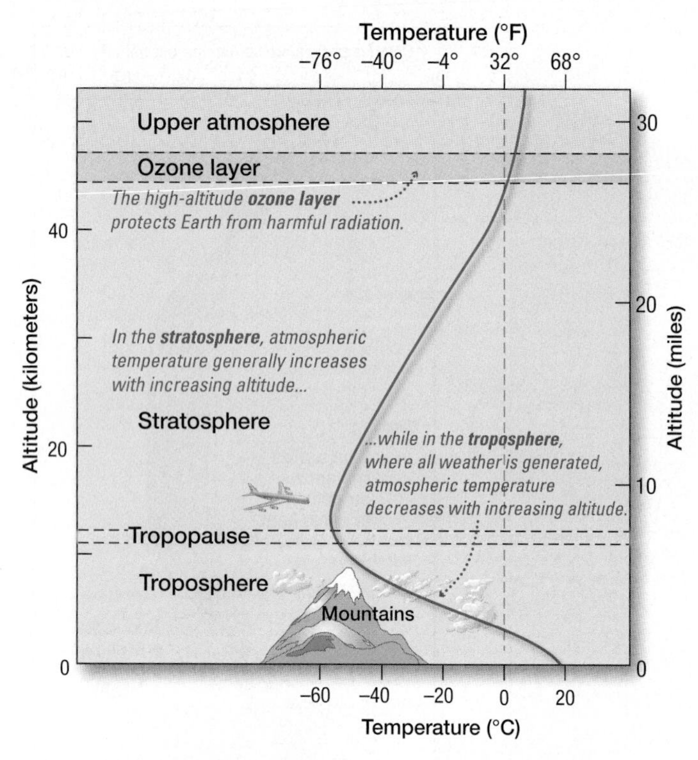


**Figure 6.6 Temperature profile of the atmosphere including the names of atmospheric layers.**

Warm air rising
Convection cell
Cold window
Cool air falling
Hot radiator

*A circular-moving loop of air (a **convection cell**) is created in this room by warm air rising and cool air sinking.*

**Figure 6.7 How a convection cell in a room is created by a hot radiator and a cold window.**

## Temperature Variation in the Atmosphere

Intuitively, it seems logical that the higher one goes in the atmosphere, the warmer it should be since it's closer to the Sun. However, as unusual as it seems, the atmosphere is actually heated from *below*. That is because Earth's atmosphere is effectively transparent to incoming radiation from the Sun, which means energy passes through Earth's atmosphere without heating it. Instead, solar radiation strikes and warms Earth's surface (both land and water), which in turn reradiates this energy back into the atmosphere as heat. This process is one of the mechanisms underlying the *greenhouse effect* and will be discussed in more detail in Chapter 16, "The Oceans and Climate Change."

Figure 6.6 shows a temperature profile of the atmosphere. The lowermost portion of the atmosphere, which extends from the surface to about 12 kilometers (7 miles), is called the **troposphere** (*tropo* = turn, *sphere* = a ball) and is where all weather is produced. The troposphere gets its name because of the abundance of mixing that occurs within this layer of the atmosphere, mostly as a result of being heated from below. Within the troposphere, temperature gets cooler with altitude to the point that at high altitudes, the air temperature is well below freezing. If you have ever flown in an airplane, for instance, you may have noticed that any water on the wings or inside your window freezes during a high-altitude flight.

## Density Variation in the Atmosphere

It may seem surprising that air has density, but since air is composed of molecules, it certainly does. Temperature has a dramatic effect on the density of air. At higher temperatures, for example, air molecules move more quickly and the distance between molecules increases, causing the volume of air to increase and its density to decrease. Thus, the general relationship between density and temperature is as follows:

- Warm air is less dense, so it rises; this is commonly expressed as "heat rises."
- Cool air is more dense, so it sinks.

Figure 6.7 shows how a radiator (heater) uses convection to heat a room. The heater warms the nearby air and causes it to expand. This expansion makes the air less dense, causing it to rise. Conversely, a cold window cools the nearby air and causes it to contract, thereby becoming more dense, which causes it to sink. A **convection cell** (*con* = with, *vect* = carried) forms, composed of the rising and sinking air moving in a circular fashion, similar to the convection in Earth's mantle discussed in Chapter 2.

## Atmospheric Water Vapor Content

The amount of water vapor in air depends in part on the air's temperature. Warm air, for instance, can hold more water vapor than cold air because the air molecules are moving more quickly and come into contact with more water vapor. Thus, warm air is typically moist, and, conversely, cool air is typically dry. This is why a warm day is best for drying your laundry outside, and evaporation is especially quick if it is also a breezy day.

Water vapor influences the density of air. The addition of water vapor decreases the density of air because water vapor has a lower density than air. Thus, humid air is less dense than dry air.

## Atmospheric Pressure

Atmospheric pressure is 1.0 atmosphere (14.7 pounds per square inch) at sea level and decreases with increasing altitude. (The *atmosphere* is a unit of pressure; 1.0

atmosphere is the average pressure exerted by the overlying atmosphere at sea level and is equivalent to 760 millimeters of mercury, 101,300 Pascal, or 1013 millibars.) Atmospheric pressure depends on the weight of the column of air above. For instance, a tall column of air produces higher atmospheric pressure than a short column of air. An analogy to this is water pressure in a swimming pool: the taller the column of water above, the higher the water pressure. Thus, the highest pressure in a pool is at the bottom of the deep end.

Similarly, the tall column of air at sea level means air pressure is high at sea level and decreases with increasing elevation. When sealed bags of potato chips or pretzels are taken to a high elevation, there is a shorter column air overhead and the atmospheric pressure is much lower than where the bags were sealed. This may cause the bags to swell and sometimes burst. You may also have experienced this change in pressure when your ears "popped" during the takeoff or landing of an airplane, or while driving on steep mountain roads.

Changes in atmospheric pressure cause air movement as a result of changes in the molecular density of the air. The general relationship is shown in **Figure 6.8**, which indicates that:

- A column of cool, dense air causes high pressure at the surface, which will lead to sinking air (movement *toward* the surface and compression).
- A column of warm, less dense air causes low pressure at the surface, which will lead to rising air (movement *away from* the surface and expansion).

In addition, sinking air tends to warm because of its compression, while rising air tends to cool due to expansion. Note that there are complex relationships between air composition, temperature, density, water vapor content, and pressure.

**CREATURE FEATURE 6.1**

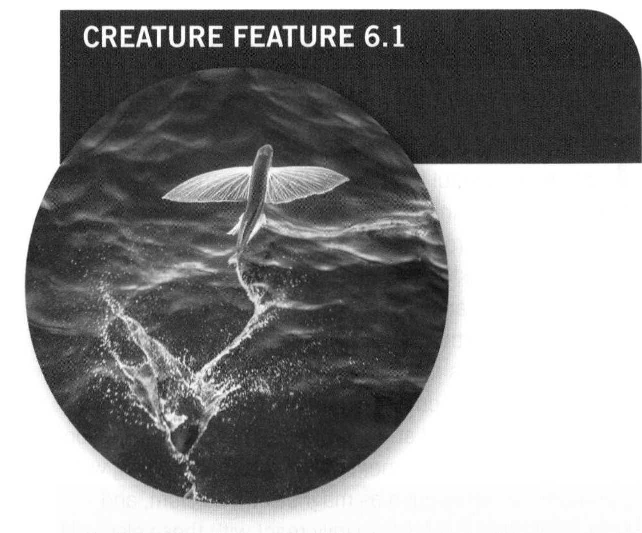

## I can fly!

Several species of **flying fish** make powerful, self-propelled leaps out of water into air, where their long, wing-like fins enable the fish to glide through the air for great distances.

Flying fish take advantage of the property of air being less dense than water, which makes air easier to travel through and helps them avoid predators; they sometimes hit boats as they soar just above the water's surface.

## Movement of the Atmosphere

Air *always* moves from high-pressure regions toward low-pressure regions. This moving air is called **wind**. If a balloon is inflated and let go, what happens to the air inside the balloon? It rapidly escapes, moving from a high-pressure region inside the balloon (caused by the balloon pushing on the air inside) to the lower-pressure region outside the balloon.

**EXPLORING DATA ▼**

**1) For cool, sinking air, what atmospheric pressure conditions are created at Earth's surface? Explain this in terms of molecular packing and the density of air.**

**2) For warm, rising air, what atmospheric pressure conditions are created at Earth's surface? Explain this in terms of molecular packing and the density of air.**

## An Example: A Nonspinning Earth

Imagine for a moment that Earth is not spinning on its axis but that the Sun rotates around Earth, with the Sun directly above Earth's equator at all times (**Figure 6.9**). Because more solar radiation is received along the equator than at the poles, the air at the equator in contact with Earth's surface is warmed. This warm, moist air rises, creating low pressure at the surface. This rising air cools (see Figure 6.6) and releases its moisture as rain. Thus, a zone of low pressure and much precipitation occurs along the equator.

As the air along the equator rises, it reaches the top of the troposphere and begins to move toward the poles. Because the temperature is much lower at high altitudes, the air cools, and its density increases. This cool, dense air sinks at the poles, creating high pressure at the surface. The sinking air is quite dry because cool air cannot hold much water vapor. Thus, the poles experience high pressure and clear, dry weather.

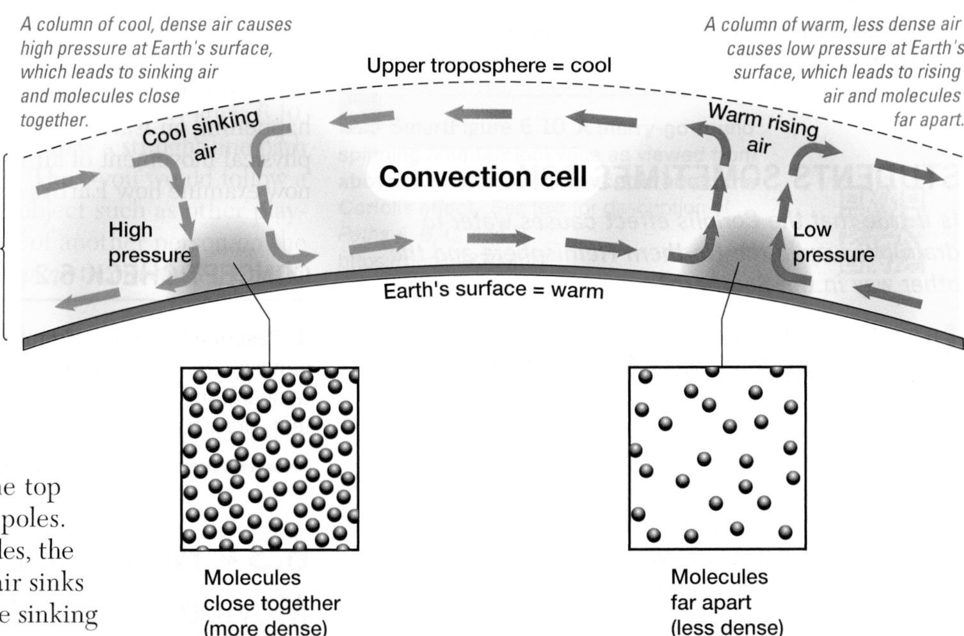


**Figure 6.8 Characteristics of high and low atmospheric pressure zones.**

**SmartFigure 6.20 Typical North Atlantic hurricane storm track and detail of internal structure.**
https://goo.gl/DcEaEV

60 Km
100 Mi

(a) Satellite photo of Hurricane Andrea off the U.S. East Coast in 2007.

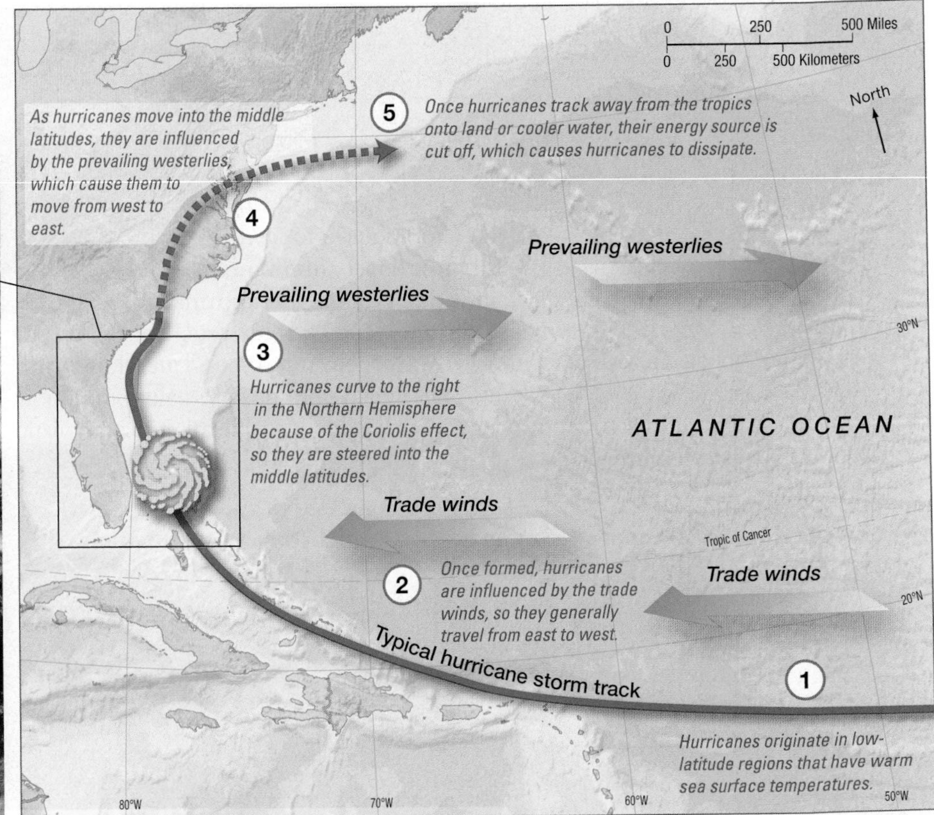
0 250 500 Miles
0 250 500 Kilometers
North
As hurricanes move into the middle latitudes, they are influenced by the prevailing westerlies, which cause them to move from west to east.
5
Once hurricanes track away from the tropics onto land or cooler water, their energy source is cut off, which causes hurricanes to dissipate.
4
Prevailing westerlies
Prevailing westerlies
30°N
3
Hurricanes curve to the right in the Northern Hemisphere because of the Coriolis effect, so they are steered into the middle latitudes.
ATLANTIC OCEAN
Trade winds
Tropic of Cancer
Trade winds
2
Once formed, hurricanes are influenced by the trade winds, so they generally travel from east to west.
20°N
Typical hurricane storm track
1
Hurricanes originate in low-latitude regions that have warm sea surface temperatures.
80°W
70°W
60°W
50°W

(b) Map showing a typical hurricane storm track, including the steps involved in its origin, movement, and dissipation.

**Animation**
Hurricanes
http://goo.gl/99HYop

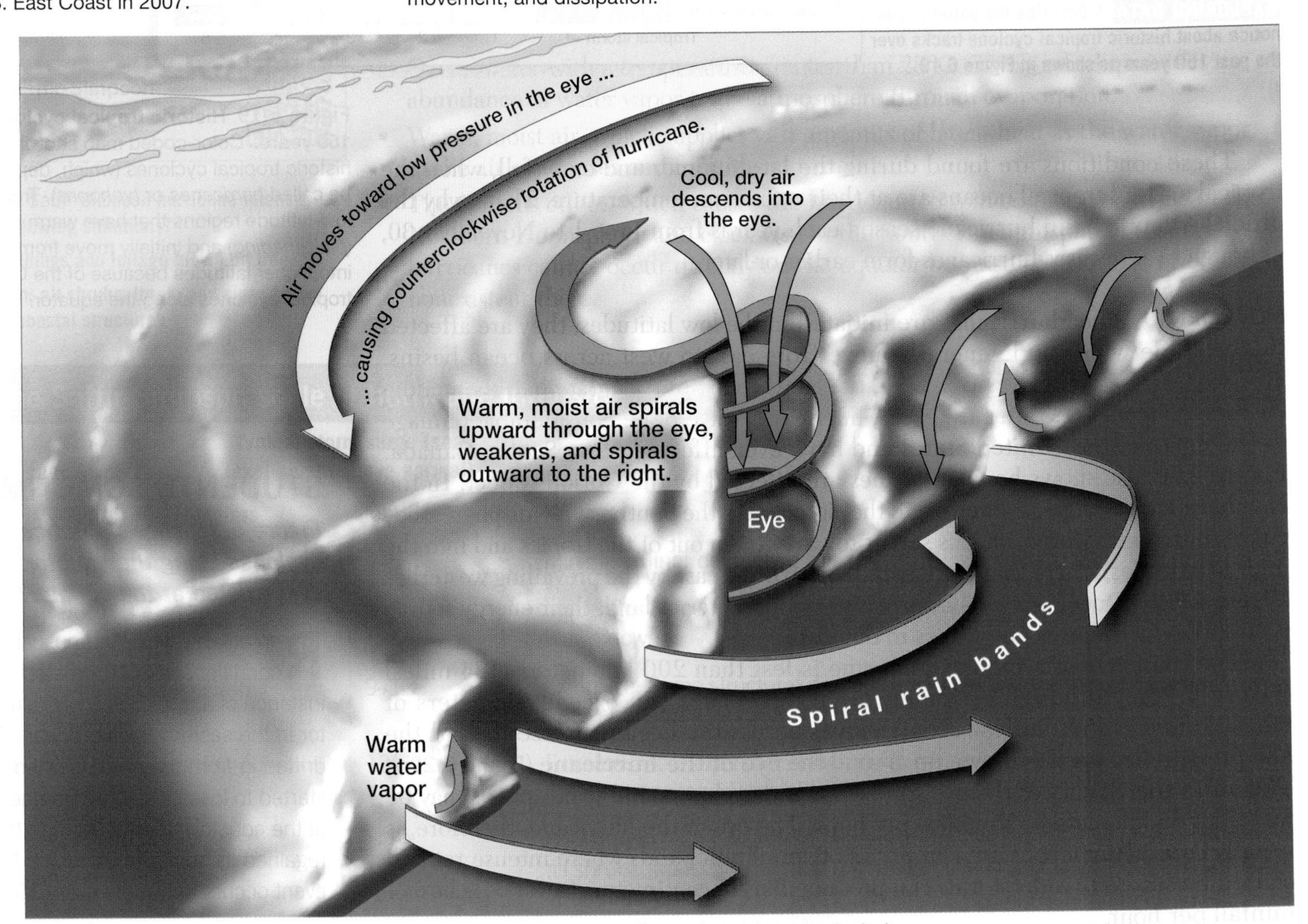
Air moves toward low pressure in the eye ...
... causing counterclockwise rotation of hurricane.
Cool, dry air descends into the eye.
Warm, moist air spirals upward through the eye, weakens, and spirals outward to the right.
Eye
Spiral rain bands
Warm water vapor

(c) Enlarged cut-away view of a hurricane showing its components, internal structure, and winds.

Climate Connection

**IMPACT OF OTHER FACTORS** A host of factors influences the development and strength of hurricanes. For example, warmer sea surface temperatures tend to favor the development of hurricanes, while strong wind shear in the upper atmosphere can ventilate heat away from a developing hurricane and thus interfere with hurricane formation. Other factors that can either enhance or disrupt the development and intensification of hurricanes include the amount of atmospheric convective instability, air humidity, the degree of rotation of spinning winds, and even El Niño/La Niña events (which are discussed in Chapter 7, "Ocean Circulation").

New research combined with careful analysis of historical data shows an out-of-phase relationship between tropical cyclone variability in the North Atlantic and the eastern North Pacific, meaning that when one basin has high storm occurrence, the other has low storm occurrence. The recognition of this pattern has helped improve hurricane forecasting.

Human-caused climate change has been linked to the documented warming of ocean surface waters, which fuels hurricanes. As a result, several recent studies show that hurricane severity is expected to increase. In fact, climate models suggest that, while typical tropical storms will likely decrease in number, the risk of Category 4 and 5 storms making landfall will likely increase because the warming of coastal waters is especially significant for adding to the strength of hurricanes making landfall. Since the late 1970s in the northwest Pacific Ocean, for example, the annual number of Category 4 and 5 typhoons–the strongest–has increased by 40%. For more details about climate change and its impact on hurricanes, see Chapter 16, "The Oceans and Climate Change."

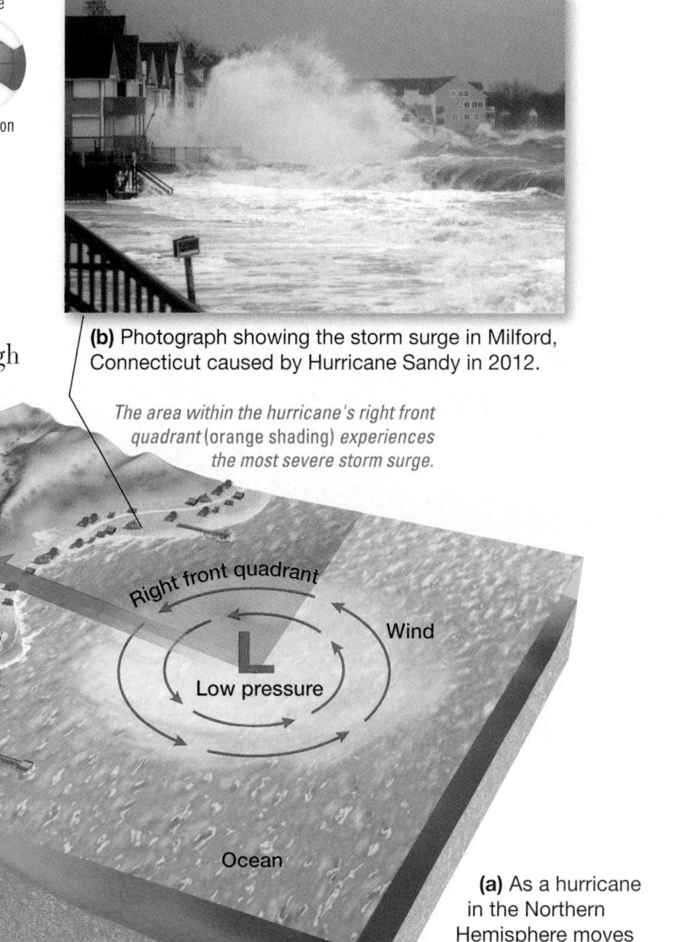


**(b)** Photograph showing the storm surge in Milford, Connecticut caused by Hurricane Sandy in 2012.

**(a)** As a hurricane in the Northern Hemisphere moves ashore, the low-pressure center around which the storm winds blow, combined with strong onshore winds, produces a high-water storm surge that floods the coast.

Figure 6.21 **A hurricane's storm surge batters the coast.**

**TYPES OF DESTRUCTION** Destruction from hurricanes is caused by high winds and flooding from intense rainfall. **Storm surge**, however, causes the majority of a hurricane's coastal destruction. In fact, storm surge is responsible for 90% of the deaths associated with hurricanes.

When a hurricane develops over the ocean, its low-pressure center produces a low "hill" of water (**Figure 6.21**). As the hurricane migrates across the open ocean, the hill moves with it. As the hurricane approaches shallow water nearshore, the portion of the hill over which the wind is blowing shoreward produces a mass of elevated, wind-driven water. This mass of water—the storm surge—can be as high as 12 meters (40 feet), resulting in a dramatic increase in sea level at the shore, large storm waves, and tremendous destruction to low-lying coastal areas (particularly if it occurs at high tide). In addition, the area of the coast that is hit with the right front quadrant of the hurricane—where onshore winds further pile up water—experiences the most severe storm surge (Figure 6.21). Note that as the mass of water traveling with the storm grows, water is withdrawn from surrounding areas, sometimes leaving shallow coastal areas dry. Table 6.3 shows typical storm surge heights associated with Saffir-Simpson hurricane intensities.

**HISTORIC DESTRUCTION ON THE U.S. MAINLAND** Periodic destruction from hurricanes occurs along the East Coast and the Gulf Coast regions of the United States. In fact, the deadliest natural disaster in U.S. history was caused by a hurricane that struck Galveston Island, Texas, in September 1900. Galveston Island is a thin strip of sand called a *barrier island* located in the Gulf of Mexico off Texas (**Figure 6.22**). In 1900, it was a popular beach resort that averaged only 1.5 meters (5 feet) above sea level. At least 6000 people in and around Galveston were killed when the hurricane's 6-meter (20-foot) high storm surge completely submerged the island, accompanied by heavy rainfall and winds of 160 kilometers (100 miles) per hour.

Category 4 hurricanes, like the one in 1900 that devastated Galveston, have been surpassed by Category 5 hurricanes making landfall only three times in the United States: (1) in 1935, an unnamed hurricane flattened the Florida Keys; (2) in 1969, Hurricane Camille struck Mississippi; and (3) in 1992, Hurricane Andrew came ashore in southern Florida, with winds as high as 258 kilometers (160 miles) per hour, ripping down every tree in its path as it crossed the Everglades. (Note that prior to 1950, Atlantic hurricanes were not named, but the unnamed 1935 hurricane is often referred to as the "Labor Day Hurricane" because it came ashore on that date. Today, hurricanes are named by forecasters using an alphabetized list of both female and male names.) Hurricane Andrew did more than $26.5 billion of damage in Florida and along the Gulf Coast. In the aftermath of Hurricane Andrew, more than 250,000 people were left homeless and although most people heeded the warnings to evacuate, 54 were killed.

**Figure 6.22 Destruction from the Galveston hurricane of 1900.** Photo showing destruction from the 1900 hurricane at Galveston (*above*) and location map of Galveston, Texas (*right*). At least 6000 people died as a result of the Galveston hurricane, which completely submerged Galveston Island and still stands as the single deadliest U.S. natural disaster.

In October 1998, Hurricane Mitch proved to be one of the most devastating tropical cyclones to affect the Western Hemisphere. At its peak, it was estimated to have winds of 290 kilometers (180 miles) per hour—a strong Category 5 hurricane. It hit Central America with winds of 160 kilometers (100 miles) per hour and as much as 130 centimeters (51 inches) of total rainfall, causing widespread flooding and mudslides in Honduras and Nicaragua that destroyed entire towns. The hurricane resulted in more than 11,000 deaths, left more than 2 million homeless, and caused more than $10 billion in damage across the region.

In September 2008, Hurricane Ike reached Category 4 in the Gulf of Mexico and made landfall near Galveston in low-lying Gilchrist, Texas, as a Category 2 hurricane. Ike resulted in 146 deaths and $24 billion in damages, making it the third costliest U.S. hurricane of all time, behind only Hurricane Katrina (2005) and Hurricane Andrew (1992). In August 2011, Hurricane Irene achieved Category 3 status in the Caribbean and wreaked havoc as it moved along the eastern seaboard from Florida to New England. In all, Irene caused severe flooding that was responsible for 56 deaths and more than $10 billion in damages.

In October 2012, Hurricane Sandy, a large Category 1 storm, affected the Caribbean and the eastern seaboard from Florida to Maine. Hurricane Sandy was the largest Atlantic hurricane on record, with winds spanning an enormous area that was over 1800 kilometers (1100 miles) wide. When Hurricane Sandy came ashore in the United States, its peak wave heights and storm surge coincided with peak high tides, with the largest waves and storm surge focused along the heavily populated New York and New Jersey coasts. Hurricane Sandy caused extensive wave damage (**Figure 6.23**), severe coastal erosion, and extreme flooding

**Figure 6.23 Damage to a New Jersey amusement pier from Hurricane Sandy in 2012.** Hurricane Sandy, which was the widest Atlantic hurricane on record, caused flooding and destruction from the Caribbean to the U.S. East Coast from Florida to Maine, particularly in New York and New Jersey.

that destroyed thousands of homes and left millions without electric service throughout the Mid-Atlantic states. In all, the storm was responsible for 233 deaths and more than $68 billion in damages, making it the second costliest hurricane in U.S. history, behind only Hurricane Katrina.

**THE RECORD-BREAKING 2005 ATLANTIC HURRICANE SEASON: HURRICANES KATRINA, RITA, AND WILMA** Although the official Atlantic hurricane season extends each year from June 1 to November 30, the 2005 Atlantic hurricane season persisted into January 2006 and was the most active season on record, shattering numerous records. For example, a record 27 named tropical storms formed, of which a record 15 became hurricanes. Of these, seven strengthened into major hurricanes, a record-tying five became Category 4 hurricanes and a record four reached Category 5 strength, the highest categorization for hurricanes on the Saffir-Simpson Scale of hurricane intensity (see Table 6.3). For the first time ever, NOAA's National Hurricane Center, which oversees the naming of Atlantic hurricanes, ran out of the usual names for storms and resorted to naming storms using the Greek alphabet.

The most notable storms of the 2005 season were the five Category 4 and Category 5 hurricanes: Dennis, Emily, Katrina, Rita, and Wilma. These storms made a combined 12 landfalls as major hurricanes (Category 3 strength or higher) throughout Cuba, Mexico, and the Gulf Coast of the United States, causing over $180 billion in damages and more than 2000 deaths.

For example, Hurricane Katrina, the sixth-strongest Atlantic hurricane ever recorded, was the costliest and one of the deadliest hurricanes in U.S. history. Katrina formed over the Bahamas on August 23 and crossed southern Florida as a moderate Category 1 hurricane before passing over the warm Loop Current and strengthening rapidly in the Gulf of Mexico, becoming one of the strongest hurricanes ever recorded in the Gulf. The storm weakened considerably before making its second landfall as a Category 3 storm on the morning of August 29 in southeast Louisiana (**Figure 6.24a**). Still, Katrina was the largest hurricane of its strength to make landfall in the United States in recorded history; its sheer size caused devastation over a radius of 370 kilometers (230 miles). Katrina's 9-meter (30-foot) storm surge—the highest ever recorded in the United States—caused severe damage along the coasts of Mississippi, Louisiana, and Alabama.

As forecasters watched these events unfold, they recognized a potential catastrophe—Katrina was on a collision course with New Orleans. This scenario was considered particularly disastrous because nearly all of the New Orleans metropolitan area is below sea level along Lake Pontchartrain. Even without a direct hit, the storm surge from Katrina was forecast to be greater than the height of the levees protecting New Orleans. This risk of devastation was well known; several previous studies warned that a direct hurricane strike on New Orleans could lead to massive flooding, which would lead to thousands of drowning deaths, as well as many more suffering from disease and dehydration after the hurricane passed. Although Katrina passed to the east of New Orleans, levees separating Lake Pontchartrain from New Orleans were breached by Katrina's high winds, storm surge, and heavy rains, ultimately flooding roughly 80% of the city and many neighboring areas (**Figure 6.24b**). Damages from Katrina exceeded $100 billion, easily making it the costliest single hurricane in U.S. history. The storm also left hundreds of thousands homeless and killed approximately 1800 people, making it the deadliest U.S. hurricane since the 1928 Okeechobee Hurricane, which killed as many as 2500 people. Responders in the aftermath of Katrina attributed many deaths to drowning because of rising water that trapped residents in the attics of single-story homes.

**(a)** Satellite view of Hurricane Katrina coming ashore along the Gulf Coast on August 29, 2005, showing the hurricane's counterclockwise direction of spin and prominent central eye. Hurricane Katrina, which had a diameter of about 670 kilometers (415 miles), was the largest hurricane of its strength to make landfall in the United States in recorded history.

**(b)** Hurricane Katrina breached levees and flooded New Orleans, Louisiana, causing damages of more than $75 billion and claiming at least 1600 lives.

Figure 6.24 **Hurricane Katrina, the most destructive hurricane in U.S. history.**

Figure 6.25 **Withdrawal of water from a coastal region during Hurricane Irma in 2017.** As the bulge of water from a storm surge forms offshore, water from coastal regions is drained away, leaving the sea floor exposed, such as here along the west coast of Florida.

**THE HISTORIC 2017 ATLANTIC HURRICANE SEASON: HURRICANES HARVEY, IRMA, AND MARIA** The 2017 Atlantic hurricane season was a hyperactive, deadly, and extremely destructive season, featuring 17 named storms, ranking alongside 1936 as the fifth-most active season since records began in 1851. The season also featured the highest number of major hurricanes since 2005. All ten of the season's hurricanes occurred in a row, the greatest number of consecutive hurricanes in the satellite era, and tied for the greatest number of consecutive hurricanes ever observed in the Atlantic basin since records began in 1851. In addition, it was by far the costliest season on record, with damages of approximately $281.14 billion U.S. dollars, which is about $100 billion higher than the damages of the record-breaking 2005 season. Essentially all of the season's damage was caused by three of the season's most destructive major hurricanes—Harvey, Irma, and Maria—a deadly trio, the last of which had especially damaging effects in Puerto Rico and other parts of the Caribbean. This season is also one of only six years on record to feature multiple Category 5 hurricanes, and only the second after 2007 to feature two hurricanes making landfall at that intensity.

Of the 2017 hurricanes, Hurricane Irma was the first Category 5 hurricane of the season and the strongest storm on record to exist in the open Atlantic Ocean. Irma caused widespread and catastrophic damage throughout its long lifetime, particularly in the northeastern Caribbean and the Florida Keys. It was also the most intense hurricane to strike the continental United States since Katrina in 2005 and the first major hurricane to make landfall in Florida since Wilma in the same year. In addition, strong offshore winds from Hurricane Irma blew water away from shore and into the offshore storm surge bulge, causing large areas of the coastal sea floor to be bared along the west coast of Florida and in the Bahamas (**Figure 6.25**). This phenomenon appears similar to what happens just before a large tsunami arrives, where seawater can sometimes withdrawal from the coastal region (for more information about tsunami, see Chapter 8, "Waves and Water Dynamics").

## STUDENTS SOMETIMES ASK . . .

***What is the strongest tropical cyclone to ever make landfall?***

On November 8, 2013, Typhoon Haiyan slammed into the Philippines with sustained winds of 305 to 314 kilometers (190 to 195 miles) per hour, making it the strongest Category 5 tropical cyclone on record to hit land. Three prior tropical cyclones, the earliest of which was in 1958, had higher wind speeds when out at sea but they all weakened somewhat before making landfall. Typhoon Haiyan is blamed for more than 6000 deaths and for destroying or damaging the homes of more than 6 million people.

**HISTORIC DESTRUCTION IN OTHER REGIONS** The majority of the world's tropical cyclones are formed in the waters north of the equator in the western Pacific Ocean. These storms, called *typhoons*, do enormous damage to coastal areas and islands in Southeast Asia (see Figure 6.19).

Other areas of the world such as Bangladesh, which borders the Indian Ocean, experience tropical cyclones on a regular basis. Bangladesh is particularly vulnerable because it is a highly populated and low-lying country, much of it only 3 meters (10 feet) above sea level. In 1970, a 12-meter (40-foot)-high storm surge from a tropical cyclone killed an estimated 1 million people. Another tropical cyclone hit the area in 1972 and caused up to 500,000 deaths. In 1991, Hurricane Gorky's winds of 233 kilometers (145 miles) per hour and large storm surge caused extensive damage, killing over 200,000 people.

Even islands near the centers of ocean basins can be struck by hurricanes. The Hawaiian Islands, for example, were hit hard by Hurricane Dot in August 1959 and by Hurricane Iwa in November 1982. Hurricane Iwa hit very late in the hurricane season and produced winds up to 130 kilometers (81 miles) per hour. Damage of

## PROCESS OF SCIENCE 6.1
## Ocean Soundscapes Interrupted by Hurricanes

### BACKGROUND

Earth's oceans are not vast, silent landscapes—waves breaking, marine life sounds, and human-induced noises all add a "voice" to the seas. Scientists have studied ocean sounds, such as whale calls, for decades, but only recently has technology advanced to the point that long-term sound studies can be conducted. Passive acoustic monitoring systems, also called hydrophones, can now record sound on land and underwater for months at a time, giving scientists a unique opportunity to study the sounds of the ocean.

Coral reef ecosystems are teeming with life, and those organisms that live on the reef make quite a bit of noise! The life of a coral reef uses sound to communicate, find prey, and avoid predation. For example, studies have found that many coral reef fish sing "choruses" to locate one another. And snapping shrimp, the noisiest of all coral reef life, snap their claws to threaten and stun their prey. Could the sounds of a reef tell us more about their health?

### FORMING A HYPOTHESIS

Studying the biology, physics, and chemistry of a coral reef is a challenging task. Some reefs are in remote locations, making access difficult. Moreover, for a whole-scale study of a reef ecosystem, scientists often need expensive technology. Recently, a group of scientists hypothesized that hydrophones could provide long-term data on the overall health and stability of coral reef ecosystems. With an understanding of the various sounds of a reef, and their short- and long-term variations, scientists could develop a baseline for reef sounds and monitor the effects of disturbances.

### DEVISING AN EXPERIMENT

As part of a long-term study, researchers set up hydrophones in coral reefs near Puerto Rico's southwest coast to continuously record how sounds change over time.

In September 2017, Hurricanes Irma and Maria provided these scientists with a surprising study of how the soundscapes change in response to large environmental disturbances.

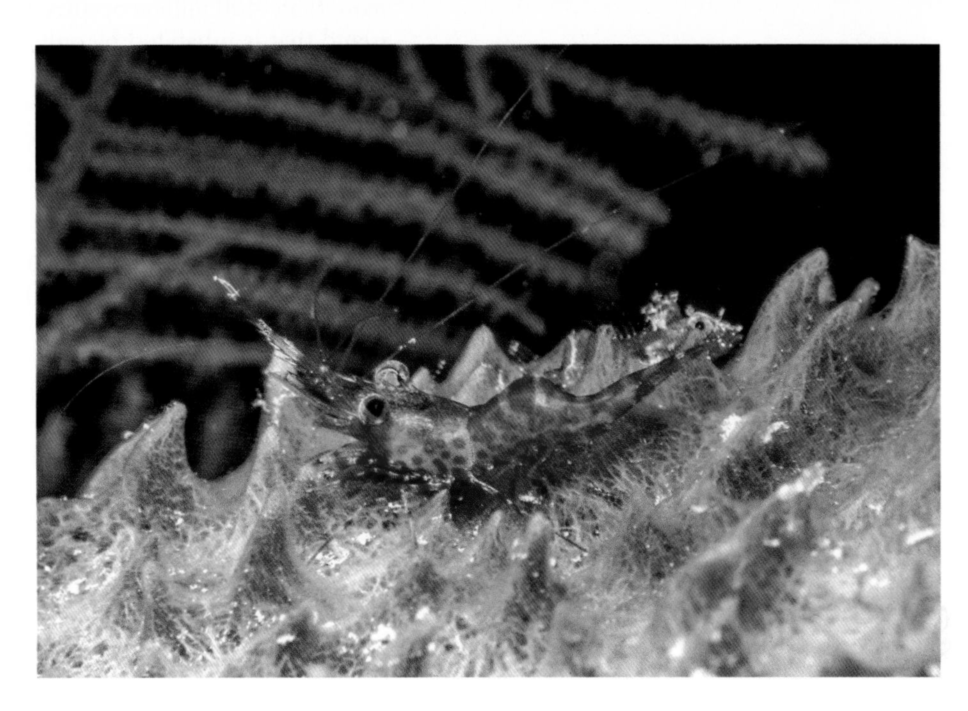

**Figure 6B** **A Caribbean snapping shrimp.**

Notably, the researchers discovered that the storms had short-term effects on the sounds of fish and the activity of snapping shrimp. Following the hurricanes, the intensity of the nightly choruses of fish increased for several nights before returning to baseline levels. And the noise of snapping shrimp decreased significantly during the hurricanes and stayed at reduced values for several days after the storms passed.

### INTERPRETING THE RESULTS

Why did the sounds of the coral reef change during and after Hurricanes Irma and Maria? Researchers think ocean turbidity is the cause—as the storms passed near Puerto Rico, large waves and strong currents churned up sediment into the water, and runoff from land deposited sediment into the reefs.

In this murky water, the fish had to make louder sounds to locate one another, which explains their higher-than-normal intensity of choruses. Scientists also think the snapping shrimp may have needed time to clean out their burrows, causing a reduction in searching for prey. Another idea is that the shrimp could not see their prey in the turbid waters and thus had no reason to snap. This surprising study of sound gave scientists an interesting look into how a coral reef ecosystem responds to significant environmental disturbances.

### THINKING LIKE A SCIENTIST: WHAT'S NEXT?

Acoustic sensors have applicability in a variety of studies. Think of a human-induced disturbance that affects oceanic ecosystems. Could we use hydrophones to understand the effects of such a disturbance? Formulate a hypothesis for such a study.

### REFERENCES

Acoustic Monitoring Reveals Hurricane Maria's Impact on Puerto Rico's Coral Reefs https://coastalscience.noaa.gov/news/acoustic-monitoring-reveals-resilient-natural-marine-soundscapes/ (accessed October 11, 2019).

Hurricanes Irma and Maria temporarily altered choruses of land and sea animals. https://news.agu.org/press-release/hurricanes-irma-and-maria-temporarily-altered-choruses-of-land-and-sea-animals/ (accessed October 10, 2019).

Martinez, F.; Gottesman, B.; Appeldoorn, R.; Mason, D.; Olson, J.; Pijanowski, B.C.; Ruberg, S. A.; Weil, E. Altered Soundscapes Help Reveal Hurricane Maria's Impact on Three Coral Reef Habitats off the Coast of Puerto Rico. Presented at 2018 Ocean Sciences Meeting, Portland, Oregon, February 16, 2018; AI53B-05.

Ogden, L.E. Sounds Good? The Acoustic Monitoring of Coral Reef Health. *BioScience* **2018**, *68* (1), 48.

more than $100 million occurred on the islands of Kauai and Oahu. Niihau, a small island that is inhabited by only a few hundred native Hawaiians, was directly in the path of the storm and suffered severe property damage but no serious injuries. Hurricane Iniki roared across the islands of Kauai and Niihau in September 1992, with 210-kilometer (130-mile)-per-hour winds. It was the most powerful hurricane to hit the Hawaiian Islands in the past 100 years, with property damage that approached $1 billion.

**FUTURE THREAT TO LIFE AND PROPERTY** Each year, tropical cyclones and hurricanes leave millions homeless worldwide and account for, on average, over $100 billion of damage in the United States alone. Hurricanes will continue to be a threat to life and property around the globe. Because of increasingly accurate forecasts and prompt evacuation, however, the loss of life has been decreasing. Property damage, on the other hand, has been increasing because increasing coastal populations have resulted in more and more construction along the coast. Inhabitants of areas subject to a hurricane's destructive force must be made aware of the danger so that they can be prepared for its eventuality. The impact of human-caused climate change on the inevitable economic losses from tropical cyclones is a major concern, too. In fact, new research shows that human-caused climate change may double the global economic losses caused by tropical cyclones and hurricanes.

Climate Connection

**RECAP Hurricanes are intense—and sometimes destructive—tropical storms that form where water temperatures are high; where there is an abundance of warm, moist air; and where the Coriolis effect influences their spin.**

## The Ocean's Climate Patterns

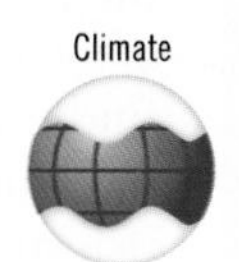


Just as land areas have climate patterns, so do regions of the oceans. The open ocean is divided into climatic regions that run generally east–west (parallel to lines of latitude) and have relatively stable boundaries that are somewhat modified by ocean surface currents (**Figure 6.26**).

The **equatorial** region spans the equator, which gets an abundance of solar radiation. As a result, the major air movement is upward because heated air rises. Surface winds, therefore, are weak and variable, which is why this region is called the *doldrums*. Surface waters are warm, and the air is saturated with water vapor. Daily rain showers are common, which keeps surface salinity relatively low. The equatorial regions just north or south of the equator are also the breeding grounds for tropical cyclones.

**Tropical** regions extend north or south of the equatorial region up to the Tropic of Cancer and the Tropic of Capricorn, respectively. They are characterized by strong trade winds, which blow from the northeast in the Northern Hemisphere and from the southeast in the Southern Hemisphere. These winds push the equatorial currents and create moderately rough seas. Relatively little precipitation falls at higher latitudes within tropical regions, but precipitation increases toward the equator. Once tropical cyclones form, they gain energy here as large quantities of heat are transferred from the ocean to the atmosphere.

Beyond the tropics are the **subtropical** regions. Belts of high pressure are centered there, so the dry, descending air produces little precipitation and a high rate of evaporation, resulting in the highest surface salinities in the open ocean (see the surface salinity map in Chapter 5, Figure 5.22). Winds are weak and currents are sluggish, typical of the horse latitudes. However, strong boundary currents (along the boundaries of continents) flow north and south, particularly along the western margins of the subtropical oceans.

The **temperate** regions (also called the *middle latitudes* or *midlatitudes*) are characterized by strong westerly winds (the prevailing westerlies) that blow from the southwest in the Northern Hemisphere and from the northwest in the Southern Hemisphere (see Figure 6.12). Severe storms are common, especially during winter, and precipitation is heavy. In fact, the North Atlantic is noted for fierce storms, which have claimed many ships and numerous lives over the centuries.

The **subpolar** region experiences extensive precipitation due to the subpolar low. Sea ice covers the subpolar ocean in winter, but it melts away, for the most part,

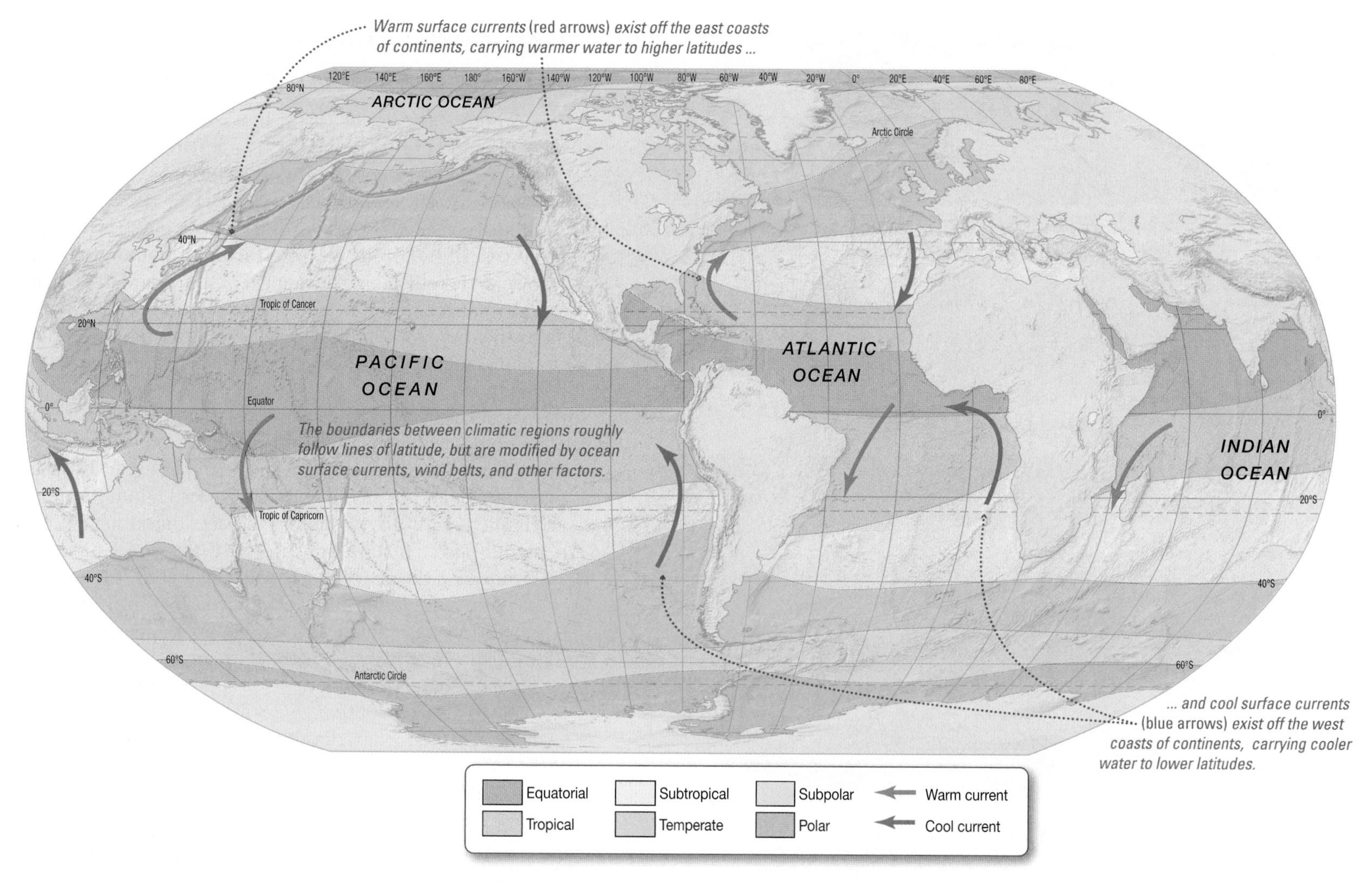
Warm surface currents (red arrows) exist off the east coasts of continents, carrying warmer water to higher latitudes ...
ARCTIC OCEAN
Arctic Circle
Tropic of Cancer
PACIFIC OCEAN
ATLANTIC OCEAN
Equator
The boundaries between climatic regions roughly follow lines of latitude, but are modified by ocean surface currents, wind belts, and other factors.
INDIAN OCEAN
Tropic of Capricorn
Antarctic Circle
... and cool surface currents (blue arrows) exist off the west coasts of continents, carrying cooler water to lower latitudes.
Equatorial
Tropical
Subtropical
Temperate
Subpolar
Polar
Warm current
Cool current

Figure 6.26 **The ocean's climatic regions.**

in summer. Icebergs are common, and the surface temperature seldom exceeds 5°C (41°F) in the summer months.

Surface temperatures remain at or near freezing in the **polar** regions, which are covered with ice throughout most of the year. The polar high pressure dominates the area, which includes the Arctic Ocean and the ocean adjacent to Antarctica. There is no sunlight during the winter and constant daylight during the summer.

## CONCEPT CHECK 6.5 ▸ Describe how the ocean influences global weather phenomena and climate patterns.

1 Describe the difference between cyclonic and anticyclonic flow and show how the Coriolis effect is important in producing both clockwise and counterclockwise flow patterns.

2 How do sea breezes and land breezes form? During a hot summer day, which one would be most common and why?

3 Name the polar and tropical air masses that affect U.S. weather. Describe the pattern of movement across the continent and patterns of precipitation associated with warm and cold fronts.

4 What are the conditions needed for the formation of a tropical cyclone? Why do most middle latitude areas only rarely experience a hurricane? Why are there no hurricanes at the equator?

5 Describe the types of destruction caused by hurricanes. Which one causes the majority of fatalities and destruction?

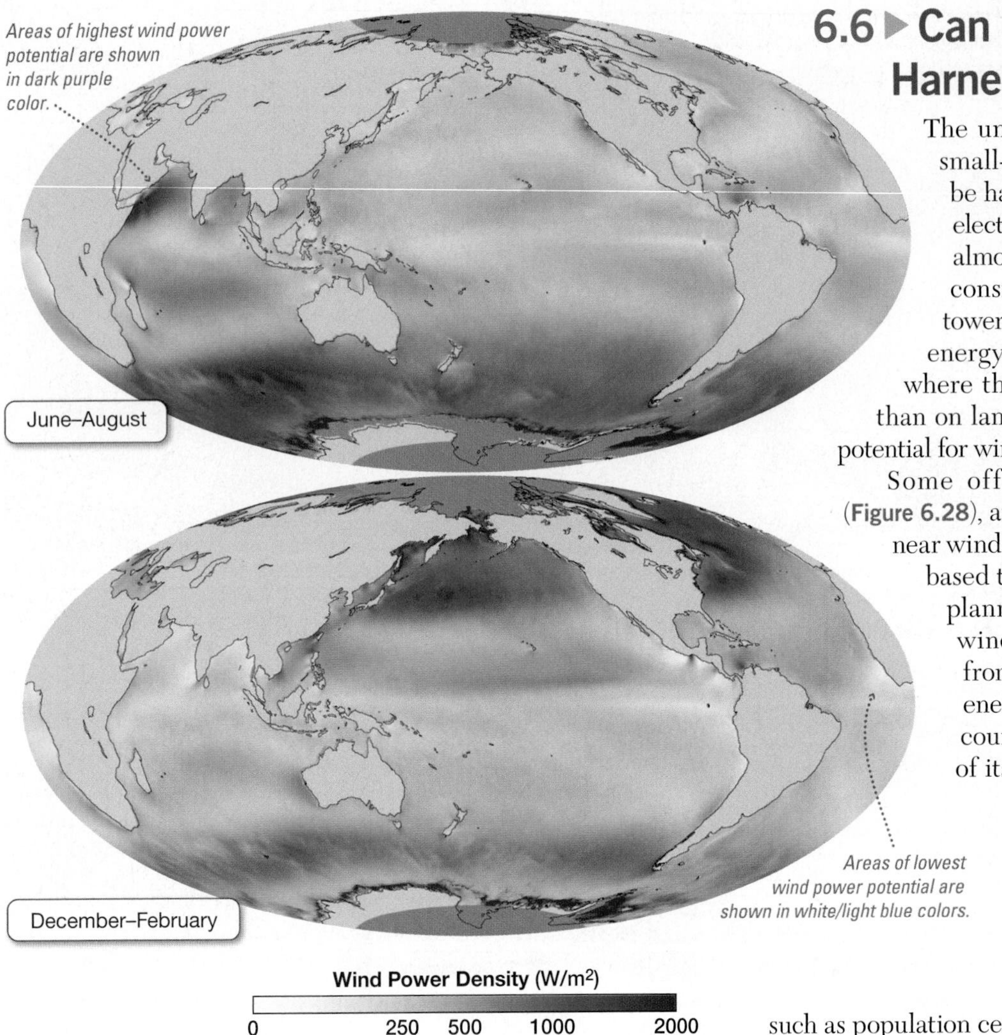
Areas of highest wind power potential are shown in dark purple color.
June–August
December–February
Areas of lowest wind power potential are shown in white/light blue colors.
Wind Power Density (W/m2)
0
250
500
1000
2000

**Figure 6.27 Global ocean wind energy potential.** Average ocean wind intensity maps during 2000–2007 for June–August (*top*) and December–February (*bottom*).

## 6.6 ▸ Can Power from Wind Be Harnessed as a Source of Energy?

The uneven heating of Earth by the Sun drives various small- and large-scale winds. These winds, in turn, can be harnessed to turn windmills or turbines that generate electricity. At various places on land where the wind blows almost constantly, wind farms have been constructed that consist of hundreds of large turbines mounted on tall towers, thereby taking advantage of this renewable, clean energy source. Similar facilities could be built offshore, where the wind generally blows harder and more steadily than on land. Figure 6.27 shows the offshore areas where the potential for wind farms exist.

Some offshore **wind farms** have already been built (Figure 6.28), and many more are being planned. In the North Sea, near windswept northern Europe, for example, about 100 sea-based turbines are already operating, with hundreds more planned. In 2015, offshore wind accounted for 24% of wind-power installations in the European Union, up from 13% the year before. And since 2011, the overall energy capacity installed off the coasts of 11 European countries has tripled. In fact, Denmark generates 18% of its power by wind—more than any other country—and hopes to increase its proportion of wind power to 50% by 2030.

One major disadvantage of wind power is that wind strength varies and sometimes it's not windy at all, which is particularly problematic when a high demand for electricity exists. There is also the problem of getting the energy to viable markets such as population centers. On paper, wind and solar power could supply the United States and some other countries with all the electricity they require. In practice, however, both sources are too erratic to supply more than about 20% of a region's total energy capacity, according to the U.S. Department of Energy. What are needed are cheap and efficient ways of storing the generated

**Figure 6.28 Offshore wind farm.** Offshore wind turbines form part of a wind farm that harnesses wind energy off the west coast of Scotland in the United Kingdom.

power, then tapping those supplies when needed. Some of the best solutions to the storage problem include pumping water to high areas and using the water to turn turbines later, using pumps to store underground air at high pressures, and storing energy in advanced batteries.

**RECAP** **There is vast potential for developing wind power as a renewable source of energy, although the erratic nature of wind energy is problematic. Several offshore wind farms currently exist.**

**CONCEPT CHECK 6.6** ▶ **Evaluate the advantages and disadvantages of harnessing winds as a source of energy.**

1 Discuss the advantages and disadvantages of building an offshore wind farm.

2 Describe the location, timeframe, and power generating capability of America's first offshore wind farm.

# ESSENTIAL CONCEPTS REVIEW

## 6.1 ▶ What causes variations in solar radiation on Earth?

- *The atmosphere and the ocean act as one interdependent system*, linked by complex feedback loops. There is a close association between most atmospheric and oceanic phenomena.
- *The Sun heats Earth's surface unevenly due to the change of seasons (caused by the tilt of Earth's rotational axis*, which is 23.5 degrees from vertical) and the daily cycle of sunlight and darkness (Earth's rotation on its axis). *Differences in latitude also cause changes in the amount of solar radiation* received on Earth.

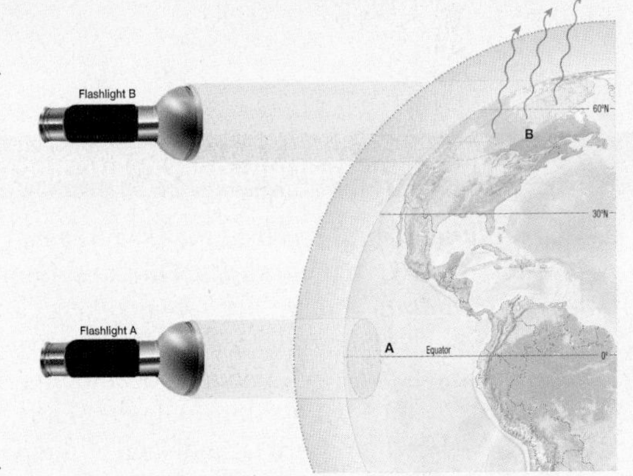

### Selected Key Terms

Use the **glossary** at the end of this book to discover the meanings of these Selected Key Terms: **vernal equinox, summer solstice, autumnal equinox, winter solstice, albedo**.

### Critical Thinking Question

Earth's axis of rotation is angled 23.5 degrees from perpendicular relative to the plane of its ecliptic. Specify how the tilt of Earth's axis affects the change of seasons, the length of day, and the angle of sunlight over the timespan of a year using as an example a location from both hemispheres.

### Active Learning Exercise

With another student in class, make a list of changes that would occur if Earth's axis of rotation was vertical (not tilted). For example, would seasons still exist?

## 6.2 ▶ What physical properties does the atmosphere possess?

- The *uneven distribution of solar energy on Earth* influences most of the physical properties of the atmosphere (such as *temperature, density, water vapor content,* and *pressure differences*) that produce *atmospheric movement.*

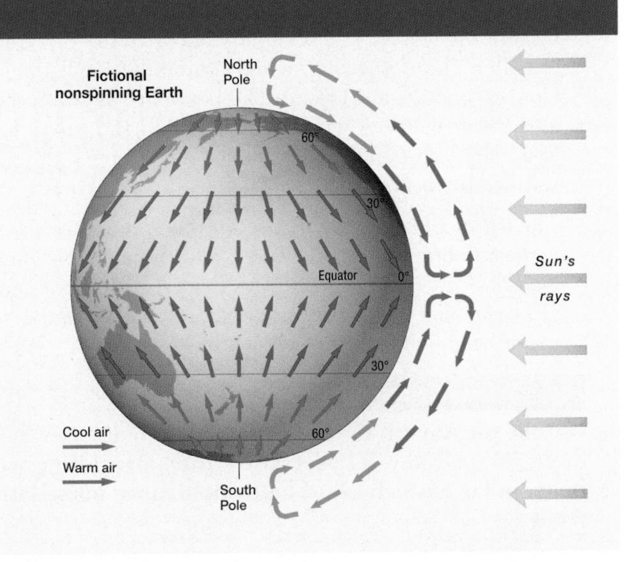


### Selected Key Terms

Use the **glossary** at the end of this book to discover the meanings of these Selected Key Terms: **troposphere, convection cell, wind**.

### Critical Thinking Question

In a nonspinning Earth, describe the basic atmospheric circulation pattern that would exist.

### Active Learning Exercise

With another student in class, use the Internet to research the ancient Greek story of Icarus. Based on the temperature profile of the atmosphere shown in Figure 6.6, is the tragedy that befalls Icarus based in physical fact? Explain.

## 6.3 ▶ How does the Coriolis effect influence moving objects?

- *The Coriolis effect influences the paths of moving objects on Earth and is caused by Earth's rotation.* Because Earth's surface rotates at different velocities at different latitudes, *objects in motion tend to veer to the right in the Northern Hemisphere and to the left in the Southern Hemisphere.* The Coriolis effect is *nonexistent at the equator but increases with latitude,* reaching a maximum at the poles.

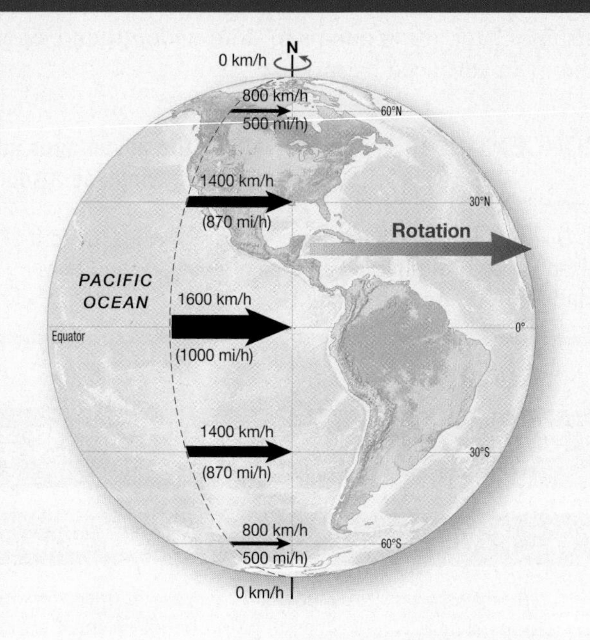


### Selected Key Terms

Use the **glossary** at the end of this book to discover the meanings of this Selected Key Term: **Coriolis effect**.

### Critical Thinking Question

How does the Coriolis effect influence the *direction* of moving objects? How does it affect the *speed* of moving objects? Explain.

### Active Learning Exercise

With another student in class, explain why the Coriolis effect is strongest at the poles. Then, switch roles and explain why the Coriolis effect is nonexistent at the equator.

## 6.4 ▶ What global atmospheric circulation patterns exist?

- *More solar energy is received than is radiated back into space at low latitudes than at high latitudes.* On the spinning Earth, this creates *three circulation cells in each hemisphere*: a *Hadley cell* between 0 and 30 degrees latitude, a *Ferrel cell* between 30 and 60 degrees latitude, and a *polar cell* between 60 and 90 degrees latitude. High-pressure regions, where dense air descends, are located at about 30 degrees north or south latitude and at the poles. Belts of low pressure, where air rises, are generally found at the equator and at about 60 degrees latitude.
- *The movement of air within the circulation cells produces the major wind belts of the world.* The air at Earth's surface that is moving away from the subtropical highs produces *trade winds* moving toward the equator and *prevailing westerlies* moving toward higher latitudes. The air moving along Earth's surface from the polar high to the subpolar low creates the *polar easterlies*.
- *Calm winds characterize the boundaries between the major wind belts of the world.* The boundary between the two trade wind belts is called the *doldrums*, which coincides with the Intertropical Convergence Zone (ITCZ). The boundary between the trade winds and the prevailing westerlies is called the *horse latitudes*. The boundary between the prevailing westerlies and the polar easterlies is called the *polar front*.
- The *tilt of Earth's axis of rotation*, the *lower heat capacity of rock material* compared to seawater, and the *distribution of continents modify the wind and pressure belts of the idealized three-cell model.* However, the three-cell model closely matches the pattern of the major wind belts of the world.

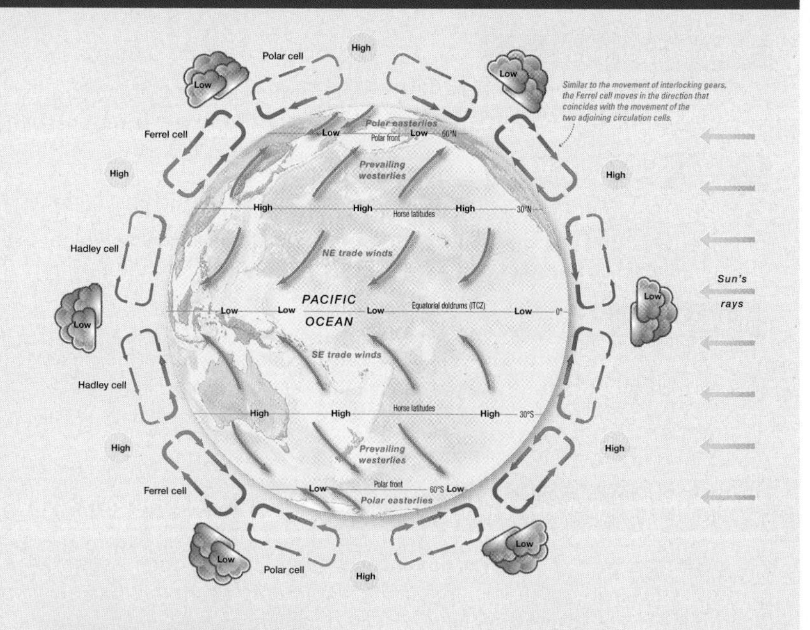


### Selected Key Terms

Use the **glossary** at the end of this book to discover the meanings of these Selected Key Terms: **trade winds, prevailing westerly wind belts, polar easterly wind belts, doldrums, horse latitudes, polar front**.

### Critical Thinking Question

To help reinforce your knowledge of atmospheric circulation patterns, draw from memory the pattern of surface wind belts on Earth, showing atmospheric circulation cells in the upper/lower atmosphere, zones of high and low atmospheric pressure, the names of the wind belts, and the names of the boundaries between the wind belts.

### Active Learning Exercise

With another student in class, discuss how the world's idealized wind belts shown in Figure 6.12 are modified on the real Earth (Figure 6.13). Use specific examples including wind belts, boundaries, high and low atmospheric pressures, and differences between ocean and land. Report your findings to the class.

## 6.5 ▶ How does the ocean influence global weather phenomena and climate patterns?

- *Weather describes the conditions of the atmosphere at a given place and time, while climate is the long-term average of weather.* Atmospheric motion (*wind*) *is always from high-pressure regions toward low-pressure regions*. In the Northern Hemisphere, therefore, there is a *counterclockwise cyclonic movement* of air around low-pressure cells and a *clockwise anticyclonic movement* around high-pressure cells. Coastal regions commonly experience *sea and land breezes*, due to the daily cycle of heating and cooling.
- *Many storms are due to the movement of air masses*. In the middle latitudes, cold air masses from higher latitudes meet warm air masses from lower latitudes and create *cold and warm fronts* that move from west to east across Earth's surface. *Tropical cyclones (hurricanes) are large, powerful storms that mostly affect tropical regions of the world.* Destruction caused by hurricanes is caused by storm surge, high winds, and intense rainfall.
- *The ocean's climate patterns are closely related to the distribution of solar energy and the wind belts of the world*. Ocean surface currents somewhat modify oceanic climate patterns.

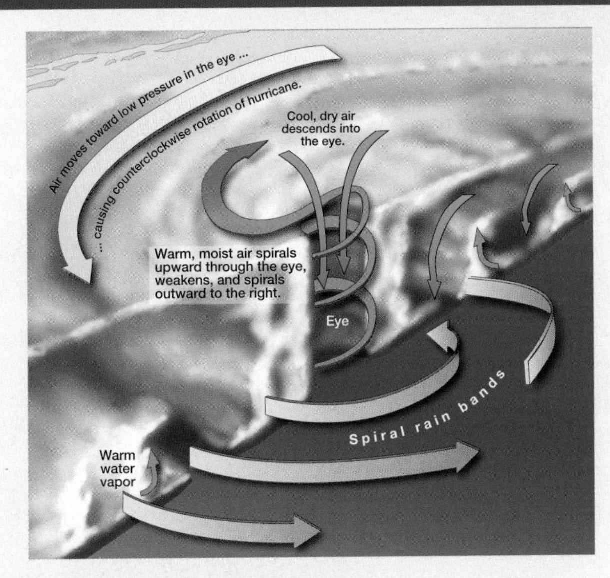
Air moves toward low pressure in the eye ...
... causing counterclockwise rotation of hurricane.
Cool, dry air descends into the eye.
Warm, moist air spirals upward through the eye, weakens, and spirals outward to the right.
Eye
Spiral rain bands
Warm water vapor

### Selected Key Terms

Use the **glossary** at the end of this book to discover the meanings of these Selected Key Terms: **weather, climate, air mass, tropical cyclone, hurricane, typhoon, cyclone, storm surge**.

### Critical Thinking Question

Specify differences between *weather* and *climate*. Then answer this question: When it rains in a region that experiences an arid climate, does it mean the region's climate has changed from dry to wet? Explain.

### Active Learning Exercise

Pair up with another student in class. Using the Internet, have each student determine the latitude and today's offshore surface water temperature for one of the following two locations: San Diego, California, and Charleston, South Carolina. Working together, mark each location on the map shown as Figure 6.26. Using information from Figure 6.26, explain why the surface water temperature is so different at these two locations.

## 6.6 ▶ Can power from wind be harnessed as a source of energy?

- *Winds can be harnessed as a source of power.* There is vast potential for developing this clean, renewable resource, but *problems exist* related to generating power when needed, supplying it to consumers, and storing it. Several offshore wind farm systems currently exist.

### Selected Key Terms

Use the **glossary** at the end of this book to discover the meanings of this Selected Key Term: **wind farm**.

### Critical Thinking Question

Specify some negative environmental factors that might inhibit the development of large offshore wind farms.

### Active Learning Exercise

Working as a group and using Figure 6.27, identify a specific location on Earth that would allow wind turbines to work at maximum capacity year-round (not just during a particular season). Present your findings to the class, including the location and an explanation of your reasoning.

## Mastering Oceanography™

www.masteringoceanography.com

Looking for additional review and test prep materials? With individualized coaching on the toughest topics of the course, Mastering Oceanography offers a wide variety of ways for you to move beyond memorization and deeply grasp the underlying processes of how the oceans work. Visit the Study Area in **www.masteringoceanography.com** to find practice quizzes, study tools, and multimedia that will improve your understanding of this chapter's content. Sign in today to access the following features: Self Study Quizzes, SmartFigures, Oceanography Videos and Animations, Squidtoons, Dynamic Study Modules, and an optional Pearson eText with embedded videos.

# Ocean Circulation

# 7

## ESSENTIAL LEARNING CONCEPTS

At the end of this chapter, you should be able to:

- ☐ **7.1** Demonstrate an understanding of how ocean currents are measured.
- ☐ **7.2** Explain the origin of ocean surface currents and how surface circulation patterns are organized globally.
- ☐ **7.3** Describe the conditions that produce upwelling.
- ☐ **7.4** Specify the main surface circulation patterns in each ocean basin.
- ☐ **7.5** Describe how sea ice and icebergs form.
- ☐ **7.6** Explain the origin and characteristics of deep-ocean currents.
- ☐ **7.7** Evaluate the advantages and disadvantages of harnessing currents as a source of energy.

↑ *Check when completed*

Ocean **currents** are masses of ocean water that flow from one place to another. The amount of water can be large or small, currents can be at the surface or at great depth in the ocean, and the phenomena that create them can be simple or quite complex. Simply put, currents are *water masses in motion.*

Huge current systems dominate the surfaces of the major oceans. These currents transfer heat from warmer to cooler areas on Earth, just as the major wind belts of the world do. Wind belts transfer about two-thirds of the total amount of heat from the tropics to the poles; ocean surface currents transfer the other third. Ultimately, energy from the Sun drives surface currents, and they closely follow the pattern of the world's major wind belts. As a result, the movement of currents has aided the travel of prehistoric people across ocean basins. Ocean currents also influence the abundance of life in surface waters by affecting the growth of microscopic algae, which are the basis of most oceanic food webs.

More locally, surface currents affect the climates of coastal continental regions. Cold currents flowing toward the equator on the western sides of continents produce arid conditions. Conversely, warm currents flowing poleward on the eastern sides of continents produce warm, humid conditions. Ocean currents, for example, contribute to the mild climate of northern Europe and Iceland, whereas conditions at similar latitudes along the Atlantic coast of North America (such as Labrador, Canada) are much colder. In addition, water sinks in high-latitude regions, initiating deep currents that help regulate the planet's climate.

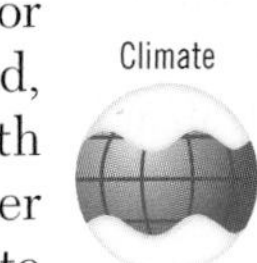


*"The ocean, far from a steady circulation, is a lovely confusion of swirling, turbulent, rapidly evolving motion incredibly complex and varying in space and time."*

*— Jennifer MacKinnon, Physical oceanographer (2017)*

## 7.1 ▸ How Are Ocean Currents Measured?

Ocean currents are either *wind driven* or *density driven*. Moving air masses—particularly the major wind belts of the world—set wind-driven currents in motion. Wind-driven currents move water horizontally and occur primarily in the ocean's surface waters, so these currents are called **surface currents**. Density-driven circulation, on the other hand, moves water vertically and accounts for the thorough mixing of the deep masses of ocean water. Some surface waters become high in density—through low temperature and/or high salinity—and so sink beneath the surface. This dense water sinks and spreads slowly beneath the surface, so these currents are called **deep currents**.

◂ **Patterns of ocean currents seen from space.** This composite SeaWiFS/SeaStar satellite view during the austral summer highlights ocean circulation patterns. The deep blue color represents low chlorophyll (phytoplankton) concentrations, and the orange and red colors represent high chlorophyll (phytoplankton) concentrations. Note the wavy pattern of eddies between Africa and Antarctica, where the Agulhas Current meets the Antarctic Circumpolar Current and is turned to the east, creating the Agulhas Retroflection. Off the west coast of Africa, coastal upwelling is shown in bright red colors.

1 m
3.3 ft

(a) A **drift current meter** afloat in the ocean.

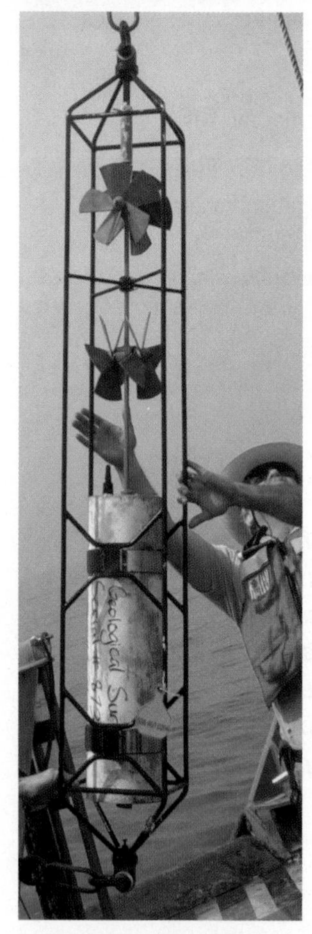

(b) A **propellor-type flow meter** being brought back aboard a research vessel.

**Figure 7.1 Examples of direct-method current-measuring devices.**

## Surface Current Measurement

Because surface currents are driven by the wind, they rarely flow in the same direction and at the same rate for very long, so measuring average flow rates can be difficult. Some consistency, however, exists in the *overall* surface current pattern worldwide. Surface currents can be measured directly or indirectly.

**DIRECT METHODS** Two main methods are used to measure surface currents *directly*. In one, a floating device is released into the current and its position is tracked through time. Typically, radio-transmitting float bottles or other devices are used (**Figure 7.1a**), but other accidentally released items also make good drift meters (**Diving Deeper 7.1**). The other method uses a current-measuring device, such as the propeller flow meter shown in **Figure 7.1b**, that is deployed from a fixed position, such as a pier or a stationary ship. A propeller device can also be towed behind a ship, with the ship's speed subtracted to determine the current's true flow rate.

**INDIRECT METHODS** Three different methods can be used to measure surface currents *indirectly*. The first method involves *pressure gradients*, which are the slopes caused by large-scale bulges and depressions in the ocean's surface (note that pressure gradients are also used to determine the movement of winds based on high and low atmospheric pressure; see, for example, the weather map shown in Figure 6.15). Water flows parallel to a pressure gradient (that is, downhill), so this method determines the internal distribution of density and the corresponding pressure gradient across an area of the ocean. A second method uses radar altimeters—such as those launched aboard Earth-observing satellites today—to determine the lumps and bulges at the ocean surface, which are a result of the shape of the underlying sea floor (note that this technique is described in Chapter 3, Section 3.1: "Using Satellites to Map Ocean Properties from Space.") as well as current flow. From these data, *dynamic topography* maps can be produced that show the speed and direction of surface currents (**Figure 7.2**). A third method uses a *Doppler flow meter* to transmit low-frequency sound signals through the water. The flow meter remains

**Figure 7.2 Satellite view of ocean dynamic topography.** Map of TOPEX/Poseidon radar altimeter data showing variation of sea surface height, in centimeters, from September 1992 to September 1993.

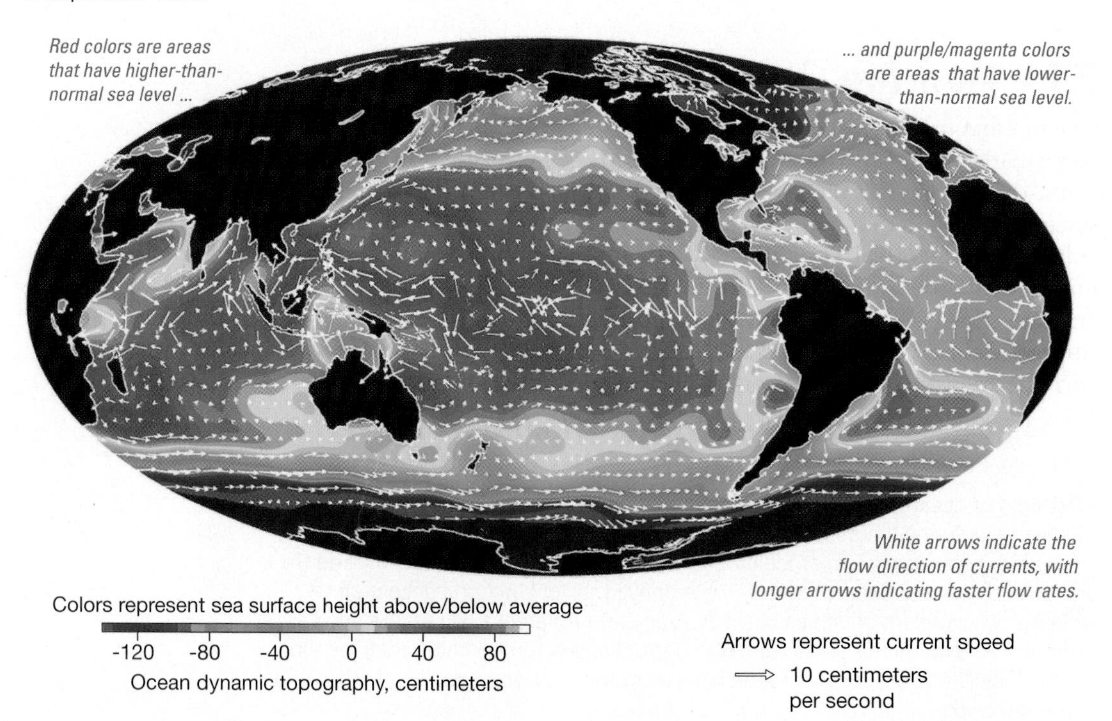
Red colors are areas that have higher-than-normal sea level ...
... and purple/magenta colors are areas that have lower-than-normal sea level.
White arrows indicate the flow direction of currents, with longer arrows indicating faster flow rates.
Colors represent sea surface height above/below average
-120 -80 -40 0 40 80
Ocean dynamic topography, centimeters
Arrows represent current speed
10 centimeters per second

stationary and measures the shift in frequency between the sound waves emitted and those backscattered by particles in the water to determine current movement.

## Deep Current Measurement

The great depth at which deep currents exist makes them even more difficult to measure than surface currents. Most often, they are mapped using underwater floats that are carried within deep currents. One such unique oceanographic program that began in 2000, called **Argo**, is a global array of free-drifting profiling floats (**Figure 7.3b**) that move vertically and measure the temperature, salinity, and other water characteristics of the upper 2000 meters (6600 feet) of the ocean. Once deployed, each float sinks to a particular depth, drifts for up to 10 days collecting data, then resurfaces and transmits data on its location and ocean variables, which are made publically available within hours. Each float then sinks back down to a programmed depth and drifts for up to another 10 days, collecting more data, before resurfacing and repeating the cycle. In 2007, the goal of the program was achieved with the launch of the 3000th Argo float; currently, nearly 4000 floats that report publically-available data are operating worldwide (**Figure 7.3a**). The program will allow oceanographers to develop a forecasting system for the oceans analogous to weather forecasting on land, and also to track changes in ocean properties as a result of human-caused climate change.

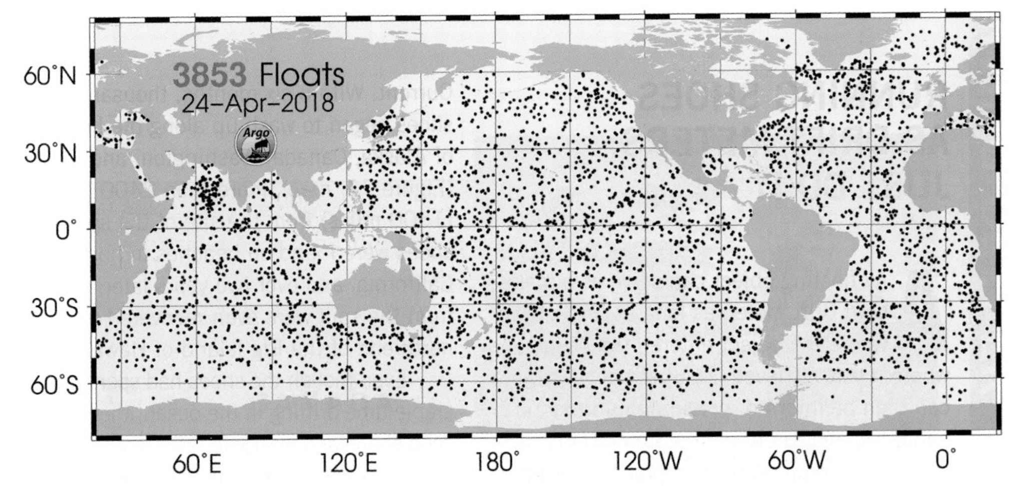


**(a)** Map showing the location of Argo floats, which can dive to 2000 meters (6600 feet) and collect data on the ocean properties before resurfacing and transmitting their data.

**(b)** Argo floats are deployed from research or cargo vessels.

Figure 7.3 **The Argo system of free-drifting submersible floats.**

While the present Argo array provides observations of the upper ocean, this is only about one-half of the total oceanic volume. In 2014, the first of an array of 25 deep-water floats known as *Deep Argo* was released to collect data on seawater temperature and salinity down to 6000 meters (19,700 feet), allowing oceanographers to study ocean circulation and long-term climate trends.

Climate Connection

Other techniques used for measuring deep currents include identifying the distinctive temperature and salinity characteristics of a deep-water mass and tracking telltale chemical tracers. Some tracers are naturally present in seawater, while others are intentionally added. Some useful tracers that have inadvertently been added to seawater include tritium (a radioactive isotope of hydrogen produced by nuclear bomb tests in the 1950s and early 1960s) and chlorofluorocarbons (Freon and other gases that deplete Earth's ozone layer).

**RECAP** **Wind-induced surface currents are measured with floating objects, by satellites, and through other techniques. Density-induced deep currents are measured using submerged floats, water properties, and chemical tracers.**

### CONCEPT CHECK 7.1 ▸ Demonstrate an understanding of how ocean currents are measured.

1 Compare the forces that are directly responsible for creating horizontal and deep vertical circulation in the oceans. What is the ultimate source of energy that drives both circulation systems?

2 Describe the different ways in which currents are measured.

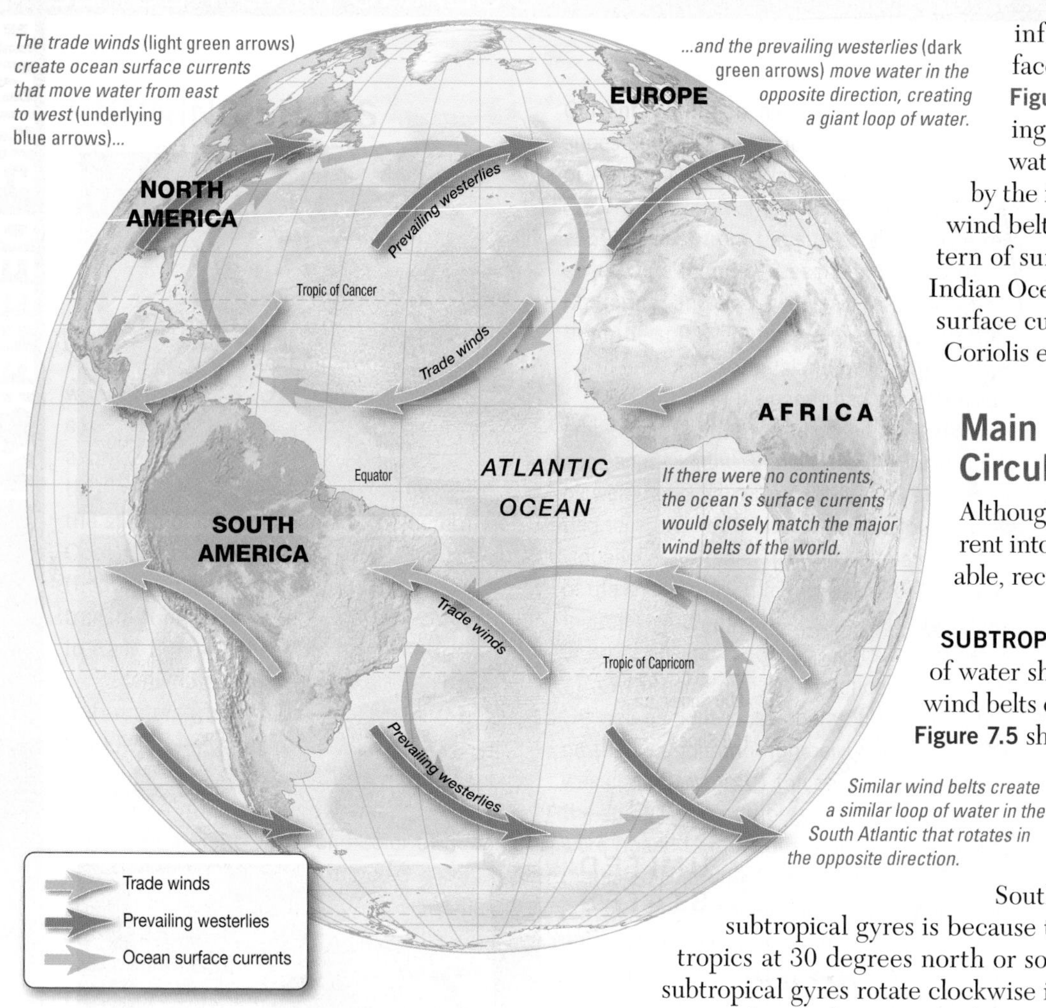

The trade winds (light green arrows) create ocean surface currents that move water from east to west (underlying blue arrows)...
...and the prevailing westerlies (dark green arrows) move water in the opposite direction, creating a giant loop of water.
EUROPE
NORTH AMERICA
Prevailing westerlies
Tropic of Cancer
Trade winds
AFRICA
Equator
ATLANTIC OCEAN
If there were no continents, the ocean's surface currents would closely match the major wind belts of the world.
SOUTH AMERICA
Trade winds
Tropic of Capricorn
Prevailing westerlies
Similar wind belts create a similar loop of water in the South Atlantic that rotates in the opposite direction.
Trade winds
Prevailing westerlies
Ocean surface currents

**Figure 7.4 How major wind belts affect the movement of surface currents in the Atlantic Ocean.**

influences the nature and direction of flow of surface currents in each ocean basin. As an example, **Figure 7.4** shows how the trade winds and prevailing westerlies create large circular-moving loops of water in the Atlantic Ocean basin, which is bounded by the irregular shape of continents. These same global wind belts affect the other ocean basins, so a similar pattern of surface current flow also exists in the Pacific and Indian Oceans. As we shall see, other factors that influence surface current patterns include gravity, friction, and the Coriolis effect.

## Main Components of Ocean Surface Circulation

Although ocean water continuously flows from one current into another, ocean surface currents have a predictable, recurring pattern within each ocean basin.

**SUBTROPICAL GYRES** The large, circular-moving loops of water shown in Figure 7.4 that are driven by the major wind belts of the world are called **gyres** (*gyros* = a circle). **Figure 7.5** shows the world's five **subtropical gyres**: (1) the *North Pacific Gyre*, (2) the *South Pacific Gyre*, (3) the *North Atlantic Gyre*, (4) the *South Atlantic Gyre*, and (5) the *Indian Ocean Gyre* (which is mostly within the Southern Hemisphere). The reason they are called subtropical gyres is because the center of each gyre coincides with the subtropics at 30 degrees north or south latitude. As shown in Figures 7.4 and 7.5, subtropical gyres rotate clockwise in the Northern Hemisphere and counterclockwise in the Southern Hemisphere. Studies of floating objects (see Diving Deeper 7.1) indicate that the average drift time in a smaller subtropical gyre, such as the North Atlantic Gyre, is about three years, whereas in larger subtropical gyres, such as the North Pacific Gyre, it is about six years.

**Animation**
Ocean Circulation
http://goo.gl/bZqQBD

Generally, *each subtropical gyre is composed of four main currents* that flow progressively into one another (**Table 7.1**). The North Atlantic Gyre, for instance, is composed of the North Equatorial Current, the Gulf Stream, the North Atlantic Current, and the Canary Current (Figure 7.5). Let's examine each of the four main currents that comprise subtropical gyres.

***Equatorial Currents*** The trade winds, which blow from the southeast in the Southern Hemisphere and from the northeast in the Northern Hemisphere, set in motion the water masses between the tropics. The resulting currents are called **equatorial currents**, which travel westward along the equator and form the equatorial boundary current of subtropical gyres (Figure 7.5). They are called *north equatorial currents* or *south equatorial currents*, depending on their position relative to the equator.

***Western Boundary Currents*** When equatorial currents reach the western portion of an ocean basin, they must turn because they cannot cross land. The Coriolis effect deflects these currents away from the equator as **western boundary currents**, which comprise the western part of all subtropical gyres. Western boundary currents are so named because they travel along the western boundaries of their respective ocean basins. Notice that western boundary currents are off the *eastern* coasts of adjoining continents. It's easy to be confused about this because we have a land-based perspective. From an *oceanic perspective*, however, the western side

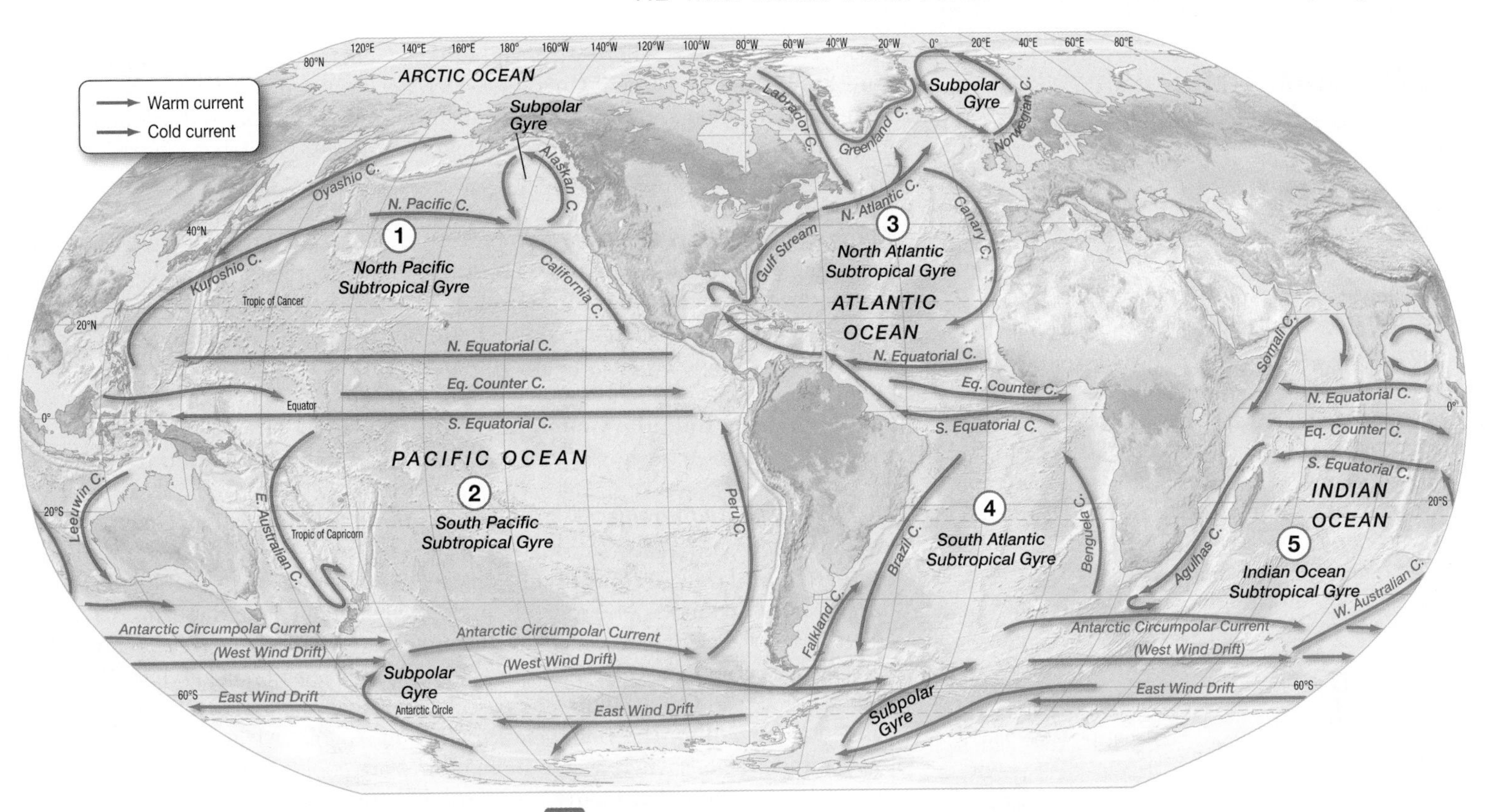


**SmartFigure 7.5 Wind-driven surface currents.** Major wind-driven surface currents of the world's oceans during February–March. The five major subtropical gyres are: (1) the North Pacific Gyre, (2) the South Pacific Gyre, (3) the North Atlantic Gyre, (4) the South Atlantic Gyre, and (5) the Indian Ocean Gyre. Smaller subpolar gyres exist in higher latitudes and rotate in the reverse direction of their adjacent subtropical gyres. https://goo.gl/Gnj9zw

**SmartTable 7.1 Subtropical gyres and surface currents** https://goo.gl/DHJGM6

## SmartTable 7.1 Subtropical gyres and surface currents

| Pacific Ocean | Atlantic Ocean | Indian Ocean |
|---|---|---|
| **1. North Pacific Gyre** | **3. North Atlantic Gyre** | **5. Indian Ocean Gyre** |
| North Pacific Current | North Atlantic Current | South Equatorial Current |
| California Current[a] | Canary Current[a] | Agulhas Current[b] |
| North Equatorial Current | North Equatorial Current | West Wind Drift |
| Kuroshio (Japan) Current[b] | Gulf Stream[b] | West Australian Current[a] |
| **2. South Pacific Gyre** | **4. South Atlantic Gyre** | **Other Major Currents** |
| South Equatorial Current | South Equatorial Current | Equatorial Countercurrent |
| East Australian Current[b] | Brazil Current[b] | North Equatorial Current |
| West Wind Drift | West Wind Drift | Leeuwin Current |
| Peru (Humboldt) Current[a] | Benguela Current[a] | Somali Current |
| **Other Major Currents** | **Other Major Currents** | |
| Equatorial Countercurrent | Equatorial Countercurrent | |
| Alaskan Current | Florida Current | |
| Oyashio Current | East Greenland Current | |
| | Labrador Current | |
| | Falkland Current | |

[a]Denotes an eastern boundary current of a gyre, which is relatively *slow*, *wide*, and *shallow* (and is also a *cold-water* current).
[b]Denotes a western boundary current of a gyre, which is relatively *fast*, *narrow*, and *deep* (and is also a *warm-water* current).

## STUDENTS SOMETIMES ASK . . .

***What is the name of the current that's mentioned in the movie* Finding Nemo?**

It's the East Australian Current, which is called the "EAC" in the 2003 Disney animated film and is even geographically correct (if only all movies could do the same!). The EAC is a western intensified ocean surface current that helps Nemo's dad and Dory travel from the Great Barrier Reef to Sydney Harbor along the east coast of Australia. While being swept along in the EAC, they meet Crush the sea turtle, who famously asks, "What brings you to the EAC?" In the storyline, Crush and other sea turtles help Nemo's dad and Dory navigate the EAC and after many close calls ( . . . spoiler alert . . . ), Nemo is successfully rescued. In fact, a sequel called *Finding Dory* was released in 2016 that continues the adventures of Nemo, Dory, and friends.

**Video**
Perpetual Ocean by NASA
https://goo.gl/q1Vo4i

of the ocean basin is where western boundary currents reside. For example, the Gulf Stream and the Brazil Current, which are shown in Figure 7.5, are western boundary currents. They come from equatorial regions, where water temperatures are warm, so they carry warm water to higher latitudes. Note that Figure 7.5 shows warm currents as red arrows.

***Northern or Southern Boundary Currents*** Between 30 and 60 degrees latitude, the prevailing westerlies blow from the northwest in the Southern Hemisphere and from the southwest in the Northern Hemisphere. These winds direct ocean surface water in an easterly direction across an ocean basin (see the North Atlantic Current and the Antarctic Circumpolar Current [West Wind Drift] in Figure 7.5). In the Northern Hemisphere, these currents comprise the northern parts of subtropical gyres and are called **northern boundary currents**; in the Southern Hemisphere, they comprise the southern parts of subtropical gyres and are called **southern boundary currents**.

***Eastern Boundary Currents*** When currents flow back across the ocean basin, the Coriolis effect and continental barriers turn them toward the equator, creating **eastern boundary currents** of subtropical gyres along the eastern boundary of the ocean basins. Examples of eastern boundary currents include the Canary Current and the Benguela Current, which are shown in Figure 7.5. (Note that ocean surface currents are often named for a prominent geographic location near where they pass. For instance, the Canary Current passes the Canary Islands, and the Benguela Current is named for the Benguela Province in Angola, Africa.) They come from high-latitude regions where water temperatures are cool, so they carry cool water to lower latitudes. Note that Figure 7.5 shows cold currents as blue arrows.

**EQUATORIAL COUNTERCURRENTS** A large volume of water is driven westward by the north and south equatorial currents and piles up water on the western side of an ocean basin near the equator, creating higher sea level there. As a result, this bulge of water flows downhill toward the east under the influence of gravity. This current, called the **equatorial countercurrent**, is a narrow, easterly flow of water that occurs *counter to* and *between* the adjoining equatorial currents.

Figure 7.5 shows that an equatorial countercurrent is particularly apparent in the Pacific Ocean. This is because of the large equatorial region that exists in the Pacific Ocean and because of a dome of equatorial water that becomes trapped in the island-filled embayment between Australia and Asia. Continual influx of water from equatorial currents builds the dome and creates an eastward countercurrent that stretches across the Pacific toward South America. The equatorial countercurrent in the Atlantic Ocean, on the other hand, is not nearly as well defined because of the shapes of the adjoining continents, which limit the equatorial area that exists in the Atlantic Ocean. The presence of an equatorial countercurrent in the Indian Ocean is strongly influenced by the monsoons, which will be discussed later in this chapter.

**SUBPOLAR GYRES** Northern or southern boundary currents that flow eastward as a result of the prevailing westerlies eventually move into subpolar latitudes (about 60 degrees north or south latitude). Here, they are driven in a westerly direction by the polar easterlies, producing **subpolar gyres** that rotate opposite the adjacent subtropical gyres. Subpolar gyres are smaller and fewer than subtropical gyres. Two examples include the subpolar gyre in the Atlantic Ocean between Greenland and Europe and in the Weddell Sea off Antarctica (Figure 7.5).

**RECAP** **The principal ocean surface current pattern on Earth consists of large subtropical gyres and smaller subpolar gyres, both of which are big, circular-moving loops of water powered by the major wind belts of the world.**

## Other Factors Affecting Ocean Surface Circulation

Several other factors influence circulation patterns in subtropical gyres, including Ekman spiral and Ekman transport, geostrophic currents, and western intensification of subtropical gyres.

**EKMAN SPIRAL AND EKMAN TRANSPORT** During the voyage of the *Fram* (see Mastering Oceanography Web Diving Deeper 7.1), Norwegian explorer Fridtjof Nansen (1861–1930) observed that Arctic Ocean ice moved 20 to 40 degrees to the right of the wind blowing across its surface (**Figure 7.6**). Not only ice but surface waters in the Northern Hemisphere were observed to move to the right of the wind direction; in the Southern Hemisphere, surface waters move to the *left* of the wind direction. Why does surface water move in a direction different than the wind? **V. Walfrid Ekman** (1874–1954), a Swedish physicist, developed an ocean circulation model in 1905 called the **Ekman spiral** (**Figure 7.7**) that explains Nansen's observations as a balance between the Coriolis effect, which causes objects to curve from their intended path, and frictional effects, which reduce the water's speed with depth.

The Ekman spiral describes the speed and direction of flow of surface waters at various depths. Ekman's model assumes that a uniform column of water is set in motion by wind blowing across its surface (Figure 7.7, *large green wind arrow*). Under ideal conditions in the Northern Hemisphere, the Coriolis effect causes surface water in contact with the wind to move in a direction 45 degrees to the right of the wind direction (Figure 7.7, *purple arrow*). In the Southern Hemisphere, where Coriolis curvature is to the left, the surface layer moves 45 degrees to the left of the wind direction. The surface water moves as a thin layer on top of deeper layers of water. As the surface layer moves, other layers beneath it are set in motion, thus passing the energy of the wind down through the water column. This is similar to the way a deck of cards can be fanned out by pressing on and rotating only the top card in the deck.

Current speed decreases with increasing depth, however, and the Coriolis effect increases curvature to the right (like a spiral). Thus, each successive layer of water is set in motion at a progressively slower speed and in a direction progressively to the right of the one above it. In Figure 7.7, for example, the purple surface current arrow sets the water in motion beneath it, which is curved more

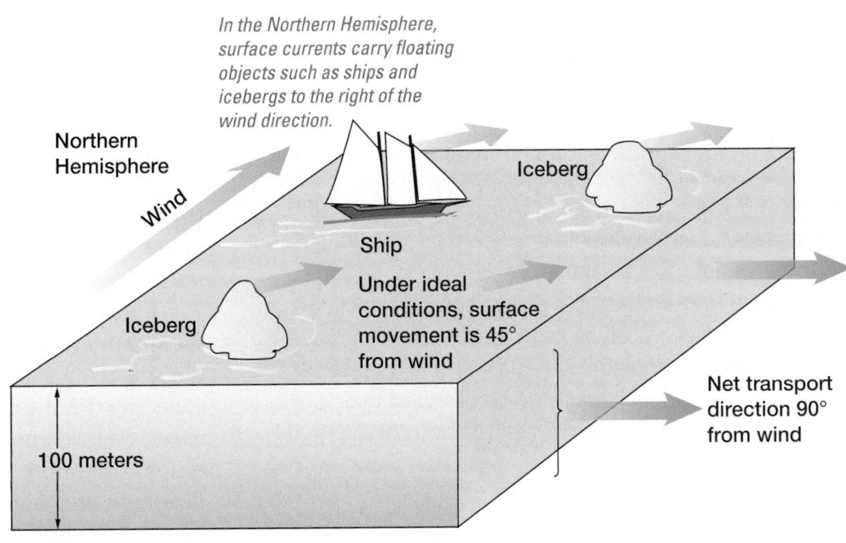


**Figure 7.6 Transport of floating objects is to the right of the wind direction in the Northern Hemisphere.**

**Animation**
Ekman Spiral and Ekman Transport
http://goo.gl/awYLBb

**SmartFigure 7.7 The Ekman spiral produces Ekman transport.** **(a)** Perspective view and **(b)** top view of the same area in the Northern Hemisphere showing how the Ekman spiral produces Ekman transport.
https://goo.gl/Tn7Upl

**EXPLORING DATA** ▼ **Describe how the Ekman spiral and Ekman transport are related to one another.**

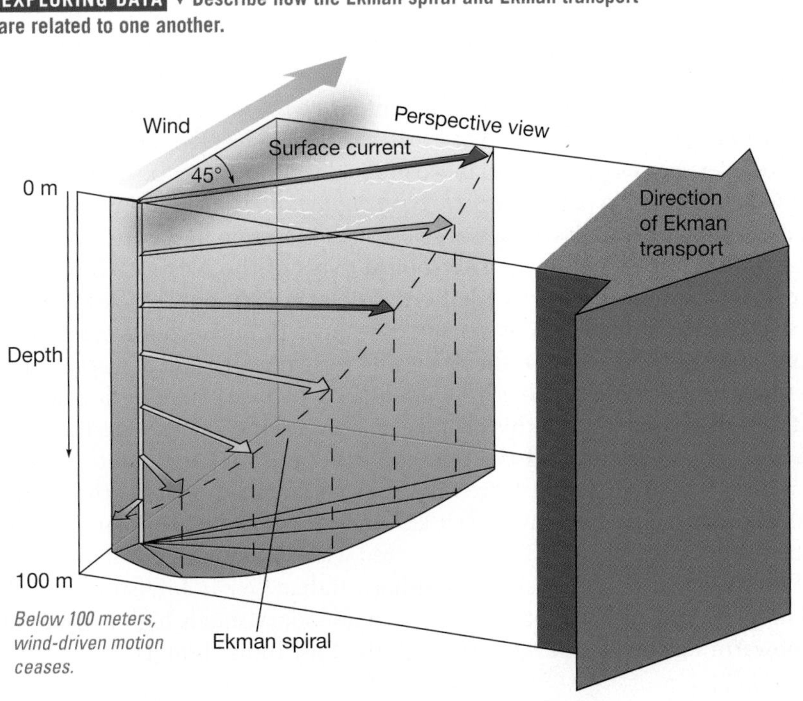

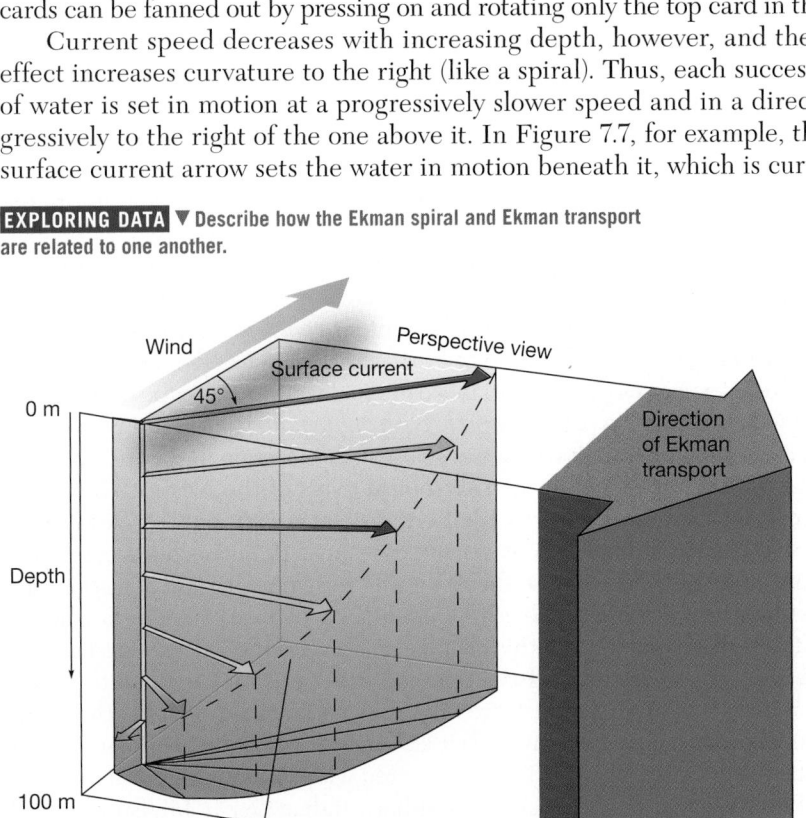

**(a)** Wind drives surface water in a direction 45 degrees to the right of the wind in the Northern Hemisphere. Deeper water continues to deflect to the right and moves at a slower speed with increased depth, causing the Ekman spiral.

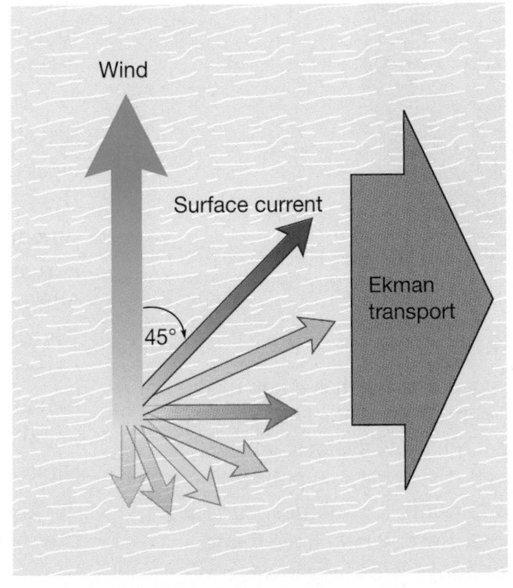


**(b)** Ekman transport, which is the average water movement for the entire column, is at a right angle (90 degrees) to the wind direction.

## STUDENTS SOMETIMES ASK . . .

***What does an Ekman spiral look like at the surface? Is it strong enough to disturb ships?***

The Ekman spiral creates different layers of surface water that move in slightly different directions at slightly different speeds. It is too weak to create eddies or whirlpools (vortices; *vortex* = to turn) at the surface and so presents no danger to ships. In fact, the Ekman spiral is unnoticeable at the surface. It can be observed, however, by lowering oceanographic equipment over the side of a vessel. At various depths, the equipment can be observed to drift at various angles from the wind direction according to the Ekman spiral.

to the right but moves more slowly and is represented by the shorter pink arrow. In turn, the pink arrow sets the water in motion beneath it, which is curved more to the right but moves more slowly and is represented by the even shorter gray arrow, and so on down the water column. Deeper in the ocean, a layer of water actually exists that moves in a direction *exactly opposite from the wind direction that initiated it*! (See Figure 7.7, *small orange arrow*.) If the water is deep enough, friction will consume the energy imparted by the wind, and no motion will occur below that depth. Although it depends on wind speed and latitude, this stillness normally occurs at a depth of about 100 meters (330 feet).

Figure 7.7 shows the spiral nature of this movement with increasing depth from the ocean's surface. The length of each colored arrow in Figure 7.7 is proportional to the speed of each individual thin layer, and the direction of each colored arrow indicates the direction it moves. (Note that the name Ekman *spiral* refers to the spiral observed by connecting the tips of the arrows shown in Figure 7.7.) Under ideal conditions, therefore, the surface layer should move at an angle of 45 degrees from the direction of the wind (Figure 7.7, *purple arrow*). All the layers combine, however, to create a net water movement that is 90 degrees from the direction of the wind. This average movement, called **Ekman transport**, is 90 degrees to the *right* in the Northern Hemisphere and 90 degrees to the *left* in the Southern Hemisphere.

"Ideal" conditions rarely exist in the ocean, so the actual movement of surface currents deviates slightly from the angles shown in Figure 7.7. As Nansen originally observed, surface currents generally move at an angle somewhat less than 45 degrees from the direction of the wind; Ekman transport in the open ocean is typically about 70 degrees from the wind direction. In shallow coastal waters, Ekman transport may be very nearly the same direction as the wind.

**GEOSTROPHIC CURRENTS** Ekman transport deflects surface water to the right in the Northern Hemisphere, so a clockwise rotation develops within an ocean basin and produces the **Subtropical Convergence** of water in the middle of the gyre, causing water literally to pile up in the center of the subtropical gyre. Thus, there is a hill of water within all subtropical gyres that is as much as 2 meters (6.6 feet) high.

Surface water in the Subtropical Convergence tends to flow downhill in response to gravity. The Coriolis effect opposes gravity, however, deflecting the water to the right in a curved path (**Figure 7.8a**) into the hill again. When these two factors balance, the net effect is a **geostrophic current** (*geo* = earth, *strophio* = turn) that moves in a circular path around the hill and is shown in Figure 7.8a as the *path of ideal geostrophic flow*. (The term *geostrophic* for these currents is appropriate, since the currents behave as they do because of Earth's rotation.) Friction between water molecules, however, causes the water to move gradually down the slope of the hill as it flows around it. This is the path of *actual geostrophic flow* labeled in Figure 7.8a.

If you reexamine the satellite image of sea surface elevation in Figure 7.2, you'll see that the hills of water within the subtropical gyres of the Atlantic Ocean are clearly visible. The hill in the North Pacific is visible as well, but the elevation of the equatorial Pacific is not as low as expected because the map shows conditions during a moderate El Niño event, so there is a well-developed region of warm water across the equatorial Pacific that has an anomalously high sea surface height. (El Niño events are discussed later in this chapter, under "Pacific Ocean Circulation.") Figure 7.2 also shows very little distinction between the North and South Pacific Gyres. Moreover, the South Pacific Gyre hill is less pronounced than in other gyres because (1) it covers such a large area, (2) it lacks confinement by continental barriers along its western margin, and (3) it is interfered with by numerous islands (really the tops of tall sea floor mountains). The southern Indian Ocean hill is rather well developed in the figure, although its northeastern boundary stands high because of the influx of warm Pacific Ocean water through the East Indies islands.

**WESTERN INTENSIFICATION OF SUBTROPICAL GYRES** Figure 7.8a shows that the apex (top) of the hill formed within a rotating gyre is closer to the western boundary

## CREATURE FEATURE 7.1

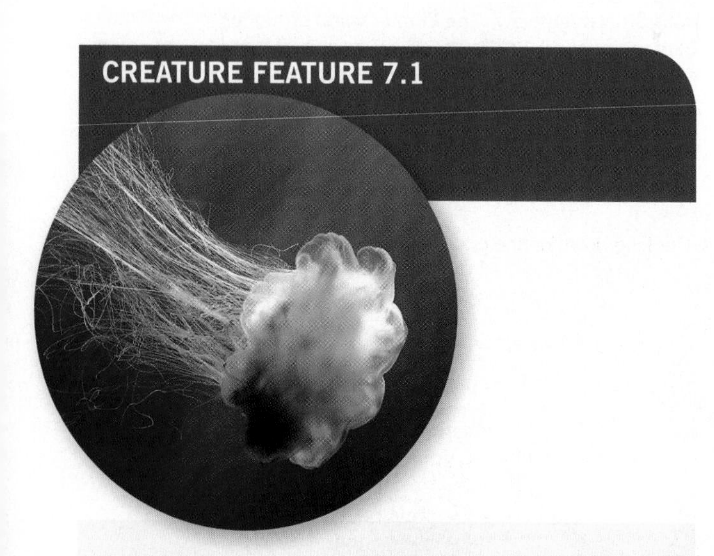

### Caution: I can sting you!

The **lion's mane jelly** (*Cyanea capillata*) has a bell (body) up to 2.3 meters (7.5 feet) across and tentacles 37 meters (120 feet) long, making it the world's largest jelly.

Jellies can't swim well (they can only pulsate) and so are carried by ocean currents; sometimes, currents bring them close to shore. Researchers have found that the best remedy for stings is a vinegar wash followed by a heat pack.

than the geographic center of the gyre. As a result, the western boundary currents of the subtropical gyres are faster, narrower, and deeper than their eastern boundary current counterparts. For example, the Kuroshio Current (a western boundary current) of the North Pacific Gyre is up to 15 times faster, 20 times narrower, and 5 times deeper than the California Current (an eastern boundary current). This phenomenon is called **western intensification**, and currents affected by this phenomenon are said to be *western intensified*. Note that the western boundary currents of *all* subtropical gyres are western intensified, *even in the Southern Hemisphere*.

A number of factors cause western intensification, including the Coriolis effect. The Coriolis effect increases toward the poles, so eastward-flowing high-latitude water turns toward the equator more strongly than westward-flowing equatorial water turns toward higher latitudes. This causes a wide, slow, and shallow flow of water toward the equator across most of each subtropical gyre, leaving only a narrow band between land and the western margin of the ocean basin through which the poleward flow can occur. If a constant volume of water rotates around the apex of the hill in **Figure 7.8b**, then the velocity of the water along the western margin will be much greater than the velocity around the eastern side. A good analogy for this phenomenon is a funnel: in the narrow end of a funnel, the flow rates are speeded up (such as in western intensified currents); in the wide end, the flow rates are sluggish (such as in eastern boundary currents). In Figure 7.8b, the lines are close together along the western margin, indicating the faster flow. The end result is a high-speed western boundary current that flows along the hill's steeper westward slope and a slow drift of water toward the equator along the more gradual eastern slope. **Table 7.2** summarizes the differences between western and eastern boundary currents of subtropical gyres.

Northern Hemisphere Subtropical Gyre

**(a)** Perspective view of a subtropical gyre showing how water literally piles up in the center, forming a hill up to 2 meters (6.6 feet) high. Ideally, gravity and the Coriolis effect balance each other to create an ideal geostrophic current that flows in equilibrium around the hill. However, friction makes the current gradually run downslope *(path of actual geostrophic flow)*.

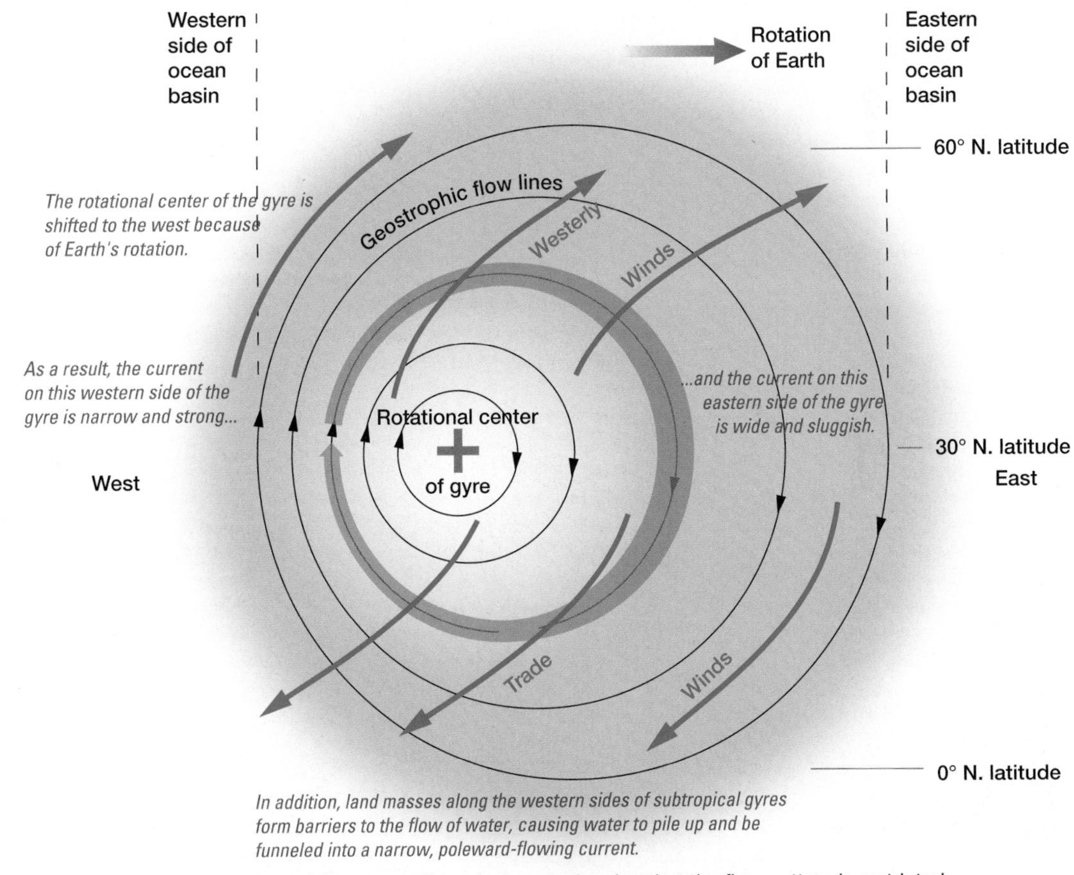


**(b)** Corresponding map view of the same subtropical gyre, showing that the flow pattern is restricted (lines are closer together) on the western side of the gyre, resulting in western intensification.

Figure 7.8 **Geostrophic current and western intensification.** **(a)** Perspective view and **(b)** map view of a subtropical gyre showing how the center of rotation of the gyre is shifted to the west by Earth's rotation and creates the western intensification of currents.

## Ocean Currents and Climate

Ocean surface currents directly influence the climate of adjoining landmasses. For instance, warm ocean currents warm the nearby air. This warm air can hold a large amount of water vapor, which puts more moisture (high humidity) in the atmosphere. When this warm, moist air travels over a continent, it releases its water vapor in the form

Climate
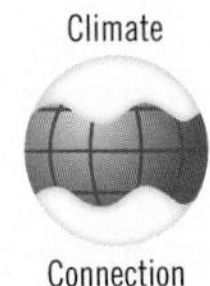
Connection

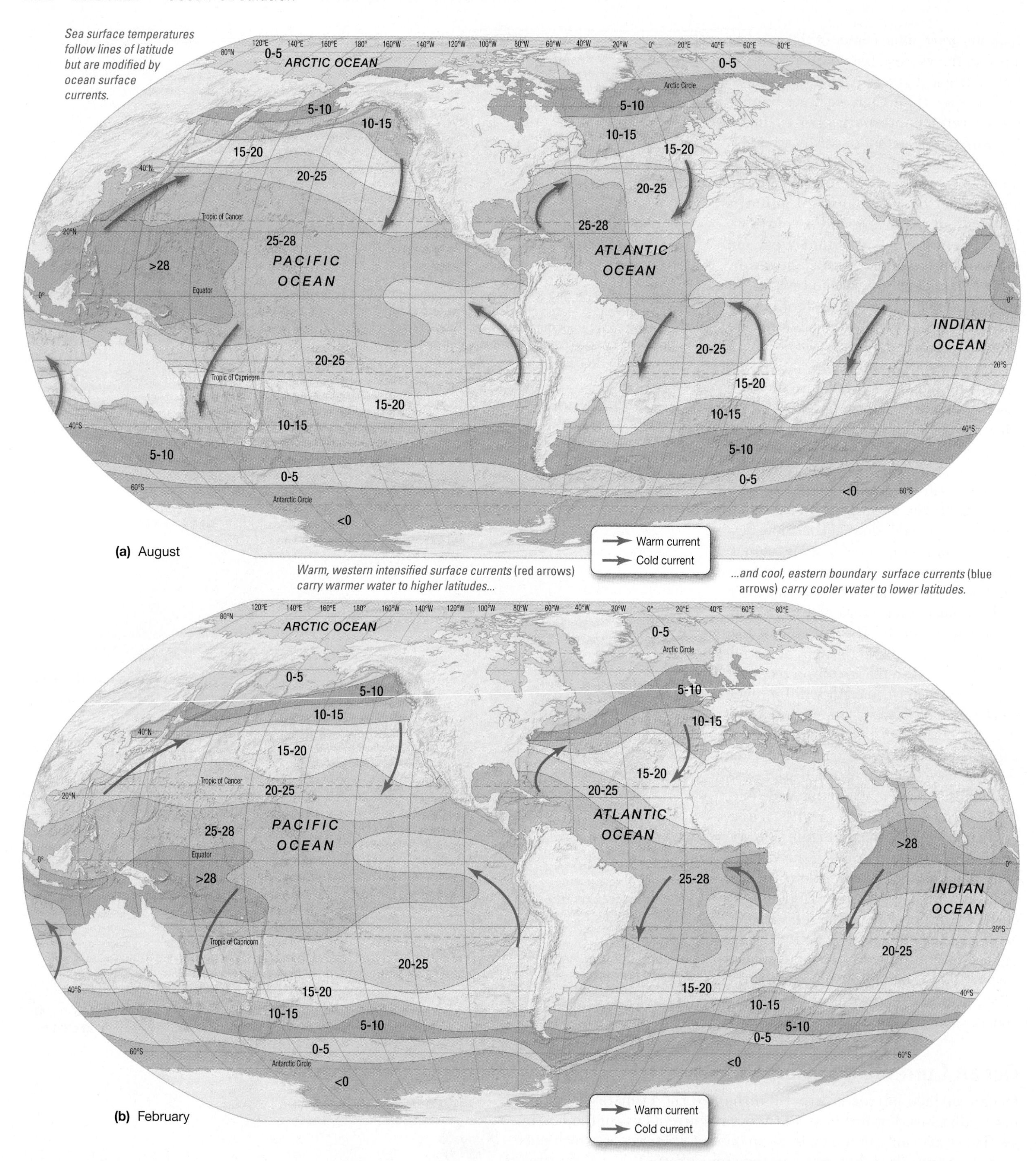
Sea surface temperatures follow lines of latitude but are modified by ocean surface currents.
ARCTIC OCEAN
PACIFIC OCEAN
ATLANTIC OCEAN
INDIAN OCEAN
Arctic Circle
Tropic of Cancer
Equator
Tropic of Capricorn
Antarctic Circle
0-5
5-10
10-15
15-20
20-25
25-28
>28
<0
(a) August
Warm current
Cold current
Warm, western intensified surface currents (red arrows) carry warmer water to higher latitudes...
...and cool, eastern boundary surface currents (blue arrows) carry cooler water to lower latitudes.
(b) February

Figure 7.9 **Sea surface temperature of the world ocean.** Paired maps showing average sea surface temperature in degrees centigrade for **(a)** August and **(b)** February. A comparison of the maps shows that sea surface temperatures migrate north–south with the seasons.

**EXPLORING DATA ▲ Explain why ocean temperatures, which are represented by different colors in the figure, don't follow lines of latitude that continue horizontally across the maps.**

**SmartTable 7.2 Characteristics of western and eastern boundary currents of subtropical gyres**

| Current type | Examples | Width | Depth | Speed | Transport volume (millions of cubic meters per second[a]) | Comments |
|---|---|---|---|---|---|---|
| **Western boundary current** | Gulf Stream, Brazil Current, Kuroshio Current | *Narrow:* usually less than 100 kilometers (60 miles) | *Deep:* to depths of 2 kilometers (1.2 miles) | *Fast:* hundreds of kilometers per day | *Large:* as much as 100 Sv[a] | Waters derived from low latitudes and are warm; little or no upwelling |
| **Eastern boundary current** | Canary Current, Benguela Current, California Current | *Wide:* up to 1000 kilometers (600 miles) | *Shallow:* to depths of 0.5 kilometer (0.3 mile) | *Slow:* tens of kilometers per day | *Small:* typically 10 to 15 Sv[a] | Waters derived from middle latitudes and are cool; coastal upwelling common |

[a]One million cubic meters (35.3 million cubic feet) per second is a flow rate equal to one Sverdrup (Sv).

of precipitation. Continental margins that have warm ocean currents offshore (**Figure 7.9**, *red arrows*) typically have a humid climate. The presence of a warm current off the East Coast of the United States helps explain why the area experiences such high humidity, especially in the summer.

Conversely, cold ocean currents cool the nearby air, which is more likely to have low water vapor content. When the cool, dry air travels over a continent, it results in very little precipitation. Continental margins that have cool ocean currents offshore (Figure 7.9, *blue arrows*) typically have a dry climate. The presence of a cold current off California is part of the reason the climate there is so arid.

**SmartTable 7.2 Characteristics of western and eastern boundary currents of subtropical gyres.** https://goo.gl/P2G56w

**RECAP** **Western intensification is a result of Earth's rotation and causes the western boundary currents of all subtropical gyres to be fast, narrow, and deep.**

**CONCEPT CHECK 7.2 ▸ Explain the origin of ocean surface currents and how surface circulation patterns are organized globally.**

1 How many subtropical gyres exist worldwide? How many main currents exist within each subtropical gyre?

2 On a base map of the world, plot and label the major currents involved in the surface circulation gyres of the oceans. Use colors to represent warm versus cool currents and indicate which currents are western intensified. On an overlay, superimpose the major wind belts of the world on the gyres and describe the relationship between wind belts and currents.

3 Explain why the subtropical gyres in the Northern Hemisphere move in a clockwise fashion while the subpolar gyres rotate in a counterclockwise pattern.

4 Diagram and discuss how Ekman transport produces the "hill" of water within subtropical gyres that causes geostrophic current flow. As a starting place on the diagram, use the wind belts (the trade winds and the prevailing westerlies). What causes the apex of the geostrophic "hills" to be offset to the west of the center of the ocean gyre systems?

5 Describe western intensification, including the characteristics of western and eastern boundary currents of subtropical gyres.

## 7.3 ▸ What Causes Upwelling and Downwelling?

**Upwelling** is the upward movement of cold, deep, nutrient-rich water to the surface; **downwelling** is the downward movement of surface water to deeper parts of the ocean. Upwelling hoists chilled water to the surface. This cold water, rich in nutrients, creates high **productivity** (an abundance of microscopic algae), which establishes the base of the food web and, in turn, supports incredible numbers of larger marine life like fish and whales. Downwelling, on the other hand, is associated with much lower amounts of surface productivity but carries necessary dissolved oxygen to those organisms living on the deep-sea floor.

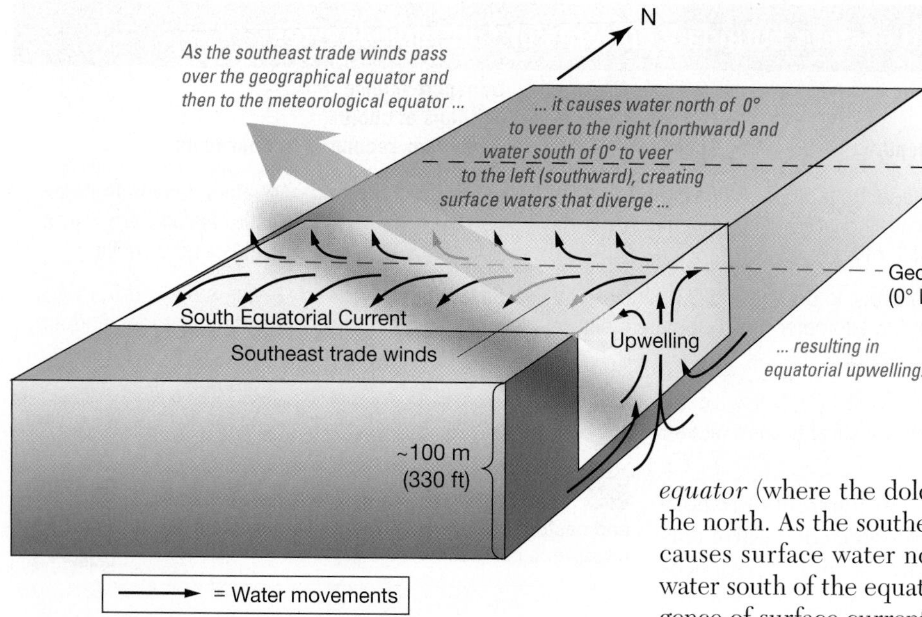


**Figure 7.10 Equatorial upwelling is produced by the southeast trade winds, which cause surface waters to diverge.**

Upwelling and downwelling provide important mixing mechanisms between surface and deep waters, and are the result of a variety of processes.

## Diverging Surface Water

Current divergence occurs when surface waters move *away from* an area on the ocean's surface, such as along the equator. As shown in **Figure 7.10**, the South Equatorial Current occupies the area along the *geographical equator* (most notably in the Pacific Ocean; see Figure 7.5), while the *meteorological equator* (where the doldrums exist) typically occurs a few degrees of latitude to the north. As the southeast trade winds blow across this region, Ekman transport causes surface water north of the equator to veer to the right (northward) and water south of the equator to veer to the left (southward). The net result is a divergence of surface currents along the geographical equator, which causes upwelling of cold, nutrient-rich water. Because this type of upwelling is common along the equator—especially in the Pacific—it is called **equatorial upwelling**, and it creates areas of high productivity that are some of the most prolific fishing grounds in the world.

## Converging Surface Water

Current convergence occurs when surface waters move *toward* each other. In the North Atlantic Ocean, for instance, the Gulf Stream, the Labrador Current, and the East Greenland Current all come together in the same region of the ocean. When currents converge, water stacks up and has no place to go but downward. Under the extra weight of the water that has piled up, the surface water slowly sinks downward in a process called *downwelling* (**Figure 7.11**). Unlike upwelling, areas of downwelling are not associated with prolific marine life because the necessary nutrients are not continuously replenished from cold, nutrient-rich deep water below. Consequently, downwelling areas have low productivity.

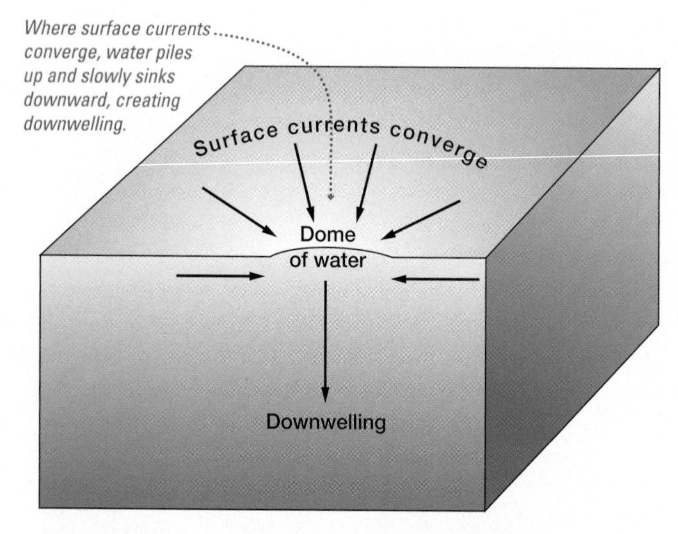


**Figure 7.11 Downwelling caused by convergence of surface currents.**

## Coastal Upwelling and Downwelling

Coastal winds can cause upwelling or downwelling due to Ekman transport. **Figure 7.12** shows a coastal region along the west coast of a continent in the Northern Hemisphere with winds moving parallel to the coast. If the winds are from the north (Figure 7.12b), Ekman transport moves the coastal water to the right of the wind direction, causing the water to flow *away from* the shoreline. Water rises from below to replace the water moving away from shore in a process called **coastal upwelling**. Areas where coastal upwelling occurs, such as the West Coast of the United States, are characterized by high concentrations of nutrients, resulting in high productivity and rich marine life. This coastal upwelling also creates low water temperatures in areas such as San Francisco, providing a natural form of air conditioning (and much cool weather and fog) in the summer.

If the winds are from the south, Figure 7.12b shows that Ekman transport still moves the coastal water to the right of the wind direction but, in this case, the water

**Animation**
Ekman Spiral and Coastal Upwelling/Downwelling
http://goo.gl/UIFbuF

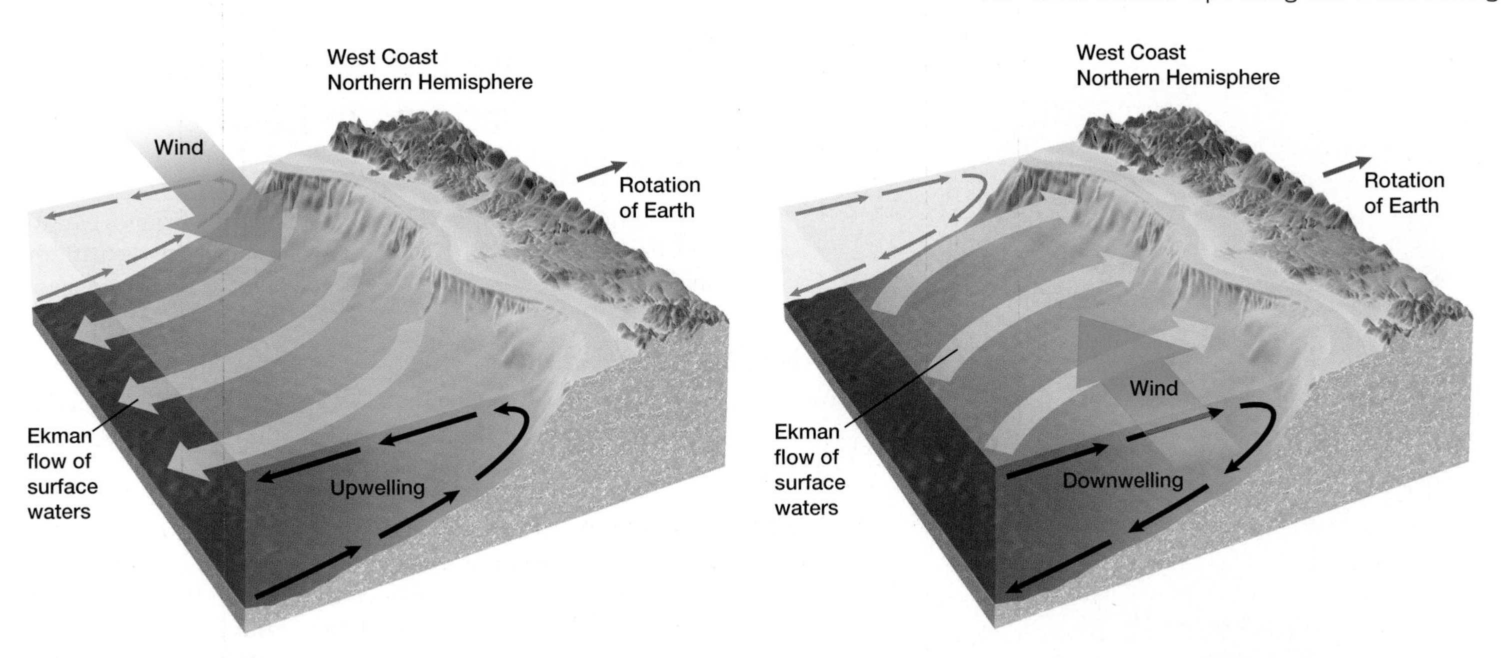


SmartFigure 7.12 **Coastal upwelling and downwelling.** Coastal upwelling and downwelling can be produced by winds that blow parallel to the coast. Similar situations exist for coastal winds and upwelling/downwelling in the Southern Hemisphere, except that Ekman transport is to the *left* of the wind direction. https://goo.gl/gVYzp5

**EXPLORING DATA** ▲ **Draw similar diagrams for the west coasts of the Southern Hemisphere, showing the same wind directions that parallel the coast, and describe differences in Ekman transport and upwelling/downwelling.**

flows *toward* the shoreline. The water stacks up along the shoreline and has nowhere to go but down, in a process called **coastal downwelling**. Areas where coastal downwelling occurs have low productivity and a lack of marine life. If the winds reverse, areas that are typically associated with coastal downwelling can experience upwelling.

A similar situation exists for coastal winds and upwelling/downwelling in the Southern Hemisphere, except that Ekman transport is to the *left* of the wind direction.

## Other Causes of Upwelling

Figure 7.13 shows how upwelling can be created by offshore winds, sea floor obstructions, or a sharp bend in a coastline. Upwelling also occurs in high-latitude regions, where there is no pycnocline (a layer of rapidly changing density). The absence of a pycnocline allows significant vertical mixing between high-density cold surface water and high-density cold deep water below. Thus, both upwelling and downwelling are common in high latitudes.

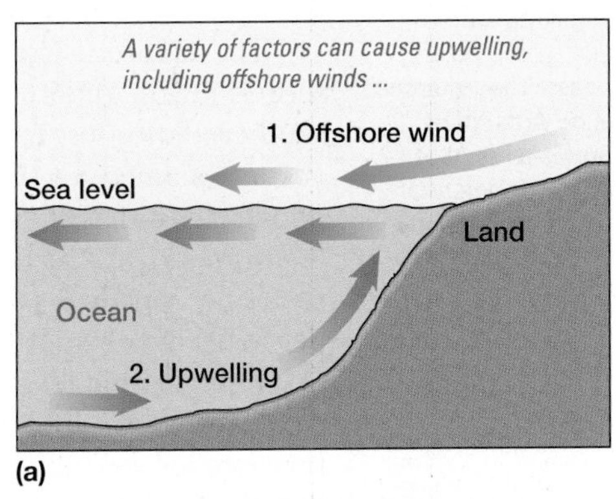


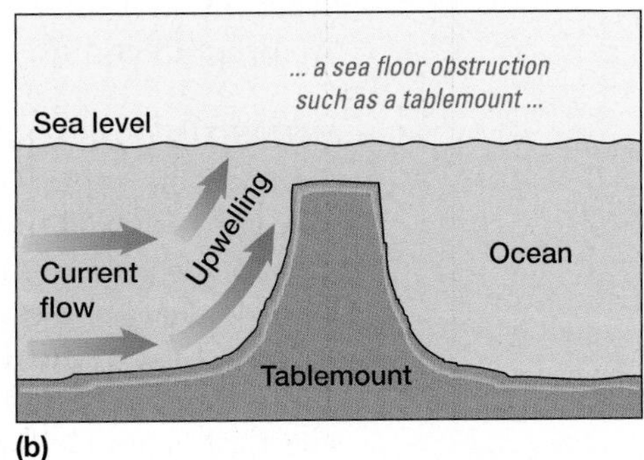


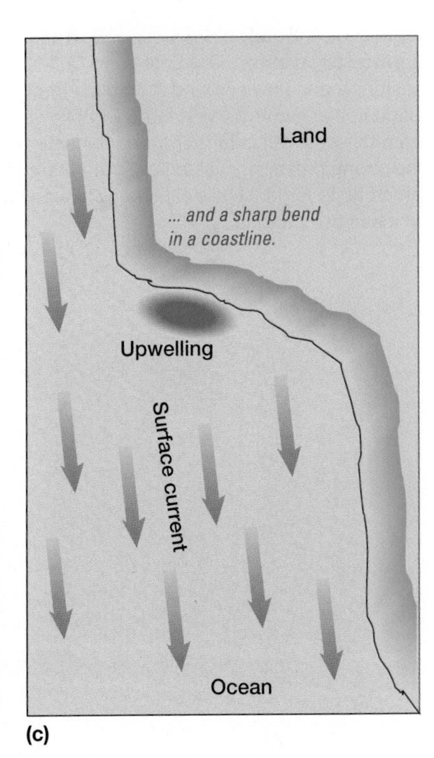


Figure 7.13 **Examples of other types of upwelling.**

**CONCEPT CHECK 7.3 ▸ Describe the conditions that produce upwelling.**

1 Draw and describe several different oceanographic conditions that produce upwelling.

2 Explain why upwelling areas are associated with an abundance of marine life.

**RECAP Upwelling and downwelling cause vertical mixing between surface and deep water. Upwelling brings cold, deep, nutrient-rich water to the surface, which results in high productivity.**

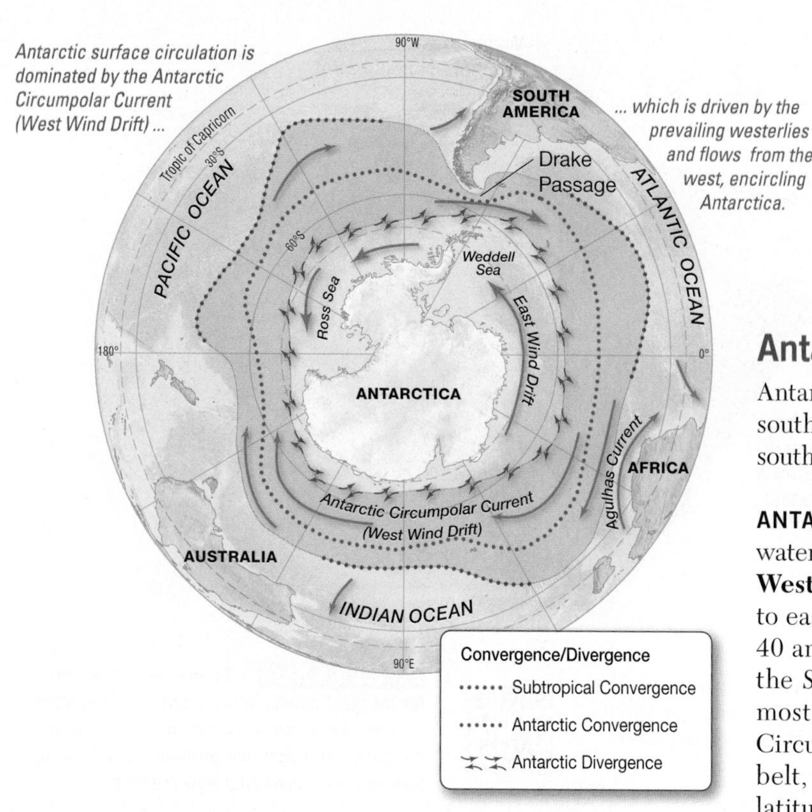


**Figure 7.14 South polar view of Earth showing Antarctic surface circulation.** The East Wind Drift is driven by the polar easterlies and flows around Antarctica from the east. The Antarctic Circumpolar Current (West Wind Drift) flows around Antarctica from the west but is further from the continent and is a result of the strong prevailing westerlies. The Antarctic Convergence and Antarctic Divergence are caused by interactions at the boundaries of these two currents.

# 7.4 ▸ What Are the Main Surface Circulation Patterns in Each Ocean Basin?

The specific pattern of surface currents varies from ocean to ocean, depending on the geometry of the ocean basin, the pattern of major wind belts, seasonal factors, and other periodic changes.

## Antarctic Circulation

Antarctic circulation is dominated by the movement of water masses in the southern Atlantic, Indian, and Pacific Oceans south of about 50 degrees south latitude.

**ANTARCTIC CIRCUMPOLAR CURRENT** The main current in Antarctic waters is the **Antarctic Circumpolar Current**, which is also called the **West Wind Drift**. This current encircles Antarctica and flows from west to east at approximately 50 degrees south latitude but can vary between 40 and 65 degrees south latitude. At about 40 degrees south latitude is the *Subtropical Convergence* (**Figure 7.14**), which forms the northernmost boundary of the Antarctic Circumpolar Current. The Antarctic Circumpolar Current is driven by the powerful prevailing westerly wind belt, which creates winds so strong that these Southern Hemisphere latitudes have been called the "Roaring Forties," "Furious Fifties," and "Screaming Sixties."

The Antarctic Circumpolar Current is the only current that completely circumscribes Earth, and it is allowed to do so because of the lack of land barriers at high southern latitudes. It meets its greatest restriction as it passes through the *Drake Passage* (named for the famed English sea captain and ocean explorer Sir Francis Drake [1540–1596]) between the Antarctic Peninsula and the southern islands of South America, which is about 1000 kilometers (600 miles) wide. Although the current is not speedy (its maximum surface velocity is about 2.75 kilometers [1.65 miles] per hour), it transports on average about 130 Sverdrups (130 million cubic meters [4.6 billion cubic feet] per second), which is more than any other surface current. Note that one million cubic meters (35.3 million cubic feet) per second is a useful flow rate for describing ocean currents, so it has become a standard unit, named the **Sverdrup (Sv)**, after Norwegian meteorologist and physical oceanographer Harald Sverdrup (1888–1957).

**ANTARCTIC CONVERGENCE AND DIVERGENCE** The **Antarctic Convergence** (Figure 7.14), or *Antarctic Polar Front*, is where colder, denser Antarctic waters converge with (and sink sharply below) warmer, less-dense sub-Antarctic waters at about 50 degrees south latitude. The Antarctic Convergence marks the northernmost boundary of the Southern, or Antarctic, Ocean.

The **East Wind Drift**, a surface current propelled by the polar easterlies, moves from an easterly direction around the margin of the Antarctic continent. The East Wind Drift is most extensively developed to the east of the Antarctic Peninsula in the Weddell Sea region and in the area of the Ross Sea (Figure 7.14). As the East Wind Drift and the Antarctic Circumpolar Current (West Wind Drift) flow around Antarctica in opposite directions, they create a surface divergence. Recall that the Coriolis effect deflects moving masses to the left in the Southern Hemisphere, so the East Wind Drift is deflected toward the continent, and the Antarctic Circumpolar Current is deflected away from it. This creates a divergence of currents along a boundary called the **Antarctic Divergence**. The Antarctic Divergence has abundant marine life in the Southern Hemisphere summer because of the mixing of these two currents, which supplies nutrient-rich water to the surface through upwelling.

## Atlantic Ocean Circulation

**Figure 7.15** shows Atlantic Ocean surface circulation, which consists of two large subtropical gyres: the North Atlantic Gyre and the South Atlantic Gyre.

**THE NORTH AND SOUTH ATLANTIC SUBTROPICAL GYRES** The **North Atlantic Subtropical Gyre** rotates clockwise, and the **South Atlantic Subtropical Gyre** rotates counterclockwise, due to the combined effects of the trade winds, the prevailing westerlies, and the Coriolis effect. Figure 7.15 shows that each gyre consists of a poleward-moving warm current (*red*) and an equatorward-moving cold "return" current (*blue*). The two gyres are partially offset by the shapes of the surrounding continents, and the **Atlantic Equatorial Countercurrent** moves in between them.

In the South Atlantic Gyre, the **South Equatorial Current** reaches its greatest strength just below the equator, where it encounters the coast of Brazil and splits in two. Part of the South Equatorial Current moves off along the northeastern coast of South America toward the Caribbean Sea and the North Atlantic. The rest is turned southward as the **Brazil Current**, which ultimately merges with the Antarctic Circumpolar Current (West Wind Drift) and moves eastward across the South Atlantic. The Brazil Current is much smaller than its Northern Hemisphere counterpart, the Gulf Stream, due to the splitting of the South Equatorial Current. The **Benguela Current**, slow moving and cold, flows toward the equator along Africa's western coast, completing the gyre.

Outside the gyre, the *Falkland Current* (Figure 7.15), which is also called the *Malvinas Current*, moves a significant amount of cold water along the coast of Argentina as far north as 25 to 30 degrees south latitude, wedging its way between the continent and the southbound Brazil Current.

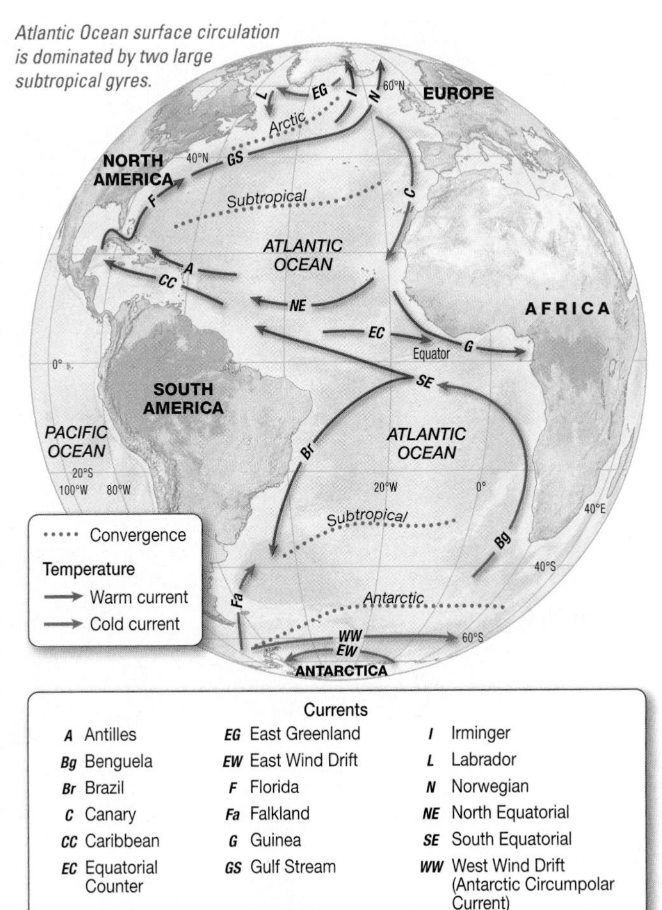


**Figure 7.15 Atlantic Ocean surface currents.**

**THE GULF STREAM** The **Gulf Stream** is the world's best studied ocean current. It moves northward along the East Coast of the United States, warming coastal states and moderating winters in these and northern European regions.

**Figure 7.16** shows the network of currents in the North Atlantic Ocean that contribute to the flow of the Gulf Stream. The **North Equatorial Current** moves parallel to the equator in the Northern Hemisphere, where it is joined by the portion of the South Equatorial Current that turns northward along the South American coast. This flow then splits into the **Antilles Current**, which passes along the Atlantic side of the West Indies, and the **Caribbean Current**, which passes through the Yucatán Channel into the Gulf of Mexico. These water masses reconverge to form the **Florida Current**.

The Florida Current flows close to shore over the continental shelf at a rate that at times exceeds 35 Sverdrups. As it moves off North Carolina's Cape Hatteras and flows across the deep ocean in a northeasterly direction, it is called the *Gulf Stream*. The Gulf Stream is a western boundary current, so it is subject to western intensification. Thus, it is only 50 to 75 kilometers (31 to 47 miles) wide, but it reaches depths of 1.5 kilometers (1 mile) and speeds from 3 to 10 kilometers (2 to 6 miles) per hour, making it the world's fastest ocean current.

The western boundary of the Gulf Stream is usually abrupt, but it periodically migrates closer to and farther away from the shore. Its eastern boundary is very difficult to identify because it is usually masked by circular and meandering water flows that continuously change their positions.

***The Sargasso Sea*** The Gulf Stream gradually merges eastward with the water of the **Sargasso Sea**. The Sargasso Sea is the water that circulates around the North Atlantic Gyre's center of rotation, which is shifted to the west because of Earth's rotation. The Sargasso Sea can be thought of as the stagnant eddy on the western side of the North Atlantic Gyre. Its name is derived from a type of floating marine alga called *sargassum* (*sargassum* = grapes) that abounds on its surface.

## STUDENTS SOMETIMES ASK . . .

### *Is the Gulf Stream rich in life?*

The Gulf Stream *itself* isn't rich in life, but its *boundaries* often are. The oceanic areas that have abundant marine life are typically associated with cool water—either in high-latitude regions, or in any region where upwelling occurs. These areas are constantly resupplied with oxygen- and nutrient-rich water, which results in high productivity. Warm-water areas develop a prominent thermocline that isolates the surface water from colder, nutrient-rich water below. Nutrients used up in warm waters tend not to be resupplied. The Gulf Stream, which is a western intensified, warm-water current, is therefore associated with low productivity and an absence of marine life. New England fishers knew about the Gulf Stream (see **Diving Deeper 7.2**) because they sought their catch along the sides of the current, where mixing and upwelling occur.

Actually, all western intensified currents are warm and are associated with low productivity. The Kuroshio Current in the North Pacific Ocean, for example, is named for its conspicuous absence of marine life. In Japanese, *Kuroshio* means "black current," in reference to its clear, lifeless waters.

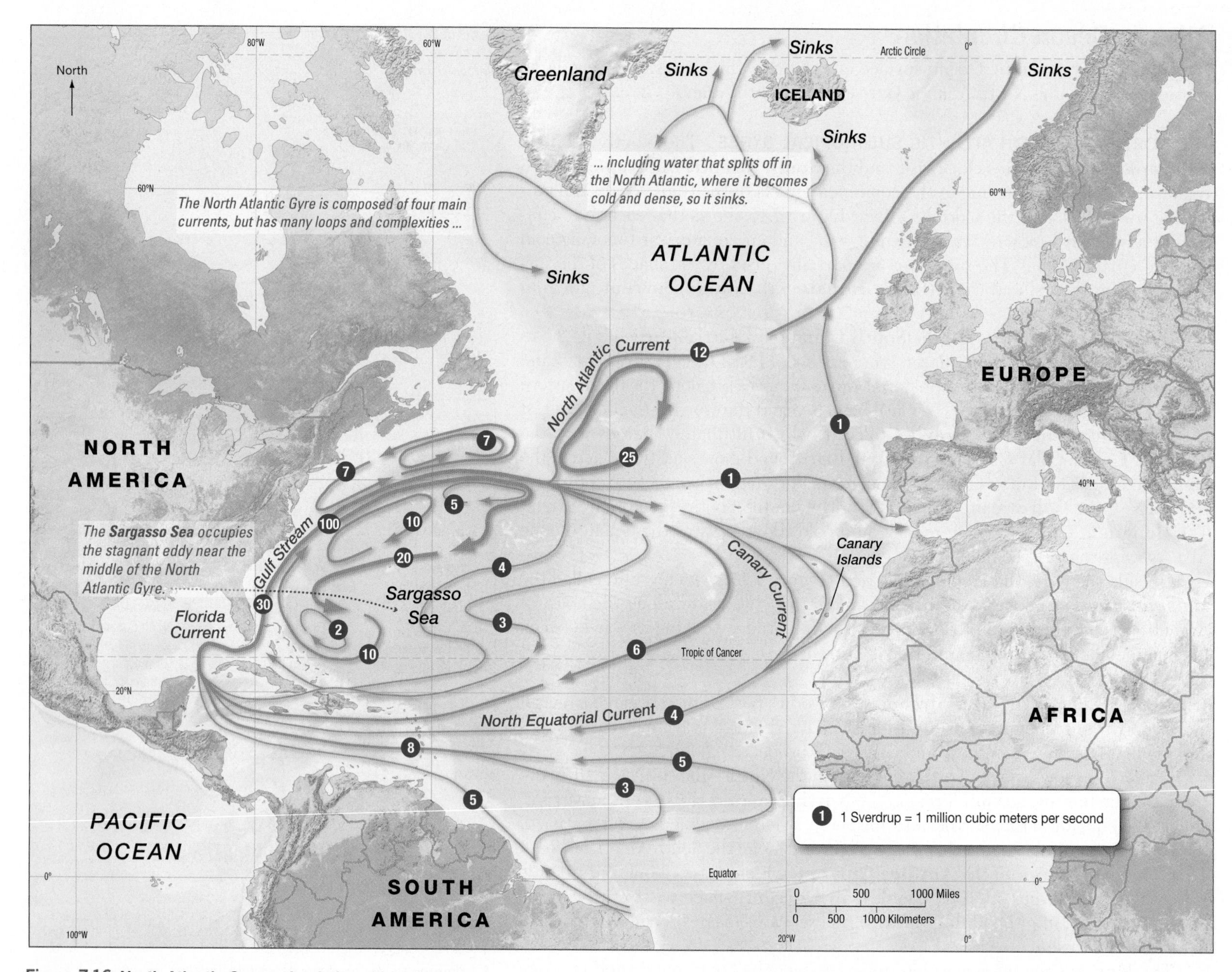


Figure 7.16 **North Atlantic Ocean circulation.** Map of major currents in the North Atlantic Gyre, showing average flow rates in *Sverdrups* (1 Sverdrup = 1 million cubic meters [35.3 million cubic feet] per second). The four major currents include the western intensified Gulf Stream, the North Atlantic Current, the Canary Current, and the North Equatorial Current, but there are many complex flow patterns.

**EXPLORING DATA ▲**

**1. According to the figure, what is the maximum flow rate in Sverdrups for the Gulf Stream? What is the flow rate in Sverdrups for the Canary Current? (Hint: estimate!)**

**2. According to the figure, what happens to surface waters once they reach the vicinity of Iceland? Using the physical properties of water, explain why the water does this.**

The transport rate of the Gulf Stream off Chesapeake Bay is about 100 Sverdrups, which suggests that a large volume of water from the Sargasso Sea has combined with the Florida Current to produce the Gulf Stream. (The Gulf Stream's flow of 100 Sverdrups equates to a volume of about 100 major league sport stadiums passing by the southeast U.S. coast *each second* and is more than 100 times greater than the combined flow of *all* the world's rivers!) By the time the Gulf Stream nears Newfoundland, however, the transport rate is only 40 Sverdrups, which suggests that a large volume of water has rejoined the diffuse southerly flow of the Sargasso Sea.

***Warm- and Cold-Core Rings*** The mechanisms that produce the dramatic loss of water as the Gulf Stream moves northward are yet to be determined. Meanders, however, may cause much of it. **Meanders** (*Menderes* = a river in Turkey that has a very sinuous course) are snake-like bends in the current that often disconnect from the Gulf Stream and form large rotating masses of water called *vortices* (*vortex* = to turn), which are more commonly known as *eddies* or *rings*. **Figure 7.17** shows several of these rings, which are noticeable near the center of each image.

The figure also shows that meanders along the north boundary of the Gulf Stream pinch off and trap warm Sargasso Sea water in eddies that rotate clockwise, creating **warm-core rings** (*yellow*) surrounded by cooler (*blue and green*) water. These warm rings contain shallow, bowl-shaped masses of warm water about 1 kilometer (0.6 mile) deep, with diameters of about 100 kilometers (60 miles). Warm-core rings remove large volumes of water as they disconnect from the Gulf Stream.

Cold nearshore water spins off to the south of the Gulf Stream as counterclockwise-rotating **cold-core rings** (*green*) surrounded by warmer (*yellow* and *red-orange*) water (Figure 7.17). The cold rings consist of spinning cone-shaped masses of cold water that extend over 3.5 kilometers (2.2 miles) deep. These rings may exceed 500 kilometers (310 miles) in diameter at the surface. The diameter of the cone increases with depth and sometimes reaches all the way to the sea floor, where cones have a tremendous impact on sea floor sediment. Cold rings move southwest at speeds of 3 to 7 kilometers (2 to 4 miles) per day toward Cape Hatteras, where they often rejoin the Gulf Stream.

Because both warm- and cold-core rings maintain unique temperature characteristics, they often contain distinct populations of marine life. For example, studies of rings have found they are isolated habitats for either warm-water organisms in a cold ocean or, conversely, cold-water organisms in a warmer ocean. The organisms can survive as long as the ring does; in some cases, rings have been documented to last up to two years. In addition, cold-core rings are typically associated with high nutrient levels and an abundance of marine life, while warm-core rings are zones of downwelling that lack nutrients and are deficient in marine life.

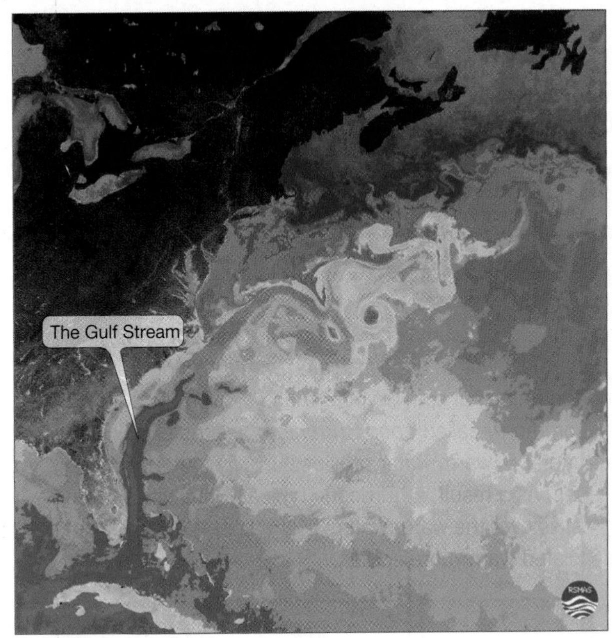


**(a)** The Gulf Stream is shown flowing along the U.S. East Coast in this NOAA satellite false-color image of sea surface temperature (warm waters = red and orange; cool waters = green, blue, purple, pink).

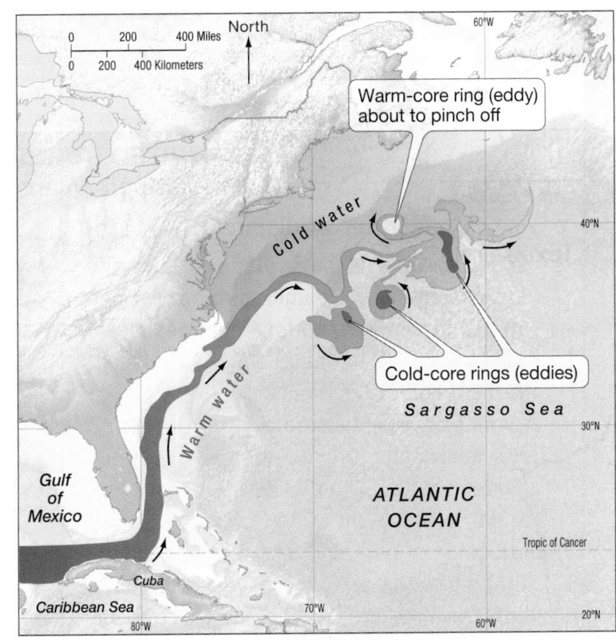


**(b)** Matching map of the same area as part a. As the Gulf Stream flows northward, some of its meanders pinch off and form either warm-core or cold-core rings.

Figure 7.17 **The Gulf Stream and sea surface temperatures. (a)** A NOAA satellite false-color image of sea surface temperature and **(b)** a matching map of the same area, showing the development of warm- and cold-core rings (eddies).

**OTHER NORTH ATLANTIC CURRENTS** Southeast of Newfoundland, the Gulf Stream continues in an easterly direction across the North Atlantic (Figure 7.16). Here the Gulf Stream breaks into numerous branches, many of which become cold and dense enough to sink beneath the surface. As shown in Figure 7.15, one major branch combines the cold water of the **Labrador Current** with the warm Gulf Stream, producing abundant fog in the North Atlantic. This branch eventually breaks into the **Irminger Current**, which flows along Iceland's west coast, and the **Norwegian Current**, which moves northward along Norway's coast. The other major branch crosses the North Atlantic as the **North Atlantic Current** (also called the *North Atlantic Drift*, emphasizing its sluggish nature), which turns southward to become the cool **Canary Current**. The Canary Current is a broad, diffuse, southward flow that eventually joins the North Equatorial Current, thus completing the gyre.

**CLIMATIC EFFECTS OF NORTH ATLANTIC CURRENTS** The warming effects of the Gulf Stream are far ranging. The Gulf Stream moderates temperatures not only along the East Coast of the United States but also in northern Europe (in conjunction with heat transferred by the atmosphere). Thus, the temperatures across the Atlantic at different latitudes are much higher in Europe than in North America because of the effects of heat transfer from the Gulf Stream to Europe. For example, Spain and Portugal have warm climates, even though they are at the same latitude as the New England states, which are known for severe winters. The warming that northern Europe experiences because of the Gulf Stream is as much as 9°C (20°F), which is enough to keep high-latitude Baltic ports ice-free throughout the year.

Climate Connection

The warming effects of western boundary currents in the North Atlantic Ocean can be seen on the average sea surface temperature map for February shown in Figure 7.9b. Off the east coast of North America from latitudes 20 degrees north (the latitude of Cuba) to 40 degrees north (the latitude of Philadelphia), for example, there is a 20°C (36°F) difference in sea surface temperatures. On the eastern side of the North Atlantic, on the other hand, there is only a 5°C (9°F) difference in temperature between the same latitudes, indicating the moderating effect of the Gulf Stream.

**Animation**
The Gulf Stream: Its Meanders and Cold- and Warm-Core Eddies
https://goo.gl/LwqnV4

**EXPLORING DATA** ▼

1. **Carefully compare normal conditions (part *a*) with El Niño conditions (part *b*). Describe at least eight separate items that are different between the two diagrams.**
2. **Carefully compare normal conditions (part *a*) with La Niña conditions (part *c*). Describe at least eight separate items that are different between the two diagrams.**

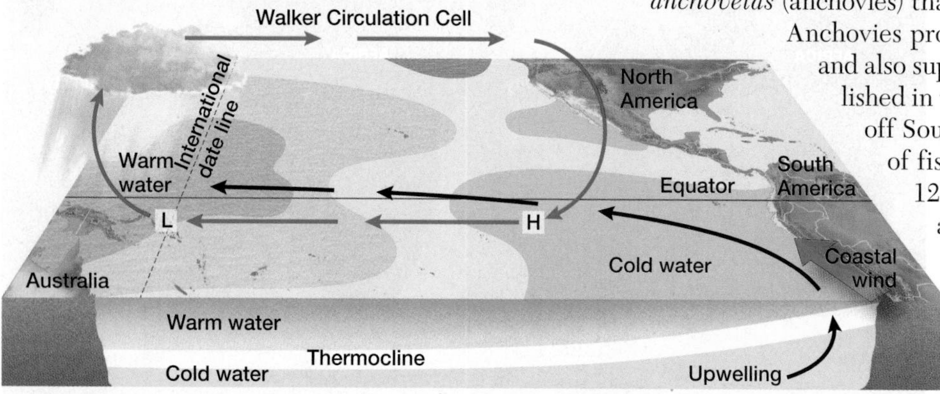


**(a)** Normal conditions

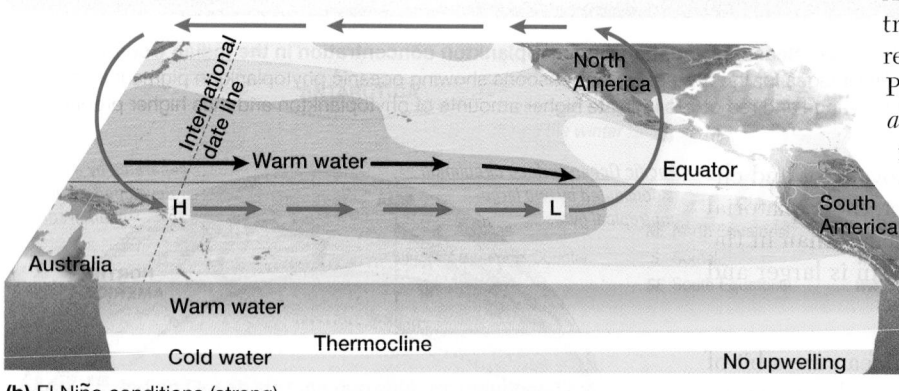

**(b)** El Niño conditions (strong)

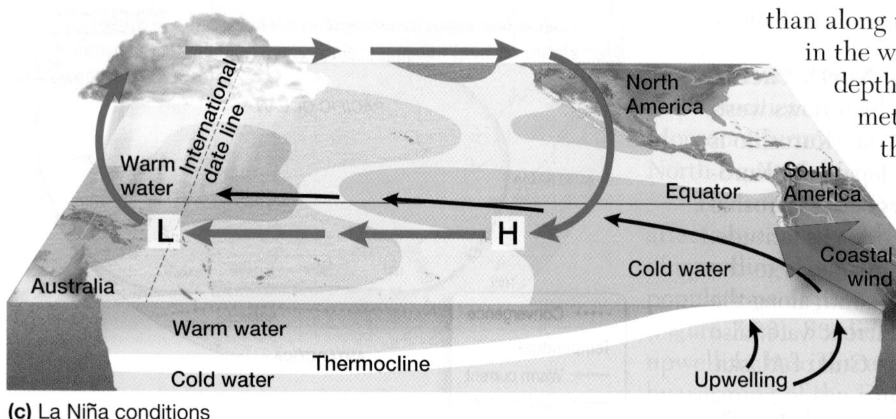


**(c)** La Niña conditions

**SmartFigure 7.22 Normal, El Niño, and La Niña conditions.** Perspective views of oceanic and atmospheric conditions in the equatorial Pacific Ocean. **(a)** Normal conditions; **(b)** El Niño (ENSO warm phase) conditions (strong); and **(c)** La Niña (ENSO cool phase) conditions. https://goo.gl/FhkBIE

**FISHERIES AND THE PERU CURRENT** The cool waters of the Peru Current have historically been one of Earth's richest fishing grounds. What conditions produce such an abundance of fish? **Figure 7.22a** shows that along the west coast of South America, coastal winds create Ekman transport that moves water away from shore, causing upwelling of cool, nutrient-rich water. This upwelling increases productivity and results in an abundance of marine life, including small silver-colored fish called *anchovetas* (anchovies) that become particularly plentiful near Peru and Ecuador. Anchovies provide a food source for many larger marine organisms, and also supply Peru's commercial fishing industry, which was established in the 1950s. Anchovies had been so abundant in the waters off South America that by 1970 Peru was the largest producer of fish from the sea in the world, with a peak production of 12.3 million metric tons (27 billion pounds), accounting for about one-quarter of *all* fish from the sea worldwide.

**WALKER CIRCULATION** Figure 7.22a shows that high pressure and sinking air dominate the coastal region of South America, resulting in clear, fair, and dry weather. On the other side of the Pacific, a low-pressure region and rising air create cloudy conditions with plentiful precipitation in Indonesia, New Guinea, and northern Australia. This pressure difference causes the strong southeast trade winds to blow across the equatorial South Pacific. The resulting atmospheric circulation cell in the equatorial South Pacific Ocean is named the **Walker Circulation Cell** (*green arrows*) after Sir Gilbert T. Walker (1868–1958), the British meteorologist who first described the effect in the 1920s.

**PACIFIC WARM POOL** The southeast trade winds set ocean water in motion across the Pacific from east to west. The water warms as it flows in the equatorial region and creates a wedge of warm water on the western side of the Pacific Ocean, called the **Pacific Warm Pool** (see Figure 7.9). Due to the movement of equatorial currents to the west, the Pacific Warm Pool is thicker along the western side of the Pacific than along the eastern side. The *thermocline* beneath the Warm Pool in the western equatorial Pacific occurs below 100 meters (330 feet) depth. In the eastern Pacific, however, the thermocline is within 30 meters (100 feet) of the surface. The difference in depth of the thermocline can be seen by the sloping boundary between the warm surface water and the cold deep water in Figure 7.22a.

**EL NIÑO–SOUTHERN OSCILLATION (ENSO) CONDITIONS** Peru's residents have known for generations that every few years, a current of warm water reduces the population of anchovies in coastal waters. The decrease in anchovies causes a dramatic decline not only in the fishing industry, but also in marine life such as sea birds, sea lions, and seals that depend on anchovies for food. The warm current also brings about changes in the weather—usually intense rainfall—and even brings such interesting items as floating coconuts from tropical islands near the equator. At first, these events were called *años de abundancia* (years of abundance) because the additional rainfall dramatically increased plant growth on the normally arid land. What was once thought of as a joyous event, however, soon became associated with the ecological and economic disaster that is now a well-known consequence of the phenomenon.

**Animation**
El Niño and La Niña
https://goo.gl/yqatJZ

The Pacific warming was first described in the late 1880s by a Peruvian Navy captain who reported on an unusually warm *"corriente del Niño"* (ocean current of the Christ Child), so named because it appeared around Christmas time. Thus, this warm-water current was given the name **El Niño**, Spanish for "the child," in reference to baby Jesus. In the 1920s, Walker was the first to recognize that an east–west atmospheric pressure seesaw accompanied the warm current and he called the phenomenon the **Southern Oscillation**. Today, the combined oceanic and atmospheric effects are called **El Niño–Southern Oscillation (ENSO)**, which periodically alternates between warm and cold phases and causes dramatic environmental changes.

***ENSO Warm Phase (El Niño)*** **Figure 7.22b** shows the atmospheric and oceanic conditions during an ENSO warm phase, which is known as *El Niño*. The high pressure along the coast of South America weakens, reducing the difference between the high- and low-pressure regions of the Walker Circulation Cell. This, in turn, causes the southeast trade winds to diminish. In very strong El Niño events, the trade winds actually blow in the *reverse* direction.

Without the trade winds, the Pacific Warm Pool that has built up on the western side of the Pacific begins to flow back across the ocean toward South America, creating a band of warm water that stretches across the equatorial Pacific Ocean (**Figure 7.23a**). The warm water usually begins to move in September of an El Niño year and reaches South America by December or January. During strong to very strong El Niños, the water temperature off Peru can be up to 10°C (18°F) higher than normal. In addition, the average sea level can increase as much as 20 centimeters (8 inches), simply due to thermal expansion of the warm water along the coast.

As the warm water increases sea surface temperatures across the equatorial Pacific, temperature-sensitive corals are decimated in Tahiti, the Galápagos, and other tropical Pacific islands. In addition, many other organisms are affected by the warm water (see **Process of Science 7.1**). Once the warm water reaches South America, it moves north and south along the west coast of the Americas, increasing average sea level and the number of tropical hurricanes formed in the eastern Pacific.

The flow of warm water across the Pacific also causes the sloped thermocline boundary between warm surface waters and the cooler waters below to flatten out and become more horizontal (Figure 7.22b). Near Peru, upwelling brings warmer, nutrient-depleted water to the surface instead of cold, nutrient-rich water. In fact, *downwelling* can sometimes occur as the warm water stacks up along coastal South America. Productivity diminishes and most types of marine life in the area are dramatically reduced.

As the warm water moves to the east across the Pacific, the low-pressure zone also migrates. In a strong to very strong El Niño event, the low pressure can move across the entire Pacific and remain over South America. The low pressure substantially increases precipitation along coastal South America. Conversely, high pressure replaces the Indonesian low, bringing dry conditions or, in strong to very strong El Niño events, drought conditions to Indonesia and northern Australia.

***ENSO Cool Phase (La Niña)*** In some instances, conditions opposite of El Niño prevail in the equatorial South Pacific; these events are known as *ENSO cool phase* or **La Niña** (Spanish for "the female child"). **Figure 7.22c** shows La Niña conditions, which are similar to normal conditions but more intensified because there is a larger pressure difference across the Pacific Ocean. This larger pressure difference creates stronger Walker Circulation and stronger trade winds, which in turn cause more upwelling, a shallower thermocline in the eastern Pacific, and a band of cooler than normal water that stretches across the equatorial South Pacific (**Figure 7.23b**).

La Niña conditions commonly occur following an El Niño. For instance, the 1997–1998 El Niño was followed by several years of persistent La Niña conditions. The alternating pattern of El Niño–La Niña conditions since 1950 is shown by the

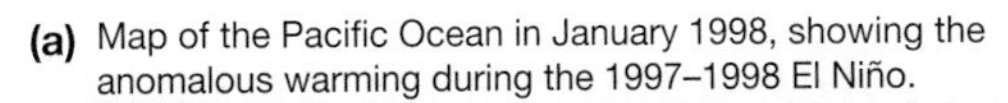
**(a)** Map of the Pacific Ocean in January 1998, showing the anomalous warming during the 1997–1998 El Niño.

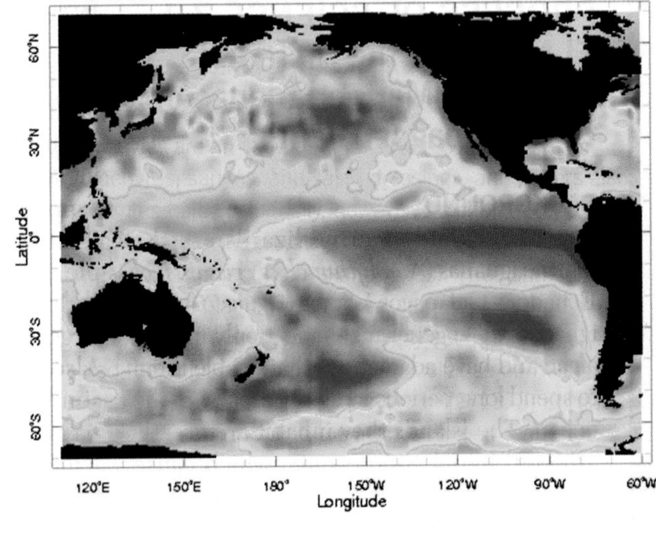


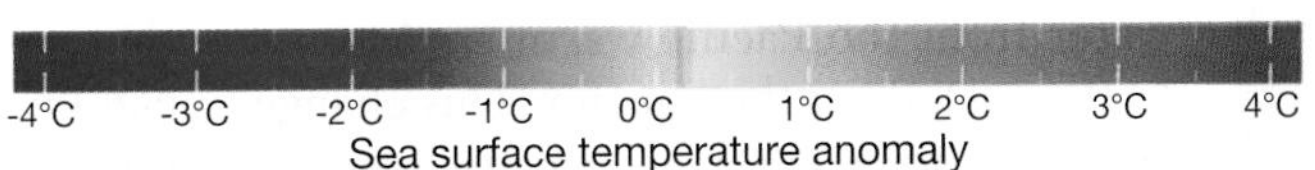


**(b)** Map of the same area in January 2000, showing cooling in the equatorial Pacific related to La Niña.

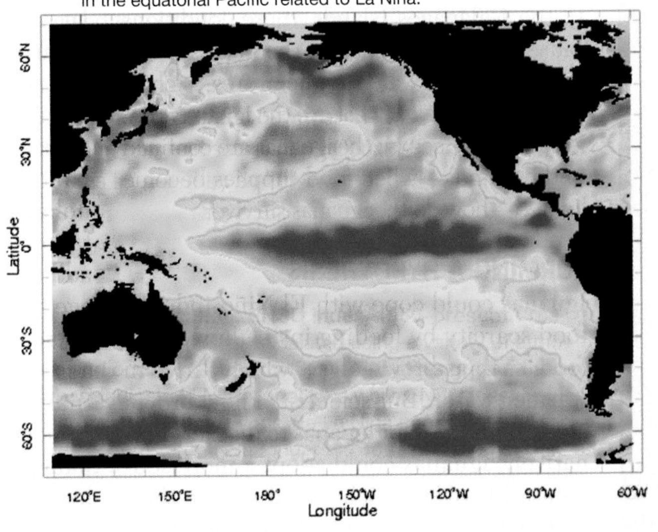


Figure 7.23 **Sea surface temperature anomaly maps during El Niño and La Niña.** Maps showing satellite-derived sea surface temperature anomalies, which represent departures from normal conditions. Values are in °C; red colors indicate water warmer than normal and blue colors represent water cooler than normal.

**Video**
Weekly Pacific Sea Surface Temperature Anomalies
http://goo.gl/9jBBnr

appears to influence Pacific sea surface temperatures. Analysis of satellite data suggests that the Pacific Ocean was in the warm phase of the PDO from 1977 to 1999 and that it is now in its cool phase, which may suppress the initiation of El Niño events during the next few decades.

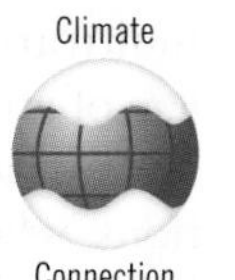


***Effects of El Niños and La Niñas*** Mild El Niño events influence only the equatorial South Pacific Ocean, while strong to very strong El Niño events can influence worldwide weather patterns. Typically, stronger El Niños alter the atmospheric jet stream and produce unusual weather in most parts of the globe. Sometimes the weather is drier than normal; at other times, it is wetter. The weather may also be warmer or cooler than normal. It is still difficult to predict exactly how a particular El Niño will affect any region's weather.

**Figure 7.25** shows how very strong El Niño events can result in flooding, erosion, droughts, fires, tropical storms, and effects on marine life worldwide. These weather perturbations also affect the production of corn, cotton, and coffee. More locally, the satellite images in **Figure 7.26** show that sea surface temperatures off western North America are significantly higher during an El Niño year.

Even though severe El Niños are typically associated with vast amounts of destruction, they can be beneficial in some areas. Tropical hurricane formation, for instance, is generally suppressed in the Atlantic Ocean because of greater wind shear in the upper atmosphere, some desert regions receive much-needed rain, and organisms adapted to warm-water conditions thrive in the Pacific.

La Niña events are associated with sea surface temperatures and weather phenomena opposite those of El Niño. Indian Ocean monsoons, for instance, are typically drier than usual in El Niño years but wetter than usual in La Niña years.

***Examples from Recent El Niños*** Recent El Niños provide an indication of the variability of the effects of El Niño events. For instance, in the winter of 1976, a

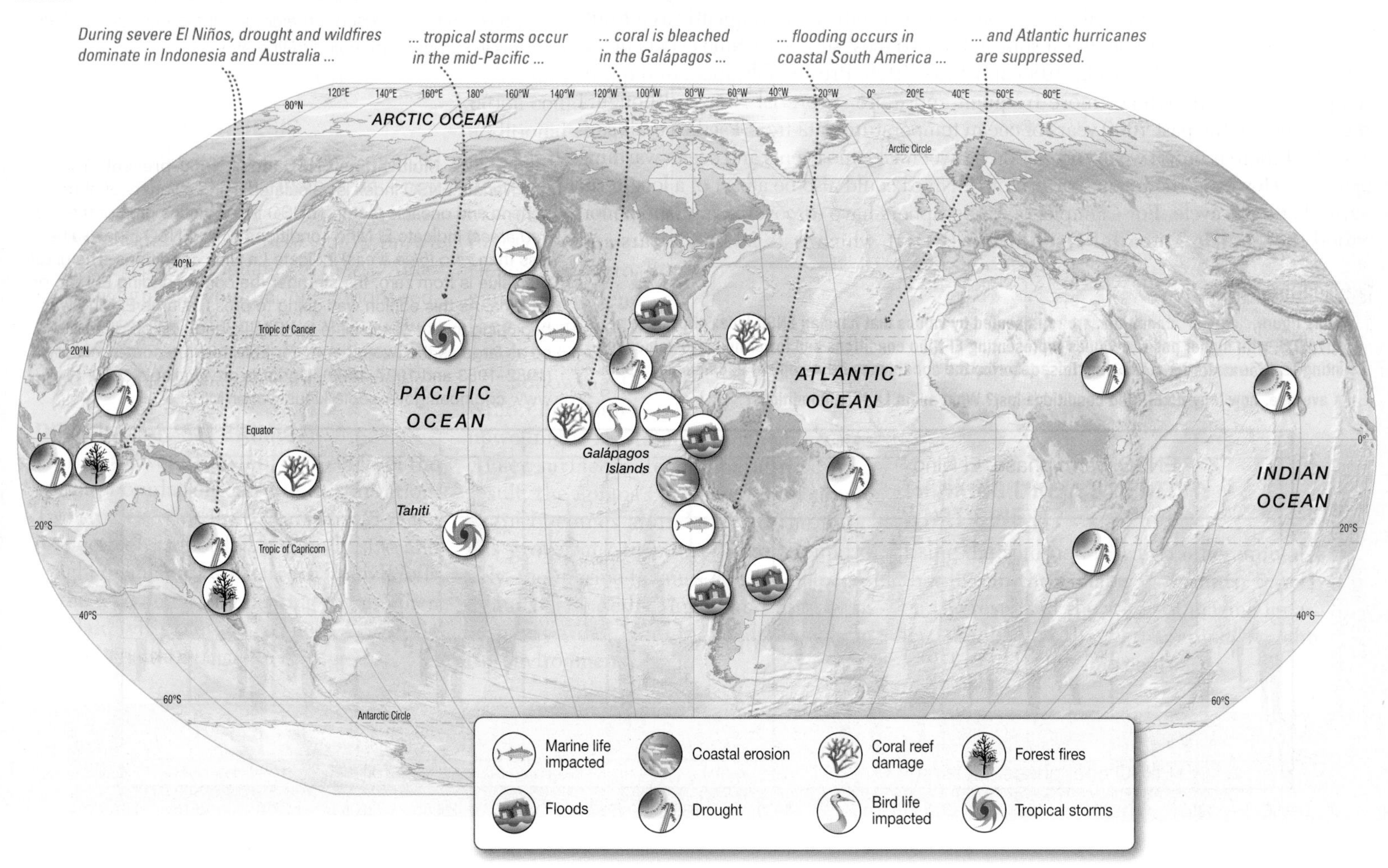


**Figure 7.25 Effects of severe El Niños.** Map showing the locations of flooding, erosion, droughts, fires, tropical storms, and effects on marine life that are associated with severe El Niño events.

moderate El Niño event coincided with northern California's worst drought of the 20th century, showing that El Niño events don't always bring torrential rains to the western United States. During that same winter, the eastern United States experienced record cold temperatures.

**THE 1982–1983 EL NIÑO** The 1982–1983 El Niño was the strongest ever recorded, causing far-ranging effects around the globe. Not only was there anomalous warming in the tropical Pacific, but the warm water also spread along the west coast of North America, influencing sea surface temperatures as far north as Alaska. Sea level was higher than normal (due to thermal expansion of the water), which, when high surf was experienced, caused damage to coastal structures and increased coastal erosion. In addition, the jet stream swung much farther south than normal across the United States, bringing a series of powerful storms that resulted in three times normal rainfall across the southwestern United States. The increased rainfall caused severe flooding and landslides as well as higher-than-normal snowfall in the Rocky Mountains. Alaska and western Canada had a relatively warm winter, and the eastern United States had its mildest winter in 25 years.

The full strength of El Niño was experienced in western South America. Normally arid Peru was drenched with more than 3 meters (10 feet) of rain, causing extreme flooding and landslides. Sea surface temperatures were so high for so long that temperature-sensitive coral reefs across the equatorial Pacific were decimated. Marine mammals and sea birds, which depend on the food normally available in the highly productive waters along the west coast of South America, went elsewhere or died. In the Galápagos Islands, for example, over half of the island's fur seals and sea lions died of starvation during the 1982–1983 El Niño.

French Polynesia had not experienced a hurricane in 75 years; in 1983, it endured six. The Hawaiian Island of Kauai also experienced a rare hurricane. Meanwhile, in Europe, severe cold weather prevailed. Elsewhere, droughts occurred in Australia, Indonesia, China, India, Africa, and Central America. Worldwide, more than 2000 deaths and at least $10 billion in property damage ($2.5 billion in the United States) were attributed to the 1982–1983 El Niño event.

**THE 1997–1998 EL NIÑO** The 1997–1998 El Niño event began several months earlier than normal and peaked in January 1998. The amount of Southern Oscillation and sea surface warming in the equatorial Pacific was initially as strong as in the 1982–1983 El Niño, which caused a great deal of concern. However, the 1997–1998 El Niño weakened in the last few months of 1997 before reintensifying in early 1998. The impact of the 1997–1998 El Niño was felt mostly in the tropical Pacific, where surface water temperatures in the eastern Pacific averaged more than 4°C (7°F) warmer than normal, and, in some locations, reached up to 9°C (16°F) above normal (see Figure 7.23a). High pressure in the western Pacific brought drought conditions that caused wildfires to burn out of control in Indonesia. Also, the warmer than normal water along the west coast of Central and North America increased the number of hurricanes off Mexico.

In the United States, the 1997–1998 El Niño caused killer tornadoes in the Southeast, massive blizzards in the upper Midwest, and flooding in the Ohio River Valley. Most of California received twice the normal rainfall, which caused flooding and landslides in many parts of the state. The lower Midwest, the Pacific Northwest, and the eastern seaboard, on the other hand, had relatively mild weather. In all, the 1997–1998 El Niño caused 2100 deaths and $33 billion in property damage worldwide.

**THE 2015–2016 EL NIÑO** The 2015–2016 El Niño was a particularly strong event, comparable to the extreme 1997–1998 El Niño which was dubbed "the climate

**(a)** Sea surface temperature anomaly map for January 1998, an El Niño year.

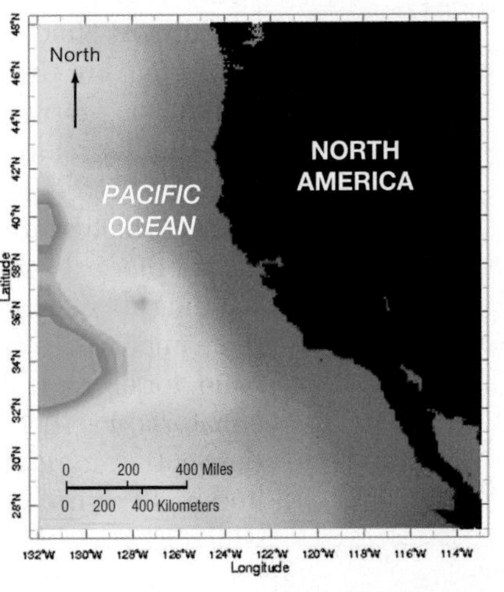


**(b)** Sea surface temperature anomaly map for the same region a year later (January 1999), during a La Niña event.

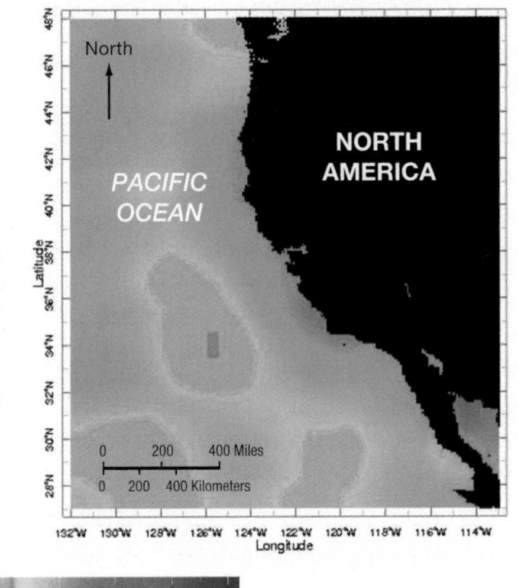


-4°C -3°C -2°C -1°C 0°C 1°C 2°C 3°C 4°C
Sea Surface Temperature Anomaly

Figure 7.26 **Sea surface temperatures off western North America during El Niño and La Niña.** Satellite-derived sea surface temperature anomaly maps (in °C) along the west coast of North America. Red color represents water that is warmer than normal; blue color represents water that is colder than normal.

## STUDENTS SOMETIMES ASK . . .

### *Do El Niño events occur in other ocean basins?*

Yes, the Atlantic and Indian Oceans both experience events similar to the Pacific's El Niño. These events are not nearly as strong, however, nor do they influence worldwide weather phenomena to the same extent as those that occur in the equatorial Pacific Ocean. The great width of the Pacific Ocean in equatorial latitudes is the main reason that El Niño events occur more strongly in the Pacific.

In the Atlantic Ocean, this phenomenon is related to the North Atlantic Oscillation (NAO), which is a periodic change in atmospheric pressure between Iceland and the Azores Islands. This pressure difference determines the strength of the prevailing westerlies in the North Atlantic, which in turn affects ocean surface currents there. The Atlantic Ocean periodically experiences NAO events, which sometimes cause intense cold in the northeast United States, unusual weather in Europe, and heavy rainfall along the normally arid coast of southwest Africa.

event of the 20th century." It developed rapidly and had spectacular weather impacts worldwide, including drought conditions in Venezuela, Australia, and a number of Pacific islands, while significant flooding was also recorded elsewhere. It was hoped that this El Niño would bring some relief from five years of drought conditions in California, but the event did not bring extensive rainfall to the region and failed to end the long-term dryness.

Another reason why the event was significant is that scientists had experienced only two extreme El Niños in modern times: the 1982–1983 and the 1997–1998 events, which shared similar key characteristics with each other. Moreover, observing another extreme event was very valuable for testing the current state of understanding about extreme El Niños. And it turned out that the 2015–2016 El Niño had some important differences from its extreme predecessors. During the 2015–2016 El Niño, for example, the highest sea surface temperatures peaked toward the central equatorial Pacific, whereas the 1982–1983 and 1997–1998 events peaked closer to the South American coast. Consequently, the 2015–2016 El Niño exhibited higher rainfall in the central Pacific than in the eastern Pacific, which is opposite to the 1982–1983 and 1997–1998 patterns. The latest event also provided a glimpse into how background ocean warming trends, which have become more prominent since the last extreme El Niño in 1997–1998, can affect the development and impacts of El Niño.

***Predicting El Niño Events*** The 1982–1983 El Niño event was not predicted, nor was it recognized until it was near its peak. Because it affected weather worldwide and caused such extensive damage, the **Tropical Ocean–Global Atmosphere (TOGA)** program was initiated in 1985 to study how El Niño events develop. The goal of the TOGA program was to monitor the equatorial South Pacific Ocean during El Niño events to enable scientists to model and predict future El Niño events. The 10-year program studied the ocean from research vessels, analyzed surface and subsurface data from radio-transmitting sensor buoys, monitored oceanic phenomena by satellite, and developed computer models.

These models have made it possible to predict El Niño events since 1987 as much as one year in advance. After the completion of TOGA, the **Tropical Atmosphere and Ocean (TAO)** project (sponsored by the United States, Canada, Australia, and Japan) has continued to monitor the equatorial Pacific Ocean with a series of 70 moored buoys, providing real-time information about the conditions of the tropical Pacific that is available on the Internet. Although monitoring has improved, the trigger mechanisms of El Niño events are still not fully understood.

**RECAP** **El Niño is a combined oceanic–atmospheric phenomenon that occurs periodically in the tropical Pacific Ocean, bringing warm water to the east. La Niña describes conditions opposite those of El Niño.**

## CONCEPT CHECK 7.4 ▸ Specify the main surface circulation patterns in each ocean basin.

1 Explain why Gulf Stream eddies that develop northeast of the Gulf Stream rotate clockwise and have warm-water cores, whereas those that develop to the southwest rotate counterclockwise and have cold-water cores.

2 Describe changes in atmospheric pressure, precipitation, winds, and ocean surface currents during the two monsoon seasons of the Indian Ocean.

3 Describe changes in atmospheric and oceanographic phenomena that occur during El Niño/La Niña events, including changes in atmospheric pressure, winds, Walker Circulation, weather, equatorial surface currents, coastal upwelling/downwelling and the abundance of marine life, sea surface temperature and the Pacific Warm Pool, sea surface elevation, and position of the thermocline.

4 How often do El Niño events occur? Using Figure 7.24, determine how many years since 1950 have been El Niño years. Has the pattern of El Niño events occurred at regular intervals?

5 How is La Niña different from El Niño? Describe the pattern of La Niña events in relation to El Niños since 1950 (see Figure 7.24).

6 Describe the global effects of severe El Niños.

# 7.5 ▶ How Do Sea Ice and Icebergs Form?

Ocean circulation is not just affected by wind patterns and atmospheric pressure changes. In high-latitude regions, for example *low temperatures* cause a permanent or nearly permanent ice cover on the sea surface, and this in turn affects both deep circulation and surface conditions. The term **sea ice** is used to distinguish such masses of frozen seawater from **icebergs**, which are also found at sea but originate by breaking off (*calving*) from glaciers that originate on land. Sea ice is found throughout the year around the margin of Antarctica, within the Arctic Ocean, and in the extreme high-latitude region of the North Atlantic Ocean.

## Formation of Sea Ice

*Sea ice* is ice that forms directly from seawater (**Figure 7.27**). It begins as small, needle-like, hexagonal (six-sided) crystals, which eventually become so numerous that a partially frozen/partially liquid *slush* develops. As the slush begins to form into a thin sheet, it is broken by wind stress and wave action into disk-shaped pieces called **pancake ice** (Figure 7.27a). As further freezing occurs, the pancakes merge together to form larger **ice floes** (*flo* = layer; Figure 7.27b). Over time, ice floes merge and create large sheets of ice, which are moved by winds and currents to produce **pressure ridges** along their margins (Figure 7.27c).

The rate at which sea ice forms is closely tied to temperature conditions. Large quantities of ice form in relatively short periods when the temperature falls to extremely low levels (such as temperatures below −30°C [−22°F]). Even at these low temperatures, the rate of ice formation slows as sea ice thickens because the ice (which has poor heat conduction) effectively insulates the underlying water from freezing. In addition, calm water enables pancake ice to join together more easily, which aids the formation of sea ice.

The process of sea ice formation tends to be a self-perpetuating process. As sea ice forms at the surface, only a small percentage of the dissolved components can be accommodated into the crystalline structure of ice. As a result, most of the dissolved substances remain in the surrounding seawater, which causes its salinity to increase. Recall from Chapter 5 that increasing the amount of dissolved materials decreases the freezing point of water, which doesn't appear to enhance ice formation. However, also recall that increasing the salinity of water increases its density and its tendency to sink. As it sinks below the surface, it is replaced by lower-salinity (and lower-density) water from below, which will freeze more readily than the high-salinity water it replaced, thereby establishing a circulation pattern that enhances the formation of sea ice.

Recent satellite analyses of the extent of Arctic Ocean sea ice shows that it has decreased dramatically in the past few decades. This accelerated melting appears to be linked to shifts in Northern Hemisphere atmospheric circulation patterns that have caused the region to experience anomalous warming. For more on this topic, see Chapter 16, "The Oceans and Climate Change."

Climate
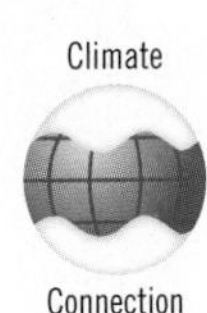
Connection

## Formation of Icebergs

An *iceberg* is a body of floating ice that has broken away from a glacier (**Figure 7.28a**) and so is quite distinct from sea ice. Icebergs are formed by vast ice sheets on land, which grow from the accumulation of snow and slowly flow outward to the sea. Once at sea, the ice either breaks up and produces icebergs there or, because it is less dense than water, floats on top of the water, often extending a great distance away from shore before breaking up under the stress of current, wind, and wave action. Most calving occurs during the summer months, when temperatures are highest.

In the Arctic, icebergs originate primarily by calving from glaciers that extend to the ocean along the western coast of Greenland (**Figure 7.28b**). Icebergs are also produced by glaciers along the eastern coasts of Greenland, Ellesmere Island, and other Arctic islands. In all, about 10,000 or so icebergs are calved off these glaciers each year, and the number of icebergs has been increasing recently. Many of these

**Figure 7.27 Stages of formation of sea ice.** Sea ice, which forms directly from the freezing of seawater, begins as **(a)** pancake ice. As further freezing occurs, pancake ice thickens into **(b)** ice floes, eventually forming large sheets of ice that contain **(c)** pressure ridges.

**(a)** Pancake ice, which is the initial stage of sea ice formation, is a frozen slush that is broken by wind stress and wave action into disk-shaped pieces.

**(b)** Ice floes form with additional freezing, causing pancake ice to merge together and become thicker.

**(c)** Ridged ice is created over time when large sheets of sea ice collide, forming thick pressure ridges.

icebergs are carried by currents in and around the Labrador Sea (Figure 7.28b, *blue arrows*) into North Atlantic shipping lanes, where they become navigational hazards. In recognition of this fact, the area is called *Iceberg Alley*; it is here that the luxury liner RMS *Titanic* hit an iceberg and sank (see *black x* on Figure 7.28b; see also Mastering Oceanography **Web Diving Deeper 7.2**). Because of their large size, some of these icebergs take several years to melt, and, in that time, they may be carried as far south as 40 degrees north latitude, which is the same latitude as Philadelphia, Pennsylvania.

**SHELF ICE** In Antarctica, where glaciers cover nearly the entire continent, the edges of glaciers form thick floating sheets of ice called **shelf ice** that break off and produce extensive plate-like icebergs (**Figures 7.28c** and **7.28d**). In March 2000, for example, a Connecticut-sized iceberg—11,000 square kilometers (4250 square miles)—known as B-15 and nicknamed "Godzilla" broke loose from the Ross Ice Shelf into the Ross Sea. Since its origin, B-15 has broken up into smaller icebergs, but several of its parts still exist. In July 2017, a Delaware-sized iceberg—5200 square kilometers (2000 square miles) and estimated to weigh more than 1.1 trillion tons—known as A-68 broke off from the Larsen Ice Shelf on the opposite side of Antarctica (Figure 7.28d). Even larger icebergs have also been observed in Antarctic waters. For instance, the largest iceberg ever reported was an incredible 32,500 square kilometers (12,500 square miles)—nearly three times the size of B-15 or about the same size as Connecticut and Massachusetts combined. (Note that this record has been questioned by researchers, who now believe that the U.S. Navy icebreaker USS *Glacier* incorrectly measured the iceberg in 1956, and so the berg was far smaller.)

Icebergs from shelf ice have flat tops that may stand as much as 200 meters (650 feet) above the ocean surface, although most rise less than 100 meters (330 feet) above sea level, and as much as 90% of their mass is below the waterline. Once icebergs are created, ocean currents driven by strong winds carry the icebergs north, where they eventually melt. Because this region is not a major shipping route, the icebergs pose little serious navigation hazard except to supply ships traveling to Antarctica. Officers aboard ships sighting these gigantic bergs have, in some cases, mistaken them for land!

The rate at which Antarctica is producing icebergs—especially large icebergs—has recently increased, most likely as a result of warming ocean and atmospheric temperatures. In fact, these increases in temperature have been implicated as a factor in earlier disintegrations of ice shelves elsewhere on the Antarctic Peninsula, most notably Larsen A in 1995, Larsen B in 2002, and Pine Island Glacier in the past few decades. In addition, newly discovered channels on the sea floor where the bedrock on which glaciers rest meets the sea could funnel warmer ocean water to the base of glaciers, thereby accelerating their melting. For more information about Antarctic warming and its relationship to climate change, see Chapter 16, "The Oceans and Climate Change."

**(a)** Icebergs, such as this small North Atlantic berg, are formed when pieces of ice calve from glaciers that extend to the sea.

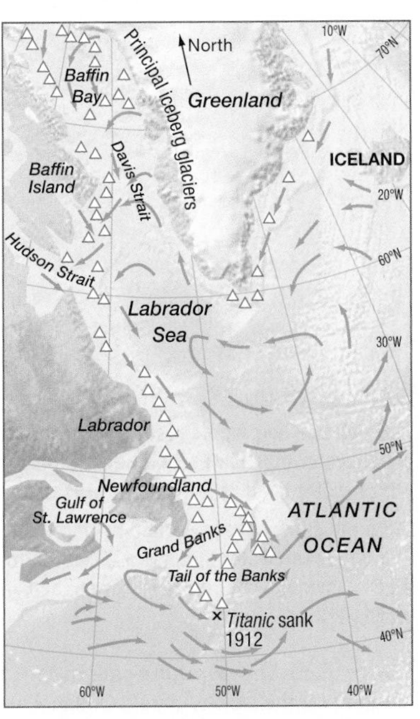


**(b)** Map showing North Atlantic currents (*blue arrows*), typical iceberg distribution (*white triangles*), and the site of the 1912 *Titanic* sinking (*black x*).

**(c)** Aerial view of part of a large tabular Antarctic iceberg, which extends beyond the horizon.

**(d)** Satellite view of iceberg A-68, which broke off from Antarctica's Larsen C Ice Shelf in July 2017 and was as big as the state of Delaware.

**Figure 7.28 Icebergs.** Icebergs form when land-based freshwater glacial ice breaks off into the sea. The world's two main iceberg-producing regions are Greenland (*a* and *b*) and Antarctica (*c* and *d*).

## CONCEPT CHECK 7.5 ▸ Specify how sea ice and icebergs form.

1 Why does the formation of sea ice tend to be a self-perpetuating process?

2 Describe differences between sea ice, icebergs, and shelf ice, including how each is formed.

**RECAP Sea ice is created when seawater freezes; icebergs form when chunks of ice break off from coastal glaciers that reach the sea.**

## 7.6 ▶ How Do Deep-Ocean Currents Form?

Deep currents occur in the deep zone below the pycnocline, so they influence about 90% of all ocean water. Density differences create deep currents. Although these density differences are usually small, they are large enough to cause denser waters to sink. Deep-water currents move larger volumes of water and are much slower than surface currents. Typical speeds of deep currents range from 10 to 20 kilometers (6 to 12 miles) per year. Thus, it takes a deep current an *entire year* to travel the same distance that a western intensified surface current can move in *one hour.*

Because the density variations that cause deep-ocean circulation are the result of differences in temperature and salinity, deep-ocean circulation is also referred to as **thermohaline circulation** (*thermo* = heat, *haline* = salt).

**EXPLORING DATA ▼**

1. On an oceanographic research cruise, you take a sample of deep-ocean water that has a temperature of 4.5°C and a salinity of 34.1‰. Which water mass is it from, and what is its density?

2. On the same oceanographic research cruise, you take another sample of deep-ocean water that has a temperature of 4°C and a salinity of 35.0‰. Which water mass is it from, and what is its density? (Hint: estimate!)

Figure 7.29 **Temperature–salinity (T–S) diagram.** A density T–S diagram for the North Atlantic Ocean. Lines of constant density are in grams/cm$^3$. After various deep-water masses sink below the surface and spread out, they can be identified in any ocean basin based on their characteristic temperature (*vertical axis*), salinity (*horizontal axis*), and resulting density (*blue curving lines*).

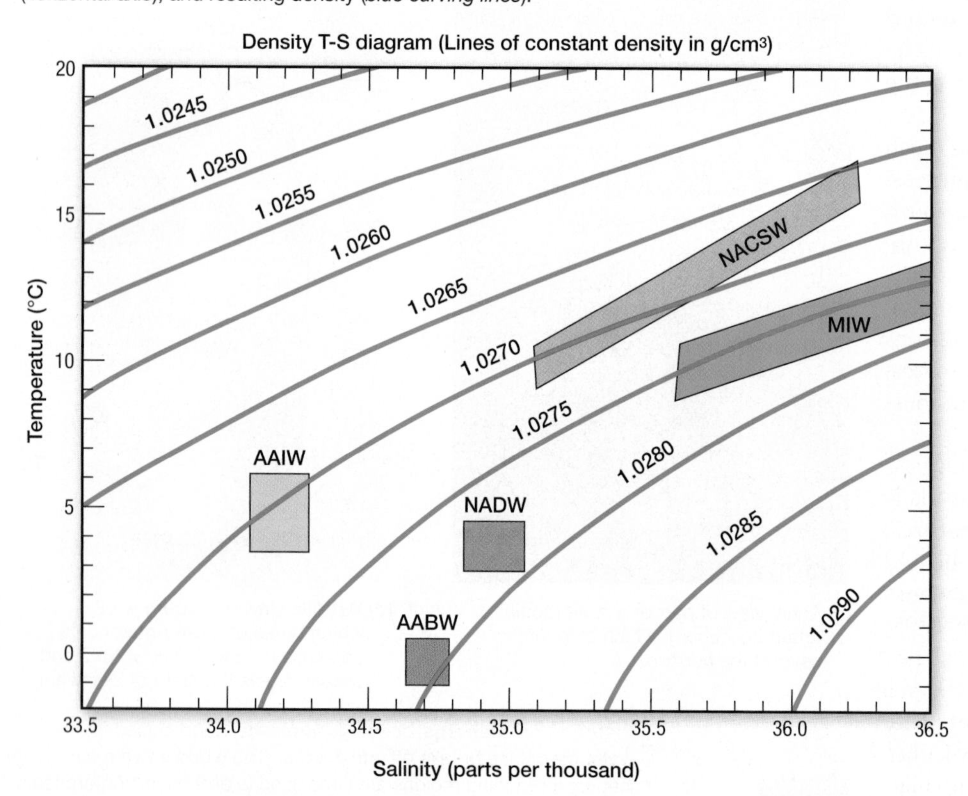


### Origin of Thermohaline Circulation

Recall from Chapter 5 that an increase in seawater density can be caused by a *decrease* in temperature or an *increase* in salinity. Temperature, though, has the greater influence on density. Density changes due to salinity are important only in very high latitudes, where water temperature remains low and relatively constant.

Most water involved in deep-ocean currents (thermohaline circulation) originates in high latitudes *at the surface*. In these regions, surface water becomes cold and its salinity increases as sea ice forms. When this surface water becomes dense enough, it sinks, initiating deep-ocean currents. Once this water sinks, it is removed from the physical processes that increased its density in the first place, so its temperature and salinity remain largely unchanged for the duration it spends in the deep ocean. Thus, a **temperature–salinity (T–S) diagram** can be used to identify deep-water masses based on their characteristic temperature, salinity, and resulting density. **Figure 7.29** shows a density T–S diagram for the North Atlantic Ocean.

As these surface water masses become dense and are sinking (downwelling) in high-latitude areas, deep-water masses are also rising to the surface (upwelling). Because the water temperature in high-latitude regions is the same at the surface as it is down below, the water column is isothermal, there is no thermocline or associated pycnocline (see Chapter 5), and upwelling and downwelling can easily occur.

### Sources of Deep Water

In southern subpolar latitudes, huge masses of deep water form beneath sea ice along the margins of the Antarctic continent. Here, rapid winter freezing produces very cold, high-density water that sinks down the continental slope of Antarctica and becomes **Antarctic Bottom Water**, the densest water in the open ocean (**Figure 7.30**). Antarctic Bottom Water slowly sinks beneath the surface and spreads into all the world's ocean basins, eventually returning to the surface perhaps 1000 years later.

In the northern subpolar latitudes, large masses of deep water form in the Norwegian Sea. From there, the deep water flows as a subsurface current into the North Atlantic, where it becomes part of the **North Atlantic Deep Water**. North Atlantic Deep

Water also comes from the margins of the Irminger Sea off southeastern Greenland, the Labrador Sea, and the dense, salty Mediterranean Sea. Like Antarctic Bottom Water, North Atlantic Deep Water spreads throughout the ocean basins. It is less dense, however, so it layers on top of the Antarctic Bottom Water (Figure 7.30).

Surface water masses converge within the subtropical gyres and in the Arctic and Antarctic. Subtropical Convergences do not produce deep water, however, because the density of warm surface waters is too low for them to sink. Major sinking does occur, however, along the **Arctic Convergence** and Antarctic Convergence (Figure 7.30, *inset*). The deep-water mass formed from sinking at the Antarctic Convergence is called the **Antarctic Intermediate Water** mass (Figure 7.30), which remains one of the world's most poorly studied water masses.

Figure 7.30 also shows that the highest-density water is found along the ocean bottom, with less-dense water above. In low-latitude regions, the boundary between the warm surface water and the deeper cold water is marked by a prominent thermocline and corresponding pycnocline that prevent vertical mixing. There is no pycnocline in high-latitude regions, so substantial vertical mixing (upwelling and downwelling) occurs.

This same general pattern of layering based on density occurs in the Pacific and Indian Oceans as well. These oceans have no source of Northern Hemisphere deep water, however, so they lack a deep-water mass. In the northern Pacific Ocean, the low salinity of surface waters prevents them from sinking into the deep ocean. In the northern Indian Ocean, surface waters are too warm to sink. **Oceanic Common Water**, which is created when Antarctic Bottom Water and North Atlantic Deep Water mix, lines the bottoms of these basins.

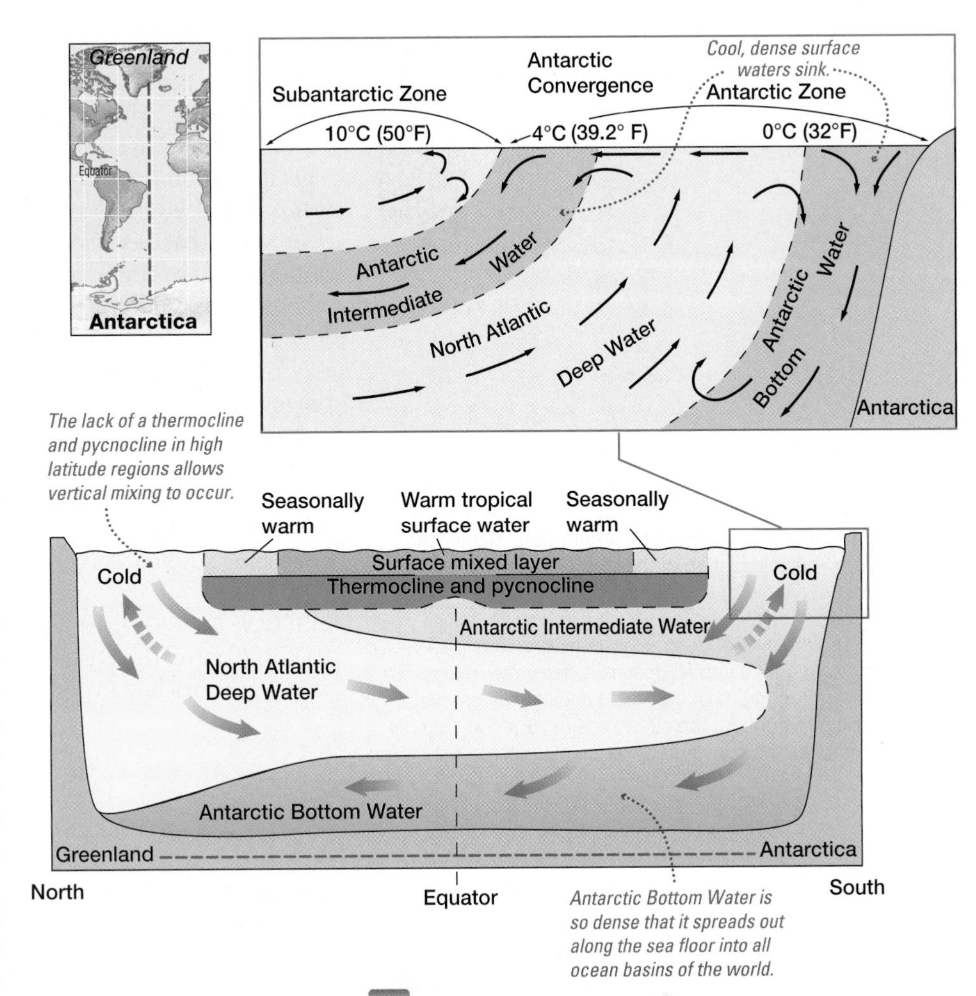


**SmartFigure 7.30 Atlantic Ocean subsurface water masses.** Schematic diagram of the various water masses in the Atlantic Ocean. Similar but less distinct layering based on density occurs in the Pacific and Indian Oceans as well. Upwelling and downwelling occur in the North Atlantic and near Antarctica (*inset*), creating deep-water masses. https://goo.gl/ow7wr6

**EXPLORING DATA ▲ In almost all ocean basins, if you go deep enough, you will eventually encounter Antarctic Bottom Water. Explain why this is true.**

## Worldwide Deep-Water Circulation

For every liter of water that sinks from the surface into the deep ocean, a liter of deep water must return to the surface somewhere else. However, it is difficult to identify specifically *where* this upward flow to the surface is occurring. It is generally believed that it occurs as a gradual, uniform upwelling throughout the ocean basins and that it may be somewhat greater in low-latitude regions, where surface temperatures are higher. Alternatively, scientific studies on turbulent mixing rates between deep-ocean and surface waters in the Southern Ocean suggest that deep water traveling across rugged bottom topography is a major factor in producing the upwelling that returns deep water toward the surface.

**CONVEYOR-BELT CIRCULATION** An integrated model combining deep thermohaline circulation and surface currents is shown in **Figure 7.31**. Because the overall circulation pattern resembles a large conveyor belt, the model is called **conveyor-belt circulation**. Beginning in the North Atlantic, surface water carries heat to high latitudes via the Gulf Stream. During the cold winter months, this heat is transferred to the overlying atmosphere, warming northern Europe.

**Animation**
North Atlantic Deep-Water Circulation
https://goo.gl/ojdtTq

**Animation**
Deep-Ocean Conveyor-Belt Circulation
http://goo.gl/jzGmDh

Cooling in the North Atlantic increases the density of this surface water to the point where it sinks to the bottom and flows southward, initiating the lower limb of the "conveyor." Here, seawater flows downward at a rate equal to 100 Amazon Rivers and begins its long journey into the deep basins of all the world's oceans. This limb extends all the way to the southern tip of Africa, where it joins the deep water that encircles Antarctica. The deep water that encircles Antarctica includes deep water that descends along the margins of the Antarctic continent. This mixture of deep waters flows northward into the deep Pacific and Indian Ocean basins, where it eventually surfaces and completes the conveyor belt by flowing west and then north again into the North Atlantic Ocean.

Does this simple conveyor-belt model of ocean circulation adequately reflect the movement of both surface and deep-ocean currents? Satellite and deep sampling of the oceans confirms the basic conveyor-belt movement of warm waters poleward at the surface and cold waters equatorward at depth. However, the conveyor-belt model ignores some crucial complex components of the ocean's circulation system—including small-scale eddies and oceanic fronts—that affect climate change and are now being studied at ever-finer scales.

**DISSOLVED OXYGEN IN DEEP WATER** Recall from Chapter 5, Section 5.8 ("What Factors Control the Distribution of Carbon and Oxygen in the Ocean?") that surface waters are enriched in dissolved oxygen because of photosynthesis by phytoplankton.

Figure 7.31 **Idealized conveyor-belt circulation.** Schematic map of conveyor-belt circulation of the world ocean. Source areas for deep water (*purple ovals*) exist in high-latitude regions where surface water cools, becomes high density, and sinks. These source areas feed the flow of deep, high-density waters (*blue bands*), which slowly drifts into all oceans. Deep water returns to the surface in localized areas of upwelling and also as gradual, uniform upwelling throughout the ocean basins. Surface currents (*red bands*) complete the conveyer by returning water to the source areas.

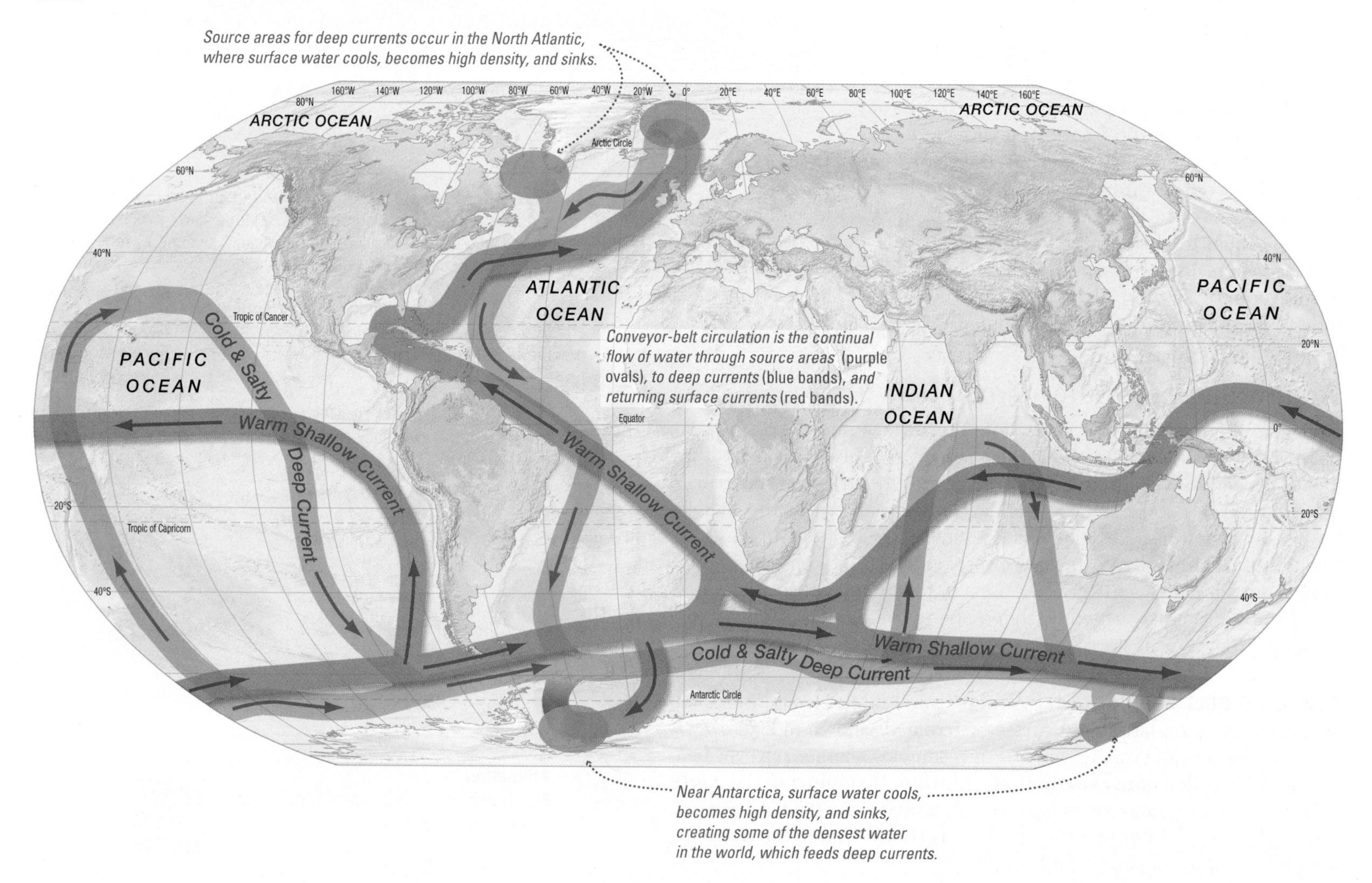

Also, remember that cold water can dissolve more oxygen than warm water. Thus, deep-water circulation brings dense, cold, oxygen-enriched water from the surface in high latitudes to the deep ocean. This means that in many regions of the ocean, deep water has higher dissolved oxygen content than water much closer to the surface (see Figure 5.29). In addition, during its time in the deep ocean, deep water becomes enriched in nutrients, caused mainly by the decomposition of dead organisms, but also by the lack of living organisms using up nutrients there.

Climate
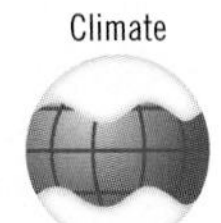
Connection

At various times in the geologic past, warmer water probably constituted a larger proportion of deep oceanic waters. As a result, the oceans had a lower oxygen concentration than today's oceans because warm water cannot hold as much oxygen. Moreover, the oxygen content of the oceans has probably fluctuated widely throughout time.

To summarize, if high-latitude surface waters did not sink and eventually return from the deep sea to the surface, the distribution of life in the sea would be considerably different. There would be very little life in the deep ocean, for instance, because there would be no oxygen for organisms to breathe. In addition, life in surface waters might be significantly reduced without the upwelling of deep water that brings nutrients to the surface.

**CONVEYOR-BELT CIRCULATION AND CLIMATE CHANGE** Conveyor-belt circulation is an important part of global ocean circulation; as such, it can profoundly affect global climate. For example, the part of the conveyor in the North Atlantic that includes both surface and deep currents is called the **Atlantic meridional overturning circulation (AMOC)**. Studies of the AMOC reveal that it is weaker now than it has been in the past 1000 years because glacial ice from Greenland has melted, thereby releasing low-density freshwater into the North Atlantic, which acts as a cap and prohibits deep water from sinking. Scientists who study the phenomenon predict that the AMOC will continue to weaken in the next few decades as global warming causes more glacial ice to melt in Greenland, which would increase the size of the freshwater wedge in the North Atlantic and may cause the complete shutdown of the AMOC. If this happens, the formation of deep water will be prevented, and global circulation patterns will be reorganized. For example, the freshwater wedge may cause the Gulf Stream in the North Atlantic to loop back towards the equator instead of reaching higher latitudes. If the warming waters of the Gulf Stream don't reach high latitudes, climate would likely be dramatically altered in Europe, North America, and many other places around the world. For more details about how the reorganization of ocean circulation patterns can alter Earth's climate, see Chapter 16, "The Oceans and Climate Change."

Climate Connection

## STUDENTS SOMETIMES ASK . . .

*The movie* The Day After Tomorrow *depicts a freezing superstorm as a result of alterations in deep-water circulation. Could this really happen?*

Although Hollywood movies are well known for dramatization, what is interesting about this classic 2004 sci-fi film is that its story line is based on scientific findings: the ocean's deep-water circulation helps drive ocean currents around the globe and is important to world climate. In fact, strong evidence suggests that North Atlantic deep-water circulation has already weakened and may further weaken during this century, resulting in unpleasant effects on our climate—but certainly not as rapid or as cataclysmic as the situation portrayed in the film. Computer models suggest that the continued weakening of North Atlantic deep-water circulation will result in some long-term cooling, particularly over parts of northern Europe.

**RECAP Thermohaline circulation describes the movement of deep currents, which is initiated when surface water in high latitudes becomes cold and dense, and as a result, sinks.**

## CONCEPT CHECK 7.6 ▸ Explain the origin and characteristics of deep-ocean currents.

1 Discuss the origin of thermohaline vertical circulation. Why do deep currents form only in high-latitude regions?

2 Using Figure 7.29, notice that AAIW and NACSW fall on the same density line. What single measurement would allow you to tell them apart? Explain.

3 What are the two major deep-water masses? Where do they form at the ocean's surface?

4 Explain why no matter where you are in the ocean, if you go deep enough, you will encounter Oceanic Common Water.

5 Describe how the distribution of life in the ocean would be different if there were very little dissolved oxygen in deep-water currents.

**Figure 7.32 Power from ocean currents.** A prototype ocean current power system in Strangford Lough, Ireland, which generates power with underwater blades that can be raised for maintenance. As currents flow past the tower, they turn the 16-meter (52-foot)-long propellers, which rotate internal rotors and generate electricity. The propellers can rotate in either direction, so this system can generate power from bidirectional tidal flows.

## 7.7 ▶ Can Power from Currents Be Harnessed as a Source of Energy?

The movement of ocean currents has often been considered to be capable of providing a source of renewable, clean energy similar to that of wind farms (see Chapter 6)—but underwater. Even though currents move much more slowly than winds, currents carry much more energy because water has about 800 times the density of air. As a result, currents have the potential to generate even more power than wind farms do. In theory, oceans could power the entire globe without adding any pollution to the atmosphere. Power from currents could also provide a more dependable source of electricity than the wind or Sun because ocean currents flow all day and night.

One location that has received much consideration as a site for harnessing power from ocean currents is the Florida–Gulf Stream Current System, which you may recall is a fast, western intensified surface current that runs along the East Coast of the United States. In fact, researchers have determined that at least 2000 megawatts of electricity could be recovered from this ocean current system along the southeastern coast of Florida alone. (Note that each megawatt of electricity is enough to serve the energy needs of about 800 average U.S. homes.)

Various devices have been proposed to extract energy from ocean currents. All of them involve some sort of mechanism for converting the movement of water into electrical energy. **Figure 7.32** shows one solution. Here, underwater turbines similar to windmills (but anchored to the ocean floor) are placed in waters that experience strong currents. As these currents flow past the tower, they turn the propellers, which rotate internal rotors and generate electricity. These systems must also work in locations where bi-directional tidal flows occur, so in some cases the propellers can rotate in either direction; in other cases, the entire turbine can swivel on its anchor to face oncoming currents. Such a system of six turbines has been successfully tested in the East River near New York City. Once this system is expanded to its capacity of 300 turbines, it will have the ability to generate about 10 megawatts of electricity.

Systems that generate power from currents have some significant hurdles to overcome. For example, current systems are expensive, difficult to maintain, and potentially hazardous to ship traffic. Further, the moving devices that harness current power can potentially disturb, injure, or kill marine life, although detailed environmental studies suggest that marine organisms are not harmed by the devices. (Interestingly, the same environmental studies suggest that these sea floor devices may, in fact, act as makeshift marine protected areas because fishing practices do not occur at these locations. See the Afterword near the end of this book for more details about marine protected areas.) Also, underwater deployment of complex machines with moving parts exposed to seawater for long periods of time presents challenges associated with corrosion and *biofouling*, which is the accumulation of algae and other sea life on machinery. In addition, the variability of current flow causes irregular power generation, which is problematic. Still, similar turbine systems powered by currents are in use in Strangford Lough, Ireland (Figure 7.32), and are planned for Canada's Bay of Fundy and offshore South Korea.

**RECAP** **In spite of problems with placing mechanical devices in the ocean, there is vast potential for developing power from currents as a renewable source of energy.**

### CONCEPT CHECK 7.7 ▶ Evaluate the advantages and disadvantages of harnessing currents as a source of energy.

1 Explain why ocean currents have the potential to generate even more power than wind farms, which are much more common today.

2 Discuss the advantages and disadvantages of building an offshore current power system.

# ESSENTIAL CONCEPTS REVIEW

## 7.1 ▶ How are ocean currents measured?

- *Ocean currents are masses of water that flow* from one place to another and can be divided into *surface currents that are wind driven* and *deep currents that are density driven*. Currents can be measured directly or indirectly by various methods.

### Selected Key Terms

Use the **glossary** at the end of this book to discover the meanings of these Selected Key Terms: **surface currents, deep currents.**

### Critical Thinking Question

Compare and contrast the characteristics and origins of surface currents and deep currents.

### Active Learning Exercise

Working with another student in class, compile a list of the ways in which both surface and deep currents are measured.

## 7.3 ▶ What causes upwelling and downwelling?

- *Upwelling and downwelling help vertically mix deep and surface waters*. Upwelling—the movement of *cold, deep, nutrient-rich water to the surface*—stimulates productivity and creates a large amount of marine life. Upwelling and downwelling can occur in a variety of ways.

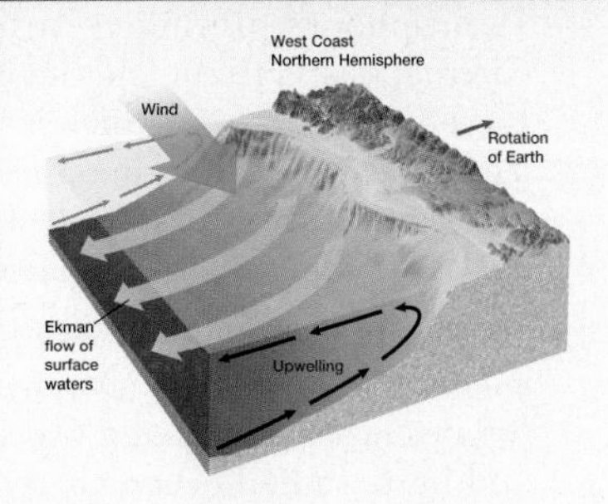


### Selected Key Terms

Use the **glossary** at the end of this book to discover the meanings of these Selected Key Terms: **upwelling, downwelling, productivity.**

### Critical Thinking Question

From a practical standpoint, give several reasons why upwelling is a much more intensely studied process than downwelling. For example, why are the world's major fisheries associated with areas of upwelling?

### Active Learning Exercise

Working with another student in class, discuss this question: If Earth's rotation on its axis were to suddenly stop, how would it affect the processes that cause both upwelling and downwelling?

## 7.2 ▶ What creates ocean surface currents and how are they organized?

- *Surface currents occur within and above the pycnocline*. They consist of circular-moving loops of water called gyres, set in motion by the major wind belts of the world. They are modified by the positions of the continents, the Coriolis effect, seasonal changes, and other factors. There are *five major subtropical gyres in the world*; they rotate *clockwise in the Northern Hemisphere* and *counterclockwise in the Southern Hemisphere*. Water is pushed toward the center of the gyres, forming low "hills" of water.
- The *Ekman spiral* influences shallow surface water and is *caused by winds and the Coriolis effect*. The average net flow of water affected by the Ekman spiral causes the water to move at *90-degree angles to the wind direction*. At the center of a gyre, the Coriolis effect deflects the water so that it tends to move into the hill, whereas gravity moves the water down the hill. When gravity and the Coriolis effect balance, a *geostrophic current* flowing parallel to the contours of the hill is established.
- As a result of Earth's rotation, *the apex (top) of the hill is located to the west of the geographical center of the gyre*. A phenomenon called *western intensification* occurs in which *western boundary currents of subtropical gyres are faster, narrower, and deeper* than their eastern boundary counterparts.

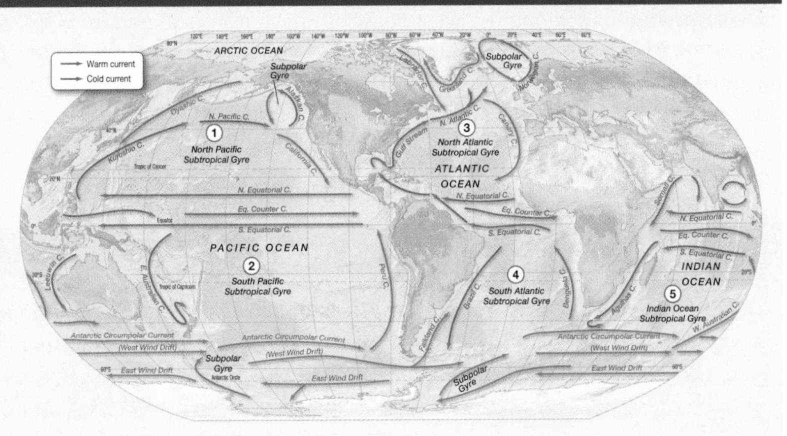


### Selected Key Terms

Use the **glossary** at the end of this book to discover the meanings of these Selected Key Terms: **subtropical gyre, subpolar gyre, Ekman spiral, Ekman transport, geostrophic current, western intensification.**

### Critical Thinking Question

From memory, construct a map showing the world's five subtropical gyres, including nearby continents. For each gyre, name the main currents involved, including any that experience western intensification.

### Active Learning Exercise

With another student in class, describe what the pattern of ocean surface currents would look like if there were no continents on Earth.

## 7.4 ▶ What are the main surface circulation patterns in each ocean basin?

- *Antarctic circulation is dominated by a single large current, the Antarctic Circumpolar Current* (West Wind Drift), which flows in a clockwise direction around Antarctica and is driven by the Southern Hemisphere's prevailing westerly winds. Between the Antarctic Circumpolar Current and the Antarctic continent is a current called the *East Wind Drift*, which is powered by the polar easterly winds. The two currents flow in opposite directions, and the Coriolis effect deflects them away from each other, creating the *Antarctic Divergence*, an area of abundant marine life due to upwelling and current mixing.
- *The North Atlantic Gyre and the South Atlantic Gyre dominate circulation in the Atlantic Ocean*. A poorly developed equatorial countercurrent separates these two subtropical gyres. The highest-velocity and best-studied ocean current is the *Gulf Stream*, which carries warm water along the southeastern U.S. Atlantic coast. Meanders of the Gulf Stream produce *warm- and cold-core rings*. The warming effects of the Gulf Stream extend along its route and reach as far away as northern Europe.
- *The Indian Ocean consists of one gyre, the Indian Ocean Gyre*, which exists mostly in the Southern Hemisphere. The *monsoon wind system*, which changes direction with the seasons, dominates circulation in the Indian Ocean. The monsoons blow from the northeast in the winter and from the southwest in the summer.
- *Circulation in the Pacific Ocean consists of two subtropical gyres: the North Pacific Gyre and the South Pacific Gyre*, which are separated by a well-developed equatorial countercurrent.
- *A periodic disruption of normal sea surface and atmospheric circulation patterns in the Pacific Ocean is called El Niño–Southern Oscillation (ENSO)*. The *warm phase of ENSO (El Niño)* is associated with the eastward movement of the Pacific Warm Pool, halting or reversal of the trade winds, a rise in sea level along the equator, a decrease in productivity along the west coast of South America, and, in very strong El Niños, worldwide changes in weather. El Niños fluctuate with the *cool phase of ENSO (La Niña conditions)*, which are associated with cooler than normal water in the eastern tropical Pacific.

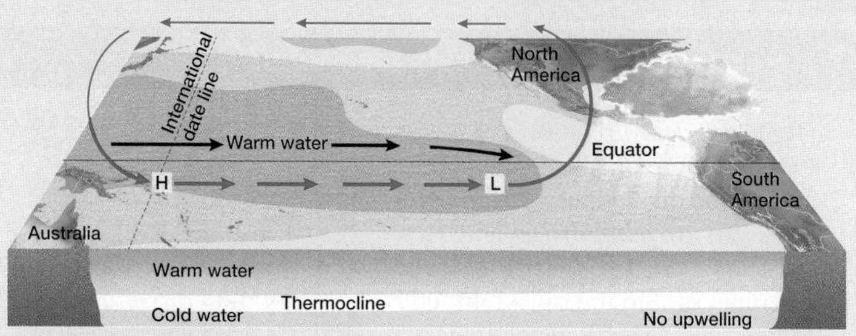

El Nino conditions (strong)

### Selected Key Terms

Use the **glossary** at the end of this book to discover the meanings of these Selected Key Terms: **Antarctic Circumpolar Current (West Wind Drift), North Atlantic Subtropical Gyre, South Atlantic Subtropical Gyre, Gulf Stream, warm-core/cold-core ring, monsoon, Indian Ocean Subtropical Gyre, North Pacific Subtropical Gyre, South Pacific Subtropical Gyre, El Niño–Southern Oscillation (ENSO), La Niña.**

### Critical Thinking Question

Using Figure 7.22, specify the atmospheric and oceanic changes that occur between normal and El Niño conditions. Also, compare normal and La Niña conditions.

### Active Learning Exercise

During flood stage, the largest river in the world—the mighty Amazon River—dumps 200,000 cubic meters of water into the Atlantic Ocean each second. With another student in class, compare its flow rate with the volume of water transported by (1) the West Wind Drift and (2) the western intensified Gulf Stream. How many times larger than the Amazon is each of these two ocean currents?

## 7.5 ▶ How do sea ice and icebergs form?

- *In high latitudes, low temperatures freeze seawater and produce sea ice*, which forms as a slush and breaks into pancakes that ultimately grow into ice floes and, over time, large sheets of ice with *pressure ridges*. *Icebergs form when chunks of ice break off glaciers* that form on Antarctica, Greenland, and some Arctic islands. Floating sheets of ice called *shelf ice* near Antarctica produce the largest icebergs.

### Selected Key Terms

Use the **glossary** at the end of this book to discover the meanings of these Selected Key Terms: **sea ice, iceberg, pancake ice, ice floe, pressure ridge, shelf ice.**

### Critical Thinking Question

Specify the places in the world where most icebergs form. What hazards do icebergs present? Explain.

### Active Learning Exercise

Using the Internet and working with another student in class, find three instances within the past 10 years where icebergs have been involved in oceanic shipping or transportation accidents (or near accidents). Specify where the incidents took place.

## 7.6 ▶ How do deep-ocean currents form?

- *Deep currents occur below the pycnocline*. They affect much larger amounts of ocean water and move much more slowly than surface currents. Changes in temperature and/or salinity at the surface create slight increases in density, which set deep currents in motion. Deep currents, therefore, are called *thermohaline circulation*.
- *The deep ocean is layered based on density. Antarctic Bottom Water*, the densest deep-water mass in the oceans, forms near Antarctica and sinks along the continental shelf into the South Atlantic Ocean. Farther north, at the Antarctic Convergence, the low-salinity *Antarctic Intermediate Water* sinks to an intermediate depth dictated by its density. Sandwiched between these two masses is the *North Atlantic Deep Water*, which has high nutrient levels after remaining below the surface for hundreds of years. Layering in the Pacific and Indian Oceans is similar, except there is no source of Northern Hemisphere deep water.
- *Worldwide circulation models that include both surface and deep currents resemble a conveyor belt*. Deep currents carry oxygen into the deep ocean, which is extremely important for life on the planet.

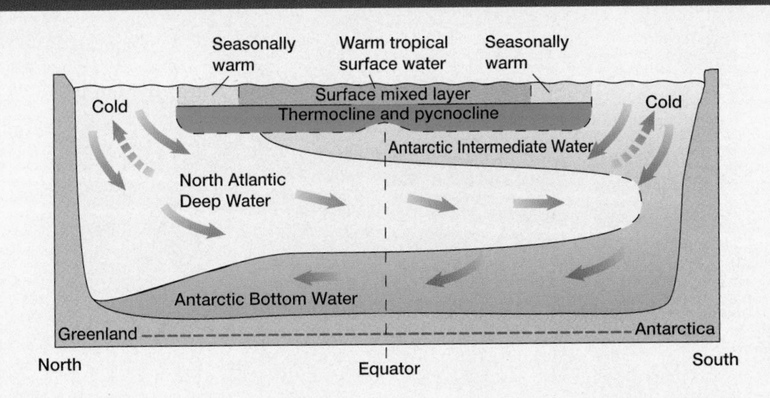


### Selected Key Terms

Use the **glossary** at the end of this book to discover the meanings of these Selected Key Terms: **thermohaline circulation, Antarctic Bottom Water, North Atlantic Deep Water, Antarctic Intermediate Water, conveyor-belt circulation.**

### Critical Thinking Question

Antarctic Intermediate Water can be identified throughout much of the South Atlantic based on its temperature, salinity, and dissolved oxygen content. Why is it colder and less salty—and why does it contain more oxygen—than the surface water mass above it and the North Atlantic Deep Water below it?

### Active Learning Exercise

The density T–S graph shown as Figure 7.29 shows that the Antarctic Intermediate Water (AAIW) has about the same density as the North Atlantic Central Surface Water (NACSW). With another student in class, discuss how you can differentiate the two water masses based on physical properties (not just their locations). As a group, explain how it is possible that these two distinct water masses have nearly the same density.

## 7.7 ▶ Can power from currents be harnessed as a source of energy?

- *Ocean currents can be harnessed as a source of power.* Although there is vast potential for developing this clean, renewable resource, significant problems must be overcome to make this a practical source of energy.

### Critical Thinking Question

What are some environmental factors that could inhibit the development of offshore ocean current power systems?

### Active Learning Exercise

With another student in class, discuss the advantages of current power devices that are attached to the sea floor and capable of operating in two or more directions.

## Mastering Oceanography www.masteringoceanography.com

Looking for additional review and test prep materials? With individualized coaching on the toughest topics of the course, Mastering Oceanography offers a wide variety of ways for you to move beyond memorization and deeply grasp the underlying processes of how the oceans work. Visit the Study Area in **www.masteringoceanography.com** to find practice quizzes, study tools, and multimedia that will improve your understanding of this chapter's content. Sign in today to access the following features: Self Study Quizzes, SmartFigures, Oceanography Videos and Animations, Squidtoons, Dynamic Study Modules, and an optional Pearson eText with embedded videos.

# Waves and Water Dynamics

# 8

**ESSENTIAL LEARNING CONCEPTS**

At the end of this chapter, you should be able to:

- ☐ **8.1** Demonstrate an understanding of how waves are generated and how they move.
- ☐ **8.2** Describe the characteristics of waves.
- ☐ **8.3** Discuss how wind-generated waves develop.
- ☐ **8.4** Explain how waves change in the surf zone.
- ☐ **8.5** Specify the origin and characteristics of tsunami.
- ☐ **8.6** Evaluate the advantages and disadvantages of harnessing waves as a source of energy.

↑ *Check when completed*

***"Can ye fathom the ocean, dark and deep, where the mighty waves and the grandeur sweep?"***

***—Poet Fanny Crosby (1820–1915)***

What combination of oceanographic factors causes waves to reach extreme heights at places such as Praia do Norte beach, just west of the town of Nazaré along the central Portuguese coast? One factor is that this premier big wave surf spot is located offshore of a prominent point of land, which, as will be explained in this chapter, tends to concentrate wave energy due to *wave refraction*. Another factor is that the point juts directly into the North Atlantic Ocean, which is known for its wintertime storms and giant waves. Still another factor is the presence of an offshore submarine canyon that focuses wave energy toward the shore. Usually, open-ocean waves approaching the coastline lose some of their energy when they make contact with the shallower ocean floor, but at Nazaré, the energy of ocean waves gets funneled through this deep submarine canyon, enabling the waves to become gigantic in size right at the shoreline. These factors make the site challenging to even the most skilled surfers, who brave these waters for the thrill of catching some of the world's most extreme waves. In fact, Praia do Norte is listed in *Guinness World Records* as having the biggest waves ever surfed. In November 2011, American surfer Garrett McNamara was towed via Jet Ski into a giant wave that was officially recorded as 23.8 meters (78 feet) high, which set a then-world record for the largest wave ever surfed. In November 2017, Brazilian big wave surfer Rodrigo Koxa surfed an even larger wave at Praia do Norte, which judges determined was 24.4 meters (80 feet) high.

Most waves are driven by the wind and are relatively small, so they release their energy gently. However, ocean storms can build up waves to extreme heights, and when these waves come ashore, they often produce devastating effects—or, in the case at Praia do Norte beach, a wild ride. Waves are *moving energy* traveling along the interface between ocean and atmosphere, often transferring energy from a storm far out at sea over distances of several thousand kilometers. That's why, even on calm days, the ocean is in continual motion as waves travel across its surface.

## 8.1 ▶ How Are Waves Generated, and How Do They Move?

All waves begin as disturbances, and ocean waves form as the result of a **disturbing force**. A rock thrown into a still pond, for example, creates a disturbance that generates waves radiating out in all directions. Ocean waves are caused by a similar transfer of energy to the surface of the ocean.

◂ **Surfer riding a giant wave at Nazaré, Portugal.** Big wave surfers are drawn to locations producing the largest waves, and in 2017 a new world's record was set by Brazilian surfer Rodrigo Koxa, who successfully rode a 24.4-meter (80-foot) breaking wave at Praia do Norte beach near Nazaré, Portugal. The giant waves at this site are produced by a unique combination of oceanographic factors (see text for details).

(a) A "desktop ocean." As the base tips the glass chamber back and forth, an internal wave develops along the boundary between the two different colored fluids that do not mix.

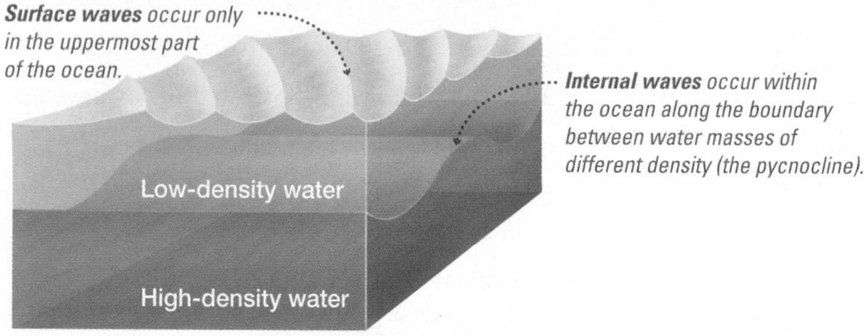
Surface waves occur only in the uppermost part of the ocean.
Internal waves occur within the ocean along the boundary between water masses of different density (the pycnocline).
Low-density water
High-density water

(b) Block diagram showing the differences between surface waves and internal waves.

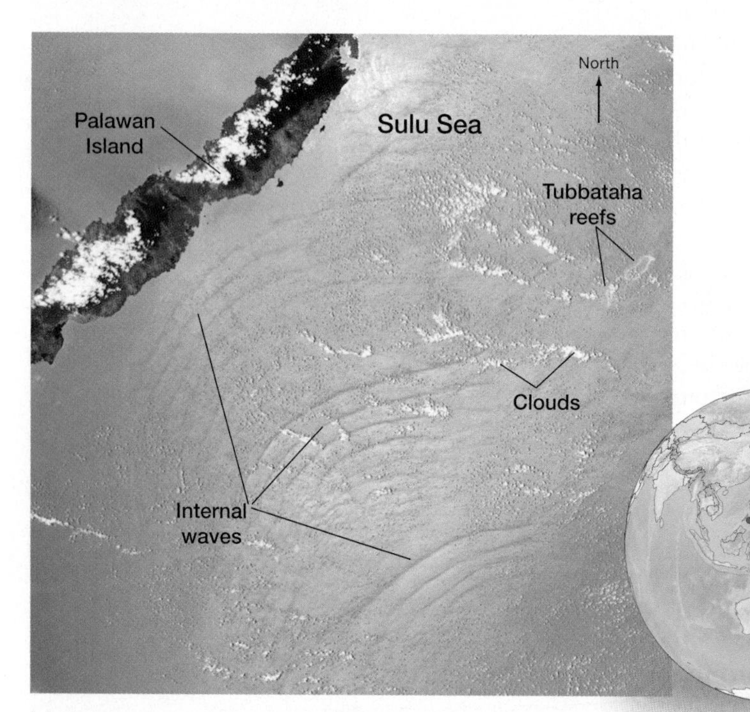
North
Palawan Island
Sulu Sea
Tubbataha reefs
Clouds
Internal waves

(c) View from space of internal waves highlighted by sunglint (bright areas) in the Sulu Sea between the Philippines and Malaysia. Image taken April 8, 2003, by the Moderate Resolution Imaging Spectroradiometer (MODIS) instrument aboard the Aqua satellite.

**Figure 8.1 Internal waves.**

## Disturbances Generate Ocean Waves

Wind blowing across the surface of the ocean generates most ocean waves. The waves radiate out in all directions, just as when a rock is thrown into the pond, but on a much larger scale.

Waves can also be created by the movement of fluids with different densities. These waves travel along the interface (boundary) between two different fluids. Both the air and the ocean are fluids, so waves can be created along interfaces *between* and *within* these fluids as follows:

- Along an *air–water interface*, the movement of air across the ocean surface creates **ocean waves** (simply called *waves*).
- Along an *air–air interface*, the movement of different air masses creates **atmospheric waves**, which are often represented by ripple-like clouds in the sky. Atmospheric waves are especially common when cold fronts (high-density air) move into an area.
- Along a *water–water interface*, the movement of water of different densities creates **internal waves**, as shown by the "desktop ocean" in Figure 8.1a and the block diagram of the ocean in Figure 8.1b. Because these waves travel along the boundary between waters of different density, they are associated in the ocean with a *pycnocline*, which, as discussed in Chapter 5, is a layer of rapidly changing density. Internal waves can be much larger than surface waves, with heights exceeding 100 meters (330 feet). Tidal movement, turbidity currents, wind stress, and even passing ships at the surface create internal waves, which can sometimes be observed from space (Figure 8.1c). Another clue to the presence of internal waves below is when lines of debris at the water's surface form parallel rows of slicks.

Internal waves can also be a hazard for submarines: If a submarine is caught in an internal wave while testing its depth limits, the submarine can inadvertently be carried to depths exceeding its designed pressure strength. In addition, internal waves are involved in ocean mixing and thus the transfer of heat, so understanding them is crucial to developing accurate global climate models.

Climate
Connection

Mass movement into the ocean, such as coastal landslides or large icebergs that fall from coastal glaciers, also creates waves. These waves are commonly called *splash waves* (see Mastering Oceanography **Web Diving Deeper 8.1** for a description of a large splash wave).

Another way in which large waves are created involves the uplift or downdropping of large areas of the sea floor or other sudden geological events, all of which can transfer large amounts of energy to the entire water column (compared to wind-driven waves, which affect only surface water). Examples include underwater avalanches (turbidity currents), volcanic eruptions, and fault slippage. The resulting waves are called *seismic sea waves* or *tsunami*, which are discussed later in this chapter. Fortunately, tsunami occur infrequently. When they do, however, they can flood coastal areas and cause large amounts of destruction.

The tides are also a type of wave, in this case caused mainly by the gravitational pull of the Moon and, to a lesser extent, the Sun. The tides are actually a wide-ranging and very predictable type of wave that will be discussed in Chapter 9, "Tides."

Human activities also cause ocean waves. When ships travel across the ocean, they leave behind a *wake*, which is a wave. In fact, smaller boats are often carried along in the wake of larger ships, and marine mammals sometimes play there.

In all cases, though, some type of energy transfer creates waves. Figure 8.2 shows the distribution of energy in waves, indicating that most ocean waves are wind generated.

## Wave Movement

A good way to think about waves is that waves are simply *energy in motion*. All types of waves transmit energy—or propagate—by means of cyclic movement through matter. The medium itself (solid, liquid, or gas) does not actually travel in the direction of the wave that is passing through it. The particles in the medium simply oscillate—or cycle—back and forth, and up and down, transferring energy from one particle to another. If you thump your fist on a table, for example, the energy travels through the table as waves that someone sitting at the other end can feel, but the whole table itself does not move its position. Another useful analogy is a field of grain: if you have ever seen wind pass across it, you know that the waveform moves as the grains sway, but the individual plants stay in the same location.

Different types of waves move in a variety of ways. Simple *progressive waves* (Figure 8.3) are waves that oscillate uniformly and progress or travel without breaking. Progressive waves are divided into *longitudinal*, *transverse*, or a combination of the two motions, called *orbital*.

In **longitudinal waves** (also known as push–pull waves), the particles that vibrate "push and pull" in the same direction that the wave is traveling, like a spring whose coils are alternately compressed and expanded. The shape of the wave (called a *waveform*) moves through the medium by compressing and decompressing as it goes. Sound, for instance, travels as longitudinal waves. Clapping your hands initiates a percussion that compresses and decompresses the air as the sound moves through a room. Energy can be transmitted through all states of matter—gaseous, liquid, or solid—by this longitudinal movement of particles.

In **transverse waves** (also known as side-to-side waves), energy travels at right angles to the direction of the vibrating particles. If one end of a rope is tied to a doorknob while the other end is moved up and down (or side to side) by hand, for example, a waveform progresses along the rope and energy is transmitted from the motion of the hand to the doorknob. The waveform moves up and down (or side to side) with the hand, but the motion is at right angles to the direction in which energy is transmitted (from the hand to the doorknob). Generally, transverse waves transmit energy only through solids, because the particles in solids are bound to one another strongly enough to transmit this kind of motion.

Longitudinal and transverse waves are called *body waves* because they transfer energy through a body of matter. Ocean waves are considered *surface waves* because they travel along the interface between two different fluids (air and water). The movement of particles in ocean waves involves components of *both* longitudinal and transverse waves, so particles move in circular orbits. Thus, waves at the ocean surface are **orbital waves** (also called *interface waves*).

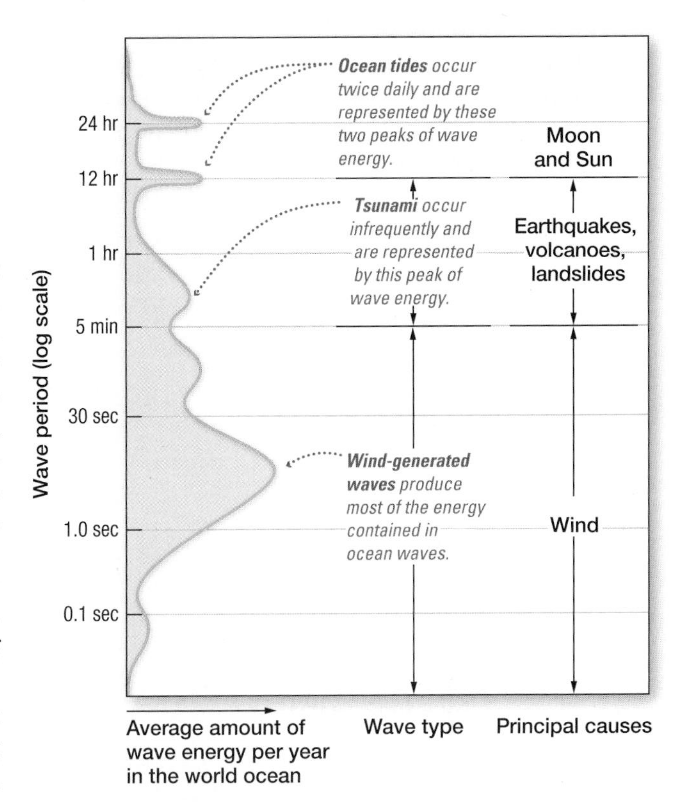


Figure 8.2 **Global distribution of energy in ocean waves.** Graph showing the average annual energy, types, and principal causes of waves in the world ocean. Note that the wave period is defined as the time it takes one full wave—one wavelength—to pass a fixed position.

### STUDENTS SOMETIMES ASK . . .

***Can internal waves break?***

Internal waves do not break in the way that surface waves break in the surf zone because the density difference across an interface within the water column is much smaller than that between water and air (the ocean–atmosphere interface). When internal waves approach the edges of continents, however, they undergo physical changes similar to those of waves in the surf zone. This causes the waves to buildup and expend their energy with much turbulent motion—in essence "breaking" against the continent.

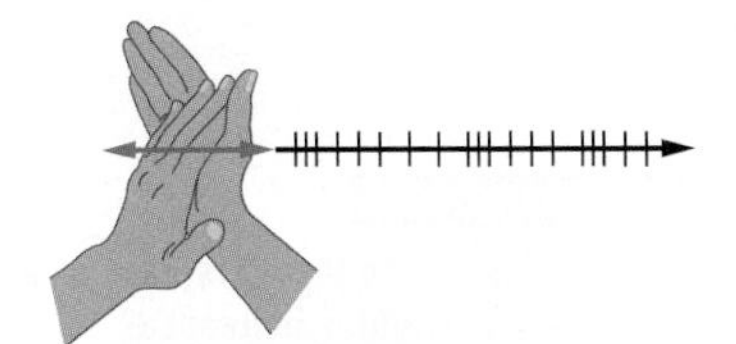

① LONGITUDINAL WAVE
*Hands clapping or thumping a table.* Particles (*blue color*) move back and forth in the direction of energy transmission. These waves transmit energy through all states of matter.

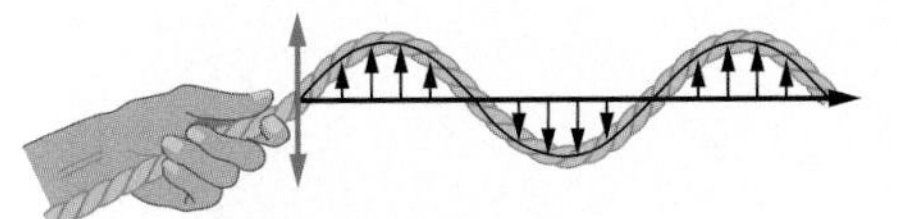

② TRANSVERSE WAVE
*A rope attached to a wall.* Particles (*blue color*) move back and forth at right angles to the direction of energy transmission. These waves transmit energy only through solids.

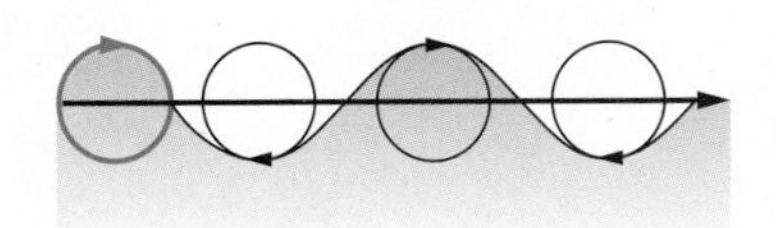

③ ORBITAL WAVE
*The movement of water waves.* Particles (*blue color*) move in a circular path. These waves transmit energy along interface between two fluids of different density (liquids and/or gases).

Figure 8.3 **Types of progressive waves.** Diagrammatic view of the three types of progressive waves. Examples include: ① longitudinal waves, ② transverse waves, and ③ orbital waves.

**RECAP** **Most ocean waves are caused by wind, but many other types of waves, including internal waves, splash waves, tsunami, tides, and human-induced waves, are created by transfer of energy in the ocean.**

**CONCEPT CHECK 8.1 ▶ Demonstrate an understanding of how waves are generated and how they move.**

1 Discuss several different ways in which waves form. How are most ocean waves generated?

2 Why is the development of internal waves likely within the pycnocline?

3 Describe the three types of motion in which progressive waves move. Which type includes waves at the ocean surface?

## 8.2 ▶ What Characteristics Do Waves Possess?

Figure 8.4a shows the characteristics of an idealized ocean wave. The simple, uniform, moving waveform transmits energy from a single source and travels along the ocean–atmosphere interface. These waves are also called *sine waves* because their uniform shape resembles the oscillating pattern expressed by a sine curve. Even though idealized waveforms do not exist in nature, they help us understand wave characteristics.

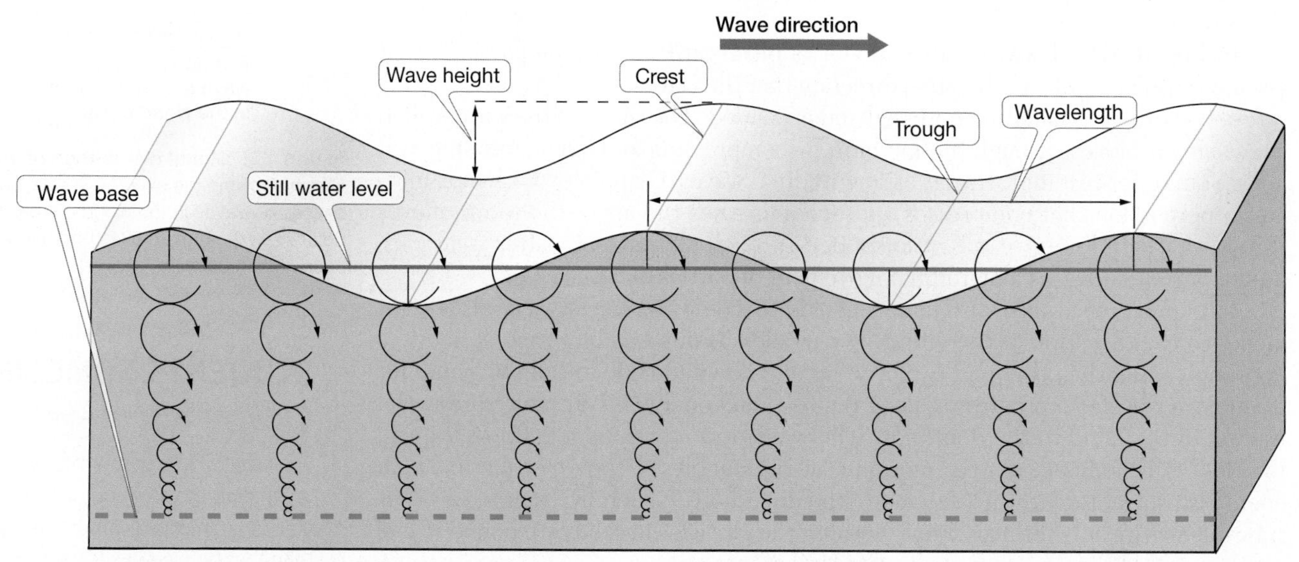


(a) Diagrammatic view of an idealized progressive ocean wave showing wave characteristics and terminology.

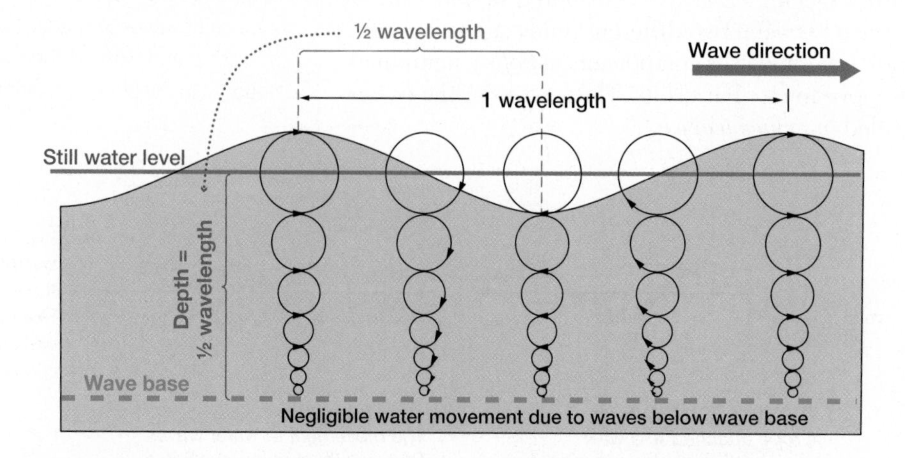


(b) Detail showing the decreasing size of orbital motion of water particles in waves with depth. Note that wave base (the depth at which orbital motion ceases) is at a depth of one-half the wavelength, measured from still water level.

**EXPLORING DATA ▲**

**1. What happens to the wave base as a wave's wave height is increased?**

**2. What happens to the wave base as a wave's wavelength is increased?**

Figure 8.4 **Characteristics and terminology of a typical progressive wave.**

## Wave Terminology

As an idealized wave passes a permanent marker (such as a pier piling), a succession of high parts of the waves, called **crests**, alternate with low parts, called **troughs**. Halfway between the crests and the troughs is the **still water level**, or *zero energy level*. This is the level of the water if there were no waves. The **wave height**, designated by the symbol $H$, is the vertical distance between a crest and a trough.

The horizontal distance between any two corresponding points on successive waveforms, such as from crest to crest or from trough to trough, is the **wavelength**, $L$. **Wave steepness** is the ratio of wave height to wavelength:

$$\text{Wave steepness} = \frac{\text{Wave height } (H)}{\text{Wave length } (L)} \tag{8.1}$$

If the wave steepness exceeds 1/7, the wave *breaks* (spills forward) because the wave is too steep to support itself. A wave can break anytime the 1:7 ratio is exceeded, either along the shoreline or out at sea. This ratio also dictates the maximum height of a wave. For example, a wave 7 meters long can only be 1 meter high; if the wave is any higher than that, it will break.

The time it takes one full wave—one wavelength—to pass a fixed position (such as a pier piling) is the **wave period**, $T$. Typical wave periods range between 6 and 16 seconds. The **frequency** ($f$) is defined as the number of wave crests passing a fixed location per unit of time and is the inverse of the period:

$$\text{Frequency } (f) = \frac{1}{\text{Period } (T)} \tag{8.2}$$

For example, consider waves with a period of 12 seconds. These waves have a frequency of 1/12 or 0.083 waves per second, which converts to 5 waves per minute.

## Circular Orbital Motion

Waves can travel great distances across ocean basins. In one classic study conducted in 1963 by a team of researchers led by Dr. Walter Munk at the Scripps Institution of Oceanography, waves generated near Antarctica were tracked at various points as they traveled through the Pacific Ocean basin. After more than 10,000 kilometers (over 6000 miles), the waves finally expended their energy a week later along the shoreline of the Aleutian Islands of Alaska. The result of this study and many others indicate that water itself doesn't travel the entire distance, but *the waveform does*. As the wave travels, the water passes the energy along by moving in a circle. This movement is called **circular orbital motion**.

Observation of an object floating in the waves reveals that it moves not only up and down but also slightly forward and backward with each successive wave. **Figure 8.5** shows that a floating object moves up and backward as the crest approaches, up and forward as the crest passes, down and forward after the crest, down and backward as the trough approaches, and up and backward again as the next crest advances. When the movement of the rubber ducky, shown in Figure 8.5, is traced as a wave passes, it can be seen that the ducky moves in a circle and returns close to its original position. (Actually, the circular orbit does not quite return the floating object to its original position because the half of the orbit accomplished in the trough is slower than the crest half of the orbit. This results in a slight forward movement [net mass transport], which is called *wave drift*). Nonetheless, this circular orbital motion allows a waveform (the wave's shape) to move forward through the water while the individual water particles that transmit the wave move around in a circle and return to essentially the same place. Wind moving across a field of wheat causes a similar phenomenon: the wheat itself doesn't travel across the field, but the waves do.

The circular orbits of an object floating at the surface have a diameter equal to the wave height (Figure 8.4a). **Figure 8.4b** shows that circular orbital motion dies out

### STUDENTS SOMETIMES ASK . . .

***Since the water is actually going around in a circle and returns to just about the same place, what actually is being transmitted as waves move?***

In a word: energy. The energy of waves has been known to wreak havoc on coastal structures, lift submerged obstacles that weigh many tons, and even change the shape of entire beaches. If you have ever watched really big waves breaking at a beach, you'll witness the impressive amount of energy waves can transmit.

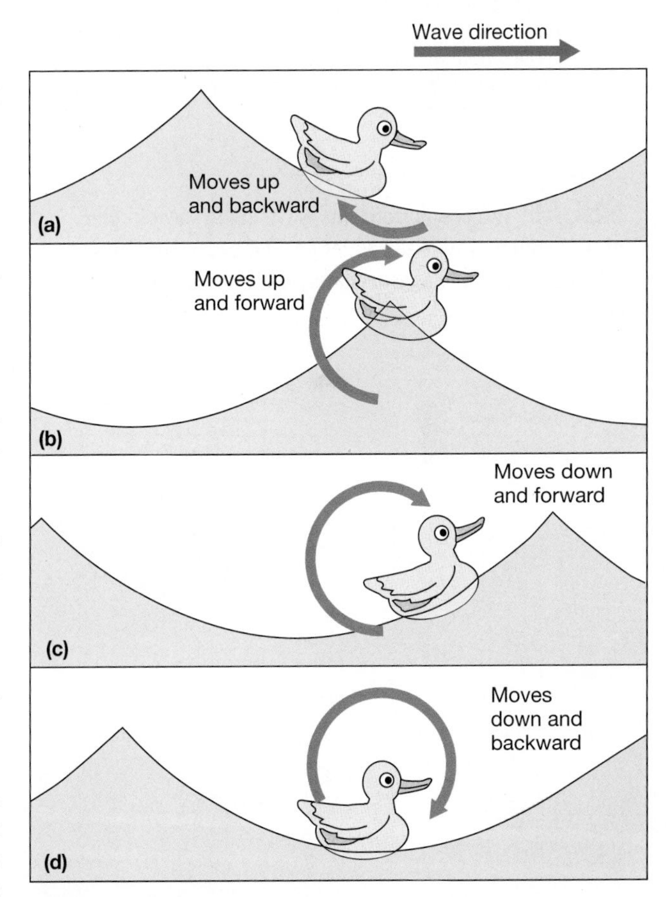


**SmartFigure 8.5 A floating rubber ducky shows circular orbital motion.** As a wave passes from left to right, the motion of a floating rubber ducky resembles that of a circle, which is known as circular orbital motion. https://goo.gl/zvZ6Mf

**Animation**
Wave Circular Orbital Motion
http://goo.gl/1wEScd

**Figure 8.6 Floating airport runway in waters off Japan.** This floating airport runway, called "Mega-Float," was built offshore Yokosuka in Tokyo Bay, Japan. It is the world's largest floating runway, with a length of 1000 meters (3300 feet). Floating runways and bridges use submerged pontoons (not shown) to make them stable at the surface; note that the majority of the structure's mass is below wave base.

quickly below the surface. At some depth below the surface, the circular orbits become so small that movement is negligible. This depth is called the **wave base**, and it is equal to one-half the wavelength ($L/2$) *measured from still water level*. Only wavelength controls the depth of the wave base, so the longer the wave, the deeper the wave base.

The decrease of orbital motion with depth has many practical applications. For instance, submarines can avoid large ocean waves simply by submerging below the wave base. Even the largest storm waves will go unnoticed if a submarine submerges to only 150 meters (500 feet). Floating bridges and floating oil rigs are constructed so that most of their mass is below wave base, so they will be unaffected by wave motion. In fact, offshore floating airport runways have been designed using similar principles (**Figure 8.6**). In addition, seasick scuba divers find relief when they submerge into the calm, motionless water below wave base. Finally, as you walk from the beach into the ocean, you reach a point where it is easier to dive under an incoming wave than to jump over it. It is easier to swim through the smaller orbital motion below the surface than to fight the large waves at the surface.

**RECAP** **The ocean transmits wave energy by circular orbital motion, where the water particles move in circular orbits and return to approximately the same location.**

## Deep-Water Waves

If the water depth ($d$) is greater than the wave base ($L/2$), the waves are called **deep-water waves** (**Figure 8.7a**). Deep-water waves have no interference with the ocean bottom, so they include all wind-generated waves in the open ocean, where water depths far exceed wave base.

**Wave speed** ($S$) is the rate at which a wave travels. Numerically, it is the distance traveled divided by the travel time; for a wave, it can be calculated as:

$$\text{Wave speed } (S) = \frac{\text{Wavelength } (L)}{\text{Period } (T)} \tag{8.3}$$

Wave speed is more correctly known as *celerity* ($C$), which is different from the traditional concept of speed. Celerity is used only in relation to waves where no mass is in motion, just the waveform.

According to the equations that govern the movement of progressive waves, the speed of deep-water waves is dependent upon (1) wavelength and (2) several other variables (such as gravitational attraction) that remain constant on Earth. So, by filling in the constants with numbers, the equation for wave speed of deep-water waves varies only with wavelength and becomes (in meters per second):

$$S(\text{in meters per second}) = 1.25\sqrt{L(\text{in meters})} \tag{8.4}$$

or in feet per second:

$$S(\text{in feet per second}) = 2.26\sqrt{L(\text{in feet})} \tag{8.5}$$

We can also determine wave speed if we know only the period ($T$) because wave speed ($S$) is defined in Equation 8.3 as $L/T$. Filling in the known variables with numbers gives (in meters per second):

$$S(\text{in meters per second}) = 1.56 \times T \tag{8.6}$$

or in feet per second:

$$S(\text{in feet per second}) = 5.12 \times T \quad (8.7)$$

The graph in **Figure 8.8** uses the above equations to relate the wavelength, period, and speed of deep-water waves. Of the three variables, the wave period is usually easiest to measure. Since all three variables are related, the other two can be determined by using Figure 8.8. For example, the vertical red line in Figure 8.8 shows that a wave with a period of 8 seconds has a wavelength of 100 meters. Thus, the speed of the wave is shown by the horizontal red line in Figure 8.8, which is:

$$\text{Speed } (S) = \frac{L}{T} = \frac{100 \text{ meters}}{8 \text{ seconds}} = 12.5 \text{ meters per second} \quad (8.8)$$

In summary, the general relationship shown by Equations 8.3 through 8.8 (and shown in Figure 8.8) for deep-water waves is *the longer the wavelength, the faster the wave travels*. A fast wave does not necessarily have a large wave height, however, because wave speed depends *only* on wavelength.

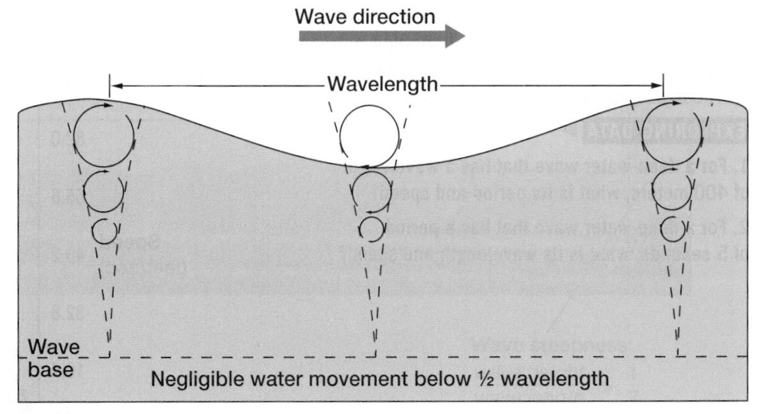


**(a) Deep-water wave:** Circular orbits diminish in size with increasing depth. Water depth is greater than $^1/_2$ wavelength.

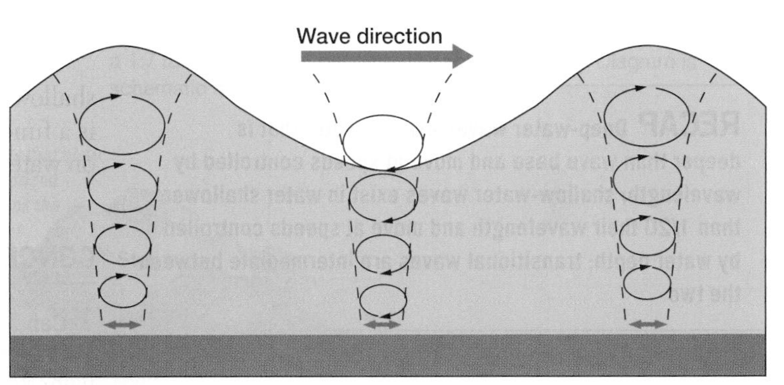


**(b) Transitional wave:** Intermediate between deep-water and shallow-water waves. Water depth is greater than $^1/_{20}$ wavelength, but less than $^1/_2$ wavelength.

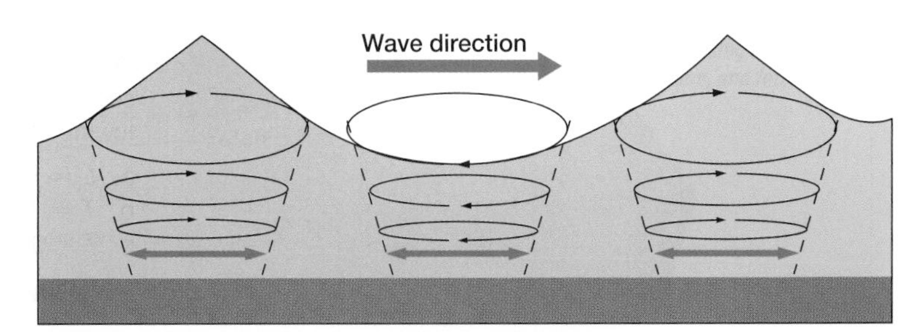


**(c) Shallow–water wave:** The ocean floor interferes with circular orbital motion, causing the orbits to become progressively flattened. Water depth is less than $^1/_{20}$ wavelength.

Figure 8.7 **Characteristics of deep-water, transitional, and shallow-water waves.** Diagrammatic views showing the characteristics of **(a)** deep-water waves, **(b)** transitional waves, and **(c)** shallow-water waves. Diagrams are not to scale.

## Shallow-Water Waves

Waves in which depth ($d$) is less than 1/20 of the wavelength ($L$/20) are called **shallow-water waves**, or *long waves* (**Figure 8.7c**). Shallow-water waves are said to *touch bottom* or *feel bottom* because they touch the ocean floor, which interferes with the wave's orbital motion. Note that for shallow-water waves, orbital motion is present throughout the entire water column, so scuba divers will find no relief from the effects of circular orbital motion because they can't descend below wave base.

The speed of shallow-water waves is influenced only by gravitational acceleration ($g$) and water depth ($d$). Since gravitational acceleration remains constant on Earth, the equation for wave speed becomes (in meters per second):

$$S(\text{in meters per second}) = 3.13\sqrt{d(\text{in meters})} \quad (8.9)$$

or in feet per second:

$$S(\text{in feet per second}) = 5.67\sqrt{d(\text{in feet})} \quad (8.10)$$

Equations 8.9 and 8.10 show that wave speed in shallow-water waves is determined only by water depth, where *the deeper the water, the faster the wave travels*.

Examples of shallow-water waves include wind-generated waves that have moved into shallow nearshore areas; *tsunami* (seismic sea waves), generated by earthquakes in the ocean floor; and the *tides*, which are a type of wave generated by the gravitational attraction of the Moon and the Sun. Although tsunami and tides travel across the deepest ocean basins, they are considered shallow-water waves because their wavelengths far exceed the depth of the ocean.

Particle motion in shallow-water waves is in a very flat elliptical orbit that approaches horizontal (back-and-forth) oscillation. The vertical component of particle motion decreases with increasing depth below sea level, causing the orbits to become even more flattened.

## Transitional Waves

Waves that have some characteristics of shallow-water waves and some of deep-water waves are called **transitional waves**. The wavelengths of transitional waves are between 2 and 20 times the water depth (**Figure 8.7b**). The wave speed of

## SmartTable 8.1 Beaufort wind scale and the state of the sea

| Beaufort number | Descriptive term | Wind speed km/hr | Wind speed mi/hr | Photo | Appearance of the sea |
|---|---|---|---|---|---|
| 0 | Calm | <1 | <1 | | Like a Mirror |
| 1 | Light air | 1-5 | 1-3 | | Ripples with the appearance of scales, no foam crests |
| 2 | Light breeze | 6-11 | 4-7 | | Small wavelets; crests of glassy appearance, no breaking |
| 3 | Gentle breeze | 12-19 | 8-12 | | Large wavelets; crests begin to break, scattered whitecaps |
| 4 | Moderate breeze | 20-28 | 13-18 | | Small waves, becoming longer; numerous whitecaps |
| 5 | Fresh breeze | 29-38 | 19-24 | | Moderate waves, taking longer form; many whitecaps, some spray |
| 6 | Strong breeze | 39–49 | 25–31 | | Large waves begin to form, whitecaps everywhere, more spray |
| 7 | Near gale | 50–61 | 32–38 | | Sea heaps up and white foam from breaking waves begins to be blown in streaks |
| 8 | Gale | 62–74 | 39–46 | | Moderately high waves of greater length, edges of crests begin to break into spindrift, foam is blown in well-marked streaks |
| 9 | Strong gale | 75–88 | 47–54 | | High waves, dense streaks of foam and sea begins to roll, spray may affect visibility |
| 10 | Storm | 89–102 | 55–63 | | Very high waves with overhanging crests; foam is blown in dense white streaks, causing the sea to appear white; the rolling of the sea becomes heavy; visibility reduced |
| 11 | Violent storm | 103–117 | 64–72 | | Exceptionally high waves (small and medium-sized ships might for a time be lost from view behind the waves), the sea is covered with white patches of foam, everywhere the edges of the wave crests are blown into froth, visibility further reduced |
| 12 | Hurricane | 118+ | 73+ | | The air is filled with foam and spray, sea completely white with driving spray, visibility greatly reduced |

**SmartTable 8.1 Beaufort wind scale and the state of the sea.**
https://goo.gl/R9PVQf

**EXPLORING DATA ▲**

**1. What is the Beaufort wind scale number and descriptive term for a sea where small and medium-sized ships might for a time be lost from view behind waves?**

**2. What is the Beaufort wind scale number and descriptive term for a sea that has dense streaks of foam?**

Figure 8.11 is a map based on satellite data of average wave heights during October 3–12, 1992. The waves in the Southern Hemisphere are particularly large because the prevailing westerlies between 40 and 60 degrees south latitude reach the highest average wind speeds on Earth, creating the latitudes called the "Roaring Forties," "Furious Fifties," and "Screaming Sixties."

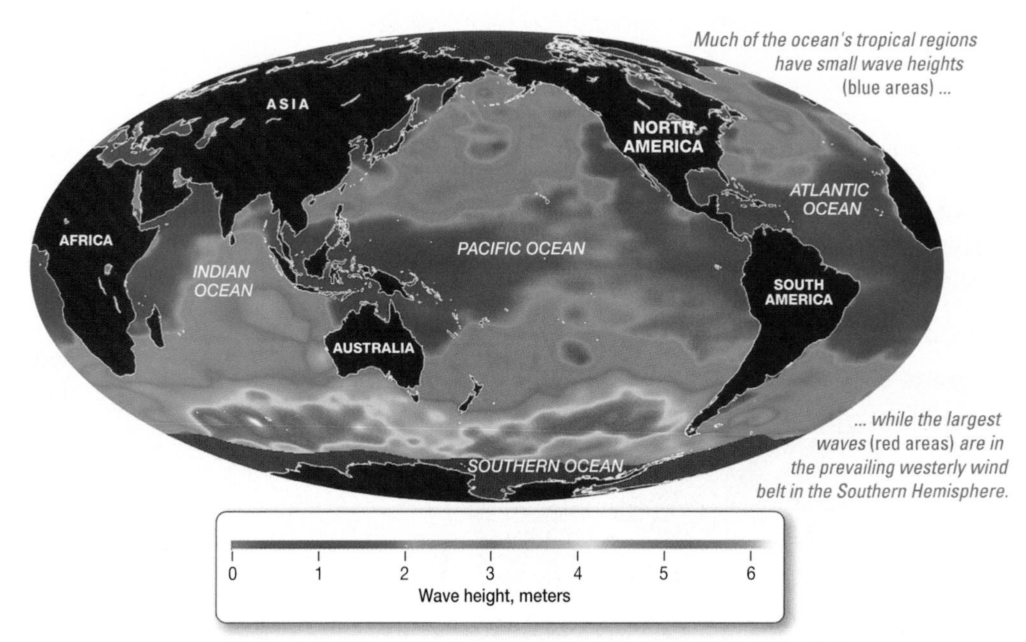


**Figure 8.11 Map of satellite-derived wave height.** While in operation, the TOPEX/Poseidon satellite received a return of stronger radar signals from calm seas and weaker radar signals from seas with large waves. Based on these data, a map of average wave height can be produced. Data was collected during October 3–12, 1992; scale is in meters.

**HOW HIGH CAN WAVES BE?** According to a U.S. Navy Hydrographic Office bulletin published in the early 1900s, the theoretical maximum height of wind-generated waves should be no higher than 18.3 meters (60 feet); this became known as the "60-foot rule." Although there were some isolated eyewitness accounts of larger waves, the U.S. Navy considered any sightings of waves over 60 feet to be exaggerations. Certainly, embellishment of reported wave height under conditions of extremely rough seas would be understandable. For many years, the "60-foot rule" was accepted as fact.

Careful observations made aboard the 152-meter- (500-foot-) long U.S. Navy tanker USS *Ramapo* in 1933 proved the "60-foot rule" incorrect. The ship was caught in a typhoon in the western Pacific Ocean and encountered 108-kilometer-per-hour (67-mile-per-hour) winds en route from the Philippines to San Diego. The resulting waves were symmetrical, uniform, and had a period of 14.8 seconds. Because the *Ramapo* was traveling with the waves, the vessel's officers were able to measure the waves accurately. The officers used the dimensions of the ship, including the *eye height* of an observer on the ship's bridge (Figure 8.12). Geometric relationships revealed that the waves were 34 meters (112 feet) high, which is taller than an 11-story building! These waves proved to be a record that still stands today for the largest authentically recorded wind-generated waves, shattering the "60-foot rule." Although the *Ramapo* was largely undamaged, other ships traveling in rough seas aren't always so lucky (Figure 8.13). In fact, several large ships disappear at sea every year simply due to enormous waves.

**FULLY DEVELOPED SEA** For a given wind speed, Table 8.2 lists the minimum fetch and duration of wind beyond which the waves cannot grow. Waves cannot grow because an equilibrium condition, called a **fully developed sea**, has been achieved. Waves can grow no further in a fully developed sea because they lose as much energy breaking as whitecaps under the force of gravity as they receive from the wind. Table 8.2 also lists the average characteristics of waves resulting from a fully developed sea, including the height of the highest 10% of the waves.

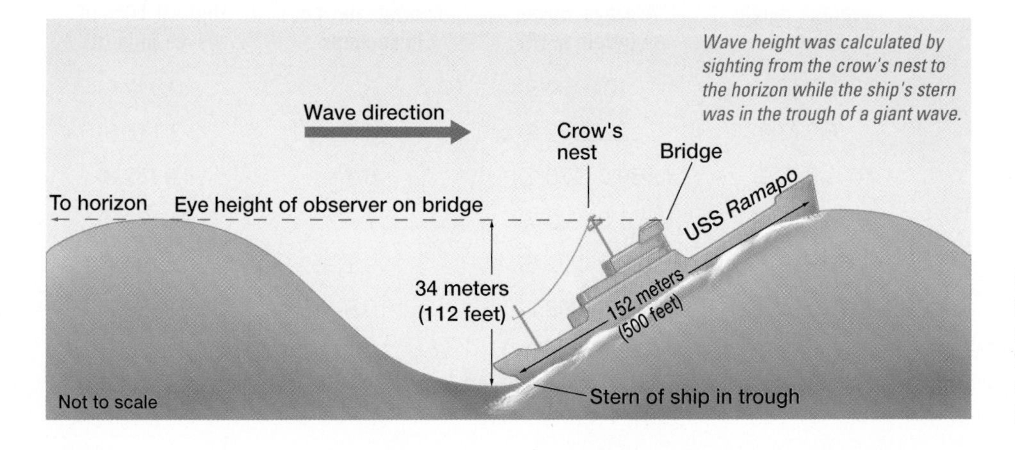


**Figure 8.12 The U.S. Navy tanker USS *Ramapo* measured the largest authentically recorded waves (1933).** Diagrammatic view of the record-setting wave height of 34 meters (112 feet) that was calculated based on geometric relationships of the vessel in the giant waves. Note that the record has not been exceeded since.

**Figure 8.13 Wave damage on the aircraft carrier *Bennington* (1945).** The *Bennington* returns from heavy seas encountered in a typhoon off Okinawa in 1945, with part of its reinforced steel flight deck bent down over its bow. Damage to the flight deck, which is 16.5 meters (54 feet) above still water level, was caused by large waves.

**SWELL** As waves generated in a sea area move toward its margins, wind speeds diminish and the waves eventually move faster than the wind. When this occurs, wave steepness decreases and waves become long-crested waves called **swells** (*swellan* = swollen), which are uniform, symmetrical waves that have traveled out of their area of origination. Swells move with little loss of energy over large stretches of the ocean surface, transporting energy away from one sea area and depositing it in another. The movement of swells to distant areas is the reason why there can be waves at a shoreline even though there is no wind.

Waves with longer wavelengths travel faster and thus leave the sea area first. They are followed by slower, shorter **wave trains**, or groups of waves. The progression from long, fast waves to short, slow waves illustrates the principle of **wave dispersion** (*dis* = apart, *spargere* = to scatter)—the sorting of waves by their wavelength. Waves of many wavelengths are present in the generating area. Wave speed depends on wavelength in deep water (see Figure 8.8), however, so the longer waves "outrun" the shorter ones. This has important implications regarding which waves reach a beach first when there is a distant storm out at sea; as most surfers know, the long wavelength waves arrive before the shorter, choppy waves come ashore. The distance over which waves change from a choppy "sea" to uniform swell is called the **decay distance**, which can be up to several hundred kilometers.

As a group of waves leaves a sea area and becomes a *swell wave train*, the leading wave keeps disappearing. However, the same number of waves always remains in the group because as the leading wave disappears, a new wave replaces it at the back of the group (**Figure 8.14**). For example, if four waves are generated, the lead wave keeps dying out as the wave train travels, but one is created in the back, so the

## STUDENTS SOMETIMES ASK . . .

***I know that swell is what surfers hope for. Is swell always big?***

Not necessarily. Swell is defined as waves that have moved out of their area of origination, so these waves do not have to be a certain wave height to be classified as swell. It is true, however, that the uniform and symmetrical shape of most swell delights surfers.

**Table 8.2 Conditions necessary to produce a fully developed sea at various wind speeds and the characteristics of the resulting waves**

| These conditions . . . | | | . . . produce these waves | | | |
|---|---|---|---|---|---|---|
| **Wind speed in km/hr (mi/hr)** | **Fetch in km (mi)** | **Duration in hours** | **Average height in m (ft)** | **Average wave-length in m (ft)** | **Average period in seconds** | **Highest 10% of waves in m (ft)** |
| 20 (12) | 24 (15) | 2.8 | 0.3 (1.0) | 10.6 (34.8) | 3.2 | 0.8 (2.5) |
| 30 (19) | 77 (48) | 7.0 | 0.9 (2.9) | 22.2 (72.8) | 4.6 | 2.1 (6.9) |
| 40 (25) | 176 (109) | 11.5 | 1.8 (5.9) | 39.7 (130.2) | 6.2 | 3.9 (12.8) |
| 50 (31) | 380 (236) | 18.5 | 3.2 (10.5) | 61.8 (202.7) | 7.7 | 6.8 (22.3) |
| 60 (37) | 660 (409) | 27.5 | 5.1 (16.7) | 89.2 (292.6) | 9.1 | 10.5 (34.4) |
| 70 (43) | 1093 (678) | 37.5 | 7.4 (24.3) | 121.4 (398.2) | 10.8 | 15.3 (50.2) |
| 80 (50) | 1682 (1043) | 50.0 | 10.3 (33.8) | 158.6 (520.2) | 12.4 | 21.4 (70.2) |
| 90 (56) | 2446 (1517) | 65.2 | 13.9 (45.6) | 201.6 (661.2) | 13.9 | 28.4 (93.2) |

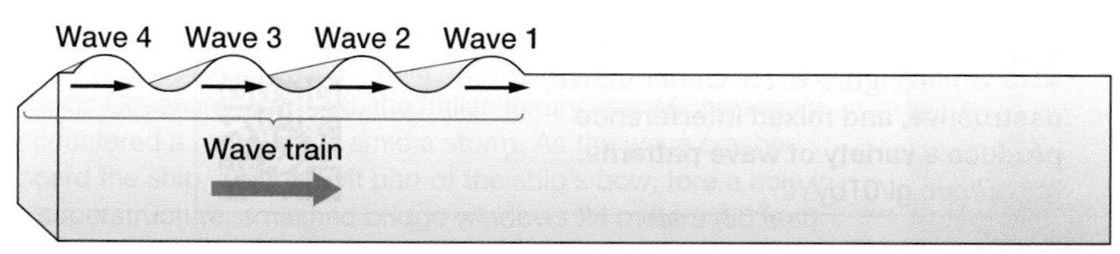


**(a)** Energy in the leading waves (Waves 1 and 2) is transferred into circular orbital motion.

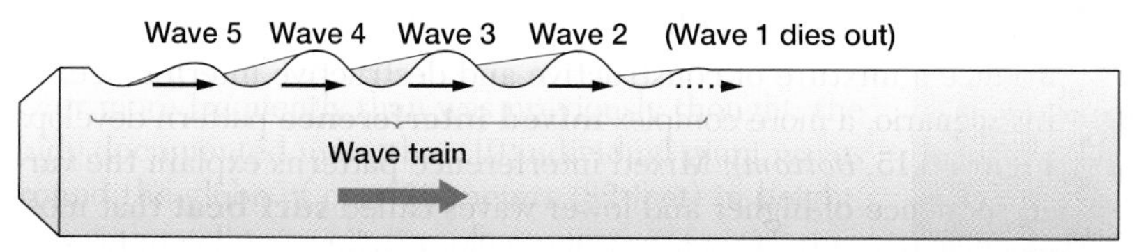


**(b)** Wave 1 dies out and is replaced by Wave 2; note new Wave 5 behind.

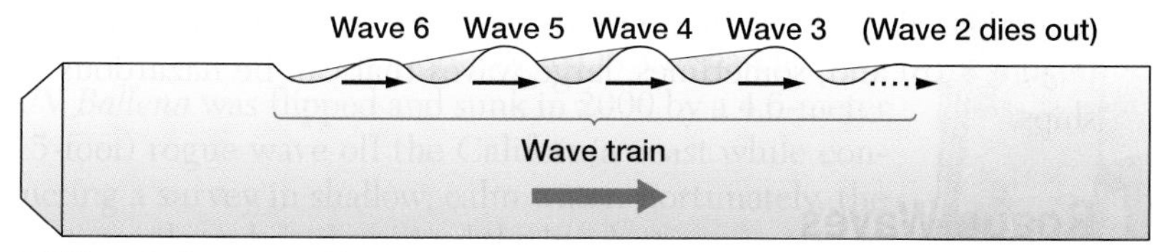


**(c)** Wave 2 dies out and is replaced by Wave 3; note new Wave 6 behind.

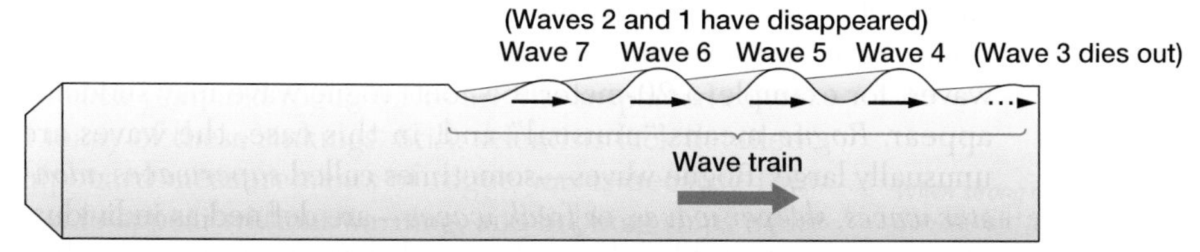


**(d)** Wave 3 dies out and is replaced by Wave 4; note new Wave 7 behind. Even though new waves take up the lead, the length of the wave train and the total number of waves remain the same. This causes the group speed to be one-half that of the individual wave.

Figure 8.14 **Movement of a wave train.** Diagrammatic sequence showing the movement of a wave train. Note how waves die out in the front but are replaced in the back, causing the length of the wave train and the total number of waves to remain the same even as the wave train progresses.

wave train stays four waves. Because of the progressive dying out and creation of new waves, the group moves across the ocean surface at only *half* the velocity of an individual wave in the group.

## Interference Patterns

When swells from different storms run together, the waves clash, or *interfere* with one another, giving rise to **interference patterns**. An interference pattern produced when two or more wave systems collide is the sum of the disturbance that each wave would have produced individually. **Figure 8.15** shows that the result may be a larger or smaller trough or crest, depending on conditions.

**CONSTRUCTIVE INTERFERENCE** **Constructive interference** occurs when wave trains having the same wavelength come together *in phase*, meaning crest to crest and trough to trough. If the displacements from each wave are added together, the interference pattern results in a wave with the same wavelength as the two overlapping wave systems but with a wave height equal to the sum of the individual wave heights (Figure 8.15, *top*).

**DESTRUCTIVE INTERFERENCE** **Destructive interference** occurs when wave trains having the same wavelength come together *out of phase*, meaning the crest from one wave coincides with the trough from a second wave. If the waves have identical heights, the sum of the crest of one and the trough of another is zero, so the energies of these waves cancel each other (Figure 8.15, *middle*).

## STUDENTS SOMETIMES ASK . . .

### *What is the difference between "groundswell" and "wind swell"?*

The term *groundswell* was originally a sailor's word (and is now a surfer's term) for deep-ocean swell, such as might be generated by a distant storm or earthquake. In its original sense, groundswell referred to waves so huge that their troughs bared the "ground" of the sea bottom. Groundswell is essentially the same thing as *wind swell,* although groundswell often refers to very large waves from a distant origin, whereas wind swell refers to smaller, locally produced waves.

## 8.4 ▶ How Do Waves Change in the Surf Zone?

Most waves generated in the sea area by storm winds move across the ocean as swell. These waves then release their energy along the margins of continents in the **surf zone**, which is the zone of breaking waves. Breaking waves exemplify power and persistence, sometimes moving objects weighing several tons. In doing so, energy from a distant storm can travel thousands of kilometers until it is finally expended along a distant shoreline in a few wild moments.

### Physical Changes as Waves Approach Shore

As deep-water waves of swell move toward continental margins over gradually **shoaling** (*shold* = shallow) water, they eventually encounter water depths that are less than one-half of their wavelength (**Figure 8.19**) and become transitional waves. Actually, any shallowly submerged obstacle (such as a coral reef, sunken wreck, or sand bar) will cause waves to release some energy. Navigators have long known that breaking waves indicate dangerously shallow water.

Many physical changes occur to a wave as it encounters shallow water, becomes a shallow-water wave, and eventually breaks. The progressively shallower depths interfere with water particle movement at the base of the wave, so the *wave speed decreases*. As one wave slows, the following waveform, which is still moving at its original speed, moves closer to the wave that is being slowed, causing a *decrease in wavelength*. Although some wave energy is lost due to friction, the wave energy that remains must go somewhere, so *wave height increases*. This increase in wave height combined with the decrease in wavelength causes an *increase in wave steepness* ($H/L$). When the wave steepness reaches the 1:7 ratio, the *waves break as surf* (Figure 8.19).

If the surf is swell that has traveled from distant storms, breakers will develop relatively near shore, in shallow water. The horizontal motion characteristic of shallow-water waves moves water alternately toward and away from the shore as an oscillation. The surf will be characterized by parallel lines of relatively uniform breakers.

If the surf consists of waves generated by local winds, the waves may not have been sorted into swell. The surf may be mostly unstable, deep-water, high-energy waves with steepness already near the 1:7 ratio. In this case, the waves will break shortly after feeling bottom some distance from shore, and the surf will be rough, choppy, and irregular.

When the water depth is about one and one-third times the wave height, the crest of the wave breaks, producing surf. (Using this relationship provides a handy way of estimating water depth in the surf zone: the depth of the water where waves are breaking is one and one-third times the breaker height.) When the water depth becomes less than $^1/_{20}$ the wavelength, waves in the surf zone begin to behave like shallow-water waves (see Figure 8.7). Particle motion is greatly

### STUDENTS SOMETIMES ASK . . .

***How much more energy does a 3-foot wave pack compared to a 1-foot wave?***

Interestingly, the energy that a wave produces is proportional to the height of the wave *squared*. Therefore, a 3-foot wave is nine—not three—times more powerful than a 1-foot wave. When onshore breaking waves reach 5 feet, the surf is generally too dangerous for swimming, and anyone who has been tumbled around by waves can attest to their incredible power. Although experienced surfers relish big waves, good surfing beaches are generally not safe for swimming.

**Animation**
Wave Motion and Wave Refraction
http://goo.gl/55FMau

**EXPLORING DATA** ▼

1. **What happens to a wave when its steepness exceeds the critical 1/7 ratio?**
2. **Assume that there is a shallowly-submerged object that extends above the wave base on the left side of the figure. What will happen to waves there? Can a wave break twice?**

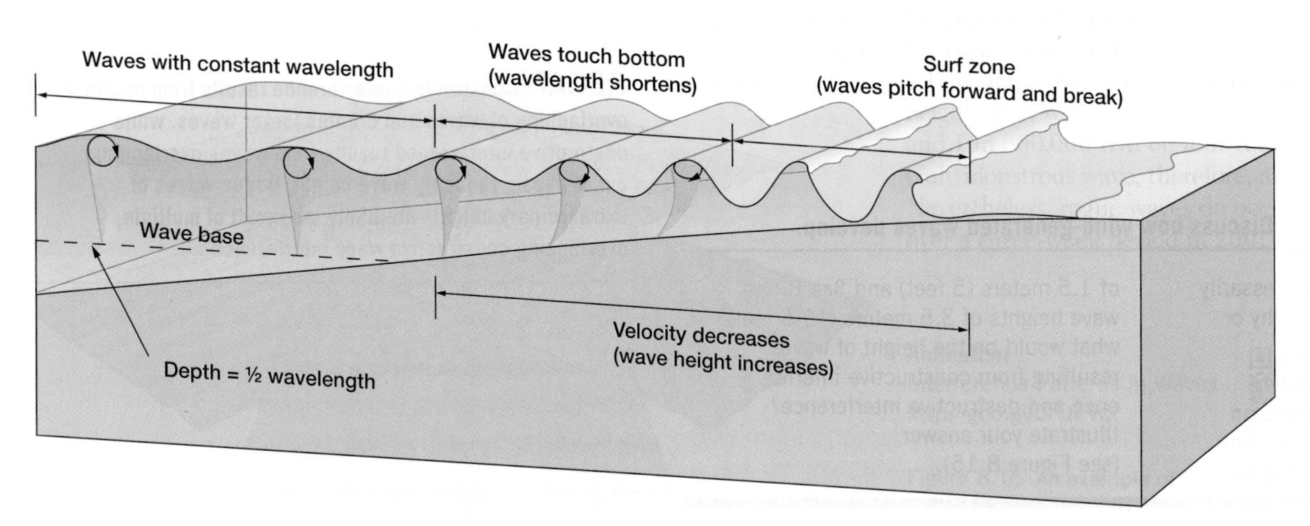


**Figure 8.19 Physical changes of a wave in the surf zone.** As waves approach the shore and encounter water depths of less than one-half wavelength, the waves "feel bottom." The *wave speed decreases*, and waves stack up against the shore, causing the *wavelength to decrease*. This results in an *increase in wave height* to the point where the wave *steepness is increased* beyond the 1:7 ratio, causing the wave to *pitch forward and break* in the surf zone.

impeded by the bottom, and a significant transport of water toward the shoreline occurs (Figure 8.19).

Waves break in the surf zone because particle motion near the bottom of the wave is severely restricted, slowing the waveform. At the surface, however, individual orbiting water particles have not yet been slowed because they have no contact with the bottom. In addition, the wave height increases in shallow water. The difference in speed between the top and bottom parts of the wave causes the top part of the wave to overrun the lower part, which results in the wave toppling over and breaking. Breaking waves are analogous to a person who leans too far forward. If you don't catch yourself, you may also "break" something when you fall.

## Breakers and Surfing

There are three main types of breakers. Figure 8.20a shows a **spilling breaker**, which is a turbulent mass of air and water that runs down the front slope of the wave as it breaks. Spilling breakers result from a gently sloped ocean bottom, which gradually extracts energy from the wave over an extended distance and produces breakers with low overall energy. As a result, spilling breakers have a longer life span and give surfers a long—but somewhat less exciting—ride than other breakers. In addition, spilling breakers are also the norm for open-ocean waves when gusty wind conditions cause waves to form whitecaps offshore.

Figure 8.20b shows a **plunging breaker**, which has a curling crest that moves over an air pocket. The curling crest occurs because the particles in the crest literally outrun the wave, and there is nothing beneath them to support their motion. Plunging breakers form on moderately steep beach slopes and are the best waves for surfing (see the chapter-opening photo).

When the ocean bottom has an abrupt slope, the wave energy is compressed into a shorter distance, and the wave will surge forward, creating a **surging breaker** (Figure 8.20c). These waves build up and break right at the shoreline, so board surfers tend to avoid them. For body surfers, however, these waves present the greatest challenge.

**Surfing** is analogous to riding a gravity-operated water sled by balancing the forces of gravity and buoyancy. The particle motion of ocean waves (see Figure 8.4) shows that water particles move up into the front of the crest. This force, along with the buoyancy of the surfboard, helps maintain a surfer's position in front of a breaking wave. The trick is to perfectly balance the force of gravity (directed downward) with the buoyant force (directed perpendicular to the wave face) to enable a surfer to be propelled forward by the wave's energy. A skillful surfer, by positioning the board properly on the wave front, can regulate the degree to which the propelling gravitational forces exceed the buoyancy forces, and the surfer can obtain speeds up to 40 kilometers (25 miles) per hour while moving along the face of a breaking wave. When the wave passes over water that is too shallow to allow the upward movement of water particles to continue, the wave has expended its energy, and the ride is over.

## Wave Refraction

Waves seldom approach a shore at a perfect right angle (90 degrees). Instead, some segment of the wave will "feel bottom" first and will slow before the rest of the wave. This results in **wave refraction** (*refringere* = to break up), or the *bending* of each *wave crest* (also called a *wave front*) as waves approach the shore.

Figure 8.21a shows how waves coming toward a straight shoreline are refracted and tend to align themselves *nearly* parallel to the shore. This explains why all waves come almost straight in toward a beach, no matter what their original orientation was. Figure 8.21b shows how waves coming toward an irregular shoreline refract so that they, too, nearly align with the shore. Figure 8.21c shows a classic example of wave refraction around Rincon Point in California.

The refraction of waves along an irregular shoreline distributes wave energy unevenly along the shore. To help illustrate how this works, notice the long black arrows in Figure 8.21b, which are called **orthogonal lines**

**(a)** Spilling breaker, resulting from a gradual beach slope.

**(b)** Plunging breaker, resulting from a steep beach slope; these are the best waves for surfing.

**(c)** Surging breaker, resulting from an abrupt beach slope.

Figure 8.20 **Types of breakers.** Photos of the three types of breakers, which are a result of different beach slopes.

**Animation**
Three Types of Breakers
https://goo.gl/7D8t9S

## PROCESS OF SCIENCE 8.1: A Wave's Sweet Spot Revealed

### BACKGROUND

Finding the most powerful part of a wave that propels a surfer forward—often called the "sweet spot"—is part of the thrill and challenge of surfing. Although surfers have an intuitive feel for where the "sweet spot" of a wave is as they ride a wave, it is exceedingly difficult to identify it by studying real, moving, breaking waves. So no one exactly knew where a wave's "sweet spot" is located until recently, when a physical oceanographer, who is also an avid surfer, decided to take on the challenge.

Figure 8A **A wave's "sweet spot" revealed.**

### FORMING A HYPOTHESIS

In examining surfing waves, the researcher noted two things: first, to catch a wave at all, a surfer paddling a surfboard must achieve a speed near the speed of the underlying wave. Second, different sections of the wave move at different speeds, so the location where a surfer positions a board on a wave is of utmost importance. A key component to success in surfing is positioning a board to take best advantage of the wave's *horizontal acceleration*, which is what propels a surfer forward. So, the question the researcher tried to answer is: "Where on a breaking wave do you find the maximum horizontal acceleration, which is necessary to get the fastest speeds and the best ride out of a wave?"

### DEVISING AN EXPERIMENT

Comparing theoretical models of breaking waves with sophisticated mathematical models of real-world waves, the researcher studied various breaking waves and determined their maximum horizontal acceleration. The analysis is based on the physics of how air and water interact, and how a wave's energy is transferred to particles touching its surface, such as those in a solid surfboard.

### INTERPRETING THE RESULTS

The study reveals that the mathematical "sweet spot" of a wave that provides maximum horizontal acceleration is the steep part right inside the curl of the wave just below the wave's breaking crest (**Figure 8A**). Surfers can use this information to get maximum acceleration and an enhanced ride out of a wave.

In addition, the study has a practical application for understanding weather and climate by revealing how mixing by breaking waves occurs. Although breaking waves aren't as common out at sea as they are close to the shore, as waves break they create currents in the water as well as produce water droplets in the form of sea spray that is ejected into the atmosphere. The interactions of these tiny events can help to predict storms and hurricanes as well as long-term changes in climate, so the benefits of the research could eventually go far beyond the surfing community.

### THINKING LIKE A SCIENTIST: WHAT'S NEXT?

What happens to a surfer if he/she is moving too slow relative to the underlying wave being surfed? What about when a surfer is moving faster than the underlying wave being surfed? And, for both cases, based on the research presented here, what should the surfer do about it to enhance their ride?

(*ortho* = straight, *gonia* = angle), or *wave rays*. Orthogonal lines are always oriented perpendicular to the wave crests, so they indicate the direction that waves travel. More importantly, orthogonals are spaced so that the energy between lines is equal at all times and can thus be used to show variations in wave energy. Far from shore, for example, notice how the orthogonals in Figure 8.21b are evenly spaced, which indicates that all parts of the wave have the same amount of energy. As the waves approach the shore and refract, however, notice how the orthogonals *converge* on headlands that jut into the ocean and *diverge* in bays. This means that wave

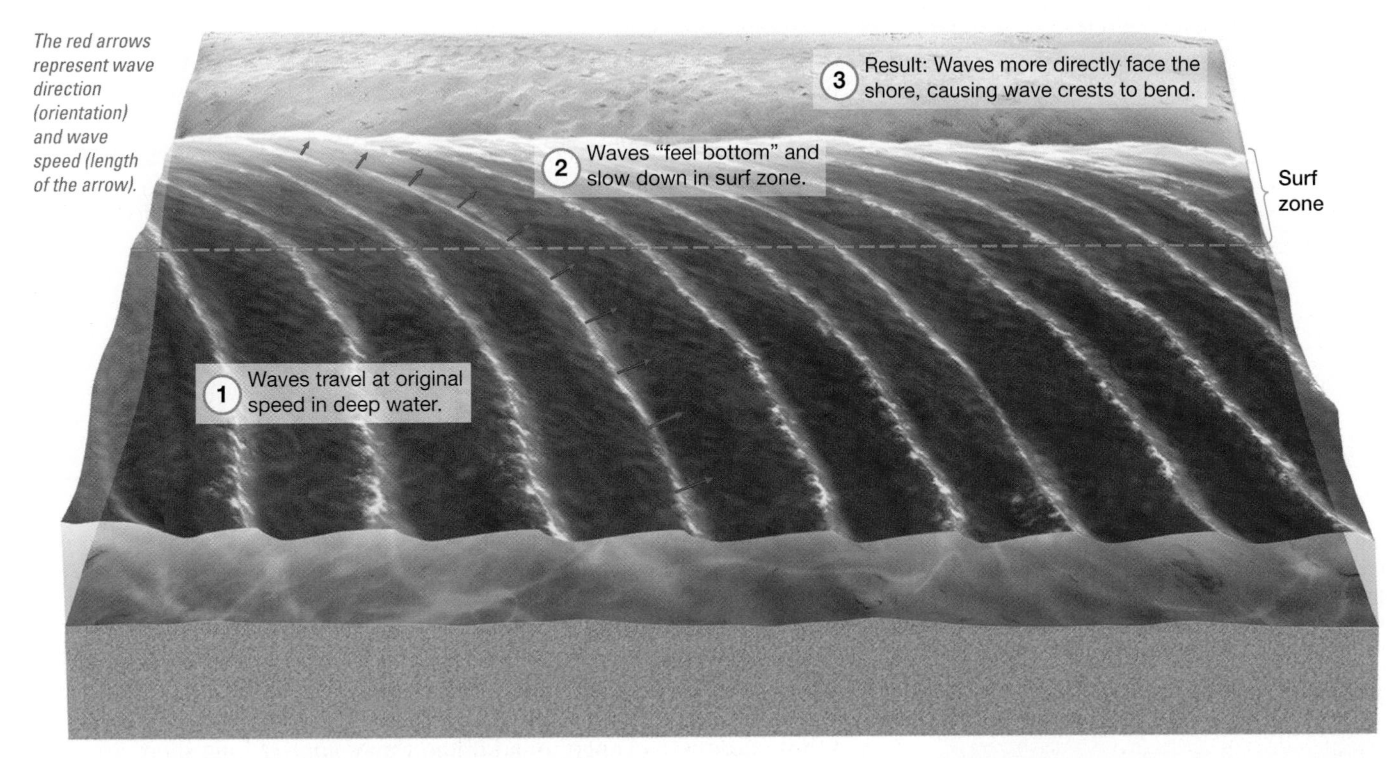
The red arrows represent wave direction (orientation) and wave speed (length of the arrow).
3 Result: Waves more directly face the shore, causing wave crests to bend.
2 Waves "feel bottom" and slow down in surf zone.
Surf zone
1 Waves travel at original speed in deep water.

**(a)** Aerial view of wave refraction along a straight shoreline.

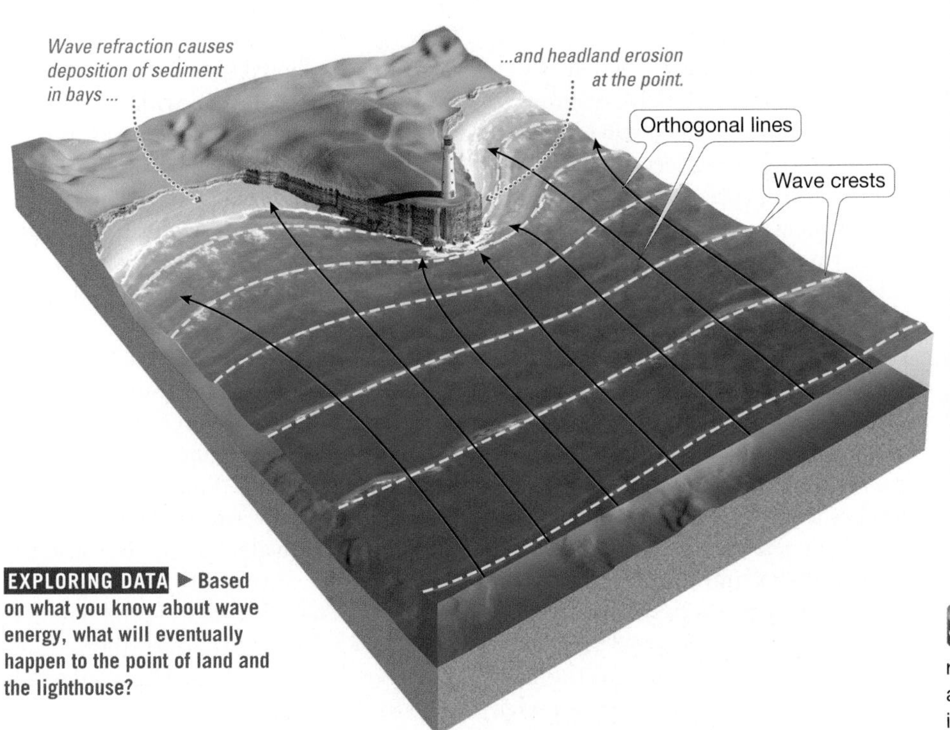
Wave refraction causes deposition of sediment in bays ...
...and headland erosion at the point.
Orthogonal lines
Wave crests

**EXPLORING DATA** ▶ **Based on what you know about wave energy, what will eventually happen to the point of land and the lighthouse?**

**(b)** Perspective view of wave refraction along an irregular shoreline.

Original wave angle
Waves refract (bend) around point

**(c)** Photo of wave refraction at Rincon Point, California (looking west).

**SmartFigure 8.21 Wave refraction.** Wave refraction is the bending of waves, as shown here along **(a)** a straight shoreline and **(b** and **c)** an irregular shoreline.
https://goo.gl/mJrwgb

energy is focused against the headlands but dispersed in bays. As a result, large waves occur at headlands, which are areas of good surfing and sites of erosion. (Sailors have long known that "the points draw the waves." Surfers also know how wave refraction causes good "point breaks.") Conversely, smaller waves occur in bays, which often provide areas for boat anchorages and are also associated with sediment deposition. Interestingly, the same waves break against both headlands and in nearby bays, but their energy is different because wave refraction causes the spacing of orthogonals

**Animation**
Wave Motion and Wave Refraction
http://goo.gl/55FMau

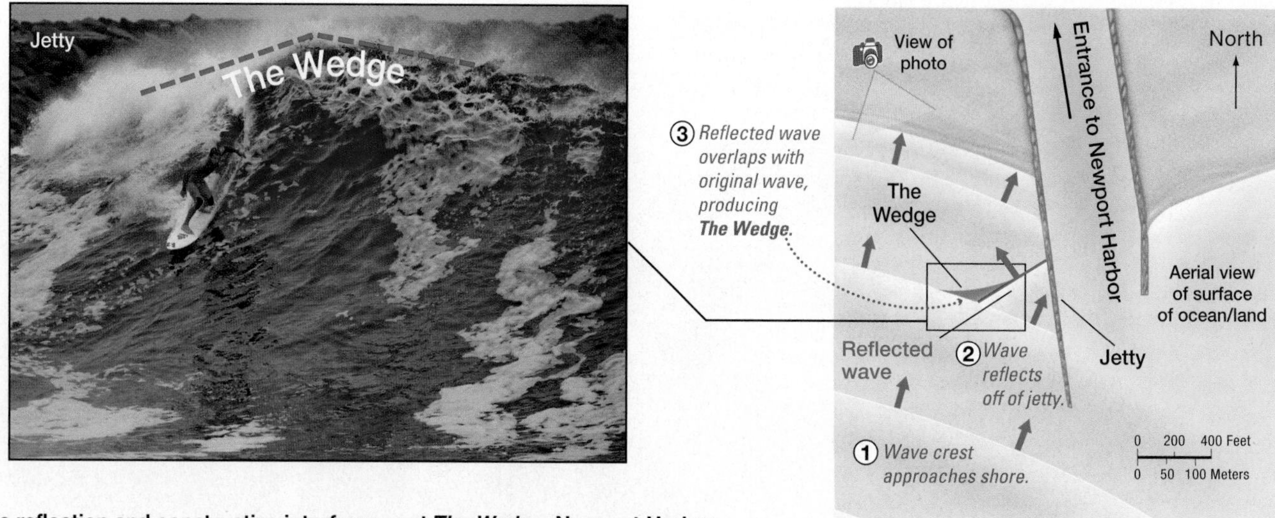


SmartFigure 8.22 **Wave reflection and constructive interference at *The Wedge*, Newport Harbor, California.** As waves approach the shore, (1) some of the wave energy is reflected off the long jetty at the entrance to the harbor. (2) The reflected wave overlaps and constructively interferes with the original wave, (3) resulting in a wedge-shaped wave (*dark blue triangle*) that may reach heights exceeding 8 meters (26 feet). Photo (*inset*) shows a surfer riding *The Wedge*; note the jetty in the background. https://goo.gl/iFmS1c

(and thus energy) to change. In addition, waves approaching shore are also influenced by sea floor features such as shallow banks or submarine canyons.

## Wave Reflection

Not all wave energy is expended as waves rush onto the shore. A vertical barrier, such as a seawall or a rock ledge, can reflect waves back into the ocean with little loss of energy—a process called **wave reflection** (*reflecten* = to bend back), which is similar to how a mirror reflects (bounces) back light. If the incoming wave strikes the barrier at a right (90-degree) angle, for example, the wave energy is reflected back parallel to the incoming wave, often interfering with the next incoming wave and creating unusual waveforms. More commonly, waves approach the shore at an angle, causing wave energy to be reflected at an angle equal to the angle at which the wave approached the barrier.

**THE WEDGE: A CASE STUDY OF WAVE REFLECTION AND CONSTRUCTIVE INTERFERENCE** An outstanding example of wave reflection and constructive interference occurs in an area called *The Wedge*, which develops west of the jetty that protects the harbor entrance at Newport Harbor, California (**Figure 8.22**). The jetty is a solid human-made object that extends into the ocean 400 meters (1300 feet) and has a near-vertical side facing the waves. As incoming waves strike the vertical side of the jetty at an angle, they are reflected at an equivalent angle. Because the original waves and the reflected waves have the same wavelength, a constructive interference pattern develops, creating plunging breakers that may exceed 8 meters (26 feet) in height (Figure 8.22, *inset*). Too dangerous for board surfers, these waves present a fierce challenge to the most experienced body surfers. The Wedge has crippled and even killed many who have come to challenge it.

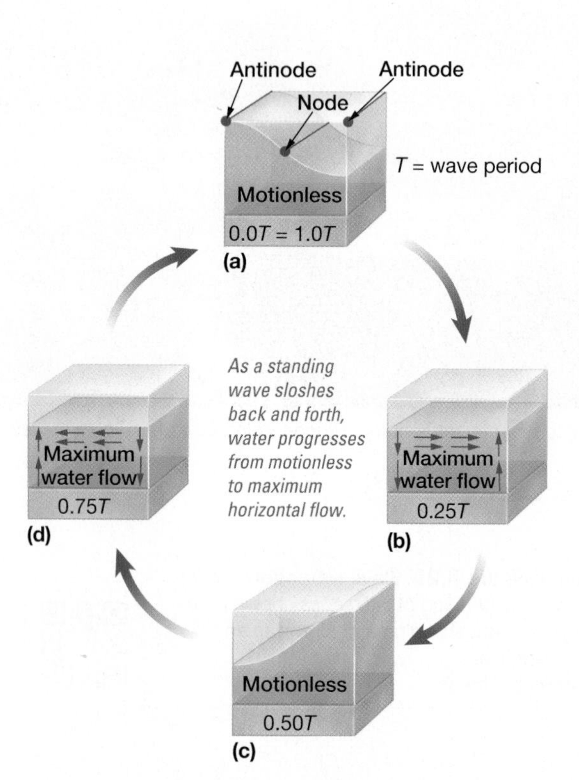


**Figure 8.23 Sequence of motion in a standing wave.** In a standing wave, water is motionless when antinodes reach maximum displacement (**a** and **c**). Water movement is at a maximum (*blue arrows*) when the water level is horizontal (**b** and **d**). Movement is vertical beneath the antinodes, and maximum horizontal movement occurs beneath the node.

**STANDING WAVES, NODES, AND ANTINODES** **Standing waves** (also called *stationary waves*) can be produced when waves are reflected at right angles to a barrier. Standing waves are the sum of two waves with the same wavelength moving in opposite directions, resulting in no net movement. Although the water particles continue to move vertically and horizontally, there is none of the circular motion that is characteristic of a progressive wave.

**Figure 8.23** shows the movement of water during the wave cycle of a standing wave. Lines along which there is no vertical movement are called *nodes* (*nodus* = knot), or *nodal lines*. *Antinodes*, which are crests that alternately become troughs, are the points of greatest vertical movement within a standing wave.

When water sloshes back and forth in a basin, the maximum vertical displacements are at the antinodes. When water moves from crest to trough at an antinode, the displaced volume of water has to move horizontally to raise the water level of the adjacent antinode from trough to crest. As a result, there is no vertical motion at the node; instead, there is only horizontal motion. At the antinodes, however, the movement of water particles is entirely vertical.

We'll consider standing waves further when tidal phenomena are discussed in Chapter 9, "Tides." Under certain conditions, the development of standing waves significantly affects the tidal character in coastal regions.

**RECAP** **Wave refraction is the bending of waves caused when waves slow in shallow water; wave reflection is the bouncing back of wave energy caused when waves strike a hard barrier.**

## CONCEPT CHECK 8.4 ▸ Explain how waves change in the surf zone.

1 What is the 1:7 ratio? What happens to a wave when the 1:7 ratio is exceeded?

2 Describe the physical changes that occur to a wave's wave speed (*S*), wavelength (*L*), height (*H*), and wave steepness (*H*/*L*) as the wave moves across progressively shallower water to break on the shore.

3 Describe the three different types of breakers and indicate the slope of the beach that produces each of the three types. How is the energy of the wave distributed differently within the surf zone by the three types of breakers?

4 Using examples, explain how wave refraction is different from wave reflection.

5 Using orthogonal lines, illustrate how wave energy is distributed along a shoreline with headlands and bays. Identify areas of high- and low-energy release.

## STUDENTS SOMETIMES ASK . . .

***Why is surfing so much better along the West Coast of the United States than along the East Coast?***

There are three main reasons the U.S. West Coast has better surfing conditions:

1. The waves are generally bigger in the Pacific. The Pacific is larger than the Atlantic, so the fetch is larger, allowing bigger waves to develop in the Pacific.
2. The beach slopes are generally steeper along the West Coast. Along the East Coast, the gentle slopes often create spilling breakers, which are not as favorable for surfing. Steeper beach slopes along the West Coast result in plunging breakers, which are better for surfing.
3. The wind is more favorable. Most of the United States is influenced by the prevailing westerlies, which blow toward shore and enhance waves along the West Coast. Along the East Coast, the wind generally blows away from shore.

# 8.5 ▸ How Are Tsunami Created?

The Japanese term for the large, sometimes destructive waves that occasionally roll into their harbors is **tsunami** (*tsu* = harbor, *nami* = wave[s]). Tsunami originate from sudden changes in the topography of the sea floor caused by such events as slippage along underwater faults, underwater landslides such as those that form turbidity currents or from the collapse of large oceanic volcanoes, and underwater volcanic eruptions. Many people mistakenly call them "tidal waves," but tsunami are unrelated to the tides. The mechanisms that trigger tsunami are typically seismic events, so tsunami are more accurately called *seismic sea waves*.

The majority of tsunami are caused by vertical *fault movement*. Underwater fault movement displaces Earth's crust, generates earthquakes, and, if it ruptures the sea floor, produces a sudden change in water level at the ocean surface (**Figure 8.24**). Faults that produce *vertical* displacements (the uplift or downdropping of ocean floor) change the volume of the ocean basin, which affects the entire water column and generates tsunami. Conversely, faults that produce *horizontal* displacements (such as the lateral movement associated with transform faulting) generally do not generate tsunami because the side-to-side movement of these faults does not change the volume of the ocean basin. Much less common events, such as underwater avalanches triggered by shaking or underwater volcanic eruptions—which create the largest waves—also produce tsunami. In addition, large objects that splash into the ocean, such as above-water coastal landslides (see Mastering Oceanography **Web Diving Deeper 8.1**) or meteorite impacts, produce **splash waves**, which are a type of tsunami.

Figure 8.24 **How a tsunami differs from wind-generated waves.**

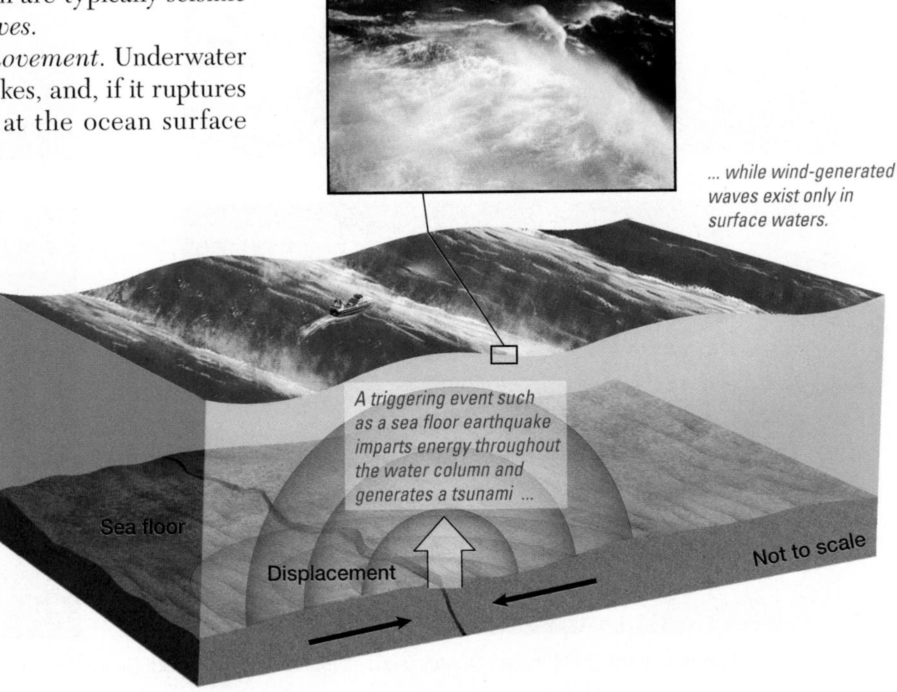
... while wind-generated waves exist only in surface waters.
A triggering event such as a sea floor earthquake imparts energy throughout the water column and generates a tsunami ...
Sea floor
Displacement
Not to scale

SmartFigure 8.25 **Tsunami generation, propagation, and destruction.**
https://goo.gl/X4dwgA

**EXPLORING DATA ▼ Will a tsunami likely be generated if the offshore fault has a horizontal (side-to-side) motion? Explain.**

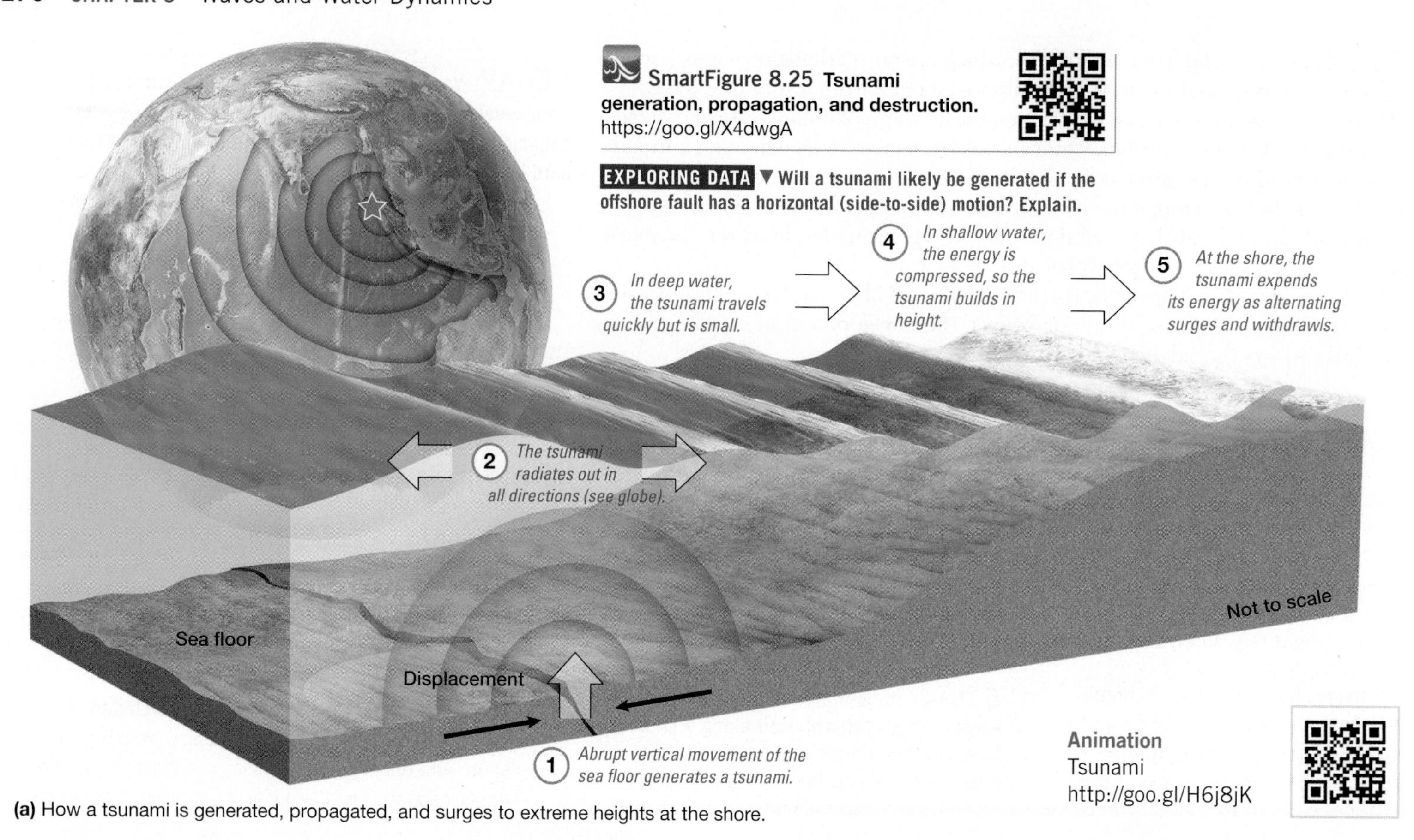

1 Abrupt vertical movement of the sea floor generates a tsunami.
2 The tsunami radiates out in all directions (see globe).
3 In deep water, the tsunami travels quickly but is small.
4 In shallow water, the energy is compressed, so the tsunami builds in height.
5 At the shore, the tsunami expends its energy as alternating surges and withdrawls.
Sea floor
Displacement
Not to scale

**(a)** How a tsunami is generated, propagated, and surges to extreme heights at the shore.

**Animation**
Tsunami
http://goo.gl/H6j8jK

1
2
3
4

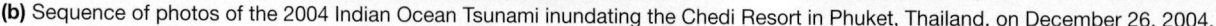

**(b)** Sequence of photos of the 2004 Indian Ocean Tsunami inundating the Chedi Resort in Phuket, Thailand, on December 26, 2004.

Because tsunami generally originate from energy imparted to the ocean basin, they affect the entire water column of the ocean, which on average is 4 kilometers (2.5 miles) deep and in the deepest trenches the depth can exceed 11 kilometers (7 miles). Wind-generated surface waves, on the other hand, affect only the topmost waters of the ocean (Figure 8.24) and do not contain nearly as much energy as tsunami. When tsunami come into shallow water, the energy from the entire oceanic water column is compressed into a shallow region and tsunami expend their energy at the shore as a series of strong alternating surges and withdrawals of seawater that can cause considerable damage (**Figure 8.25**).

During the 20th century, 498 measurable tsunami occurred worldwide, 66 of them resulting in fatalities. The source events were earthquakes (86%), volcanic activity (5%), landslides (4%), and combinations of those processes (5%). Tsunami generated by meteorite or asteroid impacts occur much less frequently.

The wavelength of a typical tsunami exceeds 200 kilometers (125 miles), so it is a shallow-water wave everywhere in the ocean. (Recall that the depth of the wave base is equal to one-half a wave's wavelength. Thus, theoretically tsunami could feel the ocean bottom to depths of 100 kilometers [62 miles], which is deeper than even the deepest ocean trenches.) Because tsunami are shallow-water waves, their speed is determined only by water depth. In the open ocean, tsunami move at well over 700 kilometers (435 miles) per hour—they can easily keep pace with a jet airplane—and have heights of only about 1 meter (3.3 feet) or less. Even though they are fast, tsunami are small in the open ocean and pass unnoticed in deep water until they reach shore, where they slow in the shallow water and undergo physical changes (just as wind-generated waves do) that cause tsunami to increase greatly in wave height.

**Figure 8.26 Tsunami damage in Hilo, Hawaii.** Flattened parking meters in Hilo, Hawaii, caused by the 1946 tsunami that resulted in more than $25 million in damage and 159 deaths.

## Coastal Effects

Contrary to popular belief, a tsunami does not form a huge breaking wave at the shoreline. Instead, it is a strong flood or surge of water that causes the ocean to advance (or, in certain cases, retreat) dramatically. In fact, a tsunami resembles a sudden, *extremely* high tide, which is why they are misnamed "tidal waves." It may take many minutes for the crest of the tsunami to arrive on shore, during which time sea level can rise up to 40 meters (131 feet) above normal, with normal waves superimposed on top of the higher sea level. The strong surge of water can rush into low-lying areas and cause extensive destruction (Figure 8.25b), including loss of life.

As the trough of the tsunami approaches the shore, the water rapidly drains off the land. In coastal areas, it looks like a sudden and *extremely* low tide, where sea level is many meters lower than even the lowest low tide. Because tsunami are typically a series of waves, there is often an alternating series of dramatic surges and withdrawals of water that are widely separated in time. The first surge is only rarely the largest; instead, the third, fourth—or even seventh—surge may be the largest and can occur several hours later.

Depending on the geometry of sea floor motion that creates a tsunami, the trough can sometimes arrive at a coast first. For example, on the side of a fault where the sea floor drops down, the first part of the wave to propagate outward will be the trough, followed by the crest. Conversely, on the side of the fault that moves upward, the crest leads the trough. At a coast where the trough arrives first, the water drains out and exposes parts of the lowermost shoreline that are rarely seen. For people at the shoreline, the temptation is to explore these newly exposed areas and catch stranded organisms. Within a few minutes, however, a strong surge of water (the crest of the tsunami) is due to arrive.

The alternating surges and retreats of water by tsunami can severely damage coastal structures and can injure and kill people as well. The speed of the advance—up to 4 meters (13 feet) per second—is faster than any human can run. Those who are caught in a tsunami are often drowned or crushed by floating debris (**Figure 8.26**).

### STUDENTS SOMETIMES ASK . . .

***What is the record height of a tsunami?***

Japan, which is in close proximity to several subduction zones and endures more tsunami than any other place on Earth (followed by Chile and Hawaii), holds the record. The largest documented high water from a tsunami occurred in the Ryukyu Islands of southern Japan in 1971, when normal sea level was raised by an astonishing 85 meters (278 feet). In low-lying coastal areas, such an enormous vertical rise can send seawater many kilometers inland, causing flooding and widespread damage.

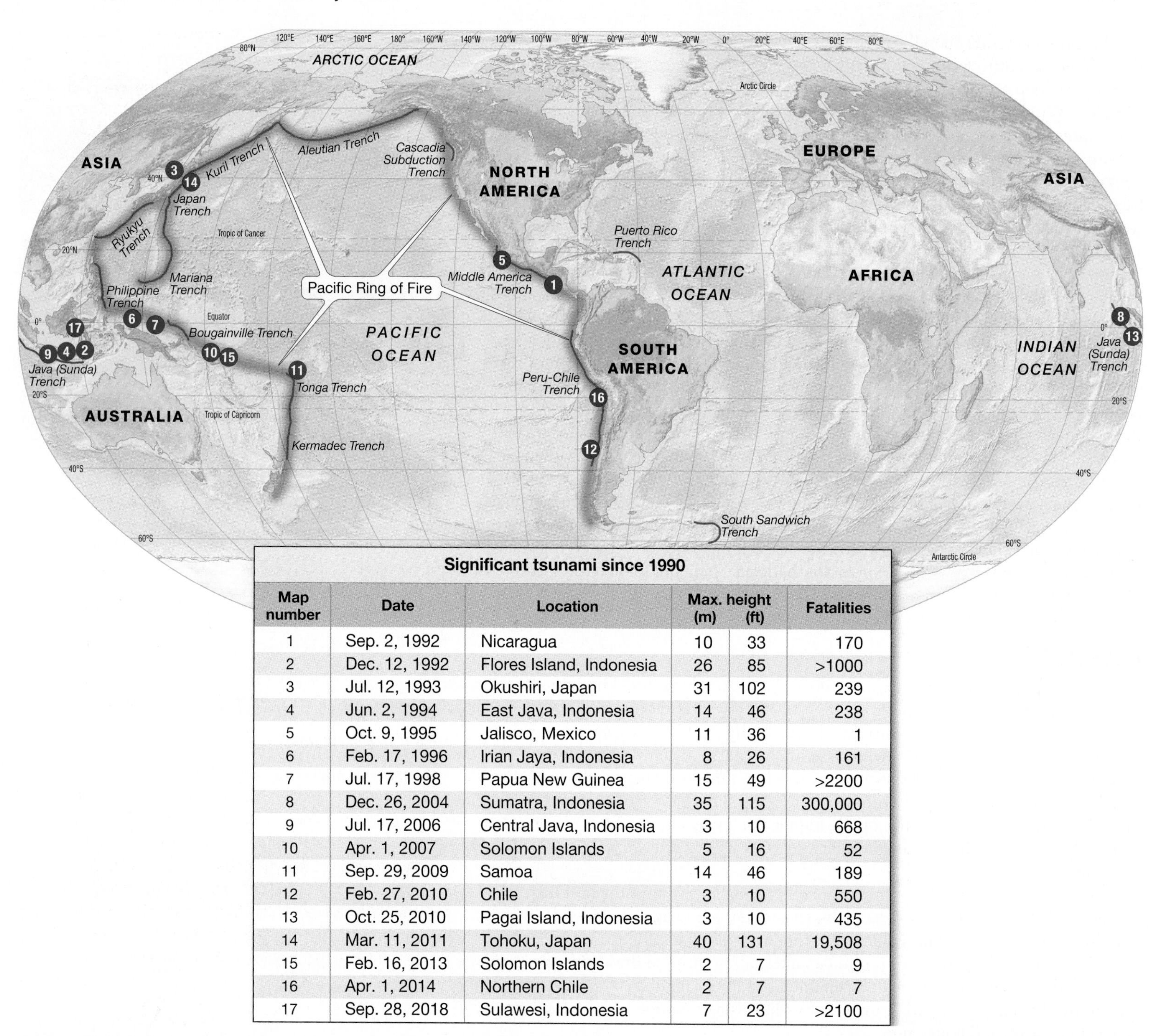


**Significant tsunami since 1990**

| Map number | Date | Location | Max. height (m) | Max. height (ft) | Fatalities |
|---|---|---|---|---|---|
| 1 | Sep. 2, 1992 | Nicaragua | 10 | 33 | 170 |
| 2 | Dec. 12, 1992 | Flores Island, Indonesia | 26 | 85 | >1000 |
| 3 | Jul. 12, 1993 | Okushiri, Japan | 31 | 102 | 239 |
| 4 | Jun. 2, 1994 | East Java, Indonesia | 14 | 46 | 238 |
| 5 | Oct. 9, 1995 | Jalisco, Mexico | 11 | 36 | 1 |
| 6 | Feb. 17, 1996 | Irian Jaya, Indonesia | 8 | 26 | 161 |
| 7 | Jul. 17, 1998 | Papua New Guinea | 15 | 49 | >2200 |
| 8 | Dec. 26, 2004 | Sumatra, Indonesia | 35 | 115 | 300,000 |
| 9 | Jul. 17, 2006 | Central Java, Indonesia | 3 | 10 | 668 |
| 10 | Apr. 1, 2007 | Solomon Islands | 5 | 16 | 52 |
| 11 | Sep. 29, 2009 | Samoa | 14 | 46 | 189 |
| 12 | Feb. 27, 2010 | Chile | 3 | 10 | 550 |
| 13 | Oct. 25, 2010 | Pagai Island, Indonesia | 3 | 10 | 435 |
| 14 | Mar. 11, 2011 | Tohoku, Japan | 40 | 131 | 19,508 |
| 15 | Feb. 16, 2013 | Solomon Islands | 2 | 7 | 9 |
| 16 | Apr. 1, 2014 | Northern Chile | 2 | 7 | 7 |
| 17 | Sep. 28, 2018 | Sulawesi, Indonesia | 7 | 23 | >2100 |

**EXPLORING DATA ▲ Using information on this figure, explain why the Atlantic Ocean experiences very few tsunami.**

Figure 8.27 **Significant tsunami since 1990.** Global locations and corresponding table showing significant tsunami (those with large tsunami height and/or responsible for a large number of fatalities) since 1990. Tsunami have claimed more than 300,000 lives worldwide since 1990. These killer waves are most often generated by earthquakes along colliding tectonic plates of the Pacific Rim, although the most deadly tsunami in history was the 2004 Indian Ocean Tsunami (*number 8*). Locations of ocean trenches are shown by dark red lines; Pacific Ring of Fire is shown by red shading.

## Some Examples of Historic and Recent Tsunami

Although many small tsunami are created each year, most go unnoticed. On average, about 50 noticeable tsunami occur every decade, with a large tsunami occurring somewhere in the world every 2 to 3 years and an extremely large and damaging one occurring every 15 to 20 years. Remarkably, two of the largest and most deadly tsunami have occurred within the past decade.

Where do most tsunami occur? About 86% of all great waves are generated in the Pacific Ocean because large-magnitude earthquakes occur along the series of trenches that ring its ocean basin where oceanic plates are subducted along convergent plate boundaries. Volcanic activity is also common along the *Pacific Ring of Fire*, and the large earthquakes that occur along its margin are capable of producing extremely large tsunami. **Figure 8.27** shows significant tsunami that have struck since 1990—with the majority occurring along the Pacific Ring of Fire.

**THE CASCADIA EARTHQUAKE AND TSUNAMI (1700)** A large earthquake occurred along the Cascadia Subduction Zone offshore of the U.S. Pacific Northwest on January 26, 1700, with an estimated $M_w$ = 8.7–9.2. The megathrust earthquake involved the Juan de Fuca Plate from mid-Vancouver Island all the way to Northern California. The length of the fault rupture was about 1000 kilometers (620 miles), with an average slip of 20 meters (66 feet).

The earthquake caused what we now see as drowned "ghost forests"—once-living forests that were suddenly dropped below sea level in the Pacific Northwest, and then covered by tsunami deposits. In addition, based on detailed historical records in Japan that match the timing of this earthquake, the tsunami also struck the coast of Japan, was 4.9 meters (16 feet) tall, and did considerable damage. According to seismologists, the historic occurrence of a large, damaging earthquake along the Cascadia Subduction Zone is on average about once every 300-500 years.

**THE ERUPTION OF KRAKATAU (1883)** One of the most destructive historical tsunami ever generated came from the eruption of the volcanic island of Krakatau on August 27, 1883. (Note that the island Krakatau, which is *west* of Java, is also called Krakatoa.) Approximately the size of a small Hawaiian island in what is now Indonesia, Krakatau exploded with the greatest release of energy from Earth's interior ever recorded in historic times. The island, which stood 450 meters (1500 feet) above sea level, was nearly obliterated. The sound of the explosion was heard throughout the Indian Ocean, up to 4800 kilometers (3000 miles) away, and remains the loudest noise on human record. Dust from the explosion ascended into the atmosphere and circled Earth on high-altitude winds, producing brilliant red sunsets worldwide for nearly a year.

Not many people were killed by the outright explosion of the volcano because the island was uninhabited. However, the displacement of water from the energy released during the explosion was enormous, creating a tsunami that exceeded 35 meters (116 feet)—as high as a 12-story building. It devastated the coastal region of the Sunda Strait between the nearby islands of Sumatra and Java, drowning more than 1000 villages and taking more than 36,000 lives. The energy carried by this wave reached every ocean basin and was even detected by tide-recording stations as far away as London and San Francisco.

Like most of the other approximately 130 active volcanoes in Indonesia, Krakatau was formed along the Sunda Arc, a 3000-kilometer (1900-mile) curving chain of volcanoes associated with the subduction of the Australian Plate beneath the Eurasian Plate. Where these two sections of Earth's crust meet, earthquakes and volcanic eruptions are common.

**INDIAN OCEAN (2004)** Although most tsunami occur in the Pacific Ocean, they do sometimes occur in other ocean basins. On December 26, 2004, an enormous earthquake struck in a subduction zone off the west coast of Sumatra in Indonesia on the shores of the Indian Ocean. Known as the Sumatra–Andaman Earthquake, it was the second biggest earthquake recorded during the past century and the largest to be recorded by modern seismograph equipment. The earthquake was so large, in fact, that it triggered small earthquakes as far away as Alaska, altered Earth's rotation, and by some calculations may have even changed Earth's gravity field. Initially, it was deemed a magnitude $M_w$ = 9.0, but it was subsequently upgraded to magnitude $M_w$ = 9.2. The earthquake occurred about 30 kilometers (19 miles) beneath the sea floor, near the Sunda Trench, where the Indian Plate is being subducted beneath the Eurasian Plate. Moreover, about 1200 kilometers (750 miles) of sea floor was ruptured along the interface of these two tectonic plates, thrusting sea floor upward and generating about 10 meters (33 feet) of vertical displacement. This abrupt vertical movement of the sea floor is what generated the deadliest tsunami in history.

Once generated, the tsunami spread out at jetliner speeds across the Indian Ocean. Only 15 minutes after the earthquake, the tsunami hit the shores of Sumatra with a series of alternating rapid withdrawals and strong surges up to

## STUDENTS SOMETIMES ASK . . .

***If a tsunami warning is issued, what is the best thing to do?***

The *smartest* thing to do is to stay out of coastal areas, but people often want to observe the tsunami firsthand. For instance, after the February 2010 magnitude $M_w$ = 8.8 Chilean earthquake, tsunami warnings were issued in 53 countries. Across the Pacific Ocean in Australia, widespread media broadcasts warned people of the impending danger, yet live television coverage showed hundreds of people standing on low-lying beaches waiting for the tsunami to arrive. Worse, some of them were deliberately swimming into the incoming tsunami, despite the efforts of volunteer lifeguards to prevent this.

If you *must* go to the beach to observe a tsunami, expect crowds, road closings, and general mayhem. It would be a good idea to stay at least 30 meters (100 feet) above sea level. If you happen to be at a remote beach and notice the water suddenly withdrawing, evacuate immediately to higher ground (**Figure 8.28**). And, if you happen to be at a beach when an earthquake occurs and shakes the ground so hard that you can't stand up, then *RUN*—don't walk—for high ground as soon as you *can* stand up!

After the first surge of the tsunami, stay out of low-lying coastal areas for several hours because several more surges (and withdrawals) can be expected. There are many documented cases of curious people being killed when they were caught in the third or fourth (. . . or ninth) surge of a tsunami.

**Figure 8.28 Tsunami warning sign.** This tsunami warning sign in coastal Oregon advises residents to evacuate low-lying areas during a tsunami.

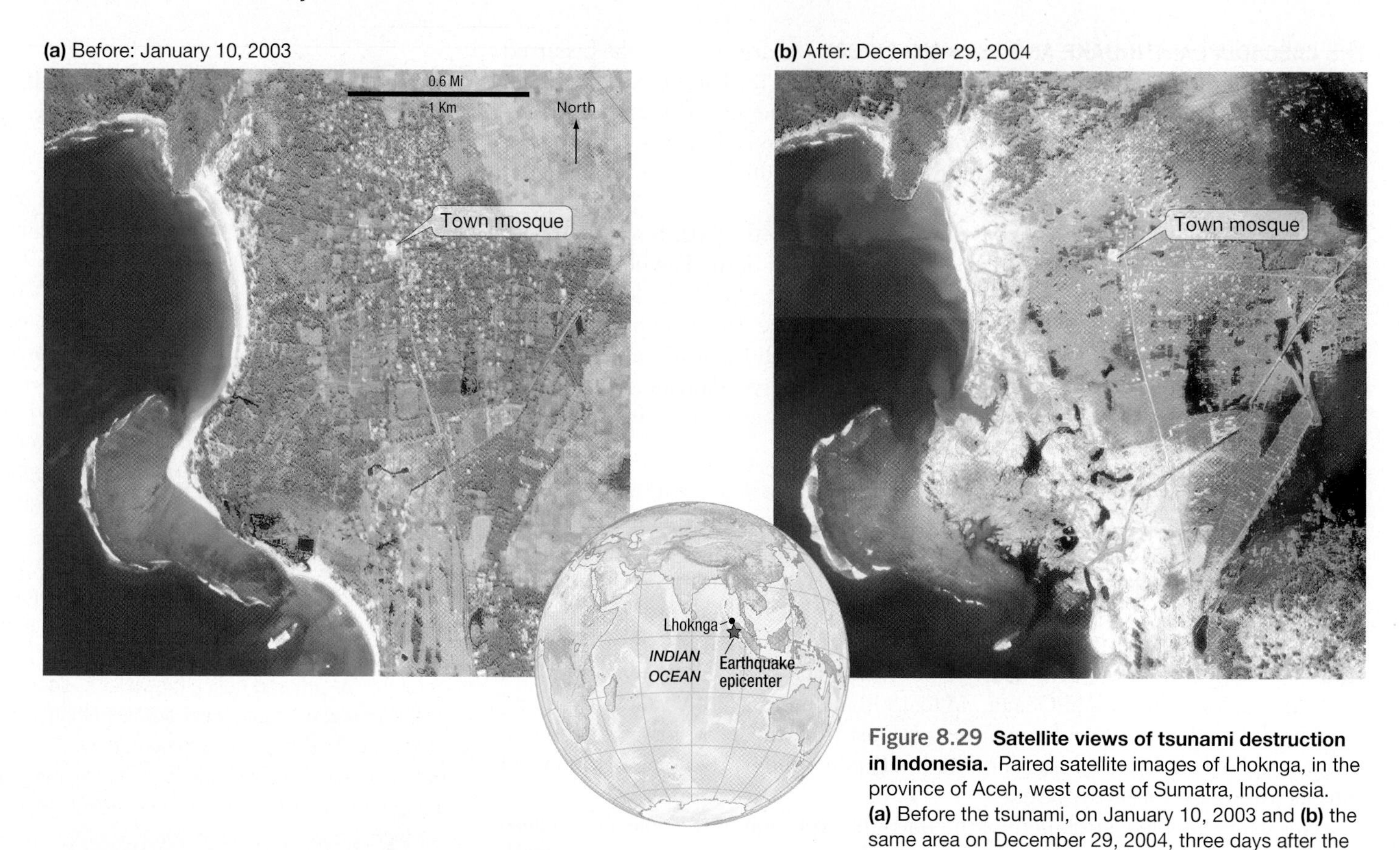


**Figure 8.29 Satellite views of tsunami destruction in Indonesia.** Paired satellite images of Lhoknga, in the province of Aceh, west coast of Sumatra, Indonesia. **(a)** Before the tsunami, on January 10, 2003 and **(b)** the same area on December 29, 2004, three days after the tsunami that inundated the city with a 15-meter (50-foot) surge of seawater. The town mosque, indicated in both images, was one of the few buildings that remained standing.

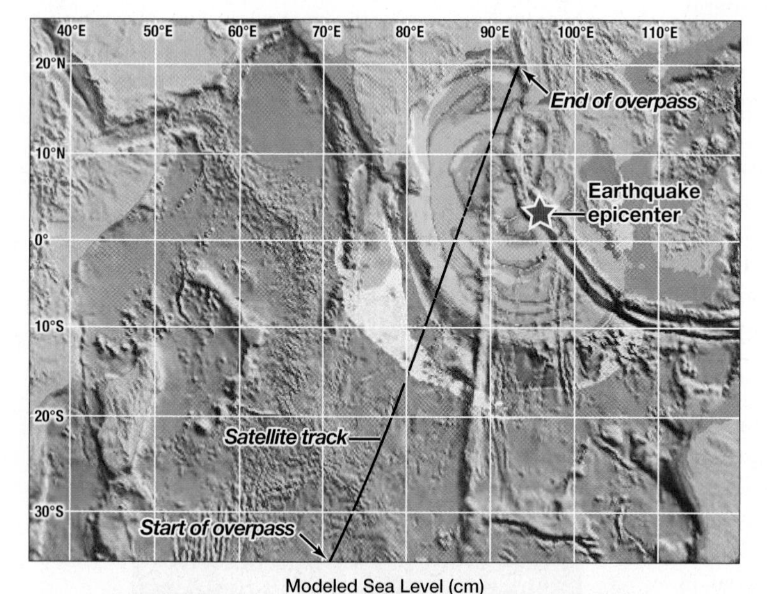


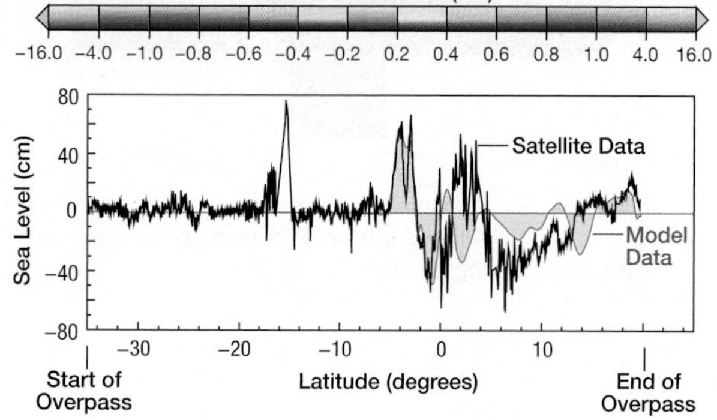


35 meters (115 feet) high. Many coastal villages were completely washed away (**Figure 8.29**), causing several hundred thousand deaths. Sites farther from the quake zone experienced smaller but nevertheless deadly waves. Particularly affected were Thailand (see Figure 8.25b), which was struck by the tsunami about 75 minutes after the quake occurred, and Sri Lanka and India, which were pounded by devastating waves about three hours after the earthquake. After seven hours and at a distance of more than 5000 kilometers (3000 miles), the tsunami hit the east coast of Africa, where it still had enough power to kill more than a dozen people. Although much smaller, the tsunami was also detected in the Atlantic, Pacific, and Arctic Oceans.

In a remarkable coincidence, the Jason-1 satellite happened to be passing over the Indian Ocean two hours after the tsunami originated (**Figure 8.30**). The satellite's radar altimeter, which is designed to accurately measure the elevation of the ocean surface (see Diving Deeper 3.1), was able to detect the crests and troughs of the tsunami as

**Figure 8.30 Jason-1 satellite detects the Indian Ocean Tsunami.** The Indian Ocean Tsunami was initiated by a large earthquake offshore Sumatra on December 26, 2004 (*red star*). By a fortuitous circumstance, the Jason-1 satellite passed over the Indian Ocean (*black line*) two hours after the tsunami was generated. Its radar altimeter detected the crests and troughs of the tsunami (*colors*), which showed a wave height of about 1 meter (3.3 feet) in the open ocean. The graph of the satellite's overpass (*below*) shows the difference between the measured sea level from satellite data (*black line*) and the modeled wave height (*blue curve*).

it radiated out across the Indian Ocean with a wavelength of about 500 kilometers (300 miles). Although this sighting occurred about an hour before the first waves struck Sri Lanka and India, the satellite data couldn't have been used to warn tsunami victims because scientists needed several hours to analyze the information. However, these satellite data are particularly valuable because they have allowed scientists to check the accuracy of open-ocean tsunami travel models, which are based on seismic data, bathymetric information, and coastal tide gauges.

In all, between 230,000 and 300,000 people in 11 countries were killed by the tsunami, ranking this tragedy as the deadliest tsunami in recorded history. The tsunami also caused billions of dollars of damage throughout the region and left millions homeless. Why was there such a large loss of life during the tsunami? One factor was the lack of a warning system in the Indian Ocean, like the network of buoys and deep-sea instruments that monitors earthquake and wave activity in the Pacific Ocean, where most tsunami have historically occurred. Another is the lack of an emergency response system to warn high-population-density coastal communities and beachgoers. Still another was the lack of good public awareness in recognizing the signs of an impending tsunami, such as when a rapid withdrawal of water is observed at the shore, which is caused by the trough of the tsunami arriving first and indicates that an equally strong surge of water will soon follow.

Studies of the effect of the tsunami on different coastlines suggest that those areas lacking protective coral reefs or mangroves received stronger surges. In many cases, the shape of the coastline as well as the geometry of the offshore sea floor affected the height of the tsunami and the amount of coastal destruction. Sediment cores reveal that several other large tsunami have occurred in the region during the past 1000 years.

On July 17, 2006, another strong earthquake—this time an $M_w = 7.7$ earthquake associated with faulting in the Java Trench off the southern coast of Java, Indonesia—triggered a 3-meter (10-foot) tsunami that killed 668 people, injured at least 600, and destroyed many coastal structures. This destruction near the area, which experienced even greater damage from the 2004 tsunami, emphasized the fact that the region still lacked a comprehensive tsunami warning system. In 2010, however, the Indian Ocean Tsunami Warning and Mitigation System became fully operational, with its network of deep-ocean pressure sensors, buoys, land seismographs, tidal gauges, data centers, and communications upgrades. The system will continue to be upgraded with more detection buoys in the coming years. With this new warning system in place and enhanced public education about what to do in the event of a tsunami, all countries bordering the Indian Ocean will be much better prepared for the destructive power of future tsunami.

**JAPAN (2011)** On March 11, 2011, an $M_w = 9.0$ earthquake—the fourth largest ever to be recorded—occurred in the Japan Trench offshore of northeastern Japan. Known as the Tohoku Earthquake, it uplifted the sea floor by several meters, moved the island of Japan itself 4 meters (13 feet), and created a large, devastating tsunami that was felt throughout the Pacific Ocean (**Diving Deeper 8.1**). In all, damages have been estimated at $235 billion, making it the most expensive natural disaster in world history.

**TSUNAMI WARNING SYSTEM** In response to the tsunami that struck Hawaii in 1946, a tsunami warning system was established throughout the Pacific Ocean. It led to what is now the **Pacific Tsunami Warning Center (PTWC)**, which coordinates information from 25 Pacific Rim countries and is headquartered in Ewa Beach (near Honolulu), Hawaii. The tsunami warning system uses seismic waves—some of which travel through Earth at speeds 15 times faster than tsunami—to forecast destructive tsunami. In addition, oceanographers have recently established a network of sensitive pressure sensors on the deep-ocean floor of the Pacific. The program, called **Deep-ocean Assessment and Reporting of Tsunamis (DART)**,

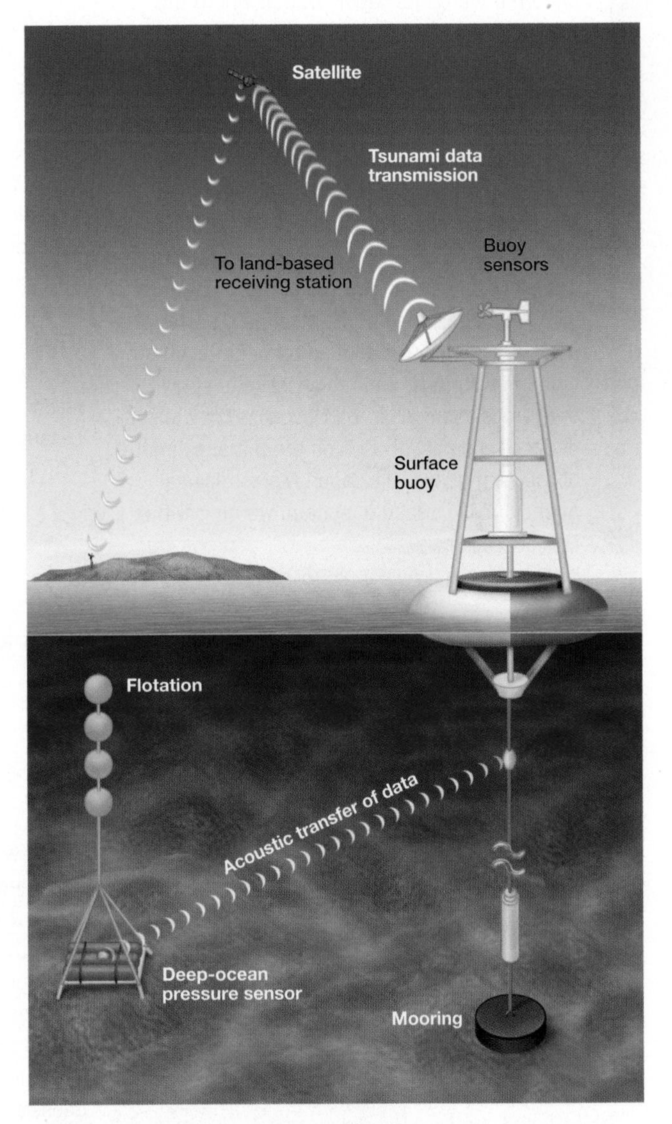


Figure 8.31 **Deep-ocean Assessment and Reporting of Tsunamis (DART).** The DART system consists of a deep-ocean pressure sensor that can detect a tsunami passing above. The pressure sensor relays information to a buoy at the surface that transmits the data via satellite, allowing oceanographers to detect the passage of a tsunami in the open ocean.

# WAVES OF DESTRUCTION: THE 2011 JAPANESE TSUNAMI

On March 11, 2011, the nation of Japan and seismologists around the world received a terrible surprise: A huge earthquake, significantly stronger than people had anticipated or prepared for in the region, struck off the northeastern shore of Japan. The great 2011 $M_w = 9.0$ Tohoku Earthquake—named for the region it struck—occurred along the Japan Trench where the Pacific Plate is being subducted underneath Japan (**Figure 8B**). It shifted the coast of Japan up to 8 meters (26 feet) eastward and also uplifted the sea floor by as much as 5 meters (16.4 feet) over an area comparable in size to the state of Connecticut; this sudden vertical movement of the sea floor imparted energy into the Pacific Ocean and generated one of the largest and best-studied tsunami in history.

Although Japan prides itself in disaster preparedness, many residents did not get accurate tsunami warnings, and many chose to stay in dangerous locations in part because they misunderstood the risks. Based on the initial estimates of the earthquake, the Japan Meteorological Agency (JMA) issued a tsunami warning within three minutes from the start of the shaking. The agency's warning predicted a tsunami surge of only 3 to 6 meters (10 to 20 feet), and unfortunately, many people did not take immediate action because of their belief that Japan's extensive network of tsunami walls would protect them. In addition, some residents ignored the warning because of previous false alarms. Within the next few minutes, JMA reissued its warning for a 10-meter (33-foot) tsunami, but many people did not receive the updated warning because of the destruction of the regional power supply system.

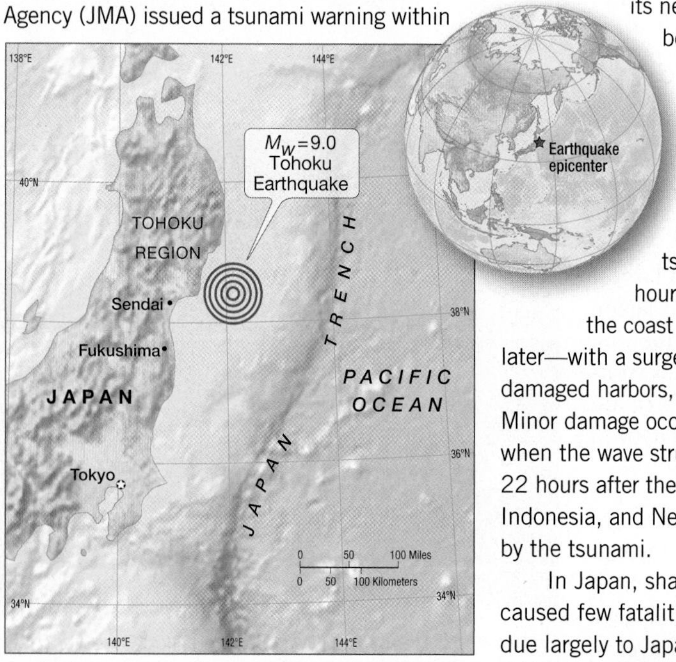

**Figure 8B The great 2011 Tohoku Earthquake.** Location of the $M_w = 9.0$ Tohoku Earthquake, which was the most powerful earthquake ever to have hit Japan. It occurred along the Japan Trench, uplifted the sea floor, and created a large, devastating tsunami.

About 20 minutes after the earthquake began, the tsunami swept onto the coastline of northeastern Japan, where most fishing villages and towns were built at the ends of long, narrow bays because these bays offer protection from wind waves. However, these long narrow bays amplified the height of the tsunami. The initial surge of the tsunami reached heights of about 15 meters (49 feet) and easily overtopped harbor-protecting tsunami walls and coastal margins (**Figure 8C**), penetrating as far as 10 kilometers (6 miles) inland. In one location, the tsunami reached a record height of 40 meters (131 feet) because of amplification by offshore topography.

The Pacific Tsunami Warning Center, with its network of buoys that had been greatly expanded since the 2004 Indian Ocean tsunami, sent out tsunami warnings across the Pacific and accurately predicted its arrival, allowing coastal communities to prepare. The tsunami struck Hawaii seven hours after the earthquake—and the coast of California three hours later—with a surge up to 2 meters (7 feet) that damaged harbors, marinas, and coastal resorts. Minor damage occurred in Peru and Chile when the wave struck their coastlines 20 to 22 hours after the event. The Philippines, Indonesia, and New Guinea were also affected by the tsunami.

In Japan, shaking from the earthquake caused few fatalities and did minimal damage due largely to Japan's strict building codes, thorough earthquake preparation, and advance warning systems. However, the tsunami destroyed many small towns and villages, killed 19,508 people, and initially displaced nearly half a million people; nearly two years later, thousands were still living in temporary shelters such as high school gymnasiums. The tsunami also disrupted the power supply needed to maintain water circulation for cooling the reactors at the Fukushima Daiichi nuclear power plant. Three reactors exploded, releasing radioactivity that continues to plague central Japan. Radioactivity was also released into the ocean, and studies are now under way to determine its effect on marine life.

**Figure 8C The 2011 tsunami surges over a protective seawall in Miyako, northeastern Japan.**

The destructive power of the tsunami also generated a huge amount of floating debris, including boats, cars, building parts, household items, and even entire houses. Japanese authorities estimate that 5 million metric tons (11 billion pounds) of debris was swept out to sea by the tsunami, with about 70% sinking to the sea floor and 1.5 million metric tons (3 billion pounds) left floating. The tsunami debris is being monitored as it slowly spreads out and moves across the Pacific Ocean, transported by currents in the North Pacific Gyre. Although some of the floating debris will eventually sink, disperse, or biodegrade, the wreckage has been found in Hawaii and all the way across the Pacific. Along the U.S. West Coast, for example, tsunami debris such as a derelict Japanese fishing vessel, wood, plastics, buoys, toys, and refrigerators began washing ashore months after the tsunami, with additional floating tsunami debris arriving even years later.

## WHAT DID YOU LEARN?

1. Why did many Japanese residents not heed the tsunami warnings that were issued immediately after the 2011 Tohoku Earthquake?
2. How much floating debris was released into the Pacific Ocean by the tsunami? Where did this debris end up?

utilizes sea floor sensors that are capable of picking up the small yet distinctive pressure pulse from a tsunami passing above. The pressure sensors relay information to a buoy at the surface that transmits the data via satellite, allowing oceanographers to detect the passage of a tsunami in the open ocean (**Figure 8.31**). DART buoys, which are essential components of tsunami warning systems, have now been deployed in all oceans.

**TSUNAMI WATCHES AND TSUNAMI WARNINGS** When a seismic disturbance occurs beneath the ocean surface that is large enough to be tsunamigenic (capable of producing a tsunami), a *tsunami watch* is issued. At this point, a tsunami may or may not have been generated, but the potential for one exists.

The PTWC is linked to a series of sea floor pressure sensors, ocean buoys, and tide-measuring stations throughout the Pacific, so the recording stations nearest the earthquake are closely monitored for any indication of unusual wave activity. If unusual wave activity is verified, the tsunami watch is upgraded to a *tsunami warning*. Generally, earthquakes smaller than magnitude $M_w = 6.5$ are not typically tsunamigenic because they lack the duration of ground shaking necessary to initiate a tsunami. In addition, transform faults do not usually produce tsunami because lateral movement does not offset the ocean floor and impart energy to the water column in the same way that vertical fault movements do.

When a tsunami is detected, warnings are sent to all the coastal regions that might encounter the destructive wave, along with its estimated time of arrival. This warning, usually just a few hours in advance of the tsunami, makes it possible to evacuate people from low-lying areas and remove ships from harbors before the waves arrive. If the disturbance is nearby, however, there is not enough time to issue a warning because a tsunami travels so rapidly. Unlike hurricanes, whose high winds and waves threaten ships at sea and send them to the protection of a coastal harbor, a tsunami washes ships from their coastal moorings into the open ocean or onto shore. The best strategy during a tsunami warning is to move ships out of coastal harbors and into deep water, where tsunami are not easily felt.

**EFFECTIVENESS OF TSUNAMI WARNINGS** Since the PTWC was established in 1948, it has effectively prevented loss of life due to tsunami when people have heeded the evacuation warnings. Property damage, however, has increased as more buildings have been constructed close to shore. To combat the damage caused by tsunami, countries that are especially prone to tsunami, such as Japan, have invested in shoreline barriers, seawalls, and other coastal fortifications.

Perhaps one of the best strategies to limit tsunami damage and loss of life is to restrict construction projects in low-lying coastal regions where tsunami have frequently struck in the past. However, the long time interval between large tsunami—typically 200 years or more—often causes people to let down their guard about remembering past disasters.

## CONCEPT CHECK 8.5 ▸ Specify the origin and characteristics of tsunami.

1 Why is it more likely that a tsunami will be generated by the vertical movement of sea floor faults rather than the horizontal movement of sea floor faults?

2 Describe the differences between wind-generated waves and tsunami. Which waves transmit more energy?

3 Explain what it would look like at the shoreline when the trough of a tsunami arrives there first. What is the impending danger?

4 Explain how the tsunami warning system in the Pacific Ocean works. Why must the tsunami be verified at the closest tide recording station?

## STUDENTS SOMETIMES ASK . . .

***Where and when will the next major tsunami hit?***

Although most tsunami occur in the Pacific Ocean, a tsunami can affect any coastal region worldwide. Tsunami prediction is still in its infancy, yet seismologists who examine the history of large earthquakes (*paleoseismologists*) can produce forecasts of future tsunami based on evidence such as sediment records that indicate the size and frequency of past tsunami. Using this technique, paleoseismologists have issued forecasts for large, damaging tsunami in two regions: (1) in the Indian Ocean, where in 2004 the deadliest tsunami in history killed 300,000 people, a similar tsunami has been forecast to occur sometime within the next 30 years and (2) off the U.S. Pacific Northwest coast, where there is a subduction zone just offshore which generated a large tsunami in 1700; predictions suggest that this region may be overdue for another large tsunami. For both these locations, it's only a matter of time.

**RECAP** **Most tsunami are generated by underwater fault movement, which transfers energy to the entire water column. When these fast and long waves surge ashore, they can do considerable damage.**

## CREATURE FEATURE 8.1

### I'm an exotic invader!

**Asian amur sea stars** (*Asterias amurensis*) and nearly 300 other species from Japan have been recently discovered along the North American coast, having ferried across the Pacific Ocean on floating rafts of plastics and other floatables.

In 2011, during Japan's great Tohoku Earthquake and resulting tsunami, 5 million metric tons of debris was washed out to sea, providing a habitat for exotic species to hitch a ride on and be transported alive across an entire ocean basin.

## 8.6 ▸ Can Power from Waves Be Harnessed as a Source of Energy?

Moving water has a huge amount of energy, which is why there are so many hydroelectric power plants on rivers. Even greater energy exists in ocean waves, but significant problems must be overcome for the power to be harnessed efficiently. For example, a serious obstacle to the use of any device to harness wave energy is the monumental engineering problem of preventing the devices from being destroyed by the wave force they are built to harness. Also, deployment of complex machines with exposed moving parts to the marine environment presents challenges associated with corrosion and *biofouling*, which is the accumulation of algae and other sea life on machinery.

Another key disadvantage of wave energy is that the system produces significant power only when large storm waves break against it, so the system can't continuously provide reliable power and thus would serve only as a power supplement. In addition, a series of 100 or more of these structures along the shore would be required. Structures of this type could have a significant impact on the environment, with negative effects on marine organisms that rely on wave energy for dispersal, transporting food supplies, or removing wastes. Also, harnessing wave energy might alter the transport of sand along the coast, causing erosion in areas deprived of sediment.

Still, the immense power contained in waves could be used for generating electricity. Offshore wave-generating plants would be able to tap into the higher wave energy found offshore, but they are more likely to be damaged in large waves and more difficult to maintain. The most promising locations for coastal power generation from waves are where waves refract (bend) and converge, such as at headlands, which tend to focus wave energy (see Figure 8.21b). Using this advantage, an array of wave power plants might extract up to 10 megawatts of power per 1 kilometer (0.6 mile) of shoreline. (Note that each megawatt of electricity is enough to serve the energy needs of about 250 average U.S. homes.)

Internal waves are a potential source of energy, too. Along shores that have favorable sea floor shape for focusing wave energy, internal waves may be effectively concentrated by wave refraction and thus could power an energy-conversion device that generates electricity.

### Wave Power Plants and Wave Farms

In 2000, the world's first commercial-scale wave power plant began generating electricity. Built by Voith Hydro Wavegen and called **LIMPET 500** (*L*and *I*nstalled *M*arine *P*owered *E*nergy *T*ransformer), the plant is located on Islay, a small island off the west coast of Scotland that experiences high wave energy potential. The plant consists of a partially submerged chamber facing the sea (**Figure 8.32a**). As a wave approaches, the water level inside the structure rises, compressing air in the top of the chamber that is expelled through a turbine, which rotates to generate electricity (**Figure 8.32b**). As the wave recedes, the water level falls in the chamber and air is drawn back into the structure through the turbine. Even though there are two directions of air flow through the turbine, it is designed to continually produce power throughout the wave cycle. LIMPET 500 was constructed as a research and test facility that was originally configured with a turbo-generator of only 500 kilowatts (capable of supplying the energy needs of about 125 average U.S. homes) but has since been upgraded with a more efficient turbine that can generate more power and is continually being improved.

Wavegen's technology has recently been deployed in the 16-unit Mutriku breakwater power plant. Located in the Basque Country in northern Spain and developed by the Basque Energy Board EVE, this power plant is the world's only commercially operated wave energy plant. Another wave power plant using Wavegen technology will also be installed in Siadar on the Isle of Lewis off the west coast

(a) Photo of the exterior of LIMPET 500, the world's first commercial wave power plant.

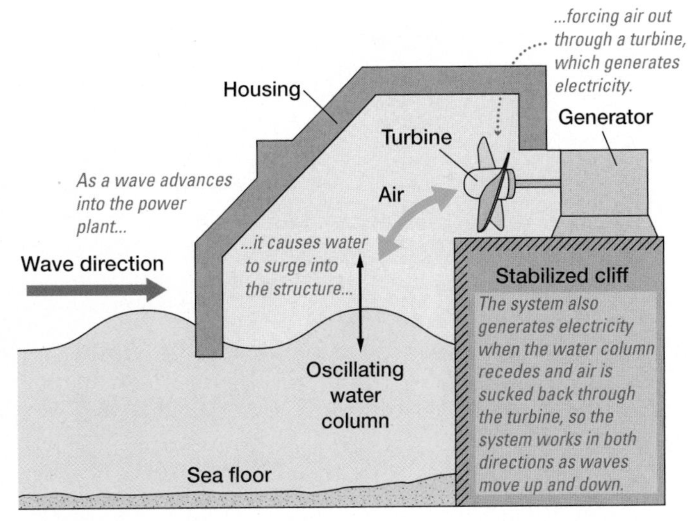


(b) Schematic view of the interior of a wave power plant showing how it generates electricity from waves.

**SmartFigure 8.32 How a wave power plant works.**
https://goo.gl/Vx6PJT

of Scotland. This installation will have a peak operating capacity of 30 megawatts and will be one of the world's first large-scale developments of wave energy.

In 2008, Pelamis Wave Power completed the world's first wave farm off the coast of northern Portugal. This project uses three 150-meter-long (500-foot-long) devices that resemble giant segmented snakes and float half-submerged in the ocean (**Figure 8.33**). As each segment surges up or down with the crest of an oncoming wave, its hydraulic power plant pumps a biodegradable hydraulic fluid through a turbine, thus generating electricity. These wave-energy devices are already supplying up to 2.25 megawatts of power to Portugal's electrical grid, and up to 26 more devices are planned for the future.

Currently, about 50 wave-energy projects are in development at various sites around the world. These projects use different methods to harness wave power, including floats or submerged pistons that move up and down with each passing wave, tethered paddles that oscillate back and forth, and collection of water from breaking waves that overtops coastal structures and then using the weight of that water to turn turbines as it moves downhill and returns to the ocean. In 2013, the London-based consultancy Bloomberg New Energy Finance projected that up to 22 tidal projects and 17 wave projects generating more than one megawatt of electricity—enough to power around 250 average U.S. homes—could be installed by 2020.

Figure 8.33 **Harnessing the power of ocean waves.** This wave energy device resembles a large segmented floating snake and is designed to flex as waves pass, generating electricity in the process. A wave farm of three of these devices is currently generating electricity off northern Portugal, and up to 26 more are planned.

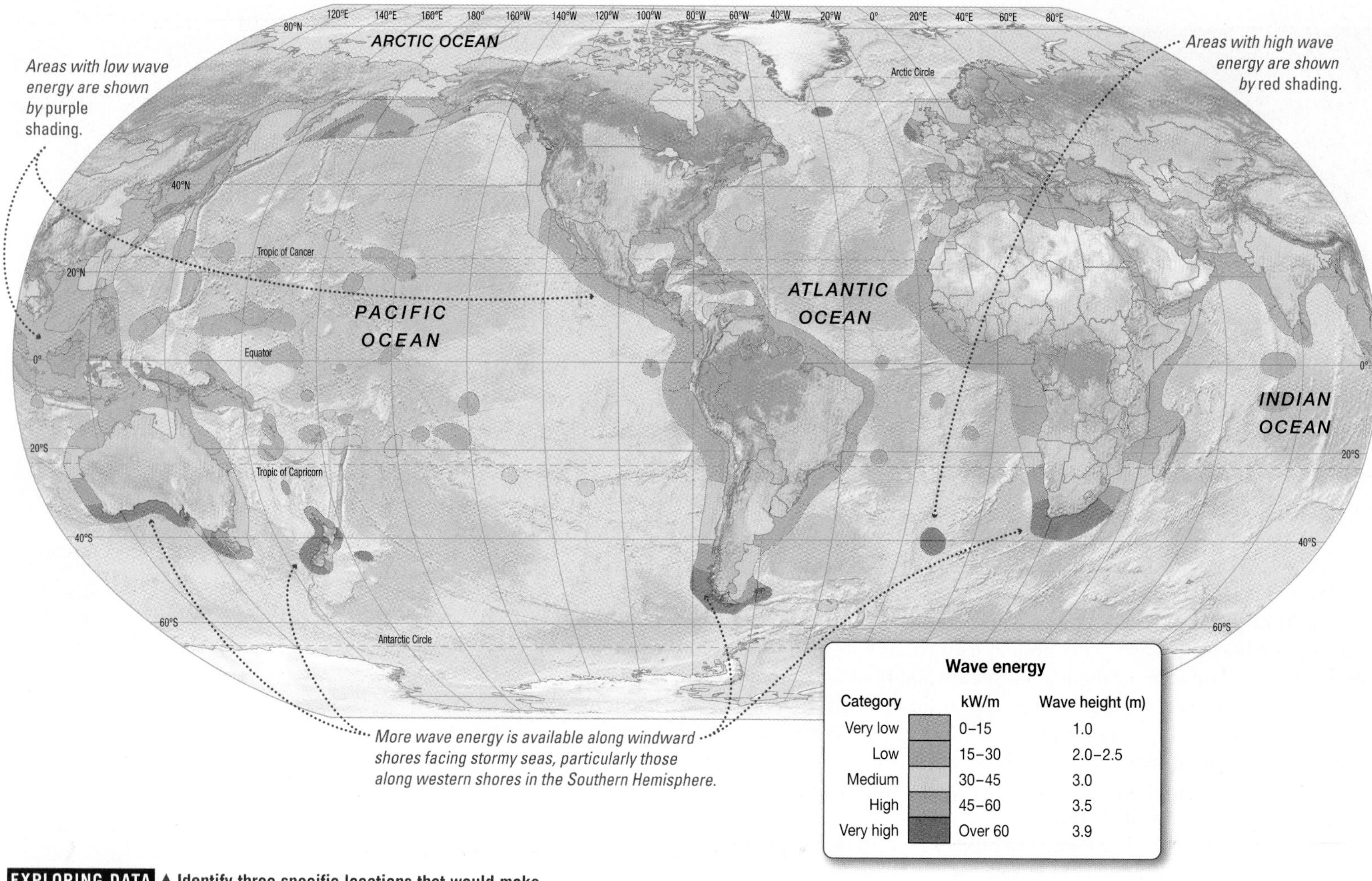


**EXPLORING DATA** ▲ **Identify three specific locations that would make good wave energy generation sites.**

Figure 8.34 **Global coastal wave energy resources.** In this map of coastal wave energy, areas of highest wave energy are shown in red. kW/m is kilowatts per meter (for example, every meter of "red" shoreline has the potential of generating over 60 kilowatts of electricity); corresponding average wave height is in meters.

**RECAP Ocean waves produce large amounts of energy. Although significant problems exist in harnessing wave energy effectively, several types of devices are extracting wave energy today.**

## Global Coastal Wave Energy Resources

Leading estimates suggest that the global resource for wave energy lies between 1 and 10 terawatts; the world currently produces about 12 terawatts from all sources. Where are the best places to develop additional wave power plants and wave farms? **Figure 8.34** shows the average wave heights experienced along coastal regions and indicates the sites most favorable for wave-energy generation (*red areas*). The map shows that west-to-east movement of storm systems in the middle latitudes between 30 and 60 degrees north or south latitude causes the western coasts of continents to be struck by larger waves than eastern coasts. Thus, more wave energy is generally available along western than eastern shores. Furthermore, some of the largest waves (and greatest potential for wave power) are associated with the prevailing westerly wind belt in the Southern Hemisphere middle latitudes.

### CONCEPT CHECK 8.6 ▸ Evaluate the advantages and disadvantages of harnessing waves as a source of energy.

1 Discuss some problems that might result from developing facilities for conversion of wave energy to electrical energy.

2 Describe the locations and power-generating capability of existing offshore wave power plants and wave farms.

3 Worldwide, where are the best locations for new wave farms? Explain the oceanographic conditions that make these locations favorable.

# ESSENTIAL CONCEPTS REVIEW

## 8.1 ▶ How are waves generated, and how do they move?

- *All ocean waves begin as disturbances that release energy into the ocean.* Energy sources include wind, the movement of fluids of different densities (which create internal waves), catastrophic releases of energy into the ocean by landslides and meteor impacts, underwater sea floor movements, the gravitational pull of the Moon and the Sun on Earth, and human activities in the ocean.
- Once initiated, *waves transmit energy through matter by setting up patterns of oscillatory motion* in the particles that make up the matter. Progressive waves are longitudinal, transverse, or orbital, depending on the pattern of particle oscillation. Particles in ocean waves move primarily in orbital paths.

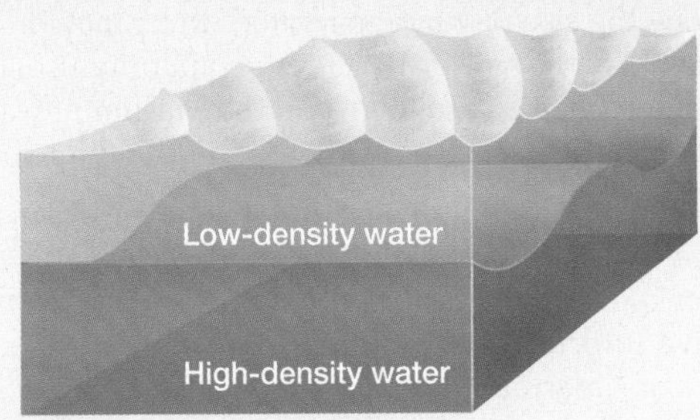

Block diagram showing the differences between surface waves and internal waves.

### Selected Key Terms

Use the **glossary** at the end of this book to discover the meanings of these Selected Key Terms: **ocean wave, atmospheric wave, internal wave, orbital wave.**

### Critical Thinking Question

Compare and contrast the properties of the three types of progressive waves: longitudinal, transverse, and orbital waves.

### Active Learning Exercise

Working in groups of three, have each student in class pick one of the following types of progressive waves: (1) longitudinal wave, (2) transverse wave, and (3) orbital wave. For each type of wave, use your own words to explain how the wave moves and whether the wave can transmit through the three states of matter: (a) solid, (b) liquid, and (c) gas. Then recombine as a group and have each student take turns explaining their wave type.

## 8.2 ▶ What characteristics do waves possess?

- *Waves are described according to their wavelength* (L), *wave height* (H), *wave steepness* (H/L), *wave period* (T), *frequency* (f), *and wave speed* (S). As a wave travels, the water passes the energy along by moving in a circle, called *circular orbital motion*. This motion advances the waveform, not the water particles themselves. *Circular orbital motion decreases with depth*, ceasing entirely at wave base, which is equal to one-half the wavelength measured from still water level.
- If water depth is greater than one-half the wavelength, a progressive wave travels as a *deep-water wave with a speed that is directly proportional to wavelength*. If water depth is less than $^1/_{20}$ wavelength (L/20), the wave moves as a *shallow-water wave with a speed that is directly proportional to water depth. Transitional waves* have wavelengths between deep- and shallow-water waves, with speeds that depend on both wavelength and water depth.

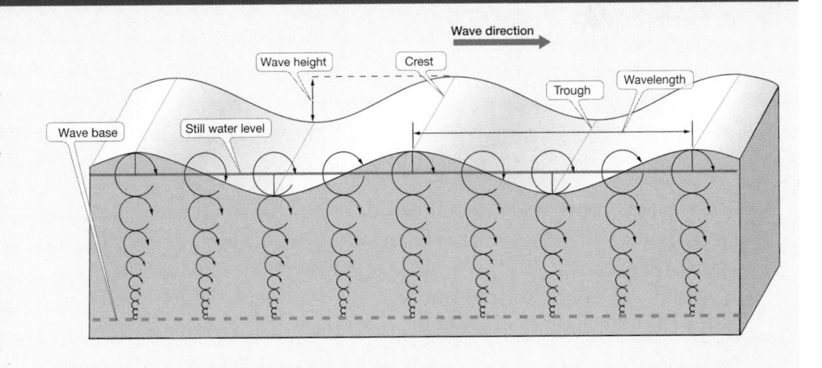


### Selected Key Terms

Use the **glossary** at the end of this book to discover the meanings of these Selected Key Terms: **crest, trough, still water level, wave height, wavelength, wave steepness, wave period, circular orbital motion, wave base, deep-water wave, wave speed, shallow-water wave.**

### Critical Thinking Question

To help reinforce your knowledge of wave terminology, draw a diagram of a simple progressive wave from memory. Include wave orbitals and label the crest, trough, wavelength, wave height, wave base, and still water level.

### Active Learning Exercise

Working in pairs, have each student in class use Figure 8.8 to consider one of the following two deep-water waves: (1) a wave with a wavelength of 200 meters and (2) a wave with a wavelength of 400 meters. Determine the corresponding wave period and wave speed for each wave and compare your results. Then discuss how you determined your answers and what general relationships exist between wavelength, wave period, and wave speed.

## 8.3 ▶ How do wind-generated waves develop?

- As wind-generated waves form in a sea area, *capillary waves* with rounded crests and wavelengths less than 1.74 centimeters (0.7 inch) *form first. As the energy of the waves increases, gravity waves form,* with increased wave speed, wavelength, and wave height. Factors that influence the size of wind-generated waves include wind *speed, duration* (time), and *fetch* (distance). An equilibrium condition called a *fully developed sea* is reached when the maximum wave height is achieved for a particular wind speed, duration, and fetch.
- *Energy is transmitted from the sea area across the ocean by uniform, symmetrical waves called swell.* Different wave trains of swell can create either *constructive, destructive,* or *mixed interference* patterns. Constructive interference produces unusually large waves called rogue waves or superwaves.

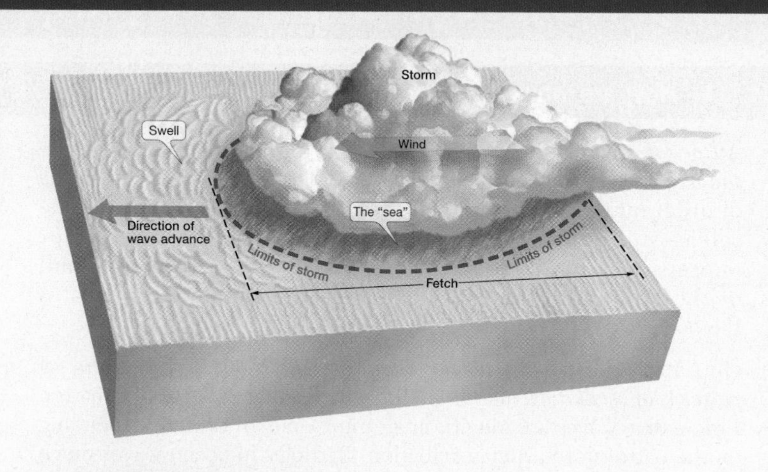


### Selected Key Terms

Use the **glossary** at the end of this book to discover the meanings of these Selected Key Terms: **capillary wave, gravity wave, fully developed sea, swell, wave train, wave dispersion, constructive interference, destructive interference, mixed interference, surf beat, rogue wave.**

### Critical Thinking Question

Using the information about the record-breaking waves experienced by the USS *Ramapo* in 1933, determine the waves' wavelength and speed.

### Active Learning Exercise

Working with another student in class, use Table 8.2 to determine the characteristics of the average height, wavelength, period, and highest 10% of waves created by the following two conditions: (1) a wind speed of 40 kilometers (25 miles) per hour, and (2) a wind speed of 80 kilometers (50 miles) per hour. With a doubling of wind speed, are waves created that have double the properties? Explain.

## 8.4 ▶ How do waves change in the surf zone?

- *As waves approach shallow water near the shore, they undergo many physical changes.* Waves release their energy in the surf zone when their steepness exceeds a 1:7 ratio and break. If waves break on a relatively flat surface, they produce *spilling breakers*. The curling crests of *plunging breakers*, which are the best for surfing, form on steep slopes, and abrupt beach slopes create *surging breakers*.
- When swell approaches the shore, *segments of the waves that first encounter shallow water are slowed* whereas other segments of the wave in deeper water move at their original speed, *causing each wave to refract, or bend.* Refraction concentrates wave energy on headlands, while low-energy breakers are characteristic of bays.
- *Reflection of waves off seawalls or other barriers* can cause an interference pattern called a *standing wave*. The crests of standing waves do not move laterally as in progressive waves but alternate with troughs at antinodes. Between the antinodes are nodes, where there is no vertical movement of the water.

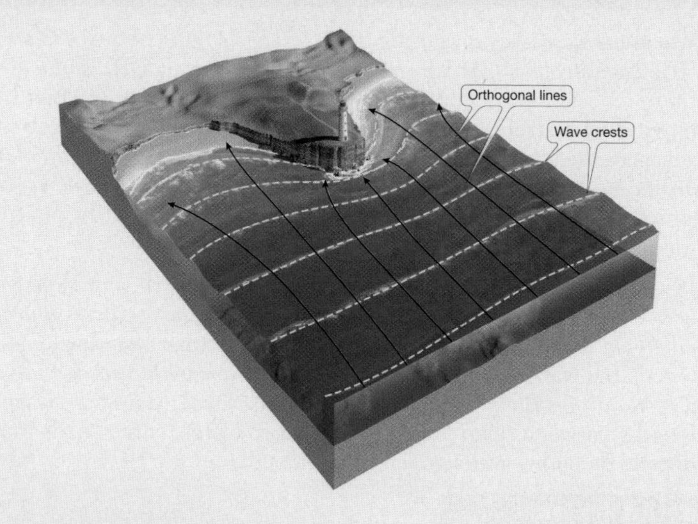


### Selected Key Terms

Use the **glossary** at the end of this book to discover the meanings of these Selected Key Terms: **spilling breaker, plunging breaker, surging breaker, surfing, wave refraction, wave reflection.**

### Critical Thinking Question

As a wave comes into shallow water, it experiences five physical changes that cause the wave to break at the shore. Name each physical change and explain why each occurs.

### Active Learning Exercise

Working with another student in class and using the Internet, explain the basic physics of surfing. Your answer should include important properties of waves as well as important properties of surfboards.

## 8.5 ▶ How are tsunami created?

- *Sudden changes in the elevation of the sea floor, such as from fault movement or volcanic eruptions, generate tsunami, or seismic sea waves*. These waves often have lengths exceeding 200 kilometers (125 miles) and travel across the open ocean with undetectable heights of about 0.5 meter (1.6 feet) at speeds in excess of 700 kilometers (435 miles) per hour. Upon approaching shore, a tsunami produces a *series of rapid withdrawals and surges*, some of which may increase the height of sea level by 40 meters (131 feet) or more.
- *Most tsunami occur in the Pacific Ocean*, where they have caused billions of dollars of coastal damage and taken tens of thousands of lives. A recent example is the *2011 Japanese Tohoku Earthquake and resulting tsunami*. However, the *2004 Indian Ocean tsunami killed as many as 300,000*, making it the deadliest tsunami in history. The *Pacific Tsunami Warning Center (PTWC)* has dramatically reduced fatalities by successfully predicting tsunami using real-time seismic information and a network of deep-ocean pressure sensors. A new tsunami warning system has become operational in the Indian Ocean.

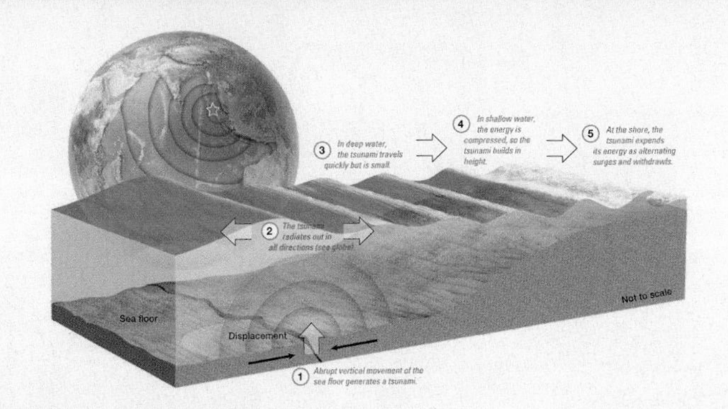

1 Abrupt vertical movement of the sea floor generates a tsunami.
2 The tsunami radiates out in all directions (see globe)
3 In deep water, the tsunami travels quickly but is small.
4 In shallow water, the energy is compressed, so the tsunami builds in height.
5 At the shore, the tsunami expends its energy as alternating surges and withdrawls.
Sea floor
Displacement
Not to scale

### Selected Key Terms

Use the **glossary** at the end of this book to discover the meanings of these Selected Key Terms: **tsunami, splash wave, Pacific Tsunami Warning Center (PTWC), Deep-ocean Assessment and Reporting of Tsunamis (DART).**

### Critical Thinking Question

What ocean depth would be required for a tsunami with a wavelength of 220 kilometers (136 miles) to travel as a deep-water wave? Is it possible that such a wave could become a deep-water wave anyplace in the world ocean? Explain.

### Active Learning Exercise

Working with another student in class, analyze this scenario: while shopping in a surf shop, you overhear some surfing enthusiasts mention that they would really like to ride the curling wave of a tidal wave at least once in their life because it is a single breaking wave of enormous height. What would you say to these surfers?

## 8.6 ▶ Can power from waves be harnessed as a source of energy?

- *Ocean waves can be harnessed to produce hydroelectric power*, but significant problems must be overcome to make this a practical source of energy. Some of these problems include the *constant battering* of any structure that exists in such a high-energy environment, *corrosion and/or biofouling* of machinery exposed to ocean waters, and the *varying amount of wave energy* at different times associated with natural waves.

### Selected Key Terms

Use the **glossary** at the end of this book to discover the meaning of this Selected Key Term: **LIMPET 500.**

### Critical Thinking Question

What are some negative environmental factors that could inhibit the development of ocean wave power systems? What are your ideas for how these issues could be overcome?

### Active Learning Exercise

Working as a group and using Figure 8.34, identify a specific location on Earth that has high wave power potential *and* is close to a major population center so that it is easy to distribute the generated electricity. Present your findings to the class, including the location and an explanation of your reasoning.

## Mastering Oceanography

Looking for additional review and test prep materials? With individualized coaching on the toughest topics of the course, Mastering Oceanography offers a wide variety of ways for you to move beyond memorization and deeply grasp the underlying processes of how the oceans work. Visit the Study Area in **www.masteringoceanography.com** to find practice quizzes, study tools, and multimedia that will improve your understanding of this chapter's content. Sign in today to access the following features: Self Study Quizzes, SmartFigures, Oceanography Videos and Animations, Squidtoons, Dynamic Study Modules, and an optional Pearson eText with embedded videos.

BIG BLUE
SANDRA LEANNE
BIG BLUE
SANDRA LEANNE

# Tides

# 9

## ESSENTIAL LEARNING CONCEPTS

At the end of this chapter, you should be able to:

- ☐ **9.1** Demonstrate an understanding of the forces that cause ocean tides.
- ☐ **9.2** Explain how tides vary during a monthly tidal cycle.
- ☐ **9.3** Specify what tides look like in the ocean.
- ☐ **9.4** Compare the characteristics and coastal locations of the three types of tidal patterns.
- ☐ **9.5** Describe tidal phenomena that occur in coastal regions.
- ☐ **9.6** Evaluate the advantages and disadvantages of harnessing tides as a source of energy.

↑ *Check when completed*

**Tides** are the periodic raising and lowering of sea level that occurs daily throughout the ocean. As sea level rises and falls each day, the edge of the sea slowly shifts landward and seaward; as it rises, it often destroys sand castles that were built during low tide. Knowledge of tides is important in many coastal activities, including tide pooling, shell collecting, surfing, fishing, navigation, and preparing for storms. Tides are so important that accurate records have been kept at nearly every port for several centuries, and there are many examples of the term *tide* in everyday vocabulary (for instance, "to tide someone over," "to go against the tide," or to wish someone "good tidings").

Although the original inhabitants of coastal regions most certainly noticed the tides, the first written record of tides doesn't appear until about 450 B.C.E. Even the earliest sailors knew the Moon had some connection with the tides because both followed a similar pattern. For example, high tides were associated with either a full or new moon, and lower tides were associated with quarter moons. However, it wasn't until **Isaac Newton** (1642–1727) developed the universal law of gravitation that the tides could adequately be explained.

Although the study of tides can be complex, tides are fundamentally very long and regular shallow-water waves. As we shall see, their wavelengths are measured in thousands of kilometers, and their heights range to more than 15 meters (50 feet).

***"I derive from the celestial phenomena the forces of gravity with which bodies tend to the sun and several planets. Then from these forces, by other propositions which are also mathematical, I deduce the motions of the planets, the comets, the moon, and the sea."***

**—*Sir Isaac Newton*, Philosophiae Naturalis Principia Mathematica (Philosophy of Natural Mathematical Principles) *(1686)***

## 9.1 ▸ What Causes Ocean Tides?

Simplistically, the gravitational attraction of the Sun and Moon on Earth creates ocean tides. In a more complete analysis, tides are generated by forces imposed on Earth that are caused by a combination of *gravity* and *motion* among Earth, the Moon, and the Sun.

### Tide-Generating Forces

Newton's work on quantifying the forces involved in the Earth–Moon–Sun system led to the first understanding of the underlying forces that keep bodies in orbit around each other. It is well known that gravity is the force that interconnects the Sun, its planets, and their moons and keeps them in relatively fixed orbits. For example, most of us are taught that "the Moon orbits Earth," but it is not quite this simple. The two bodies actually rotate around a common center of mass called the **barycenter** (*barus* = heavy, *center* = center), which is the *balance point* of the system, located about 1700 kilometers (1050 miles) beneath Earth's surface (**Figure 9.1a**).

◂ **Extreme tidal variation.** High and low tides in a small harbor near Blomidon Provincial Park, Nova Scotia, Canada, demonstrate the dramatic change in sea level experienced daily in the Bay of Fundy. This region experiences the largest tidal range in the world, up to 17 meters (56 feet).

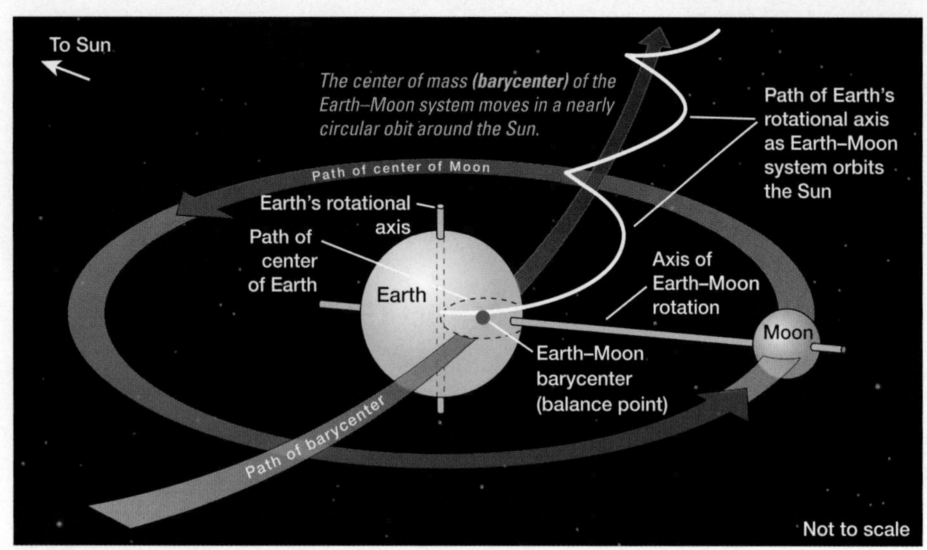


**(a)** Schematic diagram showing the movement of the barycenter (balance point) of the Earth–Moon system around the Sun.

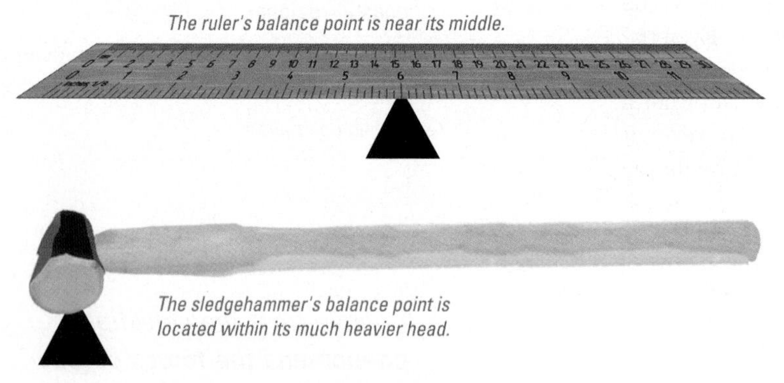


**(b)** Balance points of a ruler *(above)* and a sledgehammer *(below)*.

**Figure 9.1 Earth–Moon system barycenter.** The Earth–Moon barycenter is similar to a sledgehammer flung into space, with Earth represented by the hammer's head and the Moon represented by the end of the handle.

Why isn't the barycenter halfway between the two bodies? It's because Earth's mass is so much greater than that of the Moon. This can be visualized by imagining Earth and its Moon as ends of an object that is much heavier on one end than the other. Visualize a sledgehammer, for example, which has a lighter handle and a much heavier head. If you were to position your finger to balance the sledgehammer with its handle sticking out to one side, you'd find that its balance point is within the head of the hammer (**Figure 9.1b**). Now imagine that the sledgehammer is flung into space, tumbling slowly end over end about its balance point. This is exactly the situation that describes the movement of the Earth–Moon system. The purple arrow in Figure 9.1a shows the smooth, nearly circular path of the Earth–Moon barycenter around the Sun.

If the Moon and Earth are attracted to one another, why don't the two collide? The Earth–Moon system is in a mutually stable orbit held together by a balance between centripetal (gravitational) and inertial (motion) forces, which prevents the Moon and Earth from colliding or flying apart. This is how orbits are established that keep objects at more or less fixed distances.

Newton's work also allowed an understanding of why the tides behave as they do. Just as gravity and motion serve to keep bodies in mutual orbits, they also exert an influence on every particle of water in the oceans, thus creating the tides.

**GRAVITATIONAL AND CENTRIPETAL FORCES IN THE EARTH–MOON SYSTEM** To understand how *tide-generating forces* influence the oceans, let's examine how *gravitational forces* and *centripetal forces* affect objects on Earth within the Earth–Moon system. (We'll ignore the influence of the Sun for the moment.)

The **gravitational force** is derived from **Newton's law of universal gravitation**, which states that *every object that has mass in the universe is attracted to every other object*. An object can be as small as a sub-atomic particle or as large as a sun. The basic equation for this relationship is:

$$F_g = \frac{Gm_1m_2}{r^2} \quad (9.1)$$

This equation states that the gravitational force ($F_g$) is directly proportional to the product of the masses of the two bodies ($m_1$, $m_2$) and is inversely proportional to the square of the distance between the two masses ($r^2$). Note that $G$ is the gravitational constant, so it does not change.

Let's simplify Newton's law of universal gravitation and examine the effect of both mass and distance on the gravitational force, which can be expressed with arrows (up arrow = increase, down arrow = decrease):

If mass increases (↑), then gravitational force increases (↑).

A practical example of this can be seen in an object with a large mass (such as the Sun), which produces a large gravitational attraction (**Figure 9.2a**).

Looking at how distance influences gravitational force, the relationship is as follows:

If distance increases (↑), then gravitational force greatly decreases (↓↓).

Equation 9.1 shows that the gravitational attraction varies with the *square* of distance, so even a small *increase* in the distance between two objects significantly *decreases* the gravitational force between them—hence the double arrows in the distance relationship illustrated above. This means that when an object is twice as far away, the gravitational attraction is only one-quarter as strong. As a practical example, this is why astronauts can experience zero gravity (weightlessness) in space if they get far enough away from Earth's gravitational pull (**Figure 9.2b**). In summary, then, the *greater* the mass of the objects and (especially) the *closer* they are together, the greater their gravitational attraction.

**Figure 9.3** shows how gravitational forces for points on Earth (caused by the Moon) vary depending on their distances from the Moon. The greatest gravitational attraction (the longest arrow) is at Z, the **zenith** (*zenith* = a path over the head), which is the point closest to the Moon. The gravitational attraction is weakest at *N*, the **nadir** (*nadir* = opposite the zenith), which is the point farthest from the Moon. The direction of the gravitational attraction between most particles and the center of the Moon is at an angle relative to a line connecting the center of Earth and the Moon (Figure 9.3). This angle causes the force of gravitational attraction between each particle and the Moon to be slightly different.

The **centripetal force** (*centri* = the center, *pet* = seeking) required to keep planets in their orbits is provided by the gravitational attraction between the center-of-mass of each of the planets and the Sun. (Note that the *centripetal force* is not to be confused with the so-called *centrifugal force* [*centri* = the center, *fug* = flee], which is an apparent or fictitious force that is oriented outward.) Centripetal force connects the center-of-mass of an orbiting body to its parent, pulling the object *inward* toward the parent, "seeking the center" of its orbit. For example, if you tie a string to a ball and swing the ball around your head (**Figure 9.4**), the string pulls the ball toward your hand. The string exerts a *centripetal force* on the ball, forcing the ball to *seek the center* of its orbit. If the string breaks, the force is gone, and the ball can no longer maintain its circular orbit. In this case, the ball will continue along a straight-line path, obeying Newton's first law of motion (the *law of inertia*), which states that moving objects follow straight-line paths until they are compelled

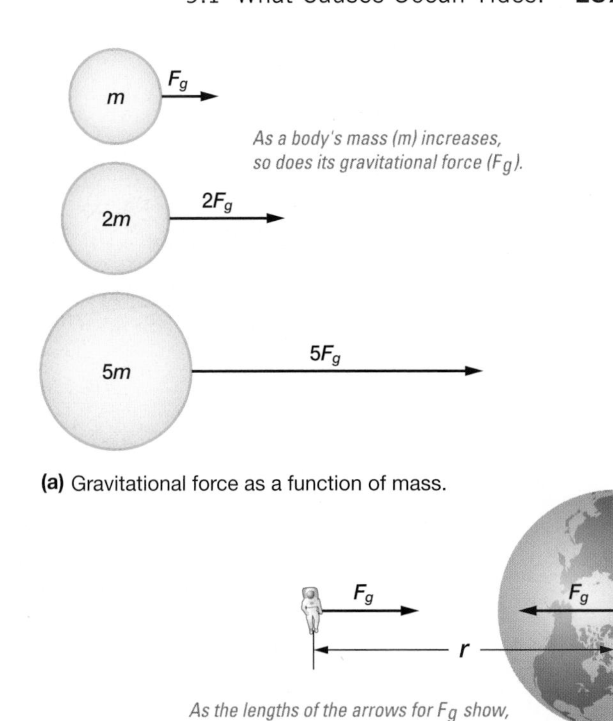


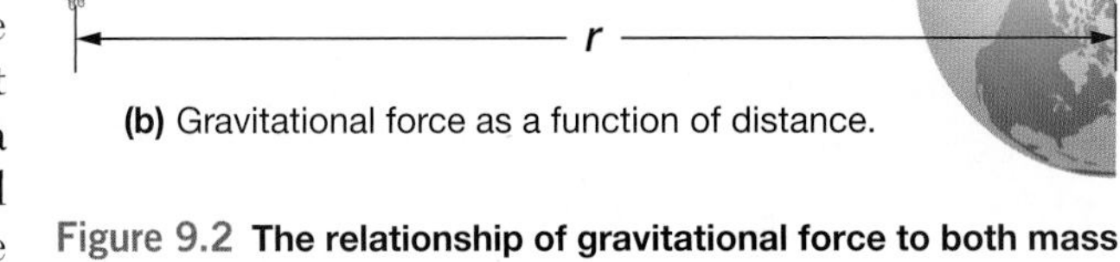


**Figure 9.2 The relationship of gravitational force to both mass (a) and distance (b).**

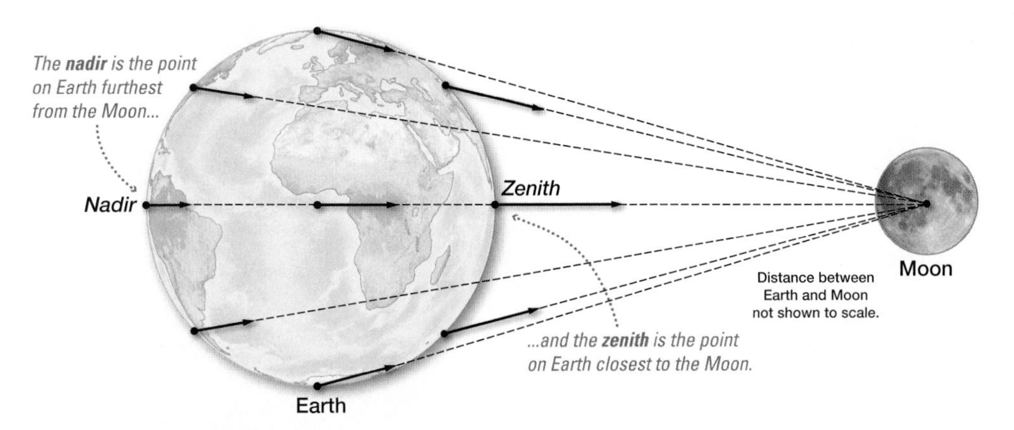

The **nadir** is the point on Earth furthest from the Moon...
Nadir
Zenith
Moon
Distance between Earth and Moon not shown to scale.
...and the **zenith** is the point on Earth closest to the Moon.
Earth

**Figure 9.3 Gravitational forces on Earth due to the Moon.** Black arrows represent the gravitational forces on objects located at different places on Earth due to the Moon. The length and orientation of the arrows indicate the strength and direction of the gravitational force. Notice the length and angular differences of the arrows for different points on Earth.

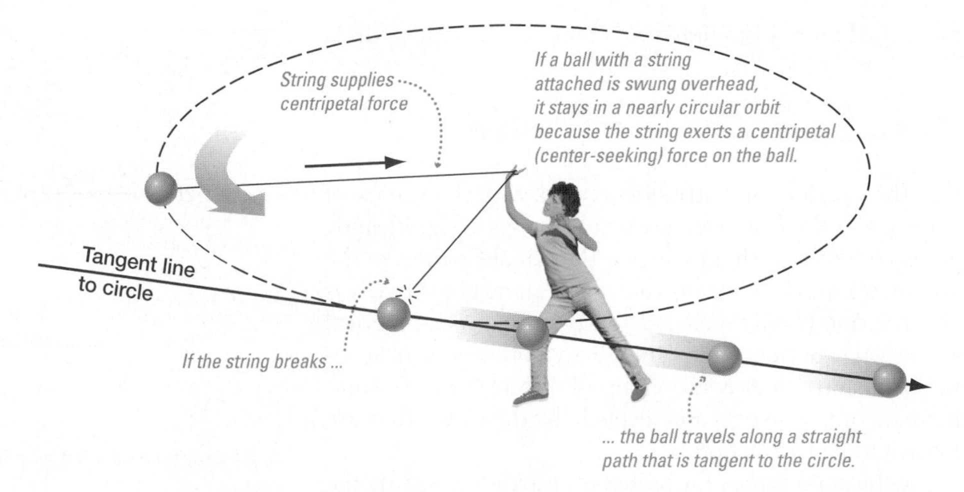
String supplies centripetal force
If a ball with a string attached is swung overhead, it stays in a nearly circular orbit because the string exerts a centripetal (center-seeking) force on the ball.
Tangent line to circle
If the string breaks ...
... the ball travels along a straight path that is tangent to the circle.

Figure 9.4 **Centripetal force.** The string on a rotating ball exerts a centripetal force that holds the rotating ball in a nearly circular orbit. This is similar to how Earth's gravity exerts a centripetal force on the Moon to hold the Moon in a nearly circular orbit.

to change that path by other forces. As a result, the ball flies off in a *straight* line that is *tangent* (*tangent* = touching) to the circle it was on (Figure 9.4).

Earth and the Moon are interconnected, too, but by gravity rather than by a string. Gravity provides the centripetal force that holds the Moon in its orbit around Earth. If all gravity in the solar system could be shut off, centripetal force would vanish, and the momentum of the celestial bodies would send them flying off into space along straight-line paths, tangent to their orbits.

**RESULTANT FORCES** Particles of identical mass rotate in identical-sized paths due to the Earth–Moon rotation system (Figure 9.5). Each particle requires an identical centripetal force to maintain it in its circular path. Gravitational attraction between the particle and the Moon supplies the centripetal force, but the *supplied* force is different than the *required* force (because gravitational attraction varies with distance from the Moon) except at the center of Earth. This difference creates tiny **resultant forces**, which are the mathematical difference between the two sets of arrows shown in Figures 9.3 and 9.5.

Figure 9.6 combines Figures 9.3 and Figure 9.5 to show that resultant forces are produced by the difference between the required centripetal ($C$) and supplied

## STUDENTS SOMETIMES ASK . . .

***Are there also tides in other objects, such as lakes and swimming pools?***

The Moon's and Sun's gravity act on all objects, so anything that has the ability to flow will exhibit measurable tides. For example, there are tides in lakes, wells, and swimming pools. In fact, there are even extremely tiny tidal bulges in a glass of water! However, tidal effects become negligible the smaller the body of water is, and other effects dominate them, so they are not observable for the most part. On the other hand, tides in the atmosphere and the "solid" Earth are much larger. Tides in the atmosphere—called *atmospheric tides*—can be miles high and are also affected by solar heating. The tides inside Earth's interior—called *solid-body tides*, or *Earth tides*—are up to 50 centimeters (1.6 feet) high and cause the daily stretching and contraction of Earth. Interestingly, studies have shown that Earth tides add to the stress along faults, which makes it more likely that major earthquakes will strike at these times.

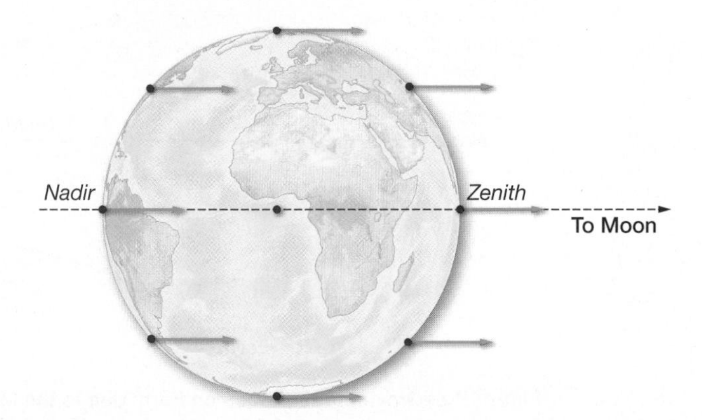
Nadir
Zenith
To Moon

Figure 9.5 **Required centripetal (center-seeking) forces.** Red arrows represent the centripetal forces required to keep identical-sized particles in identical-sized orbits as a result of the rotation of the Earth–Moon system around its barycenter. Notice that the arrows are all the same length and are oriented in the same direction for all points on Earth.

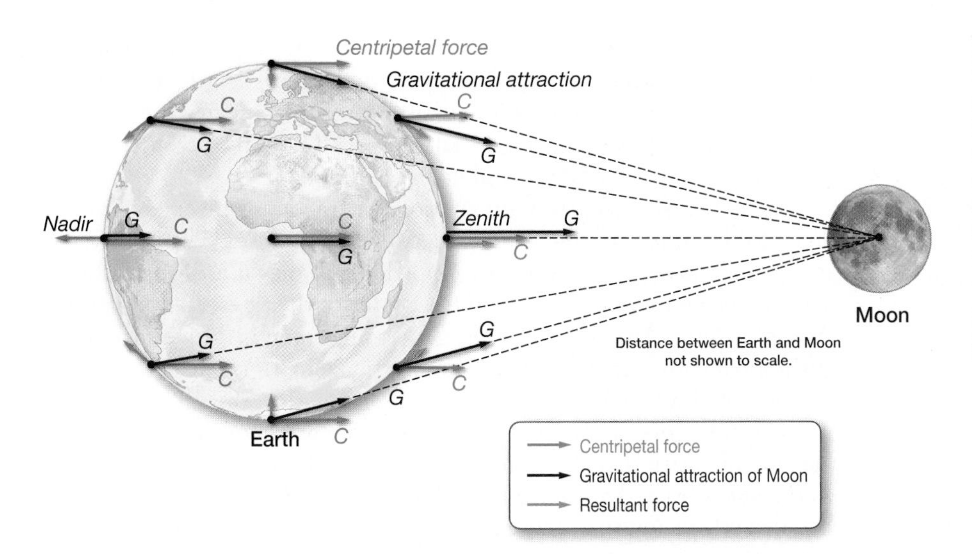


**SmartFigure 9.6 Resultant forces.** Red arrows indicate centripetal forces (*C*), which are not equal to the black arrows that indicate gravitational attraction (*G*). The small blue arrows show resultant forces, which are established by constructing an arrow from the tip of the centripetal (*red*) arrow to the tip of the gravity (*black*) arrow and located where the red and black arrows begin. https://goo.gl/eBCMYZ

**EXPLORING DATA ◄ Describe two key differences between the centripetal forces (red arrows) and the gravitational forces (black arrows).**

gravitational (*G*) forces. However, do not think that both of these forces are being applied to the points, because *C* is a force that would be required to keep the particles in a perfectly circular path, while *G* is the force actually provided for this purpose by gravitational attraction between the particles and the Moon. The resultant forces (*blue arrows*) are established by constructing an arrow from the tip of the centripetal (*red*) arrow to the tip of the gravity (*black*) arrow and located where the red and black arrows begin.

**TIDE-GENERATING FORCES** Resultant forces are small, averaging about one-millionth the magnitude of Earth's gravity. Moreover, there is no tide-generating force where the resultant forces either point straight upward toward the sky or straight downward into Earth. In these cases, the resultant force is oriented *vertically* relative to Earth's surface (**Figure 9.7**), and no tides occur here. These conditions occur at three locations on Earth: (1) at the zenith, (2) at the nadir, and (3) along an "equator" connecting all points halfway between the zenith and nadir. However, if the resultant force has a significant *horizontal component*—that is,

**RECAP The tides are caused by an imbalance between the required centripetal and the provided gravitational forces acting on Earth. This difference produces residual forces, the horizontal component of which creates two equal tidal bulges on opposite sides of Earth.**

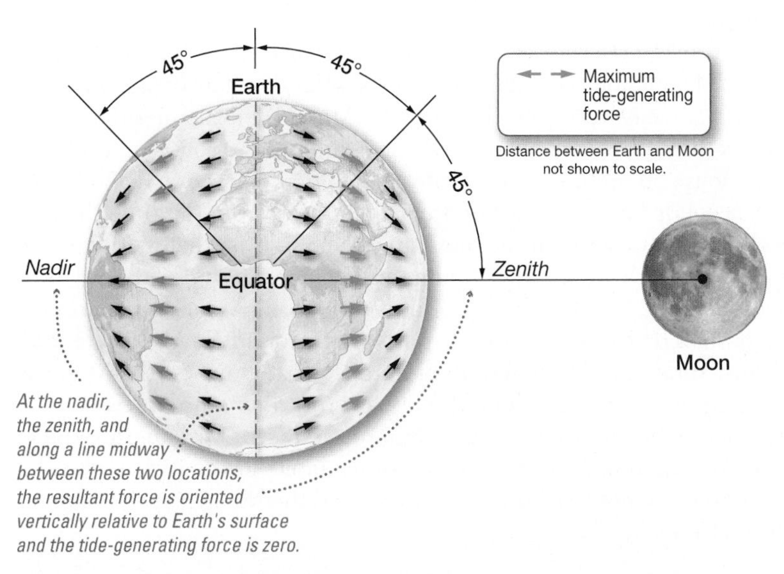


**Figure 9.7 Tide-generating forces.** Schematic diagram showing the tide-generating forces on Earth (*arrows*) that are caused by the Moon. The maximum tide-generating forces (*blue arrows*) occur where the resultant forces have a significant horizontal component. See the text for a description.

**Animation**
Tidal Forces
https://goo.gl/DKLLdk

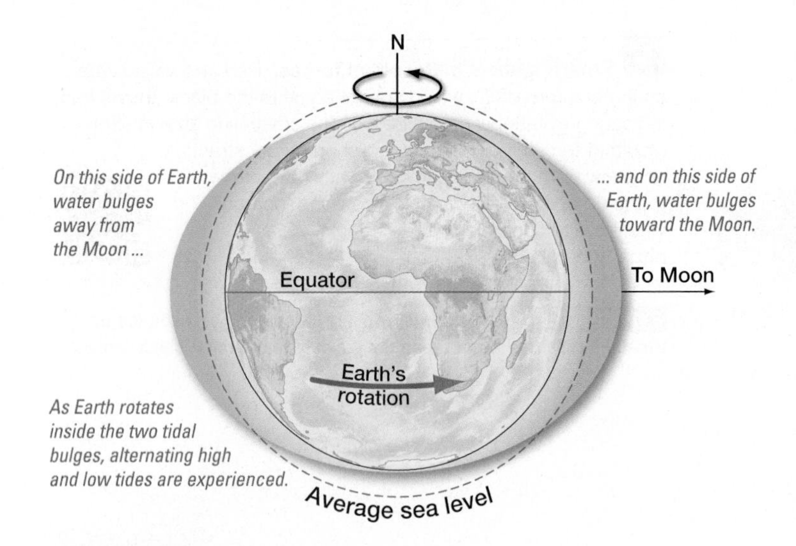
N
On this side of Earth, water bulges away from the Moon ...
... and on this side of Earth, water bulges toward the Moon.
Equator
To Moon
Earth's rotation
As Earth rotates inside the two tidal bulges, alternating high and low tides are experienced.
Average sea level

**EXPLORING DATA** ▲ **Why does Earth have not just one tidal bulge on the side of Earth facing the Moon but two equal tidal bulges on both sides of Earth?**

Figure 9.8 **Idealized tidal bulges.** In an idealized case, the Moon creates two bulges in the ocean surface: one that extends *toward* the Moon and the other *away from* the Moon. As Earth rotates, it carries various locations into and out of the two tidal bulges so that all points on its surface (except the poles) experience two alternating high and low tides daily.

tangential or parallel to Earth's surface—it produces tidal bulges on Earth, creating what are known as the **tide-generating forces**. These tide-generating forces are quite small but reach their maximum value at points on Earth's surface at a "latitude" of 45 degrees relative to the "equator" between the zenith and nadir (Figure 9.7).

As previously discussed, gravitational attraction is inversely proportional to the *square* of the distance between two masses. The tide-generating force, however, is inversely proportional to the *cube* of the distance between each point on Earth and the *center* of the tide-generating body (Moon or Sun). Although the tide-generating force is derived from the gravitational force, it is not linearly proportional to it. As a result, distance is a more highly weighted variable for tide-generating forces.

The tide-generating forces create two simultaneous bulges: one on the side of Earth directed *toward* the Moon (the zenith) and the other on the side directed *away from* the Moon (the nadir) (**Figure 9.8**). On the side directly facing the Moon, the bulge is created because the provided gravitational force is greater than the required centripetal force. Conversely, on the side facing away from the Moon, the bulge is created because the required centripetal force is greater than the provided gravitational force. Although the forces are oriented in opposite directions on the two sides of Earth, the resultant forces are equal in magnitude, so the bulges are equal, too.

## STUDENTS SOMETIMES ASK . . .

***I understand how the tidal bulge is created on the side of Earth facing the Moon, but how does the second bulge form on the far side of Earth, away from the Moon?***

Yes, it's easier to see how the tidal bulge forms on the side of Earth facing the Moon. After all, those parts of Earth are physically closer to the Moon, and gravitational attraction is strongly related to distance. The other bulge is created because of the *motion* of Earth and the Moon. In essence, the Moon is accelerating toward Earth and the Earth is accelerating toward the Moon, so fluid objects such as the oceans on Earth get left behind, hence the second identical bulge on the far side of Earth. It's a little more difficult to grasp, but the second bulge is a result of equal but opposite forces caused by motion of the mutual Earth–Moon system.

## Tidal Bulges: The Moon's

It is easier to understand how tides on Earth are created if we consider an ideal Earth and an ideal ocean. The ideal Earth has two tidal bulges, one toward the Moon and one away from the Moon (called the **lunar bulges**), as shown in Figure 9.8. The ideal ocean has a uniform depth, with no friction between the seawater and the sea floor. Newton made these same simplifications when he first explained Earth's tides.

If the Moon is stationary and aligned with the ideal Earth's equator, the maximum bulge will occur on the equator on opposite sides of Earth. If you were standing on the equator, you would experience two high tides each day. The time between high tides, which is the **tidal period**, would be 12 hours. If you moved to any latitude north or south of the equator, you would experience the same tidal period, but the high tides would be less high because you would be at a lower point on the bulge.

In most places on Earth, however, high tides occur every 12 hours 25 minutes because tides depend on the lunar day, not the solar day. The **lunar day** (also called a *tidal day*) is measured from the time the Moon is on the meridian of an observer—that is, directly overhead—to the next time the Moon is on that meridian and is 24 hours 50 minutes long. (Note that a lunar day is exactly 24 hours, 50 minutes, 28 seconds long.) The **solar day** is measured from the time the Sun is on the meridian of an observer to the next time the Sun is on that meridian and is 24 hours long. Why is the lunar day 50 minutes longer than the solar day? During the 24 hours it takes Earth to make a full rotation, the Moon has continued moving another 12.2 degrees to the east in its orbit around Earth (**Figure 9.9**). Thus, Earth must rotate an additional 50 minutes to "catch up" to the Moon so that the Moon is again on the meridian (directly overhead) of our observer.

**RECAP** **A solar day (24 hours) is shorter than a lunar day (24 hours and 50 minutes). The extra 50 minutes is the result of the Moon's movement in its orbit around Earth.**

The difference between a solar day and a lunar day can be seen in some of the natural phenomena related to the tides. For example, alternating high tides are normally 50 minutes *later* each successive day, and the Moon rises 50 minutes *later* each successive night.

## Tidal Bulges: The Sun's Effect

The Sun affects the tides, too. Like the Moon, the Sun produces tidal bulges on opposite sides of Earth, one oriented *toward* the Sun and one oriented *away from*

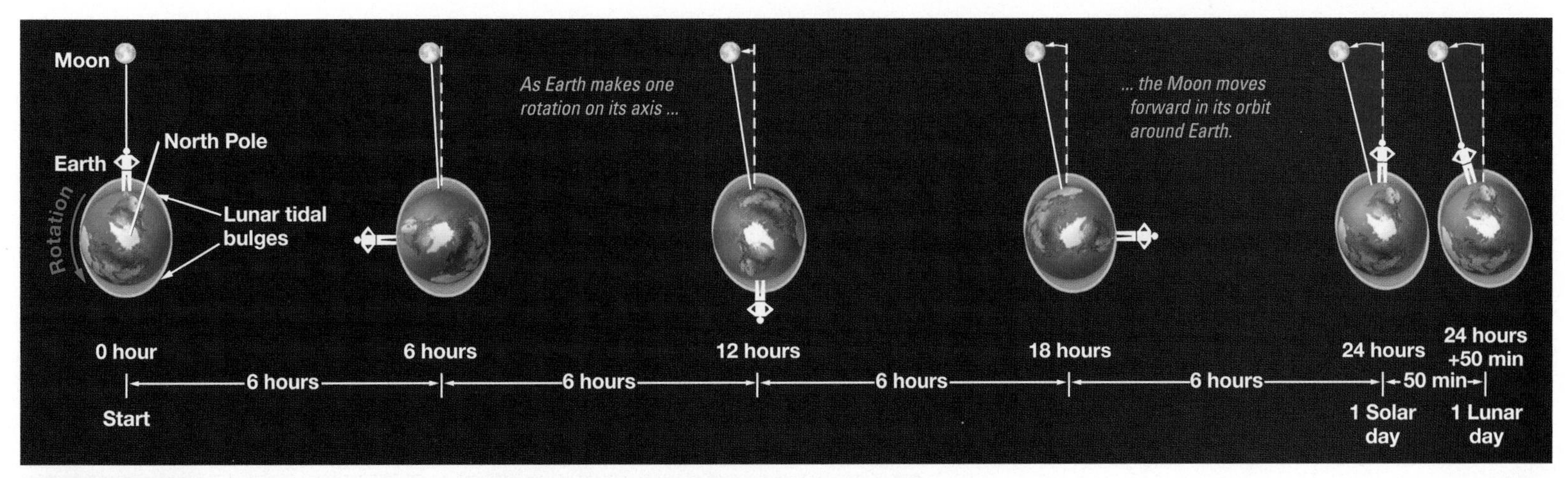


**EXPLORING DATA** ▲ **1) Describe the differences between a solar day and a lunar day.**

**2) You are photographing a sunset at 6:00 p.m. and notice the full Moon rising in the east behind you. You decide to return the next evening to photograph the moonrise. Based on your knowledge of the lunar day, what time would moonrise be?**

Animation
The Lunar Day
https://goo.gl/ElkpUH

SmartFigure 9.9 **The lunar day.** A lunar day is the time that elapses between when the Moon is directly overhead and the next time the Moon is directly overhead. During one complete rotation of Earth (the 24-hour solar day), the Moon moves eastward 12.2 degrees, and Earth must rotate an additional 50 minutes for the Moon to be in the exact same position overhead. Thus, a lunar day is 24 hours 50 minutes long.
https://goo.gl/bbYGx0

the Sun. These **solar bulges**, however, are only about half the size of the lunar bulges. Although the Sun is 27 million times more massive than the Moon, its tide-generating force is not 27 million times greater than the Moon's. This is because the Sun is 390 times farther from Earth than the Moon (**Figure 9.10**). Moreover, tide-generating forces vary inversely as the *cube* of the distance between objects. Thus, the tide-generating force is reduced by the cube of 390, or about 59 million times compared with that of the Moon. As a consequence, the Sun's tide-generating force is only $^{27}/_{59}$ that of the Moon, or 46% (about one-half), and the solar bulges are 46% the size of the lunar bulges. For simplicity, it is easy just to remember that the Moon exerts over two times the gravitational pull of the Sun on the tides.

Even though the Moon exerts over two times the gravitational pull of the Sun on Earth's tides, note that the Sun does not exert a smaller gravitational force on Earth as compared to the Moon. In fact, the Sun's total "pull" on all points on Earth is much greater than that of the Moon, but the *difference* across Earth is small because the diameter of Earth is very small in relation to the distance from the Sun. In contrast, the diameter of Earth is quite large in relation to the distance to the center of the Moon. In summary, the reason the Moon controls tides far more than the Sun is that the Moon is much closer to Earth, although it is much smaller in size and mass compared to the Sun.

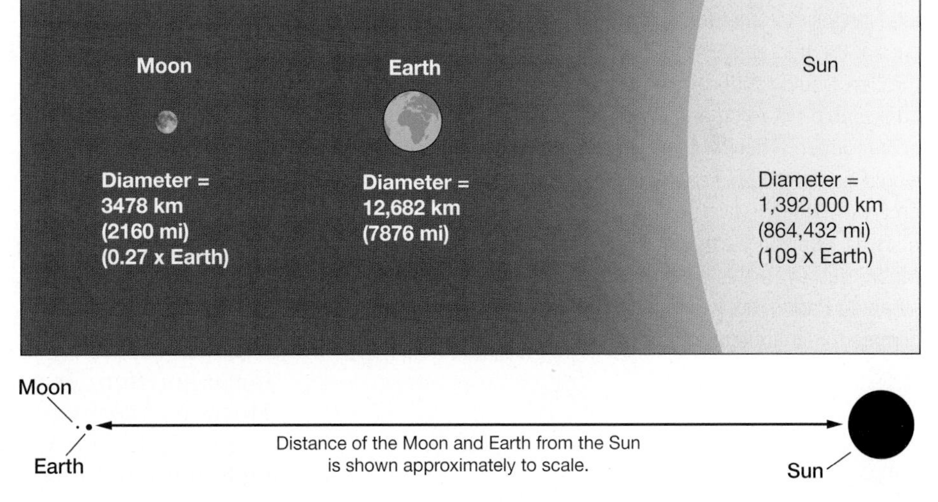


Figure 9.10 **Relative sizes and distances of the Moon, Earth, and Sun.** *Top*: The relative sizes of the Moon, Earth, and Sun are shown to scale. Note that the diameter of the Moon is roughly one-fourth that of Earth, while the diameter of the Sun is 109 times the diameter of Earth. *Bottom*: The relative distances of the Moon, Earth, and Sun are shown to scale.

**RECAP Although the Moon is much smaller than the Sun, it controls tides far more than the Sun because the Moon is much closer to Earth. As a result, the Moon creates lunar tidal bulges that are twice the size of the solar tidal bulges.**

## Earth's Rotation and the Tides

The tides appear to move water in toward shore (the **flood tide**) and to move water away from shore (the **ebb tide**). However, according to the nature of the idealized tides presented so far, *Earth's rotation carries various locations into and out of*

**RECAP** **In an idealized case, the rise and fall of the tides are caused by Earth's rotation carrying various locations into and out of the tidal bulges.**

*the tidal bulges*, which are in fixed positions relative to the Moon and the Sun. In essence, alternating high and low tides are created as Earth constantly rotates inside fluid bulges that are supported by the Moon and the Sun.

**CONCEPT CHECK 9.1 ▸ Demonstrate an understanding of the forces that cause ocean tides.**

1 Why are there tidal bulges on both sides of Earth (for example, not just the side of Earth that faces the Moon or the Sun)?

2 Explain why the Sun's influence on Earth's tides is only 46% that of the Moon, even though the Sun is so much more massive than the Moon.

3 Why is a lunar day 24 hours 50 minutes long, while a solar day is 24 hours long?

4 If Earth did not have the Moon orbiting it, would there still be tides? Why or why not?

### STUDENTS SOMETIMES ASK . . .

***What would Earth be like if the Moon didn't exist?***

For starters, Earth would spin faster, and days would be much shorter because the tidal forces that act as slow brakes on Earth's rotation wouldn't exist. In fact, geologists have evidence that an Earth day was originally five or six hours long in the distant geologic past; it might be just a little longer than that today if the Moon didn't exist. In the ocean, the tidal range would be much smaller because only the Sun would produce relatively small tidal bulges. Spring tides would not exist, and coastal erosion would be markedly reduced. There would be no moonlight, and nighttime would be much darker, which would affect nearly all life on Earth. There is even some speculation that life would not exist at all on Earth without the stabilizing effect of the Moon: during Earth's early development, the Moon's gravity steadied Earth's tilt, which in turn helped stabilize Earth's climate for life to flourish.

## 9.2 ▸ How Do Tides Vary during a Monthly Tidal Cycle?

The monthly tidal cycle is 29-and-a-half days because that's how long it takes the Moon to complete an orbit around Earth. (The 29-and-a-half-day monthly tidal cycle is also called a *lunar cycle*, a *lunar month*, or a *synodic* [*synod* = meeting] *month*. In fact, the word "month" is derived from the word "moon.") During its orbit around Earth, the Moon's changing position influences tidal conditions on Earth.

### The Monthly Tidal Cycle

During the monthly tidal cycle, the phase of the Moon changes dramatically. When the Moon is between Earth and the Sun, it cannot be seen at night; this phase is called **new moon**. When the Moon is on the side of Earth opposite the Sun, its entire disk is brightly visible; this phase is called **full moon**. A **quarter moon**—a moon that is half lit and half dark as viewed from Earth—occurs when the Moon is at right angles to the Sun relative to Earth.

**Figure 9.11** shows the positions of Earth, the Moon, and the Sun at various points during the 29-and-a-half-day lunar cycle. When the Sun and Moon are aligned, either with the Moon between Earth and the Sun (new moon; Moon in *conjunction*) or with the Moon on the side opposite the Sun (full moon; Moon in *opposition*), the tide-generating forces of the Sun and Moon combine (Figure 9.11, *top*). At this time, the **tidal range** (the vertical difference between high and low tides) is large (very *high* high tides and quite *low* low tides) because there is *constructive interference* between the lunar and solar tidal bulges. (As mentioned in Chapter 8, *constructive interference* occurs when two waves [or, in this case, two tidal bulges] overlap crest to crest and trough to trough.) The maximum tidal range is called a **spring tide** (*springen* = to rise up) because the tide is extremely large, or "springs forth." Note that spring tides have nothing to do with the spring season; they happen twice a month during all times of the year. When the Earth–Moon–Sun system is aligned, the Moon is said to be in **syzygy** (*syzygia* = union).

When the Moon is in either the first- or third-quarter phase (Figure 9.11, *bottom*), the tide-generating force of the Sun is working at right angles to the tide-generating force of the Moon. (Note that the third-quarter moon is often called the last-quarter moon, which is not to be confused with certain sports that have a fourth or last quarter.) The tidal range is small (*lower* high tides and *higher* low tides) because there is *destructive interference* between the lunar and solar tidal bulges. (Recall from Chapter 8 that *destructive interference* occurs when two waves

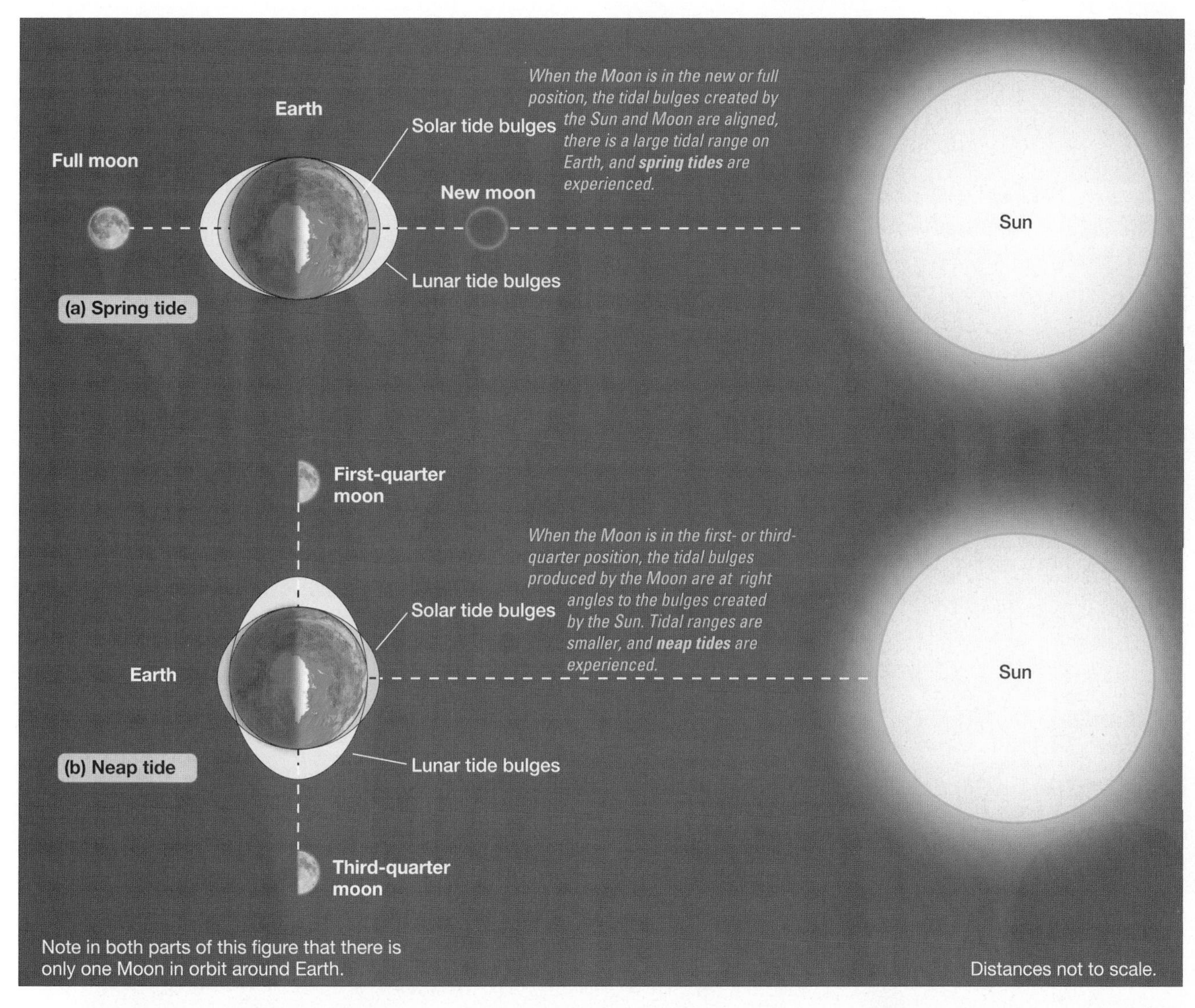

When the Moon is in the new or full position, the tidal bulges created by the Sun and Moon are aligned, there is a large tidal range on Earth, and **spring tides** are experienced.
Earth
Solar tide bulges
Full moon
New moon
Sun
Lunar tide bulges
(a) Spring tide
First-quarter moon
When the Moon is in the first- or third-quarter position, the tidal bulges produced by the Moon are at right angles to the bulges created by the Sun. Tidal ranges are smaller, and **neap tides** are experienced.
Solar tide bulges
Earth
Sun
(b) Neap tide
Lunar tide bulges
Third-quarter moon
Note in both parts of this figure that there is only one Moon in orbit around Earth.
Distances not to scale.

**EXPLORING DATA** ◄

1) Explain why during some weeks tides are higher than during other weeks over the course of a month.

2) Explain why there is a pattern of two high tides and two low tides daily during a spring tide. Your answer must include the terms "tidal bulges" and "Earth's rotation."

**SmartFigure 9.11 Earth–Moon–Sun positions and the tides.** Schematic diagram showing **(a)** spring tide and **(b)** neap tide. Note that the phase of the Moon is shown as it appears from Earth. https://goo.gl/DmuidP

[or, in this case, two tidal bulges] match up crest to trough and trough to crest.) This is called a **neap tide** (*nep* = scarcely or barely touching), and the Moon is said to be in **quadrature** (*quadra* = four). To help you remember a *neap tide*, think of it as one that has been *"nipped in the bud,"* thus indicating a small tidal range.

The time between successive spring tides (full moon and new moon) or neap tides (first quarter and third quarter) is one-half the monthly lunar cycle, which is about two weeks. The time between a spring tide and a successive neap tide is one-quarter the monthly lunar cycle, which is about one week.

Figure 9.12 shows the appearance of the Moon as it moves through its monthly cycle. As the Moon progresses from new moon to first-quarter phases, the Moon is a **waxing crescent** (*waxen* = to increase; *crescere* = to grow). In between the first-quarter and full moon phases, the Moon is a **waxing gibbous** (*gibbus* = hump). Between the Moon's full and third-quarter phases, it is a **waning gibbous** (*wanen* = to decrease). And, in between the third-quarter and new moon phases, the Moon is a **waning crescent**. The Moon has identical periods of rotation on its axis and orbit around Earth (a property called *synchronous rotation*). What this means is that during the Moon's early history, it became "locked" in Earth's gravitational "tractor beam." As a result, the same side of the Moon always faces Earth.

Why does the Moon appear to change phase? The answer is that, unlike the Sun and other stars, the Moon emits no light of its own. Instead, it shines brightly only because it reflects light from the Sun. As shown in Figure 9.12, the Moon's

**RECAP** Spring tides occur during the full moon and new moon phases, when the lunar and solar tidal bulges constructively interfere, producing a large tidal range. Neap tides occur during the quarter moon phases, when the lunar and solar tidal bulges destructively interfere, producing a small tidal range.

**Animation**
Monthly Tidal Cycle
http://goo.gl/cg3NoK

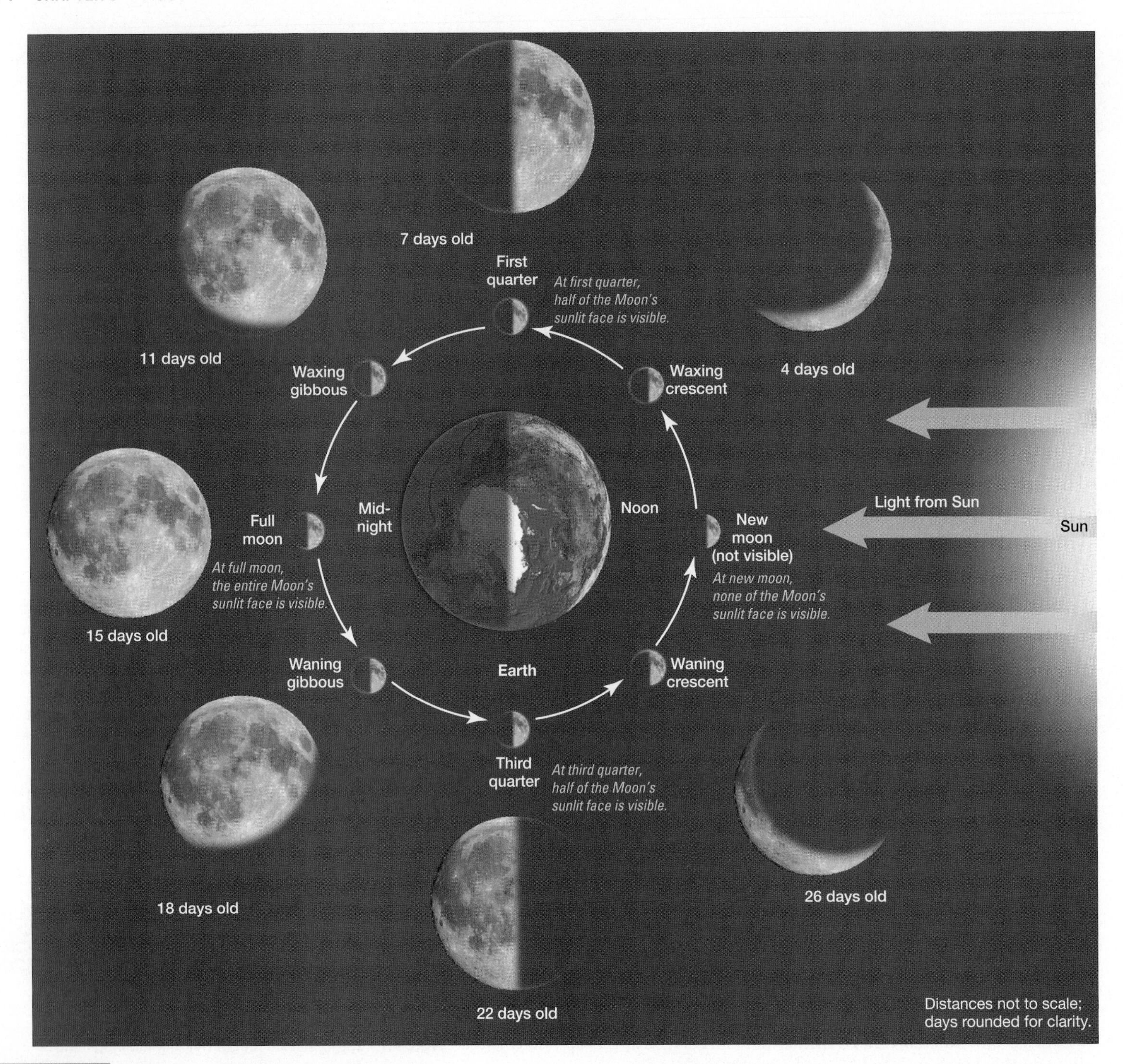


**EXPLORING DATA** ▲ **How many spring tides occur in the time it takes the Moon to make one complete orbit around Earth? Describe what the Moon looks like from Earth during each occurrence.**

Figure 9.12 **Phases of the Moon.** As the Moon moves around Earth during its 29-and-a-half-day lunar cycle, its phase changes depending on its position relative to the Sun and Earth. Names of Moon phases are shown (*inner circle*) along with what the Moon looks like as viewed from Earth (*outer circle*).

circular disk is always present but only half of the Moon's surface is illuminated by the Sun at all times. However, not all of the Moon's sunlit face can be seen because of the Moon's position with respect to Earth and the Sun. When the Moon is in full moon phase, for example, we see the entire "daylit" face because the Sun and the Moon are in opposite directions from Earth in the sky (although they do not exist on the same plane; as a result, sunlight is not blocked). In the case of a new moon, the Moon and the Sun are in almost the same part of the sky, and the sunlit side of the Moon is oriented away from our view (Figure 9.12).

## Complicating Factors

Besides Earth's rotation and the relative positions of the Moon and the Sun, there are many other factors that influence tides on Earth. Two of the most prominent of these factors are the declination of the Moon and Sun and the elliptical shapes of Earth's and the Moon's orbits. Let's examine both of these factors.

**DECLINATION OF THE MOON AND SUN** Up to this point, we have assumed that the Moon and Sun have remained directly overhead at the equator, but this is not usually the case. Most of the year, in fact, they are either north or south of the equator. The angular distance of the Sun or Moon above or below Earth's equatorial plane is called **declination** (*declinare* = to turn away).

Earth orbits the Sun along an invisible ellipse in space. The imaginary plane that contains this ellipse is called the **ecliptic** (*ekleipein* = to fail to appear). Recall from Chapter 6 that Earth's axis of rotation is tilted 23.5 degrees with respect to the ecliptic, and that this tilt causes Earth's seasons. It also means the maximum declination of the Sun relative to Earth's equator is 23.5 degrees.

To complicate matters further, the plane of the Moon's orbit is tilted 5 degrees with respect to the ecliptic. Thus, the maximum declination of the Moon's orbit relative to Earth's equator is 28.5 degrees (5 degrees plus the 23.5 degrees of Earth's tilt). The declination changes from 28.5 degrees south to 28.5 degrees north and back to 28.5 degrees south of the equator during the multiple lunar cycles within one year. As a result, tidal bulges are rarely aligned with the equator. Instead, they occur mostly north and south of the equator. The Moon affects Earth's tides more than the Sun, so tidal bulges follow the Moon, ranging from a maximum of 28.5 degrees north to a maximum of 28.5 degrees south of the equator (**Figure 9.13**).

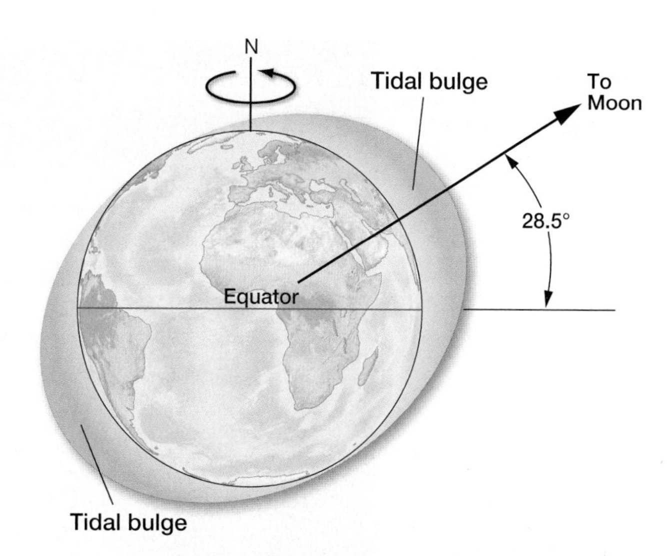


**Figure 9.13 Maximum declination of tidal bulges from the equator.** The center of the tidal bulges may lie at any latitude from the equator to a maximum of 28.5 degrees on either side of the equator, depending on the season of the year (solar angle) and the Moon's position.

**EFFECTS OF ELLIPTICAL ORBITS** Earth orbits the Sun in an elliptical orbit (**Figure 9.14**) such that Earth is 148.5 million kilometers (92.2 million miles) from the Sun during the Northern Hemisphere winter and 152.2 million kilometers (94.5 million miles) from the Sun during summer. Thus, the distance between Earth and the Sun varies by 2.5% over the course of a year. Tidal ranges are largest when Earth is near its closest point, called **perihelion** (*peri* = near, *helios* = Sun) and small-

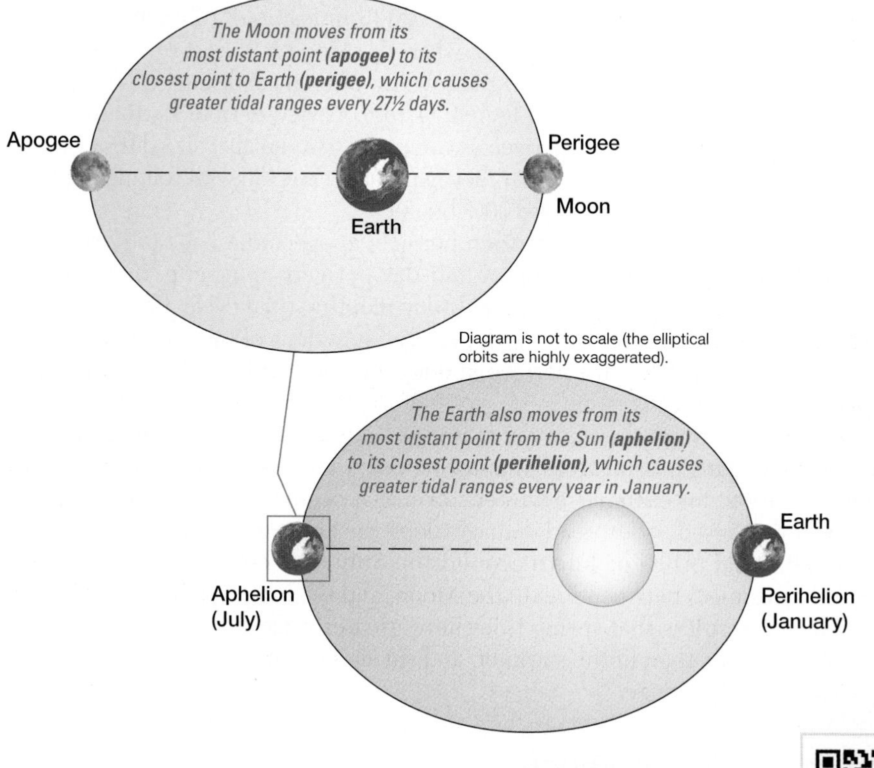

The Moon moves from its most distant point **(apogee)** to its closest point to Earth **(perigee)**, which causes greater tidal ranges every 27½ days.
Apogee
Perigee
Moon
Earth
Diagram is not to scale (the elliptical orbits are highly exaggerated).
The Earth also moves from its most distant point from the Sun **(aphelion)** to its closest point **(perihelion)**, which causes greater tidal ranges every year in January.
Earth
Aphelion (July)
Perihelion (January)

**Figure 9.14 Effects of elliptical orbits.**

**Animation**
Effects of Elliptical Orbits
https://goo.gl/w0Sc4Q

## STUDENTS SOMETIMES ASK . . .

***Do the phases of the Moon influence human behavior?***

All kidding about werewolves aside, tidal and gravitational effects have been used to explain the apparent increase in people's bizarre behavior when the Moon is full. The belief is that since the phases of the Moon affect ocean tides, they may have similar effects on the bodily fluids of human beings, whose composition is about 65% water. A number of scientists have tried to determine whether the Moon is in fact related to human behavior by calculating the precise number of births, crimes, and unusual behaviors that purportedly occur during various phases of the Moon. A few such investigators have indeed found an association between a full moon and unintentional poisonings, absenteeism, and crime. More often, however, researchers have found no relationship between lunar cycles and human biology or behavior. And it would seem that since tidal forces are so similar during full and new moon phases, there would be reports of similar unusual behavior during both full and new moons. It may simply be that certain people expect to be influenced by a full moon, so they may be more attentive and responsive to their internal sensations or desires at that time. It's no wonder these people are called *lunatics*!

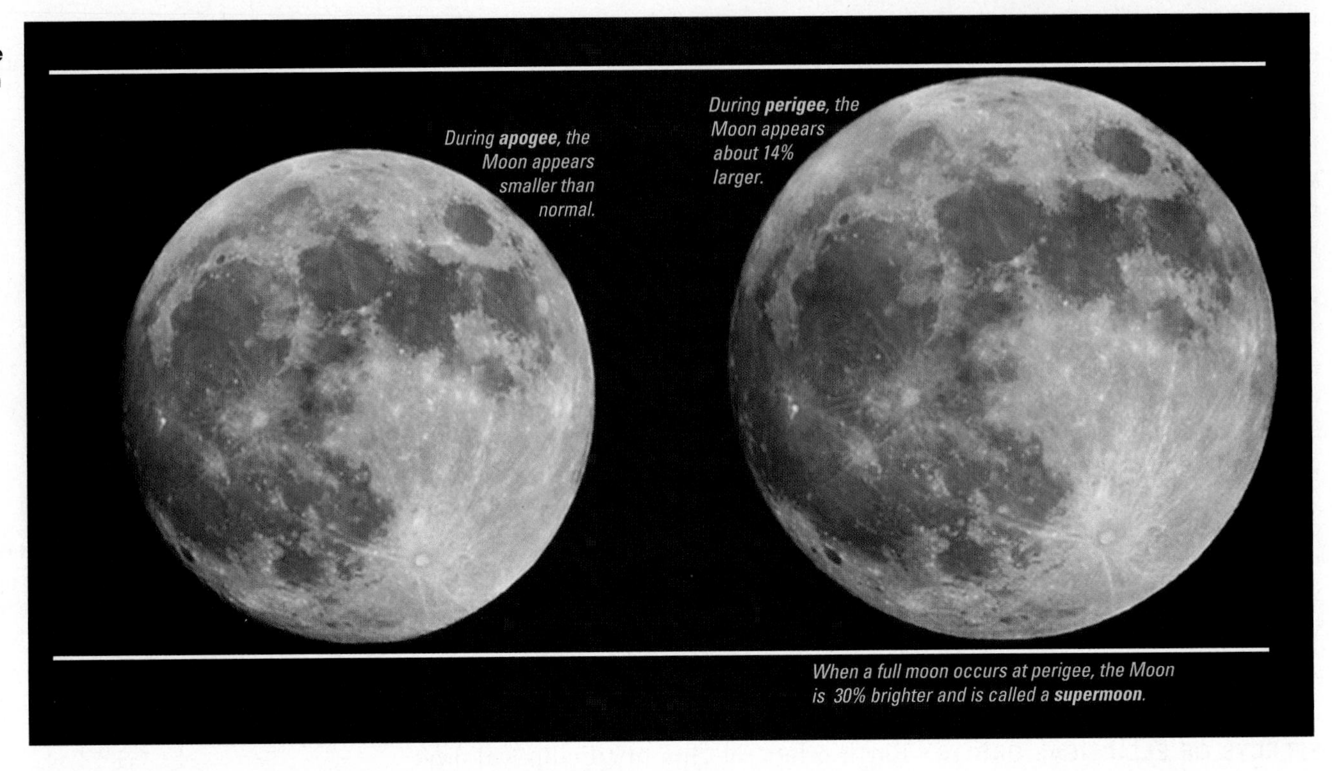

During apogee, the Moon appears smaller than normal.
During perigee, the Moon appears about 14% larger.
When a full moon occurs at perigee, the Moon is 30% brighter and is called a supermoon.

Figure 9.15 **A comparison of the size of the Moon as it appears from Earth during apogee and perigee.**

## STUDENTS SOMETIMES ASK . . .

***I've heard of a blue moon. Is the Moon really blue then?***

No. "Once in a blue moon" is a phrase that has gained popularity and is synonymous with a rather infrequent occurrence. A blue moon is the second full moon of any calendar month, which occurs when the 29-and-a-half-day lunar cycle falls entirely within a 30- or 31-day month. Because the divisions between our calendar months were determined arbitrarily, a blue moon has no special significance aside from the fact that it occurs only once every 2.72 years (about 33 months). At that rate, it's certainly less common than a month of Sundays!

The origin of the term *blue moon* is not exactly known, but it probably has nothing to do with color—although large forest fires or volcanic eruptions can put enough soot and ash particles in the atmosphere to cause the Moon to appear blue. One likely explanation involves the Old English word *belewe*, meaning "to betray." Thus, the Moon is belewe because it betrays the usual perception of one full moon per month. Another explanation links the term to a 1946 article in *Sky and Telescope* that tried to correct a misinterpretation of the term *blue moon*, but the article itself was misinterpreted to mean the second full moon in a given month. Apparently, the erroneous interpretation was repeated so often that it eventually stuck.

est near its most distant point, called **aphelion** (*apo* = away from, *helios* = Sun). Thus, the greatest tidal ranges typically occur in January each year.

The Moon orbits Earth in an elliptical orbit, too. The Earth–Moon distance varies by 8% (between 375,000 kilometers [233,000 miles] and 405,800 kilometers [252,000 miles]). Tidal ranges are largest when the Moon is closest to Earth, called **perigee** (*peri* = near, *geo* = Earth), and smallest when most distant, called **apogee** (*apo* = away from, *geo* = Earth) (Figure 9.14, *top*). As viewed from Earth, the Moon appears about 14% larger during perigee as compared to its smaller size during apogee (**Figure 9.15**); these larger moons, when they happen to coincide with full moon phase, are called *supermoons*, which are also 30% brighter.

Note that the Moon cycles between perigee, apogee, and back to perigee every 27-and-a-half days. (This 27-and-a-half-day perigee–apogee–perigee cycle is not to be confused with the 29-and-a-half-day monthly tidal cycle; both cycles occur simultaneously and are ongoing but are independent of one another.) Spring tides happen to coincide with perigee about every one-and-a-half years, producing **proxigean** (*proximus* = nearest, *geo* = Earth), or "closest of the close moon" tides. During this time, the tidal range is especially large and often results in the flooding of low-lying coastal areas; if a storm occurs simultaneously, damage can be extreme. In 1962, for example, a winter storm that occurred at the same time as a proxigean tide caused widespread damage along the entire U.S. East Coast.

The elliptical orbits of Earth around the Sun and the Moon around Earth change the distances between Earth, the Moon, and the Sun, thus affecting Earth's tides. The net result is that spring tides have greater ranges during the Northern Hemisphere winter than in the summer, and spring tides have greater ranges when they coincide with perigee.

## Idealized Tide Prediction

The declination of the Moon determines the position of the tidal bulges. The example illustrated in **Figure 9.16** shows that the Moon is directly overhead at 28 degrees

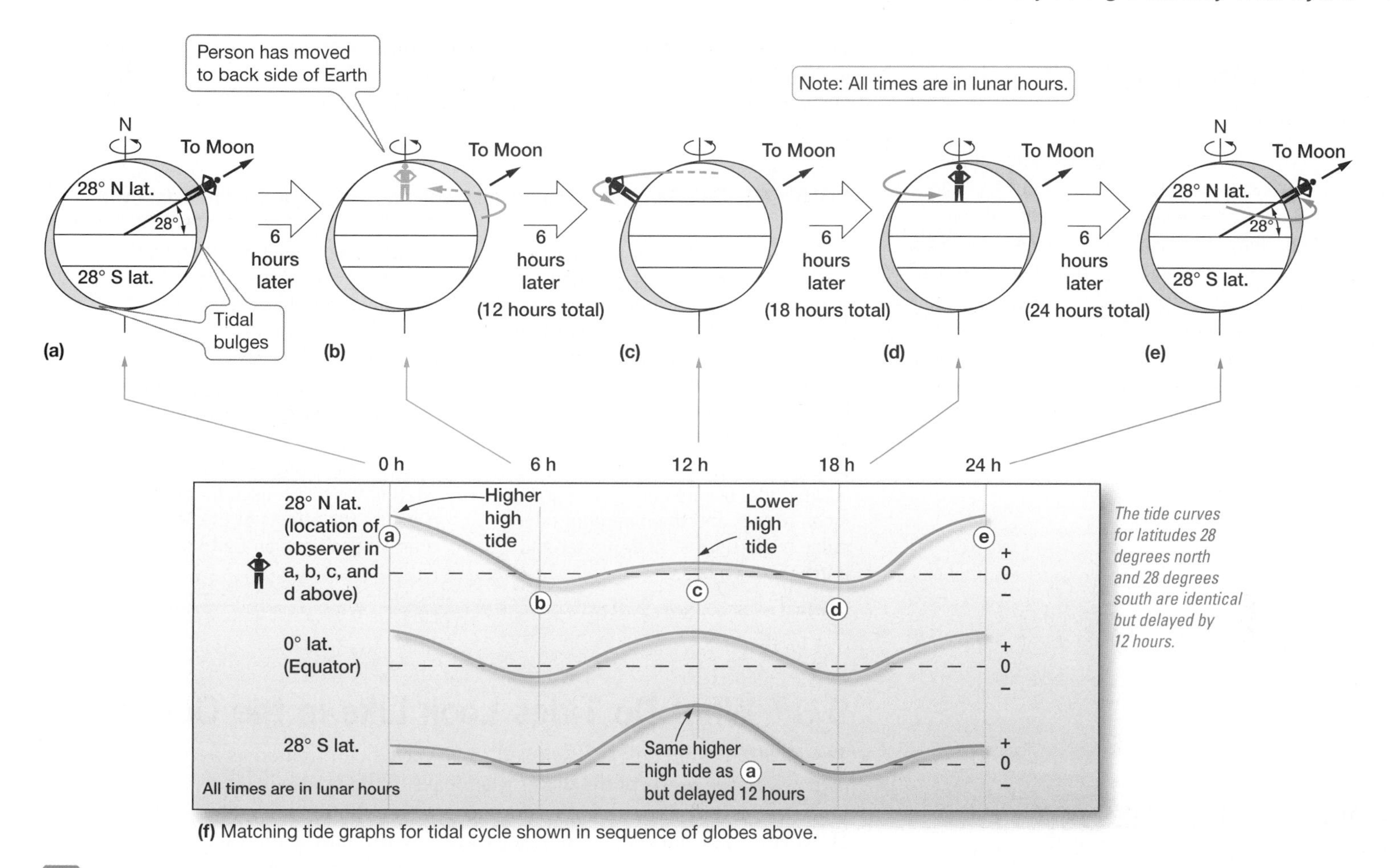


(f) Matching tide graphs for tidal cycle shown in sequence of globes above.

*The tide curves for latitudes 28 degrees north and 28 degrees south are identical but delayed by 12 hours.*

**SmartFigure 9.16 Predicted idealized tides. (a)–(e)** Sequence showing the tide experienced every 6 lunar hours at 28 degrees north latitude when the declination of the Moon is 28 degrees north. **(f)** Corresponding tide curves for the lunar day shown in the sequence above for the latitudes 28 degrees north, 0 degrees (equator), and 28 degrees south.
https://goo.gl/CFdvNW

north latitude when its declination is 28 degrees north of the equator. Imagine standing at 28 degrees north latitude and experiencing tidal conditions during a day, which is the sequence shown in Figure 9.16a–e:

- With the Moon directly overhead, the tidal condition experienced will be high tide (Figure 9.16a).
- Low tide occurs 6 lunar hours later (6 hours 12½ minutes solar time) (Figure 9.16b).
- Another high tide, but one much lower than the first, occurs 6 lunar hours later (Figure 9.16c).
- Another low tide occurs 6 lunar hours later (Figure 9.16d).
- Six lunar hours later, at the end of a 24-lunar-hour period (24 hours 50 minutes solar time), you will have passed through a complete lunar-day cycle of two high tides and two low tides (Figure 9.16e).

The graphs in Figure 9.16f show the heights of the tides observed during the same lunar day at 28 degrees north latitude, the equator, and 28 degrees south latitude when the declination of the Moon is 28 degrees north of the equator. Tide curves for 28 degrees north and 28 degrees south latitude have identically timed highs and lows, but the *higher* high tides and *lower* low tides occur 12 hours later.

The reason they occur out of phase by 12 hours is because the bulges in the two hemispheres are on opposite sides of Earth in relation to the Moon. Mastering Oceanography **Web Table 9.1** summarizes the characteristics of the tides on the idealized Earth.

**CONCEPT CHECK 9.2 ▸ Explain how tides vary during a monthly tidal cycle.**

1 To help reinforce your knowledge of how tides work, draw the positions of the Earth–Moon–Sun system during a complete monthly tidal cycle from memory. Indicate the tide conditions experienced on Earth, the phases of the Moon, the time between those phases, and syzygy and quadrature.

2 Explain why the maximum tidal range (spring tide) occurs during new and full moon phases and the minimum tidal range (neap tide) at first-quarter and third-quarter moons.

3 What is declination? Discuss the degree of declination of the Moon and Sun relative to Earth's equator. What are the effects of declination of the Moon and Sun on the tides?

4 Diagram the Earth–Moon system's orbit about the Sun. Label the positions on the orbit at which the Moon and Sun are closest to and farthest from Earth, stating the terms used to identify them. Discuss the effects of the Moon's and Earth's positions on Earth's tides.

## STUDENTS SOMETIMES ASK . . .

***What are tropical tides?***

Differences between successive high tides and successive low tides occur each lunar day (see, for example, Figure 9.16f). Because these differences occur within a period of one day, they are called *diurnal (daily) inequalities*. These inequalities are at their greatest when the Moon is at its maximum declination relative to the equator, and such tides are called *tropical tides* because at maximum declination the Moon is over one of Earth's tropics. When the Moon is over the equator (*equatorial tides*), the difference between successive high tides and low tides is minimal.

# 9.3 ▸ What Do Tides Look Like in the Ocean?

The idealized tidal bulges discussed so far have been treated as *freely propagating waves* with the crests of the waves (the peaks of the tidal bulges) separated by a distance of one-half Earth's circumference—about 20,000 kilometers (12,420 miles). However, the tidal bulges themselves are continually pulled on by astronomical forces, so they actually exist on Earth as *forced waves*. What this means is they are continuously pulled along by a driving force, which is the tidal force caused mostly by the Moon. As a result, the tidal bulges move across Earth's oceans at about 1600 kilometers (1000 miles) per hour to keep up with the tidal force of the Moon. Note that it's the tidal bulge waveform—not the water itself—that moves at such great speeds.

The idealized tidal bulge model also assumes no continents on Earth, oceans that are infinitely deep, and no frictional losses. These conditions clearly don't exist on Earth, so the idealized tidal bulge model can only take us so far in explaining the tides. In reality, a variety of conditions on the real Earth cause ocean tides to break up into distinct, large circulation units in each ocean basin called *cells*.

## STUDENTS SOMETIMES ASK . . .

***How often are conditions right to produce the maximum tide-generating force?***

Maximum tides occur when Earth is closest to the Sun (at perihelion), the Moon is closest to Earth (at perigee), and the Earth–Moon–Sun system is aligned (at syzygy) with both the Sun and Moon at zero declination. This rare condition—which creates an absolute *maximum* spring tidal range—occurs once every 1600 years. Fortunately, the next occurrence is predicted for the year 3300.

However, there are other times when conditions produce large tide-generating forces and produce what is often called a *king tide*, which is simply the very highest tide experienced only a few times a year. During early 1983, for example, large, slow-moving low-pressure cells developed in the North Pacific Ocean and caused strong northwest winds. In late January, the winds produced a near fully developed 3-meter (10-foot) swell that affected the U.S. West Coast from Oregon to Baja California. The large waves would have been trouble enough under normal conditions, but there were also unusually high spring tides of 2.25 meters (7.4 feet) because Earth was near perihelion at the same time that the Moon was at perigee. In addition, a strong El Niño had raised sea level by as much as 20 centimeters (8 inches). When the waves hit the coast during these unusual conditions, they caused more than $100 million in damage, including the destruction of 25 homes, damage to 3500 others, the collapse of several commercial and municipal piers, and at least a dozen deaths.

## Amphidromic Points and Cotidal Lines

In the open ocean, the crests and troughs of the tide wave rotate around an **amphidromic point** (*amphi* = around, *dromus* = running) near the center of each cell. There is essentially no tidal range at amphidromic points, but radiating from each point are **cotidal lines** (*co* = with, *tidal* = tide), which connect all nearby locations where high tide occurs simultaneously. The labels on the cotidal lines in **Figure 9.17** indicate the time of high tide in hours as they rotate around the cell.

The times in Figure 9.17 indicate that the tide wave in each cell generally rotates counterclockwise in the Northern Hemisphere and clockwise in the Southern Hemisphere. The tide wave in each cell must complete one rotation during the tidal period (usually 12 lunar hours), so this limits the size of the cells.

Low tide occurs six hours after high tide in an amphidromic cell. If high tide is occurring along the cotidal line labeled "10," for example, then low tide is occurring along the cotidal line labeled "4."

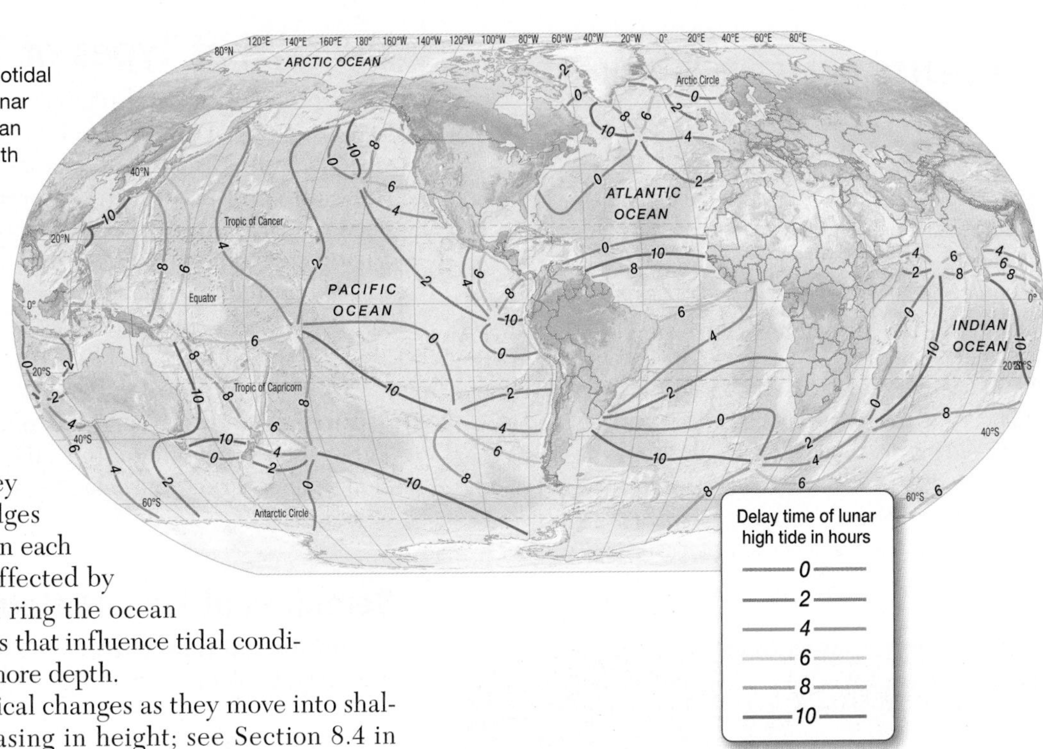

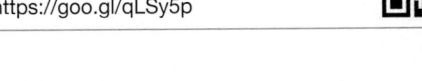

**SmartFigure 9.17 Cotidal map of the world.** Cotidal lines indicate times of the main lunar daily high tide in lunar hours after the Moon has crossed the Greenwich Meridian (0 degrees longitude). Tidal ranges generally increase with increasing distance along cotidal lines away from the amphidromic points (center of the cell). Where cotidal lines terminate at both ends in amphidromic points, maximum tidal range will be near the midpoints of the lines.
https://goo.gl/qLSy5p

## Effect of the Continents

The continents affect tides, too, because they interrupt the free movement of the tidal bulges across the ocean surface. Tides are expressed in each ocean basin as freestanding waves that are affected by the position and shape of the continents that ring the ocean basin. In fact, two of the most important factors that influence tidal conditions along a coast are coastline shape and offshore depth.

Just like surface waves that undergo physical changes as they move into shallow water (such as slowing down and increasing in height; see Section 8.4 in Chapter 8), tides experience similar physical changes as they enter the shallow water of continental shelves. These changes tend to amplify the tidal range as compared to the deep ocean, where the maximum tidal range is only about 45 centimeters (18 inches).

In addition, increased turbulent mixing rates in deep water over areas of rough bottom topography (as discussed in Chapter 7) are associated with internal waves created by tides breaking on this rough topography and against continental slopes. These tide-generated internal waves have recently been observed along the chain of Hawaiian Islands, have heights of up to 300 meters (1000 feet), and contribute to increased turbulence and mixing, which in turn strongly affect the tides.

## Other Considerations

A detailed analysis of all the variables that affect the tides at any particular coast reveals that nearly 400 factors are involved—far more than can adequately be addressed here. The combination of all these factors creates some conditions that are unexpected based on a simple tidal model. For example, high tide rarely occurs when the Moon is at its highest point in the sky. Instead, the time between the Moon crossing the meridian and a corresponding high tide varies from place to place.

Because of the complexity of the tides, a completely mathematical model of the tides is beyond the limits of marine science. Instead, a combination of mathematical analysis and observation is required to adequately model the tides. Moreover, successful models must take into account at least 37 independent factors related to tides (the two most important are forces caused by the Moon and the Sun). Tidal models that take into account the most dominant tide-producing factors are usually quite successful in predicting future tides.

### STUDENTS SOMETIMES ASK . . .

***Why don't all areas of the world experience the same type of tidal pattern?***

If Earth were a perfect sphere without large continents, all areas on the planet would experience two equally proportioned high and low tides every lunar day (a semidiurnal tidal pattern). The large continents on the planet, however, block the westward passage of the tidal bulges as Earth rotates. Unable to move freely around the globe, the tides instead establish complex patterns within each ocean basin that often differ greatly from tidal patterns of adjacent ocean basins or even other regions of the same ocean basin.

**RECAP Many factors influence the tides, and tidal prediction is more complicated than what is predicted based on a simple lunar and solar tidal bulge model.**

### CONCEPT CHECK 9.3 ▸ Specify what tides look like in the ocean.

1 Are tides considered deep-water waves anywhere in the ocean? Why or why not?

2 What are amphidromic points and cotidal lines?

3 Discuss reasons why the tides don't follow a simple tidal bulge model.

## STUDENTS SOMETIMES ASK . . .

***I noticed that Figure 9.18 shows negative tides. How can there ever be a negative tide?***

Negative tides occur because the *datum*—the starting point or reference point from which tides are measured—is an average of the tides over many years. Along the West Coast of the United States, for instance, the datum is mean lower low water (MLLW), which is the average of the *lower* of the two low tides that occur daily in a mixed tidal pattern. Because the datum is an average, there will be some days when the tide is less than the average (similar to the distribution of exam scores, some of which will be below the average). These lower-than-average tides are given negative values, occur only during spring tides, and are often the best times to visit local tide pool areas.

## CREATURE FEATURE 9.1

### No water? No problem, I can breathe air!

Several species of **mudskippers** are amphibious fish that inhabit coastal regions of Africa and Asia and have the ability to breathe air if they are trapped in small pools when the tide goes out.

Mudskippers are found in coastal habitats and use their fins to walk on land. Mudskippers breathe through their skin, the lining of their mouth, and their throat; they also use their enlarged gill chambers to retain a bubble of air for breathing if the water they are submerged in becomes depleted of dissolved oxygen.

**RECAP A diurnal tidal pattern exhibits one high and one low tide each lunar day, a semidiurnal tidal pattern exhibits two high and two low tides daily of about the same height, and a mixed tidal pattern usually has two high and two low tides of different heights daily but may also exhibit diurnal characteristics.**

# 9.4 ▸ What Types of Tidal Patterns Exist?

In theory, most coastal regions on Earth should experience two high tides and two low tides of unequal heights during a lunar day. In practice, however, the various depths, sizes, and shapes of ocean basins modify tides so they exhibit three different patterns in different parts of the world. The three types of tidal patterns, which are illustrated in **Figure 9.18**, are *diurnal* (*diurnal* = daily), *semidiurnal* (*semi* = half, *diurnal* = daily), and *mixed semidiurnal*, which is often simply referred to as *mixed*.

## Diurnal Tidal Pattern

A **diurnal tidal pattern** has one high tide and one low tide each lunar day. These tides are common in shallow inland seas such as the Gulf of Mexico and along the coast of Southeast Asia. Diurnal tides have a tidal period of 24 hours 50 minutes.

## Semidiurnal Tidal Pattern

A **semidiurnal tidal pattern** has two high tides and two low tides each lunar day. The heights of successive high tides and successive low tides are approximately the same. (Because tides are always growing higher or lower at any location due to the spring tide–neap tide sequence, successive high tides and successive low tides can never be *exactly* the same at any location.) Semidiurnal tides are common along the Atlantic coast of the United States. The tidal period is 12 hours 25 minutes.

## Mixed Tidal Pattern

A **mixed tidal pattern** may have characteristics of both diurnal and semidiurnal tides. Successive high tides and/or low tides will have significantly different heights, a condition called *diurnal inequality*. Mixed tides commonly have a tidal period of 12 hours 25 minutes, but they may also exhibit diurnal periods. Mixed tidal patterns are commonly found along the Pacific coast of North America.

**Figure 9.19** shows examples of monthly tidal curves for various coastal locations. Even though a tide at any particular location follows a single tidal pattern, it still may pass through stages of one or both of the other tidal patterns. Typically, however, the tidal pattern for a location remains the same throughout the year. Also, the tidal curves in Figure 9.19 clearly show the weekly switching of the spring tide–neap tide cycle.

**CONCEPT CHECK 9.4 ▸ Compare the characteristics and coastal locations of the three types of tidal patterns.**

1 What do the terms *diurnal*, *semidiurnal*, and *mixed* mean, as related to tidal patterns?

2 Describe the number of high and low tides in a lunar day, the period, and any inequality of the following tidal patterns: diurnal, semidiurnal, and mixed.

3 Using the Internet, find a coastal tidal prediction for your birthday this year (or, if your birthday has already happened, find next year's tide prediction). What tidal pattern is displayed?

4 Of the three tidal patterns, which one is most common along the U.S. East Coast? The U.S. West Coast? Worldwide?

**Animation**
Tidal Patterns
http://goo.gl/tdDbEx

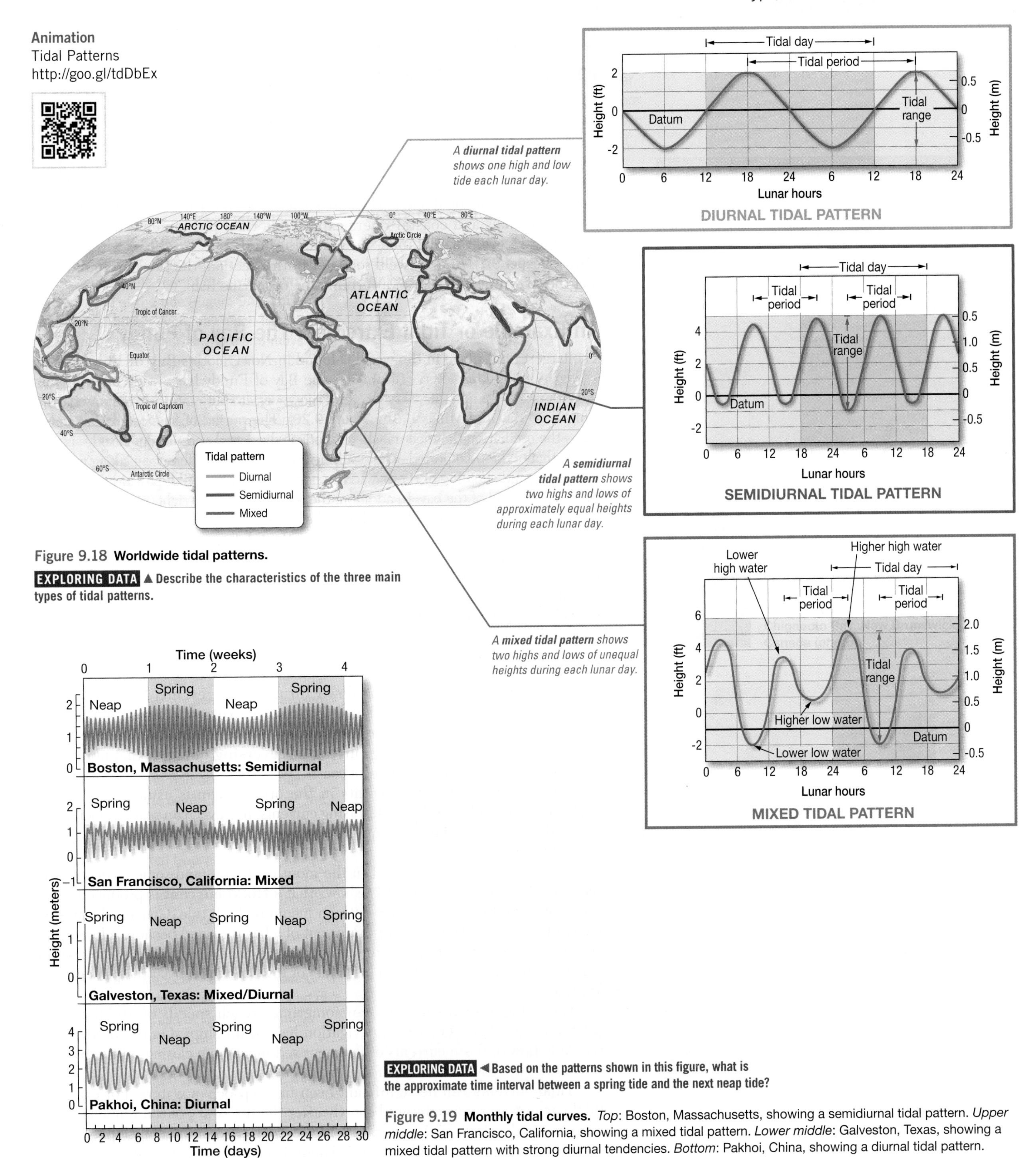


Figure 9.18 **Worldwide tidal patterns.**

**EXPLORING DATA** ▲ **Describe the characteristics of the three main types of tidal patterns.**

**EXPLORING DATA** ◀ **Based on the patterns shown in this figure, what is the approximate time interval between a spring tide and the next neap tide?**

Figure 9.19 **Monthly tidal curves.** *Top*: Boston, Massachusetts, showing a semidiurnal tidal pattern. *Upper middle*: San Francisco, California, showing a mixed tidal pattern. *Lower middle*: Galveston, Texas, showing a mixed tidal pattern with strong diurnal tendencies. *Bottom*: Pakhoi, China, showing a diurnal tidal pattern.

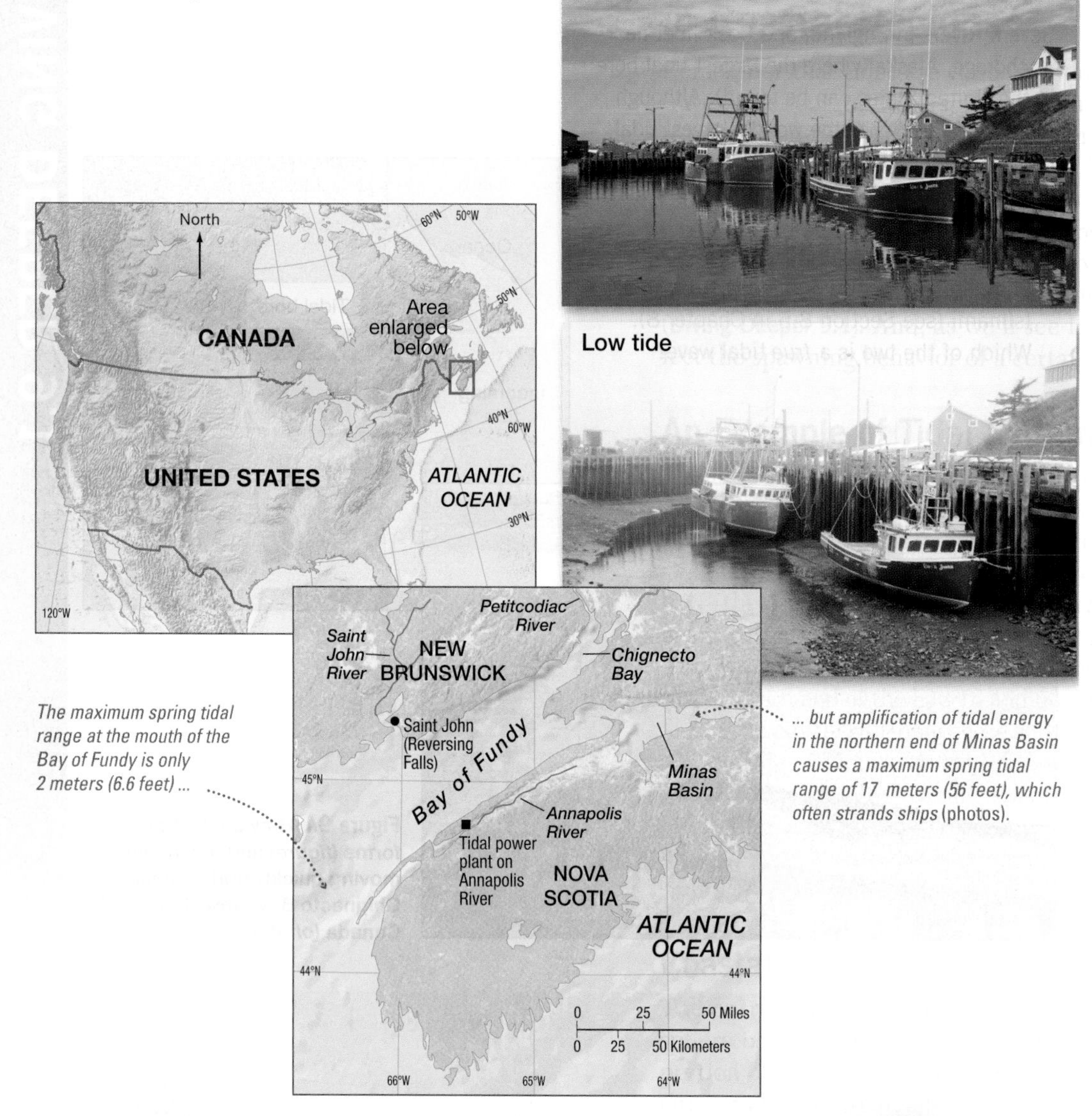


**Figure 9.20 The Bay of Fundy, site of the world's largest tidal range.**

forced researchers to abandon the use of the camera-equipped, tethered, remotely operated vehicle *Jason Jr.*

## Whirlpools: Fact or Fiction?

A **whirlpool**—a rapidly spinning body of water, which is also termed a *vortex* (*vertere* = to turn)—can be created in some restricted coastal passages due to reversing tidal currents. Whirlpools most commonly occur in shallow passages connecting two large bodies of water that have different tidal cycles. The different tidal heights of the two bodies cause water to move vigorously through the passage. As water rushes through the passage, it is affected by the shape of the shallow sea floor, causing turbulence, which, along with spin due to opposing tidal currents, creates whirlpools. The larger the tidal difference between the two bodies of water and the smaller the passage, the greater the vortex caused by the tidal currents. Because whirlpools can have high flow rates of up to 16 kilometers (10 miles) per hour, they can cause ships to spin out of control for a short time.

One of the world's most famous whirlpools is the *Maelstrom* (*malen* = to grind in a circle, *strom* = stream), which occurs in a passage off the west coast of Arctic Norway (**Figure 9.22**). This and another famous whirlpool in the Strait of Messina, which separates mainland Italy from Sicily, are probably the source of ancient legends of huge churning funnels of water that destroy ships and carry mariners to their deaths, although they are not nearly as deadly as legends suggest. Other notable whirlpools occur off the west coast of Scotland, in the Bay of Fundy at the border between Maine and the Canadian province of New Brunswick, and off Japan's Shikoku Island.

## Grunion: Doing What Comes Naturally on the Beach

From March through September, shortly after the maximum spring tide has occurred, **grunion** (*Leuresthes tenuis*) come ashore along sandy beaches of Southern California and Baja California to bury their eggs. Grunion—slender, silvery fish up to 15 centimeters (6 inches) long—are the only marine fish in the world that come completely out of water to spawn. The name *grunion* comes from the Spanish *gruñón*, which means "grunter" and refers to the faint noise they make during spawning.

A mixed tidal pattern occurs along Southern California and Baja California beaches. On most lunar days (24 hours and 50 minutes), there are two high tides and two low tides. There is usually a significant difference in the heights of the two high tides that occur each day. During the summer months, the higher high tide occurs at night. The night high tide becomes higher each night as the maximum spring tide

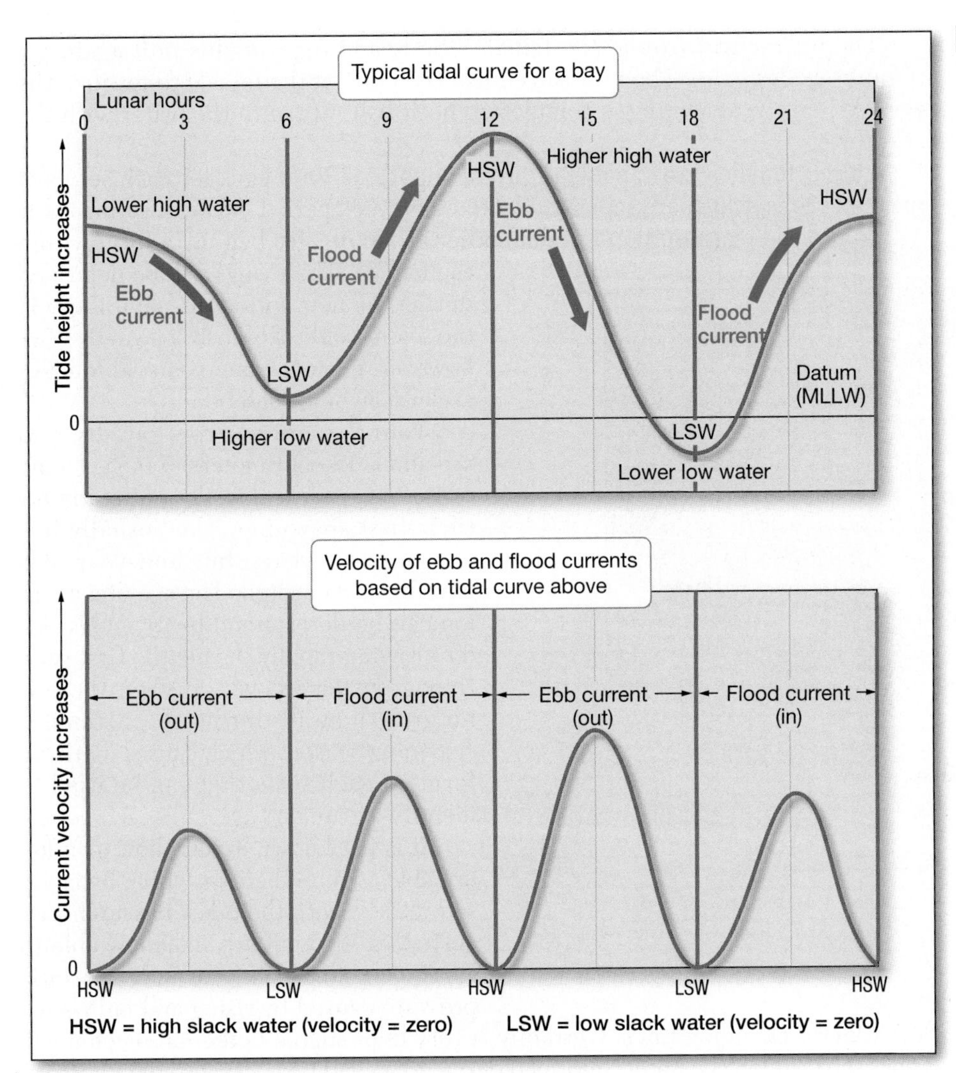


**EXPLORING DATA** ◄ **Based on the information in this figure, when would be the best time to navigate a ship into a small, rocky harbor? When would be the best time to set sail and leave the harbor?**

**SmartFigure 9.21 Reversing tidal currents in a bay.** *Top*: Typical tidal curve for a bay that experiences a mixed tidal pattern, showing alternating ebb currents (approaching low tide) and flood currents (approaching high tide). No currents occur during either high slack water (*HSW*) or low slack water (*LSW*). The datum *MLLW* stands for *mean lower low water*, which is the average of the lower of the two low tides that occur daily in a mixed tidal pattern. *Bottom*: Corresponding chart showing the velocity of ebb and flood currents based on the tidal curve above. https://goo.gl/g2XtZD

range is approached, causing sand to be eroded from the beach (**Figure 9.23**, *graph*). After the maximum spring tide has occurred, the night high tide diminishes each night. As neap tide is approached, sand is deposited on the beach.

Grunion spawn only after each night's higher high tide has peaked on the three or four nights following the night of the highest spring high tide. This ensures that their eggs will be covered deeply in sand deposited by the receding higher high tides each succeeding night. The fertilized eggs buried in the sand are ready to hatch about 10 days after spawning. By this time, another spring tide is approaching, so the night high tide is getting progressively higher each night again. The beach sand is eroding again, too, which exposes the eggs to the waves that break ever higher on the beach. The eggs hatch about three minutes after being freed in the water. Tests done in laboratories have shown that the grunion eggs will not hatch until agitated in a manner that simulates that of the eroding waves.

The spawning begins as the grunion come ashore immediately following an appropriate high tide, and it may last from one to three hours. Spawning usually peaks about an hour after it starts and may last an additional 30 minutes to an hour. During this time, the beach may be littered with thousands of fish. Females, which are larger than males, catch waves and swim high onto the beach. If no males are near, a female may return to the water without depositing her eggs. In the presence of males, she drills her tail into the semifluid sand until only her head is visible. The female continues to twist, depositing her eggs 5 to 7 centimeters (2 to 3 inches) below the surface.

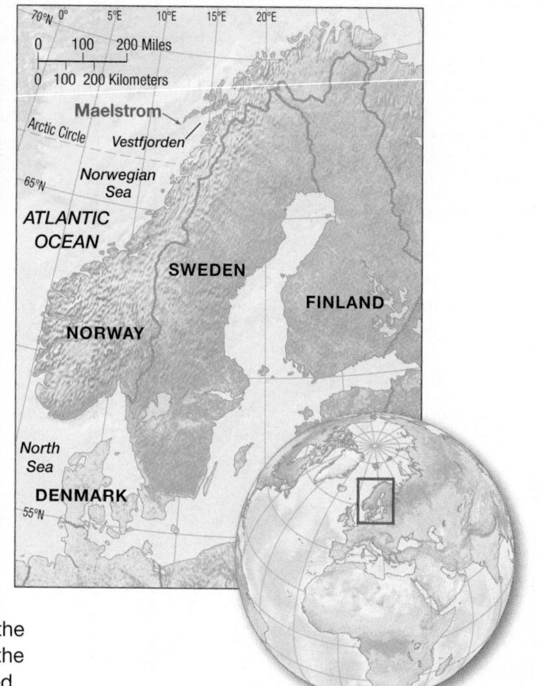
0 100 200 Miles
0 100 200 Kilometers
Maelstrom
Arctic Circle
Vestfjorden
Norwegian Sea
ATLANTIC OCEAN
SWEDEN
FINLAND
NORWAY
North Sea
DENMARK

**Figure 9.22 The Maelstrom.** The Maelstrom, located off the west coast of Norway, is one of the strongest whirlpools in the world and can cause ships to spin out of control. It is created by tidal currents that pass through a narrow, shallow passage between Vestfjorden and the Norwegian Sea.

The male curls around the female's body and deposits his milt against it (Figure 9.23, *photo*). The milt runs down the body of the female to fertilize the eggs. When the spawning is completed, both fish return to the water with the next wave.

Larger females are capable of producing up to 3000 eggs for each series of spawning runs, which are separated by the two-week period between spring tides. As soon as the eggs are deposited, another group of eggs begins to form within the female. These eggs will be deposited during the next spring tide run. Early in the season, only older fish spawn. By May, however, even the one-year-old females are in spawning condition.

Young grunion grow rapidly and are about 12 centimeters (5 inches) long when they are a year old and ready for their first spawning. They usually live two or three years, but four-year-olds have been recovered. The age of a grunion can be determined by its scales. After growing rapidly during the first year, they grow very slowly. In fact, there is no growth at all during the six-month spawning season, which causes marks to form on each scale that can be used to identify a grunion's age.

It is not known exactly how grunion are able to time their spawning behavior so precisely with the tides. Research suggests that grunion are somehow able to sense very small changes in hydrostatic pressure caused by rising and falling sea level due to changing tides. Certainly, a very dependable detection mechanism keeps the grunion accurately informed of the tidal conditions because their survival depends on a spawning behavior precisely tuned to the tides.

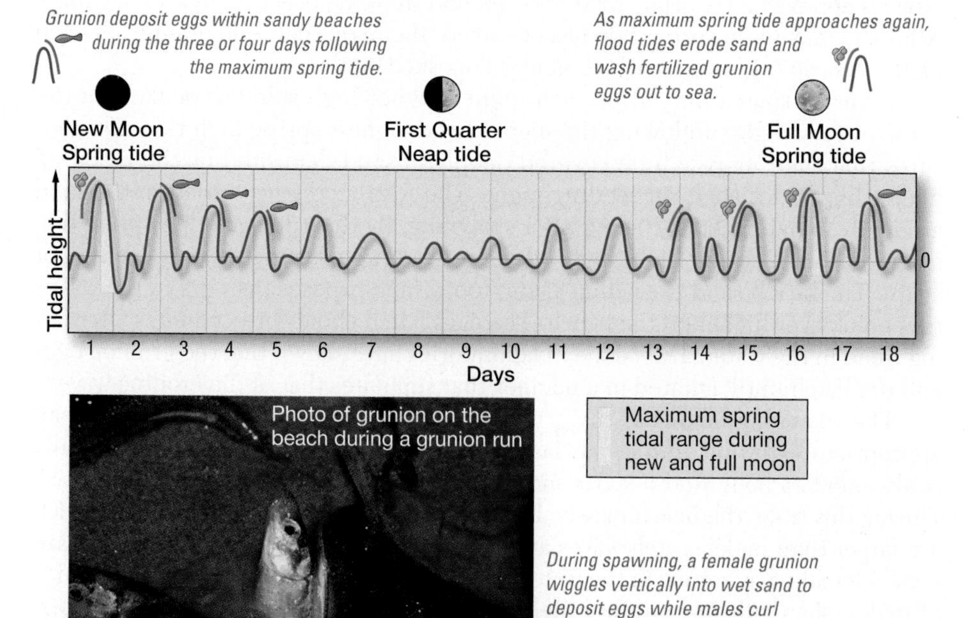
Grunion deposit eggs within sandy beaches during the three or four days following the maximum spring tide.
As maximum spring tide approaches again, flood tides erode sand and wash fertilized grunion eggs out to sea.
New Moon
Spring tide
First Quarter
Neap tide
Full Moon
Spring tide
Tidal height
0
1 2 3 4 5 6 7 8 9 10 11 12 13 14 15 16 17 18
Days
Photo of grunion on the beach during a grunion run
Maximum spring tidal range during new and full moon
During spawning, a female grunion wiggles vertically into wet sand to deposit eggs while males curl around her body, releasing milt and fertilizing the eggs.

**EXPLORING DATA** ▶ **Describe the pattern of grunion spawning in relation to the tides.**

**Figure 9.23 The tidal cycle and spawning grunion.** During summer months and for 3 or 4 days after the highest spring tides (*graph*), grunion deposit their eggs on sandy beaches (*photo*). The successively lower high tides during the approaching neap tide conditions won't wash the eggs from the sand until they are ready to hatch about 10 days later. As the next spring tide is approached, successively higher high tides wash the eggs free and allow them to hatch. The spawning cycle begins a few days later, after the peak of spring tide conditions, with the next cycle of successively lower high tides.

**CONCEPT CHECK 9.5 ▸ Describe tidal phenomena that occur in coastal regions.**

1 Discuss factors that help produce the world's largest tidal range in the Bay of Fundy.

2 Specify the differences between rotary and reversing tidal currents.

3 Of flood current, ebb current, high slack water, and low slack water, when is the best time to enter a bay by boat? When is the best time to navigate in a shallow, rocky harbor? Explain.

4 Describe the spawning cycle of grunion, indicating the relationship among tidal phenomena, where grunion lay their eggs, and the movement of sand on the beach.

**RECAP Coastal tidal phenomena include large tidal ranges (the largest of which—17 meters [56 feet]—occurs in the Bay of Fundy), tidal currents, rapidly spinning vortices called whirlpools, and grunion, which time their spawning cycles with the tides.**

## 9.6 ▸ Can Tidal Power Be Harnessed as a Source of Energy?

Throughout history, ocean tides have been used as a source of power. During high tide, water can be trapped in a basin and then harnessed to do work as it flows back to the sea. In the 12th century, for example, water wheels driven by the tides were used to power gristmills and sawmills. During the 17th and 18th centuries, much of Boston's flour was produced at a tidal mill.

Today, tidal power is considered a clean, renewable resource with vast potential. The initial cost of building a tidal power electricity-generating plant may be higher than the cost of building a conventional thermal power plant, but the operating costs are lower because it does not use fossil fuels or radioactive substances to generate electricity.

One disadvantage of tidal power, however, is the periodicity of the tides, which allows power to be generated only during a portion of a 24-hour day. People operate on a solar period, but tides operate on a lunar period, so the energy available from the tides would coincide with need only part of the time. Power would have to be distributed to the point of need at the moment it was generated, which could be a great distance away, resulting in an expensive transmission problem. The power could be stored, but even this alternative presents a large and expensive technical problem.

To generate electricity effectively, electrical turbines (generators) need to run at a constant speed, which is difficult to maintain when generated by the variable flow of tidal currents in two directions (flood tide and ebb tide). Specially designed turbines that allow both advancing and receding water to spin their blades are necessary to solve the problem of generating electricity from the tides.

Other potential disadvantages of tidal power include environmental concerns such as change in habitat and harm to wildlife. For example, most tidal power plants involve a dam that alters the ecology of estuaries where they are located. Also, tidal power plants disturb the normal flow of tidal currents and negatively affect marine organisms that depend on these currents for bringing food or for aiding migrations. Another example is the harm done to marine organisms entrapped or injured by moving tidal power devices. Underwater turbines produce noise, too, that could disturb marine organisms or have long-term negative effects. Lastly, a tidal power plant would likely interfere with many traditional human uses of estuaries, such as transportation and fishing.

### Tidal Power Plants

Tidal power can be harnessed in one of two ways: (1) tidal water trapped behind coastal barriers in bays and estuaries during high tide can be released at a later time to turn turbines and generate electrical energy and (2) tidal currents that pass through narrow channels can be used to turn underwater pivoting turbines, which produce energy (see Section 7.7, "Can Power from Currents Be Harnessed as a Source of Energy?", in Chapter 7). Although the first type is much more commonly employed, Norway, the United Kingdom, and the United States have recently installed offshore turbines that harness swift coastal tidal currents and plan to expand these devices into tidal energy farms.

One example of a successful tidal power plant that uses water trapped behind a coastal barrier and has been in operation since 1966 is at Saint-Malo, France, on the estuary of the La Rance River (**Figure 9.24**). The La Rance estuary has a surface area of approximately 23 square kilometers (9 square miles) and its tidal range is 13.4 meters (44 feet). A general rule of tidal power is that usable tidal energy increases as the area of the basin increases and as the tidal range increases.

The power-generating barrier was built across the La Rance estuary a little over 3 kilometers (2 miles) upstream to protect it from storm waves. The barrier is 760 meters (2500 feet) wide and supports a two-lane road (Figure 9.24). Water passing through the barrier powers 24 electricity-generating units that operate beneath the power plant. At peak operating capacity, each unit can generate 10 megawatts of electricity, for a total of 240 megawatts. (Each megawatt of electricity is enough to serve the energy needs of about 250 average U.S. homes.)

To generate electricity, the La Rance plant needs a sufficient water height between the estuary and the ocean—which occurs only about half of the time. Annual power production of about 540 million kilowatt-hours can be increased to 670 million kilowatt-hours by using the turbine generators as pumps to move water into the estuary at proper times.

Worldwide, only six other tidal power plants exist besides the La Rance plant, and half of them have a relatively small power generation of less than 2 megawatts. However, larger tidal power plants with increased efficiency of operation are being considered. For example, a tidal power plant across the Bay of Fundy at Passamaquoddy Bay near the U.S.–Canadian border has often been proposed, in large part because the Bay of Fundy has the largest tidal range in the world and its tidal flow volume is over 100 times greater than at the La Rance plant. Still, a tidal power plant has never been built across the Bay of Fundy because of engineering difficulties, great costs, and environmental concerns. In 1984 the Canadian province of Nova Scotia constructed a small tidal power plant in a nearby location that can generate 20 megawatts of electricity. The plant is built on the Annapolis River estuary, an arm of the Bay of Fundy (see Figure 9.20), where maximum tidal range is 8.7 meters (26 feet). Canada's Bay of Fundy and its potential to produce tidal power continues to be an item of keen interest in the region, with plans to install several new turbine systems similar to windmills but anchored to the sea floor.

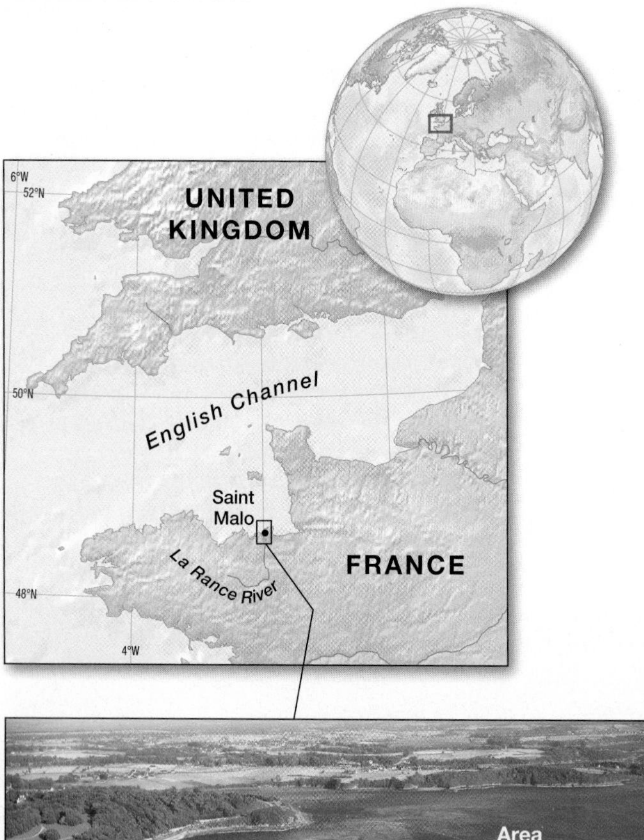

**Figure 9.24 La Rance tidal power plant at Saint-Malo, France.** Electricity is generated at the La Rance tidal power plant at Saint-Malo, France, when **(a)** water from a falling tide flows out of the estuary and turns turbines; electricity is also generated when **(b)** high water from a rising tide enters the estuary and turns turbines in the other direction.

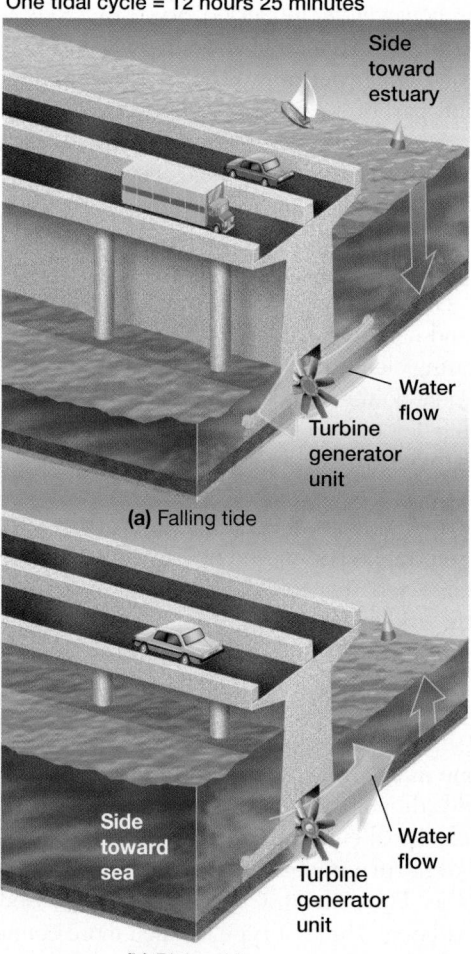

Many countries recognize the benefits of carbon-free renewable energy that tidal power provides, so tidal power plants are being constructed worldwide. In 2011, for example, South Korea installed a tidal power plant at Sihwa Lake capable of producing 254 megawatts of electricity. Other countries with proposed tidal projects include Russia, the Philippines, India, and the United Kingdom. In 2013, the London-based consultancy Bloomberg New Energy Finance projected that up to 22 tidal projects, each capable of generating more than 1 megawatt of electricity, could be installed worldwide by 2020.

One of the most high-profile tidal projects is in the United Kingdom, where there are plans to build a tidal power barrage across the Severn Estuary that separates England and Wales. The Severn River has the second-largest tidal range in the world and is a prime target for producing tidal power. If completed, it would be the world's largest tidal power plant, with a 12-kilometer (7.5-mile)-long dam that could produce 8.6 gigawatts of energy, or about 5% of the electricity currently used in the United Kingdom.

**RECAP** **The daily change in water level as a result of ocean tides can be harnessed as a source of energy. Despite significant drawbacks, several tidal power plants in coastal estuaries successfully extract tidal energy today.**

## CONCEPT CHECK 9.6 ▸ Discuss advantages and disadvantages of harnessing tides as a source of energy.

1 Discuss at least two positive and two negative factors related to tidal power generation.

2 Explain how a tidal power plant works, using as an example an estuary that has a mixed tidal pattern. Why does potential for usable tidal energy increase with an increase in the tidal range?

# ESSENTIAL CONCEPTS REVIEW

## 9.1 ▶ What causes ocean tides?

- *Gravitational attraction of the Moon and Sun create ocean tides*, which are fundamentally very long and regular shallow-water waves. According to a *simplified model of tides*, which assumes an ocean of uniform depth and ignores the effects of friction, small horizontal forces (the tide-generating forces) tend to push water into *two bulges on opposite sides of Earth*. One bulge is directly facing the tide-generating body (the Moon and the Sun), and the other is directly opposite.
- Despite its vastly smaller size, *the Moon has about twice the tide-generating effect of the Sun* because the Moon is so much closer to Earth. The tidal bulges due to the Moon's gravity (the lunar bulges) dominate, so lunar motions dominate the periods of Earth's tides. However, the changing position of the solar bulges relative to the lunar bulges modifies tides. According to the simplified idealized tide theory, *Earth's rotation carries locations on Earth into and out of the various tidal bulges*.



### Selected Key Terms

Use the **glossary** at the end of this book to discover the meanings of these Selected Key Terms: **tides, barycenter, gravitational force, centripetal force, resultant force, tide-generating force, lunar bulge, solar bulge, flood tide, ebb tide.**

### Critical Thinking Question

Specify the forces involved in creating ocean tides.

### Active Learning Exercise

Two students are talking about what causes the tides. Student A says that the tide comes in and goes out based on differences in the height of the sea level. Student B says that the changing tides are caused by Earth rotating into and out of the tidal bulges. Working with another student in class, discuss the explanations of the tides by Student A and Student B. Which explanation is more technically correct? Explain your reasoning.

## 9.2 ▶ How do tides vary during a monthly tidal cycle?

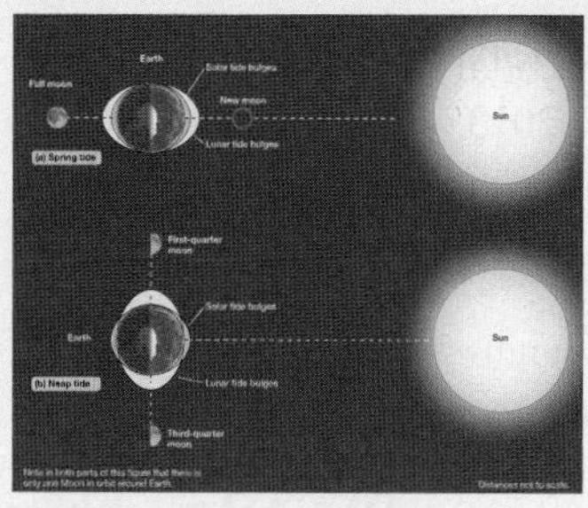


- Tides would be easy to predict if Earth were a uniform sphere covered with an ocean of uniform depth. For most places on Earth, *the time between successive high tides would be 12 hours 25 minutes (half a lunar day)*. The *29-and-a-half-day monthly tidal cycle* would consist of tides with maximum tidal range (spring tides) and minimum tidal range (neap tides). *Spring tides would occur each new moon and full moon, and neap tides would occur each first- and third-quarter phases of the Moon.*
- The *declination of the Moon* varies between 28.5 degrees north or south of the equator during the lunar month, and the *declination of the Sun* varies between 23.5 degrees north or south of the equator during the year, so *the location of tidal bulges usually creates two high tides and two low tides of unequal height per lunar day.* Tidal ranges are greatest when Earth is nearest the Sun and Moon.

### Selected Key Terms

Use the **glossary** at the end of this book to discover the meanings of these Selected Key Terms: **tidal range, spring tide, neap tide, perihelion, aphelion, perigee, apogee.**

### Critical Thinking Question

Specify differences between these two astronomical cycles that occurs simultaneously: the 27-and-a-half-day perigee–apogee–perigee cycle and the 29-and-a-half-day monthly tidal cycle.

### Active Learning Exercise

This is a take-home assignment. Observe the Moon from a reference location every night at about the same time for two weeks. Keep track of your observations about the shape (phase) of the Moon and its position in the sky. Then compare your observations to the reported tides in your area (or a coastal area you visit frequently) and report your findings to the class.

## 9.3 ▶ What do tides look like in the ocean?

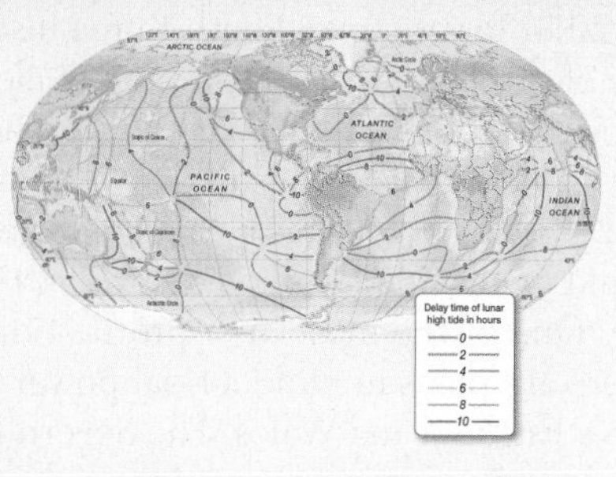


- *When friction and the true shape of ocean basins are considered, the dynamics of tides become more complicated.* Moreover, the two bulges on opposite sides of Earth cannot exist because they cannot keep up with the rotational speed of Earth. Instead, the bulges are broken up into *several tidal cells that rotate around an amphidromic point*—a point of zero tidal range. Rotation is counterclockwise in the Northern Hemisphere and clockwise in the Southern Hemisphere. *Many other factors influence tides on Earth*, too, such as the positions of the continents, the varying depth of the ocean, and coastline shape.

### Selected Key Terms

Use the **glossary** at the end of this book to discover the meanings of these Selected Key Terms: **amphidromic point, cotidal line.**

### Critical Thinking Question

Explain amphidromic points and cotidal lines. How do they influence tides in the open ocean?

### Active Learning Exercise

Assume that the tides are free waves and are therefore governed by the conditions and equations for ocean waves in Chapter 8. Consider the tidal bulges as wave crests with a wavelength of one-half Earth's circumference—about 20,000 kilometers (12,420 miles)—and assume that the average height of the lunar bulges is 3 meters (10 feet). Working with another student in class and using information from Chapter 8, determine if tides are deep-water or shallow-water waves. Note that the average depth of the oceans is 3.7 kilometers (2.3 miles) and the deepest ocean depth is 11 kilometers (6.8 miles).

## 9.4 ▶ What types of tidal patterns exist?

- The *three types of tidal patterns* observed on Earth are *diurnal* (a single high and low tide each lunar day), *semidiurnal* (two high and two low tides each lunar day), and *mixed* (characteristics of both). Mixed tidal patterns usually consist of semidiurnal periods with significant diurnal inequality. Mixed tidal patterns are the most common type in the world.

### Selected Key Terms

Use the **glossary** at the end of this book to discover the meanings of these Selected Key Terms: **diurnal tidal pattern, semidiurnal tidal pattern, mixed tidal pattern.**

### Critical Thinking Question

Imagine that there are two moons in orbit around Earth that are on the same orbital plane but always on opposite sides of Earth and that each moon is the same size and mass of our Moon. How would this affect the tidal range during spring and neap tide conditions? Also, how would this affect the tidal pattern observed?

### Active Learning Exercise

Working with another student in class, draw a monthly tidal curve for a fictional location. On the horizontal axis, use time in hours to show the length of two days; on the vertical axis, use tidal height. Also label the high and low tides. Present it to the class and indicate if it represents diurnal, semidiurnal, or mixed tide conditions.

## 9.5 ▶ What tidal phenomena occur in coastal regions?

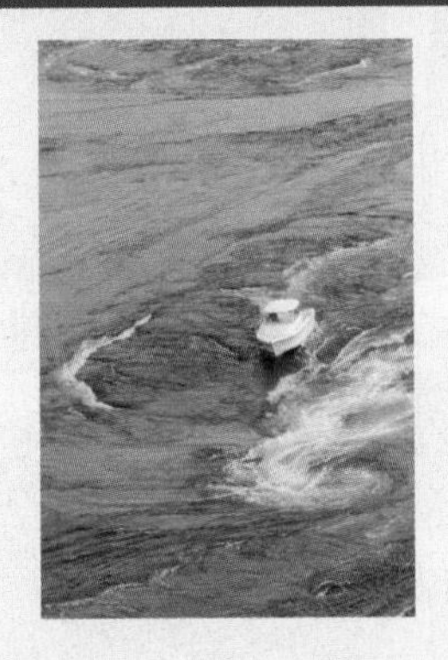

- There are many types of *observable tidal phenomena in coastal areas. Tidal bores are true tidal waves* (a wave produced by the tides) that occur in certain rivers and bays due to an incoming high tide. The effects of constructive interference together with the shoaling and narrowing of coastal bays creates the *largest tidal range in the world—17 meters (56 feet)—at the northern end of Nova Scotia's Bay of Fundy. Tidal currents follow a rotary pattern* in open-ocean basins but are converted to *reversing currents* along continental margins. The maximum velocity of reversing currents occurs during flood and ebb currents, when the water is halfway between high slack water and low slack water. *Whirlpools* can be created in some restricted coastal passages due to reversing tidal currents. *The tides are also important to many marine organisms.* For instance, *grunion*—small silvery fish that inhabit waters along the West Coast of North America—time their spawning cycle to match the pattern of the tides.

### Selected Key Terms

Use the **glossary** at the end of this book to discover the meanings of these Selected Key Terms: **flood current, ebb current, high slack water, low slack water, whirlpool, grunion.**

### Critical Thinking Question

Examine a current monthly tide calendar for a coastal location. When is the best time for a field trip to the tide pools? If this were a grunion spawning month, when would the conditions be right for the next grunion spawning run?

### Active Learning Exercise

Working with another student in class and using the Internet, find a monthly tidal chart for the Bay of Fundy. Discuss the tidal information that is displayed, including spring and neap tide conditions, the corresponding phases of the Moon, and the amount of tidal range.

## 9.6 ▶ Can tidal power be harnessed as a source of energy?

- *Tides can be used to generate power* without need for fossil or nuclear fuel. There are some *significant drawbacks*, however, to creating successful tidal power plants. Still, many sites worldwide have the *potential for tidal power generation*.

### Critical Thinking Question

What are some negative environmental factors that could inhibit the development of ocean tidal power systems?

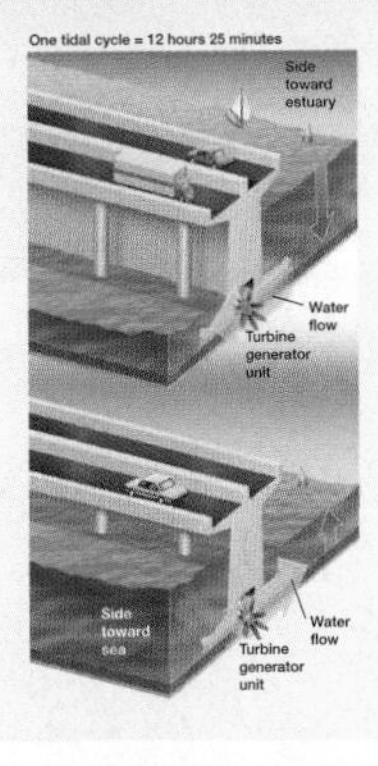


### Active Learning Exercise

Split the class into groups of two. In each group, have each student pick one of the two methods in which tidal power is harnessed. Debate the relative advantages and disadvantages of each method. For example, which method has fewer negative environmental concerns?

## Mastering Oceanography

Looking for additional review and test prep materials? With individualized coaching on the toughest topics of the course, Mastering Oceanography offers a wide variety of ways for you to move beyond memorization and deeply grasp the underlying processes of how the oceans work. Visit the Study Area in **www.masteringoceanography.com** to find practice quizzes, study tools, and multimedia that will improve your understanding of this chapter's content. Sign in today to access the following features: Self Study Quizzes, SmartFigures, Oceanography Videos and Animations, Squidtoons, Dynamic Study Modules, and an optional Pearson eText with embedded videos.

# Beaches, Shoreline Processes, and the Coastal Ocean

# 10

## ESSENTIAL LEARNING CONCEPTS

At the end of this chapter, you should be able to:

- ☐ **10.1** Use appropriate beach terminology to specify how coastal regions are defined.
- ☐ **10.2** Explain how sand moves on the beach.
- ☐ **10.3** Describe the characteristic features of erosional and depositional shores.
- ☐ **10.4** Discuss how changes in sea level produce emerging and submerging shorelines.
- ☐ **10.5** Describe the types of hard stabilization and evaluate various alternatives.
- ☐ **10.6** Compare the various types of coastal waters.
- ☐ **10.7** Specify the issues that face coastal wetlands.

↑ *Check when completed*

***"The waves which dash upon the shore are, one by one, broken, but the ocean conquers nevertheless. It overwhelms the Armada, it wears out the rock."***

***—Lord Byron (1821)***

The coastal ocean is a very busy place. Humans have always been attracted to the coastal regions of the world for their mild climate, abundance of seafood, ease of transportation, abundant recreational opportunities, and other commercial benefits. Population studies, for example, reveal that about 50% of the world's population—more than 3.5 billion people—live along coasts, and more than 80% of all Americans live within an hour's drive from an ocean or the Great Lakes. In the future, these figures are expected to increase. In the United States, for example, 8 of the 10 largest cities are in coastal environments, and about 3600 people move to the coast every day, often in pursuit of job opportunities in these urban areas. By 2025, as much as 75% of the global population is expected to live at the coast. Although coastal regions are desirable places to live, this rapid increase in human population negatively impacts coastal environments.

The coastal ocean is also filled with marine life. About 95% of all fish caught in the ocean are obtained within 320 kilometers (200 miles) of the shore. In addition, coastal waters support about 95% of the total mass of life in the oceans. Further, coastal estuary and wetland environments are among the most biologically productive ecosystems on Earth and serve as nursery grounds for many species of marine organisms that inhabit the open ocean. Coastal wetlands also serve as a vital natural cleanser for storm runoff, removing pollutants before they reach the ocean.

The coastal region is constantly changing because waves crash along most shorelines about 10,000 times a day, releasing their energy from distant storms. Waves cause erosion in some areas and deposition in others, resulting in changes that occur hourly, daily, weekly, monthly, seasonally, and yearly.

In this chapter, we'll examine the major features of beaches and shorelines, including the processes that modify them. We'll also discuss ways people interfere with these processes, creating hazards to themselves and to the environment. And finally, we'll examine the characteristics and types of coastal waters, including how human activities have impacted those regions.

## 10.1 ▶ How Are Coastal Regions Defined?

The **shore** is a zone that lies between the lowest tide level (low tide) and the highest elevation on land that is affected by storm waves. The **coast** extends inland from the shore as far as ocean-related features can be found (**Figure 10.1**). The width of the shore varies between a few meters and hundreds of meters. The width of the coast may vary from less than 1 kilometer (0.6 mile) to many tens of kilometers. The **coastline** marks the boundary between the shore and the coast. It is the landward limit of the effect of the highest storm waves on the shore.

◂ **Beaches of the world.** Beaches are composed of whatever material is locally available and differ depending on a variety of oceanographic factors. Beach locations shown include California; Oregon; Hawaii; the Galápagos Islands, Ecuador; and Baja California, Mexico.

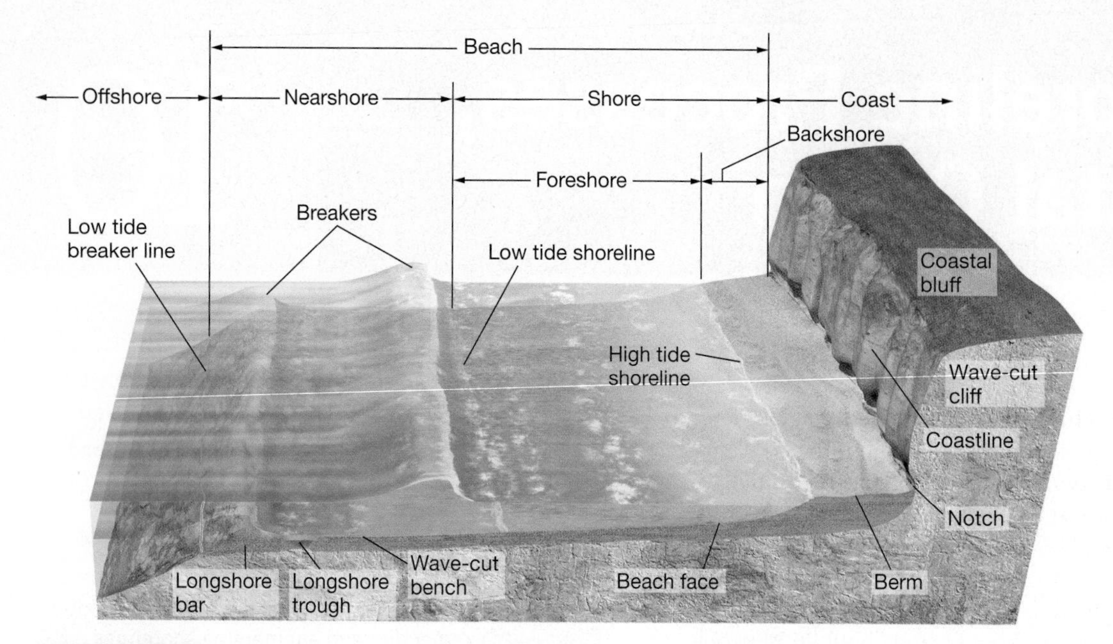


**EXPLORING DATA ▲ Examine Figure 10.1 and compare it to the beaches in your local area. Identify differences and/or similarities.**

Figure 10.1 **Diagrammatic view of a cliffed coastal region showing beach terminology and landforms.** The beach is defined as the entire active area of a coastline that is affected by waves; it extends from the low tide breaker line offshore (*left*) to the far end of the berm (*right*). Although most of these beach features can be found at any beach, not every beach has coastal cliffs.

## Beach Terminology

The beach profile in Figure 10.1 shows features characteristic of a cliffed shoreline. The shore is divided into the **backshore** and the **foreshore**. (The foreshore is often referred to as the *intertidal zone*, or *littoral [litoralis* = the shore] *zone*.) The backshore is above the high tide shoreline and is covered with water only during storms. The foreshore is the portion exposed at low tide and submerged at high tide. The **shoreline** migrates back and forth with the tide and is the water's edge. The **nearshore** extends seaward from the low tide shoreline to the low tide breaker line. It is never exposed to the atmosphere, but it is affected by waves that touch bottom. Beyond the low tide breakers is the **offshore** zone, which is deep enough that waves rarely affect the bottom.

A **beach** is a deposit of the shore area (**Figure 10.2**). It consists of wave-worked sediment that moves along the **wave-cut bench** (a flat, wave-eroded surface). A beach may continue from the coastline across the nearshore region to the line of breakers. Thus, the beach is the entire active area of a coast that experiences changes due to breaking waves. The area of the beach above the shoreline is often called the *recreational beach*.

The **berm** is the dry, gently sloping, slightly elevated margin of the beach that can be found at the foot of coastal cliffs or sand dunes. Because the berm is normally composed of dry sand (Figure 10.2), it is a favorite place of beachgoers for activities such as lying in the sun, beach volleyball, barbecues, and bonfires. Moving offshore, the **beach face** is the wet, sloping surface that extends from the berm to the shoreline. It is more fully exposed during low tide and is also known as the *low-tide terrace*. The beach face is a favorite place for runners because the sand is wet and hard packed. Offshore beyond the beach face are one or more **longshore bars**—sand bars that parallel the coast. A longshore bar may not always be present throughout the year, but when one is, it may be exposed during extremely low tides. Longshore bars can "trip" waves as they approach shore and cause them to begin breaking. Separating the longshore bar from the beach face is a **longshore trough**.

Figure 10.2 **Photo of a typical beach.** Many of the features of a typical beach are seen in this photo, such as the *berm*, which is the dry, flat area close to land, and the *beach face*, which is the wet, gently-sloping area (*left*).

## Beach Composition

Beaches are composed of whatever material is locally available (for example: sand, sea shells, pebbles, cobbles, and rocks; see the chapter-opening photos). When this material—sediment—comes from the erosion of beach cliffs or nearby coastal mountains, beaches are composed of mineral particles from these rocks and may be relatively coarse in texture. When the sediment comes primarily from rivers that drain lowland areas, beaches are finer in texture. Often, mud flats develop along the shore because only tiny clay-sized and silt-sized particles are emptied into the ocean. Such is the case for muddy coastlines such as along the coast of Suriname in South America and the Kerala coast of southwest India.

Other beaches have a significant biological component. For example, in low-relief, low-latitude areas such as southern Florida, there are no mountains or other sources of rock-forming minerals nearby. As a result, beaches in these areas are generally composed of shell fragments, broken coral, and the remains of organisms that live in coastal waters. Many beaches on volcanic islands in the open ocean are composed of black or green fragments of the basaltic lava that comprises the islands, or of coarse debris from coral reefs that develop around islands in low latitudes.

Regardless of the composition, though, the material that comprises the beach does not stay in one place. Instead, the waves that crash along the shoreline are constantly moving it. Thus, beaches can be thought of as *material in transit along the shoreline.*

**RECAP** **The beach is the coastal area affected by breaking waves and includes the berm, beach face, longshore trough, and longshore bar.**

**CONCEPT CHECK 10.1 ▸ Use appropriate beach terminology to specify how coastal regions are defined.**

1 Explain the difference between the shore and the coast.

2 What specific features are included in a typical beach?

3 How does the berm differ from the beach face?

4 Why do beaches reflect the composition of locally available materials? Include examples in your answer.

## 10.2 ▸ How Does Sand Move on the Beach?

The movement of sand on the beach occurs both perpendicular to the shoreline (*toward* shore and *away from* shore) and parallel to the shoreline (often referred to as *upcoast* and *downcoast*).

### Movement Perpendicular to the Shoreline

Sand on the beach moves perpendicular to the shoreline as a result of breaking waves.

**MECHANISM** As each wave breaks, water rushes up the beach face toward the berm. Some of this **swash** soaks into the beach and eventually returns to the ocean. However, most of the water drains away from shore as **backwash**, though usually not before the next wave breaks and sends its swash over the top of the previous wave's backwash.

While standing in ankle-deep water at the shoreline, you can see that swash and backwash transport sediment up and down the beach face perpendicular to the shoreline. Whether swash or backwash dominates determines whether sand is deposited or eroded from the berm.

**LIGHT VERSUS HEAVY WAVE ACTIVITY** During *light wave activity* (characterized by less energetic waves), much of the swash soaks into the beach, so backwash is reduced. The swash dominates the transport system, therefore causing a net movement of the sand up the beach face toward the berm, which creates a wide, well-developed berm.

During *heavy wave activity* (characterized by high-energy waves), the beach is saturated with water from previous waves, so very little of the swash soaks into the beach. Backwash dominates the transport system, therefore causing a net movement of sand down the beach face, which erodes the berm. When a wave breaks, moreover, the incoming swash comes *on top of* the previous wave's backwash, effectively protecting the beach from the swash and adding to the eroding effect of the backwash.

During heavy wave activity, where does the sand from the berm go? The orbital motion in waves is too shallow to move the sand very far offshore. Thus, the sand accumulates just beyond where the waves break and forms one or more offshore sand bars (the longshore bars).

**SUMMERTIME AND WINTERTIME BEACHES** Light and heavy wave activity alternate seasonally at most beaches, so the characteristics of the beaches they produce change, too (**Table 10.1**). For example, light wave activity produces a wide sandy berm and an overall steep beach face—a **summertime beach**—at the expense of

**CREATURE FEATURE 10.1**

### Catch me if you can: I live within the beach face!

**Sand crabs** (*Emerita* sp.) are small crustaceans with tough exoskeletons that live within sandy beaches and can bury themselves in saturated sand in just under two seconds.

Sand crabs have a barrel-shaped body and can hold their paddle-like appendages close to their bodies, allowing them to roll in the swash zone of waves. Their feathery antennae filter plankton and other detritus from receding waves, and although the crabs are most often hidden just below the sand, their antennae protrude above and can be seen creating a V-shape in the backwash.

**RECAP** **Smaller, low-energy waves move sand up the beach face toward the berm and create a summertime beach; larger, high-energy waves scour sand from the berm and create a wintertime beach.**

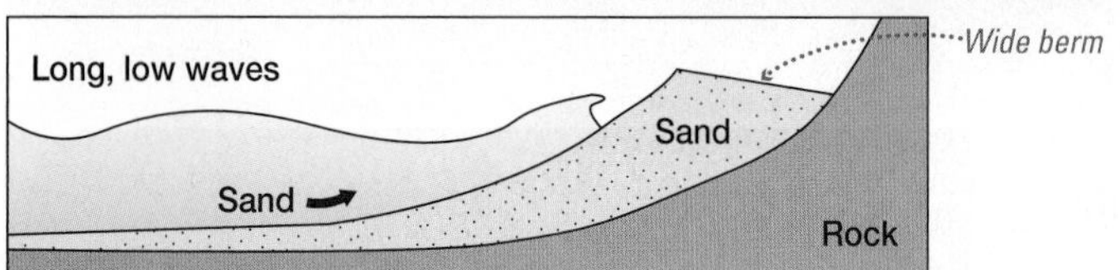


(a) Summertime beach (fair weather)

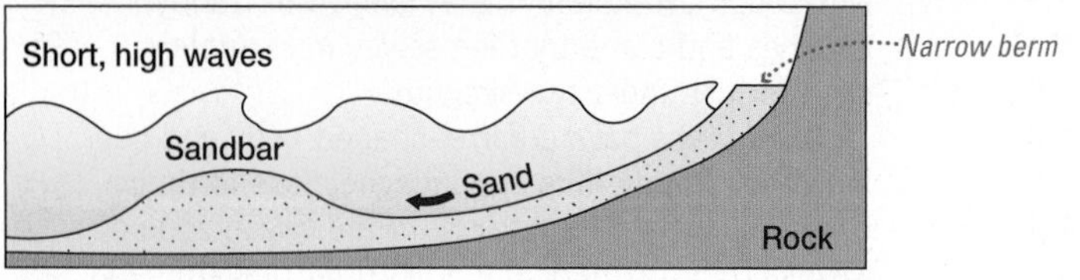


(b) Wintertime beach (storm)

Figure 10.3 **Summertime and wintertime beach conditions.** Dramatic differences occur between **(a)** summertime and **(b)** wintertime beach conditions at Boomer Beach in La Jolla, California.

**Animation**
Summertime/Wintertime Beach Conditions
http://goo.gl/XsoVZk

**Table 10.1 Characteristics of beaches affected by light and heavy wave activity**

| | Light wave activity (small waves) | Heavy wave activity (large waves) |
|---|---|---|
| Berm/longshore bars | Berm is built at the expense of the longshore bars | Longshore bars are built at the expense of the berm |
| Wave energy | Low wave energy (non-storm conditions) | High wave energy (storm conditions) |
| Time span | Long time span (weeks or months) | Short time span (hours or days) |
| Characteristics | Creates summertime beach: sandy, wide berm, steep beach face | Creates wintertime beach: rocky, narrow berm, flattened beach face |

the longshore bar (**Figure 10.3a**). Conversely, heavy wave activity produces a narrow rocky berm and an overall flattened beach face—a **wintertime beach**—and builds prominent longshore bars (**Figure 10.3b**). A wide berm that takes several months to build can be destroyed in just a few hours by high-energy wintertime storm waves.

## Movement Parallel to the Shoreline

At the same time that movement occurs perpendicular to shore, movement parallel to the shoreline also occurs.

**MECHANISM** Recall from Chapter 8 that within the surf zone, waves *refract* (bend) and line up *nearly* parallel to shore. With each breaking wave, the swash moves up onto the exposed beach at a slight angle; gravity then pulls the backwash down the beach face perpendicular to the shore. As a result, water moves in a zigzag fashion along the shore.

**LONGSHORE CURRENT AND LONGSHORE DRIFT (LONGSHORE TRANSPORT)** The zigzag movement of water along the shore is called a **longshore current** (**Figure 10.4**). Longshore currents have speeds up to 4 kilometers (2.5 miles) per hour. Speeds increase as beach slope increases, as the angle at which breakers arrive at the beach increases, as wave height increases, and as wave frequency increases.

Swimmers can inadvertently be carried by longshore currents and find themselves carried far from where they initially entered the water. This demonstrates that longshore currents are strong enough to move people as well as a vast amount of sand in a zigzag fashion along the shore.

**LONGSHORE DRIFT** (also called **longshore transport**, *beach drift*, or *littoral drift*) is the movement of *sediment* in a zigzag fashion caused by the longshore current (Figure 10.4b). Both longshore currents and longshore transport occur only within the surf zone and not farther offshore because the water is too deep there.Recall from Chapter 8 that the depth of a wave's *wave base* is one-

(a) Waves approaching the beach at a slight angle along the Oregon coast, producing a longshore current moving toward the right of the photo.

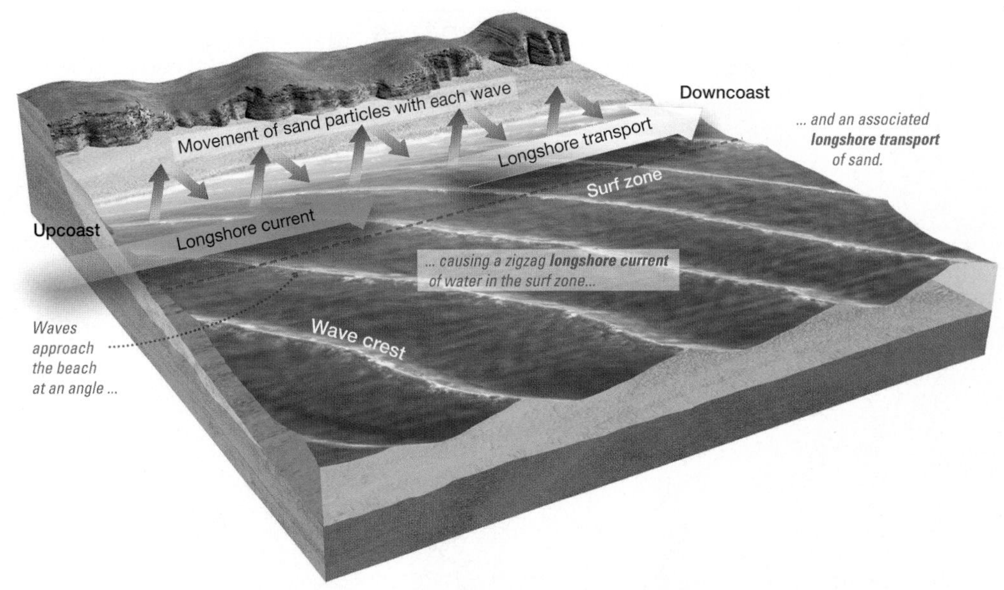

(b) A longshore current, caused by refracting waves, moves water in a zigzag fashion along the shoreline. This causes a net movement of sand grains (longshore drift) from upcoast to downcoast ends of a beach.

**EXPLORING DATA** ▲ **How would your position relative to the beach be affected if you weren't paying much attention and playing around in the waves within the surf zone?**

SmartFigure 10.4 **Longshore current and longshore drift.** https://goo.gl/SqWFnp

**Animation**
Longshore Current and Longshore (Beach) Drift
http://goo.gl/UlwJjR

half its wavelength, measured from still water level. Below this depth, waves don't touch bottom, and they don't refract; as a result, longshore currents can't form.

**THE BEACH: A RIVER OF SAND** By various processes, both rivers and coastal regions transport water and sediment from one area (*upcoast* or *upstream*) to another (*downcoast* or *downstream*). As a result, the beach has often been referred to as a "river of sand." There are, however, differences between how beaches and rivers transport sediment. For example, a longshore current moves in a zigzag fashion, while rivers flow mostly in a turbulent, swirling fashion. In addition, the direction of flow of longshore currents along a shoreline can change, whereas rivers always flow in the same basic direction (downhill). The longshore current can change direction because the waves that approach the beach typically come from different directions in different seasons. Nevertheless, the longshore current generally flows *southward along both the Atlantic and Pacific coasts of the United States* (**Figure 10.5**).

## STUDENTS SOMETIMES ASK . . .

***How much sand is moved along coasts by longshore drift?***

Very impressive amounts! For example, longshore drift rates are typically in the range of 75,000 to 230,000 cubic meters (100,000 to 300,000 cubic yards) per year. To help you visualize how much sand this is, think of a typical garbage truck, which has a volume of about 14 cubic meters (18 cubic yards). In essence, longshore drift carries the equivalent of many thousands of full garbage trucks along coastal regions each year. And a few coastal regions have longshore drift rates as high as 765,000 cubic meters (1,000,000 cubic yards) per year.

## CONCEPT CHECK 10.2 ▸ Explain how sand moves on the beach.

1 Describe differences between summertime and wintertime beaches. Explain why these differences occur.

2 What variables affect the speed of longshore currents?

3 What is longshore drift, and how is it related to a longshore current?

4 Why does the direction of longshore current sometimes reverse in direction? What is the primary direction of longshore current along the Pacific coast? Along the Atlantic coast? (See Figure 10.5 on page 318.)

**RECAP** **Longshore currents are produced by waves approaching the beach at an angle and create longshore drift, which transports sand along the coast in a zigzag fashion.**

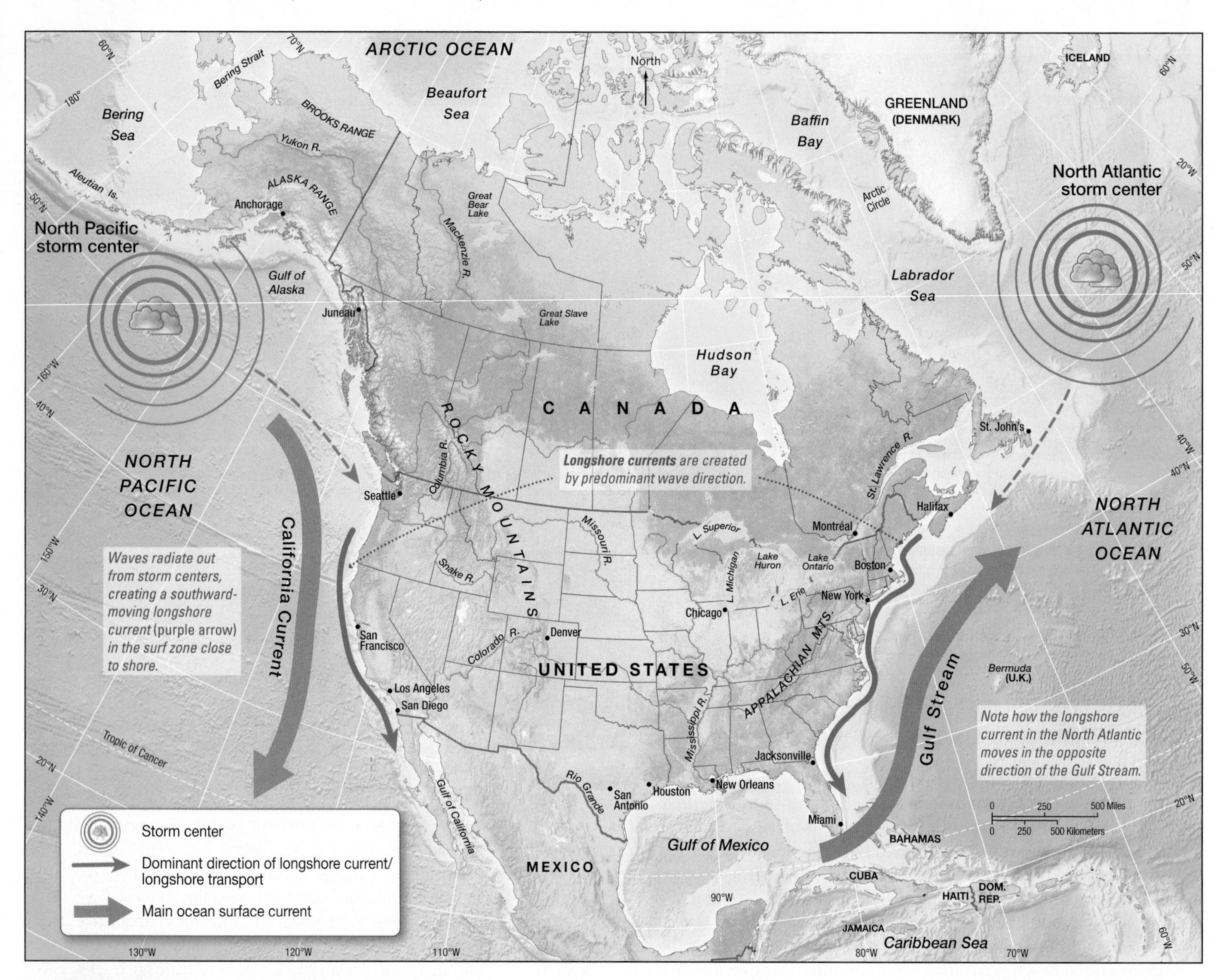


**EXPLORING DATA** ▲ Explain how the dominant direction of longshore current/longshore transport can go in the *opposite* direction of the main ocean surface current offshore. Along which U.S. coast does this occur?

Figure 10.5 **Major storm centers and the development of longshore current and longshore transport along U.S. coasts.** Map showing the regions of major storm centers in the North Pacific and North Atlantic Oceans. Waves generated from these storm centers radiate out and create dominant southward-moving longshore currents and longshore transport along *both* the Pacific and Atlantic coasts. For comparison, the main offshore surface currents are also shown.

## 10.3 ▸ What Features Are Typical of Erosional and Depositional Shores?

Sediment eroded from the beach is transported along the shore and deposited in areas where wave energy is low. Even though all shores experience some degree of both erosion and deposition, shores can often be identified primarily as one type or the other. **Erosional shores** typically have well-developed cliffs and are in areas where tectonic uplift of the coast occurs, such as along the U.S. Pacific coast.

The U.S. southeastern Atlantic coast and the Gulf coast, on the other hand, are primarily **depositional shores**. Sand deposits and offshore barrier islands are common there because the shore is gradually subsiding (sinking). Erosion can still be a major problem on depositional shores, especially when human-made coastal structures interfere with natural coastal processes (as discussed later in this chapter). For more details about the rates of erosion and deposition along U.S. coasts, see Mastering Oceanography Web Diving Deeper 10.2).

## Features of Erosional Shores

As discussed in Chapter 8, *wave refraction* (the bending of waves) causes a concentration of wave energy on **headlands** that jut out from the landmass, while the amount of energy reaching the shore in bays is reduced (see Figure 8.21). As a result, headlands are eroded more quickly, which causes the shoreline to retreat and often leaves erosional shoreline features. Some of these erosional features are shown in Figure 10.6.

Waves pound relentlessly away at the base of headlands, undermining the upper portions, which eventually collapse to form **wave-cut cliffs**. The waves may form **sea caves** at the base of the cliffs.

As waves continue to pound the headlands, the caves may eventually erode through to the other side, forming openings called **sea arches** (Figure 10.7). Some sea arches are large enough to allow a boat to maneuver safely through them. With continued erosion, the tops of sea arches eventually crumble to produce **sea stacks** (Figure 10.7). Waves also erode the bedrock of the bench. Uplift of the wave-cut bench creates a gently sloping **marine terrace** above sea level (Figure 10.8). In some regions that experience episodic uplift, such as the islands offshore of Southern California, a whole series of progressively older marine terraces—the oldest at the top—exist above sea level (Figure 10.9).

Rates of coastal erosion are influenced by the degree of exposure to waves, the amount of tidal range, and the composition of the coastal bedrock. Regardless of the erosion rate, all coastal regions follow the same developmental path. As long as there is no change in the elevation of the landmass relative to the ocean surface, the cliffs will continue to erode and retreat until the beaches widen sufficiently to prevent waves from reaching them. The eroded material is carried from high-energy areas and deposited in low-energy areas.

### STUDENTS SOMETIMES ASK . . .

***Along the East Coast of the United States, how can the longshore current move to the south when the strong Gulf Stream current is moving to the north?***

Longshore currents and major ocean surface currents are different things and completely independent of one another. For one thing, longshore currents occur only within the surf zone, while ocean surface currents are much wider and occur farther from shore. In addition, longshore currents are caused by waves coming into shore at an angle (and so can reverse), while ocean surface currents are caused by the major wind belts of the world and modified by the Coriolis effect (and so rarely reverse). And remember that waves (which cause longshore currents) can move in the opposite direction from ocean surface currents.

Along the U.S. East Coast, the reason that the longshore current goes to the south is because the major storm centers that create waves occur in the stormy northern part of the North Atlantic Ocean. As waves radiate southward from these storm centers, a southward-moving longshore current is produced along the East Coast. A similar situation occurs in the North Pacific Ocean, creating a southward-moving longshore current along the West Coast, which just happens to move in the same direction as the California Current (see Figure 10.5).

## Features of Depositional Shores

Coastal erosion of sea cliffs produces large amounts of sediment. Additional sediment, which is carried to the shore by rivers, comes from the erosion of inland rocks. Waves then distribute all this sediment along the continental margin.

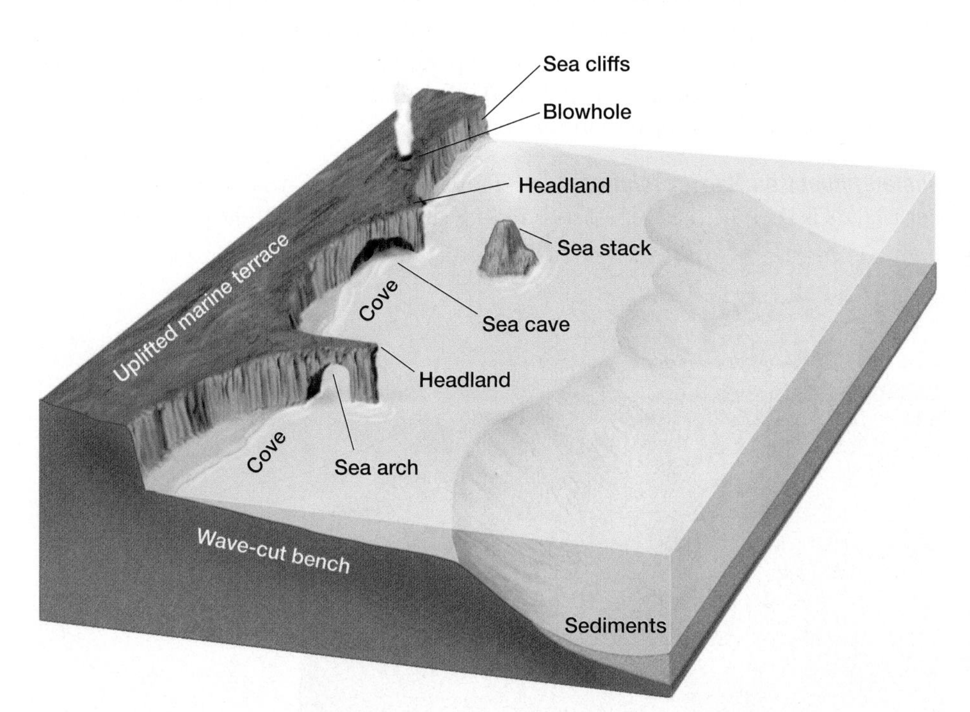


Figure 10.6 **Features of erosional rocky coasts.** Diagrammatic view of features characteristic of erosional rocky coasts.

Figure 10.7 **Sea arches and sea stack at Praia da Marinha Beach, near Armacao de Pera, Algarve, Portugal.** When the roof of a sea arch (*behind*) collapses, a sea stack (*middle*) is formed.

**Figure 10.8 Wave-cut bench and marine terrace.** A wave-cut bench is exposed at low tide near Kaikoura, New Zealand (*right*). An elevated wave-cut bench, called a marine terrace, is shown on the left.

Figure 10.10 shows some of the features of depositional coasts. These features are primarily deposits of sand moved by longshore drift but are also modified by other coastal processes. Some are partially or wholly separated from the shore.

A **spit** (*spit* = spine) is a linear ridge of sediment that extends in the direction of longshore drift from land into the deeper water near the mouth of a bay. The end of the spit normally curves into the bay due to the movement of currents.

Tidal currents or currents from river runoff are usually strong enough to keep the mouth of the bay open. If not, the spit may eventually extend across the bay and connect to the mainland, forming a, **bay barrier**, or **bay-mouth bar** (Figure 10.11a), which cuts off the bay from the open ocean. Although a bay barrier is a buildup of sand usually less than 1 meter (3.3 feet) above average sea level, permanent buildings are often constructed on them.

A **tombolo** (*tombolo* = mound) is a sand ridge that connects an island or a sea stack to the mainland (Figure 10.11b). Tombolos can also connect two adjacent islands. Tombolos form in the wave-energy shadow of an island and as a result are usually oriented perpendicular to the average direction of incoming waves.

## STUDENTS SOMETIMES ASK . . .

***What is the difference between a rip current and a rip tide? Are they the same thing as an undertow?***

Like tidal waves (tsunami), *rip tides* are a misnomer and have nothing to do with the tides. Rip tides are more correctly called *rip currents*. Perhaps rip currents have incorrectly been called rip tides because they occur suddenly (like an incoming tide). The origin of rip currents and their associated dangers are discussed in Diving Deeper 10.1.

An *undertow* is similar to a rip current, in that it is a flow of water away from shore. An undertow is much wider, however, and is usually more concentrated along the ocean floor. An undertow is really a continuation of backwash that flows down the beach face and is strongest during heavy wave activity. Undertows can be strong enough to knock people off their feet, but they are confined to the immediate floor of the ocean and only within the surf zone.

**BARRIER ISLANDS** Extremely long offshore deposits of sand that are parallel to the coast are called **barrier islands** (Figure 10.12). They form a first line of defense against rising sea level and high-energy storm waves, which would otherwise exert their full force directly against the shore. The origin of barrier islands is complex, but many appear to have developed during the worldwide rise in sea level that is associated with the melting of glaciers at the end of the most recent ice age, about 18,000 years ago.

According to a recent study of global satellite images, 2149 barrier islands have been identified worldwide in every climate and every tide–wave combination. Nearly 300 barrier islands ring the Atlantic and Gulf coasts of the United States (Figure 10.13). They exist from Massachusetts to eastern Florida in a nearly continuous line; they also occur discontinuously in the Gulf of Mexico from western Florida into Mexico. Barrier islands may exceed 100 kilometers (60 miles) in length, have widths of several kilometers, and are separated from the mainland by a lagoon. Notable barrier islands include Fire Island off the New York coast, North Carolina's Outer Banks, and Padre Island off the coast of Texas.

***Human Impact on Barrier Islands*** One human-related environmental issue of barrier islands is their attractiveness as building sites because of their proximity to the

**Figure 10.9 Marine (wave-cut) terraces.** The gently sloping surface shown near the top of this photo is a marine terrace near Cape Blanco, Oregon. It was created by wave activity at sea level and was subsequently elevated to its present position by tectonic uplift.

## WARNING: RIP CURRENTS . . . DO YOU KNOW WHAT TO DO?

The backwash from breaking waves usually returns to the open ocean as a flow of water across the ocean bottom, so it is commonly referred to as "sheet flow." Some of this water, however, can return to the ocean in strong, narrow surface currents called **rip currents** that flow away from shore and are generally oriented perpendicular to the beach.

Rip currents are between 15 and 45 meters (15 and 150 feet) wide and can attain velocities of 7 to 8 kilometers (4 to 5 miles) per hour—faster than most people can swim for any length of time. In fact, it is useless to swim for long against a current stronger than about 2 kilometers (1.2 miles) per hour. Rip currents can travel hundreds of meters from shore before they break up. If a light-to-moderate swell is breaking, numerous rip currents that are moderate in size and velocity may develop. A heavy swell usually produces fewer, more concentrated, and stronger rips. They can often be recognized by the way they interfere with incoming waves, by their characteristic brown color caused by suspended sediment, or by their foamy and choppy appearance (**Figure 10A**).

The rip currents that occur during heavy swell are a significant hazard to coastal swimmers. In fact, rip currents cause an estimated 70 to 100 drownings annually in the United States, and 80% of rescues at beaches by lifeguards involve people who are trapped in rip currents. What is the best thing to do if you are caught in a rip current? Swimmers can escape rip currents by swimming parallel to the shore for a short distance (simply swimming out of the narrow rip current) and then riding the waves in toward the beach. However, even excellent swimmers who panic or try to fight the current by swimming directly into it are eventually overcome by exhaustion and may drown. Even though most beaches have warnings posted and are frequently patrolled by lifeguards, many people tragically lose their lives each year because of rip currents.

### WHAT DID YOU LEARN?

1. This may save your life one day: what is the best strategy to ensure that you won't drown if you are caught in a rip current?
2. How do rip currents change the shape of the offshore portion of a beach?

Video
Rip Current
http://goo.gl/FqwlWb

**Figure 10A Rip current and warning sign.** A rip current (*red arrow*), which extends outward from shore and interferes with incoming waves, and a warning sign (*inset*).

ocean. For example, the population densities of barrier islands are three times higher than those of adjoining beaches along non-barrier island coastlines. In addition, the overall population on barrier islands increased 14% from 1990 to 2000 and continues to increase. Although it seems unwise to build a coastal structure on a narrow, low-lying, shifting strip of sand, many large buildings have been constructed on barrier islands (see Figure 10.12c). Some of these structures have either fallen into the ocean or have needed to be moved (see Mastering Oceanography **Web Diving Deeper 10.1**).

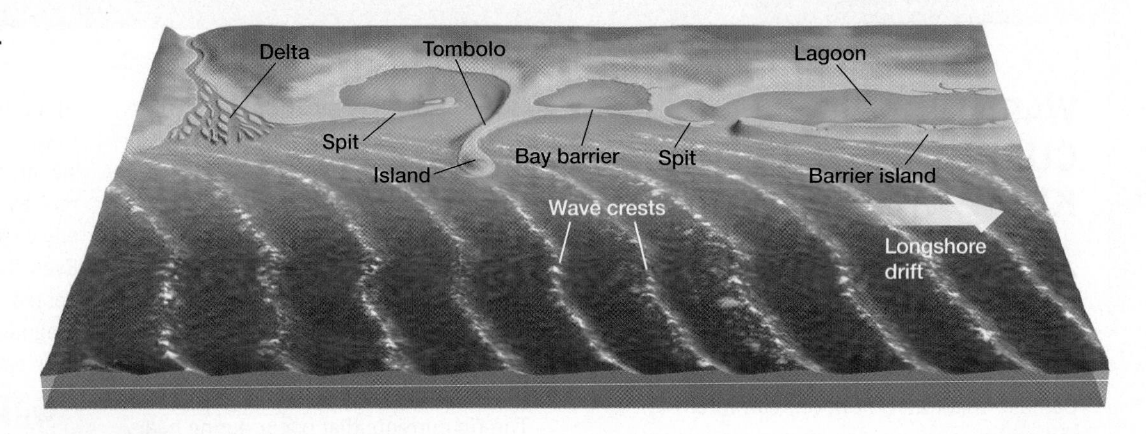


**Figure 10.10 Features of depositional coasts.** Diagrammatic view of features characteristic of depositional coasts.

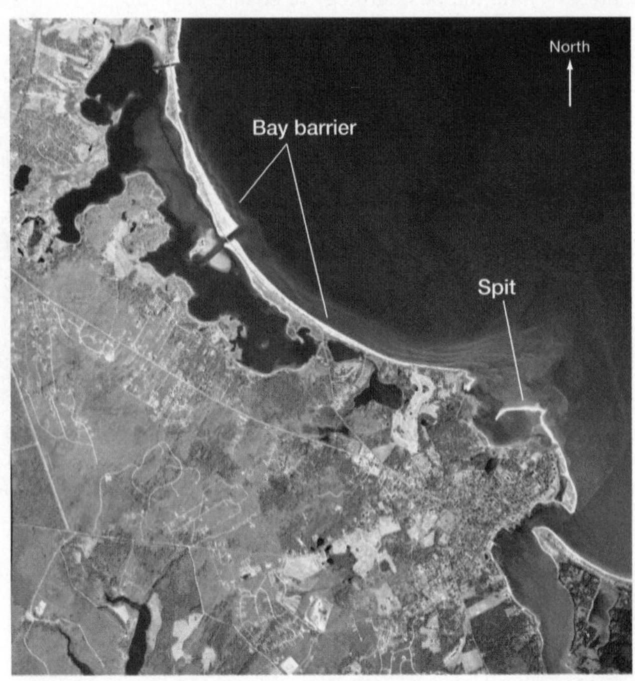


**(a)** Barrier coast, spit, and bay barrier along the coast of Martha's Vineyard, Massachusetts.

**(b)** Tombolo at Goat Rock Beach, California, looking north.

**Figure 10.11 Coastal depositional features.** Photos showing various coastal depositional features, including a bay barrier, spit, and tombolo.

***Features of Barrier Islands*** A typical barrier island has the physiographic features shown in **Figure 10.14**. In order, moving from the ocean toward land, they are:

- ocean beach
- dunes
- barrier flat
- high salt marsh
- low salt marsh, and
- lagoon between the barrier island and the mainland.

During the summer, gentle waves carry sand to the *ocean beach*, so it widens and becomes steeper. During the winter, higher-energy waves carry sand offshore and produce a narrow, gently sloping beach.

Winds blow sand inland during dry periods to produce coastal *dunes*, which are stabilized by dune grasses. These plants can withstand salt spray and burial by sand. Dunes protect the lagoon against excessive flooding during storm-driven high tides. Numerous passes exist through the dunes, particularly along the southeastern Atlantic coast, where dunes are less well developed than to the north.

The *barrier flat* forms behind the dunes from sand driven through the passes during storms. Grasses quickly colonize these flats, and seawater washes over them during storms. If storms wash over the barrier flat infrequently enough, the plants undergo natural biological succession, with the grasses successively replaced by thickets, woodlands, and eventually forests.

*Salt marshes* typically lie inland of the barrier flat. They are divided into the *low marsh*, which extends from about mean sea level to the high neap-tide line, and the *high marsh*, which extends to the highest spring-tide line. The low marsh is by far the most biologically productive part of the salt marsh.

New marshland is formed as overwash carries sediment into the lagoon, filling portions so they become intermittently exposed by the tides. Marshes may be poorly developed on parts of the island that are far from floodtide inlets. Their development is greatly restricted on barrier islands, where people perform artificial dune enhancement and fill inlets in an attempt to prevent overwashing and flooding.

***Barrier Island Migration*** The gradual sea level rise experienced along the eastern North American coast is causing barrier islands to migrate landward. As *Sequence 1–4* in Figure 10.14 shows, the movement of the barrier island is similar to a slowly moving tractor tread, with the entire island rolling over itself, impacting structures built on these islands. *Peat deposits*, which are formed by the accumulation of organic matter in marsh environments, provide further evidence of barrier island migration. As the island slowly

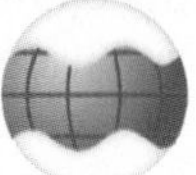

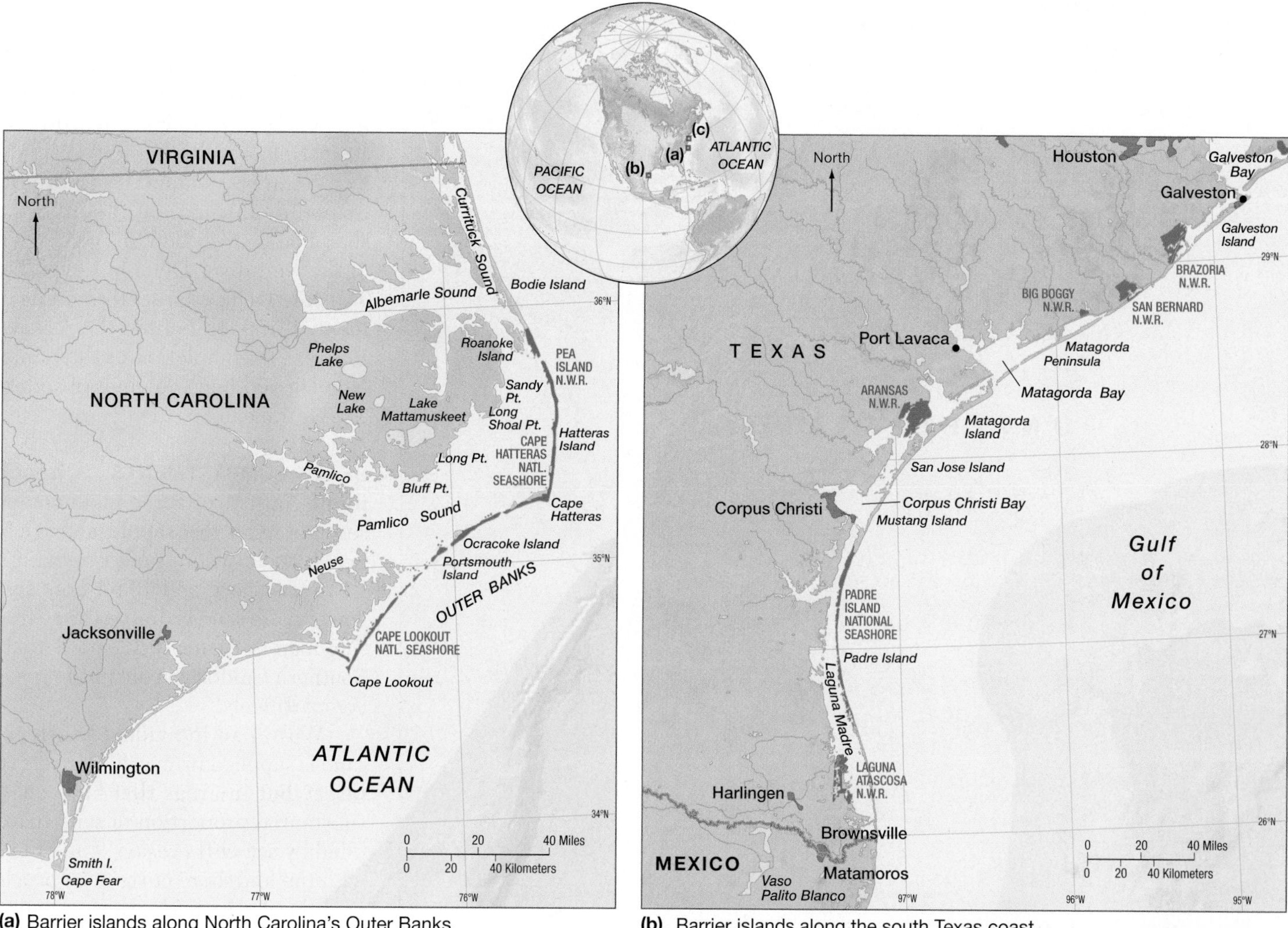


(a) Barrier islands along North Carolina's Outer Banks.

(b) Barrier islands along the south Texas coast.

rolls over itself and migrates toward land, it buries ancient peat deposits. These peat deposits can be found beneath the island and may even be exposed on the ocean beach when the barrier island has moved far enough.

**DELTAS** Some rivers carry more sediment to the ocean than longshore currents can distribute. These rivers develop a **delta** (*delta* = triangular) deposit at their mouths. The Mississippi River, which empties into the Gulf of Mexico (**Figure 10.15a**), forms one of the largest deltas on Earth. Deltas are fertile, flat, low-lying areas that are subject to periodic flooding.

Delta formation begins when a river has filled its mouth with sediment. The delta then grows through the formation of *distributaries*, which are branching channels that deposit sediment as they radiate out over the delta in finger-like extensions (Figure 10.15a). When the fingers get too long, they become choked with sediment. At this point, a flood

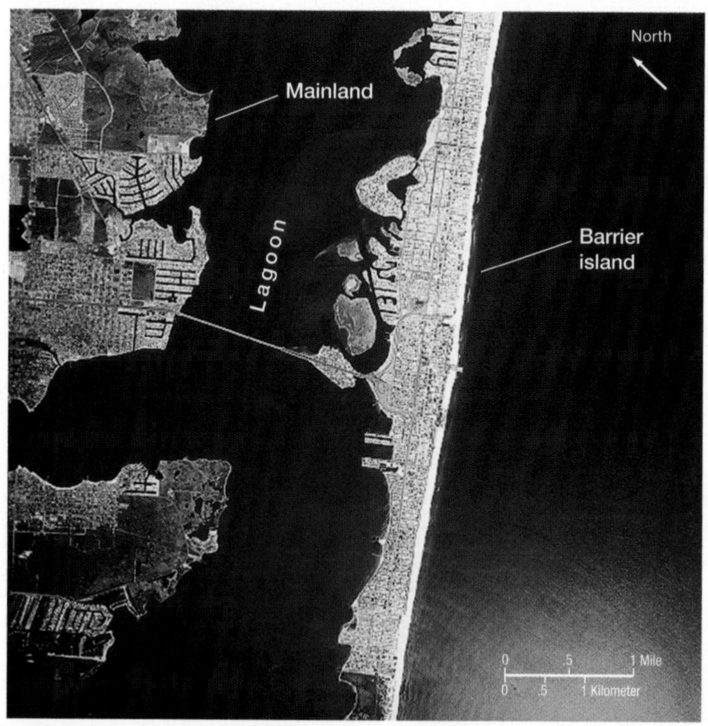


(c) A portion of a heavily developed barrier island near Tom's River, New Jersey.

**Figure 10.12 Examples of barrier islands.** Maps and aerial photo of barrier islands in **(a)** North Carolina, **(b)** Texas, and **(c)** New Jersey.

**Figure 10.13 Locations of barrier islands along the U.S. Atlantic and Gulf coasts.** Barrier islands occur from Maine to eastern Florida along the Atlantic coast and from western Florida to Mexico along the Gulf coast. Barrier islands do not occur along the Pacific coast.

may easily shift the distributary's course and provide sediment to low-lying areas between the fingers. When depositional processes exceed coastal erosion and transportation processes, a branching "bird's foot" delta similar to the Mississippi River Delta results.

When erosion and transportation processes exceed deposition, on the other hand, a delta shoreline is smoothed to a gentle curve, like that of the Nile River Delta in Egypt (**Figure 10.15b**). The Nile Delta is presently eroding because sediment is trapped behind the Aswan High Dam. Before the completion of the dam in 1964, the Nile carried huge volumes of sediment into the Mediterranean Sea.

**BEACH COMPARTMENTS** A **beach compartment** consists of three main components: (1) a series of rivers that supply sand to a beach; (2) the beach itself, where sand is moving due to longshore transport; and (3) offshore submarine canyons, where sand is drained away from the beach. The map in **Figure 10.16** shows that the coast of Southern California contains four separate beach compartments.

Within an individual beach compartment, sand is supplied primarily by rivers (Figure 10.16, *inset*), but in areas that have coastal bluffs, a substantial proportion of sand may also be supplied by sea cliff erosion. The sand moves south with the longshore current, so beaches are wider near the southern (*downcoast*) end of each beach compartment. Although some sand is washed offshore along the way or blows further inland to produce coastal sand dunes, most sand eventually moves near the head of a submarine canyon. Surprisingly, many submarine canyons come very close to shore. This allows sand to be drained off away from the beach and onto the ocean floor, lost from the beach forever. To the south of this beach compartment, the beaches are typically thin and rocky, without much sand. The process begins all over again at the upcoast end of the next beach compartment, where rivers add their sediment. Farther downcoast, the beach widens and has an abundance of sand until that sand is also diverted down a submarine canyon.

***Beach Starvation*** Human activities have altered the natural system of beach compartments. When a dam is built along one of the rivers that feed into a beach compartment, it deprives the beach of sand. Lining rivers with concrete for flood control further reduces the sediment load delivered to coastal regions. Longshore transport continues to sweep the shoreline's sand into the submarine canyons, so the beaches become narrower and experience **beach starvation**. If all the rivers are blocked, the beaches may nearly disappear.

SmartFigure 10.14
**Physiographic features of barrier islands and migration of a barrier island in response to sea level rise.**
https://goo.gl/arbvRf

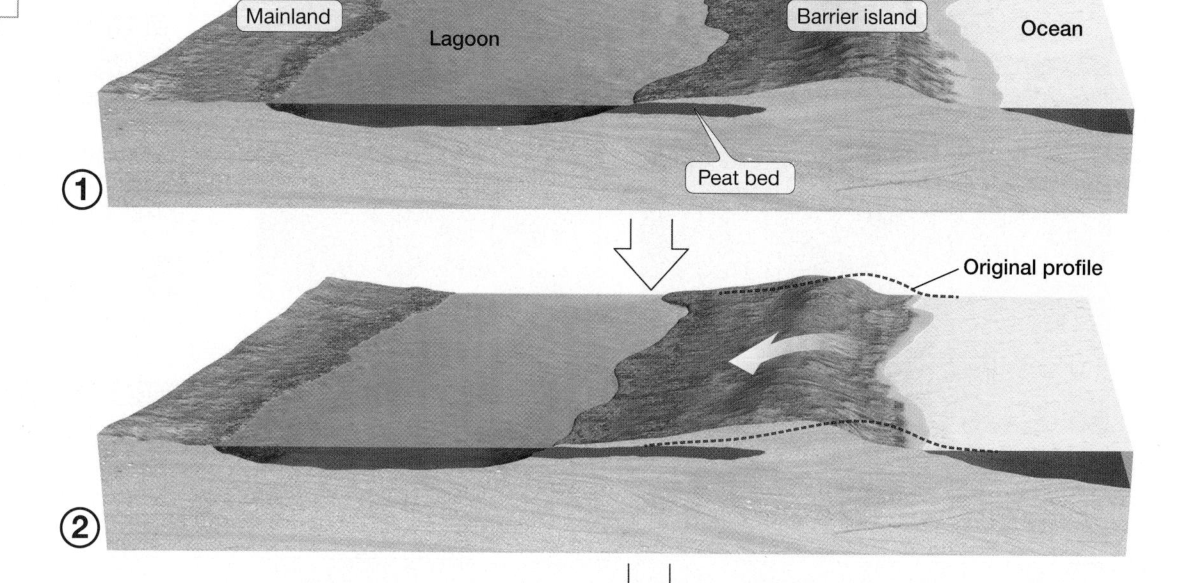


**Animation**
Movement of a Barrier Island in Response to Rising Sea Level
http://goo.gl/KUPJcl

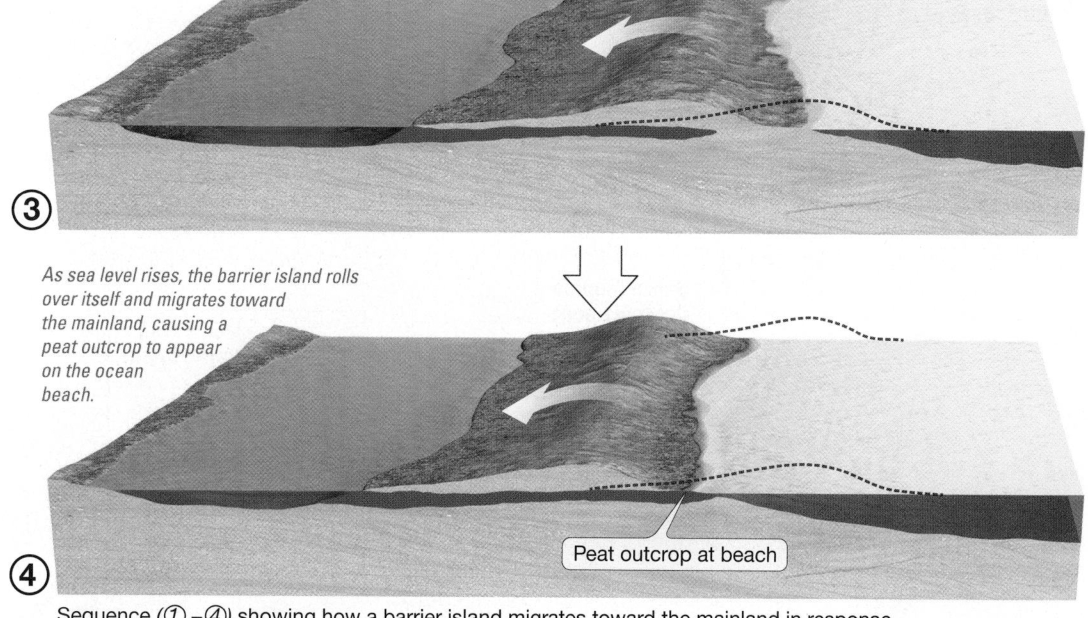

3
As sea level rises, the barrier island rolls over itself and migrates toward the mainland, causing a peat outcrop to appear on the ocean beach.
Peat outcrop at beach
4

Sequence (①–④) showing how a barrier island migrates toward the mainland in response to rising sea level and exposes peat deposits that have been covered by the island.

What can be done to prevent beach starvation in beach compartments? One obvious solution is to eliminate the dams, which would allow rivers to supply sand to the beach and return beach compartments to a natural balance. However, most

Figure 10.15 **Examples of deltas.** Aerial photos showing **(a)** the Mississippi River Delta and **(b)** the Nile River Delta.

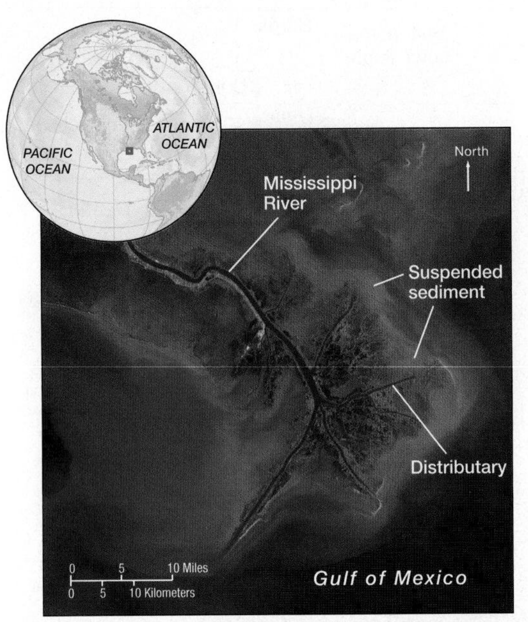


**(a)** Satellite image of the branching "bird's foot" structure of the Mississippi River Delta, which flows into the Gulf of Mexico and shows suspended sediment in the water.

**(b)** Photograph from the Space Shuttle of Egypt's Nile River Delta, which has a smooth, curved shoreline as it extends into the Mediterranean Sea.

**Animation**
Movement of Sand in a Beach Compartment
goo.gl/u8fkDn

**SmartFigure 10.16 Beach compartments.**
https://goo.gl/sm2k0K

**EXPLORING DATA ▼ Explain why a beach compartment is narrow and rocky near its upstream end, and is wide and sandy near its downstream end.**

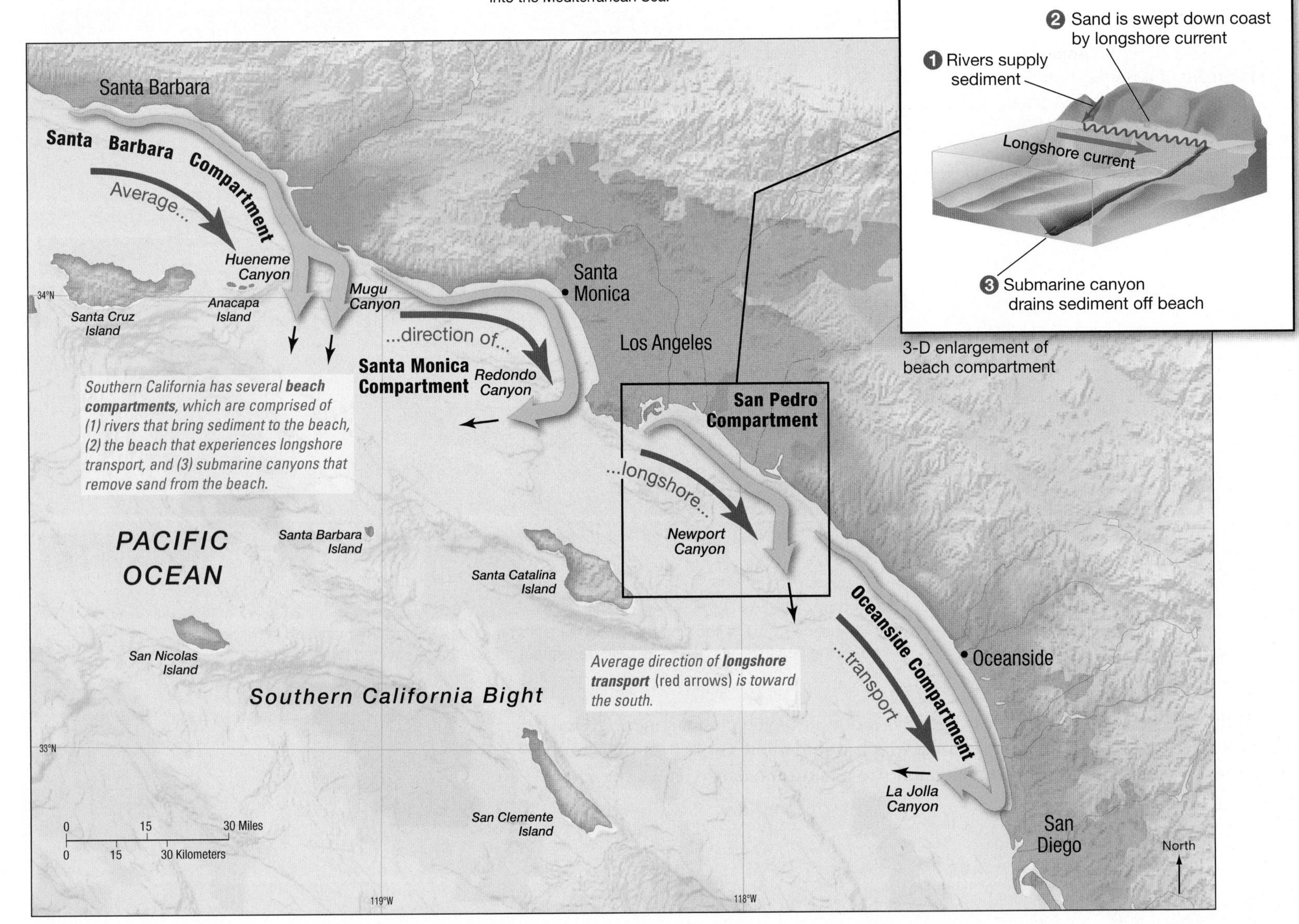

dams are built for flood protection, water storage, and the generation of hydropower, so it is unlikely that many will be removed. Another option, called *beach nourishment*, is discussed later in this chapter.

**CONCEPT CHECK 10.3 ▶ Describe the characteristic features of erosional and depositional shores.**

1 Discuss the formation of such erosional features as wave-cut cliffs, sea caves, sea arches, sea stacks, and marine terraces.

2 Describe the origin of these depositional features: spit, bay barrier, tombolo, and barrier island.

3 Describe the response of a barrier island to a rise in sea level. Why do some barrier islands develop peat deposits running through them from the ocean beach to the salt marsh?

4 Discuss why some rivers have deltas and others do not. What are the factors that determine whether a "bird's foot" delta (like the Mississippi Delta) or a smoothly curved delta (like the Nile Delta) will form?

5 Describe the three parts of a beach compartment. What happens when dams are built across all the rivers that supply sand to the beach?

**RECAP** **Erosional shores are characterized by erosional features such as cliffs, sea arches, sea stacks, and marine terraces. Depositional shores are characterized by depositional features such as spits, tombolos, barrier islands, deltas, and beach compartments.**

**STUDENTS SOMETIMES ASK . . .**

***Can submarine canyons fill with sediment?***

Yes. In many beach compartments, the submarine canyons that drain sand from the beach empty into deep basins offshore. However, given several million years and tons of sediments per year sliding down submarine canyons, offshore basins begin to fill up and can eventually be exposed above sea level. In fact, the Los Angeles basin in California was filled in by sediment derived from local mountains in this manner during the geologic past.

# 10.4 ▶ How Do Changes in Sea Level Produce Emerging and Submerging Shorelines?

In addition to being described as primarily erosional or depositional, shorelines can also be classified based on their position relative to sea level. *Sea level, however, has changed throughout time*, intermittently exposing large regions of continental shelf and then plunging them back under the sea. Sea level can change because the level of the land changes, the level of the sea changes, or a combination of the two. Shorelines that are rising above sea level are called **emerging shorelines**, and those sinking below sea level are called **submerging shorelines**.

## Features of Emerging Shorelines

*Marine terraces* (**Figure 10.17**; see also Figures 10.8 and 10.9) are one feature characteristic of emerging shorelines. Marine terraces are flat platforms backed by cliffs, which form when a wave-cut bench is exposed above sea level. **Stranded beach deposits** and other evidence of marine processes such as ancient sea cliffs may exist many meters above the present shoreline, indicating that the former shoreline has risen above sea level (Figure 10.17).

## Features of Submerging Shorelines

Features characteristic of submerging shorelines include wave-cut benches below sea level that contain **drowned beaches** (Figure 10.17). Other features of submerging shorelines include **submerged dune topography** and **drowned river valleys** along the present shoreline.

## Changes in Sea Level

What causes the changes in sea level that produce submerging and emerging shorelines? One mechanism is the raising or lowering of the land surface relative to sea level through the movement of Earth's crust. Another mechanism is the alteration of the level of the sea itself through worldwide changes in sea level.

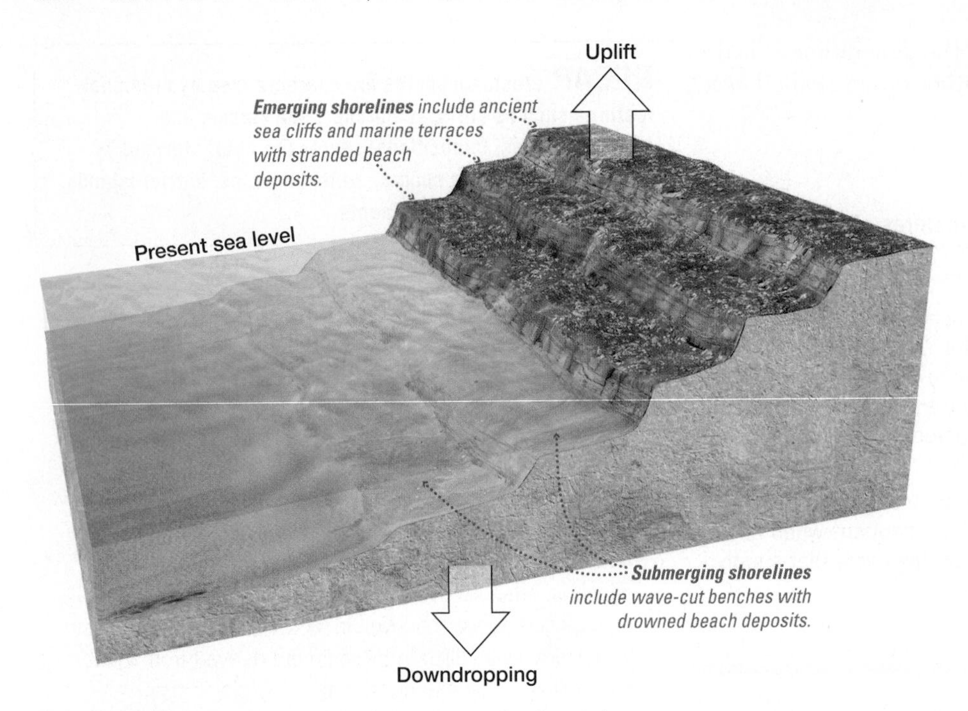
Uplift
Emerging shorelines include ancient sea cliffs and marine terraces with stranded beach deposits.
Present sea level
Submerging shorelines include wave-cut benches with drowned beach deposits.
Downdropping

Figure 10.17 **Features of ancient emerging and submerging shorelines.**

## STUDENTS SOMETIMES ASK . . .

***Because of plate motions, I know that the continents have not always remained in the same geographic positions. Has the movement of the continents ever affected sea level?***

Remarkably, yes. When plate motion moves large continental masses into polar regions, thick continental glaciation can occur (such as in Antarctica today). Glacial ice forms from water vapor in the atmosphere (in the form of snow), which is ultimately derived from the evaporation of seawater. Thus, water is removed from the oceans when continents assume positions close to the poles that provide a platform for large land-based ice accumulation, thereby lowering sea level worldwide.

**MOVEMENT OF EARTH'S CRUST** The elevation of Earth's crust relative to sea level can be affected by tectonic movements and by isostatic adjustment. (Recall that isostatic adjustment of Earth's crust is also discussed in Chapter 1.) These are called *changes in relative sea level*, because it's the land that has changed, not the sea.

***Tectonic Movements*** The most dramatic changes in sea level during the past 3000 years have been caused by *tectonic movements*, which affect the elevation of the land. These changes include uplift or subsidence of major portions of continents or ocean basins, as well as localized folding, faulting, or tilting of the continental crust.

Most of the U.S. Pacific coast, for example, is an emerging shoreline because continental margins where plate boundaries occur are tectonically active, producing earthquakes, volcanoes, and mountain chains paralleling the coast. Most of the U.S. Atlantic coast, on the other hand, is a submerging shoreline. When a continent moves away from a spreading center (such as the Mid-Atlantic Ridge), its trailing edge subsides because of cooling and the additional weight of accumulating sediment. Passive margins experience only a low level of tectonic deformation, earthquakes, and volcanism, making the Atlantic coast far more quiet and stable than the Pacific coast.

***Isostatic Adjustment*** Earth's crust also undergoes *isostatic adjustment*: it sinks under the accumulation of heavy loads of ice, vast piles of sediment, or outpourings of lava, and it rises when heavy loads are removed (**Figure 10.18**).

For example, at least four major accumulations of glacial ice—and dozens of smaller ones—have occurred in high-latitude regions over the past 3 million years. Although Antarctica is still covered by a very large, thick ice cap, much of the ice that once covered northern Asia, Europe, and North America has melted.

The weight of ice sheets as much as 3 kilometers (2 miles) thick caused the crust beneath to sink (Figure 10.18). Today, these areas are still slowly rebounding, 18,000 years after the ice began to melt. The floor of Hudson Bay, for example, which is now about 150 meters (500 feet) deep, will be close to or above sea level by the time it stops isostatically rebounding. Another example is the Gulf of Bothnia (between Sweden and Finland), which has isostatically rebounded 275 meters (900 feet) during the past 18,000 years.

Generally, tectonic and isostatic changes in sea level are confined to a segment of a continent's shoreline. For a *worldwide* change in sea level, there must be a change in seawater volume or ocean basin capacity.

**WORLDWIDE (EUSTATIC) CHANGES IN SEA LEVEL** Changes in sea level that are experienced worldwide due to changes in seawater volume or ocean basin capacity are called **eustatic sea level changes** (*eu* = good, *stasis* = standing). (The term *eustatic* refers to a highly idealized situation in which all of the continents remain static [in *good standing*], while only the sea rises or falls.) The formation or destruction of large inland lakes, for example, causes small eustatic changes in sea level. When lakes form, they trap water that would otherwise run off the land into the ocean, so sea level is lowered worldwide. When lakes are drained and release their water back to the ocean, sea level rises.

Another example of a eustatic change in sea level is through changes in sea floor spreading rates, which can change the capacity of the ocean basin and affect sea level worldwide. Fast spreading, for instance, produces larger rises, such as the East Pacific Rise, which displace more water than slow-spreading ridges such as the Mid-Atlantic Ridge. Thus, fast spreading raises sea level, whereas slower spreading lowers sea level worldwide. Significant changes in sea level due to changes in spreading rate typically take hundreds of thousands to millions of years and may have changed sea level by 1000 meters (3300 feet) or more in the geologic past.

**Changes to Sea Level during Ice Ages** Ice ages cause eustatic sea level changes, too. As glaciers form, they tie up vast volumes of water on land, eustatically lowering sea level. An analogy to this effect is a sink of water representing an ocean basin. To simulate an ice age, some of the water from the sink is removed and frozen, causing the water level of the sink to be lower. In a similar fashion, worldwide sea level is lower during an ice age. During interglacial stages (such as the one we are in at present), the glaciers melt and release great volumes of water that drain to the sea, eustatically raising sea level. This would be analogous to putting a frozen chunk of ice on the counter near the sink and letting the ice melt, causing the water to drain into the sink and raise "sink level."

During the *Pleistocene Epoch*, glaciers advanced and retreated many times on land in middle- to high-latitude regions, causing sea level to fluctuate considerably. (The Pleistocene Epoch of geologic time, which is also called the "Ice Age," occurred 2.6 million to 10,000 years ago [see the Geologic Time Scale, Figure 1.30]). The thermal contraction and expansion of the ocean as its temperature decreased and increased, respectively, affected sea level, too. The thermal contraction and expansion of seawater work much like a mercury thermometer: As the mercury inside the thermometer warms, it expands and rises up the thermometer; as it cools, it contracts. Similarly, cooler seawater contracts and occupies less volume, thereby eustatically *lowering* sea level. Warmer seawater expands, eustatically *raising* sea level.

Although it is difficult to state definitely the range of shoreline fluctuation during the Pleistocene, evidence suggests that it was at least 120 meters (400 feet) below the present shoreline (**Figure 10.19**). It is also estimated that if *all* the remaining glacial ice on Earth were to melt, sea level would rise another 70 meters (230 feet). Thus, the maximum sea level change during the Pleistocene would have been on the order of 190 meters (630 feet), most of which was due to the capture and release of Earth's water by land-based glaciers and polar ice sheets.

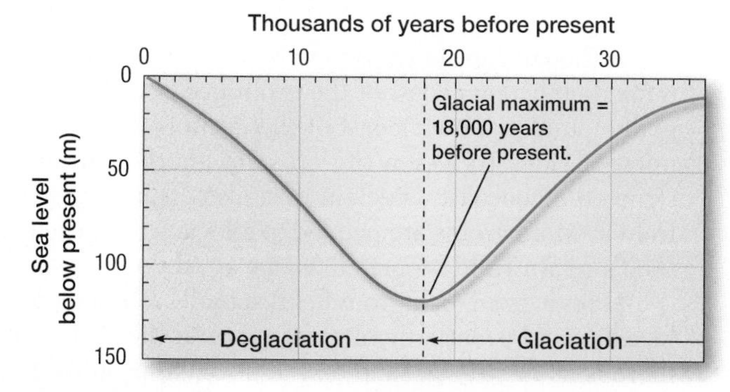
Thousands of years before present
0
10
20
30
0
50
100
150
Sea level below present (m)
Glacial maximum = 18,000 years before present.
Deglaciation
Glaciation

**Figure 10.19 Sea level change during the most recent advance and retreat of Pleistocene glaciers.** Graph showing how sea level dropped worldwide by about 120 meters (400 feet) as the last glacial advance removed water from the oceans and transferred it to continental glaciers. Sea level began to rise about 18,000 years ago as the glaciers melted and water returned to the oceans.

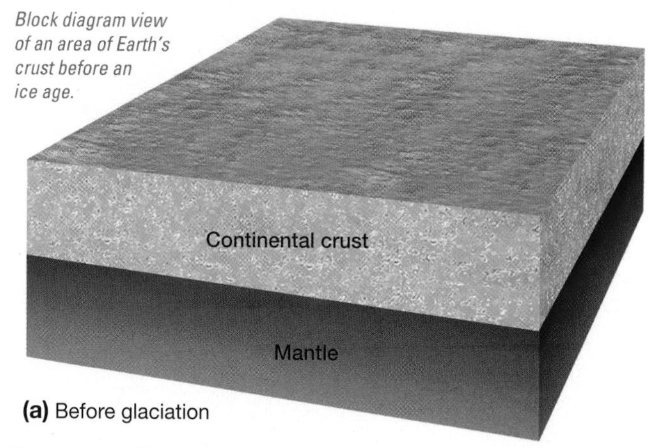
Block diagram view of an area of Earth's crust before an ice age.
Continental crust
Mantle
(a) Before glaciation

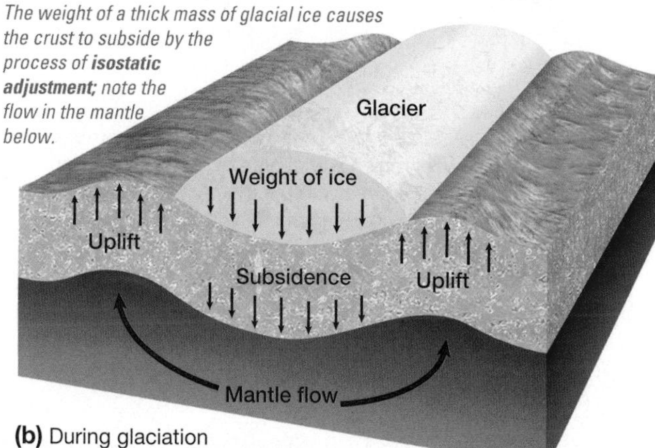
The weight of a thick mass of glacial ice causes the crust to subside by the process of isostatic adjustment; note the flow in the mantle below.
Glacier
Weight of ice
Uplift
Subsidence
Uplift
Mantle flow
(b) During glaciation

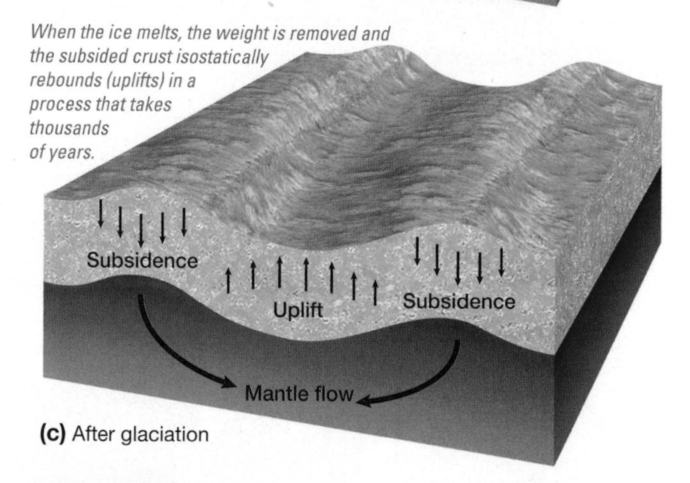
When the ice melts, the weight is removed and the subsided crust isostatically rebounds (uplifts) in a process that takes thousands of years.
Subsidence
Uplift
Subsidence
Mantle flow
(c) After glaciation

**EXPLORING DATA** ▲ **Identify at least two specific places in the world where would you expect to see evidence of isostatic adjustment.**

**Figure 10.18 Isostatic adjustment caused by glacial ice.**

**Animation**
Glacial Isostasy
http://goo.gl/vz3ZDT

The combination of tectonic and eustatic changes in sea level is very complex, so it is difficult to classify coastal regions as purely emergent or submergent. In fact, most coastal areas show evidence of *both* submergence and emergence in the recent past. Evidence suggests, however, that until recently, sea level has experienced only minor changes as a result of melting glacial ice during the past 3000 years.

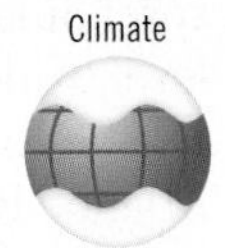


More recently, there has been a documented sea level rise as a result of human-caused climate change. This topic is discussed in Chapter 16, "The Oceans and Climate Change."

**RECAP Sea level is affected by the movement of land and changes in seawater volume or ocean basin capacity. Sea level has changed dramatically in the past because of changes in Earth's climate.**

**CONCEPT CHECK 10.4 ▸ Discuss how changes in sea level produce emerging and submerging shorelines.**

1 Compare the causes and effects of tectonic versus eustatic changes in sea level.

2 List the two basic processes by which coasts advance seaward and list their counterparts that lead to coastal retreat.

3 Describe how an ice age affects sea level.

## 10.5 ▸ How Does Hard Stabilization Affect Coastlines?

Coastal residents continually modify coastal sediment erosion/deposition in attempts to improve or preserve their property. Structures built to protect a coast from erosion or to prevent the movement of sand along a beach are known as **hard stabilization**, or *armoring of the shore.* Hard stabilization can take many forms and often results in predictable yet unwanted outcomes.

### Groins and Groin Fields

One type of hard stabilization is a **groin** (*groin* = ground). Groins are built perpendicular to a coastline and are specifically designed to trap sand moving along the coast in longshore transport (**Figure 10.20**). They are constructed of many types of material, but the most common is large blocks of rocky material called **rip-rap**. Sometimes groins are even constructed of sturdy wood pilings (similar to a fence built out into the ocean).

Although a groin traps sand on its *upcoast side*, erosion occurs immediately downcoast of the groin because the sand that is normally found just downcoast of the groin is trapped on the groin's upcoast side. To lessen the erosion, another groin can be constructed downcoast, which in turn also creates erosion downcoast from it. More groins are needed to alleviate the beach erosion, and soon a **groin field** is created (**Figure 10.21**).

Does a groin (or a groin field) actually retain more sand on the beach? Sand eventually migrates around the end of the groin, so there is no additional sand on the beach; it is only *distributed differently*. With proper engineering and by taking into account the regional sand transport budget and seasonal wave activity, an equilibrium may be reached that allows sufficient sand to move along the coast before excessive erosion occurs downcoast from the last groin. However, some serious erosional problems have developed in many areas because of attempts to stabilize sand on the beach through the excessive use of groins.

SmartFigure 10.20 **Interference of sand movement.** Aerial view of a coastal region with a type of hard stabilization called a groin, which interferes with the movement of sand along the beach. Note how the groin modifies the shape of the beach by causing deposition of sand upcoast of the groin and erosion immediately downcoast. https://goo.gl/9xaSdT

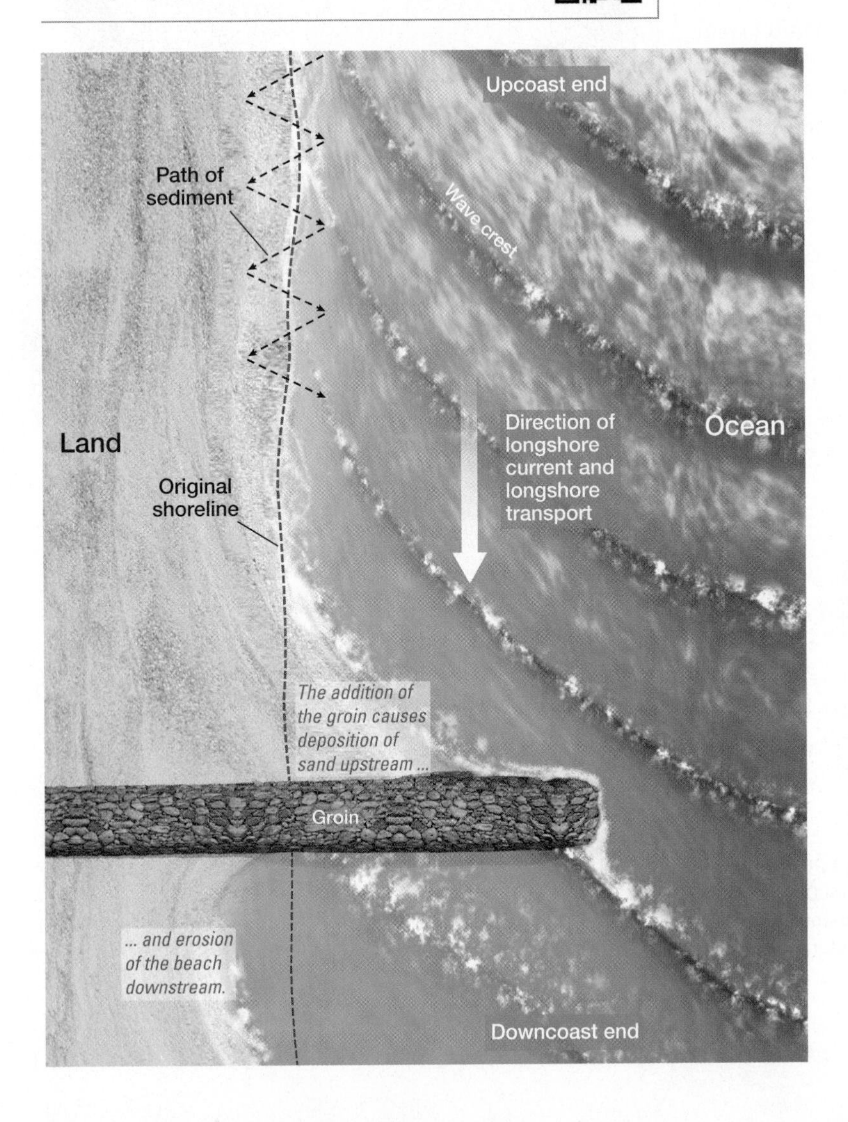

## Jetties

Another type of hard stabilization is a **jetty** (*jettee* = to project outward). A jetty is similar to a groin in that it is built perpendicular to the shore and is usually constructed of rip-rap. The purpose of a jetty, however, is to protect harbor entrances from waves, and only secondarily does it trap sand (**Figure 10.22**). Because jetties are usually built in closely spaced pairs and can be quite long, they can cause more pronounced upcoast deposition and downcoast erosion than groins (**Figure 10.23**).

## Breakwaters

A **breakwater** is a form of hard stabilization that is built parallel to a shoreline (**Figure 10.24**). **Figure 10.25** shows a breakwater that was constructed to create the harbor at Santa Barbara, California. California's longshore drift is predominantly southward, so the breakwater on the western side of the harbor accumulated sand that had migrated eastward along the coast. The beach to the west of the harbor continued to grow until finally the sand moved around the breakwater and began to fill in the harbor (Figure 10.25).

While abnormal deposition occurred to the west, erosion proceeded at an alarming rate east of the harbor. The waves east of the harbor were no greater than before, but the sand that had formerly moved down the coast was now trapped behind the breakwater.

A similar situation occurred in Santa Monica, California, where a breakwater was built to provide a boat anchorage. A bulge in the beach soon formed behind (inshore of) the breakwater, and severe erosion occurred downcoast (**Figure 10.26**). The breakwater interfered with the natural transport of sand by blocking the waves that used to keep the sand moving. If something was not done to put energy back into the system, the breakwater would soon be attached by a tombolo of sand, and further erosion downcoast might destroy coastal structures.

In Santa Barbara and Santa Monica, dredging was used to compensate for erosion downcoast from the breakwater and to keep the harbor or anchorage from filling with sand. Sand dredged from behind the breakwater is pumped down the coast so it can re-enter the longshore drift and replenish the eroded beach.

**EXPLORING DATA** ▲ **Based on evidence in the photo, how can the primary direction of longshore current be determined?**

Figure 10.21 **Groin field.** A series of groins has been built along the shoreline near the English seaside resort town of Brighton, southern England, in an attempt to trap sand, altering the distribution of sand on the beach. The view is toward the northwest, and the primary direction of longshore current is toward the bottom right of the photo (toward the east).

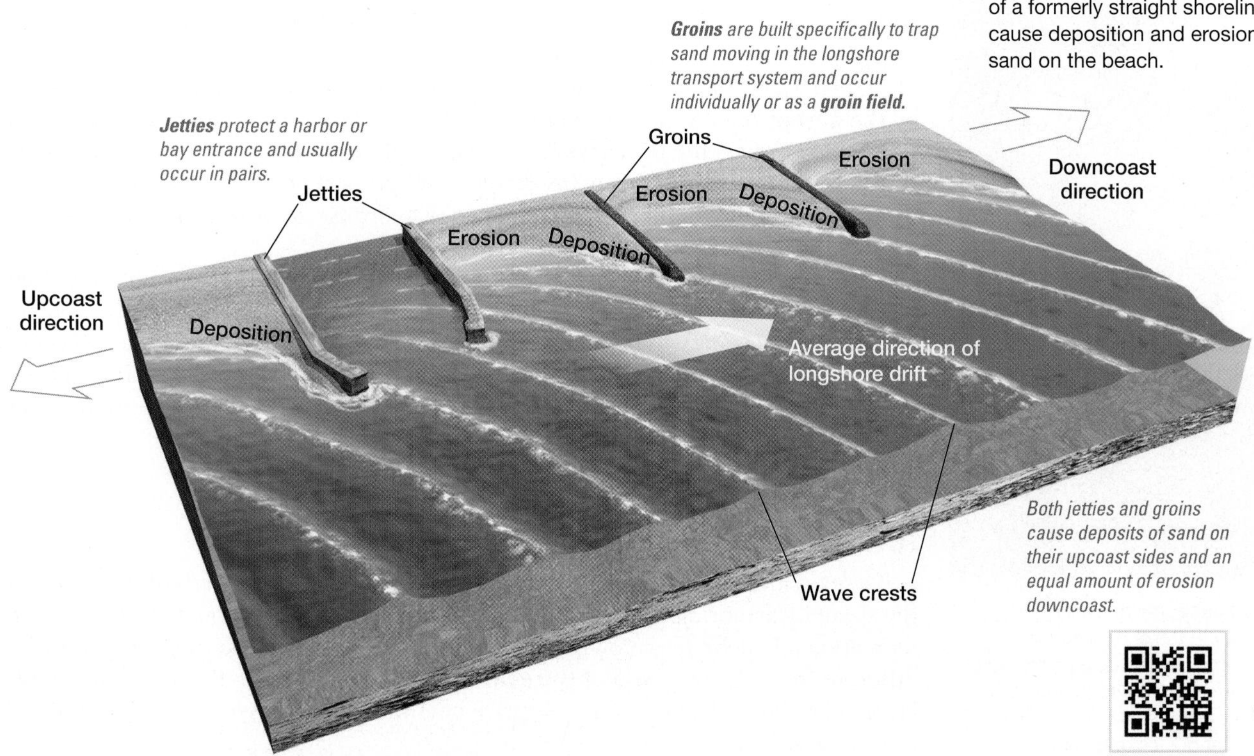
*Groins are built specifically to trap sand moving in the longshore transport system and occur individually or as a groin field.*
*Jetties protect a harbor or bay entrance and usually occur in pairs.*
Jetties
Groins
Erosion
Deposition
Erosion
Deposition
Erosion
Deposition
Upcoast direction
Downcoast direction
Average direction of longshore drift
Wave crests
*Both jetties and groins cause deposits of sand on their upcoast sides and an equal amount of erosion downcoast.*

Figure 10.22 **Effect of jetties and groins.** Block diagram of a formerly straight shoreline showing how jetties and groins cause deposition and erosion and thus alter the distribution of sand on the beach.

**Animation**
Coastal Stabilization Structures
https://goo.gl/Fb2StT

Figure 10.23 **Jetties at Santa Cruz Harbor, California.** These jetties protect the inlet to Santa Cruz Harbor and interrupt the flow of sand, which is toward the right (southward). Notice the buildup of sand to the left (upcoast) of the jetties and the corresponding erosion to the right (downcoast).

The dredging operation has stabilized the situation in Santa Barbara, but at considerable (and ongoing) expense. In Santa Monica, dredging was conducted until the breakwater was largely destroyed during winter storms in 1982–1983. Shortly thereafter, wave energy was able to move sand along the coast again, and the system was restored to normal conditions. When people interfere with natural processes in the coastal region, they must provide the energy needed to replace what they have misdirected through modification of the shore environment.

Figure 10.24 **Breakwater at Nea Fokea, northern Greece.** Breakwaters are built parallel to the shore and are composed of rocky, blocky material that is piled up a meter or so (several feet) above sea level. They are designed to reduce wave energy, thus creating a protected area of quiet water inshore of (behind) the breakwater. View is from shore looking north.

## Seawalls

One of the most destructive types of hard stabilization is a **seawall** (Figure 10.27), which is built parallel to the shore, along the landward side of the berm. The purpose of a seawall is to armor the coastline and protect landward developments from ocean waves.

Once waves begin breaking against a seawall, however, turbulence generated by the abrupt release of wave energy quickly erodes the sediment on its seaward side, which can eventually cause it to collapse into the surf (Figure 10.27). In many cases where seawalls have been used to protect property on barrier islands, the seaward slope of the island beach has steepened and the rate of erosion has increased, causing the destruction of the recreational beach.

A well-designed seawall may last for many decades, but the constant pounding of waves eventually takes its toll (Figure 10.28). In the long run, the cost of repairing or replacing seawalls will be more than the property is worth, and the sea will claim more of the coast through the natural processes of erosion. It's just a matter of time for homeowners who live too close to the coast, many of whom are gambling that their houses won't be destroyed in their lifetimes.

## Alternatives to Hard Stabilization

Is it a good idea to preserve the houses of a few people who have built too close to the shore by armoring the coast with hard stabilization even though it destroys the recreational beach? If you own coastal property, your response would probably be different from the response of the general beachgoing public. Because hard stabilization has been shown to have negative environmental consequences, alternatives have been sought.

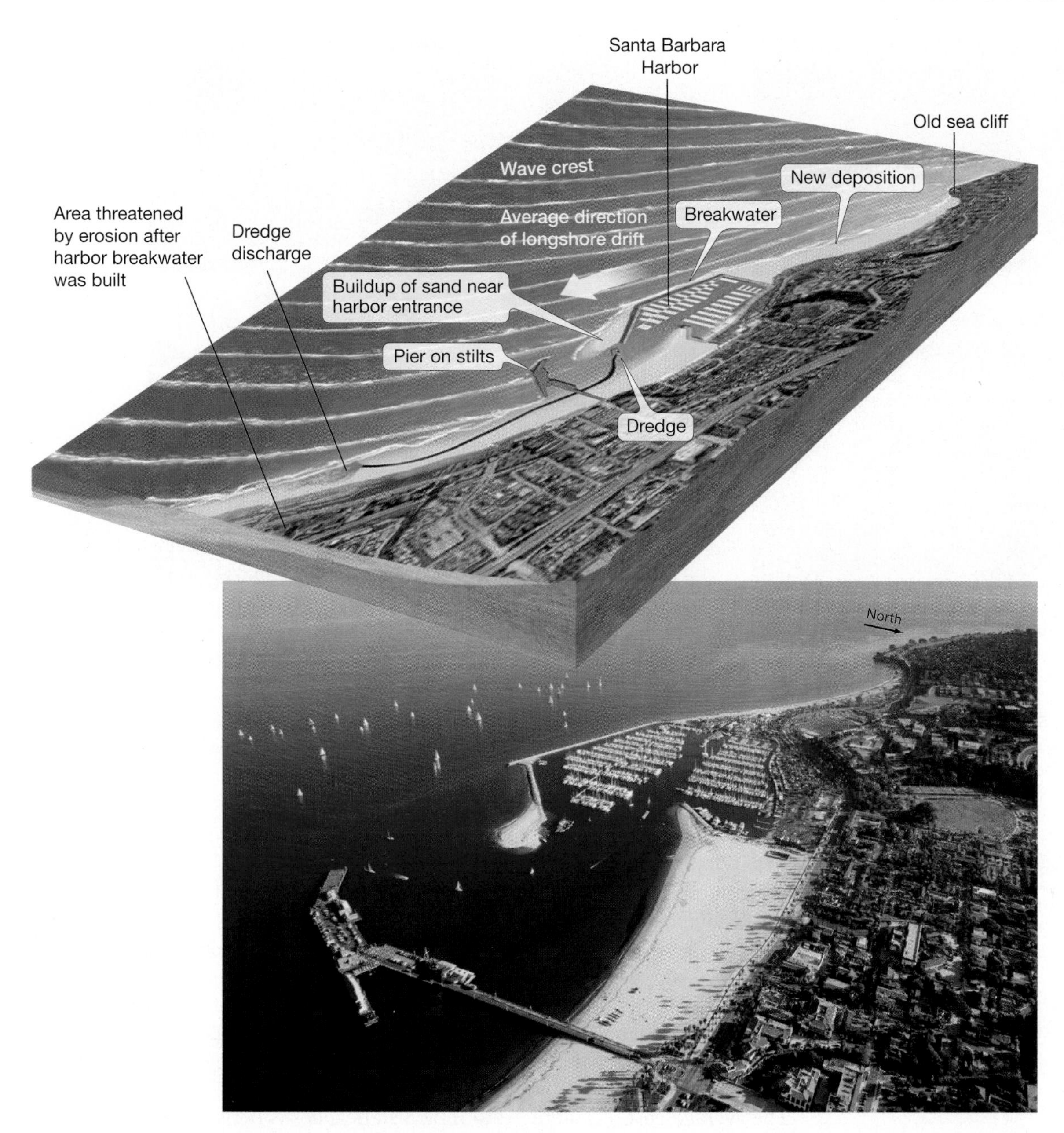


Figure 10.25 **Breakwater at Santa Barbara Harbor, California.**
**(a)** Illustration showing an aerial view of Santa Barbara Harbor and its shore-connected breakwater, which interferes with longshore drift and has created a broad beach. As the beach extended around the breakwater into the harbor, the harbor was in danger of being closed off by accumulating sand. As a result, dredging operations were initiated to move sand from the harbor downcoast, where it helped reduce coastal erosion. **(b)** Photograph of Santa Barbara Harbor from the air, looking west.

**CONSTRUCTION RESTRICTIONS** One of the simplest alternatives to the use of hard stabilization is to restrict construction in areas prone to coastal erosion. Unfortunately, this is becoming less and less an option as coastal regions experience population increases and governments increase the risk of damage and injuries because of programs like the *National Flood Insurance Program* (*NFIP*). Since its inception in 1968, NFIP has paid out billions of dollars in federal subsidies to repair or replace high-risk coastal structures. As a result, NFIP has actually *encouraged* construction in exactly the unsafe locations it was designed to prevent! However, changes in regulations of the Federal Emergency Management Agency (FEMA), which oversees NFIP, are intended to curb this practice. In addition, many homeowners spend large amounts of money rebuilding structures and fortifying their property.

**BEACH REPLENISHMENT** Another alternative to hard stabilization is **beach replenishment** (also called **beach nourishment**), in which sand is added to the beach to replace lost sediment (**Figure 10.29**). Although rivers naturally supply sand to

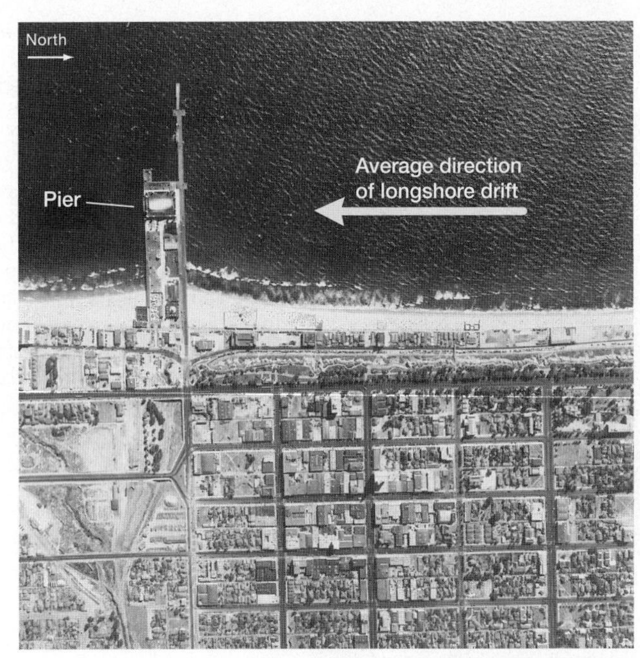


(a) The shoreline and pier at Santa Monica as it appeared in September 1931, before the breakwater was constructed in 1933. Note that the pier is on stilts and thus does not affect longshore transport.

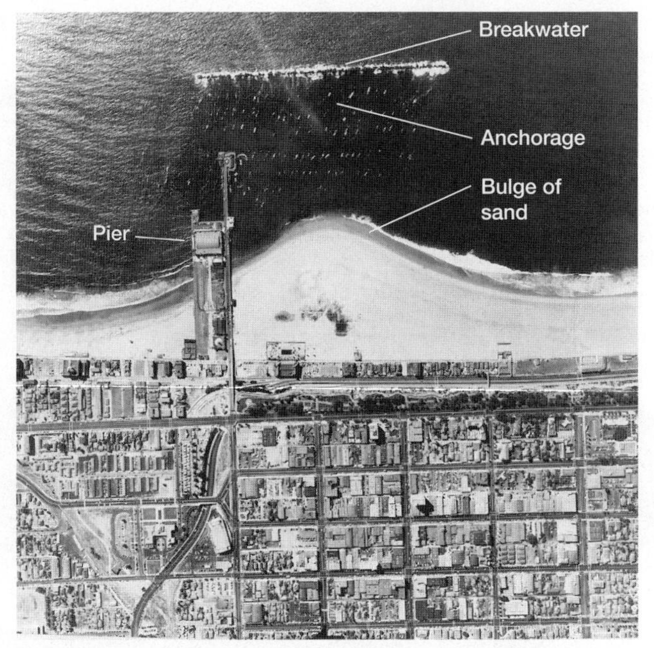


(b) The same area in 1949, showing that the construction of the breakwater to create a boat anchorage disrupted the longshore transport of sand and caused a bulge of sand on the beach in the wave shadow behind the breakwater.

Figure 10.26 **Breakwater at Santa Monica, California.** Paired aerial photos of the Santa Monica pier and shoreline **(a)** before a breakwater was built and **(b)** after the breakwater, showing how the breakwater caused a bulge of sand on the beach. After the breakwater was destroyed by waves in 1983, the bulge disappeared and the shoreline returned to a straight shoreline.

most beaches, dams on rivers restrict the sand supply that would normally arrive at beaches. When inland dams are built, their effects on beaches far downstream are rarely considered. It's not until beaches begin disappearing that the rivers are seen as parts of much larger systems that operate along the coast.

Beach replenishment is expensive, however, because huge volumes of sand must be continually supplied to the beach. The cost of beach replenishment depends on the type and quantity of material placed on the beach, how far the material must be transported, and how it is to be distributed on the beach. Most sand used for replenishment comes from offshore areas, but sand that is dredged from nearby rivers, drained dams, harbors, and lagoons is also used.

The average cost of sand used to replenish beaches is between $5 and $10 per 0.76 cubic meter (1 cubic yard). In comparison, a typical top-loading trash dumpster holds about 2.3 cubic meters (3 cubic yards) of material, and a typical dump truck holds a volume of about 45 cubic meters (60 cubic yards). The drawbacks of beach replenishment projects are that a huge volume of sand must be moved, and that new sand must be supplied on a regular basis. These problems often cause replenishment projects to exceed the monetary limits of what can be reasonably accomplished. For example, a small beach replenishment project of several hundred cubic meters can cost around $10,000 per year. Larger projects—requiring several thousand cubic meters of sand—cost several million dollars per year.

**RELOCATION** U.S. coastal policy has recently shifted from defending coastal property in high-hazard areas to removing structures and letting nature reclaim the beach. This approach, called **relocation**, involves moving structures to safer locations as they become threatened by erosion. One example of the successful use of this technique is the relocation of the Cape Hatteras Lighthouse in North Carolina (see Mastering Oceanography **Web Diving Deeper 10.1**). Relocation, if done wisely, can allow humans to live in balance with the natural processes that continually modify beaches.

## STUDENTS SOMETIMES ASK . . .

***I have the opportunity to live in a house at the edge of a coastal cliff where there is an incredible view along the entire coast. Is it safe from coastal erosion?***

Based on what you've described, most certainly not! Geologists have long known that cliffs are naturally unstable. Even if cliffs appear to be stable (or have been stable for a number of years), they can be severely damaged during just one significant storm.

The most common cause of coastal erosion is direct wave attack, which undermines the support and causes the cliff to fail. You might want to check the base of the cliff and examine the local bedrock to determine for yourself if you think it will withstand the pounding of powerful storm waves that can move rocks weighing several tons. Other dangers include drainage runoff, weaknesses in the bedrock, slumps and landslides, seepage of water through the cliff, and even burrowing animals. Although all states enforce a setback from the edge of the cliff for all new buildings, sometimes that isn't enough because large sections of "stable" cliffs can fail all at once. For instance, several city blocks of real estate have been eroded from the edge of cliffs during the past 100 years in some areas of Southern California. Even though the view sounds outstanding, you may find out the hard way that the house is built a little *too* close to the edge of a cliff!

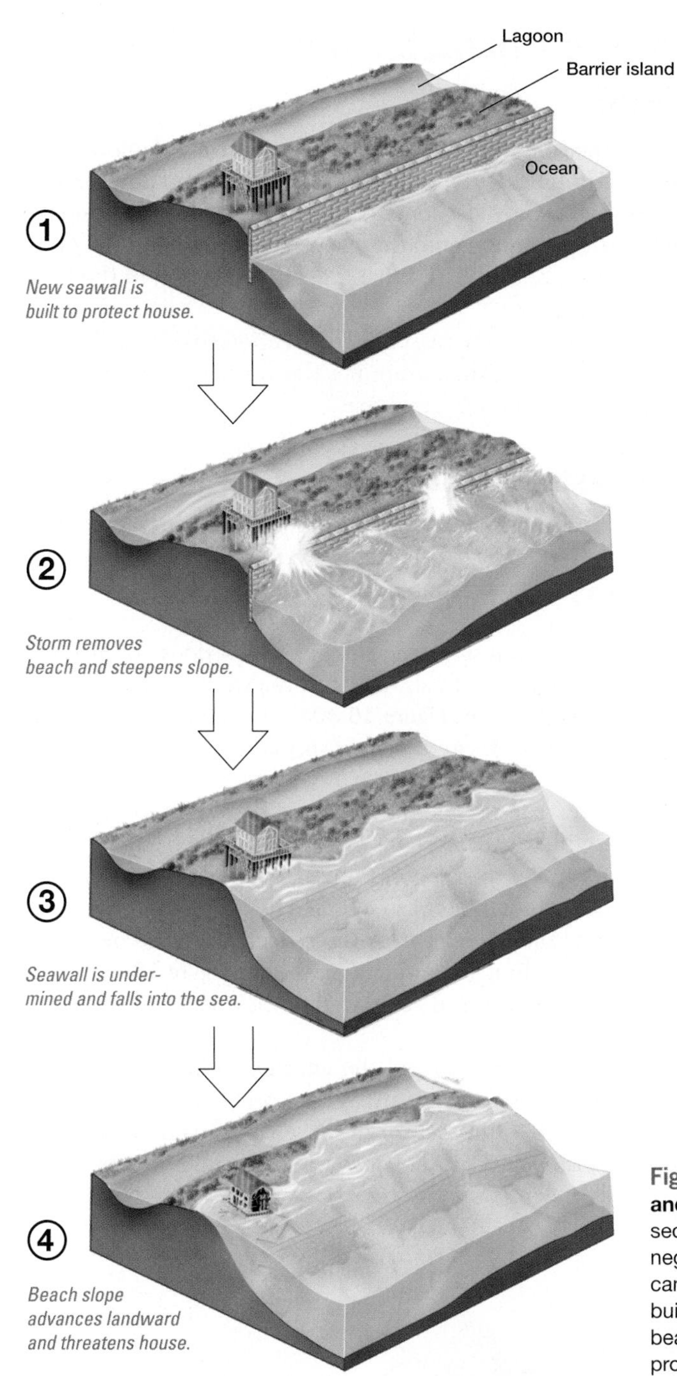
Lagoon
Barrier island
Ocean
①
New seawall is built to protect house.
②
Storm removes beach and steepens slope.
③
Seawall is undermined and falls into the sea.
④
Beach slope advances landward and threatens house.

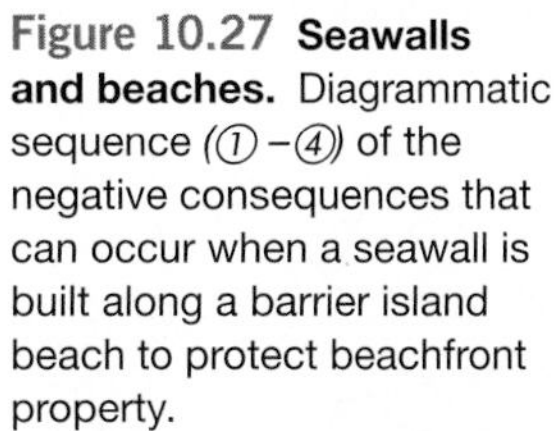

Figure 10.27 **Seawalls and beaches.** Diagrammatic sequence (①–④) of the negative consequences that can occur when a seawall is built along a barrier island beach to protect beachfront property.

Figure 10.28 **Seawall damage.** A seawall in Solana Beach, California, that has been damaged by waves and needs repair. Although seawalls appear to be sturdy, they can be destroyed by the continual pounding of high-energy storm waves. In addition, high energy waves sometimes carry driftwood or logs that pound seawalls with the strength of battering rams.

Dredged sand exiting pipe

Figure 10.29 **Beach replenishment.** Beach replenishment projects, such as this one in Carlsbad, California, are used to widen beaches. Beach replenishment involves dredging sand from offshore or coastal locations, pumping the wet slurry through a pipe (*lower right*), and spreading it across the beach.

## CONCEPT CHECK 10.5 ▸ Describe the types of hard stabilization and evaluate various alternatives.

1 List the types of hard stabilization and describe what each is intended to do.

2 Overall, does a groin add any additional sand to the beach? Explain.

3 Why do groins often multiply to form a groin field?

4 When a breakwater was built in Santa Monica, what unexpected problem occurred? What was done to alleviate the problem (before the breakwater was destroyed by waves)?

5 Describe alternatives to hard stabilization, including potential drawbacks of each.

**RECAP** **Hard stabilization includes groins, jetties, breakwaters, and seawalls, all of which alter the coastal environment and result in changes in the shape of the beach. Alternatives to hard stabilization include construction restrictions, beach replenishment, and relocation.**

## STUDENTS SOMETIMES ASK . . .

***I've seen reports of severe beach erosion along various U.S. coasts and that beaches are virtually nonexistent in some places. Is this a normal cycle that will self-correct, or are these beaches permanently gone from sea level rise or other environmental causes?***

Based on U.S. Environmental Protection Agency estimates, 80%–90% of the sandy beaches along America's coastlines are eroding—and have been for decades. In many of these cases, individual beaches may be losing only a few inches per year, but in some cases the problem is much worse. For example, the outer coast of Louisiana, which is referred to by experts as the erosion "hotspot" of the United States, is losing some 15 meters (50 feet) of beach every year.

Once a beach is lost, it doesn't automatically get replaced—hence the abundance of expensive beach replenishment projects. Of particular concern is the effect of climate change, which not only causes sea level to rise but also increases the severity and possibly the frequency of harsh storms, all of which cause even more beach erosion. While sea level rise sets the conditions for landward displacement of the shore, coastal storms supply the energy to do the work of moving sand offshore and/or downshore from a beach.

# 10.6 ▶ What are the Characteristics and Types of Coastal Waters?

Just offshore of beaches are **coastal waters**, which are the relatively shallow-water areas that adjoin continents or islands. If the continental shelf is broad and shallow, coastal waters can extend several hundred kilometers from land. If the continental shelf has significant relief or drops rapidly onto the deep-ocean basin, on the other hand, coastal waters will occupy a relatively thin band near the margin of the land. Beyond coastal waters lies the *open ocean*.

Coastal waters are important for many reasons. This section first describes the unique characteristics of coastal waters, then examines the various types of coastal waters, including estuaries, lagoons, and marginal seas.

## Characteristics of Coastal Waters

Because of their proximity to land, coastal waters are directly influenced by processes that occur on or near land. River runoff and tidal currents, for example, have a far more significant effect on coastal waters than on the open ocean.

**SALINITY** Freshwater is less dense than seawater, so river runoff does not mix well with seawater along the coast. Instead, the freshwater forms a wedge at the surface, which creates a well-developed **halocline** (**Figure 10.30a**). (Recall that a halocline [*halo* = salt, *cline* = slope] is a layer of rapidly changing salinity, as discussed in Chapter 5.)

Other processes can create strong haloclines in coastal regions as well. For example, prevailing offshore winds can increase the salinity in coastal regions through evaporation. As winds travel over a continent, they usually lose most of their moisture. When these dry winds reach the ocean, they typically evaporate considerable amounts of water as they move across the sea surface. The increased evaporation rate increases surface salinity, creating a halocline (**Figure 10.30b**). The gradient of the halocline, however, is reversed compared to the one developed from the input of freshwater (Figure 10.30a).

In general, however, freshwater runoff from the continents lowers the salinity of coastal regions compared to the open ocean. Where precipitation on land is mostly rain, river runoff peaks in the rainy season. Where runoff is due mainly to melting snow and ice, on the other hand, runoff always peaks in summer. If the

**SmartFigure 10.30 Salinity variation in the coastal ocean.** Diagrammatic views showing various factors that affect the salinity of coastal oceans. Red curves represent vertical salinity profiles. https://goo.gl/eVkzha

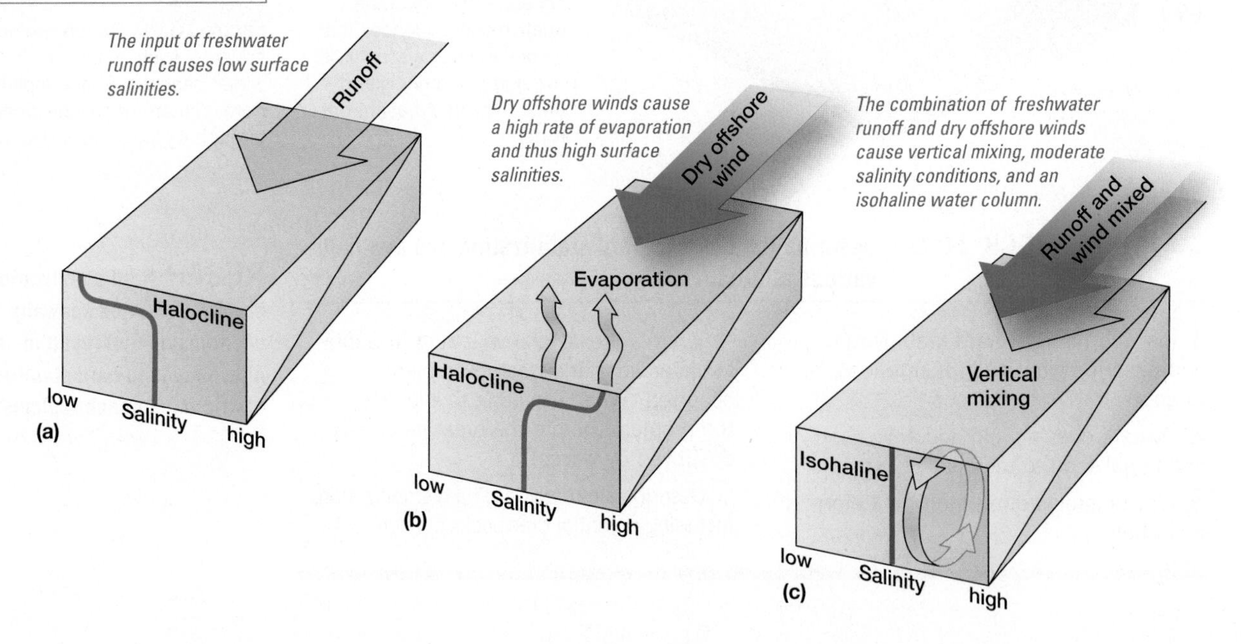
The input of freshwater runoff causes low surface salinities.
Runoff
Halocline
low
Salinity
high
(a)
Dry offshore winds cause a high rate of evaporation and thus high surface salinities.
Dry offshore wind
Evaporation
Halocline
low
Salinity
high
(b)
The combination of freshwater runoff and dry offshore winds cause vertical mixing, moderate salinity conditions, and an isohaline water column.
Runoff and wind mixed
Vertical mixing
Isohaline
low
Salinity
high
(c)

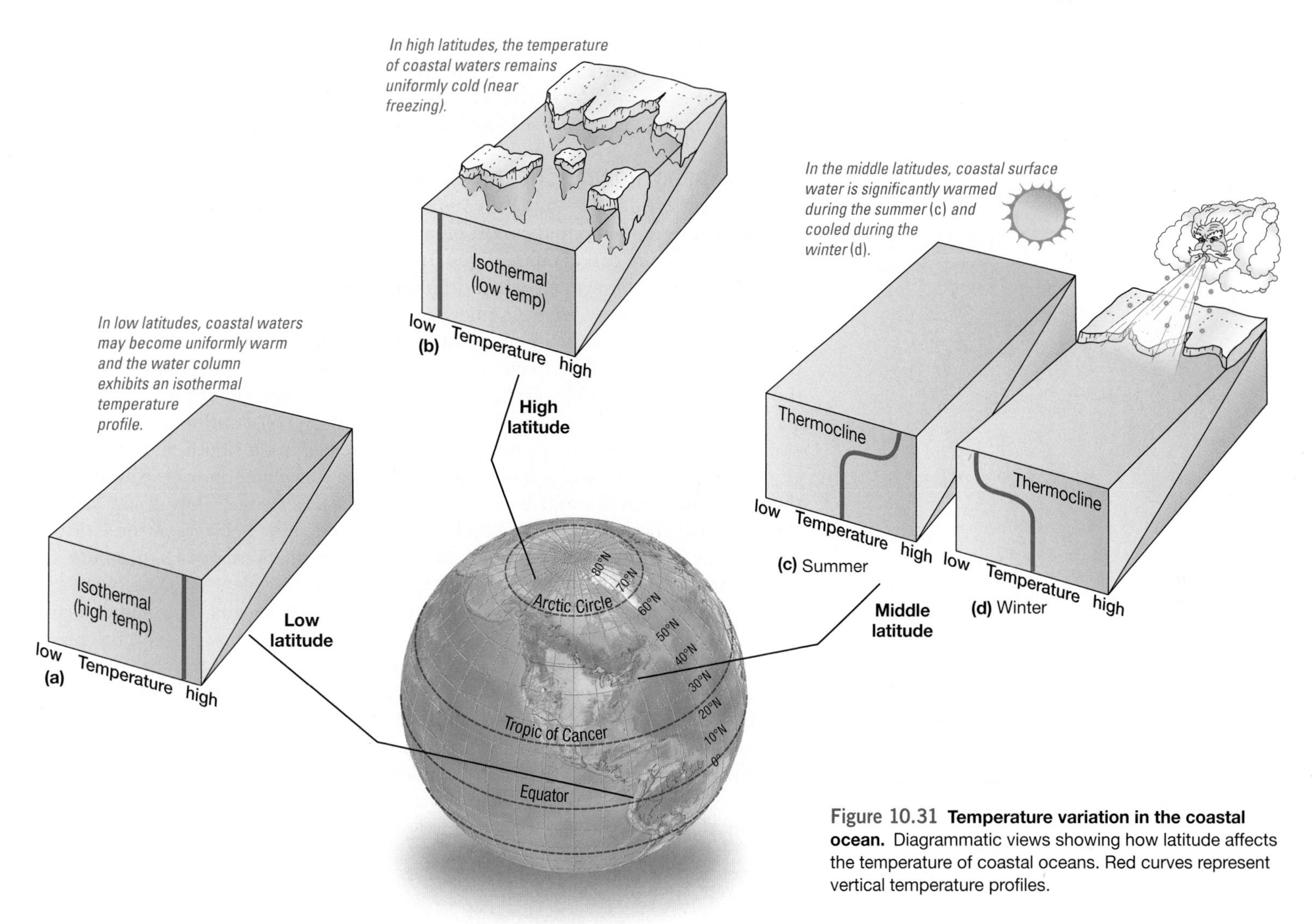


Figure 10.31 **Temperature variation in the coastal ocean.** Diagrammatic views showing how latitude affects the temperature of coastal oceans. Red curves represent vertical temperature profiles.

water is shallow enough, tidal mixing causes freshwater to mix with seawater, thus reducing the salinity of the water column (**Figure 10.30c**). In this case, there is no halocline; instead, the water column is **isohaline** (*iso* = same, *halo* = salt).

**TEMPERATURE** In low-latitude coastal regions, where circulation with the open ocean is restricted, surface waters are prevented from mixing thoroughly, so sea surface temperatures may approach 45°C (113°F) (**Figure 10.31a**). Alternatively, sea ice forms in many high-latitude coastal areas where water temperatures are uniformly cold—generally lower than −2°C (28.4°F) (**Figure 10.31b**). In both low- and high-latitude coastal waters, **isothermal** (*iso* = same, *thermo* = heat) conditions prevail.

Surface temperatures in middle-latitude coastal regions are coolest in winter and warmest in late summer. A strong **thermocline** may develop from surface water being warmed during the summer (**Figure 10.31c**) and cooled during the winter (**Figure 10.31d**). (Recall that a thermocline [*thermo* = heat, *cline* = slope] is a layer of rapidly changing temperature, as discussed in Chapter 5.) In summer, very high-temperature surface water may form a relatively thin layer. Vertical mixing reduces the surface temperature by distributing the heat through a greater volume of water, thus pushing the thermocline deeper and making it less pronounced. In winter, cooling increases the density of surface water, which causes it to sink.

Prevailing offshore winds can significantly affect surface water temperatures. These winds are relatively warm during the summer, so they increase the ocean surface temperature and seawater evaporation. During winter, they are much cooler than the ocean surface, so they absorb heat and cool surface water near shore. Mixing from strong winds may drive the thermoclines in Figure 10.31c and 10.31d deeper and even mix the entire water column, producing isothermal conditions. Tidal currents can also cause considerable vertical mixing in shallow coastal waters.

**COASTAL GEOSTROPHIC CURRENTS** Recall from Chapter 7 that *geostrophic* (*geo* = earth, *strophio* = turn) *currents* move in a circular path around the middle of a current gyre. Wind and runoff create geostrophic currents in coastal waters, too, where they are called **coastal geostrophic currents**.

Where winds blow in a certain direction parallel to a coastline, they transport water toward the coast, where it piles up along the shore. Gravity eventually pulls this water back toward the open ocean. As it runs downslope away from the shore, the Coriolis effect causes it to curve to the right in the Northern Hemisphere and to the left in the Southern Hemisphere. Thus, in the Northern Hemisphere, the coastal geostrophic current curves *northward* on the western coast and *southward* on the eastern coast of continents. These currents are reversed in the Southern Hemisphere.

A high-volume runoff of freshwater produces a surface wedge of freshwater that thins away from the shore (**Figure 10.32**). This causes a surface flow of low-salinity water toward the open ocean, which curves to the right because of the Coriolis effect in the Northern Hemisphere and to the left in the Southern Hemisphere.

Coastal geostrophic currents are variable because they depend on the wind and the amount of runoff for their strength. If the wind is strong and the volume of runoff is high, then the currents are relatively strong. They are bounded on the ocean side by the steadier eastern or western boundary currents of subtropical gyres.

An example of a coastal geostrophic current is the **Davidson Current**, which develops along the coast of Washington and Oregon (Figure 10.32). Although the current is present year-round, it is more strongly developed during the rainy winter season when high volumes of runoff combine with strong southwesterly winds to produce a relatively strong northward-flowing current. It flows between the shore and the southward-flowing California Current.

**RECAP The shallow coastal ocean adjoins land and experiences changes in salinity and temperature that are more dramatic than the open ocean. Coastal geostrophic currents can also develop.**

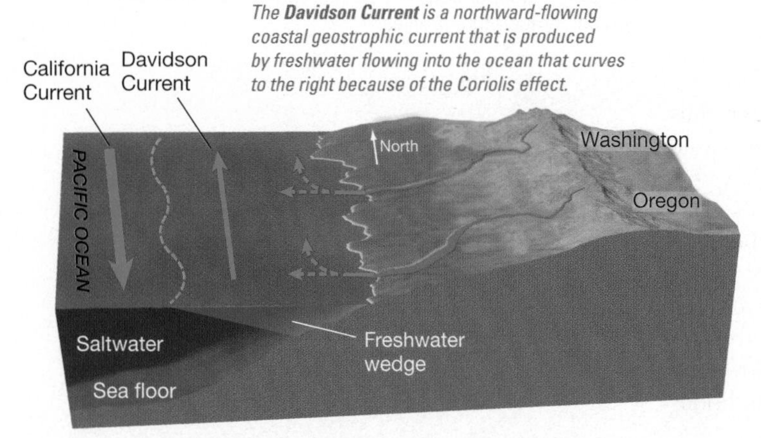


**Figure 10.32 Davidson coastal geostrophic current.** Block diagram of the Pacific Northwest coast showing that runoff from Oregon and Washington produces a freshwater wedge (*light blue*) that thins away from shore. This causes a surface flow of low-salinity water toward the open ocean, which is acted upon by the Coriolis effect and curves to the right, producing the Davidson Current that flows close to shore and in the opposite direction of the California Current. Note that the Davidson Current is more strongly developed during the winter rainy season when there is a large amount of river runoff.

## Estuaries

An **estuary** (*aestus* = tide) is a partially enclosed coastal body of water in which freshwater runoff from a river dilutes the input of salty ocean water. Estuaries are marine environments whose pH, salinity, temperature, and water levels vary, depending on the mixing between the river that feeds the estuary and the ocean from which it derives its salinity. The most common example of an estuary is a river mouth, where a river empties into the sea. Other coastal bodies of water such as bays, inlets, gulfs, and sounds are considered estuaries, too.

The mouths of large rivers form the most economically significant estuaries because many are seaports, centers of ocean commerce, and important commercial fisheries. Examples include Baltimore, New York, San Francisco, Buenos Aires, London, Tokyo, and many others.

**ORIGIN OF ESTUARIES** The estuaries of today exist because sea level has risen approximately 120 meters (400 feet) since major continental glaciers began melting about 18,000 years ago. As described in Section 10.4, these glaciers covered portions of North America, Europe, and Asia during the Pleistocene Epoch, which is also referred to as the *Ice Age*.

Four major types of estuaries can be identified based on their geologic origin (**Figure 10.33**):

1. A **coastal plain estuary** forms as sea level rises and floods existing river valleys. These estuaries, such as the Chesapeake Bay in Maryland and Virginia, are called *drowned river valleys* (Figure 10.33a).

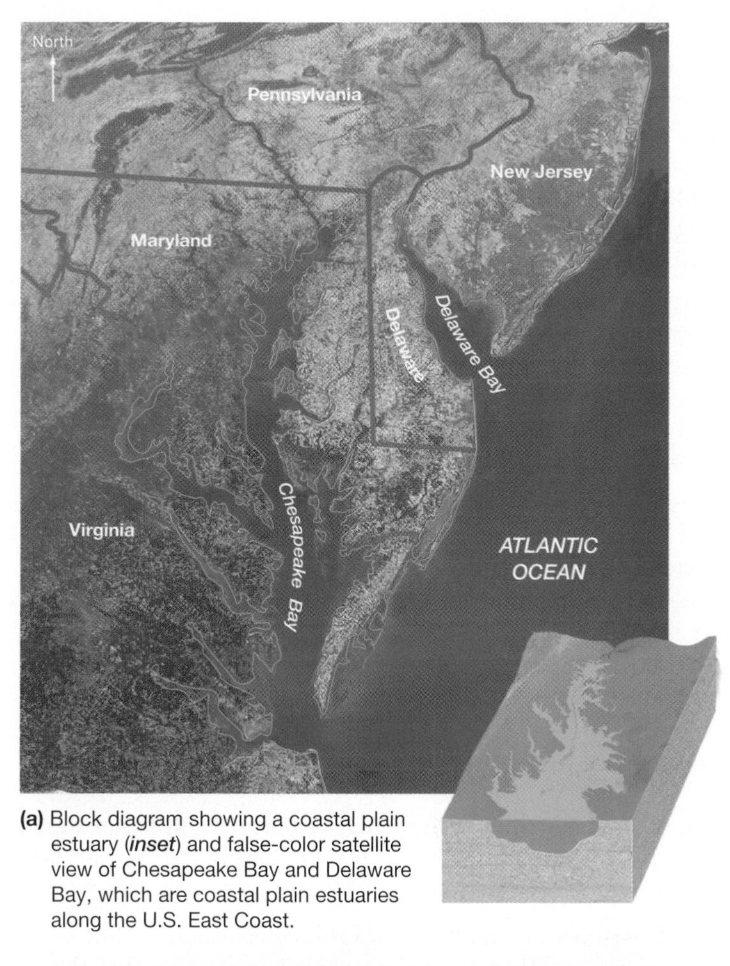


**(a)** Block diagram showing a coastal plain estuary (*inset*) and false-color satellite view of Chesapeake Bay and Delaware Bay, which are coastal plain estuaries along the U.S. East Coast.

**(b)** Block diagram showing a glacially carved fjord (*inset*) and aerial view of an Alaskan fjord with an active glacier that extends into the upper part of the estuary. Fjords are steep-sided, deep, glacially formed estuaries that are flooded by the sea.

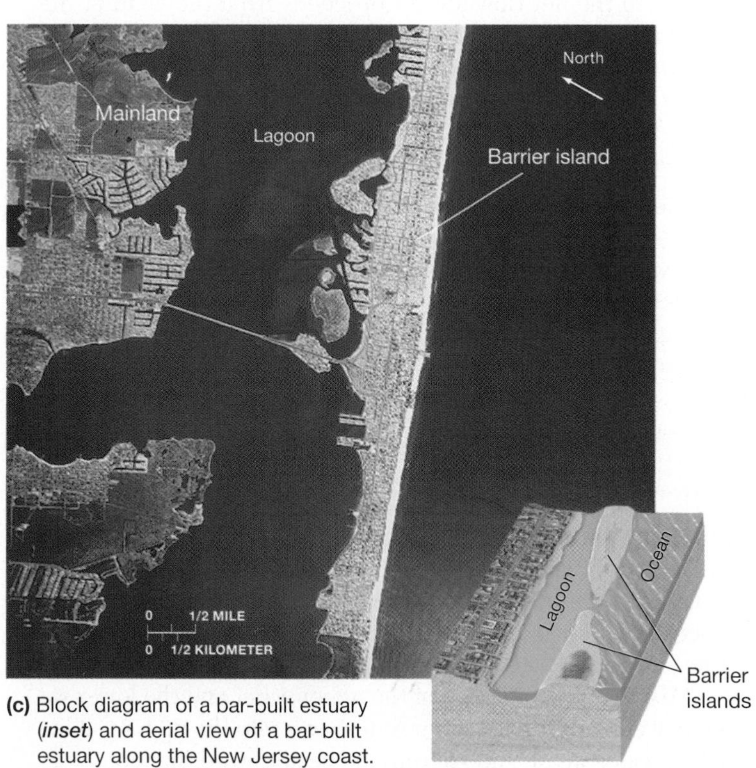


**(c)** Block diagram of a bar-built estuary (*inset*) and aerial view of a bar-built estuary along the New Jersey coast.

**(d)** Block diagram of a tectonic estuary (*inset*) and aerial view of California's San Francisco Bay, which was created by downdropping between two faults (*red lines*).

Figure 10.33 **Classifying estuaries by geologic setting.** Block diagrams and corresponding photos of the four types of estuaries, based on geologic setting.

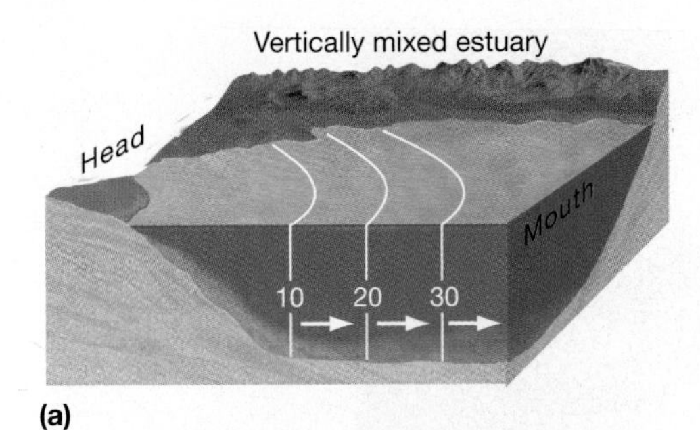


(a)

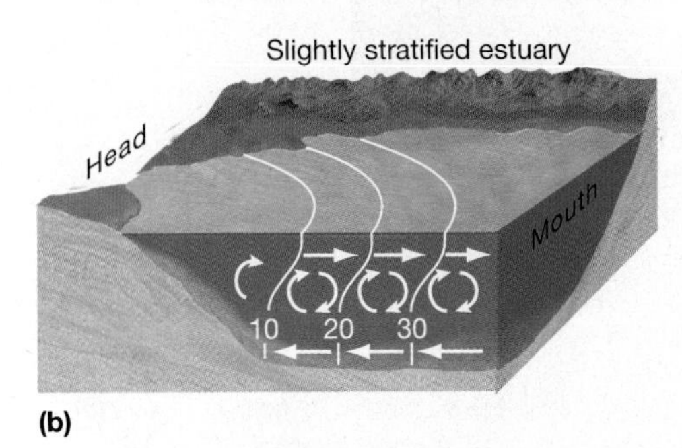


(b)

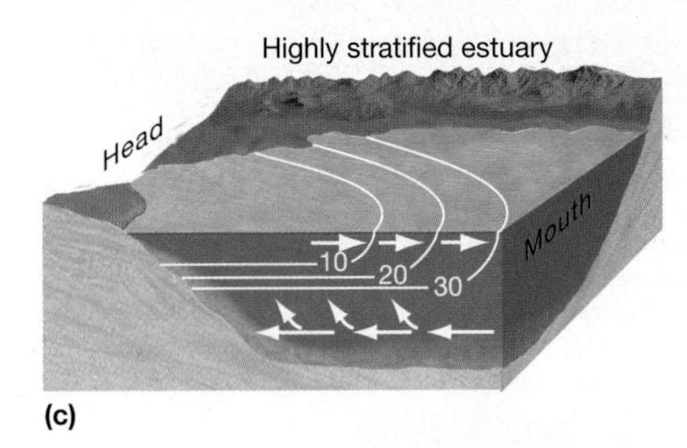


(c)

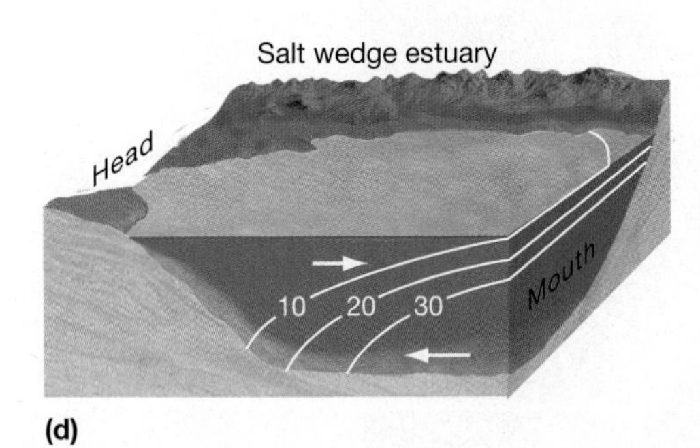


(d)

**SmartFigure 10.34 Classifying estuaries by mixing.** Block diagrams of the four types of estuaries, based on mixing. Numbers represent salinity in ‰; arrows indicate flow directions. https://goo.gl/EsQKKT

2. A **fjord** forms as sea level rises and floods a glaciated valley. (The Norwegian term *fjord* is pronounced "FEE-yord" and means a long, narrow sea inlet bordered by steep glacially-carved cliffs.) Water-carved valleys have V-shaped profiles, but fjords are U-shaped valleys with steep walls. Commonly, a shallowly submerged glacial deposit of debris (called a *moraine*) is located near the ocean entrance, marking the farthest extent of the glacier. Fjords are common along the coasts of Alaska, Canada, New Zealand, Chile, and Norway (Figure 10.33b).
3. A **bar-built estuary** is shallow and is separated from the open ocean by sand bars that are deposited parallel to the coast by wave action. Lagoons that separate *barrier islands* from the mainland are bar-built estuaries. They are very common along the U.S. Gulf Coast and East Coast (see Figure 10.13). Examples include Laguna Madre in Texas and Pamlico Sound in North Carolina (Figure 10.33c).
4. A **tectonic estuary** forms when faulting or folding of rocks creates a restricted downdropped area into which the sea has flooded. California's San Francisco Bay is in part a tectonic estuary (Figure 10.33d), formed by movement along faults, including the San Andreas Fault.

**WATER MIXING IN ESTUARIES** Because freshwater from a river is less dense than seawater, the basic flow pattern in an estuary is a surface flow of less dense freshwater toward the ocean and an opposite flow below the surface of salty seawater into the estuary. Mixing takes place where these two water masses are in contact with one another.

Based on the physical characteristics of the estuary and the resulting mixing of freshwater and seawater, marine estuaries are classified into one of four main types, as shown in **Figure 10.34**:

1. A **vertically mixed estuary** or a *well-mixed estuary* is a shallow, low-volume estuary where the net flow always proceeds from the head of the estuary toward its mouth. Salinity at any point in the estuary is uniform from surface to bottom because river water mixes evenly with ocean water at all depths. Salinity simply increases from the head to the mouth of the estuary, as shown in Figure 10.34a. Salinity lines curve at the edge of the estuary because the Coriolis effect influences the inflow of seawater.
2. A **slightly stratified estuary** or a *partially mixed estuary* is a somewhat deeper estuary in which salinity increases from the head to the mouth at any depth, as in a vertically mixed estuary. However, two water layers can be identified. One is the less-saline, less-dense upper water from the river, and the other is the more-saline, more-dense deeper water from the ocean. These two layers are separated by a zone of mixing. The circulation that develops in slightly stratified estuaries is a net surface flow of low-salinity water toward the ocean and a net subsurface flow of seawater toward the head of the estuary (Figure 10.34b), which is called an **estuarine circulation pattern**.
3. A **highly stratified estuary** or a *fjord circulation estuary* is a deep estuary in which upper-layer salinity increases from the head to the mouth, reaching a value close to that of open-ocean water. The deep-water layer has a rather uniform open-ocean salinity at any depth throughout the length of the estuary. An estuarine circulation pattern is well developed in this type of estuary (Figure 10.34c). Mixing at the interface of the upper water and the lower water creates a net movement from the deep-water mass into the upper water. Less-saline surface water simply moves from the head toward the mouth of the estuary, growing more saline as water from the deep mass mixes with it. Relatively strong haloclines develop at the contact between the upper and lower water masses.

4. A **salt wedge estuary** or *highly stratified estuary* is an estuary in which a wedge of salty water intrudes from the ocean beneath the river water. This kind of estuary is typical of the mouths of deep, high-volume rivers. No horizontal salinity gradient exists at the surface because surface water is essentially fresh throughout the length of—and even beyond—the estuary (Figure 10.34d). There is, however, a *horizontal* salinity gradient at depth and a very pronounced vertical salinity gradient (a halocline) at any location throughout the length of the estuary. This halocline is shallower and more highly developed near the mouth of the estuary.

In addition, there is one other type of marine estuary sometimes recognized by oceanographers: A **reversing estuary** or *variable-salinity estuary*, such as the type of estuary that exists at Laguna Madre, Texas, and is characterized by fluctuating salinity depending on the season (Laguna Madre is discussed in the next section under Lagoons). Within all estuaries, the predominant mixing pattern may vary with location, season, or tidal conditions. In addition, mixing patterns in real estuaries are rarely as simple as the models presented here.

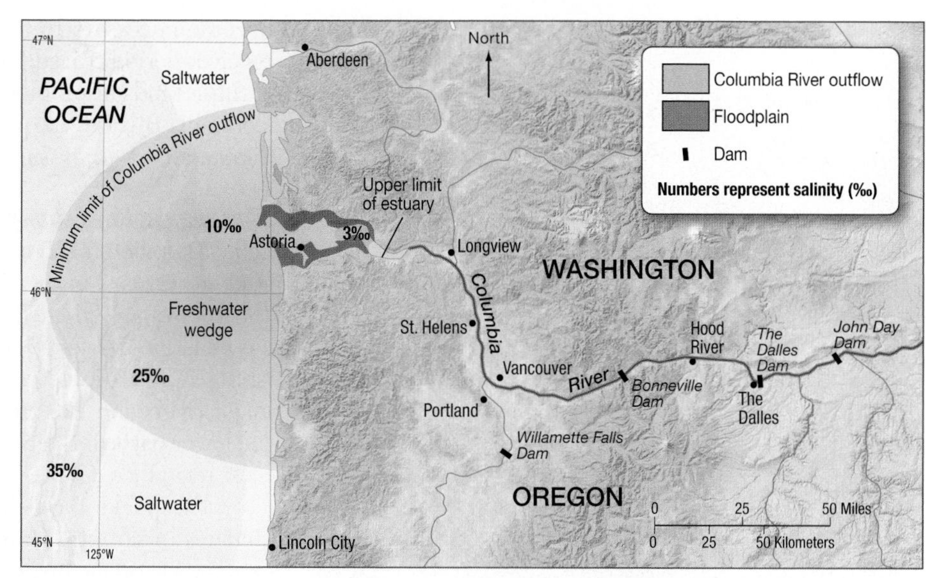


**Figure 10.35 Columbia River estuary.** The long estuary at the mouth of the Columbia River has been severely affected by interference with floodplains that have been diked, by logging activities, and—most significantly—by the construction of hydroelectric dams. The tremendous outflow of the Columbia River creates a large wedge of low-density freshwater that remains traceable far out at sea.

**RECAP Estuaries were formed by the rise in sea level after the last ice age. They can be classified based on geologic origin as coastal plain, fjord, bar-built, or tectonic estuaries. Estuaries can also be classified based on mixing as vertically mixed, slightly stratified, highly stratified, or salt wedge.**

**ESTUARIES AND HUMAN ACTIVITIES** Estuaries are important breeding grounds and protective nurseries for many marine animals, so the ecological well-being of estuaries is vital to fisheries and coastal environments worldwide. Nevertheless, estuaries support shipping, logging, manufacturing, waste disposal, and other activities that can potentially damage the environment.

Estuaries are most threatened where human population is large and expanding, but they can be severely damaged where populations are still modest, too. Development in the sparsely populated Columbia River estuary, for example, demonstrates how human activities can damage an estuary.

***Columbia River Estuary*** The Columbia River, which forms most of the border between Washington and Oregon, has a long salt wedge estuary at its entrance to the Pacific Ocean (**Figure 10.35**). The strong flow of the river and tides drive a salt wedge as far as 42 kilometers (26 miles) upstream and raise the river's water level more than 3.5 meters (12 feet). When the tide falls, the huge flow of freshwater (up to 28,000 cubic meters [1,000,000 cubic feet] per second) creates a freshwater wedge that can extend hundreds of kilometers into the Pacific Ocean.

Most rivers create floodplains along their lower courses, which have rich soil that can be used for growing crops. In the late 19th century, farmers moved onto the floodplains to establish agriculture along the Columbia River. Eventually, protective dikes were built to prevent agricultural damage done by annual flooding. Flooding brings new nutrients, however, so the dikes deprived the floodplain of the nutrients necessary to sustain agriculture.

The river has been the principal conduit for the logging industry, which dominated the region's economy through most of its modern history. Fortunately, the river's ecosystem has largely survived the additional sediment caused by clear cutting by the logging industry. The construction of more than 250 dams along the river and its tributaries, on the other hand, has permanently altered the river's ecosystem. Many of these dams, for example, do not have salmon ladders, which help fish "climb" in short vertical steps around the dams to reach their spawning grounds at the headwaters of their home streams.

Even though the dams have caused a multitude of problems, they do provide flood control, electrical power, and a dependable source of water, all of which have

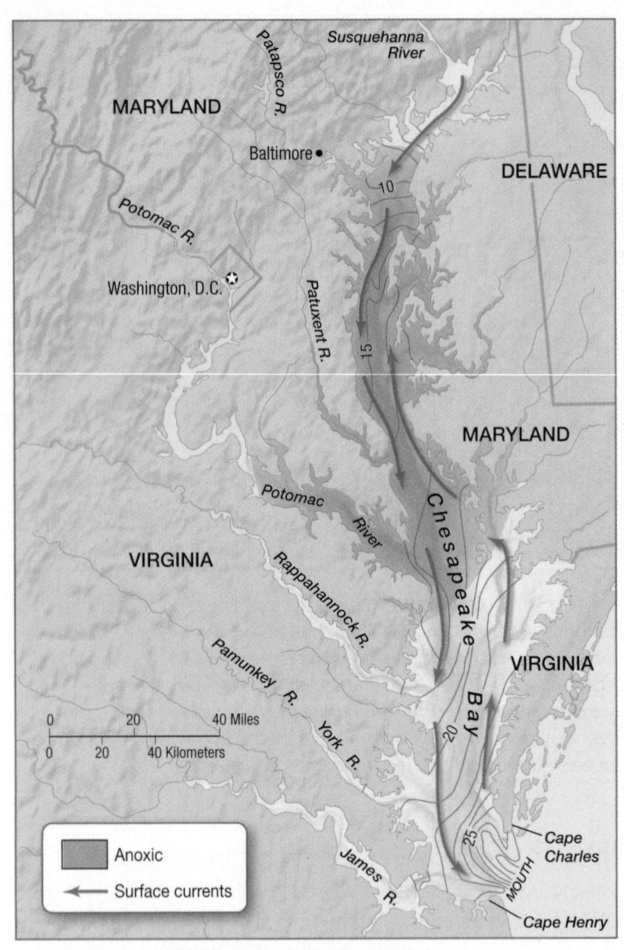


**(a)** Map of Chesapeake Bay, showing average surface salinity *(blue lines)* in ‰. The red area in the middle of the bay represents anoxic (oxygen-depleted) waters.

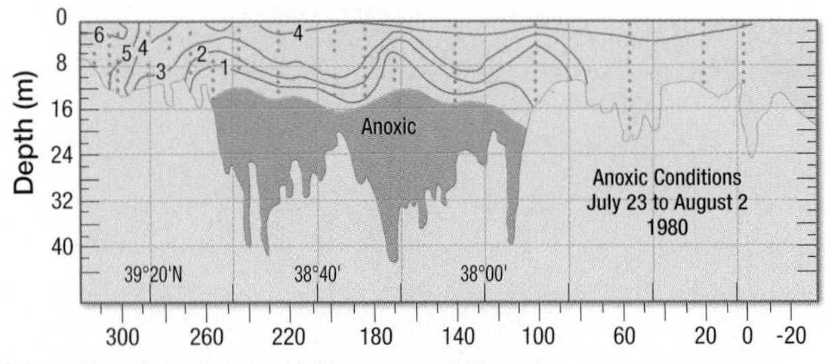


**(b)** Profile along length of Chesapeake Bay showing dissolved oxygen concentration (in ppm) during July–August 1980, indicating deep anoxic waters *(dark red)*.

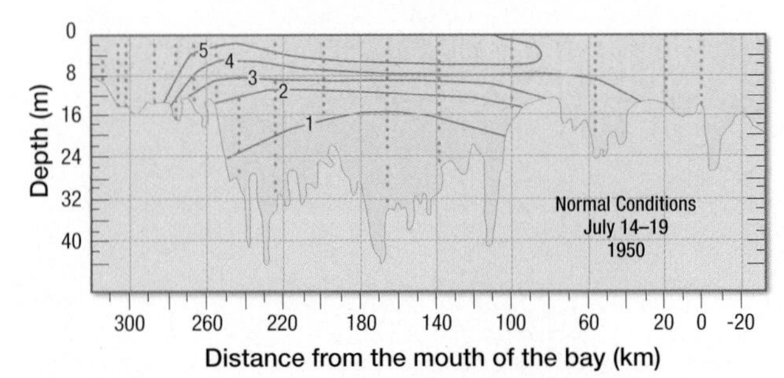


**(c)** Comparison profile showing normal dissolved oxygen concentration (in ppm) during July 1950.

**Figure 10.36 Chesapeake Bay salinity and dissolved oxygen.**

become necessary to the region's economy. To aid shipping operations, the river receives periodic dredging of sediment, which brings an increased risk for pollution. If these kinds of problems have developed in such sparsely populated areas as the Columbia River estuary, then larger environmental effects must exist in more highly populated estuaries, such as the Chesapeake Bay.

***Chesapeake Bay Estuary*** Chesapeake Bay is about 320 kilometers (200 miles) long and 56 kilometers (35 miles) wide at its widest point, making it the largest (and best studied) estuary in the United States (**Figure 10.36**). It drains a watershed of about 166,000 square kilometers (64,000 square miles) spread over six states that includes a population of over 15 million people. The length of the bay's shoreline is an astonishing 17,700 kilometers (11,000 miles) because of all the inlets created by the 19 major rivers and 400 creeks and tributaries that flow into it. The bay formed when the lower parts of the Susquehanna River were drowned by rising sea level after the most recent ice age.

Chesapeake Bay is a slightly stratified estuary that experiences large seasonal changes in salinity, temperature, and dissolved oxygen. Figure 10.36a shows the estuary's average surface salinity, which increases oceanward. The salinity lines are oriented virtually north–south in the middle of the bay because of the Coriolis effect. Recall that the Coriolis effect causes flowing water to curve to the right in the Northern Hemisphere, so seawater entering the bay tends to hug the bay's *eastern* side, and freshwater flowing through the bay toward the ocean tends to hug its *western* side.

With maximum river flow in the spring, a strong halocline (and *pycnocline*) develops, preventing the fresh surface water and saltier deep water from mixing. (Recall that a *pycnocline* [*pycno* = density, *cline* = slope] is a layer of rapidly changing density, as discussed in Chapter 5. A pycnocline is caused by a change in temperature and/or salinity with depth.) Beneath the pycnocline, which can be as shallow as 5 meters (16 feet), waters may become **anoxic** (*a* = without, *oxic* = oxygen) from May through August, as dead organic matter decays in the deep water (Figure 10.36b). Major kills of commercially important blue crab, oysters, and other bottom-dwelling organisms occur during this time.

The degree of stratification and extent of mortality of bottom-dwelling animals have increased since the early 1950s. Increased nutrients from sewage and agricultural fertilizers have been added to the bay during this time, too, which has increased the productivity of microscopic algae (algal blooms). When these organisms die, their remains accumulate as organic matter at the bottom of the bay and promote the development of anoxic conditions. In drier years with less river runoff, however, anoxic conditions aren't as widespread or severe in bottom waters (Figure 10.36c) because fewer nutrients are supplied.

## Lagoons

Landward of barrier islands lie protected, shallow bodies of water called **lagoons** (see Figure 10.33c). Lagoons form in a bar-built type of estuary. Because of restricted circulation between lagoons and the ocean, three distinct zones can usually be identified within lagoons (**Figure 10.37**): (1) A *freshwater zone* that lies near the head of the lagoon where rivers enter, (2) a *transitional zone* of brackish water, which has a salinity between that of freshwater and seawater and occurs near the middle of the lagoon, and (3) a *saltwater zone* that lies close to the lagoon's mouth.

Salinity within a lagoon is highest near the entrance and lowest near the head (Figure 10.37b). In latitudes that have seasonal variations in temperature and precipitation, ocean water flows through the entrance during a warm, dry summer to compensate for the volume of water lost through evaporation, thus increasing the salinity in the lagoon. Lagoons may actually become hypersaline (excessively salty) in arid regions, where evaporation rates are extremely high. Even though water flows into the lagoon from the open ocean to replace water lost by evaporation, the dissolved components do not evaporate and sometimes accumulate to extremely high levels. During the rainy season, the lagoon becomes much less saline as freshwater runoff increases.

Tidal effects are greatest near the entrance to the lagoon (Figure 10.37c) and diminish inland from the saltwater zone until they are nearly undetectable in the freshwater zone.

**LAGUNA MADRE** Laguna Madre is located along the Texas coast between Corpus Christi and the mouth of the Rio Grande (**Figure 10.38**). This long, narrow body of water is protected from the open ocean by Padre Island, a barrier island 160 kilometers (100 miles) long. The lagoon probably formed about 6000 years ago as sea level approached its present height.

The tidal range of the Gulf of Mexico in this area is about 0.5 meter (1.6 feet). The inlets at each end of Padre Island are quite narrow (Figure 10.38), so there is very little tidal interchange between the lagoon and the open sea.

Laguna Madre is a hypersaline lagoon, and much of it is less than 1 meter (3.3 feet) deep. As a result, there are large seasonal changes in temperature and salinity. Water temperatures reach 32°C (90°F) in the summer and can dip below 5°C (41°F) in winter. Salinities range from 2‰ when infrequent local storms provide large volumes of freshwater to over 100‰ during dry periods. High evaporation generally keeps salinity well above 50‰. (For comparison, recall that normal salinity in the open ocean averages 35‰.)

Because even salt-tolerant marsh grasses cannot withstand such high salinities, the marsh has been replaced by an open sand beach on Padre Island. At the inlets, ocean water flows in as a surface wedge *over* the denser water of the lagoon and water from the lagoon flows out as a *subsurface* flow, which is the exact opposite of a typical estuarine circulation pattern.

## Marginal Seas

At the margins of the ocean are relatively large, semi-isolated bodies of water called **marginal seas**. Most of these seas result from tectonic events that have isolated low-lying pieces of ocean crust between continents, such as the Mediterranean Sea, or are created behind volcanic island arcs, such as the Caribbean Sea. These waters are shallower than and have varying degrees of exchange with the open ocean, depending on climate and geography; as a result, salinities and temperatures are substantially different from those of typical open ocean seawater.

**A CASE STUDY: THE MEDITERRANEAN SEA** The **Mediterranean Sea** (*medi* = middle, *terra* = land) is actually a number of small seas connected by narrow necks of water into one larger sea. It is the remnant of the ancient Tethys Sea that existed when all the continents were combined about 200 million years ago. It is more than 4300 meters (14,100 feet) deep and is one of the few inland seas in the world underlain by oceanic crust. Thick salt deposits and other evidence on the floor of the Mediterranean suggest that it nearly dried up about 6 million years ago, only to refill with a large saltwater waterfall (see Mastering Oceanography **Web Diving Deeper 4.1**).

The Mediterranean is bounded by Europe and Asia Minor on the north and east and Africa on the south (**Figure 10.39a**). It is surrounded by land except for very shallow and narrow connections to the Atlantic Ocean through the Strait of Gibraltar (about 14 kilometers [9 miles] wide), and to the Black Sea through the Bosporus (roughly 1.6 kilometers [1 mile] wide). In addition, the Mediterranean Sea has a human-made passage to the Red Sea via the Suez Canal, a waterway 160 kilometers (100 miles) long that was completed in 1869. The Mediterranean Sea has a very

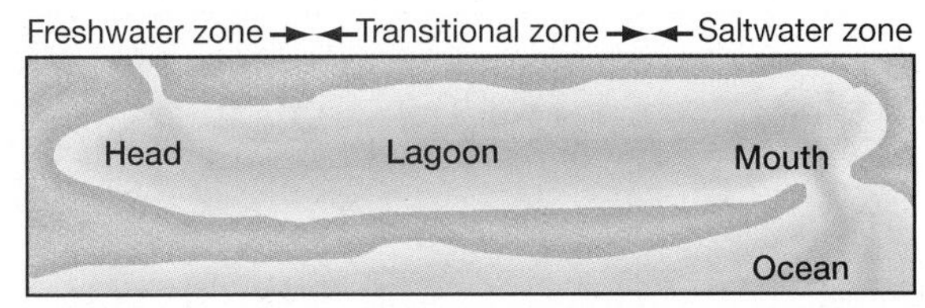


**(a)** Geometry and features of a typical lagoon as seen from above.

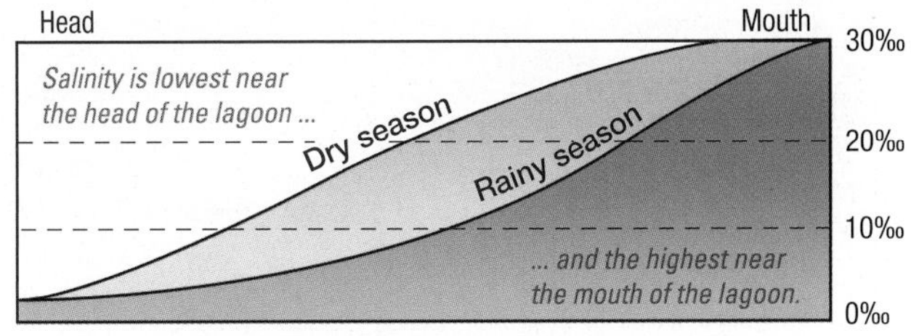


**(b)** Salinity profile of a typical lagoon, which is affected by seasonal changes in freshwater input.

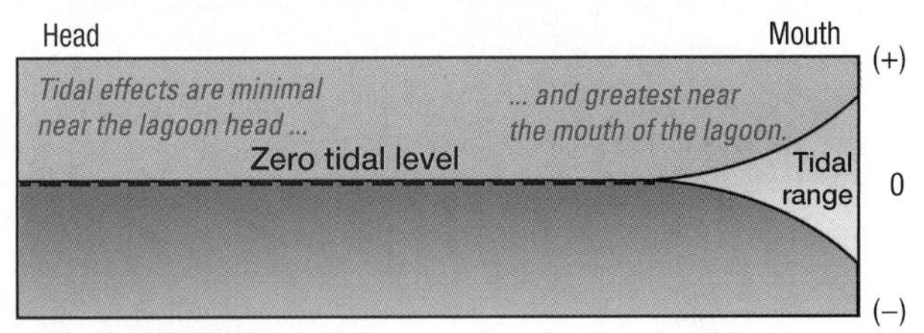


**(c)** Tidal effects from the head to the mouth of a typical lagoon.

**Figure 10.37 Lagoons.** Diagrammatic representations of the general features of a typical lagoon.

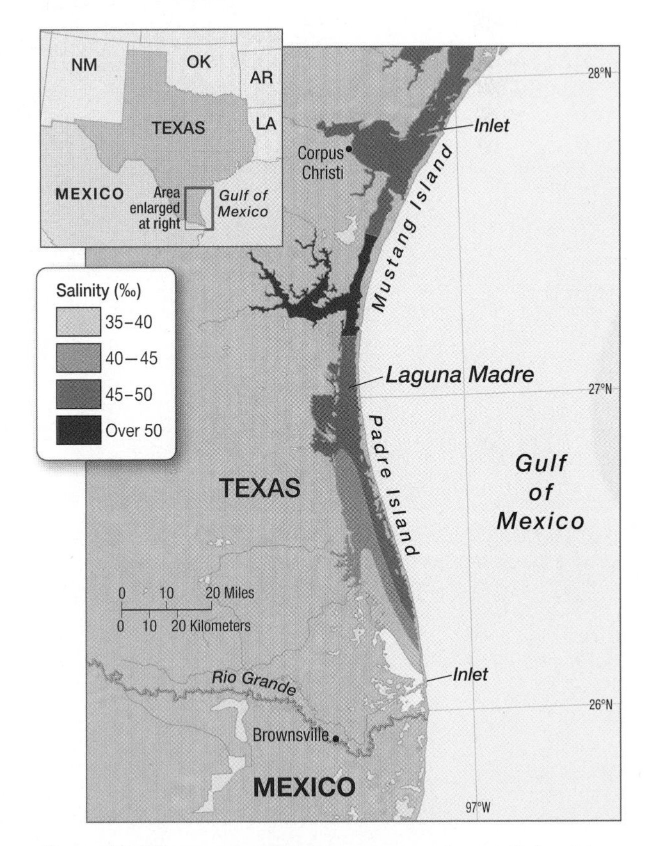


**Figure 10.38 Laguna Madre summer surface salinity.** Map showing geometry of Laguna Madre, Texas, and typical summer surface salinity (in ‰).

Figure 10.39 **Mediterranean Sea bathymetry and circulation.**

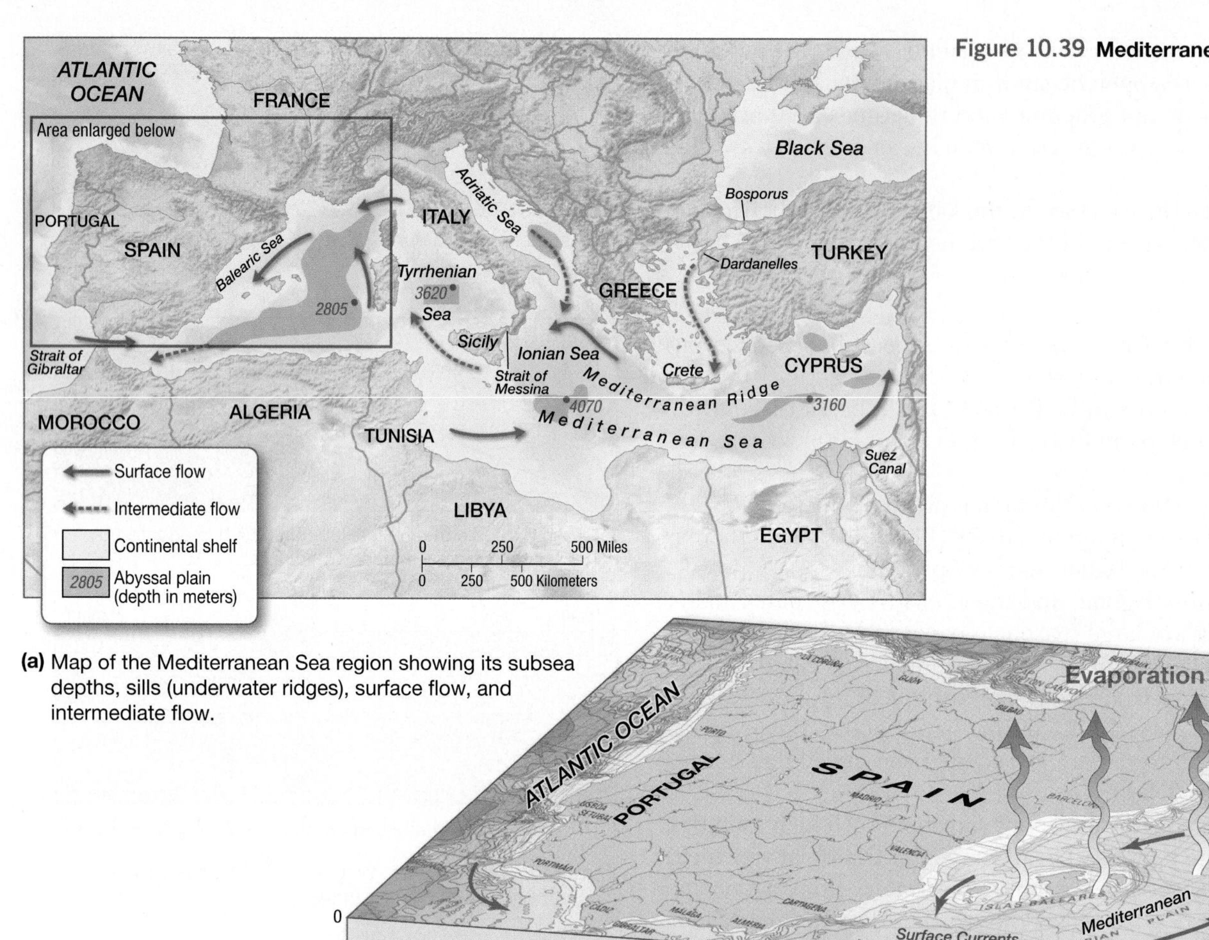


**(a)** Map of the Mediterranean Sea region showing its subsea depths, sills (underwater ridges), surface flow, and intermediate flow.

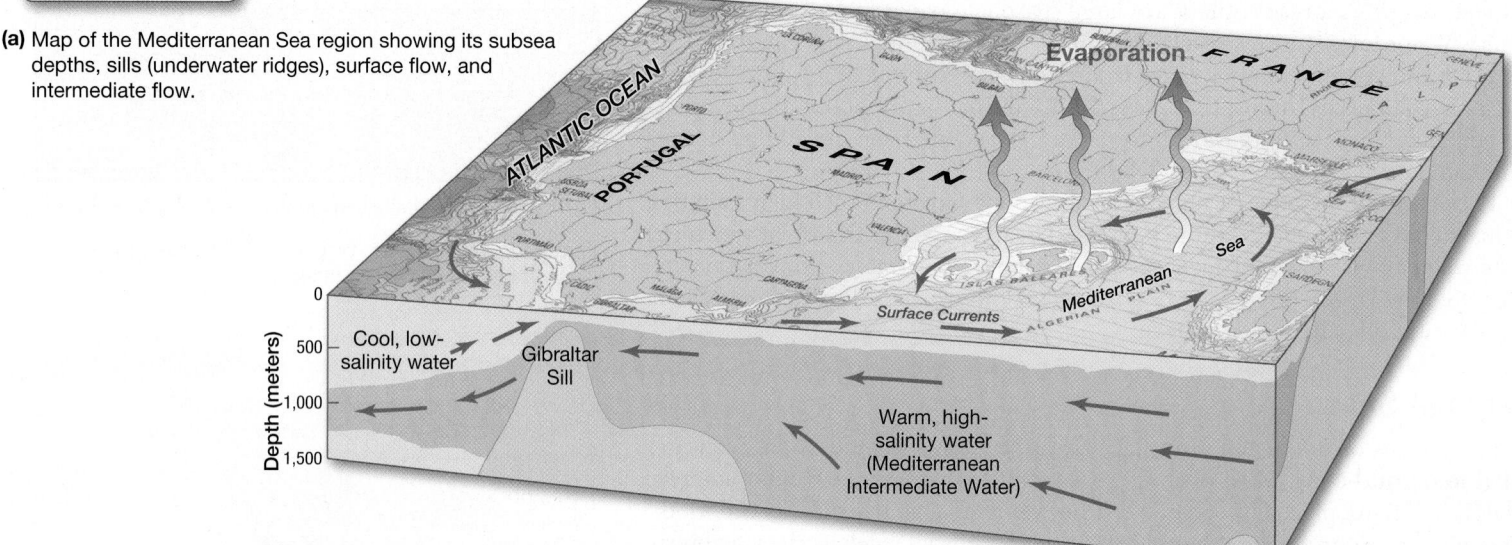

**(b)** Diagrammatic view of Mediterranean circulation in the Gibraltar Sill area.

irregular coastline, which divides it into smaller subseas such as the Aegean Sea and Adriatic Sea, each of which has a separate circulation pattern.

An underwater ridge called a **sill**, which extends from Sicily to the coast of Tunisia at a depth of 400 meters (1300 feet), separates the Mediterranean into two major basins. This sill restricts the flow between the two basins, resulting in strong currents that run between Sicily and the Italian mainland through the Strait of Messina (Figure 10.39a).

***Mediterranean Circulation*** The Mediterranean Sea has a unique circulation pattern. This circulation is caused by the dry, intense heat of the Middle East, where a huge volume of water evaporates from the eastern Mediterranean and causes a tremendous surface inflow of Atlantic Ocean water through the Strait of Gibraltar to replace the evaporated water. In fact, the water level in the eastern Mediterranean is generally 15 centimeters (6 inches) lower than at the Strait of Gibraltar. The surface flow follows the northern coast of Africa throughout the length of the Mediterranean and spreads northward across the sea (Figure 10.39a).

The remaining Atlantic Ocean water continues eastward to Cyprus. During winter, it sinks to form what is called the *Mediterranean Intermediate Water*, which has a temperature of 15°C (59°F) and a salinity of 39.1‰. This water flows westward at a depth of 200 to 600 meters (660 to 2000 feet) and returns to the North Atlantic as a *subsurface* flow through the Strait of Gibraltar (**Figure 10.39b**). During World War II, German submarines routinely escaped detection when crossing through the Strait of Gibraltar by switching off their engines and taking advantage of the currents flowing into and out of the Mediterranean Sea. The submarine captains would adjust the buoyancy of the submarine so that the sub would be transported silently, either into the Mediterranean Sea with its surface current or out of the sea within its intermediate waters.

By the time the Mediterranean Intermediate Water passes through Gibraltar, its temperature has dropped to 13°C (55°F) and its salinity to 37.3‰. It is still denser than even Antarctic Bottom Water and much denser than water at this depth in the Atlantic Ocean, so it moves down the continental slope. While descending, it mixes with Atlantic Ocean water and becomes less dense. At a depth of about 1000 meters (3300 feet) (Figure 10.39b), its density equals that of the surrounding Atlantic Ocean, so it spreads in all directions, sometimes forming deep-ocean eddies that last for more than two years and can be detected by satellite as far north as Iceland.

This circulation pattern, which is called **Mediterranean circulation**, is opposite that of most estuaries, which experience estuarine circulation where freshwater flows at the surface into the open ocean and salty water flows below the surface into the estuary. In estuaries, however, freshwater input exceeds water loss to evaporation, whereas evaporation exceeds input in the Mediterranean.

## STUDENTS SOMETIMES ASK . . .

***How can Mediterranean Intermediate Water sink if it's so warm?***

While it is true that warm water has low density, remember that *both* salinity and temperature affect seawater density. In the case of the Mediterranean Intermediate Water, it has high enough salinity to increase its density despite being warm. Once its density increases enough, it sinks beneath the surface and retains its temperature and salinity characteristics as it flows out through the Strait of Gibraltar into the North Atlantic.

Circulation between the Mediterranean Sea and the Atlantic Ocean is typical of closed, restricted basins where evaporation exceeds precipitation. Low-latitude restricted basins such as this always rapidly lose water to evaporation, so surface flow from the open ocean must replace it. Evaporation of inflowing water from the open ocean increases the sea's salinity to very high values. This denser water eventually sinks and returns to the open ocean as a subsurface flow.

**RECAP High evaporation rates in the Mediterranean Sea cause it to have a shallow inflow of surface seawater and a subsurface high-salinity outflow—a circulation pattern opposite that of most estuaries.**

### CONCEPT CHECK 10.6 ▸ Compare the various types of coastal waters.

1 For coastal oceans where deep mixing does not occur, describe the effect that offshore winds and freshwater runoff have on salinity distribution. How will the winter and summer seasons affect the temperature distribution in the water column?

2 Describe how coastal runoff of low-salinity water produces a coastal geostrophic current and give a specific location where a coastal geostrophic current can be found.

3 Describe the four main types of estuaries, based on geologic origin.

4 Describe the difference between vertically mixed and salt wedge estuaries in terms of salinity distribution, depth, and volume of river flow. Which displays the more classical estuarine circulation pattern?

5 Discuss factors that cause the surface salinity of Chesapeake Bay to be greater along its east side. Also, why are periods of summer anoxia in Chesapeake Bay's deep water becoming increasingly worse?

6 What factors lead to a wide seasonal range of salinity in Laguna Madre?

7 Describe the circulation between the Atlantic Ocean and the Mediterranean Sea, and explain how and why it differs from typical estuarine circulation.

# 10.7 ▸ What Issues Face Coastal Wetlands?

**Wetlands** are ecosystems in which the water table is close to the surface, so they are typically saturated most of the time. Wetlands can border either freshwater or coastal environments. Coastal wetlands occur along the margins of coastal waters such as estuaries, lagoons, and marginal seas; they include swamps, tidal flats, coastal marshes, and bayous.

## Types of Coastal Wetlands

The two most important types of coastal wetlands are **salt marshes** and **mangrove swamps**. Both experience intermittent submergence by ocean water and contain

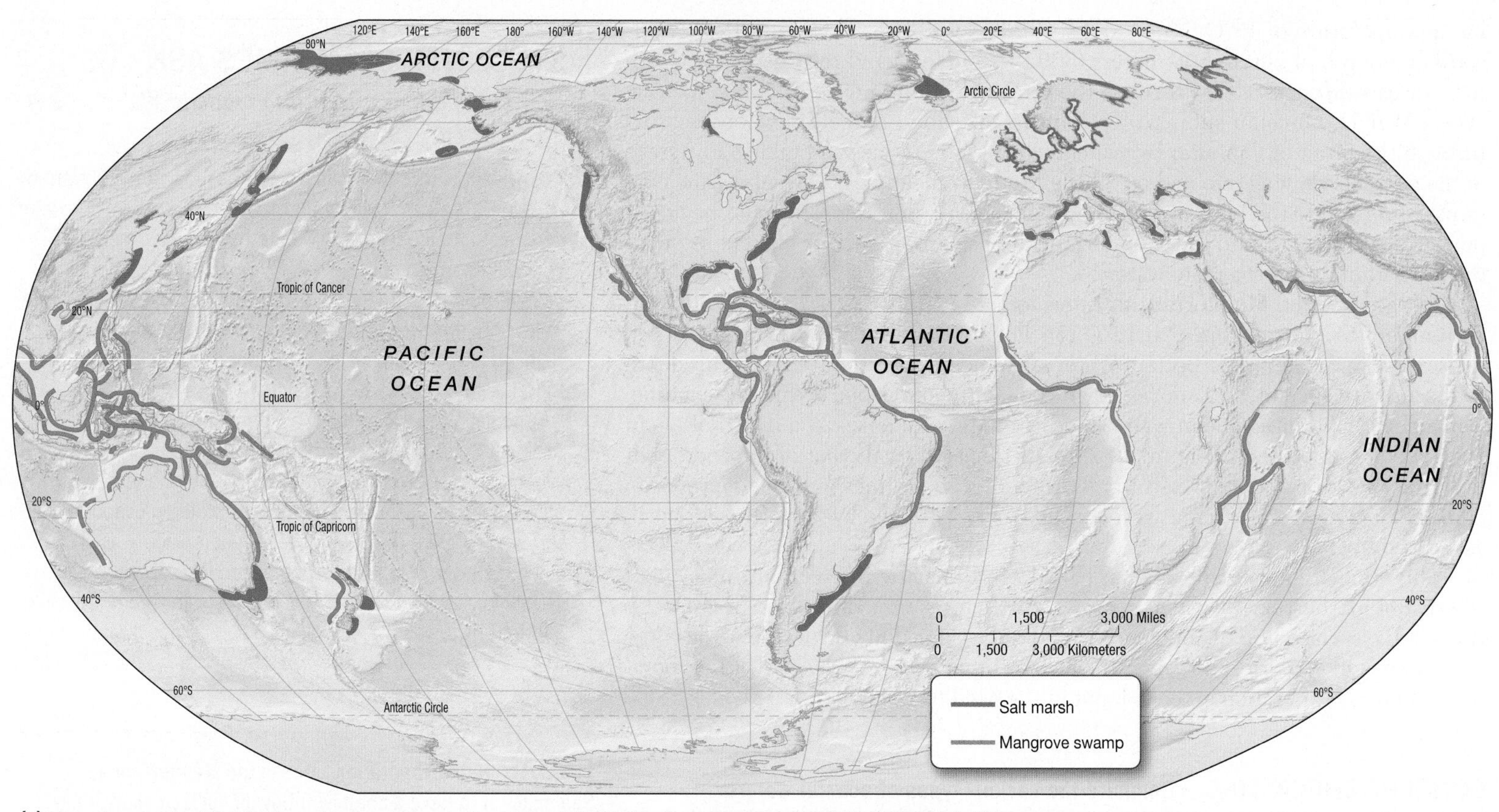


**(a)** Map showing the distribution of salt marshes in the higher latitudes and mangrove swamps in the lower latitudes.

**(b)** A typical salt marsh in Morro Bay, California.

**(c)** A dense mangrove swamp bordering a seaway in the Florida Keys.

Figure 10.40 **Salt marshes and mangrove swamps. (a)** Map showing the distribution of salt marshes and mangrove swamps, with accompanying photos (**b** and **c**).

salt-adapted plants, oxygen-depleted muds, and accumulations of organic matter called *peat deposits*.

Salt marshes generally occur between about 30 and 65 degrees latitude (**Figures 10.40a** and **10.40b**) and support a variety of salt-tolerant grasses and other low-lying plants that are termed *halophytic* (*halo* = salt, *phyto* = plant). Examples of halophytic grasses include cordgrass and salt-meadow cordgrass, both of which belong to the genus *Spartina* and have the ability to get rid of excess salt by producing exterior salt crystals. Other plants that live in this habitat, like pickleweed (*Salicornia*), accumulate salts in their tissues and dispose of excess salts by breaking off the tissues once they become highly salty. Well-developed salt marsh habitats are found along most coasts of the continental United States and also along the coasts of Europe, Japan, and eastern South America.

Mangrove swamps are restricted to tropical regions (below 30 degrees latitude; **Figures 10.40a** and **10.40c**) and support various species of salt-tolerant mangrove trees, shrubs, and palms. To live in these salty conditions, some mangroves produce tall tripod-like root systems to stay above the salty water; others crystalize excess salt on their leaves. Mangrove swamps occur throughout the Caribbean and Florida; the most extensive mangroves in the world are found throughout Southeast Asia.

Figure 10.41 **Marine wetlands provide habitat and protection for many species of fish.** Both salt marshes and mangrove swamps are types of marine wetlands that are important nursery areas for many species of fish, such as these Atlantic silversides (*Menidia menidia*) that seek protection within mangrove roots.

## Characteristics of Coastal Wetlands

Wetlands are home to a diverse assortment of plants and animals and are some of the most highly productive ecosystems on Earth. When left undisturbed, wetlands provide enormous economic benefits. Salt marshes, for example, serve as nurseries for more than half the species of commercially important fishes in the southeastern United States (**Figure 10.41**). Other fish, such as flounder and bluefish, use marshes for feeding and protection during the winter. Fisheries of oysters, scallops, clams, eels, and smelt are located directly in marshes, too. Mangrove ecosystems are important nursery areas and habitats for commercially valuable shrimp, prawn, shellfish, and fish species. Both marshes and mangroves also serve as important stopover points for many species of waterfowl and migrating birds.

Wetlands also soak up the nutrients that run off farmlands and down rivers, which, if they reached coastal waters, could fuel harmful algal blooms and create marine oxygen-free dead zones. In essence, wetlands are amazingly efficient at cleansing polluted water; that's why they are often referred to as "nature's kidneys." Just 0.4 hectare (1 acre) of wetlands, for example, can filter up to 2,760,000 liters (730,000 gallons) of water each year, cleaning agricultural runoff, toxins, and other pollutants long before they reach the ocean. Wetlands remove inorganic nitrogen compounds (from sewage and fertilizers) and metals (from groundwater polluted by land sources), which become attached to clay-sized particles in wetland mud. Some nitrogen compounds trapped in sediment are decomposed by bacteria that release the nitrogen to the atmosphere as gas, and many of the remaining nitrogen compounds fertilize plants, further increasing the productivity of wetlands. As marsh plants die, their remains either accumulate as peat deposits or are broken up to become food for bacteria, fungi, and fish.

Another important attribute of coastal wetlands is that they trap carbon in their soils, making them an important sink for atmospheric carbon. In fact, studies have shown that existing mangrove ecosystems alone could store in their soils as much as 20 billion metric tons (22 billion short tons) of carbon—equivalent to more than two years of global carbon emissions—much of which would stay in the atmosphere and contribute to additional global warming if the mangrove trees were destroyed.

Climate Connection

In addition, wetlands protect shorelines from erosion and serve as a first line of defense against hurricanes and tsunami by dissipating wave energy and absorbing excess water, much like a natural sponge. The 2004 Indian Ocean tsunami, for example, devastated some coastal regions, yet others with protective offshore coral reefs or coastal mangroves experienced much less damage. As another example,

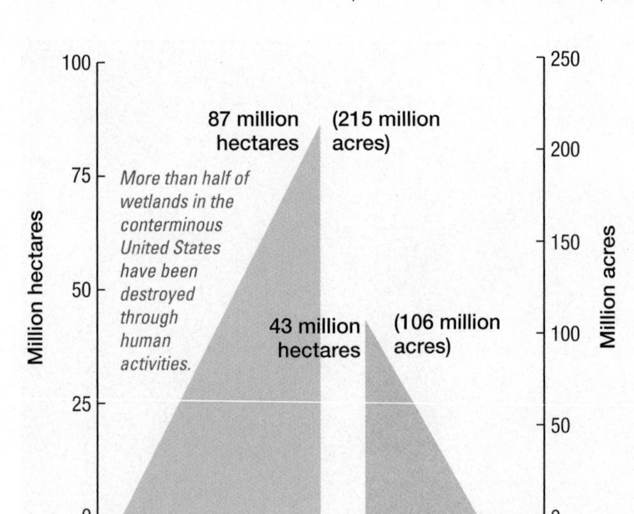


**Figure 10.42 Wetland loss in the conterminous United States.** A comparison between the original areal extent of wetlands (*green triangle*) versus today (*orange triangle*) in the conterminous United States (excluding Alaska and Hawaii).

the loss of protective coastal wetlands in the Mississippi River Delta contributed to the extensive flooding associated with the storm surge caused by Hurricane Katrina in 2005 (see Section 6.5 "How Does the Ocean Influence Global Weather Phenomena and Climate Patterns?" in Chapter 6). During Hurricane Sandy in 2012, the complete lack of protective wetlands around New York City caused much more severe flooding there than in neighboring regions that retained just remnants of wetlands.

## Serious Loss of Valuable Wetlands

Despite all the benefits wetlands provide, more than half of the nation's wetlands have vanished. Of the original 87 million hectares (215 million acres) of wetlands that once existed in the conterminous United States, only about 43 million hectares (106 million acres) remain (**Figure 10.42**). Wetlands have been filled in and developed for agriculture such as rice paddies and palm plantations, industry such as fish and shrimp farms, or tourism and real-estate projects. In spite of the numerous benefits of wetlands, many people view wetlands as unproductive, useless land that harbors diseases that would be better served as being developed. In many places, wetland loss is compounded by the lack of fresh sediment from regular river floods. Instead, flooding rivers and their sediment are channeled away from wetland areas.

Louisiana's coastal wetlands, for example, are among those that are steadily disappearing. Over time, the soil in wetlands naturally compresses under its own weight, in a process called *subsidence*. Normally, the growth of plants and the infusion of fresh sediment from river floodwaters offset subsidence. With these factors reduced or eliminated, many wetlands are sinking into the ocean faster than they are building up. For example, scientists estimate that subsidence of the Mississippi River Delta—along with rising sea level—will cause about 10% of Louisiana to sink beneath the ocean surface by the end of the century.

Other countries have experienced similar losses of wetlands, too. In fact, scientists estimate that 50% of wetlands worldwide have been destroyed in the past century. Mangroves, for example, are already critically endangered or approaching extinction in 26 out of the 120 countries that have mangroves. Indonesia has lost over 50% of its mangroves in the past three decades, and the Philippines has reported losing 70% of its original mangrove cover. Worldwide, 3.6 million hectares (8.9 million acres) of mangroves have been lost since 1980. Of the mangrove swamps that remain, many are in critical condition or seriously damaged. At the current rate of mangrove loss, there is increasing concern that all mangrove ecosystems worldwide will be destroyed within the next 100 years.

To help prevent the loss of remaining wetlands, the U.S. Environmental Protection Agency (EPA) established an Office of Wetlands Protection (OWP) in 1986. At that time, wetlands were being lost to development at a rate of 121,000 hectares (300,000 acres) per year! In 1997, the rate of coastal wetland loss had slowed to about 8100 hectares (20,000 acres) per year. The goal of the OWP is to reduce the loss of wetlands in the United States to zero by actively enforcing regulations against wetlands pollution and identifying the most valuable wetlands to be protected or restored.

In spite of these global, long-term trends, recent documentation of U.S. wetlands shows that there has been an overall *increase* in wetlands during this century. In fact, a study from 1998 to 2004 revealed that the conterminous United States gained an estimated 13,000 hectares (32,000 acres) of wetlands each year of the study. This gain—although small—was primarily due to an increase in freshwater wetlands; coastal wetlands were still decreasing, but at a slower rate than previously reported. The fact that coastal regions were losing wetlands despite the national trend of a net gain in wetlands points to the need for more research on the natural and human forces behind these trends and to an expanded effort on conservation of wetlands, particularly in coastal areas.

## PROCESS OF SCIENCE 10.1: Recycling Christmas Trees to Save Louisiana's Disappearing Coast

### BACKGROUND

Coastal wetland erosion has been a growing issue for decades along the Louisiana coast, but now a novel approach is being used to help. Flood control measures, as well as oil and gas exploration, have eroded about 3900 square kilometers (1500 square miles) of valuable coastal wetlands in Louisiana since 1930, and another 65 square kilometers (25 square miles) are lost every year. What could be done in a cost-effective way to prevent the continued loss of coastal wetlands?

### FORMING A HYPOTHESIS

Researchers at Louisiana State University studied how other countries protect their low-lying coastal wetlands from wave erosion. In a visit to the Netherlands, they witnessed how the Dutch government was tying together willow trees and other shrubs in bundles to create thousands of hectares of wetlands along the country's northern coast. Back home, the researchers noticed the surplus of cut, recyclable Christmas trees after the Christmas season. With tens of thousands of Christmas trees destined for the dump every January, repurposing old Christmas trees would be a feel-good way for residents to contribute to coastal restoration at a fraction of the cost of government-sponsored dredging or diversion projects. In addition, the natural sap inside the trees would help prevent them from breaking down too quickly in seawater.

Putting this all together, the researchers proposed a low-cost program that over the past 20 years has blossomed into a statewide initiative involving 16 coastal communities.

**Figure 10B** Recycled Christmas trees are stacked in offshore bins to prevent coastal erosion in Louisiana.

### DEVISING AN EXPERIMENT

Each year, discarded Christmas trees are collected by various agencies and placed in offshore wooden bins called "cribs" that parallel the coast for kilometers (**Figure 10B**). Tens of thousands of trees are stacked into the wooden cribs, which shield fragile shorelines from pounding waves and trap the vital sediment needed to rebuild marshes. In effect, the tree branches create something akin to an underwater sand bar that stabilizes sediment and keeps it from being washed out to sea. In addition, the stillness of this protected environment allows native aquatic plants to take root and grow more quickly, which also reduces coastal erosion by fortifying the wetlands. Each year new trees are stacked on top of the previous years' trees to maintain the cribs.

### INTERPRETING THE RESULTS

The Christmas tree cribs appear to be slowing the rate of coastal wetland erosion and in some cases have even been responsible for *reversing* the destruction of coastal wetlands. In Goose Bayou of Jefferson Parish, for example, which in some places had experienced as much as 100 meters (330 feet) of erosion each year, only 1.2 hectares (3 acres) of land was lost in the first year of Christmas tree crib protection. For comparison, in a nearby area without Christmas tree cribs that served as a control, 9.3 hectares (23 acres) was lost. Over the past 16 years in Jefferson Parish, the program has been responsible for restoring 100 hectares (250 acres) of coastal wetlands, compared with the area's previous wetland loss of 6500 hectares (16,000 acres). And because the Christmas tree barriers facilitate the regrowth of wetland plants, the project has resulted in renewed habitat for native fish and wildlife.

### THINKING LIKE A SCIENTIST: WHAT'S NEXT?

Other than Christmas trees, what other materials would you suggest be stacked into the wooden offshore cribs to prevent coastal wetland erosion?

Future sea level rise is predicted to exacerbate the loss of wetlands. Even using a conservative estimate of sea level rise over the next 100 years of 50 centimeters (20 inches), it is estimated that as much as 61% of existing U.S. coastal wetlands will be lost. Human-caused global warming could cause additional sea level rise and thus even more wetland loss. Some of this wetland loss, however, would be partially offset by new wetland formation on former upland areas, although even under ideal circumstances, not all lost wetlands would be replaced.

Climate

Connection

**RECAP Coastal wetlands such as salt marshes and mangrove swamps are highly productive areas that serve as important nurseries for many marine organisms, act as filters for polluted runoff, trap atmospheric carbon in their soils, and help prevent coastal erosion.**

**CONCEPT CHECK 10.7 ▶ Specify the issues that face coastal wetlands.**

1 Name the two types of coastal wetland environments and the latitude ranges where each will likely develop.

2 How do wetlands contribute to the biology of the oceans and the cleansing of polluted river water?

3 From the information in Figure 10.42, determine how large an area of wetlands has been lost in the conterminous United States. What percentage of original wetlands still remains? What efforts are being made in the United States to reverse this trend?

# ESSENTIAL CONCEPTS REVIEW

## 10.1 ▶ How are coastal regions defined?

- *The coastal region changes continuously*. The *shore* is the region of contact between the oceans and the continents, lying between the lowest low tides and the highest elevation on the continents affected by storm waves. The *coast* extends inland from the shore as far as marine-related features can be found. The *coastline* marks the boundary between the shore and the coast. The shore is divided into the *foreshore*, extending from low tide to high tide, and the *backshore*, extending beyond the high tide line to the coastline. Seaward of the low tide shoreline are the *nearshore* zone, extending to the breaker line, and the *offshore* zone beyond.
- *A beach is a deposit of the shore area*, consisting of wave-worked sediment that moves along a wave-cut bench. It includes the *recreational beach, berm, beach face, low-tide terrace, longshore trough*, and one or more *longshore bars*. Beaches are composed of whatever material is locally available.

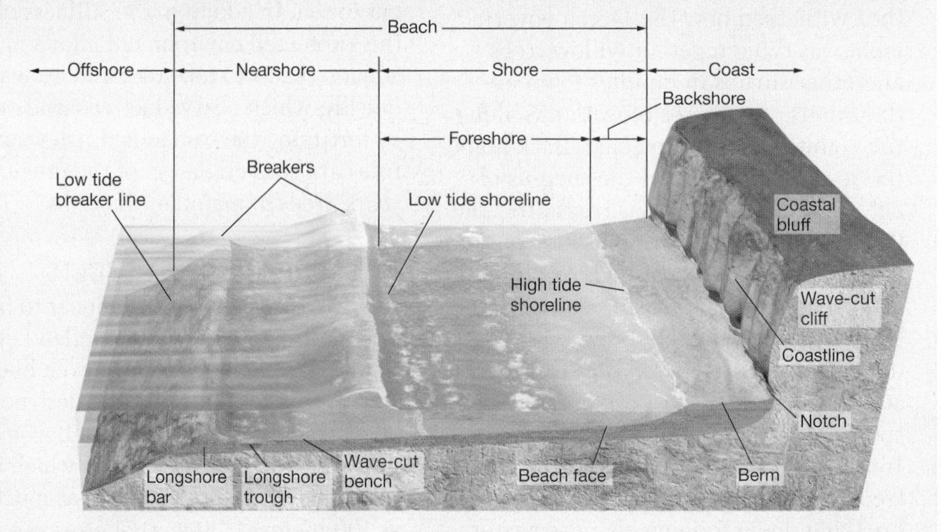


### Selected Key Terms

Use the **glossary** at the end of this book to discover the meanings of these Selected Key Terms: **shore, coast, coastline, beach, berm, beach face, longshore bar, longshore trough.**

### Critical Thinking Question

To help reinforce your knowledge of beach terminology, construct and label your own diagram similar to Figure 10.1 from memory.

### Active Learning Exercise

Working with another student in class, determine the technical term for the beach area where people go to sunbathe or to have a barbeque. Also, what is the technical term for the area of the beach where most people run in the sand? Include characteristics of each area.

## 10.2 ▶ How does sand move on the beach?

- *Waves that break at the shore move sand perpendicular to shore* (toward and away from shore). In *light wave activity, swash dominates the transport system*, and sand is moved up the beach face toward the berm. In *heavy wave activity, backwash dominates the transport system*, and sand is moved down the beach face, away from the berm and toward longshore bars. In a natural system, there is a *balance between light and heavy wave activity*, alternating between sand piled on the berm (*summertime beach*) and sand stripped from the berm (*wintertime beach*), respectively.
- *Sand is moved parallel to the shore, too*. Waves breaking at an angle to the shore create a *longshore current that results in a zigzag movement of sediment called longshore drift* (*longshore transport*). Each year, *millions of tons of sediment are moved from upcoast to downcoast* ends of beaches. Most of the year, *longshore drift moves southward along both the Pacific and Atlantic coasts* of the United States.

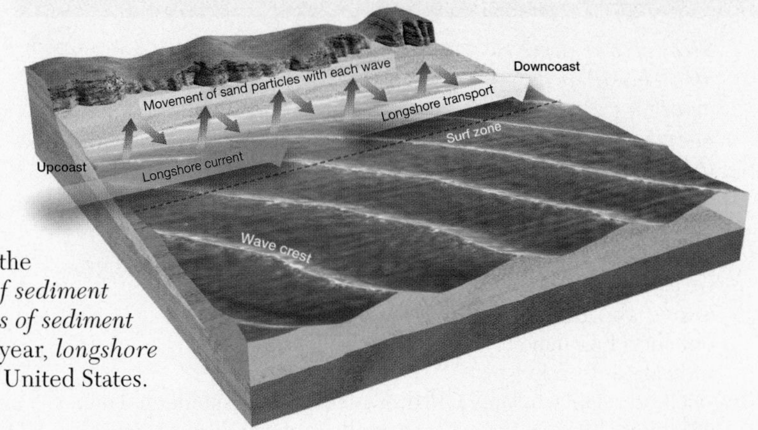


### Selected Key Terms

Use the **glossary** at the end of this book to discover the meanings of these Selected Key Terms: **swash, backwash, summertime beach, wintertime beach, longshore current, longshore drift/longshore transport.**

### Critical Thinking Question

Imagine you are floating at the beach within the surf zone. You notice that the longshore current and resulting longshore transport change direction (for example, from northward to southward) throughout the day. Based on your knowledge of shoreline processes, explain how this is possible.

### Active Learning Exercise

Working with another student in class, discuss how the flow of water in a stream is similar to that of a longshore current. In addition, discuss how the two are different.

## 10.3 ▶ What features are typical of erosional and depositional shores?

- *Erosional shores are characterized by headlands, wave-cut cliffs, sea caves, sea arches, sea stacks, and marine terraces* (caused by uplift of a wave-cut bench). Wave erosion increases as more of the shore is exposed to the open ocean, tidal range decreases, and bedrock weakens.
- *Depositional shores are characterized by beaches, spits, bay barriers, tombolos, barrier islands, deltas, and beach compartments*. Viewed from ocean side to lagoon side, barrier islands commonly have an ocean beach, dunes, barrier flat, and salt marsh. Deltas form at the mouths of rivers that carry more sediment to the ocean than the longshore current can carry away. *Beach starvation* occurs in beach compartments and other areas where the sand supply is interrupted.

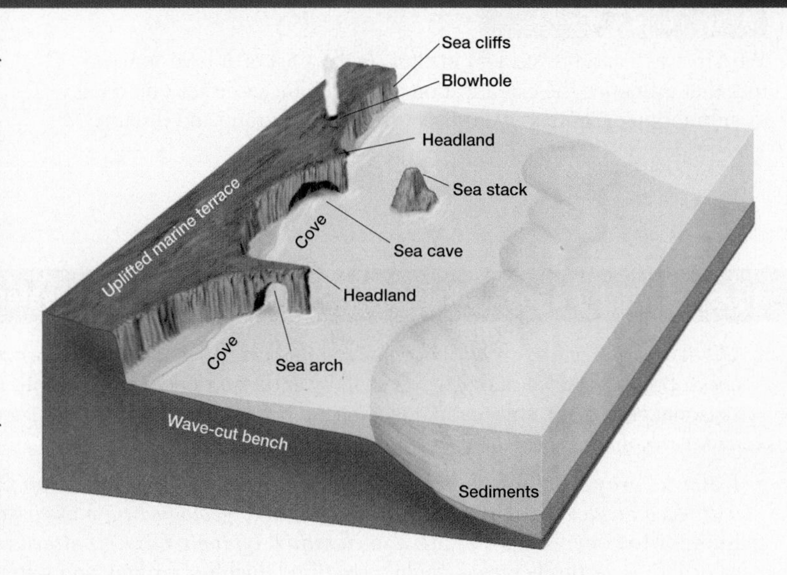


### Selected Key Terms

Use the **glossary** at the end of this book to discover the meanings of these Selected Key Terms: **erosional shore, depositional shore, sea cave, sea arch, sea stack, marine terrace, spit, bay-mouth bar, tombolo, barrier island, delta, beach compartment, beach starvation.**

### Critical Thinking Question

Specify the characteristics and coastal features that differentiate erosional and depositional shores.

### Active Learning Exercise

Working with another student in class, make a list of at least four factors that influence the classification of a coast as either erosional or depositional. Compare your lists with another group and discuss.

## 10.4 ▶ How do changes in sea level produce emerging and submerging shorelines?

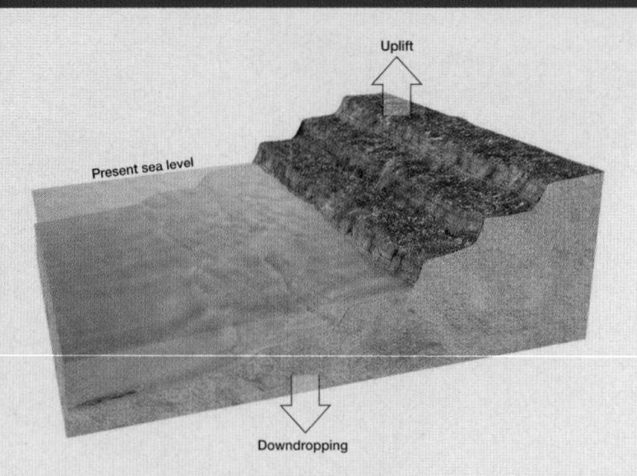


- *Shorelines are also classified as emerging or submerging, based on their position relative to sea level.* Ancient wave-cut cliffs and stranded beaches well above the present shoreline may indicate a drop in sea level relative to land. Old drowned beaches, submerged dunes, wave-cut cliffs, or drowned river valleys may indicate a rise in sea level relative to land. *Changes in sea level may result from tectonic processes causing local movement of the landmass or from eustatic processes changing the amount of water in the oceans or the capacity of ocean basins.* Melting of continental ice caps and glaciers during the past 18,000 years has caused a *eustatic rise in sea level of about 120 meters (400 feet).*

### Selected Key Terms

Use the **glossary** at the end of this book to discover the meanings of these Selected Key Terms: **emerging shoreline, submerging shoreline, stranded beach deposit, drowned beach, eustatic sea level change.**

### Critical Thinking Question

Specify the characteristics and coastal features that differentiate emerging and submerging shorelines.

### Active Learning Exercise

Working with another student in class, make a list of at least four factors that influence the classification of a shoreline as either emerging or submerging. Compare your lists with another group and discuss.

## 10.5 ▶ How does hard stabilization affect coastlines?

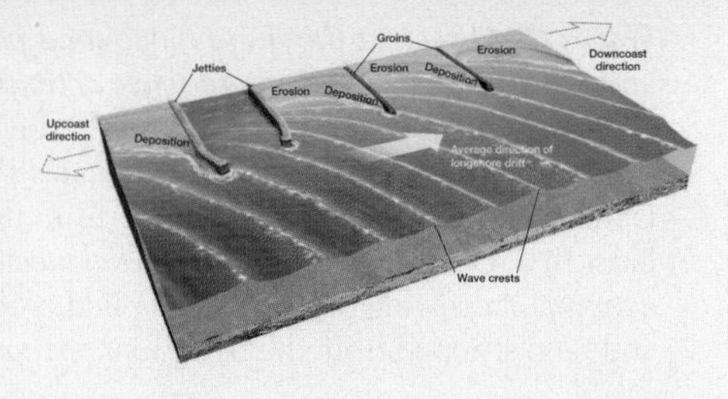


- *Hard stabilization—such as groins, jetties, breakwaters, and seawalls—is often used in an attempt to stabilize a shoreline. Groins* (built to trap sand) and *jetties* (built to protect harbor entrances) widen the beach by trapping sediment on their upcoast side, but erosion usually becomes a problem downcoast. Similarly, *breakwaters* (built parallel to a shore) trap sand behind the structure but cause unwanted erosion downcoast. *Seawalls* (built to armor a coast) often cause loss of the recreational beach. Eventually, the constant pounding of waves destroys all types of hard stabilization.
- *Alternatives to hard stabilization* include *construction restrictions* in areas prone to coastal erosion, *beach replenishment* (*beach nourishment*), which is an expensive and temporary method of reducing beach starvation, and *relocation*, which is a technique that has been successfully used to protect coastal structures.

### Selected Key Terms

Use the **glossary** at the end of this book to discover the meanings of these Selected Key Terms: **hard stabilization, groin, rip-rap, groin field, jetty, breakwater, seawall, beach replenishment/ beach nourishment, relocation.**

### Critical Thinking Question

Draw an aerial view of a shoreline to show the effect on erosion and deposition caused by constructing a groin, a jetty, a breakwater, and a seawall within the coastal environment.

### Active Learning Exercise

Working with another student in class, evaluate the alternatives to hard stabilization and select the one method that your group thinks would work best. Be sure to include your rationale as you present your answer to the class.

## 10.6 ▶ What are the characteristics and types of coastal waters?

- *The temperature and salinity of the coastal ocean vary over a greater range than the open ocean* because the coastal ocean is shallow and experiences river runoff, tidal currents, and seasonal changes in solar radiation. *Coastal geostrophic currents are produced from freshwater runoff and coastal winds.*
- *Estuaries are partially enclosed bodies of water where freshwater runoff from the land mixes with ocean water.* Estuaries are *classified by their geologic origin* as coastal plain, fjord, bar built, or tectonic. Estuaries are also *classified by their mixing patterns* of freshwater and saltwater as vertically mixed, slightly stratified, highly stratified, and salt wedge. *Typical circulation in an estuary consists of a surface flow of low-salinity water toward its mouth and a subsurface flow of marine water toward its head.*
- *Estuaries provide important breeding and nursery areas for many marine organisms* but often suffer from human population pressures. The *Columbia River Estuary*, for example, has degraded from agriculture, logging, and the construction of dams upstream. In the *Chesapeake Bay*, an anoxic zone occurs during the summer that kills many commercially important species.

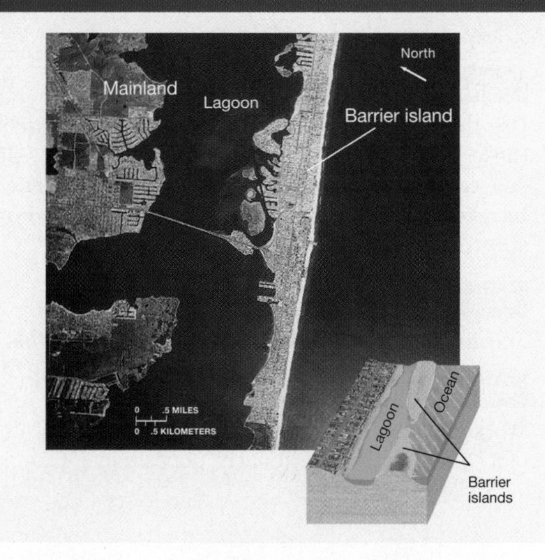


(Continued)

## 10.6 ▶ What are the characteristics and types of coastal waters? (*Continued*)

- *Long offshore sand deposits called barrier islands protect marshes and lagoons*. Some lagoons have restricted circulation with the ocean, so water temperatures and salinity may vary widely with the seasons.
- *Circulation in the Mediterranean Sea is characteristic of restricted bodies of water in areas where evaporation greatly exceeds precipitation*. Called *Mediterranean circulation*, it is the reverse of estuarine circulation.

### Selected Key Terms

Use the **glossary** at the end of this book to discover the meanings of these Selected Key Terms: **coastal geostrophic current, Davidson Current, coastal plain estuary, fjord, bar-built estuary, tectonic estuary, vertically mixed estuary, slightly stratified estuary, estuarine circulation pattern, highly stratified estuary, salt wedge estuary, reversing estuary, lagoon, marginal sea, Mediterranean circulation.**

### Critical Thinking Question

Based on their geologic origin, draw and describe the four major types of estuaries and give an example of where each one occurs.

### Active Learning Exercise

With another student in class, describe the temperature variation of the coastal ocean in (1) the low latitudes, (2) the high latitudes, and (3) the middle latitudes. In each of the three locations, discuss if a thermocline exists (and explain why), or if the water column is isothermal (and explain why).

## 10.7 ▶ What issues face coastal wetlands?

- *Wetlands are some of the most biologically productive regions on Earth. Salt marshes* and *mangrove swamps* are important *examples of coastal wetlands. Wetlands are ecologically important* because they *provide critical habitat* for many seagoing species, *remove land-derived pollutants* from water before they reach the ocean, *trap atmospheric carbon* in their soils, and *help prevent coastal erosion*. Nevertheless, *human activities continue to destroy wetlands*.

### Selected Key Terms

Use the **glossary** at the end of this book to discover the meanings of these Selected Key Terms: **salt marsh, mangrove swamp.**

### Critical Thinking Question

Specify the reasons wetlands are being destroyed worldwide in spite of their many benefits.

### Active Learning Exercise

Working with another student in class, discuss the steps you would recommend be taken to bring back lost coastal wetlands.

## Mastering Oceanography

Looking for additional review and test prep materials? With individualized coaching on the toughest topics of the course, Mastering Oceanography offers a wide variety of ways for you to move beyond memorization and deeply grasp the underlying processes of how the oceans work. Visit the Study Area in www.masteringoceanography.com to find practice quizzes, study tools, and multimedia that will improve your understanding of this chapter's content. Sign in today to access the following features: Self Study Quizzes, SmartFigures, Oceanography Videos and Animations, Squidtoons, Dynamic Study Modules, and an optional Pearson eText with embedded videos.

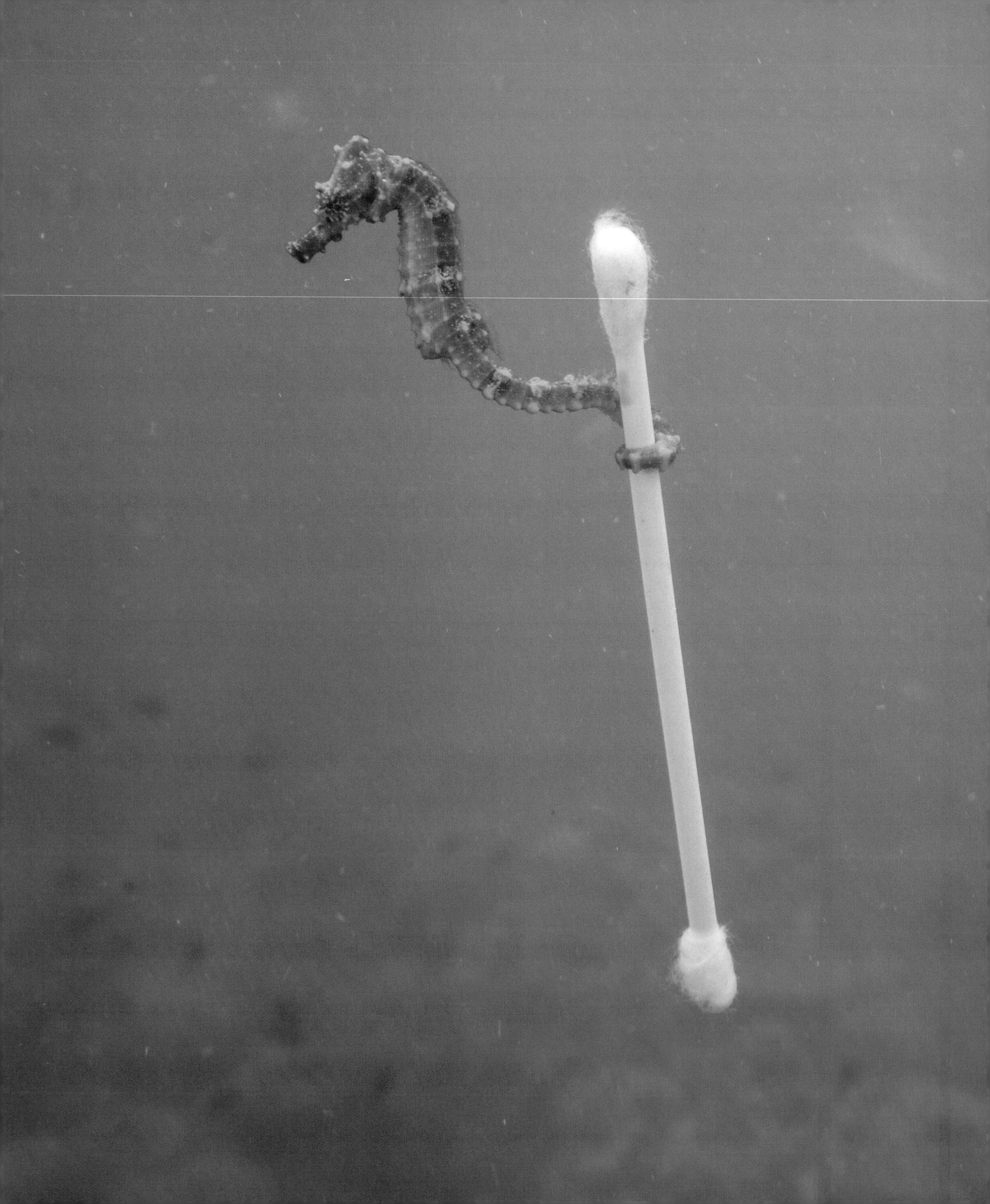

# Marine Pollution

# 11

## ESSENTIAL LEARNING CONCEPTS

At the end of this chapter, you should be able to:

- ☐ **11.1** Explain how pollution is defined.
- ☐ **11.2** Specify the marine environmental problems associated with petroleum pollution.
- ☐ **11.3** Specify the marine environmental problems associated with non-petroleum chemical pollution.
- ☐ **11.4** Specify the marine environmental problems associated with non-point source pollution, including trash.
- ☐ **11.5** Describe actions you can take to help prevent marine pollution.
- ☐ **11.6** Specify the environmental problems associated with biological pollution.

↑ *Check when completed*

***"Most people think of oceans as so immense and bountiful that it's difficult to imagine any significant impact from human activity. Now we've begun to recognize how much of an impact we do have."***

***—Jane Lubchenco, marine ecologist (2002)***

When one looks out over the vast expanse of the ocean, it's hard to believe that it could be in serious trouble because of human activities. Throughout human history, the oceans have been known to possess a tremendous ability to assimilate waste materials, yet there is a limit to the amount of society's waste the oceans can hold. Earth's rapidly expanding human population has increased the amount of marine pollution, which puts an ever-increasing stress on the marine environment. As a result, negative effects in the marine environment are now being noticed worldwide, particularly in the coastal ocean and enclosed bodies of water where water circulation and vertical mixing are limited. These negative effects, along with impacts on land, have become large enough for humans to finally acknowledge our role in altering our planet's ecosystems on a global scale. In the United States, for example, comprehensive reports such as those from the U.S. Commission on Ocean Policy and the Pew Oceans Commission have identified an emerging national crisis regarding damage being done to the oceans, calling for a plan of action to restore vital ocean habitats.

Marine pollution comes from a variety of sources including marine transportation (see Diving Deeper 7.1), ocean mining operations, fishing practices, sewage sludge disposed at sea, large volumes of polluted runoff from land, and increasing use of seawater for human benefit (for example, using seawater as a coolant in coastal power plants). Marine pollution also comes from accidental spills of petroleum, industrial waste, toxic chemicals (such as dichloro-diphenyl-trichloroethane [DDT], polychlorinated biphenyls [PCBs], and mercury), and trash from just about everybody. These pollutants—both alone or in combination with each other—often have severe deleterious effects on individual organisms and may even affect entire marine ecosystems. In addition, there is a growing realization that heat and carbon dioxide injected into the ocean as a result of increases in atmospheric $CO_2$ is a type of marine pollution that is having a profound impact on not only marine ecosystems but also on physical processes in the ocean. This topic will be discussed in Chapter 16, "The Oceans and Climate Change."

In this chapter, we'll first examine a working definition of marine pollution. Then we'll explore the various types of marine pollution, including what can be done to reduce or eliminate pollutants from ocean waters. Finally, we'll examine the environmental problems of marine biological pollution.

## 11.1 ▸ What Is Pollution?

The ocean supplies humans with many commodities that are valuable in today's society, such as recreational opportunities, a source of water, an inexpensive form of transportation, extensive biological resources, and vast geologic resources within

◂ ***Sewage surfer*: A seahorse latches onto a plastic cotton swab.** To ride currents, seahorses clutch drifting seagrass or other debris. In the polluted waters off the Indonesian island of Sumbawa, this seahorse latched onto a plastic cotton swab—"a photo I wish didn't exist," says award-winning wildlife photographer Justin Hofman, who captured the image.

FOCUS ON THE ENVIRONMENT

DIVING DEEPER 11.1

## THE 2010 GULF OF MEXICO *DEEPWATER HORIZON* OIL SPILL

On April 20, 2010, the *Deepwater Horizon*, an offshore drilling platform operating in the Gulf of Mexico by British Petroleum (BP), was completing the final stages of drilling a deep oil well known as Macondo. The floating platform was located about 80 kilometers (50 miles) off the coast of Louisiana and in 1500 meters (5000 feet) of water beyond the edge of the continental shelf. Unexpectedly, it received a "kick" from a large bubble of natural gas that was under high pressure at the bottom of the well, 4 kilometers (2.5 miles) beneath the sea floor. The blowout preventer on the sea floor failed, and the natural gas bubble rushed to the drilling platform, which exploded and caught fire (**Figure 11A**), killing 11 crew members. Two days after the explosion, the *Deepwater Horizon* sank, leaving the well gushing crude oil at the seabed at a rate of about 9 million liters (2.4 million gallons) per day. Three months later, an underwater robotic vehicle was finally able to cap the well, but the amount of oil released became the world's second-largest oil spill (the world's largest accidental spill into the ocean) and by far the largest oil spill in U.S. waters.

In all, nearly 795 million liters (210 million gallons) of oil were released from the well. A deluge of chemical dispersants—detergent-like solvents designed to break up the oil but considered more toxic to marine life than the oil itself—were applied both at the surface and in deep water. Scientists estimate that BP removed about a quarter of the oil, most of it recovered directly from the well, burned at sea, or skimmed by boats. Another quarter of the oil evaporated or dissolved into scattered molecules. And a third quarter was either naturally dispersed in the water as small droplets (which might still be toxic to some organisms) or chemically dispersed. But the last quarter—around five times the amount released by the *Exxon Valdez*—hasn't been fully accounted for; this oil is called *residual oil*. Some scientists suggest that the majority of it formed slicks or sheens on the water (Figure 11A), washed onto local beaches and marshes, and accumulated as tar balls on the sea floor. Other scientists suggest that some oil never made it to the surface; instead, the oil formed diffuse plumes more than 1000 meters (3300 feet) below the surface and likely dispersed into the ocean, was biodegraded by microbes, or sunk to the sea floor as tar mats. Even years later, it's still not clear what happened to the missing oil.

Fortunately, most of the oil remained offshore, where wave energy combined with oxygen, sunlight, and the Gulf's abundant oil-eating bacteria were able to naturally biodegrade it. Some of the oil, however, washed onto local beaches and salt marshes, fouling the shore and killing marine animals. The oil that sank to the bottom and became entrained in low-oxygen sediments, like those on the sea floor—or a marsh—can hang around for decades, degrading the environment. The cost of the cleanup to BP and its partners, which includes settlement fees and environmental restoration, has exceeded $62 billion.

Birds, sea turtles, marine mammals, fish, and shellfish were some of the most affected marine organisms. Local fisheries were shut down after the spill but have been reopened since. It is unknown how this vast release of oil and chemical dispersants will affect marine organisms and the Gulf ecosystem, especially over time. As a tragic example, more than 1300 bottlenose dolphins (*Tursiops truncatus*) washed up dead in the Gulf of Mexico between 2010 and 2012 with lung and adrenal-gland lesions that are consistent with exposure to petroleum compounds. Long-term studies are under way that will reveal how much damage the spill has done to the Gulf's marine life.

### WHAT DID YOU LEARN?

Describe what happened to the spilled oil from the 2010 Gulf of Mexico *Deepwater Horizon* oil spill. Why was it fortunate that the location of the oil spill occurred offshore?

The *Deepwater Horizon* oil drilling platform on fire in the Gulf of Mexico.

Aerial view of a ship passing though an oil slick.

Map showing the extent of the 2010 Gulf of Mexico *Deepwater Horizon* oil spill.

Oil washing ashore at Fourchon Beach, Louisiana.

**Figure 11A The 2010 Gulf of Mexico oil spill.**

biodegradable, many marine pollution experts consider oil to be among the *least* damaging pollutants introduced into the ocean! Certainly oil spills appear ghastly as the oil coats the water, the shore, and helpless marine organisms, including seabirds, and it can cause grievous short-term damage. But oil dissipates and breaks down, becoming food for various microbes. As an example, vast quantities of oil were spilled in the Pacific and, especially, the Atlantic during World War II. In fact, some U.S. East Coast beaches were coated by oil several centimeters deep, yet not a trace of those spills remains today. In a broader view, some natural undersea oil seeps have occurred for millions of years, and the ocean ecosystem seems unaffected—or even *enhanced*—by them (because oil is a source of energy). As we'll see, other types of pollutants last far longer and can do much more damage than an oil spill.

Data from the *Exxon Valdez* oil spill are a case in point. The oil spill released almost 44 million liters (11.6 million gallons) of oil into a pristine wilderness area in Alaska. The affected waters were expected to have a long, slow recovery, but the fisheries that closed in 1989 bounced back with record takes in 1990. A study conducted 10 years after the spill revealed that several key species had rebounded to the point where their numbers were greater than before the spill (**Figure 11.7**). Scientific studies on the long-term effects of the spill, however, reveal that a significant amount of poorly weathered oil has seeped into intertidal sediments and remains below the surface.

Figure 11.6 **Firefighting boats attempt to put out the blaze on the Iranian oil tanker *Sanchi* in 2018 in the East China Sea.** The oil tanker *Sanchi* collided with another ship in the East China Sea off the eastern coast of China. The tanker, which was carrying 570,000 liters (150,000 gallons) of petroleum distillate, produced a large oil slick, caught fire, and eventually sank.

**OTHER CONCERNS ABOUT OIL IN THE OCEAN** Oil is a complex mixture of various hydrocarbons and other substances, including the elements oxygen, nitrogen, sulfur, and various trace metals. When this complex chemical mixture combines with seawater—another complex chemical mixture that also contains organisms—the results are usually devastating for marine organisms. Studies have revealed, for example, that crude oil in seawater at concentrations of only 0.7 parts per billion kills or damages certain fish eggs. In addition, many organisms are killed outright when they are coated by oil, rendering their insulating feathers or fur useless (**Figure 11.8**).

A **toxic compound** (*toxicum* = poison) is defined as a poisonous substance that has the capability of causing injury or death, especially by chemical means. Toxic compounds in crude oil vary, but the most worrisome are polycyclic aromatic hydrocarbons (PAHs), such as napthalenes, benzene, toluene, and xylenes. Even in small doses, these compounds can sicken humans, animals, and plants. Further, these hydrocarbons are particularly dangerous if inhaled or ingested by animals because they can be transformed into even more toxic products, which can cause mutations in an organisms' DNA.

There is also concern about the long-term effects of oil spills, such as chronic, delayed, or indirect impacts, many of which are often difficult to document and link to the spill because of the considerable time lag. For example, exposure to even tiny concentrations of the chemicals present in oil can cause harmful biological effects that usually go unnoticed. Scientific studies have shown that fish exposed to PAHs in crude oil exhibited changes in gene expression that are linked to developmental abnormalities, decreased embryo survival, and lower reproductive success that wouldn't be noticed until years

**EXPLORING DATA** ▼ **For the four key organisms shown in the figure that were affected by the *Exxon Valdez* oil spill in Alaska, how did each respond in the years after the spill?**

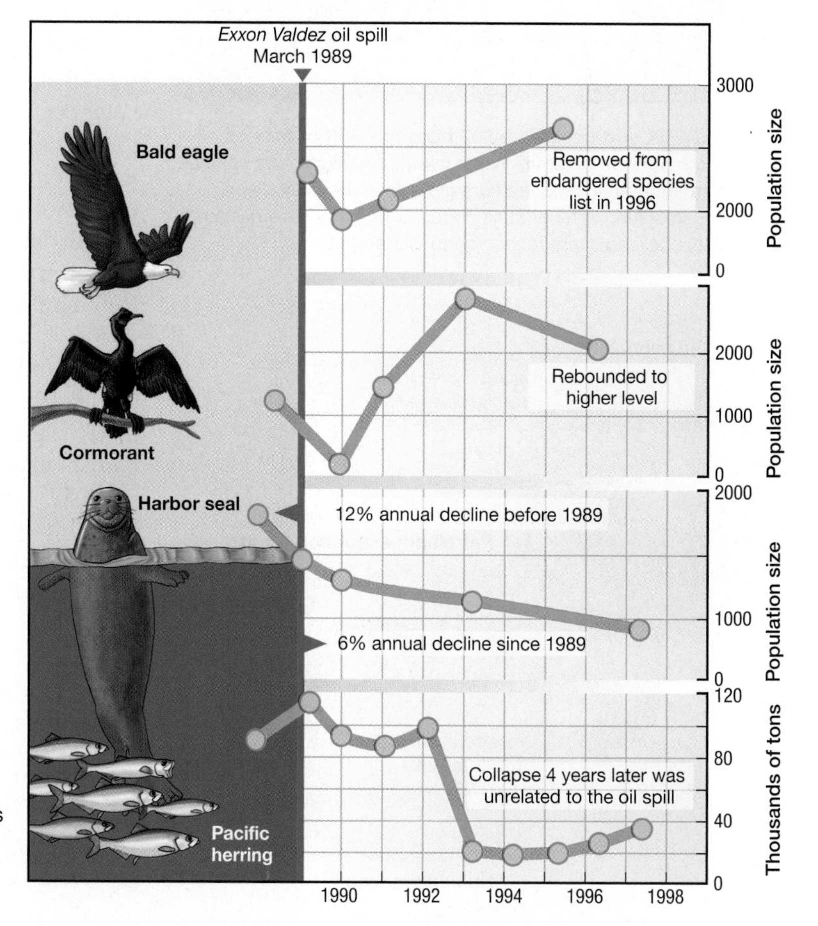


Figure 11.7 **Recovery of organisms affected by the *Exxon Valdez* oil spill.** The populations of several key organisms in the Prince William Sound area of Alaska have rebounded since the 1989 *Exxon Valdez* oil spill. The bald eagle is so numerous that it was removed from the endangered species list in 1996. The collapse of the Pacific herring population four years after the spill was caused by disease unrelated to the oil spill.

Figure 11.8 **A bird covered by oil from the Gulf of Mexico *Deepwater Horizon* oil spill.** When marine organisms are covered by oil from an oil spill, their feathers or fur lose their insulation properties, resulting in high fatality rates. Some marine organisms, such as this pelican, were rescued and cleaned of oil.

later. Another example is from the oil tanker that ran aground in the Galápagos Islands in 2001, spilling roughly 3 million liters (800,000 gallons) of diesel and bunker oil. The oil spread westward and was dispersed by strong currents, so only a few marine animals were immediately killed. Marine iguanas on a nearby island, however, suffered a massive 62% mortality in the year after the accident, due to a small amount of residual oil contamination in the sea.

In addition, spills from oil tankers may release a wide variety of petroleum products (not just crude oil), each of which has a different level of toxicity and behaves differently in the environment. For example, refined oil, such as fuel oil, is rich in compounds that are much more toxic to the environment than crude oil.

Although spills from oil tankers receive much media attention, they are not the primary source of oil to the oceans. Remarkably, the overwhelming majority of the petroleum that enters the oceans due to human activity is the result of small but frequent and widespread releases of oil related to activities that consume petroleum. Figure 11.9 shows that, although 47% of worldwide oil to the oceans is caused by underwater *natural oil seeps* (many of which occur in U.S. waters), of the remaining 53% attributed to human sources, 72% comes from *petroleum consumption*, which includes non-tank vessels, runoff from increasingly paved urban areas, and individual car, boat, and watercraft owners; 22% comes from *petroleum transportation*, including refining and distribution activities; and only 6% comes from *petroleum extraction*, which is associated with oil and gas exploration and production.

**CLEANING OIL SPILLS** When oil enters the ocean, it initially floats because oil is less dense than water and forms a slick at the surface, where it starts to break down through natural processes (**Figure 11.10**). The volatile, lighter components of crude oil evaporate over the first few days, leaving behind a more viscous substance that aggregates into tar balls and eventually sinks. The tarry oil also coats suspended particles, which settle to the sea floor, too.

If the floating oil hasn't dispersed, it can be collected with specially designed skimmers or absorbent materials. The collected oil (or oiled materials), however, must still be disposed of elsewhere. Waves, winds, and currents serve to further disperse an oil slick and mix the remaining oil with water to make a frothy emulsion called *mousse*. In addition, bacteria combined with the process of photo-oxidation by sunlight act to break down the oil into compounds that dissolve in water.

Microorganisms such as bacteria and fungi naturally biodegrade oil, so they can be used to help clean oil spills—a method called **bioremediation** (*bio* = biologic, *remedium* = to heal again). Virtually all marine ecosystems harbor naturally occurring bacteria that degrade hydrocarbons. Although certain types of bacteria and fungi can break down particular kinds of hydrocarbons, none is effective against all forms. In 1980, however, microbiologists discovered a microorganism capable of breaking down nearly two-thirds of the hydrocarbons in most crude oil spills.

Releasing bacteria directly into the marine environment is one form of bioremediation. For example, a strain of oil-degrading bacteria was released into the Gulf of Mexico to test its effectiveness in cleaning up about 15 million liters

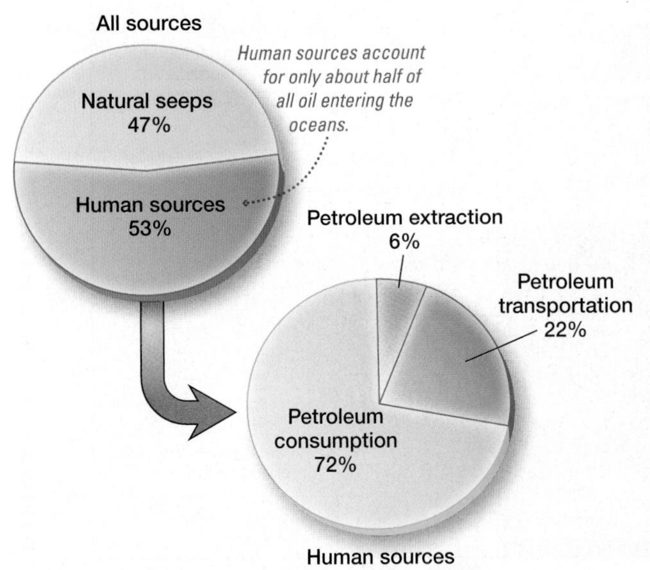


Figure 11.9 **Sources of oil to the oceans.** Of worldwide oil to the oceans, 47% comes from natural seeps, while 53% comes from human sources. Of all human sources, 72% comes from petroleum consumption activities, such as individual car and boat owners, non-tank vessels, and runoff from increasingly paved urban areas. Surprisingly, combined petroleum transportation and extraction account for only 28% of all human-caused spilled oil.

**EXPLORING DATA ◄ Describe the activities that contribute to marine pollution in each of the three categories of human sources of oil to the ocean. Of the three, which is the greatest source of oil to the ocean?**

**SmartFigure 11.10 Processes acting on oil spills.** When oil enters the ocean, it is acted upon by various natural processes that break up the oil. The lighter components evaporate, while the heavier components form tar balls or coat suspended particles and sink. The remaining dispersed oil photo-oxidizes or can mix with water, creating a frothy substance called "mousse."
https://goo.gl/8kHbgP

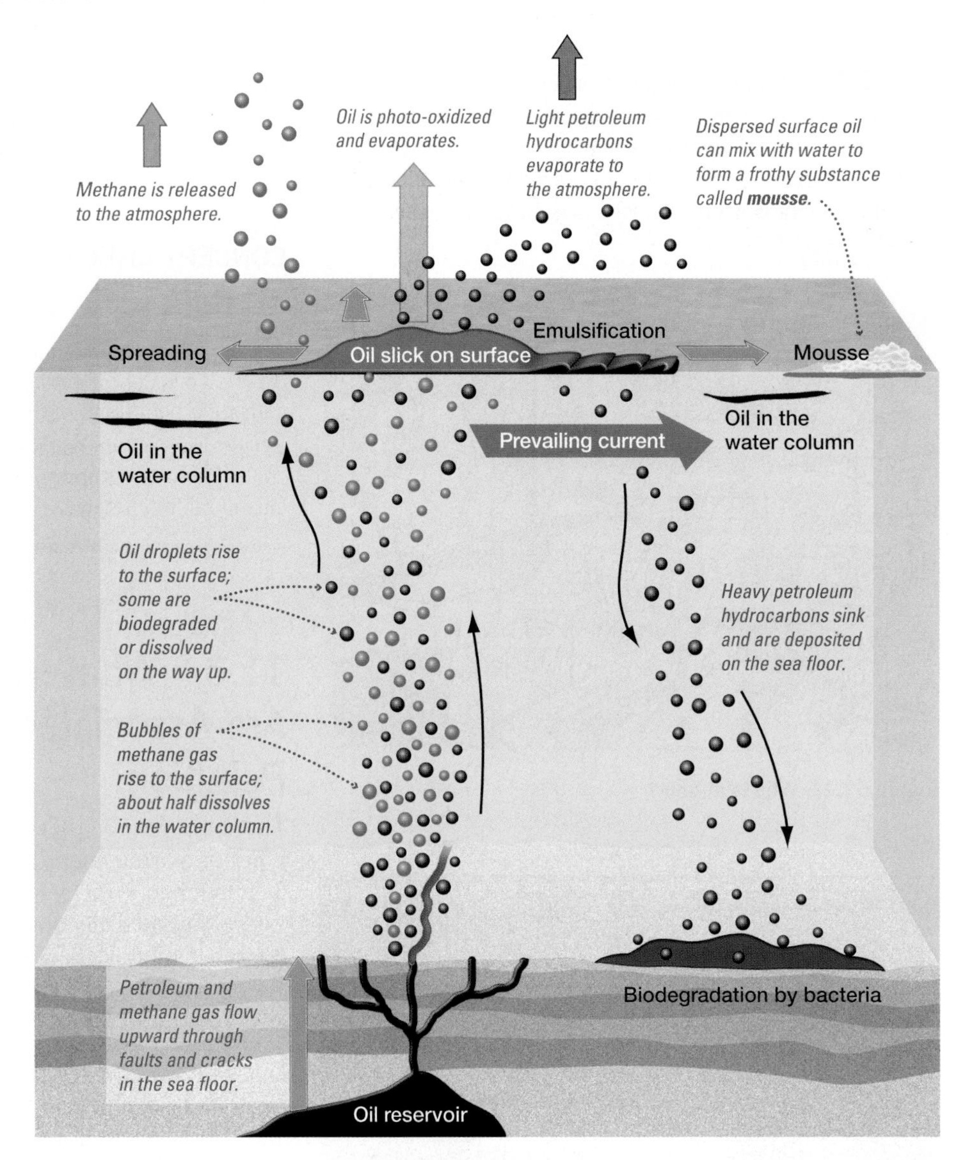


(4 million gallons) of crude oil spilled after an explosion disabled the tanker *Mega Borg* in 1990. Results indicate that the bacteria reduced the amount of oil and had no negative effects on the area's ecology.

Providing conditions that stimulate the growth of naturally occurring oil-degrading bacteria is another form of bioremediation. Exxon, for example, spent $10 million to spread fertilizers rich in phosphorus and nitrogen on Alaskan shorelines to boost the growth of indigenous oil-eating bacteria after the *Exxon Valdez* spill. The resulting cleanup rate was more than twice the rate that occurs under natural conditions.

**PREVENTING OIL SPILLS** One of the best ways to protect areas from oil spills is to prevent spills from occurring in the first place. Because our society relies on petroleum products, however, oil spills are a likely occurrence in the future (**Figure 11.11**), especially as petroleum reserves beneath the continental shelves of the world are increasingly exploited.

After the *Exxon Valdez* oil spill in 1989, Congress enacted the Oil Pollution Act of 1990, which defined responsibility for fiscal damage and cleanup. The act also phased out single-hulled oil tankers traveling in U.S. waters and mandated double-hulled construction by 2015. Currently, single-hulled tankers are barred from U.S. ports, and European countries, such as France and Spain, do not allow them within 320 kilometers (200 miles) of their coasts. A double hull houses two layers; an inner hull can prevent oil spillage if damage occurs to the outer hull. Studies of hull designs during groundings and collisions indicate that double-hull designs are more effective overall at reducing oil spills. However, analysis of the *Exxon Valdez* spill suggests that even a double-hulled tanker would not have prevented the disaster. Tanker designs are also being modified to limit the amount of oil spilled in the event of a hull rupture.

In February 1999, the Japanese-owned freighter M/V *New Carissa* ran aground just offshore of Coos Bay, Oregon, with nearly 1.5 million liters (400,000 gallons) of tar-like fuel oil aboard, which began leaking through cracks in its hull. When the ship washed into the surf zone and an approaching storm threatened to tear it apart, federal and state authorities decided to ignite the vessel and its fuel rather than risk a larger oil spill (**Figure 11.12**). This was the first time that oil on a ship in U.S. waters was intentionally burned to prevent an oil spill. Eventually, the ship split in two, and about half of its oil burned, limiting the amount of oil spilled into the ocean. Most of the remaining oil was sunk with the wrecked ship a month later, when it was towed offshore and sunk in water 3 kilometers (1.9 miles) deep by U.S. Naval gunfire and a torpedo.

## STUDENTS SOMETIMES ASK . . .

### *What is the best way to clean oiled animals?*

Detergents, dispersants, and de-greasers have all been tried by rescue workers to clean the fur and feathers of oiled animals. Remarkably, experts say that Dawn® dishwashing liquid works best because of its chemical surfactants, which allow it to cut grease but not harm the animal's skin. Although its exact formula is a trade secret, Dawn® does contain petroleum products that aid its cleaning ability.

The process of cleaning an oiled animal is quite complex. For example, it takes three people up to an hour to clean an oiled pelican. The workers start by rubbing the bird with cooking oil to loosen the sticky petroleum. After spraying the pelican with dish liquid, the crew then scrubs the feathers to work the soap into the feathers. The final step is thoroughly rinsing the bird to wash away the mixture of oil and soap.

Survival rates of cleaned birds vary between 50–80% but can be higher, based on species, toxicity of the oil, how rapidly a bird is treated, and the bird's prior health.

**RECAP** **Although transportation accidents and other oil spills pose a problem, natural seeps and ordinary human activities such as boating and auto use are a far greater source of petroleum into the marine environment. Microbes such as bacteria naturally biodegrade oil in the ocean.**

Figure 11.11 **Who is at fault?**

Figure 11.12 **The *New Carissa* on fire off the Oregon coast.** When the freighter M/V *New Carissa* ran aground in 1999 in shallow water offshore Coos Bay, Oregon, and began leaking oil, it was intentionally set on fire to prevent further oil from spilling into the ocean.

In addition to the chemicals that comprise petroleum, a variety of other chemical compounds—including sewage sludge—is released into the ocean. In sufficient quantities, these materials are often responsible for marine pollution incidents. We'll explore these substances next.

**CONCEPT CHECK 11.2 ▸ Specify the marine environmental problems associated with petroleum pollution.**

1 Explain why many marine pollution experts consider oil among the *least* damaging pollutants in the ocean.

2 Discuss techniques used to clean oil spills. Why is it important to begin the cleanup immediately?

3 Describe the world's largest accidental and intentionally released oil spills. How many times larger was each than the *Exxon Valdez* oil spill?

## 11.3 ▸ What Marine Environmental Problems Are Associated with Non-Petroleum Chemical Pollution?

There are many other types of pollution besides petroleum that are considered types of marine chemical pollution. Examples include sewage sludge, DDT, PCBs, mercury, and even chemicals contained in prescription and non-prescription drugs. Let's examine some of these types of chemical pollutants, including how these substances get into the ocean.

### Sewage Sludge

One of the main types of marine chemical pollution is **sewage sludge**. Sewage treated at a facility typically undergoes **primary treatment**, where solids are allowed to settle and separate from the liquid, and **secondary treatment**, where it is exposed to bacteria-killing chlorine. Sewage sludge is the semisolid material that remains after such treatment. It contains a toxic brew of human waste, oil, zinc, copper, lead, silver, mercury, pesticides, and other chemicals. Since the 1960s, at least 500,000 metric tons (1.1 billion pounds) of sewage sludge have been dumped into the coastal waters of Southern California, and more than 8 million metric tons (18 billion pounds) of sewage sludge has been dumped into the New York Bight between Long Island and the New Jersey shore.

Although the Clean Water Act of 1972 prohibited the dumping of sewage into the ocean after 1981, the high cost of treating and disposing of sewage sludge on land resulted in extension waivers being granted to many municipalities. In the summer of 1988, however, non-biodegradable debris including medical waste—probably carried by heavy rains into the ocean through storm drains—washed up on Atlantic coast beaches and adversely affected the tourist business. Although this event was completely unrelated to sewage disposal at sea, it focused public awareness on ocean pollution and helped pass legislation that makes it illegal to dump sewage sludge at sea.

**NEW YORK'S SEWAGE SLUDGE DISPOSAL AT SEA** Sewage sludge from New York and Philadelphia has traditionally been transported offshore by barge and dumped in the ocean at sites totaling 150 square kilometers (58 square miles) within the New York Bight sludge site and the Philadelphia sludge site (**Figure 11.13**).

The water depth is about 29 meters (95 feet) at the New York Bight sludge site and about 40 meters (130 feet) at the Philadelphia sludge site. The water column in such shallow water is relatively uniform, so even the smallest sludge particles reach the bottom without undergoing much horizontal transport, and the ecology of the dump site can be severely affected. At the very least, such a concentration of organic and inorganic matter seriously disrupts the chemical cycling of nutrients. Greatly reduced species diversity results, and in some locations, the result is an overabundance of algae that causes dissolved oxygen to be reduced to very low levels. (Water with very low dissolved oxygen creates *hypoxic* [*hypo* = under, *oxic* = oxygen] conditions that can kill marine life; for details, see Section 13.2 in Chapter 13, "Biological Productivity and Energy Transfer.")

In 1986, the shallow-water sites were abandoned, and sewage was subsequently transported to a deep-water site 171 kilometers (106 miles) out to sea (Figure 11.13). The deep-water site is beyond the continental shelf break, so there is usually a well-developed density gradient that separates low-density, warmer surface water from high-density, colder deep water. Internal waves moving along this density gradient can horizontally transport particles at rates 100 times faster than they sink.

Local fishermen reported adverse effects on their fisheries soon after deep-water dumping began. Also, concern was expressed that the sewage could be transported great distances in eddies of the Gulf Stream (see Chapter 7), even as far as the coast of the United Kingdom. This program was terminated in 1993, and municipalities must now dispose of their sewage on land.

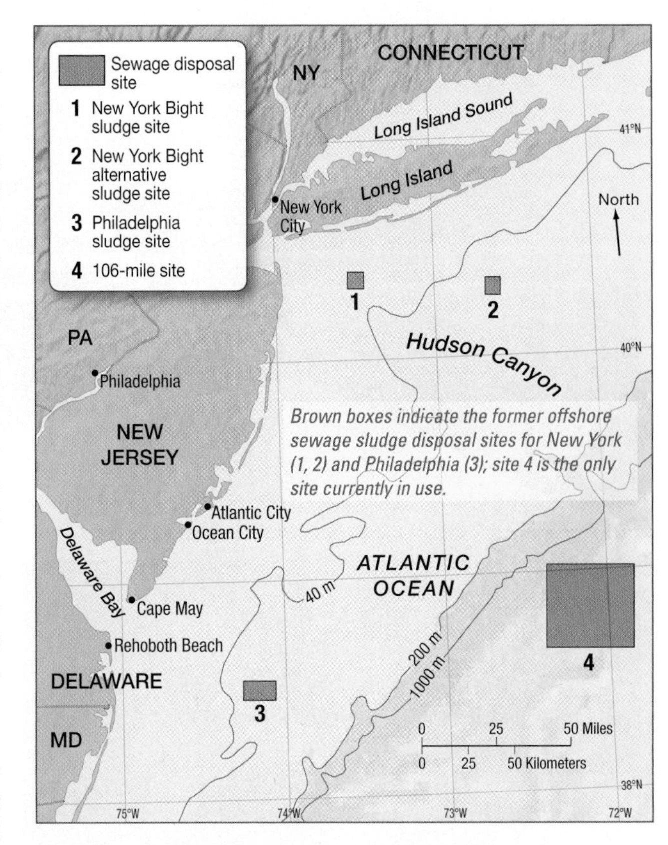


**Figure 11.13 Atlantic sewage sludge disposal sites.** Prior to 1986, more than 8,000,000 metric tons (18 billion pounds) of sewage sludge was dumped each year by barge at the New York Bight sludge sites (*1* and *2*) and the Philadelphia sludge site (*3*). In 1986, the dump site was moved to the larger and deeper-water 171-kilometer (106-mile) site (*4*).

**BOSTON HARBOR SEWAGE PROJECT** Prior to the 1980s, some 48 different communities that comprise the greater Boston area used an antiquated sewage system to dump sludge and partially treated sewage at the entrance to Boston Harbor. Tidal currents often swept the sewage back into the bay, and at other times, the system became overloaded and dumped raw sewage directly into the bay, making Boston Harbor one of the most polluted bays in the country.

A court-ordered cleanup of Boston Harbor in the 1980s resulted in the construction of a new waste treatment facility at Deer Island, which came online in 1998. The facility treats all sewage with bacteria-killing chlorine and carries it through a tunnel 15.3 kilometers (9.5 miles) long into deeper waters offshore (**Figure 11.14a**), which prevents it from returning to the bay. Since the cleanup of Boston Harbor, beaches have reopened, clammers are digging again, and marine life—including harbor seals, porpoises, and even whales—has returned. To pay for the $3.8 billion sewage system, however, the average annual sewage bill for a Boston-area household was increased to about $1200, more than five times what it had been.

Despite the benefits, at the time it was proposed, some opponents feared that the project would degrade the environment in Cape Cod Bay and Stellwagen Bank (**Figure 11.14b**), which is an important whale habitat. In 1992, six years before the facility went online, this area was designated as a U.S. National Marine Sanctuary, which restricts the amount of sewage that can be dumped there.

## DDT and PCBs

The pesticide **DDT** and the industrial chemicals called **PCBs** are now found throughout the marine environment. They are persistent, biologically active chemicals that have been introduced into the oceans entirely as a result of human activities. Because of their toxicity, persistence, and propensity for accumulating in food chains, these and other chemicals have been classified as *persistent organic pollutants (POPs)* capable of causing cancer, birth defects, and other grave harm.

DDT was widely used in agriculture during the 1950s and improved crop production throughout developing countries for several decades. However, its extreme effectiveness as an insecticide and persistence as a toxin in the environment

*The Boston Harbor sewage project includes tunnels that transport sewage to an outfall 15 kilometers (9.5 miles) offshore at a depth of 76 meters (250 feet) beneath the ocean floor.*

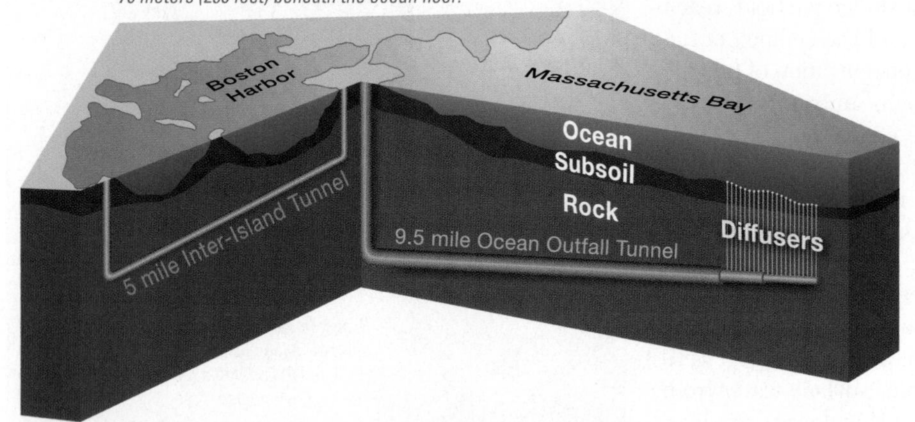


**(a)** Diagrammatic view of the Boston Harbor sewage tunnels.

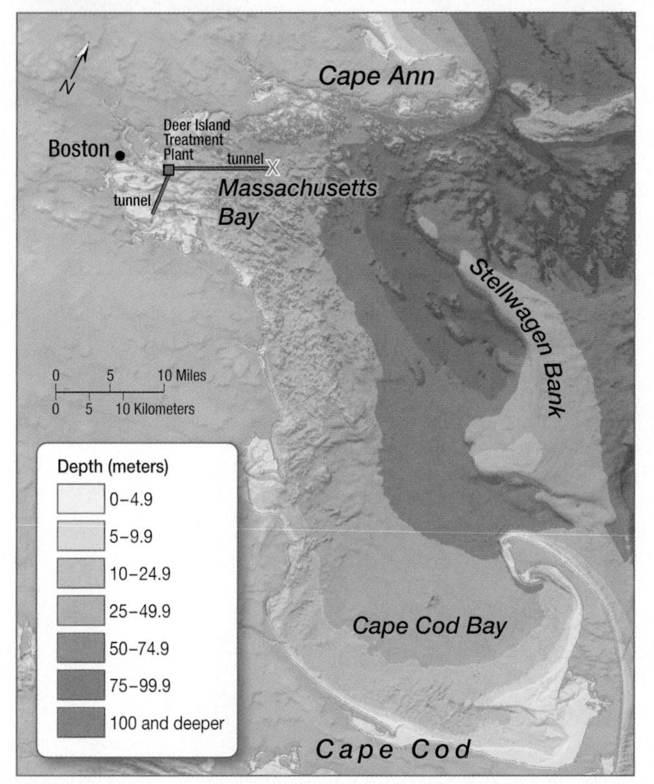


**(b)** Bathymetric map of the coastal ocean in the Boston-Cape Cod area, showing the proximity of the new sewage outfall *(red x)* to Stellwagen Bank, which is a National Marine Sanctuary.

**Figure 11.14 Boston Harbor sewage project.** (**a**) Diagrammatic view of the Boston Harbor sewage project tunnels and (**b**) a map of the sewage outfall area.

eventually resulted in a host of environmental problems, including devastating effects on marine food chains. In 1962, biologist Rachel Carson published her seminal book *Silent Spring* about the dangers of DDT and other chemicals in the environment. Her book played an important role in the environmental movement, and DDT was subsequently banned in the United States in 1972.

PCBs are industrial chemicals that were once widely used as liquid coolants and insulation in industrial equipment such as power transformers, where they were released into the environment. PCBs were also widely used in wiring, paints, caulks, hydraulic oils, carbonless copy paper, and a host of other products. PCBs have been shown to cause liver cancer and harmful genetic mutations in animals. PCBs can also affect animal reproduction; they have been indicated as causes of spontaneous abortions in sea lions and the death of shrimp in Escambia Bay, Florida.

**DDT AND EGGSHELLS** Since 1972, the EPA has banned the use of DDT in the United States. Worldwide, the pesticide is banned from agricultural use, but it continues to be used in limited quantities for public health purposes. As a result, U.S. companies continue to produce DDT and supply it to other countries.

The danger of excessive use of DDT and similar pesticides first became apparent in the marine environment when it affected marine bird populations. During the 1960s, there was a serious decline in the brown pelican population of Anacapa Island off Southern California (**Figure 11.15**). High concentrations of DDT in the fish eaten by the birds had caused them to produce eggs with excessively thin shells because DDT made it more difficult for calcium to be incorporated into the egg shells.

The osprey is a bird of prey, similar to a large hawk, that is common in coastal waters. The osprey population of Long Island Sound declined in the late 1950s and 1960s because DDT contamination caused them to produce eggs with thin shells, too. Since the ban on DDT, the osprey, brown pelican, and many other species affected by the chemical are making remarkable comebacks.

**DDT AND PCBS LINGER IN THE ENVIRONMENT** The chemicals DDT, which was banned in 1972, and PCBs, which were banned in 1977, generally enter the ocean through the atmosphere and river runoff. They are initially concentrated in the thin slick of organic chemicals at the ocean surface, and then they gradually sink to the bottom, attached to sinking particles. A study off the coast of Scotland indicated that open-ocean concentrations of DDT and PCBs are 10 and 12 times less, respectively, than in coastal waters. Long-term studies have shown that DDT residue in mollusks along the U.S. coasts peaked in 1968.

Although most countries have banned their use, DDT and PCBs are so pervasive in the marine environment that even Antarctic marine organisms contain measurable quantities of them. Because there has been no agriculture or industry in Antarctica to introduce them directly, these chemicals must have been transported to Antarctica from distant sources by winds and ocean currents.

## Mercury and Minamata Disease

The elemental metal **mercury**, a silvery metal with the rare property of being liquid at normal room temperature, has many industrial uses. For example, mercury is used in the manufacture of industrial chemicals and for electrical and electronic applications. Mercury in its gaseous form is also used in fluorescent lighting, so it is a very useful substance.

Where does the mercury in the ocean come from? Although some oceanic mercury comes from natural sources, such as volcanic eruptions and leaching from mercury-rich rocks, about two-thirds comes from human activities. The biggest single source is the burning of fossil fuels, especially coal, which releases mercury into the atmosphere, which then gets washed into the ocean through precipitation and runoff. Humans also discharge mercury-laden industrial effluent directly into rivers or the ocean. And the improper disposal of mercury batteries is a concern, too. In fact, scientists have determined that mercury levels in the ocean have increased by as much as six times since the beginning of the Industrial Revolution.

But even large discharges of mercury wouldn't pose a major threat to human health if the mercury weren't converted by bacteria in low-oxygen environments to its toxic form, *methylmercury*, which diffuses into phytoplankton and then passes up marine food chains in ever-accumulating quantities. Exposure to and consumption of methylmercury, a known neurotoxin, can cause serious human health issues.

**Figure 11.15 Survival of brown pelicans threatened by DDT.** Brown pelicans (*Pelecanus occidentalis*) that breed on Anacapa Island offshore Southern California were found to have high levels of DDT, which decreased the thickness of their eggshells. Since DDT has been banned, healthy pelicans (*photo*) have returned to these waters.

**MINAMATA DISEASE** A chemical plant built in 1938 on Minamata Bay, Japan, produced acetaldehyde, which requires mercury in its manufacture. Industrial wastewater that contained methylmercury was discharged into Minamata Bay, where the chemical was later ingested and concentrated in the tissues of marine organisms such as fish and shellfish. The first ecological changes in Minamata Bay were reported in 1950, and human effects were noted as early as 1953. The mercury poisoning that is now known as **Minamata disease** became epidemic in 1956, when the plant was only 18 years old.

Minamata disease is a degenerative neurological disorder that affects the human nervous system and causes sensory disturbances, including blindness and tremor, brain damage, birth defects, paralysis, and even death. This mercury poisoning was the first major human disaster resulting from ocean pollution. However, the Japanese government did not declare mercury as the cause of the disease until 1968. The plant was immediately shut down, but more than 100 people were known to acutely suffer from Minamata disease by 1969 (**Figure 11.16**), and almost half of them died as a result of the disease.

A second acetaldehyde plant was closed in 1965 to the north in Niigata, Japan, amid evidence that similar methylmercury discharges into the bay were poisoning people there as well. Many lessons were learned from Minamata and the investigation into the cause of the outbreak in Niigata proceeded much more smoothly than it had in Minamata.

As of 2001, 2265 Japanese victims of Minamata disease—including 1784 fatalities—had been officially recognized. Today the concentration of methylmercury in Minamata Bay is no longer unusually high, suggesting that there has been enough time for the mercury to be widely dispersed within the marine environment.

**BIOACCUMULATION AND BIOMAGNIFICATION** During the 1960s and 1970s, methylmercury contamination in seafood received considerable attention. Certain marine organisms concentrate within their tissues many substances found in minute concentrations in seawater in a process called **bioaccumulation**. When animals eat other animals, some of these substances (including toxic chemicals) move up food chains and become concentrated in the tissues of larger animals, in a process called **biomagnification** (**Figure 11.17**). Because the amount of mercury in the ocean has been increasing, some seafood such as tuna and swordfish contains unusually high amounts of mercury.

## STUDENTS SOMETIMES ASK . . .

***I've heard that some organizations want to lift the ban on DDT. Why would they want to do that?***

Since production of DDT was banned in 1972, outbreaks of malaria have dramatically increased because DDT was the most effective and readily available pesticide used to kill mosquitoes that transmit malaria. According to the World Health Organization, malaria infects up to 200 million people a year—mostly in tropical regions—and kills as many as 450,000, including at least 1 child every two minutes. In addition, drug-resistant strains of malaria have begun to show up worldwide. This resurgence of malaria has caused many health organizations to call for an exception to the ban on DDT—despite its well-documented perseverance and negative effects on the environment—so that it can be used selectively to spray houses in malaria-prone areas such as tropical Africa and Indonesia. The ambitious Global Malaria Action Plan seeks to eliminate all malaria deaths through research into vaccines and the use of bed nets, drugs that treat the disease, and spraying with DDT.

Figure 11.16 **A mother holding a child victim of Minamata disease.** Ingestion or contact with the toxic metal mercury can cause Minamata disease, which affects the human nervous system and can result in brain damage, birth defects, paralysis, and even death.

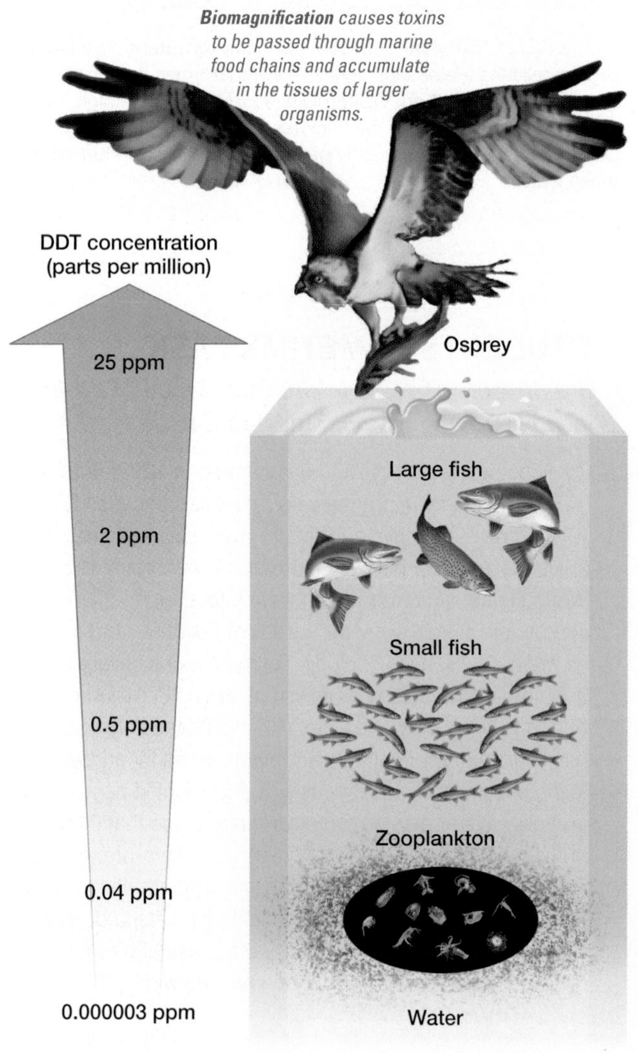


**HOW MUCH MERCURY-RICH SEAFOOD IS SAFE TO EAT?** Studies conducted on the amount of seafood consumed by various human populations have helped establish safe levels of methylmercury in marketed fish. (Note that most animals can get rid of elemental mercury quickly, and so it does not pose the same threat as methylmercury.) To establish these levels, three variables were considered:

1. The rate at which each group of people consumed fish
2. The methylmercury concentration in the fish consumed by that population
3. The minimum ingestion rate of methylmercury that induces disease symptoms

These three variables help establish a maximum allowable methylmercury concentration that will safeguard people from mercury poisoning, as long as they don't exceed the recommended intake of fish.

**Figure 11.18** shows the relative risk of contracting Minamata disease for people in the United States, Sweden, and Japan, including those from the Minamata fishing community. The graph shows that the risk increases with increased consumption of fish and that the higher the methylmercury concentration of the fish, the greater the risk.

Figure 11.18 is a complex graph that shows the methylmercury concentration in fish, fish consumption rates for various populations, and the danger levels of mercury poisoning. Scientists have determined that the minimum level of methylmercury consumption that causes poisoning symptoms is 0.3 milligrams (0.00001 ounce) per day. Using a safety factor of 10 times, the safe ingestion level is set at 0.03 milligrams (0.000001 ounce) of methylmercury per day. The graph shows that the higher the consumption of fish—especially fish contaminated with mercury—the greater the chance of mercury poisoning symptoms. For example, the graph shows that a high rate of consumption of high mercury concentration fish (such as Minamata fish) results in extreme danger of mercury poisoning. The graph also shows that people in the United States consume, on average, 17 grams (0.6 ounce) of fish per day. At this rate, the chance of mercury poisoning symptoms first occurs when methylmercury concentrations in fish exceed 20 parts per million (ppm), and the safe ingestion level is 2.0 ppm methylmercury in fish.

Using the established safe U.S. ingestion level of 2.0 ppm methylmercury in fish, the U.S. Food and Drug Administration (FDA) doubled the safety factor and established a limit of methylmercury concentration for fish at 1.0 ppm. Based on consumption rates, this limit has adequately protected the health of U.S. citizens because essentially all tuna and most swordfish fall below this concentration. Still, the FDA issued an advisory in 2001 stating that pregnant women, women of childbearing age, nursing mothers, and young children should avoid eating certain kinds of fish that may contain high levels of methylmercury, such as swordfish, sharks, king mackerel, and tilefish.

Figure 11.18 shows that for people in Sweden and Japan, the methylmercury concentration of fish deemed to be at a safe level is lower because these populations eat more fish. The graph also shows the extreme danger to which residents of Minamata were inadvertently subjected when they ate so much of the highly contaminated fish from Minamata Bay.

## Other Types of Chemical Pollutants

Non-prescription, prescription, and illegal drugs all enter municipal waste treatment systems, which separate solid materials but do not remove all the dissolved chemicals from processed water. As such, these chemicals make it to the ocean

SmartFigure 11.17 **How biomagnification concentrates toxins in higher-level organisms.**
https://goo.gl/P96fU2

**EXPLORING DATA** ◀ **How many times more concentrated is DDT at each step of the biomagnification chain?**

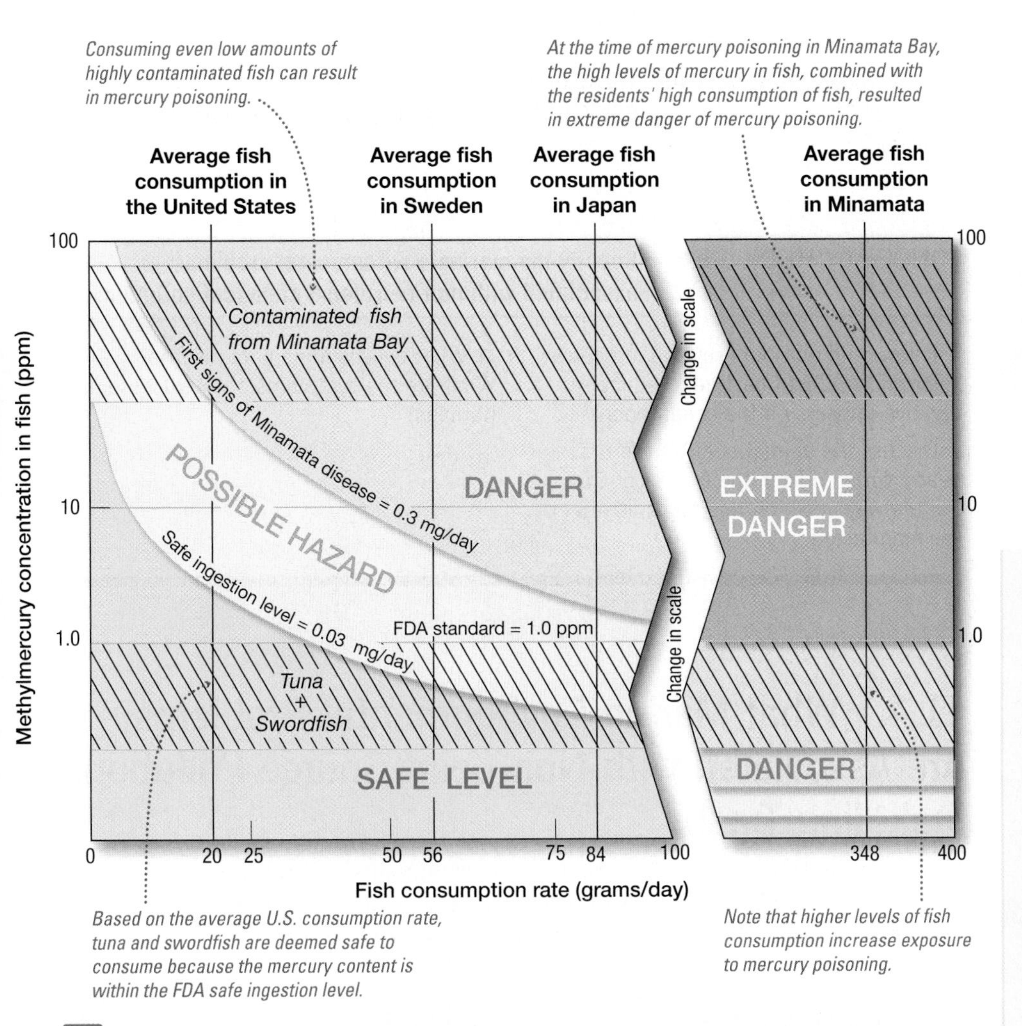


SmartFigure 11.18 **Methylmercury concentration in fish, fish consumption rates for various populations, and the danger levels of mercury poisoning.** Graph showing the relative risk of contracting Minamata disease based on the amount of fish consumed and the concentration of toxic methylmercury in fish for people in the United States, Sweden, and Japan—including Minamata. The graph shows that a high consumption rate of fish containing high concentrations of mercury results in extreme danger of mercury poisoning (*upper right*) and that tuna and most swordfish are safe to consume at average U.S. consumption rates (*lower left*). Although a typical serving portion of fish is about 170 grams (6 ounces), the graph shows that the average U.S. fish consumption rate is 17 grams (0.6 ounce) per day. https://goo.gl/YspqKE

**EXPLORING DATA**

1. **How many grams of tuna and/or swordfish would you have to consume every day in order to ingest possibly hazardous levels of mercury? Assume both fish meet the FDA standard of 1.0 ppm of mercury.**
2. **Approximately how many typical servings of tuna and/or swordfish per day does your answer to Question 1 correspond to?**

in the remaining liquid-waste stream, although normally in very low concentrations. One example is hormones from both pharmaceuticals prescribed for human use and those naturally produced by humans, which have survived sewage treatment and are discharged into the environment, eventually making their way to the ocean. Hormones employed by intensive livestock operations may also be washed into the oceans. Another example is caffeine, which is a noted stimulant that is present in a variety of popular beverages. As caffeine is passed through human bodies in urine, it becomes part of the waste disposal stream that eventually enters the oceans. Studies show that amount of caffeine in sewage effluent has been increasing over time; this upsurge has an unknown effect on marine organisms.

In addition, industrial chemicals enter the oceans via water runoff from land. Fertilizers are one example of an industrial chemical that is spread on lawns to

## STUDENTS SOMETIMES ASK . . .

***How has the radioactivity that leaked from Japan's Fukushima power plant affected marine fish? For example, is it safe to eat fish from the ocean?***

The failure of the cooling system in Japan's Fukushima Daiichi nuclear power plant as a result of the destructive 2011 tsunami caused explosions and the release of radioactive substances into the atmosphere, groundwater, and ocean surrounding the plant. Even though the ocean is large and has good mixing mechanisms that serve to dilute the radioactivity, radioactive cesium and iodine from Fukushima have been transported by ocean surface currents and subsequently detected in waters along North America's West Coast. However, the radioactivity is present in such tiny amounts that the levels are not deemed a human health hazard. Overall, a much greater human-exposure radiation issue remains for the people who live and work in the contaminated area on land in Japan.

In the ocean, marine organisms in the vicinity of Fukushima received some of the highest doses of radioactive materials, and so commercial fishing is closed in those regions. However, large ocean-going fish that eat contaminated organisms can biomagnify the radioactive materials and carry those substances throughout the oceans. For example, Pacific bluefin tuna that migrate between the waters off San Diego, California, and near Japan have been found to carry tiny amounts of Fukushima cesium **(Creature Feature 11.1)**. Even so, these fish are deemed safe to eat. Marine chemists who study probability statistics suggest that a person who eats five times the amount of fish that an average American does—and eats only contaminated fish for a year—would end up with a radiation dose that would cause an extra two cancers in 10 million people. So the risk to human health by eating contaminated fish is extremely small. In fact, the risk is hundreds of times *less* than that posed by the ingestion of polonium-210, a naturally-occurring radioactive element present in small amounts in seafood, although most people aren't concerned about it.

**RECAP** **Many other types of pollution besides petroleum are considered marine chemical pollutants, including sewage sludge, DDT, PCBs, mercury, and chemicals contained in prescription and non-prescription drugs.**

## CREATURE FEATURE 11.1

### I'm a trans-oceanic swimmer!

From the spawning grounds in the Sea of Japan where they are born, young **Pacific bluefin tuna** (*Thunnus orientalis*) embark on a journey of over 8000 kilometers (5000 miles) across the Pacific Ocean to arrive near the California coast where they spend several years feeding and growing before returning to the Sea of Japan to spawn.

Although studies have found traces of radioactive contamination from Fukushima's 2011 nuclear power plant disaster in water and tuna tissue samples collected along the California coast, the concentration of radioactive isotopes is well below levels of concern for humans or marine life. In Japan, however, coastal fisheries nearest the reactors remain closed because of concern over exposure by some species, particularly those that live on or near the sea floor.

Figure 11.19 **A labeled storm drain that leads to the ocean.** Although many people believe that storm drain runoff is processed by sewage treatment plants, any material that enters storm drains goes directly into streams or the ocean.

promote growth. When these fertilizers are washed off land and make their way to coastal waters, they can cause the overabundance of algae known as a *harmful algal bloom (HAB)* (see Section 13.2 in Chapter 13, "Biological Productivity and Energy Transfer").

**CONCEPT CHECK 11.3 ▸ Specify the marine environmental problems associated with non-petroleum chemical pollution.**

1 How would dumping sewage in deeper water off the East Coast help reduce negative impacts on the ocean floor?

2 Discuss the animal populations that clearly suffered from the effects of DDT and the way in which this negative effect was manifested.

3 What causes Minamata disease? What are the symptoms of the disease in humans?

# 11.4 ▸ What Marine Environmental Problems Are Associated with Non-point Source Pollution, Including Trash?

Through intentional and unintentional actions, humans also release a vast amount of trash and other unwanted substances into the oceans. Let's examine these substances, how they enter the ocean, and the environmental problems they cause.

## Non-point Source Pollution and Trash

**Non-point source pollution**—also called *poison runoff*—is any type of pollution entering the ocean from multiple sources rather than from a single discrete source, point, or location. In most urban areas, non-point source pollution arrives at the ocean via runoff from storm drains, many of which now have labels indicating that they lead to the ocean (**Figure 11.19**). The U.S. National Academy of Sciences estimates that 5.8 million metric tons (13 billion pounds) of litter enters the world's oceans each year.

Because non-point source pollution comes from many different locations, it is difficult to pinpoint where it originates, although the *cause* of the pollution may readily be apparent. One such example is trash that is washed down storm drains to the ocean and then washes up on beaches (**Figure 11.20**). Others include pesticides and fertilizers from agriculture and oil from automobiles that are washed to the ocean whenever it rains. In fact, the amount of road oil and improperly disposed oil regularly discharged each year into U.S. waters as non-point source pollution is as much as one-and-a-half times the amount of the *Deepwater Horizon* oil spill!

Trash enters the ocean as a result of ocean dumping, too. According to existing laws (**Figure 11.21**), certain types of trash, such as glass, metal, rags, and food, can legally be dumped in the ocean—as long as they are dumped far enough away from shore or ground up small enough. Mostly, this material sinks or biodegrades and does not accumulate at the surface. The exception is plastics.

## Plastics as Marine Debris

Globally, **plastics** (*plasticus* = molded) constitute the vast majority of marine debris. About 80% of marine debris comes from land-based sources, and the majority of that is plastics. When plastics enter the ocean, they float and are not readily biodegradable (**Figure 11.22**). As a result, plastics can remain in the marine environment almost indefinitely, affecting marine organisms through entanglement and ingestion. In fact, there are many documented cases of plastic waste such as six-pack rings or packing straps strangling fish, marine mammals, and birds that have been entangled in plastic trash (**Figure 11.23a** and **11.23b**). Marine birds have also ingested so much floating plastic trash that it fills their stomachs and they die from starvation (**Figure 11.23c**). In probably the most well-known case of lethal ingestion of plastic trash, marine turtles have been killed when they eat floating plastic bags, evidently mistaking them for jellyfish or other transparent plankton on which they typically feed. Over 700 species of marine animals—some of them endangered—have been reported to have become entangled in or eaten plastic. And a human health question remains: does the plastic ingested by marine life affect humans when we eat it? Researchers are currently studying the effect of ingestion of plastics by various types of marine life to see how plastics are passed back to humans.

Entanglement and ingestion, however, are not the worst problems caused by the ubiquitous plastic pollution in the ocean. Researchers have recently discovered that floating plastic pieces have a high affinity for non-water-soluble toxic compounds, most notably DDT, PCBs, and other oily pollutants. As a result, some plastic pieces accumulate poisons to levels as high as a million times their concentrations in seawater. When marine organisms ingest the toxic plastic pieces, they accumulate vast quantities of the toxic materials.

Despite the fact that plastics are one of the few substances that are illegal to dump anywhere in the ocean (see Figure 11.21), their properties and abundant usage in society contribute to their increasing profusion in the marine environment.

**A BRIEF HISTORY OF PLASTICS** Even though human-produced plastics made their debut in 1862 at the Great International Exhibition in London, their commercial development didn't occur until World War II, when shortages of rubber and other materials created great demand for alternative products. Plastic products are *lightweight*, *strong*, *durable*, and *inexpensive*, so they have many advantages over products made from other materials. By the 1970s, plastic products found their way into virtually every aspect of human lives, from airplane parts to zippers (**Figure 11.24**). Humans, for example, wear plastics, cook with plastics, drive in plastics, and even have internal artificial parts made of plastics. In addition, plastics save lives daily by their use in airbags, incubators, and safety equipment like helmets; one of the most useful applications is that plastic bottles help deliver clean drinking water to those in need. The convenience of plastic items intended for one-time use has also contributed to their popularity.

What was once thought of as a miracle substance, however, has been found to have several disadvantages. Disposing of plastics has already strained the capacity of land-based solid-waste disposal systems. Plastic waste is now an increasingly abundant component of oceanic flotsam. (*Flotsam* is floating refuse that can originate from either land-based or ocean-based sources.) Unfortunately, the very same properties that make plastics so advantageous make them unusually persistent and damaging when released into the marine environment:

- They are *lightweight*, so they float and concentrate at the surface.
- They are *strong*, so they entangle marine organisms.
- They are *durable*, so they don't biodegrade easily, causing them to last almost indefinitely.
- They are *inexpensive*, so they are mass-produced and used in almost everything.

## STUDENTS SOMETIMES ASK . . .

***Don't storm drains receive treatment before emptying into the ocean?***

Contrary to popular belief, water (and any other material) that goes down a storm drain does not receive any treatment before being emptied into a river or directly into the ocean. Sewage treatment plants receive enough waste to process without the additional runoff from storms, so it is important to monitor carefully what is disposed into storm drains. For instance, some people discharge used motor oil into storm drains, thinking that it will be processed by a sewage plant. A good rule of thumb is this: *Don't put anything down a storm drain that you wouldn't put directly into the ocean itself.*

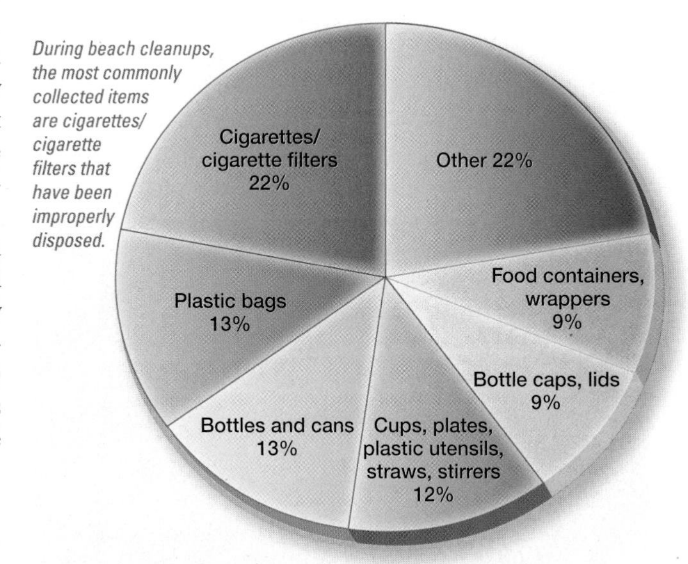


**Figure 11.20 Most commonly found items collected during beach cleanups.** Most of the trash collected during beach cleanups consists of single-use items that are improperly disposed of every day.

## STUDENTS SOMETIMES ASK . . .

### *How effective are biodegradable plastics?*

Unfortunately, biodegradable plastics don't work—at least not yet. In theory, plastics that include biodegradable additives such as the plant product cellulose should break down more readily. But according to material science experts, that's not the case. Researchers used plastic bags mixed with the leading additives that claim to break down plastics and submitted those materials to the three most common types of biodegradation: compost, landfill, and burial of the materials. After three years in those conditions, the samples were retrieved and analyzed. The results show no significant difference in biodegradation between plastics mixed with biodegradable additives and those without. The study suggests that at least several decades are needed for plastics with additives to biodegrade. In spite of this setback, packaging researchers continue to seek a solution to the plastic bag disposal problem.

**Ocean Dumping Regulations in U.S. Waters**

State and local regulations may further restrict the disposal of garbage.

| U.S. lakes, rivers, bays, sounds, and 3 miles from shore | 3 to 12 miles | 12 to 25 miles | Outside 25 miles |
|---|---|---|---|
| ILLEGAL TO DUMP: Plastic, Paper, Rags, Glass, Food, Garbage, Metal, Crockery, Dunnage (lining & packing materials that float) | ILLEGAL TO DUMP: Plastic, Dunnage (lining & packing materials that float). Also, if not ground to less than one inch: Garbage, Paper, Rags, Glass, Metal, Crockery, Food | ILLEGAL TO DUMP: Plastic, Dunnage (lining & packing materials that float) | ILLEGAL TO DUMP: Plastic |

It is illegal for any vessel to dump plastic trash anywhere in the ocean or navigable waters of the United States. Annex V of the MARPOL TREATY is a new international law for a cleaner, safer marine environment. Each violation of these requirements may result in civil penalty up to $25,000, a fine up to $50,000, and imprisonment up to 5 years.

**EXPLORING DATA** ▲ **Outside of 25 miles from shore, what is the one item that cannot legally be dumped at sea? Why is there such tight regulation for this substance?**

SmartFigure 11.21 **International laws regulate ocean dumping in U.S. waters.** Many types of trash can legally be dumped in the ocean, as long as it is ground fine enough and not composed of plastic. In fact, plastic is the only substance that cannot be dumped anywhere in the ocean.
https://goo.gl/tb04AG

**PLASTIC NURDLES IN THE MARINE ENVIRONMENT** Today, nearly all plastic products are produced from small pre-production plastic pellets called **nurdles** that range in size from a BB to a pea (**Figure 11.25**). Nurdles are transported in bulk aboard commercial vessels and are found throughout the oceans, probably due to spillage at loading terminals.

Because of ocean surface currents that wash plastics ashore, plastic nurdles and other trash can be found on most beaches—even in remote areas. In one of the best-documented studies of trash on beaches, for example, researchers painstakingly counted all debris larger than a pinhead along selected transits in Orange County, California. Based on what was collected during the six-week sampling period, researchers project that each year Orange County beaches receive an astonishing 106.9 million pieces of debris, 98% (105.2 million) of which are nurdles. Other studies have shown that some Bermuda beaches contain up to 10,000 nurdles per square meter (925 nurdles per square foot) and some beaches on Martha's Vineyard in Massachusetts contain 16,000 nurdles per square meter (1480 nurdles per square foot).

Figure 11.22 **Floating plastic trash.** When plastic trash enters the ocean, it floats and is not readily biodegradable.

**MICROPLASTICS** An increasing concern in the ocean are microscopic pieces of plastic from the use of human health care products and other applications. These **microplastics** or *microbeads* are small plastic particles generally between 1 and 5 millimeters (0.04 and 0.2 inch) in diameter (**Figure 11.26**). The tiny plastic beads are used as cleaners and scrubbers in products such as hand cleaners, exfoliating facial scrubs, and toothpaste, and are also used in industrial processes such as air blasting technology. In some cosmetics, the combined microplastics in the product contain more plastic than the container they came in! Microplastics can enter the ocean via runoff from land or are washed down drains and transported unaltered through waste water treatment plants because of their microscopic size. In fact, scientific studies reveal that microplastics in the ocean have increased more than

100 times since the 1970s. Microplastics are known to transport pollutants and be eaten by fish, so environmental groups are now advocating for a complete ban on products that contain microplastics.

Microplastics are also generated as larger pieces of plastic break down over time. Studies have shown that floating plastics *photodegrade*, a process in which sunlight breaks them into progressively smaller pieces, which only serves to facilitate the ingestion of plastics by all types of marine organisms, particularly microscopic organisms that constitute lower levels of marine food webs.

The amount of microplastics in the ocean is just beginning to be studied. In 2014, for example, a study was conducted using fine mesh nets towed behind research vessels that trapped particles less than 5 millimeters (0.2 inch). Based on the results of the study, researchers estimate that the Pacific Ocean contains at least 21,000 metric tons (46 million pounds) of floating microplastic pieces. In addition, microplastics have been discovered in Arctic sea ice, which forms when seawater freezes.

**(a)** A female northern elephant seal (*Mirounga angustirostris*) with a plastic packing strap tight around her neck.

**(b)** A herring gull (*Larus argentatus*) entangled in a plastic six-pack ring.

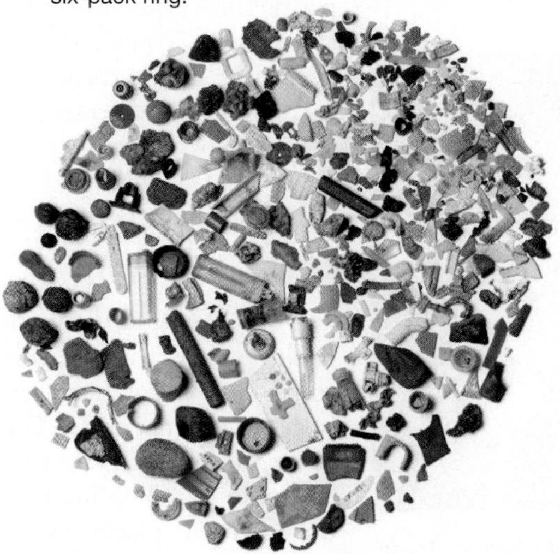

**(c)** Photo of a young Laysan albatross (*Phoebastria immutabilis*) from Midway Island at the northwestern end of the Hawaiian chain that died from a stomach full of pieces of floating plastic trash (*right*) including lighters, bottle caps, electrical connectors, and other materials unintentionally fed to it by its parents.

Figure 11.23 **Examples of how floating plastic trash endangers marine life.**

Figure 11.24 **Examples of plastic products.** A variety of everyday products, from airplane parts to zippers, are made of plastic. Many plastic products are intended to be single-use and so are soon discarded as trash.

**THE ISSUE OF PLASTICS IN THE OCEAN** Oceanographers have recently documented the accumulation of floating plastic trash in unprecedented amounts in the open ocean, particularly in the middle of the five major subtropical gyres. These gyres are far from population centers and slowly rotate, which causes floating debris to accumulate in their calm centers. From Arctic sea ice to the distant waters near Antarctica, from surface waters to sea floor sediment, from once-pristine coral reefs to remote uninhabited islands, in every marine environment where scientists have looked, they have found plastic. In fact, scientists have found plastic in the stomachs of tiny sea creatures living in Pacific Ocean trenches nearly 11 kilometers (7 miles) deep. Indeed, equipment deployed from research and commercial fishing vessels often returns entangled in various types of plastic trash. Clearly, the plastics that are being released into the marine environment are affecting it greatly.

Floating plastic trash is now so abundant that it is being used as artificial habitat in the open ocean by marine creatures, from microbes that use the trash as floating platforms, to fish that are protected within the floating debris. As such, it is beginning to alter the composition of marine ecosystems. For example, the marine water strider *Halobates sericeus* lays its eggs on floating objects in the open ocean. Studies reveal that as floating trash has increased in the ocean, so has the abundance of the water strider's eggs, which are eaten by crabs, fish, and sea birds.

Today, the facts and figures about plastics are so staggering that they're hard to comprehend. For example, half the plastic ever made was produced in the past 15 years. And a trillion plastic bags are used worldwide each year, with an average working life of just 15 minutes. Some 8.2 million metric tons (18 trillion pounds) of plastic ends up in the ocean each year. Not to mention that estimates for how long plastic endures in the environment range from 450 years to forever. Scientific studies reveal that marine plastic particles have increased by 100 times over the past 40 years. And researchers reported in 2014 that more than 5.25 trillion plastic pieces weighing nearly 270,000 metric tons (600 million pounds) are now floating in the ocean. What is even more curious is that the reported amount is less than 1% of the annual global production of plastic, so there should, in fact, be an even greater amount of floating plastic trash in the ocean, unless plastics are being removed somehow.

Where does all this floating plastic trash go? Marine pollution experts agree that some of it washes up on various beaches as a result of the movements of ocean surface currents, some of it is eaten by marine animals, and some of it becomes encrusted by marine life and sinks to the sea floor. In fact, recent studies reveal that plastic trash is pervasive throughout the sea floor, and is found even in the depths of the Mariana Trench, the deepest point of the oceans. However, a large proportion of floating plastic trash simply remains at sea, constantly breaking down into ever-smaller pieces. Microscopic examination of these bits of floating plastic reveals that they are host to an abundance of marine bacteria, which further break the plastic down into smaller bits.

In addition, oceanographers have discovered large floating regions of trash in all oceans of the world where studies reveal that there are at least six times more plastic pieces in surface waters than marine plankton. These regions are located near the center of the world's five major subtropical gyres (see Chapter 7), where trash becomes trapped in large and relatively stable calm areas caused by the convergence of major ocean surface currents. One of the best studied of these regions is the **Eastern Pacific Garbage Patch** (Figure 11.27), which is reportedly about twice the size of Texas and, according to a 2018 report, contains at least 79,000 metric tons (170 million pounds) of debris and an estimated 1.8 trillion pieces of floating ocean plastic, mostly as tiny grains.

Figure 11.25 **Plastic pellets (nurdles) found at a beach.** These pre-production plastic nurdles were found at a Southern California beach. Nurdles have been found floating in the ocean and on all beaches, even in remote areas.

**REDUCING THE AMOUNT OF PLASTICS IN THE OCEAN** What can be done to limit the amount of plastics in the marine environment? For starters, people can limit their use of single-use plastic, recycle plastic material, and dispose of their plastic trash properly, including not dumping any plastics at sea. (Currently in the United States, less than 10% of plastic bottles are recycled; for many other countries, the amount is even less.) Nearly 200 cities and counties across the United States have recently banned single-use plastic bags and polystyrene (plastic-foam) takeout containers, with mounting pressure to enforce statewide prohibition on these items. In 2014, California became the first U.S. state to sign into law a ban on the use of plastic grocery bags, and many other states have since followed. Worldwide, more than three dozen countries have banned disposable plastic grocery bags, including countries as diverse as China, Australia, France, Great Britain, Italy, South Africa, Tanzania, Kenya, Rwanda, and Bangladesh. In addition, beach cleanups are responsible for removing an amazing amount of trash from beaches (see Figures 11.1, 11.20, and 11B). In 2017, 193 nations—including the United States—signed a United Nations resolution called the Clean Seas agreement. The legislation is nonbinding, but the signees vowed to eliminate plastic pollution from the sea, establishing a priority for tackling the plastic pollution problem. All of these actions help reduce the amount of plastics in the marine environment.

In the 1980s, increasing concern about floating trash—in particular syringes and other medical waste that closed long stretches of beaches in the New York area—prompted nations of the world to come together to try to correct the problem. In 1988, the International Convention for the Prevention of Pollution from Ships, which is commonly known as **MARPOL** (short for "Marine Pollution"), proposed a treaty banning the disposal of all plastics and regulating the dumping of most other garbage at sea (see Figure 11.21). By 2005, 122 nations had ratified MARPOL. Some studies suggest that MARPOL has reduced both marine debris and entanglements by discarded fishing nets in some places, most notably in

LO #
plast
.355

**Figure 11.26 Microplastics.** Microplastic beads are used as cleaners and scrubbers in personal hygiene products such as hand cleaners, facial scrubs, and toothpaste.

## STUDENTS SOMETIMES ASK . . .

***I've heard that a microbe has been recently discovered that biodegrades certain types of plastics. Is this true?***

Not only is it true, it was discovered outside of a bottle-recycling facility in Japan! In 2016, researchers analyzed bacteria samples from soil, wastewater, and recycling plant sludge, all contaminated with polyethylene terephthalate (PET), a stiff and very stable plastic fiber that is the main ingredient in polyester clothing and disposable plastic bottles. The researchers had a hunch that in the waste they might find microorganisms that had evolved the ability to digest PET. They successfully identified a bacterium, *Ideonella sakaiensis*, that has an enzyme capable of breaking down PET. While studying the novel enzyme, the scientists modified its structure and, to their surprise, inadvertently made the enzyme even better at degrading plastic. Although bioengineered enzymes may someday help recycle PET plastics by breaking them down into their building blocks, this may not be the answer to the huge number of plastic bottles discarded every year. That's why it's still a good idea to avoid using single-use plastic containers.

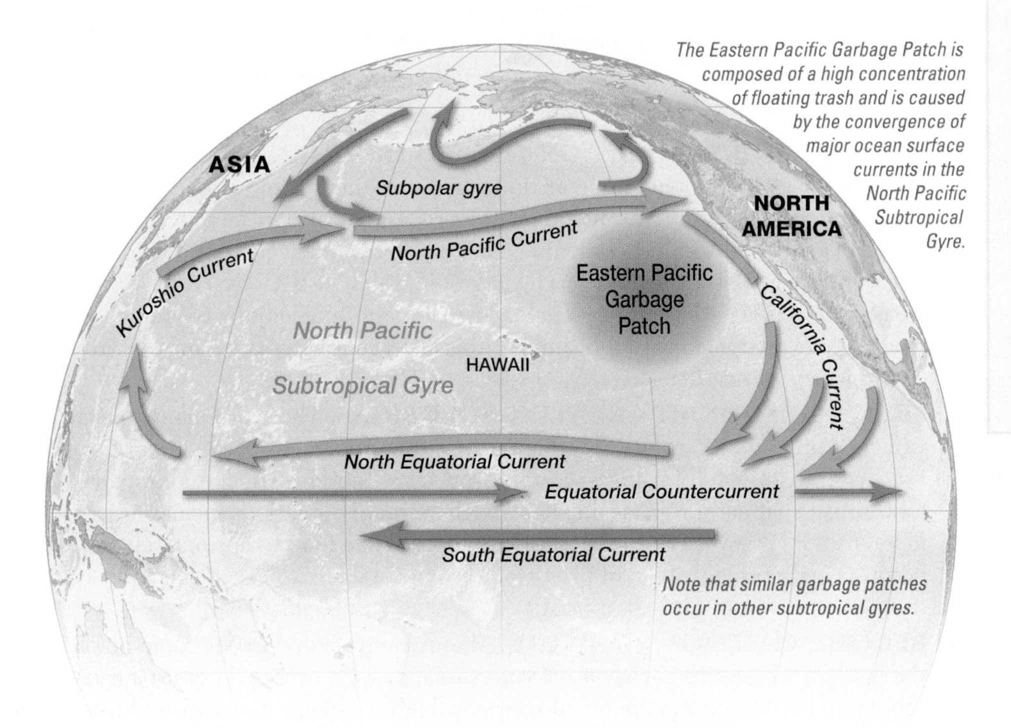
The Eastern Pacific Garbage Patch is composed of a high concentration of floating trash and is caused by the convergence of major ocean surface currents in the North Pacific Subtropical Gyre.
ASIA
Subpolar gyre
NORTH AMERICA
North Pacific Current
Kuroshio Current
Eastern Pacific Garbage Patch
California Current
North Pacific Subtropical Gyre
HAWAII
North Equatorial Current
Equatorial Countercurrent
South Equatorial Current
Note that similar garbage patches occur in other subtropical gyres.

**Figure 11.27 The Eastern Pacific Garbage Patch, which is located between Hawaii and the West Coast of North America.**

waters off Alaska and California. Other studies, however, show no improvement in areas such as the Southern Ocean, South Atlantic, and Hawaiian Islands. As with many other environmental protections and international treaties, enforcement lags far behind stated intentions. Although an annex to MARPOL requires that nations that have signed the treaty provide shore-based facilities where ships can easily dispose of their garbage, many developing countries have not been able to provide these facilities. As a result, even well-meaning captains and shippers who want to comply with existing international marine pollution laws can find it difficult to do so.

**RECAP Although some plastics wash up on beaches, the vast majority of plastics don't ever leave the ocean. Instead, they break down into smaller pieces and enter marine food webs at progressively lower levels.**

**CONCEPT CHECK 11.4 ▶ Specify the marine environmental problems associated with non-point source pollution, including trash.**

1 What is non-point source pollution, and how does it get to the ocean? What other ways does trash get into the ocean?

2 What properties contributed to plastics being considered a miracle substance? How do those same properties cause them to be unusually persistent and damaging in the marine environment?

## STUDENTS SOMETIMES ASK . . .

***What does the garbage patch look like? Can it be seen from space? How can it be cleaned up?***

One of the problems with determining the size of the garbage patch is that it lacks definitive edges, and so there are no accurate estimates of its size. Contrary to what many people have heard about it, the patch is not a floating island of garbage. Since most of the plastic pieces are about the size of a fingernail or smaller, the garbage patch cannot easily be seen from a boat, let alone from space. In fact, oceanographers use fine mesh trawl nets to determine its existence. A better way to visualize the garbage patch is to imagine a very thin brothy soup with some tiny scattered pieces of plastic.

Removing the floating garbage is not as easy as it sounds. For one thing, the trash is unevenly distributed. For another, it involves gathering small pieces that are spread across a huge area, which would require an armada of ships working continuously for many years to collect all the trash. Even if large amounts of floating trash were collected, what would be done with all of it once back on shore, especially the parts that can't be recycled? Certainly a bigger issue is that skimming the ocean's surface waters for floating trash would also remove plankton that are crucial to marine food webs. At the moment, the best approach appears to be preventing the trash from getting to the ocean in the first place.

# 11.5 ▶ What Can You Do to Help Prevent Marine Pollution?

As discussed throughout this chapter, the ocean has historically been used as a dumping ground for many of society's wastes, in large part because the ocean is vast and capable of absorbing many substances. Even today, humans are adding pollutants to the ocean at staggering rates. What can each of us do to help prevent marine pollution? Some ways to help the environment in general and the ocean in particular include the following:

- **MINIMIZE YOUR IMPACT ON THE ENVIRONMENT.** Reduce the amount of waste you generate by making wise consumer choices. Avoid single-use plastic products and Styrofoam containers. Stay clear of products with excessive packaging and support companies that have good environmental records. Use nontoxic or less hazardous products around the house. Conserve resources. Reuse and recycle items and also help close the recycling loop by buying goods made of recycled materials. Do simple things that make a positive impact on the environment (**Diving Deeper 11.2**).
- **SEEK AND SUPPORT CREATIVE SOLUTIONS.** There are many ways in which creative solutions have been applied to the problem of ocean pollution. For example, instead of using a plastic throw-away toothbrush, consider using a toothbrush that has a replaceable head. Other creative solutions include using metal reusable straws instead of plastic ones, reusable food wraps made from beeswax and cotton instead of plastic, and compostable six-pack rings made from brewery waste instead of plastic.
- **BECOME POLITICALLY AWARE.** Many ocean-related issues come before the public and require a majority of voters to approve a proposal before it is enacted into law. A common political saying is that the "majority rules," and this situation is true for local as well as national and international issues.

  Within our lifetimes, for instance, we may very well decide whether to spend large amounts of money to add finely ground iron to the ocean to reduce the amount of carbon dioxide in the atmosphere (see Chapter 16, "The Oceans and Climate Change"). Many political issues in the future will involve the ocean.

## STUDENTS SOMETIMES ASK . . .

***All this pollution is distressing. Is there anything I can do about it?***

Yes, there are many things you can do to help protect the ocean, some of which are listed in Diving Deeper 11.2. They all involve making intelligent choices and minimizing your impact on the environment. Non-point source pollution, for example, is something that the general public is directly responsible for, so one of the best methods of prevention may be educating people. Once people understand the impact their choices have on the environment, they often realize that the solution is up to *all* of us.

## TWELVE SIMPLE THINGS YOU CAN DO TO HELP PREVENT MARINE POLLUTION

*"Nobody made a greater mistake than he who did nothing because he could only do a little."*
*—Edmund Burke (circa 1790)*

There are many simple things you can do every day to help prevent marine pollution, such as the following:

1. **Avoid plastic bottles . . . and plastic straws.** Instead, invest in a refillable water bottle. Some come with filters if you're worried about water quality. Around the world, nearly a million plastic beverage bottles are sold every minute, and only about 18% are recycled (the recycling rate of plastic bottles in the United States is less than 10%). As for plastic straws, consider using paper straws. Americans toss 500 million plastic straws every day, or about 1.5 per person.
2. **Steer clear of products with excess plastic packaging.** Avoid products that come wrapped in plastic and then wrapped individually in plastic again. Buy bar soap in a paper box instead of liquid soap sold in a plastic container. Always consider buying in bulk. Avoid produce sheathed in plastic. And while you're at it, give up single-use plastic and polystyrene (plastic-foam) plates, cups, and containers, which can often find their way into the marine environment.
3. **Snip plastic six-pack rings . . . or avoid them altogether.** Plastic six-pack rings entangle many marine organisms, so snip each circle with scissors before you toss them into the garbage—or, better yet, recycle them along with the cans or bottles to which they are attached. If you find any at a beach, pick them up, snip them, and recycle them, too. Probably the best advice is to avoid purchasing cans that are bound together by them in the first place.
4. **Use lawn fertilizers, yard chemicals, and laundry detergent sparingly.** Lawn fertilizers contain nitrates and phosphates, which cause harmful algal blooms when runoff from land enters the ocean through storm drains. Detergents also contain phosphates, so read detergent labels to find one that is phosphate free and use a smaller amount than recommended. Use yard chemicals such as pesticides and weed killers only when absolutely necessary.
5. **Clean up after your pet.** Dog and cat feces have high bacteria levels. When they wash into a stream or storm drain and eventually to the ocean, they also provide nitrates and phosphates, which create harmful algal blooms.
6. **Make sure your car doesn't leak oil.** Oil that leaks from automobiles is responsible for a large percentage of the oil that gets into the ocean. Annually, the amount of oil that enters the ocean as runoff from road sources (non-point source pollution) through storm drains is greater than a major oil spill. If you change your car's oil yourself, be sure to recycle the used oil at an appropriate recycling center.
7. **Drive less and carpool more**. Reducing the amount of gasoline you use reduces the amount of oil that must be transported across the ocean, which minimizes the potential for oil spills.
8. **Take your own recyclable bags to the grocery store.** Cloth bags are the best choice. Paper bags are biodegradable, but plastic bags are not, and plastics are an increasing problem in the ocean, especially for animals like sea turtles that eat plastic bags when they mistake them for jellyfish. That's why plastic bags—as well as Styrofoam containers—have been banned in certain countries, states, and coastal communities.
9. **Don't release balloons.** Balloons that are released far from the ocean can still wind up there, where they deflate, quickly lose their color, and resemble drifting jellyfish that marine animals can ingest.
10. **Don't litter.** Any material that is carelessly discarded on land can become non-point source pollution when it washes down a storm drain, into a stream, and eventually into the ocean.
11. **Pick up trash at the beach or volunteer for an organized beach cleanup** (Figure 11B). It is important to remove trash that washes up at the beach so it can't endanger marine organisms. Trash arrives on beaches from non-point source pollution, ships, recreational boaters, beachgoers, and other sources. It is truly surprising (and also somewhat horrifying) to discover what ends up on beaches as the result of human activities.
12. **Inform and educate others.** Many people are unaware that their actions have a negative influence on the environment—especially the marine environment.

Figure 11B **Volunteers picking up trash at a beach cleanup event.**

### WHAT DID YOU LEARN?

What was the most surprising and/or important item on this list?

- **EDUCATE YOURSELF ABOUT HOW THE OCEAN WORKS.** A recent poll indicates that more than 90% of the American public consider themselves scientifically illiterate. With our society becoming more scientifically advanced, people need to understand how science operates. Science is not meant to be comprehended by an elite few. Rather, science is for everyone. By studying

**RECAP** **Although the ocean has historically been used as a dumping ground for many of society's wastes, there are informed actions each of us can take to help prevent marine pollution.**

oceanography, you have begun to understand how the ocean works. It is our hope that you will be a lifelong student of the ocean. And if you are considering participating in the scientific study of the oceans, look for more details on working in marine science in Mastering Oceanography **Appendix V, "Careers in Oceanography."**

**CONCEPT CHECK 11.5 ▸ Describe actions you can take to help prevent marine pollution.**

1 List four main ways you can help the environment in general and the ocean in particular.

2 List several specific actions you can take to help reduce marine pollution.

## 11.6 ▸ What Marine Environmental Problems Are Associated with Biological Pollution?

As human activities increase around the globe, so does the transportation of biological pollutants, otherwise known as *non-native species*. **Non-native species** (also called *exotic*, *alien*, or *invasive species*) are species that originate in a particular area but are introduced into new environments either by the deliberate or accidental actions of humans and so are classified as biological pollutants. Because non-native species inhabit new areas where they may lack predators or other natural controls, they can wreak ecological havoc by outcompeting and dominating native populations. Non-native species can also introduce new parasites and/or diseases. In some cases, non-native species completely transform ecosystems. In the United States alone, more than 7000 introduced species (not counting microorganisms) have been documented, of which about 15% cause ecological and economic damage. In fact, invasive species cause an estimated $137 billion in loss and damages in the United States each year. Let's examine a few examples of non-native marine species that are causing problems worldwide.

### The Seaweed *Caulerpa taxifolia*

The seaweed *Caulerpa taxifolia*, which is native in tropical waters, is one example of an invasive, non-native marine species. It is ideal as a decorative alga in saltwater aquariums because it is hardy, fast growing, and not edible by most fish. However, when it is introduced into suitable new habitats (most likely as a result of the dumping of household saltwater aquariums), it becomes a dominant and persistent species that displaces native seaweeds and other marine life. In 1984, a cold-tolerant clone of *Caulerpa taxifolia*, which was produced for the aquarium industry, was first introduced into the Mediterranean Sea, where it has overwhelmed aquatic ecosystems and continues to spread. In 2000, this clone was also found along the coast of New South Wales, Australia, and in two Southern California lagoons. The *Caulerpa taxifolia* outbreak in one of these lagoons, Aqua Hedionda Lagoon in Carlsbad, California, was most probably caused by an aquarium owner who improperly dumped the seaweed into a storm drain, which allowed it to enter the lagoon where the invasion was discovered. California has since passed a law forbidding the possession, sale, or transport of *Caulerpa taxifolia* within the state. A public awareness campaign was initiated that was successful in stopping the spread of this non-native species (**Figure 11.28**).

The distribution of *Caulerpa* in Aqua Hedionda Lagoon was small enough to be considered controllable, so divers covered patches of the seaweed with large tarps held down with sandbags at the edges of the infestation. Then chlorine was injected below the tarp, killing all living organisms inside. Eradication efforts in Southern California appear to have been successful. However, because *Caulerpa taxifolia* can regenerate from very small fragments, repeated surveys are being conducted in these lagoons to eliminate all remaining occurrences of the seaweed.

## Zebra Mussels

Another example of a non-native aquatic species is the European **zebra mussel** (*Dreissena polymorpha*), which was first discovered in the Great Lakes region in 1988. Zebra mussels probably entered North America in the ballast water of a freighter from Europe and have since proliferated rapidly in waters of eastern Canada and the United States. (Ballast water is taken into the hold of a ship to enhance stability and then released in a port when it is no longer needed.) In doing so, the transplanted mussels have driven out native mussels, altered the ecology of freshwater lakes and streams, and blocked the water-carrying pipes of power plants and many other industrial facilities. Although zebra mussels are exceedingly hardy organisms, researchers are working to identify predators, parasites, and infectious microbes that can kill zebra mussels but leave native populations unharmed.

Figure 11.28 **Invasion of *Caulerpa taxifolia*.** The seaweed *Caulerpa taxifolia* is thought to have been illegally dumped by an aquarium owner into a storm drain that allowed the seaweed to spread into Southern California coastal lagoons. The alga was eradicated because of good public awareness and quick action by local authorities.

## Other Notable Examples of Marine Biological Pollution

Other notable examples of harmful non-native aquatic species include the Atlantic comb jelly *Mnemiopsis leidyi*, which was transported in ballast water to the Black Sea and has done extensive damage to the region's fishing and tourist industries; the Atlantic cordgrass *Spartina alterniflora*, which has invaded soft-bottom coasts of California, Washington, and China; the water hyacinth *Eichhornia crassipes*, which infests tropical estuaries and other water bodies; the European green crab *Carcinus maenas*, which has invaded the Pacific Coast and is altering coastal food webs; and the red lionfish *Pterois volitans*, which has proliferated out of control in southeast U.S. and Caribbean coastal waters.

In 2011, during Japan's great Tohoku Earthquake and resulting tsunami (see Chapter 8), 5 million metric tons of debris was washed out to sea, providing a habitat for exotic species to hitch a ride on and be transported alive across an entire ocean basin. As a result, nearly 300 species from Japan have been recently discovered along the North American coast, having ferried across the Pacific Ocean on floating rafts of plastics and other floatables (see Creature Feature 8.1).

**RECAP** **Worldwide, there is a growing concern about the introduction of non-native species, which are organisms that are transported to regions where they have no natural predators and so their populations explode and crowd out other species that normally live there.**

**CONCEPT CHECK 11.6** ▸ **Specify the environmental problems associated with biological pollution.**

1 What are non-native species? Why can they be so damaging to ecosystems?

2 How did the invasive species *Caulerpa taxifolia* and zebra mussels get released into new environments?

# ESSENTIAL CONCEPTS REVIEW

## 11.1 ▶ What is pollution?

- *Although marine pollution seems easily defined, an all-encompassing definition is quite detailed.* It is often difficult to establish the degree to which pollution affects ocean areas. The most widely used technique for determining the effects of pollution is the *environmental bioassay*, which determines the concentration of a pollutant that causes a *50% mortality rate* among test organisms. The debate continues about whether society's wastes should be dumped in the ocean.

### Selected Key Terms

Use the **glossary** at the end of this book to discover the meanings of these Selected Key Terms: **pollution, environmental bioassay**.

### Critical Thinking Question

Why do some experts believe that the ocean can be a repository for many of society's wastes? If we do use the ocean as a disposal site, what conditions should be required?

### Active Learning Exercise

Working with another student in class, determine five common characteristics of pollution. Then write down as many different types of marine pollution as you can in two minutes. Share your answers with the class.

## 11.2 ▶ What marine environmental problems are associated with petroleum pollution?

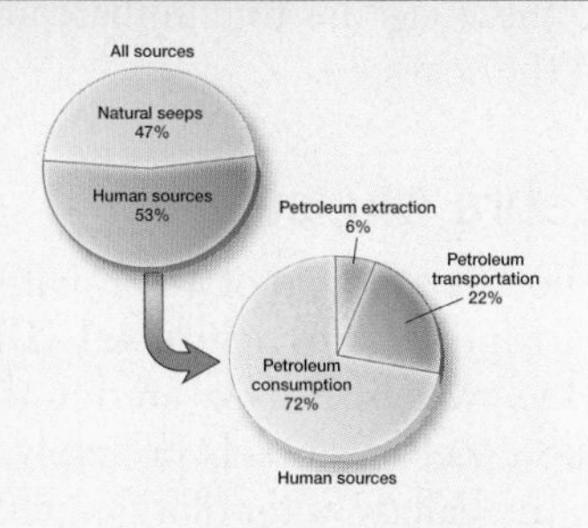


- *Oil is a complex mixture of hydrocarbons and other substances*, most of which are naturally *biodegradable*. Thus, *many marine pollution experts consider oil to be among the least damaging of all substances* introduced into the marine environment. In areas that have experienced oil spills, *recovery can be as rapid as a few years*. Still, *oil spills can cover large areas and kill many animals*. After the much-publicized *Exxon Valdez* oil spill in Alaska, *many novel approaches were used to clean the spilled oil*, including releasing oil-eating bacteria (*bioremediation*).

### Selected Key Terms

Use the **glossary** at the end of this book to discover the meanings of these Selected Key Terms: **petroleum, hydrocarbon, toxic compound**.

### Critical Thinking Question

Specify the marine environmental problems associated with petroleum pollution.

### Active Learning Exercise

Working with another student in class and using the blue circles shown in Figure 11.3, calculate the total amount of oil that has been spilled into the oceans since the 1991 Persian Gulf War from "natural seeps and regular human activities such as transportation." Using the circles in Figure 11.3 for scale, sketch circles that represent the amounts in both U.S. waters and globally. Compare these values to the other circles shown in Figure 11.3 and share your results with the class.

## 11.3 ▶ What marine environmental problems are associated with non-petroleum chemical pollution?

- *Millions of tons of sewage sludge have been dumped offshore in coastal waters.* Although 1972 legislation required an *end to dumping of sewage* in the coastal ocean by 1981, *exceptions continue to be made*. Increased public concern resulted in new legislation to prohibit sewage dumping in the ocean.
- *DDT and PCBs are persistent, biologically hazardous chemicals that have been introduced into the ocean by human activities.* DDT pollution produced a decline in the Long Island *osprey population* in the 1950s and the *brown pelican population* of the California coast in the 1960s. *Virtual cessation of DDT use* in the Northern Hemisphere in 1972 allowed the *recovery of both populations*. The DDT thinned the eggshells and reduced the number of successful hatchings. PCBs have been implicated in causing health problems in sea lions and shrimp.
- *Mercury poisoning was the first major human disaster resulting from ocean pollution.* It is now called *Minamata disease*, after the bay in Japan where it first occurred in 1953. The toxic form of mercury is *methylmercury, which bioaccumulates in the tissues of many large fish*, most notably tuna and swordfish, and *works its way up the food web* in a process called *biomagnification*. To prevent mercury poisonings in the United States, *stringent methylmercury contamination levels in fish have been established by the FDA.*

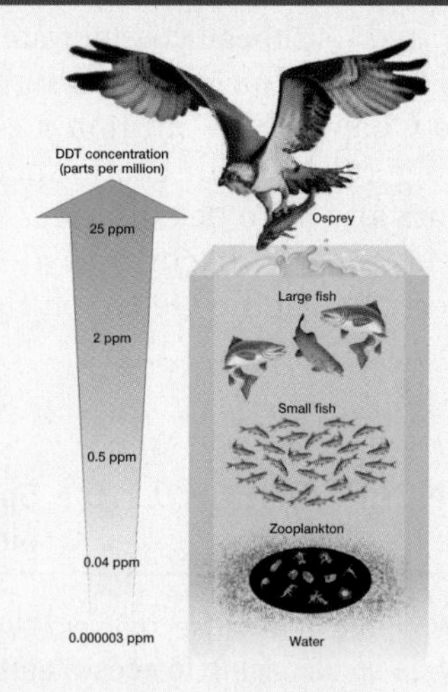


### Selected Key Terms

Use the **glossary** at the end of this book to discover the meanings of these Selected Key Terms: **sewage sludge, DDT, PCBs, mercury, Minamata disease, bioaccumulation, biomagnification**.

### Critical Thinking Question

Specify the marine environmental problems associated with non-petroleum chemical pollution.

### Active Learning Exercise

Working with another student in class, discuss what we should do with sewage sludge if we can't dump it into the ocean. What are the most feasible options?

## 11.4 ▶ What marine environmental problems are associated with non-point source pollution, including trash?

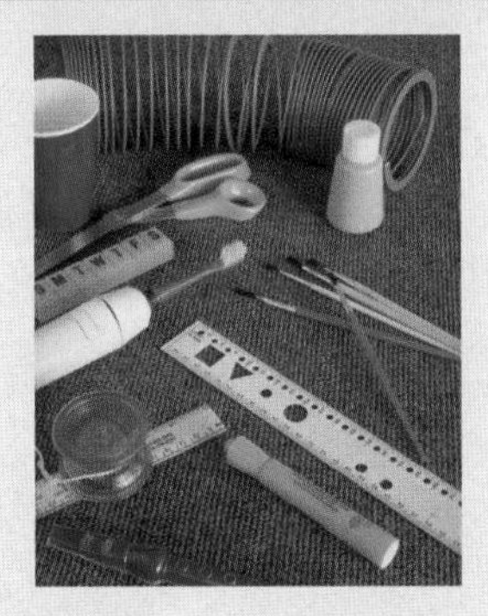

- *Non-point source pollution includes road oil and trash. Plastics* have gained popularity in modern culture because they *are lightweight, strong, durable, and inexpensive.* Unfortunately, these same properties make them a relentless *source of floating trash* in the ocean, especially because they *break down into progressively smaller pieces* over time. *The amount of plastics accumulating in the oceans has increased dramatically.* Certain types of plastics are known to be *lethal to marine mammals, birds, and turtles.* International legislation such as *MARPOL regulates the disposal of trash in the ocean* but lacks enforcement.

### Selected Key Terms

Use the **glossary** at the end of this book to discover the meanings of these Selected Key Terms: **non-point source pollution, plastics, nurdles, microplastics, Eastern Pacific Garbage Patch**.

### Critical Thinking Question

List and describe the three main problems that floating plastic trash presents to marine organisms. Of the three, which one is potentially the most serious threat? Explain.

### Active Learning Exercise

Working with another student in class, describe the two main sources of microplastics and discuss problems that microplastics create in the oceans.

## 11.5 ▶ What can you do to help prevent marine pollution?

- Although the ocean has historically been used as *a dumping ground for many of society's wastes*, there are *informed actions* each of us can take to help prevent marine pollution. Some examples include *minimize your impact on the environment, seek and support creative solutions, become politically aware*, and *educate yourself about how the ocean works.*

### Critical Thinking Question

One of the best ways of preventing marine pollution is to keep it from getting into the marine environment in the first place. Considering all the different types of marine pollution, brainstorm ideas on how this could be best accomplished.

### Active Learning Exercise

With another student in class, review the 12 suggested actions to help prevent marine pollution in Diving Deeper 11.2. Of the 12 items listed, which ones are you already doing? Also, which ones could you easily adopt and get others to adopt? Have each group report their findings to the class.

## 11.6 ▶ What marine environmental problems are associated with biological pollution?

- The introduction of *biological pollution* such as *non-native species* (including *Caulerpa taxifolia* and zebra mussels, among others) into new environments can cause *severe ecological and economic damage.*

### Selected Key Terms

Use the **glossary** at the end of this book to discover the meanings of these Selected Key Terms: **non-native species, zebra mussel**.

### Critical Thinking Question

Specify the marine environmental problems associated with biological pollution.

### Active Learning Exercise

Working with another student in class, search the Internet to find information about one of the invasive species listed in the section "Other Notable Examples of Marine Biological Pollution." Determine how the species got to its new location, what damage it is currently doing, and what measures are being done to prevent its spread. Report your findings to the class.

## Mastering Oceanography

Looking for additional review and test prep materials? With individualized coaching on the toughest topics of the course, Mastering Oceanography offers a wide variety of ways for you to move beyond memorization and deeply grasp the underlying processes of how the oceans work. Visit the Study Area in **www.masteringoceanography.com** to find practice quizzes, study tools, and multimedia that will improve your understanding of this chapter's content. Sign in today to access the following features: Self Study Quizzes, SmartFigures, Oceanography Videos and Animations, Squidtoons, Dynamic Study Modules, and an optional Pearson eText with embedded videos.

# Marine Life and the Marine Environment

# 12

## ESSENTIAL LEARNING CONCEPTS

At the end of this chapter, you should be able to:

- ☐ **12.1** Discuss the characteristics of life and how living things are classified.
- ☐ **12.2** Demonstrate an understanding of how marine organisms are classified.
- ☐ **12.3** Specify the number of marine species that exist.
- ☐ **12.4** Explain how marine organisms are adapted to the physical conditions of the ocean.
- ☐ **12.5** Compare the main divisions of the marine environment.

↑ *Check when completed*

Although most people may not be aware of it, an astonishingly wide variety of marine organisms inhabit the world's oceans. These organisms range in size from microscopic bacteria and algae to the blue whale, which is as long as three buses lined up end to end. Marine biologists have identified more than 228,000 marine species; this number is constantly increasing as new organisms are discovered.

Most marine organisms live within the sunlit surface waters of the ocean. Strong sunlight supports photosynthesis by marine algae, which either directly or indirectly provides food for the vast majority of marine organisms. All marine algae must live near the surface because they need sunlight for photosynthesis, and most marine animals live near the surface because this is where food can be obtained. In shallow-water areas close to land, sunlight reaches all the way to the ocean floor, resulting in an abundance of marine life throughout the water column.

There are advantages and disadvantages to living in the marine environment. One advantage is that there is an unlimited supply of water available, which is necessary for maintaining all types of life. One disadvantage is that maneuvering in water can be difficult because the high density of water impedes movement. The success of a species depends on an individual's ability to find food, avoid predators, reproduce, and adapt to its environment. In this chapter, we'll examine some of the unique adaptations of marine life that allow various species to thrive in the ocean environment.

***"A species is a masterpiece of evolution, a million-year-old entity encoded by five billion genetic letters, exquisitely adapted to the niche it inhabits."***

***—E. O. Wilson, biologist and global conservation advocate (2001)***

## 12.1 ▸ What Are Living Things, and How Are They Classified?

Living things are classified based on their physical characteristics, with organisms that are closely related to one another sharing a host of common traits. Recently, researchers have determined the genomic sequencing of the DNA of many organisms, including human, mouse, dog, cat, cow, elephant, honeybee, platypus, diatom, red algae, sea urchin, comb jelly, dolphin, and hundreds of bacteria and viruses, among others. This analysis has allowed genetic comparison, sometimes confirming many of the classification groupings based on structure and other times indicating some unexpected relationships. First, however, let's examine what characteristics qualify an entity as being alive.

### A Working Definition of Life

It might seem easy to differentiate that which is living from the nonliving, but the unusual nature of some life-forms makes defining *life* a challenging task. Also, both

◂ **Amazing adaptations of marine life.** The goosefish (*Lophius americanus*; also called monkfish or angler) has unique adaptations, including feathery appendages as camouflage and a modified dorsal fin that serves as a lure to attract prey. It has been described as mostly mouth with a tail attached, and it can eat prey almost as big as itself. The bottom-dwelling goosefish lives in water as deep as 800 meters (2600 feet) and ranges along the Atlantic coast from Grand Banks, Newfoundland, to Cape Hatteras, North Carolina.

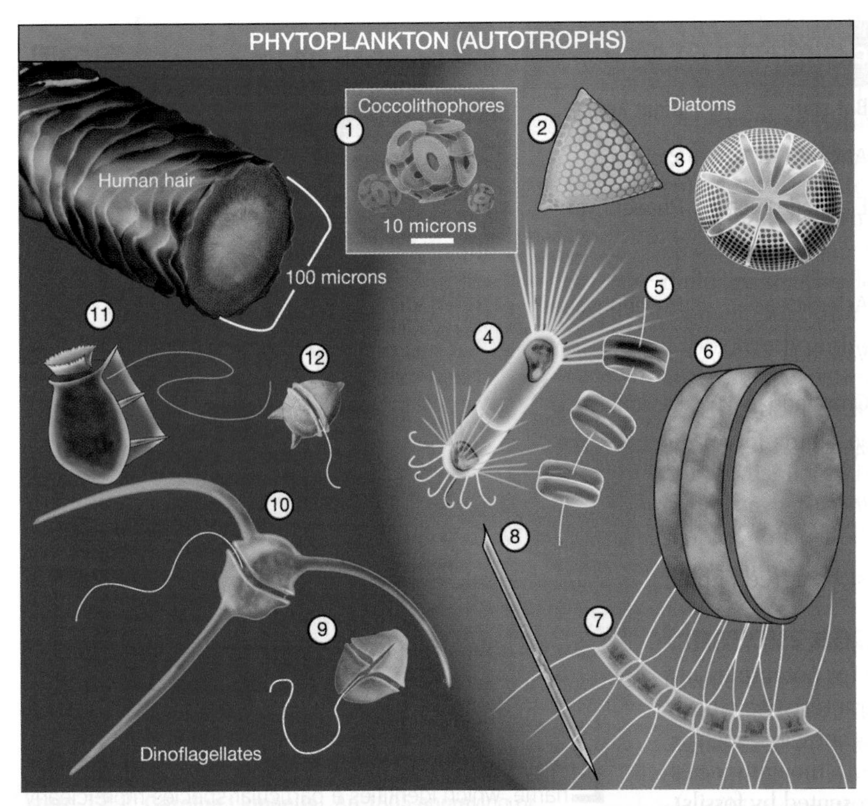


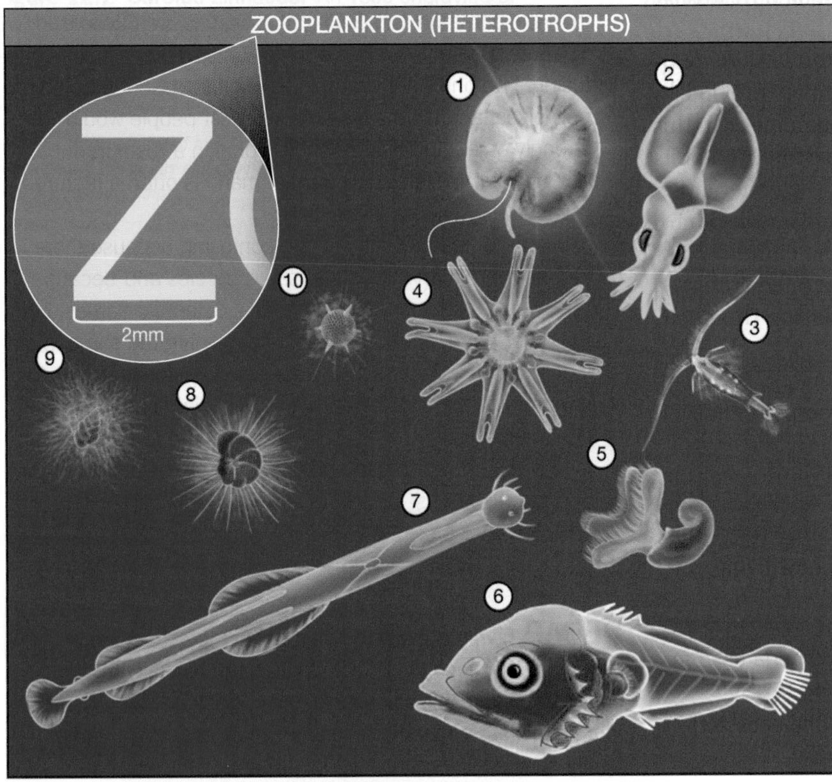


**Figure 12.4 Phytoplankton and zooplankton (floaters).** Schematic drawings of various phytoplankton (*above*) and zooplankton (*below*). **Phytoplankton:** (1) Coccolithophores; (2)–(8) Diatoms; (9)–(12) Dinoflagellates. Width of human hair = 100 microns (0.004 inch). **Zooplankton:** (1) Noctiluca, a predatory dinoflagellate; (2) Squid larva; (3) Copepod; (4) Jelly larva; (5) Snail larva; (6) Fish larva; (7) Arrowworm; (8)–(9) Foraminifers; (10) Radiolarian. Scale = 2 millimeters (0.08 inch).

## Plankton (Drifters)

**Plankton** (*planktos* = wandering) include all organisms—algae, animals, and bacteria—that drift with ocean currents. An individual organism is called a **plankter**. Just because plankters drift does not mean they are unable to swim. In fact, many plankters can swim but either move only weakly or move only vertically. As such, they cannot determine their horizontal position within the ocean.

Plankton are hugely abundant and incredibly important within the marine environment. In fact, *most of Earth's* ***biomass****—the mass of living organisms—consists of plankton adrift in the oceans.* Even though 98% of marine *species* are bottom dwelling, the vast majority of the ocean's *biomass* is planktonic.

**TYPES OF PLANKTON** Plankton can be classified based on their feeding styles. If an organism can photosynthesize and therefore produce its own food, it is termed **autotrophic** (*auto* = self, *tropho* = nourishment). Autotrophic plankton are called **phytoplankton** (*phyto* = plant, *planktos* = wandering) and can range in size from microscopic algae to larger species of drifting kelp. If an organism cannot produce its own food and relies instead on food produced by other organisms, it is termed **heterotrophic** (*hetero* = different, *tropho* = nourishment). Heterotrophic plankton are called **zooplankton** (*zoo* = animal, *planktos* = wandering), which includes drifting marine animals. Representative members of each group are shown in Figure 12.4. Plankton also include bacteria. Scientists have discovered that free-living **bacterioplankton** are much more abundant and far more widely distributed than previously thought. Having an average diameter of only one-half of a micrometer (0.00002 inch), they were missed in earlier studies because they are so incredibly small (note that one micrometer [also known as a *micron*] is one-millionth of a meter and is designated by the symbol μm). Recently, microbiologists have begun to study oceanic bacterioplankton and have even discovered an extremely small yet abundant bacterium (*Prochlorococcus* sp.) that has been estimated to constitute at least half of the ocean's total photosynthetic biomass. This tiny bacterium is likely the most abundant photosynthetic organism on Earth.

Plankton also include viruses, which are called **virioplankton**. Virioplankton are an order of magnitude smaller than bacterioplankton and are similarly little-known. Only recently through advanced sampling methods has the role of viruses in marine planktonic communities been better understood. For example, studies on marine microbial communities indicate that viruses are surprisingly abundant in marine ecosystems; in some locations, they are the most abundant biologic entities in the ocean. As such, viruses can strongly influence the structure of marine microbial assemblages by limiting the abundance of other types of plankton through infection. In fact, detailed microbial studies suggest that viruses probably infect and kill approximately 20% of total marine biomass per day. In addition, scientific studies have shown that marine viruses provide a governing force that influences patterns of oceanic nutrient and energy cycling. Marine viruses also affect gas exchange between the atmosphere and the ocean surface. As a result, viruses may play a key role in the ocean's response to human-caused climate change (For more on human-caused climate change and its effect on the ocean, see Chapter 16, "The Oceans and Climate Change").

Climate
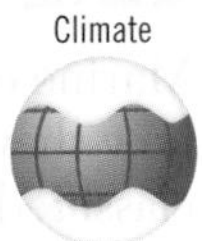
Connection

Although plankton can be classified as either phytoplankton, zooplankton, bacterioplankton, or virioplankton, they can also be classified according to the portion of their life cycle spent as plankton. Organisms that spend their entire lives as plankton are **holoplankton** (*holo* = whole, *planktos* = wandering). Many organisms that spend their adult lives as nekton or benthos spend their juvenile and/or larval stages as plankton (**Figure 12.5**). These organisms are called **meroplankton** (*mero* = a part, *planktos* = wandering). Finally, plankton can also be classified based on size. For example, large floating animals and algae, such as jellies and Sargassum (which is a floating type of brown macro marine algae commonly referred to as a seaweed that is particularly abundant in the Sargasso Sea), are called **macroplankton** (*macro* = large, *planktos* = wandering) and measure 2 to 20 centimeters (0.8 to 8 inches). Plankton also include *bacterioplankton*, which are so small that they can be removed from the water only with special microfilters. These very tiny drifters are called **picoplankton** (*pico* = small, *planktos* = wandering) and measure 0.2 to 2 microns (0.000008 to 0.00008 inch).

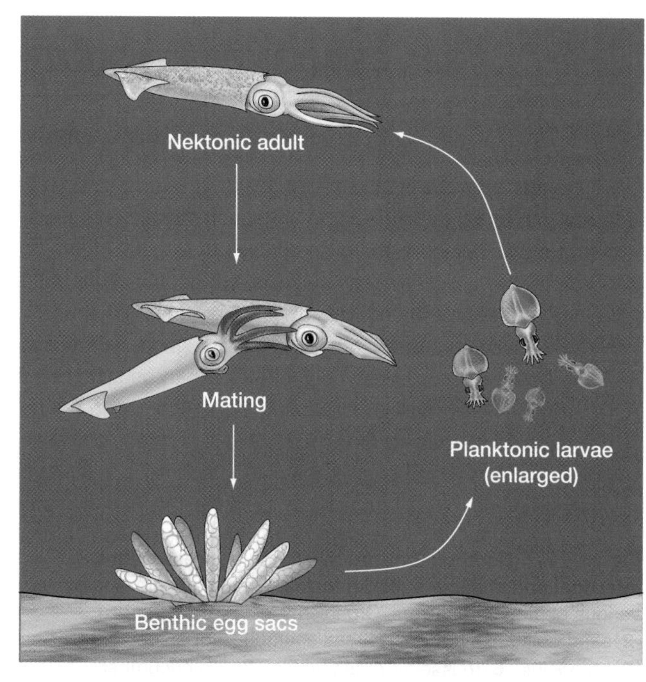


**Figure 12.5 Typical life cycle of a squid.** Squid are meroplankton because they are planktonic only during their larval stage. Adult squid are nekton, and their egg sacs are benthos.

## Nekton (Swimmers)

**Nekton** (*nektos* = swimming) include all animals capable of moving independently of the ocean currents by swimming or other means of propulsion. They are capable not only of determining their own positions within the ocean but also, in many cases, undertake long migrations. Nekton include most adult fish, marine mammals, marine reptiles, and some marine invertebrates such as squid (**Figure 12.6**). When you go ocean swimming, you become nekton, too. Although nekton move freely, many are unable to move throughout the breadth of the ocean. Gradual changes

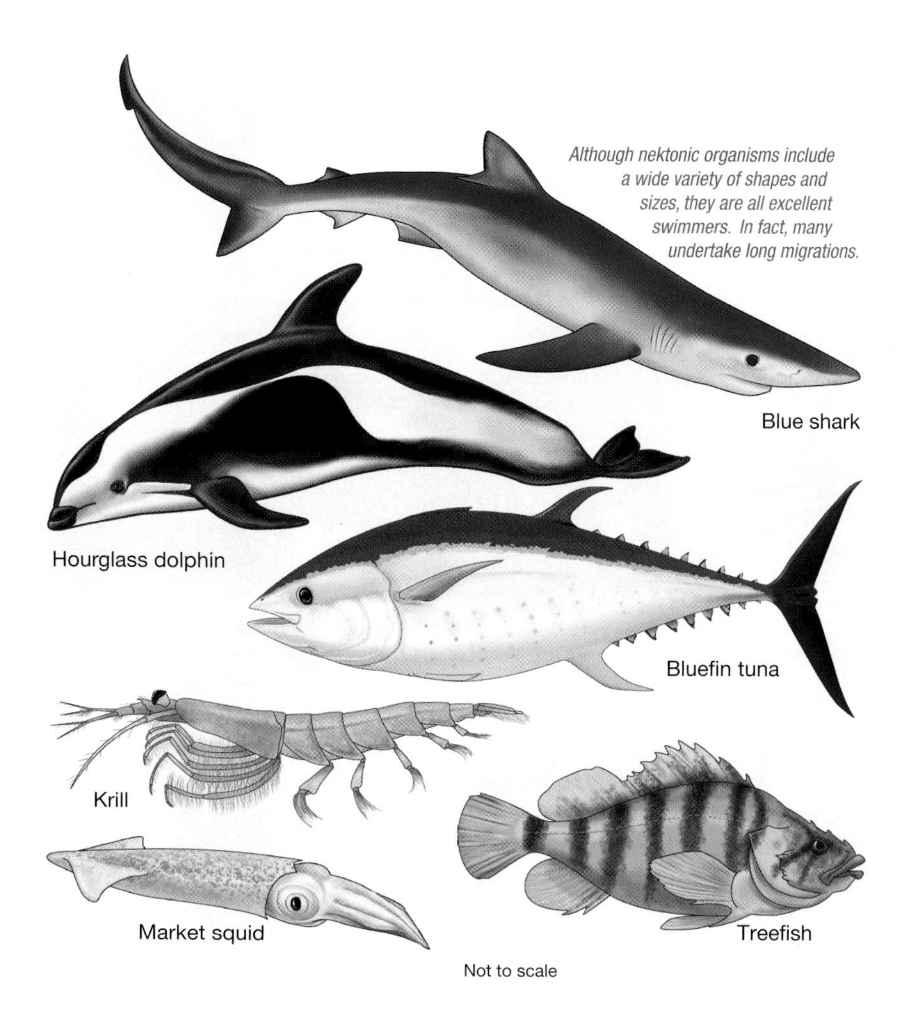


**Figure 12.6 Nekton (swimmers).** Schematic drawings of various nektonic organisms. Note that scale varies in this figure and that organisms pictured here range in size from tiny krill (5 centimeters or 2 inches) to the blue shark (4 meters or 13 feet).

*By the process of **osmosis**, water molecules (but not dissolved substances) diffuse through a semipermeable membrane that separates two liquids of different salinities.*

Semipermeable membrane

*Osmosis causes water molecules to move from the less concentrated **(hypotonic)** solution on the left into the more concentrated **(hypertonic)** solution on the right.*

Hypotonic (lower salinity)

Water molecules

Hypertonic (higher salinity)

**(a)** Two solutions of different salinities separated by a semipermeable membrane.

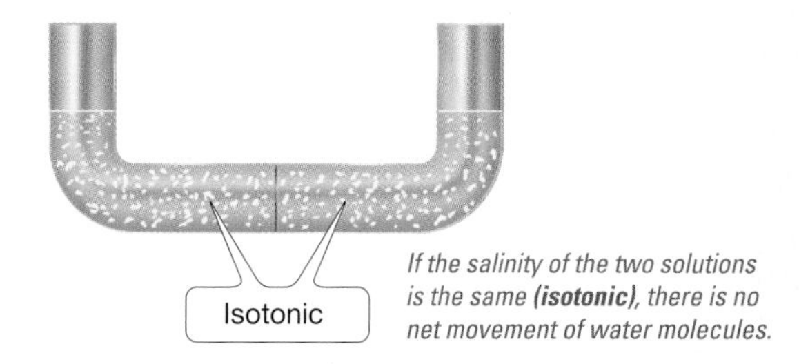
Isotonic
If the salinity of the two solutions is the same (isotonic), there is no net movement of water molecules.

**(b)** Two solutions of the same salinity separated by a semipermeable membrane.

Figure 12.17 **Osmosis.** Osmosis is the process by which water molecules move through a semipermeable membrane from the solution containing a *lower* concentration of dissolved salts (lower salinity) to a solution containing a *higher* concentration of dissolved salts (higher salinity).

**Animation**
Osmosis
https://goo.gl/Ei4Lpa

from within the cell into the surrounding fluid. The waste products are then carried away by circulating fluid that services cells in higher animals or by the water that surrounds simple one-celled organisms.

**OSMOSIS** When water solutions of unequal salinity are separated by a semipermeable membrane (such as the membrane surrounding a living cell), water molecules (but not dissolved salts) move freely across the membrane. The imbalance in the concentration of dissolved salt particles on either side of the membrane, however, causes more water to leave the low salinity side and move into the high salinity side. In the absence of any other forces, this process, called **osmosis** (*osmos* = to push), would continue to drive water across the membrane until the concentrations of dissolved particles on both sides of the membrane are equal (**Figure 12.17a**). **Osmotic pressure** is the pressure that must be applied to the more concentrated solution to prevent water molecules from passing into it. Osmosis causes water to move through an organism's tissues and affects both marine and freshwater organisms. If the salinity of an organism's body fluid equals that of the ocean, it is **isotonic** (*iso* = same, *tonos* = tension) and has equal osmotic pressure, and so no net transfer of water will occur through the membrane in either direction (**Figure 12.17b**).

If seawater has a lower salinity than the fluid within an organism's cells, water will pass through the cell membranes into the cells (toward the more concentrated solution). This organism is **hypertonic** (*hyper* = over, *tonos* = tension), which means it is saltier than the surrounding seawater.

If the salinity within an organism's cells is less than that of the surrounding seawater, water from the cells will pass through the cell membranes out into the seawater (again, toward the more concentrated solution). This organism is **hypotonic** (*hypo* = under, *tonos* = tension) relative to the water outside its body.

In essence, osmosis is a process that produces a net transfer of water molecules through a semipermeable membrane from the side with the *lower concentration* of *dissolved particles* (lower salinity) to the side with the *higher concentration* of *dissolved particles* (higher salinity).

In summary, osmosis and diffusion work in concert to move dissolved substances and water across the cell membrane. These three things can occur simultaneously:

1. Osmosis drives water molecules through the semipermeable membrane toward the side—inside or outside the cell—that has the higher concentration of dissolved particles.
2. Nutrient molecules diffuse from outside the cell, where they are more concentrated, into the cell, where they are used to maintain the cell.
3. Waste molecules diffuse from within the cell into the surrounding seawater.

The body fluids of marine invertebrates (those without backbones) such as worms, mussels, and octopuses, are nearly isotonic with the seawater in which they live. As a result, these organisms have not had to evolve special mechanisms to maintain their body fluids at a proper concentration. This gives them an advantage over their freshwater relatives, whose body fluids are hypertonic.

**AN EXAMPLE OF OSMOSIS: SALTWATER VERSUS FRESHWATER FISH** Saltwater fish have body fluids that are only slightly more than one-third as saline as ocean water, possibly because they evolved in low-salinity coastal waters. They are, therefore, hypotonic (less salty) compared to the surrounding seawater.

This salinity difference means that saltwater fish, without some means of regulation, would lose water from their body fluids into the surrounding ocean and eventually dehydrate. This loss is counteracted, however, because saltwater fish drink ocean water and excrete the salts through special chloride-releasing cells located in

## STUDENTS SOMETIMES ASK . . .

### *Why do my fingers get wrinkly when I stay in the water for a long time?*

People often assume that the wrinkling of the skin is the result of water passing into the outer layer of the skin via osmosis, making it swell up. But researchers have known since the 1930s that the effect does not occur when there is nerve damage in the fingers. This points to the change being an involuntary reaction by the body's autonomic nervous system—the system that also controls breathing, heart rate, and perspiration. In fact, researchers have demonstrated that the distinctive wrinkling is caused by the constriction of blood vessels in tissues below the skin, which is a temporary condition that goes back to normal after you leave the water. Scientific studies also show that there may be an evolutionary advantage to having wrinkly skin: Furrowed fingertips and toes make it easier to grip slick surfaces.

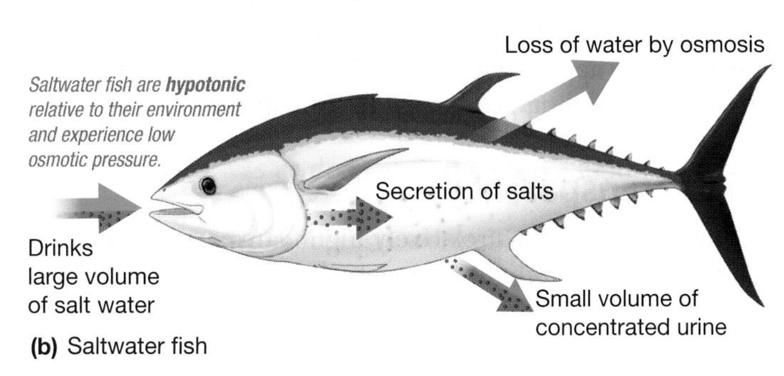


**SmartFigure 12.18 Salinity adaptations of freshwater and saltwater fish.** Osmotic processes cause freshwater and saltwater fish to have different adaptations to their environment. As a result, freshwater fish are hypertonic relative to their environment, and saltwater fish are hypotonic relative to theirs. https://goo.gl/o9YA04

**EXPLORING DATA ◀ What are the osmotic adaptation differences between hypertonic and hypotonic fish?**

their gills. Saltwater fish also help maintain their body water by discharging a very small amount of very highly concentrated urine (**Figure 12.18a**). Freshwater fish are hypertonic (internally more saline) compared to the freshwater in which they live. The osmotic pressure of the body fluids of such fish may be 20 to 30 times greater than that of the freshwater that surrounds them, so freshwater fish risk rupturing cell walls from taking in excessive quantities of water through osmosis. To prevent this, freshwater fish do not drink water, and their cells have the capacity to absorb salt. They also excrete large volumes of very dilute urine to reduce the amount of water in their cells (**Figure 12.18b**).

**RECAP Osmosis produces a net transfer of water molecules through a semipermeable membrane, from the side with the lower concentration of dissolved particles to the side with the higher concentration of dissolved particles.**

## Dissolved Gases

The amount of gases that dissolve in seawater increases as the temperature of seawater decreases, so cold water dissolves more gas than warm water. (This is an *inverse relationship*, similar to how temperature affects the density of seawater, which was discussed in Section 5.6, "How Does Seawater Density Vary with Depth?".) In addition, you may recall from Section 5.8 ("What Factors Control the Distribution of Carbon and Oxygen in the Ocean?") that cold surface waters in the high latitudes contain an abundance of dissolved gases, specifically carbon dioxide (which phytoplankton need for photosynthesis) and oxygen (which all organisms need to metabolize their food). As a result, vast phytoplankton communities develop in the high latitudes during summer, when solar energy becomes available for photosynthesis. In addition, the cold, oxygen-rich water of high-latitude regions sinks and flows along the ocean bottom, supplying deep-sea organisms with an abundant supply of dissolved oxygen.

Most animals that live in the ocean—except air-breathing marine mammals and certain fishes—must extract dissolved oxygen from seawater. How do they do this? Most marine animals have specially designed fibrous respiratory organs called **gills** that exchange oxygen and carbon dioxide directly with seawater. Most fish, for instance, take water in through their mouths (which gives them the appearance of "breathing" underwater), pass it through their gills to extract oxygen, and then expel it through the gill slits on the sides of their bodies (**Figure 12.19**). Most fish need at least 4.0 parts per million (ppm by weight) of dissolved oxygen in seawater to survive

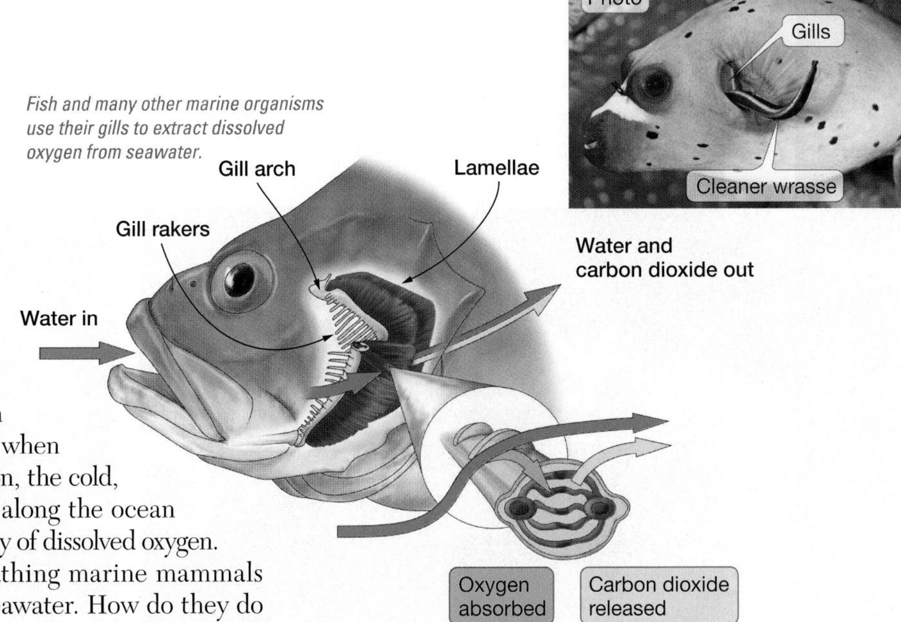


**Figure 12.19 Gills on fish.** Water is taken in through the mouth and passes through the gills, which extract dissolved oxygen. Afterward, water and carbon dioxide are expelled through the gill slits. Photo (*inset*) shows a blue-streaked cleaner wrasse (*Labroides dimidiatus*) cleaning the gills of a blackspotted puffer (*Arothron nigropunctatus*).

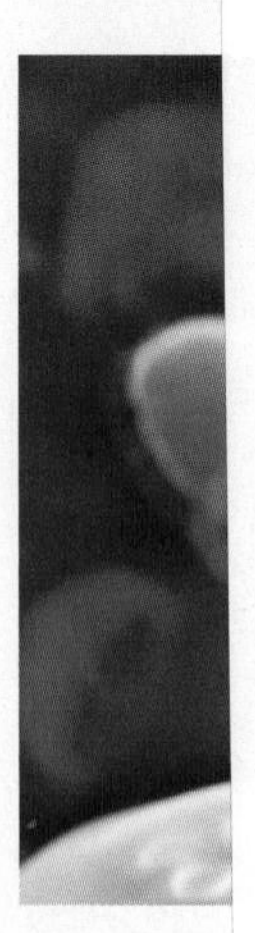

Figure 12.
jellies (*Aure*
ficult for pre
tograph, the

(a) The head

Figure 12.2



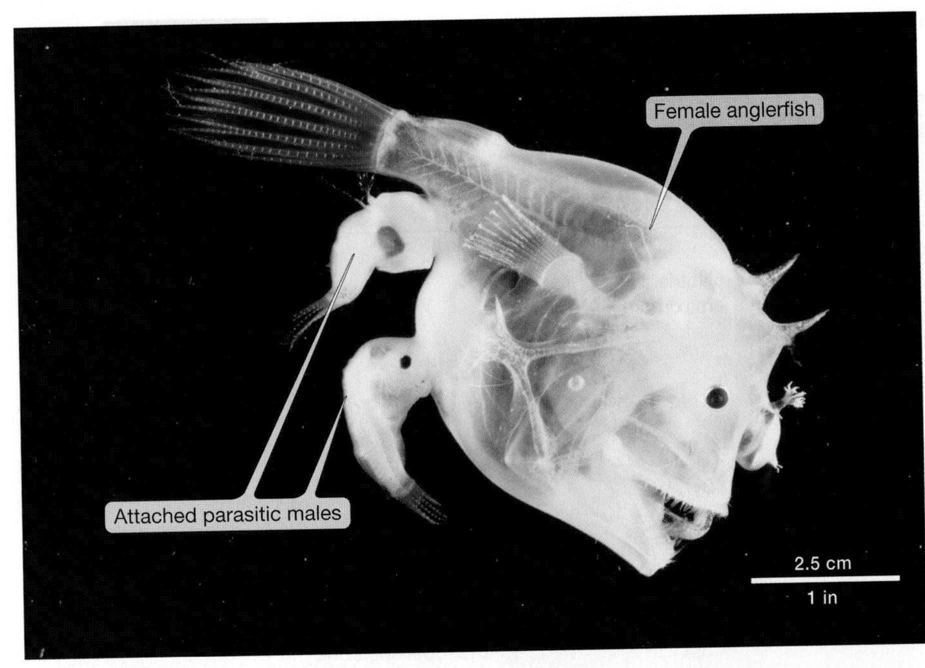


**Figure 12.27 Adaptations of deep-sea anglerfish.** A female deep-sea anglerfish (*Edriolychnus schmidti*) has a transparent body, small eyes, and sharp teeth. The feathery structure projecting from the front of its head is a bioluminescent lure, which is used to attract prey. Note the two much smaller parasitic males that are attached to the bottom of the female's body.

Many also have sharp teeth and extremely large mouths relative to their body size.

Oxygen content increases with depth below the oxygen minimum layer because it is replenished by deep currents originating in polar regions as cold surface water high in oxygen. The abyssopelagic zone is the realm of the bottom-water masses, which commonly move in the direction opposite the deep-water masses in the bathypelagic zone.

## Benthic (Sea Bottom) Environment

Similar to the way that the water column is divided into zones with different physical conditions, the sea bottom environment can also be divided into provinces that provide a variety of habitats for bottom-dwelling organisms. The transitional region from land to sea floor above the spring high tide line is called the **supralittoral zone** (*supra* = above, *littoralis* = the shore) (see Figure 12.25). Commonly called the spray zone, it is covered with water only during periods of extremely high tides and when tsunami or large storm waves break on the shore.

The rest of the benthic, or sea floor, environment is divided into two main units that correspond to the neritic and oceanic provinces of the pelagic environment (see Figure 12.25):

- The **subneritic province** extends from the spring high tide shoreline to a depth of 200 meters (660 feet), approximately encompassing the continental shelf.
- The **suboceanic province** includes the benthic environment below 200 meters (660 feet).

**SUBNERITIC PROVINCE** The subneritic province is subdivided into the littoral and sublittoral zones. The *intertidal zone* (the zone between high and low tides) coincides with the **littoral zone** (*littoralis* = the shore). The **sublittoral zone** (*sub* = below, *littoralis* = the shore), or *shallow subtidal zone*, extends from low tide shoreline out to a depth of 200 meters (660 feet).

The sublittoral zone consists of inner and outer regions. The **inner sublittoral zone** extends to the depth at which marine algae no longer grow attached to the ocean bottom (approximately 50 meters [160 feet]), so the seaward boundary varies. All photosynthesis seaward of the inner sublittoral zone is carried out by floating microscopic algae.

The **outer sublittoral zone** extends from the inner sublittoral zone out to a depth of 200 meters (660 feet) or the shelf break, which is the seaward edge of the continental shelf.

**Figure 12.28 Benthic organisms produce tracks on the ocean floor.** As benthic organisms move across or burrow through the ocean bottom, they often leave tracks in the sediment on the ocean floor. Width of view is about 0.6 meter (2 feet).

**SUBOCEANIC PROVINCE** The suboceanic province is subdivided into bathyal, abyssal, and hadal zones. The **bathyal zone** (*bathus* = deep) extends from a depth of 200 to 4000 meters (660 to 13,000 feet) and corresponds generally to the continental slope.

The **abyssal zone** (*a* = without, *byssus* = bottom) extends from a depth of 4000 to 6000 meters (13,000 to 20,000 feet) and includes more than 80% of the benthic environment. The ocean floor of the abyssal zone is covered by soft oceanic sediment, primarily abyssal clay. Tracks and burrows of animals that live in this sediment can be seen in **Figure 12.28**.

The **hadal zone** (*hades* = hell; this inhospitable high-pressure environment is, in fact, aptly named) extends below 6000 meters (20,000 feet), so it consists only of deep trenches along the margins of continents. Animal communities that are found in these deep environments have been isolated from each other, often resulting in unique adaptations.

**RECAP** **The pelagic environment includes the water column and the benthic environment includes the sea bottom. Subdivisions of pelagic and benthic environments are based on depth, which influences the amount of sunlight.**

**CONCEPT CHECK 12.5 ▸ Compare the main divisions of the marine environment.**

1 Construct a table listing the subdivisions of the pelagic and benthic environments and the physical factors used in assigning their boundaries.

2 Describe the three zones based on the availability of sunlight. Which one is where most marine life exists?

# ESSENTIAL CONCEPTS REVIEW

## 12.1 ▶ What are living things, and how are they classified?

- *A wide variety of organisms lives in the ocean*, ranging in size from microscopic bacteria and algae to blue whales. *All living things belong to one of the three major domains (branches) of life: Archaea*, simple microscopic bacteria-like creatures; *Bacteria*, simple life-forms consisting of cells that usually lack a nucleus; and *Eukarya*, complex organisms (including plants and animals) consisting of cells that have a nucleus.
- *Organisms are further divided into six kingdoms: Eubacteria*, microscopic single-celled organisms without a nucleus; *Archaebacteria*, ancient bacteria-like organisms that live in extreme environments; *Plantae*, many-celled plants; *Animalia*, many-celled animals; *Fungi*, molds and lichens, and *Protista*, single-celled and multicelled organisms with a nucleus. *Classification of organisms involves placing individuals within the kingdoms into increasingly specific groupings of phylum, class, order, family, genus, and species*, the last two of which denote an organism's scientific name. Many organisms also have one or more common names.

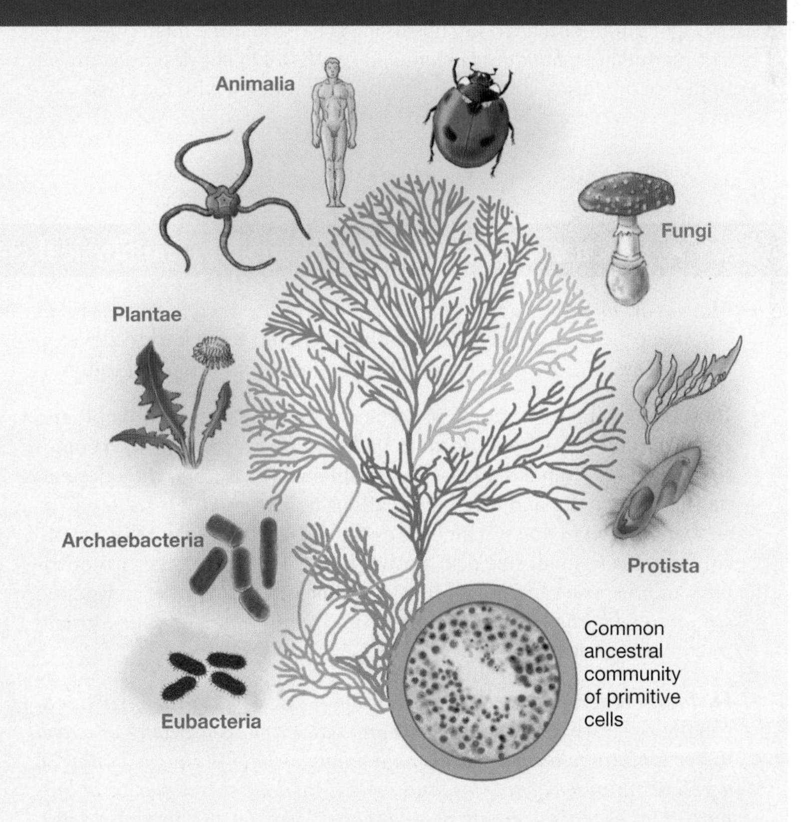


### Selected Key Terms

Use the **glossary** at the end of this book to discover the meanings of these Selected Key Terms: **Bacteria, Archaea, Eukarya, Eubacteria, Archaebacteria, Plantae, Animalia, Fungi, Protista, taxonomy, species.**

### Critical Thinking Question

Discuss why it is often difficult to differentiate between living and nonliving things.

### Active Learning Exercise

With another student in class, construct a nested box diagram similar to Figure 12.3 showing the taxonomic classification of humans. Use the information from Table 12.1 to construct your diagram. Also, use the Internet to find taxonomic information about a marine invertebrate (which lacks a backbone) not listed in Table 12.1 and construct a nested box diagram for it.

## 12.2 ▶ How are marine organisms classified?

- *Marine organisms can be classified into one of three groups, based on habitat and mobility. Plankton* are free-floating forms with little power of locomotion, *nekton* are swimmers, and *benthos* are bottom dwellers. *Most of the ocean's biomass is planktonic.*

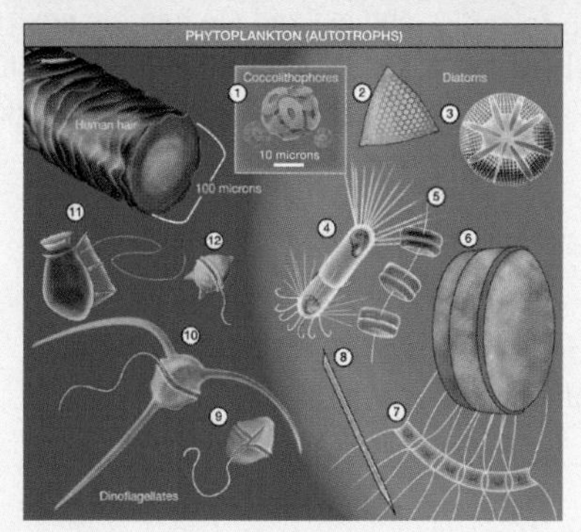


### Selected Key Terms

Use the **glossary** at the end of this book to discover the meanings of these Selected Key Terms: **plankton, phytoplankton, zooplankton, holoplankton, meroplankton, nekton, benthos, epifauna, infauna, nektobenthos.**

### Critical Thinking Question

Explain why most of the ocean's biomass is planktonic.

### Active Learning Exercise

Working with another student in class and the Internet, come up with a list of eight different marine organisms in each of the three main categories of marine life: *plankton, nekton,* and *benthos*. Do not duplicate any examples of organisms found in the text. Share your list with the class.

## 12.3 ▶ How many marine species exist?

- *Only about 13% of all known species inhabit the ocean, and more than 98% of marine organisms are benthic.* The *marine environment*—especially the pelagic environment—*is much more stable than the terrestrial environment*, so there is less pressure on marine organisms to diversify.

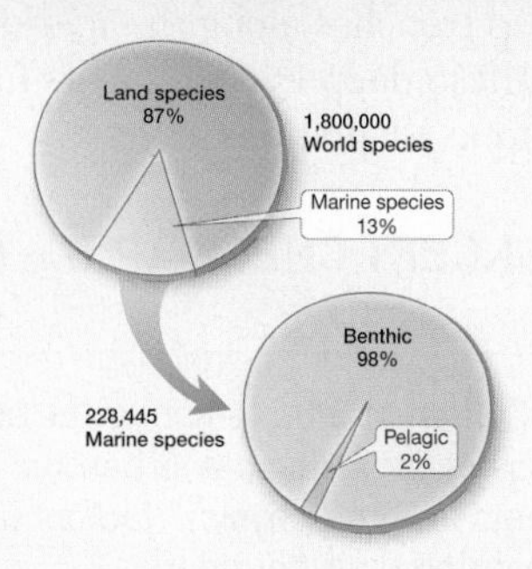


### Selected Key Terms

Use the **glossary** at the end of this book to discover the meanings of these Selected Key Terms: **pelagic environment, benthic environment.**

### Critical Thinking Question

Explain how environmental variability affects the number of species present.

### Active Learning Exercise

Working with another student in class and the Internet, go to the Census of Marine Life Website at **www.coml.org**. Explore the site and make a list of five new species that were discovered during the census. Report your findings to the class.

## 12.4 ▶ How are marine organisms adapted to the physical conditions of the ocean?

- *Marine organisms are well adapted to life in the ocean.* Those organisms that have established themselves on land have had to develop complex systems for support and for acquiring and retaining water.
- Both *algae*, which must stay in surface waters to receive sunlight, and *small animals* that feed on them *lack an effective means of locomotion.* To keep from sinking below sunlit surface waters, they depend on their *small size* and other adaptations to *increase their ratio of surface area to body mass*, which gives them high frictional resistance to sinking. Their small size also allows them to efficiently absorb nutrients and dispose of wastes. Many nektonic organisms have developed *streamlined bodies so they can overcome the viscosity of seawater* and move through it more easily.
- *Surface temperature of the world ocean does not vary* on a daily, seasonally, or yearly basis *as much as on land. Organisms living in warm water tend to be individually smaller, have ornate plumage, comprise a greater number of species, and constitute a much smaller total biomass than organisms living in cold water.* Warm-water organisms also tend to live shorter lives and reproduce earlier and more frequently than cold-water organisms.
- *Osmosis is the process by which water molecules pass through a semipermeable membrane from a less concentrated solution to a more concentrated solution.* If the cells of an organism are separated from seawater by a membrane that allows water molecules to pass through, water will move from its cells into the seawater and the organism may become severely dehydrated from osmosis. Many

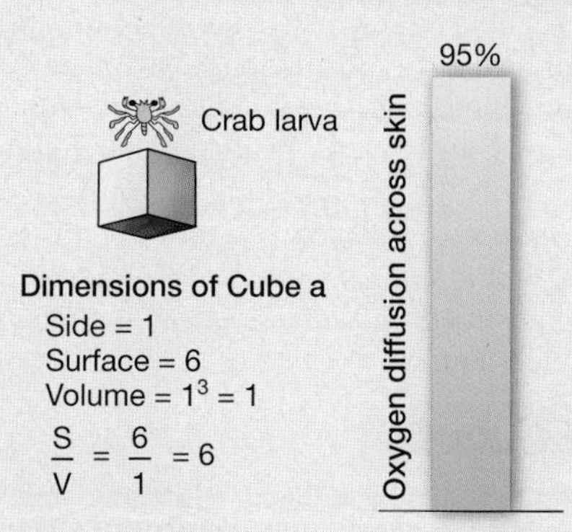


marine invertebrates are essentially *isotonic*: The salinity of their body fluids is similar to that of ocean water. Most marine vertebrates are *hypotonic*: The salinity of their body fluids is lower than that of ocean water, so they tend to lose water through osmosis. Freshwater organisms are essentially all *hypertonic*: The salinity of their body fluids is greater than the water in which they live, so they tend to gain water through osmosis.

- *Most marine animals extract oxygen through their gills.* Many marine organisms have well-developed eyesight because water is so transparent. *To avoid being seen and consumed by predators*, many marine organisms are transparent, camouflaged, countershaded, or disruptively colored. Unlike humans, most *marine organisms are unaffected by the high pressure at depth* because they do not have large internal air pockets that can be compressed.

*(continued)*

## 12.4 ▶ How are marine organisms adapted to the physical conditions of the ocean? *(continued)*

### Selected Key Terms

Use the **glossary** at the end of this book to discover the meanings of these Selected Key Terms: **viscosity, streamlining, broadcast spawning, diffusion, osmosis, hypertonic, hypotonic, gills, countershading, deep scattering layer (DSL), disruptive coloration, swim bladder.**

### Critical Thinking Question

Determine the surface-to-volume ratio of an organism whose average linear dimension is (a) 1 centimeter (0.4 inch), (b) 3 centimeters (1.1 inches), and (c) 5 centimeters (2 inches). Which one is better able to resist sinking, and why?

### Active Learning Exercise

Working with another student in class, describe how the depth of the deep scattering layer varies over the course of a day. Be sure to include reasons why it does this and which organisms comprise the DSL. Report your findings to the class.

## 12.5 ▶ What are the main divisions of the marine environment?

- *The marine environment is divided into pelagic (open sea) and benthic (sea bottom) environments.* These regions are further divided based on depth and have varying physical conditions to which marine life is superbly adapted. One of the most important layers of the pelagic environment is the *euphotic zone*, which includes the sunlit surface waters and *contains enough sunlight to support photosynthesis.*

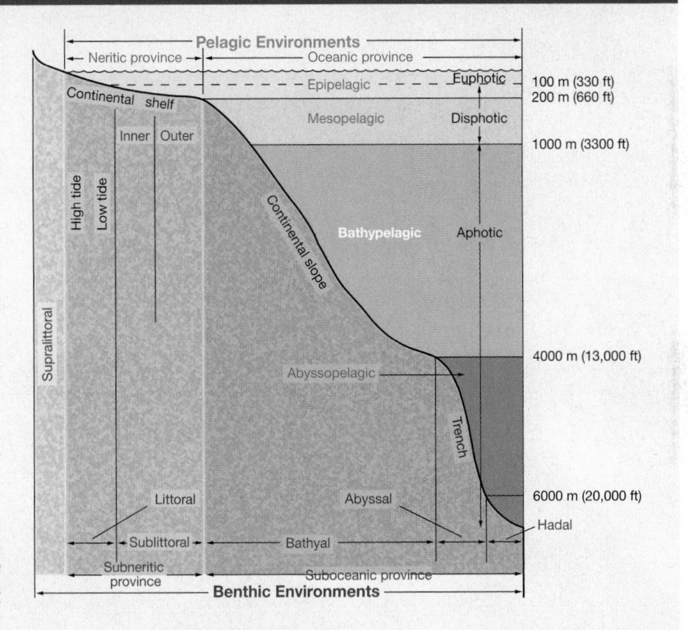


### Selected Key Terms

Use the **glossary** at the end of this book to discover the meanings of these Selected Key Terms: **neritic province, oceanic province, euphotic zone, disphotic zone, aphotic zone, oxygen minimum layer (OML), bioluminescence, detritus, bathyal zone, abyssal zone, hadal zone.**

### Critical Thinking Question

To help reinforce your knowledge of oceanic biozones, construct and label your own diagram similar to Figure 12.25 from memory.

### Active Learning Exercise

With another student in class, explain why the two curves in Figure 12.26 have the shape that they do (for example, why the oxygen minimum layer exists, and why it coincides with the nutrient maximum). Share your analysis with the class.

## Mastering Oceanography

Looking for additional review and test prep materials? With individualized coaching on the toughest topics of the course, Mastering Oceanography offers a wide variety of ways for you to move beyond memorization and deeply grasp the underlying processes of how the oceans work. Visit the Study Area in **www.masteringoceanography.com** to find practice quizzes, study tools, and multimedia that will improve your understanding of this chapter's content. Sign in today to access the following features: Self Study Quizzes, SmartFigures, Oceanography Videos and Animations, Squidtoons, Dynamic Study Modules, and an optional Pearson eText with embedded videos.

# Biological Productivity and Energy Transfer

# 13

## ESSENTIAL LEARNING CONCEPTS

At the end of this chapter, you should be able to:

- ☐ **13.1** Demonstrate an understanding of the mechanisms that control marine primary productivity.
- ☐ **13.2** Describe various kinds of photosynthetic marine organisms.
- ☐ **13.3** Explain variations in regional oceanic primary productivity.
- ☐ **13.4** Discuss how energy and nutrients are passed along in marine ecosystems.
- ☐ **13.5** Evaluate several issues that affect marine fisheries.

↑ *Check when completed*

***"Give a man a fish and he will eat today. Teach a man how to fish and he will eat for a lifetime."***

***—Ancient proverb***

Producers are organisms that synthesize their own food from carbon dioxide, water and sunlight. Through the process called *photosynthesis*, producers capture solar energy and create food in the form of sugars that sustain all other organisms in the marine biological community (except those near hydrothermal vents—which are discussed in more detail in Chapter 15, "Animals of the Benthic Environment"—where *chemosynthesis* is the major source of "food" energy). Photosynthetic producers in the ocean include plants, algae, and bacteria. As such, the ocean's producers are the foundation of the oceanic food web.

In the ocean, there are few true marine plants and large species of marine algae play only a minor role. As a result, it is the microscopic photosynthetic producers—called *phytoplankton*—that comprise the majority of organisms responsible for the conversion of solar energy. These microscopic organisms, consisting of algae, some protists, and bacteria, are mostly scattered throughout the ocean's sunlit surface waters and represent the largest community of organisms in the marine environment.

In this chapter, we'll examine primary productivity and the factors that cause it to vary as a function of latitude and depth in the water column. We will also describe various types of photosynthetic marine organisms, discuss productivity in different regions of the ocean, examine feeding relationships such as food chains and food webs, and explore environmental issues related to marine fisheries.

## 13.1 ▶ What Is Primary Productivity?

**Primary productivity** is the rate at which organisms store energy through the formation of organic matter (carbon-based compounds) from inorganic carbon (carbon dioxide), a process called **carbon fixation**. This process uses energy derived from solar radiation during **photosynthesis** (*photo* = light, *syn* = with, *thesis* = an arranging) or from chemical reactions during **chemosynthesis** (*chemo* = chemistry, *syn* = with, *thesis* = an arranging; more details about chemosynthesis can be found in Chapter 15, "Animals of the Benthic Environment"). Other organisms may then use this organic matter as food. Although chemosynthesis supports hydrothermal vent biocommunities along oceanic spreading centers, it is much less significant than photosynthesis in global marine primary production. In fact, 99.9% of the ocean's **biomass** relies either directly or indirectly on organic matter supplied by photosynthetic primary productivity as its source of food, and only 0.1% of the ocean's biomass relies on chemosynthesis (recall that *biomass* is the mass of living organisms). Therefore, the discussion of primary productivity presented here focuses on photosynthetic productivity.

◀ **Looking upward through schooling fish in sunlight.** The behavior of schooling, which is demonstrated by these bigeye trevally (*Caranx sexfasciatus*), helps protect the group from predators. Sunlight powers nearly all marine food webs, which are generally based on the passing of nutrients from one organism to another through eating.

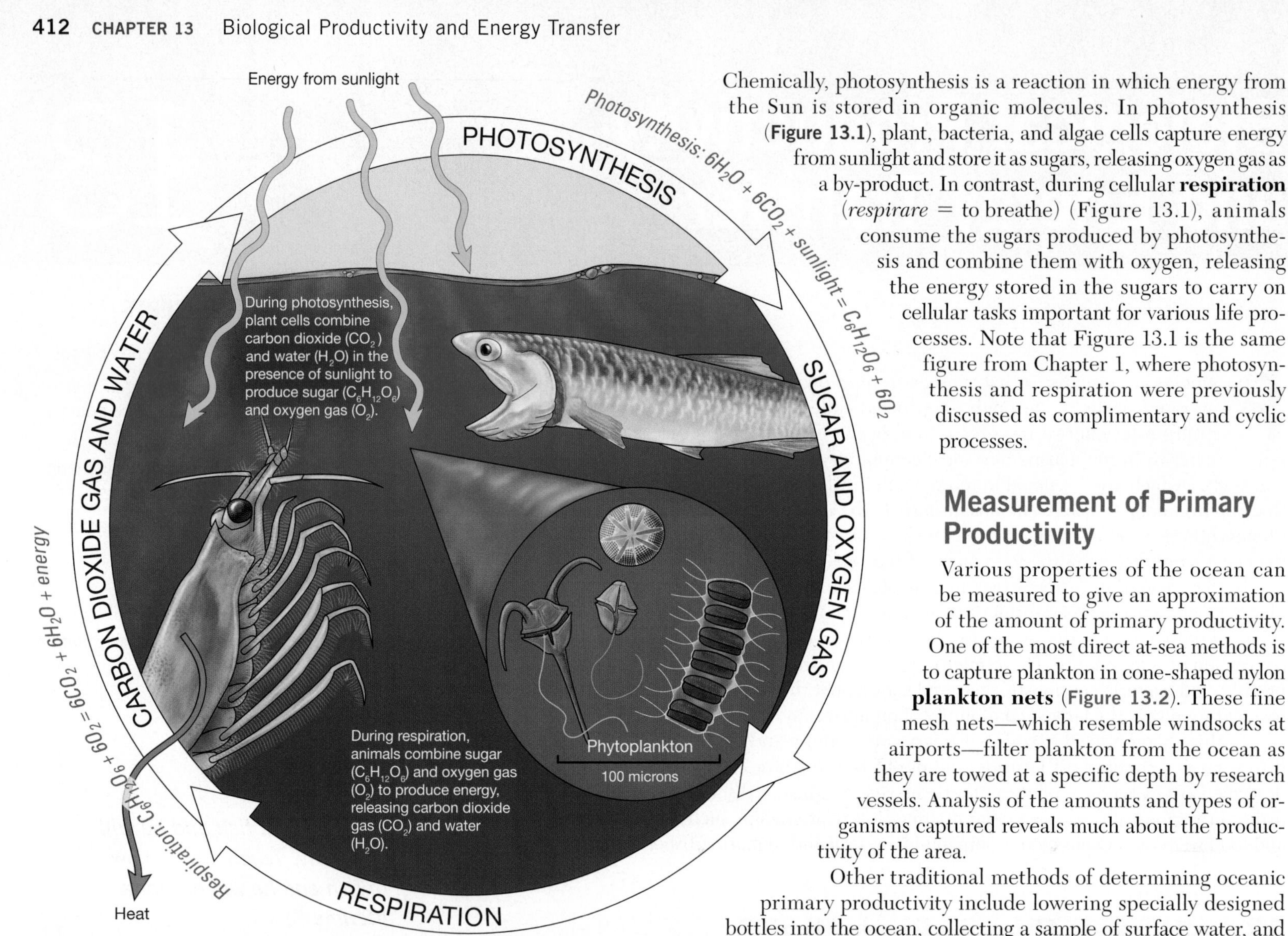

Energy from sunlight
PHOTOSYNTHESIS
Photosynthesis: $6H_2O + 6CO_2$ + sunlight = $C_6H_{12}O_6 + 6O_2$
SUGAR AND OXYGEN GAS
CARBON DIOXIDE GAS AND WATER
During photosynthesis, plant cells combine carbon dioxide ($CO_2$) and water ($H_2O$) in the presence of sunlight to produce sugar ($C_6H_{12}O_6$) and oxygen gas ($O_2$).
During respiration, animals combine sugar ($C_6H_{12}O_6$) and oxygen gas ($O_2$) to produce energy, releasing carbon dioxide gas ($CO_2$) and water ($H_2O$).
Phytoplankton
100 microns
Respiration: $C_6H_{12}O_6 + 6O_2 = 6CO_2 + 6H_2O$ + energy
RESPIRATION
Heat

**SmartFigure 13.1 Photosynthesis and respiration are cyclic and complimentary processes that are fundamental to life on Earth.** Note that this is the same image used in Chapter 1 as Figure 1.27. https://goo.gl/avxu5g

**RECAP Primary productivity is the rate at which carbon (organic matter) is produced by microbes, algae, and plants, mostly through photosynthesis; however, primary productivity also includes microbes that perform chemosynthesis.**

Chemically, photosynthesis is a reaction in which energy from the Sun is stored in organic molecules. In photosynthesis (**Figure 13.1**), plant, bacteria, and algae cells capture energy from sunlight and store it as sugars, releasing oxygen gas as a by-product. In contrast, during cellular **respiration** (*respirare* = to breathe) (Figure 13.1), animals consume the sugars produced by photosynthesis and combine them with oxygen, releasing the energy stored in the sugars to carry on cellular tasks important for various life processes. Note that Figure 13.1 is the same figure from Chapter 1, where photosynthesis and respiration were previously discussed as complimentary and cyclic processes.

## Measurement of Primary Productivity

Various properties of the ocean can be measured to give an approximation of the amount of primary productivity. One of the most direct at-sea methods is to capture plankton in cone-shaped nylon **plankton nets** (**Figure 13.2**). These fine mesh nets—which resemble windsocks at airports—filter plankton from the ocean as they are towed at a specific depth by research vessels. Analysis of the amounts and types of organisms captured reveals much about the productivity of the area.

Other traditional methods of determining oceanic primary productivity include lowering specially designed bottles into the ocean, collecting a sample of surface water, and measuring the uptake of radioactive carbon by phytoplankton in the sample. The amount of labeled carbon incorporated into the phytoplankton sample in a day can then be used to estimate the total rate of photosynthesis in a particular region of the ocean.

Global primary productivity, on the other hand, is best measured from the vantage point of space. Monitoring ocean color from Earth-orbiting satellites allows scientists to measure the concentration of **chlorophyll** (*khloros* = green, *phylum* = leaf) in surface waters, which can used to estimate **phytoplankton** (*phyto* = plant, *planktos* = wandering) abundance and, in turn, productivity. Today, ocean color is collected worldwide every two days by MODIS (Moderate Resolution Imaging Spectroradiometer) instruments aboard the *Terra* and *Aqua* satellites. MODIS measures 36 spectral frequencies of light, including ocean fluorescence, which provides a wealth of information about ocean phytoplankton productivity, health, and efficiency.

## Factors Affecting Primary Productivity

In the ocean, the two main factors that limit the amount of photosynthetic primary productivity are the availability of solar radiation and the availability of nutrients. Sometimes other variables—such as the amount of carbon dioxide—can also limit primary productivity if they become scarce in seawater. Human-caused climate change can affect marine productivity, too.

**AVAILABILITY OF NUTRIENTS** The distribution of life throughout the ocean's breadth and depth depends mainly on the availability of nutrients that phytoplankton need, such as nitrogen, phosphorus, iron, and silica. Marine populations reach their greatest concentration where the physical conditions supply large quantities of nutrients. The sources of nutrients must be considered to understand where these areas are found.

Water in the form of runoff erodes the continents, carrying material to the oceans and depositing it as sediment on the continental margins. Runoff also dissolves and transports compounds such as nitrates and phosphates, which are the main nutrients for phytoplankton. Nitrates and phosphates are also the basic ingredients in all garden and farm fertilizers. When these chemicals reach coastal areas, they can cause **eutrophication** (*eu* = good, (*tropho* = nourishment, (*ation* = action), which is the enrichment of an ecosystem with chemical nutrients. Eutrophication and its associated problems will be discussed in the next section.

The continents are the major sources of nutrients, so the greatest concentrations of marine life are found along the continental margins. The concentration of marine life decreases as the distance from the continental margins into the open sea increases. Marine life also decreases with increasing depth in the ocean because sunlight doesn't penetrate that far into the ocean, even in the clearest waters. The vast depth of the world's oceans and the great distance between the open ocean and the coastal regions where nutrients are concentrated account for these differences.

Often, the lack of certain nutrients, particularly nitrogen (as nitrates) and phosphorus (as phosphates), can limit productivity. As a result, these compounds are among the most studied in chemical oceanography.

Carbon is an important element in productivity, too, because carbon is the basic component of all organic compounds (including carbohydrates, proteins, and fats). In the ocean, however, various forms of carbon are quite abundant, so there is no scarcity of carbon for photosynthetic production. Thus, carbon does not limit productivity.

When nutrients are not limiting productivity, the ratio of carbon to nitrogen to phosphorus in the tissues of algae is in the proportion of 106:16:1 (C:N:P), which is called the *Redfield ratio*, after American oceanographer Alfred C. Redfield, who first described it in 1963. This ratio is also observed in zooplankton that feed on diatoms, and in most ocean water samples taken worldwide. Moreover, phytoplankton

*These large, cone-shaped, fine-mesh plankton nets are lowered into the water and towed behind a research vessel to collect plankton.*

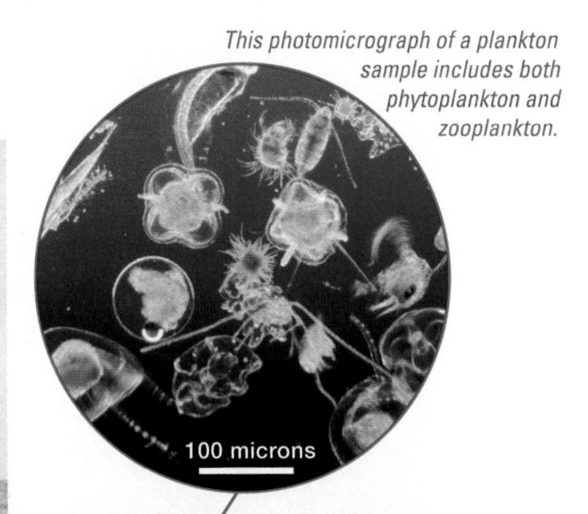


*This photomicrograph of a plankton sample includes both phytoplankton and zooplankton.*

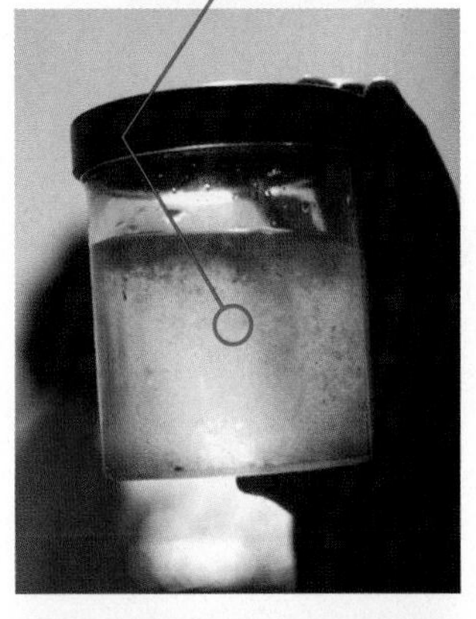

*Each speck in this plankton sample is a marine organism that can be further analyzed under a microscope.*

*Plankton accumulates in the closed end of a plankton net, where it can be transferred into a jar.*

**Figure 13.2 Plankton nets collect a plankton sample.**

take up nutrients in the ratio in which they are available in the ocean water and pass them on to zooplankton in the same ratio. When these plankton and animals die, carbon, nitrogen, and phosphorus are recycled into the water in this same ratio.

Scientific studies in the waters near Antarctica and the Galápagos Islands have revealed that photosynthetic production is low even though the concentration of all nutrients—except iron—is high. (The idea of fertilizing the ocean with iron to stimulate productivity and increase the amount of carbon dioxide gas absorbed by the ocean is discussed in Chapter 16, "The Oceans and Climate Change.") Production is high only in regions of shallow water down-current from islands or landmasses where a significant amount of iron from rocks and sediments is dissolved in water. Therefore, the lack of iron can also severely limit primary productivity.

**AVAILABILITY OF SOLAR RADIATION** Photosynthesis cannot proceed unless light energy (solar radiation) is available. Despite the atmosphere's thickness of more than 80 kilometers (50 miles), its high transparency allows sunlight to penetrate it quite readily, so land-based plants almost always have an abundance of solar radiation to conduct photosynthesis.

In the clearest ocean water, however, solar energy may be detected to depths of only about 1 kilometer (0.6 mile) and, even then, the amount reaching these depths is inadequate for photosynthesis. Photosynthesis in the ocean, therefore, is restricted to the uppermost surface waters and those areas of the sea floor where the water is shallow enough to allow light to penetrate. The water depth at which light is so limited that net photosynthesis becomes zero is called the **compensation depth for photosynthesis**.

The **euphotic zone** (*eu* = good, *photos* = light) extends from the surface down to the compensation depth for photosynthesis, which is approximately 100 meters (330 feet) in the open ocean (see Figure 12.25). Near the coast, the euphotic zone may extend to less than 20 meters (66 feet) because the water contains more suspended inorganic material (turbidity) or microscopic organisms that limit light penetration.

How do the two factors necessary for photosynthesis—the supply of nutrients and the presence of solar radiation—differ between coastal areas and the open ocean? In the open ocean (far from continental margins), solar energy extends deeper into the water column, but concentration of nutrients is low. In coastal regions, on the other hand, light penetration is much less, but the concentration of nutrients is much higher. Because the coastal zone is much more productive, nutrient availability must be the most important factor affecting the distribution of life in the oceans.

## STUDENTS SOMETIMES ASK . . .

***How is the ocean's primary productivity being affected by climate change?***

Human-caused global climate change is expected to produce significant negative impacts on the primary productivity of entire ocean ecosystems. In fact, human-caused climate change is already affecting ocean primary production in many regions. For example, because of changes in wind patterns or water temperatures, areas of the ocean that in the past exhibited strong seasonal phytoplankton blooms now experience weakened blooms or periods in which phytoplankton don't bloom at all. Human-caused climate change is also predicted to produce a decrease in the strength of ocean surface currents, altering the transportation range of fish larvae and plankton, which will ultimately negatively affect primary production.

Another major negative effect of climate change is a shift in the duration or timing of oceanic growing seasons. As explained by the "match-mismatch hypothesis," the growth and survival of predatory species depends on the synchronous production of their main food source. In essence, if a phytoplankton bloom is too early or is delayed, it changes the time when food is available for the organisms that feed on them. In a worst-case scenario, the absence of phytoplankton would cause most zooplankton to die from starvation, which, in turn, would dramatically impact ocean food webs.

Climate Connection

## Light Transmission in Ocean Water

The graph in **Figure 13.3** shows that most solar energy falls in the range of wavelengths called **visible light**. This radiant energy from the Sun powerfully affects three major components of the oceans:

1. *Ocean winds.* The major wind belts of the world, which produce ocean currents and wind-driven ocean waves, ultimately derive their energy from solar radiation. Wind belts and ocean currents strongly influence world climates.
2. *Ocean stratification.* At the ocean surface, a thin layer of water created by solar heating is warmer than the water below and overlies a great mass of cold water that fills most ocean basins. In most places, this causes the ocean's water column to be stratified into layers.
3. *Primary productivity.* Photosynthesis can occur only where sunlight penetrates the ocean water, so phytoplankton and most animals that eat them must live where the light is, in the relatively thin layer of sunlit surface water, which is the "life layer" where most marine life exists.

**THE ELECTROMAGNETIC SPECTRUM** The Sun radiates a wide range of wavelengths of electromagnetic radiation. Together they comprise the **electromagnetic spectrum**, which is shown in the upper part of Figure 13.3. Only a very narrow portion of the electromagnetic spectrum is visible to humans as visible light. We call it "visible" light because our electromagnetic sensors—our eyes—are adapted to detect only the wavelengths in the visible region. In essence, our eyes "tune into" the visible light wavelengths, just as a radio "tunes into" specific radio waves.

Visible light can be further divided by wavelength into energy levels associated with the colors red, orange, yellow, green, blue, and violet (the acronym used for the color spectrum is ROYGBV). Together, these different wavelengths produce white light. The lower energy, longer wavelengths of light to the left of visible light (for example, infrared, microwaves, and radio waves) are used for heat transfer and communication. The higher energy, shorter wavelengths of light to the right of visible light (for example, X-rays and gamma rays) are capable of damaging tissue, in high enough doses.

**THE COLOR OF OBJECTS** As Figure 13.3 shows, light from the Sun includes all the visible colors. Most of the light we see is reflected from objects. All objects absorb and reflect different wavelengths of light, and each wavelength represents a color in the visible spectrum. Vegetation, for example, absorbs most wavelengths except green and yellow, which they reflect, so most plants look green. Similarly, a red jacket absorbs all wavelengths of color except red, which is reflected.

The lower part of Figure 13.3 shows how the ocean selectively absorbs the longer-wavelength colors (red, orange, and yellow) of visible light. The true colors of objects can be observed in natural light only in the surface waters because only there can all wavelengths of the visible spectrum be found. Red light is absorbed within the upper 10 meters (33 feet) of the ocean, and yellow is completely absorbed before a depth of 100 meters (330 feet). Thus, the shorter-wavelength portion of the visible spectrum is all that can be transmitted to greater depths (mostly blue

**SmartFigure 13.3 The electromagnetic spectrum and transmission of visible light in seawater.**
https://goo.gl/Dvl14o

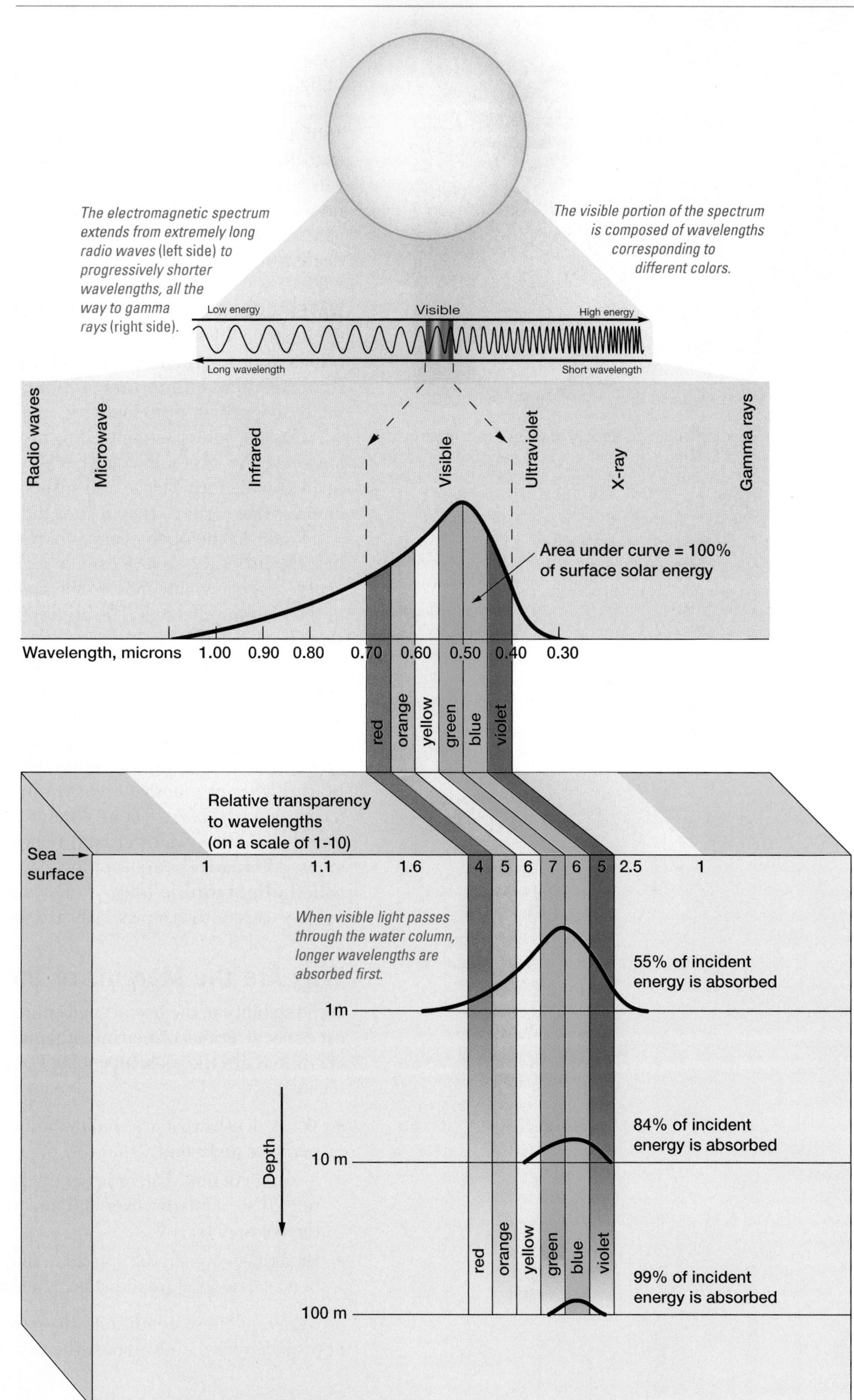
The electromagnetic spectrum extends from extremely long radio waves (left side) to progressively shorter wavelengths, all the way to gamma rays (right side).
The visible portion of the spectrum is composed of wavelengths corresponding to different colors.
Low energy
Visible
High energy
Long wavelength
Short wavelength
Radio waves
Microwave
Infrared
Visible
Ultraviolet
X-ray
Gamma rays
Area under curve = 100% of surface solar energy
Wavelength, microns 1.00 0.90 0.80 0.70 0.60 0.50 0.40 0.30
red
orange
yellow
green
blue
violet
Relative transparency to wavelengths (on a scale of 1-10)
Sea surface
1 1.1 1.6 4 5 6 7 6 5 2.5 1
When visible light passes through the water column, longer wavelengths are absorbed first.
55% of incident energy is absorbed
1m
84% of incident energy is absorbed
10 m
Depth
99% of incident energy is absorbed
100 m

Figure 13.4 **Students using a Secchi disk.** A Secchi disk is lowered from a vessel into the water with a line to measure the depth of penetration of sunlight, which indicates the clarity of the water.

light, with some violet and green wavelengths), and even then, their intensity is low. In the open ocean, sunlight strong enough to support photosynthesis occurs only within the euphotic zone to a depth of 100 meters (330 feet), and no sunlight penetrates below a depth of about 1000 meters (3300 feet).

A **Secchi** (pronounced "SECK-ee") **disk**, such as the one shown in Figure 13.4, is used to measure water transparency, and based on that, the depth of light penetration can be estimated. The Secchi disk is named after its inventor, Angelo Secchi (1818–1878), an Italian astronomer who first used the device in 1865 to measure water clarity in lakes. It consists of a disk 20 to 40 centimeters (8 to 16 inches) in diameter attached to a line that is marked off at regular intervals. As the disk is slowly lowered into the ocean, the depth at which it can last be seen indicates the water's clarity. Increased turbidity, which includes microorganisms and suspended sediment, increases the degree of light absorption, thus decreasing the depth to which visible light can penetrate into the ocean.

**WATER COLOR AND LIFE IN THE OCEANS** The color of the ocean ranges from deep indigo (blue) to yellow-green. Why are some areas of the ocean blue, whereas others appear green? Ocean color is influenced by (1) the amount of turbidity from runoff and (2) the amount of photosynthetic pigment, which increases with increasing primary productivity.

Coastal waters and upwelling areas are biologically very productive and almost always yellow-green in color because they contain large amounts of yellow-green microscopic marine algae and suspended particles. When these materials are present in surface waters, they scatter the wavelengths for greenish or yellowish light.

Water in the open ocean—particularly in the tropics—is less productive and has less turbidity, so it is usually a clear, indigo-blue color. Here, it is water molecules that contribute most to the scattering of light, and they scatter light primarily in the blue wavelengths. The atmosphere scatters blue light, too, which is why clear skies are blue.

Although photosynthetic marine algae and bacteria are microscopic, they occur in such large numbers that they can change the color of the ocean to such a degree that orbiting satellites are able to measure the changes from space. Figure 13.5, for example, shows a satellite view of ocean chlorophyll, which is an approximation for productivity. The figure shows high chlorophyll concentrations (highly productive areas) in light green color, which are called **eutrophic** (*eu* = good, *tropho* = nourishment). Generally, eutrophic waters are naturally found in shallow-water coastal regions, areas of upwelling, and high-latitude regions. Alternatively, areas of low chlorophyll concentration (low productivity) are called **oligotrophic** (*oligo* = few, *tropho* = nourishment) and are found in the open oceans of the tropics; they are shown in dark blue colors in the figure.

## Why Are the Margins of the Oceans So Rich in Life?

If the stability of the ocean environment is ideal for sustaining life, why are the richest concentrations of marine organisms in the very margins of the oceans, where conditions are the most *un*stable? For example, characteristics of the coastal ocean include:

- Water depths that are shallow, allowing much greater seasonal variations in temperature and salinity than the open ocean.
- A water column that varies in thickness in the nearshore region in response to tides that regularly cover and uncover a thin strip of land along the margins of the continents.
- Breaking waves in the surf zone that release large amounts of energy, which has been carried for great distances across the open ocean.

Each of these conditions stresses organisms. In spite of hardships, however, new species have evolved over the vast expanse of geologic time spanning billions of

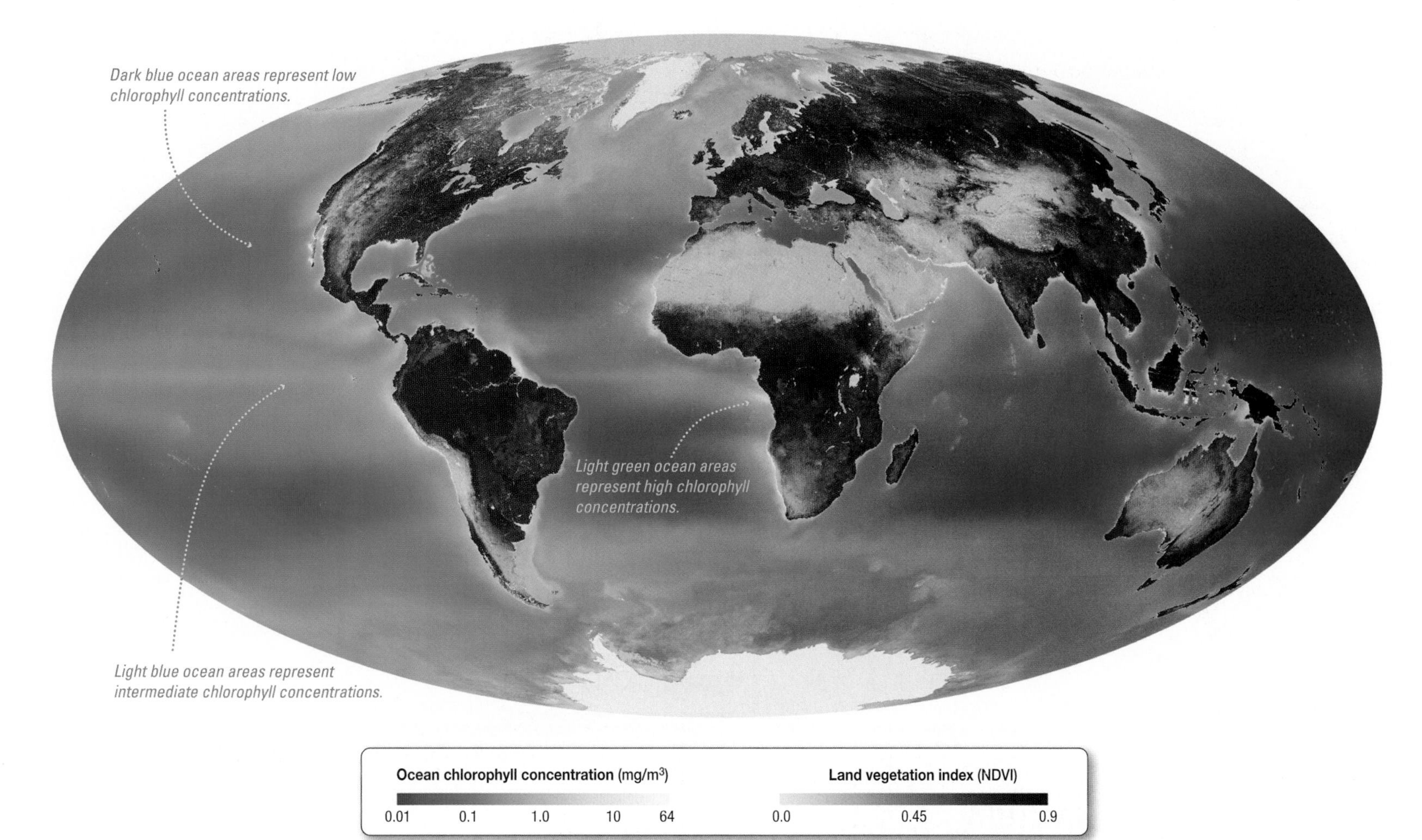
Dark blue ocean areas represent low chlorophyll concentrations.
Light green ocean areas represent high chlorophyll concentrations.
Light blue ocean areas represent intermediate chlorophyll concentrations.
Ocean chlorophyll concentration (mg/m3)
0.01
0.1
1.0
10
64
Land vegetation index (NDVI)
0.0
0.45
0.9

Figure 13.5 **Satellite image of ocean chlorophyll.** Satellite data (1998–2010) showing average ocean chlorophyll concentration, which is an approximation for productivity. Data were gathered during the entire 13-year operation of the SeaWiFS instrument aboard the *SeaStar* satellite, which detected changes in seawater color caused by changing concentrations of chlorophyll that varies with photosynthetic productivity. Ocean chlorophyll concentrations are reported in milligrams per cubic meter (mg/m$^3$). On land, data are depicted using the Normalized Difference Vegetation Index (NDVI), which shows the density of green vegetation.

years by the process of natural selection to fit every imaginable biological niche—even in environments that pose difficulties for organisms (see Diving Deeper 1.1 for a description of evolution and natural selection). In fact, many organisms have adapted to live under adverse conditions—such as coastal environments—as long as nutrients are available.

Along continental margins, some areas have more abundant life than others. What characteristics create such an uneven distribution of life? Again, only the basic requirements for the production of food need be considered. For example, areas that have the lowest water temperatures have the greatest biomass, because cold water contains higher amounts of nutrients and dissolved gases—such as oxygen and carbon dioxide—than warm water. These nutrients and gases stimulate phytoplankton growth, which profoundly affects the distribution of all other life in the oceans.

**UPWELLING AND NUTRIENT SUPPLY** As discussed in Chapter 7, **upwelling** is a flow of deep water toward the surface that brings water from depths below the euphotic zone. This deep water is rich in nutrients and dissolved gases because there are no phytoplankton at these depths to consume these compounds. When chilled water from below the surface rises, it hoists nutrients from the depths to the surface, where phytoplankton thrive and become food for larger organisms—copepods, fish, and on up to larger organisms such as sharks and whales. However, as will be discussed later, surface warming and the resulting stratification of the ocean's water column can limit upwelling and thereby inhibit primary productivity.

Where does upwelling occur in the oceans? One common location are the highly productive areas of *coastal upwelling* that are found along the western margins of continents, where surface currents are moving toward the equator

**Animation**
Ekman Spiral and Coastal Upwelling/Downwelling
https://goo.gl/Y4LOD9

ARCTIC OCEAN
ATLANTIC OCEAN
PACIFIC OCEAN
INDIAN OCEAN
Area enlarged below
Areas of upwelling resulting from winds
Wind direction

**(a)** Coastal winds *(green arrows)* cause Ekman transport, which drives surface water away from the west coasts of continents *(blue arrows)*.

NAMIBIA
ATLANTIC OCEAN
SOUTH AFRICA
Chlorophyl a concentration (mg/m$^3$)
.01 .02 .03 .05 .1 .2 .3 .5 1 2 3 5 10 15 20 30 50

**(b)** SeaWiFS image of chlorophyll concentration along the southwest coast of Africa (February 21, 2000). High chlorophyll concentrations indicate high phytoplankton biomass, which is caused by coastal upwelling. Concentration is reported in milligrams of chlorophyll a per cubic meter of seawater (mg/m$^3$).

1 *Wind causes surface water to move to the left away from shore.*
Wind
Coastal upwelling
2 *Deep, cold water upwells to replace surface water.*
Southern Hemisphere
1000 m
Phytoplankton
Zooplankton

**(c)** Block diagram showing how coastal upwelling in the Southern Hemisphere is caused by coastal winds that cause surface waters to move away from shore due to Ekman transport, thus bringing cold, nutrient-rich water to the surface.

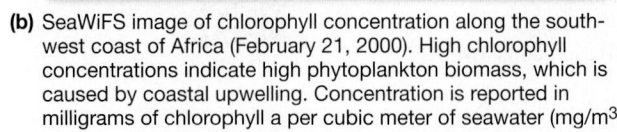

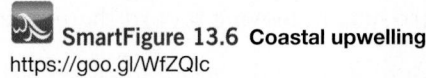

**SmartFigure 13.6 Coastal upwelling.**
https://goo.gl/WfZQlc

(**Figure 13.6**). Ekman transport (see Chapter 7) causes surface water to move away from these coasts, so nutrient-rich water from depths of 200 to 1000 meters (660 to 3300 feet) constantly rises to replace it. Another location is along the equator, where the process of *equatorial upwelling* occurs.

**RECAP** **Photosynthetic productivity is limited in the marine environment by the amount of sunlight and the supply of nutrients. Upwelling greatly enhances the conditions for life by lifting cold, nutrient-rich water to the sunlit surface waters.**

**CONCEPT CHECK 13.1 ▶ Demonstrate an understanding of the mechanisms that control marine primary productivity.**

1 Discuss chemosynthesis as a method of primary productivity. How does it differ from photosynthesis?

2 An important variable in determining the distribution of life in the oceans is the availability of nutrients. How are the following variables related: proximity to the continents, availability of nutrients, and the concentration of life in the oceans?

3 Another important determinant of productivity is the availability of solar radiation. Why is biological productivity relatively low in the tropical open ocean, where the penetration of sunlight is greatest?

4 Discuss the general characteristics of the coastal ocean where unusually high concentrations of marine life are found.

5 Explain why everything in the deep ocean below the shallowest surface water appears blue-green in color.

# 13.2 ▶ What Kinds of Photosynthetic Marine Organisms Exist?

Many types of marine organisms photosynthesize. They are primarily represented by microscopic bacteria and algae but also include larger forms of algae and some seed-bearing plants. Recall from Chapter 12 that marine algae are not plants. The main difference between the two is in their complexity. Algae are simple organisms, sometimes unicellular, and even the largest types are relatively simple in structure. True plants, on the other hand, are quite complex, with many specialized structures such as roots, stems, leaves, and flowers, all of which algae lack.

## Seed-Bearing Plants (Anthophyta)

The only members of kingdom Plantae that exist in the marine environment belong to the seed-bearing members of phylum Anthophyta (*antho* = flower, *phytum* = plant), which occur exclusively in shallow coastal areas. Eelgrass (*Zostera*), for example, is a grass-like plant with true roots that exists primarily in the quiet waters of bays and estuaries from the low-tide zone to a depth of 6 meters (20 feet). Surf grass (*Phyllospadix*) (**Figure 13.7**), which is also a seed-bearing plant with true roots, is typically found in the high-energy environment of exposed rocky coasts from the intertidal zones down to a depth of 15 meters (50 feet).

Other seed-bearing plants are found in salt marshes and include grasses (mostly of the genus *Spartina*), whereas mangrove swamps contain mostly mangroves (genera *Rhizophora*, *Avicennia*, and *Laguncularia*). All these plants are important sources of food and protection for the marine animals that inhabit coastal environments, as discussed in Chapter 10.

Figure 13.7 **Surf grass.** Green surf grass (*Phyllospadix*) and various species of brown algae are exposed during an extremely low tide in this California tide pool. When the tide rises, the anchored surf grass floats and provides a protective hiding place for many tide pool organisms.

## Macroscopic (Large) Algae

Various types of marine macro algae (the "seaweeds") are typically found in shallow waters along the ocean margins. These algae are usually attached to the bottom, but a few species float. Algae are classified in part based on the color of the pigment they contain (**Figure 13.8**). Although modern classification of algae uses more than just its color, the division of algae into groups based on color is still a useful means of describing the different types of algae.

Figure 13.8 **Examples of macroscopic algae.**

**(a)** Green alga *Codium fragile*, also known as sponge weed or dead man's fingers.

**(b)** Red algae of two different types, *Bossiella californica* (*left, center*) and *Corallina* sp. (*right*). Both have tips that show portions of their internal calcareous skeleton (*white*).

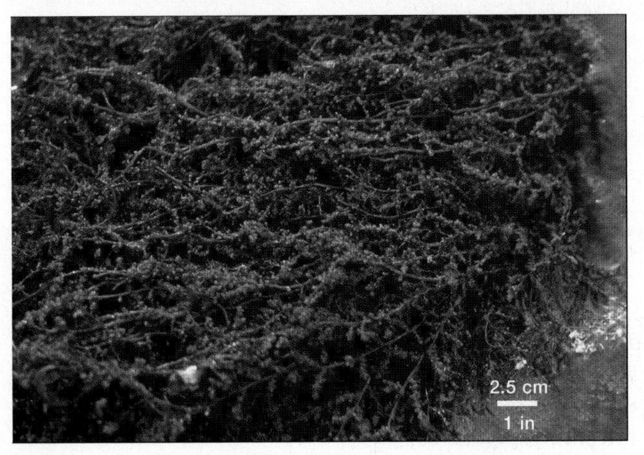


**(c)** Brown alga *Sargassum*. This attached form is similar to the floating form that is the namesake of the Sargasso Sea.

**(d)** A small strand of brown alga *Macrocystis*, which is a major component of kelp beds.

## STUDENTS SOMETIMES ASK . . .

***What type of algae was involved in the massive cleanup effort prior to the 2008 Olympic sailing events in China?***

Every year during the past several years, a bloom of floating green algae *Enteromorpha* occurs along the coast of Qingdao, China. In 2008, Olympic sailing events were scheduled to be held there but the algae covered the entire sailing venture. The algae bloom is caused by excessive nitrogen from fertilizers and from septic and sewage systems that provide nutrients for the algae, which can grow to over 0.3 meter (1 foot) in thickness. Threatened by having to cancel the sailing competition, the Chinese government initiated one of the world's largest algae-cleanup efforts (**Figure 13.9**), which involved tens of thousands of citizens and defense personnel, thousands of fishing boats, and hundreds of dump trucks. The massive cleanup effort was able to collect enough of the floating algae that China was able to successfully host the Olympic sailing events as planned. In all, about 682 million metric tons (1.5 trillion pounds) of algae was removed from coastal waters at an estimated cost of $87.3 million. According to news reports, much of the algae was transported to farms as fertilizer or feed for livestock.

**GREEN ALGAE** Although green algae of phylum Chlorophyta (*khloros* = green, *phytum* = plant) are common in freshwater environments, they are not well represented in the ocean. Most marine species are intertidal or grow in shallow bay waters. They contain the pigment chlorophyll which gives them their green color. They grow only to moderate size, seldom exceeding 30 centimeters (12 inches) in the largest dimension. Forms range from finely branched filaments to thin sheets.

Various species of sea lettuce (*Ulva*), a thin membranous sheet only two cell layers thick, are widely scattered throughout cold-water areas. Sponge weed (*Codium*), a two-branched form more common in warm waters, can exceed 6 meters (20 feet) in length (**Figure 13.8a**).

**RED ALGAE** Red algae of phylum Rhodophyta (*rhodos* = red, *phytum* = plant) are the most abundant and widespread of marine macroscopic algae. Over 4000 species occur from the very highest intertidal levels to the outer edge of the inner sublittoral zone. Many are attached to the bottom, either as branching forms (**Figure 13.8b**) or as forms that encrust surfaces. They are very rare in freshwater. Red algae range from being just barely visible to the unaided eye to being 3 meters (10 feet) long. Red algae are found in both warm and cold waters, but the warm-water varieties are relatively small.

The color of red algae varies considerably, depending on its depth in the intertidal or inner sublittoral zones. In upper, well-lighted areas, it may be green to black or purplish. In deeper water zones, where less light is available, it may be brown to pinkish-red.

The vast majority of marine photosynthetic productivity occurs within the surface layer of the ocean to a depth of 100 meters (330 feet), which corresponds to the depth of the euphotic zone. At this depth, the amount of light is reduced to 1% of that available at the surface. Remarkably, some deep-water species can survive on the very faint amount of sunlight that exists below the euphotic zone. For example, a species of red alga has been documented growing at a depth of 268 meters (880 feet) on a seamount near San Salvador in the Bahamas, where available light is only 0.0005% of that available at the surface.

**BROWN ALGAE** Brown algae of the phylum Phaeophyta (*phaeo* = dusky, *phytum* = plant) include the largest members of the attached (that is, not free-floating) species of marine algae. Their color ranges from very light brown to black. Brown algae occur primarily in middle latitude, cold-water areas.

The sizes of brown algae range widely. One of the smallest is *Ralfsia*, which occurs as a dark brown encrusting patch in upper and middle intertidal zones. One of the largest is bull kelp (*Pelagophycus*), which may grow in water deeper than 30 meters (100 feet) and extend to the surface. Other types of brown algae include *Sargassum* (**Figure 13.8c**) after which the Sargasso Sea is named and *Macrocystis* (**Figure 13.8d**).

Figure 13.9 **Algae cleanup in Qingdao, China, before the 2008 Olympic sailing events.** The Chinese government undertook a massive cleanup of floating green algae in order to host the 2008 Olympic sailing competition. The cleanup effort involved tens of thousands of people that removed about 682 million metric tons (1.5 trillion pounds) of algae at an estimated cost of $87.3 million.

## Microscopic (Small) Algae

Microscopic algae are either directly or indirectly the source of food for more than 99% of marine animals. Most microscopic algae are phytoplankton—photosynthetic organisms that live in the upper surface waters and drift with currents. However, other types of microscopic algae live on the bottom in the nearshore environment, where sunlight reaches the shallow ocean floor.

**GOLDEN ALGAE** The golden algae of phylum Chrysophyta (*chrysus* = golden, *phytum* = plant) contain the orange-yellow pigment *carotin*. They consist of diatoms and coccolithophores, both of which store food as carbohydrates and oils, as discussed in Chapter 4.

***Diatoms*** **Diatoms** (*diatoma* = cut in half) are a class of algae that are contained in a microscopic shell called a **test** (*testa* = shell). Diatom tests are composed of opaline silica ($SiO_2 \cdot nH_2O$) and are important geologically because they accumulate on the ocean bottom, producing **diatomaceous earth**. Some deposits of diatomaceous earth that have been elevated above sea level by tectonic forces are mined and used in filtering devices and numerous other applications (see Diving Deeper 4.3). Diatoms are the most productive group of marine algae.

Diatom tests have a variety of shapes, but all have a top half and a bottom half that fit together (**Figure 13.10a**). The single cell is contained within this test, and it exchanges nutrients and waste with the surrounding water through holes in its test.

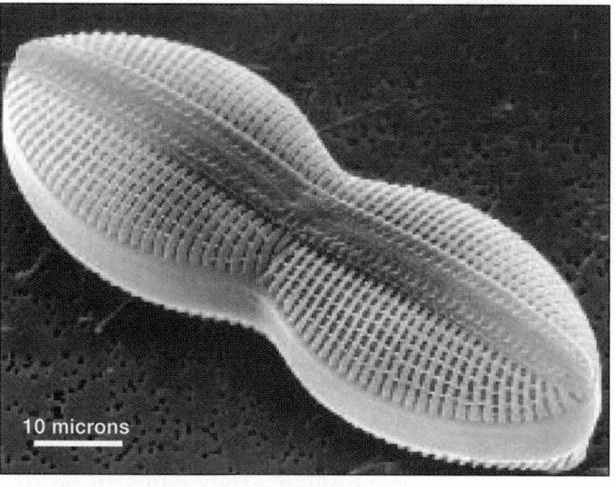


(a) Peanut-shaped diatom *Diploneis.*

(b) Coccolithophore *Emiliania huxleyi*, showing disk-shaped calcium carbonate ($CaCO_3$) plates—called coccoliths—that cover the organism.

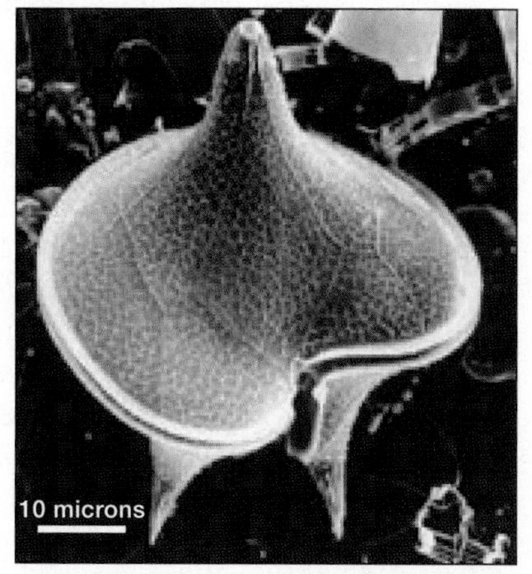


(c) Dinoflagellate *Protoperidinium divergens.*

(d) Leaflike tropical dinoflagellate *Heterodinium whittingae.*

Figure 13.10 **Examples of microscopic algae.**

**Figure 13.11 Red tide in the ocean.** When conditions are right, dinoflagellates reproduce and populate surface waters in staggering numbers, turning the water red. Although this phenomenon is called a red tide, it has nothing to do with the tides.

***Coccolithophores*** **Coccolithophores** (*coccus* = berry, *lithos* = stone, *phorid* = carrying) are covered with small calcareous plates called **coccoliths**, made of calcium carbonate ($CaCO_3$) (**Figure 13.10b**). Each individual plate is about the size of a bacterium (about 1 micron, or 0.000004 inch), and the entire organism is too small to be captured in plankton nets. Coccolithophores live in temperate and warmer surface waters and contribute significantly to calcareous sea floor deposits.

***Dinoflagellates*** **Dinoflagellates** (*dino* = whirling, *flagellum* = whip) belong to the phylum Pyrrophyta (*pyrrhos* = fire, *phytum* = plant) (**Figure 13.10c** and **d**). They possess **flagella** (small, whiplike structures) for locomotion, giving them a slight capacity to move into areas that are more favorable for photosynthetic productivity. Dinoflagellates are rarely important geologically because their tests are made of cellulose, which is biodegradable and not preserved as deposits on the sea floor.

***Red Tides*** Dinoflagellates, which contain a red pigment, sometimes exist in such great abundance that they color surface waters red, producing the phenomenon known as a **red tide** (**Figure 13.11**), which has nothing to do with tidal phenomena. When conditions are right, phytoplankton populations can grow exponentially and create a phenomenon known as a *bloom*, which can be seen in satellite images. Red tides and associated algal blooms that do not color the water red but are detrimental to marine animals, humans, or the environment are more accurately called **harmful algal blooms (HABs)**. These toxic blooms can make marine life—including manatees and other marine mammals—very sick or kill them, and can have the same effect on people who eat contaminated seafood (see details below). In addition to producing toxins, many of the 1100 species of dinoflagellates undergo bizarre structural changes in response to changes in their environment (see Mastering Oceanography **Web Diving Deeper 13.1**).

What causes red tides? Natural oceanographic conditions sometimes stimulate the productivity of certain dinoflagellates. During these times, up to 2 million dinoflagellates may be found in 1 liter (about 1 quart) of water, giving the water a reddish color (**Figure 13.11**). In other instances, red tides appear to be associated with nutrient-laden runoff from land. Red tides are by no means a new phenomenon. In fact, the Old Testament and other ancient literature make reference to waters turning blood red, which most likely influenced the colorful names of certain seas, such as the Red Sea and the Vermillion Sea.

***Dinoflagellate Toxins*** Although many red tides are harmless to marine animals and humans, they can still be responsible for mass die-offs of marine organisms. When huge numbers of dinoflagellates die, the resulting decomposition removes oxygen from seawater, and many types of marine life literally suffocate to death. In other cases, dinoflagellates that are responsible for many red tides produce neurotoxins that can spread to many different types of organisms—including humans (**Figure 13.12**). *Karenia* and *Gonyaulax*, for example, are two common genera of dinoflagellates in red tides that produce water-soluble toxins. Certain filter-feeding shellfish called bivalves—various clams, mussels, and oysters—then strain the dinoflagellates from the water for food. *Karenia* toxin kills fish and shellfish. *Gonyaulax* toxin is not poisonous to shellfish, but it concentrates in their tissues and is poisonous to humans who eat the shellfish, even after the shellfish are cooked. This malady is called *paralytic shellfish poisoning (PSP)*.

The symptoms of PSP in humans are similar to those of drunkenness—incoherent speech, uncoordinated movement, dizziness, and nausea—and can occur only 30 minutes after ingesting contaminated shellfish or swimming in the bloom waters. There is no known antidote for the toxin, which attacks the human central nervous system, but the critical period usually passes within 24 hours. At least 300 fatal and 1750 nonfatal cases of PSP have been documented worldwide.

Dinoflagellates are also associated with various types of seafood poisoning following ingestion of fish. One example is **ciguatera**, which is caused by eating

## STUDENTS SOMETIMES ASK . . .

### *Why does a red tide glow bluish green at night?*

Many of the species of dinoflagellates that produce red tides (most notably those of genus *Gonyaulax*) also have bioluminescent capabilities—that is, they can produce light organically. When the organisms are disturbed, they emit a faint bluish-green glow. When waves break during a red tide at night, the waves are often spectacularly illuminated by millions of bioluminescent dinoflagellates. During these times, one can easily observe marine animals moving through the water because their bodies are silhouetted by bioluminescent dinoflagellates that light up as they pass over the animals' bodies.

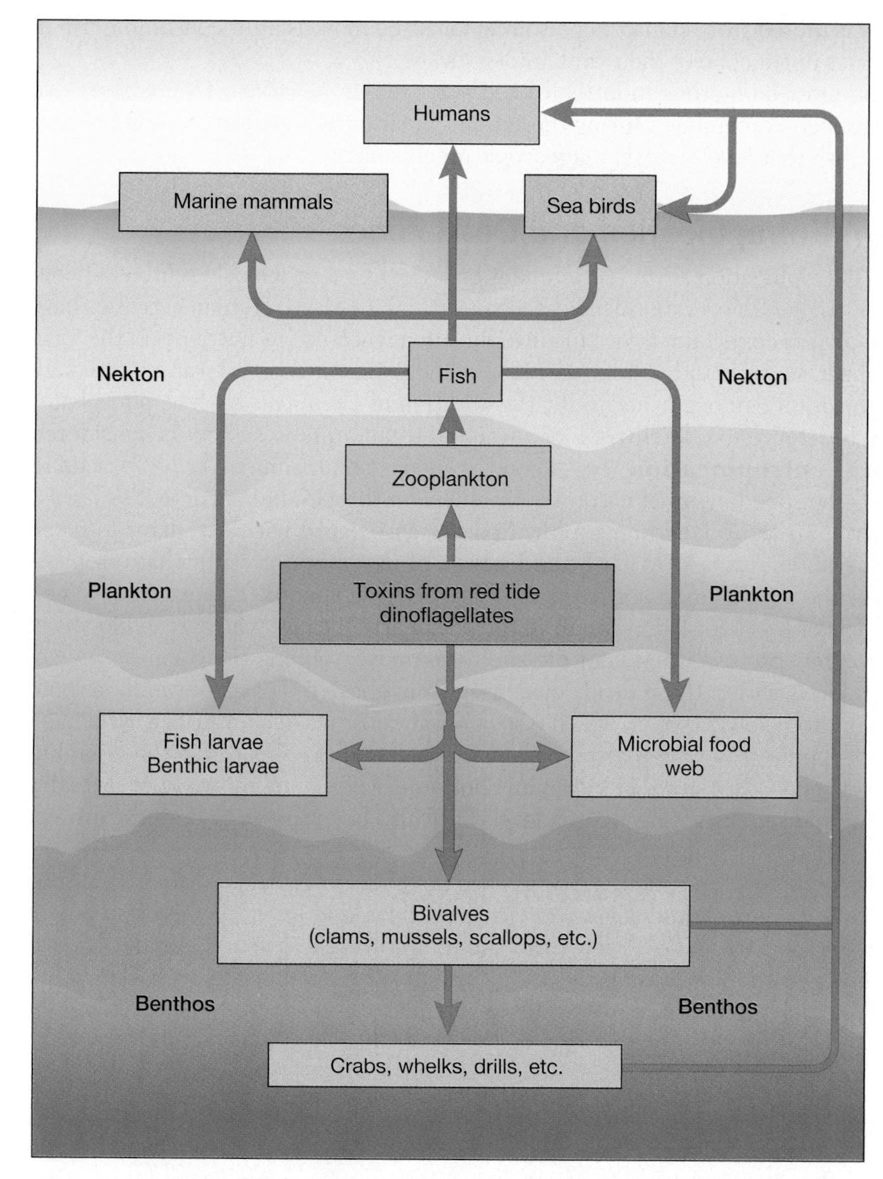


**Figure 13.12 Routes through which dinoflagellate toxins spread to marine organisms and humans.** During algal blooms, certain species of dinoflagellates and other phytoplankton produce powerful toxins, which can spread throughout marine food webs. These toxic blooms can kill marine life and even people who eat contaminated seafood.

certain tropical reef fish (most notably large predators such as barracuda, red snapper, and grouper). These reef fish accumulate high levels of naturally occurring dinoflagellate toxins in their tissues in a process called *biomagnification*, which is discussed in Section 11.3 "What Marine Environmental Problems Are Associated with Non-Petroleum Chemical Pollution?" in Chapter 11. These toxins don't affect the fish, but they do affect humans. Symptoms of ciguatera in humans usually involve a combination of gastrointestinal, neurological, and cardiovascular disorders but are rarely fatal, with symptoms usually clearing in one to four weeks. Worldwide, ciguatera causes more cases of human illness than any other form of seafood poisoning and even famous British explorer Captain James Cook (1728–1779) and

## STUDENTS SOMETIMES ASK . . .

***What other strange occurrences have been linked to the ingestion of algae toxins?***

Worldwide, a number of strange occurrences and mysterious poisonings have been traced back to toxins produced by marine microorganisms that have spread throughout the marine food web and even to humans. For instance, domoic acid—a biotoxin produced by a diatom (*Pseudo-nitzschia*)—was first recognized in 1987 as the poison that sickened more than 100 people who ate contaminated mussels from Prince Edward Island, Canada. Domoic acid poisoning causes symptoms such as confusion, disorientation, seizures, coma, and even death. In the Prince Edward Island incident, 4 of the victims died, and 10 suffered permanent short-term memory loss, which led researchers to call poisoning by domoic acid *amnesic shellfish poisoning*.

Domoic acid poisoning has also been linked to famously odd behavior of seabirds in California's Monterey Bay during the summer of 1961. A local newspaper at the time reported thousands of crazed seabirds, which flew erratically, smashed into structures, rammed cars, and pecked eight people. Studies of the gut contents of archived zooplankton samples from 1961 have found that toxin-making algae were present in 79% of zooplankton. Researchers surmise that domoic acid was concentrated in plankton-eating fish, which in turn were eaten by migratory sooty shearwaters (*Puffinus griseus*), which then sickened. The incident was reported to have inspired legendary filmmaker Alfred Hitchcock (who owned a home in the area) to describe similar events in his classic thriller *The Birds* (**Figure 13.13**), which was released two years later.

There have been other incidents involving toxic diatoms in Monterey Bay. In 1991, brown pelicans and Brandt's cormorants exhibited odd behavior—they acted drunk, swam in circles, and made loud squawking sounds. Research revealed that toxic diatoms (*Pseudo-nitzschia australis*) were to blame and caused more than 100 birds to wash up dead on the shore.

**Figure 13.13 Actress Tippi Hedren fights off a menacing seagull in Hitchcock's 1963 classic suspense film The Birds.** The inspiration for Hitchcock's classic thriller was an actual event that occurred in Monterey Bay in 1961, when birds acted erratically after eating fish contaminated with domoic acid, a biotoxin produced by diatoms.

## STUDENTS SOMETIMES ASK . . .

***Are red tides becoming a more common occurrence? If so, why?***

In many places worldwide, such as the northern Gulf of Mexico, algal blooms known as red tides have been documented in some of the earliest written records. Yet it is unclear how large a role human-caused versus natural factors play in their development. For example, the occurrence of algal blooms in some locations appears to be entirely natural, while in others the blooms appear to be a result of increased nutrients such as nitrates and phosphates from land-based runoff. In addition to human-caused coastal water nutrient loading, the recently documented increase in seawater surface temperatures has also been implicated as a contributing factor in algal blooms. Currently, it is under debate in the scientific community whether the apparent increase in frequency and severity of algal blooms in various parts of the world is in fact a real increase or whether it is simply due to an enhanced observation and reporting network.

his crew suffered from ciguatera poisoning for three months after consuming fish in the Azores during one of their exploratory voyages.

The most dangerous months for red tides in the Northern Hemisphere are April through September. During these times, quarantines exist to prohibit harvesting shellfish that feed on toxic microscopic organisms.

## Ocean Eutrophication and Dead Zones

Ocean eutrophication is the enrichment of waters by a previously scarce nutrient that can trigger an overabundance of algae such as an HAB. Human activities have been shown to contribute to ocean eutrophication when excess nutrients in the form of fertilizer, sewage, and animal waste make their way into coastal waters. Although eutrophication can occur naturally, the addition of chemicals such as phosphates (through detergents), fertilizers, or sewage into an aquatic system is considered **cultural eutrophication** (*eu* = good, *tropho* = nourishment, *ation* = action), which is the speeding up of natural eutrophication through human activities.

Large areas of ocean eutrophication are associated with extensive *hypoxic* (*hypo* = under, *oxic* = oxygen) **dead zones** of oxygen-poor water that often occur near the mouths of major rivers after large spring runoffs (**Figure 13.14**). When rivers deliver fertilizer-laden runoff to the ocean, it can cause widespread blooms of algae to grow profusely and later die and decompose, robbing the water of oxygen. Oxygen levels within these dead zones drop from above 5.0 parts per million (ppm by weight) to below 2.0 ppm, which is lower than most marine animals can tolerate. Some of the most mobile marine organisms that swim well can flee the area, but the dead zone suffocates and kills many bottom-dwelling organisms such as crabs, sea stars, and sea slugs. Low oxygen levels have also been shown to limit the growth and reproduction of marine life.

The number of marine dead zones has doubled every decade from the 1960s into the 21st century, and scientists currently have documented the existence of more than 500 worldwide (**Figure 13.15**). In the future, the size and number of dead zones is expected to increase because of continuing human impacts such as polluted runoff and destruction of coastal wetland habitats.

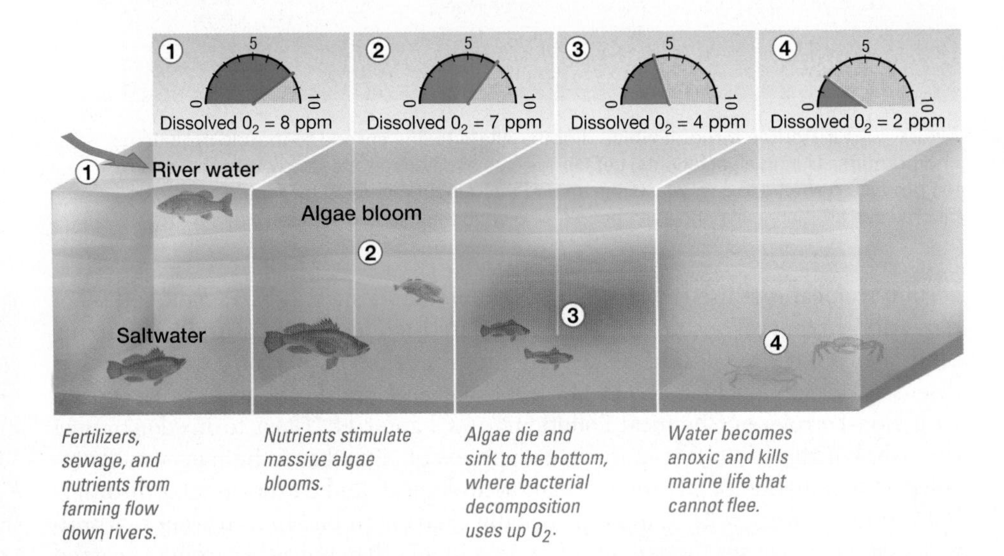

1
Dissolved $O_2$ = 8 ppm
2
Dissolved $O_2$ = 7 ppm
3
Dissolved $O_2$ = 4 ppm
4
Dissolved $O_2$ = 2 ppm
River water
Algae bloom
Saltwater
Fertilizers, sewage, and nutrients from farming flow down rivers.
Nutrients stimulate massive algae blooms.
Algae die and sink to the bottom, where bacterial decomposition uses up $O_2$.
Water becomes anoxic and kills marine life that cannot flee.

**SmartFigure 13.14 How dead zones form.** Dead zones are low-oxygen regions related to excess nutrients from agricultural runoff that stimulates algae blooms. When the algae die, decomposition consumes vast amounts of oxygen, thereby creating an anoxic dead zone that can kill marine life. Gauges shown in each panel illustrate dissolved oxygen concentration in parts per million (ppm). https://goo.gl/jxhl90

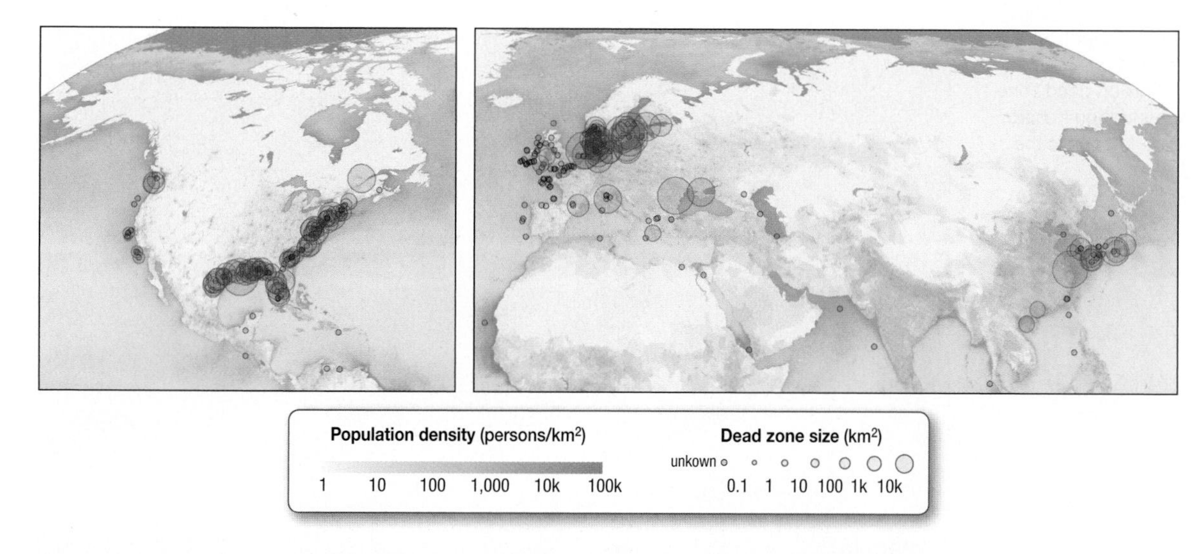


**Figure 13.15 Oceanic dead zones in the Northern Hemisphere.** Map of population density (*brown*) and documented dead zones (*red circles*, based on size) in the Northern Hemisphere. Small black circles show where dead zones have been reported, but their size is unknown. Notice that many dead zones occur where large rivers enter the sea. Not shown: the Southern Hemisphere, where few dead zones appear, likely because this region has fewer continents and fewer people.

The world's largest dead zone is in the Baltic Sea, where a combination of agricultural runoff, deposition of nitrogen from burning fossil fuels, and human waste discharge has over-fertilized the sea. The countries surrounding the Baltic Sea established a commission in 2007 to help protect the environmental health of the sea by implementing policies and measures to reduce its incoming load of nutrients.

The second-largest dead zone in the world is the one that forms each summer near the mouth of the Mississippi River in the Gulf of Mexico off Louisiana (**Figure 13.16**). In 2017, in fact, it reached a record size of 22,729 square kilometers (8800 square miles)—about the size of the state of New Jersey. Smaller dead zones have occurred each summer in the region for decades, but they dramatically increased in size after the record-breaking Midwest floods in 1993 and 2011. Over the past several years, the average size of the dead zone in the northern Gulf of Mexico has been about 17,000 square kilometers (6560 square miles), the size of Lake Ontario.

The Gulf's dead zone appears to be related to runoff of nutrients—especially nitrates from agricultural activities—that are washed down the Mississippi River and eventually reach the Gulf, where they trigger algal blooms (see Figure 13.14). Once these algae die and rain down on the sea floor, bacteria feed on them and on fecal matter, depleting the water of oxygen along the bottom. Other factors—most notably stratification of the water column that inhibits water mixing—can influence the formation of dead zones, which can persist for several months until there is strong mixing of ocean waters from a hurricane or other storm.

Proposals to combat the spread of the Gulf's dead zone include controlling nutrient runoff from agriculture, preserving and utilizing wetlands that filter runoff before it enters the Gulf, planting buffer strips of trees and grasses between farm fields and streams, altering the times when fertilizers are applied, improving crop rotation, and enforcing existing clean-water regulations. In this way, scientists, land planners, and policymakers are working to institute an action plan to shrink the yearly Gulf of Mexico dead zone. There are, however, a few naturally occurring low-oxygen zones in the world, such as the Bay of Bengal and the Atlantic coastal

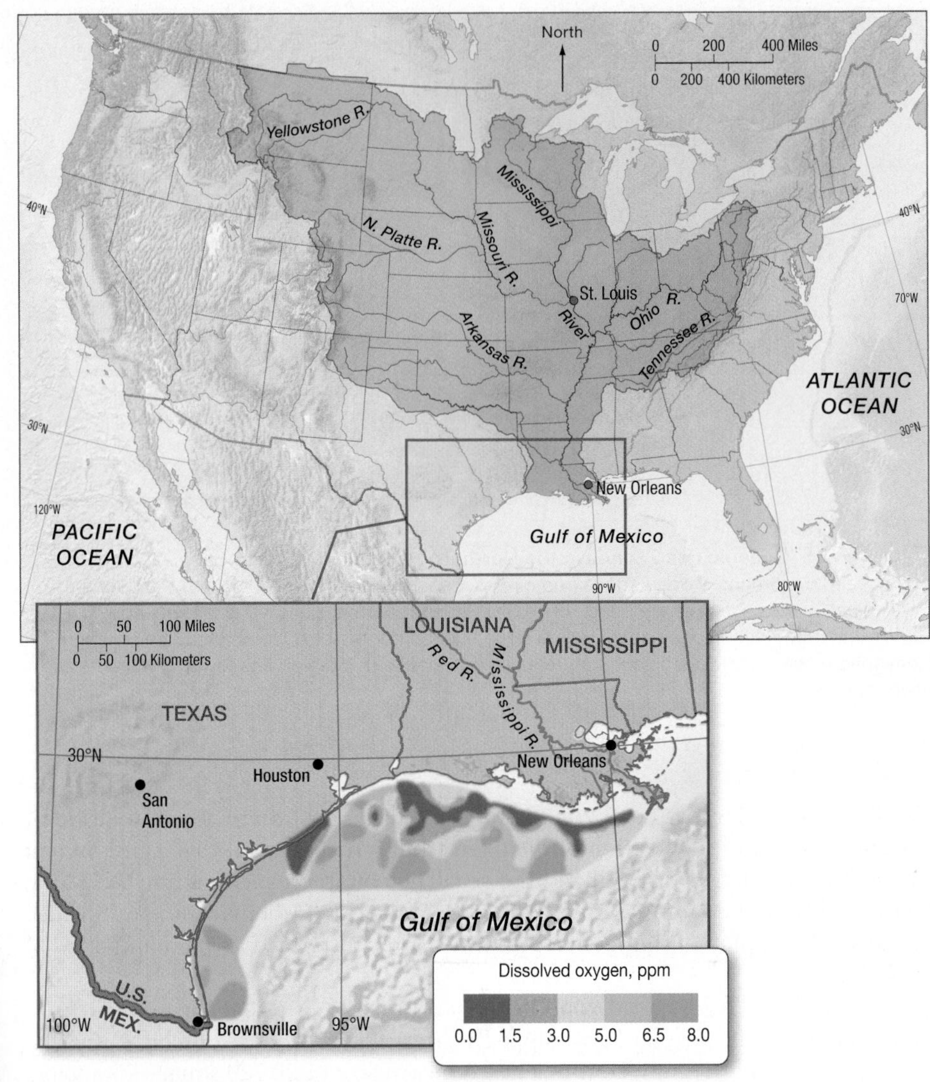


**Figure 13.16 The Gulf of Mexico dead zone.** Map (*above*) showing Mississippi River drainage basin (*gold*) and enlargement (*below*) showing the extent of the 2011 Gulf of Mexico dead zone with colors representing dissolved oxygen concentration in parts per million (ppm). The northern Gulf of Mexico dead zone is the second largest in the world, after the Baltic Sea.

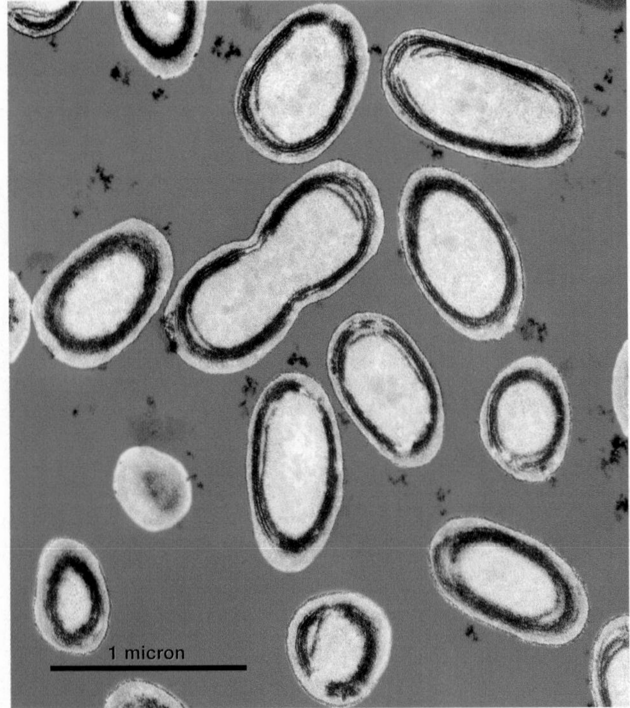


**Figure 13.17 Photosynthetic bacterium *Prochlorococcus*.** The bacterium *Prochlorococcus* is the most abundant and smallest of marine phytoplankton, reaching only about 0.6 micron (0.00002 inch) in diameter.

ocean west of South Africa, where marine organisms are adapted to the low-oxygen conditions that exist there.

## Photosynthetic Bacteria

Until recently, the role of marine bacteria in photosynthesis has been largely ignored. Because of the extremely small size of bacteria, earlier samplings of marine life completely overlooked them. Recent improvements in sampling methods for bacteria-sized organisms and genome sequencing studies have revealed bacteria's incredible abundance and importance in the oceans.

For example, one of the first types of marine photosynthetic bacteria to be identified was *Synechococcus*, which are extremely abundant in coastal and open-ocean environments, sometimes reaching densities greater than 100,000 cells per milliliter (0.03 ounce) of seawater. At certain times and places, these cells can be responsible for as much as half of the primary production of food in the ocean. More recently, microbiologists have discovered an extremely small but abundant bacteria named *Prochlorococcus* (**Figure 13.17**) that reaches concentrations many times that of *Synechococcus*. In fact, *Prochlorococcus* has been estimated to constitute at least half of the world ocean's total photosynthetic biomass, which means it is probably the most abundant photosynthetic organism on Earth.

In addition, recent large-scale gene sequencing of microbes in the Sargasso Sea has revealed a host of new types of bacteria, suggesting substantial yet previously unrecognized oceanic microbial diversity. Clearly, microbes exert a critical influence on marine ecosystems and carry significant implications for sustainability, global climate change, ocean system cycles, and human health.

**RECAP Marine photosynthetic organisms include seed-bearing plants (such as surf grass), macroscopic algae (seaweeds), microscopic algae (diatoms, coccolithophores, and dinoflagellates), and bacteria. Red tides are caused by an overabundance of dinoflagellates.**

**CONCEPT CHECK 13.2 ▸ Describe various kinds of photosynthetic marine organisms.**

1 Compare and contrast green, red, and brown algae in terms of composition, color, depth in which they grow, and size.

2 The golden algae include two classes of important phytoplankton: diatoms and coccolithophores. Compare and contrast the composition and structure of their tests, and explain their importance in the geologic fossil record.

3 What is a red tide? What conditions cause red tides?

4 How does paralytic shellfish poisoning (PSP) differ from amnesic shellfish poisoning? What types of microorganisms create each?

5 What conditions create ocean eutrophication (dead zones)? What can be done to limit their spread?

## 13.3 ▸ How Does Regional Primary Productivity Vary?

Primary photosynthetic production in the oceans varies dramatically from place to place (see Figure 13.5). Typical units of photosynthetic production are in mass of carbon gC (*grams of carbon*) per unit of area $m^2$ (*square meter*) per unit of time yr (*year*), which is abbreviated as $gC/m^2/yr$. Values range from as low as 1 $gC/m^2/yr$ in some areas of the open ocean to as much as 4000 $gC/m^2/yr$ in some highly productive coastal estuaries (**Table 13.1**). This variability is a result of the uneven distribution of nutrients and seasonal changes in the availability of solar energy (for a review of Earth's seasons, see Section 6.1 "What Causes Variations in Solar Radiation on Earth?" in Chapter 6) throughout the ocean's shallow, photosynthetic surface layer.

On average, about 90% of the organic biomass generated in the euphotic (sunlit) zone of the open ocean decomposes before descending below this zone. The remaining 10% of this material sinks into deeper water, where about 9% is decomposed. The remaining 1% of this material reaches the deep-ocean floor and accumulates there. The way in which material is removed from the euphotic zone to the sea floor is called a **biological pump** because it "pumps" carbon dioxide and nutrients from the upper ocean and concentrates them in deep-sea waters and sea floor sediments. (For a discussion about the role of the ocean's biological pump in carbon cycling and climate change, see Chapter 16, "The Oceans and Climate Change.")

Surface warming and the resulting stratification of the ocean's water column also affects primary productivity. Throughout much of the subtropical oceans, for

**SmartTable 13.1 Values of net primary productivity for various ecosystems**

| Ecosystem | Primary productivity | |
|---|---|---|
| | Range ($gC/m^2/yr$) | Average ($gC/m^2/yr$) |
| **Oceanic** | | |
| Algae beds and coral reefs | 1000–3000 | 2000 |
| Estuaries | 500–4000 | 1800 |
| Upwelling zone | 400–1000 | 500 |
| Continental shelf | 300–600 | 360 |
| Open ocean | 1–400 | 125 |
| **Land** | | |
| Freshwater swamp and marsh | 800–4000 | 2500 |
| Tropical rainforest | 1000–5000 | 2000 |
| Middle latitude forest | 600–2500 | 1300 |
| Cultivated land | 100–4000 | 650 |

**SmartTable 13.1 Values of net primary productivity for various ecosystems**
https://goo.gl/zsRF6K

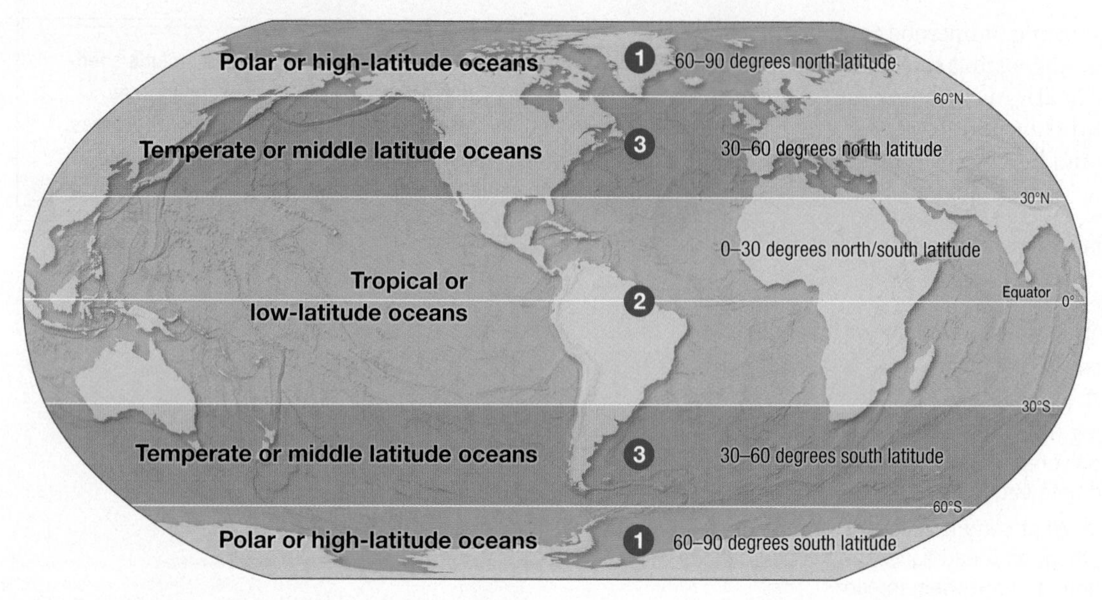


Figure 13.18 **Locations of three marine productivity areas.** Map showing the locations of three open ocean areas where yearly marine productivity patterns are examined: (1) polar or high-latitude oceans (60 degrees to 90 degrees north and south latitude); (2) tropical or low-latitude oceans (0 degrees to 30 degrees north and south latitude); and (3) temperate or middle latitude oceans (30 degrees to 60 degrees north and south latitude). See the text for full descriptions of the conditions found in each productivity area.

**RECAP** **The thermocline acts as an impenetrable lid that inhibits the movement of nutrient-rich deep water to the surface, thereby limiting primary productivity.**

example, a permanent **thermocline** (and resulting **pycnocline**) develops. (Recall that a *thermocline* is a layer of rapidly changing temperature, and a *pycnocline* is a layer of rapidly changing density; development of ocean thermoclines and pycnoclines is discussed in Chapter 5.) The thermocline forms a barrier to vertical mixing, preventing the resupply of nutrients to the sunlit surface layer. In essence, the thermocline acts as an impenetrable lid that prohibits the movement of nutrient-rich deep water to the surface, thereby inhibiting primary productivity. In middle latitude oceans, a thermocline develops only during the summer season. A thermocline does not usually develop in polar oceans because of the lack of adequate surface warming. As discussed below, the degree to which waters develop a thermocline profoundly affects the patterns of primary productivity in different latitudes.

Let's examine the yearly productivity patterns of three open ocean areas: (1) polar or high-latitude oceans, (2) tropical or low-latitude oceans, and (3) temperate or middle latitude oceans (**Figure 13.18**). Note that in the following discussion, we'll consider only open ocean areas far from land, ensuring that these locations aren't affected by runoff from the continents, which often contain high levels of nutrients that interfere with the seasonal patterns described below.

## Productivity in Polar (High-Latitude) Oceans: 60 to 90 degrees North and South Latitude

Polar oceans such as the Arctic Ocean's Barents Sea, which is off the northern coast of Europe, experience continuous darkness for about three months during winter and continuous illumination for about three months during summer. Diatom productivity peaks in the Barents Sea during May (**Figure 13.19a**), when the Sun rises high enough in the sky and there is deep penetration of sunlight into the water. As soon as diatoms develop, zooplankton—mostly small crustaceans such as copepods (**Figure 13.19b**)—begin feeding on them. The zooplankton biomass peaks in June and continues at a relatively high level until winter darkness begins in October.

In the Antarctic region—particularly at the southern end of the Atlantic Ocean—productivity is somewhat greater. This is caused by the upwelling of North Atlantic Deep Water, which forms on the opposite side of the ocean basin, where it sinks and moves southward below the surface. Hundreds of years later, it rises to the surface near Antarctica, carrying with it high concentrations of nutrients (**Figure 13.19c**). When the Sun provides sufficient solar radiation in summer, there is an explosion of biological productivity. A recent study of Antarctic waters, however, documented as much as a 12% decrease in phytoplankton productivity because of increased ultraviolet radiation as a result of the Antarctic ozone hole caused by the use of CFCs (chlorofluorocarbons). For more details about the atmospheric ozone hole and CFCs, which are also strong greenhouse gases, see Chapter 16, "The Oceans and Climate Change."

Blue whales—the largest of all whales (see Figure 14.20)—eat mostly zooplankton and time their migration through middle latitude and polar oceans to coincide with maximum zooplankton productivity. This enables the whales to develop and support calves that can exceed 7 meters (23 feet) in length at birth. The mother blue whale suckles the calf with rich, high-fat milk for six months. By the time the calf is weaned, it is over 16 meters (50 feet) long. In two years, it will be 23 meters (75 feet) long, and after about three years, it will weigh 55 metric tons

Figure 13.19 **Productivity in polar oceans.**

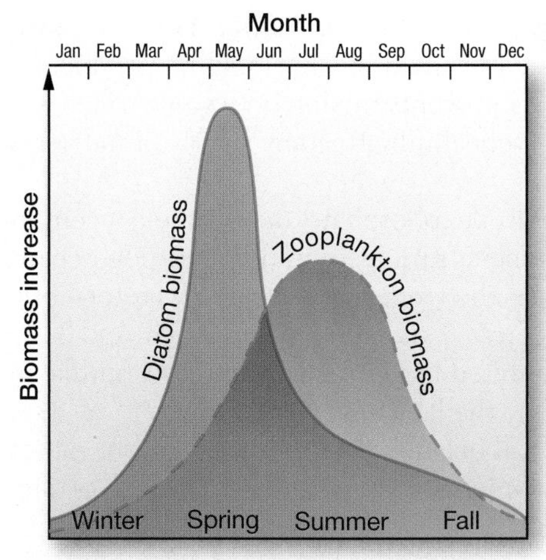


**(a)** Barents Sea productivity graph, which shows the dramatic springtime increase of diatom biomass that results in an increase in zooplankton abundance.

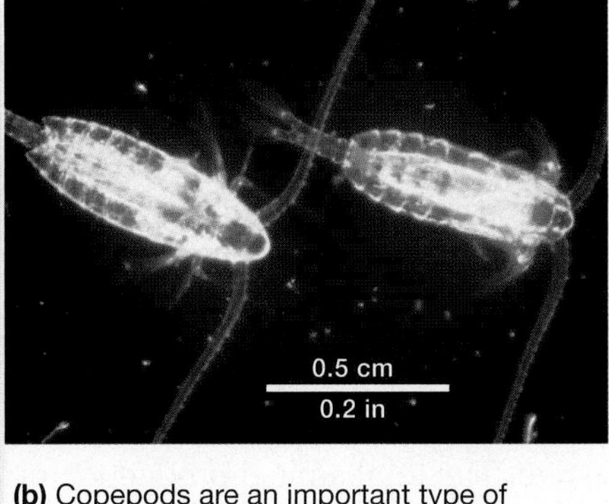


**(b)** Copepods are an important type of zooplankton in polar oceans. These copepods are of the genus *Calanus*.

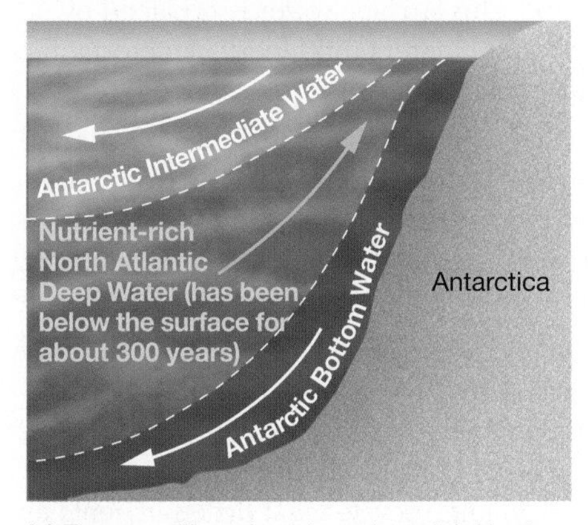


**(c)** The upwelling of cold, nutrient-rich North Atlantic Deep Water near Antarctica continually supplies Antarctic waters with nutrients.

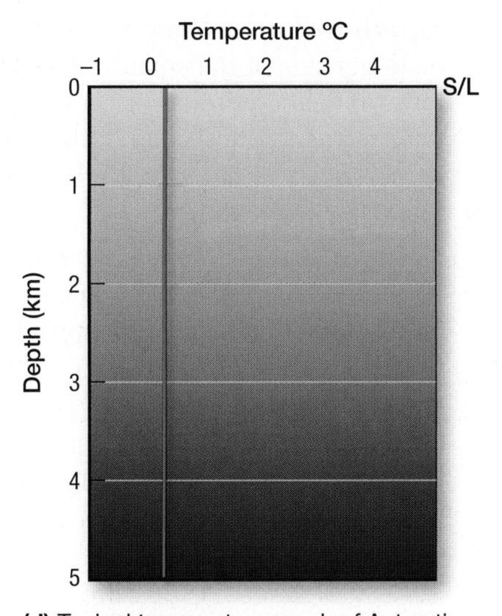


**(d)** Typical temperature graph of Antarctic waters showing a nearly uniform water temperature with depth (an isothermal water column).

(121,000 pounds)! This phenomenal growth rate gives some indication of the enormous biomass of small copepods and krill upon which these large mammals feed (as a similar size analogy, consider how many ants you would have to eat as a child to grow to adult size!).

Density and temperature change very little with depth in polar oceans (Figure 13.19d), so these waters are **isothermal** (*iso* = same, *thermo* = temperature), and there is no barrier to mixing between surface waters and deeper, nutrient-rich waters. In the summer, however, melting ice creates a thin, low-salinity layer that does not readily mix with the deeper waters. This stratification is crucial to summer primary production because it helps prevent phytoplankton from being carried into deeper, darker waters. Instead, they are concentrated in the sunlit surface waters, where they reproduce continuously.

Nutrient concentrations (mostly nitrates and phosphates) are usually adequate in high-latitude surface waters, so the availability of solar energy limits photosynthetic productivity in these areas more than the availability of nutrients.

## Productivity in Tropical (Low-Latitude) Oceans: 0 to 30 degrees North and South Latitude

Perhaps surprisingly, productivity is low in tropical oceans. Because the Sun is more directly overhead, light penetrates much deeper into tropical oceans than in middle

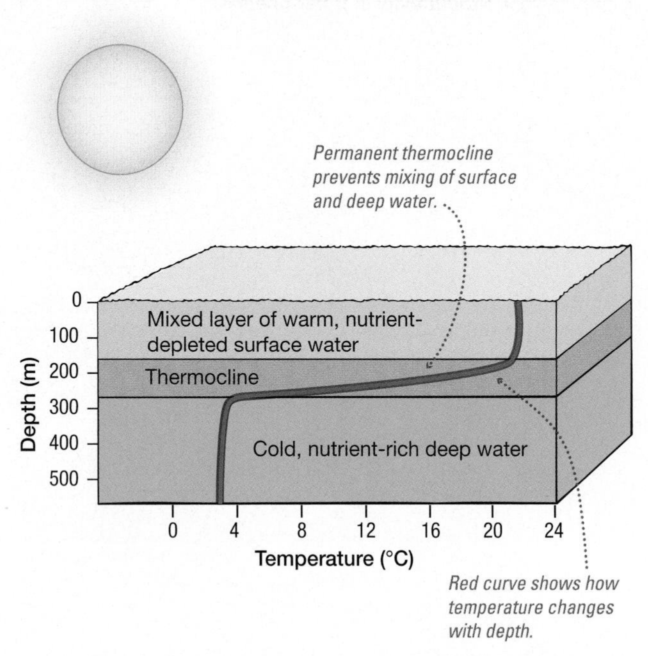


Figure 13.20 **Productivity in tropical oceans.** Although tropical regions receive adequate sunlight year-round, a permanent thermocline prevents the mixing of surface water and deep water. As phytoplankton consume nutrients in the surface layer, productivity is limited because the thermocline prevents replenishment of nutrients from deeper water. Thus, productivity remains at a steady, low level.

## STUDENTS SOMETIMES ASK . . .

***The number and variety of tropical species on land is astounding. I don't understand how the tropical oceans can have such low productivity.***

Life on land does not necessarily correspond to life in the ocean. Tropical rainforests support an amazing diversity of species and an enormous biomass. In the tropical ocean, however, a strong, permanent thermocline limits the availability of nutrients that are necessary for the growth of phytoplankton. Without abundant phytoplankton, not much else can live in the ocean. In fact, these areas are often considered biological deserts. It is ironic that the clear blue water of the tropics so prominently displayed in tourist brochures indicates seawater that is biologically quite sterile!

latitude and polar waters, and solar energy is available year-round. However, productivity is low in tropical oceans because a permanent thermocline produces a stratification (layering) of water masses. This prevents mixing between surface waters and nutrient-rich deeper waters, effectively eliminating any supply of nutrients from deeper waters below (**Figure 13.20**).

At about 20 degrees north and south latitude, phosphate and nitrate concentrations are commonly less than $\frac{1}{100}$ of their concentrations in middle latitude oceans during winter. In fact, nutrient-rich waters in the tropics lie below 150 meters (500 feet), with the highest concentrations between 500 and 1000 meters (1640 and 3300 feet). So, productivity in tropical oceans is limited by the lack of nutrients (unlike in polar oceans, where productivity is limited by the lack of sunlight).

Generally, primary production in tropical oceans occurs at a steady but rather low rate. The total annual production of tropical oceans is only about half of that found in middle latitude oceans.

Exceptions to the general pattern of low productivity in tropical oceans include the following:

1. ***Equatorial upwelling.*** Where trade winds drive westerly equatorial currents on either side of the equator, Ekman transport causes surface water to diverge toward higher latitudes (see Figure 7.10). This surface water is replaced by nutrient-rich water from depths of up to 200 meters (660 feet). Equatorial upwelling is best developed in the eastern Pacific Ocean.
2. ***Coastal upwelling.*** Where the prevailing winds blow toward the equator and along western continental margins, surface waters are driven away from the coast. They are replaced by nutrient-rich waters from depths of 200 to 900 meters (660 to 2950 feet). This upwelling promotes high primary production along the west coasts of continents (see Figure 13.6), which can support large fisheries.
3. ***Coral reefs.*** Organisms that comprise and live among coral reefs are superbly adapted to low-nutrient conditions, similar to the way certain organisms are adapted to desert life on land. Symbiotic microscopic algae living within the tissues of coral and other species allow coral reefs to be highly productive ecosystems. Coral reefs also tend to retain and recycle what little nutrients exist. Coral reef ecosystems are discussed further in Chapter 15, "Animals of the Benthic Environment."

## Productivity in Temperate (Middle Latitude) Oceans: 30 to 60 degrees North and South Latitude

As discussed above, productivity is limited by available sunlight in polar oceans and by nutrient supply in tropical oceans. In temperate or middle latitude oceans, a combination of these two limiting factors controls productivity, as shown in **Figure 13.21a** (which shows the seasonal pattern for the Northern Hemisphere; in the Southern Hemisphere, the seasons are reversed).

**WINTER** Productivity in middle latitude oceans is very low during winter, even though nutrient concentration is *highest* at this time (Figure 13.21a). During this season, the water column is isothermal, too, like in polar oceans, so nutrients are well distributed throughout the water column. **Figure 13.21b** (*winter*) shows, however, that the Sun is at its lowest position above the horizon during winter, so a high percentage of the available solar energy is reflected, leaving only a small percentage to be absorbed into surface waters. As a result, the compensation depth for photosynthesis—the depth at which net photosynthesis becomes zero—is so shallow that phytoplankton do not grow much. Moreover, the absence of a thermocline allows algal cells to be carried down beneath the euphotic zone for extended periods by turbulence associated with winter waves.

**SPRING** The Sun rises higher in the sky during spring than during winter (Figure 13.21b, *spring*), so the compensation depth for photosynthesis deepens. A

**(a)** Relationship among phytoplankton, zooplankton, amount of sunlight, and nutrient levels for surface waters in northern temperate latitudes.

Month: Jan, Feb, Mar, April, May, June, July, Aug, Sept, Oct, Nov, Dec

Increase

Phytoplankton · Sunlight · Nutrients · Zooplankton · Spring bloom · Fall bloom · Northern Hemisphere

**EXPLORING DATA**

**1. Describe the factors that limit the spring bloom of phytoplankton.**

**2. Describe the factors that limit the fall bloom of phytoplankton.**

| | Winter | Spring | Summer | Fall |
|---|---|---|---|---|
| | Dec–March | March–June | June–Sept | Sept–Dec |
| | Isothermal water column | Developing thermocline | Strong thermocline | Weakening thermocline |
| **Sunlight:** | Lowest (–) | Increasing (↑) | Highest (+) | Decreasing (↓) |
| **Nutrients:** | Highest (+) | Decreasing (↓) | Lowest (–) | Increasing (↑) |

Reflected sunlight · Absorbed sunlight · Depth · Temp

Phytoplankton · Zooplankton · Nutrient-rich water · Nutrient-poor water

*Compensation depth* (blue line) *is the depth at which net photosynthesis becomes zero.*

**(b)** The seasonal cycle of sunlight affects the presence and depth of the thermocline, which affects the availability of nutrients. This, in turn, affects the abundance of phytoplankton and other organisms such as zooplankton that rely on phytoplankton for food.

Figure 13.21 **Productivity in middle latitude (temperate) oceans in the Northern Hemisphere.**

**spring bloom** of phytoplankton occurs (Figure 13.21a) because solar energy and nutrients are available, and a seasonal thermocline develops due to increased solar heating that traps algae in the euphotic zone (Figure 13.21b). This creates a tremendous demand for nutrients in the euphotic zone, so the supply becomes limited, causing productivity to decrease sharply. Even though the days are lengthening and sunlight is increasing, productivity during the spring bloom is limited by the lack of nutrients. In most areas of the Northern Hemisphere, therefore, phytoplankton populations decrease in April due to insufficient nutrients and because their population is being consumed by zooplankton (grazers).

**Animation**
Oceanic Midlatitude Productivity
http://goo.gl/z1AWwN

**SUMMER** The Sun rises even higher in the summer than in the spring (Figure 13.21b, *summer*), so surface waters in the middle latitudes continue to warm. A strong seasonal thermocline is created at a depth of about 15 meters (50 feet). The thermocline,

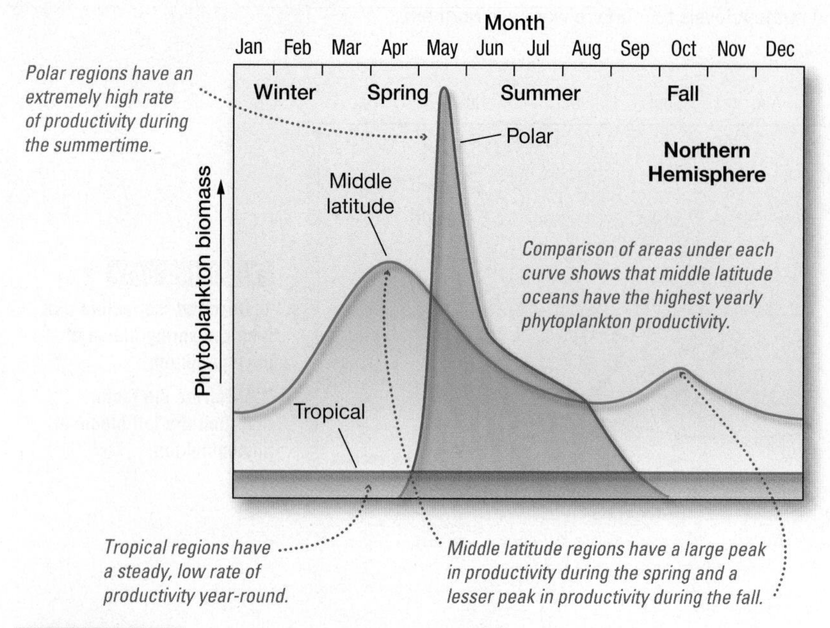
Month
Jan Feb Mar Apr May Jun Jul Aug Sep Oct Nov Dec
Winter
Spring
Summer
Fall
Polar
Northern Hemisphere
Middle latitude
Tropical
Phytoplankton biomass
Polar regions have an extremely high rate of productivity during the summertime.
Comparison of areas under each curve shows that middle latitude oceans have the highest yearly phytoplankton productivity.
Tropical regions have a steady, low rate of productivity year-round.
Middle latitude regions have a large peak in productivity during the spring and a lesser peak in productivity during the fall.

**EXPLORING DATA** ▲ **As compared to the polar and middle latitudes, why do the tropical oceans have a steady, low rate of productivity year-round?**

Figure 13.22 **Comparison of phytoplankton productivity in tropical, middle latitude, and polar oceans (Northern Hemisphere).** Graph comparing the seasonal variations in phytoplankton biomass in various oceanic regions.

**RECAP In polar oceans, productivity peaks during the summer and is limited by sunlight during the rest of the year. In tropical oceans, productivity is low year-round and is limited by nutrients. In middle latitude oceans, productivity peaks in the spring and fall and is limited by a lack of solar radiation in the winter and a lack of nutrients in the summer. Despite this, middle latitude oceans exhibit the highest overall productivity of the three regions.**

in turn, prevents vertical mixing, so nutrients depleted from surface waters cannot be replaced by those from deeper waters. Throughout summer, the phytoplankton population remains relatively low (Figure 13.21a). Even though the compensation depth for photosynthesis is at its maximum, phytoplankton can actually become scarce in late summer.

**FALL** Solar radiation diminishes in the fall, as the Sun moves lower in the sky (Figure 13.21b, *fall*), so surface temperatures drop, and the summer thermocline breaks down. Nutrients return to the surface layer as increased wind strength mixes surface waters with deeper waters. These conditions create a **fall bloom** of phytoplankton, which is much less dramatic than the spring bloom (Figure 13.21a). The fall bloom is very short-lived because sunlight (not nutrient supply, as in the spring bloom) becomes the limiting factor as winter approaches to repeat the seasonal cycle.

## Comparing Regional Productivity

Figure 13.22 compares the seasonal variation in phytoplankton biomass of tropical, north polar, and north middle latitude oceans, where the total area under each curve represents photosynthetic productivity. The graph shows the dramatic peak in productivity in polar oceans during the summer; the steady, low rate of productivity year-round in the tropical ocean; and the seasonal pattern of productivity that occurs in middle latitude oceans. It also shows that the highest overall productivity occurs in middle latitude oceans.

**CONCEPT CHECK 13.3 ▸ Explain variations in regional oceanic primary productivity.**

1 In your own words, describe how a biological pump works. What percentage of organic material from the euphotic zone accumulates on the sea floor? What are the three environments that are exceptions to this, and what factors contribute to their higher productivity?

2 Describe the yearly productivity pattern in polar oceans, including the main factor that limits productivity in polar oceans.

3 Why is the productivity in tropical oceans uniformly low year-round?

4 Describe the yearly productivity pattern in middle latitude oceans. Describe the limiting factor(s) for both the spring and fall phytoplankton blooms.

# 13.4 ▸ How Are Energy and Nutrients Passed Along in Marine Ecosystems?

A **biotic community** is an assemblage of organisms that live together within some definable area or habitat. An **ecosystem** includes the biotic community plus the **abiotic** (*a* = without, *biotic* = life) environment with which organisms exchange energy and chemical substances. A kelp forest biotic community, for instance, includes all organisms living within or near the kelp and receiving some benefit from it. A kelp forest ecosystem, on the other hand, includes all those organisms plus the surrounding seawater, the hard substrate onto which the kelp is attached, and the atmosphere where gases are exchanged.

Two of the most important commodities passed along in marine ecosystems are energy and nutrients.

## Flow of Energy in Marine Ecosystems

Energy flow in photosynthetic marine ecosystems is not a cycle but a *unidirectional flow* based on a continuous supply of solar energy. Consider an algae-supported biotic community (**Figure 13.23**), where energy enters the system and algae absorb solar radiation. Photosynthesis converts this solar energy into chemical energy (carbohydrates), which is used for the algae's respiration. This chemical energy is also passed on to the animals that consume the algae for their growth and other life functions. The animals expend mechanical and heat energy, which are progressively less recoverable forms of energy, until this residual energy becomes dissipated within the ecosystem in the form of heat, thereby increasing **entropy** (*en* = in, *trope* = transformation). In essence, the ecosystem relies on a constant input of energy in the form of sunlight.

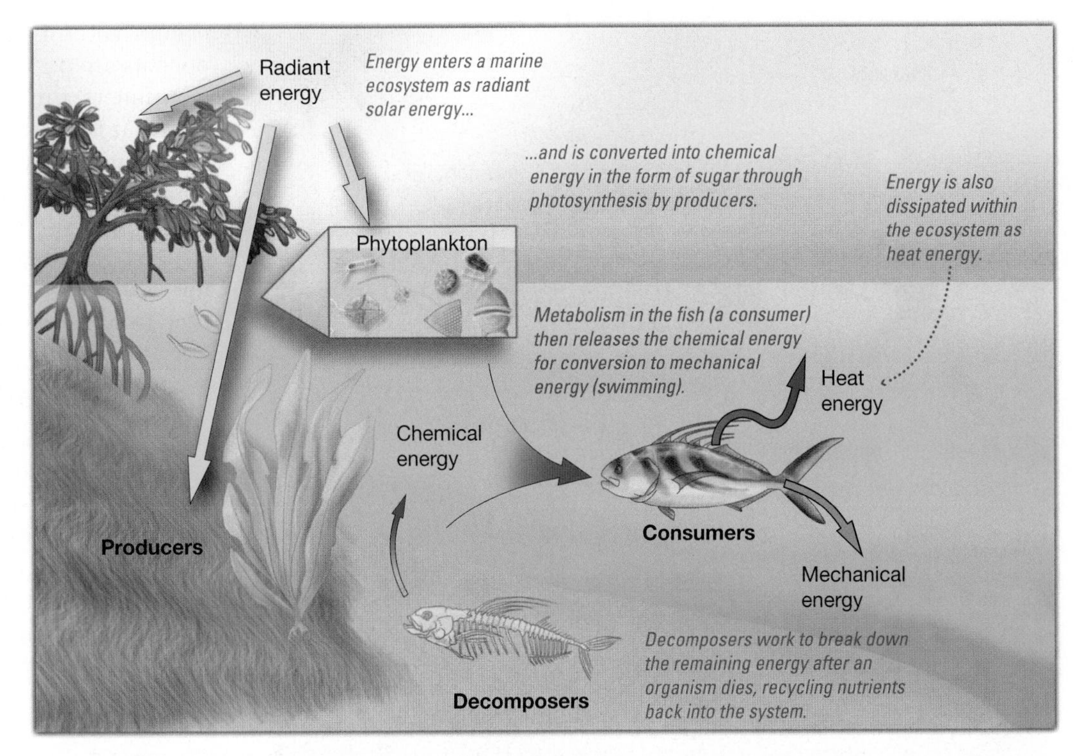


Figure 13.23 **Energy flow through a photosynthetic marine ecosystem.**

**PRODUCERS, CONSUMERS, AND DECOMPOSERS** Generally, three basic categories of organisms exist within an ecosystem: **producers**, **consumers**, and **decomposers** (Figure 13.23).

Producers can nourish themselves through either photosynthesis or chemosynthesis. Examples of producers include algae, plants, archaea, and photosynthetic bacteria, all of which are called **autotrophic** (*auto* = self, *tropho* = nourishment) organisms. Consumers and decomposers, on the other hand, are called **heterotrophic** (*hetero* = different, *tropho* = nourishment) organisms because they depend, either directly or indirectly, on the organic compounds produced by autotrophs for their energy supply.

Consumers eat other organisms and are categorized as either **herbivores** (*herba* = grass, *vora* = eat), which feed directly on plants or algae; **carnivores** (*carni* = meat, *vora* = eat), which feed only on other animals; **omnivores** (*omni* = all, *vora* = eat), which feed on both; or **bacteriovores** (*bacterio* = bacteria, *vora* = eat), which feed only on bacteria.

Decomposers such as bacteria break down organic compounds that comprise **detritus** (*detritus* = to lessen)—dead and decaying remains and waste products of organisms—for their own energy requirements. In the decomposition process, compounds are released and recycled as they become available again as nutrients for use by autotrophs.

## Flow of Nutrients in Marine Ecosystems

Unlike the noncyclic, unidirectional flow of energy through a biotic community, the flow of nutrients depends on **biogeochemical cycles**, which are so named because they involve *bio*logical, *geo*logical (Earth processes), and *chemical* components. Moreover, matter does not dissipate and becomes lost as energy does, but it is *cycled* from one chemical form to another by the various members of the biotic community.

**BIOGEOCHEMICAL CYCLING** **Figure 13.24** shows the biogeochemical cycling of matter within the marine environment. The chemical components of organic matter enter the biological system through photosynthesis (or, less commonly, through

### CREATURE FEATURE 13.1

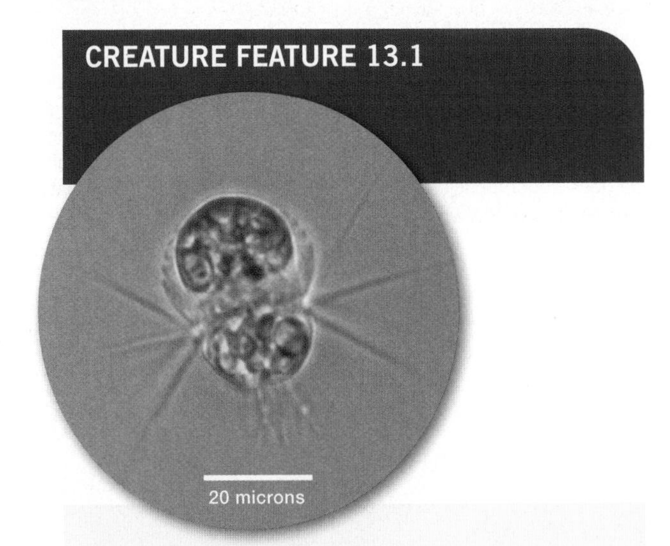


### I'm a hybrid organism: both animal and algae!

The single-celled protozoan ***Mesodinium chamaeleon*** uses its hair-like appendages to move around rapidly in the oceans. It hunts and eats other organisms, which makes it an animal.

Oddly, this tiny creature can also photosynthesize. When it engulfs certain algae, it digests everything but the algae's green photosynthetic machinery, which it puts to work to produce food. This hybrid strategy of eating like an animal and photosynthesizing like algae make it a *mixotroph* (*mixo* = mixing, *tropho* = nourishment), blurring the line between animal and algae.

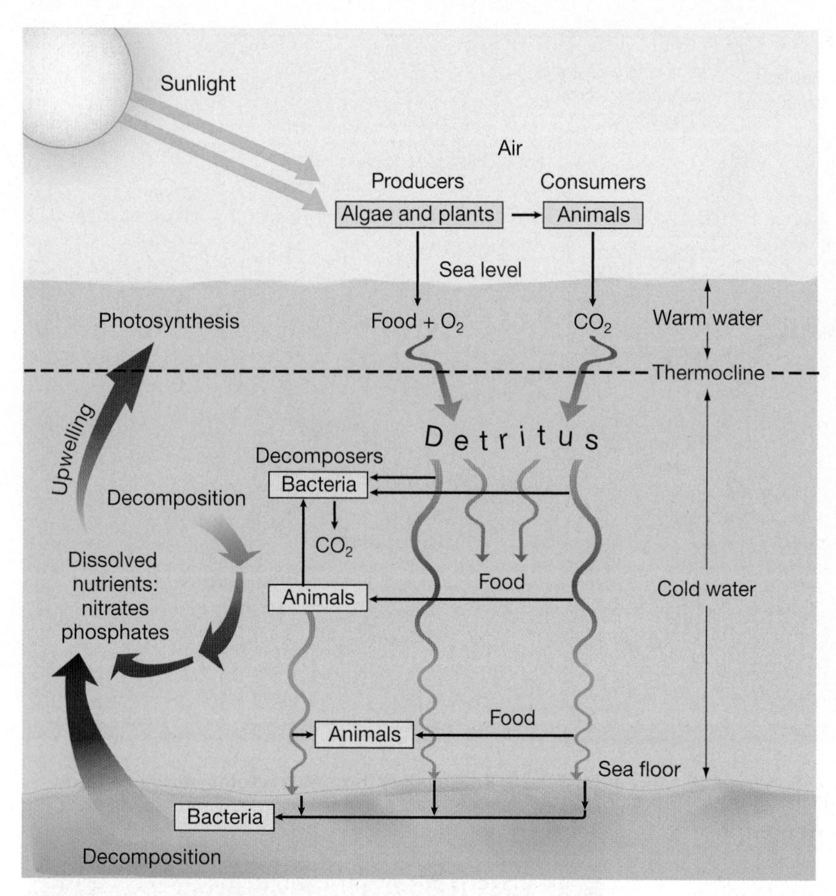


**Figure 13.24 Biogeochemical cycling of matter.** The chemical components of organic matter enter the biological system through photosynthesis and are passed on to consumers through feeding. Detritus sinks and feeds organisms living below the surface or undergoes decomposition, which returns nutrients to the water that can be hoisted to the surface by upwelling.

chemosynthesis at hydrothermal vents). These chemical components are passed from the producer on to animal populations (consumers) through feeding. When organisms die, some of the material is used and reused within the euphotic zone, while some sinks as detritus. Some of this detritus feeds organisms living in deep water or on the sea floor, while some undergoes bacterial or other decomposition processes that convert organic remains into usable nutrients (nitrates and phosphates). When upwelling hoists these nutrients to the surface again, they can be used by algae and plants to begin the cycle anew.

## Oceanic Feeding Relationships

As producers make food (organic matter) available to the consuming animals of the ocean, the food passes from one feeding population to the next. Only a small percentage of the energy (on average about 10%) taken in at any level is passed on to the next because energy is consumed and lost, mostly as heat, at each level. As a result, the producers' biomass in the ocean is many times greater than the mass of the top consumers, such as sharks or whales.

**FEEDING STRATEGIES** For most marine animals, obtaining food occupies the majority of their time. Some animals are streamlined, fast, and agile, and are able to obtain food through active predation. Other animals move more leisurely, or not at all, and instead obtain food by filtering it from seawater or locating it as deposits on the sea floor. **Figure 13.25** shows several modes of feeding in animals that live along sediment-covered shores.

In **suspension feeding**, which is also called **filter feeding**, organisms use specially designed structures to filter plankton from seawater. For example, barnacles, which are sessile crustaceans (**Figure 13.26**), attach to hard surfaces and use their legs to strain passing food particles from the water. Clams (Figure 13.25) bury themselves in sediment and extend siphons up to the surface. They pump in overlying water and filter from it suspended plankton and other organic matter.

In **deposit feeding**, organisms feed on food items that occur as deposits. These deposits include detritus—dead and decaying organic matter and waste products—and the sediment itself, which is coated with organic matter. Some deposit feeders, such as the amphipod *Orchestoidea* (Figure 13.25), feed on more concentrated deposits of organic matter (detritus) on the sediment surface. Others, such as the segmented worm *Arenicola* (Figure 13.25) feed by ingesting sediment and extracting organic matter from it.

In **carnivorous feeding**, organisms directly capture and eat other animals. This predation can be either passive or active. Passive predation involves waiting for prey items and ensnaring them; an example of this is the way a sea anemone feeds. Active predation involves seeking prey. Examples include sharks and the sand star *Astropecten* (Figure 13.25), which burrows rapidly into sandy beaches and feeds voraciously on crustaceans, mollusks, worms, and other echinoderms.

**TROPHIC LEVELS** Chemical energy stored in the mass of the ocean's algae (the "grass of the sea") is transferred to the animal community mostly through feeding. Most zooplankton are *herbivores*, like cows, so they eat diatoms and other microscopic marine algae (in fact, zooplankton could be considered the miniature drifting "cows of the sea"). Larger herbivores feed on the larger algae and marine plants that grow attached to the ocean bottom near the shore.

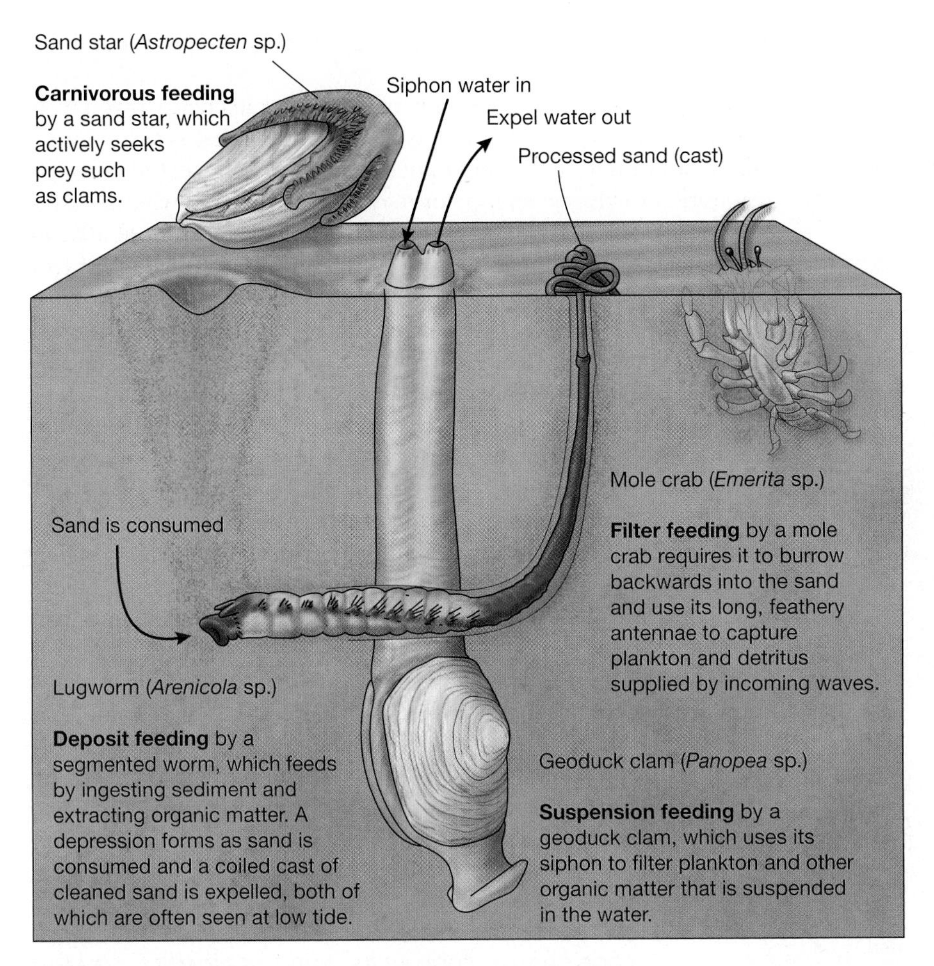
Sand star (*Astropecten* sp.)
**Carnivorous feeding** by a sand star, which actively seeks prey such as clams.
Siphon water in
Expel water out
Processed sand (cast)
Mole crab (*Emerita* sp.)
**Filter feeding** by a mole crab requires it to burrow backwards into the sand and use its long, feathery antennae to capture plankton and detritus supplied by incoming waves.
Sand is consumed
Lugworm (*Arenicola* sp.)
**Deposit feeding** by a segmented worm, which feeds by ingesting sediment and extracting organic matter. A depression forms as sand is consumed and a coiled cast of cleaned sand is expelled, both of which are often seen at low tide.
Geoduck clam (*Panopea* sp.)
**Suspension feeding** by a geoduck clam, which uses its siphon to filter plankton and other organic matter that is suspended in the water.

Figure 13.25 **Modes of feeding along sediment-covered shores.**

1.0 cm
0.4 in

Figure 13.26 **Barnacles filter feeding.** Barnacles are shelled organisms that attach to hard surfaces and filter feed by using their feathery legs to strain microscopic food particles from seawater.

Herbivores are then eaten by larger animals—*carnivores*—that, in turn, are eaten by other populations of larger carnivores, and so on, in a progression that may encompass four or five steps. Each of these feeding stages is called a **trophic level** (*tropho* = nourishment).

Generally, individual members of a feeding population are larger—but not too much larger—than the organisms they eat. There are conspicuous exceptions, however, such as the blue whale. At 30 meters (100 feet) long, it is possibly the largest animal that has ever existed on Earth, yet it feeds on **krill** (*kril* = young fry of fish), a small crustacean, which has a maximum length of only 6 centimeters (2.4 inches).

The transfer of energy from one population to another is a continuous *flow* of energy. Small-scale recycling and storage interrupt the flow. This, in turn, slows the conversion of potential (chemical) energy to kinetic energy, then to heat energy, and, finally, the energy is so dissipated in the form of heat it becomes useless.

**TRANSFER EFFICIENCY** The transfer of energy between trophic levels is very inefficient, especially at the lowest level. The efficiencies of different algal species vary, but the average is only about 2%, which means that only 2% of the light energy available in the sunlit surface waters is ultimately synthesized into food by algae and made available to herbivores.

The **gross ecological efficiency** at any trophic level is the ratio of energy passed on to the next higher trophic level divided by the energy received from the trophic level below. The ecological efficiency of herbivorous anchovies, for example, would be the energy consumed by carnivorous tuna that feed on the

## STUDENTS SOMETIMES ASK . . .

### *How likely am I to see a whale during a whale-watching trip?*

It depends on the area and the time of the year, but generally your chances are quite low because large marine organisms comprise such a small proportion of marine life. In fact, it has been estimated that large nektonic organisms such as whales comprise only *one-tenth of 1%* of all biomass in the sea! With your knowledge of food pyramids, it should be no surprise that the majority of the ocean's biomass is comprised of phytoplankton. Perhaps commercial boat operators should conduct *plankton*-watching trips! Everyone would be guaranteed to see dozens of different species—and all that would be required would be a plankton net and a microscope.

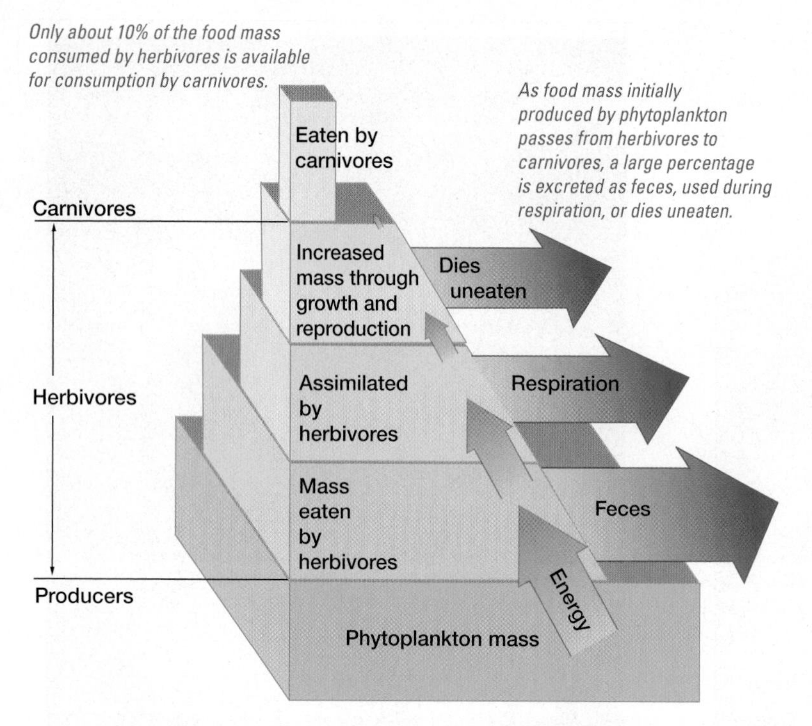
Only about 10% of the food mass consumed by herbivores is available for consumption by carnivores.
As food mass initially produced by phytoplankton passes from herbivores to carnivores, a large percentage is excreted as feces, used during respiration, or dies uneaten.
Eaten by carnivores
Carnivores
Increased mass through growth and reproduction
Dies uneaten
Herbivores
Assimilated by herbivores
Respiration
Mass eaten by herbivores
Feces
Producers
Energy
Phytoplankton mass

**Figure 13.27 Passage of energy through trophic levels.**

anchovies divided by the energy contained in the phytoplankton that the anchovies consumed.

**Figure 13.27** shows that some of the chemical energy taken in as food by herbivores is excreted as feces, and the rest is assimilated. Of the assimilated chemical energy, much is converted through respiration to kinetic energy for maintaining life, and what remains is available for growth and reproduction. Thus, only about 10% of the food mass consumed by herbivores is available to the next trophic level.

**Figure 13.28** shows the passage of energy between trophic levels through an entire ecosystem, from the solar energy assimilated by phytoplankton through all trophic levels to the ultimate carnivore—humans. Because energy is lost at each trophic level, it takes thousands of smaller marine organisms to produce a *single* fish that is so easily consumed during a meal!

The efficiency of energy transfer between trophic levels depends on many variables. Young animals, for example, have a higher growth efficiency than older animals. In addition, when food is plentiful, animals expend more energy in digestion and assimilation than when food is scarce.

Most ecological efficiencies in natural ecosystems average about 10% but range between 6% and 15%. There is some evidence, however, that ecological efficiencies in populations important to present fisheries may run as high as 20%. The true value of this efficiency is of practical importance because it determines the size of the fish harvest that can be taken safely from the oceans without damaging the ecosystem.

### FOOD CHAINS, FOOD WEBS, AND THE BIOMASS PYRAMID

The loss of energy between each feeding population limits the number of feeding populations in an ecosystem. If there were too many levels, there would not be enough energy to support the organisms in higher and higher trophic levels. In addition, each feeding population must necessarily have less mass than the population it eats. As a result, individual members of a feeding population are generally *larger in size* and *less numerous* than their prey.

***Food Chains*** A **food chain** is a sequence of organisms through which energy is transferred, starting with an organism that is the primary producer, then an herbi-

**EXPLORING DATA** ◀ **For a 10-kilogram (22 pound) salmon at trophic level 4, how many kilograms of phytoplankton at trophic level 1 are needed to support it? Assume the relative efficiencies at each step as shown in the figure.**

SmartFigure 13.28 **Ecosystem energy flow and efficiency.** For every 500,000 units of radiant energy input available to producers (phytoplankton) at trophic level 1, only 1 unit of equivalent mass is added to the fifth trophic level. For trophic level 1, the average transfer efficiency is 2% (98% loss); all other trophic levels average 10% efficiency (90% loss). https://goo.gl/n2SDWQ

vore, then one or more carnivores, and finally culminating with the "top carnivore," which is not usually preyed upon by any other organism.

Because energy transfer between trophic levels is inefficient, it is advantageous for fishers to target populations that feed as close to primary producers as possible. An example of such a target population would be a primary consumer (herbivore) or a secondary consumer (carnivore). This increases the biomass available for food and the number of individuals available to be taken by the fishery. Newfoundland herring, for example, are an important fishery that usually represents the third trophic level in a food chain. Newfoundland herring feed primarily on small crustaceans (copepods) that, in turn, feed upon diatoms (**Figure 13.29**).

***Food Webs*** Feeding relationships are rarely as simple as that of the Newfoundland herring. More often, top carnivores in a food chain feed on a number of different animals, each of which has its own simple or complex feeding relationships or food chains. This constitutes a **food web** of several interconnected food chains, as shown in Figure 13.29b for North Sea herring.

Animals that feed through a food web rather than a food chain are more likely to survive because they have alternative foods to eat should one of their food sources diminish in quantity or even disappear. Newfoundland herring, which feed through a food chain, eat only copepods, so the disappearance of copepods would catastrophically affect their population. Conversely, Newfoundland herring are more likely to have a larger biomass to eat, because they are only two steps removed from the producers, whereas North Sea herring are three steps removed in some of the food chains within their web.

***Biomass Pyramid*** The ultimate effect of energy transfer between trophic levels can be seen in the oceanic **biomass pyramid** shown in **Figure 13.30**. It shows that for the survival of each large marine organism, many levels of progressively larger populations of smaller-sized organisms must exist to support those higher on the pyramid. In essence, the *number of individuals* and *total biomass* decrease at successive trophic levels because the amount of available energy decreases. The figure also shows that organisms *increase in size* at successive trophic levels up the pyramid.

In some marine environments, inverted biomass pyramids can occur. An inverted biomass pyramid can exist, for example, where a smaller population of

**RECAP The transfer of energy between various trophic levels operates at low efficiencies, averaging only 2% for marine algae and 10% for most consumer levels.**

## STUDENTS SOMETIMES ASK . . .

***I've heard that harvesting krill could be the next big fishery. Is this true?***

Perhaps. Fishing at lower trophic levels certainly makes sense because of the abundance of biomass. In fact, krill is remarkably abundant in Antarctic waters. Accordingly, the Commission for the Conservation of Antarctic Marine Living Resources (CCAMLR), a consortium formed in 1982 by the United States, the United Kingdom, Australia, South Africa, New Zealand, Chile, the European community, Germany, and Japan, permits as much as 620,000 metric tons (1.4 billion pounds) of krill to be harvested annually. The krill is used as fish food for fish farms, cattle fodder, dog food, Omega 3 oils, enzymes for medical applications, and cosmetic additives. Some of it is even used to make food products for humans (krill burger, anyone?).

Unfortunately, as is the case for many fisheries worldwide, the harvesting of so much krill creates a reduction of food supply for other marine species. For example, the availability of krill during the short Antarctic summer is critical to the reproductive success of a wide range of Antarctic species, including seabirds, penguins, seals, and blue whales. Even so, industrial krill fishing has been increasing recently in Antarctic waters and vessels often use foraging penguins and other predators to locate large schools of krill.

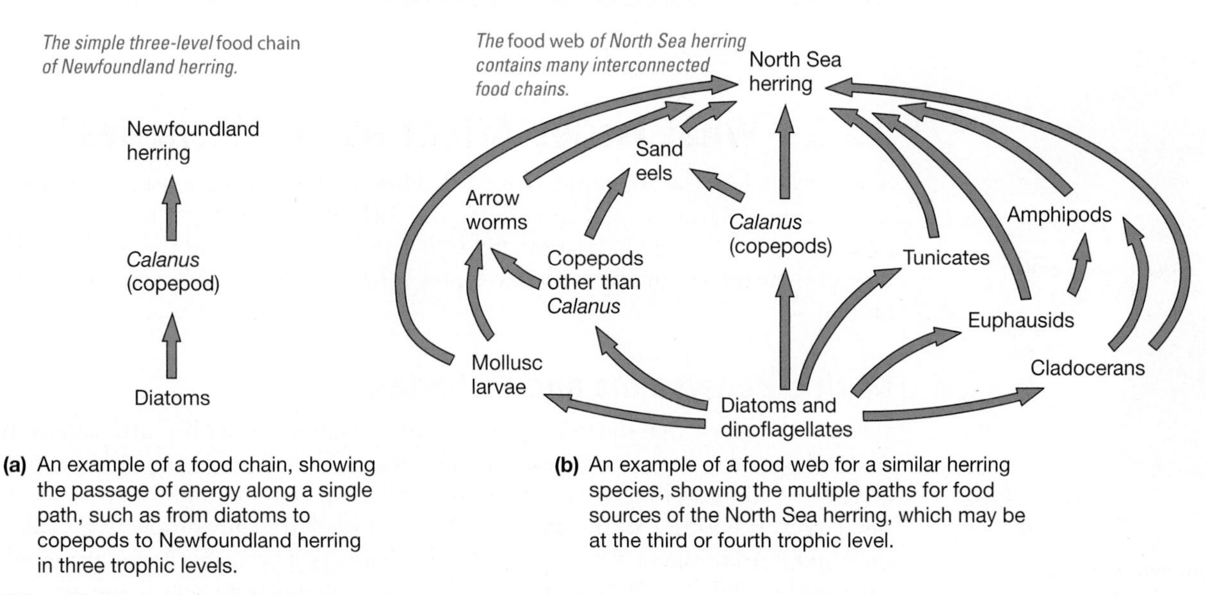


**(a)** An example of a food chain, showing the passage of energy along a single path, such as from diatoms to copepods to Newfoundland herring in three trophic levels.

**(b)** An example of a food web for a similar herring species, showing the multiple paths for food sources of the North Sea herring, which may be at the third or fourth trophic level.

**Figure 13.29 Comparison between a food chain and a food web.**

**SmartFigure 13.30 Oceanic biomass pyramid.**
https://goo.gl/BV7E5Q

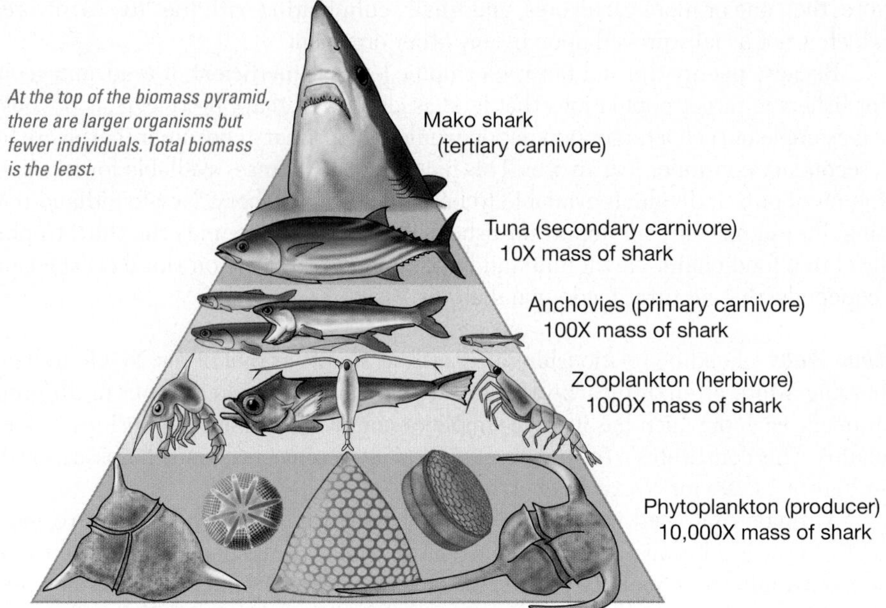


**RECAP A food chain is a linear feeding relationship among producers and one or more consumers, while a food web is a network of interconnected food chains of feeding relationships among many different organisms. The oceanic biomass pyramid shows energy transfer between trophic levels.**

phytoplankton with a very high turnover rate supports a larger population of zooplankton. As a result, oceanic biomass pyramids can have a variety of shapes, depending on the turnover rates of different trophic levels.

**CONCEPT CHECK 13.4 ▸ Discuss how energy and nutrients are passed along in marine ecosystems.**

1 Describe the flow of energy through the biotic community and include the forms into which solar radiation is converted. How does this flow differ from the manner in which matter is moved through the ecosystem?

2 Describe the three types of feeding strategies utilized by marine organisms.

3 Describe the advantage a top carnivore gains by eating from a food web, as compared to a single food chain.

4 Describe trends that occur in number of individuals, total biomass, and organism size at successive trophic levels going up a biomass pyramid. What conditions produce an inverted biomass pyramid?

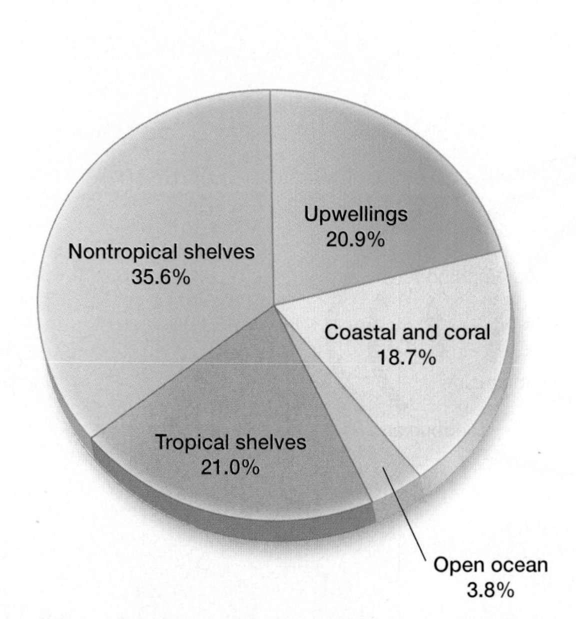


**Figure 13.31 Marine fishery ecosystems.** Pie diagram showing the relative contribution of various ecosystems to the total world marine fishery.

# 13.5 ▸ What Issues Affect Marine Fisheries?

Since well before the beginning of recorded history, humans have used the sea as a source of food. Over the past several decades, **fisheries** (fish caught from the ocean by commercial fishers) have provided billions of people around the world nearly 20% of their protein intake, with developing relying on fish for 27% of their dietary protein.

## Marine Ecosystems and Fisheries

**Figure 13.31** shows that the world marine fishery is drawn from five ecosystems. In descending order, they are (1) nontropical continental shelves, (2) tropical continental shelves, (3) upwelling areas, (4) coastal and coral systems, and (5) open ocean. The largest proportion of the marine fishery is found in highly productive shallow shelf and coastal waters, whereas low-productivity open ocean areas comprise only 3.8% of the total. Nearly 21% of the world's total catch is from very highly productive upwelling areas, which represent only about 0.1% of the ocean surface area.

## Overfishing

Fisheries harvest from the **standing stock** of a population, which is the mass present in an ecosystem at a given time. Successful fisheries leave enough individuals from the standing stock to repopulate the ecosystem after fisheries have made their harvest.

**Overfishing** occurs when harvesting of fish stocks takes place so rapidly that the majority of the population is sexually immature and is therefore unable to reproduce. Overfishing can occur in any size or body of water including ponds, rivers, lakes, or oceans when fish or shellfish are harvested beyond their sustainable level. Predictably, overfishing results in the decline of marine fish populations and the overall size of fish in a population. Fisheries biologists determine the sustainable levels of a fishery by calculating its **maximum sustainable yield (MSY)**, which is the maximum amount of fish biomass that can be removed yearly from a stock and still allow the population to be sustained indefinitely. The MSY must be determined yearly for each fishing stock and, in addition to harvesting, is affected by several factors including the number of predators, food availability, the fish's reproductive success, and water temperature (which is affected by human-caused climate change; see details below). It is important to accurately estimate a fish stock's MSY and to avoid exceeding this value during harvesting.

According to a report by the United Nations Food and Agriculture Organization (FAO), 80% of the 523 world marine fish stocks for which assessment information is available are classified as fully exploited, overexploited, or depleted/recovering from depletion (**Figure 13.32**). Although these numbers may seem discouraging, these trends can be reversed. For example, of the 85 fish stocks in U.S. waters declared overfished or approaching being overfished between the years 1997 and 2012, 41 stocks are no longer classified as overfished as a result of proper management.

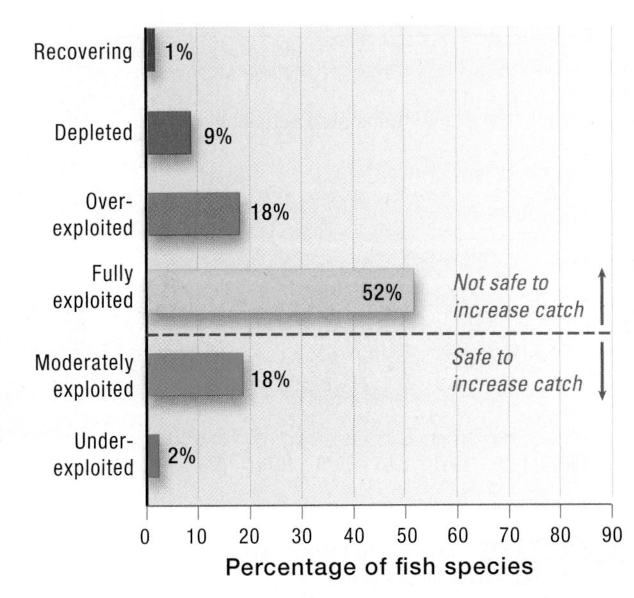


**Figure 13.32 Exploitation status of world marine fisheries.** Bar chart showing the current status of world fishery stocks. The only catch safe to increase are those fisheries that are moderately exploited and underexploited (*green*).

**ECOSYSTEM EFFECTS OF OVERFISHING LARGE PREDATORS** Scientific studies show that large predatory fish, called *keystone species* or top predators, are a vital part of healthy marine ecosystems. For instance, they prevent smaller fish from overrunning and potentially destroying ocean ecosystems. They also increase ecosystem health by culling sick and old populations of marine herbivores. Yet modern fishing practices have harvested 90% of large predatory fish species from the world ocean (**Diving Deeper 13.1**).

Removing large predators from the marine environment often has unintended consequences. One example is the recent population explosion of the American (Maine) lobster (*Homarus americanus*) in the North Atlantic Ocean. In the mid-1990s, the lobster's main predator—cod—was overfished and its numbers greatly reduced. As a result, the American lobster population increased to record numbers, which at first seems like a good thing; however, too many scavenging lobsters can disturb the ocean's ecological balance. Another example is the overfishing of large predatory sharks (such as bull, great white, dusky, and hammerhead sharks), which has led to an explosion of their prey species: rays, skates, and small sharks. In 2004, for instance, the overabundance of rays, which feed primarily on scallops and crustaceans, decimated the populations of these species in some areas and caused the permanent closure of North Carolina's century-old bay scallop fishery.

Coral reef ecosystems are being affected by the removal of large predators, too. For example, the removal of large reef residents such as sharks reduces predation and competition, which can clear the way for smaller fish to proliferate. The removal of large fish like parrotfish and surgeonfish—both herbivores that eat algae from reefs—increases the number of smaller nonherbivore fish. As a result, algae can overgrow and cover coral reefs, which then blocks sunlight, denies nutrients for corals, and ultimately suppresses coral growth.

With the reduction of large fish, the fishing industry has concentrated on smaller fish that occupy successively lower trophic levels. In fact, low-trophic level species now account for over 30% of global fisheries production. The danger here is that removing large numbers of individuals from lower trophic levels can cause

### STUDENTS SOMETIMES ASK . . .

***Is it true that the Maine lobster fishery is experiencing a plentiful supply of lobsters?***

Yes, but it may not last long. In 2014, for example, Maine lobsterfishers hauled ashore 56 million kilograms (124 million pounds) of lobsters, six times more than what they'd caught in 1984. But scientists haven't been able to agree what is causing the unexpected uptick in lobster population. A rise in sea temperatures due to climate change, which has sped up lobster growth and opened up new coastal habitats for baby lobsters, is one likely reason. Another is that by plundering cod and other big fish in the Gulf of Maine, fishers thinned out the predators that had long kept lobster numbers in check. There is also a growing consensus that the trend is the result of the long history of effective regulations that the lobster industry abides by, such as returning to the sea juvenile and egg-bearing female lobsters. However, a recent study of the optimum conditions for the waters that support lobsters predicts that the Gulf of Maine lobster population will shrink by as much as 62% over the next 30 years because of additional ocean temperature increase.

Climate Connection

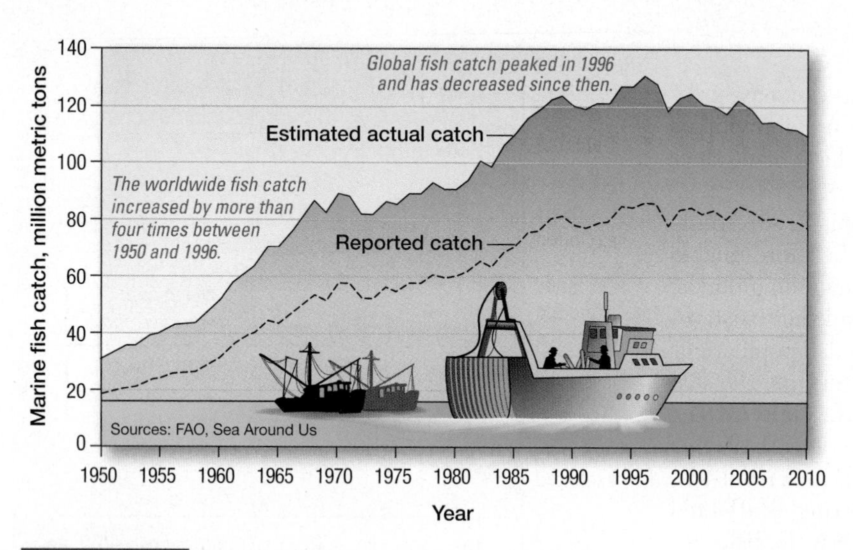


**EXPLORING DATA ▲ Explain why there is such a large discrepancy between the two curves on this graph.**

Figure 13.33 **World total marine fish production 1950–2010.** Graph comparing the FAO reported world marine fish catch (*dashed line*) and that calculated using a model that included estimates of unreported catch (*solid line*).

## STUDENTS SOMETIMES ASK . . .

### *As the world's oceans are overfished and degraded, are they being taken over by jellies?*

There's an increasing concern that "jellies gone wild" is, in fact, happening now. As predators, jellies appear to be slow and passive. Unable to see or chase their prey, most drift along, creating tiny eddies that guide food particles toward their tendrils. Yet in many parts of the ocean, jellies are thriving as many of their competitors are eliminated by overfishing and other human impacts, creating environmental conditions that favor jellies. All around the world, jellies are creating havoc by reproducing in astonishing numbers and congregating where they've rarely been seen before. For example, jellies have shut down nuclear power plants that use seawater as a coolant, disabled a military aircraft carrier by blocking its seawater intake system, halted sea floor mining operations by clogging dredges, capsized fishing vessels by their sheer numbers caught in nets, endangered bathers, and even stung and killed hordes of farmed fish. Although some studies show the apparent increase in the size and frequency of jelly blooms in coastal waters and estuaries around the world during the past few decades is part of a natural cycle, jellies' impact on marine food webs is likely to increase in the future, leading to speculation that "the rise of slime" may lead to a "gelatinous ocean."

an unwanted cascade of effects on other parts of marine ecosystems, particularly for fish, seabirds, and marine mammals that rely on these small prey items as a food source.

**THE END OF FISH?** It's difficult to imagine the oceans without fish, but scientists are predicting exactly that scenario. Overall, the selective removal of larger fish by commercial and recreational fishers leaves fish populations with fewer and smaller individuals. When smaller populations of individuals reproduce, their offspring have less genetic variability, which can cause genetic changes that cut the average growth rate, time to mature, and size of fish from generation to generation. This phenomenon has severe implications for the health and sustainability of fisheries. In fact, scientists have determined that the negative effects of pollution, habitat loss, and overfishing will deplete current marine fishing stocks by 2048. If these wasteful practices continue, scientists predict that within only a few decades, fish will no longer live in the oceans. Not only will the loss of seafood threaten humans' food supplies, the loss of fish will seriously damage entire marine ecosystems.

The outlook is grim, but there are actions that are being undertaken to avert this fate. For example, Seikai National Fisheries Research Institute in Japan and the University of Maryland in the United States have started to raise bluefin tuna in large tanks to reduce fishing pressure on wild tuna populations while at the same time supplying tuna for sushi. Tuna are also being raised in enclosed ocean pens (an enclosure much like a corral on land) in coastal waters off Mexico. In Carlsbad, California, white seabass (*Atractoscion nobilis*) are being raised in a hatchery, and millions have already been released into the ocean to repopulate local waters.

**RECREATIONAL FISHING** Although it is widely assumed that recreational fishing has a minimal influence on most species of fish, recreational fishing has been shown to have an impact on fish populations, too. A detailed analysis of multiple U.S. fisheries records shows that for some threatened sportfishing species—such as red drum, bocaccio rockfish, and red snapper—recreational fishing actually poses more of a threat than commercial fishing because of the size and number of fish taken by recreational fishers. Thus, new fishing regulations that target recreational fishers are needed to complement existing catch restrictions on commercial fishers. Interestingly, commercial fishers blame recreational fishing for the decline in some of the species, and vice-versa. However, many recreational fishers are actively involved in the practice of *catch and release*, which uses circle hooks and returns caught fish to the ocean, thereby helping to sustain fish populations.

**WORLD FISH PRODUCTION** Since 1950, the FAO has monitored the amount of wild marine fish caught in the world, and it has based its conclusions on the accuracy of numbers submitted by participating countries. In 2016, fisheries researchers suggested that numbers published by the FAO over the past several decades underreported the world fish catch. Using a sophisticated analytical model, which included the reported amount of fish caught plus estimates of unreported catch, the researchers reconstructed what they believed to be a more accurate world fish catch. That study concluded the world marine fisheries catch in 1950 was about 25 million metric tons (55 billion pounds) per year and rose rapidly until 1996, when it hit a peak of 130 million metric tons (287 billion pounds) per year (**Figure 13.33**). The world catch has been declining steadily since then and in 2010 was about 108 million metric tons (238 billion pounds). One of the main reasons why the worldwide catch has decreased is that many marine fish stocks have been severely depleted by *overfishing*, as discussed earlier in this chapter. As a partial offset to this reduction of fish, worldwide marine aquaculture has increased five-fold since the 1980s and now accounts for more than 25 million metric tons (55 billion pounds) of fish.

## FISHING DOWN THE FOOD WEB: SEEING IS BELIEVING

Marine scientists can measure the health of an ecosystem by determining its original fish abundance, diversity, and size. But how can these variables be assessed effectively for periods before extensive scientific field surveys were conducted? Recently, scientists have been using novel ways to estimate these parameters. A new field of research called *historical marine ecology* uses items such as old photographs, newspaper accounts, ships' logs, cannery records, and even old restaurant menus to estimate the type and quantity of fish that used to live in the sea.

Striking evidence of past fish abundance and size comes from old photographs. Certainly, people have always enjoyed having their picture taken with the fish they catch. Loren McClenachan, a graduate student at Scripps Institute of Oceanography at the University of California, San Diego, found a historical archive of old photographs in the Monroe County Public Library in Key West, Florida. The archive allowed her to examine fish caught by day-trippers at coral reef sites aboard a series of vessels called *Gulf Stream* and *Greyhound*, which have operated out of Key West for the past 50 years.

By comparing old photographs with more recent ones (**Figure 13A**) of fishing in the same area, it is apparent there has been a pronounced decrease through the years in both fish abundance and size. Research suggests that chronic overfishing by both commercial and recreational fishers is the likely source of this decline. In fact, some species completely disappear over time, reflecting a loss of the largest fish from the coral reef environment. For example, in the 1950s, fishers caught huge grouper and sharks. In the 1970s, they landed a few grouper but more jack. Today's main catch lacks grouper and instead consists of small snapper, which at one time weren't even deemed worthy of a photograph; instead, they were just piled up below the hanging racks. Historical records from other regions throughout the world also reveal astonishing declines in most fish stocks. Scientists use the phrase "fishing down the food web" to describe how key large species in healthy functioning ecosystems have been replaced by progressively smaller, less valuable species that occupy lower levels of the food web.

Worldwide, fishers typically catch the biggest animals first, whether turtles, whales, cods, or groupers. Then they catch whatever is left—including animals so young they haven't yet reproduced—until, in some cases, the animals are gone. To change human habits, it is important to gain a clearer picture of what has been lost. In the Florida Keys, historical photographs provide such a window into a more pristine coral reef ecosystem that existed a half-century ago.

### WHAT DID YOU LEARN?

Describe the change over time that took place in the populations of fish in response to fishing pressure off Key West, Florida.

**Figure 13A Decreasing size of fish catch.** Historical photos from chartered fishing vessels that concentrated on the fish of the coral reefs surrounding Key West, Florida, show decreasing catch size and abundance from 1958 (*above*), 1980s (*middle*), and 2007 (*below*).

Figure 13.34 **Fishing bycatch.** Photograph of dead or dying fish caught in a fishing net as bycatch in the North Atlantic Ocean. In commercial fishing operations, about one-fourth of all catch is discarded as unwanted bycatch, which includes birds, turtles, sharks, dolphins, and many species of noncommercial fish.

## Incidental Catch

**Incidental catch**, or **bycatch**, includes any marine organisms caught incidentally by fishers seeking commercial species. On average, close to one-fourth of the catch is discarded, although for some fisheries, such as shrimp, the incidental catch may be up to eight times larger than the catch of the target species. Incidental catch includes birds, turtles, sharks, and dolphins, as well as many species of noncommercial fish (**Figure 13.34**). In most cases, these animals die before they are thrown back overboard, even though some of them are protected by U.S. and international law. According to a recent study, bycatch peaked at 19 million metric tons (42 billion pounds) per year in 1989 and have since declined to around 10 million metric tons (22 billion pounds) per year, which still accounts for about 10% of the world's total marine fish catch.

**TUNA AND DOLPHINS** Schools of yellowfin tuna are commonly found swimming beneath spotted and spinner dolphins in the eastern Pacific Ocean (**Figure 13.35**). Fishers commonly use these dolphins to locate tuna and set a *purse seine net* around the entire school. When an underwater line is drawn tight, the net traps the tuna underwater as well as the dolphins at the surface. Unfortunately, dolphins, which are marine mammals, get caught in the net below the water's surface and can't reach the surface to breathe air, so they often drown. **Figure 13.36** illustrates how a purse seine net works; the figure also shows the variety of methods and fishing gear used in modern-day commercial fishing practices.

In 1988, biologist Samuel F. La Budde presented the problem of dolphin deaths caused by tuna fishing through graphic video footage taken of dolphins struggling in tuna fishing nets. In 1990, under intense public outcry and a boycott on tuna, the U.S. tuna canning industry declared it would not buy or sell tuna caught using methods that kill or injure dolphins. In 1992, a special addendum was added to the **Marine Mammals Protection Act**, further protecting dolphins. As a result of these measures, purse seine nets were modified so that dolphins could be released alive. In spite of reducing dolphin mortality as bycatch, dolphin populations have not rebounded accordingly. Research suggests that tuna fishing operations are still having a negative effect on dolphin populations by reducing dolphin survival and birth rates.

Figure 13.35 **Spotted dolphin (*Stenella attenuata*), which are commonly associated with yellowfin tuna.** Eastern Pacific spotted dolphins are often found swimming above yellowfin tuna. This relationship is thought to exist because the dolphins use tuna to help find schools of prey items (squid and small fish), which are eaten by both animals.

**DRIFTNETS** Another means of netting tuna and other species is by use of **driftnets**, or **gill nets**, which involves capturing fish by their gills (Figure 13.36). Driftnets are made of crisscrossed monofilament fishing line that is virtually invisible and cannot be detected by most marine animals as they swim into them. Depending on the size of the holes in the net, it is highly effective at catching anything large enough to become entangled in it. As a result, driftnets often have high amounts of bycatch.

Up until 1993, Japan, Korea, and Taiwan had the largest driftnet fleets, deploying as many as 1500 fishing vessels into the North Pacific and setting over 48,000 kilometers (30,000 miles) of driftnets in one day. Although driftnetting was supposed to be restricted to specific fisheries, some fishers who claimed to be fishing for squid were involved in illegally taking large quantities of salmon and steelhead trout. Driftnetters were also targeting immature tuna in the South Pacific, which could result in the reduced abundance of South Pacific tuna. In addition, tens of thousands of birds, turtles, dolphins, and other species were killed annually in these nets as bycatch.

In an effort to reduce the wasteful practice of driftnet fishing, the United States signed an international treaty in 1989 that prohibits driftnets greater than 2.5 kilometers (1.5 miles) in the South Pacific and includes a clause that prohibits the import of any fish caught in driftnets. Although long driftnets are banned in international waters, the United States allows limited use of shorter driftnets in

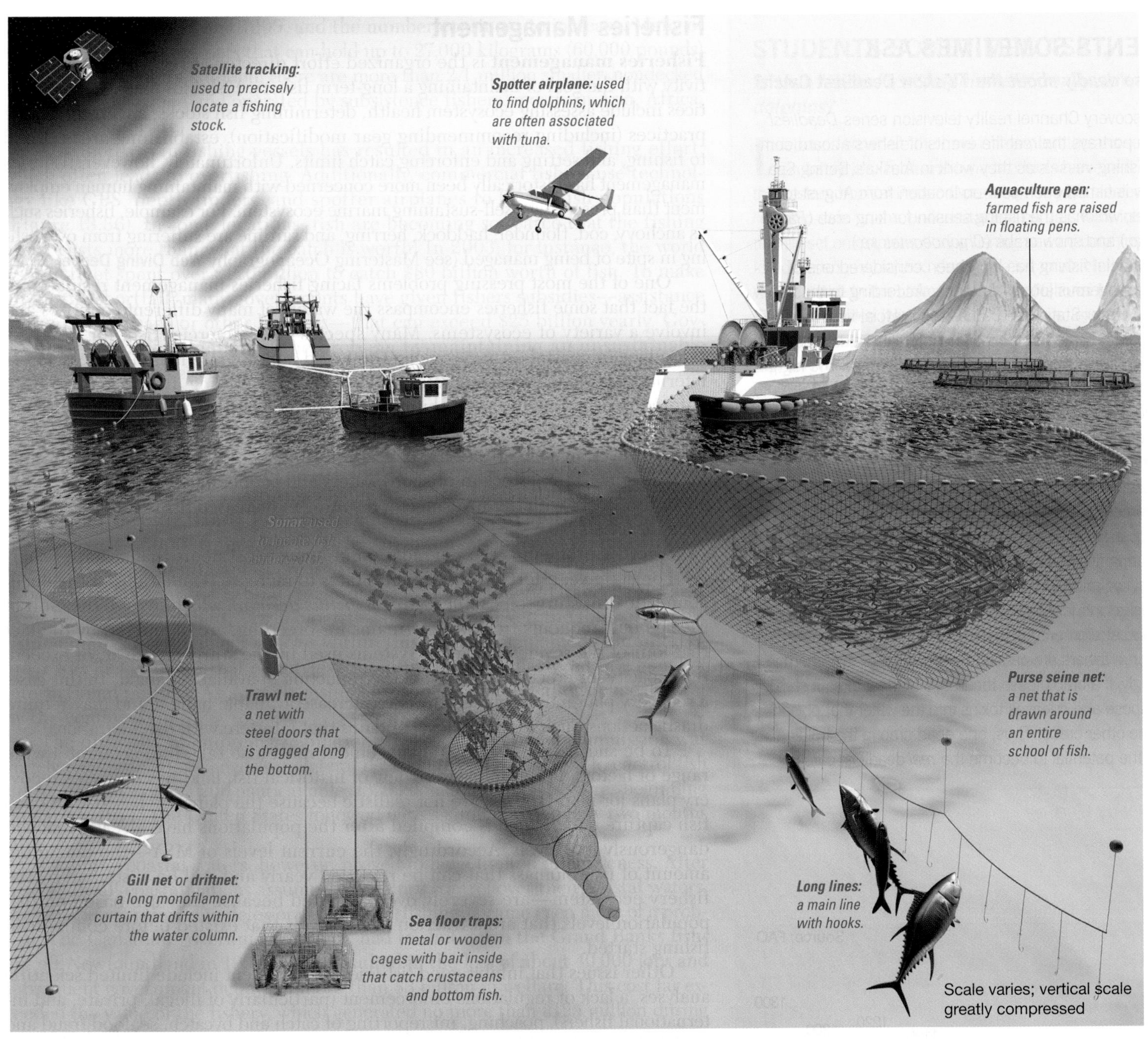

Satellite tracking: used to precisely locate a fishing stock.
Spotter airplane: used to find dolphins, which are often associated with tuna.
Aquaculture pen: farmed fish are raised in floating pens.
Sonar: used to locate fish underwater.
Trawl net: a net with steel doors that is dragged along the bottom.
Purse seine net: a net that is drawn around an entire school of fish.
Gill net or driftnet: a long monofilament curtain that drifts within the water column.
Sea floor traps: metal or wooden cages with bait inside that catch crustaceans and bottom fish.
Long lines: a main line with hooks.
Scale varies; vertical scale greatly compressed

**SmartFigure 13.36 Methods and gear used in commercial fishing.**
https://goo.gl/tc03yN

some rivers, lakes, and bays. Additionally, some fishers continue to use long driftnets illegally in international waters.

**GHOST FISHING** Another area of concern for fisheries is called **ghost fishing**, which describes any lost or discarded fishing gear that continues to catch fish, marine mammals, or other organisms after it has been abandoned. Examples of ghost fishing gear include longlines, gill nets, entangling nets, trammel nets, traps, and even crab and lobster pots. Ghost fishing is detrimental to the environment because anything caught by ghost fishing is killed and wasted. One of the reasons ghost fishing is so deadly is that the abandoned fishing gear continues to entangle and kill marine organisms as long as it remains intact. One solution to this problem, particularly for crab and lobster pots, is the use of biodegradable panels that decompose within a few months after the pots are lost, allowing the captured animals to escape.

# Animals of the Pelagic Environment

# 14

## ESSENTIAL LEARNING CONCEPTS

At the end of this chapter, you should be able to:

- ☐ **14.1** Compare the various methods by which marine organisms are able to avoid sinking through the water column.
- ☐ **14.2** Specify adaptations that pelagic organisms possess for seeking prey.
- ☐ **14.3** Specify adaptations that pelagic organisms possess to avoid becoming prey.
- ☐ **14.4** Differentiate between the main groups of marine mammals based on their physical characteristics.

↑ *Check when completed*

Pelagic organisms live suspended in seawater (not on the ocean floor) and comprise the vast majority of the ocean's **biomass**, which as you may recall is the mass of living organisms. Phytoplankton and other photosynthesizing microbes live within the sunlit surface waters of the ocean and are the food source for nearly all other marine life. As a result, many marine animals live in surface waters so they can be close to their food supply. One of the most important challenges facing many marine organisms is to stay afloat and not sink below surface waters into the immense depth of the oceans.

Phytoplankton and other photosynthesizing microbes depend primarily on their small size to provide a high degree of frictional resistance to sinking. Most animals, however, are more dense than ocean water and have less surface area per volume of body mass (they have a smaller surface-area-to-volume ratio). Therefore, they tend to sink more rapidly than phytoplankton.

To remain in surface waters where the food supply is greatest, pelagic marine animals must increase their buoyancy or swim continually. Animals apply one or both of these strategies in amazing ways using a variety of adaptations.

***"The whale rose even closer. It had a distinct hazel eye that looked directly at me. It studied my hair, looked at my beard, passed its gaze over my nose, and then looked deeply into my eyes. It looked past all those biology classes, between the volumes of whale literature I had studied, and beyond the thousands of gray whales in my memory. It looked into my soul."***

***—Lindblad Expeditions Naturalist Robert "Pete" Pederson, describing a close encounter with a gray whale (1999).***

## 14.1 ▸ How Are Marine Organisms Able to Stay above the Ocean Floor?

Some animals increase their buoyancy to remain in near-surface waters. They may have internal structures containing gas, which significantly reduce their average density, or they may have soft bodies void of hard, high-density parts. Larger animals often have the ability to swim, but if their bodies are denser than seawater, they must exert more energy to propel themselves through the water.

### Use of Gas Containers

Air is almost 800 times less dense than water at sea level, so even a small amount of air inside an organism can dramatically increase its buoyancy. Generally, animals use either an internal, rigid gas container or a swim bladder to achieve *neutral buoyancy*, using the amount of air in their bodies to regulate their density, so they can remain at a particular depth without expending energy to do so.

◂ **A superbly streamlined shark glides through the ocean.** Sharks like this silky shark (*Carcharhinus falciformis*) have unique adaptations that make them efficient ocean predators.

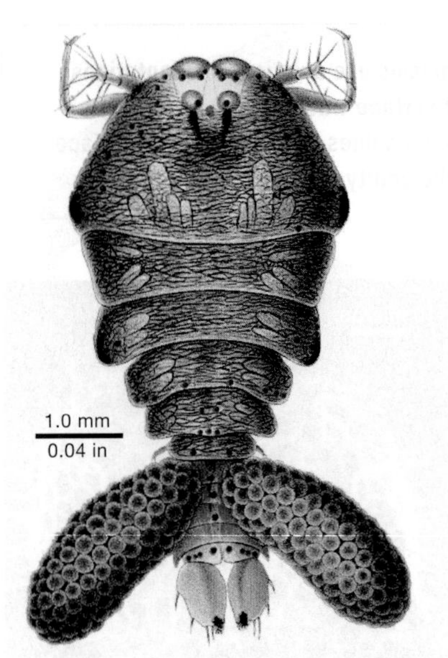


**(a)** The adult female *Sapphirina auranitens* carries a pair of lobe-like egg sacs *(blue)*.

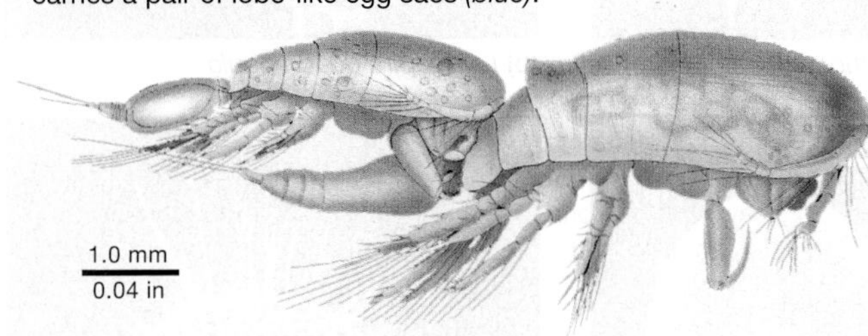


**(b)** Copulating pair of *Oncaea conifera.*

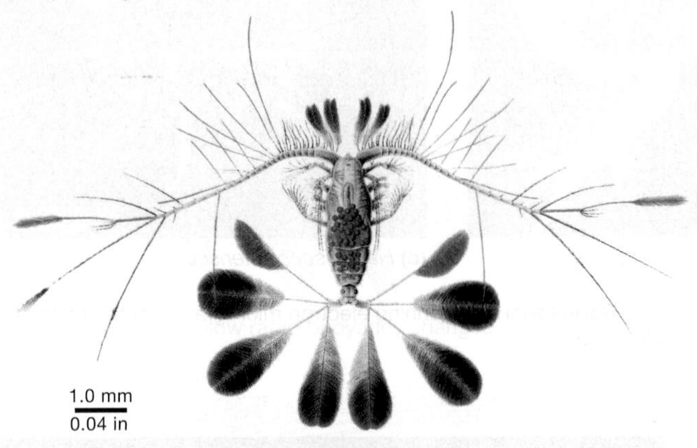


**(c)** *Calocalanus pavo* showing elaborate feathery appendages that are characteristic of warm-water species.

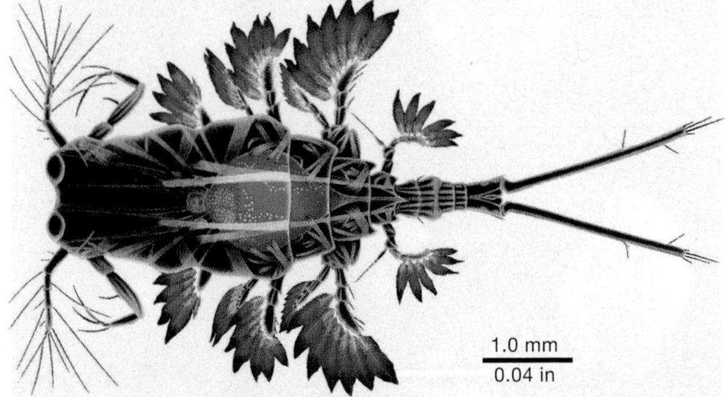


**(d)** *Copilia vitrea* uses its appendages to cling to large particles in the water column or to larger zooplankton.

**Figure 14.5 Copepods.** Line drawings of various copepods from Wilhelm Giesbrecht's 1892 book on the flora and fauna of the Gulf of Naples (Mediterranean Sea).

organisms on the planet. As such, copepods comprise the majority of the ocean's zooplankton biomass and are a vital link in many marine food webs between phytoplankton (producers) and larger species such as plankton-eating fish.

**EXAMPLES OF MACROSCOPIC ZOOPLANKTON** Many types of zooplankton are large enough to be seen without the aid of a microscope yet don't swim well. Two of the most important groups of macroscopic zooplankton are krill and various types of cnidarians.

**Krill**, which means "young fry of fish" in Norwegian, are actually in the subphylum Crustacea (genus *Euphausia*) and resemble mini-shrimp or large copepods (**Figure 14.6**). There are more than 1500 species of krill, most of which achieve a length no longer than 5 centimeters (2 inches). They are abundant near Antarctica and form a critical link in the food web there, supplying food for many organisms from sea birds to the largest whales in the world.

**Cnidarians** (*cnid* = nematocyst [*nemato* = thread, *cystis* = bladder]), which were formerly known as *coelenterates* (*coel* = hollow, *enteron* = intestines), have soft bodies that are more than 95% water and tentacles armed with stinging cells called *nematocysts*. Macroscopic zooplankton that are cnidarians fall into one of two basic groups: the *hydrozoans* and the *scyphozoans* (jellies).

**Hydrozoan** (*hydro* = water, *zoa* = animal) *cnidarians* are represented in all oceans by the "Portuguese man-of-war" (genus *Physalia*) and the "by-the-wind sailor" (genus *Velella*). Their gas chambers, called *pneumatophores*

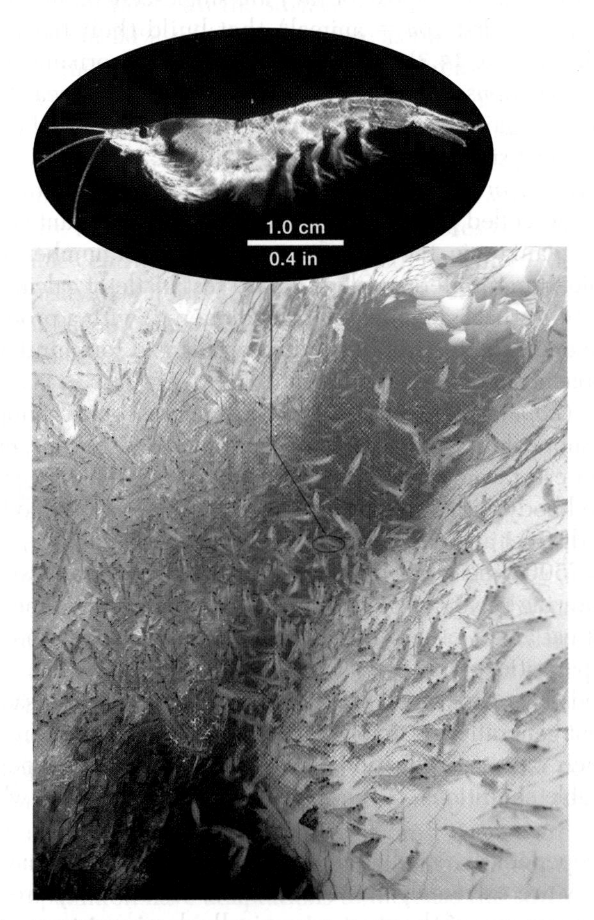


**Figure 14.6 Krill.** Shrimp-like krill in surface waters near Antarctica and close-up view of krill *Meganyctiphanes norvegica* (*inset*).

to swim in broad circles-
while swimming.

Caudal fin shapes are
forked, 4) lunate, and 5) l
illustrated and explained

Caudal fin shape and descripti

**Rounded:** a rounded cauda
blue-face angel *(Pomacanth*
is flexible and useful in acce
maneuvering at slow speed

**Truncated:** a truncated cau
gray angelfish *(Pomacanthu*
moderately flexible for gooc
also used for maneuvering.

**Forked:** a forked caudal fin
doctorfish *(Acanthurus coer*
of faster fish; it is moderatel
propulsion and is useful for

**Lunate:** a lunate caudal fin
marlin *(Kajikia audax)* is four
fish; the fin is very rigid, whi
for maneuvering, but very e
propulsion.

**Heterocercal:** a heterocerc
gray reef shark *(Carcharhinu*
very rigid and produces trer
However, sharks sacrifice m
are usually seen swimming

Figure 14.11 **Caudal fin sh**
rounded, truncate, forked, lun
maneuverability and propulsic

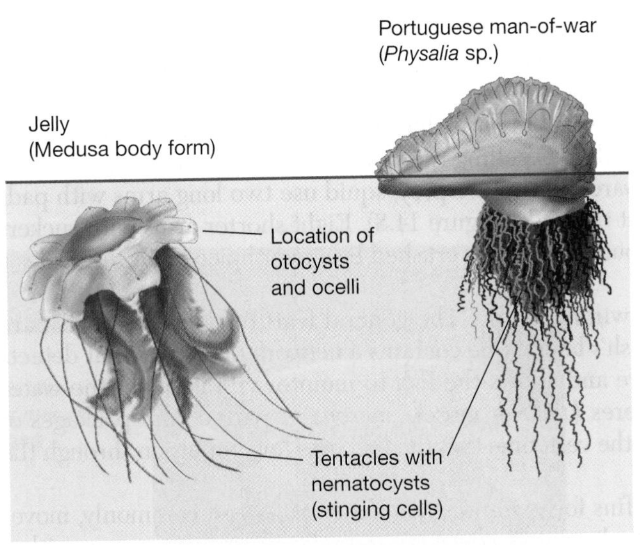


**(a)** Line drawings of a typical medusa jelly (*left*) and a Portuguese man-of-war (*Physalia*) (*right*).

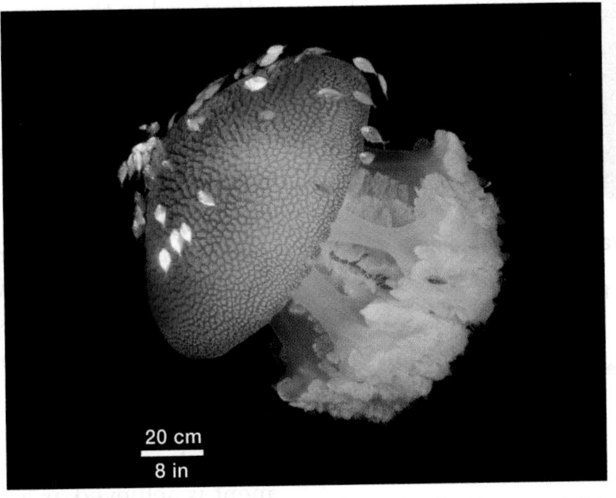


**(b)** An Australian spotted medusa jelly (*Phyllorhiza punctata*) provides habitat on its bell for juvenile fish known as blue trevally (*Carangoides ferdau*).

Figure 14.7 **Planktonic cnidarians. (a)** Line drawings and **(b)** a photo of various species of planktonic cnidarians.

(*pneumato* = breath, *phoros* = bearing), serve as floats and sails that allow the wind to push them across the ocean surface (**Figure 14.7a**, *right*). Sometimes the wind pushes large numbers of these organisms toward a beach, where they can wash ashore and die. In the living organism, an entire colony of other small organisms that rely on the hydrozoan for habitat can be found within and beneath the float. Portuguese man-of-war tentacles may be many meters long and, because they are armed with nematocysts, they have been known to inflict a painful and occasionally dangerous neurotoxin poisoning in humans.

Jellies, or **scyphozoan** (*skyphos* = cup, *zoa* = animal) *cnidarians*, have a bell-shaped body with a fringe of tentacles and a mouth at the end of a clapper-like extension hanging beneath the bell-shaped float (Figure 14.7a, *left*, and **Figure 14.7b**). Ranging in size from nearly microscopic to 2 meters (6.6 feet) in diameter, most jellies are less than 0.5 meter (1.6 feet) in diameter. The largest jellies can have tentacles as long as 60 meters (200 feet).

Jellies move by muscular contraction of their bell. Water enters the cavity under the bell and is forced out when muscles that circle the bell contract, slowly jetting the animals in random directions as they pulsate. (Recall from Chapter 12 that jellies are considered plankton [floaters] and not nekton [swimmers] because they can't control where they move.) More importantly, the motion swooshes water that contains tiny floating food particles toward their tentacles. To allow the animal to orient itself generally in an upward direction, light-sensitive or gravity-sensitive organs exist around the outer edge of the bell. The ability to orient is important because jellies feed by swimming to the surface and sinking slowly through the rich surface waters. Like hydrozoans, jellies are particularly important in the open ocean, where they provide habitat for a variety of other organisms ranging from juvenile fish to worms and crabs.

Other types of macroscopic zooplankton include tunicates and salps (barrel-shaped, often colonial organisms), ctenophores (comb jellies or sea gooseberries), and chaetognaths (arrowworms).

**EXAMPLES OF SWIMMING ORGANISMS** Swimming, or *nektonic*, organisms include invertebrate squid, fish, sea turtles, and marine mammals.

Swimming squid include the common squid (genus *Loligo*), flying squid (*Ommastrephes*), and giant squid (*Architeuthis*), all of which are active predators of fish. Most squid possess long, slender bodies with paired fins (**Figure 14.8**) and must

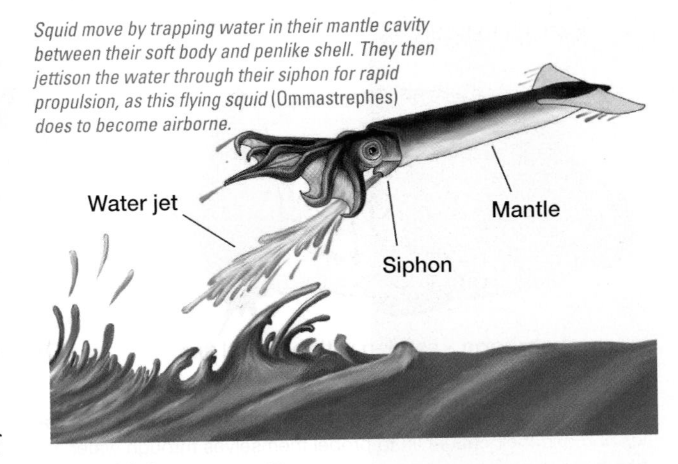

Squid move by trapping water in their mantle cavity between their soft body and penlike shell. They then jettison the water through their siphon for rapid propulsion, as this flying squid (Ommastrephes) does to become airborne.
Water jet
Mantle
Siphon

Figure 14.8 **Squid motility.**

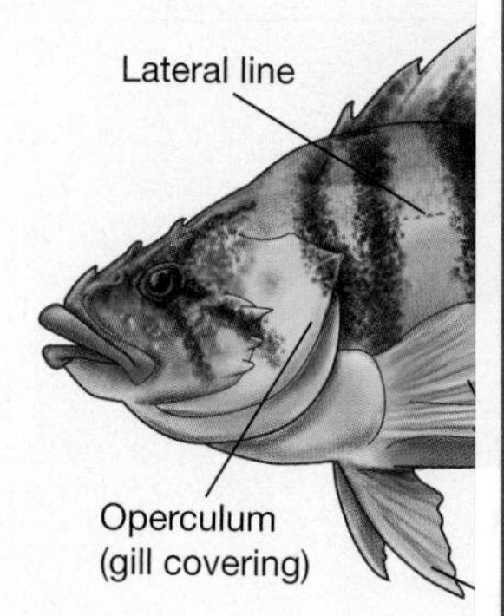


SmartFigure 14.9 Ge
of a fish
https://goo.gl/hPA484

Thunniform - caudal f

The fin that
used to pro
fish is the c


Fish not to scale.

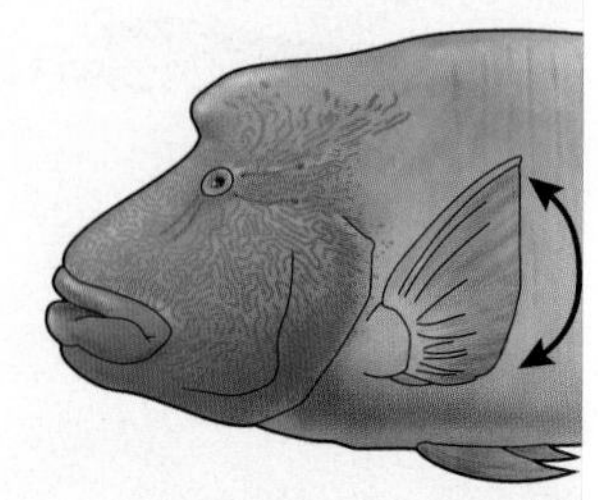

Labriform - scullin

Figure 14.10 Fish locomoti
ies and use their caudal fin to
fish for locomotion.



marine life is bioluminescent. The vast majority of bioluminescent organisms use light-producing organs called **photophores** (*photo* = light, *phoros* = bearing), which can be simple luminous spots or may be quite complex and equipped with lenses, shutters, color filters, and reflectors.

Bioluminescent light is produced from compounds during the digestion of prey, from specialized cells in the organism, or associated with symbiotic bacteria that is cultured and lives inside the organism. The light is produced when molecules of the biological pigment *luciferin* (*lucifer* = light bringing) are excited and emit photons of light in the presence of oxygen (this is like the chemical reaction that occurs when you snap a glow stick). The light-production process is remarkably efficient: only a 1% loss of energy is required to produce this illumination. Some marine animals control light production in unique ways. For example, scientific research suggests that some sharks control their bioluminescence by using hormones.

In a world of darkness, the ability to bioluminesce is useful for a variety of purposes, including:

- Searching for food items in the dark
- Attracting prey (for example, the female deep-sea anglerfish shown in Figures 14.13f and 14.13g use their specially modified dorsal fin as a bioluminescent lure)
- Staking out territory by constantly patrolling an area
- Communicating or seeking a mate by sending signals
- Escaping from predators by using a flash of light to temporarily blind or distract them
- Avoiding predators by use of a "burglar alarm" by attracting unwanted attention with brilliant displays of bioluminescence
- Camouflaging by using belly lights to match the color and intensity of dim filtered sunlight from above and obliterate a telltale shadow to become effectively invisible; this is known as **counterillumination**

To take advantage of bioluminescent light, many deep-sea fish species have large and sensitive eyes—perhaps 100 times more sensitive to light than human

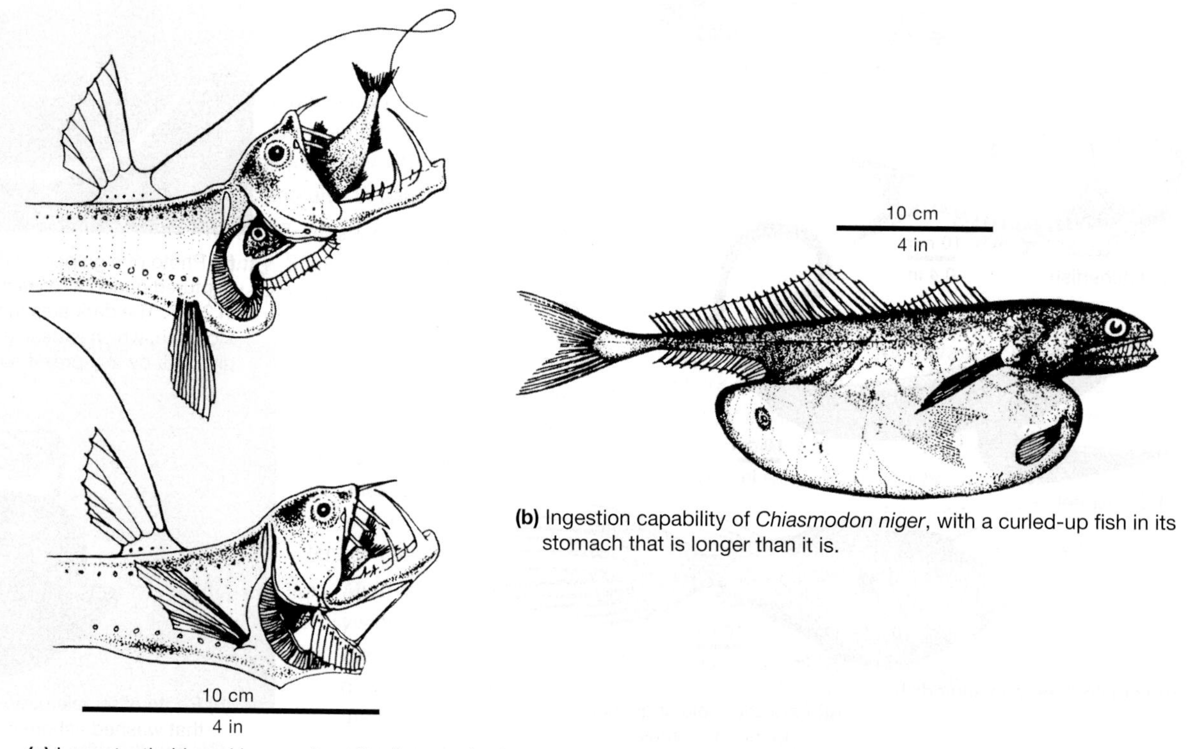


**Figure 14.15 Adaptations of deep-sea fish.** Line drawings showing various adaptions of deep-sea fish.

**(a)** Large teeth, hinged jaw, and swallowing mechanism of the deep-sea viperfish *Chauliodus sloani*.

**(b)** Ingestion capability of *Chiasmodon niger*, with a curled-up fish in its stomach that is longer than it is.

eyes—that enable them to see potential prey. To avoid becoming prey, most species are dark in color so that they blend into the environment. Still other species are blind and rely on senses such as smell to track down prey.

Other adaptations that various deep-sea fish species possess include large sharp teeth, expandable bodies to accommodate large food items, hinged jaws that can unlock to open widely, and mouths that are huge in proportion to their bodies (**Figure 14.15**). These adaptations allow deep-sea fish to ingest species that are larger than they are and to process food efficiently whenever it is captured.

**RECAP** **Adaptations of pelagic organisms for seeking prey include mobility (lunging versus cruising), high swimming speed, and high body temperature. Deep-water nekton exhibit a host of unusual adaptations—including bioluminescence—that allow them to survive in deeper waters.**

**CONCEPT CHECK 14.2 ▸ Specify adaptations that pelagic organisms possess for seeking prey.**

1 Are most fast-swimming fish cold-blooded or warm-blooded? What advantage does this provide?

2 What are the two food sources of deep-water nekton? List several adaptations of deep-water nekton that allow them to survive in their environment.

3 Describe the mechanism by which bioluminescence is accomplished in deep-sea organisms. What is bioluminescence useful for in the marine environment?

# 14.3 ▸ What Adaptations Do Pelagic Organisms Possess to Avoid Becoming Prey?

Many animals have unique adaptations to avoid being captured and eaten. Examples of adaptations that organisms use to enhance their survival include schooling and symbiosis.

## Schooling

The term **school** refers to large numbers of fish, squid, or shrimp that form well-defined social groupings. Although vast populations of phytoplankton and zooplankton may be highly concentrated in certain areas of the ocean, they are not usually referred to as schools.

The number of individuals in a school can vary from a few larger predaceous fish (such as bluefin tuna) to hundreds of thousands of small filter feeders (such as anchovies). Within the school, individuals move in the same direction and are evenly spaced. Spacing is probably maintained through visual contact and, in the case of fish, by use of the lateral line system (see Figure 14.9) that detects vibrations of swimming neighbors. Similar to a flock of birds, the school can turn abruptly in spectacular fashion or even reverse direction as individuals at the head or rear of the school assume leadership positions (**Figure 14.16**). Research suggests that each fish decides where to move not based on the behavior of its nearest neighbors, as is often assumed, but on a synthesis of where all the fish in its field of view are headed.

**Figure 14.16** **Schooling.** A school of blue-lined snapper near a reef in the Maldives, Indian Ocean. Schooling increases chances of survival, and more than half of all fish species are known to join schools during at least a portion of their lives.

What are the advantages of schooling? One advantage is that during spawning, schooling ensures that there will be males to release sperm to fertilize the eggs released into the water or deposited on the bottom by females. Another advantage is that schools of smaller

**(a)** Commensalism occurs when an organism benefits without harming its host, such as these remoras attached to a lemon shark.

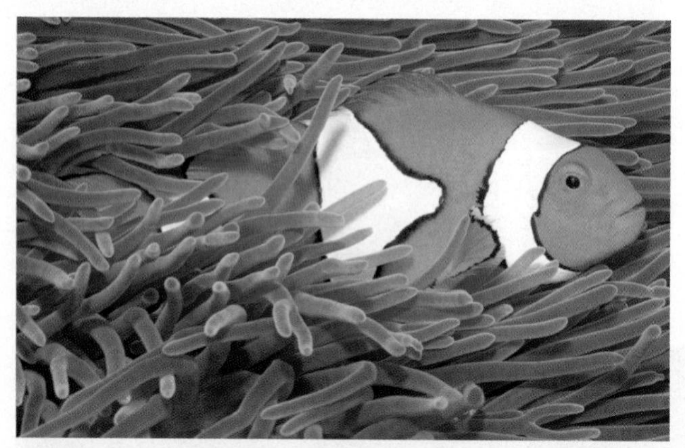

**(b)** Mutualism occurs when both participants benefit, such as this clown anemonefish and sea anemone.

**(c)** Parasitism occurs when one participant benefits at the expense of the other, such as this isopod that has attached itself to the head of a whitetip soldierfish.

Figure 14.17 **Types of symbiosis.** The three main types of symbiosis are **(a)** commensalism, **(b)** mutualism, and **(c)** parasitism.

fish can invade the territory of larger aggressive species and feed there because the "owner" of the territory can never chase away the whole school. The most important function of schooling in small fish, however, is protection from predators.

It may seem illogical that schooling would be protective. For instance, schooling creates tighter groupings of organisms so that any predator lunging into a school would surely catch something, just as land predators run a herd of grazing animals until one weakens and becomes a meal. So, aren't the smaller fish making it easier for the predators by forming a large target? Scientists who study fish behavior suggest that schooling does indeed serve to protect a group of organisms based on strategies that give them "safety in numbers." In many parts of the marine environment, such as the open ocean where there is no place to hide, schooling has the following advantages:

1. When members of a species form schools, they reduce the percentage of ocean volume in which a cruising predator might find one of their kind.
2. When a predator encounters a large school, it is less likely that every fish in the school is consumed, compared to encounters with an individual or even a small school.
3. The school may appear as a single large and dangerous opponent to a potential predator and prevent some attacks.
4. Predators may find the continually changing position and direction of movement of fish within the school confusing, making attack particularly difficult for predators, which can attack only one fish at a time.

In addition, the fact that more than half of all fish species join schools during at least a portion of their lives suggests that schooling enhances survival of species, especially for those with no other means of defense. Schooling may also help fish swim greater distances than individuals because each schooling fish gets a boost from the vortex created by the fish swimming in front of it.

Recently, a new ocean predator has developed a method to take advantage of the schooling behavior of many species of fish. Human fishers have developed nets large enough to encircle whole schools of fish. These nets are very efficient at catching fish, thereby leading to the decline of many fish stocks (see Section 13.5 in Chapter 13).

## Symbiosis

Many marine organisms seek relationships with other organisms to help them survive. One such relationship is **symbiosis** (*sym* = together, *bios* = life), which occurs when two or more organisms associate in a way that benefits at least one of them. There are three main types of symbiotic relationships: commensalism, mutualism, and parasitism.

In **commensalism** (*commensal* = sharing a meal, *ism* = process), a smaller or less dominant participant benefits without harming its host, which affords subsistence or protection to the other. A remora, for example, attaches itself to a shark or another fish to obtain food and transportation, generally without harming its host (**Figure 14.17a**).

In **mutualism** (*mutuus* = borrowed, *ism* = process), both participants benefit. For example, the stinging tentacles of the sea anemone protect the clown fish (**Figure 14.17b**), and the clown fish, which is small but aggressive, chases away any fish that tries to feed on the anemone itself. In addition, the clown fish helps clean the anemone and may even supply scraps of food. Remarkably, clown fish are not stung by the anemone because clown fish have a protective agent in the mucus that coats their bodies.

In **parasitism** (*parasitos* = a person who eats at someone else's table, *ism* = process), one participant (the parasite) benefits at the expense of the other (the host). Many fish are hosts to isopods, which attach to the fish and derive their nutrition from the body fluids of the fish, thereby robbing the host of some of its energy supply (**Figure 14.17c**). Usually, the parasite does not rob enough energy to kill the host because if the host dies, so does the parasite.

Recently, symbiosis had been discovered to be an important component that drives evolution. For example, the sequencing of the genome of a species of diatom reveals that it apparently acquired new genes by engulfing microbial neighbors. The research suggests that early on in the evolution of diatoms, the most significant acquisition was an algal cell that provided the diatom with photosynthetic machinery. In fact, scientists have discovered that planktonic *mixotrophs* (*mixo* = mixing, *tropho* = nourishment), which eat other organisms like animals and photosynthesize like algae by stealing the photosynthetic organelles of their digested victims, possess an evolutionary advantage that may make them one of the most prevalent types of plankton in the ocean (see Creature Feature 13.1).

**RECAP** **Many pelagic species (especially fish) school, engage in symbiosis, or have other adaptations to increase their chances of survival by avoiding predators.**

## Other Adaptations

Marine animals exhibit a variety of behaviors that serve as defensive mechanisms to help them ward off predators—or to be more successful predators themselves. These include using speed, secreting poisons, and mimicking other poisonous or distasteful species. Others use transparency, camouflage, or countershading, as discussed in Chapter 12.

**CONCEPT CHECK 14.3 ▸ Specify adaptations that pelagic organisms possess to avoid becoming prey.**

1 What are several benefits of schooling?

2 What are the three types of symbiosis, and how do they differ?

3 Besides schooling and symbiosis, what other adaptations do pelagic animals possess to avoid becoming prey?

### STUDENTS SOMETIMES ASK . . .

***What are the world's largest and smallest fish?***

The world's largest fish is the whale shark (*Rhincodon typus*), which reaches lengths of up to 15 meters (50 feet) and can weigh up to 13.6 metric tons (30,000 pounds). Its mouth is an enormous 1.5 meters (5 feet) wide. It is a slow-moving, wide-ranging, filter-feeding animal that exists almost entirely on plankton. Alternatively, the smallest known fish is *Paedocypris progenetica*, a relative of carp and minnows. Its adult length is only 7.9 millimeters (0.31 inch), which is about as thick as a pencil. It lives exclusively in Indonesian swamps and, remarkably, was only discovered in 2005. There is, however, one other fish that is smaller: the male deep-sea anglerfish *Photocorynus spiniceps*, which has an adult length of only 6.2 millimeters (0.24 inch). It is generally not considered the world's smallest fish because it is not a self-sustaining organism. Instead, the parasitic males bite into larger females and fuse for life (see Figure 14.13g).

# 14.4 ▸ What Characteristics Do Marine Mammals Possess?

Marine mammals include some of the largest, best known, and most charismatic animals in the sea, such as seals, sea lions, manatees, porpoises, dolphins, and whales.

Although all marine mammals have an aquatic existence, their ancestors were land animals. For example, a series of striking fossil discoveries of ancient whales in Pakistan, India, and Egypt provide strong evidence that whales evolved from mammals on land about 50 million years ago. Some whale ancestors had small, unusable hind legs, suggesting that the land mammal predecessor had no need for its hind legs when it developed a large paddle-shaped structure for a tail used to swim through water. Other fossils show a remarkable progression of skeletal adaptations for an increasingly aquatic existence, such as the migration of the blowhole (nostrils) toward the top of the head, upper vertebrae that become increasingly fused, shrinking hip and ankle structures, and jawbone and ear components adapted for underwater hearing. Additional lines of evidence—including DNA analysis of modern whales and a host of anatomical similarities between land and marine mammals—confirm the evolution of whales from a hippopotamus-like land-dwelling ancestor.

The geologic record shows that life on land evolved from marine organisms millions of years ago. Why would a land mammal migrate back to the sea? One hypothesis suggests that they may have returned to the sea because of more abundant food sources. Another is that the extinction of many large marine predators that occurred at the same time as the demise of the dinosaurs allowed mammals to expand into a new environment: the sea. Interestingly, recent research suggests that the rise of diatoms as dominant marine primary producers at a time of global cooling on Earth associated with the most recent Ice Age were key factors that influenced the evolution of modern whales.

### STUDENTS SOMETIMES ASK . . .

***Why do whales move their tails up and down when they swim while sharks propel themselves forward by moving their rear fin side-to-side?***

The answer has to do with evolution. Whales move their tail flukes up and down because they evolved from land mammals about 50 million years ago. When four-legged land mammals run, their spine flexes up and down; whales retained this flexible spine, allowing them to "gallop" underwater with the familiar up-and-down motion of their tail flukes. On the other hand, because sharks are fish, they move their caudal (rear) fin back-and-forth as all fish do. Even when fish first ventured onto land, they kept this musculature and slithered in a side-to-side motion. In fact, many modern reptiles (such as lizards and snakes) still move in a side-to-side shimmy. But as ancient animals evolved, the bones and muscles of locomotion changed, and some animals developed movement up-and-down. Which propulsion method is more efficient? Apparently both are equally effective, otherwise evolutionary changes over time would cause one propulsion method to be favored over the other.

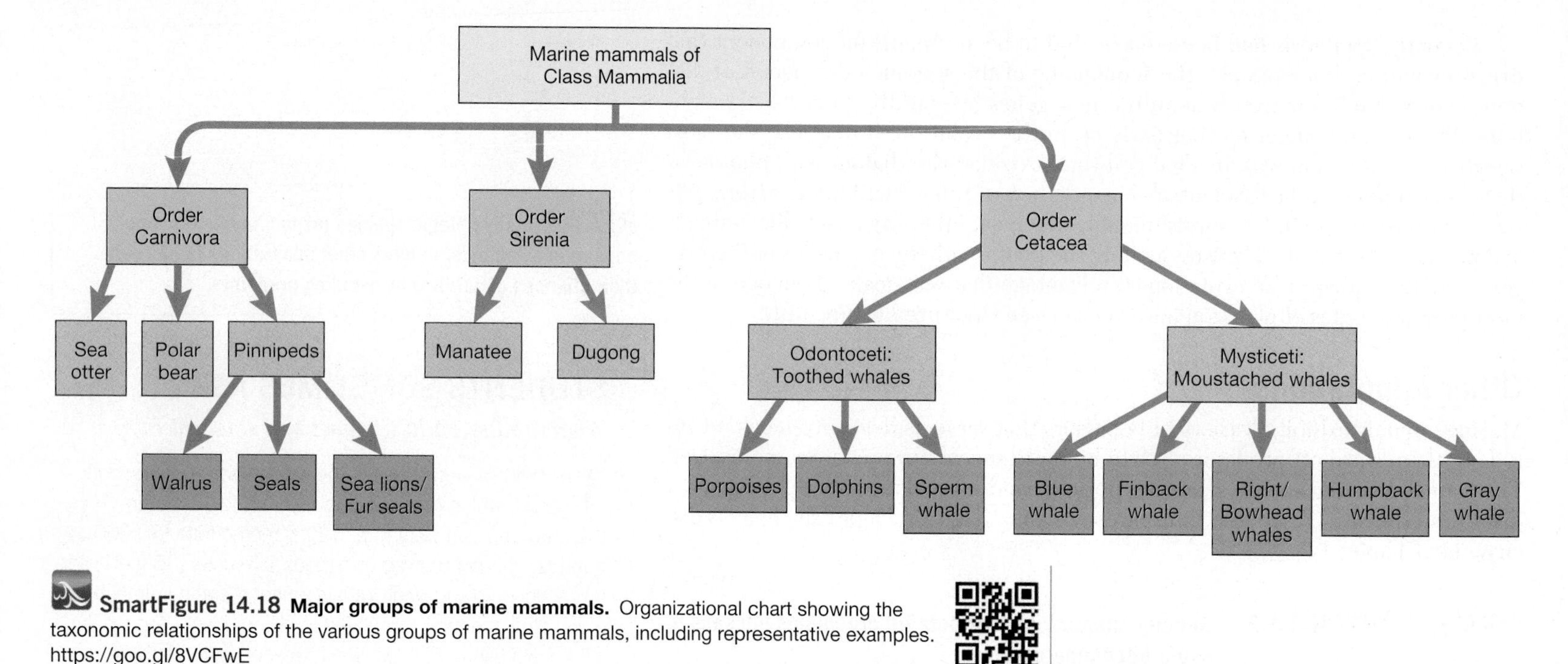


**SmartFigure 14.18 Major groups of marine mammals.** Organizational chart showing the taxonomic relationships of the various groups of marine mammals, including representative examples. https://goo.gl/8VCFwE

## STUDENTS SOMETIMES ASK . . .

***What is the fat content of the milk of marine mammals?***

Marine mammals produce milk with some of the highest fat content in the animal kingdom. For example, the milk of most dolphins contains about 14% fat (by comparison, whole cow milk contains 4% fat and human breastmilk contains 4.5% fat). Polar bears have milk with 31% fat and baleen whales have milk with 35–41% fat. Pinnipeds have even a higher percentage of milkfat. In fact, the gray seal (*Halichoerus grypus*) produces milk with 53% fat content and the world record for milkfat by animals is the hooded seal (*Cystophora cristata*), which produces milk containing 61% fat! The high fat content in milk helps baby marine animals quickly build up a layer of blubber that insulates them from cold water.

## Mammalian Characteristics

All organisms in class Mammalia (including marine mammals) share the following characteristics:

- They are warm-blooded.
- They breathe air.
- They have hair (or fur) during at least some stage of their development.
- They bear live young (this is true except for a few egg-laying monotreme mammals of Australia from the subclass Prototheria, which includes the duck-billed platypus and the spiny anteater [echidna]).
- The females of each species have mammary glands that produce milk for their young.

Marine mammals include at least 117 species within the orders Carnivora, Sirenia, and Cetacea. The major groups of marine mammals are shown in **Figure 14.18** and described below.

## Order Carnivora

All animals within order **Carnivora** (*carni* = meat, *vora* = eat)—such as the familiar cat and dog families on land—have prominent canine teeth. Marine representatives of order Carnivora include sea otters, polar bears, and the **pinnipeds** (*pinna* = feather, *ped* = foot), which include walruses, seals, sea lions, and fur seals. The name *pinniped* describes these organisms' prominent skin-covered flippers, which are well adapted for propelling them through water.

**Sea otters** (**Figure 14.19a**) inhabit kelp beds in coastal waters of the eastern North Pacific Ocean. They are some of the smallest marine mammals, with adults reaching lengths up to 1.2 meters (4 feet). Sea otters lack an insulating layer of blubber but have extremely dense fur, which is extremely luxurious and was highly sought for pelts; as a result, they were hunted to the brink of extinction in the late 1800s. Fortunately, they have made a remarkable comeback and now inhabit most areas where they were formerly hunted. Sea otters seem particularly playful because of their habit of continually scratching themselves, which serves to clean their fur and adds an insulating layer of air. Because they lack the insulative benefit of blubber, they have high caloric requirements and are voracious eaters.

Sea otters eat more than 50 kinds of marine life, including sea urchins, crabs, lobsters, sea stars, abalone, clams, mussels, octopuses and fish. They are one of the few types of animals known to use tools. During dives for prey items, they keep a tool—usually a rock—tucked under one arm while they use their dexterous hands to obtain food. When they return to the surface, they use the tool to break open the shells of their food while floating on their backs.

**Polar bears** (**Figure 14.19b**) are a type of marine mammal with massive webbed paws that make them excellent swimmers. The polar bear's fur is thick, and each hair is hollow to trap air for better insulation; the hairs also function like fiber-optic cables by channeling sunlight to the animal's dark/black skin that absorbs heat from sunlight. Polar bears also have large teeth and sharp claws, which they use for prying and killing. Their diet consists mainly of seals, which they often capture at holes in the Arctic ice when the seals come up for a breath of air. The topic of shrinking Arctic sea ice and its effect on polar bear populations is discussed in Chapter 16, "The Oceans and Climate Change."

Climate Connection

**Walruses** have large bodies, and adults—both male and female—have ivory tusks up to 1 meter (3 feet) long (**Figure 14.19c**). Their tusks are used for territorial fighting, for hauling themselves onto icebergs, and sometimes for stabbing their prey.

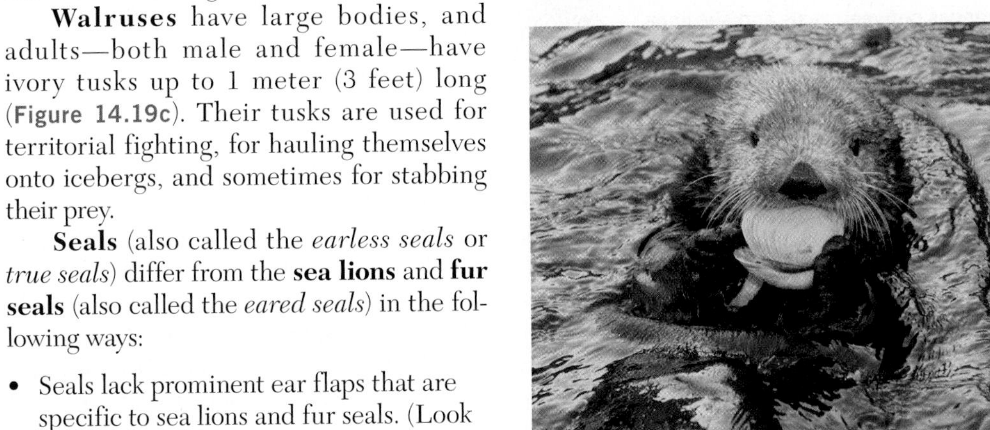

**Seals** (also called the *earless seals* or *true seals*) differ from the **sea lions** and **fur seals** (also called the *eared seals*) in the following ways:

- Seals lack prominent ear flaps that are specific to sea lions and fur seals. (Look closely and compare **Figures 14.19d** and **14.19e**.)
- Seals have smaller and less-prominent front flippers (called *fore flippers*) than sea lions and fur seals.
- Seals have prominent claws that extend from their fore flippers that sea lions and fur seals lack (**Figure 14.20**).
- Seals have a different hip structure than sea lions and fur seals. As a result, seals cannot move their rear flippers underneath their bodies in the way that sea lions and fur seals do (Figure 14.20).
- Seals, with their smaller front flippers and different hip structure, do not move around on land very well and can only slither along like caterpillars. Sea lions and fur seals, on the other hand, use their large front flippers and their rear flippers, which they can turn under their bodies, to walk easily on land and can even ascend steep slopes, climb stairs, and do other acrobatic tricks.
- Seals propel themselves through the water by using a back-and-forth motion of their rear flippers (similar to a wagging tail), whereas sea lions and fur seals flap their large front flippers.

**(a)** Sea otter (*Enhydra lutris*).

**(b)** Polar bear (*Ursus maritimus*).

**(c)** Walrus (*Odobenus rosmarus*).

**(d)** Harbor seals (*Phoca vitulina*).

**(e)** California sea lions (*Zalophus californianus*).

Figure 14.19 **Marine mammals of order Carnivora.** Photos of various types of marine mammals of order Carnivora, including **(a)** sea otter and **(b)** polar bear. Pinnipeds include **(c)** walrus, **(d)** harbor seals, and **(e)** California sea lions.

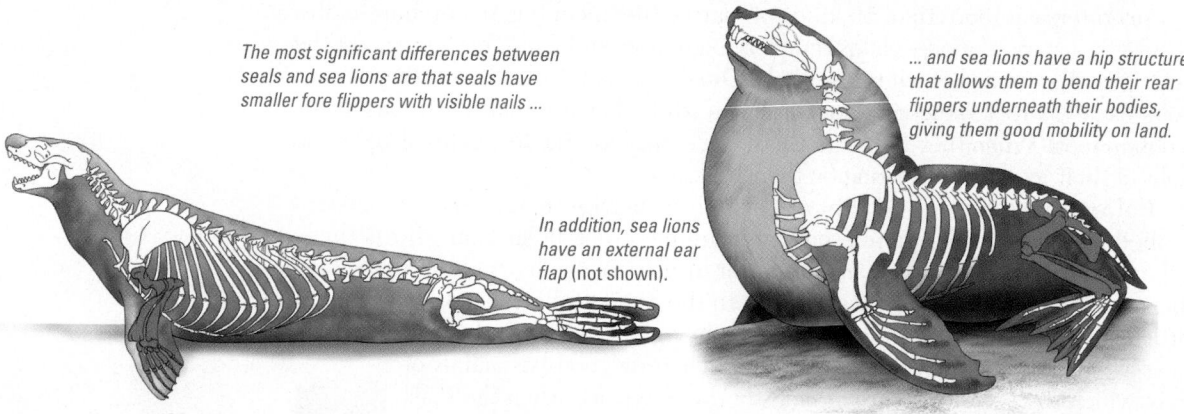

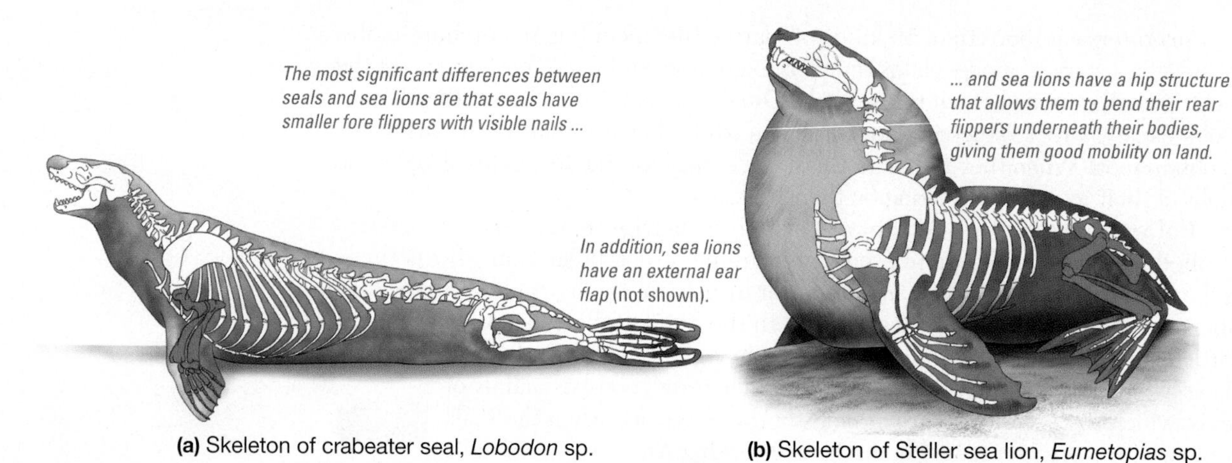

The most significant differences between seals and sea lions are that seals have smaller fore flippers with visible nails ...
... and sea lions have a hip structure that allows them to bend their rear flippers underneath their bodies, giving them good mobility on land.
In addition, sea lions have an external ear flap (not shown).


**EXPLORING DATA** ▶ Describe the skeletal and morphological differences between seals and sea lions.

**(a)** Skeleton of crabeater seal, *Lobodon* sp.

**(b)** Skeleton of Steller sea lion, *Eumetopias* sp.

SmartFigure 14.20 **Skeletal and morphological differences between seals and sea lions.** https://goo.gl/87e3M9

## Order Sirenia

Animals of order **Sirenia** (*siren* = a mythical mermaid-like creature with an enticing voice) include the *manatees* and *dugongs*, collectively known as "sea cows." (Note that this group also includes the cold-water Steller's sea cow [*Hydrodamalis gigas*], which was driven to extinction in 1768 by early whalers only 27 years after its discovery.) Manatees are concentrated in coastal areas of the tropical Atlantic Ocean, while dugongs populate the tropical regions of the Indian and Western Pacific Oceans.

Both manatees and dugongs have a paddle-like tail and rounded front flippers (Figure 14.21). Their bodies are covered with sparse hairs, which are concentrated around the mouth. They are large animals that can reach lengths of up to 4.3 meters (14 feet) and weigh more than 1360 kilograms (3000 pounds). The land-dwelling ancestors of sirenians were elephant-like; in fact, the front flippers of manatees have prominent nails that bear a striking resemblance to the nails on elephant feet.

Sirenians eat only shallow-water coastal grasses and are thus the only vegetarian marine mammals. They spend most of their lives in coastal waters that are heavily used by humans for commerce, recreation, development, and waste disposal, and so a major concern for sirenian survival is habitat destruction. A scientific study of biologically sensitive seagrass habitat—which is vital for sirenians—indicates that it is being destroyed so rapidly that it is one of the most threatened ecosystems on Earth. The rate of seagrass loss, in fact, is similar to that of other endangered ecosystems, such as mangroves, coral reefs, and rainforests.

Figure 14.21 **Marine mammals of order Sirenia.** Photos of representative marine mammals of order Sirenia, including **(a)** West Indian manatees and **(b)** Indian Ocean dugong.

**(a)** West Indian manatees, which have a rounded tail fin and nails on their front flippers.

**(b)** Indian Ocean dugong, which has a fluked tail fin similar to that of a whale and has no visible nails on its front flippers.

In addition, there have been many accidents with motorboats that run over these slow-moving animals. In 2017, for example, the Florida Fish and Wildlife Conservation Commission reported the deaths of 538 Florida manatees, of which 20% were attributed to boat collisions and another 34% were either undetermined or unrecovered. Scientists have recently developed a forward-looking sonar system called a "manatee finder" to help boaters locate and avoid the animals. The continuing decline in populations of manatees and dugongs, however, has led to their being classified as endangered species.

## Order Cetacea

The order **Cetacea** (*cetus* = whale) includes the whales, dolphins, and porpoises (Figure 14.22). The cetacean body is more or less cigar-shaped and insulated with a thick layer of blubber. Cetacean forelimbs are modified into flippers that move only at the "shoulder" joint. The hind limbs are vestigial (rudimentary), not attached to the rest of the skeleton, and are usually not visible externally. All cetaceans share the following characteristics:

- An elongated (telescoped) skull
- Blowholes on top of the skull
- Very few hairs
- A horizontal tail fin called a *fluke* (*flok* = to be flat) that is used for propulsion by vertical movements

These characteristics make cetaceans' bodies very streamlined, allowing them to be excellent swimmers.

**MODIFICATIONS TO INCREASE SWIMMING SPEED** Cetaceans' muscles are not a great deal more powerful than those of other mammals, so their ability to swim fast must result from modifications that reduce frictional drag. The muscles of a small dolphin, for example, would need to be five times stronger than they are to swim at 40 kilometers (25 miles) per hour in turbulent flow.

In addition to a streamlined body, cetaceans improve the flow of water around their bodies with a specialized skin structure. Their skin consists of a soft outer layer that is 80% water and has narrow canals filled with spongy material, and a stiffer inner layer composed mostly of tough connective tissue. The soft layer decreases the pressure differences at the skin–water interface by compressing when pressure is high and expanding when pressure is low, reducing turbulence and drag.

**MODIFICATIONS TO ALLOW DEEP DIVING** Humans can free-dive to a maximum depth of 130 meters (428 feet) and hold their breath in rare instances for up to six minutes. In contrast, sperm whales (*Physeter macrocephalus*) dive deeper than 2800 meters (9200 feet), and northern bottlenose whales (*Hyperoodon ampullatus*) can stay submerged for up to two hours. These remarkable feats require unique adaptations, such as special structures that allow them to use oxygen efficiently, muscular adaptations, and an ability to resist nitrogen narcosis, all of which are discussed below.

***Oxygen Usage*** Figure 14.23 shows the internal structures that allow cetaceans to remain submerged for extended periods. Inhaled air finds its way to tiny terminal chambers, the *alveoli* (*alveus* = small hollow). Alveoli are lined with a thin alveolar membrane that is in contact with a dense bed of capillaries. The exchange of gases between the inhaled air and the blood (oxygen in, carbon dioxide out) occurs across the alveolar membrane. Some cetaceans have an exceptionally large concentration of capillaries surrounding the alveoli (Figure 14.23b), which have muscles that move air against the membrane by repeatedly contracting and expanding.

Cetaceans take from one to three breaths per minute while resting, compared with about 15 in humans. Because they hold the inhaled breath much longer, and because of the large capillary mass in contact with the alveolar membrane and the

CREATURE FEATURE 14.1

### I'm an acrobat!

The **humpback whale** (*Megaptera novaeangliae*) is one of the most acrobatic whales, known for breaching and slapping the water with its fins. Once hunted to the brink of extinction, the population of humpback whales decreased by an estimated 90% before a whaling moratorium was established in 1966.

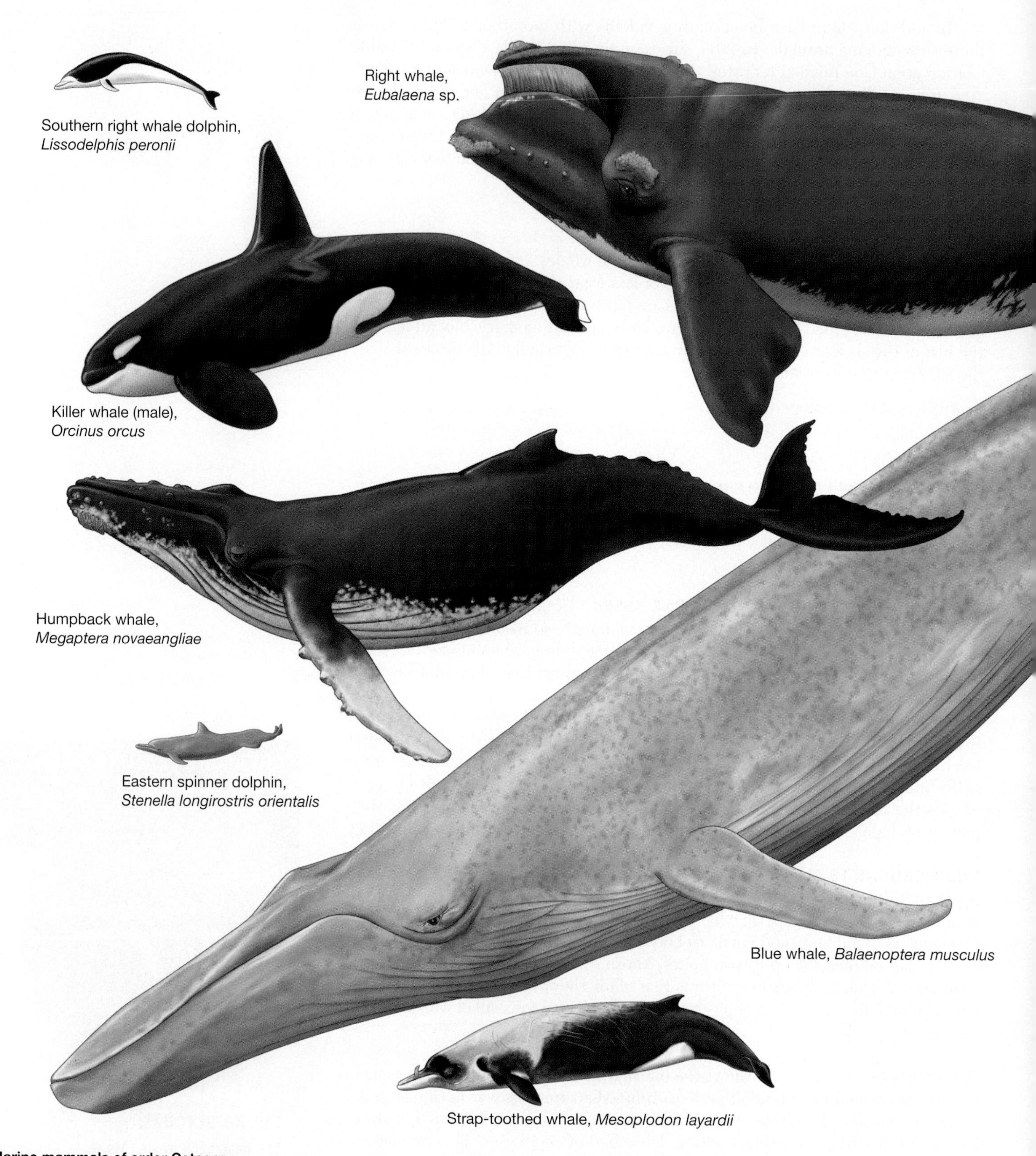


**Figure 14.22 Marine mammals of order Cetacea.** A composite drawing of representatives of the two cetacean suborders, Odontoceti (toothed whales) and Mysticeti (baleen whales). All whales and dolphins are drawn to relative scale; also note the human diver in the upper right.

circulation of the air by muscular action, cetaceans can extract almost 90% of the oxygen in each breath, whereas terrestrial mammals extract only 4–20%.

To use oxygen efficiently during long dives, cetaceans store it and limit its use. The storage of so much oxygen is possible because these prolonged divers have such a large blood volume per unit of body mass.

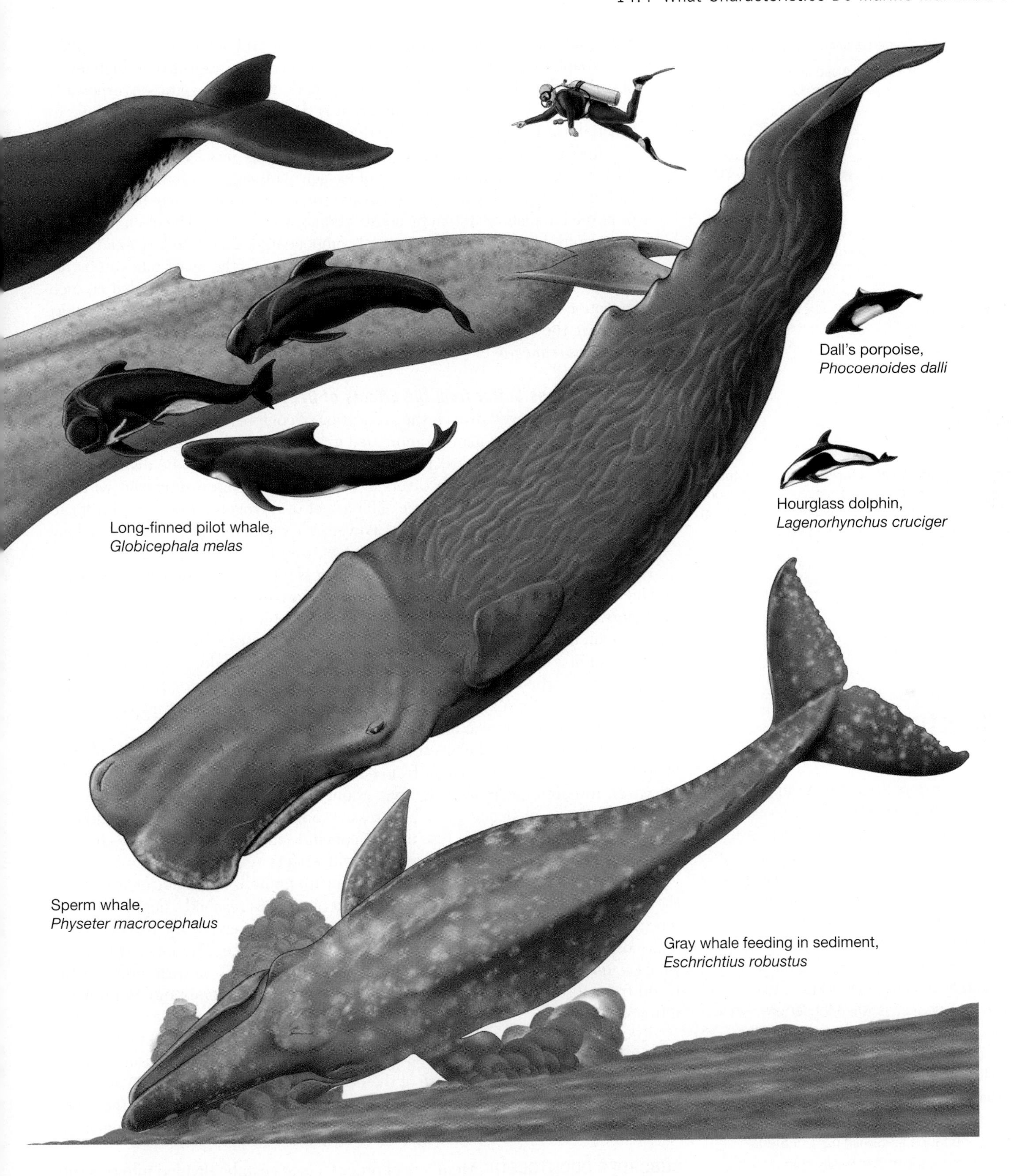


Some cetaceans have twice as many red blood cells per unit of blood volume and up to nine times as much myoglobin in their muscle tissue as terrestrial animals. As a result, large supplies of oxygen can be stored chemically in **hemoglobin** (*hemo* = blood, *globus* = sphere) within red blood cells and in myoglobin within muscles.

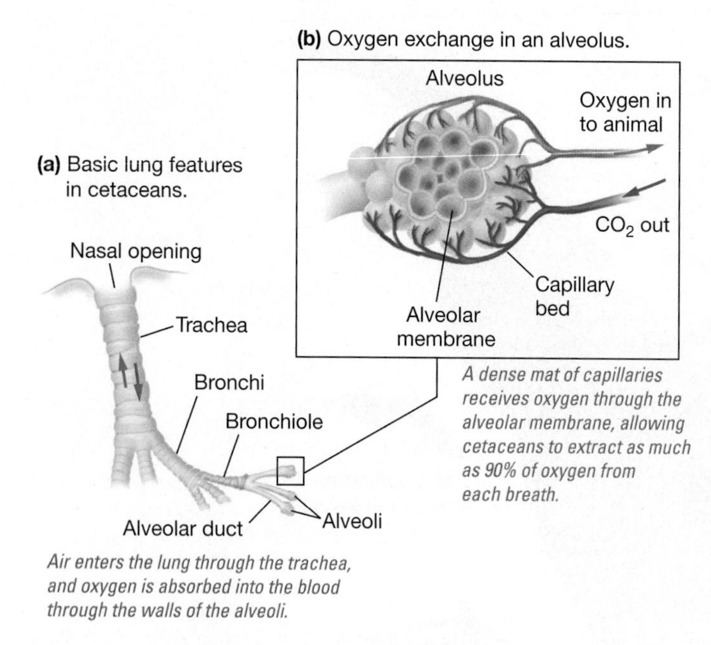
(b) Oxygen exchange in an alveolus.
Alveolus
Oxygen in to animal
$CO_2$ out
Capillary bed
Alveolar membrane
A dense mat of capillaries receives oxygen through the alveolar membrane, allowing cetaceans to extract as much as 90% of oxygen from each breath.
(a) Basic lung features in cetaceans.
Nasal opening
Trachea
Bronchi
Bronchiole
Alveolar duct
Alveoli
Air enters the lung through the trachea, and oxygen is absorbed into the blood through the walls of the alveoli.

Figure 14.23 **Internal adaptations of cetaceans that facilitate prolonged submergence.** Cetaceans possess internal adaptations that enable them to stay underwater for extended periods, including **(a)** basic lung features and **(b)** oxygen exchange in the alveolus.

## STUDENTS SOMETIMES ASK . . .

### *How do whales, you know, DO IT?!*

Whale sex is truly an incredible achievement. Imagine trying to copulate as whales do when you have nothing with which to grasp a mate and, in the ocean, you have nothing to push against for support! Copulation in large whales has been infrequently studied and witnessed in only a remarkably few cases. In fact, the world's largest whale, the blue whale, has never been observed copulating.

From coastal-dwelling whale species such as the gray whale, here's what is known: Whales are one of the few animals besides humans and bonobo chimpanzees that copulate belly-to-belly. Mating begins when the female is in a forward position horizontally near the water surface. She turns toward the male, who emerges from below and enters her as they spiral on their long axis through the water, which helps push them together. As with many other mammals, mating takes only a few seconds. Both a whale's penis and its testes are internal to enhance streamlining, but the tapered fibro-elastic tip of the penis is especially suited for the tricky act of copulation with no limbs. Once it is extruded outside the body for mating, it is dexterous enough to maneuver itself into a female's genital slit to deposit sperm.

***Muscular Adaptations*** Cetaceans' muscles are well adapted for deep dives. One adaptation is that their muscle tissue is relatively insensitive to high levels of carbon dioxide, which builds up in the body through respiration, especially during deep dives. Another adaptation is that their muscles can continue to function through anaerobic respiration when oxygen becomes depleted.

Research has shown that cetaceans' swimming muscles can still function during a dive, even in the absence of oxygen. This suggests that these muscles and other organs, such as the digestive tract and kidneys, may be sealed off from the circulatory system by constriction of key arteries. The circulatory system would then service only essential components, such as the heart and brain. Because of the decreased circulatory requirements, the heart rate can be reduced by 20–50% of normal. Other research has shown, however, that no such reduction in heart rate occurs during dives by the common dolphin (*Delphinus delphis*), the white whale (*Delphinapterus leucas*), or the bottlenose dolphin (*Tursiops truncatus*).

***Do Cetaceans Suffer from the Effects of Deep Diving?*** One difficulty with deep and prolonged dives is the absorption of compressed gases into the blood. When humans dive using compressed air—which includes nitrogen and oxygen—the higher pressure at depth causes more nitrogen to be dissolved in a diver's body and can cause divers to experience **nitrogen narcosis**, which is also called *rapture of the deep*. The effect of nitrogen narcosis is similar to drunkenness and can occur when a diver either goes too deep or stays too long at depths greater than 30 meters (100 feet) (see Diving Deeper 12.1).

Another difficulty can occur when divers return to the surface too rapidly, in which case they may experience **decompression sickness**, which is also called *the bends*. During a rapid ascent, the lungs cannot remove excess gases from the bloodstream fast enough, and the reduced pressure causes small bubbles of nitrogen to form in a diver's blood and tissue. This process is analogous to the bubbles that form in a carbonated beverage when the container is opened. The bubbles interfere with blood circulation and can cause bone damage, excruciating pain, severe physical debilitation, or even death.

Until recently, it was believed that cetaceans and other marine mammals had adaptations that prevented them from suffering from the effects of deep diving. However, research involving a rigorous examination of sperm whale skeletons reveals that sperm whales acquire progressive bone damage caused by recurring decompression sickness. The researchers conclude that sperm whales are neither anatomically nor physiologically immune to the effects of deep diving.

Still, the debilitating effects of deep diving are minimized in cetaceans because they have collapsible rib cages. By the time a cetacean has reached a depth of 70 meters (230 feet), its rib cage has collapsed under the 8 kilograms per square centimeter (8 atmospheres, or 118 pounds per square inch) of pressure. The lungs within the rib cage also collapse, removing all air from the alveoli. This, in turn, prevents the blood from absorbing additional gases across the alveolar membrane, minimizing nitrogen narcosis.

In addition, cetaceans may be naturally resilient to the buildup of nitrogen gas in their bodies. In a study where enough nitrogen was put into the tissue of a dolphin to give a human a severe case of the bends, for instance, the dolphin appeared to suffer no ill effects. This suggests that dolphins (as well as other marine mammals) may have simply evolved an insensitivity to the buildup of excess nitrogen gas.

**SUBORDER ODONTOCETI** Members of order Cetacea can be divided into two suborders: Odontoceti (the *toothed whales*) and Mysticeti (the *baleen whales*). Suborder **Odontoceti** (*odonto* = tooth, *cetus* = whale) includes the dolphins, porpoises, killer whales, and sperm whales. (Interestingly, genetic analyses suggest that sperm whales are more closely related to baleen whales than to other toothed whales, another example of how genetic analysis is allowing scientists to rethink evolutionary relationships.)

***Characteristics of Toothed Whales*** All toothed whales have prominent teeth that are used to hold and position fish and squid to facilitate swallowing them whole. Killer whales, however, are known to feed on a variety of larger animals, including other whales. The toothed whales form complex and long-lived social groups. A toothed whale has one external nasal opening (blowhole), while a baleen whale has two. Although both toothed and baleen whales can emit and receive sounds, the ability to use sound is best developed in toothed whales (particularly sperm whales, which are the most vocal cetaceans).

***Differences between Dolphins and Porpoises*** Dolphins and porpoises are small toothed whales of suborder Odontoceti. They have similarities in appearance, behavior, and range and so are easily confused. For instance, both dolphins and porpoises (as well as seals, sea lions, and fur seals) can exhibit a behavior known as "porpoising," which is leaping out of the water while swimming. However, there are several morphological differences between dolphins and porpoises.

Porpoises are somewhat smaller and have a more stout (bulky and robust) body shape than dolphins, which are more elongated and streamlined. Generally, porpoises have a blunt snout (rostrum), while dolphins have a longer rostrum. Porpoises have a smaller and more triangular (or, on one species, no) dorsal fin, whereas a dolphin's dorsal fin is sickle-shaped, or **falcate** (*falcatus* = sickle), and appears hooked and curved backward in profile view.

Dolphins and porpoises also have differences in the shape of their teeth, although it is often difficult to get close enough to see them. The teeth of dolphins end in points, while the teeth of porpoises are blunt or flat (shovel shaped) and resemble our incisors (front teeth). Killer whales have teeth that end in points (**Figure 14.24**), confirming that they are indeed members of the dolphin family.

***Echolocation and Hearing in Toothed Whales*** All marine mammals have good vision, but ocean conditions often limit its effectiveness. In coastal waters (where suspended sediment and dense plankton blooms make the water turbid) and in deeper waters (where light is limited or absent), the use of sound—which travels well in water—has many advantages over sight.

Despite their lack of vocal cords, toothed whales can produce a variety of sounds—some of which are within the range of human hearing. Speculations about the sounds' purpose range from **echolocation**—using sound to determine the direction and distance of objects— (clearly true) to a highly developed language (doubtful). In fact, what marine biologists know about cetaceans' use of sound is limited.

To locate an object and determine its distance, toothed whales send sound signals through the water, some of which are reflected from various objects and are returned to the animal and interpreted (**Figure 14.25**). Because sound penetrates objects, echolocation can produce a three-dimensional image of the object's internal structure and density (which is more than eyesight alone can do). Scientific studies suggest that some toothed whales may also use a sharp burst of sound to stun their prey before they close in for the kill.

Sound generation by toothed whales is a remarkably complex process, and is different in different species. In sperm whales (**Figure 14.26a**), sounds are generated when the whale forces air through its right nasal passage to a special structure called the *museau du singe* ("monkey's muzzle"), which resembles a pair of giant lips that snaps shut to produce a percussive sound. This resulting *click!* travels through the **spermaceti** (*sperm* = seed, *cetus* = whale) **organ** to the top part of the skull, which is shaped like a large, curving radar dish. (Note that because the spermaceti organ is composed of a white, waxy substance, early whalers noticed its similarity to human sperm, hence the name. The spermaceti organ, however, has

## STUDENTS SOMETIMES ASK . . .

***I've heard of the problem with military sonar causing whale strandings. How does sonar harm cetaceans?***

Some (but not all) scientific evidence links mass strandings of cetaceans—predominately beaked whales, a group of deep-diving whales that have been very poorly studied—with the deployment of mid-frequency sonar used for detecting submarines during military exercises. For example, researchers have documented gas-bubble lesions in stranded cetaceans consistent with rapid decompression that is normally associated with decompression sickness. It is hypothesized that mid-frequency sonar may be responsible for the formation of these bubbles by causing the cetaceans to ascend to the surface too rapidly or by frightening them into undertaking dangerously deep dives; it is also possible that the sonar itself has physical effects on cetaceans' nitrogen-saturated tissues. In fact, scientific studies have shown that beaked whales stop their echolocations and flee the area when sonar is in use, suggesting that these whales may be more sensitive than other species to sound. Other strandings, however, have occurred during times when military sonar is not in use, but stranded animals are still found to have similar bubbles in their tissues. So, at least some of the strandings appear to be unrelated to military sonar; instead, they may be just an unfortunate coincidence. Clearly, more research is needed in this important area.

Documented whale strandings have occurred since as far back as the time of Aristotle, which implies that many strandings are a natural phenomenon. For example, strandings due to pneumonia and trauma after storms are common reasons. Other reasons for strandings are obvious, such as shark attacks or even assaults by members of the same species. In other cases, both human-derived pollutants and natural toxins, such as biotoxins from algae (see Section 13.2 "What Kinds of Photosynthetic Marine Organisms Exist?" in Chapter 13), are implicated in mass strandings. Parasites and pathologies—including bacterial and viral infections—are sometimes to blame, too. Even using diagnostic techniques common in advanced human medicine, such as CT and MRI scanning and molecular studies, there are still many instances of whale strandings with no single, clear answer. For example, of the 55 cetacean strandings in the United States since 1991, scientists have classified 29 as "undetermined."

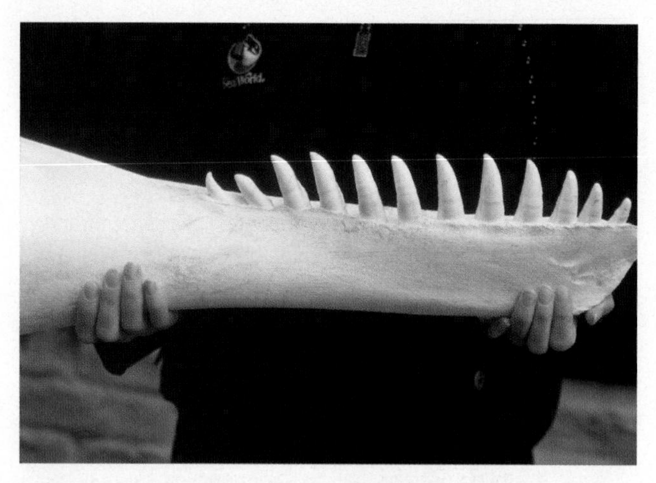

Figure 14.24 **Jawbone of a killer whale.** The lower jawbone of a killer whale (*Orcinus orca*), showing large teeth that end in points. Thus, killer whales are members of the dolphin family.

nothing to do with reproduction but instead is used for echolocation.) From there, the reflected sound is focused forward through another organ called *the junk* and amplified out into the watery world. Research suggests that sperm whales are able to manipulate the shape of both the spermaceti organ and the junk, allowing them to focus the sound and perhaps even to aim their clicks.

In small toothed whales such as dolphins and porpoises (**Figure 14.26b**), sounds are emitted from phonic lips near the blowhole. Contractions of muscles produce a wide variety of complex sounds such as clicks, buzzes, and whistles. These sounds are concentrated as they pass through an organ that sits atop the skull called the **melon**, which can be manipulated by the animal and acts as an acoustical lens to focus the sounds before leaving the body.

Using lower-frequency clicks at great distance and higher frequency at closer range, the bottlenose dolphin (*Tursiops truncatus*) can detect a school of fish at distances exceeding 100 meters (330 feet). It can pick out an individual fish 13.5 centimeters (5.3 inches) long at a distance of 9 meters (30 feet). Sperm whales can detect their main prey—squid—from distances of up to 400 meters (1300 feet).

How do toothed whales hear underwater? Toothed whales have specialized fats associated with their jaws that efficiently convey sound waves from the ocean to their ears. These structures insulate the inner ear housing from the rest of the skull, thereby allowing them to differentiate the sounds they hear underwater. In many toothed whales, sounds are picked up by the thin, flaring jawbone and passed to the inner ear via the connecting, fat-filled body. (To simulate this, try pushing the end of a vibrating tuning fork into your chin; the sound is transmitted through your jaw directly to your ear.) The signals are then sent to the brain, where the sounds are interpreted (Figure 14.26).

There is growing concern in the scientific community that noise pollution in the ocean is affecting cetaceans. This increased noise comes from the greater number of ships plying the world's oceans as well as the larger size, higher speeds, and enhanced propulsion power of individual ships. For example, scientific studies show that the world's commercial marine trade fleet has contributed to a doubling of low-frequency ship noise every decade for the past 60 years in some of the most intensely used parts of the ocean. The impact of this increased underwater noise on cetacean hearing, behavior, and communication is unknown.

***How Intelligent Are Toothed Whales?*** The question of cetacean intelligence is a topic of much debate. Although there may not be a definitive answer, the following facts about toothed whales imply a certain level of intelligence:

- They can communicate with each other by using sound.
- They have large brains relative to their body size; in fact, sperm whales have the largest brain of any animal on the planet—it weighs up to 9 kilograms (20 pounds), which is over six times the weight of a typical human brain.
- Their brains are highly convoluted—a characteristic shared by many organisms that are considered to have highly developed intelligence (such as humans and other primates).
- Some wild dolphins have been reported to assist drowning humans in the ocean.
- Some dolphins have been trained to respond to hand signals and do tricks on command (such as retrieve objects).

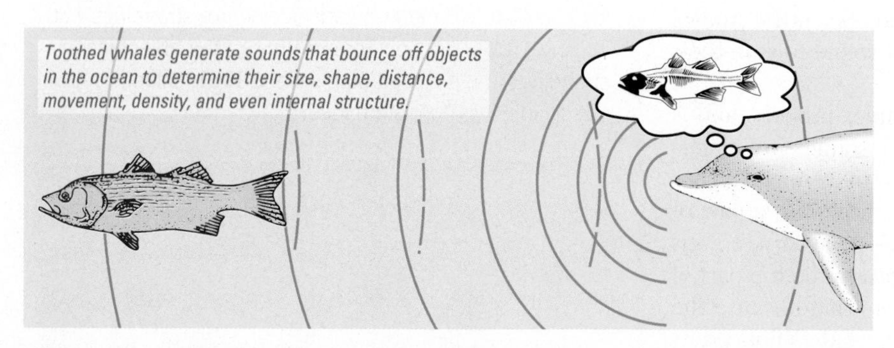

Toothed whales generate sounds that bounce off objects in the ocean to determine their size, shape, distance, movement, density, and even internal structure.

Figure 14.25 **How echolocation works in toothed whales.**

Although toothed whales have remarkable abilities, this does not necessarily imply high intelligence. Pigeons, for instance, which are not known for being highly intelligent, have also been trained to retrieve objects by using hand signals. Even crows have demonstrated the problem-solving ability to craft tools to pick locks. Perhaps many of us would like to

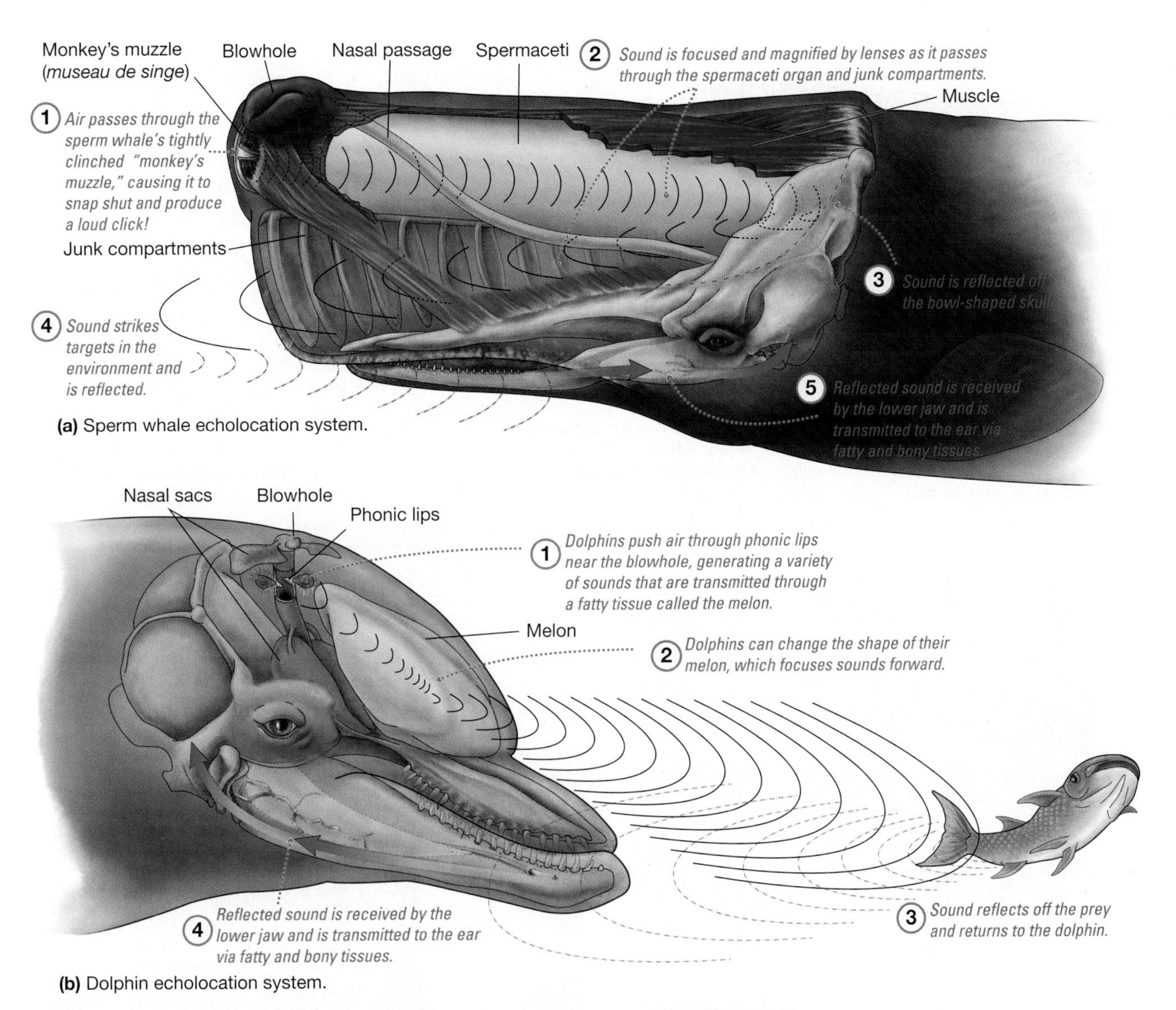

Monkey's muzzle (museau de singe)
Blowhole
Nasal passage
Spermaceti
2 Sound is focused and magnified by lenses as it passes through the spermaceti organ and junk compartments.
Muscle
1 Air passes through the sperm whale's tightly clinched "monkey's muzzle," causing it to snap shut and produce a loud click!
Junk compartments
3 Sound is reflected off the bowl-shaped skull
4 Sound strikes targets in the environment and is reflected.
5 Reflected sound is received by the lower jaw and is transmitted to the ear via fatty and bony tissues.
Nasal sacs
Blowhole
Phonic lips
1 Dolphins push air through phonic lips near the blowhole, generating a variety of sounds that are transmitted through a fatty tissue called the melon.
Melon
2 Dolphins can change the shape of their melon, which focuses sounds forward.
4 Reflected sound is received by the lower jaw and is transmitted to the ear via fatty and bony tissues.
3 Sound reflects off the prey and returns to the dolphin.

**(a)** Sperm whale echolocation system.

**(b)** Dolphin echolocation system.

**SmartFigure 14.26 Cutaway views comparing the echolocation systems of a sperm whale and a dolphin.** https://goo.gl/0GhVHb

**EXPLORING DATA ▲ Describe the differences in the echolocation systems between sperm whales and dolphins.**

think that whales and dolphins are more intelligent than they really are because humans feel an attachment to these charismatic, seemingly ever-smiling, air-breathing creatures. It is interesting to note that even experts in the field of animal intelligence disagree on how to assess *human* intelligence accurately, let alone that of a marine mammal. In fact, experts on dolphin behavior suggest that public enthusiasm for dolphin smarts has outpaced the evidence for dolphin intelligence.

If the large brain that toothed whales possess is not an indication of intelligence, then why is their brain so large? Leading whale researchers don't exactly know, but it might be because toothed whales need a large brain to process the wealth of information they receive from the sound echoes they transmit. Because intelligence is difficult to measure, perhaps it is best to say that animals of suborder Odontoceti are tremendously well adapted to the marine environment.

**SUBORDER MYSTICETI** Suborder **Mysticeti** (*mystic* = moustache, *cetus* = whale)—also known as the *baleen whales*—includes the world's largest

## STUDENTS SOMETIMES ASK . . .

### *In a battle between a killer whale and a great white shark, which one would win?*

Although many people who are fascinated with large and powerful wild animals have often wondered which of the two would win such a fight, there was little evidence to settle the dispute until recently. A remarkable video was taken in waters off northern California in 1997, documenting a battle between a 6-meter (20-foot) juvenile killer whale (*Orcinus orca*) and a 3.6-meter (12-foot) adult great white shark (*Carcharodon carcharias*). The video clearly shows the killer whale biting and completely severing the shark's head! If this is representative of the way these two animals interact in the wild, then the killer whale is the top carnivore in the ocean. It is believed that the killer whale's superior maneuverability and use of echolocation helped it defeat the shark.

**Figure 14.27 Whale baleen.** Diagrammatic drawings of **(a)** a cutaway view through the head of a typical baleen whale and **(b)** how a baleen whale feeds. Photos show **(c)** a rack of gray whale baleen and **(d)** an individual piece of right whale baleen.

**EXPLORING DATA ▶ Describe how baleen whales use their baleen to feed.**

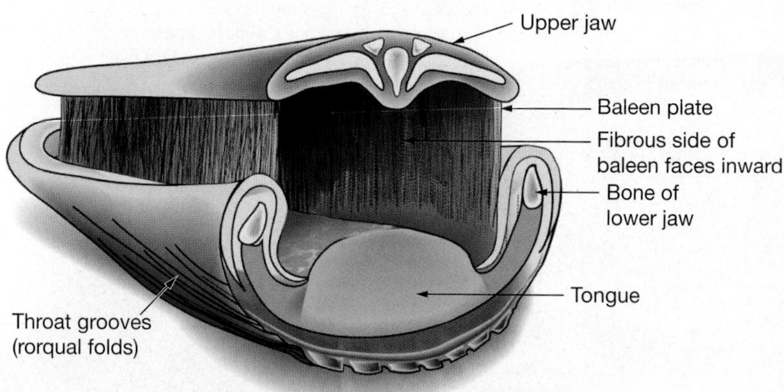


**(a)** Cross-sectional view through the head of a baleen whale, showing the baleen plates hanging from the upper jaw, which form a sieve that allows these whales to concentrate and eat large quantities of smaller organisms.

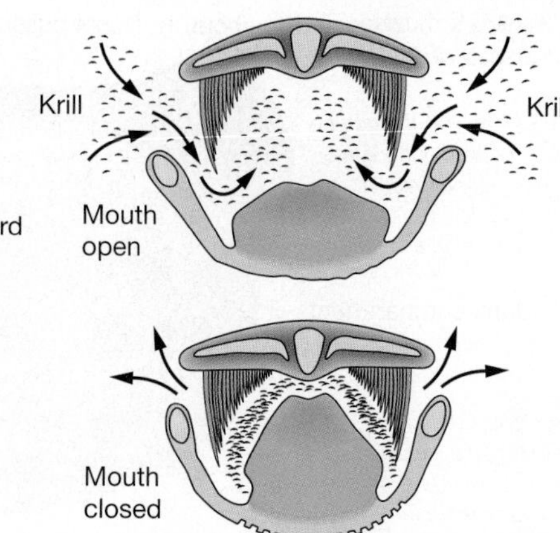


**(b)** A baleen whale takes krill and other small prey into its mouth (*above*) and then closes its mouth to force the water through its baleen (*below*), capturing its prey.

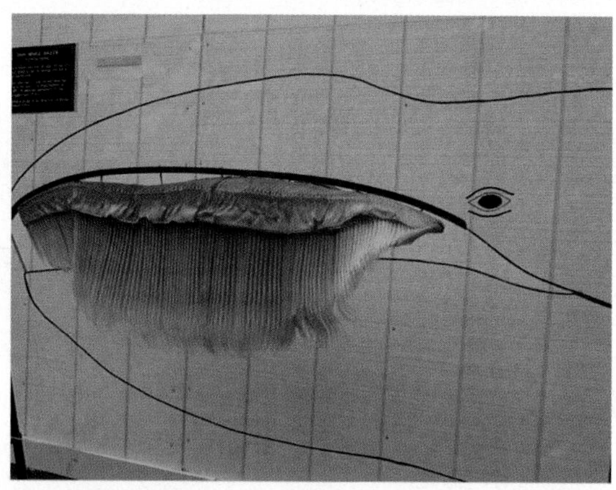

**(c)** A rack of baleen from a bottom-feeding gray whale (*Eschrichtius robustus*).

**(d)** Individual slat of baleen from a surface-skimming North Atlantic right whale (*Eubalaena glacialis*).

**Animation**
Feeding in Baleen Whales
https://goo.gl/cnolYJ

whales (the blue whale, finback whale, and humpback whale) and the gray whale (a bottom feeder).

Baleen whales are generally much larger than toothed whales because of differences in food sources. Baleen whales eat lower on the food web (including zooplankton, such as krill and small nektonic organisms), which are relatively abundant in the marine environment. How are the largest whales in the world able to survive on eating such small prey, especially when these smaller organisms are widely dispersed in the marine environment?

***Use of Baleen*** To concentrate small prey items and separate them from seawater, baleen whales have parallel rows of **baleen** (*balaena* = whale) plates in their mouths instead of teeth (**Figure 14.27a**). These baleen plates hang from the

**Figure 14.28 Feeding by rorqual whales.** Rorqual whales such as this Bryde's whale (*Balaenoptera edeni*) off Baja California, Mexico, feed by engulfing an entire patch of prey with their huge mouths. They fill their mouths with a body's weight of water and prey that expands their pleated throat; they then sift their prey from seawater by expelling seawater through their baleen plates.

whale's upper jaw and, when the whale opens its mouth, the baleen resembles a moustache (except that it is on the *inside* of their mouths), which is why these whales are sometimes called *moustached whales* (**Figures 14.27c** and **14.28**). Baleen is made of flexible keratin—the same as human nails and hair—and can be up to 4.3 meters (14 feet) long (**Figure 14.27d**). (Note that baleen [also called *whalebone*] was used for such items as buggy whips, umbrella ribs, and corset stays before synthetic materials—mostly plastics—were substituted.)

To feed, the largest baleen whales fill their mouths with a body-weight of water and prey (**Figure 14.27b**), allowing their pleated lower jaw to balloon in size (Figure 14.28). The whales force the water out between the fibrous plates of baleen, trapping small fish, krill, and other plankton inside their mouths. For example, humpback whales feed at or near the surface, sometimes working together in large groups to produce a circular curtain of bubbles to concentrate prey and then surfacing as a group within the *bubblenet* in a coordinated vertical lunge (**Figure 14.29**). The gray whale, on the other hand, has short baleen slats and feeds by straining benthic organisms such as amphipods and shellfish from bottom sediment.

***Baleen Whales and Sounds*** Baleen whales produce sound, too, but at much lower frequencies than toothed whales. For example, gray whales produce pulses (possibly for echolocation) and moans that may help them maintain contact with other gray whales. Rorqual whales produce moans that last from one to many seconds. These sounds are extremely low in frequency and are probably used to communicate over distances of up to 50 kilometers (31 miles). Blue whales have been documented to produce sounds that may travel along the SOFAR channel across entire ocean basins. (For more details about the SOFAR channel, see Mastering Oceanography **Web Diving Deeper 16.1**.) Songs of humpback whales are thought to be a form of sexual display, but it is unclear whether their main purpose is to repel other males or to attract females.

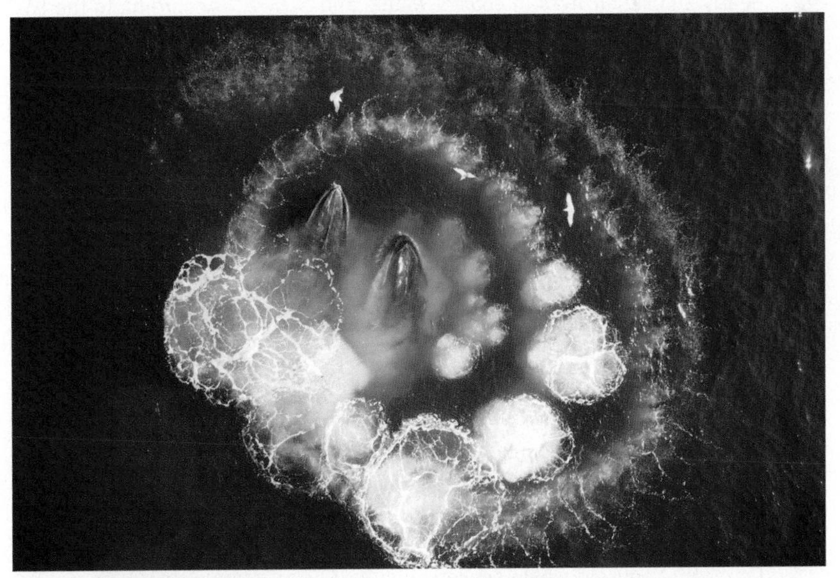

**(a)** Aerial view of bubblenet feeding by a group of humpback whales.

**(b)** Vertical lunge by a group of feeding humpback whales.

**Figure 14.29 Humpback whale bubblenet feeding in Alaskan waters. (a)** Humpback whales (*Megaptera novaeangliae*) often feed by swimming in a circle underwater and emitting a curtain of bubbles to corral prey. **(b)** Once the prey is gathered, the group of whales swims through the concentrated prey with their mouth open and surfaces in a vertical lunge. Baleen plates are used to filter prey items from the water and occur as parallel rows on the upper jaw that are separated by the roof of the mouth (*pink*).

## ARE WHALES STILL HUNTED?

There are several reasons why whales are still being hunted today, but *far* fewer whales are taken nowadays as compared to the huge numbers hunted in the last century. A detailed scientific study published in 2015 reveals that in the period after 1900, nearly 3 million whales were killed by the whaling industry. The most commonly killed whales during this time were hundreds of thousands of fin, sperm, and blue whales, but thousands of right, sei, humpback, and minke whales were also taken. In addition, gray whales were hunted to the brink of extinction in the mid-1800s mostly because gray whales spend nearly all of their lives in coastal waters and so they were easy targets for early whalers. During this time, whalers successfully traced gray whales to their birthing and breeding lagoons in Mexico, where they were hunted mainly for their oil. A common strategy was to harpoon a calf, which then lured its vigilant mother to the whalers, too. During this time, gray whales were known as "devilfish" because the adults would often capsize small whaling boats when they came to the aid of their young. By the late 1800s, the number of gray whales had diminished to the point that they were difficult to find during their annual migration. Fortunately, these low numbers also made it difficult for whalers to hunt them successfully.

The slaughter of gray whales was one factor that spurred whale conservation measures in the 1900s. In 1938, the International Whaling Treaty banned the taking of gray whales, which were thought to be nearly extinct. This protection has allowed gray whales to steadily increase in number to this day and to become the first marine creature to be removed from endangered status. Their repopulation is truly one of the most impressive success stories of how protecting animals can ensure their continued survival. What is perhaps most surprising is how friendly gray whales are now toward people in boats (**Figure 14B**) in the same lagoons where they were hunted to near extinction over 150 years ago. Devilfish, indeed!

In 1946, the **International Whaling Commission (IWC)** was established in an effort to manage the subsistence and commercial hunting of large whales. In 1986, the 72 member nations passed a ban on commercial whaling in order to allow whales to recover from overhunting and to give researchers time to develop methods for assessing whale populations. Today, the ban on whaling is still in effect, although some countries such as Japan, Norway, and Iceland have killed more than 35,000 whales since the 1986 whaling moratorium. Despite concerns about the practice of killing whales, some nations have proposed to end the ban on whaling and establish annual whale kill quotas. The ban on killing whales can be reversed only by a three-quarters majority vote of IWC member nations.

According to the IWC, there are currently three ways to engage in legal hunting of whales:

1. *Whaling by objection.* Countries can continue to hunt whales legally under an objection to the IWC ban on commercial whaling. (*Notable countries doing this:* Norway, Iceland.)
2. *Scientific whaling.* Countries may decide to kill whales for scientific research. Under the IWC convention, meat from whales hunted for this research may be sold commercially. (*Notable country doing this:* Japan. In fact, Japan is permitted to kill up to 1000 humpback, fin, and minke whales for scientific purposes each year. In 2014, the International Court of Justice found that Japan's whaling program was not for scientific purposes and even discovered whale meat being illegally sold in Japanese markets. As a result, Japan was ordered to halt its scientific whaling program; even so, in 2018 Japan took 333 minke whales from Antarctic waters.)
3. *Aboriginal subsistence whaling.* Subsistence whaling by native cultures is permitted where there is a long-standing cultural tradition of whaling and where whale meat satisfies the nutritional needs of the native population. (*Notable countries doing this*: Greenland, Russia, the United States, and the Caribbean nation of Saint Vincent and the Grenadines.).

### WHAT DID YOU LEARN?

1. Why did the International Whaling Commission (IWC) invoke a ban on commercial whaling?
2. What three ways exist to legally hunt whales? Which countries are doing each?

**Figure 14B Gray whale friendly behavior.** Gray whales (*Eschrichtius robustus*) exhibit friendly behavior by approaching a boat and initiating contact in Scammon's Lagoon, Baja California, Mexico.

***Baleen Whale Families*** Baleen whales are grouped into the following three families:

1. The **gray whale**, which has short, coarse baleen, no dorsal fin, and only two to five ventral grooves on its lower jaw. The gray whale is a bottom-feeder.
2. The **rorqual whales**, which have short baleen, many ventral grooves, and feed by gulping large mouthfuls of water containing small food items. (Note that the term *rorqual* refers to the longitudinal grooves on the lower jaw called rorqual folds; *rorqual* means "furrow whale" in Norwegian.) The rorqual whales are divided into two subfamilies:
    a. The *balaenopterids*, which have long, slender bodies, small sickle-shaped dorsal fins, and flukes with smooth edges (minke, Bryde's, sei, fin, and blue whales).
    b. The *megapterids*, or humpback whales, which have a more robust body, long flippers, flukes with uneven trailing edges, tiny dorsal fins, and nodules or tubercles (bumps that each hold a single long hair) on the head.
3. The **right whales**, which have long, fine baleen, broad triangular flukes, no dorsal fin, and no ventral grooves. (The name of the right whales comes from the fact that they are so oil-rich, slow, and valuable that early whalers considered them the "right" whales to kill.) Of the four species of right whales, the North Atlantic right whale and the North Pacific right whale are among the most critically endangered whales in the world. A third species of right whale, the Southern right whale, inhabits the Southern Ocean. The fourth species of right whale is the bowhead whale, which lives near the edge of the Arctic pack ice. Right whales feed by moving through surface waters with their mouths open, skimming for small surface-dwelling prey.

## STUDENTS SOMETIMES ASK . . .

***Are gray whales an endangered species?***

Not any more. The North Pacific gray whale (*Eschrichtius robustus*) was removed from the endangered species list in 1993, when their numbers exceeded 20,000, which surpassed the estimated size of their population before whaling. Out of nearly 1400 species listed as endangered since 1973, gray whales have become one of only 13 species to be removed from the endangered species list. However, other populations of gray whales were not so fortunate. For instance, the gray whales that used to inhabit the North Atlantic Ocean were hunted to extinction several centuries ago, and the gray whales that live in the Western Pacific near Japan are critically endangered (less than 130 whales, including only about 30 reproductive females). Recent research indicates that there is some intermixing between the populations of Western Pacific and Eastern Pacific gray whales, which might ultimately benefit the gene pools of both species.

**RECAP** **Marine mammals include orders Carnivora (sea otters, polar bears, and pinnipeds—walrus, seals, sea lions, and fur seals), Sirenia (manatees and dugongs), and Cetacea (whales, dolphins, and porpoises).**

## CONCEPT CHECK 14.4 ▸ Differentiate between the main groups of marine mammals based on their physical characteristics.

1 What common characteristics do all organisms in class Mammalia share?

2 Describe marine mammals within the order Carnivora, including their adaptations for living in the marine environment.

3 How can true seals be differentiated from the eared seals (sea lions and fur seals)?

4 Describe the marine mammals within the order Sirenia, including their distinguishing characteristics.

5 How can dolphins be differentiated from porpoises?

6 Describe differences between cetaceans of the suborder Odontoceti (toothed whales) with those of the suborder Mysticeti (baleen whales). Be sure to include examples from each suborder.

7 Compare echolocation systems in sperm whales and dolphins. Evaluate the differences and similarities between the two systems.

8 Describe the mechanism by which baleen whales feed.

## PROCESS OF SCIENCE 14.1: Why Do Gray Whales Migrate?

### BACKGROUND

Many marine organisms—such as fish, squid, sea turtles, and marine mammals—undertake seasonal migrations. One of the best-studied and longest migration of any mammal is that of the Pacific (or California) gray whale *Eschrichtius robustus*. These medium-sized, slow-moving coastal whales have a mottled gray appearance, can live 60 years, grow to 15 meters (50 feet) in length, and weigh up to 36 metric tons (79,000 pounds). Gray whales feed primarily on sea floor organisms during the summer in the cold, highly productive waters of the continental shelves of the Arctic Ocean and the far northern Pacific Ocean. They breed and give birth in warm tropical lagoons along the west coast of Baja California and mainland Mexico (**Figure 14C**). Peak calving season is mid-January, and although the destination may seem enviable, this lifestyle demands a yearly, round-trip journey of 22,000 kilometers (13,700 miles). Why do these whales undertake such a long, arduous migration?

### FORMING A HYPOTHESIS

Several hypotheses have been put forward to explain gray whales' migration:

1. One hypothesis states that whales must travel north to feed because without the bountiful food they consume during the highly productive Arctic summer, they could not sustain themselves during the mating and calving season, when feeding is minimal.
2. A second hypothesis suggests that gray whales migrate so far south to give birth because the physical environment of their cold-water feeding grounds does not meet the needs of young gray whales.
3. A third hypothesis is that the migration is a relic from the Ice Age, when sea level was lower and the shallow, high-latitude feeding grounds that are so productive today did not exist. Unable to feast on abundant food, the gray whales might have given birth to smaller calves, which could not survive in the cold water. This necessitated the migration to warmer water regions, which continues to this day despite the abundant food supply available.
4. A fourth hypothesis is that gray whales may leave colder Arctic waters to avoid killer whales, which are more numerous there and are a major threat to young whales.

### DEVISING AN EXPERIMENT

Gray whales have been radio-tagged to understand the timing of their migration and details of their journey. This is what researchers have determined:

The migration usually begins in September, after the high-latitude summer bloom in productivity has peaked. When Arctic pack ice begins to form over their feeding grounds along the continental shelf, the whales begin to move south.

Pregnant females are the first to leave on the two-month journey south. They are followed in steady succession by nonpregnant mature females, immature females, mature males, and then immature males. Traveling about 200 kilometers (125 miles) per day, most reach the lagoons of Baja California by December and early January.

In these warm-water lagoons, the pregnant females give birth to calves that are about 4.6 meters (15 feet) long and weigh about 1 metric ton (2200 pounds). The calves nurse on milk that is almost half butterfat and has the consistency of cheese, allowing the calves to put on weight quickly during the next two months. While the calves are nursing, the mature males breed with the mature females that did not bear calves. The gestation period is about 1 year, but producing large offspring and providing them with fat-rich milk for several months requires enormous amounts of energy, so it is not uncommon for females to give birth only once every two or three years.

In late February, gray whales head back north, beginning with males and females without new calves. Pregnant females and nursing mothers with their newborns are the last to depart, leaving late March to mid-April. Often, a few mothers linger in the protective lagoons with their young calves well into May. Most of the whales are back in their high-latitude feeding grounds by the end of June, when shallow continental shelf waters are ice-free and the high-latitude summer productivity bloom has begun.

How do the data match up to the four hypotheses?

1. Although gray whales rarely eat during their migration, they have been observed to feed briefly during their migration when the opportunity presents itself.
2. Research on the physiology of newborn gray whales indicates that gray whale calves can actually survive in much colder water.

Figure 14C **Gray whale migration route.** Gray whales (*Eschrichtius robustus*) undertake as much as a 22,000-kilometer (13,700-mile) annual migration, the longest of any mammal. Also shown is the Pacific Coast Feeding Group, which spends the summer in coastal waters from Canada to California.

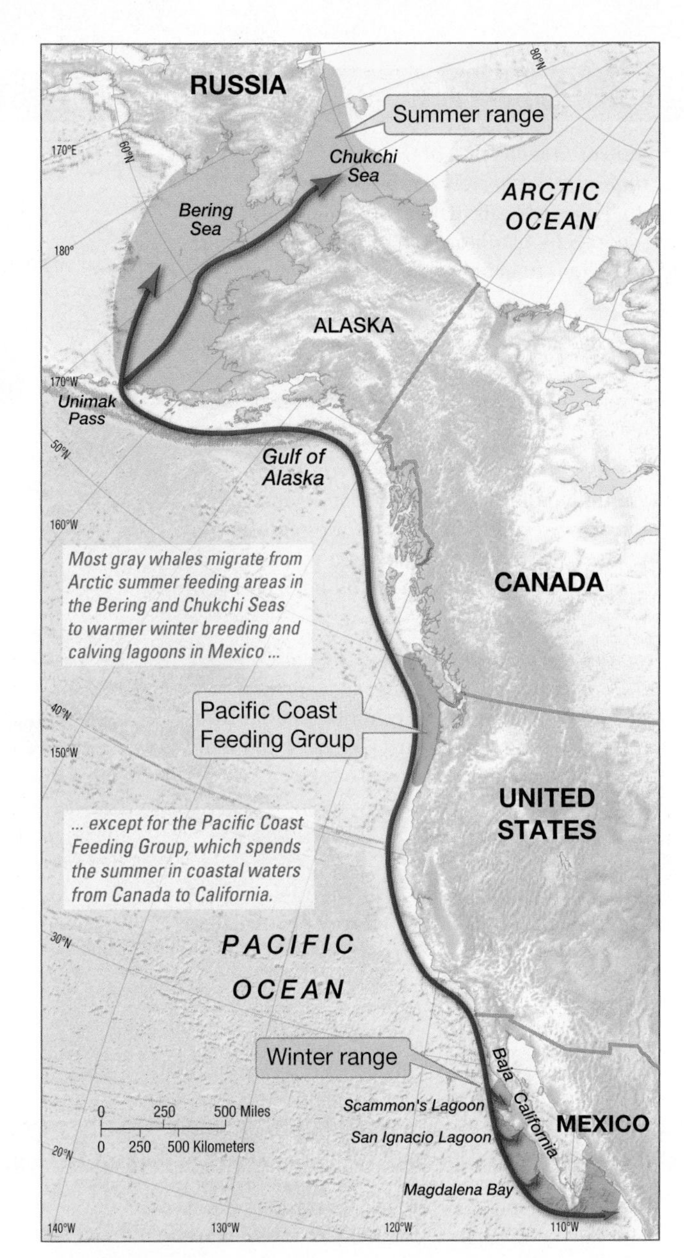


**EXPLORING DATA** ▲ **Using a calendar and tying your answer to productivity considerations, explain the seasonal cycle of migration for gray whales.**

3. Gray whales appear to be highly adaptable and sometimes give birth to their young on the way to their birthing grounds.
4. Although only those lagoons in Mexico with shallow entrances are used for calving, killer whales have been seen near the birthing lagoons and have also been observed to feed in extremely shallow water (see Mastering Oceanography **Web Diving Deeper 14.1**). In fact, killer whales take an estimated 30% of gray whale calves born each year.

## INTERPRETING THE RESULTS

This is one of those times when more data is needed! It is still not known which of the hypotheses about gray whale migration is the main reason why gray whales migrate. It could be that all of the hypotheses or a combination of some of them are the reasons behind their migration. Only through continued research will this mystery of the mammal's great migration be solved.

## THINKING LIKE A SCIENTIST: WHAT'S NEXT?

What experiments can you devise to help determine which of the four hypotheses put forward explains why gray whales migrate?

# ESSENTIAL CONCEPTS REVIEW

## 14.1 ▶ How are marine organisms able to stay above the ocean floor?

- *Pelagic animals that comprise the majority of the ocean's biomass remain mostly within the upper surface waters of the ocean*, where their primary food source exists. Those animals that are not planktonic (floating forms such as microscopic zooplankton) depend on *buoyancy* or their *ability to swim* to help them remain in food-rich surface waters.
- *The rigid gas containers in some cephalopods and the expandable swim bladders in some fish help increase buoyancy*. Other organisms maintain their positions near the surface with *gas-filled floats* (such as those of the Portuguese man-of-war) and *soft bodies that lack high-density hard parts* (such as the jellies).
- *Nekton—squid, fish, and marine mammals—are strong swimmers* that depend on their swimming ability to avoid predators and obtain food. Squid swim by trapping water in their body cavities and forcing it out through a siphon. Most fish swim by creating a wave of body curvature that passes from the front of the fish to the back and provides a forward thrust.
- *The caudal (rear) fin provides the most thrust, while the paired pelvic and pectoral (chest) fins are used for maneuvering. The dorsal (back) and anal fins serve primarily as stabilizers*. A rounded caudal fin is flexible and can be used for maneuvering at slow speeds. The lunate fin is rigid and is of little use in maneuvering but produces thrust efficiently for fast swimmers such as tuna.

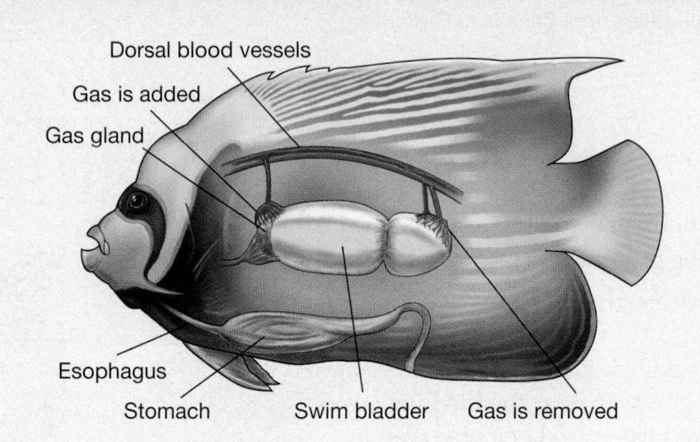


**Selected Key Terms**

Use the **glossary** at the end of this book to discover the meanings of these Selected Key Terms: **swim bladder, radiolarian, foraminifer, copepod, krill, cnidarian, hydrozoan, scyphozoan.**

**Critical Thinking Question**

Explain how a swim bladder works.

**Active Learning Exercise**

Pair up with another student in class. For each pair, have each person decide on which swim bladder adaptation they would like to describe: (1) adaptations that allow for rapid buoyancy changes or (2) adaptations that allow for slow buoyancy changes. Be sure to include how the buoyancy changes are achieved and determine which method is more advantageous. Share your analysis with each other and then report out to the class

## 14.2 ▶ What adaptations do pelagic organisms possess for seeking prey?

- *Fish can be categorized as lungers* (such as groupers) or *cruisers* (such as tuna). Lungers sit motionless and lunge at passing prey. They have mostly white muscle tissue, which fatigues more quickly than red muscle tissue. Cruisers swim constantly in search of prey and possess mostly red, myoglobin-rich muscle tissue.
- *Fish swim slowly when cruising, fast when hunting for prey, and fastest when trying to escape from predators*. Although *most fish are cold-blooded*, the fast-swimming tuna *Thunnus* is homeothermic, meaning that it maintains its body temperature well above water temperature.
- *Deep-water nekton have special adaptations*—such as *good sensory devices* and *bioluminescence*—that allow them to survive in this still and completely dark environment. *Bioluminescence*—the ability to organically produce light—*has many uses in the deep ocean.*

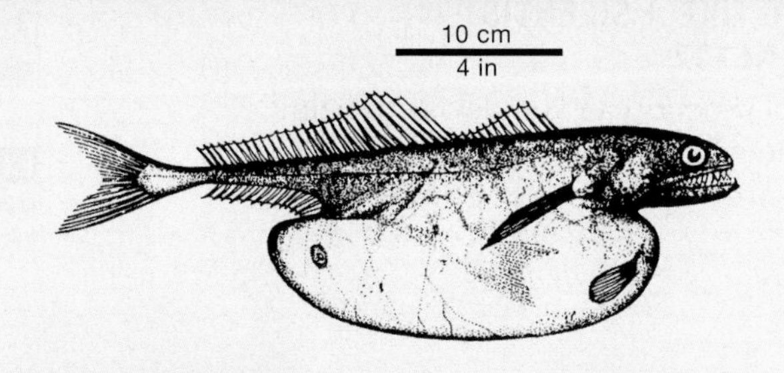


**Selected Key Terms**

Use the **glossary** at the end of this book to discover the meanings of these Selected Key Terms: **lunger, cruiser, myoglobin, cold-blooded (poikilothermic), warm-blooded (homeothermic), detritus, deep-sea fish, bioluminescence, counterillumination.**

**Critical Thinking Question**

What are the major structural and physiological differences between fast-swimming cruisers and lungers that patiently lie in wait for their prey?

**Active Learning Exercise**

Working with another student in class, list the advantages that marine fish have by being warm-blooded as compared to cold-blooded fish. Then compare your list with the list from another group of students in class.

## 14.3 ▶ What adaptations do pelagic organisms possess to avoid becoming prey?

- *Many marine organisms such as fish, squid, and crustaceans exhibit schooling*, probably because it increases their chances of avoiding predation compared to swimming alone and serves to preserve the species. Some organisms live closely together in *symbiotic relationships*.

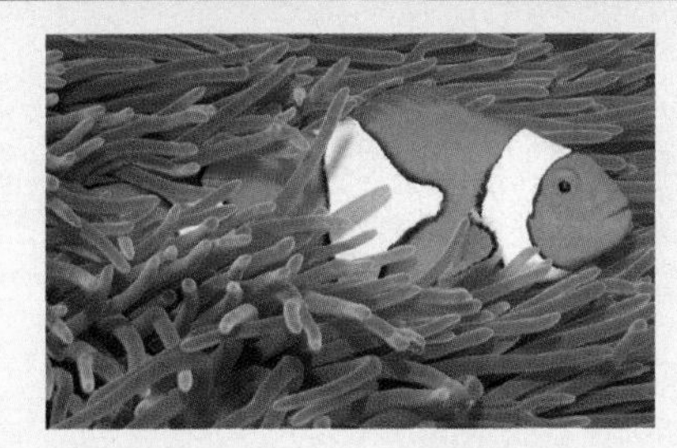

### Selected Key Terms

Use the **glossary** at the end of this book to discover the meanings of these Selected Key Terms: **school, symbiosis, commensalism, mutualism, parasitism.**

### Critical Thinking Question

Evaluate the advantages and disadvantages for a brightly colored yellow fish schooling within a large group of gray-colored fish.

### Active Learning Exercise

Working with another student in class, list the advantages of schooling. Then compare your list with the list from another group of students in class.

## 14.4 ▶ What characteristics do marine mammals possess?

- *Good fossil evidence shows that marine mammals evolved from land-dwelling animals* about 50 million years ago. *Marine mammals are warm-blooded, breathe air, have hair or fur, bear live young, and the females have mammary glands*. Marine mammals belong to orders *Carnivora*, *Sirenia*, and *Cetacea*.
- *Marine mammals in order Carnivora* have prominent canine teeth and include *sea otters*, *polar bears*, and the *pinnipeds* (*walruses*, *seals*, *sea lions*, and *fur seals*). *Marine mammals of order Sirenia*, which include *manatees* and *dugongs* ("sea cows"), have toenails (manatees only) and sparse hairs covering their bodies, and they are vegetarians.
- The mammals best adapted to life in the open ocean are those of the *order Cetacea*, which includes *whales*, *dolphins*, and *porpoises*. Cetaceans have *highly streamlined bodies* so that they are fast swimmers. Other adaptations—such as being able to absorb 90% of the oxygen they inhale, storing large quantities of oxygen, reducing the use of oxygen by noncritical organs, and having collapsible ribs and lungs)—*allow them to dive deeply* and minimize the effects of nitrogen narcosis and decompression sickness, although research suggests that they are not immune to it.
- *Cetaceans are divided into suborder Odontoceti (the toothed whales) and suborder Mysticeti (the baleen whales). Odontocetes use echolocation* to find their way through the ocean and locate prey. They emit clicking sounds and can determine the size, shape, internal structure, and distance of the objects from the nature of the returning signals and the time elapsed.
- *Mysticetes*, which include the largest whales in the world, *separate their small prey from seawater using their baleen plates as a strainer*. Baleen whales include the *gray whale*, the *rorqual whales*, and the *right whales*.

### Selected Key Terms

Use the **glossary** at the end of this book to discover the meanings of these Selected Key Terms: **Carnivora, pinniped, sea otter, polar bear, walrus, seal, sea lion, fur seal, Sirenia, Cetacea, hemoglobin, Odontoceti, falcate, echolocation, melon, Mysticeti, baleen, gray whale, rorqual whale, right whale.**

### Critical Thinking Question

List the modifications that are thought to give some cetaceans the ability to (a) increase their swimming speed, (b) dive to great depths without suffering the bends, and (c) stay submerged for long periods.

### Active Learning Exercise

Pair up with another student in class. For each pair, have each person decide on which table they would like to construct: (1) a table on the physical characteristics that differentiate true seals from eared seals (sea lions and fur seals) or (2) a table on the physical characteristics that differentiate dolphins from porpoises. Be sure to include the names of representative organisms. Share your table with each other and then compare with other students' tables that were created in class.

## Mastering Oceanography™

Looking for additional review and test prep materials? With individualized coaching on the toughest topics of the course, Mastering Oceanography offers a wide variety of ways for you to move beyond memorization and deeply grasp the underlying processes of how the oceans work. Visit the Study Area in **www.masteringoceanography.com** to find practice quizzes, study tools, and multimedia that will improve your understanding of this chapter's content. Sign in today to access the following features: Self Study Quizzes, SmartFigures, Oceanography Videos and Animations, Squidtoons, Dynamic Study Modules, and an optional Pearson eText with embedded videos.

# Animals of the Benthic Environment

# 15

## ESSENTIAL LEARNING CONCEPTS

At the end of this chapter, you should be able to:

- ☐ **15.1** Specify characteristics of the communities that exist along rocky shores.
- ☐ **15.2** Specify characteristics of the communities that exist along sediment-covered shores.
- ☐ **15.3** Specify characteristics of the communities that exist on the shallow offshore ocean floor.
- ☐ **15.4** Specify characteristics of the communities that exist on the deep-ocean floor.

↑ *Check when completed*

***"The deep sea is like a continent not yet discovered."***

***—Thomas Dahlgren, marine ecologist (2006)***

Of the approximately 230,000 known species that inhabit the marine environment, more than 98% live in or on the ocean floor. Ranging from the rocky, sandy, and muddy intertidal zone to the muddy deposits of the deepest ocean trenches, the ocean floor provides a tremendously varied environment that is home to a diverse group of specially adapted organisms.

As shown in **Figure 15.1**, the distribution of *benthic biomass* (the mass of living organisms on the sea floor), closely matches the distribution of chlorophyll in surface waters (*inset*, which is also Figure 13.5). This is because chlorophyll is an approximation of primary productivity, and life on the ocean floor depends on the productivity of the ocean's surface waters. As a result, great abundances of benthic life are found beneath areas of high primary productivity.

The vast majority of known benthic species live on the continental shelf, where the water is often shallow enough to allow sunlight to penetrate to the ocean bottom and support photosynthesis. Scientific expeditions to the deep-ocean floor, however, suggest that a large number of species yet undescribed by science live there.

The number of benthic species found at similar latitudes on opposite sides of an ocean basin depends on how ocean surface currents affect coastal water temperature—one of the most important variables affecting species diversity. The Gulf Stream, for example, is a warm surface current that flows east across the Atlantic, warming the European coast from Spain to the northern tip of Norway. This gives rise to more than three times the number of benthic species than are found in similar high latitudes of the Atlantic coast of North America, where the cold-water Labrador Current cools the water as far south as Cape Cod, Massachusetts.

Living at or near the interface of the ocean floor and seawater, an organism's success is closely related to its ability to cope with the physical conditions of the water, the ocean floor, and the other organisms that inhabit its environment. In this chapter, we'll examine a variety of benthic communities and the animals that inhabit them, starting at the shallow shore and progressing to the deep-ocean floor.

## 15.1 ▸ What Communities Exist along Rocky Shores?

If you visit any rocky shoreline around the world, you're likely to find an abundance of marine organisms that live on the surface of the ocean floor. These organisms, called **epifauna** (*epi* = upon, *fauna* = animal) are either permanently attached to the bottom (for example, marine algae) or move over it (for example, crabs).

◂ **Sea stars cling to a rocky ledge just below the ocean surface.** In the intertidal zone, waves crash, oxygen and salinity values fluctuate, predators seek food, drying out is a constant threat, and attachment sites are fought over and defended. Despite all these environmental challenges, many types of bottom-dwelling marine organisms, including these sea stars, abound in rocky intertidal zones such as the one shown here.

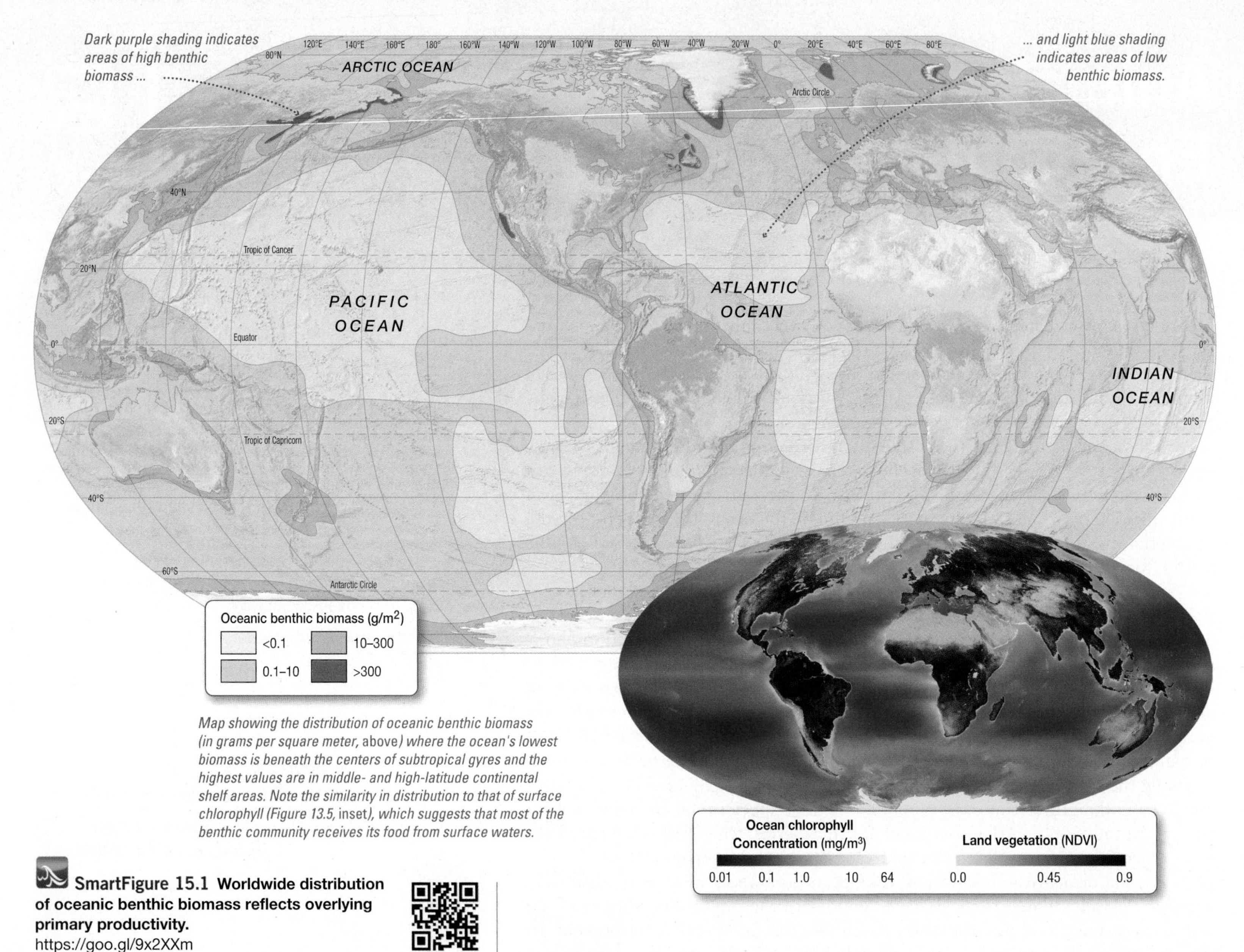


*Map showing the distribution of oceanic benthic biomass (in grams per square meter, above) where the ocean's lowest biomass is beneath the centers of subtropical gyres and the highest values are in middle- and high-latitude continental shelf areas. Note the similarity in distribution to that of surface chlorophyll (Figure 13.5, inset), which suggests that most of the benthic community receives its food from surface waters.*

**SmartFigure 15.1 Worldwide distribution of oceanic benthic biomass reflects overlying primary productivity.**
https://goo.gl/9x2XXm

**Table 15.1** lists some of the special adaptations these organisms have that enable them to withstand the rigors of life on rocky shores.

## Intertidal Zonation

Most shorelines exhibit *intertidal zonation*, which describes the natural organization of ecosystems relative to sea level that are caused by varying environmental conditions. For example, a typical rocky shore (**Figure 15.2a**) can be divided into a **spray zone**, which is above the spring high tide line and is covered by water only during storms, and an **intertidal zone**, which lies between the high and low tidal extremes. Along most shores, the intertidal zone can be clearly separated into the following subzones (Figure 15.2a):

- The **high tide zone**, which is relatively dry and is covered only during the highest high tides
- The **middle tide zone**, which is alternately covered by all high tides and exposed during all low tides
- The **low tide zone**, which is usually wet but is exposed during the lowest low tides

**SmartTable 15.1 Adverse conditions of rocky intertidal zones and organism adaptations**

| Adverse conditions of rocky intertidal zones | Organism adaptations | Examples of organisms |
|---|---|---|
| Drying out during low tide | • Ability to seek shelter or withdraw into shells<br>• Thick exterior or exoskeleton to prevent water loss<br>• External surfaces covered with rock or shell fragments to prevent water loss<br>• Physiologically adapted to periodic drying out | Sea slugs, snails, crabs, sea anemones, kelp |
| Strong wave activity | • In algae: strong holdfasts to prevent being washed away<br>• In animals: seeking shelter or employing strong attachment threads, biological adhesives, a muscular foot, multiple legs, or hundreds of tube feet to allow them to attach firmly to the bottom<br>• In both: hard structures adapted to withstand wave energy; clustering closely together | Kelp, snails, sea stars, mussels, sea urchins |
| Predators occupy area during low tide/high tide | • Firm attachment of body parts, including a hard shell<br>• Stinging cells<br>• Camouflage<br>• Inking response<br>• Ability to break off body parts and regrow them later (regenerative capability) | Mussels, sea anemones, sea slugs, octopuses, sea stars |
| Difficulty finding mates for attached species | • Release of large numbers of eggs/sperm into the water column during reproduction<br>• Long organs to reach others for sexual reproduction | Abalones, sea urchins, barnacles |
| Rapid changes in temperature, salinity, pH, and oxygen content | • Ability to withdraw into shells to minimize exposure to rapid changes in environmental conditions<br>• Ability to exist in varied temperature, salinity, pH, and low-oxygen environments for extended periods | Snails, limpets, mussels, barnacles |
| Lack of space or attachment sites | • Overtake another organism's space<br>• Attach to other organisms<br>• Planktonic larval forms that inhabit new areas, which limits parental and offspring competition for the same space | Bryozoans, coral, barnacles, limpets |

**SmartTable 15.1 Adverse conditions of rocky intertidal zones and organism adaptations**
https://goo.gl/7NqmQQ

Organisms living within these intertidal zones have different environmental conditions that they must be adapted to. For example, physical stress (such as drying out) is much more important in higher tide zones. Wave energy and predation by other marine organisms are larger stresses in lower tide zones. And competition between intertidal organisms for attachment space is the biggest stress in middle tide zones. As a result, intertidal organisms have evolved specific adaptations to cope with the environmental conditions they face. Not surprisingly, then, the subzones of the intertidal zone can be delineated based on characteristic populations of benthic organisms found within each zone.

Although the zonation of marine organisms along rocky shores produces some of the most finely delineated biozones in the marine environment, the intertidal zone can have remarkably different characteristics from place to place. For example, some of the physical characteristics that change in a matter of just a few vertical centimeters are the amount of wave energy, exposure to the atmosphere, and changes in temperature and salinity. Overall, the intertidal zone is a difficult place to live, but the organisms that inhabit the intertidal zone have specific adaptations to overcome the challenges of living there. Let's examine some of these organisms and their adaptations.

One common organism that is found in the spray zone well above the high tide line is called rock louse or sea roach (isopods of the genus *Ligia*) (**Figure 15.2c**), which lives on exposed rocks or is found among the cobbles and boulders that typically cover the floors of sea caves. These scavengers reach lengths of 3 centimeters (1.2 inches) and scurry about at night, feeding on organic debris. During the day, they mostly hide in crevices.

A distant relative of the periwinkle snail, the limpet is also found in the spray zone (genus *Acmaea*; **Figure 15.2d**). Both limpets and periwinkle snails feed on marine algae. The limpet has a flattened conical shell and a muscular foot with which it clings tightly to rocks.

## The High Tide Zone: Organisms and Their Adaptations

Like animals within the spray zone, most animals that inhabit the high tide zone have a protective covering to prevent them from drying out. For example, both the striped shore crab (*Pachygrapsus crassipes*) (**Figure 15.2e**) and periwinkles have a protective shell and can move between the spray zone and the high tide zone. Buckshot barnacles (**Figure 15.2f**) are another type of crustacean that has a protective shell, too, but they cannot live above the high-tide shoreline because they are attached, filter-feed from seawater, and their larval form is planktonic.

The most conspicuous algae in the high to middle tide zone are rock weeds, members of the genus *Fucus* that live in colder latitudes (**Figure 15.2h**) and *Pelvetia* that live in warmer latitudes. Both have thick cell walls to reduce water loss during periods of low tide.

Observations of newly created rocky shorelines or other recently disturbed regions of the coast indicate that rock weeds are among the first organisms to colonize a rocky shore. Later, **sessile** (*sessilis* = sitting on) animal forms—those attached to the bottom, such as barnacles and mussels—begin to establish themselves, competing for attachment sites with the rock weeds.

## The Middle Tide Zone: Organisms and Their Adaptations

Seawater constantly bathes the middle tide zone, so more types of marine algae and soft-bodied animals can live there. The total biomass is much greater than in the high tide zone, so there is much greater competition for rock space among sessile forms.

Shelled organisms inhabiting the middle tide zone include mussels (genera *Mytilus* and *Modiolus*) (**Figure 15.2j**); gooseneck barnacles (*Pollicipes*) (Figure 15.2j), which attach themselves to rocks with a long, muscular stalk; and acorn barnacles (*Balanus*) (**Figure 15.2i**). Mussels attach to bare rock, algae, or barnacles during their planktonic stage and remain in place by means of strong *byssus* threads.

Mussels are often grouped together into a distinctive *mussel bed* that appears as a pronounced band or layer (**Figure 15.3a**) and can often be one of the most recognizable features of middle tidal zones along rocky coasts. The mussel bed thickens toward the bottom until it reaches an abrupt bottom limit, where physical conditions restrict mussel growth. Often protruding from the mussel bed are numerous gooseneck barnacles; crowded in with the mussels are less conspicuous species, such as acorn barnacles, other crustaceans, marine worms, rock-boring clams, sea stars, and algae.

Carnivorous snails and sea stars (such as genera *Pisaster* and *Asterias*) feed upon mussels in the mussel beds. To pry open the mussel shell, sea stars pull on either side with hundreds of tube-like feet. The mussel eventually becomes fatigued and can no longer hold its shell halves closed. When the shell opens ever so slightly, the sea star turns its stomach inside out, slips it through the crack in the mussel shell, and digests the edible tissue inside (**Figure 15.3b**).

Mussel bed

**(a)** The distinctive black band covering the rock surface is a bed of mussels (*Mytilus*), which is a common feature of middle tide zones along rocky shores such as this one in Glacier Bay, Alaska. Note how the diffuse band can also be seen on the beach to the left.

**(b)** An ochre sea star (*Pisaster*), feeds on a mussel by first enveloping a mussel with its tube feet and prying apart the mussel's shell. The star can then digest the mussel's soft tissues with its own everted stomach.

Figure 15.3 **Middle tide zone mussel bed and sea star.**

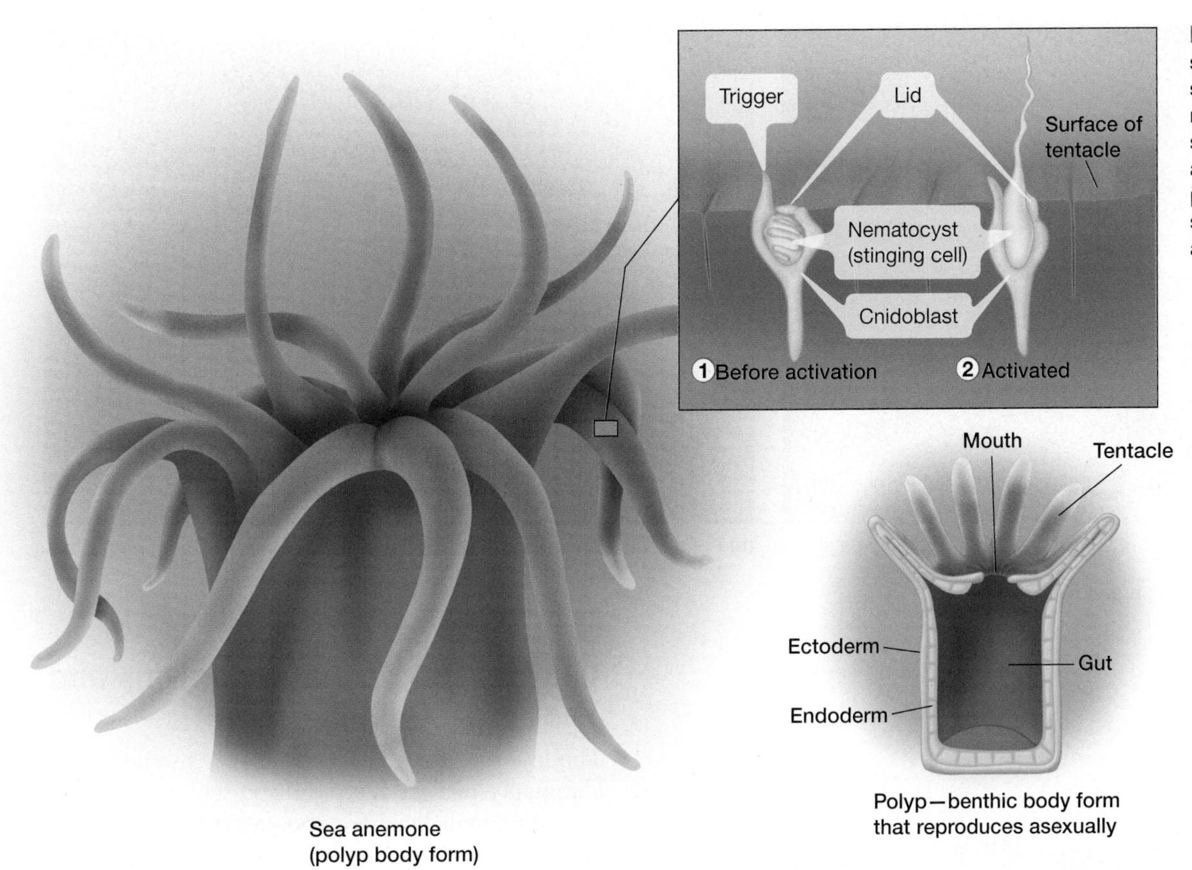


**Figure 15.4 Sea anemone structure and operation of its stinging cells.** Sea anemone morphology (*left*), with detail of its stinging nematocysts (*inset*), which are used to sting and incapacitate prey. An internal view of the basic sea anemone body form is shown at right.

Where the rock surface flattens out within the middle tidal zone, tide pools trap water as the tide goes out. These pools support microecosystems containing a wide variety of organisms. The most conspicuous member of this community is often the sea anemone (see **Figure 15.2k**), which is a relative of jellies.

Shaped like a sack, an anemone has a flat foot disk that provides a suction attachment to the rock surface. Directed upward, the open end of the sack is the mouth, which leads directly to the gut cavity and is surrounded by rows of tentacles (**Figure 15.4**). The tentacles are covered with stinging needle-like cells called **nematocysts** (*nemato* = thread, *cystis* = bladder) (Figure 15.4, *inset*), which inject the victim with a potent neurotoxin. When an organism brushes against a sea anemone's tentacles, the nematocysts are automatically released (except for certain organisms such as clownfish that live in a symbiotic relationship with a sea anemone).

Hermit crabs (*Pagurus*) inhabit tide pools, too. They have a well-armored pair of claws and upper body but a soft, vulnerable abdomen, which they protect by inhabiting an abandoned snail shell (**Figure 15.5a**). They can often be seen scurrying around the tide pool area or fighting with other hermit crabs for new shells. Their abdomen has even evolved a curl to the right to make it fit properly into snail shells. Once in the snail shell, the crab can further protect itself by closing off the shell's opening with its large claws.

In tide pools near the lower limit of the middle tide zone, sea urchins may be found feeding on algae (**Figure 15.5b**). A sea urchin has a five-toothed mouth centered on the bottom side of its hard, spherical shell, consisting of fused calcium carbonate plates perforated to allow tube feet and water to pass through. Resembling a pincushion, the shell of a sea urchin has numerous spines used for protection and to scrape out protective holes in rocks.

## STUDENTS SOMETIMES ASK . . .

***I've been at a tide pool and seen sea anemones. When I put my finger on one, it tends to gently grab my finger. Why does it do that?***

The sea anemone is trying to kill you with poison and devour you whole (seriously!). Disguised as a harmless flower, the sea anemone is actually a vicious predator that will attack any unsuspecting animal (even a human) that its stinging tentacles entrap. Fortunately, the skin on your hands is thick enough to resist the stinging nematocyst and its neurotoxin. A couple of unsuspecting people, however, were interested in finding out if the sea anemone grabbed other things with its tentacles, so they put their *tongues* into a sea anemone. After a short time, their throats swelled almost completely closed, and they had to be rushed to a hospital. They lived, but the moral of this story is: *NEVER* put your tongue into a sea anemone!

(a) Hermit crab (*Pagurus*) that has taken up residence in a *Maxwellia gemma* shell.

(b) Purple sea urchins (*Echinus*) that have burrowed into the bottom of a rocky tide pool in the middle tide zone.

Figure 15.5 **Middle tide zone hermit crab and sea urchins.**

Figure 15.6 **Surf grass.** Green surf grass (*Phyllospadix*) and various species of brown algae are exposed during an extremely low tide in this California tide pool. When the tide rises, the anchored surf grass floats and provides a protective hiding place for many low tide zone organisms.

**RECAP** **Rocky shores are divided into the spray zone and high, middle, and low tide (intertidal) zones. Many shelled organisms inhabit the upper zones, while more soft-bodied organisms and algae inhabit the lower zones.**

## The Low Tide Zone: Organisms and Their Adaptations

The low tide zone is almost always submerged, and an abundance of algae is typically present. A diverse community of animals live here, too, but they are hidden by the great variety of marine algae and surf grass (*Phyllospadix*) (**Figure 15.6**). Various types of encrusting red algae (such as *Lithophyllum* and *Lithothamnium*), which are also seen in middle-zone tide pools, become very abundant in lower tide pools. In temperate latitudes, moderate-sized red and brown algae provide a drooping canopy beneath which much of the animal life can hide during low tide.

Scampering from crevice to crevice and in and out of tide pools across the full range of intertidal zones are various species of shore crabs (Figure 15.2e and **Figure 15.7**). These scavengers help keep the shore clean. Shore crabs spend most of the day hiding in cracks or beneath overhangs. At night, they eat algae as rapidly as they can tear them from the rock surface with their large front claws, called *chelae* (*khele* = claw). Their hard exoskeleton (shell) prevents them from drying out too quickly, so they can spend long periods of time out of water.

**CONCEPT CHECK 15.1 ▸ Specify characteristics of the communities that exist along rocky shores.**

1 What are some adverse conditions of rocky intertidal zones? What are some organisms' adaptations for those adverse conditions? Which conditions seem to be most important in controlling the distribution of life?

2 One of the most noticeable features of the middle tide zone along rocky coasts is a mussel bed. Describe general characteristics of mussels and include a discussion of other organisms that are associated with mussels.

3 In which intertidal zone of a rocky shore would you typically find each of the following organisms: sea anemones, sea lettuce, rock lice, abalones, brittle stars, and buckshot barnacles?

(a) Sally lightfoot crab (*Grapsus grapsus*), which is a brightly colored and speedy crab found in tropical regions.

(b) Striped shore crab (*Pachygrapsus crassipes*). This female is carrying eggs (orange mass under her abdomen).

Figure 15.7 **Shore crabs.**

## 15.2 ▶ What Communities Exist along Sediment-Covered Shores?

Even though most sediment-covered shores have intertidal zones similar to rocky shores, life on and in sediment-covered shores requires very different adaptations than on rocky shores. Sediment-covered shores, for example, are composed of unconsolidated materials that often change shape and so require specific adaptations for organisms. In addition, there is much less species diversity in sediment-covered shores, but the organisms are usually found in great numbers. In the low tide zone of some beaches and on mud flats, for example, as many as 5000 to 8000 burrowing clams have been counted in only 1 square meter (10.8 square feet).

Nearly all large organisms that inhabit sediment-covered shores are called **infauna** (*in* = inside, *fauna* = animal) because they can burrow into the sediment. Sediment-covered shores also contain large numbers of microbial communities, particularly in quiet environments such as salt marshes and mud flats that tend to accumulate organic matter.

### Physical Environment of the Sediment

Sediment-covered shores include *coarse boulder beaches*, *sand beaches*, *salt marshes*, and *mud flats*, which represent progressively lower-energy environments and are consequently composed of progressively finer sediment. The energy level that a shore experiences is related to the strength of waves and longshore currents. Along shores that experience low energy levels, particle size becomes smaller, the sediment slope decreases, and overall sediment stability increases. Thus, the sediment in a fine-grained mud flat is more stable than that of a high-energy sandy beach.

Along high-energy sandy beaches, a large quantity of water from breaking waves rapidly sinks into the sand and brings a continual supply of nutrients and oxygen-rich water for the animals that live there. This supply of oxygen also enhances bacterial decomposition of dead tissue. The sediment in salt marshes and mud flats, on the other hand, is not nearly as rich in oxygen, however, so decomposition occurs more slowly, and these areas tend to develop a characteristic "rotten egg" smell.

### Intertidal Zonation

The intertidal zone of the sediment-covered shore consists of supratidal, high tide, middle tide, and low tide zones, as shown in Figure 15.8. These zones are best developed on steeply sloping, coarse-sand beaches and are less distinct on gentler sloping, fine-sand beaches. On mud flats, the tiny, clay-sized particles form a deposit with essentially no slope, so zonation is not possible in this protected, low-energy environment.

### STUDENTS SOMETIMES ASK . . .

***What is the most venomous marine organism?***

Many types of marine organisms are armed with potent venom. For example, certain species of jellies, stingrays, stonefish, snails, and octopuses all have venom strong enough to kill humans. Experts agree that the world's most venomous marine organism is the Australian box jelly (*Chironex fleckeri*), which has tentacles covered with poisonous darts called nematocysts that can inject poison into a person and kill them in minutes.

Coming in a close second is the cone snail, a small tropical snail with over 700 species and elaborately decorated shells that lives in coral debris among coral reefs. Cone snails immobilize their prey by using a modified radular tooth with barbs that is loaded with highly venomous compounds and is launched out of its mouth like a harpoon. The venom causes muscle paralysis and respiratory failure in its victim, which can lead to death. Because cone snails produce a mixture of toxins, there is currently no antivenom available. Researchers are analyzing the cone snail's complex venom in the hope of producing medicines to treat diabetes and a wide variety of human neurological diseases such as Parkinson's, Alzheimer's, and alcoholism.

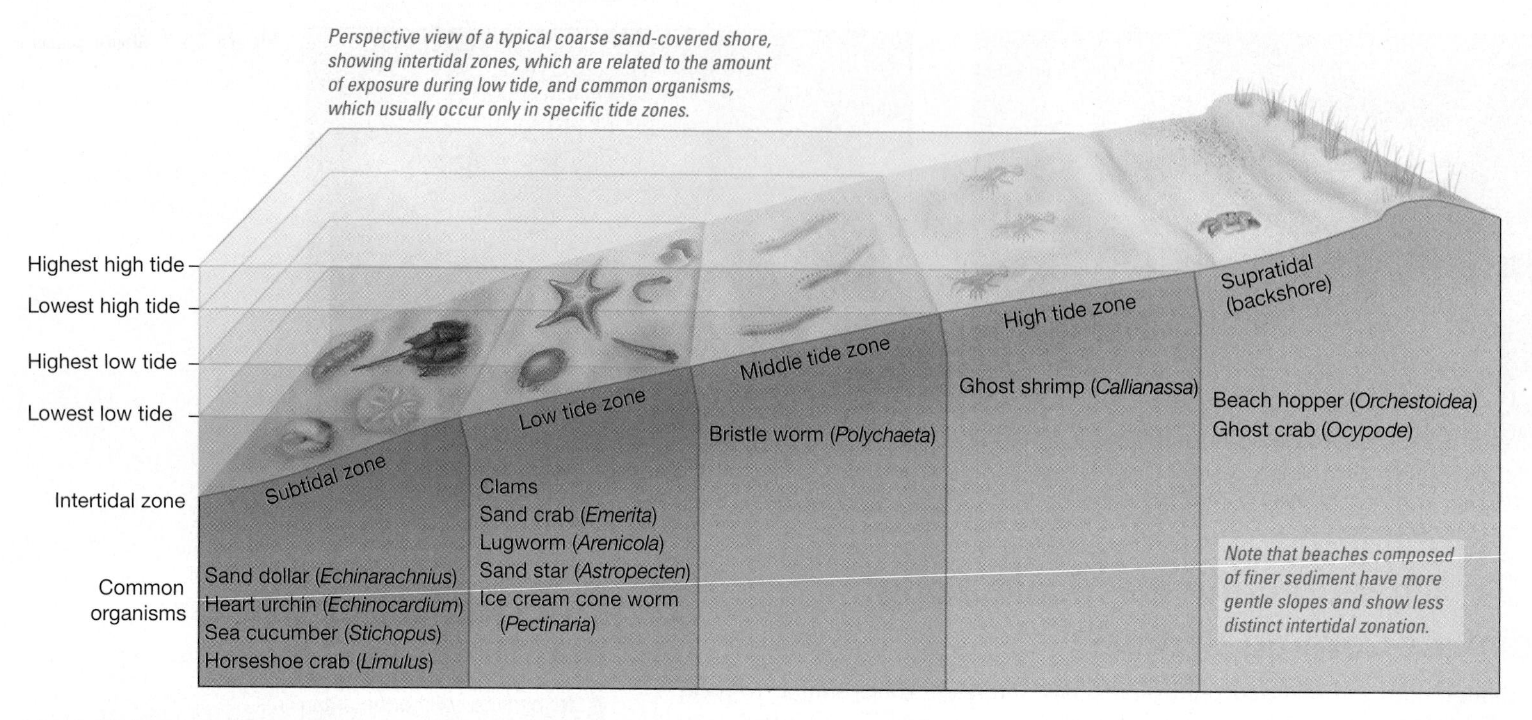


**Figure 15.8 Intertidal zonation and common organisms of sediment-covered shores.**

The species of animals differ from zone to zone. As in intertidal rocky shores, however, the maximum *number* of species and the greatest *biomass* in intertidal sediment-covered shores are found near the low tide shoreline, and both diversity and biomass decrease toward the high tide shoreline.

## Sandy Beaches: Organisms and Their Adaptations

Most animals at the beach burrow into the sand because there is no stable, fixed surface (as on rocky shores) to which they can attach. As a result, organisms are much less obvious at beaches than in other environments. By burrowing only a few centimeters beneath the surface, they encounter a much more stable environment where they are not affected by fluctuations of temperature and salinity or the threat of drying out.

**BIVALVE MOLLUSKS** A **bivalve** (*bi* = two, *valva* = a valve) is an animal that has two hinged shells, such as a clam or a mussel. A **mollusk** is a member of the phylum Mollusca (*molluscus* = soft), characterized by a soft body and either an internal or external hard calcium carbonate shell.

Bivalve mollusks are well adapted to living within sediment. A single foot digs into the sediment to pull the creature down into the sand. The method by which clams bury themselves is shown in Figure 15.9. How deeply a bivalve can bury itself depends on the length of its siphons, which must reach above the sediment surface to pull in water for food (plankton) and oxygen. Indigestible matter is forced back out the siphon periodically by quick muscular contractions.

The greatest biomass of clams is burrowed into the low-tide region of sandy beaches and decreases where the sediment becomes muddier.

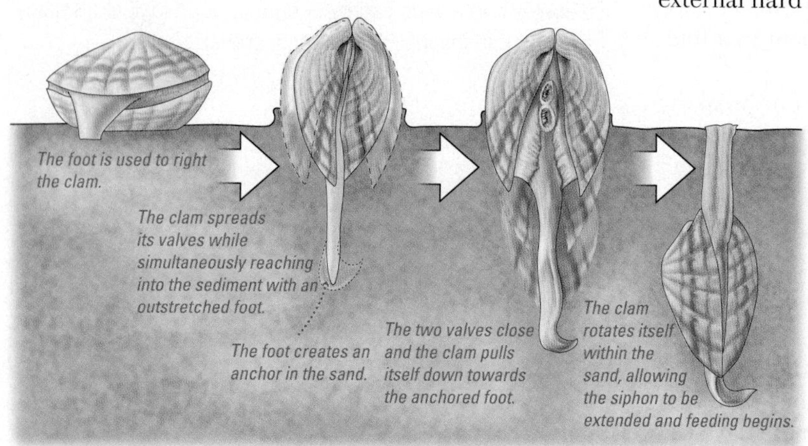

The foot is used to right the clam.
The clam spreads its valves while simultaneously reaching into the sediment with an outstretched foot.
The foot creates an anchor in the sand.
The two valves close and the clam pulls itself down towards the anchored foot.
The clam rotates itself within the sand, allowing the siphon to be extended and feeding begins.

**Figure 15.9 How a clam burrows.**

**ANNELID WORMS** A variety of **annelids** (*annelus* = ring)—worms—are well adapted to life in the sediment. The lugworm (*Arenicola*), for example, constructs a vertical, U-shaped burrow, the walls of which are strengthened with mucus. The worm feeds by extending its proboscis (snout) up into the shaft of the burrow to loosen sand with quick, pulsing movements. A cone-shaped depression forms at the surface over the head end of the burrow as sand continually slides into the burrow and is ingested by the worm. As the sand passes through the worm's digestive tract, the sand's *biofilm* (coating of organic matter) is digested, and the processed sand is deposited back at the surface.

**Figure 15.10 Sand crab.** A sand crab (*Emerita*) emerges from the sand. Sand crabs can often be found just beneath the surface of sandy beaches within the lower intertidal zone.

**CRUSTACEANS** **Crustaceans** (*crusta* = shell)—such as crabs, lobsters, shrimps, and barnacles—include predominately aquatic animals that are characterized by a segmented body, a hard exoskeleton, and paired, jointed limbs. On most sandy beaches, numerous crustaceans called *beach hoppers* feed on kelp cast up by storm waves or high tides. A common genus is *Orchestoidea*, which is only 2 to 3 centimeters (0.8 to 1.2 inches) long but can jump more than 2 meters (6.6 feet) high. Laterally flattened, beach hoppers usually spend the day buried in the sand or hidden in kelp. They become particularly active at night, when large groups may hop at the same time and form clouds above the piles of kelp on which they feed.

Sand crabs (*Emerita*) (**Figure 15.10**) are a type of crustacean common to many sandy beaches. Ranging in length from 2.5 to 8 centimeters (1 to 3 inches), they move up and down the beach near the shoreline. They bury their bodies in the sand and leave their long, curved, V-shaped antennae pointing up the beach slope. These little crabs filter food particles from the water and can be located by looking in the lower intertidal zone for a V-shaped pattern in the swash as it runs down the beach face (see Creature Feature 10.1).

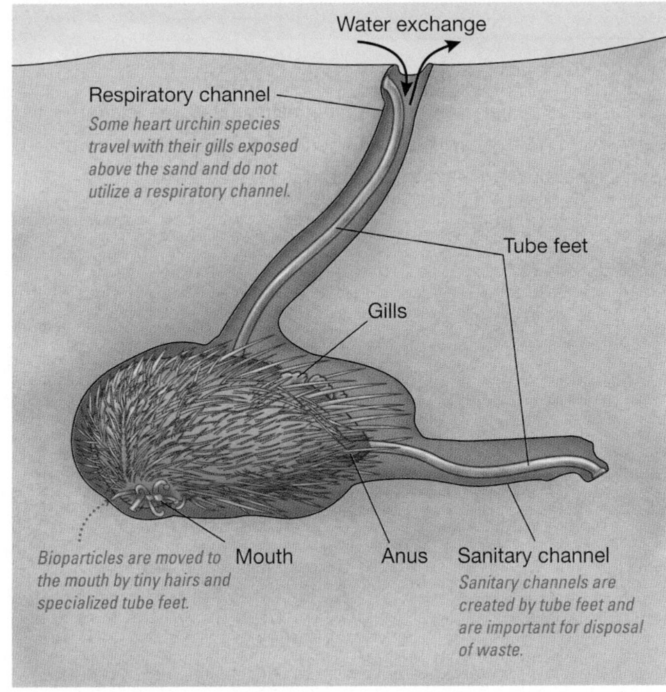


**Figure 15.11 Heart urchin.** Feeding and respiratory structures of a heart urchin (*Echinocardium*), which feeds on the biofilm of organic matter that covers sand grains.

**ECHINODERMS** **Echinoderms** (*echino* = spiny, *derma* = skin) that live in beach deposits include the sand star (*Astropecten*) and heart urchin (*Echinocardium*). Sand stars prey on invertebrates that burrow into the low-tide region of sandy beaches. The sand star is well designed for moving through sediment: It has a smooth back and five tapered legs with spines.

More flattened and elongated than the sea urchins of the rocky shore, heart urchins live buried in the sand near the low tide line (**Figure 15.11**). They gather sand grains into their mouths, where the biofilm of organic matter that coats sand grains is scraped off and ingested.

**MEIOFAUNA** **Meiofauna** (*meio* = lesser, *fauna* = animal) are small marine organisms that live in the spaces between sediment particles. These organisms, generally only 0.1 to 2 millimeters (0.004 to 0.08 inch) long, feed primarily on bacteria attached to the surface of sediment particles. Meiofauna include polychaetes, mollusks, arthropods, and nematodes (**Figure 15.12**) and live in sediment from the intertidal zone to deep-ocean trenches.

**(a)** Head of a nematode; the pit on its left side and numerous projections are sensory structures.

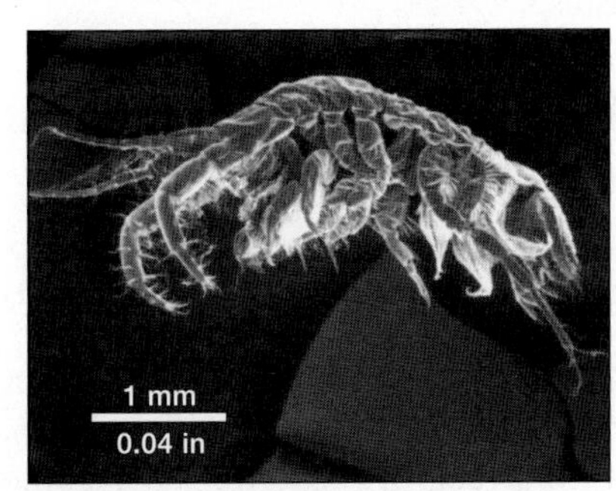


**(b)** Amphipod, which is a crustacean related to shrimp and krill.

**(c)** Polychaete worm, shown with its proboscis (mouth) extended (*left*).

**Figure 15.12 Scanning electron micrographs of meiofauna.** Various examples of meiofauna, which are small marine organisms that live in the spaces between sediment particles.

**Figure 15.13 Fiddler crab.** A male fiddler crab (*Uca*), which uses its large claw for protection and for attracting mates.

**RECAP Sediment-covered shores—including sandy beaches and mud flats—have an intertidal zonation similar to that of rocky shores but also contain many organisms that live within the sediment (infauna).**

## Mud Flats: Organisms and Their Adaptations

Eelgrass (*Zostera*) and turtle grass (*Thalassia*) are widely distributed in the low tide zone of mud flats and the adjacent shallow coastal regions. Numerous openings at the surface of mud flats attest to a large population of bivalve mollusks and other invertebrates.

Fiddler crabs (*Uca*) dig burrows within mud flats to live in; these burrows may extend 1 meter (3 feet) or more below the surface. Relatives of the shore crabs, they usually measure no more than 2 centimeters (0.8 inch) in width. A male fiddler crab has one small claw and one oversized claw, which is up to 4 centimeters (1.6 inches) long (**Figure 15.13**). Fiddler crabs get their name because they wave around their large claws as if they were playing imaginary fiddles. The females have two normal-sized claws. The large claw of the male is used to court females and to fight competing males.

**CONCEPT CHECK 15.2 ▸ Specify characteristics of the communities that exist along sediment-covered shores.**

1 Describe how sandy and muddy shores differ in terms of energy level, particle size, sediment stability, and oxygen content.

2 How does the diversity of species on sediment-covered shores compare with that of the rocky shore? Suggest at least one reason why this occurs.

3 In which intertidal zone of a steeply sloping, coarse-sand beach would you typically find each of the following organisms: clams, beach hoppers, ghost shrimp, sand crabs, and heart urchins?

# 15.3 ▸ What Communities Exist on the Shallow Offshore Ocean Floor?

The shallow offshore ocean floor extends from the spring low-tide shoreline to the seaward edge of the continental shelf. It is mainly sediment covered, but rocky bottom exposures may occur locally near shore. Rocky exposures feature many types of marine algae, which have adaptations (such as gas-filled floats) for reaching from the shallow sea floor to near the sunlit surface waters.

The sediment-covered shelf has moderate to low species diversity. Surprisingly, the diversity of benthic organisms is *lowest* beneath upwelling regions. This is because upwelling waters that are rich in nutrients cause high pelagic production, so large amounts of dead organic matter are produced. When this matter rains down on the bottom and decomposes, it consumes oxygen, so the oxygen supply can be locally depleted, thereby limiting benthic populations. However, kelp beds associated with rocky bottoms are a specialized shallow-water community with a higher diversity of benthic organisms.

## STUDENTS SOMETIMES ASK . . .

***What is an urchin barren?***

An urchin barren is created when the population of sea urchins goes unchecked and the urchins devour entire areas of giant brown bladder kelp (*Macrocystis*), one of the main types of algae in kelp forests. The urchins chew through the holdfast structure that holds the kelp in place and set it adrift. In California, the severe reduction of animals that prey on urchins (such as the wolf eel and sea otter) has upset the natural balance of ocean food webs. Consequently, sea urchins have proliferated, and urchin barrens now exist where there were once lush kelp beds.

## Rocky Bottoms (Subtidal): Organisms and Their Adaptations

A rocky bottom within the shallow inner **subtidal zone** (*sub* = under, *tidal* = the tides) is usually covered with various types of marine macro algae.

**KELP AND KELP FORESTS** Along the North American Pacific coast, the giant brown bladder **kelp** (*Macrocystis*) uses a root-like anchor called a *holdfast* (**Figure 15.14a**) to attach to rocky bottoms as deep as 30 meters (100 feet). The holdfast is so strong that only large storm waves can break the algae free. The *stipes* and *blades* of the algae are supported by gas-filled floats called *pneumatocysts* (*pneumato* = breath, *cystis* = bladder), which allow the algae to grow upward and

Figure 15.14 **Kelp and kelp forests.**

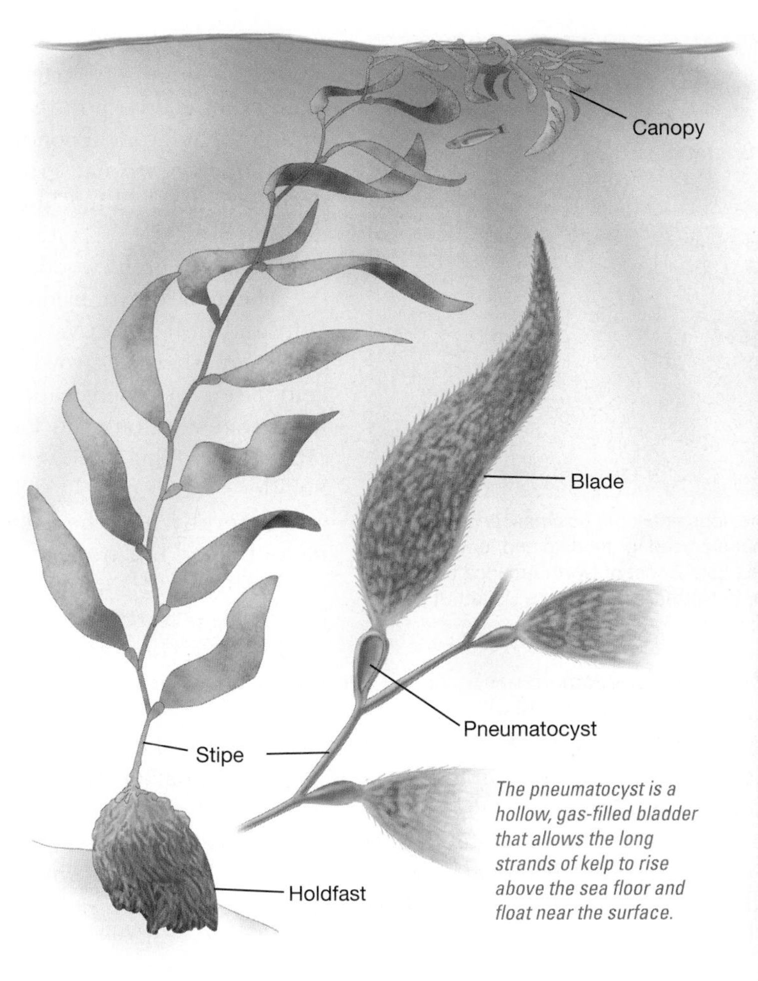

Canopy
Blade
Pneumatocyst
Stipe
Holdfast
The pneumatocyst is a hollow, gas-filled bladder that allows the long strands of kelp to rise above the sea floor and float near the surface.

**(a)** Structure and features of the giant brown bladder kelp (*Macrocystis*), a common species in kelp forests.

**(b)** Underwater photo of a kelp forest, which provides food, shelter, living space, spawning grounds, and nursery areas for many marine organisms.

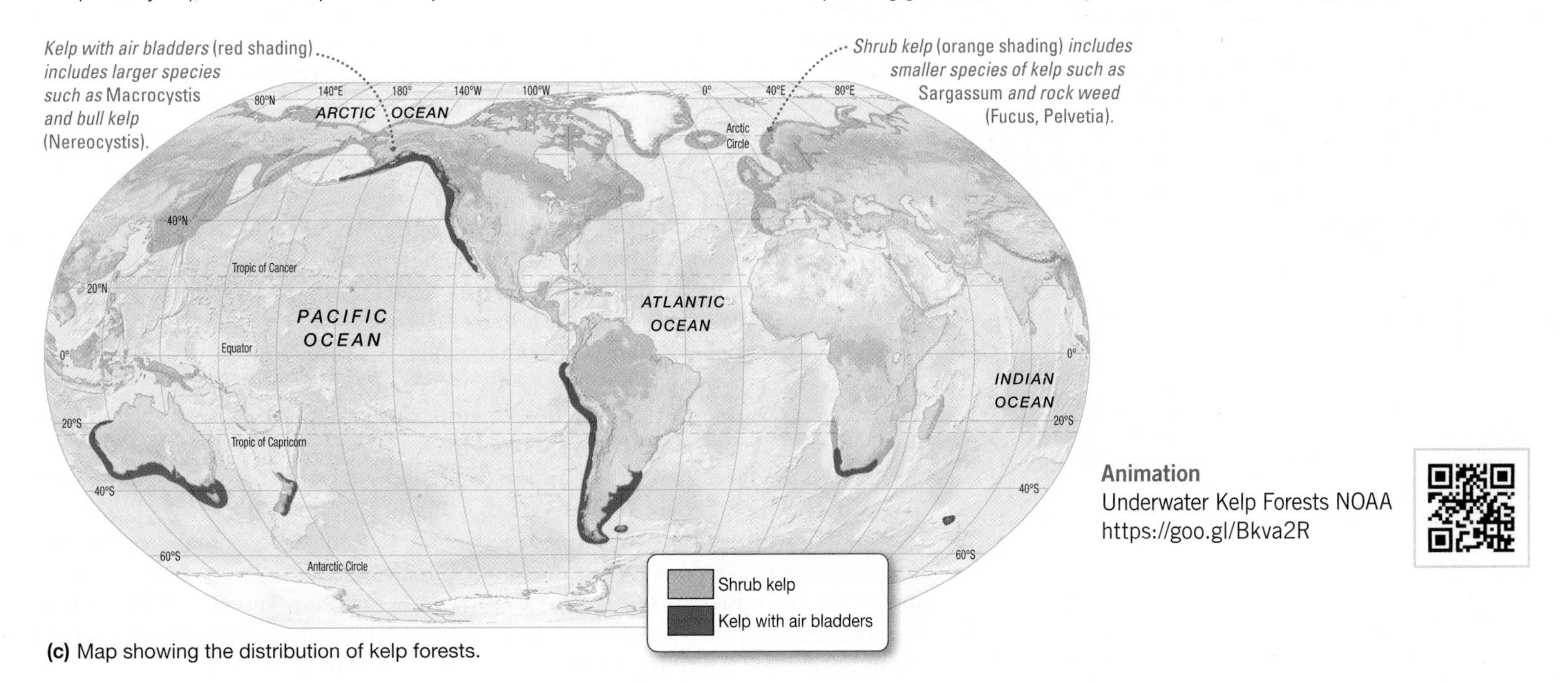


**(c)** Map showing the distribution of kelp forests.

**Animation**
Underwater Kelp Forests NOAA
https://goo.gl/Bkva2R

(a) Spiny lobster (*Panulirus interruptus*), which lacks large claws but has long spiny antennae. It is found along rocky bottoms in the Caribbean and along the West Coast of North America.

(b) American, or Maine, lobster (*Homarus americanus*), which has large claws that are used for feeding and for defense. It is found along the East Coast of North America from Labrador, Canada, to North Carolina in the United States.

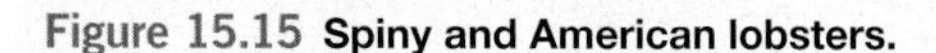

**Figure 15.15 Spiny and American lobsters.**

extend for another 30 meters (100 feet) along the surface to allow for good exposure to sunlight. Under ideal conditions, *Macrocystis* can grow up to 0.6 meter (2 feet) per day, making it the fastest-growing algae in the world.

The giant brown bladder kelp and bull kelp (*Nereocystis*), another fast-growing kelp, often form beds called **kelp forests** along the Pacific coast (**Figure 15.14b**). Smaller species of kelp that are generally less than 0.6 meter (2 feet) tall are known as shrub kelp. Examples of brown algae shrub kelp include *Sargassum* and rock weed (*Fucus, Pelvetia*). Even smaller tufts of red and brown algae are found on the bottom; these also live on the kelp blades.

Kelp forests are highly productive ecosystems that provide shelter for a wide variety of organisms living within or directly upon the kelp as epifauna. These organisms are an important food source for many of the animals living in and near the kelp forest, including mollusks, sea stars, fishes, octopuses, lobsters, and marine mammals. Surprisingly, very few animals feed directly on the living kelp. Among those that do are sea urchins and a large sea slug called the sea hare (*Aplysia*). The distribution of kelp forests is shown in **Figure 15.14c**.

## STUDENTS SOMETIMES ASK . . .

### *How big was the largest lobster ever caught?*

The largest specimen on record of the larger American lobster, *Homarus americanus*, measured 1.1 meters (3.5 feet) from the end of its tail fan to the tip of its large claw and weighed 20.1 kilograms (44.4 pounds). It was caught off Nova Scotia, Canada, in 1977 and was later sold to a New York restaurateur.

**LOBSTERS** Large crustaceans—including lobsters and crabs—are common along rocky bottoms. The spiny lobsters are named for their spiny covering and have two very large, spiny antennae (**Figure 15.15a**). These antennae serve as feelers and are equipped with noise-making devices near their base that are used for protection. The genus *Panulirus*, which reaches lengths to 50 centimeters (20 inches), is considered a delicacy and lives in water deeper than 20 meters (65 feet) along the European coast. The Caribbean lobster (*Panulirus argus*) sometimes exhibits a remarkable behavior of migrating single file across the sea floor in lines that are several kilometers long.

*Panulirus interruptus* is the spiny lobster of the American West Coast. All spiny lobsters are taken for food, but none are as highly regarded as the so-called true lobsters (genus *Homarus*), which include the American (Maine) lobster, *Homarus americanus* (**Figure 15.15b**). Although they are scavengers like their spiny relatives, the true lobsters also feed on live animals, including mollusks, crustaceans, and other lobsters.

**OYSTERS** Oysters are thick-shelled organisms that grow best where there is a steady flow of clean water to provide plankton and oxygen.

Oysters are food for sea stars, fishes, crabs, and snails that bore through the shell and rasp away the soft tissue inside (**Figure 15.16**). In fact, this is likely one of the main reasons that oysters have such a thick shell. (Note that this is an example of *coevolution*, where a weapon possessed by one species creates evolutionary pressure in another species for a defense to defeat it. This, in turn, causes other evolutionary traits for new weapons, creating a co-evolutionary "arms race.") Oysters also have great commercial importance to humans throughout the world as a food source.

Oyster beds are composed of empty shells of many previous generations that are cemented to a hard substrate or to one another, with the living generation on top. Each female produces many millions of eggs each year, which become planktonic larvae when fertilized. After a few weeks as plankton, the larvae attach themselves to the bottom. As a material upon which to anchor, oyster larvae prefer (in order) live oyster shells, dead oyster shells, and rock.

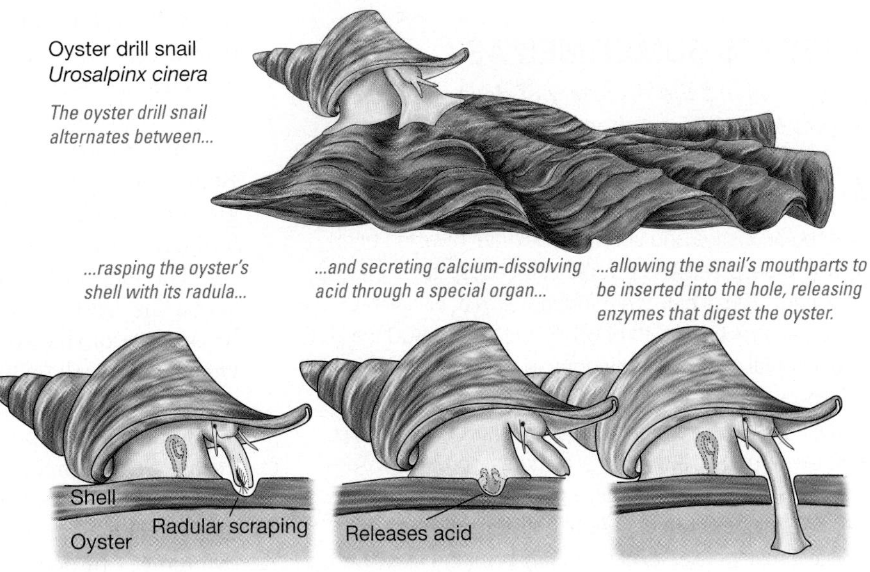

Oyster drill snail
*Urosalpinx cinera*
The oyster drill snail alternates between...
...rasping the oyster's shell with its radula...
...and secreting calcium-dissolving acid through a special organ...
...allowing the snail's mouthparts to be inserted into the hole, releasing enzymes that digest the oyster.
Shell
Oyster
Radular scraping
Releases acid


**Figure 15.16 How an oyster drill snail penetrates an oyster's shell.** An oyster drill snail (*Urosalpinx cinera*) alternates the use of its rasp-like mouthpiece called a radula and calcium-dissolving acid to penetrate an oyster's shell and feed on the oyster's soft body.

## Coral Reefs: Organisms and Their Adaptations

**Coral reefs** are wave-resistant structures produced by corals and other organisms. Corals are composed of colonies of individual **polyps** (*poly* = many, *pous* = foot), which are small benthic marine animals that feed with stinging tentacles and are related to jellies. Most species of corals are about the size of an ant, live in large colonies, and construct hard calcium carbonate structures for protection. Coral species are found throughout the ocean (and even in cold, deep water), but significant coral reef development is restricted to shallow warmer water regions.

**CONDITIONS NECESSARY FOR CORAL REEF DEVELOPMENT** Corals are very temperature sensitive and require warm water to survive. In fact, corals need water where the average monthly temperature exceeds 18°C (64°F) throughout the year (**Figure 15.17**). Furthermore, water that is too warm can kill corals: They cannot survive for long

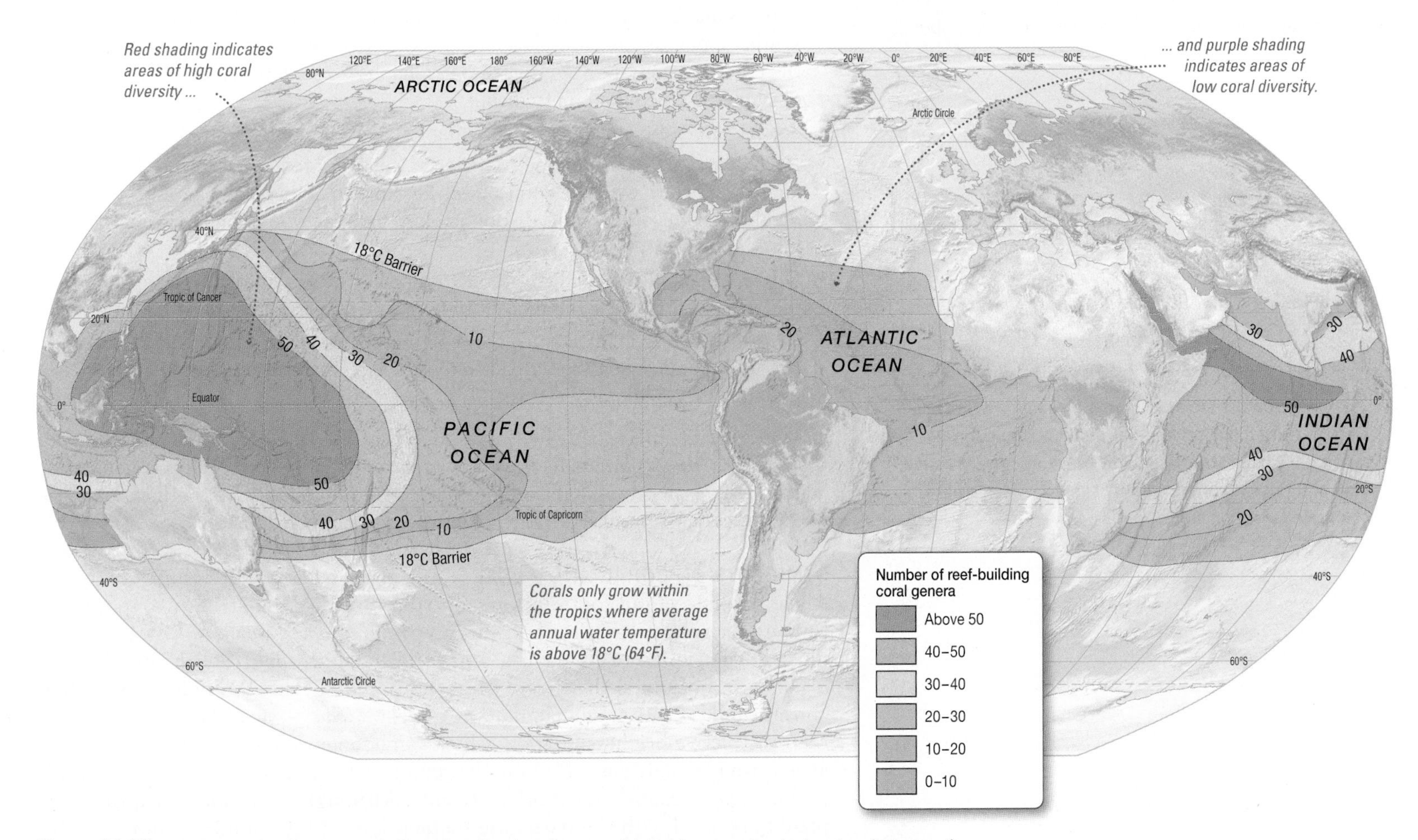


**Figure 15.17 Coral reef distribution and diversity.** Coral reefs are restricted to warm tropical waters between the two 18°C (64°F) temperature barriers. On the western side of each ocean basin, the coral reef belt is wider and the diversity of coral genera is greater, which is a result of ocean surface circulation patterns and the presence of numerous tropical islands that favor speciation.

## STUDENTS SOMETIMES ASK . . .

***I've heard about the discovery of deep-water corals. How are they different from shallow-water corals?***

Although coral reefs are generally associated with shallow tropical seas, recent deep-ocean exploration using advanced acoustics and submersibles has revealed unexpectedly widespread and diverse coral ecosystems in deep, cold waters. These corals are found below sunlit surface waters—at a record depth of 6328 meters (20,800 feet)—on continental shelves, continental slopes, seamounts, and mid-ocean ridge systems around the world. Because the majority of these corals don't live in particularly deep water, the term *cold-water coral* is more appropriate. They lack the symbiotic zooxanthellae algae that their shallow-water cousins have but can be brightly colored and use their stinging tentacles to capture tiny plankton or detritus concentrated by ocean currents. They build their skeletons of calcium carbonate as shallow corals do and create large reef structures or coral mounds that provide a habitat for many other species and may be thousands of years old. What is remarkable about them is that they have remained unnoticed for so long. New cold-water coral species, in fact, have recently been discovered in waters off highly populated Southern California!

**RECAP Corals are small colonial animals with stinging cells that are found primarily in shallow tropical waters. To thrive, they need strong sunlight, wave or current action, lack of turbidity, normal salinity seawater, and a hard substrate for attachment.**

when the water temperature exceeds 30°C (86°F). That's why warmer-than-normal sea surface temperatures—such as those experienced during severe El Niños or other warming episodes—tend to stress corals and are linked to outbreaks of coral bleaching and other diseases, as discussed later in this section.

Water warm enough to support coral growth is found primarily in the tropics. Reefs also grow as far north and south as 35 degrees latitude on the western margins of ocean basins, however, where warm-water currents raise average sea surface temperatures (Figure 15.17).

The map in Figure 15.17 also shows the greater diversity of reef-building corals on the western side of ocean basins. More than 50 genera of corals thrive in a broad area of the western Pacific Ocean and a narrow belt of the western Indian Ocean. Fewer than 30 genera, however, occur in the Atlantic Ocean, with the greatest diversity occurring in the Caribbean Sea. This pattern is related to the positions of the continents prior to about 30 million years ago, when the warm equatorial Tethys Sea connected the world's tropical oceans and provided a highway for the worldwide distribution of coral species and reef-associated organisms. With time, tectonic changes in landmass position closed the Tethys Sea and were accompanied by changes in ocean currents and climate that reduced coral reef biodiversity in areas such as the Atlantic. Further, the presence of numerous tropical islands in the western Pacific provides a variety of habitats that favors coral speciation.

Besides warm water, other environmental conditions necessary for coral growth include the following:

- *Strong sunlight.* The sunlight isn't needed by the corals themselves, which are animals and can exist in deeper water, but it is required by the symbiotic photosynthetic microscopic dinoflagellate algae called **zooxanthellae** that live within the coral's tissues. (See Section 14.3 in Chapter 14 for a discussion of the types of symbiosis, including mutualism.)
- *Strong wave or current action* to bring nutrients and oxygen.
- *Lack of turbidity.* Suspended particles in the water tend to absorb radiant energy, interfere with the coral's filter-feeding capability, and can even bury corals, so corals are not usually found close to areas where major rivers drain into the sea.
- *Salt water.* Corals die if the water is too fresh, which is another reason coral reefs do not form near the mouths of freshwater rivers.
- A *hard substrate* for attachment. Corals cannot attach to a muddy bottom, so they often build upon the hard skeletons of their ancestors, creating coral reefs that are several kilometers thick.

Once corals are established in an area that has the conditions necessary for their growth, they continue to grow upward layer by layer, with each new generation attached to the skeletons of its predecessors. Over millions of years, a thick sequence of coral reef deposits may develop if conditions remain favorable.

**SYMBIOSIS OF CORAL AND ALGAE** "Coral reefs" are more than just coral. Algae, mollusks, and foraminifers contribute to the reef structure, too. Reef-building corals are called **hermatypic** (*herma* = pillar, *typi* = type) and most hermatypic corals have a *mutualistic relationship* with microscopic algae (zooxanthellae) that live within the tissue of the coral polyp. (Note that zooxanthellae [*zoo* = animal, *xanthos* = yellow, *ella* = small] algae give corals their distinctive bright coloration, which can be many colors besides yellow.) The algae provide their coral hosts with a continual supply of food, and the corals provide the zooxanthellae with nutrients. Although coral polyps capture tiny planktonic food with their stinging tentacles, most reef-building corals receive up to 90% of their nutrition from symbiotic zooxanthellae algae. In this way, corals are able to survive in the nutrient-poor waters that are characteristic of tropical oceans. As we'll see, the mutualistic relationship between coral and

(a) Close-up of coral polyps, which are nourished by internal zooxanthellae algae and also by extending their tentacles to capture tiny planktonic organisms from the surrounding water.

(b) The blue-gray sponge *Niphates digitalis* (*below*) and the brown sponge *Agelas* (*above*), both of which host symbiotic algae or bacteria.

(c) A giant clam (*Tridacna gigas*), which depends on symbiotic algae living within its mantle tissue.

Figure 15.18 **Coral reef inhabitants that rely on symbiotic algae.**

algae is very sensitive to subtle changes in the environment, such as increased ocean temperature, salinity, and light.

Other reef animals also have a symbiotic relationship with various types of marine algae. Those that derive part of their nutrition from their algae partners are called **mixotrophs** (*mixo* = mixing, *tropho* = nourishment) and include coral, foraminifers, sponges, and mollusks (**Figure 15.18**). The algae not only nourish the coral but may also contribute to its calcification by extracting carbon dioxide from the coral's body fluids.

Coral reefs actually contain up to three times as much algal biomass as animal biomass. Zooxanthellae, for example, account for up to 75% of the biomass of reef-building coral. Nevertheless, zooxanthellae account for less than 5% of the reef's overall algal mass; most of the algae associated with coral reefs is filamentous green algae.

**RECAP** Corals are able to survive in nutrient-depleted warm water by living symbiotically with zooxanthellae algae, which live within the coral's tissues, provide it with food, and give the coral its color.

## Coral Reef Development

If coral reefs need bright sunlight to grow, what can account for the fact that typical reef deposits can be many kilometers thick? What seems odd is that the older coral structures now found at the base of these deposits had to have been in surface waters when they formed. What could explain this discrepancy? On his voyage aboard HMS *Beagle,* the famous naturalist **Charles Darwin** noticed a progression of stages in coral reef development, and hypothesized that the origin of coral reefs depended on the subsidence (sinking) of volcanic islands (**Figure 15.19**). (For more information about Charles Darwin and the voyage of HMS *Beagle*, see Diving Deeper 1.1 in Chapter 1.) Darwin published the concept in *The Structure and Distribution of Coral Reefs* in 1842. What Darwin's hypothesis lacked was a mechanism for how volcanic islands subside. Much later, advances in plate tectonic theory and samples of the deep structure of coral reefs provided evidence to help support Darwin's hypothesis.

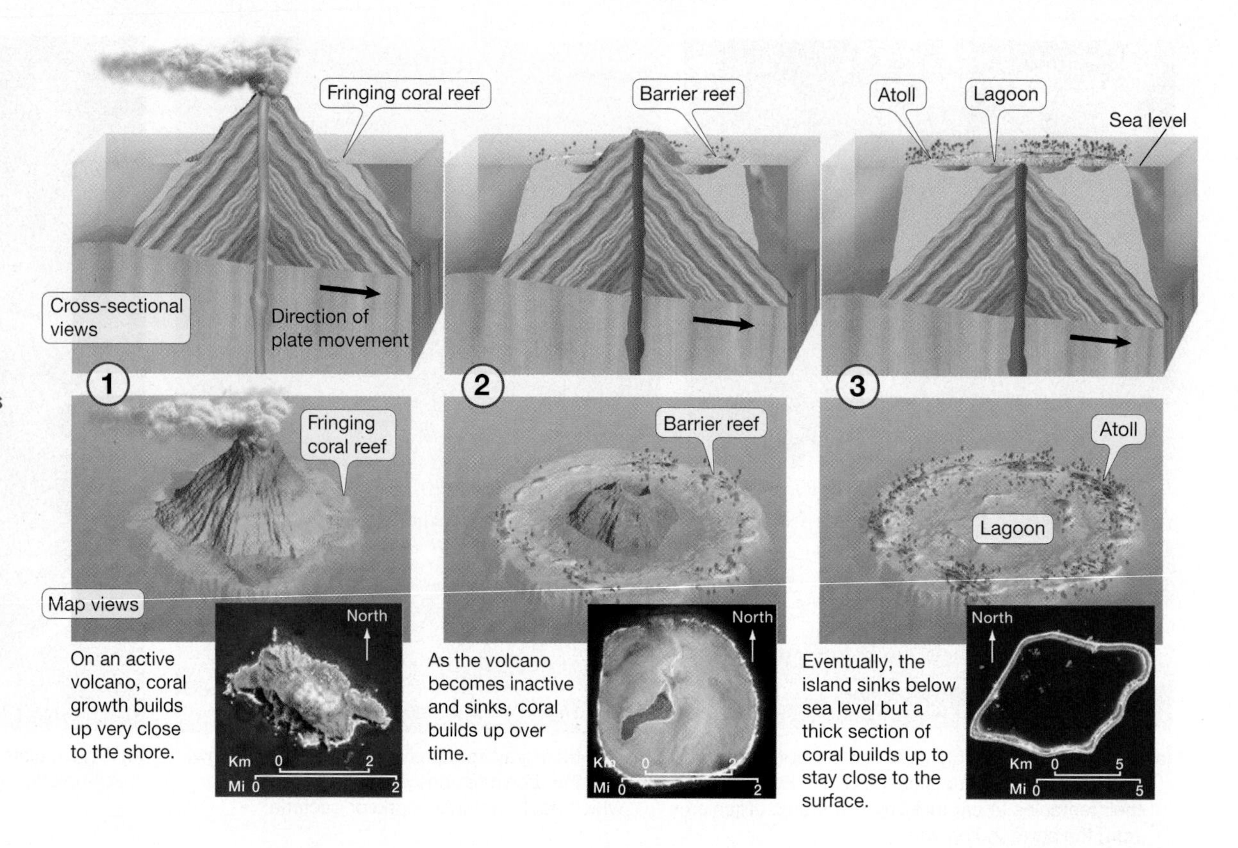


Figure 15.19 **Stages of development in coral reefs.** Cross-sectional view (*above*) and map view/aerial photographs (*below*) of (**a**) a fringing reef, (**b**) a barrier reef, and (**c**) an atoll. With the right conditions for coral growth and enough time, a coral reef progresses from fringing reef to barrier reef to atoll.

**EXPLORING DATA ▶ Besides a volcano sinking, what other process might occur in the ocean that would cause a coral reef to build up to be thousands of meters thick, with coral material that is millions of years old at the bottom?**

The three stages of development in coral reefs are called fringing, barrier, and atoll. **Fringing reefs** (Figure 15.19a) initially develop along the margin of a landmass (an island or a continent), where the temperature, salinity, and turbidity (cloudiness) of the water are suitable for reef-building corals. Often, fringing reefs are associated with active volcanoes whose lava flows run down the flanks of the volcano and kill the coral. Thus, these fringing reefs are not very thick or well developed. Because of the close proximity of the landmass to the reef, runoff from the landmass can carry so much sediment that the reef is buried. The amount of living coral in a fringing reef at any given time is relatively small, with the greatest concentration in areas protected from sediment and salinity changes. If sea level does not rise or the land does not subside, the process stops at the fringing reef stage.

The **barrier reef** stage follows the fringing reef stage. Barrier reefs are linear or circular reefs separated from the landmass by a well-developed lagoon (Figure 15.19b). As the landmass subsides, the reef maintains its position close to sea level by growing upward. Studies of reef growth rates indicate that most have grown 3 to 5 meters (10 to 16 feet) per 1000 years during the recent geologic past. Evidence suggests that some fast-growing reefs in the Caribbean have grown more than 10 meters (33 feet) per 1000 years. Note that if the landmass subsides at a rate faster than coral can grow upward, the coral reef will be submerged in water too deep for it to live.

The **atoll** (*atar* = crowded together) stage (Figure 15.19c) comes after the barrier reef stage. As a barrier reef around a volcano continues to subside, coral builds up toward the surface. After millions of years, the volcano becomes completely submerged, but the coral reef continues to grow. If the rate of subsidence is slow enough for the coral to keep up, a circular reef called an atoll is formed. The atoll encloses a lagoon usually not more than 30 to 50 meters (100 to 165 feet) deep. The reef generally has many channels that allow circulation between the lagoon and the open ocean. Buildups of crushed-coral debris often form narrow islands that encircle the central lagoon and are large enough to allow human habitation.

Alternatively, a new theory has been put forward to explain the origin of coral atolls. The theory suggests that glacial cycles cause sea level to fluctuate, leading to episodes of reef exposure and dissolution when global sea level is lower during ice ages, alternating with coral reef submergence and deposition when sea level is higher during interglacial stages. Instead of the slow growth of ring-shaped coral above a sinking volcanic island, this alternating cycle may be responsible for the formation of coral atolls. Sea level change is discussed further in Chapter 10 and Chapter 16, "The Oceans and Climate Change."

**RECAP** **Coral reefs grow over time and undergo three stages of development: fringing reef, barrier reef, and atoll.**

**CORAL REEF ZONATION** On many large coral reefs, there is a well-developed vertical and horizontal zonation of the reef slope (**Figure 15.20**), which is caused by changes in sunlight, wave energy, salinity, water depth, temperature, and other factors. These zones can readily be identified by the types of coral present and the assemblages of other organisms found in and near the reef.

Starting in the deepest water along the *reef slope*, active coral can only grow to a maximum depth of about 150 meters (500 feet) because algae within the coral's tissues need sunlight for photosynthesis. Water motion is less at these depths, so relatively delicate plate corals can live on the outer slope of the reef from 150 meters (500 feet) up to about 50 meters (165 feet) below the surface, where light intensity is as low as 4% of the surface intensity (Figure 15.20).

From 50 meters (164 feet) to about 20 meters (66 feet) below the surface, water motion from breaking waves increases on the side of the reef facing into the prevailing current flow. Correspondingly, the mass of coral growth and the strength of the coral structure supporting it increase toward the top of this zone, where light intensity is as low as 20% of the surface value.

At the surface on the seaward side of the coral reef, a *reef crest* includes a *buttress zone* that protects the *reef flat* from incoming waves (Figure 15.20). The reef flat may have a water depth of a few centimeters to a few meters at low tide, so it has at least 60% of the surface light intensity. Many species of colorful reef fish inhabit the calm, shallow water, as well as sea cucumbers, worms, and mollusks. In the protected water of the reef lagoon live gorgonian coral, anemones, crustaceans, mollusks, and echinoderms (**Figure 15.21**).

**THE IMPORTANCE OF CORAL REEFS** Coral reefs are some of the largest structures created by living creatures on Earth. (Australia's Great Barrier Reef, for example, is more than

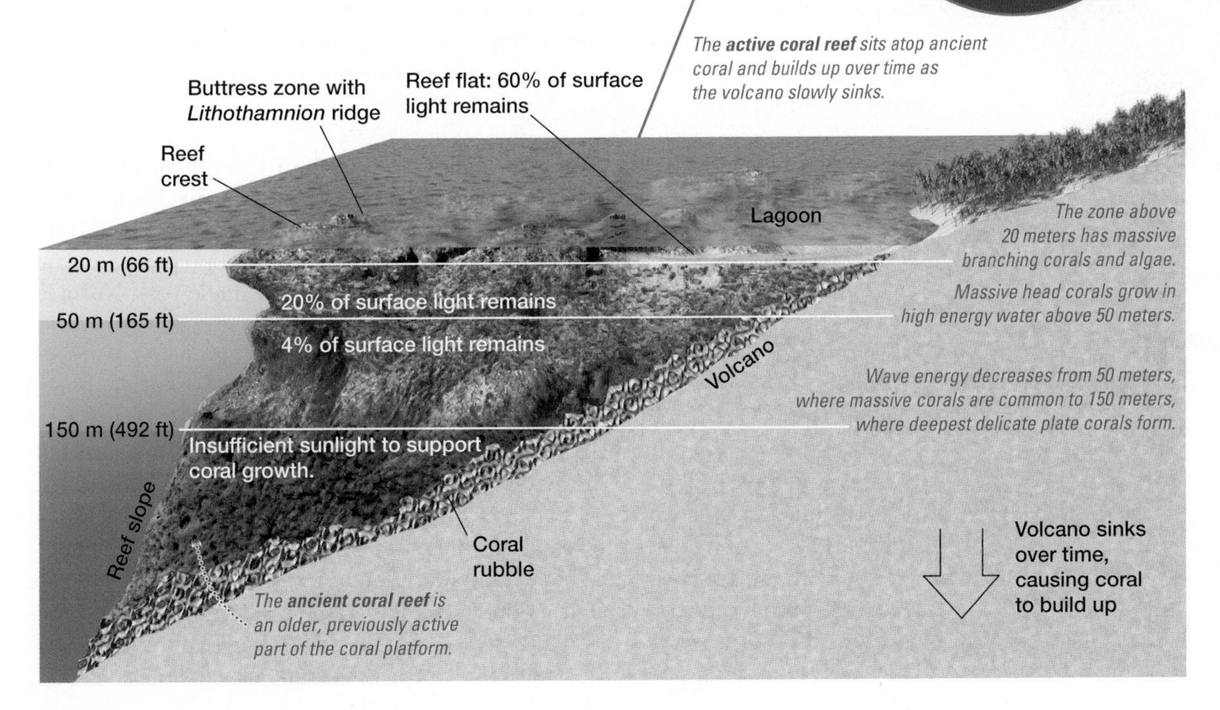

The **active coral reef** sits atop ancient coral and builds up over time as the volcano slowly sinks.
Buttress zone with *Lithothamnion* ridge
Reef flat: 60% of surface light remains
Reef crest
Lagoon
The zone above 20 meters has massive branching corals and algae.
20 m (66 ft)
20% of surface light remains
Massive head corals grow in high energy water above 50 meters.
50 m (165 ft)
4% of surface light remains
Volcano
Wave energy decreases from 50 meters, where massive corals are common to 150 meters, where deepest delicate plate corals form.
150 m (492 ft)
Insufficient sunlight to support coral growth.
Reef slope
Coral rubble
Volcano sinks over time, causing coral to build up
The **ancient coral reef** is an older, previously active part of the coral platform.

**EXPLORING DATA** ◀ **How would the coral reef structure change if sea level suddenly rose by 20 meters (66 feet)? What if sea level dropped by 20 meters (66 feet)?**

**SmartFigure 15.20 Coral reef structure and zonation.** Profile view of a typical coral reef, which exhibits zonation because of the decrease in both wave energy and sunlight intensity with increasing ocean depth. https://goo.gl/YpBzBP

**(a)** Coral reefs provide food, shelter, living space, spawning grounds, and nursery areas for many fishes, including this Guineafowl (spotted) puffer (*Arothron meleagris*).

**(b)** Unlike reef-building corals, some corals do not secrete a hard calcium carbonate structure, such as this soft gorgonian coral (*Nicella schmitti*), which has feeding polyps (*light purple*) extending from its branches.

Figure 15.21 **Nonreef-building inhabitants.**

2000 kilometers [1250 miles] long.) Although coral reefs cover less than half a percent of the ocean's surface area, they are home to 25% of all marine species, including almost one-third of the world's estimated 20,000 species of marine fish. Coral reefs provide shelter, food, and breeding grounds that attract numerous other species, including sea anemones, sea stars, crabs, sea slugs, clams, sponges, sea turtles, marine mammals, and sharks. In fact, coral reefs foster a diversity of species that surpasses even that of tropical rainforests, making coral reefs the most diverse communities in the marine environment.

Coral reefs provide many benefits to people. Over 100 million people worldwide are employed in the multibillion-dollar tropical reef tourism industry that depends on healthy coral reefs. In fact, many tropical countries that have coral reefs receive more than 50% of their gross national product as tourism related to reefs. Fisheries associated with reefs supply more than one-sixth of all fish from the sea. In addition, pharmacologists and marine chemists have discovered a storehouse of new medical compounds from coral reef inhabitants that fight maladies such as cancer and infections. In addition, reefs help prevent shoreline erosion and protect coastal communities from storm waves and tsunami. The hard calcium carbonate skeletons of coral have even been used in some human bone grafts.

**CORAL REEFS AND NUTRIENT LEVELS** When human populations increase on land adjacent to coral reefs, the reefs deteriorate. Fishing, trampling, boat collisions with the reef, sediment increase due to development, and removal of reef inhabitants by visitors all damage a reef. One of the subtler effects is the inevitable increase in the nutrient levels of the reef waters from sewage discharge and farm fertilizers.

As nutrient levels increase in reef waters, the dominant benthic community changes in the following ways:

- At low nutrient levels, hermatypic corals and other reef animals that contain algal symbiotic partners thrive.
- At moderate nutrient levels, fleshy benthic plants and algae are favored.
- At high nutrient levels, the phytoplankton mass exceeds the benthic algal mass, so benthic populations tied to the phytoplankton food web dominate. For example, high nutrient levels favor suspension feeders such as clams.

Increased phytoplankton biomass reduces the clarity of the water, which interferes with the coral's filter-feeding capability. The fast-growing members of the phytoplankton-based ecosystem destroy the reef structure by overgrowing the slow-growing coral and through *bioerosion*, which is erosion of the reef by organisms. Bioerosion by sea urchins and sponges is particularly damaging to many coral reefs.

**THE CROWN-OF-THORNS PHENOMENON** The crown-of-thorns (*Acanthaster planci*) is a sea star (**Figure 15.22**) that has greatly proliferated and destroyed living coral on many reefs throughout the western Pacific Ocean since 1962 and more recently in the Indian Ocean and Red Sea. The sea star moves across reefs and eats the coral polyps, devouring as much as 13 square meters (140 square feet) of coral in a single year. Normally, however, the coral can grow back if it has enough time and protection to do so.

Though a natural part of reef ecosystems, the sea stars can cause extensive damage because, for reasons not fully understood, a small group can suddenly multiply into millions. For example, vast numbers of crown-of-thorns sea stars have decimated many coral communities, particularly on the Great Barrier Reef. Initially, divers were employed to smash the crown-of-thorns, but sea stars (which have tremendous regenerating capabilities) can easily produce new individuals from various body parts, so it only made the problem worse.

Some investigators suggest that the proliferation of the crown-of-thorns sea star is a modern phenomenon brought about by human-caused activities. For example, the crown-of-thorns sea star has multiplied during the same time that large reef fish, which are the natural predators of crown-of-thorns sea stars, have been removed by overfishing. In addition, the larvae of crown-of-thorns sea stars, which are released into the ocean by the millions during spawning, tend to thrive in polluted waters. However, other studies show that during the past 80,000 years, the crown-of-thorns sea star has been even more abundant than it is today. Thus, the sea star may be an integral part of the reef ecology in this region, and its increase may be part of a long-term natural cycle rather than a destructive event triggered by human actions.

**Figure 15.22 Crown-of-thorns sea star.** Crown-of-thorns sea stars (*Acanthaster planci*) have plagued many tropical coral reefs such as Australia's Great Barrier Reef.

**CORAL BLEACHING AND OTHER DISEASES** **Coral bleaching** describes the loss of color in corals that causes them to turn white, as if bleach has been poured on the reef (**Figure 15.23**). Corals are very temperature sensitive, and coral bleaching is usually associated with elevated water temperatures that cause the coral's colorful symbiotic partner—zooxanthellae algae—to be expelled. Scientific studies suggest that reactive oxygen builds up in the coral's tissues and becomes toxic when temperatures are excessively warm, causing the coral to shed its normally beneficial algae. Coral reefs around the Galápagos Islands, for example, thrive in ocean water at or below 27°C (81°F). If the water is even 1°C or 2°C (2°F or 4°F) warmer for an extended period, then corals often become bleached. Once bleached, the coral no longer receives nourishment from the algae, and if the coral does not regain its symbiotic algae within a matter of weeks, it will die. Bleaching often occurs in surface waters—the top 2 to 3 meters (7 to 10 feet)—but has also been observed at depths of 30 meters (100 feet) and can occur as quickly as overnight.

Damaged corals are capable of regeneration if water temperatures return to normal and water quality remains good, but the frequency and intensity of bleaching outbreaks are now such that the percentage of reef loss from coral bleaching has been increasing dramatically. For example, Florida's coral reefs have experienced at least eight widespread bleachings since the early 1900s, and at least 70% of the corals along the Pacific Central American coast died due to bleaching associated with the severe El Niño event of 1982–1983. The warming during the 1982–1983 El Niño was so severe and long-lived that it affected all coral reefs in the eastern Pacific and two species of Panamanian coral became extinct. The bleaching episode of 1987 affected coral reefs worldwide, especially those in Florida and throughout the Caribbean. Since then, widespread bleachings have occurred with increasing frequency and intensity. For instance, the El Niño event of 1997–1998 raised water temperatures several degrees higher than normal and has been blamed for the most geographically widespread bleaching ever recorded, including in the equatorial eastern Pacific Ocean, the Yucatán coast, the Florida Keys, and the Netherlands Antilles. And yet another El Niño event in 2001–2002 caused severe bleaching of coral reefs worldwide. In the first part of the 21st century, bleaching has been observed in every ocean basin with increased frequency and severity. In 2010, for example, reefs in southeastern Asia bleached at record levels, part of the Hawaiian Islands experienced mild bleaching, and significant to severe bleaching occurred throughout the Caribbean. Most recently, a multiyear coral bleaching event during 2014–2017, caused by unusually warm sea surface temperatures, affected more than 70% of the world's coral reefs. Australia's Great Barrier Reef was particularly hard hit during this bleaching event: two-thirds of its corals were left severely bleached or dead.

Scientific studies suggest that other factors besides abnormally high sea surface temperatures (such as the warming of surface waters experienced during El Niño events described above) can cause coral bleaching, too.

**Figure 15.23 Normal and bleached coral.** Paired photographs of the same underwater region in American Samoa showing **(a)** normal coral in December 2014 and **(b)** bleached coral in February 2015 after exceptionally high sea surface temperatures, which caused extensive coral bleaching.

**(a)** Normal coral **(b)** Bleached coral

These factors include elevated ultraviolet radiation levels (especially in areas like Australia, close to the hole in Earth's protective ozone layer), a decrease in sunlight-blocking particles high in the atmosphere, marine pollution, salinity changes, invasion of disease, or a combination of factors. Still, the strong correlation between coral bleaching and elevated water temperature concerns marine scientists because sea surface temperatures are becoming warmer as a result of human-induced climate change (see Chapter 16, "The Oceans and Climate Change"). For example, the Western Pacific Warm Pool is predicted to widen and deepen as additional heat from human-caused climate change is added to the oceans, in essence making every year a candidate to become a widespread bleaching event that is normally associated with El Niño years. Scientists who study corals agree that coral reefs, which are particularly vulnerable to increases in heat and sunlight, are in serious trouble worldwide.

Climate

Connection

James Porter, a coral reef ecologist at the University of Georgia, and his colleagues study diseases that affect corals. They have been monitoring the health of corals in the Florida Keys since 1995 and have discovered the reappearance of *white plague disease* as well as a dozen new diseases, such as *white band disease, white pox, black band disease, yellow band (yellow blotch) disease, patchy necrosis,* and *rapid wasting disease.*

The cause of most of these diseases is still being investigated, and it is not known whether the new diseases are from the invasion of microorganisms—bacteria, viruses, or fungi—or related to environmental stress, as coral bleaching is. As human population has increased along the Florida Keys, the coral reefs of the Keys have begun to show signs of stress, thus making them more susceptible to a host of diseases. The increased nutrient levels and water turbidity resulting from soil runoff and improper sewage disposal in the Keys may contribute to the problem, too.

**CORAL REEFS IN DECLINE** Studies concerning the health of coral reefs worldwide show that they are in rapid decline due to various human and environmental factors. A recent survey of coral reef ecosystems, for example, shows that only 30% are healthy now, down from 41% in 2000. Another study estimated that more than one-third of the major reef-building coral species are currently at high risk of extinction, with severe consequences for entire reef ecosystems. In the Caribbean, the area of sea floor covered by live hard coral has decreased by a staggering 80% in the past 30 years. Even the Great Barrier Reef, widely regarded as one of the world's most pristine coral reefs, has lost over half of its coral cover during the past 40 years. In addition, individual coral species are disappearing. In 2014, for example, NOAA listed 20 new coral species as "threatened" under the U.S. Endangered Species Act.

The most serious threat to coral reefs—even more so than natural cataclysms such as hurricanes, floods, and tsunami—is human activity. Overfishing, for example, has depleted the populations of many of the fishes that graze on algae and if algae grow unchecked, it smothers the reefs by preventing corals from being bathed in oxygenated seawater. Runoff laden with sediment and pollutants further fuels the growth of algae and spreads harmful bacteria. A detailed study of 159 coral reefs in the Asia-Pacific region revealed that billions of pieces of plastic trash are now entangling reefs, stressing the corals through light deprivation, toxin buildup, and suffocation from oxygen depletion, thus leaving corals vulnerable to starvation and disease. Even more threatening to coral reefs are increased levels of human-caused atmospheric carbon dioxide, which is absorbed into the ocean and increases ocean acidity, making it more difficult for corals to create and maintain their calcium carbonate skeletons. A recent study, for example, artificially increased the acidity of seawater flowing over a natural coral reef community in Australia's Great Barrier Reef and found that coral calcification was reduced by about one-third. (For more information on the recent increase in ocean acidity and other climate change issues, see Chapter 16, "The Oceans and Climate Change.") In addition, global warming as a result of human activities has been shown to have increased ocean surface temperatures, affecting the growth rate of

## STUDENTS SOMETIMES ASK . . .

***How do corals synchronize their spawning?***

Although some corals spawn internally, most coral species employ a technique called *broadcast spawning*, in which sacs containing both eggs and sperm are released directly into seawater synchronously with neighboring colonies. The corals appear to use three cues to begin their mass spawning: (1) the full moon, (2) sunset, which they sense through photoreceptors, and (3) a chemical that allows them to "smell" each other spawning. For night divers lucky enough to witness coral spawning, it's a phenomenal experience! Researchers are trying to determine what will happen to fertilization rates as depleted coral colonies become fewer and farther between.

## PROCESS OF SCIENCE 15.1: Coral Reef Fishes—Adapt or Move?

### BACKGROUND

Many coral reef fishes are stenothermal—that is, they are able only to tolerate a narrow temperature range. Current estimates of ocean warming predict sea surface temperature increases of 2.0–4.8°C by 2100. Past research suggests that stenothermal reef fishes can acclimate to warmer temperatures, but at the expense of their metabolic and reproduction rates. Scientists have also documented a number of stenothermal tropical fishes shifting poleward away from warming waters. This has led researchers to the question: When exposed to warming temperatures, will stenothermal fishes adapt (stay in place) or move poleward?

### FORMING A HYPOTHESIS

Researchers focused on a stenothermal coral reef fish, the blue-green damselfish, to study their question. Scientists hypothesized that at higher temperatures, tropical stenothermal coral reef fish such as the blue-green damselfish would move poleward to waters within their optimal temperature range rather than acclimate to the warm temperatures. The reason, they hypothesized, was that the fish could not maintain their optimal metabolic and energy functions in warmer waters, thus affecting their ability to survive.

### DEVISING AN EXPERIMENT

To test their hypothesis, the researchers collected, weighed, and tagged 72 adult blue-green damselfish of about the same size from the northern portion of the Great Barrier Reef. They transported the fish to controlled aquarium tanks with temperatures in their preferred range, as well as tanks with temperatures in the range predicted for Sea Surface Temperature (SST) warming by 2100.

In one experiment, the researchers tested the fishes' oxygen consumption as an indicator of stress after being chased by hand in several different water temperatures. The researchers also subjected the fish to increasing temperatures until they lost the ability to right themselves. These experiments tested the hypothesis that fish cannot acclimate to warmer temperatures without considerable stress.

In another experiment, fish were placed in tanks with two chambers, one warmer and one cooler, between which the fish could freely move. Researchers increased and decreased the temperatures of the two chambers until the fish moved to the opposite chamber, or until maximum temperatures (those predicted for SST by 2100) were reached. This experiment tested the hypothesis that fish would want to move to more optimal temperatures when subjected to warmer-than-preferred temperatures.

### INTERPRETING THE RESULTS

The study found that the fish subjected to higher temperatures were unable to acclimate, as evidenced by the increased stress (oxygen consumption) demonstrated by the fish in the first experiments. The researchers also found that the fish preferred to seek out an optimal temperature rather than remain in a warmer-than-normal temperature of water. This study provides one explanation for why so many stenothermal organisms are moving poleward with warming ocean temperatures—they cannot adapt to the changes without considerable stress.

The researchers concluded that stenothermal fishes exposed to prolonged higher temperatures, such as those predicted with climate change, will experience stressors that impact their sustainability as a species and will prefer to move to cooler waters within their optimal temperature range.

### THINKING LIKE A SCIENTIST: WHAT'S NEXT?

What if coral reefs, the habitat of the fishes, are not able to shift poleward in warming ocean temperatures? Hypothesize the impact to the reef fishes if their habitat, coral reefs, is not able to shift poleward at the same rate the fish are moving. Design an experiment that tests your hypothesis.

### REFERENCE

Habary, A.; Johansen, J.L.; Nay, T.J.; Steffensen, J.F.; Rummer, J.L. Adapt, move or die—how will tropical coral reef fishes cope with ocean warming? *Global Change Biology* **2017**, *23*, 566–577.

**Figure 15A** **Blue-green damselfish in a healthy coral reef environment.**

temperature-sensitive corals worldwide and making them more prone to disease and bleaching episodes. The predicted rise in sea level as a result of global warming may also wreak havoc on corals by submerging them more deeply, in effect reducing the amount of sunlight they receive. The future for healthy coral reefs looks bleak unless immediate and dramatic conservation measures are enacted to preserve them.

**RECAP Coral reefs face many environmental threats and are in decline worldwide.**

**CONCEPT CHECK 15.3 ▸ Specify characteristics of the communities that exist on the shallow offshore ocean floor.**

1 Discuss the dominant species of kelp, their epifauna, and animals that feed on kelp in Pacific coast kelp forests.

2 Describe the environmental conditions required for development of coral reefs.

3 Draw and describe each of the three stages of coral reef development. How does this sequence tie into plate tectonics?

4 Describe the zones of the reef slope, the characteristic coral types, and the physical factors related to its zonation.

5 What is coral bleaching? How does it occur? What other diseases affect corals?

## 15.4 ▸ What Communities Exist on the Deep-Ocean Floor?

The vast majority of the ocean floor lies submerged below several kilometers of water. Less is known about life in the deep ocean than about life in any of the shallower nearshore environments because investigating the deep sea is difficult and expensive. For example, just obtaining samples from the deep-ocean floor requires a specially designed submersible or a properly equipped research vessel that has a spool of high-strength cable at least 12 kilometers (7.5 miles) long. In the past, the inaccessibility of ocean depths fueled much debate about the existence of life in the deep ocean (see Mastering Oceanography **Web Diving Deeper 15.1**).

Even today, collecting samples by submersible or with a biological dredge is a time-consuming process. Because the supply of oxygen is limited, submersibles with humans can stay down for only about 12 hours, and it may take 8 of those hours to descend and ascend. To send a dredge from a ship to the deep-ocean floor and retrieve it from the depths takes about 24 hours.

Robotics and remotely operated vehicles (ROVs) are making it possible to observe and sample even the deepest reaches of the ocean more easily. Because they are unmanned, ROVs are cheaper to operate and can stay beneath the surface for months, if necessary. These technological advancements should lead to further discoveries in one of Earth's least-known habitats.

### The Physical Environment

As discussed in Section 12.5 in Chapter 12, the deep-ocean floor includes bathyal, abyssal, and hadal zones. Here, the physical environment is much different than at the surface; it is quite stable and uniform. Light is present in only the lowest concentrations down to a maximum of 1000 meters (3300 feet) and absent below this depth. The temperature rarely exceeds 3°C (37°F) at the deep-ocean floor, and it falls as low as −1.8°C (28.8°F) in the high latitudes. Salinity remains at slightly less than 35‰; recall from Chapter 5 that average surface seawater salinity is 35 parts per thousand (‰). Oxygen content is constant and relatively high. Pressure exceeds 200 kilograms per square centimeter (200 atmospheres [2940 pounds per square inch]) on the oceanic ridges, exceeds 300 to 500 kilograms per square centimeter (300 to 500 atmospheres [4410 to 7350 pounds per square inch]) on the deep-ocean abyssal plains, and exceeds 1000 kilograms per square centimeter (1000 atmospheres [14,700 pounds per square inch]) in the deepest trenches. (At the ocean surface, the pressure is 1 atmosphere [1 kilogram per square centimeter (14.7 pounds per square inch)] and it increases by 1 atmosphere for each 10 meters [33 feet] of depth. Thus, a pressure of 1000 atmospheres is 1000 times that at the ocean's surface.) Bottom currents are generally slow but more variable than once believed. For example, **abyssal storms** created by warm- and cold-core eddies of surface currents affect certain areas, lasting several weeks and causing bottom currents to reverse and/or increase in speed.

A thin layer of sediment covers much of the deep-ocean floor. On abyssal plains and in deep trenches, sediment is composed of mud-like abyssal clay deposits. The accumulation of oozes—composed of dead planktonic organisms that have sunk through the water column—occurs on the flanks of oceanic ridges and rises. On the continental rise, there may be some coarse sediment from nearby land sources. Sediment may be absent on steep areas of the continental slope. It may also be absent near the crest of the mid-ocean ridge and along the slopes of seamounts and oceanic islands, where it has not had enough time to accumulate on newly formed ocean floor.

## Food Sources and Species Diversity

Because of the lack of light at the deep-ocean floor, photosynthetic primary production cannot occur. Except for the chemosynthetic productivity that occurs around hydrothermal vents, all benthic organisms receive their food from the surface waters above. Only about 1 to 3% of the food produced in the euphotic (sunlit) zone reaches the deep-ocean floor, so the scarcity of food that drifts down from the sunlit surface waters—not low temperature or high pressure—limits deep-sea benthic biomass. However, some variability in the supply of food is caused by seasonal phytoplankton blooms at the surface. **Figure 15.24** shows the food sources for deep-sea organisms.

Many of the organisms that inhabit the deep sea have special adaptations to help them detect food using chemical clues. Once food is found, these organisms are efficient at consuming it (**Diving Deeper 15.1**).

For many years, it was believed that the species diversity of the deep-ocean floor was quite low compared with shallow-water communities. Researchers studying sediment-dwelling animals in the North Atlantic, however, discovered an unexpectedly large diversity of species. An area of 21 square meters (225 square feet) contained 898 species, of which 460 were new to science. After analyzing 200 samples, it was realized that new species were being discovered at a rate that suggested the existence of millions of deep-sea species!

It turns out that deep-sea species diversity—especially for small infaunal deposit feeders—rivals that of tropical rainforests. It also appears, however, that the distribution of deep-sea life is patchy and depends to a large degree on the presence of certain microenvironments.

**RECAP** **The deep-ocean floor is a stable environment of darkness, cold water, and high pressure, but it still supports life. The food for most deep-sea organisms comes from sunlit surface waters.**

Figure 15.24 **Food sources for deep-sea organisms.**

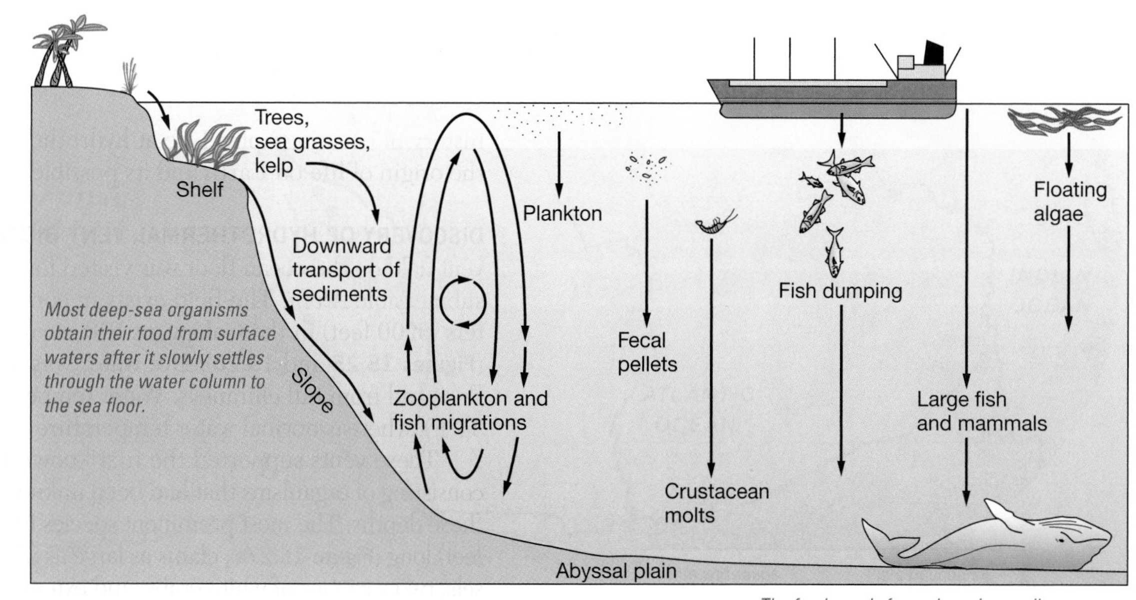


*The food supply from above is usually limited, except when large fish or mammals (such as whales) sink to the bottom.*

## Deep-Sea Hydrothermal Vent Biocommunities: Organisms and Their Adaptations

The surprising discovery of deep-sea **hydrothermal** (*hydro* = water, *thermo* = heat) **vents** and their biocommunities has been one of the most important finds in the modern

**(a)** Tubeworms (*Riftia*) up to 1 meter (3.3 feet) long are found at the Galápagos Rift and other deep-sea hydrothermal vents.

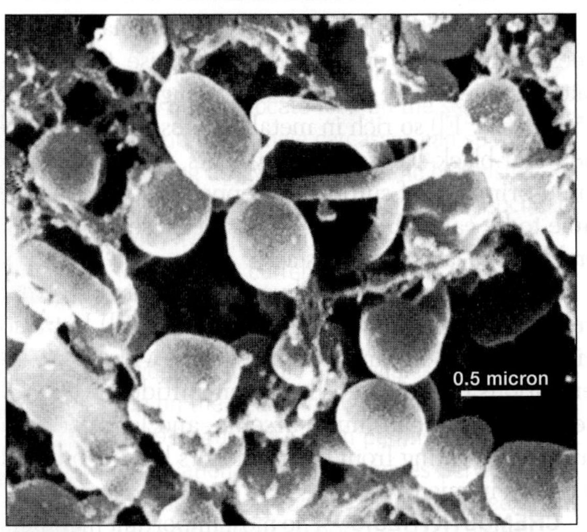


**(b)** Sulfur-oxidizing archaea that live symbiotically within the tissue of tubeworms, clams, and mussels found at hydrothermal vents.

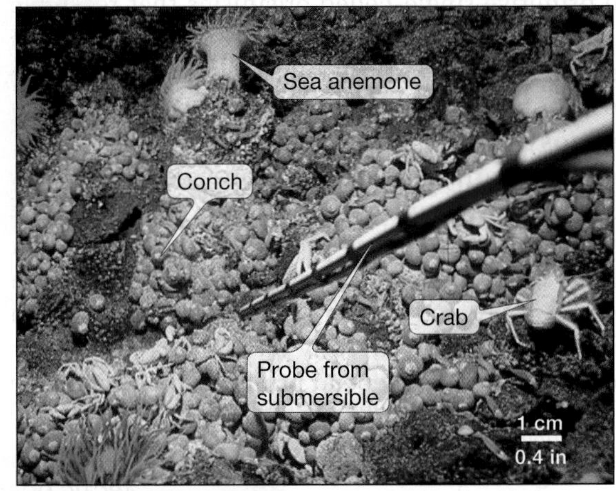


**(c)** A hydrothermal vent biocommunity from the Mariana Back-Arc Basin, which included a new genus and species of sea anemone (*Marianactis bythios*); the gastropod *Alviniconcha hessleri*, which is the first known conch to contain chemosynthetic bacteria; and the galatheid crab (*Munidopsis marianica*).

Figure 15.27 **Chemosynthetic life.**

**EXPLORING DATA ▶ By examining this figure, describe the chemical differences and similarities between photosynthesis and chemosynthesis.**

SmartFigure 15.28 **Comparing chemosynthesis (top panel) and photosynthesis (bottom panel).** The process of chemosynthesis, which is accomplished by archaea in the absence of sunlight, is represented by chemical formulas in the top panel. The process of photosynthesis, which is accomplished by plants and algae in the presence of sunlight, is shown by chemical formulas in the bottom panel. https://goo.gl/kTa8xx

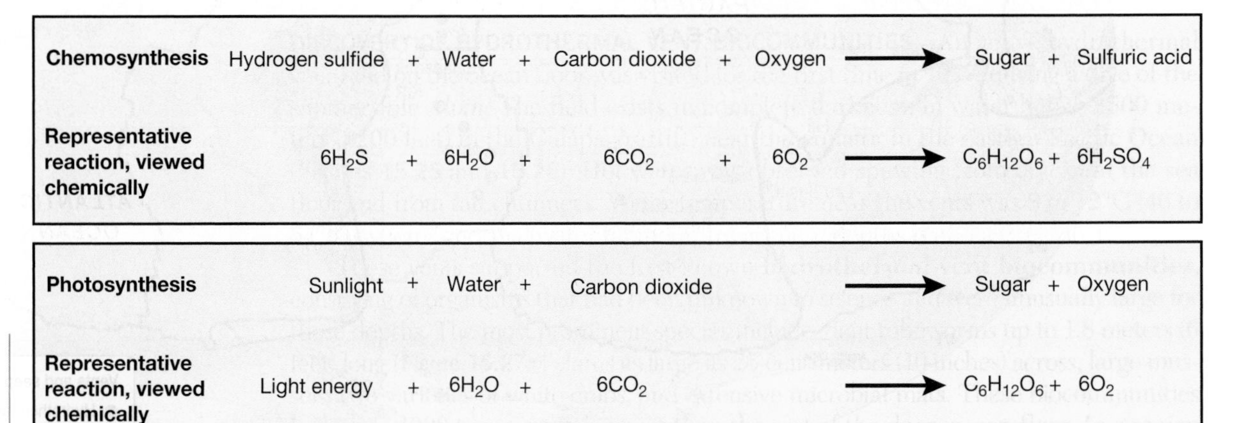

relationship with archaea. Tubeworms and giant clams, for instance, depend entirely on sulfur-oxidizing archaea that live symbiotically within their tissues. The microbes inside the tubeworms are provided habitat and bathe in hydrogen sulfide, which they chemosynthesize into sugars. The tubeworms, in return, get a steady supply of food and grow prolifically, sometimes gaining 80 centimeters (31 inches) in a year.

Analysis of the genome of the symbiotic archaea that live inside tubeworms reveals that these microbes are remarkably versatile. For example, researchers found that archaea can use two different methods to metabolize carbon dioxide and can switch back and forth to accommodate fast-changing environmental conditions. Such metabolic flexibility is a valuable asset in deep-sea vent habitats characterized by fluctuating flows of hot fluids.

**DISCOVERY OF OTHER HYDROTHERMAL VENT FIELDS?** In 1981, humans in a submersible first visited the Juan de Fuca Ridge biocommunity offshore of Oregon. Although vent fauna at this site are less abundant than at the Galápagos Rift and on the East Pacific Rise, the metallic sulfide deposits from the vents aroused much interest because they are the only active hydrothermal vent deposits in U.S. waters.

In 1982, the first hydrothermal vents beneath a thick layer of sediment were discovered during a submersible dive in the Guaymas Basin of the Gulf of California. In this region, a spreading center is actively working to rift apart the sea floor as it is being covered with sediment. Sediment samples recovered in this region were high in sulfide and saturated with hydrocarbons, which may have entered the food chain through bacteria. The abundance and diversity of life discovered here may exceed those of the Galápagos Rift and along the East Pacific Rise.

Like the Guaymas Basin, the Mariana Basin of the western Pacific has a small spreading center beneath a sediment-filled basin. A research dive via submersible in 1987 revealed many new species of hydrothermal vent organisms (**Figure 15.27c**). Subsequent exploration has revealed numerous hydrothermal vent biocommunities in other parts of the Pacific Ocean (see Figure 15.26) and additional new species. A 2017 study indicates that the local geologic setting and chemistry of the vent fluids determine the makeup of particular vent communities, and to date, more than 400 new species have been found at vent sites worldwide.

In 1985, the first active hydrothermal vents with associated biocommunities in the Atlantic Ocean were discovered at depths below 3600 meters (11,800 feet), near the axis of the Mid-Atlantic Ridge between 23 and 26 degrees north latitude. The predominant fauna of these vents consists of shrimp that have no eye lens but can detect levels of light emitted by the black smoker chimneys that are invisible to the human eye (**Figure 15.29**).

In 1993, a hydrothermal vent community was discovered on a flat-topped volcano rising to 1525 meters (5000 feet)—well above the walls of the rift valley along the Mid-Atlantic Ridge. Called the "Lucky Strike" vent field, it is about 1000 meters (3300 feet) shallower than most other sites. It is the only Mid-Atlantic Ridge site known to contain the mussels that are common at many other vent sites and is the only location where a new species of pink sea urchin has been found.

In 2000, Japanese investigators discovered the Indian Ocean's first known hydrothermal vent biocommunity, which is associated with black smokers spewing water up to 365°C (689°F). The vents are covered by shrimp similar to those found in Atlantic fields, while sea anemones mark the ambient temperature boundaries beyond. In between are clusters of animals similar to those found at other hydrothermal vents.

Also in 2000, researchers discovered the "Lost City" hydrothermal vent field about 15 kilometers (9 miles) west of the Mid-Atlantic Ridge. Remarkably, it has much lower temperatures (about 90°C [194°F]), a pH between 9 and 11 (much more alkaline than most black smokers), and tall chimneys made of calcium carbonate (instead of metal sulfides), and it releases methane and hydrogen into the surrounding water (not hydrogen sulfide or dissolved metals, which are the major outputs of volcanic black smokers). Unlike the activity at black smokers, which is driven by water superheated by shallow magma, Lost City vent activity is driven by *serpentinization*, which is a process caused

**CREATURE FEATURE 15.1**

### I'm the largest worm in the world!

The **giant deep-sea tubeworm** (*Riftia pachyptila*) lives in very deep water near black smokers and can grow to 2.5 meters (8 feet) in length and reach a diameter of 10 centimeters (4 inches); its blood-red color is due to hemoglobin. The giant deep-sea tubeworm has no digestive tract. Instead, symbiotic bacteria—which may comprise as much as half of the worm's body weight—perform chemosynthesis by converting hydrogen sulfide and other chemicals into organic molecules on which their host worm feeds.

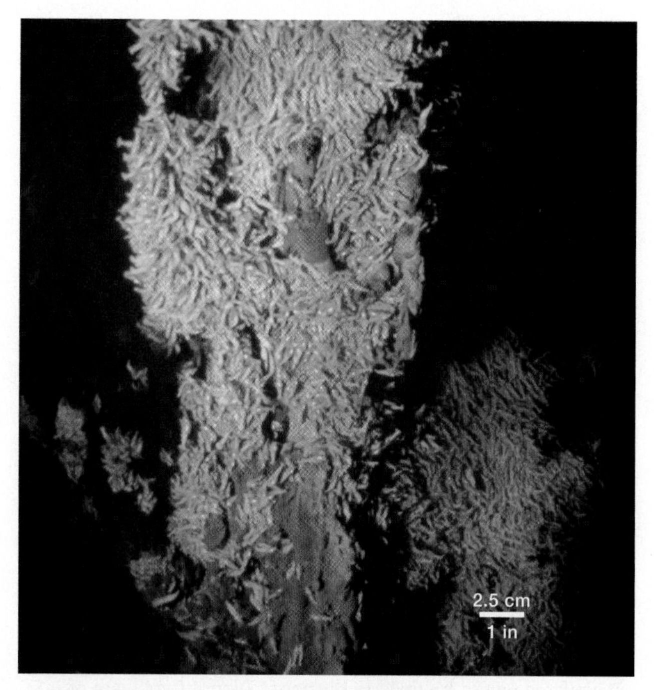


Figure 15.29 **Atlantic Ocean hydrothermal vent organisms.** Swarm of particulate-feeding shrimp, the predominant animals observed at hydrothermal vents near 26 degrees north latitude on the Mid-Atlantic Ridge.

by seawater circulating through underground hydrothermal systems that geochemically alters the underlying mantle rock and produces characteristic serpentine-group minerals. Scientists revisited the site in 2003 to collect samples, confirming that Lost City supports vastly different life-forms than black smokers—including a variety of microorganisms that live in, on, and around the vents. Scientists have suggested that life on Earth may have originated in warm, alkaline settings similar to those found today in Lost City.

Vents often differ dramatically from each other in chemical and geological characteristics. However, even vents that are physically similar can host distinctly different communities of organisms. For example, giant tubeworms are found only in Pacific vents. In the North Atlantic, meanwhile, shrimp and mussels dominate. And, on the deep-sea floor near Antarctica, the yeti crab (see Figure 12.8) is found in large numbers. Researchers are currently studying dispersal patterns and determining biogeographic relationships between vent sites to help explain this phenomenon.

Today, researchers continue to use submersibles to study deep-sea hydrothermal vents. Although there is direct evidence of about 300 hydrothermal vents worldwide, researchers estimate that another 700 or so await discovery. Every visit to a vent site—even a repeat visit—reveals new information about how vents work and the uniquely adapted microbes and other organisms that live there. In fact, the research effort is so intense at some sites that researchers have recently adopted a self-enforced code of conduct to ensure that humans do not alter vent ecosystems.

**LIFE SPAN OF HYDROTHERMAL VENTS** Because the hot-water plumbing of the sea floor is controlled by sporadic volcanic activity associated with mid-ocean ridge spreading centers, a vent may remain active for only limited periods—years or sometimes decades. For instance, a hydrothermal vent field called the Coaxial Site along the Juan de Fuca Ridge offshore of Washington that had been active was revisited a few years later and found to be inactive. Inactive sites such as this one can be identified by an accumulation of large numbers of dead hydrothermal vent organisms. When the vent becomes inactive and the hydrogen sulfide that serves as the source of energy for the community is no longer available, organisms of the community die if they cannot move elsewhere.

Other sites indicate an increase in volcanic activity. For example, at a site along the East Pacific Rise known as Nine-Degrees North, a large number of tubeworms were cooked by lava flowing into their midst, in what has been described as a "tubeworm barbecue." The discovery of newly formed and ancient vent areas along spreading centers indicates that hydrothermal vents can suddenly appear or cease to operate. Moreover, areas of active venting may lie tens or even hundreds of kilometers apart.

Hydrothermal vent organisms are well adapted to the temporary nature of hydrothermal vents. Most have high metabolic rates, for example, which cause them to mature rapidly so they can reproduce while the vent is still active.

Studies of several hydrothermal vent sites suggest that species diversity is low. Many species, however, are common to widely separated hydrothermal vent fields. Although hydrothermal vent animals typically release drifting larvae into the water, it is not clear how the larvae—which can survive a few months at most—are able to survive the journey to hydrothermal vents that lie at such great distances from one another.

One idea, called the *dead whale hypothesis*, suggests that when large animals die, they may sink to the deep-ocean floor, decompose, and provide an energy source in stepping-stone fashion for the larvae of hydrothermal vent organisms. The organisms settle and grow where a large dead animal rests on the deep-ocean floor; then they breed and release their own larvae, some of which make it to the next hydrothermal vent field. Other studies suggest that the rift valleys of mid-ocean ridges act as passageways that drifting larvae traverse to inhabit new vent fields. Still other scientific research that involves dispersing and then tracking chemical tracers indicates that deep-ocean currents are strong enough to transport drifting larvae to new sites. By whatever means they travel, they colonize new hydrothermal vents soon after the vents are created. In 2006, for example, a well-studied hydrothermal

## STUDENTS SOMETIMES ASK . . .

***Are the mussels found at hydrothermal vents edible?***

Not by humans. The microbes that form the base of the food web use hydrogen sulfide gas (which has a characteristic "rotten egg" odor) as a source of energy. To most organisms, sulfide is a deadly poison, even at low levels, and tends to concentrate in tissues. Although organisms within the hydrothermal biocommunities can ingest sulfide and have mechanisms of getting rid of it, hydrothermal vent organisms would be toxic to humans. Even if they were edible, though, mussels found at hydrothermal vents would be expensive to harvest because they live at such great depths.

vent community along the East Pacific Rise had been wiped out by seafloor eruptions. In 2007 and 2008, researchers returned to the area and discovered that tubeworms and other life-forms had already established themselves.

**HYDROTHERMAL VENTS AND THE ORIGIN OF LIFE** Life is thought to have begun in the oceans, and environments similar to those of the hydrothermal vents must have been present in the early history of the planet. The uniformity of conditions and abundant energy of the vents, therefore, have led some scientists to propose that hydrothermal vents would have provided an ideal habitat for the origin of life. In fact, hydrothermal vents may represent one of the oldest life-sustaining environments because hydrothermal activity occurs wherever there is both volcanic activity and water. The presence at vents of bacteria-like archaea, which have ancient genetic makeup, helps support this idea.

The ancient status of deep-sea life was strengthened by the discovery of deep-sea microbes with identical genes to those of microbes found in the human body. Researchers isolated two previously unknown bacterial species from hydrothermal vents near Japan and compared the new species' genomes to the genomes of two common gut pathogens, one that causes ulcers and another that causes diarrhea. The comparison shows that despite eons of evolutionary change, the deep-sea species and the pathogens share genes that enable them to colonize animal hosts. According to the researchers, genes that likely help deep-sea bacteria maintain symbiotic relationships with other vent-dwelling organisms assist their gut-dwelling relatives in evading their hosts' immune systems. The researchers suggest that the human-harming microbes evolved from deep-sea ancestors and, as life moved on to land, the microbes came with it, later acquiring more virulence factors while living in symbiosis with animals.

Although there is ample evidence to support the idea that life might have begun at hydrothermal vents, these vents tend to be unstable and short-lived, so some scientists question whether they could have adequately spawned life. The process of serpentinization may hold part of the answer to the question of whether life on Earth emerged from the deep sea. Serpentinization—the chemical alteration of mantle rock by deep circulation of seawater—releases heat, hydrogen, methane, and mineral components, all of which are important for chemosynthetic life. In fact, some scientists have proposed that serpentinization at places like Lost City hydrothermal field or in ocean trenches could have more readily fueled Earth's first life because it occurs across much larger areas and can be sustained for much longer periods of geologic time, thus providing a suitable habitat for the development of early chemosynthetic lifeforms.

**Figure 15.30 Hypersaline seep biocommunity at the base of the Florida Escarpment.**

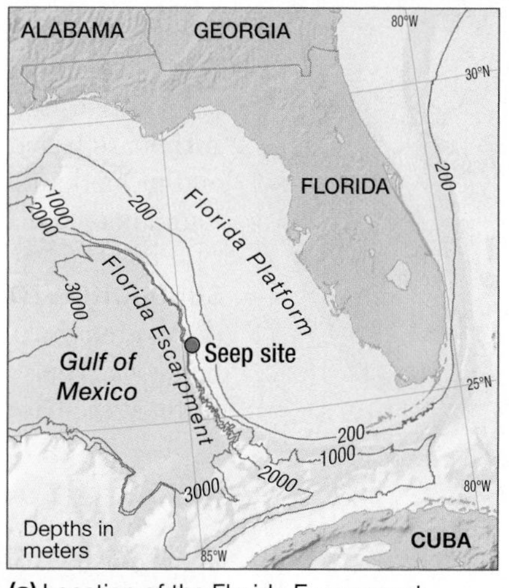


**(a)** Location of the Florida Escarpment hypersaline seep.

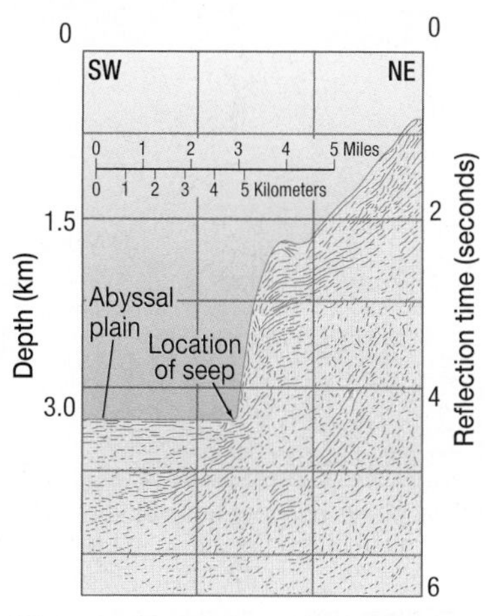


**(b)** Seismic reflection profile of the Florida Escarpment, showing the location of the hypersaline seep at its base.

**(c)** Florida Escarpment seep biocommunity of dense mussel beds. White dots are small gastropods on mussel shells. Tubeworms *(lower right)* are covered with hydrozoans and galatheid crabs.

## Low-Temperature Seep Biocommunities: Organisms and Their Adaptations

Three additional submarine seep environments—locations where water trickles out of the sea floor—have been found that chemosynthetically support biocommunities similar to hydrothermal vent communities. Because they exist at much lower temperatures, these environments are known collectively as **cold seeps**.

**HYPERSALINE SEEPS** In 1984, a hypersaline seep was studied in water depths below 3000 meters (9800 feet) at the base of the Florida Escarpment in the Gulf of Mexico (**Figure 15.30a**). The water from this seep had a salinity of 46.2‰, but its temperature was about the same as the surrounding seawater.

Researchers had discovered a **hypersaline seep biocommunity** similar in many respects to a hydrothermal vent biocommunity. The seeping water appears to flow from fractures at the base of a limestone escarpment (**Figure 15.30b**) and move out across the clay deposits of the abyssal plain at a depth of about 3200 meters (10,500 feet).

The hydrogen sulfide–rich waters of these hypersaline seeps support a number of white microbial growths called *mats*, which conduct chemosynthesis in a fashion similar to archaea at hydrothermal vents. These and other chemosynthetic microbes may provide most of the sustenance for a diverse community of animals, including sea stars, shrimp, snails, limpets, brittle stars, anemones, tubeworms, crabs, clams, mussels, and a few species of fish (**Figure 15.30c**).

**HYDROCARBON SEEPS** Also observed in 1984 were dense biological communities associated with oil and gas seeps on the Gulf of Mexico continental slope (**Figure 15.31**). Trawls at depths of between 600 and 700 meters (2000 and 2300 feet) recovered fauna similar to those observed at hydrothermal vents and at the hypersaline seep in the Gulf of Mexico. Subsequent investigations have identified nearly 100 seeps on the continental slope that have the potential of hosting chemosynthetic biocommunities; 10 of these have been visited with submersibles to depths of 2775 meters (9100 feet), where chemosynthetic bacteria and a host of other organisms were discovered.

Carbon-isotope analysis indicates that these **hydrocarbon seep biocommunities** are based on chemosynthesis that derives its energy from hydrogen sulfide and/or methane. Microbial oxidation of methane produces calcium carbonate slabs found here and at other hydrocarbon seeps (see Figure 15.26).

**SUBDUCTION ZONE SEEPS** In 1984, a **subduction zone seep biocommunity** was discovered during one of *Alvin*'s dives to study folding of the sea floor in a subduction zone. The seep is located near the Cascadia subduction zone of the Juan de Fuca Plate, at the base of the continental slope off the coast of Oregon (**Figure 15.32a**). The

Figure 15.31 **Hydrocarbon seeps on the continental slope in the Gulf of Mexico.**

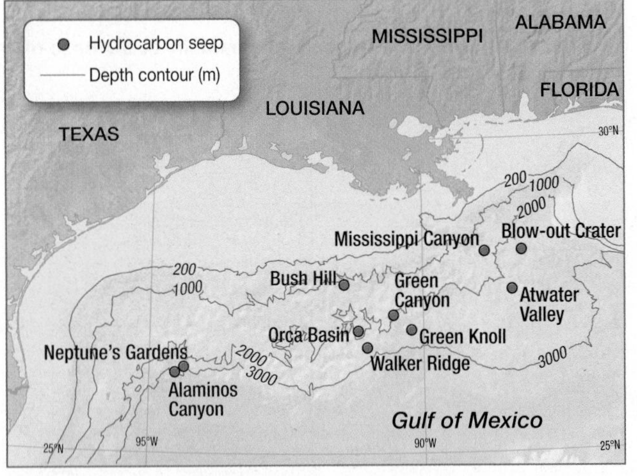


**(a)** Map showing locations of known hydrocarbon seeps that contain biocommunities in the Gulf of Mexico.

**(c)** Close-up photo of chemosynthetic mussels and tubeworms from the Bush Hill seep.

**(b)** Photo showing abundant mussels and tubeworms at Neptune's Gardens (Alaminos Canyon site).

trench is filled with sediments, which are folded into a ridge at the seaward edge of the slope. At the crest of this ridge, water slowly flows from the 2-million-year-old folded sedimentary rocks into a thin overlying layer of soft sediment on the sea floor. Eventually, the water is released from the sediment through seeps on the ocean floor.

At a depth of 2036 meters (6678 feet), the seeps produce water that is only about 0.3°C (0.5°F) warmer than seawater at that depth. The vent water contains methane that is probably produced by decomposition of organic material in the sedimentary rocks. Microbes oxidize the methane, chemosynthçetically producing food for themselves and the rest of the community, which contains many of the same genera found at other vent and seep sites (**Figure 15.32b**).

Since the detection of subduction zone seeps, similar communities have been discovered in other subduction zones, including the Japan Trench and the Peru–Chile Trench. All these subduction zone seeps are located on the landward side of the trenches, at depths from 1300 to 5640 meters (4300 to 18,500 feet).

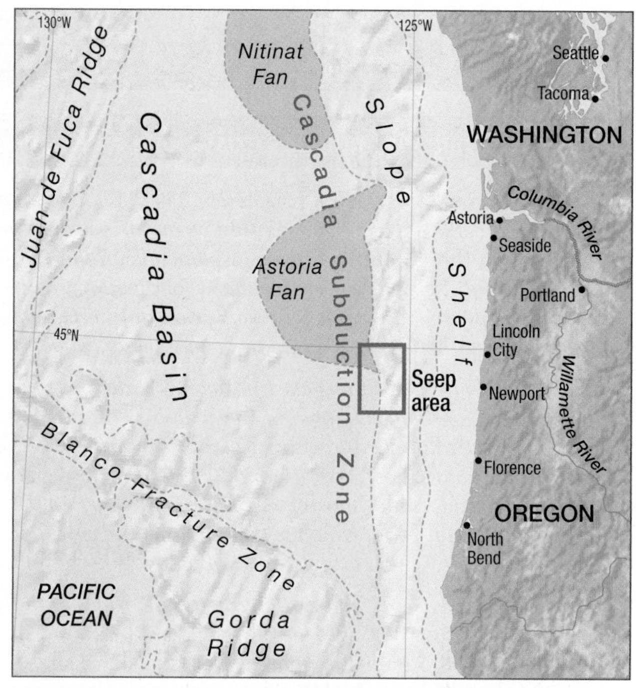


**(a)** Map showing sea floor features and the location of vent biocommunities off the coast of Oregon. These communities are associated with the Cascadia subduction zone, where sediment filling the trench is folded into a ridge with vents at its crest.

**(b)** Photo of giant white clams *(Calyptogena soyoae)* half buried in methane-rich mud at a depth of 1100 meters (3600 feet) in Sagami Bay, near the Japan Trench. The clams host sulfide-oxidizing microorganisms, which supply the clams with food.

Figure 15.32 **Subduction zone seep biocommunities.**

## The Deep Biosphere: A New Frontier

The discovery of the rich microbe communities at hydrothermal vents has resulted in the exploration of the **deep biosphere**, an environment that exists within the sea floor itself. Only recently have scientists even considered that microbial life might exist deep within Earth, and in 2002, researchers made the first expedition to study life in this environment. Cores were drilled up to 420 meters (1380 feet) deep into the sea floor off Peru, in water depths between 150 and 5300 meters (490 and 17,400 feet). At these depths, researchers discovered a host of diverse and active microbial communities living within circulating fluids that pass through the porous sea floor. Other subsequent research confirms the abundance and diversity of microbes in deep-sea floor rocks and associated sediments that rival the rich microbial ecosystems found in soil.

These studies suggest that as much as two-thirds of Earth's entire bacterial biomass might exist in the deep biosphere. Moreover, deep biosphere microbes fuel their metabolisms by taking advantage of the chemical energy stored in various minerals. In addition, the findings raise new questions about the role of the deep-sea floor in the evolution of life on Earth. Intriguingly, other bodies in the solar system have similar subsurface conditions, so they might harbor microbes, too. Earth's deep biosphere will therefore continue to be an area of active research.

**CONCEPT CHECK 15.4** ▸ **Specify characteristics of the communities that exist on the deep-ocean floor.**

1 Where does the food come from to supply organisms living on the deep-ocean floor? How does this affect benthic biomass?

2 Describe the characteristics of hydrothermal vents. What evidence suggests that hydrothermal vents have short life spans?

3 What is the "dead whale hypothesis"? What other ideas have been suggested to help explain how organisms from hydrothermal vent biocommunities populate new vent sites?

4 What are the major differences between the conditions and biocommunities of hydrothermal vents and cold seeps? How are they similar?

5 Discuss changes in the physical environment that occur as one moves from the shoreline to the deep-ocean floor.

**RECAP** **Hydrothermal vent biocommunities occur near black smokers and rely on chemosynthetic archaea for food. Other deep-sea cold seep biocommunities that depend on chemosynthesis exist around hypersaline, hydrocarbon, and subduction zone seeps.**

# ESSENTIAL CONCEPTS REVIEW

## 15.1 ▶ What communities exist along rocky shores?

- *More than 98% of the approximately 230,000 known marine species live in diverse environments within or on the ocean floor. Species diversity* of these benthic organisms *depends on their ability to adapt to the conditions of their environment,* particularly temperature. With few exceptions, the *biomass of benthic organisms closely matches that of photosynthetic productivity in surface waters above.*
- *Many adverse conditions exist in the intertidal zone of rocky shores, but organisms have adapted* so they can densely populate these environments. Influenced by the tides, rocky shores can be divided into a *high tide zone* (mostly dry), a *middle tide zone* (equally wet and dry), and a *low tide zone* (mostly wet). The *intertidal zone* is bounded by the *supratidal zone,* which is covered only by storm waves, and the *subtidal zone,* which extends below the low-tide shoreline.
- *Each of these zones contains characteristic types of marine life.* Periwinkle snails, rock lice, and limpets can be found in the subtidal zone. Sessile organisms, such as buckshot barnacles, can be found in the high tide zone. *Algae become more abundant in the middle tide zone, and the diversity and abundance of the flora and fauna increase toward the lower intertidal zone.* Acorn barnacles, gooseneck barnacles, mussels, and sea stars are commonly found in the middle tide zone, as are sea anemones, fishes, hermit crabs, and sea urchins. The low tide zone in temperate latitudes has a variety of moderately sized red and brown algae that provide a drooping canopy for animal life.

### Selected Key Terms

Use the **glossary** at the end of this book to discover the meanings of these Selected Key Terms: **epifauna, spray/supratidal zone, intertidal zone, high tide zone, middle tide zone, low tide zone, nematocyst.**

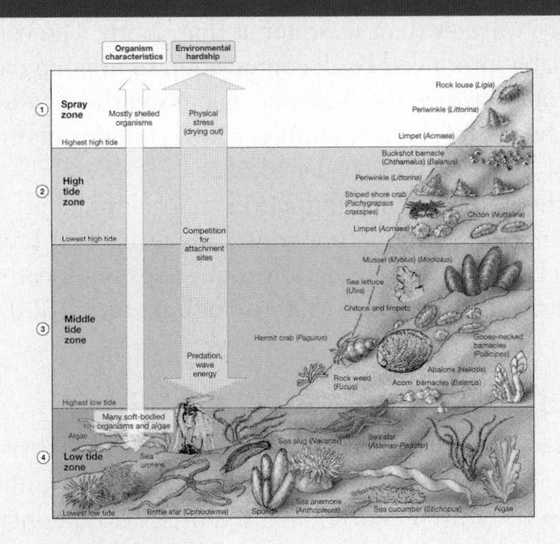


### Critical Thinking Question

From memory, construct a diagram that shows a rocky-shore intertidal region. Name each zone and include characteristic organisms that are typically found there.

### Active Learning Exercise

Divide the class into groups of four students each. Within each group, have each person select one of the four main intertidal zones along rocky shores: (1) the spray zone, (2) the high tide zone, (3) the middle tide zone, and (4) the low tide zone. Have each person select two different organisms that inhabit that intertidal zone and write down adaptations of each organism that allow it to survive there. Then compare your answers with your group and with the class.

## 15.2 ▶ What communities exist along sediment-covered shores?

- *Many varieties of burrowing infauna are common along sediment-covered shores* (that is, beaches, salt marshes, and mud flats). Compared with rocky shores, however, *sediment-covered shores have less diversity of species.* As with the rocky shore, *the diversity of species and abundance of life on the sediment-covered shore increase toward the low tide shoreline.*
- In more protected segments of the shore, wave energy is lower, so sand and mud are deposited. Sand deposits are usually well oxygenated compared to mud deposits. *The intertidal region of sediment-covered shores has high, middle, and low tide zones, similar to rocky shores.* Organisms characteristic of *sandy beaches* include bivalve (two-shelled) mollusks, lugworms, beach hoppers, sand crabs, sand stars, and heart urchins. Organisms characteristic of *mud flats* include eelgrass, turtle grass, bivalve mollusks, and fiddler crabs.

Highest high tide
Lowest high tide
Highest low tide
Lowest low tide
Intertidal zone
Common organisms
Subtidal zone
Sand dollar (Echinarachnius)
Heart urchin (Echinocardium)
Sea cucumber (Stichopus)
Horseshoe crab (Limulus)
Low tide zone
Clams
Sand crab (Emerita)
Lugworm (Arenicola)
Sand star (Astropecten)
Ice cream cone worm (Pectinaria)
Middle tide zone
Bristle worm (Polychaeta)
High tide zone
Ghost shrimp (Callianassa)
Supratidal (backshore)
Beach hopper (Orchestoidea)
Ghost crab (Ocypode)

### Selected Key Terms

Use the **glossary** at the end of this book to discover the meanings of these Selected Key Terms: **infauna, bivalve, mollusk, annelid, crustacean, echinoderm, meiofauna.**

### Critical Thinking Question

From memory, construct a diagram that shows a sand-covered intertidal region. Name each zone and include characteristic organisms that are typically found there.

### Active Learning Exercise

Divide the class into groups of five students each. Within each group, have each person select one of the five main intertidal zones along sediment-covered shores: (1) the supratidal (backshore) zone, (2) the high tide zone, (3) the middle tide zone, (4) the low tide zone, and (5) the subtidal zone. Have each person select one organism that inhabits that intertidal zone and write down its adaptations that allow it to survive there. Then compare your answers with your group and with the class.

## 15.3 ▶ What communities exist on the shallow offshore ocean floor?

- *Attached to the rocky subtidal bottom just beyond the shoreline is a band of algae that often creates kelp forests.* Kelp forests are home to many organisms, including other varieties of algae, mollusks, sea stars, fishes, octopuses, lobsters, marine mammals, sea hares, and sea urchins.
- *Spiny lobsters* are common to rocky bottoms in the Caribbean and along the West Coast, and the *American (Maine) lobster* is found from Labrador, Canada, to Cape Hatteras, North Carolina. Oyster beds found in estuarine environments consist of individuals that attach themselves to the bottom or to the empty shells of previous generations.
- *Coral reefs consist of large colonies of coral polyps and many other species that need warm water and strong sunlight to live.* Coral reefs are usually *found in nutrient-poor tropical waters.* Reef-building corals and other mixotrophs are *hermatypic,* containing *symbiotic algae (zooxanthellae)* in their tissues. Coral reefs exhibit three stages of development: *fringing reef, barrier reef,* and *atoll.* Delicate varieties of coral are found at 150 meters (500 feet), and they become more massive near the surface, where wave energy is higher. *The potentially lethal "bleaching" of coral reefs is caused by the removal or expulsion of symbiotic algae,* probably under the stress of elevated sea surface temperatures.

### Selected Key Terms

Use the **glossary** at the end of this book to discover the meanings of these Selected Key Terms: **subtidal zone, kelp forest, coral reef, polyp, zooxanthellae, hermatypic, mixotroph, fringing reef, barrier reef, atoll, coral bleaching.**

### Critical Thinking Question

Discuss the relationship between corals and algae and explain how corals benefit from having internal algae.

### Active Learning Exercise

Working with another student in class, answer this question: What in particular about the environment where corals are found makes the symbiotic relationship between corals and algae indispensable to both? Then share your answer with the class.

## 15.4 ▶ What communities exist on the deep-ocean floor?

- *The physical conditions of the deep-ocean floor* are much different from those of shallow water. There is *no light,* and the *water is uniformly cold. The primary food source is from the surface waters above, which limits biomass.* Species diversity in the deep ocean, however, is much higher than was previously thought.
- *Primary production in hydrothermal vent communities near black smokers is due to chemosynthesis.* Some evidence suggests that *hydrothermal vents may have been some of the first regions where life became established on Earth,* despite the short life span of individual vents. Chemosynthesis has also been identified in low-temperature seep biocommunities near *hypersaline, hydrocarbon,* and *subduction zone seeps.* Studies of the *deep biosphere* below the ocean floor *reveal a host of microbes.*

### Selected Key Terms

Use the **glossary** at the end of this book to discover the meanings of these Selected Key Terms: **abyssal storm, hydrothermal vent biocommunity, black smoker, archaea, chemosynthesis, cold seep, hypersaline seep biocommunity, hydrocarbon seep biocommunity, subduction zone seep biocommunity, deep biosphere.**

### Critical Thinking Question

Using Figure 15.28, discuss the major differences between chemosynthesis and photosynthesis from a chemical standpoint.

### Active Learning Exercise

Working with another student in class, create a table that compares the temperature of fluids emitted, characteristic features, and example locations of the following types of hydrothermal vents: (1) black smokers and (2) areas like Lost City in the Atlantic Ocean. Which one is more likely to have created an environment for the development of early life on Earth?

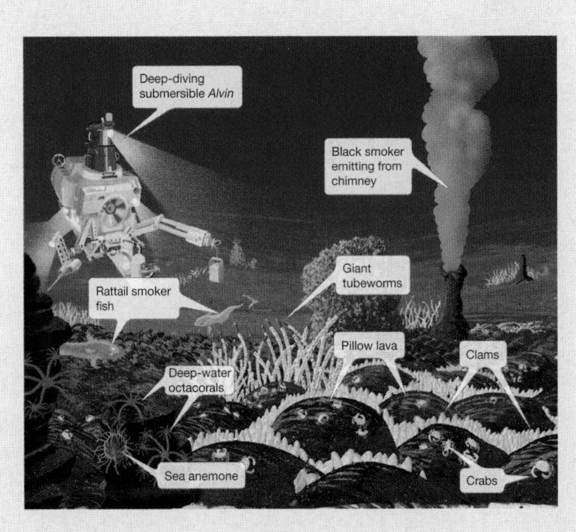


## Mastering Oceanography™

www.masteringoceanography.com

Looking for additional review and test prep materials? With individualized coaching on the toughest topics of the course, Mastering Oceanography offers a wide variety of ways for you to move beyond memorization and deeply grasp the underlying processes of how the oceans work. Visit the Study Area in **www.masteringoceanography.com** to find practice quizzes, study tools, and multimedia that will improve your understanding of this chapter's content. Sign in today to access the following features: Self Study Quizzes, SmartFigures, Oceanography Videos and Animations, Squidtoons, Dynamic Study Modules, and an optional Pearson eText with embedded videos.

# The Oceans and Climate Change

# 16

## ESSENTIAL LEARNING CONCEPTS

At the end of this chapter, you should be able to:

- ☐ **16.1** Specify the components of Earth's climate system, and the role of the carbon cycle.
- ☐ **16.2** Examine the evidence that shows how Earth's recent climate change is caused by human activities, not a natural cycle.
- ☐ **16.3** Demonstrate an understanding of how the atmosphere's greenhouse effect works.
- ☐ **16.4** Specify the changes that are occurring in the oceans as a result of global warming.
- ☐ **16.5** Evaluate options for what can be done to reduce greenhouse gases.

↑ *Check when completed*

Climate Connection

News related to climate change and global warming, which includes sea level rise, severe weather, heat waves, and drought-driven wildfires, seems to be reported daily. This has spurred intense debate on whether climate change is natural or human-caused, and what climate changes are likely to occur in the future. These topics have also become the subject of numerous international conferences and have spurred complicated discussions among journalists, policy makers, and scientists. Human-caused climate change continues to be one of the most studied aspects of climate science.

Throughout its long history, Earth has experienced both warmer and cooler global climates, as compared to today's climate. In fact, evidence from fossils, sea floor sediments, and rocks on land suggest that many places on Earth have experienced dramatic swings in climate over geologic time. For example, fossils such as corals and coal deposits that are indicative of warm temperatures are known to have formed at high latitudes. Alternatively, regions that are known to have been located at low latitudes have sediments likely to have been produced by glaciation. Even after the movement of tectonic plates over time is taken into account, evidence of past climate change remains.

In the past, climate changes have occurred as the result of natural causes. According to numerous climate science studies, today's climate changes are caused by human activities. What has convinced scientists that modern climate change is dominated by human influences—not natural variability—is that the observed changes are so large and occurring so rapidly that they exceed the bounds of any natural factors that influence Earth's climate. Moreover, these changes are likely to continue for at least the next 1000 years. Climate changes can be very disruptive, not only to humans but also to many other life-forms as well, especially if they occur as rapidly as some scientists predict. The fact that human-caused emissions are beginning to alter the basic chemistry of the oceans, for example, has implications for the health of many marine ecosystems.

In this chapter, we will examine Earth's climate system, the role of the carbon cycle, the science that explains Earth's recent and dramatic climate change, how the greenhouse effect works, what changes are occurring in the oceans today, and what can be done about this urgent problem.

***"Human-induced climate change is a reality, not only in remote polar regions and in small tropical islands, but everyplace around the country, in our own backyards. It's happening. It's happening now. It's not just a problem for the future. We are beginning to see its impacts in our daily lives. More than that, humans are responsible for the changes that we are seeing, and our actions now will determine the extent of future change and the severity of the impacts."***

***—Jane Lubchenco, marine ecologist and NOAA chief administrator (2009)***

## 16.1 ▶ What Comprises Earth's Climate System?

**Climate** is defined as the conditions of Earth's atmosphere—including temperature, precipitation, and wind—that characteristically prevail in a particular region over extended time spans.

◂ **A massive iceberg floating off the coast of Newfoundland, Canada, towers over spectators in the town of Ferryland.** Every spring, chunks of Greenland's ice sheet break off and enter shipping lanes near Labrador and Newfoundland. Known as "Iceberg Alley," these ice-strewn corridors are becoming increasingly more dangerous to navigation as global warming accelerates the melting of Greenland's ice cap.

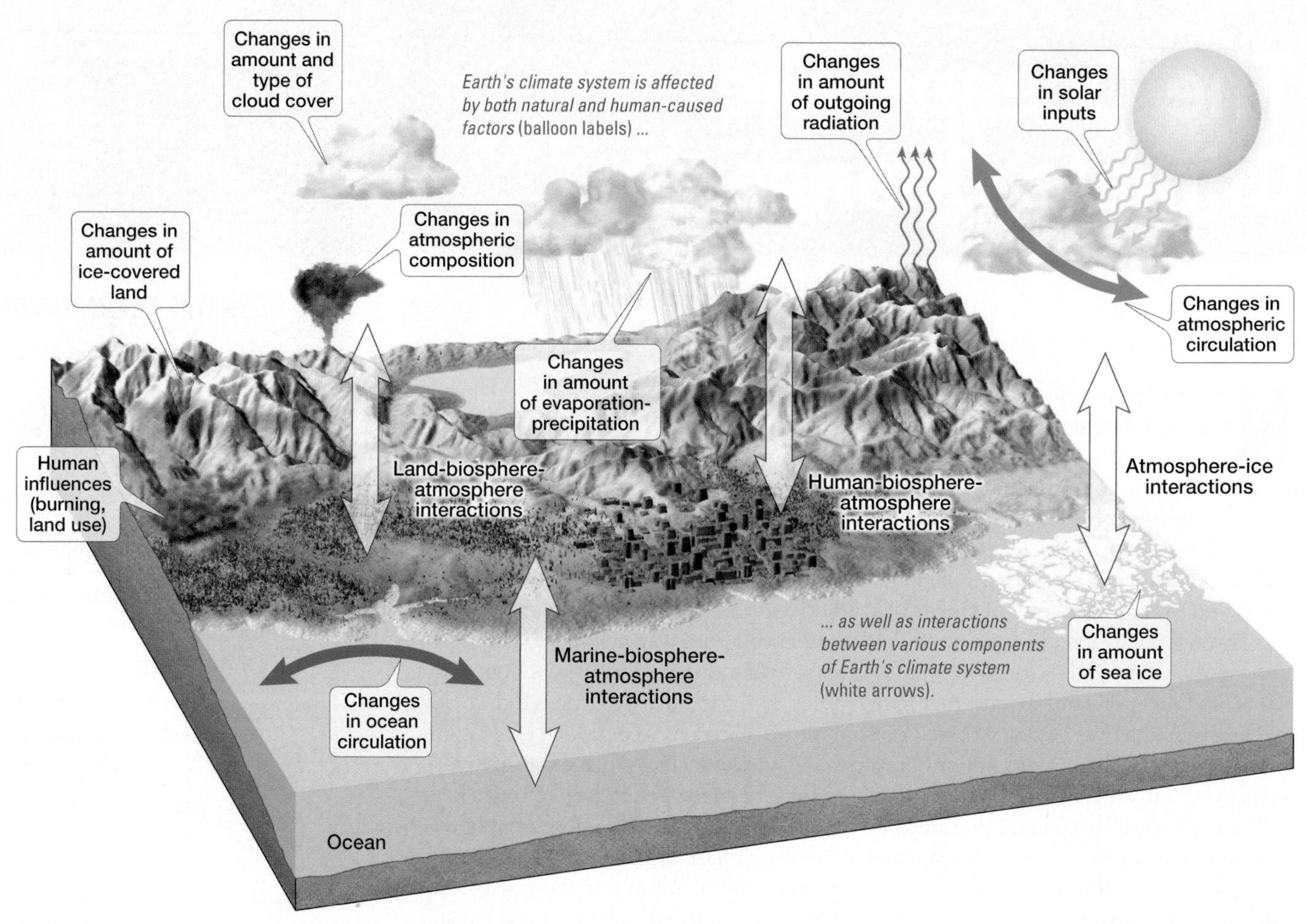


SmartFigure 16.1 **Major components of Earth's climate system.** Schematic view of the major components of Earth's climate system, including their interactions.
https://goo.gl/Vfr9hO

Obtaining a full understanding of Earth's climate involves studying more than just the atmosphere. Earth's climate is a complex and interacting system that includes five spheres: the atmosphere, hydrosphere, lithosphere, biosphere, and cryosphere. (The cryosphere [*kruos* = icy cold, *sphere* = globe] refers to the ice and snow that exists at Earth's surface.) Earth's **climate system** involves the exchanges of energy and moisture that occur among the five spheres. These exchanges link the atmosphere to the other spheres so that the entire system functions as an interactive unit. Changes to the climate system do not occur in isolation. Rather, when one part of it changes, the other components also react. The major components of Earth's climate system are shown in **Figure 16.1**. Note that the oceans are the most massive part of Earth's climate system.

Changes in Earth's climate system are large scale, complex, and involve many **feedback loops**, which are processes that modify initial changes. For example, **Figure 16.2** (*left*) shows that warmer surface temperatures increase evaporation rates. This in turn increases water vapor in the atmosphere. Like carbon dioxide, water vapor is a greenhouse gas, and it, too, absorbs heat radiated from Earth's surface. Accordingly, the more water vapor in the air, the less heat escapes back into space and the warmer the planet becomes. This type of feedback loop is called a **positive-feedback loop**, or *reinforcing loop*, because it has an amplifying effect that fortifies or adds to an initial change. That is, A produces B, which in turn produces more of A, which in turn produces more of B, and so on.

Alternatively, a **negative-feedback loop**, or *balancing loop*, has a diminishing effect that tends to counteract or subtract from an initial change. One such example is the formation of clouds (Figure 16.2, *right*). A probable result of a

## STUDENTS SOMETIMES ASK . . .

### *What's the difference between weather and climate?*

*Weather* describes the conditions of the atmosphere at a given place and time, whereas *climate* is the long-term average of weather. For example, the expected weather conditions on a particular day will help you determine if you'll wear shorts or thermal underwear that day; the ratio of shorts to thermal underwear in your drawer reflects the climate of the region. Or, as Mark Twain purportedly said about the difference between the two, "Climate is what you expect, weather is what you get."

global temperature rise is an accompanying increase in cloud cover due to the higher moisture content of the atmosphere. Most clouds are good reflectors of incoming solar energy, thus diminishing the amount of solar energy available to heat Earth's surface and warm the atmosphere. In this way, clouds can cause a decrease in overall air temperature. In this instance, A produces B, which in turn produces *less* of A.

These two examples of increased water vapor in the atmosphere show that it can be both a positive-feedback loop and a negative-feedback loop (Figure 16.2). Which effect, if either, is stronger? Studies show that the negative effect of higher reflectivity is dominant, so clouds have an overall cooling effect. Therefore, the net result of an increase in atmospheric moisture should be a decrease in air temperature. The magnitude of this negative-feedback loop, however, is not likely to be as great as the feedback caused by other positive-feedback loops between other parts of Earth's climate system. Thus, although increases in atmospheric moisture and cloud cover may partly offset a global temperature increase, climate models show that the overall effect will still be a temperature increase. In fact, climate models backed by measurements of Earth's climate indicate that increasing levels of human-caused emissions will lead to a warmer planet with a different distribution of climate patterns than what currently exist on Earth today. (Note that something created by or the result of humans is also known by the term *anthropogenic* [*anthro* = human, *generare* = to produce].)

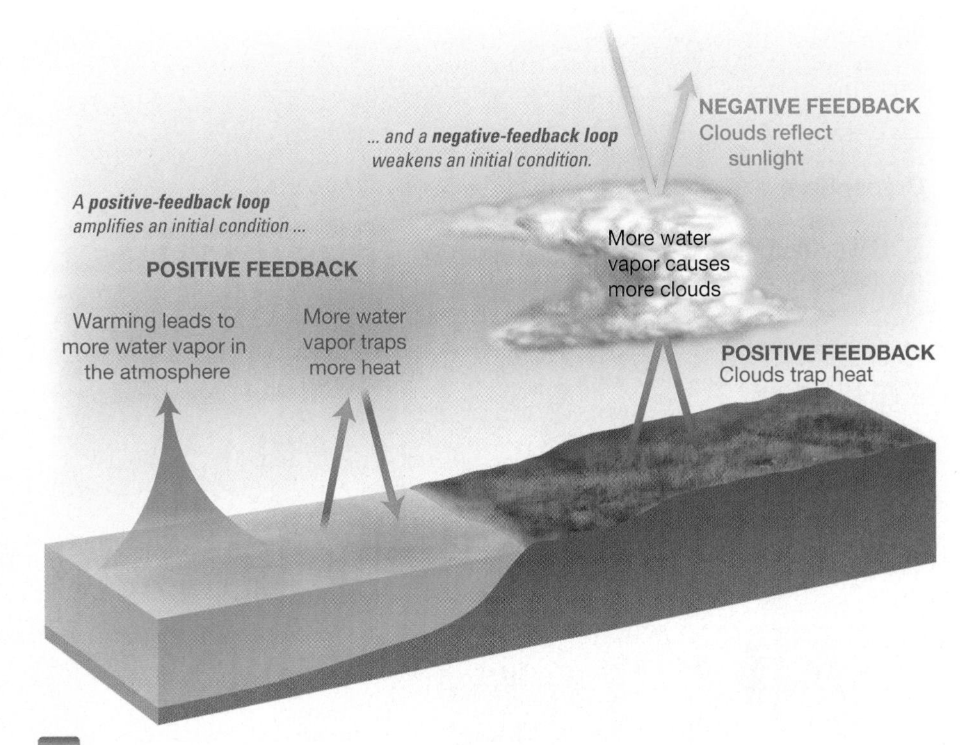


**SmartFigure 16.2 Examples of climate feedback loops.** Schematic drawing of a positive-feedback loop (*left*) that amplifies an initial change and a negative-feedback loop (*right*) that reduces an initial change. Note that clouds can be parts of both positive- and negative-feedback loops.
https://goo.gl/URQBXr

Other feedback loops may have even stronger effects on future climate. For example, rising temperatures have caused Arctic sea ice to melt, and as a result, the lack of highly reflective ice cover means that Arctic Ocean waters will absorb far more of the Sun's incoming radiation. This creates a positive-feedback loop whereby warming the ocean leads to further melting of sea ice.

## The Carbon Cycle

Central to any discussion about climate is the role that carbon plays in Earth's climate system. The **carbon cycle** describes the flow of carbon between the atmosphere, hydrosphere, lithosphere, biosphere, and cryosphere, and it affects every aspect of climate, from greenhouses gases in the atmosphere to carbonate-containing sediments on the sea floor.

Let's start with carbon dioxide, which exists in all of Earth's previously mentioned "spheres." Carbon dioxide is the product of the combustion of **fossil fuels** (oil, natural gas, and coal) and is the most abundant greenhouse gas in the atmosphere, but it is also a product of biological respiration and is the form of carbon fixed during photosynthesis (biosphere). In the oceans (hydrosphere), carbon dioxide participates in the carbonate buffering reactions and so helps maintain seawater pH. It is removed from seawater by shell-forming marine organisms in the form of calcium carbonate minerals. When these organisms die and their hard tests settle to the sea floor, the shells form carbonate-containing sediments and, ultimately, rock (lithosphere). Sea floor rocks are recycled during subduction at ocean

### STUDENTS SOMETIMES ASK . . .

***In grade school, many of us were taught that humans exhale $CO_2$ but plants absorb $CO_2$ and return oxygen to the air (keeping the carbon for fiber). Is this still valid? If so, why hasn't plant life turned the higher levels of atmospheric $CO_2$ back into oxygen?***

Plants are indeed growing more vigorously because of increased carbon dioxide in the atmosphere (about a quarter of human-produced greenhouse gas emissions are absorbed by plants), but it's not enough to offset the increases in carbon dioxide caused by the burning of fossil fuels. What has changed the balance you learned about in grade school is that the burning of fossil fuels by humans has introduced carbon that was previously locked away underground for millions of years. This amount of carbon introduced in such a short period of time simply exceeds the ability of the biosphere to absorb it. Humanity's breaths are insignificant in comparison, and have a negligible effect on atmospheric greenhouse gas concentrations, in spite of the huge number of people on Earth (see the **Afterword** at the end of this book).

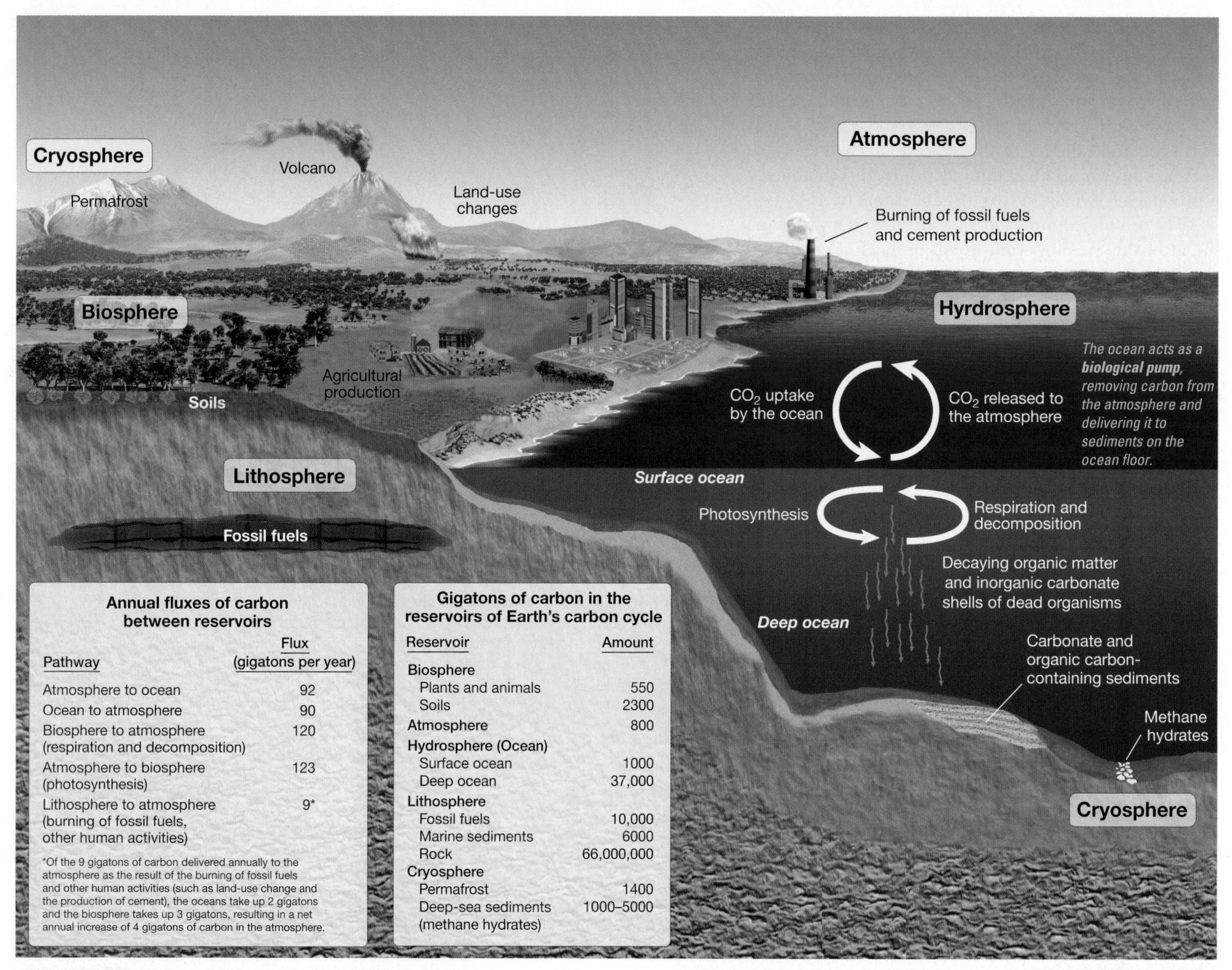


**Annual fluxes of carbon between reservoirs**

| Pathway | Flux (gigatons per year) |
|---|---|
| Atmosphere to ocean | 92 |
| Ocean to atmosphere | 90 |
| Biosphere to atmosphere (respiration and decomposition) | 120 |
| Atmosphere to biosphere (photosynthesis) | 123 |
| Lithosphere to atmosphere (burning of fossil fuels, other human activities) | 9* |

*Of the 9 gigatons of carbon delivered annually to the atmosphere as the result of the burning of fossil fuels and other human activities (such as land-use change and the production of cement), the oceans take up 2 gigatons and the biosphere takes up 3 gigatons, resulting in a net annual increase of 4 gigatons of carbon in the atmosphere.

**Gigatons of carbon in the reservoirs of Earth's carbon cycle**

| Reservoir | Amount |
|---|---|
| **Biosphere** | |
| Plants and animals | 550 |
| Soils | 2300 |
| **Atmosphere** | 800 |
| **Hydrosphere (Ocean)** | |
| Surface ocean | 1000 |
| Deep ocean | 37,000 |
| **Lithosphere** | |
| Fossil fuels | 10,000 |
| Marine sediments | 6000 |
| Rock | 66,000,000 |
| **Cryosphere** | |
| Permafrost | 1400 |
| Deep-sea sediments (methane hydrates) | 1000–5000 |

**EXPLORING DATA ▲ Consider the annual fluxes of carbon between reservoirs. How much carbon would go into the atmosphere each year if the ocean could only absorb half the amount of $CO_2$ it absorbs now? Assume everything else stays the same.**

**SmartFigure 16.3 The carbon cycle.** Carbon moves through Earth's biosphere, atmosphere, hydrosphere, lithosphere, and cryosphere. https://goo.gl/36wvXR

trenches, where they melt and the $CO_2$ they contain is injected back into the atmosphere during volcanic eruptions.

Organic carbon, on the other hand, is the food and substance of living organisms (biosphere). When these organisms die, their soft remains decay and release $CO_2$ and methane (natural gas) to the atmosphere, or they are buried and over time form deposits of coal, oil, and natural gas (geosphere). These are only some of the ways carbon interacts with the five spheres of Earth's climate system; **Figure 16.3** shows in more detail how carbon flows through these spheres and its resulting effects.

**THE OCEAN'S BIOLOGICAL PUMP** An important concept when considering the role of carbon in Earth's climate system is the idea of the ocean as a **biological pump**. What does that mean? When carbon dioxide enters the ocean, most of it is incorporated into organisms through photosynthesis and through their secretion of carbonate shells. Moreover, carbon dioxide is cycled very effectively from the atmosphere to the ocean. In fact, more than 99% of the carbon dioxide added to the atmosphere in the geologic past by volcanic activity has been removed by the ocean and deposited in marine sediments as biogenous calcium carbonate and fossil fuels. Thus, the ocean acts as a *repository* (or *sink*) for carbon dioxide, soaking it up and removing it from the environment as sea floor deposits. The way in which material is removed from the euphotic zone to the sea floor is called a biological pump because it "pumps" carbon dioxide and nutrients from the upper ocean and concentrates them in deep-sea waters and sea floor sediments (Figure 16.3). As we shall learn, however, the capacity of this biological pump to remove carbon dioxide will be reduced as the oceans become warmer and more acidic.

In summary, the global climate system contains many feedback loops, such as the effect of clouds at different altitudes, the presence of fine atmospheric particles called *aerosols*, the shading effect from air pollution, the addition of water vapor in the atmosphere from ocean warming, the reflectivity of ice, and heat uptake by the oceans. (Note that *aerosols* come from both human-made sources, such as coal-burning power plants and biomass burning, and natural sources, such as sea spray, dust, and volcanoes.) Many of these feedback loops influence other feedbacks. In addition, like interlocking gears, the carbon cycle participates in all levels of Earth's climate system. Even using some of the world's most powerful computers, one of the biggest scientific challenges today is the successful modeling of Earth's climate, its feedback loops, and the critical role played by carbon.

**RECAP** **Earth's climate system consists of exchanges of energy, moisture, and carbon between the atmosphere, hydrosphere, lithosphere, biosphere, and cryosphere. The global climate system contains many complex feedback loops.**

**CONCEPT CHECK 16.1 ▸ Specify the components of Earth's climate system, and the role of carbon.**

1 List the five parts of Earth's climate system.

2 What are the two types of climate feedback loops? Give an example of each type.

3 An example of a feedback loop occurs when melting snow exposes more dark ground, which in turn absorbs additional heat and causes more snow to melt. Is this a positive-feedback loop or a negative-feedback loop? Explain your reasoning.

4 Describe how carbon enters and leaves the five "spheres" of Earth's climate system by drawing a schematic diagram that lists the carbon inputs and outputs in each sphere.

5 Methane, a greenhouse gas, exists in a solid form as frozen methane hydrates at the near-freezing temperatures found deep in the ocean. If global warming causes deep water to heat up, thus causing methane hydrates to melt and release methane gas that would bubble up into the atmosphere, would that create a positive- or negative-feedback loop? Explain.

## 16.2 ▸ Earth's Recent Climate Change: Is It Natural or Caused by Human Activities?

Records of past climate change reveal that natural events have influenced climate throughout Earth's history. Skeptics of today's global climate change point out that because Earth's climate has fluctuated in the past, the recent climate change observed on Earth could be a natural event. How can scientists tell whether this is true? An essential part of the process of science is to *consider and test alternative explanations*—in other words, scientists do not simply collect evidence that supports their point of view, they also aggressively seek evidence that would *falsify their hypothesis*. That means that in order to demonstrate that human activities are causing the recent global climate change, scientists must show that no other explanation can account for the data.

Figure 16.4 **Tree-growth rings as climate proxies.** Tree rings are wider when conditions favor growth and narrower when times are difficult. By combining multiple tree-ring studies with other climate proxy records, scientists have been able to estimate regional and global climates for hundreds to thousands of years in the past.

Enlargement shows a slice of an ice core with its entrapped air bubbles.
In this Antarctic ice core, annual layers of snow packed in glacial ice preserve a record of temperature and atmospheric conditions that stretches back hundreds of thousands of years.

Figure 16.5 **Researchers extract an ice core from its drilling tube.**

# PROCESS OF SCIENCE:
## Are Natural Processes Causing the Climate Change We See Today?

### FORMING A HYPOTHESIS

If human activities are at the root of today's changing climate, scientists must rule out natural causes, and that starts with understanding how climate has changed in the past.

**PROXY DATA AND PALEOCLIMATOLOGY** Instrumental records go back only a couple centuries at most, and the further back in time, the less complete and more unreliable the data become. To overcome the lack of direct measurements in the past, scientists must decipher and reconstruct Earth's previous climates using indirect evidence. Such **proxy** (*proxum* = nearest) data come from natural recorders of climate variability such as sea floor sediments (see Chapter 4), tree-growth rings (**Figure 16.4**), trapped air bubbles in the annual layers of glacial ice (**Figure 16.5**), fossil pollen, coral reefs, cave deposits, and even historical documents. The data are cross-checked between the various methods and also matched with recent instrumental measurements (where overlap exists) to ensure accuracy. Scientists who analyze proxy data to reconstruct past climates are engaged in the study of **paleoclimatology** (*paleo* = ancient, *climate* = climate, *ology* = the study of). The main goal of such work is to understand Earth's past climate in order to gain insight into Earth's current climate and future climate. For example, climatologists have identified both warmer and cooler periods in Earth's recent past, such as the Medieval Warm Period (approximately 950–1250 C.E.) and the Little Ice Age (approximately 1400–1850 C.E.). As we will see, climatologists have used proxy data to construct a detailed history of Earth's climate that extends back in time over the past several hundred thousand years. If climate changes seen today are consistent with climate changes observed in the past, when human-generated emissions were not a factor, then the hypothesis that human activities are driving climate change would be falsified. Is this the case?

### DEVISING AN EXPERIMENT

Natural factors that affect Earth's climate include changes in solar energy, variations in Earth's orbit, volcanic eruptions, and even the movement of Earth's tectonic plates. To understand how each of these factors affects global climate and determine if they alone can account for all the changes we see today, climate scientists (1) look at geologic evidence and other records of Earth's past climates to see how these factors may have changed climate in the past, (2) build computer models that allow them to understand how Earth's climate system works, and (3) closely monitor Earth's current vital signs with an array of instruments, including weather balloons, thermometers at surface stations, deep-sea thermometers, Earth-orbiting satellites, and many other types of observing systems that monitor Earth's weather and climate. By careful measurement and calculation, researchers are able to determine the contribution—if any—each of these factors makes to today's changing climate.

**CHANGES IN SOLAR ENERGY** Among the most persistent hypotheses of climate change have been those based on the idea that the Sun's output of energy varies through time such that increases in solar output cause global warming, while reductions in solar energy result in global cooling. This notion is appealing because solar activity is known to influence Earth's temperature, and solar variation can be used to explain climate change of any length or intensity. However, an increase in solar output cannot explain Earth's recent warming. Earth-orbiting satellites have been making precise measurements of the Sun's output since the 1980s, and while the Sun's luminosity has increased by a small amount (0.04%), the observed changes were not large enough to account for the warming recorded during the same period. Even proxy data of solar brightness over the past 1000 years do not show a correlation with changes in climate.

Researchers have also examined solar variability related to **sunspots**, which are cooler dark areas that occur periodically on the Sun's surface (**Figure 16.6a**). **Faculae** (*facula* = bright torch), which are bright spots on the Sun, are also more abundant when sunspots are most active. Sunspots and faculae are associated with huge magnetic storms that extend from the Sun's interior to its surface and cause the Sun's ejection of particles (Figure 16.6b). These particles do not warm Earth's surface like visible and ultraviolet light from the Sun but they can disrupt satellite communications and also produce the *aurora* (*Aurora* = Roman goddess of dawn), which is a phenomenon caused by charged solar particles that interact with Earth's magnetic field and produce lights in the sky. In the Northern Hemisphere, these lights are known as the *aurora borealis*, or *northern lights*, and they have a matching component in the Southern Hemisphere, called the *aurora australis*, or *southern lights*.

When solar activity increases, as it does every 11 years or so, both sunspots and faculae become more numerous. But during the peak of a cycle, the faculae brighten the Sun more than sunspots dim it. The number of sunspots and faculae thus affects the amount of *total solar irradiance*, which is a measure of the amount of sunlight striking Earth (**Figure 16.7**, *blue curve*).

The graph in Figure 16.7 also shows the lack of correlation between solar activity (*blue curve*) and average Earth temperature since 1880

## STUDENTS SOMETIMES ASK . . .

### *In lay terms, what is a proxy?*

A proxy is a substitution or an approximation for something else. A good example of a proxy is an avatar, which is a cyber-representation of you that competes in video games. There are even proxies at school. For example, if you send someone to attend class in your absence, that person is acting as a proxy for you. And it works both ways—a substitute teacher is a proxy who conducts your class when your regular teacher can't be there.

**(a)** Visual wavelength image of the Sun from the Helioseismic and Magnetic Imager (HMI), showing sunspots.

**(b)** Matching extreme ultraviolet wavelength image of the Sun from the Atmospheric Imaging Assembly (AIA), showing the ejection of solar particles.

**Figure 16.6 Sunspots.** Matching images of the Sun, showing **(a)** sunspots and **(b)** accompanying ejection of particles taken on March 5, 2012.

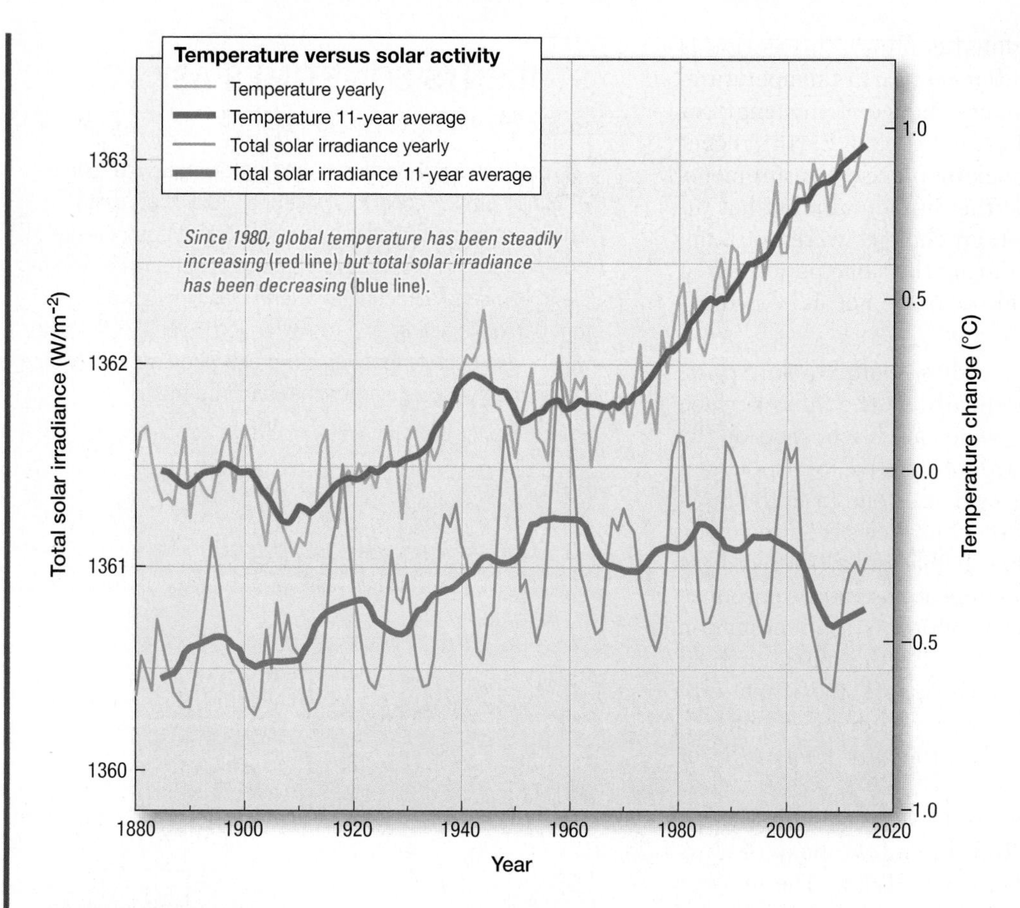


**EXPLORING DATA ▲ What does this graph tell you about the correlation between average solar irradiance and average temperatures on Earth between 1880 and 2020?**

Figure 16.7 **Earth temperature and solar activity since 1880.** Graph showing average Earth temperature (*red curve*) and total solar irradiance (*blue curve*), which is a measure of the amount of sunlight striking Earth. The red and blue curves show smoothed data on an 11-year average. Note the 11-year cycle of total solar irradiance (*green curve*), which is affected by sunspots. Total solar irradiance is in watts per square meter.

(*red curve*). Since 1980, in fact, average solar activity has been decreasing. In addition, many studies have shown that there is no significant correlation between solar activity and climate on such short timescales.

**VARIATIONS IN EARTH'S ORBIT** Another natural mechanism of climate change involves changes in Earth's orbit. Changes in (1) the shape of the orbit (*eccentricity*), (2) the variations in the angle that Earth's axis makes with the plane of its orbit (*obliquity*), and (3) the wobbling of Earth's rotational axis (*precession*) cause fluctuations in the seasonal and latitudinal distribution of solar radiation reaching Earth (**Figure 16.8**). These variations have cycles of about 100,000 years, 41,000 years, and 23,000 years, respectively; when they coincide with one another, they tend to amplify each other and cause climate variations on Earth. This is especially true for the Northern Hemisphere, which contains more landmasses and thus has a greater influence on the development of continental ice ages. For example, taken together, the following orbital conditions likely to initiate an ice age are: (1) a slightly more elliptical orbit that takes Earth further from the Sun during a Northern Hemisphere winter (eccentricity), (2) a maximum tilt (obliquity) that causes the Northern Hemisphere to be tilted further away from the Sun, and (3) the wobbling of Earth's axis causing the Northern Hemisphere summer to coincide with perihelion, which causes warmer summers but cooler winters (precession). (For a discussion of Earth's orbital cycle and perihelion, which is Earth's closest approach to the Sun, see Section 9.2 "How Do Tides Vary during a Monthly Tidal Cycle?" in Chapter 9.) When acting in unison, these factors diminish the solar radiation received in Earth's Northern Hemisphere and tend to produce major glaciations on northern landmasses that persist for tens of thousands of years or longer. This idea, first developed by Serbian astrophysicist Milutin Milankovitch, is called the *Milankovitch cycle*. It is now well established that these variations have contributed to the alternating glacial and interglacial episodes that characterize the most recent ice age, which occurred during the past few million years (**Figure 16.9**).

**VOLCANIC ERUPTIONS** Explosive volcanic eruptions emit huge quantities of gases and fine-grained debris into the atmosphere (**Figure 16.10**). The largest eruptions are sufficiently powerful to inject material high into the atmosphere, where it spreads around the globe and remains aloft for many months or even years. As was seen with historic eruptions such as Mount Tambora in Indonesia (1815), Krakatoa in Indonesia (1883), El Chichón in Mexico (1982), and Mount Pinatubo in the Philippines (1991), volcanic material ejected into the atmosphere filters out a portion of the incoming solar radiation, which in turn cools the planet. For example, the year after the 1815 eruption of Mount Tambora became widely known as the *Year without Summer* because of its effect on North American and European weather. However, the gases emitted during a volcanic eruption react with other components of the climate system and the volcanic dust eventually settles out. Thus, the cooling effect of a single eruption, no matter how large, is relatively small and short-lived.

Which emits more heat-trapping carbon dioxide: Earth's volcanoes or human activities? Analysis of worldwide human and volcanic emissions over the past several decades indicate that human activities release into the atmosphere at least 130 times more heat-trapping carbon dioxide than volcanoes do (**Figure 16.11**). As discussed later in this chapter, carbon dioxide emissions make the greatest human contribution to warming Earth's atmosphere.

If volcanism is to have a pronounced impact over an extended period, many great eruptions closely spaced in time would need to occur. If this happened, the upper atmosphere would be loaded with enough gases to alter the composition of the atmosphere and enough volcanic dust to seriously diminish the amount of solar radiation reaching the surface. Because no such period of explosive volcanism is known to have occurred in the past few hundred years, volcanism is unlikely to be responsible for the recent changes in climate. In the distant past, however, large and long-lasting volcanic eruptions may have been influential in contributing to Earth's climate shifts. For example, the Deccan Traps—enormous lava flows covering nearly 500,000 square kilometers (193,000 square miles) in India—were created by extensive volcanism that started about 66 million years ago and may have caused global climatic changes that contributed to the demise of the dinosaurs.

**MOVEMENT OF EARTH'S TECTONIC PLATES** As described in Chapter 2, Earth's tectonic plates have moved great distances. During the geologic past, plate movements have accounted for many dramatic climate changes as landmasses shifted in relation to one another and moved to different latitudinal positions. As landmasses have moved, they have changed ocean circulation, altering the transport of heat and moisture and consequently the climate. For example, the opening of the Drake Passage between South America and Antarctica about 41 million years ago caused a fundamental reorganization of ocean currents in the Southern Hemisphere, leading to the isolation of Antarctica, which caused it to become much cooler and develop a permanent ice cap. However, the rate of plate movement is very slow—only a few centimeters per year—and so appreciable changes in the positions of continents occur only over great spans of geologic time. Thus, climate changes triggered by shifting plates are extremely gradual and happen on a scale of millions of years.

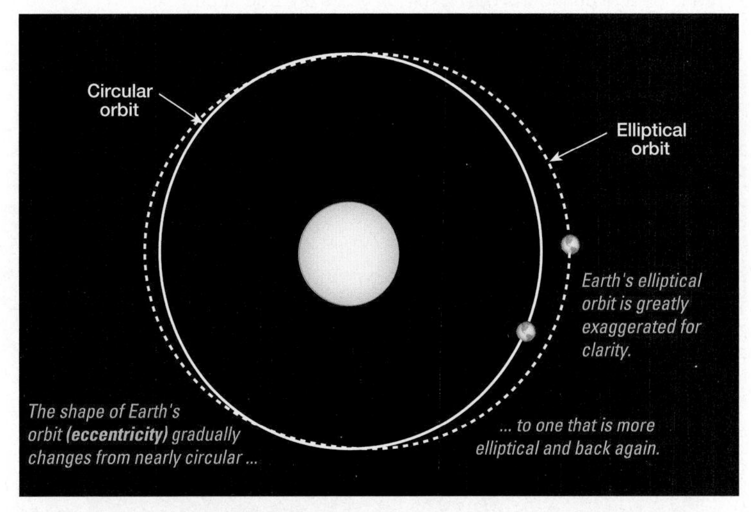


**(a)** Eccentricity cycle: 100,000 years.

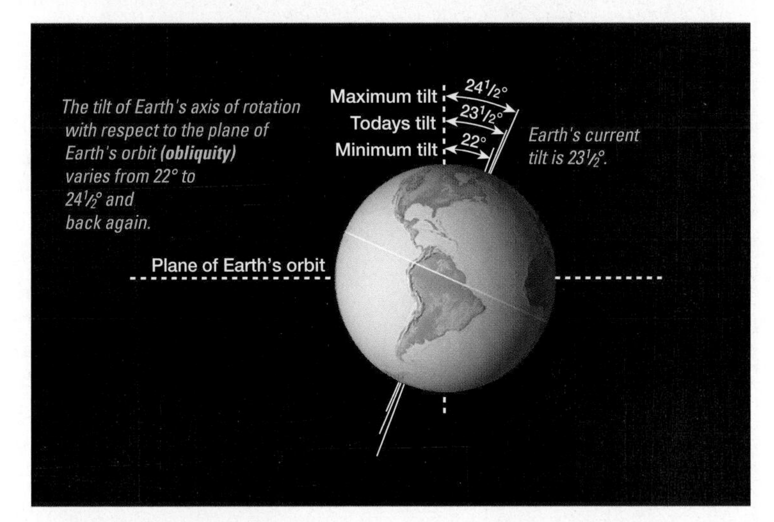


**(b)** Obliquity cycle: 41,000 years.

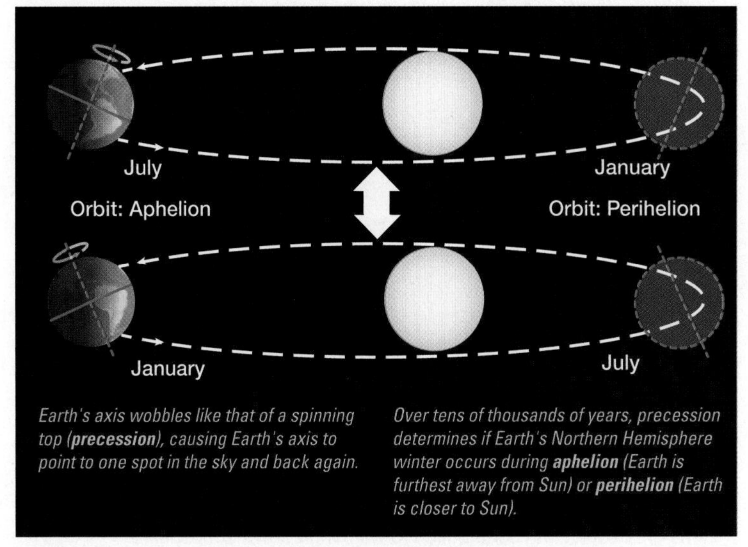


**(c)** Precession cycle: 23,000 years.

**SmartFigure 16.8 Variations in Earth's orbit.** Diagram showing the three factors that cause variations in Earth's orbit and occur over tens to hundreds of thousands of years. https://goo.gl/wF3aoa

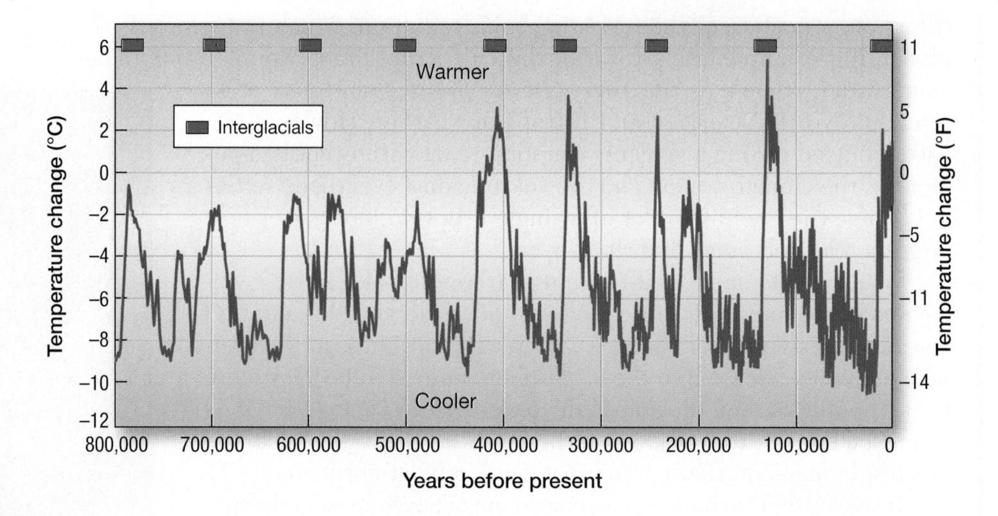


**Figure 16.9 Pleistocene Ice Age.** The alternating glacial and interglacial periods that characterize Earth's most recent ice age during the past 2 million years are due largely to cyclic changes in Earth's orbital parameters, referred to as Milankovitch cycles.

**Figure 16.10 Volcanic eruptions spew volcanic debris and gases into the atmosphere.** This 1991 eruption of Mount Pinatubo in the Philippines shows that volcanoes have the ability to inject into the atmosphere large quantities of volcanic dust and gases, which can circle the globe and block incoming solar radiation, thereby cooling the planet.

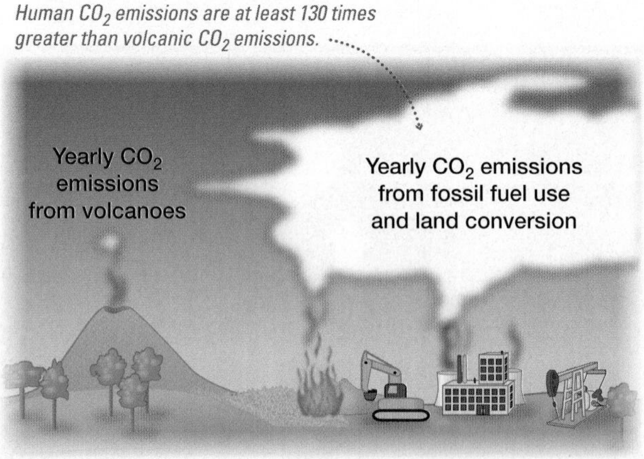

*Human $CO_2$ emissions are at least 130 times greater than volcanic $CO_2$ emissions.*

Yearly $CO_2$ emissions from volcanoes

Yearly $CO_2$ emissions from fossil fuel use and land conversion

**Figure 16.11 A comparison of human and volcanic carbon dioxide emissions.** Over the past several decades, human carbon dioxide emissions have been at least 130 times greater than volcanic emissions. Note that human-caused carbon dioxide emissions make the greatest human contribution to warming Earth's atmosphere.

## INTERPRETING THE RESULTS

In addition to the long-term climate changes that happen over the course of millions of years, evidence suggests that warmer and cooler periods have also occurred in Earth's more recent past and that these temperature fluctuations have occurred in the absence of human-caused climate change. Examples are the Pleistocene Ice Age, the Medieval Warm Period, and the Little Ice Age. How do these examples compare with the changes observed today? It is worth noting that during the past million years, the biggest temperature swings on Earth have been during glacial-interglacial transitions (Figure 16.9). When ice ages have ended, it has taken about 5000 years for the planet to warm between 4°C and 7°C (7.2°F and 12.6°F). For comparison, in the 20th century alone, Earth's average temperature climbed about 0.7°C (1.3°F), which is roughly eight times faster than any previous warming. In fact, studies show that humanity is altering Earth's climate 5000 times faster than the pace of the most rapid natural warming episode in our planet's past. Simply put, the global warming occurring today is unusual and unprecedented.

To summarize, over the past century, scientists from all over the world have been collecting data on natural factors that influence climate—such as changes in the Sun's brightness, variations in Earth's orbit, major volcanic eruptions, and other factors. These observations have failed to show any long-term changes that could fully account for the recent rapid warming of Earth's temperature. Scientists agree that the observed warming in recent decades is occurring more quickly and with greater magnitude than can be explained by any natural factors. In essence, the release of human-caused emissions is the only viable explanation for the documented and observable recent climate changes—including the increased average temperature of Earth's surface.

## THINKING LIKE A SCIENTIST: WHAT'S NEXT?

Suppose you want to predict what Earth's average temperature might be in 20 years, and you have yearly average temperature data from the last 200 years. Which data would best show the trends that would help you make your prediction? The first 100 years of data? The last 100 years of data? Or the last 50 years of data? Explain.

## Is There Scientific Consensus About Human-Caused Climate Change?

Today, there is a clear scientific consensus that human-induced emissions are responsible for the observed warming on Earth. In fact, the warming trends observed since 1950 cannot be explained without accounting for human-caused emissions.

Human-caused climate change has received broad support by major scientific organizations, including the U.S. National Academy of Sciences (NAS), the U.S. National Research Council (NRC), the U.S. National Science Foundation (NSF), the American Meteorological Society (AMS), the National Center for Atmospheric Research (NCAR), the American Association for the Advancement of Science (AAAS), and the U.S. federal government agencies NOAA and NASA. In fact, more than 200 worldwide scientific organizations support the position that humans are causing climate change.

Scientific consensus that human activities are causing climate change was clearly established more than 25 years ago (some would point out that the case was made as far back as the 1950s) and has not changed since then. However, a recent poll (**Figure 16.12**) shows that even though the vast majority of scientists—especially climatologists—accept human-caused climate change, social acceptance of the idea by the U.S. general public lags far behind. Figure 16.12 also shows that the most highly trained experts in climate science (those who are active publishers on climate change) have the highest level of confidence that human activities are causing the recent changes in Earth's climate.

Scientific peer-reviewed publications on human-caused climate change echo the consensus. For example, a recent study examined all scientific publications on climate change during 2013 and 2014, and concluded that only four of 69,406 authors of peer-reviewed articles on global warming (0.0058%, or 1 in 17,352) rejected human-caused climate change. Thus, the consensus on human-caused climate change among published scientists is above 99.99%, verging on unanimity. In essence, the peer-reviewed scientific literature contains no convincing evidence against human-caused climate change.

Although it makes interesting news stories to suggest that there is disagreement among scientists about the reality of human-caused climate change, there is overwhelming scientific consensus that human emissions are causing changes in Earth's climate. The existing scientific disagreements are, in fact, about specific future impacts and consequences of these climate changes.

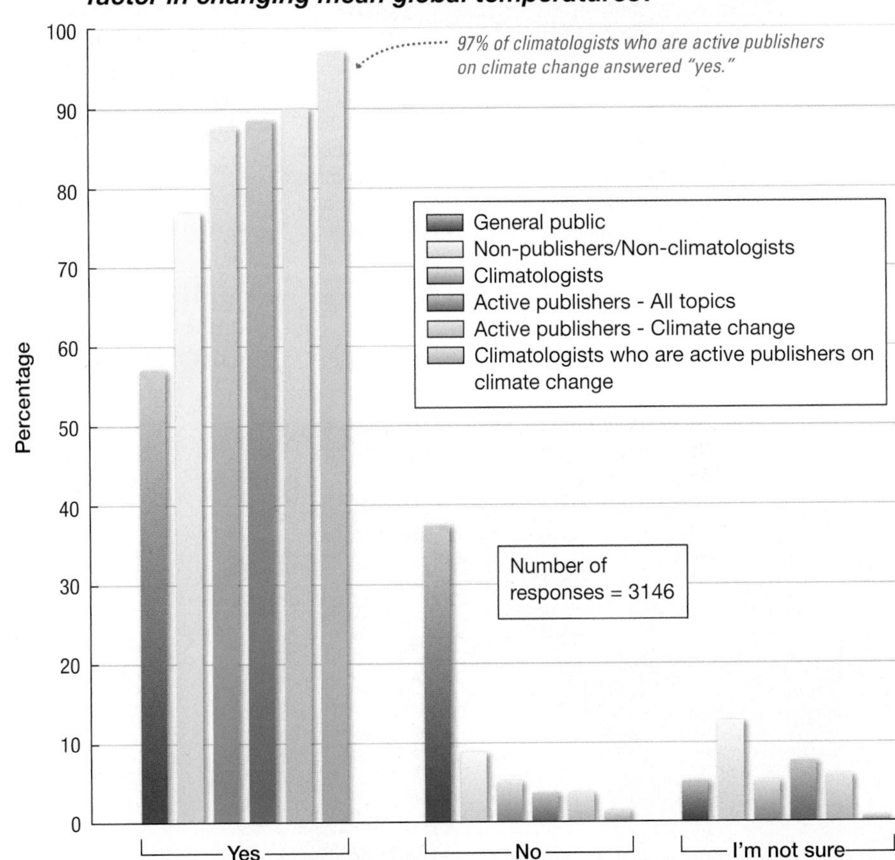


**EXPLORING DATA ▲ What is the main take-home message this graph is telling you?**

**Figure 16.12 Scientific consensus on human-caused climate change.** Bar graph showing various groups' responses to the question "Do you think human activity is a significant contributing factor in changing mean global temperatures?" Note that climatologists who are active publishers on climate change had the highest percentage of affirmative response.

## The IPCC: Documenting Human-Caused Climate Change

In 1988, the United Nations Environment Programme and the World Meteorological Organization sponsored the **Intergovernmental Panel on Climate Change (IPCC)**, a worldwide group of atmospheric and climate scientists that began studying the human effects on climate change and global warming. The IPCC utilizes peer-reviewed literature to analyze all aspects of climate change—including science, impacts, adaptation, and mitigation—to provide independent scientific advice about climate change. Since 1990, the group has published a series of assessment

Figure 16.13 **The sixth IPCC assessment report.** As this book goes to press, the IPCC released its 2018 report, which documents that these human-caused changes on Earth are happening sooner–and with even more dire consequences–than previously thought.

reports (**Figure 16.13**) that are highly regarded by both scientists and policymakers and have sparked international movement on climate change.

**THE IPCC ASSESSMENT REPORTS** The IPCC's first assessment report, released in 1990, became the basis for the United Nations Framework Convention on Climate Change, an international treaty in which signatories agreed to the idea of reducing concentrations of greenhouse gases in the atmosphere. The IPCC's second assessment report, published in 1995, states that *"the balance of evidence suggests a discernible human influence on global climate"* and that global warming *"is unlikely to be entirely due to natural causes."*

Between 2001 and 2018, the IPCC published four more reports, under the guidance of hundreds of scientists, which expressed increasing levels of confidence that climate change was due to human activities. For example, the fifth IPCC assessment report, published in 2014, included the work of 831 experts from scientific fields such as meteorology, physics, oceanography, engineering, and ecology. The report summarized 9200 recently published, peer-reviewed scientific studies on climate change and determined that warming of the atmosphere and ocean system is "unequivocal." The report states that many of the associated impacts, such as sea level rise and increased global temperatures, have occurred since 1950 at rates unprecedented in the historical record. The report also states that there is a clear human influence on climate and that it is extremely likely that human activities have been the dominant cause of observed warming since 1950, with the level of confidence raised to 95–100% certainty. In addition, the report points out that the longer humans wait to reduce our emissions, the more expensive it will become to counteract the resulting climate changes.

The IPCC assessment reports provide strong documentation of the planet's human-induced climate changes, which poses substantial risks to societies and ecosystems on all continents. In recognition of that fact, the IPCC was named a co-recipient of the 2007 Nobel Peace Prize (**Figure 16.14**), along with former U.S. Vice President Al Gore, Jr., for his work on the documentary film *An Inconvenient Truth*. When it bestowed the award, the Nobel Committee noted, *"Through the scientific reports it has issued over the past two decades, the IPCC has created an ever-broader informed consensus about the connection between human activities and global warming."*

In summary, the IPCC's six in-depth reports spanning over two decades have documented the broad scientific consensus of leading climate experts from around the globe, affirming the reality of human-caused climate change—now no longer as a prediction but as an observational reality. With new data contained in each subsequent assessment report, the message becomes increasingly certain: Humans are altering Earth's climate and actions need to be taken to reduce human-caused emissions.

Figure 16.14 **The IPCC and Al Gore, Jr. share the 2007 Nobel Peace Prize.** The Norwegian Nobel Committee awarded the 2007 Nobel Peace Prize to the Intergovernmental Panel on Climate Change (IPCC) and Albert Arnold (Al) Gore, Jr., for "their efforts to build up and disseminate greater knowledge about man-made climate change, and to lay the foundations for the measures that are needed to counteract such change."

**OTHER SCIENTIFIC REPORTS CONFIRM IPCC FINDINGS** A host of subsequent reports have confirmed the findings of the IPCC. In 2009, for example, the U.S. Global Change Research Program issued a 190-page interagency report entitled *Global Climate Change Impacts in the United States*. The report states that *"global warming is unequivocal and primarily human-induced.* The report also notes that *global average temperature has risen by about 1.5°F [0.8°C] since 1900. By 2100, it is projected to rise another 2°F to 11.5°F [1.1°C to 6.4°C]. Increases at the lower end of this range are more likely if global heat-trapping gas emissions are cut substantially. If emissions continue to rise at or near current rates, temperature increases are more likely to be near the upper end of the range."* The report warns

Figure 16.15 **Climate Science Special Report: Fourth U.S. National Climate Assessment (2017).** This report by U.S. climate science experts provides an assessment of the physical impacts of climate change in the United States. The report, which is issued every four years, notes that climate change is already affecting the American people in far-reaching ways.

## STUDENTS SOMETIMES ASK . . .

***I've heard news reports that there was a pause in the rise of Earth's average global temperature in the early part of this century, which contradicts global warming. Are the reports true?***

No; in actuality, a reexamination of the data indicates that there was never a slowdown. The presumed pause (or false pause) in the rise of Earth's average global surface temperature in the past few decades—called *the global warming hiatus*—has been shown to be a result of measurement bias. The apparent hiatus resulted from an increase in the use of buoys for measuring sea surface temperatures during the past few decades. According to experts at NOAA, buoys tend to give cooler reading than measurements taken from the sensors in the engine intake ports of ships. Uncorrected, those discrepancies suggested that there was a pause in the long-observed rise of Earth's average surface temperature and an apparent slowdown of global warming since 1998.

Now that the measurement bias has been corrected, climate change researchers are able to say with confidence that global warming since the year 2000 is the strongest it has been since the latter half of the 20th century, and the rate of warming is increasing. This of course is not good news, but it is another example of the process of science, in which researchers question, retest, and look for alternative explanations for the data.

that climate change will have numerous impacts on water resources, ecosystems, agriculture, coastal areas, human health, and other sectors.

In 2011, the U.S. National Research Council issued the final report of a five-volume series about climate change, titled *America's Climate Choices*. The reports, which were requested by the U.S. Congress, reaffirm the preponderance of scientific evidence and points to human activities—especially the release of carbon dioxide and other greenhouse gases into the atmosphere—as the most likely cause of the vast majority of global warming that has occurred over the past several decades. The reports also reiterate the pressing need for substantial action to limit the magnitude of climate change and to prepare to adapt to its impacts. The reports indicate that actions taken now *"can reduce the risk of major disruptions to human and natural systems; inaction could serve to increase these risks, especially if the rate or magnitude of climate change is particularly large."*

In 2017, the *Climate Science Special Report: Fourth National Climate Assessment* (**Figure 16.15**) was published at the behest of the U.S. Congress to provide an assessment of the state of science relating to climate change and its physical impacts. The report, which is issued every four years, was authored by teams of scientists from 13 federal departments and agencies that conduct research in global climate change and was extensively reviewed by the public and scientific experts. The report affirms the findings of previous reports and states: *"[t]his assessment concludes, based on extensive evidence, that it is extremely likely that human activities, especially emissions of greenhouse gases, are the dominant cause of the observed warming since the mid-20th century. For the warming over the last century, there is no convincing alternative explanation supported by the extent of the*

**RECAP** **Earth's climate has changed in the past due to natural causes such as changes in the Sun's output, variations in Earth's orbit, volcanic activity, and the movement of tectonic plates. Multiple lines of evidence show that the rapid climate changes Earth is currently experiencing are due primarily to human activities that release heat-trapping gases into the atmosphere.**

*observational evidence.*" The report also states that the effects of climate change are already impacting Americans and are expected to become increasingly disruptive across the nation throughout this century and beyond.

In the next two sections, we'll examine the science of climate and see what changes are occurring in the oceans as a result of global warming.

**CONCEPT CHECK 16.2 ▸ Examine the evidence that shows how Earth's recent climate change is caused by human activities, not a natural cycle.**

1 What are proxy data? List several examples. Why are such data necessary for paleoclimatology studies?

2 List several examples of natural climate change. Do natural climate change mechanisms account for the recent climate changes that Earth is experiencing? Explain.

3 Is there scientific consensus about human-caused climate change? Explain.

4 What is the IPCC? What role does it have in documenting human-caused climate change?

5 What other recent reports have been published supporting human-caused climate change?

**Animation**
Global Warming
http://goo.gl/Uz8rmm

## 16.3 ▸ What Causes the Atmosphere's Greenhouse Effect?

Numerous scientific studies indicate that human-caused emissions are responsible for the recent and dramatic climate changes experienced on Earth, including the increase in average worldwide temperature, which is called **global warming**. Although the **greenhouse effect** is a natural process that influences the temperature of Earth's surface and atmosphere, it is now being altered by human emissions, a phenomenon that is often referred to as the *anthropogenic greenhouse load* or the *enhanced greenhouse effect*.

The greenhouse effect gets its name because it keeps Earth's surface and lower atmosphere warm similar to the way a greenhouse keeps plants warm enough to grow, regardless of outside conditions (**Figure 16.16**). Energy radiated by the Sun covers the full electromagnetic spectrum, but most of the energy that reaches Earth's surface is short wavelengths, in and near the visible portion of the spectrum. In a greenhouse, shortwave sunlight passes through the glass or plastic covering, where it strikes the plants, the floor, and other objects inside and is re-radiated as longer-wavelength infrared radiation (heat). Some of this heat energy escapes from the greenhouse and some is trapped for a while by the glass or plastic covering, which keeps the greenhouse nice and snug—much like what happens in Earth's atmosphere. (Note that recent studies have indicated that an additional factor in keeping a greenhouse warm is that the greenhouse covering prevents mixing of warm air inside with cooler air outside. Although this is different from how the atmosphere works, the term *greenhouse effect* is still commonly used to describe the atmosphere's warming process.)

*Atmospheric gases such as water, carbon dioxide, and methane act just like the glass of a greenhouse, allowing sunlight to pass through but trapping outgoing heat.*

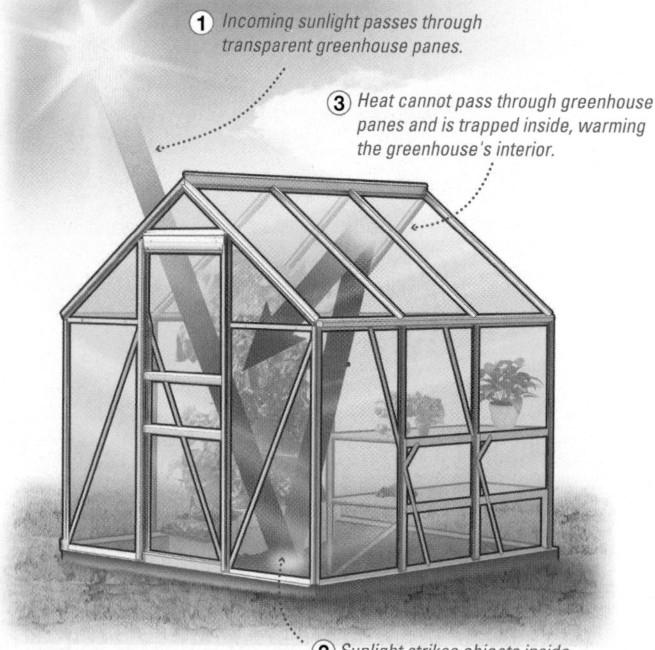

① Incoming sunlight passes through transparent greenhouse panes.
③ Heat cannot pass through greenhouse panes and is trapped inside, warming the greenhouse's interior.
② Sunlight strikes objects inside greenhouse, loses energy, and is converted to heat.

**Figure 16.16 How a greenhouse works.**

### Earth's Heat Budget and Changes in Wavelength

**Figure 16.17** shows the various components of Earth's **heat budget**, which describes all the ways in which heat is added to and subtracted from Earth. Although Earth's atmosphere blocks some forms of solar radiation, it is transparent to most wavelengths of visible light, which is able to pass through the atmosphere like sunlight coming through greenhouse glass. However, only about 47% of the solar radiation that is directed toward

Earth reaches Earth's surface and is absorbed by the oceans and continents. Of the 53% of solar radiation that isn't absorbed by land or water, about 23% is absorbed by molecules in the atmosphere, dust, and clouds, and about 30% is reflected back into space by atmospheric backscatter, clouds, and reflective regions of Earth's surface.

**Figure 16.18** shows that most of the energy coming to Earth from the Sun is within the visible part of the spectrum and peaks at a wavelength of 0.48 micron (0.00002 inch; note that a micron [$\mu$], which is actually a micrometer, is one-millionth of a meter). When this radiation is absorbed by water and rocks at Earth's surface, these materials absorb some of the energy and warm up and then emit radiation away from Earth's surface toward space as longer-wavelength infrared (heat) radiation, with a peak at a wavelength of 10 microns (0.0004 inch). Atmospheric gases such as water vapor, carbon dioxide, and other gases absorb radiation at these longer wavelengths, effectively intercepting the heat radiation that attempts to leave the planet, thus heating the atmosphere. This trapping of heat radiation and heating of the atmosphere is known as the *greenhouse effect*.

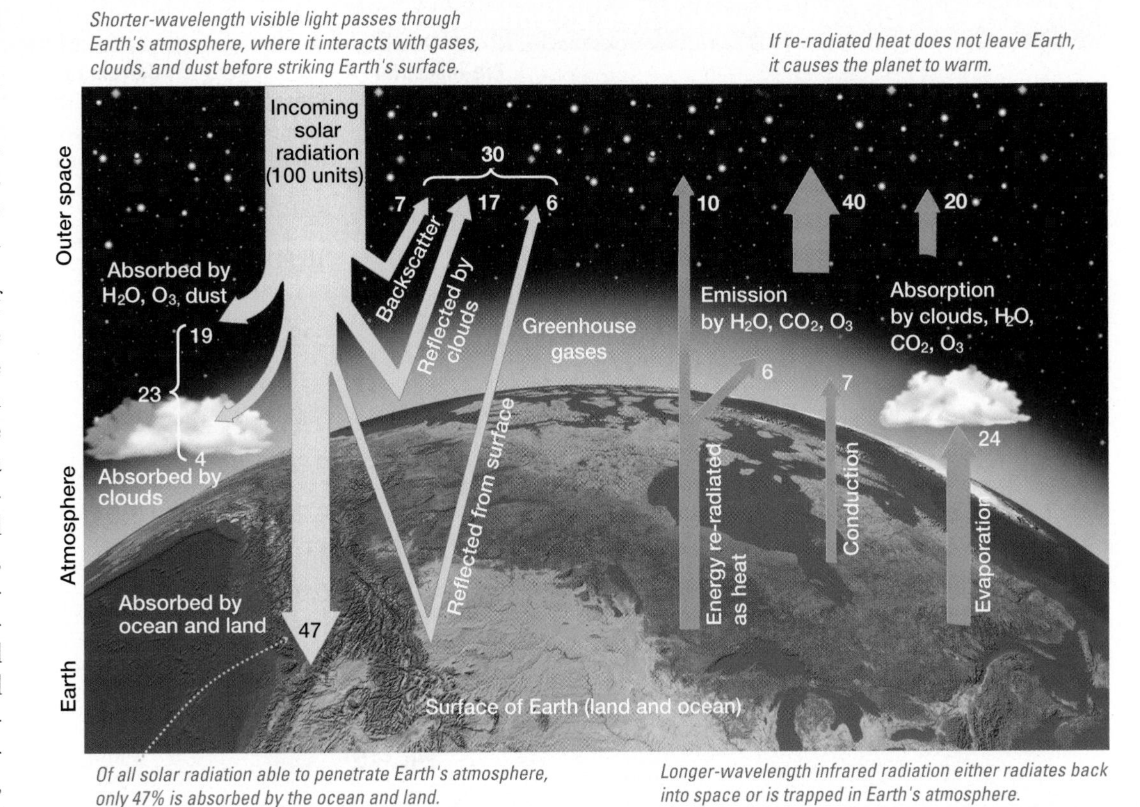


**Figure 16.17 Earth's heat budget.** In this example, 100 units of solar radiation from the Sun (mostly shorter-wavelength visible light) are reflected, scattered, and absorbed by various components of the Earth–atmosphere system. The absorbed energy is re-radiated back into space from Earth as longer-wavelength infrared radiation (heat). If this infrared radiation does not leave Earth, global warming will occur.

In summary, most of the solar radiation that is not reflected back to space passes through the atmosphere and is absorbed at Earth's surface. Earth's surface, in turn, re-emits longer-wavelength infrared radiation (heat). A portion of this energy is absorbed by certain heat-trapping gases in the atmosphere, thus producing the greenhouse effect. Thus, *the change of wavelengths from visible to infrared at Earth's surface is the key to understanding how the greenhouse effect works.*

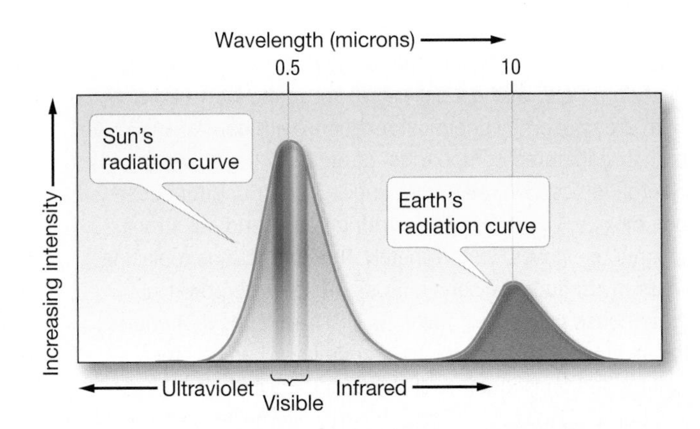


**Figure 16.18 Energy radiated by the Sun and Earth.** The intensity of energy radiated by the Sun peaks at a wavelength of 0.48 microns (0.00002 inch), which is in the visible part of the spectrum. Some of this energy is absorbed or reflected while some reradiates from Earth in the infrared (heat) range at a wavelength of 10 microns (0.0004 inch).

## Which Gases Contribute to the Greenhouse Effect?

Earth's greenhouse effect is caused by an array of atmospheric gases, many of which have both natural and human-caused sources. Take, for example, water vapor, which contributes more to the greenhouse effect than any other gas. In fact, water vapor is the single most important absorber of heat—its contribution to the greenhouse effect is between 36% and 66% of the greenhouse effect, and together with clouds, it comprises about 75% of the greenhouse effect.

Water vapor enters the atmosphere mostly through evaporation and other natural processes. Although atmospheric water vapor concentrations fluctuate regionally, studies suggest that human activity does not significantly affect water vapor concentrations except at local scales, such as near irrigated fields. And even then, the water vapor does not stay in the atmosphere very long, because it enters the rapidly changing hydrological cycle of condensation and precipitation. In essence, human activities do not directly affect the amount of water vapor in the atmosphere on a global scale.

**Animation**
Atmospheric Energy Balance
http://goo.gl/imdsCN

**SmartTable 16.1 Human-generated greenhouse gases and their contribution to increasing the greenhouse effect**

| Atmospheric gas | Human-caused sources of gas | Pre-industrial (circa 1750) concentration (ppbv[a]) | Present concentration (ppbv[a]) | Current rate of increase or decrease (% per year) | Relative contribution to increasing the greenhouse effect (%) | Infrared radiation absorption per molecule (number of times greater than $CO_2$) |
|---|---|---|---|---|---|---|
| Carbon dioxide ($CO_2$) | Combustion of fossil fuels | 280,000 | 411,000 | +0.5 | 58 | 1 |
| Methane ($CH_4$) | Leakage, domestic cattle, rice agriculture | 722 | 1834 | +1.0 | 15 | 28 |
| Nitrous oxide ($N_2O$) | Combustion of fossil fuels, industrial processes | 270 | 328 | +0.2 | 5.9 | 265 |
| Tropospheric ozone ($O_3$) | Byproduct of combustion | 237 | 338 | +0.5 | 12 | 2000 |
| Chlorofluorocarbons (CFCs) | Refrigerants, industrial uses | 0 | 0.82 | −1.0 | 7 | 12,000–15,000 |
| Hydrochlorofluorocarbons (HCFCs) | Refrigerants, industrial uses | 0 | 0.28 | +0.5 | 1.7 | 800–2000 |
| Hydrofluorocarbons (HFCs) | Refrigerants, industrial uses | 0 | 0.08 | +8-20 | 0.4 | Up to 14,800 |
| **Total** | | | | | 100 | |

[a]ppbv = parts per billion by volume (not by weight). https://goo.gl/nfxJCS

## STUDENTS SOMETIMES ASK . . .

***Is carbon dioxide causing the hole in the ozone layer?***

Here's the short answer: definitely not! The ozone layer occurs within the atmosphere's stratosphere and is composed of ozone molecules ($O_3$) that absorb most of the Sun's ultraviolet radiation. Without it, unhealthy levels of ultraviolet radiation would reach Earth's surface, making the planet largely uninhabitable. The main ozone hole (actually, a seasonal thinning of the ozone layer) occurs above the South Pole, with a smaller one above the North Pole. Both are caused by chemical reactions with natural and human-generated compounds, particularly the now-banned chlorofluorocarbon chemicals known as CFCs, but not carbon dioxide. CFCs are used in refrigeration and are strong greenhouse gases. Unfortunately, the compounds replacing CFCs, hydrofluorocarbons (HFCs), are likewise powerful greenhouse gases (see Table 16.1). This means that efforts to safeguard Earth's protective ozone layer has had the unintended consequence of worsening Earth's greenhouse gas burden. However, climate-friendly alternatives are now being introduced to replace HFCs. With these new chemicals being introduced, along with the ban of CFCs firmly in place, scientists predict that Earth's protective ozone layer will achieve its normal thickness by the middle of the 21st century. But atmospheric chemistry is complicated! As Table 16.1 shows, tropospheric (lower atmosphere) ozone, which is a byproduct of combustion, is a potent greenhouse gas, and human-generated nitrous oxide, another product of combustion and a greenhouse gas, is now the single greatest ozone-depleting substance.

However, humans may be *indirectly* affecting the amount of water vapor in the atmosphere, since warmer air can hold more moisture. In fact, satellite measurements of atmospheric water vapor show a 4% increase in global specific humidity at sea level since 1970, which researchers have tied to human-caused climate change.

Still, atmospheric water vapor plays an important role in warming. For example, a recent study suggests that an increase in stratospheric water vapor due to natural processes between 1980 and 2000 may have amplified the rapid warming of that period by as much as 30%. Conversely, a 10% decrease in the amount of water vapor in the stratosphere since 2000 may have slowed global warming by causing global temperatures to level off despite the continued rise in human emissions.

**Table 16.1** shows the concentration of **greenhouse gases**—so called because of their heat-trapping capacity—that have been increasing as a result of human activities. Remarkably, these gases exist in very small amounts in the atmosphere, yet they have a profound effect on heating. And unlike water vapor that does not stay in the atmosphere very long, many of these gases stay in the atmosphere for long periods and continue to trap heat. Some of these greenhouse gases are released by both human and natural sources. Others, however, have no natural source and thus are clearly a product of human activities.

**CARBON DIOXIDE** Of all the human-generated gases, carbon dioxide makes the greatest relative contribution to increasing the greenhouse effect (Table 16.1). Carbon dioxide enters the atmosphere as a result of combustion of carbon compounds with oxygen. It is a colorless and odorless gas that is the same one we exhale from our lungs. The conversion of *fossil fuels* (oil, natural gas, and coal) into energy by cars, factories, and power plants accounts for the majority of the annual human contribution to carbon dioxide emissions, with industrialized nations contributing the most. As a result of human activities, atmospheric concentration of carbon dioxide has increased more than 40% over the past 250 years (**Figure 16.19**). The direct measurement of carbon dioxide concentration in the atmosphere was initiated by Charles David Keeling in 1958 and is continued today by his son, Ralph Keeling.

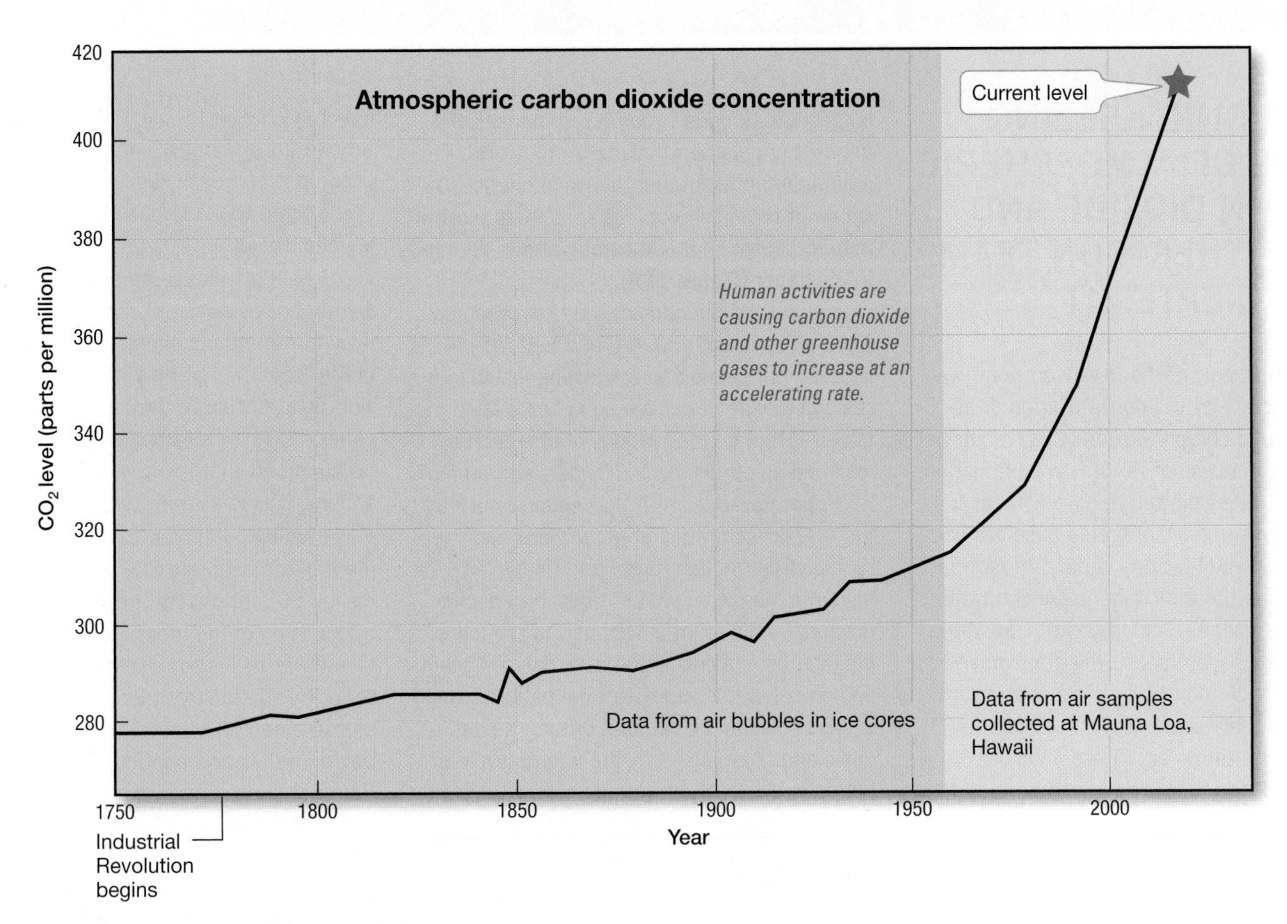


**Figure 16.19 Amount of carbon dioxide in the atmosphere since 1750.** Graph shows the dramatic increase of average worldwide atmospheric carbon dioxide since the Industrial Revolution began in the late 1700s. Values for 1958 to the present are from laboratory measurements of carbon dioxide in air samples collected at Mauna Loa Observatory in Hawaii; values prior to 1958 are from measurements of carbon dioxide in air bubbles preserved in polar ice cores.

The iconic curve showing the steady increase in atmospheric carbon dioxide now bears their name (**Diving Deeper 16.1**).

What concerns scientists is that over the past 250 years—and especially in the past 50 years—human activities have been responsible for raising the concentration of greenhouse gases in the atmosphere at an ever-increasing rate. As of 2019, the average yearly concentration of atmospheric carbon dioxide is 411 parts per million and is increasing by about 2 parts per million each year; this rate of increase is double what occurred only 50 years ago. In terms of sheer numbers, humans are now pumping into the atmosphere about 40 billion metric tons (88 trillion pounds) of carbon dioxide each year. Based on worldwide population, each person on Earth thus emits more than 5.2 metric tons (11,500 pounds) of carbon dioxide per year. In addition, this number is several times greater for those living in industrial nations (such as the United States) than for those living in developing countries (such as India). The last time that carbon dioxide levels were as high as they are today was about 15 million years ago during the Miocene warm period, when temperatures were at least 3°C to 5°C (5°F to 9°F) warmer than today.

## STUDENTS SOMETIMES ASK . . .

***Because carbon dioxide occurs in such tiny amounts in the atmosphere, how can it be responsible for global warming?***

Sometimes even tiny amounts of a substance can have dramatic effects. For example, most people are aware that even a tiny increase of cholesterol in our blood can create severe health problems. The same is true in the atmosphere: a seemingly tiny increase of heat-trapping gases can result in higher temperatures worldwide. Heat-trapping gases in the atmosphere are indeed very potent!

**METHANE** Methane is the second-most-abundant human-caused greenhouse gas (Table 16.1). It is produced by leakage from decomposing trash in landfills, by methane-belching domestic cattle, and by agriculture (particularly the cultivation of rice). Even though methane has a lower concentration in the atmosphere than carbon dioxide, it has a much greater ability to produce warming on a per-molecule basis. Since the Industrial Revolution began in about 1750, when

**RECAP The greenhouse effect is caused by gases that allow sunlight to pass through the atmosphere but trap heat energy before it is radiated back to space. Carbon dioxide is foremost in an array of gases from human activity that increase the atmosphere's ability to trap heat.**

# THE ICONIC KEELING CURVE OF ATMOSPHERIC CARBON DIOXIDE AND THE FATHER–SON TEAM WHO CREATED IT

In the first part of the 20th century it was suspected that the concentration of atmospheric carbon dioxide ($CO_2$) might be increasing because of fossil fuel combustion. However, there were relatively few measurements of this important greenhouse gas and the measurements varied widely. In an effort to determine the atmospheric concentration of carbon dioxide, Charles David (Dave) Keeling in 1958 began measuring atmospheric carbon dioxide atop Mauna Loa volcano in Hawaii, where researchers can access very clean, high altitude air. Those measurements continue to this day, representing the longest continuously monitored dataset of atmospheric carbon dioxide in the world. The graphical representation of atmospheric $CO_2$ is one of the most iconic graphs in support of human-caused global warming and is named in his honor (**Figure 16A**).

Under the direction of the U.S. Weather Bureau (now part of NOAA) and Scripps Institution of Oceanography, Keeling installed a gas analyzer at Hawaii's Mauna Loa station in March 1958; on the first day of operation it recorded an atmospheric $CO_2$ concentration of 313 ppm. In April 1958, to Keeling's surprise, the $CO_2$ concentration at Mauna Loa had risen by 1 part per million (ppm) and then even higher in May, after which it began to decline, reaching a minimum in October. In the following months, the concentration increased again and repeated the same seasonal pattern, which is now a distinct component of the Keeling curve and is explained by the natural cycle of plants in the Northern Hemisphere withdrawing $CO_2$ from the air for plant growth via photosynthesis during summer and returning it through decomposition each succeeding winter. Keeling described it as a seasonal "breathing cycle" on a global scale.

Keeling's measurements showed the first significant evidence of rapidly increasing $CO_2$ levels in the atmosphere. Many scientists credit Keeling's graph with first bringing the world's attention to the increase of atmospheric $CO_2$. In addition, researchers have documented that the steady increase in atmospheric $CO_2$ is a result of human emissions.

Keeling continued to direct the measurement of atmospheric $CO_2$ in Mauna Loa from his lab at Scripps until his death in 2005. Supervision of the measuring project was taken over by his son, Ralph Keeling, a Professor of Geochemistry at Scripps. The long-term measurement of atmospheric $CO_2$ initiated by David Keeling and continued today by his son Ralph Keeling is an important piece of climate data that clearly shows how humans are contributing to climate change. To commemorate the work done by the Keelings and to honor the importance of the curve, the American Chemical Society named the Keeling curve a National Historic Chemical Landmark in 2015.

Today, about 100 stations worldwide monitor atmospheric $CO_2$, but none of them have a longer continuous record than Keeling's Mauna Loa station. Most importantly, the data show that today's value of more than 400 ppm is the highest value of measured atmospheric $CO_2$ since long-term monitoring began in 1958.

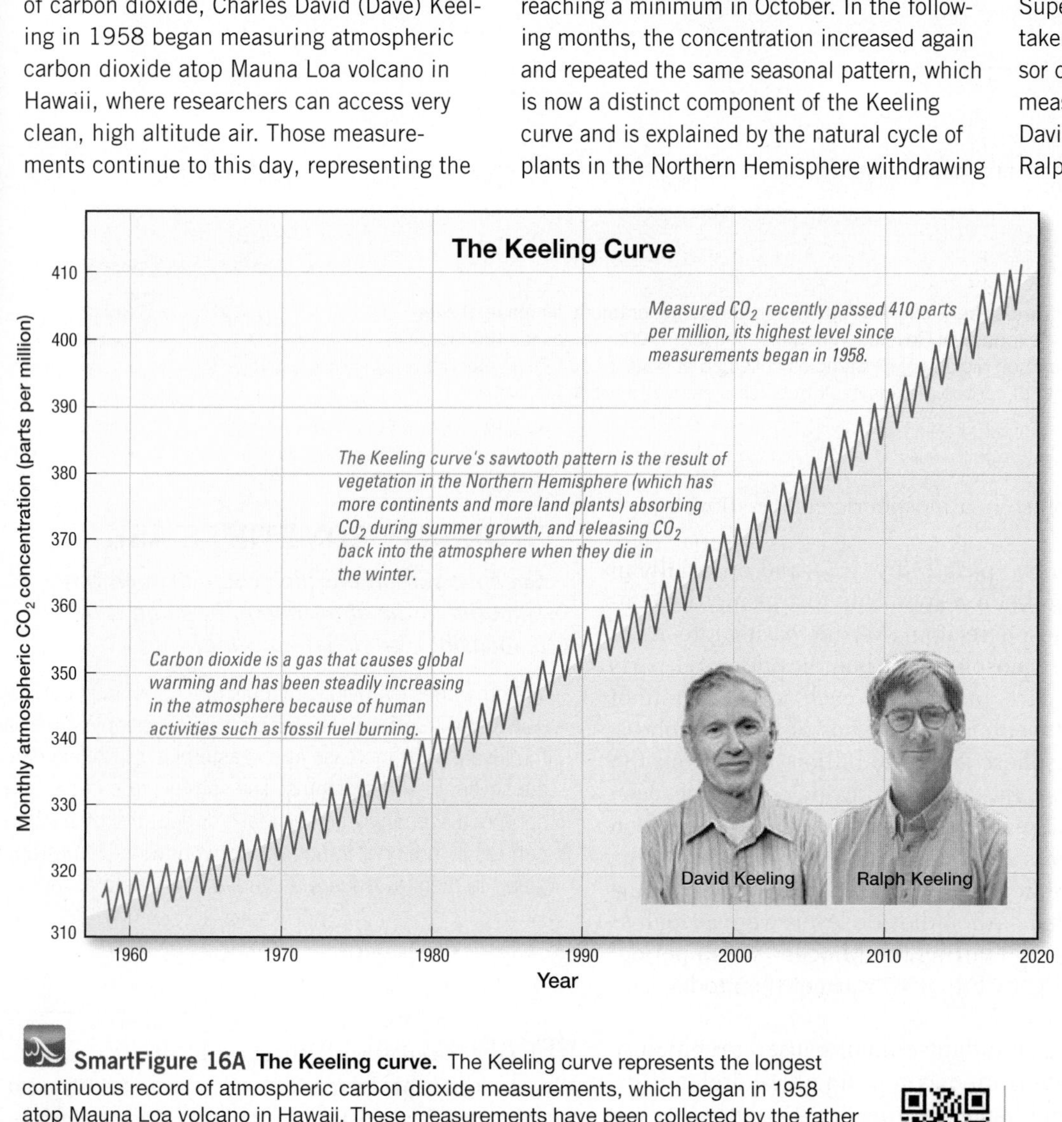

The Keeling Curve
Monthly atmospheric CO2 concentration (parts per million)
310
320
330
340
350
360
370
380
390
400
410
1960
1970
1980
1990
2000
2010
2020
Year
Measured CO2 recently passed 410 parts per million, its highest level since measurements began in 1958.
The Keeling curve's sawtooth pattern is the result of vegetation in the Northern Hemisphere (which has more continents and more land plants) absorbing CO2 during summer growth, and releasing CO2 back into the atmosphere when they die in the winter.
Carbon dioxide is a gas that causes global warming and has been steadily increasing in the atmosphere because of human activities such as fossil fuel burning.
David Keeling
Ralph Keeling

**SmartFigure 16A The Keeling curve.** The Keeling curve represents the longest continuous record of atmospheric carbon dioxide measurements, which began in 1958 atop Mauna Loa volcano in Hawaii. These measurements have been collected by the father and son team of David and Ralph Keeling (*insets*).
https://goo.gl/ffsCXD

## WHAT DID YOU LEARN?

Why does the Keeling curve, which shows the steady rise in the atmospheric concentration of carbon dioxide over the course of decades, exhibit a sawtooth seasonal pattern where atmospheric $CO_2$ concentration increases in the first part of the year, and then decreases during the following six months?

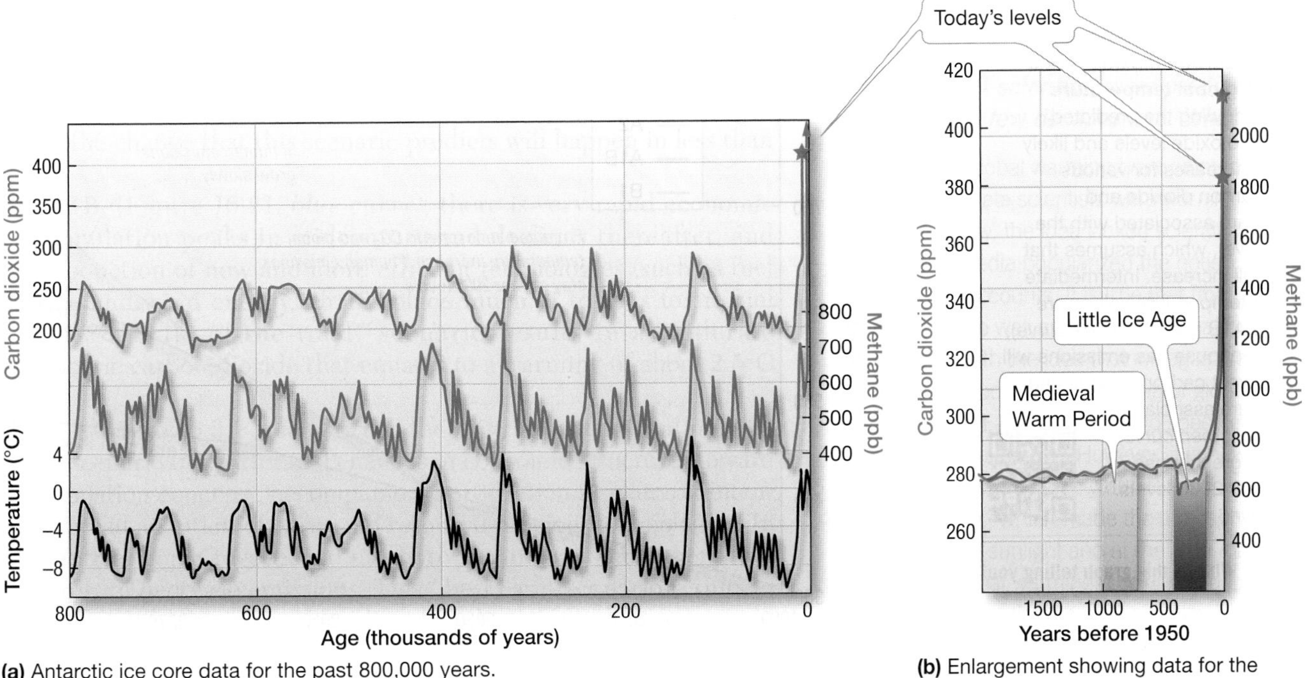


**(a)** Antarctic ice core data for the past 800,000 years.

**(b)** Enlargement showing data for the most recent 2000 years.

**EXPLORING DATA** ▲ **State the relationship between the three curves in the graph in part (a). Why do you think the Medieval Warm Period and Little Ice Age in part (b) have to be shown in an enlargement?**

**SmartFigure 16.20 Ice core data of atmospheric composition and global temperature. (a)** Antarctic ice core data for the past 800,000 years, showing atmospheric carbon dioxide (*red curve*), methane (*green curve*), and average global temperature (*black curve*). Today's carbon dioxide and methane levels are shown by the red and green stars. Atmospheric composition is from analysis of air trapped in ice bubbles; temperature reconstruction is derived from the chemical make-up of the ice. **(b)** Enlargement showing data for the most recent 2000 years. Medieval Warm Period and Little Ice Age are also shown. https://goo.gl/eyqrTr

the concentration of methane in the atmosphere was about 700 parts per billion by volume (ppbv), methane has increased in the atmosphere by more than two and a half times, to a value of 1825 ppbv. Even though methane is present in very tiny amounts in the atmosphere, scientists are concerned about this dramatic increase because, pound for pound, the comparative impact of methane on climate change is 25 times greater than that of carbon dioxide over a 100-year timescale.

**OTHER GREENHOUSE GASES** The other trace gases shown in Table 16.1—nitrous oxide, tropospheric ozone, and chlorofluorocarbons—are present in far lower concentrations than carbon dioxide and methane. Yet they are still very important because they absorb many times more infrared radiation per molecule than carbon dioxide or methane (Table 16.1, *last column*), thus making them very potent contributors to warming. Still, these gases have a smaller overall contribution to increasing the greenhouse effect because their concentrations are so low. Nevertheless, all these gases must be taken into account when considering the total amount of greenhouse warming.

**GREENHOUSE GASES: PAST AND FUTURE** In 2005, researchers recovered a nearly 3.2-kilometer (2-mile) long continuous ice core from Antarctica that contains a record of past atmospheric concentrations of carbon dioxide and methane—two important greenhouse gases—that get trapped as ice accumulates. Analysis of the core, which extends back in time 800,000 years ago (**Figure 16.20**), shows that the average level of carbon dioxide (*red curve*) naturally varied from about 180 parts per million to about 280 parts per million. During the same time, methane (Figure 16.20, *green curve*) varied from about 350 parts per billion to about 750 parts per billion, in sync with carbon dioxide levels. In addition, the chemical make-up of the ice provides a proxy of the past average temperature on Earth (Figure 16.20, *black curve*), which shows a strong correlation with atmospheric methane and carbon dioxide concentrations: when Earth experiences cooler temperatures (glacial periods), carbon dioxide and methane are low, and when Earth experiences warmer temperatures (interglacial periods), carbon dioxide and methane are high. The graph shows that our planet has passed through a cycle of glaciation and deglaciation every 100,000 years or so; the timing of

## STUDENTS SOMETIMES ASK . . .

### *What would Earth's temperature be like if there were no natural greenhouse effect?*

In a word: freezing! The worldwide average temperature of Earth and the lowermost atmosphere (troposphere) is about 15°C (59°F). If the atmosphere contained no heat-trapping greenhouse gases or warming cloud cover, the average worldwide temperature would be about −5°C (23°F). At this temperature, most of our planet would very likely have a frozen surface. Instead, the atmosphere's naturally occurring greenhouse gases help create and sustain the moderate temperatures that make Earth habitable.

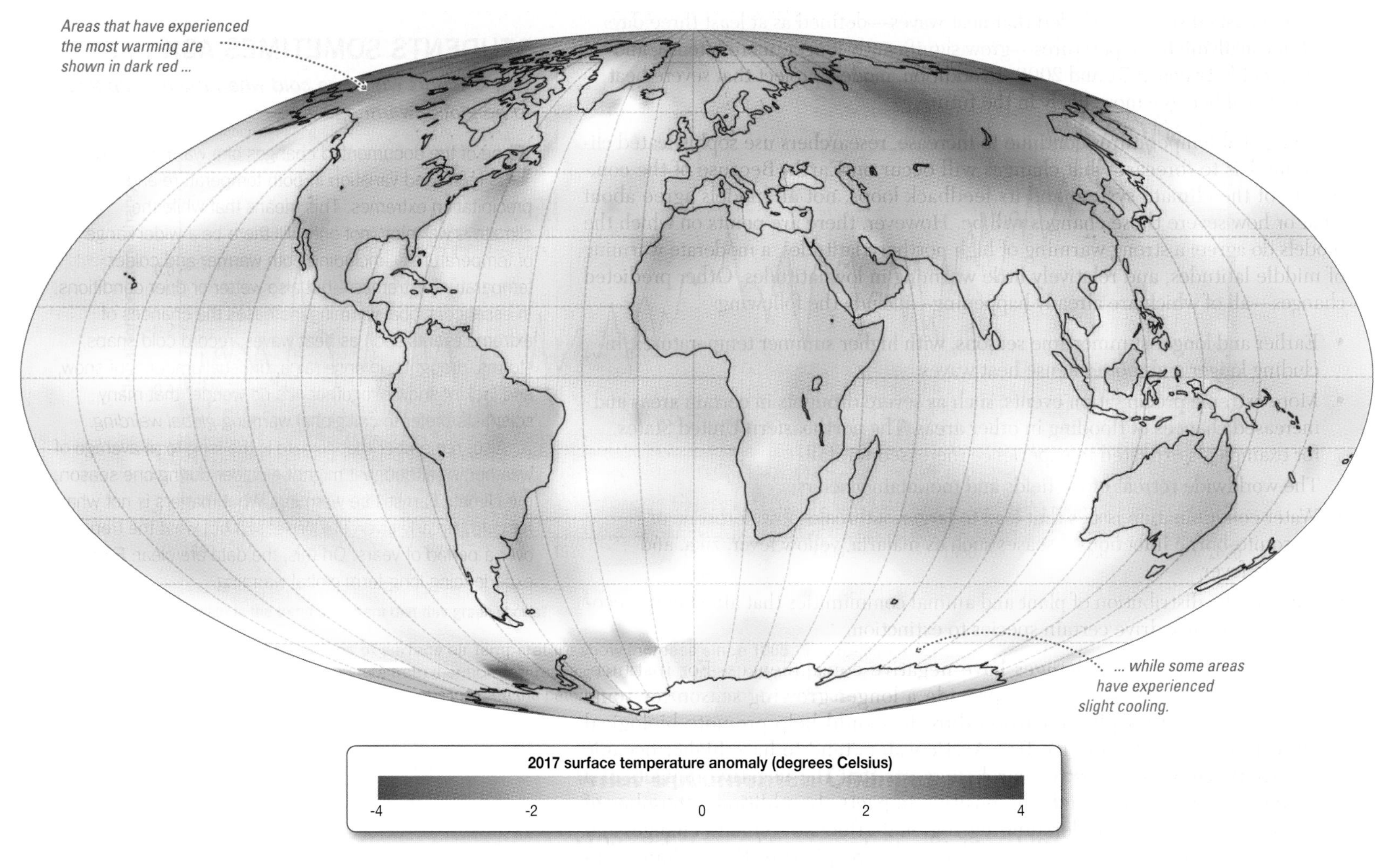

Areas that have experienced the most warming are shown in dark red ...
... while some areas have experienced slight cooling.
2017 surface temperature anomaly (degrees Celsius)
-4
-2
0
2
4

Figure 16.23 **Global temperature change.** Map showing the change in land and ocean surface temperature in 2017 compared to the 1950–1980 baseline period. Most of the globe is anomalously warm (*red areas*) as a result of greenhouse warming, with the greatest temperature increases in the Arctic Ocean, Alaska, Siberia, and West Antarctica.

**Video**
Global Temperatures 1880–2010
https://goo.gl/rmPn8E

## Increasing Ocean Temperatures

Studies have revealed that the oceans have absorbed the majority of the increased heat in the atmosphere. In fact, although the oceans absorb only 25% of the $CO_2$ being pumped into the atmosphere, it absorbs 93% of the heat. Millions of ocean temperature observations at various depths reveal that there has been an overall increase in surface temperature (**Figure 16.23**). These measurements indicate that global sea surface temperatures have risen by about 0.6°C (1.1°F) because of global warming, mainly since about 1970. However, this warming has not been uniform throughout the ocean. The greatest temperature increases have been experienced in the Arctic Ocean, near the Antarctic Peninsula, and in tropical waters. Even deep waters are showing signs of warming: warming has been documented to a depth of 2 kilometers (1.2 miles). And deeper waters are warming faster than expected. To determine the extent of warming experienced in the oceans, scientists have initiated a program to monitor changes in ocean temperature using the ocean's ability to transmit sound (Mastering Oceanography **Web Diving Deeper 16.1**).

In 2017, the *Climate Science Special Report: Fourth National Climate Assessment* (see Figure 16.15) described the change in ocean heat content from 1960 to 2015 at different ocean depths. The report stated that heating occurred at all depths in the ocean, but that surface waters warmed the most.

The impacts of a warmer ocean are far-reaching and will persist for several centuries. For example, increased seawater temperatures are likely to affect temperature-sensitive organisms such as corals. Coral reefs are already in decline, and, as discussed in Chapter 15, warmer sea surface temperatures have been implicated in widespread coral bleaching events. Warmer waters could also shift or disrupt the spawning cycles of corals. In addition, studies of coral distribution have found that some corals are already migrating with the spread of warmer waters to areas where they haven't been seen before.

It's not just the gradual warming of the oceans that is a concern, however. According to a 2018 study, marine heatwaves—prolonged periods of high ocean surface temperatures—are predicted to become more frequent, extensive, and intense as a result of global warming. Scientists report that between 1982 and 2016, the number of marine-heatwave days doubled, and this frequency is projected to increase if global temperatures continue to rise. Research has shown that marine organisms and ecosystems are particularly vulnerable to such extreme events, and so marine heatwaves pose an escalating risk to marine life.

**Figure 16.24 Flooding from Hurricane Harvey in Texas in 2017.** Hurricane Harvey, a Category 4 hurricane, dropped as much as 152 centimeters (60 inches) of rainfall in and around Houston, Texas. Flooding damage from Hurricane Harvey totaled more than $125 billion, making it the second costliest hurricane in U.S. history after Hurricane Katrina.

Other studies have documented how warmer ocean waters are changing the physical characteristics of the ocean. For example, research using the Argo program of free-drifting floats (see Section 7.1 "How Are Ocean Currents Measured?" in Chapter 7) has shown that the world's hydrologic cycle is accelerating as warming speeds up evaporation of ocean surface waters, thereby affecting ocean salinity. As mentioned earlier in this chapter, the melting of sea ice forms a positive-feedback loop that accelerates ocean warming in high latitudes because ice-free water absorbs much more solar radiation than reflective sea ice. In addition, increased seawater temperatures are likely to affect the ocean's deep-water circulation pattern, enhance warm-water El Niño events (and reduce cooler water La Niña events), and aid in the development of hurricanes.

**HAVE INCREASED OCEAN TEMPERATURES CAUSED INCREASED HURRICANE ACTIVITY?** Many scientists have suggested that warmer oceans would most certainly cause an increase in the general level of storminess because additional heat accelerates evaporation, which fuels hurricanes. In addition, the ultimate intensity of a hurricane largely depends on the temperature of deep seawater (which has also increased) that is churned upward as the storm passes overhead. The interactions between global warming and hurricane activity, however, are complex, with rising ocean temperatures, changing energy distributions, and altered atmospheric dynamics all having some effect. For example, as atmospheric temperature increases, so does atmospheric stability, a shift that limits convective transport and thus reduces the formation of tropical hurricanes.

In recent years, the frequency and severity of hurricanes—especially those in the Atlantic Ocean (see Section 6.5 in Chapter 6)—have led some to speculate that global warming has already enhanced the formation of hurricanes. For example, the recent landfall of several large, destructive Atlantic hurricanes such as Hurricane Katrina in 2005, Hurricane Sandy in 2012, and Hurricanes Harvey, Irma, and Maria in 2017 (**Figure 16.24**) has led to a general impression that hurricanes have increased, although there have been conflicting reports on the topic in the scientific literature. In fact, some research articles have attributed increases in hurricane intensity, numbers, wind speeds, and rainfall to warmer sea surface temperatures, while others have claimed that changes in data-gathering methods and instrumentation are responsible for the trends. Other studies suggest that the apparent increase in hurricanes falls within the statistical limits of normal.

Although scientists are not able to say definitively if the *number* of tropical storms has increased worldwide, the scientific consensus is that global warming

## STUDENTS SOMETIMES ASK . . .

### *Where can I find reliable information about climate change on the Web?*

Here are some recommended, scientifically accurate websites that give detailed information about climate change:

- RealClimate, which was created by real climate scientists: **www.realclimate.org**
- Skeptical Science, which explains climate science and rebuts global warming misinformation: **www.skepticalscience.com**
- NOAA's ClimateWatch, which is a portal for U.S. government climate science: **www.climate.gov/#climateWatch**
- CO2 Now, which publishes updated atmospheric carbon dioxide levels: **www.CO2Now.org**
- The Berkeley Earth Surface Temperature site, which aims to help resolve criticisms of the temperature record and lower barriers into understanding climate science: **http://berkeleyearth.org**
- This Is Climate Change, which uses visuals to educate and increase awareness of climate change among the general public: **www.thisisclimatechange.org**
- The National Center for Science Education (NCSE), which defends the teaching of evolution and has recently launched a new initiative to defend the teaching of climate change in classrooms: **http://ncse.com**

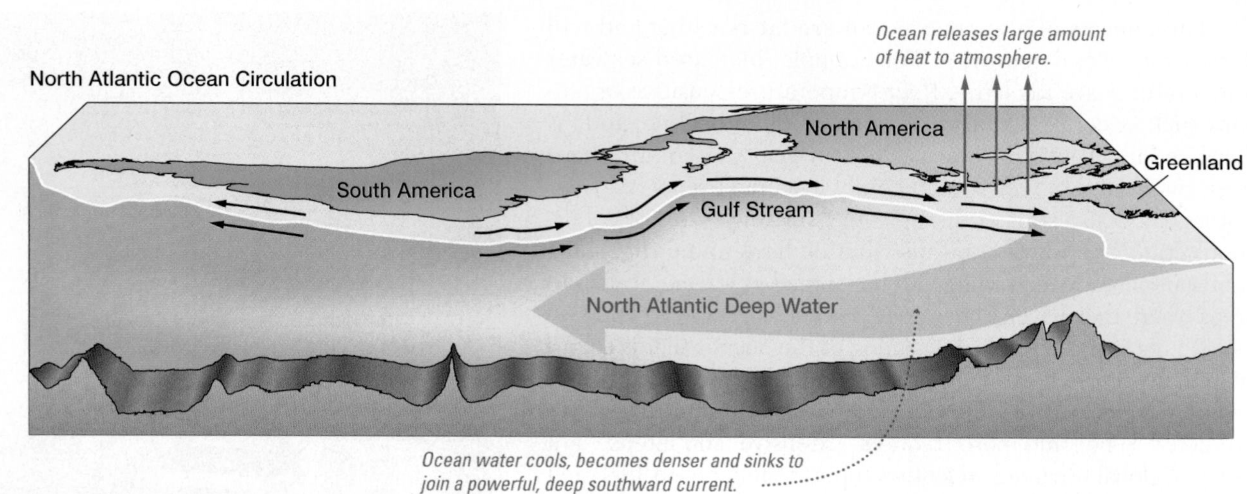
North Atlantic Ocean Circulation
Ocean releases large amount of heat to atmosphere.
North America
Greenland
South America
Gulf Stream
North Atlantic Deep Water
Ocean water cools, becomes denser and sinks to join a powerful, deep southward current.

**Figure 16.25 North Atlantic Ocean circulation.** Perspective view of circulation in the North Atlantic Ocean, showing that the Gulf Stream carries a tremendous amount of heat northward that warms the North Atlantic region. As this water cools, it generates a huge volume of cold, salty, dense water called North Atlantic Deep Water that sinks into the deep-ocean basin and flows southward. Disruption of this circulation pattern could have severe effects on global climate.

**EXPLORING DATA ▲ Near South America, the illustration shows surface currents diverging, flowing both north and south. Using what you know about ocean circulation, oceanic gyres, and global wind patterns, explain why this is so.**

**Animation**
North Atlantic Deep-Water Circulation
https://goo.gl/bXwxdY

has likely led to more *intense* hurricanes. In the most comprehensive study of recent hurricane activity to date, researchers demonstrate that there have been significant increases in tropical storm intensity and duration around the world since 1970 and that these trends are strongly related to rising sea surface temperatures. Another study of historical Atlantic hurricanes over the past 1500 years suggests that times of peak hurricane activity are related to increased sea surface temperatures and the reinforcing effects of La Niña–like climate conditions. Other research explicitly shows that the most energetic storm levels—those with Category 4 and 5 designations—have already increased significantly, particularly in the North Atlantic and northern Indian Oceans. Still other studies have detected a pronounced poleward migration in the average latitude at which tropical cyclones have achieved their lifetime-maximum intensity over the past 30 years as a result of human-caused climate change. In addition, sophisticated climate models suggest that the number of Category 4 and Category 5 storms in the western tropical Atlantic could double by the end of the century, despite a drop in the overall number of storms.

## Changes in Deep-Water Circulation

Evidence from deep-sea sediments and computer models indicates that changes in the global deep-water circulation pattern can dramatically and abruptly affect climate. Circulation in the North Atlantic Ocean, which provides an important source of deep water (**Figure 16.25**), is particularly sensitive to these changes. What drives deep-water circulation is the sinking of cold, high-salinity, dense surface waters at high latitudes, particularly in the North Atlantic. If surface waters stopped sinking because they were too warm and/or too diluted by melting ice (and thus low in density), then the oceans would absorb and redistribute heat from solar radiation much less efficiently. This would likely cause even warmer surface water temperatures and much higher land temperatures than are experienced now.

Many scientific studies suggest that the buildup of greenhouse gases in the atmosphere will change ocean circulation. One way in which this could happen is that warmer air temperatures will increase the rate at which glaciers in Greenland melt, forming a pool of fresh, low-density surface water in the North Atlantic Ocean. This freshwater could inhibit the downwelling that generates North Atlantic Deep Water, reorganizing global circulation patterns and causing a corresponding change in climate. Many climate experts warn that a large outflow of freshwater from Greenland could take the North Atlantic system of currents to a tipping point, causing rapid reorganization of deep-water currents and related changes in climate. Indeed, evidence suggests that an outburst from a North American ice-dammed lake flooded the North Atlantic with freshwater about

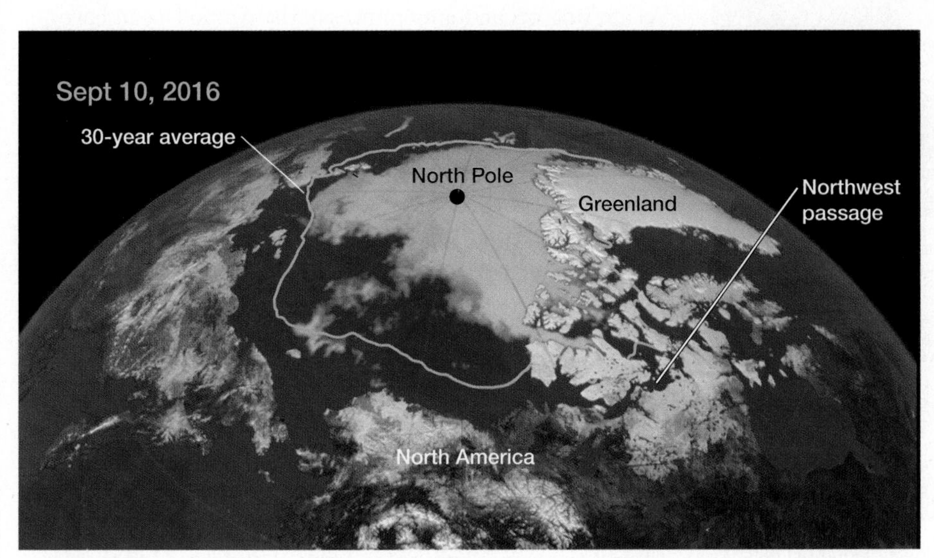


(a) Perspective view of the Arctic based on satellite data, showing the extent of Arctic sea ice in September 2016 compared with the 30-year average extent of sea ice *(yellow line)*. Note the opening of a new ice-free Northwest Passage.

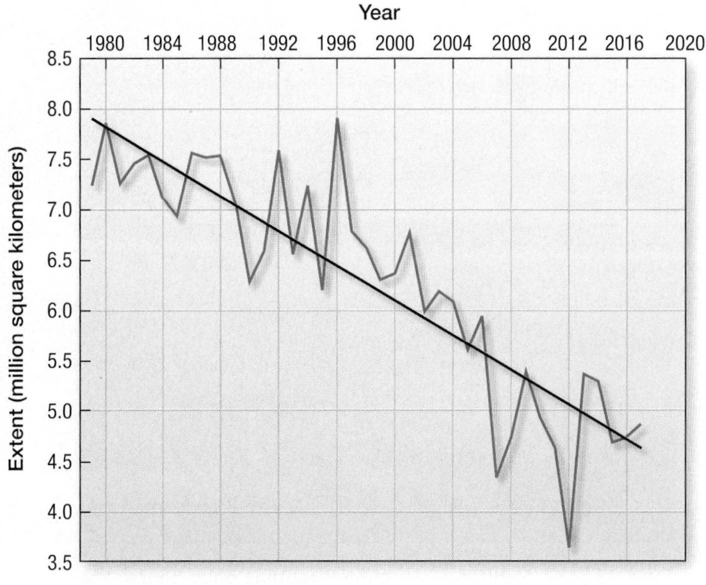


(b) Graph showing the decrease in Arctic sea ice extent based on satellite data.

**Figure 16.26 Arctic sea ice decline.** The substantial decrease in extent and thickness of Arctic Ocean sea ice is due to human-induced warming of the Arctic.

8000 years ago, causing rapid global climate change. This event may be the type of scenario that the North Atlantic could experience again because of increased precipitation and melting of ice.

## Melting of Polar Ice

Computer models have predicted that global warming will affect Earth's polar regions in a very dramatic way. One fundamental difference between the two polar regions is that in the Northern Hemisphere, the polar region is dominated by the Arctic Ocean and its cover of drifting sea ice (an ocean surrounded by land), whereas the South Pole is dominated by the continent of Antarctica and its thick ice cap, including shelf ice that extends into the ocean (land surrounded by an ocean).

The Arctic is one of the locations where the effects of global warming are being most keenly felt (see Figure 16.23) and likely will experience quite dramatic changes in the future; this phenomenon is called *Arctic amplification*. Since 1978, satellite analysis of the extent of Arctic Ocean sea ice indicates that it is dramatically shrinking and thinning at an accelerating rate (**Figure 16.26**). In the past decade alone, there has been a loss of over 2 million square kilometers (800,000 square miles) of Arctic sea ice. In fact, measurements of the ice cap in 2012 revealed that it had shrunk to its smallest size since researchers began collecting satellite measurements; summer Arctic sea ice is now about half the size it was only 30 years ago. In addition, thick multiyear ice in the Arctic has been disappearing and is being replaced by thinner first-year ice that is less likely to survive the summer melt season. As a result, Arctic ice is now unusually thin and spread out, causing wide patches of ice-free ocean during the summer—even at the North Pole. A recent study of proxy data showed that the current decline of Arctic sea ice is unprecedented for at least the past 1450 years.

Climate models are in general agreement that one of the strongest signals of greenhouse warming will be a loss of Arctic sea ice. Indeed, during the past 15 years, the decline in Arctic sea ice has occurred much faster than models had predicted. In fact, the Arctic has been warming more than twice as quickly as the Northern Hemisphere average. Cycles of natural variability are known to play a role in Arctic sea ice extent, but the sharp decline observed in the past two decades cannot be explained by natural variability alone. The accelerated Arctic sea ice melting appears to be linked to shifts in Northern Hemisphere atmospheric circulation patterns that have

**Animation**
Arctic Sea Ice Decline
https://goo.gl/ynv8dU

**Video**
Loss of Arctic Sea Ice
https://goo.gl/ziV5OU

### STUDENTS SOMETIMES ASK . . .

***I've heard that Antarctica is actually gaining ice. Is that true?***

Technically, only East Antarctica was gaining ice, and for a number of years scientists could not say for sure if these gains were sufficient to offset the large losses of ice documented in other regions of the southern continent. A new study published in 2018 analyzed 25 years of data and concluded that overall, there was a net loss of ice in Antarctica during this time period. The report concluded that melting of the southern ice sheet was "one of the largest contributors to sea level rise," and that Antarctica's contribution to sea level rise had tripled since 2012.

Figure 16.27 **Habitat destruction threatens polar bears.** Polar bears (*Ursus maritimus*) rely on floating sea ice as a feeding platform and for building dens. As a result of habitat destruction caused by shrinking Arctic sea ice, polar bears were designated a threatened species in 2008.

caused the region to experience unusually rapid warming. As a result, ocean temperatures in the Arctic Ocean have also increased, causing sea ice to melt from below. Disappearance of sea ice is likely to enhance future warming in the region because lower amounts of sea ice will reflect less of the Sun's radiation back into space, creating a positive-feedback loop and exacerbating the problem as heat is absorbed by the newly uncovered ocean. Researchers fear that the Arctic may be on the verge of a fundamental transition, or "tipping point," that will lead to the Arctic having only seasonal ice cover. Some models, for example, suggest that the Arctic could experience the complete disappearance of summer sea ice as early as 2030.

The decrease of sea ice in the Arctic has already had profound effects on Arctic ecosystems. Polar bears (*Ursus maritimus*; **Figure 16.27**), for example, are excellent swimmers but do not hunt in the water. Instead, they require a platform of floating sea ice to capture their prey, which are mainly ringed and bearded seals. As the Arctic Ocean becomes more ice-free and the ice habitat shrinks, polar bears will have more difficulty finding adequate food and making dens. As a result, polar bear breeding and survival rates may decline below the point needed to maintain the population. Their habitat destruction has been so severe that polar bears were listed as a threatened species in 2008, according to the U.S. Endangered Species Act. Studies reveal that polar bears are likely to lose nearly half of their summer sea ice habitat by the middle of the 21st century, which would in turn reduce the world's polar bear population—currently estimated at 25,000—by two-thirds.

In addition, human inhabitants of the Arctic who depend on subsistence living are being affected by the lack of sea ice. This is because some of the inhabitants' food sources are more difficult to obtain now that marine species live further from shore to be near the ice edge. Arctic residents have also claimed that their weather is changing, and a recent study supports their observations. The study determined that years with the least amount of Arctic sea ice have had significantly stronger Arctic storms. One new opportunity that now exists, however, involves the creation of a new "Northwest Passage" shipping lane linking the North Pacific and North Atlantic Oceans through the largely ice-free portions of the Arctic Ocean (Figure 16.26a).

While all these changes are taking place in the Arctic, different but equally dramatic ones are taking place in Antarctica, particularly in the western part of Antarctica that includes the Antarctic Peninsula. As discussed in Chapter 7, Antarctica produces many icebergs from glaciers on land. The rate at which Antarctica is producing icebergs—especially large icebergs the size of small U.S. states—has increased in recent years. For example, the Antarctic Peninsula's Larsen Ice Shelf has decreased by more than 40% over the past decade, including the release of a Delaware-sized iceberg—5200 square kilometers (2000 square miles) and estimated to weigh more than 1 trillion metric tons (1.1 trillion short tons)—known as A-68, which broke off in July 2017 (see Figure 7.28d). In the past 30 years, other Antarctic ice shelves have experienced similar

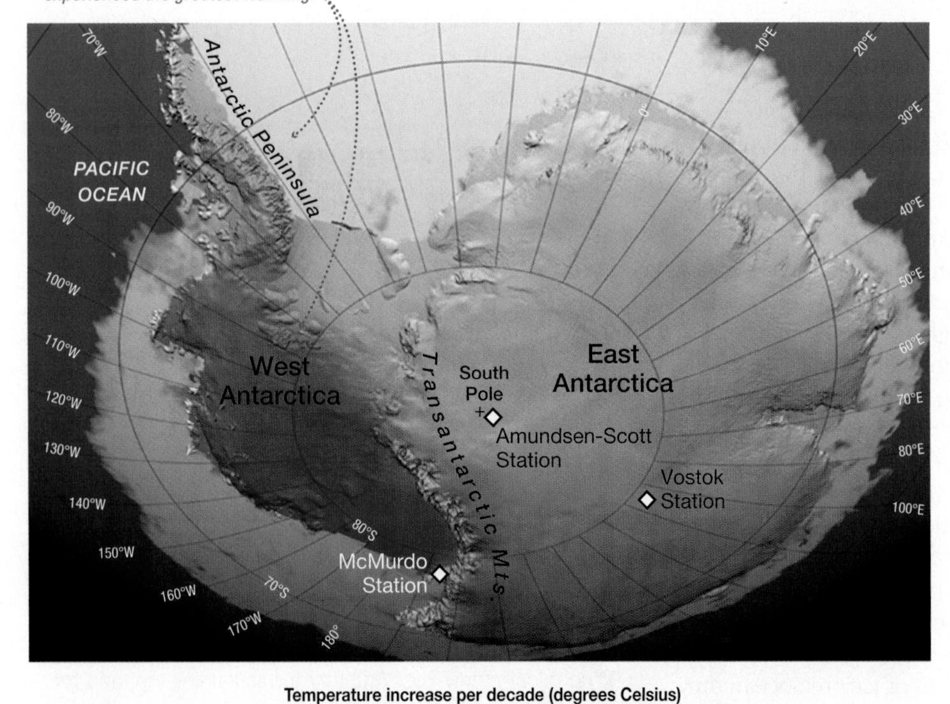


**EXPLORING DATA** ▲ **Compare the relative proportions of land and water in East and West Antarctica. Using what you know about the continental effect on temperature, explain why West Antarctica is warming up faster than East Antarctica.**

Figure 16.28 **Antarctic warming trends.** This satellite image shows the amount of warming that Antarctica has experienced since 1957. The data are from satellites, which are calibrated with weather station measurements.

collapses—including the disappearance of the ice shelves knows as Jones, Larsen A, Muller, and Wordie—after some 400 years of relative stability. Scientists attribute this catastrophic retreat to warming in Antarctica; West Antarctica and the Antarctic Peninsula have experienced some of the greatest warming worldwide (see Figure 16.23). In fact, Antarctica has warmed at a rate of about 0.12°C (0.22°F) per decade since 1957, for a total average temperature rise of 0.5°C (1.0°F) (**Figure 16.28**). Some areas of West Antarctica are warming several times faster. For example, at Byrd Station in West Antarctica, analysis of temperature records that have been kept since 1958 reveal that it has warmed by 2.4°C (4.0°F), making it one of the fastest-warming places on Earth. In addition, scientists have discovered that many Antarctic ice shelves are thinning and flowing more quickly and so could experience rapid disintegration because they are now in contact with warmer seawater that melts the ice from below.

## Ocean Acidification

The human-induced increase in the amount of carbon dioxide in the atmosphere has some severe implications for ocean chemistry and for marine life. Recent studies show that a little less than half of the carbon dioxide released by the burning of fossil fuels stays in the atmosphere and about one-quarter currently ends up in the

**(a)** Coccolithophores, which are a type of phytoplankton.

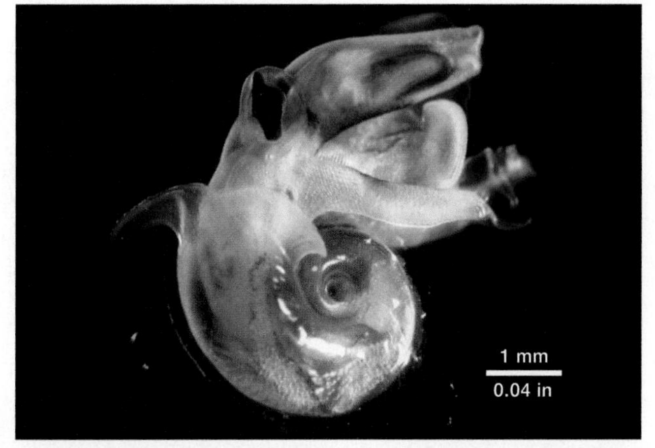


**(b)** Pteropod, which is a type of zooplankton that is a small swimming snail with a shell.

**(c)** Sea urchins, which crawl across the sea floor.

**(d)** Corals, which grow in tropical regions.

SmartFigure 16.29 **Examples of marine organisms that are affected by increased ocean acidity.** Various organisms make their skeletons or shells out of easily-dissolved calcium carbonate. As ocean acidity increases, these and many other types of organisms will have a more difficult time building and maintaining calcified hard parts.
https://goo.gl/ywiOoA

## CREATURE FEATURE 16.1

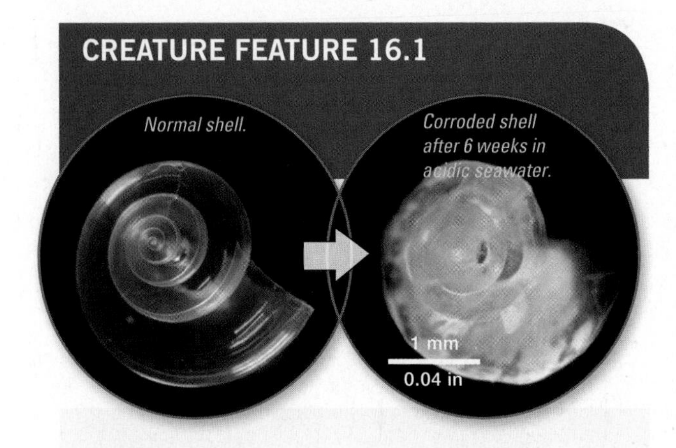


### Ocean acidification is dissolving my shell!

Sea butterflies, known as **pteropods** (family *Thecosomata*), are tiny free-floating gastropod snails that live in surface waters and build their shells out of calcium carbonate.

Laboratory experiments reveal that pteropods living in increasingly acidic seawater due to ocean acidification experience corrosion of their shells, and so have less energy to put into their metabolism and reproduction.

oceans, readily dissolving into seawater at the ocean surface. This "sink" reduces global warming—but at the expense of acidifying the sea.

When large amounts of carbon dioxide enter the ocean, it overwhelms the ocean's natural ability to buffer itself. (For a discussion of the ocean's buffering system and the pH scale, see Section 5.7 in Chapter 5.) This absorbed carbon dioxide in seawater forms carbonic acid, which is a relatively weak acid (people drink it all the time in carbonated beverages). But when carbonic acid forms in seawater, it lowers the ocean's pH (that is, increases its acidity in a process called **ocean acidification**) and changes the balance of carbonate and bicarbonate ions. In fact, the oceans have already absorbed enough carbon dioxide for surface waters to have experienced a pH decrease of 0.1 pH unit since pre-industrial times. Although 0.1 pH unit sounds like a small amount, pH is a logarithmic scale (like the moment magnitude scale $M_W$, which records the size of earthquakes), so a difference of 0.1 pH unit represents about a 30% increase in hydrogen ion concentration; every decrease of 1 pH unit is equivalent to a *tenfold* increase in hydrogen ion concentration. Another study has confirmed a 0.04 pH unit decrease in the North Pacific Ocean during just the past two decades.

Moreover, this shift toward increased acidity and the ensuing changes in ocean chemistry makes it more difficult for certain marine creatures to build and maintain hard parts out of easily dissolved calcium carbonate. (Chemically, more acidic waters have more hydrogen ions that bind with carbonates. That leaves fewer carbonates available for carbonate-forming marine organism to make their shells.) The decline in pH thus threatens a diverse assortment of calcifying organisms—creatures that grow calcium carbonate skeletons or shells—such as coccolithophores, foraminifers, pteropods, calcareous algae, sea urchins, mollusks, and corals (**Figure 16.29**). These organisms provide essential food and habitat to many other species, so their demise could affect entire ocean ecosystems. For example, research shows that the projected future level of ocean acidity prevents Antarctic krill eggs from hatching, which would affect entire Antarctic food webs. Other studies have shown that in the past 20 years, ocean acidification has already caused a 15% decrease in the growth rate of corals in Australia's Great Barrier Reef and artificially increasing seawater acidity flowing over a natural coral reef community in the Great Barrier Reef reduced coral calcification by about one-third. And, laboratory experiments that expose organisms to conditions that mimic the ocean's future acidity clearly show the negative impact on the shells of calcite-secreting organisms (**Creature Feature 16.1**).

Ocean acidification affects more than just organisms that make hard parts out of calcium carbonate. Studies show that increasing ocean acidity can interfere with basic bodily functions for all marine animals, shelled or not. For example, research has shown that the larvae of Atlantic cod (*Gadus morhua*) exposed to acidic waters experience severe tissue damage in many internal organs, thus increasing their mortality rate. Other studies of reef-dwelling clownfish and damselfish report that they are exposed to $CO_2$-infused waters, the fish exhibit learning problems and odd behavioral issues, including a tendency to seek out predators' odors. Still other studies of marine organisms exposed to acidic waters show a decrease in reproductive success. By disrupting processes as fundamental as growth, behavior, and reproduction, ocean acidification threatens marine animals' health and even the survival of species.

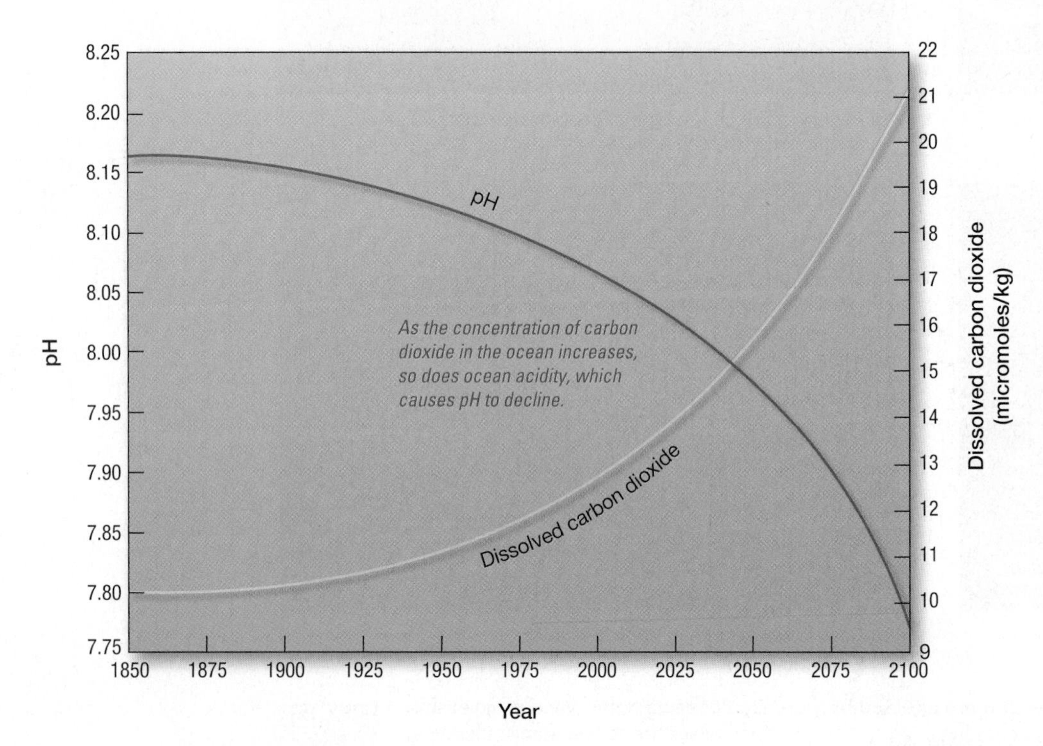


Figure 16.30 **Historical and projected dissolved carbon dioxide and ocean pH.**

The graph in **Figure 16.30** shows the projected increase of carbon dioxide in the ocean and the resulting decrease in ocean pH (rise in acidity). The graph shows that if the current trend of human-induced carbon dioxide emissions continues, by 2100, the ocean will experience a pH decrease of at least 0.3 pH unit; some studies indicate that pH could decrease by as much as 0.6 pH unit. Even at the lower value of 0.3 pH unit, this reduction of ocean pH represents an increase in hydrogen ion concentration of 100% above what it was in pre-industrial times. An additional concern is that deep currents will eventually transmit this increase in acidity to the deep-ocean floor, where it will likely affect deep-sea marine organisms, whose stable environment leaves them ill-equipped to adapt to change.

Geologic evidence shows that comparable ocean pH changes wiped out many species of sea life, especially bottom-dwelling organisms. Scientific studies show that today's alteration of ocean chemistry is unprecedented on Earth: an analysis of the past 300 million years of Earth history fails to find a time when ocean chemistry is changing as fast as it is today.

Several processes influence the amount of carbon dioxide that is absorbed by the ocean and how much remains in the atmosphere. The ocean and atmosphere are two reservoirs for carbon dioxide storage, but carbon dioxide is not equally divided between the two. This is because carbon dioxide gas is readily absorbed by the ocean and, as a result, the ocean contains a much larger amount of carbon dioxide than the atmosphere. (Of the three places where carbon dioxide resides—atmosphere, ocean, and land biosphere—approximately 93% is found in the ocean. The atmosphere, in contrast, contains the smallest amount.) The amount of carbon dioxide from the atmosphere that is dissolved in the ocean varies with the chemistry of seawater but is also controlled by positive-feedback loops. For example, one positive-feedback loop is that as the ocean approaches saturation of carbon dioxide, it will absorb less of the gas, which means that more will remain in the atmosphere. Another positive-feedback loop is that as the ocean warms, it will further reduce the amount of carbon dioxide that goes into the oceans because warmer water can't hold as much dissolved gas. Yet another positive-feedback loop involves the rate at which deep waters mix with surface waters: the more rapid the mixing, the more it facilitates the uptake of carbon dioxide from the atmosphere. If deep-water circulation slows as predicted, this will also slow the uptake of carbon dioxide from the atmosphere. In addition, another positive-feedback loop is that ocean acidification inhibits the ability of marine organisms to make calcium carbonate hard parts, so less carbon dioxide in the form of calcium carbonate can be stored by organisms and effectively removed from the environment. All of these positive-feedback loops cause increasing amounts carbon dioxide to stay in the atmosphere, which in turn will likely create additional warming.

## Rising Sea Level

Analysis of worldwide tide records indicates that there has been a rise in global sea level of between 10 and 25 centimeters (4 and 10 inches) over the past 100 years. At certain tide-recording stations where data go back well into the 19th century, there has been an increase in relative sea level of as much as 40 centimeters (16 inches) over the past 150 years (**Figure 16.31**). More recently, satellite altimeter data since 1993 indicate a global increase in sea level of about 3 millimeters (0.1 inch) per year (**Figure 16.32**). Studies have shown that the current rate of sea level rise is occurring faster than at any other time over the past 4000 years, and the rate is expected to increase with additional warming.

Two main factors contribute to the global rise in sea level: (1) thermal expansion of ocean water as it warms and (2) an increase in the amount of

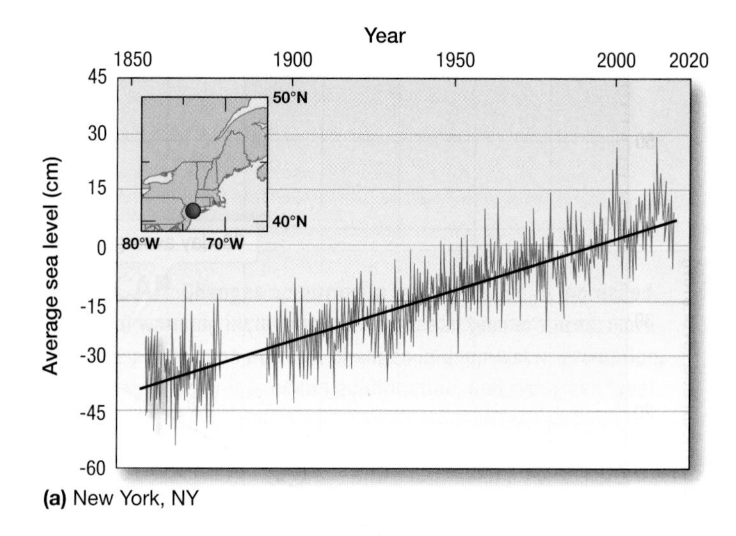


(a) New York, NY

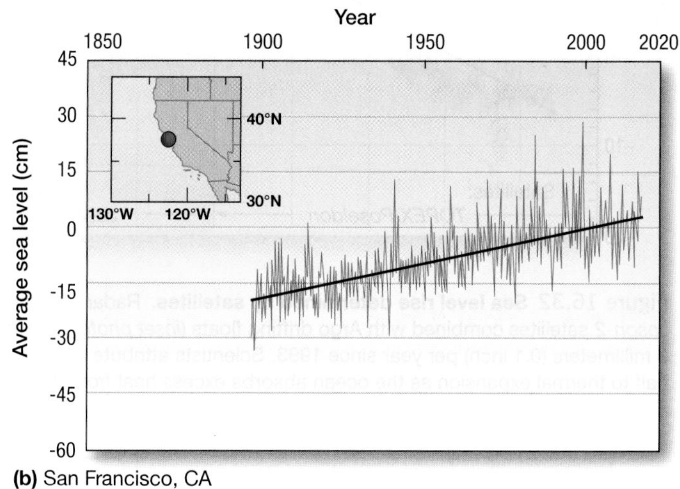


(b) San Francisco, CA

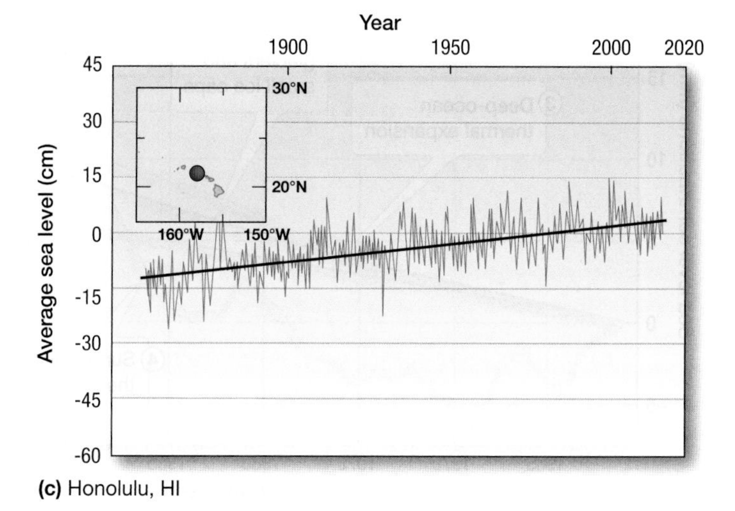


(c) Honolulu, HI

Figure 16.31 **Measured relative sea level rise from tide gauges.** Sea level data from **(a)** New York, New York, **(b)** San Francisco, California, and **(c)** Honolulu, Hawaii, all of which show an increase in sea level. While some of the documented rise is due to local effects (such as changes in the elevation of Earth's crust caused by tectonic movement or isostatic adjustment), the majority is caused by the addition of water from the melting of continental ice caps and glaciers as well as from thermal expansion of warmer ocean water.

# AFTERWORD

*"In the end, we will conserve only what we love. We love only what we understand. We will understand only what we are taught."*

—*Baba Dioum, Senegalese conservationist (1968)*

At the end of our journey through this book together, it seems fitting to examine people's perceptions of the ocean. Many people describe the ocean as "powerful," "awe-inspiring," "moving," "serene," "abundant," and "majestic." Others call it "vast," "infinite," or "boundless." These are all appropriate descriptions because even though humans have been exploring and studying the ocean for centuries, the ocean still holds many secrets. The oceans are constantly surprising researchers with unusual features, newly discovered species, and geologic wonders that exist within its watery world. In fact, less than 5% of the ocean has been seen by humans, and even less has been scientifically explored.

Despite the ocean's impressive size, it is beginning to feel the effects of human activities. For instance, every ocean contains large areas of floating plastic, and even remote beaches are littered with trash. Organisms living in the ocean also feel the effect of humankind's use of the ocean. Whaling in the 19th and 20th centuries, for example, pushed many great whale populations to near extinction. Restrictions on whaling and the development of substitutes for whale products helped whales survive this threat, but now they face destruction of their feeding and breeding grounds. Overfishing has degraded entire marine ecosystems, and human-caused climate change is altering marine chemistry on a global scale. The oceans are approaching irreversible, potentially catastrophic change, suggesting that the ocean is not quite as "vast," "infinite," or "boundless" as most people believe.

Humans have become one of the most important agents of change on the planet. This is largely due to our rapidly expanding human population, which is increasing at an exponential rate (**Figure Aft.1**). Today, more than 7.6 billion people occupy the planet, and births now exceed deaths by about two-and-a-half times. At the present rate of population increase—which sounds small enough, at about 1.1%, but produces truly staggering numbers—there are more than 2 new people on the planet each second, or about 9000 more people each hour. Each year there are about 77 million more of us, a total equal to the combined populations of the six largest cities in the world (Mumbai, Shanghai, Karachi, Delhi, Istanbul, and São Paulo). A recent study showed that there is an 80% probability that world population will increase to between 9.6 and 12.3 billion people in 2100. Some scientists, in fact, have proposed calling today's time period the *Anthropocene* (*anthro* = human, *cene* = new), or a "sixth mass extinction," in recognition of human activities that are degrading vital Earth systems and the evidence of those impacts that will be preserved in the geologic record. Clearly, our burgeoning human population is the greatest environmental threat of all.

Human-induced changes in the marine environment are broad and far-reaching. Examples include pollution, shoreline development, overfishing, introduction of non-native species, biodiversity decline, ecosystem degradation, and perhaps most serious of all, climate change. A 2008 study of the cumulative impact of 17 of the most urgent land- and ocean-based threats to worldwide marine ecosystems shows that no areas of the ocean are untouched by human activities. The study revealed that fully one-third of the oceans are strongly affected by multiple factors and that the most heavily impacted ecosystems are continental shelves, rocky reefs, coral reefs, seagrass beds, and deep-sea seamounts. In addition, a 2012 study reported that one-fifth of all invertebrate species worldwide are at risk of extinction, including almost one-third of reef-building corals.

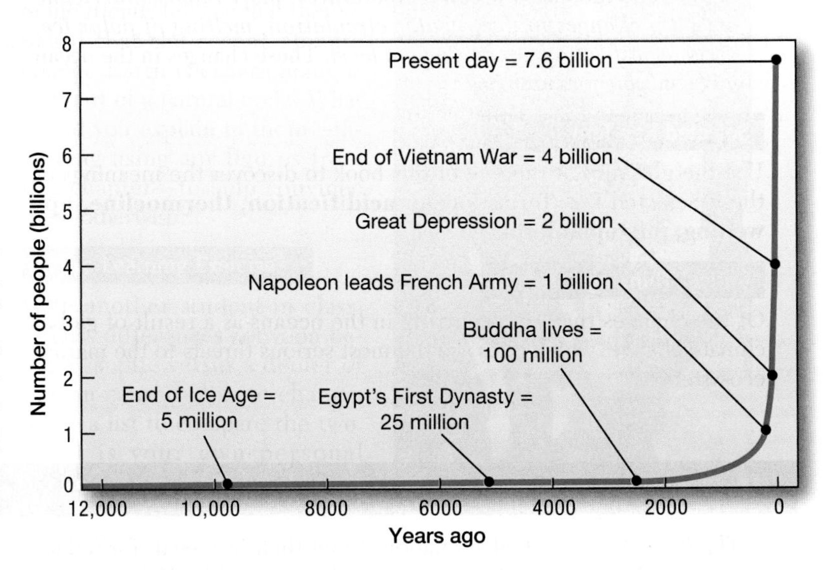


**Figure Aft.1 Human population growth.** Graph showing world population, which has experienced rapid growth in recent decades. It took 4 million years for humanity to reach the 2 billion mark but only 50 years to double that total. The current world population recently surpassed 7.6 billion people and is increasing at a rate of more than 1% per year.

Several commissions have suggested ways to protect the marine environment. In 2003, for example, the Pew Oceans Commission recommended significant changes aimed at guiding the way in which the federal government should manage American's marine environment. In 2004, the U.S. Commission on the Oceans submitted 212 recommendations for a coordinated and comprehensive national ocean policy. In 2008, the U.S. Joint Ocean Commission issued a report card assessing the nation's collective progress in ocean policy during 2007 (its overall grade: a "C," with some categories receiving a "D"). A 2012 study of the health and benefits of the global ocean produced an index of 10 diverse public goals for a healthy coupled human–ocean system, including an integrative assessment of factors such as food provision, carbon storage, tourism value, and biodiversity. The study gave the world ocean a score of 60 out of 100, with developed countries generally performing better than developing countries. (The United States achieved a score of 63.) These reports urge an integrated research and education ecosystem-based management approach to support ocean environmental health and resource sustainability. As these conclusions show, in the end, the fate of the oceans may rest on the education of policymakers and individuals alike, as we can only protect what we understand.

# GLOSSARY

## A

**Abiotic** Pertaining to nonliving factors, not life.

**Abiotic environment** The nonliving components of an ecosystem including chemical and physical factors (such as food, water, air, shelter, and temperature) that are essential for life.

**Abyssal clay** Deep-ocean (oceanic) deposits containing less than 30% biogenous sediment. Often oxidized and red in color, thus commonly termed red clay.

**Abyssal hill** A volcanic peak rising less than 1 kilometer (0.6 mile) above the ocean floor.

**Abyssal hill province** A deep-ocean region, particularly in the Pacific Ocean, where oceanic sedimentation rates are so low that abyssal plains do not form and the ocean floor is covered with abyssal hills.

**Abyssal plain** A flat depositional surface extending seaward from the continental rise or oceanic trenches.

**Abyssal storm** Storm-like occurrences of rapid current movement affecting the deep-ocean floor. They are believed to be caused by warm- and cold-core eddies of surface currents.

**Abyssal zone** The benthic environment between 4000 and 6000 meters (13,000 and 20,000 feet).

**Abyssopelagic zone** The open-ocean (oceanic) environment below 4000 meters (13,000 feet) in depth.

**Acid** A substance that releases hydrogen ions ($H^+$) in solution.

**Acoustic Thermometry of Ocean Climate (ATOC)** The measurement of ocean-wide changes in water properties such as temperature by transmitting and receiving low-frequency sound signals.

**Active margin** A continental margin marked by a high degree of tectonic activity, such as those typical of the Pacific Rim. Types of active margins include convergent active margins (marked by plate convergence) and transform active margins (marked by transform faulting).

**Adiabatic** Pertaining to a change in the temperature of a mass resulting from compression or expansion. It requires no addition of heat to or loss of heat from the substance.

**Aerosol** A type of suspended particle in the atmosphere that can affect the atmosphere's reflectivity and its ability to trap heat.

**Age of Discovery** The 30-year period from 1492–1522 when Europeans explored the continents of North and South America and the globe was circumnavigated for the first time.

**Agulhas Current** A warm surface current that carries Indian Ocean water around the southern tip of Africa and into the Atlantic Ocean.

**Air mass** A large volume of air that has a definite region of origin and distinctive characteristics.

**Alaskan Current** A cool surface current that carries water in a counterclockwise fashion in the Gulf of Alaska.

**Albedo** The fraction of incident electromagnetic radiation reflected by a surface.

**Algae** Primarily aquatic, eukaryotic, photosynthetic organisms that have no root, stem, or leaf systems. Can be microscopic or macroscopic. Singular: alga.

**Alkaline** Having a pH greater than 7, indicating a substance that releases an excess of hydroxide ions ($OH^-$) into solution. Also called basic.

**Alveoli** A tiny, thin-walled, capillary-rich sac in the lungs where the exchange of oxygen and carbon dioxide takes place.

**Amino acid** One of more than 20 naturally occurring compounds that contain $NH_2$ and COOH groups. They combine to form proteins.

**Amnesic shellfish poisoning** Partial or total loss of memory resulting from poisoning caused by eating shellfish with high levels of domoic acid, a toxin produced by a diatom.

**Amphidromic point** A nodal, or "no-tide," point in the ocean or sea around which the crest of the tide wave rotates during one tidal period.

**Amphipoda** A crustacean order containing laterally compressed members such as the "beach hoppers."

**Anadromous** Pertaining to a species of fish that spawns in freshwater and then migrates into the ocean to grow to maturity.

**Anaerobic** Requiring or occurring in the absence of free oxygen ($O_2$).

**Anaerobic respiration** Respiration carried on in the absence of free oxygen ($O_2$). Some bacteria and protozoans carry on respiration this way.

**Anchovy (anchoveta)** A small silvery fish (*Engraulis ringens*) that swims through the water with its mouth open to catch its planktonic food.

**Andesite** A gray, fine-grained volcanic rock composed chiefly of plagioclase feldspar.

**Anhydrite** A colorless, white, gray, blue, or light purple evaporite mineral (anhydrous calcium sulfate, $CaSO_4$) that usually occurs as layers associated with gypsum deposits.

**Animalia** A kingdom of many-celled animals.

**Anion** An atom that has gained one or more electrons and has an electrical negative charge.

**Annelida** A phylum of elongated segmented worms.

**Anomalistic month** The time required for the Moon to go from perigee to perigee, 27.5 days.

**Anoxic** Without oxygen.

**Antarctic Bottom Water** A water mass that forms in the Weddell Sea, sinks to the ocean floor, and spreads across the bottom of all oceans.

**Antarctic Circle** The latitude 66.5 degrees south.

**Antarctic Circumpolar Current** The eastward-flowing current that encircles Antarctica and extends from the surface to the deep-ocean floor, it is the largest volume current in the world. Also called the West Wind Drift.

**Antarctic Convergence** The zone of convergence along the northern boundary of the Antarctic Circumpolar Current where the southward-flowing boundary currents of the subtropical gyres converge on the cold Antarctic waters.

**Antarctic Divergence** The zone of divergence separating the westward-flowing East Wind Drift and the eastward-flowing Antarctic Circumpolar Current.

**Antarctic Intermediate Water** Antarctic zone surface water that sinks at the Antarctic convergence and flows north at a depth of about 900 meters (2950 feet) beneath the warmer upper-water mass of the South Atlantic Subtropical Gyre.

**Antarctic Ocean** The ocean that surrounds the continent of Antarctica and is located south of about 50 degrees south latitude; also called the Southern Ocean.

**Anthophyta** Seed-bearing plants.

**Anticyclone** An atmospheric system characterized by the rapid, outward circulation of air masses about a high-pressure center that is associated with sinking air. Anticyclones circulate clockwise in the Northern Hemisphere and counterclockwise in the Southern Hemisphere and are usually accompanied by dry, clear, fair weather.

**Anticyclonic flow** The flow of air around a region of high pressure clockwise in the Northern Hemisphere.

**Antilles Current** A warm current that flows north seaward of the Lesser Antilles from the North Equatorial Current of the Atlantic Ocean to join the Florida Current.

**Antinode** A zone of maximum vertical particle movement in standing waves where crest and trough formation alternate.

**Aphelion** The point in the orbit of a planet or comet where it is farthest from the Sun.

**Aphotic zone** A zone without light. The ocean is generally in this state below 1000 meters (3280 feet).

**Apogee** The point in the orbit of the Moon or an artificial satellite that is farthest from Earth.

**Aragonite** A form of $CaCO_3$ that is less common and less stable than calcite. Pteropod shells are usually composed of aragonite.

**Archaea** One of the three major domains of life. The domain consists of simple microscopic bacteria-like creatures (including methane producers and sulfur oxidizers that inhabit deep-sea vents and seeps) and other microscopic life-forms that prefer environments of extreme conditions of temperature and/or pressure.

**Archaebacteria** A kingdom of organisms that do not have nuclear material confined within a sheath but spread throughout the cell. Includes archaea.

**Archipelago** A large group of islands.

**Arctic Circle** The latitude 66.5 degrees north.

**Arctic Convergence** A zone of converging currents similar to the Antarctic Convergence but located in the Arctic.

**Arctic Ocean** The ocean located in the Northern Hemisphere polar region; the smallest ocean in the world.

**Argo** A global array of free-drifting profiling floats that move vertically and measure the temperature, salinity, and other water characteristics of the upper 2000 meters (6600 feet) of the ocean.

**Aspect ratio** The index of propulsive efficiency obtained by dividing the square of fin height by fin area.

**Asthenosphere** A plastic layer in the upper mantle 80 to 200 kilometers (50 to 124 miles) deep that may allow lateral movement of lithospheric plates and isostatic adjustments.

**Atlantic Ocean** The ocean located between South America, North America, Europe, and Africa; the second largest ocean in the world.

**Atlantic meridional overturning circulation (AMOC)** The part of conveyor-belt circulation in the North Atlantic that includes both surface and deep currents.

**Atlantic-type margin** See *passive margin*.

**Atmospheric wave** A type of wave that occurs in the atmosphere. See *wave*.

**Atoll** A ring-shaped coral reef growing upward from a submerged volcanic peak that usually has low-lying islands composed of coral debris.

**Atom** A unit of matter, the smallest unit of an element, having all the characteristics of that element and consisting of a dense, central, positively charged nucleus surrounded by a system of electrons.

**Atomic mass** The mass of an atom, usually expressed in atomic mass units.

**Atomic number** The number of protons in an atomic nucleus.

**Autotroph** Algae, plants, and bacteria that can synthesize organic compounds from inorganic nutrients.

**Autumnal equinox** The passage of the Sun across the equator as it moves from the Northern Hemisphere into the Southern Hemisphere, approximately September 23. During this time, all places in the world experience equal lengths of night and day. Also called fall equinox.

## B

**Backshore** The inner portion of the shore, lying landward of the mean spring tide high water line. Acted upon by the ocean only during exceptionally high tides and storms.

**Backwash** The flow of water down the beach face toward the ocean from a previously broken wave.

**Bacteria** One of the three major domains of life. The domain includes unicellular, prokaryotic microorganisms that vary in terms of morphology, oxygen and nutritional requirements, and motility.

**Bacterioplankton** Bacteria that live as plankton.

**Bacteriovore** An organism that feeds on bacteria.

**Baleen** A fibrous substance made of keratin found in parallel rows of long plates that hang down from the upper jaw of baleen whales.

**Bar-built estuary** A shallow estuary (lagoon) separated from the open ocean by a bar deposit such as a barrier island. The water in these estuaries usually exhibits vertical mixing.

**Barrier flat** An area that lies between the salt marsh and dunes of a barrier island and that is usually covered with grasses or forests if protected from overwash for a sufficient length of time.

**Barrier island** A long, narrow, wave-built island separated from the mainland by a lagoon.

**Barrier reef** A coral reef separated from the nearby landmass by open water.

**Barycenter** The center of mass of a system.

**Basalt** A dark-colored volcanic rock characteristic of the ocean crust. Contains minerals with relatively high iron and magnesium content.

**Base** A substance that releases hydroxide ions ($OH^-$) in solution. See *alkaline*.

**Bathyal zone** The benthic environment between the depths of 200 and 4000 meters (660 and 13,000 feet). It includes mainly the continental slope and the oceanic ridges and rises.

**Bathymetry** The measurement of ocean depth.

**Bathypelagic zone** The pelagic environment between the depths of 1000 and 4000 meters (3300 and 13,000 feet).

**Bathyscaphe** A specially designed deep-diving submersible.

**Bathysphere** A specially designed deep-diving submersible that resembles a sphere.

**Bay barrier** A marine deposit attached to the mainland at both ends and extending entirely across the mouth of a bay, separating the bay from the open water. Also known as a bay-mouth bar.

**Beach** Sediment seaward of the coastline through the surf zone that is in transport along the shore and within the surf zone.

**Beach compartment** A series of rivers, beaches, and submarine canyons involved in the movement of sediment to the coast, along the coast, and down one or more submarine canyons.

**Beach face** The wet, sloping surface that extends from the berm to the shoreline. Also known as the low tide terrace.

**Beach replenishment** The addition of beach sediment to replace lost or missing material. Also called beach nourishment.

**Beach starvation** The interruption of sediment supply and resulting narrowing of beaches.

**Beaufort Wind Scale** A standardized wind scale that describes the appearance of the sea surface from dead calm conditions to hurricane-force winds.

**Benguela Current** The cold eastern boundary current of the South Atlantic Subtropical Gyre.

**Benthic** Pertaining to the ocean bottom.

**Benthic environment** The ocean floor environment, which is divided into the subneritic province (water depth 0 to 200 meters or 660 feet) and the suboceanic province (water depth greater than 200 meters or 660 feet).

**Benthos** The forms of marine life that live on the ocean bottom.

**Berm** The dry, gently sloping region on the backshore of a beach at the foot of the coastal cliffs or dunes.

**Berm crest** The area of a beach that separates the berm from the beach face. The berm crest is often the highest portion of a berm.

**Bicarbonate ion ($HCO_3^-$)** An ion that contains the radical group ($HCO_3^-$).

**Bioaccumulation** The accumulation of a substance, such as a toxic chemical, in various tissues of a living organism.

**Bioerosion** Erosion of reef or other solid bottom material by the activities of organisms.

**Biofilm** Coating of organic matter such as that found on sand grains.

**Biogenous sediment** Sediment containing material produced by plants or animals, such as coral reefs, shell fragments, and housings of diatoms, radiolarians, foraminifers, and coccolithophores; components can be either macroscopic or microscopic.

**Biogeochemical cycle** The natural cycling of compounds among the living and nonliving components of an ecosystem.

**Biological pump** The movement of carbon dioxide that enters the ocean from the atmosphere through the water column to the sediment on the ocean floor by biological processes—photosynthesis, secretion of shells, feeding, and dying.

**Bioluminescence** Light organically produced by a chemical reaction. Found in bacteria, phytoplankton, and various fishes (especially deep-sea fish).

**Biomagnification** Concentration of impurities as animals are eaten and the impurity is passed through food chains.

**Biomass** The total mass of a defined organism or group of organisms in a particular community or in the ocean as a whole.

**Biomass pyramid** A representation of trophic levels that illustrates the progressive decrease in total biomass at successive higher levels of the pyramid.

**Bioremediation** The technique of using microbes to assist in cleaning toxic spills.

**Biotic community** The living organisms that inhabit an ecosystem.

**Biozone** A region of the environment that has distinctive biological characteristics.

**Bivalve** A mollusk, such as an oyster or a clam, that has a shell consisting of two hinged valves.

**Black smoker** A hydrothermal vent on the ocean floor that emits a black cloud of hot water filled with dissolved metal particles.

**Body wave** A longitudinal or transverse wave that transmits energy through a body of matter.

**Boiling point** The temperature at which a substance changes state from a liquid to a gas at a given pressure.

**Bore** A steep-fronted tide crest that moves up a river in association with an incoming high tide.

**Boundary current** The northward- or southward-flowing currents that form the western and eastern boundaries of each subtropical gyre.

**Brackish** Low-salinity water caused by the mixing of freshwater and saltwater.

**Brazil Current** The warm western boundary current of the South Atlantic Subtropical Gyre.

**Breaker zone** A region where waves break at the seaward margin of the surf zone.

**Breakwater** An artificial structure constructed roughly parallel to shore and designed to protect a coastal region from the force of ocean waves.

**Brittle** Descriptive term for a substance that is likely to fracture when force is applied to it.

**Bryozoa** A phylum of colonial animals that often share one coelomic cavity. Encrusting and branching forms secrete a protective housing (zooecium) of calcium carbonate or chitinous material.

**Buffering** The process by which a substance minimizes a change in the acidity of a solution when an acid or base is added to the solution.

**Buoyancy** The ability or tendency to float or rise in a liquid.

**Bycatch** Marine organisms that are caught incidentally by fishers seeking commercial species.

## C

**Calcareous** Containing calcium carbonate.

**Calcite** A mineral with the chemical formula $CaCO_3$.

**Calcite compensation depth (CCD)** The depth at which the amount of calcite ($CaCO_3$) produced by the organisms in the overlying water column is equal to the amount of calcite the water column can dissolve. No calcite deposition occurs below this depth, which, in most parts of the ocean, is at a depth of 4500 meters (15,000 feet).

**Calcium carbonate ($CaCO_3$)** A chalk-like substance secreted by many organisms in the form of coverings or skeletal structures.

**California Current** The cold eastern boundary current of the North Pacific Subtropical Gyre.

**Calorie** A unit of heat, defined as the amount of heat required to raise the temperature of 1 gram of water 1°C.

**Calving** The process by which a glacier breaks at an edge, so that a portion of the ice separates and falls from the glacier.

**Canary Current** The cold eastern boundary current of the North Atlantic Subtropical Gyre.

**Capillarity** The action by which a fluid, such as water, is drawn up in small tubes as a result of surface tension.

**Capillary wave** An ocean wave whose wavelength is less than 1.74 centimeters (0.7 inch). The dominant restoring force for such waves is surface tension.

**Carapace** (1) A chitinous or calcareous shield that covers the cephalothorax of some crustaceans. (2) The dorsal portion of a turtle shell.

**Carbohydrate** An organic compound containing the elements carbon, hydrogen, and oxygen with the general formula $(CH_2O)_n$.

**Carbon cycle** A biogeochemical cycle by which carbon is exchanged among Earth's atmosphere, hydrosphere, geosphere, biosphere, and cryosphere.

**Carbon fixation** The conversion of inorganic carbon in the form of $CO_2$ into organic carbon in the form of sugars and other organic compounds useful to living organisms, usually through photosynthesis and rarely through chemosynthesis.

**Carbon footprint** The amount of carbon dioxide and other carbon compounds emitted into the atmosphere due to the consumption of fossil fuels by a particular person or group.

**Carbonate ion ($CO_3^{2-}$)** An ion that contains the radical group $CO_3^{2-}$.

**Caribbean Current** The warm current that carries equatorial water across the Caribbean Sea into the Gulf of Mexico.

**Carnivora** The order of marine mammals that includes the sea otter, polar bear, and pinnipeds.

**Carnivore** An animal that depends on other animals solely or chiefly for its food supply.

**Carnivorous feeding** The process by which an organism feeds solely or chiefly on other animals as food items.

**Carotin** An orange-yellow pigment found in plants.

**Cation** An atom that has lost one or more electrons and has an electrical positive charge.

***Caulerpa taxifolia*** A tropical seaweed; a cold-water clone was introduced into the aquarium industry and found its way into coastal waters of the Mediterranean and Southern California. It continues to spread in the Mediterranean, but has been eradicated in Southern California.

**Centigrade temperature scale** A temperature scale based on the freezing point (0°C = 32°F) and boiling point (100°C = 212°F) of pure water. Also known as the Celsius scale after its founder.

**Centripetal force** A center-seeking force that tends to make rotating bodies move toward the center of rotation.

**Cephalopoda (cephalopod)** A class of the phylum Mollusca whose members have a well-developed pair of eyes and a ring of tentacles surrounding the mouth. The shell is absent or internal on most members. The class includes the squid, octopus, and nautilus.

**Cetacea (cetacean)** An order of marine mammals that includes the whales, dolphins, and porpoises.

**Chalk** A soft, compact form of calcite, generally gray-white or yellow-white in color and derived chiefly from microscopic fossils.

**Chemical energy** A form of potential energy stored in the chemical bonds of compounds.

**Chemosynthesis** A process by which bacteria or archaea synthesize organic molecules from inorganic nutrients using chemical energy released from the bonds of a chemical compound (such as hydrogen sulfide) by oxidation.

**Chloride ion ($Cl^-$)** A chlorine atom that has become negatively charged by gaining one electron.

**Chlorinity** The amount of chloride ion and ions of other halogens in ocean water expressed in parts per thousand (‰) by weight.

**Chlorophyll** A group of green pigments that make it possible for plants to carry on photosynthesis.

**Chlorophyta** Green algae. Characterized by the presence of chlorophyll and other pigments.

**Chloroplast** Photosynthetic structures containing chlorophyll found in plants, algae, and other photosynthetic organisms.

**Chondrite** A stony meteorite composed primarily of silicate rock material and containing chondrules (spheroidal granules). They are the most commonly found meteorites.

**Chronometer** An exceptionally precise timepiece.

**Chrysophyta** An important phylum of planktonic algae that includes the diatoms. The presence of chlorophyll is masked by the pigment carotene that gives the plants a golden color.

**Ciguatera** A type of seafood poisoning caused by ingestion of certain tropical reef fish (most notably barracuda, red snapper, and grouper) that have high levels of naturally occurring dinoflagellate toxins.

**Cilium** A short, hair-like structure common on single-celled or simple multicelled animals. Beating in unison, cilia may create water currents that carry food toward the mouth of an animal or may be used for locomotion.

**Circadian rhythm** Behavioral and physiological rhythms of organisms related to the 24-hour day. Sleeping and waking patterns are an example.

**Circular orbital motion** The motion of water particles caused by a wave as the wave is transmitted through water.

**Circumpolar Current** An eastward-flowing current that extends from the surface to the ocean floor and encircles Antarctica.

**Clay** (1) A particle size between silt and colloid. (2) Any of various hydrous aluminum silicate minerals that are plastic, expansive, and have ion exchange capabilities.

**Climate** The meteorological conditions, including temperature, precipitation, and wind that characteristically prevail in a particular region; the long-term average of weather.

**Climate system** Refers to all factors that influence the climate of a region, including both natural and human-induced climate changes and their interactions.

**Cnidaria** A phylum that contains some 10,000 species of predominantly marine animals with a sack-like body and stinging cells on tentacles that surround the single opening to the gut cavity. There are two basic body forms. The medusa is a pelagic form represented by the jelly. The polyp is a predominantly benthic form found in sea anemones and corals. Previously named Coelenterata.

**Cnidoblast** A stinging cell of phylum Cnidaria that contains the stinging mechanism (nematocyst) used in defense and capturing prey.

**Coast** A strip of land that extends inland from the coastline as far as marine influence is evidenced in the landforms.

**Coastal geostrophic current** See *geostrophic current*.

**Coastal plain estuary** An estuary formed by rising sea level flooding a coastal river valley.

**Coastal upwelling** The movement of deeper nutrient-rich water into the surface water mass as a result of windblown surface water moving offshore.

**Coastal waters** The relatively shallow-water areas that adjoin continents or islands.

**Coastal wetland** A biologically productive region bordering estuaries and other protected coastal areas; typically as salt marshes in regions greater than 30 degrees latitude and as mangrove swamps in lower latitudes.

**Coastline** The landward limit of the effect of the highest storm waves on the shore.

**Coccolith** Tiny calcareous discs averaging about 3 microns (0.00012 inch) in diameter that form the cell wall of coccolithophores.

**Coccolithophore** A microscopic planktonic form of algae encased by a covering composed of calcareous discs (coccoliths).

**Coelenterata** A phylum of radially symmetrical animals that includes two basic body forms, the medusa and the polyp. Includes jellies (medusoid) and sea anemones (polypoid). Preferred name is now Cnidaria.

**Cohesion** The intermolecular attraction by which the elements of a body are held together.

**Cold-blooded** See *poikilothermic* and *ectothermic*.

**Cold-core ring** See *ring*.

**Cold front** A weather front in which a cold air mass moves into and under a warm air mass. It creates a narrow band of intense precipitation.

**Colonial animal** An animal that lives in groups of attached or separate individuals. Groups of individuals may serve special functions.

**Columbia River estuary** An estuary at the border between the states of Washington and Oregon that has been most adversely affected by the construction of hydroelectric dams.

**Comb jelly** Common name for members of the phylum Ctenophora. See also *ctenophore*.

**Commensalism** A symbiotic relationship in which one party benefits and the other is unaffected.

**Compensation depth for photosynthesis** The depth at which net photosynthesis becomes zero; below this depth, photosynthetic organisms can no longer survive. This depth is greater in the open ocean (up to 100 meters or 330 feet) than near the shore due to increased turbidity that limits light penetration in coastal regions.

**Compound** A substance containing two or more elements combined in fixed proportions.

**Condensation** The conversion of water from the vapor to the liquid state. When it occurs, the energy required to vaporize the water is released into the atmosphere. This is about 585 calories per gram of water at 20°C.

**Condensation point** The point at which condensation occurs.

**Conduction** The transmission of heat by the passage of energy from particle to particle.

**Conjunction** The apparent closeness of two heavenly bodies. During the new moon phase, Earth and the Moon are in conjunction on the same side of the Sun.

**Constant proportions, principle of** A principle which states that the major constituents of ocean-water salinity are found in the same relative proportions throughout the ocean, independent of salinity.

**Constructive interference** A form of wave interference in which two waves come together in phase, for example, crest to crest, to produce a greater displacement from the still-water line than that produced by either of the waves alone.

**Consumer** An animal within an ecosystem that consumes the organic mass produced by the producers.

**Continent** About one-third of Earth's surface that rises above the deep-ocean floor to be exposed above sea level. Continents are composed primarily of granite, an igneous rock of lower density than basaltic oceanic crust.

**Continental accretion** Growth or increase in size of a continent by gradual external addition of crustal material.

**Continental arc** An arc-shaped row of active volcanoes produced by subduction that occurs along convergent active continental margins.

**Continental borderland** A highly irregular portion of the continental margin that is submerged beneath the ocean and is characterized by depths greater than those characteristic of the continental shelf.

**Continental drift** A term applied to early theories supporting the idea that the continents move across Earth's surface through time.

**Continental effect** Describes areas that are less affected by the sea and therefore having a greater range of temperature differences (both daily and yearly).

**Continental margin** The submerged area next to a continent comprising the continental shelf, continental slope, and continental rise.

**Continental rise** A gently sloping depositional surface at the base of the continental slope.

**Continental shelf** A gently sloping depositional surface extending from the low water line to the depth of a marked increase in slope around the margin of a continent or island.

**Continental slope** A relatively steeply sloping surface lying seaward of the continental shelf.

**Convection** Heat transfer in a gas or liquid by the circulation of currents from one region to another.

**Convection cell** A circular-moving loop of matter involved in convective movement.

**Convergence** The act of coming together from different directions. There are polar, tropical, and subtropical regions of the oceans where water masses with different characteristics come together. Along these lines of convergence, the denser masses sink beneath the others.

**Convergent plate boundary** A lithospheric plate boundary where adjacent plates converge, producing ocean trench–island arc systems, ocean trench–continental volcanic arcs, or folded mountain ranges.

**Conveyor-belt circulation** An integrated deep-water and surface current circulation pattern that resembles a large conveyor belt.

**Copenhagen Accord** A non-binding agreement that endorses the continuation of the Kyoto Protocol and was signed in 2009 at the United Nations Climate Change Conference in Copenhagen, Denmark.

**Copepoda** An order of microscopic to nearly microscopic crustaceans that are important members of zooplankton in temperate and subpolar waters.

**Coral** A group of benthic anthozoans that exist as individuals or in colonies and secrete $CaCO_3$ external skeletons. Under the proper conditions, corals may produce reefs composed of their external skeletons and the $CaCO_3$ material secreted by varieties of algae associated with the reefs.

**Coral bleaching** The loss of color in coral reef organisms that causes them to turn white. Coral bleaching is caused by the expulsion of the coral's symbiotic zooxanthellae algae in response to high water temperatures or other adverse conditions.

**Coral reef** A calcareous organic reef composed significantly of solid coral and coral sand. Algae may be responsible for more than half of the $CaCO_3$ reef material. Found in waters where the minimum average monthly temperature is 18°C or higher.

**Core** (1) The deep, central layer of Earth, composed primarily of iron and nickel. It is subdivided into a liquid outer core 2270 kilometers (1410 miles) thick and a solid inner core with a radius of 1216 kilometers (756 miles). (2) A cylinder of sediment and/or rock material usually obtained by drilling.

**Coriolis effect** An apparent force resulting from Earth's rotation that causes particles in motion to be deflected to the right in the Northern Hemisphere and to the left in the Southern Hemisphere.

**Cosmogenous sediment** Sediment derived from outer space.

**Cotidal line** A line connecting points where high tide occurs simultaneously.

**Counterillumination** Camouflaging by using bioluminescence to match the color and intensity of dim filtered sunlight from above and obliterate a telltale shadow.

**Countershading** Protective coloration in an animal or insect, characterized by darker coloring of areas exposed to light and lighter coloring of areas that are normally shaded.

**Covalent bond** A chemical bond formed by the sharing of one or more electrons, especially pairs of electrons, between atoms.

**Crepuscular** A term that describes animals that are active primarily during twilight (between dawn and dusk).

**Crest (wave)** The portion of an ocean wave that is displaced above the still-water level.

**Cruiser** Fish (such as the bluefin tuna) that constantly cruise pelagic waters in search of food.

**Crust** (1) The uppermost outer layer of Earth's structure that is composed of basaltic oceanic crust and granitic continental crust. The average thickness of the crust ranges from 8 kilometers (5 miles) beneath the ocean to 35 kilometers (22 miles) beneath the continents. (2) A hard covering or surface layer of hydrogenous sediment.

**Crustacea (crustacean)** A class of subphylum Arthropoda that includes barnacles, copepods, lobsters, crabs, and shrimp.

**Crystalline rock** Igneous or metamorphic rocks. These rocks are made up of crystalline particles with orderly molecular structures.

**Ctenophore** A member of the phylum of gelatinous organisms that are more or less spheroidal with biradial symmetry. These exclusively marine animals have eight rows of ciliated combs for locomotion, and most have two tentacles for capturing prey.

**Current** A physical movement of water; if in the ocean, called an ocean current, which can be further subdivided into surface currents and deep currents.

**Cyclone** An atmospheric system characterized by the rapid, inward circulation of air masses about a low-pressure center that is associated with rising air. Cyclones circulate counterclockwise in the Northern Hemisphere and clockwise in the Southern Hemisphere and are usually accompanied by stormy, often destructive, weather.

**Cyclonic flow** The flow of air around a region of low pressure counterclockwise in the Northern Hemisphere.

## D

**Davidson Current** A northward-flowing surface current along the Washington–Oregon coast that is driven by geostrophic effects on a large freshwater runoff.

**DDT** An insecticide (dichlorodiphenyltrichloroethane) that caused damage to marine bird populations in the 1950s and 1960s. Its use is now banned throughout most of the world.

**Dead zone** A region of hypoxic conditions that kills off most marine organisms that cannot escape. It is usually the result of eutrophication caused by runoff from land-based fertilizer applications.

**Decay distance** The distance over which waves change from a choppy "sea" to uniform swell.

**Declination** The angular distance of the Sun or Moon above or below the plane of Earth's equator.

**Decomposer** Primarily bacteria that break down nonliving organic material, extract some of the products of decomposition for their own needs, and make available the compounds needed for primary production.

**Decompression sickness** A serious condition that occurs in divers when they ascend too rapidly, causing nitrogen bubbles to form in the blood and tissue, resulting in great pain and sometimes death. Also known as the bends.

**Deep biosphere** The microbe-rich region beneath the sea floor.

**Deep boundary current** A relatively strong deep current flowing across the continental rise along the western margin of ocean basins.

**Deep current** A density-driven circulation that is initiated at the ocean surface by temperature and salinity conditions that produce a high-density water mass, which sinks and spreads slowly beneath surface waters.

**Deep-ocean Assessment and Reporting of Tsunamis (DART)** A system that utilizes sea floor sensors capable of picking up the small yet distinctive pressure pulse from a tsunami at the surface.

**Deep-ocean basin** Areas of the ocean floor that have deep water, are far from land, and are underlain by basaltic crust.

**Deep-ocean trench** See *trench*.

**Deep scattering layer (DSL)** A layer of marine organisms in the open ocean that scatter signals from an echo sounder. It migrates daily from depths of slightly over 100 meters (330 feet) at night to more than 800 meters (2600 feet) during the day.

**Deep Sea Drilling Project (DSDP)** A sea floor drilling program initiated in 1968 that was designed to obtain cores from the deep sea; it was the predecessor to the Ocean Drilling Program (ODP) and the Integrated Ocean Drilling Program (IODP).

**Deep-sea fan** A large fan-shaped deposit commonly found on the continental rise seaward of such sediment-laden rivers as the Amazon, Indus, or Ganges-Brahmaputra. Also known as a submarine fan.

**Deep-sea fish** Any of a large group of fishes that lives within the aphotic zone and has special adaptations for finding food and avoiding predators in darkness.

**Deep-sea system** The bathyal, abyssal, and hadal benthic environments.

**Deep water** The water beneath the permanent thermocline (and resulting pycnocline) that has a uniformly low temperature.

**Deep-water wave** An ocean wave traveling in water that has a depth greater than one-half the average wavelength. Its speed is independent of water depth.

**Delta** A low-lying deposit at the mouth of a river, usually having a triangular shape as viewed from above.

**Density** The mass per unit volume of a substance. Usually expressed as grams per cubic centimeter ($g/cm^3$) For ocean water with a salinity of 35‰ at 0°C, the density is 1.028 $g/cm^3$.

**Density stratification** A layering based on density, where the highest density material occupies the lowest space.

**Deposit feeding** The process by which an organism feeds on food items that occur as deposits, including detritus and various detritus-coated sediment.

**Depositional shore** A shoreline dominated by processes that form deposits (such as sand bars and barrier islands) along the shore.

**Desalination** The removal of salt ions from ocean water to produce pure water.

**Destructive interference** A form of wave interference in which two waves come together out of phase, for example, crest to trough, and produce a wave with less displacement than the larger of the two waves would have produced alone.

**Detritus** (1) Any loose material produced directly from rock disintegration. (2) Material resulting from the disintegration of dead organic remains.

**Diatom** A member of the class Bacillariophyceae of algae that possesses a wall of overlapping silica valves.

**Diatomaceous earth** A deposit composed primarily of the tests of diatoms mixed with clay. Also called diatomite.

**Diffraction** A change in the direction or intensity of a wave after passing an obstacle that cannot be interpreted as refraction or reflection.

**Diffusion** A process by which materials move through fluids by random molecular movement from areas of high concentration to areas in which they are in lower concentrations, thus becoming evenly distributed.

**Dinoflagellate** A single-celled microscopic planktonic organism that may possess chlorophyll and belong to the phylum Pyrrophyta (autotrophic) or may ingest food and belong to the class Mastigophora of the phylum Protozoa (heterotrophic).

**Dipolar** Having two poles. The water molecule possesses a polarity of electrical charge with one pole being more positive and the other more negative in electrical charge.

**Discontinuity** An abrupt change in a property such as temperature or salinity at a line or surface.

**Disphotic zone** The dimly lit zone, corresponding approximately to the mesopelagic, in which there is not enough light to support photosynthetic organisms; sometimes called the twilight zone.

**Disruptive coloration** A marking or color pattern that confuses prey.

**Dissolved oxygen** Oxygen that is dissolved in ocean water.

**Distillation** A method of purifying liquids by heating them to their boiling point and condensing the vapor.

**Distributary** A small stream flowing away from a main stream. Such streams are characteristic of deltas.

**Disturbing force** The energy that causes waves to form.

**Diurnal inequality** The difference in the heights of two successive high tides or two successive low tides during a lunar (tidal) day.

**Diurnal tidal pattern** A tidal pattern exhibiting one high tide and one low tide during a tidal day; a daily tide.

**Divergence** A horizontal flow of water from a central region, as occurs in upwelling.

**Divergent plate boundary** A lithospheric plate boundary where adjacent plates diverge, producing an oceanic ridge or rise (spreading center).

**Doldrums** A global belt of light, variable winds near the equator, resulting from the vertical flow of low-density air masses upward within this equatorial belt. Associated with much precipitation.

**Dolphin** (1) A brilliantly colored fish of the genus *Coryphaena*. (2) The name applied to the small, beaked members of the cetacean family Delphinidae.

**Dorsal** Pertaining to the back or upper surface of most animals.

**Downwelling** In the open or coastal ocean, where Ekman transport causes surface waters to converge or impinge on the coast, surface water that moves down beneath the surface.

**Drift bottle** Equipment used to study ocean surface current movement by drifting with currents.

**Driftnet** A fishing net made of monofilament fishing line that catches organisms by entanglement.

**Drifts** Thick sediment deposits on the continental rise produced where a deep boundary current slows and loses sediment when it changes direction to follow the base of the continental slope.

**Drowned beach** An ancient beach now beneath the coastal ocean because of rising sea level or subsidence of the coast.

**Drowned river valley** The lower part of a river valley that has been submerged by rising sea level or subsidence of the coast.

**Dune (coastal)** A coastal deposit of sand lying landward of the beach and deriving its sand from onshore winds that transport beach sand inland.

**Durban Platform** An agreement to establish a new treaty to limit global carbon emissions that was held in 2011 in Durban, South Africa.

**Dynamic topography** A surface configuration resulting from the geopotential difference between a given surface and a reference surface of no motion. A contour map of this surface is useful in estimating the nature of geostrophic currents.

## E

**Earthquake** A sudden release of energy inside Earth that creates seismic waves and is usually caused by either faulting (movement along a fracture in Earth's crust) or volcanic activity.

**East Australian Current** The warm western boundary surface current of the South Pacific Subtropical Gyre.

**East Pacific Rise** A fast-spreading divergent plate boundary extending southward from the Gulf of California through the eastern South Pacific Ocean.

**East Wind Drift** The coastal surface current driven in a westerly direction by the polar easterly winds blowing off Antarctica.

**Eastern boundary current** Equatorward-flowing cold, sluggish movements of surface water on the eastern side of each subtropical gyre.

**Eastern Pacific Garbage Patch** An area in the eastern part of the North Pacific Gyre that serves as an accumulation area for floating plastic and other trash; it is about twice the size of Texas.

**Ebb current** The flow of water seaward during a decrease in the height of the tide.

**Ebb tide** An outgoing, falling tide.

**Echinodermata (echinoderm)** A phylum of animals that have bilateral symmetry in larval forms and usually a five-sided radial symmetry as adults. Benthic and possessing rigid or articulating exoskeletons of calcium carbonate with spines, this phylum includes sea stars, brittle stars, sea urchins, sand dollars, sea cucumbers, and sea lilies.

**Echo sounder** A device that transmits sound from a ship's hull to the ocean floor where it is reflected back to receivers. The speed of sound in the water is known, so the depth can be determined from the travel time of the sound signal.

**Echolocation** A sensory system in odontocete cetaceans in which usually high-pitched sounds are emitted and their echoes interpreted to determine the direction and distance of objects.

**Ecliptic** The plane of the center of the Earth–Moon system as it orbits around the Sun.

**Ecosystem** All the organisms in a biotic community and the abiotic environmental factors with which they interact.

**Ectothermic** Of or relating to an organism that regulates its body temperature largely by exchanging heat with its surroundings; cold-blooded.

**Eddy** A current of any fluid forming on the side of a main current. It usually moves in a circular path and develops where currents encounter obstacles or flow past one another. See *vortex* and *whirlpool*.

**Ekman spiral** A theoretical consideration of the effect of a steady wind blowing over an ocean of unlimited depth and breadth and of uniform viscosity. The result is a surface flow at 45 degrees to the right of the wind in the Northern Hemisphere. Water at increasing depth below the surface will drift in directions increasingly more slowly and to the right, until at about 100 meters (330 feet) depth it may move in a direction opposite to that of the wind.

**Ekman transport** The net transport of surface water set in motion by surface winds and the Ekman spiral. It is theoretically in a direction 90 degrees to the right of the wind direction in the Northern Hemisphere and 90° to the left of the wind direction in the Southern Hemisphere.

**El Niño** A southerly flowing warm surface current that generally develops off the coast of Ecuador around Christmastime. Occasionally it will move farther south into Peruvian coastal waters and cause the widespread death of plankton, fish, and other organisms such as marine mammals that depend on fish for food.

**El Niño–Southern Oscillation (ENSO)** The correlation of El Niño events with an oscillatory pattern of pressure change in a persistent high-pressure cell in the southeastern Pacific Ocean and a persistent low-pressure cell over the East Indies.

**Electrical conductivity** The ability or power to conduct or transmit electricity.

**Electrolysis** A separation process by which salt ions are removed from saltwater through water-impermeable membranes toward oppositely charged electrodes.

**Electromagnetic energy** Energy that travels as waves or particles with the speed of light. Different kinds possess different properties based on wavelength. The longest wavelengths belong to radio waves, up to 100 kilometers (60 miles) in length. At the other end of the spectrum are cosmic rays with greater penetrating power and wavelengths of less than 0.000001 micron.

**Electromagnetic spectrum** The spectrum of radiant energy emitted from stars, ranging from cosmic rays with wavelengths of less than one millionth of a micron, to very long waves with wavelengths in excess of 100 kilometers (60 miles).

**Electron** A subatomic particle that orbits the nucleus of an atom and has a negative electric charge.

**Electron cloud** The diffuse area surrounding the nucleus of an atom where electrons are found.

**Electrostatic force of attraction** The force between charged particles; also called electrostatic attraction or electrostatic repulsion, depending on if the particles have different or the same charge.

**Element** One of a number of substances, each of which is composed entirely of like particles—atoms—that cannot be broken into smaller particles by chemical means.

**Emerging shoreline** A shoreline resulting from the emergence of the ocean floor relative to the ocean surface. It is usually rather straight and characterized by marine features usually found at some depth.

**ENSO** See *El Niño–Southern Oscillation (ENSO)*.

**ENSO index** An index showing the relative strength of El Niño and La Niña conditions.

**Entropy** The degree of randomness or disorder in a system.

**Environment** The sum of all physical, chemical, and biological factors to which an organism or community is subjected.

**Environmental bioassay** An environmental assessment technique that determines the concentration of a pollutant that causes 50% mortality among a specific group of test organisms.

**Epicenter** The point on Earth's surface that is directly above the focus of an earthquake.

**Epifauna** Animals that live on the ocean bottom, either attached or moving freely over it.

**Epipelagic zone** A subdivision of the oceanic province that extends from the surface to a depth of 200 meters (660 feet).

**Equator** The imaginary great circle around Earth's surface, equidistant from the poles and perpendicular to Earth's axis of rotation. It divides Earth into the Northern Hemisphere and the Southern Hemisphere.

**Equatorial** Pertaining to the equatorial region.

**Equatorial countercurrent** Eastward-flowing currents found between the North and South Equatorial Currents in all oceans but particularly well developed in the Pacific Ocean.

**Equatorial current** Westward-flowing surface currents that travel along the equator in all ocean basins, caused by the trade winds. They are called North or South Equatorial Currents, depending on their position north or south of the equator.

**Equatorial low** A band of low atmospheric pressure that encircles the globe along the equator.

**Equatorial upwelling** The movement of deeper nutrient-rich water into the surface water mass as a result of divergence of currents along the equator.

**Erosion** The group of natural processes, including weathering, dissolution, abrasion, corrosion, and transportation, by which material is worn away from Earth's surface.

**Erosional shore** A shoreline dominated by processes that form erosional features (such as cliffs and sea stacks) along the shore.

**Estuarine circulation pattern** A flow pattern in an estuary characterized by a net surface flow of low-salinity water toward the ocean and an opposite net subsurface flow of seawater toward the head of the estuary.

**Estuary** A partially enclosed coastal body of water in which salty ocean water is significantly diluted by freshwater from land runoff. Examples of estuaries include river mouths, bays, inlets, gulfs, and sounds.

**Eubacteria** A kingdom of organisms that do not have nuclear material confined within a sheath but spread throughout the cell. Includes bacteria and blue-green algae; previously known as Kingdom Monera.

**Eukarya** One of the three major domains of life. The domain includes single-celled or multicellular organisms whose cells usually contain a distinct membrane-bound nucleus.

**Euphotic zone** A layer that extends from the surface of the ocean to a depth where enough light exists to support photosynthesis, rarely deeper than 100 meters (330 feet).

**Euryhaline** A descriptive term for organisms with a high tolerance for a wide range of salinity conditions.

**Eurythermal** A descriptive term for organisms with a high tolerance for a wide range of temperature conditions.

**Eustatic sea level change** A worldwide raising or lowering of sea level.

**Eutrophic** A region of high productivity.

**Eutrophication** The enrichment of waters by a previously scarce nutrient; if caused by humans, called cultural eutrophication.

**Evaporation** The process of changing from the liquid to the vapor state at a temperature below the boiling point of a substance.

**Evaporite** A sedimentary deposit that is left behind when water evaporates; also known as evaporite minerals, which include gypsum, calcite, and halite.

**Evolution** The change of groups of organisms with the passage of time, mainly as a result of natural selection, so that descendants differ morphologically and physiologically from their ancestors.

**Exclusive economic zone (EEZ)** A coastal zone that generally extends 200 nautical miles (370 kilometers) from shore and establishes coastal nation jurisdiction including mineral resources, fish stocks, and pollution. If the continental shelf extends beyond the 200-mile EEZ, the EEZ is extended to 350 nautical miles (648 kilometers) from shore.

**Extrusive rock** Igneous rock that flows out onto Earth's surface before cooling and solidifying (lava).

**Eye** (1) An organ of vision or of light sensitivity. (2) The circular low-pressure area of relative calm at the center of a hurricane.

## F

**Fact** Something having real, demonstrable existence. A scientific fact is an occurrence that has been repeatedly confirmed.

**Faculae** A bright spot on the Sun that is associated with magnetic storms.

**Fahrenheit temperature scale (°F)** A temperature scale whereby the freezing point of water is 32° and the boiling point of water is 212°. Named after its founder Daniel Fahrenheit.

**Falcate** Curved and tapering to a point; sickle-shaped.

**Falkland Current** A northward-flowing cold surface current that occurs off the southeastern coast of South America.

**Fall bloom** A middle-latitude bloom of phytoplankton that occurs during the fall and is limited by the availability of sunlight.

**Fall equinox** See *autumnal equinox*.

**Fan** A gently sloping, fan-shaped feature normally located near the lower end of a canyon. Also known as a submarine fan.

**Fat** An organic compound formed from alcohol, glycerol, and one or more fatty acids; a lipid, it is a solid at atmospheric temperatures.

**Fathom (fm)** A unit of depth in the ocean, commonly used in countries using the English system of units. It is equal to 1.83 meters (6 feet).

**Fault** A fracture or fracture zone in Earth's crust along which displacement has occurred.

**Fault block** A crustal block bounded on at least two sides by faults. Usually elongate; if it is down-dropped, it produces a graben; if uplifted, it is a horst.

**Fauna** The animal life of any particular area or of any particular time.

**Fecal pellet** The excrement of planktonic crustaceans that assist in speeding up the descent rate of sedimentary particles by combining them into larger packages.

**Feedback loop** A self-repeating processes in which an initial change is modified. A positive-feedback loop amplifies an initial change, and a negative-feedback loop counteracts an initial change.

**Ferrel cell** The large atmospheric circulation cell that occurs between 30 and 60 degrees latitude in each hemisphere.

**Ferromagnesium** Describes minerals rich in iron and magnesium.

**Fetch** (1) Pertaining to the area of the open ocean over which the wind blows with constant speed and direction, thereby creating a wave system. (2) The distance across the fetch (wave-generating area) measured in a direction parallel to the direction of the wind.

**Filter feeding** The process by which an organism obtains its food by filtering seawater to collect floating organisms to ingest. Also known as suspension feeding.

**Fishery** Fish caught from the ocean by commercial fishers.

**Fishery management** The organized effort directed at regulating fishing activity with the goal of maintaining a long-term fishery.

**Fissure** A long, narrow opening; a crack or cleft.

**Fjord** A long, narrow, deep, U-shaped inlet that usually represents the seaward end of a glacial valley that has become partially submerged after the melting of the glacier.

**Flagellum** A small whip-likestructure used by some cells for locomotion.

**Floe** A piece of floating ice other than fast ice or icebergs. May range in maximum horizontal dimension from about 20 centimeters (8 inches) to more than 1 kilometer (0.6 mile).

**Flood current** A tidal current associated with increasing height of the tide, generally moving toward the shore.

**Flood tide** An incoming, rising tide.

**Flora** The plant life of any particular area or of any particular time.

**Florida Current** A warm surface current flowing north along the coast of Florida that merges into the Gulf Stream.

**Flotsam** Any floating debris, particularly from wrecked ships.

**Folded mountain range** A mountain range formed as a result of the convergence of lithospheric plates. Folded mountain ranges are characterized by masses of folded sedimentary rocks that formed from sediments deposited in the ocean basin that was destroyed by the convergence.

**Food chain** The passage of energy materials from producers through a sequence of a herbivore and a number of carnivores.

**Food web** A group of interrelated food chains.

**Foraminifer** An order of planktonic and benthic protozoans that possess protective coverings, usually composed of calcium carbonate. Also called forams.

**Forced wave** A wave that is generated and maintained by a continuous force such as the gravitational attraction of the Moon.

**Foreshore** The portion of the shore lying between the normal high and low water marks; the intertidal zone.

**Fossil** Any remains, print, or trace of an organism that has been preserved in Earth's crust.

**Fossil fuel** A natural fuel such as petroleum, gas, or coal that formed in the geological past from the remains of living organisms.

**Fracture zone** An extensive linear zone of unusually irregular topography of the ocean floor, characterized by large seamounts, steep-sided or asymmetrical ridges, troughs, or long, steep slopes. Usually represents ancient, inactive transform fault zones.

**Free wave** A wave created by a sudden rather than a continuous impulse that continues to exist after the generating force is gone.

**Freeze separation** The desalination of seawater by multiple episodes of freezing, rinsing, and thawing.

**Freezing** The process by which a liquid is converted to a solid at its freezing point.

**Freezing point** The temperature at which a liquid becomes a solid under any given set of conditions. The freezing point of water is 0°C at one atmosphere pressure.

**Frequency** See *wave frequency*.

**Fringing reef** A reef that is directly attached to the shore of an island or continent. It may extend more than 1 kilometer (0.6 mile) from shore. The outer margin is submerged and often consists of algal limestone, coral rock, and living coral.

**Fucoxanthin** The reddish-brown pigment that gives brown algae its characteristic color.

**Full moon** The phase of the Moon that occurs when the Sun and Moon are in opposition; that is, they are on opposite sides of Earth. During this time, the lit side of the Moon faces Earth.

**Fully developed sea** The maximum average size of waves that can be developed for a given wind speed when it has blown in the same direction for a minimum duration over a minimum fetch.

**Fungi** Any of numerous eukaryotic organisms of the kingdom Fungi, which lack chlorophyll and vascular tissue and range in form from a single cell to a body mass of branched filamentous structures that often produce specialized fruiting bodies. The kingdom includes the yeasts, molds, lichens, and mushrooms.

**Fur seal** Any of several eared seals of the genera *Callorhinus* or *Arctocephalus*, having thick, soft underfur.

## G

**Galápagos Rift** A divergent plate boundary extending eastward from the Galápagos Islands toward South America. The first deep-sea hydrothermal vent biocommunity was discovered here in 1977.

**Galaxy** One of the billions of large systems of stars that make up the universe.

**Gas hydrate** A lattice-like compound composed of water and natural gas (usually methane) formed in high-pressure and low-temperature environments such as those found in deep-ocean sediments. Also known as clathrates because of their cage-like chemical structure.

**Gaseous state** A state of matter in which molecules move by translation and only interact through chance collisions.

**Gastropoda** A class of mollusks, most of which possess an asymmetrical, spiral one-piece shell and a well-developed flattened foot. A well-developed head will usually have two eyes and one or two pairs of tentacles. Includes snails, limpets, abalone, cowries, sea hares, and sea slugs.

**Geologic time scale** A table that lists the names of the geologic time periods as well as important advances in the development of life-forms on Earth.

**Geostrophic current** A surface current that is the result of a near balance between gravitational force and the Coriolis effect.

**Ghost fishing** Any lost or discarded fishing gear that continues to catch marine organisms after it has been abandoned.

**Gill** A thin-walled projection from some part of the external body or the digestive tract used for respiration in a water environment.

**Gill net** See *driftnet*.

**Glacial deposit** A sedimentary deposit formed by a glacier and characterized by poor sorting.

**Glacier** A large mass of ice formed on land by the recrystallization of old, compacted snow. It flows from an area of accumulation to an area of wasting where ice is removed from the glacier by melting.

**Irminger Current** A warm surface current that branches off from the Gulf Stream and moves up along the west coast of Iceland.

**Iron hypothesis** A hypothesis that states that an effective way of increasing productivity in the ocean is to fertilize the ocean by adding the only nutrient that appears to be lacking—iron. Adding iron to the ocean also increases the amount of carbon dioxide removed from the atmosphere.

**Iron meteorite** A meteorite consisting essentially of iron, but it may also contain up to 30% nickel.

**Island arc** A linear arrangement of islands, many of which are volcanic, usually curved so that the concave side faces a sea separating the islands from a continent. The convex side faces the open ocean and is bounded by a deep-ocean trench.

**Island mass effect** An effect that occurs as surface current flows past an island and causes surface water to be carried away from the island on the downcurrent side. This water is replaced in part by upwelling of water on the downcurrent side of the island.

**Isohaline** Of the same salinity.

**Isopoda** An order of dorsoventrally flattened crustaceans that are mostly scavengers or parasites on other crustaceans or fish.

**Isopycnal** Of the same density.

**Isostasy** A condition of equilibrium, comparable to buoyancy, by which Earth's brittle crust floats on the plastic mantle.

**Isostatic adjustment** The adjustment of crustal material due to isostasy.

**Isostatic rebound** The upward movement of crustal material due to isostasy.

**Isotherm** A line connecting points of equal temperature.

**Isothermal** Of the same temperature.

**Isotonic** Pertaining to the property of having equal osmotic pressure. If two such fluids were separated by a semipermeable membrane that will allow osmosis to occur, there would be no net transfer of water molecules across the membrane.

**Isotope** One of several atoms of an element that has a different number of neutrons, and therefore a different atomic mass, than the other atoms, or isotopes, of the element.

## J

**Jelly (Jellyfish)** (1) Free-swimming, umbrella-shaped medusoid members of the cnidarian class Scyphozoa. (2) Also frequently applied to the medusoid forms of other cnidarians.

**Jet stream** An easterly moving air mass at an elevation of about 10 kilometers (6 miles). Moving at speeds that can exceed 300 kilometers (185 miles) per hour, the jet stream follows a wavy path in the middle latitudes and influences how far polar air masses may extend into the lower latitudes.

**Jetty** A structure built from the shore into a body of water to protect a harbor or a navigable passage from being closed off by the deposition of longshore drift material; plural is jetties.

**Juan de Fuca Ridge** A divergent plate boundary off the Oregon–Washington coast.

## K

**K–T event** An extinction event marked by the disappearance of the dinosaurs that occurred about 66 million years ago at the boundary between the Cretaceous (K) and Tertiary (T) Periods of geologic time. Because of recent changes in the geologic time scale, also known as the Cretaceous–Paleogene (K–Pg) event.

**Kelp** Large varieties of Phaeophyta (brown algae).

**Kelp forest** An extensive bed of various species of macroscopic brown algae that provides a habitat for many other types of marine organisms.

**Key** A low, flat island composed of sand or coral debris that accumulates on a reef flat.

**Kinetic energy** Energy of motion, which increases as the mass or speed of the object in motion increases.

**Knot (kt)** A unit of speed equal to 1 nautical mile per hour, approximately 1.15 statute (land) miles per hour.

**Krill** A common name frequently applied to members of crustacean order Euphausiacea (euphausiids).

**Kuroshio Current** The warm western boundary surface current of the North Pacific Subtropical Gyre.

**Kyoto Protocol** An agreement among 60 nations to voluntarily limit greenhouse gas emissions, signed in 1997 in Kyoto, Japan.

## L

**La Niña** An event where the surface temperature in the waters of the eastern South Pacific falls below average values. It often follows an El Niño event.

**Labrador Current** A cold surface current flowing south along the coast of Labrador in the northwest Atlantic Ocean.

**Lagoon** A shallow stretch of seawater partly or completely separated from the open ocean by an elongate narrow strip of land such as a reef or barrier island.

**Laguna Madre** A hypersaline lagoon located landward of Padre Island along the south Texas coast.

**Laminar flow** The manner in which a fluid flows in parallel layers or sheets such that the direction of flow at any point does not change with time. Also known as nonturbulent flow.

**Land breeze** The seaward flow of air from the land caused by differential cooling of Earth's surface.

**Langmuir circulation** A cellular circulation set up by winds that blow consistently in one direction with speeds in excess of 12 kilometers (7.5 miles) per hour. Helical spirals running parallel to the wind direction are alternately clockwise and counterclockwise.

**Larva** An embryo that has a different form before it assumes the characteristics of the adult of the species.

**Latent heat** The quantity of heat gained or lost per unit of mass as a substance undergoes a change of state (such as liquid to solid) at a given temperature and pressure.

**Latent heat of condensation** The heat energy that must be removed from 1 gram of a substance to convert it from a vapor at a given temperature below its boiling point. For water, it is 585 calories at 20°C.

**Latent heat of evaporation** The heat energy that must be added to 1 gram of a liquid substance to convert it to a vapor at a given temperature below its boiling point. For water, it is 585 calories at 20°C.

**Latent heat of freezing** The heat energy that must be removed from 1 gram of a substance at its melting point to convert it to a solid. For water, it is 80 calories.

**Latent heat of melting** The heat energy that must be added to 1 gram of a substance at its melting point to convert it to a liquid. For water, it is 80 calories.

**Latent heat of vaporization** The heat energy that must be added to 1 gram of a substance at its boiling point to convert it to a vapor. For water, it is 540 calories.

**Lateral line system** A sensory system running down both sides of fishes to sense subsonic pressure waves transmitted through ocean water.

**Latitude** Location on Earth's surface based on angular distance north or south of the equator. Equator = 0 degrees; North Pole = 90 degrees north; South Pole = 90 degrees south.

**Laurasia** An ancient landmass of the Northern Hemisphere. The name is derived from Laurentia, pertaining to the Canadian Shield of North America, and Eurasia, of which it was composed.

**Lava** Fluid magma coming from an opening in Earth's surface, or the same material after it solidifies.

**Law of gravitation** See *gravitational force*.

**Law of the sea** See *United Nations Conference on the Law of the Sea*.

**Leeuwin Current** A warm surface current that flows south out of the East Indies along the western coast of Australia.

**Leeward** The direction toward which the wind is blowing or waves are moving.

**Levee** (1) Natural low ridges on either side of river channels that result from deposition during flooding. (2) Artificial ridges built by humans to control water flow.

**Light** Electromagnetic radiation that has a wavelength in the range from about 4000 (violet) to about 7700 (red) angstroms and may be perceived by the normal unaided human eye; also called visible light.

**Limestone** A class of sedimentary rocks composed of at least 80% carbonates of calcium or magnesium. Limestones may be either biogenous or hydrogenous.

**Limpet** A mollusk of the class Gastropoda that possesses a low conical shell that exhibits no spiraling in the adult form.

**LIMPET 500** The world's first commercial wave power plant that can generate up to 500 kilowatts of power. It is located on Islay, a small island off the west coast of Scotland, and began generating electricity in November 2000.

**Linnaeus, Carolus** Latinized name of Swedish botanist Carl von Linné, the father of taxonomic classification and binomial nomenclature.

**Liquid state** A state of matter in which a substance has a fixed volume but no fixed shape.

**Lithify** A process by which sediment becomes hardened into sedimentary rock.

**Lithogenous sediment** Sediment composed of mineral grains derived from the weathering of rock material and transported to the ocean by various mechanisms of transport, including running water, gravity, the movement of ice, and wind.

**Lithosphere** The outer layer of Earth's structure, including the crust and the upper mantle to a depth of about 200 kilometers (124 miles). Lithospheric plates are the major components involved in plate tectonic movement.

***Lithothamnion* ridge** A feature common to the windward edge of a reef structure, characterized by the presence of the red algae *Lithothamnion*.

**Littoral zone** The benthic zone between the highest and lowest spring tide shorelines; also known as the intertidal zone.

**Lobster** A large marine crustacean considered a delicacy. *Homarus americanus* (American or Maine lobster) possesses two large chelae (pincers) and is found offshore from Labrador to North Carolina. Various species of *Panulirus* (spiny lobsters or rock lobsters) have no chelae but possess long, spiny antennae effective in warding off predators. *P. argus* is found off the coast of Florida and in the West Indies, whereas *P. interruptus* is common along the coast of Southern California.

**Loihi** The name of the volcanically active submerged seamount that is located southeast of the island of Hawaii.

**Longitude** Location on Earth's surface based on angular distance east or west of the Prime (Greenwich) Meridian (0 degrees longitude). 180 degrees longitude is the International Date Line.

**Longitudinal wave** A wave phenomenon when particle vibration is parallel to the direction of energy propagation; also known as a push-pull wave.

**Longshore bar** A deposit of sediment that forms parallel to the coast within or just beyond the surf zone.

**Longshore current** A current located in the surf zone and running parallel to the shore as a result of waves breaking at an angle to the shore.

**Longshore drift** The load of sediment transported along the beach from the breaker zone to the top of the swash line in association with the longshore current. Also called longshore transport or littoral drift.

**Longshore trough** A low area of the beach that separates the beach face from the longshore bar.

**Lophophore** A horseshoe-shaped feeding structure bearing ciliated tentacles characteristic of the phyla Bryozoa, Brachiopoda, and Phoronidea.

**Low slack water** The period of time associated with the peak of low tide when there is no visible flow of water into or out of bays and rivers.

**Low tide terrace** See *beach face*.

**Low tide zone** The portion of the intertidal zone that lies between the lowest low tide shoreline and the highest low tide shoreline.

**Low water (LW)** The lowest level reached by the water surface at low tide before the rise toward high tide begins.

**Lower high water (LHW)** The lower of two high waters occurring during a tidal day where tides exhibit a mixed tidal pattern.

**Lower low water (LLW)** The lower of two low waters occurring during a tidal day where tides exhibit a mixed tidal pattern.

**Lunar day** The time interval between two successive transits of the Moon over a meridian, approximately 24 hours and 50 minutes of solar time. Also called a tidal day.

**Lunar month** The time it takes for one orbit of the Moon around Earth; 29.5 days. Also known as the lunar cycle or synodic month.

**Lunar tide** The part of the tide caused solely by the tide-producing force of the Moon.

**Lunger** Fish (such as grouper) that sit motionless on the ocean floor waiting for prey to appear. A lunger uses quick bursts of speed over short distances to capture prey.

**Lysocline** The level in the ocean at which calcium carbonate begins to dissolve, typically at a depth of about 4000 meters (13,100 feet). Below the lysocline, calcium carbonate dissolves at an increasing rate with increasing depth until the calcite compensation depth (CCD) is reached.

## M

**Macroplankton** Plankton larger than 2 centimeters (0.8 inch) in their smallest dimension.

**Magma** Fluid rock material from which igneous rock is derived through solidification.

**Magnetic anomaly** A distortion of the regular pattern of Earth's magnetic field resulting from the various magnetic properties of local concentrations of ferromagnetic minerals in Earth's crust.

**Magnetic dip** The dip of magnetite particles in rock units of Earth's crust relative to sea level. It is approximately equivalent to latitude. Also called magnetic inclination.

**Magnetic field** A condition found in the region around a magnet or an electric current, characterized by the existence of a detectable magnetic force at every point in the region and by the existence of magnetic poles.

**Magnetic inclination** See *magnetic dip*.

**Magnetite** The mineral form of black iron oxide, $Fe_3O_4$, that often occurs with magnesium, zinc, and manganese and is an important ore of iron.

**Magnetometer** A device used for measuring the magnetic field of Earth.

**Manganese nodule** A concretionary lump containing oxides of manganese, iron, copper, cobalt, and nickel found scattered over the ocean floor.

**Mangrove swamp** A marsh-like environment dominated by mangrove trees. They are restricted to latitudes below 30 degrees.

**Mantle** (1) The zone between the core and crust of Earth; rich in ferromagnesian minerals. (2) In pelecypods, the portion of the body that secretes shell material.

**Mantle plume** A rising column of molten magma from Earth's mantle.

**Marginal sea** A semienclosed body of water adjacent to a continent and floored by submerged continental crust.

**Mariculture** The application of the principles of agriculture to the production of marine organisms.

**Marine effect** Describes locations that experience the moderating influences of the ocean, usually along coastlines or islands.

**Marine Mammals Protection Act** An act by the U.S. Congress in 1972 that specifies rules to protect marine mammals in U.S. waters.

**Marine Protected Area (MPA)** An area of the ocean in which there is some level of restriction to protect living, nonliving, cultural, and/or historic resources.

**Marine reserve** A type of MPA in which all species and their habitat are fully protected.

**Marine sanctuary** A type of MPA in which biologic or cultural resources are protected. In some marine sanctuaries, fishing, recreational boating, and mining are allowed.

**Marine terrace** A gently seaward-sloping horizontal platform cut by waves (wave-cut bench) that has been uplifted above sea level.

**MARPOL (Marine Pollution)** An international treaty that banned the disposal of all plastics and regulated the dumping of most other garbage at sea.

**Marsh** An area of soft, wet, flat land that is periodically flooded by saltwater and common in portions of lagoons.

**Maximum sustainable yield (MSY)** The maximum fishery biomass that can be removed yearly and still be sustained by the fishery ecosystem.

**Mean high water (MHW)** The average height of all the high waters occurring over a 19-year period.

**Mean higher high water (MHHW)** The average height of the daily higher of the high waters occurring over a 19-year period where tides exhibit a mixed tidal pattern.

**Mean low water (MLW)** The average height of all the low waters occurring over a 19-year period.

**Mean lower low water (MLLW)** The average height of the daily lower of the low waters occurring over a 19-year period where tides exhibit a mixed tidal pattern.

**Mean sea level (MSL)** The mean surface water level determined by averaging all stages of the tide over a 19-year period, usually determined from hourly height observations along an open coast.

**Mean tidal range** The difference between mean high water and mean low water.

**Meander** A sinuous curve, bend, or turn in the course of a current.

**Mechanical energy** Energy manifested as work being done; the movement of a mass over some distance.

**Mediterranean circulation** Circulation characteristic of bodies of water with restricted circulation with the ocean that results from an excess of evaporation as compared to precipitation and runoff similar to the Mediterranean Sea. Surface flow is into the restricted body of water with a subsurface counterflow as exists between the Mediterranean Sea and the Atlantic Ocean.

**Medusa** A free-swimming, bell-shaped cnidarian body form with a mouth at the end of a central projection and tentacles around the periphery.

**Meiofauna** Small species of animals that live in the spaces among particles in a marine sediment.

**Melon** A fatty organ located forward of the blowhole on certain odontocete cetaceans that is used to focus echolocation sounds.

**Melting point** The temperature at which a solid substance changes to the liquid state.

**Mercury** A silvery white poisonous metallic element, liquid at room temperature and used in thermometers, barometers, vapor lamps, and batteries and in the preparation of chemical pesticides.

**Meridian of longitude** Great circles running through the North and South Poles.

**Meroplankton** Planktonic larval forms of organisms that are members of the benthos or nekton as adults.

**Mesopelagic zone** That portion of the oceanic province 200 to 1000 meters (660 to 3300 feet) deep. Corresponds approximately with the disphotic (twilight) zone.

***Mesosaurus*** An extinct, presumably aquatic, reptile that lived about 250 million years ago. The distribution of its fossil remains helps support plate tectonic theory.

**Mesosphere** The middle region of Earth below the asthenosphere and above the core.

**Metal sulfide** A compound containing one or more metals and sulfur.

**Metamorphic rock** Rock that has undergone recrystallization while in the solid state in response to changes of temperature, pressure, and chemical environment.

**Meteor** A bright trail or streak that appears in the sky when a meteoroid is heated to incandescence by friction with Earth's atmosphere. Also called a falling or shooting star.

**Meteorite** A stony or metallic mass of matter that has fallen to Earth's surface from outer space.

**Methane hydrate** A white compact icy solid made of water and methane. The most common type of gas hydrate.

**Microplankton** Plankton not easily seen by the unaided eye but easily recovered from the ocean with the aid of a silk-mesh plankton net.

**Microplastic** Microscopic pieces of plastic from the use of human health care products and other applications.

**Mid-Atlantic Ridge** A slow-spreading divergent plate boundary running north–south and bisecting the Atlantic Ocean.

**Mid-ocean ridge** A linear, volcanic mountain range that extends through all the major oceans, rising 1 to 3 kilometers (0.6 to 2 miles) above the deep-ocean basins. Averaging 1500 kilometers (930 miles) in width, rift valleys are common along the central axis. Source of new oceanic crustal material.

**Middle latitudes** The region of Earth between approximately 30 and 60 degrees north and south latitudes; also called the midlatitudes.

**Middle tide zone** The portion of the intertidal zone that lies between the highest low tide shoreline and the lowest high tide shoreline.

**Migration** Long journeys undertaken by many marine species for the purpose of successful feeding and reproduction.

**Minamata Bay, Japan** The site of the occurrence of human poisoning in the 1950s by mercury contained in marine organisms that were consumed by victims.

**Minamata disease** A degenerative neurological disorder caused by poisoning with a mercury compound found in seafood obtained from waters contaminated with mercury-containing industrial waste.

**Mineral** An inorganic substance occurring naturally on Earth and having distinctive physical properties and a chemical composition that can be expressed by a chemical formula. The term is also sometimes applied to ice or to organic substances such as coal and petroleum.

**Mixed interference** A pattern of wave interference in which there is a combination of constructive and destructive interference.

**Mixed surface layer** The surface layer of the ocean water mixed by wave and tide motions to produce relatively isothermal and isohaline conditions.

**Mixed tidal pattern** A tidal pattern exhibiting two high tides and two low tides per tidal day with a marked diurnal inequality. Coastal locations that experience such a tidal pattern may also show alternating periods of diurnal and semidiurnal patterns. Also called mixed semidiurnal.

**Mixotroph** An organism that depends on a combination of autotrophic and heterotrophic behavior to meet its energy requirements. Many coral reef species exhibit such behavior.

**Mohorovicíc discontinuity (Moho)** A sharp compositional discontinuity between the crust and mantle of Earth. It may be as shallow as 5 kilometers (3 miles) below the ocean floor or as deep as 60 kilometers (37 miles) beneath some continental mountain ranges.

**Molecular motion** Molecules move in three ways: vibration, rotation, and translation.

**Molecule** A group of two or more atoms bound together by ionic or covalent bonds.

**Mollusca (mollusk)** A phylum of soft, unsegmented animals usually protected by a calcareous shell and having a muscular foot for locomotion. Includes snails, clams, chitons, and octopi.

**Moment magnitude ($M_w$)** A scale used for measuring earthquake intensity based on energy released in creating very long-period seismic waves.

**Mononodal** Pertaining to a standing wave with only one nodal point or nodal line.

**Monsoon** A name for seasonal winds derived from the Arabic word for season, *mausim*. The term was originally applied to winds over the Arabian Sea that blow from the southwest during summer and from the northeast during winter.

**Moraine** A deposit of unsorted material deposited at the margins of glaciers. Many such deposits have become important economically as fishing banks after being submerged by the rising level of the ocean.

**Mud** Sediment consisting primarily of silt and clay-sized particles smaller than 0.06 millimeters (0.002 inch).

**Mutualism** A symbiotic relationship in which both participants benefit.

**Myoglobin** A red, oxygen-storing pigment found in muscle tissue.

**Mysticeti** The baleen whales.

## N

**Nadir** The point on the celestial sphere directly opposite the zenith and directly beneath the observer.

**Nannoplankton** Plankton less than 50 microns (0.002 inch) in length that cannot be captured in a plankton net and must be removed from the water by centrifuge or special microfilters.

**Nansen bottle** A device used by oceanographers to obtain samples of ocean water from beneath the surface.

**National Flood Insurance Program (NFIP)** A program financed by the U.S. government that was intended to prevent costly federal aid after a natural disaster but instead encourages building in risk-prone areas.

**Natural selection** The process in nature by which only the organisms best adapted to their environment tend to survive and transmit their genetic characters in increasing numbers to succeeding generations while those less adapted tend to be eliminated.

**Nauplius** A microscopic, free-swimming larval stage of crustaceans such as copepods, ostracods, and decapods. Typically has three pairs of appendages.

**Neap tide** Tides of minimal range occurring about every two weeks when the Moon is in either first- or third-quarter moon phase.

**Nearshore** The zone of a beach that extends from the low tide shoreline seaward to where breakers begin forming.

**Nebula** A diffuse mass of interstellar dust and/or gas.

**Nebular hypothesis** A model that describes the formation of the solar system by contraction of a nebula.

**Negative-feedback loop** See *feedback loop*.

**Nektobenthos** Those members of the benthos that can actively swim and spend much time off the bottom.

**Nekton** Pelagic animals such as adult squids, fish, and mammals that are active swimmers to the extent that they can determine their position in the ocean by swimming.

**Nematath** A linear chain of islands and/or seamounts that are progressively older in one direction. It is created by the passage of a lithospheric plate over a hotspot.

**Nematocyst** The stinging mechanism found within the cnidoblast of members of the phylum Cnidaria.

**Neritic province** That portion of the pelagic environment from the shoreline to where the depth reaches 200 meters (660 feet).

**Neritic sediment** Sediment composed primarily of lithogenous particles and deposited relatively rapidly on the continental shelf, continental slope, and continental rise; neritic deposits are those that are produced from neritic sediment.

**Net primary production** The primary production of producers after they have removed what is needed for their metabolism.

**Neutral** A state in which there is no excess of either the hydrogen or the hydroxide ion.

**Neutron** An electrically neutral subatomic particle found in the nucleus of atoms that has a mass approximately equivalent to that of a proton.

**New moon** The phase of the Moon that occurs when the Sun and the Moon are in conjunction; that is, they are both on the same side of Earth. During this time, the dark, unlit side of the Moon faces Earth.

**New production** Primary production supported by nutrients supplied from outside the immediate ecosystem by upwelling or other physical transport.

**Newton's law of universal gravitation** An equation that quantifies gravitational force between two bodies; it states that the gravitational force is directly proportional to the product of the masses of the two bodies and is inversely proportional to the square of the distance between the two masses.

**Niche** The ecological role of an organism and its position in the ecosystem.

**Niigata, Japan** The site of mercury poisoning of humans in the 1960s by ingestion of contaminated seafood.

**Nitrogen narcosis** A sickness that affects divers. It results from too much nitrogen gas being dissolved in the blood and reducing the flow of oxygen to tissues. The threat of this problem increases with increasing pressure (depth).

**Node** The point on a standing wave where vertical motion is lacking or minimal. If this condition extends across the surface of an oscillating body of water, the line of no vertical motion is a nodal line.

**Non-native species** Species that are introduced into waters in which they are alien and often cause severe problems by displacing native species. Also called exotic, alien, or invasive species.

**Non-point source pollution** Any type of pollution entering the ocean from multiple sources rather than from a single discrete source, point, or location. Examples include urban runoff, trash, pet waste, lawn fertilizer, and other types of pollution generated by a multitude of sources. Also called poison runoff.

**North Atlantic Current** The northernmost surface current of the North Atlantic Subtropical Gyre.

**North Atlantic Deep Water** A deep-water mass that forms primarily at the surface of the Norwegian Sea and moves south along the floor of the North Atlantic Ocean.

**North Atlantic Subtropical Gyre** The large, clockwise-flowing subtropical gyre that exists in the North Atlantic Ocean.

**North-East Pacific Time-series Undersea Networked Experiments (NEPTUNE)** A cutting-edge sea floor observatory system designed to monitor tectonic activity along the Juan de Fuca tectonic plate in the northeast Pacific Ocean.

**North Pacific Current** The northernmost surface current of the North Pacific Subtropical Gyre.

**North Pacific Subtropical Gyre** The large, clockwise-flowing subtropical gyre that exists in the North Pacific Ocean.

**Northern boundary current** The northern boundary surface current of Northern Hemisphere subtropical gyres.

**Northeast Monsoon** A northeast wind that blows off the Asian mainland onto the Indian Ocean during the winter season.

**Norwegian Current** A warm surface current that branches off from the Gulf Stream and flows into the Norwegian Sea between Iceland and the British Isles.

**Nucleus** The positively charged central region of an atom, composed of protons and neutrons and containing almost all of the mass of the atom.

**Nudibranch** A sea slug. A member of the mollusk class Gastropoda that has no protective covering as an adult. Respiration is carried on by gills or other projections on the dorsal surface.

**Nurdle** A small, rounded piece of pre-production plastic.

**Nutrient** Any of a number of organic or inorganic compounds used by primary producers. Nitrogen and phosphorus compounds are important examples.

## O

**Observation** An occurrence that can be measured with one's senses.

**Ocean** The entire body of saltwater that covers 70.8% of Earth's surface.

**Ocean acidification** The process by which the ocean's pH is lowered, which increases its acidity.

**Ocean acoustical tomography** A method by which changes in water temperature may be determined by changes in the speed of transmission of sound. It has the potential to help map ocean circulation patterns over large ocean areas.

**Ocean beach** The beach on the open-ocean side of a barrier island.

**Ocean Drilling Program (ODP)** A program that replaced the Deep Sea Drilling Project in 1983, focusing on drilling the continental margins using the drill ship *JOIDES Resolution*.

**Ocean thermal energy conversion (OTEC)** A technique that involves generating energy by using the difference in temperature between surface waters and deep waters in low latitude regions.

**Ocean wave** A type of wave that occurs in the ocean. See *wave*.

**Oceanic Common Water** Deep water found in Pacific and Indian Oceans as a result of mixing of Antarctic Bottom Water and North Atlantic Deep Water.

**Oceanic crust** A mass of rock with a basaltic composition that is about 5 kilometers (3 miles) thick.

**Oceanic province** The division of the pelagic environment where the water depth is greater than 200 meters (660 feet).

**Oceanic ridge** A portion of the global mid-ocean ridge system that is characterized by slow spreading and steep slopes.

**Oceanic rise** A portion of the global mid-ocean ridge system that is characterized by fast spreading and gentle slopes.

**Oceanic sediment** The inorganic abyssal clays and the organic oozes that accumulate slowly on the deep-ocean floor.

**Oceanography** The scientific study of the floor of the ocean, the water itself, physical processes such as waves, currents, and tides, and the organisms contained within the ocean. Also known as oceanology.

**Ocelli** A light-sensitive organ around the base of many medusoid bells.

**Odontoceti** The toothed whales.

**Offset** Separation that occurs due to movement along a fault.

**Offshore** The comparatively flat submerged zone of variable width extending from the breaker line to the edge of the continental shelf.

**Oligotrophic** Exhibiting low levels of biological production, such as the centers of subtropical gyres.

**Omnivore** An animal that feeds on both plants and animals.

**Oolite** A deposit formed of small spheres from 0.25 to 2 millimeters (0.01 to 0.08 inch) in diameter. Each oolite is composed of concentric layers of calcite.

**Ooze** A pelagic sediment containing at least 30% skeletal remains of pelagic organisms, the balance being clay minerals. Oozes are further defined by the chemical composition of the organic remains (siliceous or calcareous) and by their characteristic organisms (e.g., diatomaceous ooze, foraminifer ooze).

**Opal** An amorphous form of silica ($SiO_2 \cdot nH_2O$) that usually contains from 3 to 9% water. It forms the shells of radiolarians and diatoms.

**Ophiolite** An assemblage of mafic and ultramafic igneous rocks from Earth's oceanic crust and the underlying upper mantle that has been uplifted and exposed above sea level and often emplaced onto continental crustal rocks by plate tectonic processes. Ophiolites are generally composed of green-colored metamorphic rocks such as peridotite and serpentinite.

**Opposition** The separation of two heavenly bodies by 180 degrees relative to Earth. The Sun and Moon are in opposition during the full moon phase.

**Orbital wave** A wave phenomenon in which energy is moved along the interface between fluids of different densities. The wave form is propagated by the movement of fluid particles in orbital paths.

**Orthogonal line** A line constructed perpendicular to a wave front and spaced so that the energy between lines is equal at all times. Orthogonals are used to help determine how energy is distributed along the shoreline by breaking waves.

**Osmosis** The process by which water molecules move through a semipermeable membrane from higher water molecule concentration (lower salinity) to lower water molecule concentration (higher salinity).

**Osmotic pressure** A measure of the tendency for osmosis to occur. It is the pressure that must be applied to the more concentrated solution to prevent the passage of water molecules into it from the less concentrated solution.

**Osmotic regulation** Physical and biological processes used by organisms to counteract the osmotic effects of differences in osmotic pressures of their body fluids and the water in which they live.

**Ostracoda** An order of crustaceans that are minute and compressed within a bivalve shell.

**Otocyst** Gravity-sensitive organs around the bell of a medusa.

**Outer sublittoral zone** The continental shelf below the intersection with the euphotic zone where no plants grow attached to the bottom.

**Outgassing** The process by which gases are removed from within Earth's interior.

**Overfishing** A situation that occurs when adult fish in a population are harvested faster than their natural rate of reproduction.

**Oxygen compensation depth** The depth in the ocean at which marine plants receive just enough solar radiation to meet their basic metabolic needs. It marks the base of the euphotic zone.

**Oxygen minimum layer (OML)** A zone of low dissolved oxygen concentration that occurs at a depth of about 700 to 1000 meters (2300 to 3280 feet).

## P

**Pacific Decadal Oscillation (PDO)** A natural and cyclic pattern of ocean–atmosphere variability in the Pacific Ocean that lasts 20 to 30 years and influences sea surface temperatures.

**Pacific Ocean** The ocean located between Australia, Asia, North America, and South America; the largest ocean in the world.

**Pacific Ring of Fire** An extensive zone of volcanic and seismic activity that coincides roughly with the borders of the Pacific Ocean.

**Pacific Tsunami Warning Center (PTWC)** A tsunami warning center that was established in Hawaii after the devastating 1946 tsunami; it coordinates information from 25 Pacific Rim countries.

**Pacific Warm Pool** A large region of warm surface water on the western side of the Pacific Ocean.

**Pacific-type margin** See *active margin*.

**Paleoclimatology** The study of ancient climates on Earth.

**Paleoceanography** The study of how the ocean, atmosphere, and land have interacted to produce changes in ocean chemistry, circulation, biology, and climate.

**Paleogeography** The study of the historical changes of shapes and positions of the continents and oceans.

**Paleomagnetism** The study of Earth's ancient magnetic field.

**Pancake ice** Circular pieces of newly formed sea ice from 0.3 to 3 meters (1 to 10 feet) in diameter that form in the early fall in polar regions.

**Pangaea** An ancient supercontinent of the geologic past that contained all Earth's continents.

**Panthalassa** A large, ancient ocean that surrounded Pangaea.

**Paralytic shellfish poisoning (PSP)** Paralysis resulting from poisoning caused by eating shellfish contaminated with the toxic dinoflagellate *Gonyaulax*.

**Parasitism** A symbiotic relationship between two organisms in which one benefits at the expense of the other.

**Paris Climate Agreement** An agreement between 197 participating nations that met in 2015 in Paris, France, to establish the reduction of greenhouse gas emissions and limit the global temperature increase to 2°C (3.6°F) above pre-industrial levels.

**Parts per thousand (‰)** A unit of measurement used in reporting salinity of water equal to the number of grams of dissolved substances in 1000 grams of water. For example, 1‰ is equivalent to 0.1%, or 1000 ppm.

**Passive margin** A continental margin that lacks a plate boundary and is marked by a low degree of tectonic activity, such as those typical of the Atlantic Ocean.

**PCBs** A group of industrial chemicals (polychlorinated biphenyls) used in a variety of products; responsible for several episodes of ecological damage in coastal waters.

**Peat deposit** Partially carbonized organic matter found in bogs and marshes that can be used as fertilizer and fuel.

**Pelagic sediment** Sediment composed primarily of fine lithogenous and biogenous particles that is deposited slowly on the deep ocean floor; pelagic deposits are those that are produced from pelagic sediment.

**Pelagic environment** The open-ocean environment, which is divided into the neritic province (water depth 0 to 200 meters or 660 feet) and the oceanic province (water depth greater than 200 meters or 660 feet).

**Pelecypoda** A class of mollusks characterized by two more or less symmetrical lateral valves with a dorsal hinge. These organisms filter feed by circulating water through their bodies and over their gills through posterior siphons. Many possess a hatchet-shaped foot used for locomotion and burrowing. Includes clams, oysters, mussels, and scallops.

**Perigee** The point on the orbit of an Earth satellite (Moon) that is nearest Earth.

**Perihelion** That point on the orbit of a planet or comet around the Sun that is closest to the Sun.

**Permeability** Capacity of a porous rock or sediment for transmitting fluid.

**Peru Current** The cold eastern boundary surface current of the South Pacific Subtropical Gyre.

**Petroleum** A naturally occurring liquid hydrocarbon.

***Pfiesteria piscicida*** A species of toxic dinoflagellate that has been known to cause fish kills.

**pH scale** A measure of the acidity or alkalinity of a solution, numerically equal to 7 for neutral solutions, increasing with increasing alkalinity and decreasing with increasing acidity. The pH scale commonly in use ranges from 0 to 14.

**Phaeophyta** Brown algae characterized by the carotenoid pigment fucoxanthin. Contains the largest members of the marine algal community.

**Phosphate** Any of a number of phosphorus-bearing compounds.

**Phosphorite** A sedimentary rock composed primarily of phosphate minerals.

**Photic zone** The upper ocean in which the presence of solar radiation is detectable. It includes the euphotic and disphotic zones.

**Photophore** One of several types of light-producing organs found primarily on fishes and squids inhabiting the mesopelagic and upper bathypelagic zones.

**Photosynthesis** The process by which plants and algae produce carbohydrates from carbon dioxide and water in the presence of chlorophyll, using light energy and releasing oxygen.

**Phycoerythrin** A red pigment characteristic of the Rhodophyta (red algae).

**Phytoplankton** Algal plankton. One of the most important communities of primary producers in the ocean.

**Picoplankton** Small plankton within the size range of 0.2 to 2.0 microns (0.000008 to 0.00008 inch) in size. Composed primarily of bacteria.

**Pillow basalt** A basalt exhibiting pillow structure. See *pillow lava*.

**Pillow lava** A general term for those lavas displaying discontinuous pillow-shaped masses (pillow structure) caused by the rapid cooling of lava as a result of underwater eruption of lava or lava flowing into water.

**Ping** A sharp, high-pitched sound made by the transmitting device of many sonar systems.

**Pinniped** A group of marine mammals that have prominent flippers; includes the sea lions/fur seals, seals, and walruses.

**Plane of the ecliptic** The surface connecting all points in Earth's orbit.

**Plankton** Passively drifting or weakly swimming organisms that are not independent of currents. Includes mostly microscopic algae, protozoa, and larval forms of animals. Also *plankter* (singular).

**Plankton bloom** A very high concentration of phytoplankton, resulting from a rapid rate of reproduction as conditions become optimal during the spring in high latitude areas. Less obvious causes produce blooms that may be destructive in other areas.

**Plankton net** A plankton-extracting device that is cone shaped and typically of a silk material. It is towed through the water or lifted vertically to extract plankton down to a size of 50 microns (0.0002 inch).

**Planktonic** Having the characteristics of plankton.

**Plantae** A kingdom of many-celled plants.

**Plastic** (1) Capable of being shaped or formed. (2) Composed of plastic or plastics.

**Plate tectonics** Global dynamics having to do with the movement of semirigid sections of Earth's crust called plates, with seismic activity and volcanism occurring primarily along the margins of these plates. This movement has resulted in changes in the geographic positions of continents and the shape and size of ocean basins.

**Plume** A rising column of molten mantle material that is associated with a hotspot when it penetrates Earth's crust.

**Plunging breaker** Impressive curling breakers that form on moderately sloping beaches.

**Pneumatic duct** An opening into the swim bladder of some fishes that allows rapid release of air into the esophagus.

**Poikilothermic** An organism whose body temperature varies with and is largely controlled by its environment; cold-blooded.

**Polar** Pertaining to the polar regions.

**Polar bear** A marine mammal with white fur whose native range consists of Arctic ice and adjoining islands and landmasses mostly within the Arctic Circle. It feeds primarily on seals.

**Polar cell** The large atmospheric circulation cell that occurs between 60 and 90 degrees latitude in each hemisphere.

**Polar easterly wind belt** A global wind belt that moves away from the polar regions toward the polar front at about 60 degrees north or south latitude in each hemisphere. These winds move from a northeasterly direction in the Northern Hemisphere and from a southeasterly direction in the Southern Hemisphere.

**Polar front** The boundary between the global wind belts' prevailing westerlies and polar easterlies that is centered at about 60 degrees latitude in each hemisphere and is characterized by rising air and much precipitation.

**Polar high** The region of high atmospheric pressure that occurs at the poles in both hemispheres.

**Polar wandering path** A path that shows the change in position of a pole through time. Sometimes called polar wandering curve.

**Polarity** Intrinsic polar separation, alignment, or orientation, especially of a physical property (such as magnetic or electrical polarity).

**Pollution (marine)** The introduction of substances into the environment that result in harm to the living resources of the ocean or humans who use these resources.

**Polychaeta** A class of annelid worms that includes most of the marine segmented worms.

**Polyp** A single individual of a colony or a solitary attached cnidarian.

**Population** A group of individuals of one species living in an area.

**Porifera** A phylum of sponges. Supporting structure composed of $CaCO_3$ or $SiO_2$ spicules or fibrous spongin. Water currents created by flagella-waving choanocytes enter tiny pores, pass through canals, and exit through a larger osculum.

**Porosity** The ratio of the volume of all the empty spaces in a material to the volume of the whole.

**Positive-feedback loop** See *feedback loop*.

**Potential energy** The energy of a particle or system of particles derived from position or condition rather than motion.

**Precession** Describes the change in the attitude of the Moon's orbit around Earth as it slowly changes its direction. The cycle is completed every 18.6 years and is accompanied by a clockwise rotation of the plane of the Moon's orbit that is completed in the same time interval.

**Precipitate** A substance that is formed chemically whenever dissolved materials change from existing in the dissolved state to existing in the solid state.

**Precipitation** In a meteorological sense, the discharge of water in the form of rain, snow, hail, or sleet from the atmosphere onto Earth's surface.

**Precision depth recorder (PDR)** An early type of sonar device that was developed in the 1950s.

**Pressure ridge** A ridge produced on floating sea ice by buckling or crushing under lateral pressure by wind or movement of ice.

**Prevailing westerly wind belt** A global wind belt that moves from a subtropical high-pressure belt at about 30 degrees north or south latitude toward the polar front at about 60 degrees north or south latitude. These winds move from a southwesterly direction in the Northern Hemisphere and from a northwesterly direction in the Southern Hemisphere.

**Primary productivity** The rate at which energy is stored by organisms through the formation of organic matter (carbon-based compounds) using energy derived from solar radiation (photosynthesis) or chemical reactions (chemosynthesis). Also known simply as productivity.

**Prime meridian** The meridian of longitude 0 degrees used as a reference for measuring longitude that passes through the Royal Observatory at Greenwich, England. Also known as the Greenwich Meridian.

**Process of science** See *science, process of.*

**Producer** The autotrophic component of an ecosystem that produces the food that supports the biocommunity.

**Productivity** See *primary productivity*.

**Progressive wave** A wave in which the waveform progressively moves.

**Propagation** The transmission of energy through a medium.

**Protein** A very complex organic compound made up of large numbers of amino acids. Proteins make up a large percentage of the dry weight of all living organisms.

**Protista** A kingdom of organisms that includes any of the unicellular eukaryotic organisms and their descendant multicellular organisms. Includes single-celled and multicelled marine algae as well as single-celled animals called protozoa.

**Proto-Earth** The young, early developing Earth.

**Proton** A positively charged subatomic particle found in the nucleus of atoms that has a mass approximately equivalent to that of a neutron.

**Protoplanet** Any planet that is in its early stages of development.

**Protoplasm** The self-perpetuating living material making up all organisms, mostly consisting of the elements carbon, hydrogen, and oxygen combined into various chemical forms.

**Protozoa** A phylum of one-celled animals with nuclear material confined within a nuclear sheath. Also *protozoans* (plural).

**Proxigean** A tidal condition of extremely large tidal range that occurs when spring tides coincide with perigee. Also called "closest of the close moon" tides.

**Proxy** A substitute value or measurement that takes the place of something else.

**Pseudopodia** An extension of protoplasm in a broad, flat, or long needle-like projection used for locomotion or feeding. Typical of amoeboid forms such as foraminifers and radiolarians.

**Pteropoda** An order of pelagic gastropods in which the foot is modified for swimming and the shell may be present or absent.

**Purse seine net** A style of large fishing net that resembles a purse. It is set around a grouping of organisms (such as tuna) and the bottom is drawn tight to capture the organisms.

**Pycnocline** A layer of water in which a high rate of change in density in the vertical dimension is present.

**Pyrrophyta** A phylum of dinoflagellates that possess flagella for locomotion.

## Q

**Quadrature** The state of the Moon during the first- and third-quarter moon phases when the Sun and the Moon are at right angles relative to Earth.

**Quarter moon** First- and third-quarter moon phases, which occur when the Moon is in quadrature about one week after the new moon and full moon phases, respectively. The third-quarter moon phase is also known as the last quarter moon phase.

**Quartz** A very hard mineral composed of silica, $SiO_2$.

## R

**Radiata** A grouping of phyla with primary radial symmetry—phyla Cnidarian and Ctenophora.

**Radioactivity** The spontaneous breakdown of the nucleus of an atom resulting in the emission of radiant energy in the form of particles or waves.

**Radiolaria** An order of planktonic and benthic protozoans that possess protective coverings usually made of silica. Also *radiolarians* (plural).

**Radiometric age dating** A technique that involves the use of radioactive half-lives to determine the age of a rock.

**Ray** A cartilaginous fish in which the body is dorsoventrally flattened, eyes and spiracles are on the upper surface, and gill slits are on the bottom. The tail is reduced to a whip-like appendage. Includes electric rays, manta rays, and stingrays.

**Recreational beach** The area of a beach above shoreline, including the berm, berm crest, and the exposed part of the beach face.

**Red clay** See *abyssal clay*.

**Red muscle fiber** Fine muscle fibers rich in myoglobin that are abundant in cruiser-type fishes.

**Red tide** A reddish-brown discoloration of surface water, usually in coastal areas, caused by high concentrations of microscopic organisms, usually dinoflagellates. It normally results from increased availability of certain nutrients. Toxins produced by the dinoflagellates may kill fish directly; decaying plant and animal remains or large populations of animals that migrate to the area of abundant plants may also deplete the surface waters of oxygen and cause asphyxiation of many animals.

**Reef** A strip or ridge of rocks, sand, coral, or human-made objects that rises to or near the surface of the ocean and creates a navigational hazard.

**Reef flat** A platform of coral fragments and sand on the lagoon side of a reef that is relatively exposed at low tide.

**Reef front** The upper seaward face of a reef from the reef edge (seaward margin of reef flat) to the depth at which living coral and coralline algae become rare, 16 to 30 meters (50 to 100 feet).

**Reflection (wave)** The process in which a wave has part of its energy returned seaward by a reflecting surface.

**Refraction (wave)** The process by which the part of a wave in shallow water is slowed down, causing the wave to bend and align itself nearly parallel to the shore.

**Regenerated production** The portion of gross primary production that is supported by nutrients recycled within an ecosystem.

**Relict beach** A beach deposit laid down and submerged by a rise in sea level. It is still identifiable on the continental shelf, indicating that no deposition is presently taking place at that location on the shelf.

**Relict sediment** A sediment deposited under a set of environmental conditions that still remains unchanged although the environment has changed, and it remains unburied by later sediment. An example is a beach deposited near the edge of the continental shelf when sea level was lower.

**Relocation** The strategy of moving a structure that is threatened by being claimed by the sea.

**Residence time** The average length of time a particle of any substance spends in the ocean. It is calculated by dividing the total amount of the substance in the ocean by the rate of its introduction into the ocean or the rate at which it leaves the ocean.

**Respiration** The process by which organisms use organic materials (food) as a source of energy. As the energy is released, oxygen is used and carbon dioxide and water are produced.

**Restoring force** A force such as surface tension or gravity that tends to restore the ocean surface displaced by a wave to that of a still water level.

**Resultant force** The difference between the provided gravitational force of various bodies and the required centripetal force on Earth. The horizontal component of the resultant force is the tide-producing force.

**Reverse osmosis** A method of desalinating ocean water that involves forcing water molecules through a water-permeable membrane under pressure.

**Reversing current** The tide current as it occurs at the margins of landmasses. The water flows in and out for approximately equal periods of time separated by slack water where the water is still at high and low tidal extremes.

**Reversing estuary** A type of estuary characterized by fluctuating salinity depending on the season. Also called variable-salinity estuary.

**Rhodophyta** A phylum of algae composed primarily of small encrusting, branching, or filamentous plants that receive their characteristic red color from the presence of the pigment phycoerythrin. With a worldwide distribution, they are found at greater depths than other algae.

**Ring** A circular-moving surface flow of water that can be either a warm-core ring, which contains warm water in its center, or a cold-core ring, which contains cold water in its center.

**Rift valley** A deep fracture or break, about 25 to 50 kilometers (15 to 30 miles) wide, extending along the crest of a mid-ocean ridge.

**Rifting** The movement of two plates in opposite directions such as along a divergent boundary.

**Right whale** A surface-feeding baleen whale of the family Balaenidae that were a favorite target of early whalers.

**Rip current** A strong narrow surface or near-surface current of short duration and high speed flowing seaward through the breaker zone at nearly right angles to the shore. It represents the return to the ocean of water that has been piled up on the shore by incoming waves.

**Rip-rap** Large blocky material used to armor coastal structures.

**Rogue wave** An unusually large wave that occurs unexpectedly amid other waves of smaller size. Also known as a superwave, monster wave, sleeper wave, or freak wave.

**Rorqual whale** A large baleen whale with prominent ventral groves (rorqual folds) of the family Balaenopteridae: the minke, Baird's, Bryde's, sei, fin, blue, and humpback whales.

**Rotary current** A tidal current that is observed in the open ocean and makes one complete rotation during a tidal period.

**Rotary drilling** Drilling involving the use of a long, hollow pipe with a drill bit on its end that is rotated to crush the rock around the outside and retain a cylinder of rock (a core sample) on the inside of the pipe.

**Runoff** The draining away of water from the surface area of land such as from rainfall or ice melting.

## S

**Saffir–Simpson scale** A scale of hurricane intensity that divides tropical cyclones into categories based on wind speed and damage.

**Salinity** A measure of the quantity of dissolved solids in ocean water. Formally, it is the total amount of dissolved solids in ocean water in parts per thousand (‰) by weight after all carbonate has been converted to oxide, the bromide and iodide to chloride, and all the organic matter oxidized. It is normally computed from conductivity, refractive index, or chlorinity.

**Salinometer** An instrument that is used to determine the salinity of seawater by measuring its electrical conductivity.

***Salpa*** A genus of pelagic tunicates that are cylindrical, transparent, and found in all oceans.

**Salt** A substance that yields ions other than hydrogen or hydroxyl. Salts are produced from acids by replacing the hydrogen with a metal.

**Salt deposit** An evaporative deposit composed of precipitated salts from seawater such as halite, gypsum, and in some cases calcite.

**Salt marsh** A relatively flat area of the shore where fine sediment is deposited and salt-tolerant grasses grow. One of the most biologically productive regions of Earth.

**Salt wedge estuary** A very deep river mouth with a very large volume of freshwater flow beneath which a wedge of saltwater from the ocean invades. The Mississippi River is an example.

**San Andreas Fault** A transform fault that cuts across the state of California from the northern end of the Gulf of California to Point Arena in northern California.

**Sand** Particle size of 0.0625 to 2 millimeters (0.002 to 0.08 inch). It pertains to particles that lie between silt and granules on the Wentworth scale of grain size.

**Sargasso Sea** A region of convergence in the North Atlantic lying south and east of Bermuda where the water is a very clear, deep blue color, and contains large quantities of floating algae *Sargassum*.

***Sargassum*** A brown alga characterized by a bushy form, substantial holdfast when attached, and a yellow-brown, green-yellow, or orange color. The two dominant species of macroscopic algae in the Sargasso Sea are *S. fluitans* and *S. natans*.

**Scarp** A linear steep slope on the ocean floor separating gently sloping or flat surfaces.

**Scavenger** An animal that feeds on dead organisms.

**Schooling** A well-defined large groups of fish, squid, and crustaceans that apparently aid them in survival.

**Science, process of** The principles and empirical processes of discovery and demonstration considered characteristic of or necessary for scientific investigation, generally involving the observation of phenomena, the formulation of a hypothesis concerning the phenomena, experimentation to demonstrate the truth or falseness of the hypothesis, and a conclusion that validates or modifies the hypothesis.

**Scuba** An acronym for *s*elf-contained *u*nderwater *b*reathing *a*pparatus, a portable device containing compressed air that is used for breathing underwater.

**Scyphozoa** A class of cnidarians that includes the true jellies in which the medusoid body form predominates and the polyp is reduced or absent.

**Sea** (1) A subdivision of an ocean, generally enclosed by land and usually composed of saltwater. Two types of seas are identifiable and defined. They are the Mediterranean seas, where a number of seas are grouped together collectively as one sea, and adjacent seas, which are connected individually to the ocean. (2) A portion of the ocean where waves are being generated by wind.

**Sea anemone** A member of the class Anthozoa whose bright color, tentacles, and general appearance make it resemble flowers.

**Sea arch** An opening through a headland caused by wave erosion. Usually develops as sea caves are extended from one or both sides of the headland.

**Sea breeze** The landward flow of air from the sea caused by differential heating of Earth's surface.

**Sea cave** A cavity at the base of a sea cliff formed by wave erosion.

**Sea cow** See *Sirenia*.

**Sea cucumber** A common name given to members of the echinoderm class Holothuroidea.

**Sea floor spreading** A process producing the lithosphere when convective upwelling of magma along the oceanic ridges moves the ocean floor away from the ridge axes at rates between 2 to 12 centimeters (0.8 to 5 inches) per year.

**Sea ice** A form of ice originating from the freezing of ocean water.

**Sea lion** Any of several eared seals with a relatively long neck and large front flippers, especially the California sea lion *Zalophus californianus* of the northern Pacific. Along with the fur seals, these marine mammals are known as eared seals.

**Sea otter** A seagoing otter that has recovered from near extinction along the North Pacific coasts. It feeds primarily on abalone, sea urchins, and crustaceans.

**Sea snake** A reptile belonging to the family Hydrophiidae with venom similar to that of cobras. They are found primarily in the coastal waters of the Indian Ocean and the western Pacific Ocean.

**Sea stack** An isolated, pillar-like rocky island that is detached from a headland by wave erosion.

**Sea turtle** Any of the reptilian order Testudinata found widely in warm water.

**Sea urchin** An echinoderm belonging to the class Echinoidea possessing a fused test (external covering) and well-developed spines.

**Seabeam** The first multibeam echo sounder.

**Seaknoll** See *abyssal hill*.

**Seal** (1) Any of several earless seals with a relatively short neck and small front flippers. Also known as true seals. (2) A general term that describes any of the various aquatic, carnivorous marine mammals of the families Phocidae and Otariidae (true seals and eared seals), found chiefly in the Northern Hemisphere and having a sleek, torpedo-shaped body and limbs that are modified into paddle-like flippers.

**Seamount** An individual volcanic peak extending over 1000 meters (3300 feet) above the surrounding ocean floor.

**Seasonal thermocline** A thermocline that develops due to surface heating of the oceans in middle to high latitudes. The base of the seasonal thermocline is usually above 200 meters (660 feet).

**Seawall** A wall built parallel to the shore to protect coastal property from the waves.

**SeaWiFS** An instrument aboard the SeaStar satellite launched in 1997 that measures the color of the ocean with a radiometer and provides global coverage of ocean chlorophyll levels as well as land productivities every two days.

**Secchi disk** A light-colored disk-shaped device that is lowered into water in order to measure the water's ability to transmit light.

**Sediment** Particles of organic or inorganic origin that accumulate in loose form.

**Sediment maturity** A condition in which the roundness and degree of sorting increase and clay content decreases within a sedimentary deposit.

**Sedimentary rock** Rock resulting from the consolidation of loose sediment, or rock resulting from chemical precipitation, such as sandstone and limestone.

**Seep** An area where water of various temperatures and/or salinities trickles out of the sea floor, such as in warm seeps, cold seeps, and hypersaline seeps. Also includes locations where hydrocarbons seep from the sea floor, which create hydrocarbon seeps.

**Seiche** A standing wave of an enclosed or semienclosed body of water that may have a period ranging from a few minutes to a few hours, depending on the dimensions of the basin. The wave motion continues after the initiating force has ceased.

**Seismic** Pertaining to an earthquake or Earth vibration, including those that are artificially induced.

**Seismic reflection profile** A profile view of the structure beneath the sea floor produced by the energy generated from explosions or air guns.

**Seismic sea wave** See *tsunami*.

**Seismic surveying** The use of sound-generating techniques to identify features on or beneath the ocean floor.

**Semidiurnal tidal pattern** A tidal pattern exhibiting two high tides and two low tides per tidal day with small inequalities between successive highs and successive lows; a semidaily tide.

**Sequester** To isolate, hide away, or remove a particular item from the local environment.

**Sessile** Permanently attached to the substrate and not free to move about.

**Sewage sludge** The semisolid material precipitated by sewage treatment. See also *treatment*.

**Shallow-water wave** A wave on the surface having a wavelength of at least 20 times water depth. The bottom affects the orbit of water particles and speed is determined by water depth.

**Shelf break** The depth at which the gentle slope of the continental shelf steepens appreciably. It marks the boundary between the continental shelf and continental rise.

**Shelf ice** Thick shelves of glacial ice that push out into Antarctic seas from Antarctica. Large tabular icebergs calve at the edge of these vast shelves.

**Shoal** (1) A shallowly-submerged area. (2) To become shallow; also known as shoaling.

**Shore** The area seaward of the coast, which extends from the highest level of wave action during storms to the low water line.

**Shoreline** The line marking the intersection of water surface with the shore. Migrates up and down as the tide rises and falls.

**Side-scan sonar** A method of mapping the topography of the ocean floor along a strip up to 60 kilometers (37 miles) wide using computers and sonar signals that are directed away from both sides of the survey ship.

**Silica** Silicon dioxide ($SiO_2$).

**Silicate** A mineral whose crystal structure contains $SiO_4$ tetrahedra.

**Siliceous** A condition of containing abundant silica ($SiO_2$).

**Sill** A submarine ridge partially separating bodies of water such as fjords and seas from one another or from the open ocean.

**Silt** A particle size of 0.008 to 0.0625 millimeter (0.0003 to 0.002 inch). It is intermediate in size between sand and clay.

**Siphonophora** An order of hydrozoan cnidarians that form pelagic colonies containing both polyps and medusae. An example is *Physalia*.

**Sirenia** An order of large, vegetarian marine mammals that includes dugongs and manatees, which are also known as sea cows.

**Slack water** A situation that occurs when a reversing tidal current changes direction at high tide (high slack water) or low tide (low slack water). During slack water, the speed of the current is zero.

**Slick** A smooth patch on an otherwise rippled surface caused by a monomolecular film of organic material that reduces surface tension.

**Slightly stratified estuary** An estuary of moderate depth in which marine water invades beneath the freshwater runoff. The two water masses mix so the bottom water is slightly saltier than the surface water at most places in the estuary.

**Sodium ion** A sodium atom that has become positively charged by losing one electron.

**SOFAR channel** An acronym for the *so*und *f*ixing *a*nd *r*anging channel, which is a low-velocity sound travel zone that coincides with the permanent thermocline in low and middle latitudes.

**Solar day** The 24-hour period during which Earth completes one rotation on its axis.

**Solar distillation** A process by which ocean water can be desalinated by evaporation and the condensation of the vapor on the cover of a container. The condensate then runs into a separate container and is collected as freshwater. Also called solar humidification.

**Solar humidification** See *solar distillation*.

**Solar system** The Sun and the celestial bodies, asteroids, planets, and comets that orbit around it.

**Solar tide** The partial tide caused by the tide-producing forces of the Sun.

**Solid state** A state of matter in which the substance has a fixed volume and shape. A crystalline state of matter.

**Solstice** The time during which the Sun is directly over one of the tropics. In the Northern Hemisphere, the summer solstice occurs on June 21 or 22 as the Sun is directly above the Tropic of Cancer, and the winter solstice occurs on December 21 or 22 when the Sun is directly above the Tropic of Capricorn.

**Solubility** The amount of a substance that can dissolve in a given amount of solvent.

**Solute** A substance dissolved in a solution. Salts are the solute in saltwater.

**Solution** A state in which a solute is homogeneously mixed with a liquid solvent. Water is the solvent for the solution that is ocean water.

**Solvent** A liquid that has one or more solutes dissolved in it.

**Somali Current** A surface current that flows north along the Somali coast of Africa during the southwest monsoon season.

**Sonar** An acronym for *so*und *n*avigation *a*nd *r*anging, a method that uses sound to determine the distance of objects in the ocean.

**Sorting** A texture of sediments, where a well-sorted sediment is characterized by having great uniformity of grain sizes.

**Sounding** A measured depth of water beneath a ship.

**South Atlantic Subtropical Gyre** The large, counterclockwise-flowing subtropical gyre that exists in the South Atlantic Ocean.

**South Pacific Subtropical Gyre** The large, counterclockwise-flowing subtropical gyre that exists in the South Pacific Ocean.

**Southern boundary current** The southern boundary surface current of Southern Hemisphere subtropical gyres.

**Southern Ocean** See *Antarctic Ocean*.

**Southern Oscillation** The periodic change in the pressure differential between the Southeastern Pacific high pressure and the Western Pacific equatorial low pressure that occurs in concert with El Niño–Southern Oscillation events.

**Southwest Monsoon** A southwest wind that develops during the summer season. It blows off the Indian Ocean onto the Asian mainland.

**Southwest Monsoon Current** During the southwest monsoon season, an eastward-flowing surface current that replaces the west-flowing North Equatorial Current in the Indian Ocean.

**Space dust** Micrometeoroid space debris.

**Spawn** To reproduce, including releasing eggs and sperm directly into seawater (broadcast spawning) or by internal fertilization.

**Species** A fundamental category of taxonomic classification, ranking below a genus or subgenus and consisting of related organisms capable of interbreeding.

**Species diversity** The number or variety of species found in a subdivision of the marine environment.

**Specific gravity** The ratio of density of a given substance to that of pure water at 4°C and at 1 atmosphere pressure.

**Specific heat** The amount of heat energy required to raise the temperature of 1 gram of a substance by 1°C. Also known as specific heat capacity.

**Spermaceti organ** A large fatty organ located within the head region of sperm whales (*Physeter macrocephalus*) that is used to focus echolocation sounds.

**Spermatophyta** See *Anthophyta*.

**Spherule** A cosmogenous microscopic globular mass composed of silicate rock material (tektites) or of iron and nickel.

**Spicule** A minute, needle-like calcareous or siliceous projection found in sponges, radiolarians, chitons, and echinoderms that acts to support the tissue or provide a protective covering.

**Spilling breaker** A type of breaking wave that forms on a gently sloping beach, which gradually extracts the energy from the wave to produce a turbulent mass of air and water that runs down the front slope of the wave.

**Spit** A small point, low tongue, or narrow embankment of land commonly consisting of sand deposited by longshore currents and having one end attached to the mainland and the other terminating in open water.

**Splash wave** A long-wavelength wave created by a massive object or series of objects falling into water; a type of tsunami.

**Sponge** See *Porifera*.

**Spray zone** The shore zone lying between the high tide shoreline and the coastline. It is covered by water only during storms.

**Spreading center** A divergent plate boundary.

**Spreading rate** The rate of divergence of plates at a spreading center.

**Spring bloom** A middle-latitude bloom of phytoplankton that occurs during the spring and is limited by the availability of nutrients.

**Spring equinox** See *vernal equinox*.

**Spring tide** A tide of maximum range occurring about every two weeks when the Moon is in either new or full moon phase.

**Stack** An isolated mass of rock projecting from the ocean off the end of a headland from which it has been separated by wave erosion.

**Standard seawater** Ampules of ocean water for which the chlorinity has been determined by the Institute of Oceanographic Services in Wormley, England. The ampules are sent to laboratories all over the world so that equipment and reagents used to determine the salinity of ocean water samples can be calibrated by adjustment until they give the same chlorinity as is shown on the ampule label.

**Standing stock (crop)** The mass of fishery organisms present in an ecosystem at a given time.

**Standing wave** A wave, the form of which oscillates vertically without progressive movement. The region of maximum vertical motion is an antinode. On either side are nodes, where there is no vertical motion but maximum horizontal motion.

**Stenohaline** Pertaining to organisms that can withstand only a small range of salinity change.

**Stenothermal** Pertaining to organisms that can withstand only a small range of temperature change.

**Stick chart** A device made of sticks or pieces of bamboo that was used by early navigators at sea.

**Still water level** The horizontal surface halfway between crest and trough of a wave. If there were no waves, the water surface would exist at this level. Also known as zero energy level.

**Storm** An atmospheric disturbance characterized by strong winds accompanied by precipitation and often by thunder and lightning.

**Storm surge** A rise above normal water level resulting from wind stress and reduced atmospheric pressure during storms. Consequences can be more severe if it occurs in association with high tide.

**Strait of Gibraltar** The narrow opening between Europe and Africa through which the waters of the Atlantic Ocean and Mediterranean Sea mix.

**Stranded beach deposit** An ancient beach deposit found above present sea level because of a lowering of sea level.

**Streamlining** The shaping of an object so it produces the minimum of turbulence while moving through a fluid medium. The teardrop shape displays a high degree of streamlining.

**Stromatolite** A calcium carbonate sedimentary structure in which algal assemblages trap sediment and bind it into forms that are often dome shaped. They are known to form only in shallow-water environments.

**Subduction** The process by which one lithospheric plate descends beneath another as they converge.

**Subduction zone** A long, narrow region beneath Earth's surface in which subduction takes place.

**Subduction zone seep biocommunity** Animals that live in association with seeps of pore water squeezed out of deeper sediments. They depend on sulfur-oxidizing bacteria that act as producers for the ecosystem.

**Submarine canyon** A steep, V-shaped canyon cut into the continental shelf or slope.

**Submarine fan** See *deep-sea fan*.

**Submerged dune topography** Ancient coastal dune deposits found submerged beneath the present shoreline because of a rise in sea level or submergence of the coast.

**Submerging shoreline** A shoreline formed by the relative submergence of a landmass in which the shoreline is on landforms developed under subaerial processes. It is characterized by bays and promontories and is more irregular than a shoreline of emergence.

**Subneritic province** The benthic environment extending from the shoreline across the continental shelf to the shelf break. It underlies the neritic province of the pelagic environment.

**Suboceanic province** The benthic environments seaward of the continental shelf.

**Subpolar** Pertaining to the oceanic region that is covered by sea ice in winter. The ice melts away in summer.

**Subpolar gyre** A small, circular-moving loop of water that is centered at about 60 degrees latitude in both hemispheres. Subpolar gyres rotate in a counterclockwise orientation in the Northern Hemisphere and clockwise in the Southern Hemisphere.

**Subpolar low** A global belt of low atmospheric pressure located at about 60 degrees north or south latitude that is associated with upward vertical flow of low-density air and abundant precipitation.

**Subside** To sink to a low or lower level. As a process, to subside is known as subsidence.

**Substrate** The base on which an organism lives and grows.

**Subsurface current** A current usually flowing below the pycnocline, generally at slower speed and in a different direction from the surface current.

**Subtidal zone** That portion of the benthic environment extending from low tide to a depth of 200 meters (660 feet); considered by some to be the surface of the continental shelf.

**Subtropical** Pertaining to the oceanic region poleward of the tropics (about 30 degrees latitude).

**Subtropical Convergence** The zone of convergence that occurs within all subtropical gyres as a result of Ekman transport driving water toward the interior of the gyres.

**Subtropical gyre** A large, circular-moving loop of water that is centered at about 30 degrees latitude and is initiated by the trade winds and the prevailing westerlies. A total of five subtropical gyres exist, with rotation clockwise in the Northern Hemisphere and counterclockwise in the Southern Hemisphere.

**Subtropical high** A region of high atmospheric pressure located at about 30 degrees latitude.

**Sulfur** A yellow mineral composed of the element sulfur. It is commonly found in association with hydrocarbons and salt deposits.

**Sulfur-oxidizing bacteria** Bacteria that support many deep-sea hydrothermal vents and cold-water seep biocommunities by using energy released by oxidation to synthesize organic matter chemosynthetically.

**Summer solstice** In the Northern Hemisphere, it is the instant when the Sun moves north to the Tropic of Cancer before changing direction and moving southward toward the equator approximately June 21.

**Summertime beach** A beach that is characteristic during summer months. It typically has a wide sandy berm and a steep beach face.

**Sunspot** A dark spot on the Sun that is associated with magnetic storms.

**Superwave** See *rogue wave*.

**Supratidal zone** The splash or spray zone above the spring high tide shoreline; also called the supralittoral zone.

**Surf beat** An irregular wave pattern caused by mixed interference that results in a varied sequence of larger and smaller waves.

**Surf zone** The nearshore zone of breaking waves.

**Surface tension** The tendency for the surface of a liquid to contract owing to intermolecular bond attraction.

**Surfing** The sport of riding on the crest or along the tunnel of a wave, especially while standing or lying on a surfboard.

**Surging breaker** A compressed breaking wave that builds up over a short distance and surges forward as it breaks. It is characteristic of abrupt beach slopes.

**Suspension feeding** See *filter feeding*.

**Suspension settling** The process by which fine-grained material that is being suspended in the water column slowly accumulates on the sea floor.

**Sverdrup (Sv)** A unit of flow rate equal to 1 million cubic meters per second. Named after Norwegian oceanographer Harald Sverdrup.

**Swash** A thin layer of water that washes up over exposed beach as waves break at the shore.

**Swell** A free ocean wave by which energy put into ocean waves by wind in the sea is transported with little energy loss across great stretches of ocean to the margins of continents where the energy is released in the surf zone.

**Swim bladder** A gas-containing, flexible, cigar-shaped organ that aids many fishes in attaining neutral buoyancy.

**Symbiosis** A relationship between two species in which one or both benefit or neither or one is harmed. Examples are commensalism, mutualism, and parasitism.

**Syzygy** Either of two points in the orbit of the Moon (full or new moon phase) when the Moon lies in a straight line with the Sun and Earth.

## T

**Tablemount** A conical volcanic feature on the ocean floor resembling a seamount except that it has had its top truncated to a relatively flat surface.

**Taxonomy** The classification of organisms in an ordered system that indicates natural relationships.

**Tectonic estuary** An estuary, the origin of which is related to tectonic deformation of the coastal region.

**Tectonics** Deformation of Earth's surface by forces generated by heat flow from Earth's interior.

**Tektite** See *spherule*.

**Temperate** Pertaining to the oceanic region where pronounced seasonal change occurs (about 40 to 60 degrees latitude). Also known as the middle latitudes.

**Temperature** A direct measure of the average kinetic energy of the molecules of a substance.

**Temperature of maximum density** The temperature at which a substance reaches its highest density. For water, it is 4°C.

**Temperature–salinity (T–S) diagram** A diagram with axes representing water temperature and salinity, whereby the density of the water can be determined.

**Terrane** A distinct geologic fragment of crustal material broken off from one tectonic plate and accreted or sutured onto another plate. Different than terrain, which is the lay of the land.

**Terrigenous sediment** Sediment produced from or of the Earth. Also called lithogenous sediment.

**Territorial sea** A strip of ocean, 12 nautical miles wide, adjacent to land over which the coastal nation has control over the passage of ships.

**Test** The supporting skeleton or shell (usually microscopic) of many invertebrates.

**Tethys Sea** An ancient body of water that separated Laurasia to the north and Gondwanaland to the south. Its location was approximately that of the present Alpine–Himalayan mountain system.

**Texture** The general physical appearance of an object.

**Theory** A well-substantiated explanation of some aspect of the natural world that can incorporate facts, laws (descriptive generalizations about the behavior of an aspect of the natural world), logical inferences, and tested hypotheses.

**Thermal contraction** The reduction in size as a result of lowering of temperature.

**Thermocline** A layer of water beneath the mixed layer in which a rapid change in temperature can be measured in the vertical dimension.

**Thermohaline circulation** The vertical movement of ocean water driven by density differences resulting from the combined effects of variations in temperature and salinity; produces deep currents.

**Thermonuclear fusion** A high temperature process in which hydrogen atoms are converted to helium atoms, thereby releasing large amounts of energy.

**Thermostatic effects** Refers to the natural thermostat that regulates temperature on Earth and is largely controlled by the properties of water.

**Tidal bore** A steep-fronted wave that moves up some rivers when the tide rises in the coastal ocean.

**Tidal bulge** The theoretical mound of water found on both sides of Earth caused by the relative positions of the Moon (lunar tidal bulges) and the Sun (solar tidal bulges).

**Tidal day** See *lunar day*.

**Tidal period** The time that elapses between successive high tides. In most parts of the world, it is 12 hours and 25 minutes.

**Tidal range** The difference between high tide and low tide water levels over any designated time interval, usually one lunar day.

**Tide** The periodic rise and fall of the surface of the ocean and connected bodies of water resulting from the gravitational attraction of the Moon and Sun acting unequally on different parts of Earth.

**Tide-generating force** The magnitude of the centripetal force required to keep all particles of Earth having identical mass moving in identical circular paths required by the movements of the Earth–Moon system is identical. This required force is provided by the gravitational attraction between the particles and the Moon. This gravitational force is identical to the required centripetal force only at the center of Earth. For ocean tides, the horizontal component of the small force that results from the difference between the required and provided forces is the tide-generating force on that individual particle. These forces are such that they tend to push the ocean water into bulges toward the tide-generating body on one side of Earth and away from the tide-generating body on the opposite side of Earth.

**Tide wave** A long-period gravity wave generated by tide-generating forces and manifested as rising and falling tides.

**Tissue** An aggregate of cells and their products developed by organisms for the performance of a particular function.

**Tombolo** A sand or gravel bar that connects an island with another island or the mainland.

**Topography** The configuration of a surface. In oceanography it refers to the ocean bottom or the surface of a mass of water with given characteristics.

**Toxic compound** A poisonous substance capable of causing injury or death, especially by chemical means.

**Trade winds** A global wind belt that moves from a subtropical high-pressure belt at about 30 degrees north or south latitude toward the equatorial region. These winds move from a northeasterly direction in the Northern Hemisphere and from a southeasterly direction in the Southern Hemisphere.

**Transform fault** A fault with side-to-side motion that offsets segments of a mid-ocean ridge. Oceanic transform faults occur wholly on the ocean floor, while continental transform faults occur on land.

**Transform faulting** The process by which a transform fault moves with side-to-side motion.

**Transform plate boundary** The boundary between two lithospheric plates formed by a transform fault.

**Transitional wave** A wave moving from deep water to shallow water that has a wavelength more than twice the water depth but less than 20 times the water depth. Particle orbits are beginning to be influenced by the bottom.

**Transverse wave** A wave in which particle motion is at right angles to energy propagation; also known as a side-to-side wave.

**Treatment** In reference to sewage at a sewage treatment plant, there is primary treatment where solids are allowed to settle and separate from the liquid, and secondary treatment where sewage is exposed to bacteria-killing chlorine or other means of disinfection.

**Trench** A long, narrow, and deep depression on the ocean floor with relatively steep sides that is caused by plate convergence; often referred to as an ocean trench.

**Trophic level** A nourishment level in a food chain. Plant producers constitute the lowest level, followed by herbivores and a series of carnivores at the higher levels.

**Tropic of Cancer** The latitude 23.5 degrees north, which is the furthest location north that receives vertical rays of the Sun.

**Tropic of Capricorn** The latitude 23.5 degrees south, which is the furthest location south that receives vertical rays of the Sun.

**Tropical** Pertaining to, characteristic of, occurring in, or inhabiting the tropics.

**Tropical Atmosphere and Ocean (TAO)** A scientific program that monitors the equatorial Pacific and studies how El Niño events develop.

**Tropical cyclone** See *hurricane*.

**Tropical Ocean–Global Atmosphere (TOGA)** A scientific program that was initiated in 1985 to monitor the equatorial Pacific and study how El Niño events develop; predecessor to the TAO program.

**Tropical tide** A tide that occurs twice monthly when the Moon is at its maximum declination near the Tropic of Cancer or the Tropic of Capricorn.

**Tropics** The region of Earth's surface lying between the Tropic of Cancer and the Tropic of Capricorn. Also known as the Torrid Zone.

**Troposphere** The lowermost portion of the atmosphere, which extends from Earth's surface to 12 kilometer (7 miles). It is where all weather is produced.

**Trough (wave)** The part of an ocean wave that is displaced below the still-water level.

**Tsunami** A seismic sea wave. A long-period gravity wave generated by a submarine earthquake or volcanic event. Not noticeable on the open ocean but builds up to great heights in shallow water.

**Turbidite deposit** A sediment or rock formed from sediment deposited by turbidity currents characterized by both horizontally and vertically graded bedding.

**Turbidity** A state of reduced clarity in a fluid caused by the presence of suspended matter.

**Turbidity current** A gravity current resulting from a density increase brought about by increased water turbidity. Possibly initiated by some sudden force such as an earthquake, the turbid mass continues under the force of gravity down a submarine slope.

**Turbulent flow** Flow in which the flow lines are confused heterogeneously due to random velocity fluctuations.

**Typhoon** See *hurricane*.

## U

**Ultraplankton** Plankton for which the greatest dimension is less than 5 microns (0.0002 inch). Because of their small size, they are very difficult to separate from water.

**Ultrasonic** Sound frequencies above those that can be heard by humans (above 20,000 cycles per second).

**Ultraviolet (UV) radiation** Electromagnetic radiation shorter than visible radiation and longer than X-rays. The approximate wavelength range is 0.001 to 0.4 micron.

**United Nations Conference on the Law of the Sea** A series of meetings to establish legal rights in the sea, particularly in regards to sea floor mining.

**Upper water** An area of the ocean near the surface that includes the mixed layer and the permanent thermocline. It is approximately the top 1000 meters (3300 feet) of the ocean.

**Upwelling** The process by which deep, cold, nutrient-laden water is brought to the surface, usually by diverging equatorial currents or coastal currents that pull surface waters away from a coast.

## V

**Valence** The combining capacity of an element measured by the number of hydrogen atoms with which it will combine.

**van der Waals force** A weak attractive force between molecules resulting from the interaction of one molecule and the electrons of another.

**Vapor** The gaseous state of a substance that is liquid or solid under ordinary conditions.

**Vent** An opening on the ocean floor that emits hot water and dissolved minerals; vent type is based on water temperature and includes hydrothermal (hot water) vents and warm-water vents.

**Ventral** Pertaining to the lower or under surface.

**Vernal equinox** The passage of the Sun across the equator as it moves from the Southern Hemisphere into the Northern Hemisphere, approximately March 21. During this time, all places in the world experience equal lengths of night and day. Also known as the spring equinox.

**Vertebrata** The subphylum of chordates that includes those animals with a well-developed brain and a skeleton of bone or cartilage; includes fish, amphibians, reptiles, birds, and land animals.

**Vertically mixed estuary** Very shallow estuaries such as lagoons in which freshwater and marine water are totally mixed from top to bottom so that the salinity at the surface and the bottom is the same at most places within the estuary.

**Virioplankton** Viruses that live as plankton.

**Viscosity** A property of a substance to offer resistance to flow caused by internal friction.

**Volcanic arc** An arc-shaped row of active volcanoes directly above a subduction zone. Can occur as a row of islands (island arc) or mountains on land (continental arc).

**Vortex** A circular or spiral flow of water. See *eddy* and *whirlpool*.

## W

**Walker Circulation Cell** The pattern of atmospheric circulation that involves the rising of warm air over the East Indies low-pressure cell and its descent over the high-pressure cell in the southeastern Pacific Ocean off the coast of Chile. It is the weakening of this circulation that accompanies an El Niño event, which has led to the development of the term *El Niño–Southern Oscillation event*.

**Walrus** A large marine mammal (*Odobenus rosmarus*) of Arctic regions belonging to the order Pinnipedia and having two long tusks, tough wrinkled skin, and four flippers.

**Waning crescent** The Moon when it is between third-quarter and new moon phases.

**Waning gibbous** The Moon when it is between full moon and third-quarter phases.

**Warm-blooded** See *homeothermic*.

**Warm-core ring** See *ring*. **Warm front** A weather front in which a warm air mass moves into and over a cold air mass producing a broad band of gentle precipitation.

**Water mass** A body of water identifiable from its temperature, salinity, or chemical content.

**Wave** A disturbance that moves over the surface or through a medium with a speed determined by the properties of the medium.

**Wave base** The depth at which circular orbital motion becomes negligible. It exists at a depth of one-half wavelength, measured vertically from still-water level.

**Wave-cut bench** A gently seaward-sloping horizontal platform produced by wave erosion and extending from the base of the wave-cut cliff out under the offshore region. See also *marine terrace*.

**Wave-cut cliff** A cliff produced by landward cutting by wave erosion.

**Wave dispersion** The separation of waves as they leave the sea area by wave size. Larger waves travel faster than smaller waves and thus leave the sea area first, to be followed by progressively smaller waves.

**Wave frequency (*f*)** The number of waves that pass a fixed point in a unit of time (usually one second). A wave's frequency is the inverse of its period.

**Wave height (*H*)** The vertical distance between a crest and the adjoining trough.

**Wave period (*T*)** The elapsed time between the passage of two successive wave crests (or troughs) past a fixed point. A wave's period is the inverse of its frequency.

**Wave speed (*S*)** The rate at which a wave travels. It can be calculated by dividing a wave's wavelength ($L$) by its period ($T$).

**Wave steepness** Ratio of wave height ($H$) to wavelength ($L$). If a 1:7 ratio is ever exceeded by the wave, then the wave breaks.

**Wave train** A series of waves from the same direction. Informally known as a wave set.

**Wavelength (*L*)** The horizontal distance between two corresponding points on successive waves, such as from crest to crest.

**Waxing crescent** The Moon when it is between new moon and first-quarter phases.

**Waxing gibbous** The Moon when it is between first-quarter and full moon phases.

**Weather** The state of the atmosphere at a given time and place, with respect to variables such as temperature, moisture, wind velocity, and barometric pressure.

**Weathering** A process by which rocks are broken down by chemical and mechanical means.

**Wentworth scale of grain size** A logarithmic scale for size classification of sediment particles.

**West Australian Current** A cold surface current that forms the eastern boundary of the Indian Ocean Subtropical Gyre. It is separated from the coast by the warm Leeuwin Current except during El Niño–Southern Oscillation events when the Leeuwin Current weakens.

**West Wind Drift** See *Antarctic Circumpolar Current*.

**Western boundary current** Poleward-flowing warm surface current on the western side of each subtropical gyre.

**Western boundary undercurrent (WBUC)** A bottom current that flows along the base of the continental slope eroding sediment from it and redepositing the sediment on the continental rise. It is confined to the western boundary of deep-ocean basins.

**Western intensification** Pertaining to the intensification of warm western boundary currents of each subtropical gyre that are faster, narrower, and deeper than their corresponding eastern boundary currents.

**Wetland** See *coastal wetland*.

**Whirlpool** A rapidly rotating current of water. See *eddy* and *vortex*.

**White muscle fiber** Thick muscle fibers with relatively low concentrations of myoglobin that make up a large percentage of the muscle fiber in lunger-type fishes.

**White smoker** A hydrothermal vent feature similar to a black smoker but emitting water of a lower temperature that is white in color.

**Wilson cycle** A model that uses plate tectonic processes to show the distinctive life cycle of ocean basins during their formation, growth, and destruction.

**Wind** The movement of air, usually as a result of pressure differences.

**Wind-driven circulation** A movement of ocean water that is driven by winds. This includes most horizontal movements in the surface waters of the world's oceans.

**Windward** The direction from which the wind is blowing.

**Winter solstice** The instant the southward-moving Sun reaches the Tropic of Cancer before changing direction and moving north back toward the equator, approximately December 21.

**Wintertime beach** A beach that is characteristic during winter months. It typically has a narrow rocky berm and a flat beach face.

## Z

**Zebra mussel** A non-native species released into U.S. and Canadian waters of the Great Lakes region.

**Zenith** The point on the celestial sphere directly over the observer.

**Zooplankton** Animal plankton.

**Zooxanthellae** A form of algae that lives as a symbiont in the tissue of corals and other coral reef animals and provides varying amounts of their required food supply.

# CREDITS AND ACKNOWLEDGMENTS

All uncredited figures: International Mapping/Pearson Education, Inc.

## Chapter 1

**Opener** Goddard Space Flight Center/Reto Stöckli/NASA; **Figure 1.1** JPL/University of Arizona/NASA; **Figure 1.2** International Mapping/Pearson Education, Inc.; **Figure 1.6** Office of Naval Research, US Navy; **Figure 1.7** Mark Thiessen/ZUMA-PRESS/Newscom; **Creature Feature 1.1** aaltair/Shutterstock; **Figure 1.9** ASSOCIATED PRESS; **Figure 1.10** World History Archive/Alamy Stock Photo; **Figure 1.13** US Naval Historical Center; **Figure 1D** Masa Ushioda/AGE Fotostock; **Figure 1.17** NASA, ESA, Mohammad Heydari-Malayeri (Observatoire de Paris) et al; **Figure 1.19** NASA'S GODDARD SPACE FLIGHT CENTER CONCEPTUAL IMAGE LAB; **Figure 1.25** Stringer/Reuters; **Figure 1E left** HIP/Art Resource, NY, **right** David Parry/PA Images/Alamy Stock Photo; **Figure 1.29** Zimmer, C., 2001, How old is it? National Geographic 200:3, 92; **Figure 1.30** Based on Tarbuck, E. J. and Lutgens, F. K., *Earth: An Introduction to Physical Geology*, 7th ed. (Fig. 8.15), Prentice Hall, 2002. Data from the Geological Society of America, the U.S. Geological Survey, and the International Commission on Stratigraphy. Pearson Education, Inc.; **Quote, page 3** Loren Shriver, NASA astronaut (2008); **Quote, page 18** Neil deGrasse Tyson (2014); **Quote, page 29** Theodosius Dobzhansky "Nothing in Biology Makes Sense Except in the Light of Evolution," *American Biology Teacher* vol. 35 (March 1973); **Quote, page 29** Charles Darwin, 1859. *On the Origin of Species*. London: John Murray

## Chapter 2

**Opener** Alan P. Trujillo; **Figure 2.1** bpk Bildagentur/Art Resource, NY; **Figure 2.2** Based on Dietz, R. S. and Holden, J. C., 1970, Reconstruction of Pangaea: Breakup and dispersion of continents, Permian to present, *Journal of Geophysical Research* 75:26, 4939–4956; **Figure 2.3** Based on *Continental Drift* by Don and Maureen Tarling, 1971 by G. Bell & Sons, Ltd.; **Figure 2.4** Based on Tarbuck, E. J. and Lutgens, F. K., *Earth Science*, 8th ed. (Fig. 7.6), Prentice Hall, 1997; **Figure 2.5, 2.6** Based on Tarbuck, E. J. and Lutgens, F. K., *Earth Science*, 8th ed. (Fig. 7.4), Prentice Hall, 1997; **Figure 2.7** Reprinted by permission from Tarbuck, E. J. and Lutgens, F. K., *Earth: An Introduction to Physical Geology*, 3d ed. (Fig. 18.8 and Fig 18.9), Merrill Publishing Company, 1990; **Creature Feature 2.1** stevedunleavy.com/Getty Images; **Figure 2A** David Evison/Shutterstock; **Figure 2B** Based on Tarbuck, E. J. and Lutgens, F. K., *Earth Science*, 8th ed. (Fig. 7.17), Prentice Hall,1997; **Figure 2.11** Based on Tarbuck, E. J. and Lutgens, F. K., *Earth: An Introduction to Physical Geology*, 3d ed. (Fig. 19.16), Merrill Publishing Company, 1990. After The Bedrock Geology of the World, by R. L. Larson et al., Copyright © 1985 by W. H. Freeman; **Figure 2.13** NASA/Goddard Space Flight Center; **Figure 2.15** Based on Tarbuck, E. J. and Lutgens, F. K., *Earth Science*, 8th ed. (Fig. 7.10), Prentice Hall, 1997; **Figure 2.16** Alan P. Trujillo; **Figure 2.18 top** Copyright © 2013 Cornelis Klein and Tony Philpotts. Reprinted with the permission of Cambridge University Press, **bottom** Image courtesy NASA/JPL/NGA Shuttle Radar Topography team; **Map 2.18** Based on Tarbuck, E. J. and Lutgens, F. K., *Earth Science*, 8th ed. (Fig. 7.11), Prentice Hall, 1997; **Figure 2.19 top** Scripps Institution of Oceanography Geological Collections, UCSD, **bottom** US Geological Survey Library (USGS); **Figure 2.21** US Geological Survey Library (USGS); **Figure 2.22** Frank Bienewald/ImageBroker/AGE Fotostock; **Figure 2.24** Aurora Photos/Alamy Stock Photo; **Figure 2.30** Plate tectonic reconstructions by Christopher R. Scotese, PALEOMAP Project, University of Texas at Arlington; **Figure 2.31** Based on Dietz, R. S. and Holden, J. C., 1970, The breakup of Pangaea, *Scientific American* 223: 4, 30– 41; **Figure 2.32** Adapted from Wilson, J. T., American Philosophical Society Proceedings 112, 309– 320, 1968; Jacobs, J. A., Russell, R. D., and Wilson, J. T., *Physics and Geology*, McGraw- Hill, New York, 1971; **Quote, page 41** Wegener, Alfred. *The Origins of Continents and Oceans*. Friedrich Vieweg & Sohn, Braunschweig. 1929

## Chapter 3

**Opener** Planetary Visions, Ltd./Science Source; **Figure 3.1** Courtesy Peter A. Rona, Hudson Laboratories of Columbia University; **Figure 3.2** Erika Mackay/National Institute of Water and Atmospheric Research; **Figure 3.3** Courtesy Daniel J. Fornari, Lamont-Doherty Geological Observatory, Columbia University. Reprinted with permission of the American Geophysical Union; **Figure 3.4** After Gross, M. G., *Oceanography*, 6th ed. (Fig. 16.10), Prentice Hall, 1993; **Figure 3.5-3.7** Scripps Institution of Oceanography Geological Collections, UCSD; **Figure 3.8** After Tarbuck, E. J. and Lutgens, F. K., *Earth: An Introduction to Physical Geology*, 5th ed. (Fig. 19.3), Prentice Hall, 1996; **Figure 3.12a** Tarbuck, E. J. and Lutgens, F. K., *Earth: An Introduction to Physical Geology*, 5th ed. (Fig. 19.6), Prentice Hall, 1996; **Figure 3.12b** Martin Strmiska/Alamy Stock Photo; **Figure 3.12c** Alan P. Trujillo; **Figure 3.14b** James V. Gardner/University of New Hampshire; **Creature Feature 3.1** © 2013 MBARI, **inset** Ocean Exploration Trust/Nautilus Live; **Figure 3.20** Heinrich Berann/National Geographic Creative; **Figure 3.21a** Photo Researchers, Inc./Science Source; **Figure 3.21b** Scripps Institution of Oceanography, UCSD; **Figure 3.21c** Jeffrey Grover, Cuesta College; **Figure 3A** After Tarbuck, E. J. and Lutgens, F. K., *Earth Science*, 6th ed. (Fig. 10.2), Macmillan Publishing Company, 1991; **Figure 3B** Woods Hole Oceanographic Institution; **Figure 3.22a** OET/Nautilus Live; **Figure 3.22b** Jean-Marie Auzende/Photo Ifremer, Manaute Cruise, *EOS*, Vol. 81, No. 39, 9/26/2000; **Figure 3.25** Atsushi Taketazu/AP Images; **Quote, page 81** Matthew Fontaine Maury (1854) *The Physical Geography of the Sea*. New York: Harper & Brothers

## Chapter 4

**Opener** Alfred Wegener Institute for Polar and Marine Research; **Figure 4.1** NOAA; **Figure 4.2** Courtesy of the Ocean Drilling Program, Texas A&M University; **Figure 4.3, 4.4** Alan P. Trujillo; **Figure 4.6a** NASA; **Figure 4.6b** Marine Corps University; **Figure 4.6c, d** Alan P. Trujillo; **Figure 4.7** Clement Philippe/Arterra Picture Library/Alamy Stock Photo; **Figure 4.8** Norman Kuring/SeaWiFS Project/NASA; **Figure 4.9a** CSIRO Plant Industry; **Figure 4.9b** Scripps Institution of Oceanography, UCSD; **Figure 4.9c** Taken from Imerys Diatomite mine in Lompoc, CA. Photo courtesy of Imerys; **Figure 4A** Taken from Imerys Diatomite mine in Lompoc, CA. Photo courtesy of Imerys, **inset** Gary D. Gaugler/Science Source; **Figure 4.10a, b** CSIRO Plant Industry; **Figure 4.10c, d** Scripps Institution of Oceanography; **Figure 4.11a** Steve Gschmeissner/Science Source; **Figure 4.11b** David Hughes/Robert Harding World Imagery; **Creature Feature 4.1a** Norman Kuring, Goddard Space Flight Center/NASA, **inset** The Natural History Museum, London/Science Source; **Figure 4.12b** Jarrod Boordk/Shutterstock; **Figure 4.12c** Francois Gohier/Science Source; **Figure 4.16** After Biscaye, P. E., et al., 1976; Berger, W. H., et al., 1976; and Kolla V. and Biscaye, P. E., 1976; **Figure 4.17a, b** Charles D. Winters/Science Source; **Figure 4.17c** Institute of Oceanographic Sciences/NERC/Science Source; **Figure 4.18** Alan P. Trujillo; **Figure 4.19** Integrated Ocean Drilling Program; **Figure 4.21, 4.22** Modified after Sverdrup, H. U., et al., 1942; **Figure 4.23** Photo courtesy of Susumu Honjo, Woods Hole Oceanographic Institution; **Figure 4.24** Divins, D. L., NGDC Total Sediment Thickness of the World's Oceans & Marginal Seas, Retrieved date 4/10/2006, http://www.ngdc.noaa.gov/mgg/ http://www.ngdc.noaa.gov/mgg/sedthick/sedthick.html; **Figure 4.25** Scott Gibson/Corbis; **Figure 4.26a** NOAA Okeanos Explorer Program/2013 Northeast U.S. Canyons Expedition; **Figure 4.26b, 4.28** Alan P. Trujillo; **Figure 4.29** UC San Diego Library Special Collections and Archives; **Figure 4.30** After Cronan, D. S., 1977, Deep sea nodules: Distribution and geochemistry, in Glasby, G. P., ed., *Marine Manganese Deposits*, Elsevier Scientific Publishing Co.; **Quote, page 105** Wolf, Berger: *Oceans: Reflections on a Century of Exploration* ©2009 by the Regents of the University of California. Published by the University of California Press; **Quote, page 115** Charles Darwin, 1859. *On the Origin of Species*. London: John Murray

## Chapter 5

**Opener** Pearson Education, Inc.; **Figure 5.1** After Tarbuck, E. J. and Lutgens, F. K., *Earth: An Introduction to Physical Geology*, 5th ed. (Fig. 2.4), Prentice Hall, 1996; **Figure 5.11** After data from Jet Propulsion Laboratory, NASA; **Figure 5.13** USDA; **Figure 5.14** Martyn F. Chillmaid/Science Source; **Creature Feature 5.1** Paul Nicklen/Getty Images; **Figure 5.17** Howard Shooter/Dorling Kindersley/Getty Images; **Figure 5.18** Rafael Ben-Ari/Alamy Stock Photo; **Figure 5.21** Fleming, Richard H; Johnson, Martin W; Hamre, Anna; Sverdrup, H. U.,

*OCEANS*, 1st Ed., © 1942. Reprinted and Electronically reproduced by permission of Pearson Education, Inc., Upper Saddle River, New Jersey; **Figure 5.22** Norman Kuring, Goddard Space Flight Center/NASA; **Figure 5.23** After Pickard, G. L., *Descriptive Physical Oceanography*, Pergamon Press Ltd., 1963; **Table 5.5** Data based on the mole fraction of these gases in seawater at equilibrium, after Millero, *Chemical Oceanography*, 4th edition, 2013; **Quote, page 139** Luciano Caglioti, 1985. *The Two Faces of Chemistry*. MIT Press

### Chapter 6

**Opener** Cyber Kristiyan/Shutterstock; **Creature Feature 6.1** Anthony Pierce/Alamy Stock Photo; **Figure 6.10** Department of Atmospheric Sciences; **Figure 6A** Morgan P. Sanger/The Columbus Foundation, British Virgin Islands; **Figure 6.20a** Goddard Space Flight Center/NASA; **Figure 6.21b** Michelle McLoughlin/Reuters; **Figure 6.22** NOAA; **Figure 6.23** Mark C. Olsen/U.S. Air Force; **Figure 6.24a** NOAA; **Figure 6.24b** UPI Photo/Vincent Laforet/Pool/Newscom; **Figure 6.25** Newscom; **Figure 6B** John Anderson/Alamy Stock Photo; **Figure 6.28** Chris James/Alamy Stock Photo; **Poem, page 173** Samuel Taylor Coleridge, about ships getting stuck in the horse latitudes, *Rime of the Ancient Mariner* (1798)

### Chapter 7

**Opener** NASA Earth Observatory; **Figure 7.1a** Scripps Institution of Oceanography, UCSD; **Figure 7.1b** Dann S. Blackwood/USGS, U.S. Geological Survey Library; **Figure 7.2** TOPEX/Poseidon radar/NASA; **Figure 7.3a** Based on data collected and made freely available by the International Argo Program and the national programs that contribute to it. (http://www.argo.ucsd.edu, http://argo.jcommops.org). The Argo Program is part of the Global Ocean Observing System; **Figure 7.3b** Alan P. Trujillo; **Figure 7A left** Rachel Youdelman/Pearson Education, Inc., **top right** Handout/Reuters Pictures, **center right** Jack Sullivan/Alamy Stock Photo, **bottom right** Alan P. Trujillo; **Figure 7A** Courtesy of Eos Transactions AGU 73: 34, 361 (1992). © the American Geophysical Union; **Creature Feature 7.1** Helmut Corneli/Alamy Stock Photo; **Figure 7.17a** NOAA; **Figure 7B** NOAA, **inset** Benjamin Franklin, 1706-1790,Library of Congress Prints and Photographs Division [LC-USZ62-41888]; **Figure 7.20** NASA Earth Observatory; **Figure 7.23** International Research Institute for Climate and Society; **Figure 7C** Alan P. Trujillo; **Figure 7.24** Team, E. W. (n.d.). Physical Sciences Division. Retrieved October 29, 2018, from https://www.esrl.noaa.gov/psd/enso/mei/; **Figure 7.26** Data Source: NOAA NCEP EMC CMB GLOBAL; **Figure 7.27a, b** Alan P. Trujillo; **Figure 7.27c** Sue Flood/Nature Picture Library; **Figure 7.28a** William Bacon/Science Source; **Figure 7.28c** Josh Landis/National Science Foundation; **Figure 7.28d** Joshua Stevens/NASA; **Figure 7.32** Siemens Press Pictures; **Quote, page 207** Anonymous, but often attributed to Mark Twain; said in reference to San Francisco's cool summer weather caused by coastal upwelling

### Chapter 8

**Opener** Octavio Passos/Getty Images; **Figure 8.1a** CBS Photo Archive/Getty Images; **Figure 8.1c** Jacques Descloitres/MODIS Land Rapid Response Team/NASA; **Figure 8.4a** Based on *the Tasa Collection: Shorelines*. Published by Macmillan Publishing Co., New York. Copyright © 1986 by Tasa Graphic Arts, Inc.; **Figure 8.6** Kyodo News/AP Images; **Table 8.1** U.S. Government Printing Office, photos: NOAA; **Figure 8.11** NASA; **Figure 8.13** National Archives and Records Administration; **Figure 8.18** Project Michelangelo; **Figure 8.19** Based on *the Tasa Collection: Shorelines*. Published by Macmillan Publishing Co., New York. Copyright © 1986 by Tasa Graphic Arts, Inc.; **Figure 8.20a** EpicStockMedia/Shutterstock; **Figure 8.20b** irabel8/Shutterstock; **Figure 8.20c** SweetParadise/Alamy Stock Photo; **Figure 8A** John Seaton Callahan/Getty Images; **Figure 8.21** Rich Reid/National Geographic/Getty Images; **Figure 8.22** Mark Rightmire/Newscom; **Figure 8.24** US Geological Survey Library (USGS); **Figure 8.25b** Joanee Davis/Newscom; **Figure 8.26** FLHC 12/Alamy Stock Photo; **Figure 8.28** Klotz/123RF; **Figure 8.29** US Geological Survey Library (USGS); **Figure 8.30** CNES, NOAA, EUMETSAT/NASA; **Figure 8.31** NOAA; **Figure 8C** Mainichi Shimbun/Reuters; **Figure 8.32a** Martin Bond/Science Source; **Figure 8.33** P123; **Quote, page 249** Fanny Crosby (1820–1915)

### Chapter 9

**Opener** Laszlo Podor/Moment/Getty Images; **Figure 9.9** International Mapping/Pearson Education, Inc.; **Creature Feature 9.1** twospeeds/Shutterstock; **Figure 9.11** Based on *the Tasa Collection: Shorelines*. Published by Macmillan Publishing Co., New York. © 1986 by Tasa Graphic Arts, Inc.; **Figure 9.12** NASA; **Figure 9.15** NASA; **Figure 9A** New Brunswick Department of Tourism; **Figure 9B** STRINGER Brazil/Reuters; **Figure 9.20** Laszlo Podor/Alamy Stock Photo; **Figure 9.22** Peter McBride/Aurora/Getty Images; **Figure 9.23** Mark Conlin/VWPICS/Visual&Written SL/Alamy Stock Photo; **Figure 9.24** DEA/C Sappa/De Agostini Editore/Age Fotostock; **Quote, page 285** Sir Isaac Newton, *Philosophiæ naturalis principia mathematica*: 1687. London

### Chapter 10

**Opener** Alan P. Trujillo; **Figure 10.2** The Photolibrary Wales/Alamy Stock Photo; **Creature Feature 10.1** George Ostertag/Alamy Stock Photo; **Figure 10.3a** Julian Eales/Alamy Stock Photo; **Figure 10.3b** Alan P. Trujillo; **Figure 10.4a** Marli Miller; **Figure 10.7** Mikehoward 1/Alamy Stock Photo; **Figure 10.8, 10.9** Marli Miller; **Figure Figure 10A** Alan P. Trujillo; **Figure 10.11a** USDA/FSA/Farm Service Agency; **Figure 10.11b** Martin Beebee/Alamy Stock Photo; **Figure 10.12c** USDA/FSA/Farm Service Agency; **Figure 10.15** NASA Earth Observatory; **Figure 10.21** Adam van Bunnens/Alamy Stock Photo; **Figure 10.23** US Army Corps of Engineers; **Figure 10.24** Peter Titmuss/Alamy Stock Photo; **Figure 10.25** Kevin Steele/Aurora/Getty Images; **Figure 10.26** Fairchild Aerial Photography Collection - Whittier College Collection; **Figure 10.28, 10.29** Alan P. Trujillo; **Figure 10.33a** NASA; **Figure 10.33b** Eye Ubiquitous/Glow Images; **Figure 10.33c** USDA/FSA/Farm Service Agency; **Figure 10.33d** M-Sat Ltd/Science Source; **Figure 10.40** Alan P. Trujillo; **Figure 10.41** Ethan Daniels/WaterFrame/Getty Images; **Figure 10B** Veronica Dominach; **Quote, page 313** Lord Byron, 1922. The Works of Lord Byron: Letters and Journals. John Murray

### Chapter 11

**Opener** Justin Hofman; **Figure 11.1** Sam Chadwick/Shutterstock; **Figure 11.2b** Natalie B. Fobes/National Geographic Stock; **Figure 11.2c** NOAA Central Library Photo Collection; **Figure 11.4b** Bob Jordan/AP Images; **Figure 11.5** National Oceanic and Atmospheric Administration; **Figure 11A top left** US Coast Guard, **top right** Louisiana Governors Office/Alamy Stock Photo, **bottom right** Chuck Cook/AP Images; **Figure 11.6** Ministry of Transport/AP Images; **Figure 11.8** Charlie Riedel/AP Images; **Figure 11.11** John Trever; **Figure 11.12** Jack Smith/AP Images; **Figure 11.15** Age fotostock/SuperStock; **Figure 11.16** Associated Press; **Figure 11.19** Alan P. Trujillo; **Creature Feature 11.1** Manfred Bortoli/Getty Images; **Figure 11.22** David W. Leindecker/Shutterstock; **Figure 11.23a** National Marine Fisheries Service; **Figure 11.23b** Photoshot License Limited; **Figure 11.23c** David Liittschwager/National Geographic/Getty Images; **Figure 11.23d** Photo by Susan Middleton © 2005; **Figure 11.24** Tony Freeman/Photoedit; **Figure 11.25** Alan P. Trujillo; **Figure 11.26** Brendan Bannon/Newscom; **Figure 11B** Education & Exploration 2/Alamy Stock Photo; **Figure 11.28** Alan P. Trujillo; **Quote, page 355** Jane Lubchenco, marine ecologist (2002); **Quote, page 377** Burke, E. (1834). *The works of Edmund Burke, with a memoir*. In three volumes: Vol. I. New York: George Dearborn, publisher. Sold by Collins & Hannay, New York; Carter, Hendee &, Boston; Desilver, Jr. & Thomas, Philadelphia; and Cushing & Sons, Baltimore

### Chapter 12

**Opener** Jeff Rotman/Alamy Stock Photo; **Figure 12.2** Science History Images/Photo Researchers/Alamy Stock Photo; **Figure 12.8** IFREMER/AFP/Getty Images/Newscom; **Figure 12.13** CSIRO Plant Industry; **Figure 12.19** cbimages/Alamy Stock Photo; **Figure 12.20** aaltair/Shutterstock; **Figure 12.21a** Eugene Sim/Fotolia; **Figure 12.21b** Alan P. Trujillo; **Figure 12.23** Alan P. Trujillo; **Figure 12A** Scripps Institution of Oceanography, UCSD; **Creature Feature 12.1** © 2004 MBARI; **Figure 12.27** Peter David/The Image Bank/Getty Images; **Figure 12.28** Woods Hole Oceanographic Institute; **Quote, page 383** E. O. Wilson, 2016. Half-Earth: *Our Planet's Fight for Life*. W. W. Norton & Company

### Chapter 13

**Opener** Robinson Ed/Perspectives/Getty Images; **Figure 13.2a** Scripps Institution of Oceanography, UCSD; **Figure 13.2b** Peter Parks/Image Quest Marine; **Figure 13.4** Alan P. Trujillo; **Figure 13.5** NASA's Goddard Space Flight Center; **Figure 13.7, 13.8a, b** Alan P. Trujillo; **Figure 13.8c, d** CSIRO Plant Industry; **Figure 13.9** Imaginechina/Splash News/Newscom; **Figure 13.10a, c, d** CSIRO Plant Industry; **Figure 13.10b** Scripps Institution of Oceanography, UCSD; **Figure 13.11** Pete Atkinson/Getty Images; **Figure 13.13** Everett Collection; **Figure 13.15** NASA; **Figure 13.17** Claire Ting/Science Source; **Figure 13.19b** Scripps Institution of Oceanography, UCSD; **Creature Feature 13.1** Moestrup, Ø., Garcia-Cuetos, L., Hansen, P. J. and Fenchel, T. (2012), Studies on the Genus Mesodinium I: Ultrastructure and Description of Mesodinium chamaeleon n. sp., a Benthic Marine Species with Green or Red Chloroplasts. *J. Eukaryot. Microbiol.*,, 59: 20-39. doi:10.1111/j.1550-7408.2011.00593.x; **Figure 13.26** Walter E Harvey/Science Source/Getty Images; **Figure 13A top, center** Monroe County Public Library, **bottom** Loren McClenachan, Scripps Institute of Oceanography, UCSD; **Figure 13.34** National Marine Fisheries Service; **Figure 13.35** Stefan Jacobs/Alamy Stock Photo; **Quote, page 411** Ritchie, A. (1885). *Mrs. Dymond* (Vol. 10). Cambridge: Smith, Elder

## Chapter 14

**Opener** Image Source/Alamy Stock Photo; **Figure 14.3** Takahashi, K., 1981. *Vertical flux, ecology and dissolution of Radiolaria in tropical oceans: Implications from the silica cycle.* Ph.D. Thesis, Massachusetts Institute of Technology/Woods Hole Oceanographic Institution, W.H.O.I. 81-103, pp. 461; **Figure 14.4** Siim Sepp; **Figure 14.5** Wilhelm Giesbrecht; **Figure 14.6** National Geographic Image Collection/Alamy Stock Photo, **inset** Peter Parks/Image Quest Marine; **Figure 14.7b** Danita Delimont/Getty Images; **Figure 14.11 top to bottom** Tania Zbrodko/Fotolia, lioneldivepix/Fotolia, Stephan Kerkhofs/Fotolia, Darryl Torckler/Getty Images, Cbpix/Fotolia; **Figure 14.12a** Amar/Isabelle Guillen/Alamy Stock Photo; **Figure 14.12b** Mark Conlin/Alamy Stock Photo; **Figure 14A** Michael Patrick O'Neill/Alamy Stock Photo; **Figure 14.14a** Scripps Institution of Oceanography; **Figure 14.14b** Phil Hastings/Scripps Institution of Oceanography, UC San Diego; **Figure 14.16** Frederick McConnaughey/Science Source; **Figure 14.17a** Jonathan Bird/Getty Images; **Figure 14.17b** Cbpix/Fotolia; **Figure 14.17c** Roger Steene/Image Quest Marine; **Figure 14.19a** Nicole Duplaiz/Getty Images; **Figure 14.19b** Elvele Images Limited/Alamy Stock Photo; **Figure 14.19c** Avalon/Photoshot License/Alamy Stock Photo; **Figure 14.19d** Arco Images GmbH/Alamy Stock Photo; **Figure 14.19e** Irina Mos/Shutterstock; **Figure 14.21a** Cornforth Images/Alamy Stock Photo; **Figure 14.21b** Helmut Corneli/imageBROKER/Alamy Stock Photo; **Creature Feature 14.1** Francois Gohier/VWPics/Alamy Stock Photo; **Figure 14.24, 14.27c, 14.27d** Alan P. Trujillo; **Figure 14.28** Doug Perrine/Nature Picture Library/Alamy Stock Photo; **Figure 14.29a** Acquired under National Marine Fisheries Permit 17355-01 and NOAA Class G flight authorization 2015-ESA-4-NOAA; photo by John Durban, Southwest Fisheries Science Center, NMFS, and Michael Moore, Woods Hole Oceanographic Institution; **Figure 14.29b, 14B** Justin Hofman/Pearson Education, Inc.; **Quote, page 455** Lindblad Expeditions Naturalist Robert "Pete" Pederson (1999); **Quote, page 464** Peter Benchley, *Great White Sharks*

## Chapter 15

**Opener** David Nanuk/Getty Images; **Figure 15.2b, c, i** Harold V. Thurman; **Figure 15.2d, h, j, k, l, m** Alan P. Trujillo; **Figure 15.2f** Yuval Helfman/Alamy Stock Photo; **Figure 15.2g** Suzanne Long/Alamy Stock Photo; **Figure 15.3a** Alan P. Trujillo; **Figure 15.3b** Towlake/Getty Images; **Figure 15.5a** Itsik Marom/Alamy Stock Photo; **Figure 15.5b** Harold V. Thurman; **Figure 15.6, 15.7a** Alan P. Trujillo; **Figure 15.7b** Premaphotos/Alamy Stock Photo; **Figure 15.10** Walter Dawn/Science Source; **Figure 15.12** Howard J. Spero; **Figure 15.13** blickwinkel/Woike/Alamy Stock Photo; **Figure 15.14b** Justin Hoffman/Pearson Education, Inc.; **Figure 15.15a** Alan P. Trujillo; **Figure 15.15b** Andrew J. Martinez/Science Source; **Figure 15.18a** Peter Leahy/123Rf; **Figure 15.18b** Carson Ganci/Design Pics Inc/Alamy Stock Photo; **Figure 15.18c** Sebastian Burel/Shutterstock; **Figure 15.29** NOAA; **Figure 15.21a** Kerry L. Werry/Shutterstock; **Figure 15.21b** Charles Stirling/Alamy Stock Photo; **Figure 15.22** Franco Banfi/Steve Bloom Images/Alamy Stock Photo; **Figure 15.23a** The Ocean Agency/XL Catlin Seaview Survey; **Figure 15.23b** David Burdick/NOAA; **Figure 15A** Ryan Photographic; **Figure 15B** Robert R. Hessler; **Figure 15.27a** Dr. Ken MacDonald/Science Source; **Figure 15.27b** Woods Hole Oceanographic Institute; **Figure 15.27c** Scripps Institution of Oceanography, UCSD; **Creature Feature 15.1** NOAA; **Figure 15.29** NOAA; **Figure 15.30c** Scripps Institution of Oceanography; **Figure 15.31** Charles R. Fisher/Pennsylvania State University; **Figure 15.32b** Japan Agency for Marine-Earth Science and Technology (JAMSTEC); Thomas Dahlgren, Marine Ecologist (2006); **Quote, page 489** Thomas Dahlgren, Marine Ecologist (2006)

## Chapter 16

**Opener** Jody Martin/Rueters; **Figure 16.1** Dennis Tasa; **Figure 16.4** Alan P. Trujillo; **Figure 16.5** NASA Goddard Space Flight Center, **inset** British Antarctic Survey/Science Source; **Figure 16.6** NASA; **Figure 16.8** Dennis Tasa; **Figure 16.10** InterNetwork Media/Getty Images; **Figure 16.13** © IPCC (Artwork Cover by Alisa Singer/www.environmentalgraphiti.org); **Figure 16.14** Akademie/Alamy Stock Photo; **Quote, page 536** Intergovernmental Panel in Climate Change (IPCC). (1995); **Figure 16.15** U.S. Global Change Research Program; **Figure 16A** Scripps Institution of Oceanography, UCSD; **Figure 16.24** THOMAS B. SHEA/AFP/Getty Images; **Figure 16.26a** NASA; **Figure 16.27** Kerstin Langenberger; **Figure 16.28** NASA; **Figure 16.29a** CSIRO Plant Industry; **Figure 16.29b** Sinclair Stammers/Science Source; **Figure 16.29c** Jeff Rotman/Alamy Stock Photo; **Figure 16.29d** Dobermaraner/Shutterstock; **Creature Feature 16.1** National Geographic Creative; **Figure 16.31, 16.32** NOAA; **Figure 16.32** Alan P. Trujillo; **Figure 16.35** Woods Hole Oceanographic Institute; **Figure 16.36** tobkatrina/123RF; **Quote, page 525** Dr. Jane Lubchenco, marine ecologist and Under Secretary of Commerce for Oceans and Atmosphere and NOAA Administrator (2009), at a White House news conference announcing the release of the report Global Climate Change Impacts in the United States; **Quote, page 536** Intergovernmental Panel in Climate Change (IPCC). (1995); **Quote, page 537** Thomas J. Wilbanks et al. "America's Climate Choices: Adapting to the Impacts of Climate Change", National Reasearch Council, May 2010, National Academy of Sciences; **Quote, page 537** Wuebbles, D.J. et al. "2017: Executive summary. In: Climate Science Special Report: Fourth National Climate Assessment", Volume I. U.S. Global Change Research Program, Washington, DC, USA, pp. 12-34; **Quote, page 562** Climate-related Geoengineering and Biodiversity, https://www.cbd.int/climate/geoengineering/; **Quote, page 563** John Martin from lecture at the Woods Hole Oceanographic Institution, July 1988

# INDEX

*Note:* page numbers followed by f or t refer to Figures or Tables.

## Y



## Z